ROCK-FORMING MINERALS

ROCK-FORMING MINERALS

Volume 2A Second Edition

Single-Chain Silicates

W. A. Deer, MSc, PhD, FRS
Emeritus Professor of Mineralogy & Petrology
University of Cambridge

R. A. Howie, MS ScD, FKC
Emeritus Professor of Mineralogy
University of London

J. Zussman, MA, PhD
Emeritus Professor of Geology
University of Manchester

The Geological Society
London
1997

THE GEOLOGICAL SOCIETY

The Society was founded in 1807 as The Geological Society of London and is the oldest geological society in the world. It received its Royal Charter in 1825 for the purpose of 'investigating the mineral structure of the Earth'. The Society is Britain's national society for geology with a membership of around 8000. It has countrywide coverage and approximately 1000 members reside overseas. The Society is responsible for all aspects of the geological sciences including professional matters. The Society has its own publishing house, which produces the Society's international journals, books and maps, and which acts as the European distributor for publications of the American Association of Petroleum Geologists, SEPM and the Geological Society of America.

Fellowship is open to those holding a recognized honours degree in geology or cognate subject and who have at least two years' relevant postgraduate experience, or who have not less than six years' relevant experience in geology or a cognate subject. A Fellow who has not less than five years' relevant postgraduate experience in the practice of geology may apply for validation and, subject to approval, may be able to use the designatory letters C Geol (Chartered Geologist).

Further information about the Society is available from the Membership Manager, The Geological Society, Burlington House, Piccadilly, London W1V 0JU, UK. The Society is a Registered Charity, No. 210161.

Published by The Geological Society from:
The Geological Society Publishing House
Unit 7, Brassmill Enterprise Centre
Brassmill Lane
Bath BA1 3JN
UK
(*Orders*: Tel. 01225 445046
Fax 01225 442836)

First edition published 1963
Second edition published 1978
Reprinted with corrections 1997

British Library Cataloguing in Publication Data
A catalogue record for this book is available from the British Library.

ISBN 1-897799-85-3

Printed by The Alden Press, Osney Mead, Oxford, UK.

Distributors

USA
AAPG Bookstore
PO Box 979
Tulsa
OK 74101-0979
USA
(*Orders*: Tel. (918) 584-2555
Fax (918) 560-2652)

Australia
Australian Mineral Foundation
63 Conyngham Street
Glenside
South Australia 5065
Australia
(*Orders*: Tel. (08) 379-0444
Fax (08) 379-4634)

India
Affiliated East-West Press PVT Ltd
G-1/16 Ansari Road
New Delhi 110 002
India
(*Orders:* Tel. (11) 327-9113
Fax (11) 326-0538)

Japan
Kanda Book Trading Co.
Tanikawa Building
3-2 Kanda Surugadai
Chiyoda-Ku
Tokyo 101
Japan
(*Orders*: Tel. (03) 3255-3497
Fax (03) 3255-3495)

Contents

Single-Chain Silicates: Pyroxenes

Single-Chain Silicates: Non-Pyroxenes

Abbreviations and Symbols

The following abbreviations have been used in the text except where otherwise stated.

Å	Ångstrom units (10^{-10}m)
a	cell edge in the x direction
a_{rh}	rhombohedral cell edge
a_{hex}	hexagonal cell edge
anal.	analysis or analyst
b	cell edge in the y direction
Bx_a	acute bisectrix
c	cell edge in the z direction
calc.	calculated
D	specific gravity
D	(in association with λ) sodium (yellow) light (589 nm)
d	interplanar spacing
DTA	differential thermal analysis
H	hardness (Mohs scale)
M.A.	*Mineralogical Abstracts*
max.	maximum
min.	minimum
mol.	molecular
n	refractive index (for a cubic mineral)
O.A.P.	optic axial plane
P	pressure
R	metal ions
$r < v$ (or $r > v$)	the optic axial angle in red light is less than (or greater than) that in violet light
rh	rhombohedral
T	temperature
2V	the optic axial angle
x, y, z	the crystal axes
Z	number of formula units per unit cell
α, β, γ	least, intermediate and greatest refractive indices
α, β, γ	angles between the positive directions of the y and z, x and z, and x and y crystal axes
α, β, γ	the vibration directions of the fast, intermediate and slow ray; also these rays
δ	birefringence
ϕ	polar coordinate: azimuth angle measured clockwise from [010]
ρ	polar coordinate: polar angle measured from z
λ	wavelength
ϵ	extraordinary ray, refractive index
ω	ordinary ray, refractive index
$\simeq$	approximately equal to

Preface to First Edition

In writing these volumes the primary aim has been to provide a work of reference useful to advanced students and research workers in the geological sciences. It is hoped, however, that it will also prove useful to workers in other sciences who require information about minerals or their synthetic equivalents. Each mineral has been treated in some detail, and it has thus been necessary to restrict the coverage to a selection of the more important minerals. The principle in this selection is implied in the title *Rock-Forming Minerals*, as, with a few exceptions, only those minerals are dealt with which, by their presence or absence, serve to determine or modify the name of a rock. Some may quarrel with the inclusion or omission of particular minerals; once committed, however, to the discussion of a mineral or mineral series the less common varieties have also been considered.

Most of the information contained in this text is available in the various scientific journals. An attempt has been made to collect, summarize and group these contributions under mineral headings, and the source of information is given in the references at the end of each section. The bibliography is not historically or otherwise complete, but the omission of reference to work which has been encompassed by a later and broader study does not belittle the importance of earlier investigations; where many papers have been published on a given topic, only a limited number have been selected to illustrate the scope and results of the work they report.

The collection of data and references should bring a saving of time and labour to the research worker embarking on a mineralogical study, but it is hoped also that the presentation of the results of study from many different aspects, and in particular their correlation, will further the understanding of the nature and properties of the minerals. Determinative properties are described and tabulated, but the intended function of this work is the understanding of minerals as well as their identification, and to this end, wherever possible, correlation has been attempted, optics with composition, composition with paragenesis, physical properties with structure, and so on. For each mineral the body of well-established data is summarized, but unsolved and partially solved problems are also mentioned.

The rock-forming minerals are dealt with in five volumes. The silicates are allocated on a structural basis: vol. 1. *Ortho- and Ring Silicates*, vol. 2. *Chain Silicates*, vol. 3. *Sheet Silicates*, vol. 4. *Framework Silicates*. *Non-silicates* are grouped chemically in the various sections of volume 5.

With a few exceptions, the treatment of each mineral or mineral group is in five sub-sections. In the *Structure* section, in addition to a brief description of the atomic structure, descriptions of X-ray methods for determining chemical composition and any other applications of X-rays to the study of the mineral are given. The *Chemistry* section describes the principal variations in chemical composition and includes a table of analyses representative, wherever possible, of the range of chemical and paragenetic variation. From most analyses a structural formula has been calculated. The chemistry sections also consider the synthesis and breakdown of the minerals and the phase equilibria in relevant chemical systems, together with DTA observations and alteration products. The third section lists *Optical and Physical Properties* and discusses them in relation to structure and chemistry. The fourth section contains *Distinguishing Features* or tests by which each mineral may be recognized and in particular distinguished from those with which it is most likely to be confused. The *Paragenesis* section gives the principal rock types in which the mineral occurs and some typical mineral assemblages: possible derivations of the minerals are discussed and are related wherever possible to the results of phase equilibria studies. The five sub-sections for each mineral are preceded by a condensed table of properties together with an orientation sketch for biaxial minerals and an introductory paragraph, and are followed by a list of references to the literature. The references are comprehensive to 1959 but later additions extend the coverage for some sections to 1961. In the present text, mineral data are frequently presented in diagrams, and those which can be used determinatively have been drawn to an exact centimetre scale, thus enabling the reader to use them by direct measurement: numbers on such diagrams refer to the number of the analysis of the particular mineral as quoted in the tables.

The dependence of these volumes upon the researches and reports of very many workers will be so obvious to the reader as to need no emphasis, but we wish especially to record our indebtedness to those authors whose diagrams have served as a basis for the illustrations and thus faciliated our task. In this connection we would thank also the many publishers who have given permission to use their diagrams, and Mr H. C. Waddams, the artist who has so ably executed the versions used in the present text. *Mineralogical Abstracts* have been an indispensable starting point for bringing many papers to our attention: in by far the majority of cases reference has been made directly to the original papers; where this has not been possible the *Mineralogical Abstracts* reference is also given e.g. (M.A. 13–351). Our warmest thanks are due also to our ex-colleagues in the Department of Geology, Manchester University, who have been helpful with discussions and information, and who have tolerated, together with the publishers, repeatedly over-optimistic reports about the work's progress and completion. We wish to thank Miss J. I. Norcott who has executed so efficiently the preparation of the typescript and also Longmans, Green & Co. for their continued cooperation.

Department of Geology, The University, Manchester 13. *October 1961*

Preface to Second Edition

For this completely new edition of 'Rock-Forming Minerals' we have maintained the general principles and organization adopted for the first edition. The past fifteen years, however, have seen an enormous expansion in activity in the Earth Sciences as a whole, and the subjects of Mineralogy and Petrology have certainly not been exceptions. The terms 'literature explosion' and 'exponential growth', although almost clichés, nevertheless are very apt in the present context. Not only have the numbers of researchers and their outputs increased, but exceptional growth has occurred in three particular fields: electron microprobe analysis, experimental petrology, and the determination of crystal structures.

The facility of rapid and accurate electron probe analysis has replaced to a great extent the more laborious chemical and optical analytical methods, giving many more reliable analyses for each mineral and enabling researchers to examine more specimens and to complete a wider range of studies in a shorter time. The availability of more well-analysed material has also lead to much more significant discussion of chemical variations and their relationships with crystal structure, physical properties and, most of all, parageneses. The important phenomena of fine-scale intergrowths (exsolution, etc.) and of chemical zoning have also been much more readily investigated using electron probe and other electron-optical methods.

The study of phase equilibria at elevated pressures and temperatures has continued apace, so that the cumulative number of systems which need to be described has grown. In addition, much wider ranges of pressure and temperature have become accessible with improved techniques. At the same time, there has been a growth in the determination of thermodynamic properties of minerals, and in the experimental and theoretical approaches to element distribution within and between minerals.

The advent and growing use of automatic single-crystal diffractometers has made it possible to determine crystal structures much more quickly, so that whereas there was hitherto perhaps one published structure for a mineral or even for a mineral group, now there can be structure determinations for a mineral at each of several chemical compositions, and at a number of different temperatures.

The above, and other growth areas in mineralogy, have lead to the fact that in this new edition the average number of pages devoted to each mineral is about three times that for the first edition. The extent of growth is indicated also by the list of references for each mineral which for this volume we have attempted to bring up to date to 1976, and the early months of 1977.

W. A. Deer   R. A. Howie   J. Zussman   *July, 1977*

Acknowledgements

For permission to redraw diagrams we are grateful to the authors and the following:

Akademische Verlagsgesellschaft for diagrams from *Zeitschrift für Kristallographie*; American Association for the Advancement of Science for diagram from *Science*, copyright 1966 by the American Association for the Advancement of Science; The American Ceramic Society, Inc. for diagram from *American Ceramic Society Journal*; American Geophysical Union for diagrams from *Journal Geophysical Research*; Mineralogical Association of Canada for diagram from *Canadian Mineralogist*; Carnegie Institution of Washington (Geophysical Laboratory) for diagrams from *C.I.W.A.R.D.G.L. Yearbook*; The Chemical Society of Japan for diagram from *Bulletin Chemical Society of Japan*; Chapman & Hall Ltd for diagrams from *Journal of Material Science*; Elsevier Scientific Publishing Co. for diagrams from *Phys. of Earth & Planet Interiors*, *Earth Planetary Science Letters* and *Chemical Geology*; Geological Society of America for diagrams from *Geological Society of American Memoirs* and *Bulletin Geological Society of America*; Geological Society of Australia for diagram from *Journal Geological Society of Australia*; Geological Society of London for diagrams from *Quarterly Journal Geological Society*; Hokkaido University (Dept. of Geology & Mineralogy) for diagrams from *J. Fac. Sci. Univ. Hokkaido*; Dean of the Faculty of Science, Kyushu University for diagram from *Memoirs Series D: Geology*; Mineralogical Society for diagrams from *Mineralogical Magazine*; Mineralogical Society of America for diagrams from *Mineralogical Society of America Papers* and *American Mineralogist*; Oxford University Press for diagrams from *Journal of Petrology*; Pergamon Press Ltd for diagrams from *Geochimica Cosmochimica II Lunar Sci. Conf.*, *5th Lunar Sci. Conf.*, and *Geochimica et Cosmochimica Acta*; Scottish Academic Press Ltd for diagram from *Scottish Journal of Geology*; Societé Française de Mineralogie et de Cristallographie for diagram from *Bulletin Societé Française de Mineralogie et de Cristallographie*; E. Schweizerbart'sche Verlagsbuchhandlung for diagrams from *Neues Jahrbuch für Mineralogie ABH*; Springer-Verlag (Germany) for diagrams from *Contributions to Mineralogy & Petrology*; Springer-Verlag, New York Inc. for diagram from *Electron Microscropy in Mineralogy*; Faculty of Science, Nagoya University for diagrams from *Mineralogical Journal of Japan*; The University of Chicago Press for diagrams from *Journal of Geology*; Universitetsforlaget Oslo for diagrams from *Lithos*; Yale University (Kline Geology Laboratory) for diagrams from *American Journal of Science*.

Single-Chain Silicates: Pyroxenes

Introduction

Introduction

Pyroxenes are the most important group of rock-forming ferromagnesian silicates, and occur as stable phases in almost every type of igneous rock. They are found also in many rocks of widely different compositions formed under conditions of both regional and thermal metamorphism.

The pyroxenes include varieties which have orthorhombic symmetry: thus the orthopyroxenes consist essentially of a simple chemical series of $(Mg,Fe)SiO_3$ minerals, in contrast with the larger group of monoclinic pyroxenes which have a very wide range of chemical composition. Many pyroxenes which are present in both igneous and metamorphic rocks can be considered, as a first approximation, to be members of the four component system $CaMgSi_2O_6$–$CaFeSi_2O_6$–$Mg_2Si_2O_6$–$Fe_2Si_2O_6$. The nomenclature here adopted to describe those clinopyroxenes which can be included in this system was first put forward by Poldervaart and Hess (1951), and is illustrated in Fig. 1. The monoclinic series of $MgSiO_3$–$FeSiO_3$ (clinoenstatite–clinoferrosilite) pyroxenes is rare in terrestrial rocks and is considered here together with the orthopyroxenes. Isomorphous substitutions in the clinopyroxene minerals are not restricted to the mutual replacement of divalent cations, and the compositional fields of pyroxenes in which monovalent and trivalent ions are important constituents are shown in Fig. 2.

The name pyroxene is from the Greek, *pyro*, fire and *xenos*, stranger, and was given by Haüy to the greenish crystals found in many lavas; Haüy apparently considered that these crystals had been accidentally included in the lava.

The pyroxenes can be considered broadly in terms of three major subgroups; magnesium–iron pyroxenes in which other cations occupy less than 10 per cent of the *M*1, *M*2 sites, structural formula $[(M2)(M1)(Si,Al)_2O_6)$; calcic pyroxenes in which Ca occupies more than two-thirds of the *M*2 sites, and the sodic pyroxenes where the *M*2 site is largely occupied by sodium ions and the *M*1 positions by Al,

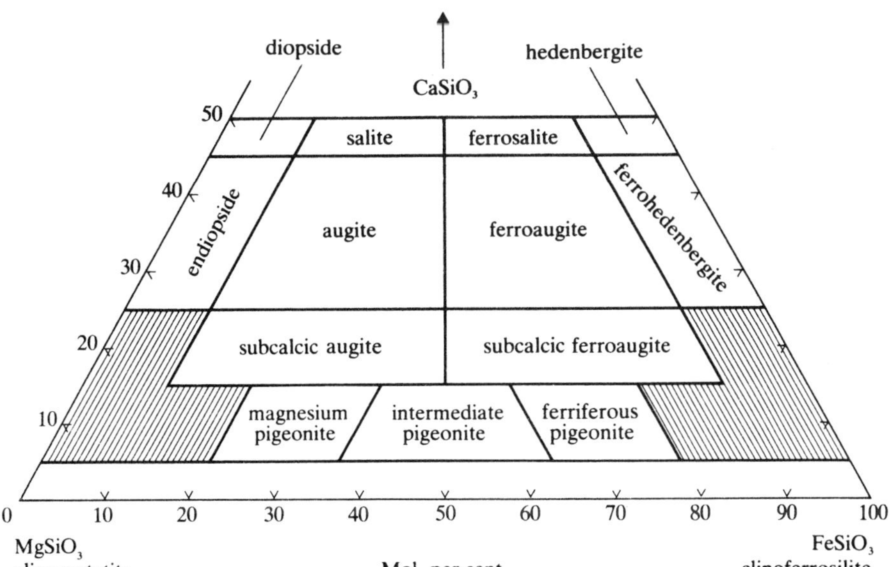

Fig. 1. Nomenclature of clinopyroxenes in the system $CaMgSi_2O_6$–$CaFeSi_2O_6$–$Mg_2Si_2O_6$–$Fe_2Si_2O_6$ (after Poldervaart and Hess, 1951).

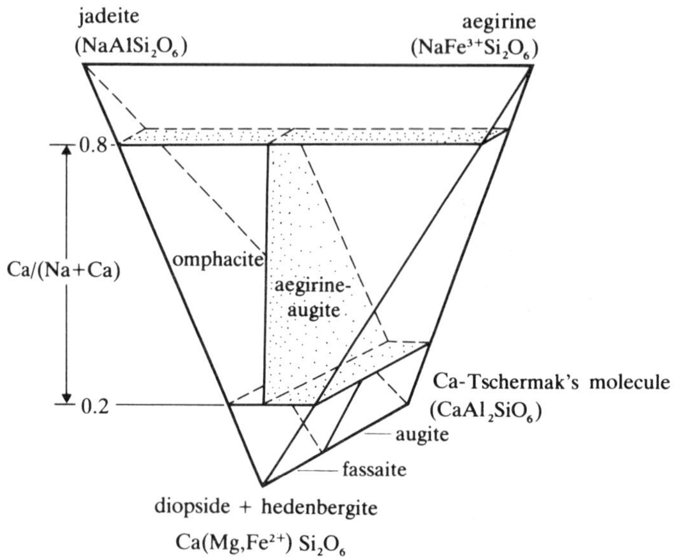

Fig. 2. Nomenclature of clinopyroxenes in the system $NaAlSi_2O_6$ (jadeite)–$NaFe^{3+}Si_2O_6$ (aegirine)–$CaMgSi_2O_6$ (diopside)–$CaFeSi_2O_6$ (hedenbergite)–$CaAl_2SiO_6$ (Tschermak's molecule) showing the compositional fields of omphacite, aegirine-augite, fassaite, and augite (after Clarke and Papike, 1968).

Fe^{3+} or Cr as in jadeite, aegirine and ureyite respectively. There are in addition two minor subgroups, the calcium–sodium pyroxenes, represented by omphacite and aegirine-augite (the latter is treated here with aegirine), and the lithium pyroxenes, represented by spodumene, $LiAlSi_2O_6$.

The compositions of the pyroxene minerals and the order in which they are considered are as follows:

Magnesium–Iron Pyroxenes
Orthopyroxenes (Enstatite–Orthoferrosilite) $(Mg,Fe^{2+})_2Si_2O_6$
Pigeonite $(Mg,Fe^{2+},Ca)(Mg,Fe^{2+})Si_2O_6$

Calcium Pyroxenes
Diopside–Hedenbergite $Ca(Mg,Fe^{2+})Si_2O_6$
Augite $(Ca,Mg,Fe^{2+}Al)_2(Si,Al)_2O_6$
Fassaite $Ca(Mg,Fe^{2+},Fe^{3+},Al)(Si,Al)_2O_6$
Johannsenite $CaMnSi_2O_6$

Calcium–Sodium Pyroxenes
Omphacite (Aegirine-augite) $(Ca,Na)(Mg,Fe^{2+},Fe^{3+},Al)Si_2O_6$

Sodium Pyroxenes
Jadeite $NaAlSi_2O_6$
Aegirine $NaFe^{3+}Si_2O_6$
Ureyite $NaCrSi_2O_6$

Lithium pyroxenes
Spodumene $LiAlSi_2O_6$

Many other names have been used to describe individual members of the pyroxene group. Most are, however, of little value and should be discarded: with few exceptions they have not been used in this text. For some pyroxenes, in order to indicate unusual amounts of a particular constituent, it is convenient to use adjectival modifiers (e.g. chromian augite), but these would not be used if the element concerned was normally present in the minerals (e.g. calcian augite). In these adjectives, the ending -ian is used for higher valence state (e.g. ferrian) whereas the ending -oan is used for the lower valence state (e.g. ferroan, titanoan), and this terminology is particularly of value in designating more precisely the members of the augite series.

A wide variety of ionic substitutions occur in members of the pyroxene group, and there is complete replacement between some of the group components, e.g. between diopside and hedenbergite, and between hedenbergite and johannsenite. The diopside–hedenbergite minerals also form a continuous series with augite and ferroaugite and subcalcic augite and ferroaugite. Because there is a continuous chemical variation between diopside, hedenbergite and augite the view has been expressed that the single name, augite, with adjectival modifiers is adequate to describe this compositional range within the pyroxene group. It is, however, desirable, particularly in petrographic descriptions, and in discussing parageneses, to convey the chemical characteristics of these pyroxenes more precisely and yet succinctly. Consequently the four divisions of the diopside–hedenbergite and augite (augite, ferroaugite, subcalcic augite and subcalcic ferroaugite) series, together with endiopside and ferrohedenbergite, both of which are associated with relatively restricted parageneses, have been retained. Similarly the name fassaite is used here to describe those Al(Fe^{3+})-rich clinopyroxenes in which most of the $M2$ positions are occupied by Ca, and the introduction of the trivalent cations into $M1$ is compensated almost entirely by the replacement of Si by Al in tetrahedral sites. In view of the long and continuing use of the name titanaugite this term has also been retained.

There is a miscibility gap between the calcium pyroxene group and the (Mg, Fe)-rich pigeonites, although there is some evidence that pyroxenes intermediate in composition between subcalcic augite and pigeonite may crystallize at high temperatures. A compositional break also occurs between pigeonites and the orthopyroxenes, the former having 0·1–0·2 Ca, and the latter less than 0·1 Ca ions to six oxygens.

Although the sodium pyroxenes, jadeite and aegirine, commonly contain more than 90 per cent of the $NaAlSi_2O_6$ or $NaFe^{3+}Si_2O_6$ component respectively, both minerals may show extensive solid solution with diopside and hedenbergite leading to omphacitic or aegirine-augite compositions (calcium–sodium pyroxenes). The arbitrary divisions between these two series, and nomenclature in the wider field defined by jadeite, aegirine, diopside–hedenbergite and $CaAl_2SiO_6$ (Ca-Tschermak's molecule), are illustrated in Fig. 2. Pyroxenes in which the jadeite, aegirine and diopside–hedenbergite components are present in approximately equal proportions, sometimes described as chloromelanites, are considered in the omphacite section (see Fig. 200).

Structure

The first of the pyroxene structures to be determined was that of diopside (Warren and Bragg, 1928). This work established that the essential feature of all pyroxene structures is the linkage of SiO_4 tetrahedra by sharing two out of four corners to form continuous chains of composition $(SiO_3)_n$ (Fig. 3). The repeat distance along the length of the chain is approximately 5·2 Å and this defines the *c* parameter of the unit cell. The chains are linked laterally by cations (Ca, Na, Mg, Fe, etc.) whose positions, labelled *M*1 and *M*2, are illustrated in Fig. 4, which represents the projection along the chain direction of an idealized pyroxene structure. *M*1 atoms lie principally between the apices of SiO_3 chains, whereas *M*2 atoms lie principally between their bases. *M*2 is the site occupied by Ca in diopside, and in general, if a large ion (Ca,Na) is present it will occupy *M*2 and not *M*1. The co-ordination of oxygens around *M*1 is very nearly a regular octahedron but the *M*2 site co-

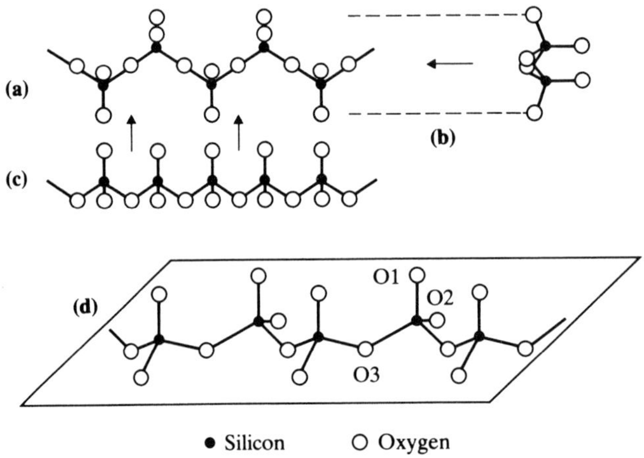

Fig. 3. Idealized illustration of a single pyroxene chain $(SiO_3)_n$ as seen in three projections (a) on (100) (b) along *z* (c) along *y*, and (d) in perspective (after Bragg, 1937; de Jong, 1959).

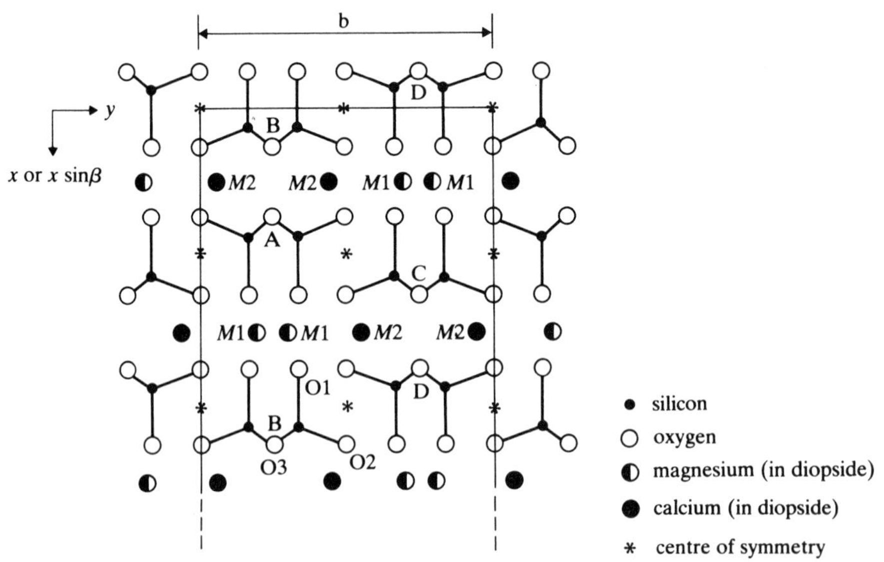

Fig. 4. Projection along *z* of an idealized pyroxene structure, showing the pyroxene chains and two distinct cation sites *M*1 and *M*2. The latter are in oxygen polyhedra which also form chains or bands parallel to *z*. In pyroxenes with space group *C*2/*c* the Si–O chains A, B, C and D are all equivalent. In those with space group $P2_1/c$ and *Pbca*, C = A and D = B but A and B are non-equivalent.

ordination is irregular and different according to the atom present, six-fold for Mg, eight-fold for Ca and Na. It should be noted also that whereas the oxygens (O1 and O2) co-ordinating *M*1 are all non-bridging (i.e. belong to only one tetrahedron of the pyroxene chain), the *M*2 atom is co-ordinated partly by oxygens which are bridging (i.e. serving to link neighbouring tetrahedra). One way of viewing the pyroxene structure is as the linkage of Si–O chains to chains or strips of *M*1–O and *M*2–O polyhedra. Alternatively, as is clear from Fig. 4, the structure can be regarded as made up of layers, parallel to (100), of tetrahedra alternating with layers of polyhedra with six-fold or higher co-ordination. The presence of distinct layers of oxygens is also very evident. In three dimensions, these oxygen layers, particularly for the Mg,Fe pyroxenes, approximate in their arrangement to that of cubic close packing.

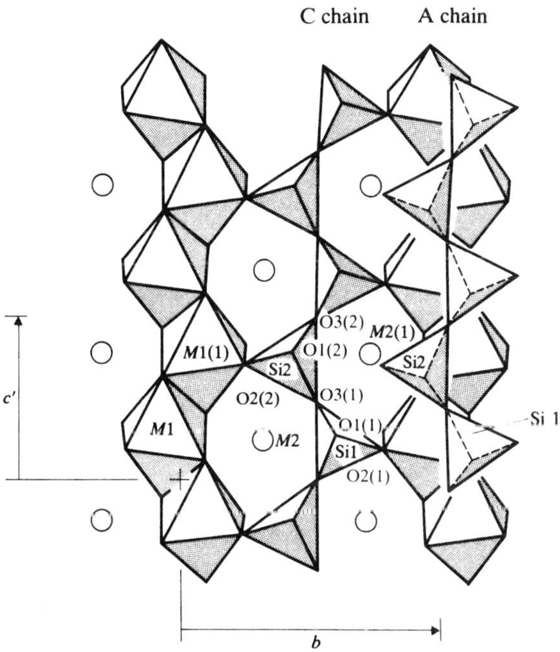

Fig. 5. View of jadeite structure along a line near the *x* direction. Cell contents are shown from 0 to *a*/2 only; origin is marked by cross near *M*1 (after Burnham *et al.*, 1967). *M*1 = Al; *M*2 = Na.

Figure 4 shows that the SiO_3 chains come back to back without displacement in the *y* direction but it does not show how they are stacked with respect to one another parallel to their length, and neither does it show any differences between chains. The latter two important features are determined largely by the nature and proportions of the different cations present, but also sometimes by the physical parameters, the resultant structure being that which is thermodynamically most favourable in the *P–T* conditions of crystallization.

There are several ways in which the structures of pyroxenes can vary:

1. The substitution of different sizes of cations can give rise to proportionate expansion or contraction in one or more of the cell parameters, with little if any change in the basic structure. This is typified by the diopside–hedenbergite series, with Ca in *M*2 and (Mg,Fe) in *M*1. The space group is *C*2/*c*. Only minor changes in the Si–O chain configuration will occur with Fe, Mg substitution.

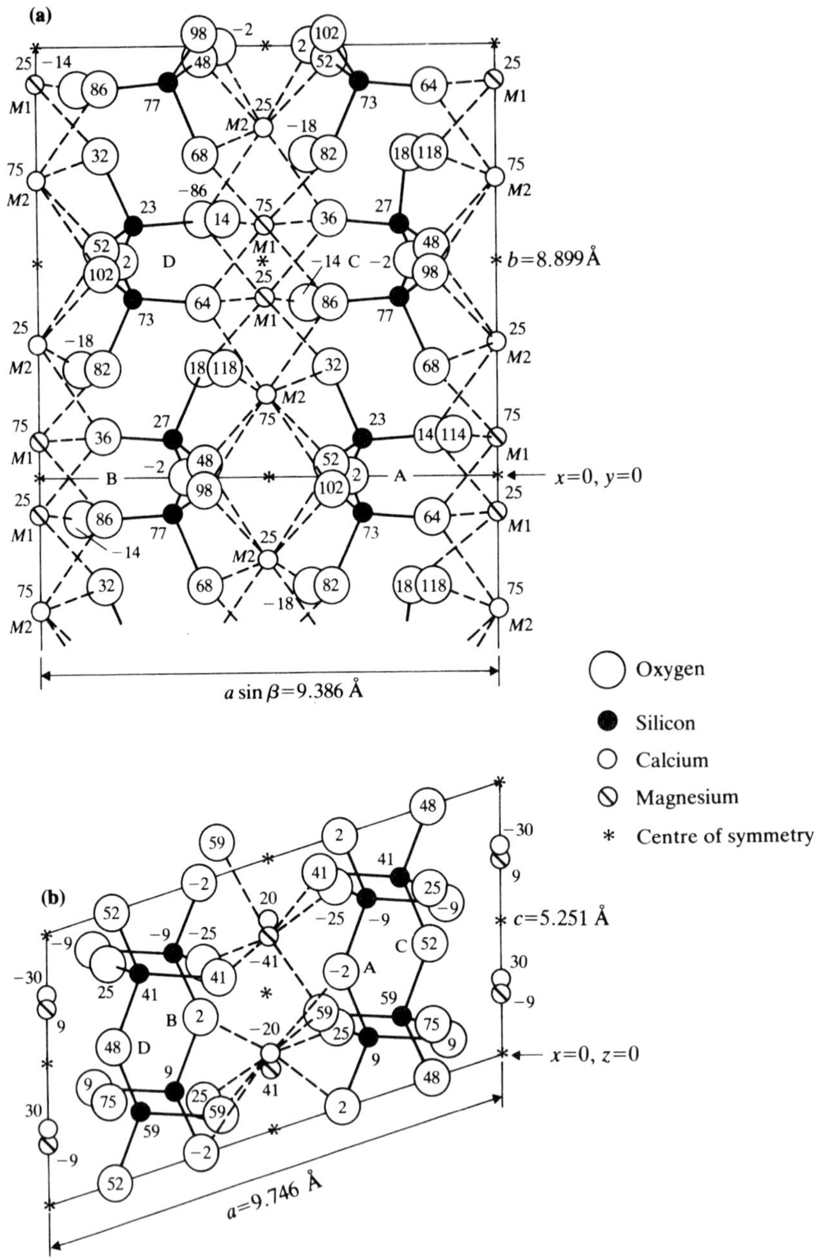

Fig. 6. Idealized structure of diopside (a) viewed along *z*, (b) along *y*. In (a) some atoms of oxygen overlying one another have been slightly displaced. In (b) the Ca and Mg atoms should overlap exactly but have been slightly displaced (after Warren and Bragg, 1928). The *z* co-ordinates in (a) and *y* in (b) are rounded off values from Clark *et al.* (1969). The choice of origin and resulting co-ordinates have been changed to conform with Burnham *et al.* (1967).

2. Greater differences in the sizes of cations as, for example, between various augites Ca (Mg,Fe), jadeite (Na,Al) and johannsenite (Ca,Mn), may be accommodated in the pyroxene structure by greater changes in the silicate chains, or in their relative positions, without a change in overall symmetry as reflected in the space group, which remains *C*2/*c*.

In each of the above kinds of structure, all silicate chains are exactly equal and the metal ions lie on diad axes. There are, however, small variations in the positions of the metal ions along the diads. Figure 5 shows a view of the structure of jadeite ($NaAlSi_2O_6$) which is typical of this kind of pyroxene. The chains of Al octahedra and Si tetrahedra are evident as are also the larger Na polyhedra. Two projections of the structure of diopside are illustrated in Fig. 6.

In a pyroxene with space group *C*2/*c* there is only one kind of *M*1 and one kind of *M*2 site. Any mixtures of atoms within either one of these sites must therefore be randomly distributed, as for example (Mg,Fe) in *M*1 in the diopside–hedenbergite series, and the substitution should be one of ideal solid solution.

The Si–O chains never have quite the 'straight' configuration shown in Fig. 3(a); they are kinked to various degrees depending upon the nature of the *M*1 and *M*2 cations, but come closest to being straight in jadeite and in a synthetic $LiFe^{3+}$ pyroxene (see Fig. 10).

The kinking of Si–O chains can be analysed in terms of rotations of the tetrahedra with respect to the *M* polyhedra. Rotations in opposite senses have been defined as *O* and *S* rotations (Thompson, 1970; see also Papike *et al.*, 1973). For spodumene the rotations are *S* but for all other *C*2/*c* pyroxenes they are *O*.

Further illustrations of the pyroxene structure are provided by Fig. 7 emphasizing the *M*1 octahedra and also the systematic labelling of the Si, O and *M* sites, and by Fig. 7 giving details of the co-ordination of *M*2 in *C*2/*c* pyroxenes.

3. When smaller atoms occur in *M*2, as for example Mg in pigeonite or clinoenstatite, in addition to further kinking of the Si–O chains and their displacement parallel to *z*, the *A* and *B* chains (Fig. 6) adopt different configurations, the *A* tetrahedra having a small *S* rotation and *B* tetrahedra a large *O* rotation. There is also displacement of the metal atoms from the diads.

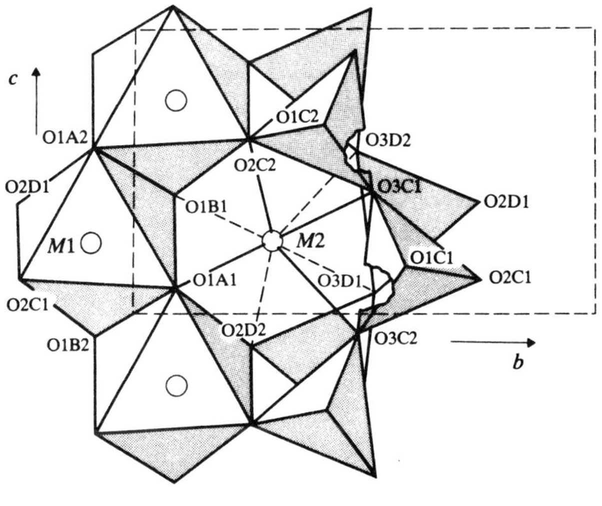

Fig. 7. The *C*2/*c* pyroxene structure projected along *a, emphasizing the co-ordination of *M*2 (Cameron *et al.*, 1973). Nomenclature after Burnham *et al.* (1967).**

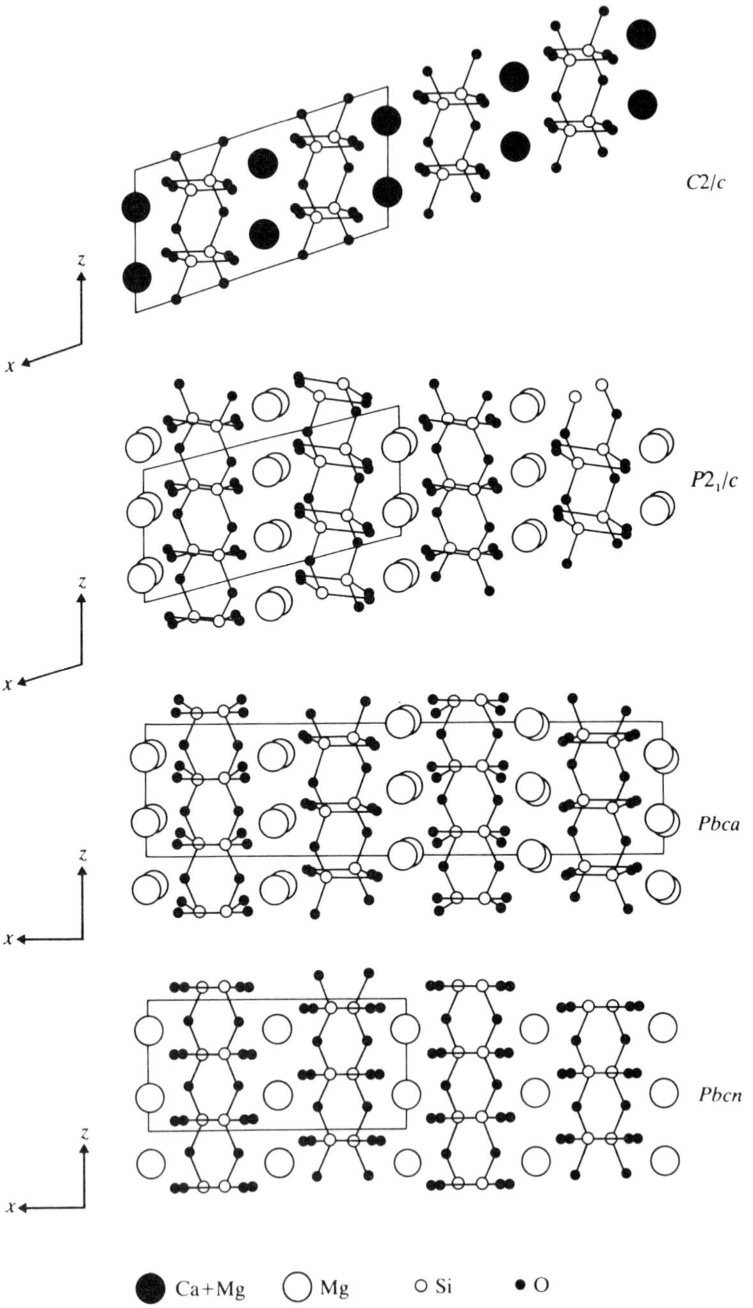

Fig. 8. Schematic illustrations of *y* axis projections of the structures of (i) diopside, (ii) low clinoenstatite, (iii) orthoenstatite, (iv) protoenstatite (after Zussman, 1968).

The loss of symmetry leads to the space group $P2_1/c$ instead of $C2/c$; compare (i) and (ii) in Fig. 8. At high temperatures the structures of these pyroxenes become more like that of diopside, and there is a transformation to the $C2/c$ space group.

4. For pyroxenes of the $MgSiO_3$–$FeSiO_3$ series, polymorphism is found, with grossly different stacking arrangements for the silicate chains in relation to each other and to the chains of octahedra. Across the series there is a monoclinic polymorph $P2_1/c$ (clinoenstatite–clinoferrosilite) and an orthorhombic polymorph *Pbca* (orthoenstatite–orthoferrosilite), and for some orthopyroxenes $P2_1ca$ has been reported. A further polymorph of $MgSiO_3$, protoenstatite, is orthorhombic *Pbcn*. In the structures of these pyroxenes the silicate chains deviate most from the straight configuration, and at the same time the oxygen layers achieve a closer approximation to the cubic close packing arrangement. This perhaps explains why polytypism and stacking disorder are common for the Mg- and Fe-rich pyroxenes and not for the others. Some of the differences between $C2/c$ and $P2_1/c$ pyroxenes and the different chain stacking arrangements are shown in Fig. 8.
5. If the *M*1 sites are occupied by two groups of cation with significantly different radii and in approximately equal proportions, for example (Mg,Fe) and Al, an ordered replacement in two distinct kinds of *M*1 site can occur, as in certain omphacites (see p. 425), which leads to a change of symmetry to monoclinic $P2/n$. There could conceivably be other ordered substitutions leading to distinct structural subgroups, but so far none has been demonstrated.

An exception to the above principles is provided by the structure of spodumene ($LiAlSi_2O_6$). Although the radius of Li in *M*2 is similar to that of Mg, the structure remains $C2/c$ rather than the $P2_1/c$ of clinoenstatite. This is attributed to the fact that Li is monovalent (see p. 528).

The most common pyroxenes are those of the Di–Hd–En–Fs quadrilateral. The more calcium-rich of these have the $C2/c$ structure, whereas the calcium-poor members are either $P2_1/c$ or one of the orthorhombic structures at room temperatures (Fig. 9); at higher temperatures they too are $C2/c$. Even at high temperatures, and although the space groups are the same, there is some gap in solid

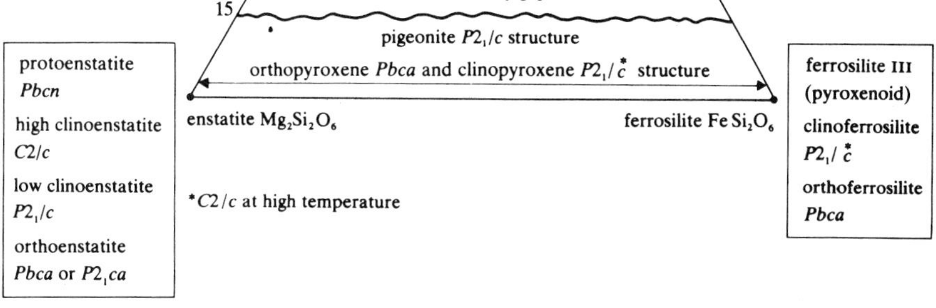

Fig. 9. Illustration of structural relationships among pyroxenes in the diopside–hedenbergite–enstatite–ferrosilite field.

solution between Ca-rich and Ca-poor pyroxenes; with decreasing temperature of crystallization, the gap widens.

Matsumoto and Banno (1970) and Brown (1972), have reviewed the possible space groups and symmetries for clinopyroxenes. Of the sixteen possible, only three (*C*2/*c*, *P*2/*n*, $P2_1/c$) have been conclusively demonstrated (others, about which there are some doubts, include $P2_1/n$ for clinoenstatite, *P*2 for omphacite, and *C*2 for spodumene).

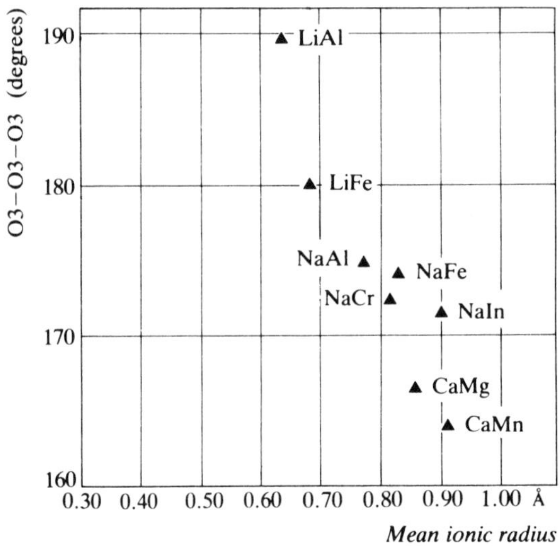

Fig. 10. Plot of tetrahedral chain angle O3–O3–O3 against mean ionic radius of *M* cations in *C*2/*c* pyroxene structures (Papike *et al.*, 1973).

Whittaker (1960) demonstrated a close relationship between the β angle and the size of the *M*2 and *M*1 cations for the clinopyroxene group as a whole. This relationship was confirmed by subsequent structure determinations (e.g. Clark *et al.*, 1969) and is illustrated by Table 1 (see also under spodumene, p. 528). It can be seen that the principal influence on β is the cation at *M*2, and that the disturbing influence of the *M*1 cation is slight.

Table 1 Relations between the ionic radii (R) of metals at the *M*1 and *M*2 sites, β-angle and space group in some important end-members of clinopyroxenes (Morimoto, 1974). Co-ordinations of *M* cations are given by Roman numerals

*M*2	R(Å)	*M*1	R(Å)	β(°)	Space group
		Mn^{2+}(VI)	0·82	105·48	
Ca^{2+}(VIII)	1·12	Fe^{2+}(VI)	0·77	105·33	
		Mg^{2+}(VI)	0·72	105·63	
		In^{3+}(VI)	0·79	107·00	*C*2/*c*
Na^{+}(VIII)	1·16	Fe^{3+}(VI)	0·65	107·42	
		Cr^{3+}(VI)	0·62	107·44	
		Al^{3+}(VI)	0·53	107·58	
Mn^{2+}(VI)	0·82	Mn^{2+}(VI)	0·82	108·22	
Fe^{2+}(VI)	0·77	Fe^{2+}(VI)	0·77	108·38	
Co^{2+}(VI)	0·735	Co^{2+}(VI)	0·735	108·45	$P2_1/c$
Mg^{2+}(VI)	0·72	Mg^{2+}(VI)	0·72	108·33	
Li^{+}(VI)	0·74	Fe^{3+}(VI)	0·65	110·16	*C*2/*c* (*C*2?)
		Al^{3+}(VI)	0·53	110·10	
Zn^{2+}(VI)	0·60	Zn^{2+}(VI)	0·75	111·42	*C*2/*c*

Clark *et al.* (1969) determined the crystal structures of eight ordered clinopyroxenes and compared them with other published structures. They conclude that the bonding in pyroxenes is largely ionic. The charge and sign of the *M*2 cation determine the structure type of most pyroxenes. Large mono- or divalent *M*2 cations give the *C*2/*c* diopside-type structure; small singly charged *M*2 cations give the *C*2 (?) spodumene structure and small divalent *M*2 cations give the $P2_1/c$ clinoenstatite or orthorhombic structures. They confirm the close relationship between β and the size of cation at *M*2. Intermediate (Na,Ca) pyroxenes have appropriately intermediate β values. Although the *c* parameter is relatively insensitive to cation substitution, Clark *et al.* draw attention to its correlation with the *M*1 cation. (c = 5·22 Å with Al in *M*1, 5·29 Å with Fe^{3+} and Mn^{2+}, and 5·37 Å with In^{3+}.)

Single-crystal studies at intervals of increasing temperature have been carried out for a number of *C*2/*c* pyroxenes (Cameron *et al.*, 1973). With increasing temperature, as for the Ca-free pyroxenes, *M* polyhedra expand regularly but Si–O tetrahedra do not change significantly. Mean thermal expansion coefficients for oxygen polyhedra increase in the order Si < Cr < Fe^{3+} < Al < Fe^{2+} < Mg < Ca < Li, as do the thermal vibration factors. Because of the differential expansion of the Si and other polyhedra a misfit occurs which is accommodated by extension of the Si chains, and distortion and tilting of the tetrahedra. Expansion of the unit cell is greatest in the *b* direction and least for *c*. Cameron and co-workers point out the importance of comparing high-temperature structures of Ca-rich and Ca-poor pyroxenes for the proper understanding of solid solution relationships. They show that an increase in solid solution is not to be explained by increased size of the pigeonite's *M*2 site to make it more like *M*2 in diopside–hedenbergite, since the latter also expands commensurately. Also, the co-ordination of *M*2 in pigeonite does not increase from six-fold to the eight-fold of diopside on heating. However, although it remains six-fold there are changes in *M*2's nearest-neighbour co-ordination which do make the structure more amenable to the introduction of Ca accompanied by small shifts of the Si–O chain. At the same time, *M*2 for the Ca-rich pyroxene also becomes nearer to six-fold, thus facilitating solid solution (see also, Takeda, 1972).

Ohashi and Burnham (1972), considering the electrostatic and repulsive energies in pyroxene structures, calculated site energies for *M*1 and *M*2 in the end-members diopside, hedenbergite, enstatite and clinoenstatite, ferrosilite and clinoferrosilite. They concluded that the site energy for a given ion (Mg, Fe, Ca) was approximately the same in both clino- and orthopyroxenes. Predicted preference of Fe for the *M*2 site was in accord with experimental observations. The way in which Fe and Mg are distributed between *M*1 and *M*2 sites is discussed in later sections (e.g. pp. 24, 166 and 295).

A discussion of the various structures exhibited by pyroxenes is presented by Papike *et al.* (1973) who analyse the differences in detail in terms of two ideal oxygen packing models, one with the cubic and the other with the hexagonal close-packed arrangement. By this means they correlate the different Si–O chain configurations with different *M*1 and *M*2 cations and suggest a reason for the existence of the $P2_1/c$ polymorph of Ca-poor pyroxenes as well as the more stable *Pbca* stacking modification.

A series with compositions between $MgSiO_3$ and $CuSiO_3$ was synthesized by Borchert and Kramer (1969) and found to have the clinopyroxene structure. Cell parameters increase with Cu content and range as follows: *a* 9·672–9·713, *b* 8·883–8·906, *c* 5·200–5·222Å, β 109°35′–110°22′.

Tauber and Kohn (1965) showed that synthetic $CoGeO_3$ had the pyroxene structure with clino- and ortho-polymorphs. Peacor (1968) determined the structure of monoclinic $CoGeO_3$. The structure is in space group *C2/c* (*a* 9·692, *b* 9·018, *c* 5·181 Å, *β* 101·2°), but because of the relative sizes of Co and Ge, the oxygens are very nearly in an ideal cubic close-packed arrangement, similar to that adopted in the $P2_1/c$ Ca-poor pyroxenes. The Co in both the *M*1 and *M*2 sites is octahedrally co-ordinated. As in the Mg, Fe, pyroxenes, the Si chain is very much kinked; the O3–O3–O3 angle is 133·4° compared with 174·7° in jadeite and 120° for an ideal cubic close packing arrangement.

At atmospheric pressure $MnSiO_3$ has the rhodonite structure but with increasing pressure this transforms to the pyroxmangite structure and then to the clinopyroxene structure (Akimoto and Syono, 1972; but see rhodonite section, p. 591). The latter has: *a* 9·864, *b* 9·179, *c* 5·198 Å, *β* 108°13′, D_{calc} 3·82 g/cm^3, and space group $P2_1/c$.

Reviews of the crystal structures of pyroxenes have been given by Prewitt and Peacor (1964), Zussman (1968), Smith (1969), Brown (1972), Papike *et al.* (1973) and Morimoto (1974).

Chemistry

The general formula for the pyroxene group proposed by Berman (1937) and later slightly modified by Hess (1949) may be expressed

$$(W)_{1-p}(X,Y)_{1+p}Z_2O_6$$

where W = Ca,Na; X = Mg,Fe^{2+},Mn,Ni,Li; Y = Al,Fe^{3+},Cr,Ti, and Z = Si, Al. In the Mg–Fe-rich pyroxene series $p \simeq 1$ and the content of *Y* ions is usually small. In the Mg–Fe-poor pyroxenes the value of *p* varies from zero (e.g. diopside in which *Y* is small, and aegirine and jadeite in which *X* is very small) to 1 (e.g. spodumene). The wide range of replacements in the (*X*,*Y*) group, commonly involving substitution of ions of different charge, necessitates compensatory replacements in either the *W* or *Z* groups and the replacements must be such that the sum of the charges in the *W*, *X*, *Y* and *Z* groups is 12. The role of Al varies greatly from one pyroxene variety to another; thus in the orthopyroxenes in which Al is usually present only in small amounts, and also in both jadeite and spodumene in which Al is one of the essential ions, the substitution of Si by Al is almost negligible. In augite, however, up to a quarter of the tetrahedral positions in the structure may be occupied by Al.

The analyses detailed in the various pyroxene sections have been recalculated on the basis of six oxygens, deficiencies in the *Z* group having been made up by adding the requisite amount of Al, and in a small number of cases also of Fe^{3+}, to Si to complete the ideal two tetrahedral ions per formula unit. An alternative method of presenting the chemical formulae of pyroxenes has been proposed by Hess (1949) in which, instead of making $Z = 2$ by adding arbitrary amounts of Al(Fe^{3+}) to Si, the cations are allotted to groups in accordance with the balance of charges on a cation for cation basis. (For a discussion and a refinement of both methods see Bown, 1964.)

As many pyroxenes show more than one type of cation substitution it is often desirable to evaluate the proportions of the different components present, and a number of different methods have been devised to present this information. For details of the various procedures reference may be made to the following: Yoder and Tilley (1962): calculation of diopside, hedenbergite, acmite, jadeite and Ca-Tschermak's molecule, and modified procedure by White (1964) (see also Coleman *et al.*, 1965); Kushiro (1962): particularly in relation to the Tschermak's components, $CaAl_2SiO_6$, $CaFe^{3+}AlSiO_6$ and $CaFe_2^{3+}SiO_6$, together with $CaTiAl_2O_6$, $NaFe^{3+}Si_2O_6$, $NaAlSi_2O_6$, $MgSiO_3$, $FeSiO_3$ and $CaSiO_3$; Carmichael (1962); method for illustrating variation of augite–aegirine-augite–aegirine in alkaline rocks in terms of (Na+K), Mg and (Fe^{2+}+Mn+(Fe^{3+})), (Fe^{3+}) being the excess, if any, of trivalent ions over (Na+K); for modification of Carmichael's procedure see Aoki (1964). Essene and Fyfe (1967): particularly in relation to calculations of the jadeite component (see also Makanjuola and Howie, 1972, and Mysen and Griffin, 1973). Edgar *et al.* (1969): determination of diopside–jadeite–acmite and diopside–hedenbergite–jadeite components in omphacitic pyroxenes. A procedure for titanium-rich pyroxenes is given by Onuma and Yagi (1971), and includes the evaluation of $Ca(Mg,Fe)Si_2O_6$, $(Mg,Fe)SiO_3$, $NaFe^{3+}Si_2O_6$, $Ca(Al,Fe^{3+})_2SiO_6$ and $CaTi(Al,Fe^{3+})_2O_6$. A more complex method, involving the estimation of twelve components has been proposed by Vogel (1966).

Analyses obtained using the electron probe do not differentiate between the valence states of iron. The usual methods of calculating end-member values and structural formulae make use of the number of oxygens per formula unit, and are not entirely satisfactory where the Fe^{2+}/Fe^{3+} ratio is not known. If the iron is regarded as being entirely divalent when an appreciable proportion is in fact trivalent, the sum of the calculated oxide percentages will be too low, the number of cations too high, and the distribution of cations between the structural sites will also be affected. In these circumstances the structural formula may be calculated, as suggested by Hamm and Vieten (1971) and Neumann (1976), on the basis of the total number of cations, rather than the total number of oxygens.

Uncertainties in the evaluation of the ferrous/ferric ratios in pyroxene analyses are discussed by Cawthorn and Collerson (1974); see also Edwards (1976). Cawthorn and Collerson have also outlined a method of recalculating analyses which reduces the significance of the content of Fe_2O_3, and permits a more precise comparison of end-member components, particularly between pyroxenes when Fe_2O_3 has been determined and those for which these data are not available.

The compositional range of the calcic and (Mg,Fe)-pyroxenes, expressed in atomic percentages of Ca, Mg and $\Sigma Fe(= Fe^{2+} + Fe^{3+} + Mn)$, listed in the relevant tables, is illustrated in Fig. 11. Although the number of analyses is not significant in relation to those available in the literature it is unlikely that the two areas which do not include compositions of natural pyroxenes would be significantly modified by the additional data. References to compilations of pyroxene analyses, particularly relating to specific parageneses, are included in the appropriate chapters. The most up-to-date and comprehensive compilation, consisting of 1309 clinopyroxene and 354 orthopyroxene analyses is that of Dobretsov *et al.* (1971); see also Dobretsov *et al.* (1972).

References

Akimoto, S. and **Syono, Y.**, 1972. High pressure transformation in $MnSiO_3$. *Amer. Min.*, **57**, 76–84.
Aoki, K-I., 1964. Clinopyroxenes from alkaline rocks of Japan. *Amer. Min.*, **49**, 1199–1223.
Berman, H., 1937. Constitution and classification of the natural silicates. *Amer. Min.*, **22**, 342–408.

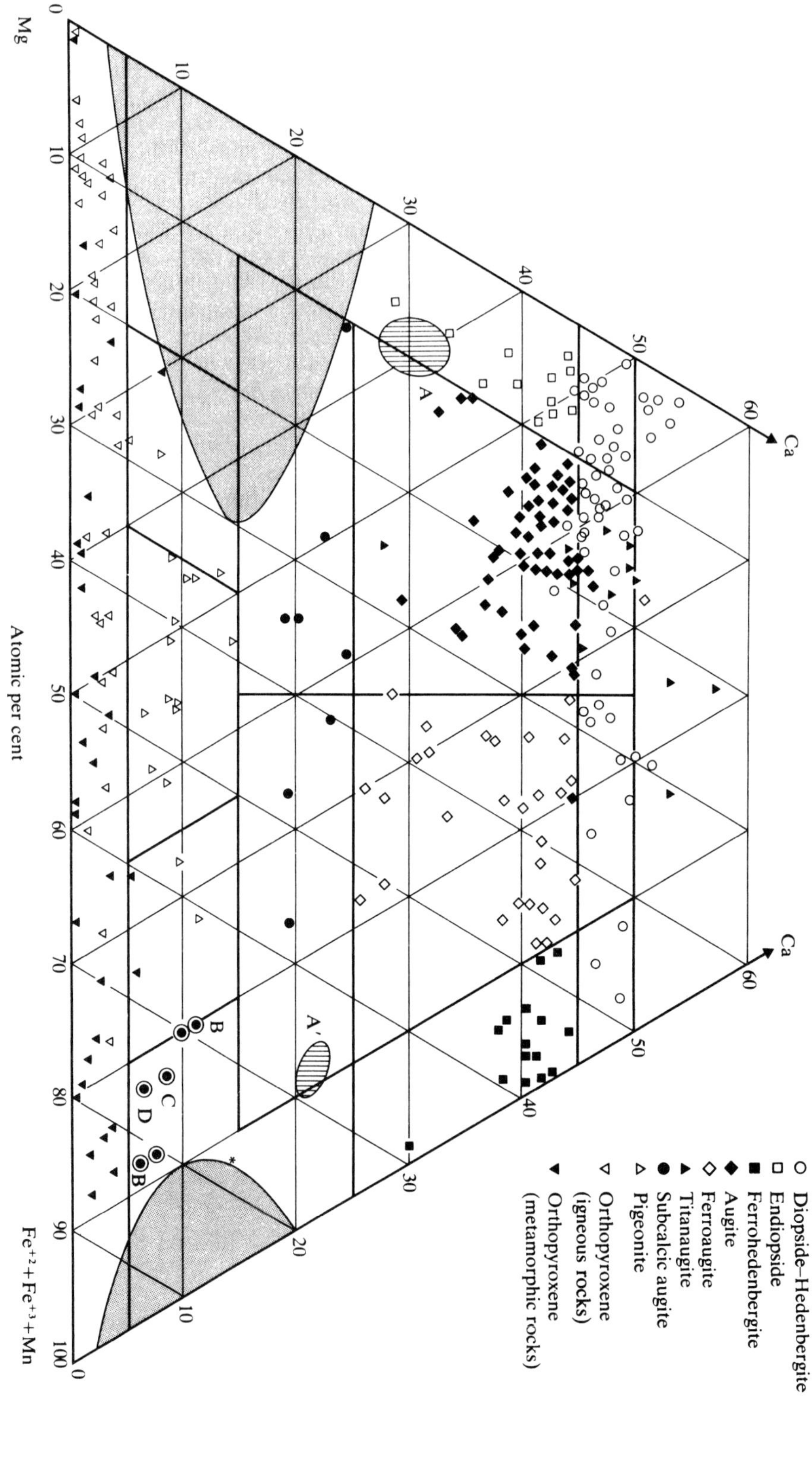
Mg
Ca
Ca
Fe+2+Fe+3+Mn
Atomic per cent
A
A′
B
C
D
B
Diopside–Hedenbergite
Endiopside
Ferrohedenbergite
Augite
Ferroaugite
Titanaugite
Subcalcic augite
Pigeonite
Orthopyroxene (igneous rocks)
Orthopyroxene (metamorphic rocks)

Borchert, W. and **Kramer, V.**, 1969. Ein synthetisches Magnesium-Kupfer-Silikat mit Klinopyroxene-struktur, $(Mg_4,Cu)_2[SiO_6]$. *Neues Jahrb. Min., Mh.*, 6–14.

Bown, M. G., 1964. Recalculation of pyroxene analyses. *Amer. Min.*, **49**, 190–194.

Bragg, W. L., 1937. *Atomic Structure of Minerals*. Cornell Univ. Press.

Brown, W. L., 1972. La symmetrie et les solutions solides des clinopyroxenes. *Bull. Soc. franç Min. Crist.*, **95**, 574–582.

Burnham, C. W., **Clark, J. R.**, **Papike, J. J.** and **Prewitt, C. T.**, 1967. A proposed crystallographic nomenclature for clinopyroxene structure. *Z. Krist.*, **125**, 109–119.

Cameron, M., **Sueno, S.**, **Prewitt, C. T.** and **Papike, J. J.**, 1973. High temperature crystal chemistry of acmite, diopside, hedenbergite, jadeite, spodumene and ureyite. *Amer. Min.*, **58**, 594–618.

Carmichael, I. S. E., 1962. Pantelleritic liquids and their phenocrysts. *Min. Mag.*, **33**, 86–113.

Cawthorn, R. G. and **Collerson, K. D.**, 1974. The recalculation of pyroxene end-member parameters and the estimation of ferrous and ferric iron content from electron microprobe analyses. *Amer. Min.*, **59**, 1203–1208.

Clark, J. R., **Appleman, D. E.** and **Papike, J. J.**, 1969. Crystal chemical characterization of clinopyroxenes based on eight new structure refinements. *Min. Soc. America, Spec. Paper*, **2**, 31–51.

Clark, J. R. and **Papike, J. J.**, 1968. Crystal-chemical characterization of omphacites. *Amer. Min.*, **53**, 840–868.

Coleman, R. G., **Lee, D. E.**, **Beatty, L. B.** and **Brannock, W. W.**, 1965. Eclogites and eclogites: their differences and similarities. *Bull. Geol. Soc. America*, **76**, 483–508.

Dobretsov, N. L., **Kochkin, Yn. N.**, **Krivenko, A. P.** and **Kutolin, V. A.**, 1971. [*Rock forming pyroxenes.*] Moscow (Akad. Nauk S.S.S.R.) (in Russian).

Dobretsov, N. L., **Khlestov, V. V.**, **Reverdatto, V. V.**, **Sobolev, N. V.** and **Sobolev, V. S.**, 1972. *The facies of metamorphism*, Volume I (Transl. from the Russian by D. A. Brown). Canberra (Australian Nat. Univ. Press), 315–331.

Edgar, A. D., **Mottana, A.** and **Macrae, N. D.**, 1969. The chemistry and cell parameters of omphacites and related pyroxenes. *Min. Mag.*, **37**, 61–74.

Edwards, A. C., 1976. A comparison of the methods for calculating Fe^{3+} contents of clinopyroxenes from microprobe analysis. *Neues Jahrb. Min., Mh.*, 508–512.

Essene, E. J. and **Fyfe, W. S.**, 1967. Omphacite in Californian metamorphic rocks. *Contr. Min. Petr.*, **15**, 1–23.

Ginzburg, I. V., 1975. [Refinement of the classification of pyroxenes from new data on their crystal chemistry.] *Zap. Vses. Min. Obshch.*, **104**, 539–546 (in Russian).

Hamm, H-M. and **Vieten, K.**, 1971. Zur Berechnung der Kristallchemischen Formel und des Fe^{3+}-Gehaltes von Klinopyroxen aus Elektronenstrahl-Mikroanalysen. *Neues Jahrb. Min., Mh.*, 310–314.

Hess, H. H., 1949. Chemical composition and optical properties of common clinopyroxenes, Part I. *Amer. Min.*, **34**, 621–666.

Jong, W. F. de., 1959. *General Crystallography*. San Francisco (Freeman).

Kushiro, I., 1962. Clinopyroxene solid solutions. Part I. The $CaAl_2SiO_6$ component. *Japan. J. Geol. Geogr.*, **33**, 213–220.

Makanjuola, A. A. and **Howie, R. A.**, 1972. The mineralogy of the glaucophane schists and associated rocks from Île de Groix, Brittany, France. *Contr. Min. Petr.*, **35**, 83–118.

Matsumoto, T. and **Banno, S.**, 1970. *P2/n*-omphacite and the possible space groups of clinopyroxenes. *IMA-IAGOD Meeting*, 1970. *Coll. Abstr.*, p. 171.

Morimoto, N., 1974. Crystal structure and fine texture of pyroxenes. *Fortsch. Min.*, **52**, 52–80.

Mysen, B. and **Griffin, W. L.**, 1973. Pyroxene stoichiometry and the breakdown of omphacite. *Amer. Min.*, **58**, 60–63.

Neumann, E-R., 1976. Two refinements for the calculation of structural formulae for pyroxenes and amphiboles. *Norsk. Geol., Tiddskr.*, **56**, 1–6.

Ohashi, Y. and **Burnham, C. W.**, 1972. Electrostatic and repulsive energies of the *M*1 and *M*2 cation sites in pyroxenes. *J. Geophys. Res.*, **77**, 5761–5766.

Onuma, K. and **Yagi, K.**, 1971. The join $CaMgSi_2O_6$–$Ca_2MgSi_2O_7$–$CaTiAl_2O_6$ in the system CaO–MgO–Al_2O_3–TiO_2–SiO_2 and its bearing on titanopyroxenes. *Min. Mag.*, **38**, 471–480.

Fig. 11. Distribution of pyroxene analyses with respect to Ca, Mg and ($Fe^{2+}+Fe^{3+}+$Mn) atoms (Tables 3–6, 13–20, 29–36). Stippled area A, pyroxenes of deformed lherzolite, griquaite and discrete nodules of the Thuba Putosa kimberlite pipe. Stippled area A′, outer zones of subcalcic ferroaugite in lunar ilmenite-rich dolerites and microgabbros, Apollo 11. B, pyroxenes of Nain adamellite and granodiorite. C, pyroxene of Bjerkrem–Sogndal anorthosite–mangerite intrusion. D, exsolved pigeonite, Skaergaard intrusion. Shaded areas, compositions not represented by natural minerals. * Ferropigeonite, outer zone of pyroxene in lunar dolerite.

Papike, J. J., **Prewitt, C. T.**, **Sueno, S.** and **Cameron, M.**, 1973. Pyroxenes: comparisons of real and ideal structural topologies. *Z. Krist.*, **69**, 254–273.
Peacor, D. R., 1968. The crystal structure of $CoGeO_3$. *Z. Krist.*, **126**, 299–306.
Poldervaart, A. and **Hess, H. H.**, 1951. Pyroxenes in the crystallization of basaltic magmas. *J. Geol.*, **59**, 472–489.
Prewitt, C. T. and **Peacor, D. R.**, 1964. Crystal chemistry of the pyroxenes and pyroxenoids. *Amer. Min.*, **49**, 1527–1542.
Smith, J. V., 1969. Crystal structure and stability of the $MgSiO_3$ polymorphs: physical properties and phase relations of Mg,Fe pyroxenes. *Min. Soc. America, Special Paper*, **2**, 3–29.
Takeda, H., 1972. Crystallographic studies of coexisting aluminian orthopyroxene and augite of high pressure origin. *J. Geophys. Res.*, **77**, 5798–5811.
Tauber, A. and **Kohn, J. A.**, 1965. Orthopyroxene and clinopyroxene polymorphs of $CoGeO_3$. *Amer. Min.*, **50**, 13–21.
Thompson, J. B., *Jr.*, 1970. Geometrical possibilities for amphibole structures. Model biopyriboles. *Amer. Min.*, **55**, 292–293.
Vogel, D. E., 1966. Nature and chemistry of the formation of clinopyroxene-plagioclase symplectite from omphacite. *Neues Jahrb. Min. Mh.*, 185–189.
Warren, B. E. and **Bragg, W. L.**, 1928. The structure of diopside $CaMg(SiO_3)_2$. *Z. Krist.*, **69**, 168–193.
White, A. J. R., 1964. Clinopyroxenes from eclogites and basic granulites. *Amer. Min.*, **49**, 883–888.
Whittaker, E. J. W., 1960. Relationships between the crystal chemistry of pyroxenes and amphiboles. *Acta Cryst.*, **13**, 741–742.
Yoder, H. S., *Jr.* and **Tilley, C. E.**, 1962. Origin of basalt magmas: an experimental study of natural and synthetic rock systems. *J. Petr.*, **3**, 342–532.
Zussman, J., 1968. The crystal chemistry of pyroxenes and amphiboles. I. Pyroxenes. *Earth-Sci. Rev.*, **4**, 39–67.

Magnesium-Iron Pyroxenes

Orthopyroxenes (Enstatite-Orthoferrosilite) $(Mg,Fe^{2+})_2Si_2O_6$

Pigeonite $(Mg,Fe^{2+},Ca)(Mg,Fe^{2+})Si_2O_6$

Orthopyroxenes

Enstatite-Ferrosilite (Mg,Fe)[SiO_3]

Orthorhombic (+) (−)

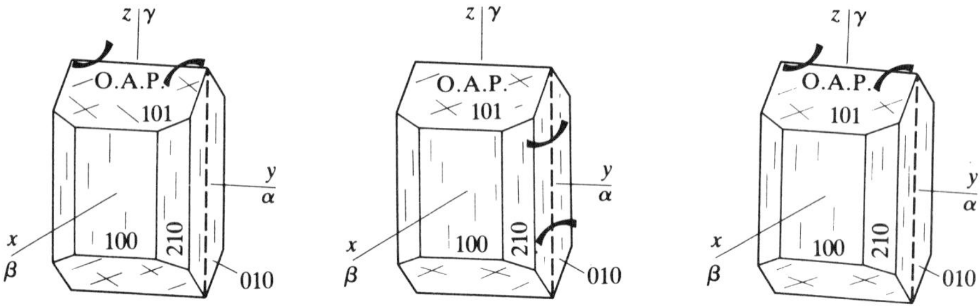

Enstatite (+) Hypersthene[1] (−) Orthoferrosilite (+)

	Enstatite	Orthoferrosilite
α	1·649–1·667	1·755–1·768
β	1·653–1·671	1·763–1·770
γ	1·657–1·680	1·772–1·788
δ	0·007–0·011	0·018–0·020
$2V_\gamma$	55°–90°[2]	55°–90°
	α = *y*, O.A.P. (100)	α = *y*, O.A.P. (100)
Dispersion:	*r* < *v*, weak to moderate (but see p. 113).	*r* < *v*, strong.
D	3·209	3·96
H	5–6	5–6
Cleavage:	{210} good; {100}, {010} parting. (210):(2$\bar{1}$0) ≃ 88°	{210} good; {100}, {010} parting. (210):(2$\bar{1}$0) ≃ 88°
Twinning:	{100} simple and lamellar.	{100}
Colour:	Colourless, grey, green, yellow, brown; colourless in thin section.	Green, dark brown; reddish or greenish in thin section.

Pleochroism: Enstatite none, for other compositions variable weak to strong:
α pale reddish brown, purple-violet, smoky brown, pink.
β pale greenish brown, pale reddish yellow, pale brown, yellow.
γ pale green, smoky green, green, grey-green, sky blue.

[1] Orthopyroxenes intermediate in composition between enstatite and ferrosilite are optically negative and have a continuous range of properties as shown in Fig. 74(a).

[2] A value of 48·5° is reported by Dobrokhotova *et al.*, 1967.

Unit cell:[1]	*a* Å 18·223	18·431
	b Å 8·815	9·080
	c Å 5·169	5·238
	Z = 16. Space group *Pbca*[2].	

Partially decomposed by HCl.

Orthopyroxenes occur commonly in basic and ultrabasic plutonic rocks, in basic and intermediate volcanic rocks and in high grade thermally and regionally metamorphosed rocks of both igneous and sedimentary origin: they are particularly characteristic constituents of norites and charnockites. Orthopyroxene is most probably an important constituent in the upper mantle, an environment in which it will be associated with magnesium-rich olivine, diopside, spinel, pyrope and possibly phlogopitic mica. A number of reactions between these phases at high pressures have been investigated and provide information on the pressure–temperature conditions in which different ultrabasic assemblages may have developed, as well as being pertinent to the understanding of the generation of basic liquids.

Orthopyroxenes form a diadochic series in which Mg and Fe^{2+} are mutually replaceable between $Mg_{100}Fe^{2+}_{0}$ and about $Mg_{10}Fe^{2+}_{90}$ at normal pressures. The polymorphism of $MgSiO_3$ (ortho-, clino-, and protoenstatite) is complex at both atmospheric and high pressures, and some uncertainties are still unresolved. Common practice has been to use the term enstatite rather than orthoenstatite for low-temperature orthorhombic $MgSiO_3$. Enstatite, and members of the enstatite–orthoferrosilite series are, with very few exceptions, the polymorphic form that occurs in both igneous and metamorphic rocks. Clinoenstatite occurs in high- and low-temperature structural forms. Protoenstatite is the stable form of $MgSiO_3$ above about 1 000°C at 1 atm.

Although Poldervaart's (1947) nomenclature for the series, viz. enstatite Fs_{0-10}, bronzite Fs_{10-30}, hypersthene Fs_{30-50}, ferrohypersthene Fs_{50-70}, eulite Fs_{70-90} and orthoferrosilite Fs_{90-100}, has been widely adopted it is not completely followed here, and the division between enstatite and bronzite, and between eulite and orthoferrosilite is placed at the change in optic sign, i.e. at Fs_{12} and Fs_{88}. Two crystallographic orientations have been used to describe the orthopyroxene minerals; in the orientation adopted here the longest axis is taken as the *a* axis, thus the optic axial plane is (100). Many orthopyroxenes, particularly those in igneous rocks, display oriented lamellae of a clinopyroxene and are inverted pigeonites. The name enstatite is from the Greek, *enstates*, opponent, in allusion to its refractory nature; hypersthene is also from the Greek *huper*, over, and *sthenos*, strength (i.e. greater hardness compared with hornblende with which it was originally confused). The theoretical iron end-member was named orthoferrosilite by Henry (1935), from the term ferrosilite suggested by Washington (1932) for the normative molecule, and is in conformity with the name clinoferrosilite for monoclinic $FeSiO_3$.

Structure

Warren and Modell (1930) investigated the crystal structure of a hypersthene (Fs_{30}) and showed that it contained the typical pyroxene $(SiO_3)_n$ chains similar to those

[1] Synthetic end-members (Turnock *et al.*, 1973).
[2] Some lunar orthopyroxenes have space group $P2_1ca$.

which had been found in diopside (p. 8). In hypersthene these chains are linked laterally by (Mg,Fe) atoms approximately in the positions of Mg and Ca in diopside. Since Mg and Fe ions are smaller than Ca, the configurations and stacking of chains are different. Two of the possible structures which can result are respectively orthorhombic, *Pbca*, for the Mg,Fe series orthoenstatite–orthoferrosilite, and monoclinic, $P2_1/c$, for the Mg,Fe series clinoenstatite–clinoferrosilite. The structures and unit cells of clinoenstatite and enstatite are compared in Fig. 8. The *a*-dimension of the orthorhombic cells is approximately double that of the monoclinic, thus:

a (orthopyroxene) $\simeq 2a\sin\beta$ (clinopyroxene) $\simeq$ 18·2 Å
b (orthopyroxene) $\simeq b$ (clinopyroxene) $\simeq$ 8·9 Å
c (orthopyroxene) $\simeq c$ (clinopyroxene) $\simeq$ 5·2 Å

The structure of almost pure enstatite (Fs_2, from Bamle, Norway: *a* 18·22, *b* 8·81, *c* 5·21Å, D 3·21) was shown by Byström (1943) to be similar to that of hypersthene. The structure of a bronzite was investigated by Ito (1935, 1950), (see also Takane, 1932) who studied the extent to which the structure of orthopyroxene could be regarded as that of diopside twinned on a unit cell scale. The parts of the twin would be related by a glide-plane (100) with translation $b/2$.

In more recent years, accurately refined crystal structures have been determined for a number of orthopyroxenes and for clinoenstatite, clinoferrosilite and pigeonites. Morimoto and Koto (1969) studied an almost pure Mg end-member orthoenstatite from the Bishopville meteorite and compared this with the previously determined structure of clinoenstatite (Morimoto *et al.*, 1960) derived by heating orthoenstatite from the same source. Other determinations of the structures of orthopyroxenes include those by Ghose (1965) – (Fs_{53}); by Burnham (1967) – pure ferrosilite, Burnham *et al.* (1971) – an orthopyroxene ($Ca_{0{\cdot}04}Mg_{0{\cdot}26}Fe_{1{\cdot}70}$) from quartz–garnet–orthopyroxene granulite; Takeda (1972) – (bronzites and an aluminian bronzite of high pressure origin), Takeda and Ridley (1972) – lunar orthopyroxenes; Smyth (1973) – hypersthene at various temperatures; Kosoi *et al.* (1974) – two orthopyroxenes from granulites; Miyamoto *et al.* (1975) – bronzite from an achondrite; and Sueno *et al.* (1976) – ferrosilite at various temperatures.

The structures of all of these orthopyroxenes show certain characteristic features. Unlike diopside, where all Si–O chains are equivalent, in orthopyroxenes there are two kinds of Si–O chain which show significantly different configurations, one (A chain) being more fully extended than the other (B chain); the (B) tetrahedra are larger than the (A) tetrahedra. The tetrahedra of both chains have *O* rotations (see p. 9). The *M*1 cation is co-ordinated in an almost regular octahedron whereas the *M*2 cation has a distorted octahedral co-ordination with two (*M*–O3) out of the six *M*–O distances greater than the other four (Fig. 12). In the Si–O chains bridging bonds (Si to an oxygen linked to another tetrahedron) are longer than non-bridging (Yin *et al.*, 1971 showed by photoelectron spectroscopy that bridging oxygens have lower binding energy than non-bridging oxygens).

The crystal structure of an orthopyroxene (hypersthene) $Mg_{0{\cdot}30}Fe_{0{\cdot}68}Ca_{0{\cdot}015}$ has been determined at 20°, 175°, 280°, 500°, 700° and 850°C by Smyth (1973). The cell parameters vary smoothly with increasing temperature and the space group remains *Pbca*. As shown by similar studies for other pyroxenes, the Si–O distances decrease with increasing temperature, while *M*1–O and *M*2–O distances expand appreciably (approximately 0·25 per cent per 100°C). To adapt to this difference in behaviour the Si–O chains become less kinked and more alike. For the (A) chain O3–O3–O3 increases from 167·2° to 172·1° and for (B) from 144·5° to 159·3°.

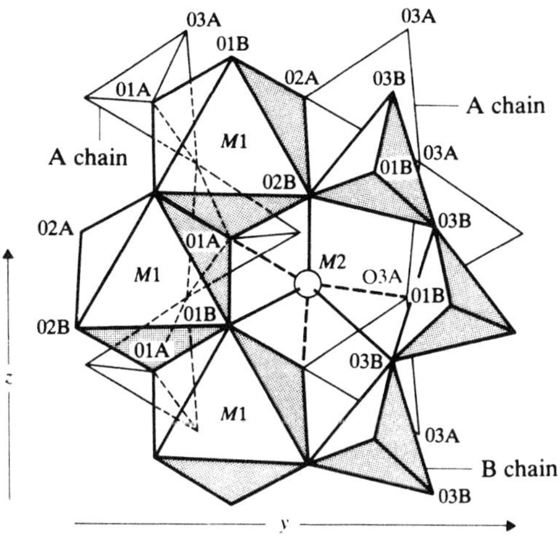

Fig. 12. Projection along *a of part of the orthoferrosilite structure, showing *M*1 and *M*2 co-ordination, typical for orthopyroxenes at room temperature (Sueno *et al.*, 1976).**

The co-ordination of the *M*2 site changes from 4+2 to 4+3; the average of the four shorter bonds remains nearly constant with increasing temperature while the average of all seven distances shows a strong expansion.

Smyth (1973) makes an interesting comparison between the behaviour of ortho- and clinopyroxenes of the same composition (Smyth and Burnham, 1972). In the $P2_1/c$ clinopyroxene the (A) tetrahedra are *S*-rotated and (B) tetrahedra are *O*-rotated (notation after Thompson, 1970). With increasing temperature the *P* structure gains symmetry; tetrahedra of the straightening A chains rotate past O3–O3–O3 − 180° to become *O*-rotated and equivalent to the (B) chains. In orthopyroxene such a displacive change to higher symmetry is not possible. Whereas *M*2 in orthopyroxene goes from six- to seven-fold co-ordination, in clinopyroxene it goes from six to seven and back to a small six-fold site again at 950°C. Smyth suggests that this inability to regain a small *M*2 site at high temperature contributes to the instability of orthopyroxene and its reconstructive inversion to high clinopyroxene.

The crystal structure of orthoferrosilite has been determined at 24°, 400°, 600°, 800°, 900° and 980°C by Sueno *et al.* (1976). Structural changes were regular and no transformation occurred. Mean Si–O distances decrease slightly with increasing temperature, more so for the B tetrahedra than for A, thus decreasing the size difference for the two kinds of tetrahedra. The Fe polyhedra both expand appreciably, and associated with the expansion of the *M*2 site, both Si–O chains become less kinked. The A chain becomes almost completely straight and the difference between the A and B chains is reduced. The variation of the O3–O3–O3 angles of six clinopyroxenes and of the A and B chains of orthoferrosilite is shown in Fig. 13.

Thermal expansion coefficients for orthoferrosilite decrease in the order $\alpha_c > \alpha_a > \alpha_b$, a different sequence as compared with $C2/c$ pyroxenes (p. 200), and $P2_1/c$ pyroxenes (p. 30).

Orthopyroxenes ($En_{86}Fs_{11}Wo_3$) with space group $P2_1ca$ have been reported by Smyth (1974b) in a lunar troctolite-granulite and by Steele (1975) ($En_{78}Fs_{19}Wo_3$) in a lunar norite. Smyth suggested that its occurrence was related to the inferred extremely slow cooling history of the rock. Another such pyroxene ($En_{90}Fs_{10}$) has been observed in a terrestrial harzburgite (Krstanovic, 1975). The space group is one which was predicted as the ideal for orthopyroxene by Thompson (1970): (see also Papike *et al.*, 1973). Other possible structure types were listed by Matsumoto (1974). In the orthopyroxene with $P2_1ca$ there are four distinct Si–O chains as compared with two in the more usual *Pbca* pyroxene.

In each member of the orthopyroxene series (other than end-members), some ordering of (Fe,Mg) as between *M*1 and *M*2 sites was observed, with Fe preferring *M*2. Ordering of this nature has been detected and measured also by Mössbauer spectroscopy (Ghose, 1965; Bancroft *et al.*, 1967; Virgo and Hafner, 1969, 1970; Dundon and Hafner, 1971; Saxena and Ghose, 1970, 1971; Burnham *et al.*, 1971; Mitra, 1976). Where the X-ray and Mössbauer methods were applied to the same specimen, consistent results were obtained. In his determination of the structures of hypersthene at various temperatures up to 850°C, Smyth (1973) found that the (Mg,Fe) distribution becomes increasingly disordered with rising temperature beyond 500°C. Distribution constants and ΔG values are given in Table 2.

The extent of disorder can be correlated with temperature of origin for naturally heated samples too. Thus the orthopyroxene studied by Ghose (1965), of low-temperature origin, was found to be almost completely ordered, as was also that

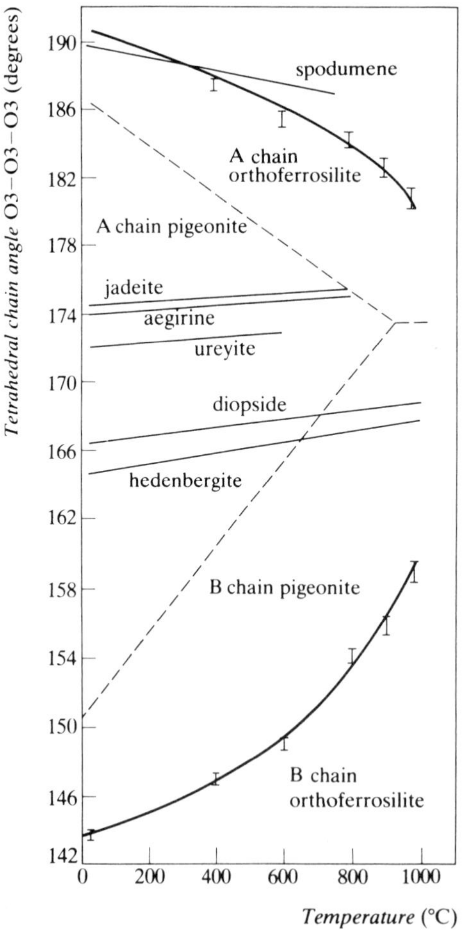

Fig. 13. Variation with increasing temperature of the O3–O3–O3 angles of six end-member clinopyroxenes and of the A and B chains in orthoferrosilite and pigeonite. The angle in the A chain for orthoferrosilite is plotted above 180° and the B chain below 180° to maintain analogy with the A and B chains in pigeonite despite the fact that both chains are *O*-rotated in the *Pbca* structure (after Sueno *et al.*, 1976).

Table 2 ***M*1, *M*2 occupancies, *k* and Δ*G* for Mg,Fe exchange in an orthopyroxene $Mg_{0.30}Fe_{0.68}Ca_{0.015}$** (Smyth, 1973)

Refinement temperature (°C)	*M*1 Mg	Fe	*M*2 Mg	Fe	Ca[a]	*k*[c]	Δ*G* (kcal/mole)
20	0·574(3)[b]	0·425	0·062	0·906	0·032		
175	0·576(4)	0·423	0·059	0·909	0·032		
280	0·574(4)	0·426	0·042	0·925	0·032		
500	0·576(5)	0·423	0·059	0·909	0·032	0·073(10)	4·01(11)
20° after 500	0·553(5)	0·447	0·083	0·885	0·032	0·109(10)	3·41(11)
700	0·512(5)	0·488	0·124	0·844	0·032	0·176(10)	3·35(11)
850	0·493(16)	0·507	0·143	0·825	0·032	0·218(30)	3·39(30)

[a] Occupancy fixed from chemical analysis.
[b] Parenthesized figures represent the standard deviation in terms of least units cited for the value to their immediate left. All Mg,Fe occupancies at a given temperature have the same standard deviation.
[c] $k = (Fe/Mg)_{M1}/(Fe/Mg)_{M2}$.

investigated by Burnham *et al.* (1971). The meteoritic bronzite studied by Miyamoto *et al.* (1975) showed a high degree of order, suggesting a slow cooling process. Mössbauer investigations showed more disorder for volcanic than for metamorphic and plutonic orthopyroxenes in the range Fs_0–Fs_{60} (Virgo and Hafner, 1970) but not as much disorder as might be expected for a mineral rapidly quenched from liquidus temperatures. However, the activation energy for the interchange of Fe and Mg between *M*1 and *M*2 is apparently low and exchange is rapid so that maximum disorder will only be found in rocks which have cooled extremely rapidly from high temperatures. The activity–composition relationship in orthopyroxenes has been calculated for various temperatures by Saxena and Ghose (1970, 1971), who found that the experimentally determined (Mg,Fe) distributions for heated pyroxenes were in better agreement with a 'simple mixture' than with an 'ideal solution' model for the sites (see also Mueller, 1970). Ideality is approached with increasing temperature and is probably achieved by 1000°C. The results show that for the more iron-rich pyroxene there is little change in site occupation between 500° and 800°C.

Snellenburg (1975) computed the ordering of Fe in orthopyroxenes with different (Fe/Mg/Ca/Mn) contents assuming a nearest neighbour site exchange mechanism and obtained reasonably good agreement with observed distributions. His results suggest that increased disorder in Fe-rich samples is to be expected through the 'blocking' action of Ca or Mn in the *M*2 site, but that Ca may not hinder ordering in Fe-poor samples.

O'Nions and Smith (1973) discussed the bonding in orthopyroxenes, and in particular the bonding to *M*2 atoms, in terms of a molecular orbital model and hold that this provides a better explanation of the extent to which site preference is shown by Fe^{2+}.

From his study of an aluminian bronzite from Takasima, North Kyushu, Japan, Takeda (1972) showed that the tetrahedral site in one of the two pyroxene chains (A) is inherently smaller than in the other. Consequently aluminium is not distributed randomly between the chains but shows preference for the (B) chain. He also showed that aluminium and ferric iron prefer the smaller of the octahedral sites (*M*1).

The structures of two synthetic high-pressure polymorphs of $ZnSiO_3$ have been determined by Morimoto *et al.* (1975). Monoclinic $ZnSiO_3$ had the space group *C*2/*c* with *a* 9·787, *b* 9·161, *c* 5·296 Å, *β* 111·42°, Z = 8; orthorhombic $ZnSiO_3$ is

Pbca with *a* 18·204, *b* 9·087, *c* 5·278 Å, Z = 16. Zn in *M*2 is co-ordinated tetrahedrally in the monoclinic polymorph and in a distorted octahedron in the orthorhombic polymorph. The monoclinic structure shows differences compared with those of most *C*2/*c* pyroxenes, being closest to that of spodumene, whereas the orthorhombic structure is very similar to that of enstatite. The unit cell twin relation between *C*2/*c* and $P2_1/c$ (Mg,Fe) pyroxenes is not applicable to $ZnSiO_3$.

The structure of synthetic $ZnMgSi_2O_6$ (*Pbca*, *a* 18·201, *b* 8·916, *c* 5·209 Å, Z = 16), and that of another (Zn,Mg) orthopyroxene were determined by Ghose *et al.* (1974). In these structures Zn shows some preference for *M*2.

Tauber *et al.* (1963) and Tauber and Kohn (1965) synthesized orthopyroxenes $MnGeO_3$ with *a* 19·29, *b* 9·25, *c* 5·48 Å, and $CoGeO_3$ with *a* 18·77, *b* 8·99, *c* 5·35 Å, both with space group *Pbca*. The latter has a clinopyroxene polymorph above 1351°C. A synthetic orthopyroxene $MgGeO_3$ with *a* 18·661, *b* 8·954, *c* 5·346 Å has also been reported (Tauber *et al.*, 1963).

In the synthetic solid solution series $MgSiO_3$–$CuSiO_3$ (Borchert and Kramer, 1969) the cell dimensions increase with increasing Cu content, *a* 9·672–9·713,

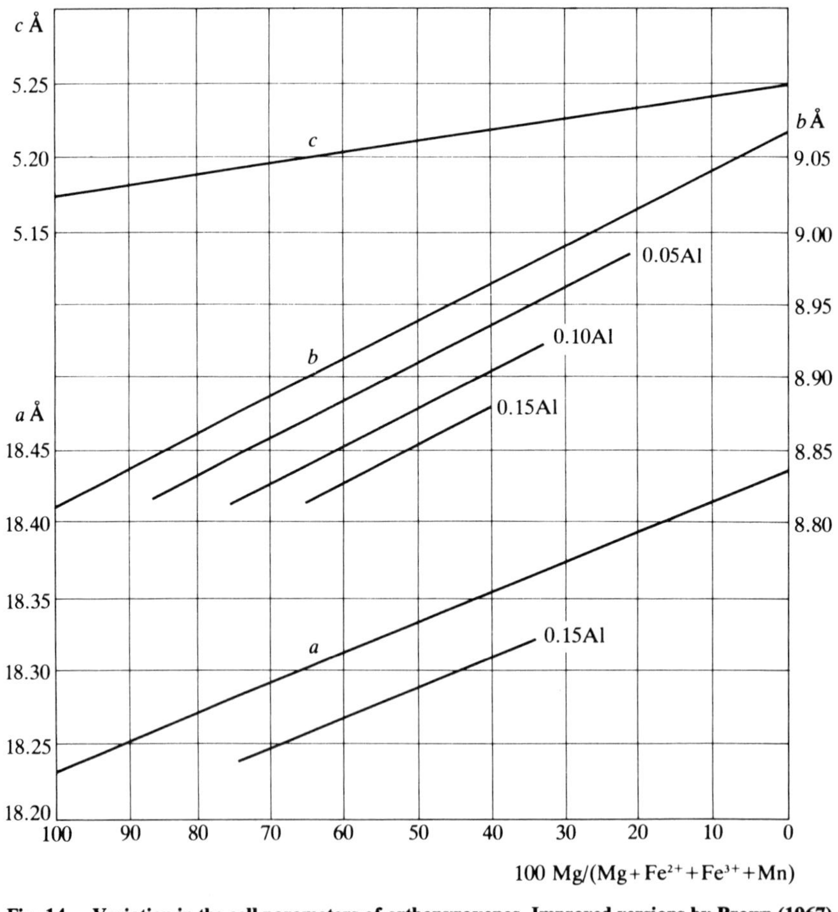

Fig. 14. Variation in the cell parameters of orthopyroxenes. Improved versions by Brown (1967) of plots given by Hess (1952) and Kuno (1954), using additional data from Howie (1963). For *a* and *b* the extra curves are for high contents of Al ions in octahedral sites (values quoted on basis of six O in the formula).

b 8·883–8·906, *c* 5·200–5·222 Å, *β* 109°35′–110°22′. X-ray powder data are given for the composition $Mg_{1\cdot31}Cu_{0\cdot69}Si_2O_6$.

Cell Parameters

The substitution of iron for magnesium in orthopyroxenes increases the cell parameters but has less effect on *c* than on *a* and *b*. The precise variation of cell parameters with (Mg,Fe) composition was first studied by Ramberg and DeVore (1951). They found that the relationship was not linear and from this and optical data they inferred that the orthopyroxenes do not form an ideal solid solution. Hess (1952) and Kuno (1954) gave straight line relationships and related deviations from the lines to the presence of aluminium or calcium. The principal effect of aluminium was to reduce *b* and of calcium to increase *a*. Howie (1963, 1964) plotted cell parameters for metamorphic orthopyroxenes of the granulite facies and confirmed the considerable effect of aluminium, as did also the work of Skinner and Boyd (1964) on synthetic enstatites. Stephenson *et al.* (1966) pointed out that the earlier work included an incorrect assignment of the 004 indices giving rise to errors in the *c* parameters. Figure 14 shows straight line plots for *a*, *b* and *c* given by Brown (1967), based upon previously published data.

Conclusive evidence of the non-ideal nature of the orthopyroxene series arising from the site preference of Fe for *M*2 has been referred to above. The different(Mg,Fe) distributions resulting from different thermal histories will be one factor therefore influencing cell parameters in addition to the presence of substituents such as calcium and aluminium. In the case of aluminium, there is the further complication that its effects on cell parameters differ according to whether it is in tetrahedral or octahedral sites.

LeFèvre (1969) showed for a suite of volcanic pyroxenes that *b* is affected by aluminium in tetrahedral as well as in octahedral co-ordination. Because of the different ways in which the aluminium content of an orthopyroxene can vary with increasing iron content, a plot of *b* against Fe will give a different straight line for different rock suites (Fig. 15). In general the Fe content cannot be determined from a

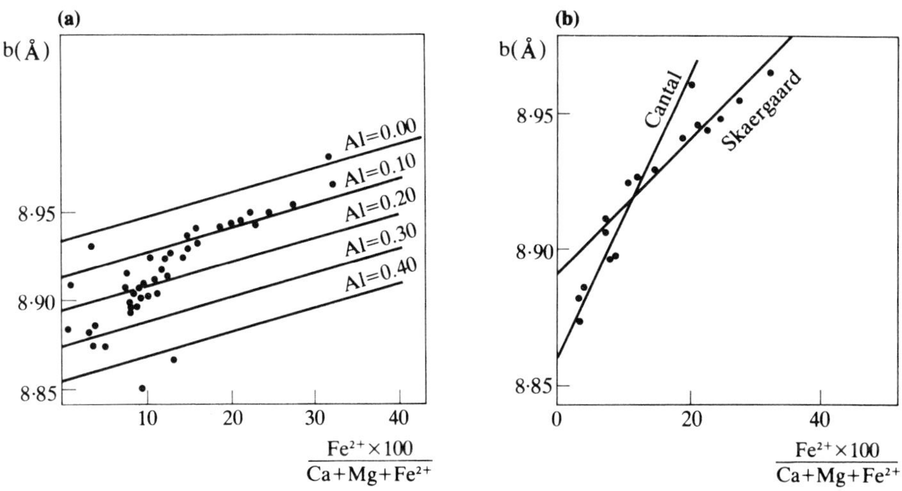

Fig. 15. (a) Values of *b* for orthopyroxenes as a function of Fe,Mg substitution and Al (atoms per six oxygens), based upon published analyses (LeFèvre, 1969). (b) Plots of *b* against Fe content by LeFèvre (1969) for orthopyroxenes of two rock suites; Cantal (LeFèvre, 1969); Skaergaard (Brown, 1960).

measurement of *b* but for a given suite of rocks where aluminium can be expected to vary systematically, a straight line plot of Fe against *b* can be used determinatively.

Winchell and Leake (1965) calculated regression formulae for the variation of cell parameters with composition based upon available analyses of natural orthopyroxenes. Smith *et al.* (1969) found, using correctly indexed X-ray data and accurate electron probe analyses of mostly metamorphic orthopyroxenes, that linear regression analysis (on Mg,Al and Ca) was only satisfactory if the two halves of the orthopyroxene series, Fs_{0-50},Fs_{50-100}, were treated separately (see also Kosoy and Shemyakin, 1971, also Optics section), and that for the series as a whole a term in $(Mg)^2$ has to be introduced. While earlier data on *a* and *b* for plutonic pyroxenes were in reasonable agreement with the newer result, for volcanic pyroxenes *a* and *b* were higher for a given composition, probably because of site preference variations associated with different petrogenetic conditions. Although qualitatively the effects of Ca and Al on cell parameters were confirmed, the regression analyses showed that the coefficients for these elements can vary considerably with Mg–Fe ratio.

It is concluded that the variables affecting cell parameters are so many that even with sophisticated regression analysis procedures, the accuracy of prediction of composition is not great. With the increased availability of direct analysis by electron probe this is clearly the best method to use. In the absence of this instrument, the straight line plots of *a* and *b* with corrections for Al and Ca (Fig. 14) are perhaps still good enough to determine the compositions of igneous orthopyroxenes in routine petrography.

A useful tabulation of cell parameters and indexed X-ray diffraction 2θ (Cu) values for natural orthopyroxenes is given by Smith *et al.* (1969), and powder patterns with intensities calculated from the known crystal structures are given by Borg and Smith (1969).

The effect of Fe–Mg ordering on the cell parameters and 2V were studied by Tarasov and Nikitina (1974). For a hypersthene, where the effect of ordering can be expected to be most marked, they show a considerable change of *a*, *b* and 2V with temperature of equilibrium (T°C). Thus $a = 18{\cdot}306 + 0{\cdot}0000413T$, $b = 8{\cdot}923 + 0{\cdot}0000286T$, $2V(-) = 54{\cdot}8 + 0{\cdot}0172T$.

The effects on cell parameters of Mg,Fe distribution alone is also shown by the work on synthetic Mg,Fe orthopyroxenes by Matsui *et al.* (1968) and by Turnock *et al.* (1973) who also studied the effect of adding Ca up to 5 per cent Wo. These show the deviation from linearity for *a*, *b* and *c*, with a smaller deviation for cell volume because of compensating effects. As deduced (though less rigorously) from natural pyroxenes, the addition of Ca significantly increases *a* but has no detectable effect on *b* and *c*. Different results are obtained by the two sets of workers and these may result from the margins of error, or may reflect a real difference in ordering due to different synthesis procedures.

Exsolution in orthopyroxene. The mechanism of exsolution of clino- in orthopyroxene has been studied by electron microscopy (e.g. Champness and Lorimer, 1973; 1974; 1976; van der Sande and Kohlstedt, 1974; Kohlstedt and van der Sande, 1976). For this exsolution the mechanism of spinodal decomposition is not possible since the two phases concerned have different structures. Figure 16 shows a typical microstructure where the orthopyroxene matrix is seen to contain coarse augite lamellae but also fine Ca-rich platelets in Guinier–Preston zones. The inversion of pigeonite to orthopyroxene has also been studied by electron microscopy (e.g. Champness and Copley, 1976). (For further discussion of exsolution in orthopyroxenes, see p. 56 and p. 115).

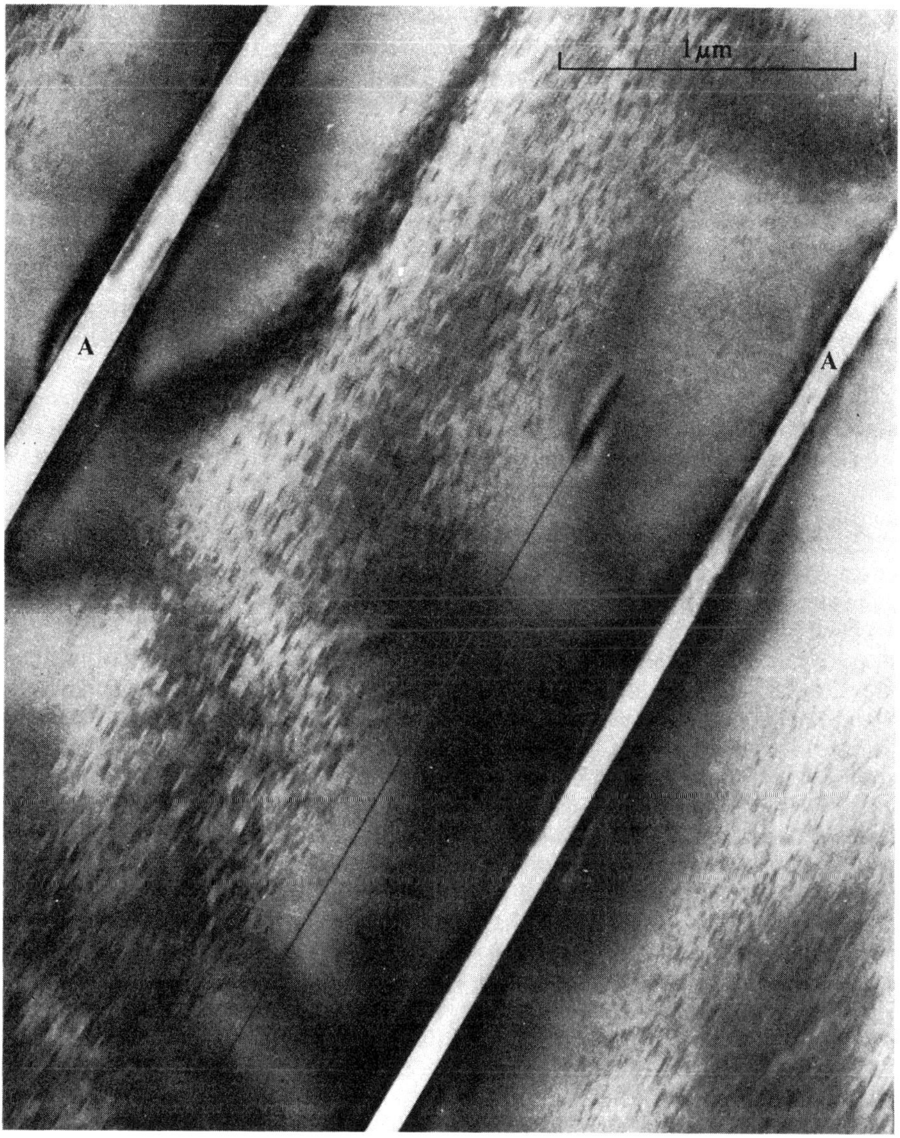

Fig. 16. Electron micrograph showing large (100) augite lamellae (A) and a fine distribution of Ca-rich platelets in orthopyroxenes (Champness and Lorimer, 1973).

Protoenstatite

Protoenstatite is an orthorhombic high-temperature form of $MgSiO_3$ which can, however, persist metastably at lower temperatures (see Kushiro 1972c; Smyth 1974a). The prediction of the structure by Atlas (1952) was shown to be correct by Smith (1959) using X-ray powder data. Lindemann (1961) gave the symmetry as monoclinic $P2_1/n$ (Pc or $P2/c$; Lindemann and Wögerbauer, 1974), but single-crystal studies at high temperature by Sadanaga *et al.* (1969) confirmed the space group as *Pbcn*. Smyth (1971) has also determined the crystal structure at high temperature and gave the space group as $P2_1cn$, a subgroup of *Pbcn*, but the specimen showed a high degree of disorder. The pyroxene chain in protoenstatite is fully extended and, as in diopside, all chains are crystallographically equivalent and have *O* rotations. The Mg–O distances in the *M*1 and *M*2 octahedra are not as different as was once thought, but the O–O distance and O–Mg–O angles show that *M*1 is approximately regular whereas the *M*2–O octahedron is distorted. Since one of the structure determinations was from limited powder data and the other from a disordered specimen, understanding of this structure can still not be regarded as complete.

Cell dimensions of a synthetic protoenstatite at room temperature were given by Smith (1959) as *a* 9·252, *b* 8·740, *c* 5·316 Å, and at 1100°C for a heated orthoenstatite by Smyth (1971) as *a* 9·304, *b* 8·902, *c* 5·351 Å.

The protoenstatite structure and unit cell are compared with those of clinoenstatite and enstatite in Fig. 8. The structure is thought to be restricted to the Mg-rich corner of the Mg–Fe–Ca triangle.

Clinoenstatite–Clinoferrosilite

There are four polymorphs of $MgSiO_3$: enstatite, protoenstatite, low and high clinoenstatite. The first two are orthorhombic and the latter two monoclinic. The structure of low clinoenstatite (Morimoto *et al.*, 1960) is in general similar to that of diopside but many differences occur because the *M*2 site is occupied by Mg rather than Ca. The space group of low clinoenstatite, like that of pigeonite (p.163) is $P2_1/c$. The screw axes of diopside's space group $C2/c$ remain in $P2_1/c$ but the diads are absent. The *M*1 and *M*2 cation sites both have six-fold co-ordination and this results in distortion of the Si–O chains (tetrahedra of A and B chains have *S* and *O* rotations respectively), displacement of *M*1 and *M*2 from the positions on the diad axes and displacement of the chains relative to the *M* sites (Fig. 17). The deviations from the diopside structure are greater in clinoenstatite than in pigeonite. The cell parameters of low clinoenstatite were given by Stephenson *et al.* (1966) as *a* 9·6065, *b* 8·8146, *c* 5·1688 Å, β 108·335°. They noted that the cell volume is smaller than for enstatite.

The existence of a high-temperature (above 980°C) monoclinic polymorph of $MgSiO_3$ was demonstrated by Perrotta and Stephenson (1965). Its cell parameters at 1 100°C are *a* 9·864, *b* 8·954, *c* 5·333 Å, β 110·03° (Smith, 1969b), and its space group is $C2/c$. The structure of a high clinoenstatite has been determined by Sadanaga and Okamura (1971).

Clinoferrosilite is isostructural with low clinoenstatite and has cell parameters *a* 9·709, *b* 9·087, *c* 5·228 Å, β 108·432° (Burnham, 1966, 1967).

The polymorphism exhibited by $MgSiO_3$ appears to extend across the Mg–Fe series. Thus, in addition to the work on $FeSiO_3$ by Burnham (1966, 1967), there has been a crystal structure determination for a clinohypersthene ($En_{31}Fs_{67}Wo_{1\cdot5}$) by Smyth and Burnham (1972) both for the low-temperature $P2_1/c$ and the high-temperature $C2/c$ polymorphs, and Smyth (1974a) has determined the structure at various temperatures below and above the inversion.

The $P2_1/c \rightarrow C2/c$ inversion is a first-order transition as shown by the discontinuity in the changes of all cell parameters with temperature occurring between 720° and 760°C, and by the behaviour of the $h+k$ odd X-ray reflections which remain sharp and observable up to the temperature of the transition determined as 725 ± 10°C. Also, the configuration of the Si–O (A) chain undergoes a sudden

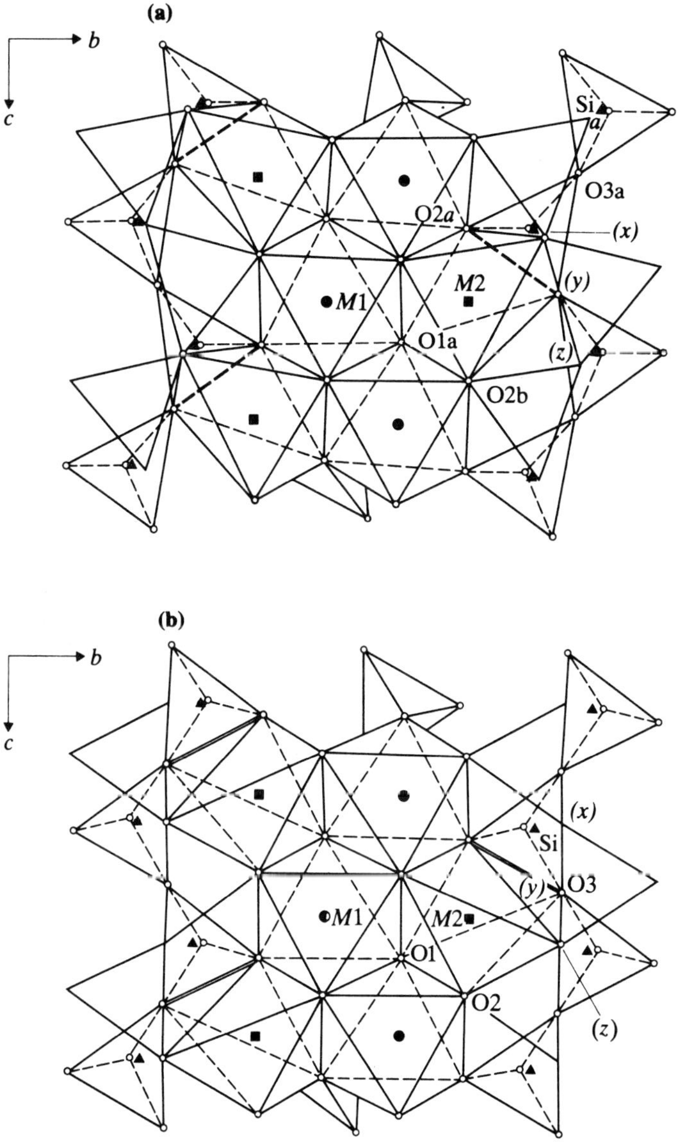

Fig. 17. (a) Projection down *a of the room-temperature structure of clinohypersthene. *M*2 is co-ordinated to O3 atoms labelled (x) and (y), and the *M*2 polyhedron shares only one edge with an adjoining Si–O tetrahedron. (b) Projection down *a** of the high-temperature structure. *M*2 is co-ordinated to O3 atoms labelled (y) and (z) rather than (x) and (y), and the *M*2 polyhedron shares two equivalent O2–O3 edges with adjoining Si–O tetrahedra (Smyth and Burnham, 1972).**

change around the transition temperature. The observation here of a sharp transition is in contrast to that found for high–low pigeonite (Prewitt *et al.*, 1971, Brown *et al.*, 1972). and for the clinohypersthene no hysteresis effect was observed (see p. 165). Smyth (1974a) suggests that the different amounts of Ca in the two pyroxenes and its distribution may be responsible for the different behaviour. The refinements of distinct crystal structures immediately above and below the transition confirms the conclusion of Brown *et al.* (1972) that the $C2/c$ structure is real and does not represent a time or space average of the two chain configurations of $P2_1/c$ structure.

The $M1$ and $M2$ sites are more similar in high clinohypersthene than in the orthopyroxene but the former nevertheless shows a slightly higher degree of (Fe,Mg) ordering with a higher ΔG of exchange. It appears that the $P2_1/c$ to $C2/c$ transition has little effect on ΔG but the effect of Ca content may be significant.

Comparison of the $P2_1/c$ and $C2/c$ structures shows that the $M2$ co-ordination does not change from six-fold to the diopside-like eight-fold co-ordination on heating; it does become more like the diopside co-ordination but remains six-fold (see Fig. 17(a) and (b)). The $M2$ polyhedron expands while Si–O tetrahedra do not, and the Si–O (B) chains become almost fully extended and equivalent to A chains. The O3–O3–O3 (B) angle, which is a measure of the straightness of the pyroxene chain changes from 145° to 166·4° while the A chain tetrahedra are rotated in the opposite sense from A tetrahedra in low clinoenstatite. Because the structure, though $C2/c$, remains different from that of the high-temperature Ca-rich pyroxenes, Smyth and Burnham (1972) suggest that the Ca-rich–Ca-poor solvus extends above the $P2_1/c$–$C2/c$ transition temperature. Nevertheless, the work of Cameron *et al.* (1973) helps to show that although the difference between sizes of the $M2$ sites in Ca-rich and Ca-poor pyroxenes remains much the same with heating, the details of co-ordination become more similar, thus increasing solid solution. Studies of the high- and low-temperature polymorphs of pigeonite yield similar results (see p.165).

Stability Fields of $MgSiO_3$ Polymorphs

The problem of the stability field of clinoenstatite has received much attention but until recently was still enigmatic. The early work includes that by Bowen and Schairer (1935), Foster (1951), Atlas (1952), Lindemann (1961), Sarver and Hummel (1962) and Brown and Smith (1963). Probably a clearer picture is emerging with the work of Smyth (1974a) and of Iijima and Buseck (1975). By single-crystal X-ray studies at high temperatures, Smyth (1974a) has shown that protoenstatite is the stable form from 1000° to 1557°C, the incongruent melting point. Below about 1000°C, orthoenstatite is the stable form down to at least 600°C and possibly to room temperature. If protoenstatite is quenched in a few seconds, twinned clinoenstatite results, but with slower cooling some enstatite is formed. Smyth suggests that if cooling is very slow a well-ordered enstatite should be produced. Smyth shows that the proto–clino and clino–proto transitions are of a rapid martensitic nature whereas proto–ortho and ortho–proto are relatively slower order/disorder-type transitions. Below 600°C it is possible that low clinoenstatite is the stable form although at these temperatures the reaction needs to be aided by shear stress. It had been suggested (Kuno, 1966) that high clinoenstatite might have a stability field above that of protoenstatite, but none was discovered in the experiment by Smyth.

Iijima and Buseck (1975) show by high resolution electron microscopy that most specimens of orthoenstatite contain regions of clinoenstatite in them, and the relation is described by them as one of twinning.

Although the evidence available is more favourable to clinoenstatite being the lower temperature form than enstatite (Boyd and England, 1965; Stephenson *et al.*, 1966; Grover, 1972), there is probably very little energy difference between the two forms. Papike *et al.* (1973) suggested that the $P2_1/c$ structure represents a metastable state because of the ease of transition from the high-temperature *C*2/*c* structure to $P2_1/c$ rather than to *Pbca* at low temperature which would involve a greater amount of bond breaking.

The relationships between $P2_1/c$ and *Pbca* structures and stabilities appear to be less uncertain for more iron-rich pyroxenes, where there is also some Ca content, in that pigeonites are clearly the high-temperature form relative to orthopyroxene.

Clinoenstatite may be produced by inversion from protoenstatite (e.g. Dallwitz *et al.*, 1966), but it can also be produced by shear deformation of enstatite, as has been shown by several workers (e.g. Turner *et al.*, 1960; Akimoto *et al.*, 1966; Riecker and Rooney, 1967; Munoz, 1968; Borg and Handin, 1966; Raleigh *et al.*, 1971). Buseck and Iijima (1975) show how features of the high resolution electron microscope image may be used to distinguish clinoenstatite of these two different geological origins. They also show that for at least one specimen it is likely that clinoenstatite was produced by transition from enstatite without shear stress.

Mechanisms of transformation between the different polymorphs of $MgSiO_3$ have been discussed by many authors (Brown *et al.*, 1961; Sadanaga *et al.*, 1969; Coe, 1970; Smyth, 1974a). They all involve relative movements of slabs of Si–O chains, breaking *M*–O but not Si–O bonds. Some movement of *M*1 and *M*2 cations is also involved, and Smyth suggests that in $MgSiO_3$ these can easily be interchanged. In all of the above processes close orientational relationships are preserved.

Coe's thermodynamic approach led him to predict that the ortho–clino transition temperature is raised by about 200°C per kilobar of stress on (100) parallel to [001], and he suggests that the rareness of clinoenstatite in naturally deformed terrestrial rocks may mean that shear stresses are rarely large enough to produce the transition. Raleigh *et al.*, found that enstatite slips and kinks rather than transforms if strain rates are low enough and temperatures are high enough. Trommsdorff and Wenk (1968) found clinoenstatite in kinked orthoenstatite. The crystallographic orientation of clinoenstatite produced by deformation has been studied by electron microscopy (e.g. Coe and Müller, 1973; Kirby and Coe, 1974; Kirby, 1976) confirming the modes of transformation outlined above.

Clinoferrosilite probably bears a similar relationship to orthoferrosilite as does clinoenstatite to enstatite (Lindsley and Munoz, 1969a).

When low clinoenstatite is produced by inversion from high clinoenstatite there is the possibility of regions of the $P2_1/c$ structure nucleating independently and therefore being out of step with each other where they meet (Morimoto and Tokonami, 1969). Such 'out of phase' boundaries have been observed (e.g. Champness *et al.*, 1971) in electron microscope studies. Similar features, but explained in terms of twinning, are described by Iijima and Buseck (1975). The latter authors also discussed the prominent parting on (100) exhibited by some enstatites, and associate this with the shearing and twinning process (Buseck and Iijima, 1975).

Stacking Disorder

The ortho- and clinoenstatite structures are a result of regular displacements of neighbouring Si–O chains, but structures disordered with respect to chain stacking are also to be expected. X-ray diffraction effects from such disorder have been described (e.g. Brown and Smith, 1963; Pollack and Ruble, 1964; Pollack, 1966, 1968; Pollack and DeCarli, 1969).

Clinoenstatite or its iron-bearing equivalents have been reported from meteorites by many workers, and are generally thought to derive from primary protopyroxene

(Mason, 1968; Reid *et al.*, 1974). Ashworth and Barber (1975) suggest that the protopyroxene itself was produced from orthopyroxene by the high temperatures produced by shock. Dundon and Hafner (1971) subjected orthoenstatite (from Bamle, Norway) to shock of various intensities in the laboratory and examined the resulting (Fe,Mg) distribution by Mössbauer spectroscopy. Between 410 and 1000 kbar increasing disorder was produced in the originally highly ordered specimen. The maximum disorder corresponded to temperatures around 1000°C, and the authors observed that natural specimens which have been subjected to intense shock will have had their previous thermal history thereby erased.

In addition to small cell regular chain sequences giving ortho-, clino- and protoenstatite, with or without disorder, large cell polytypes have been observed. Byström (1943) described an enstatite with 36 Å (approximately equal 4 × 9 Å) periodicity, and Buseck and Iijima (1975) show that regions of clinoenstatite produced by heating enstatite to 1000°C and quenching, have 27, 36 and 54 Å, repeats.

Chemistry

Analyses of orthopyroxenes, together with the numbers of ions on the basis of six oxygens, showing the range of orthopyroxene compositions in igneous and metamorphic rocks, are detailed in Tables 3 and 4. Analyses of orthopyroxenes in nodules in kimberlites, ultrabasic and basic rocks and meteorites are given in Tables 5 and 6 respectively. Although the orthopyroxenes are essentially metasilicates of magnesium and ferrous iron they invariably contain other ions, the commonest of which are Al, Ca, Mn, Fe^{3+}, Ti, Cr and Ni. In most minerals of the series, however, the total of these constituents does not exceed 10 mol. per cent (in Tables 3 and 4, the range in the content of $(Mg,Fe^{2+})SiO_3$ is 86 to 99 mol. per cent). Cr and Ni occur mainly in the magnesium-rich orthopyroxenes of igneous rocks (see Table 3). Nickel is present in only small amounts, thus the average percentage of Ni in fifteen orthopyroxenes from the peridotites and gabbros of the Kotalahti area, Finland, is 0·047 weight per cent (Häkli, 1963). Relatively high Mn contents are found in the more iron-rich orthopyroxenes of igneous rocks, and in a wide range of compositions in the minerals of metamorphic rocks (Saxén, 1925; Sundius, 1932; Bonnichsen, 1969). In general, high contents of Al are restricted to metamorphic minerals, e.g. the aluminium-rich hypersthene (Table 4, anal. 9) from a garnet-bearing granulite. Other analyses with unusually high contents of Al are given by Lokka (1943), and Rajagopalan (1946). The Mn, Cr, Ti and Ni contents of orthopyroxenes in spinel peridotites and garnet peridotites are given by Mercy and O'Hara (1967).

A number of orthopyroxenes containing substantial amounts of Ca have been reported (e.g. Groves, 1935), but these analyses are now considered to have been made on samples contaminated with calcium-rich clinopyroxenes. In most orthopyroxene analyses the content of CaO does not exceed 1·5 weight per cent (0·055 to 0·065 Ca ions per formula unit or $(Mg,Fe^{2+})_{97}Ca_3$). The content of Ca in orthopyroxenes is generally related to the temperature at which they crystallized, and from his investigation of the synthetic $MgSiO_3$–$CaMgSi_2O_6$ system, Atlas (1952) found that in enstatite the maximum number of Ca atoms on the basis of six oxygens is 0·115 at 1100°C, 0·050 at 1000°C, and 0·030 at 700°C, values with which the results of the later investigation of this system by Schairer and Boyd (1957), are in

Table 3. Orthopyroxene Analyses (igneous rocks)

	1		2		3		4		5	
SiO_2	54·01		57·73		57·10		54·76		53·26	
TiO_2	0·03		0·04		0·17		0·19		0·17	
Al_2O_3	3·95		0·95		0·70		2·24		6·59	
Cr_2O_3	0·57		0·46		0·27		0·46		0·39	
Fe_2O_3	2·07		0·42		0·60		1·77		0·97	
FeO	1·57		3·57		5·21		5·06		5·54	
MnO	0·00		0·08		0·17		0·14		0·12	
NiO	0·09		0·35		0·04		0·06		—	
MgO	35·65		36·13		34·52		34·91		31·29	
CaO	0·99		0·23		0·62		0·53		2·14	
Na_2O	0·08		—		0·07		0·11		0·07	
K_2O	0·00		—		0·03		0·05		0·02	
H_2O^+	0·59		0·52		0·64		0·00		—	
H_2O^-	—		0·04		0·06		0·05		—	
Total	99·60		100·52		100·20		100·68		100·56	
α	1·664		—		—		—		—	
β	1·667		—		—		—		—	
γ	1·672		1·6700		1·674		1·679		—	
$2V_\alpha$	118·5°		107·75°		106°		—		—	
D	3·234		3·249		—		—		—	
	Numbers of ions on the basis of six O									
Si	1·865	2·00	1·972	2·00	1·976	2·00	1·890	1·98	1·839	2·00
Al	0·135		0·028		0·024		0·091		0·161	
Al	0·026	2·02	0·010	1·99	0·004	2·00	0·000	2·05	0·107	2·01
Ti	0·001		0·001		0·004		0·005		0·004	
Cr	0·016		0·012		0·008		0·013		0·011	
Fe^{3+}	0·054		0·010		0·016		0·046		0·025	
Mg	1·835		1·839		1·780		1·796		1·611	
Ni	0·002		0·010		0·001		0·002		—	
Fe^{2+}	0·045		0·102		0·151		0·160		0·160	
Mn	0·000		0·002		0·005		0·004		0·004	
Ca	0·037		0·008		0·023		0·020		0·079	
Na	0·005		—		0·004		0·007		0·005	
K	0·000		—		0·002		0·002		0·001	
Mg	93·6		93·9		90·3		88·6		85·8	
ΣFe	5·0		5·7		8·5		10·4		10·0	
Ca	1·4		0·4		1·2		1·0		4·2	
mg*	94·9		94·2		91·2		89·5		89·5	

* mg = 100 Mg/(Mg + Fe^{2+} + Fe^{3+} + Mn).

1 Enstatite, fissure in chromite pyroxenite, Hirose mine, Tari district, Tottori Prefecture, Japan (Kitahara, 1958). Anal. J. Kitahara.
2 Enstatite, pyroxenite, Webster, North Carolina (Hess, 1952). Anal. L. C. Peck.
3 Enstatite, peridotite, Dawros, Connemara, Eire (Rothstein, 1958). Anal. E. A. Vincent.
4 Enstatite, plagioclase peridotite, Horoman, Hokkaido, Japan (Nagasaki, 1966). Anal. H. Haramura (includes P_2O_5 0·35).
5 Enstatite, peridotite, Lizard, Cornwall (Green, 1964). Anal. D. H. Green.

Table 3. Orthopyroxene Analyses (igneous rocks) *continued*

	6	7	8	9	10
SiO_2	55·70	55·94	55·02	52·51	53·51
TiO_2	0·10	0·11	0·12	0·14	0·04
Al_2O_3	1·87	1·61	2·69	3·12	2·99
Cr_2O_3	0·44	0·45	0·34	0·73	0·26
Fe_2O_3	1·13	0·97	1·81	1·46	2·66
FeO	6·47	7·15	7·18	8·61	9·87
MnO	0·15	0·19	0·19	0·20	0·23
NiO	—	0·07	—	—	0·085
MgO	32·72	32·12	32·13	30·46	29·63
CaO	0·95	1·48	0·48	1·91	0·82
Na_2O	0·05	0·00	0·02	0·10	0·07
K_2O	0·01	0·00	0·01	0·00	0·03
H_2O^+	0·26	0·09	0·21	—	—
H_2O^-	0·07	0·00	0·02	—	—
Total	99·92	100·18	100·22	99·24	100·20
α	—	—	—	—	1·677
β	—	—	—	—	1·684
γ	1·680	1·6802	1·683	—	1·690
$2V_\alpha$	97°	92°	87°	—	—
D	—	3·302	—	—	—

Numbers of ions on the basis of six O

	6		7		8		9		10	
Si	1·943	2·00	1·949	2·00	1·919	2·00	1·873	2·00	1·896	2·00
Al	0·057		0·051		0·081		0·127		0·104	
Al	0·019	2·00	0·015	2·00	0·029	2·00	0·004	2·03	0·021	2·00
Ti	0·002		0·003		0·003		0·004		0·001	
Cr	0·012		0·012		0·010		0·021		0·007	
Fe^{3+}	0·030		0·026		0·048		0·039		0·071	
Mg	1·701		1·668		1·670		1·620		1·565	
Ni	—		0·002		—		—		0·003	
Fe^{2+}	0·189		0·208		0·209		0·257		0·292	
Mn	0·004		0·006		0·006		0·006		—	
Ca	0·035		0·055		0·018		0·073		0·031	
Na	0·004		—		0·002		0·007		0·005	
K	0·001		—		0·001		0·000		0·000	
Mg	87·0		85·3		85·9		81·2		79·9	
ΣFe	11·2		11·9		13·2		15·1		18·5	
Ca	1·8		2·8		0·9		3·7		1·6	
mg*	88·4		87·4		86·4		84·3		81·2	

6 Enstatite, peridotite, Dawros, Connemara, Eire (Rothstein, 1958). Anal. E. A. Vincent.
7 Bronzite, pyroxenite, Bushveld complex (Hess, 1952). Anal. L. C. Peck.
8 Bronzite, peridotite, Dawros, Connemara, Eire (Rothstein, 1958). Anal. E. A. Vincent.
9 Bronzite, with rutile exsolution, ultramafic intrusion, Gosse Pile, central Australia (Moore, 1968). Anal. A. Moore.
10 Bronzite, websterite, Glenelg, Inverness-shire (Mercy and O'Hara, 1965). Anal. E. L. P. Mercy.

11	12	13	14	15	
51·78	55·20	54·11	53·87	52·60	**SiO_2**
0·23	0·22	0·19	0·18	0·15	**TiO_2**
6·22	1·50	1·52	1·39	0·12	**Al_2O_3**
0·01	0·07	—	0·02	0·05	**Cr_2O_3**
0·30	0·84	0·00	2·38	0·95	**Fe_2O_3**
11·78	11·86	15·73	14·97	17·14	**FeO**
0·16	0·28	0·34	0·37	0·41	**MnO**
—	0·07	—	—	0·06	**NiO**
27·49	28·14	27·03	25·33	25·65	**MgO**
1·30	1·93	1·16	1·23	2·17	**CaO**
0·12	—	—	0·07	0·88	**Na_2O**
0·00	—	—	0·02	0·00	**K_2O**
—	0·30	—	—	0·14	**H_2O^+**
—	0·06	—	0·10	—	**H_2O^-**
'99·74'	100·47	100·08	99·94	100·32	**Total**
—	—	1·6855	1·688	1·685	**α**
—	—	1·6935	1·699	1·695	**β**
—	1·6893	1·6975	1·702	1·700	**γ**
—	77°	85·2°	—	69°	**$2V_\alpha$**
—	—	—	—	3·431	**D**

11	Σ	12	Σ	13	Σ	14	Σ	15	Σ	
1·850	2.00	1·964	2·00	1·953	2·00	1·957	2·00	1·937	1·94	**Si**
0·150		0·036		0·047		0·043		0·006		**Al**
0·112		0·026		0·017		0·017		0·000		**Al**
0·006		0·006		0·006		0·005		0·004		**Ti**
0·000		0·002		—		0·001		0·002		**Cr**
0·008		0·022		—		0·065		0·026		**Fe^{3+}**
1·474	2·01	1·492	1·98	1·454	2·01	1·372	1·98	1·408	2·13	**Mg**
—		0·002		—		—		0·001		**Ni**
0·351		0·353		0·475		0·455		0·528		**Fe^{2+}**
0·005		0·008		0·011		0·011		0·013		**Mn**
0·050		0·073		0·045		0·048		0·086		**Ca**
0·008		—		—		0·005		0·062		**Na**
0·000		—		—		0·001		0·000		**K**

11	12	13	14	15	
78·1	76·9	73·6	70·3	68·8	**Mg**
19·3	19·3	24·1	27·2	27·0	**ΣFe**
2·6	3·8	2·3	2·5	4·2	**Ca**
80·2	79·6	74·9	72·0	71·3	**mg***

11 Bronzite, pyroxene–spinel symplectite, gabbro inclusion in alkali basalt, Iki Island, Japan (Aoki, 1968). Anal. K. Aoki.

12 Bronzite, norite, Stillwater complex, Montana (Hess, 1952). Anal. L. C. Peck.

13 Bronzite phenocrysts, bronzite andesite, Kokubudai, Kagawa Prefecture, Japan (Kuno, 1947b). Anal. K. Tada and M. Huzimoto.

14 Bronzite, pyroxene diorite, Feather River area, northern Sierra Nevada, California (Hietanen, 1971). Anal. S. T. Neil (includes V_2O_5 0·01; data for Ba, Be, Co, Cu, Ga, Ni, Sc, Y, Yb, Zr).

15 Bronzite, eucritic norite, Chester Pike, Delaware (Clavan *et al.*, 1954). Anal. W. Clavan.

Table 3. Orthopyroxene Analyses (igneous rocks) *continued*

	16	17	18	19	20
SiO_2	53·17	53·18	52·07	52·00	52·25
TiO_2	0·22	0·21	0·47	0·14	0·28
Al_2O_3	0·45	3·08	1·70	0·57	0·92
Cr_2O_3	—	—	—	0·003	—
Fe_2O_3	1·32	0·25	0·00	1·34	1·64
FeO	17·18	18·05	22·65	22·46	24·27
MnO	0·48	0·41	0·48	0·57	0·71
NiO	—	—	—	0·00	—
MgO	23·81	23·26	21·13	21·62	18·72
CaO	2·67	2·09	1·55	0·75	1·20
Na_2O	0·47	—	—	0·46	0·04
K_2O	0·00	—	—	0·07	0·00
H_2O^+	—	—	—	0·28	—
H_2O^-	0·21	0·20	—	—	—
Total	99·98	100·73	100·05	100·26	100·03
α	1·691	—	1·7015	1·695	—
β	1·699	—	1·7115	1·705	—
γ	1·703	1·7065	1·7145	1·710	1·719
$2V_\alpha$	63°	—	59·5°	57°	53°
D	—	—	—	3·510	—

Numbers of ions on the basis of six O

	16		17		18		19		20	
Si	1·963	1·98	1·937	2·00	1·948	2·00	1·956	1·98	1·978	2·00
Al	0·020		0·063		0·052		0·026		0·022	
Al	0·000		0·069		0·022		0·000		0·019	
Ti	0·006		0·006		0·013		0·004		0·008	
Cr	—		—		—		0·000		—	
Fe^{3+}	0·036		0·007		0·000		0·038		0·047	
Mg	1·310	2·04	1·262	1·99	1·178	2·00	1·212	2·05	1·056	1·97
Ni	—		—		—		—		—	
Fe^{2+}	0·530		0·550		0·709		0·707		0·768	
Mn	0·015		0·013		0·015		0·018		0·022	
Ca	0·106		0·082		0·062		0·030		0·049	
Na	0·033		—		—		0·034		0·003	
K	0·000		—		—		0·002		0·000	
Mg	66·1		66·4		60·4		61·0		54·4	
ΣFe	28·6		29·3		36·4		37·5		43·1	
Ca	5·3		4·3		3·2		1·5		2·5	
mg*	69·3		68·9		61·9		61·4		55·8	

16 Hypersthene, microphenocrysts, hypersthene–olivine andesite, Tengu-zawa, Hatazyuku, Hakone Volcano, Kanagawa Prefecture, Japan (Kuno and Nagashima, 1952). Anal. K. Nagashima.

17 Hypersthene, phenocrysts, augite–hypersthene andesite, Akagi Volcano, Gumma Prefecture, Japan (Ota, 1952). Anal. T. Kusida.

18 Hypersthene, phenocrysts, augite–pigeonite–hypersthene andesite, Hakone-Toge, Hakone Volcano, Kanagawa Prefecture, Japan (Kuno, 1950). Anal. K. Tada and M. Huzimoto.

19 Hypersthene, norite, Brandywine Creek, Delaware (Clavan *et al.*, 1954). Anal. W. Clavan.

20 Hypersthene, ignimbritic hypersthene–hornblende dacite, Hudson, Gosforth area. New South Wales (Wilkinson, 1971). Anal. G. I. Z. Kalocsai. (Data for Ga, Cr, V, Ni, Co, Sc, Y.)

21	22	23	24	25	
50·60	52·22	51·07	50·26	50·10	SiO_2
0·19	0·08	0·52	0·16	0·25	TiO_2
0·16	0·43	1·14	3·13	0·40	Al_2O_3
0·10	—	—	—	—	Cr_2O_3
0·97	0·70	0·62	0·65	1·85	Fe_2O_3
25·71	25·91	26·65	26·54	28·00	FeO
0·31	0·83	1·36	0·76	2·05	MnO
0·00	—	—	—	—	NiO
18·96	18·54	16·82	16·36	16·05	MgO
1·65	1·28	1·63	1·76	1·50	CaO
0·07	—	0·15	0·24	0·08	Na_2O
0·60	—	0·04	0·13	0·06	K_2O
0·44	—	—	—	—	H_2O^+
—	—	—	—	—	H_2O^-
99·76	99·99	100·00	99·99	100·64	Total
1·703	—	—	—	1·710	α
1·713	—	—	—	—	β
1·717	1·721	1·728	1·726	1·727	γ
53°	54·5°	53°	52°	50°–52°	$2V_\alpha$
3·530	—	—	—	—	D
1·955 } 1·96	1·989 } 2·00	1·963 } 2·00	1·927 } 2·00	1·950 } 1·97	Si
0·008	0·011	0·037	0·073	0·018	Al
0·000	0·009	0·015	0·069	0·000	Al
0·005	0·002	0·015	0·005	0·007	Ti
0·004	—	—	—	—	Cr
0·028	0·020	0·018	0·019	0·054	Fe^{3+}
1·092	1·053	0·964	0·935	0·931	Mg
— } 2·07	— } 1·99	— } 2·00	— } 2·00	— } 2·04	Ni
0·831	0·826	0·857	0·851	0·911	Fe^{2+}
0·010	0·027	0·044	0·025	0·068	Mn
0·068	0·052	0·067	0·072	0·063	Ca
0·004	—	0·013	0·018	0·006	Na
0·030	—	0·002	0·006	0·003	K
54·1	54·0	49·5	49·8	45·9	Mg
42·5	43·4	47·1	46·4	51·0	ΣFe
3·4	2·6	3·4	3·8	3·1	Ca
55·7	54·7	51·2	51·1	47·4	mg*

21 Hypersthene, hybridized norite, Norristown, Delaware (Clavan *et al.*, 1954). Anal. W. Clavan.
22 Hypersthene, phenocrysts, augite-bearing hypersthene dacite, Hirogawara, Yugawara Prefecture, Japan. (Kuno, 1954). Anal. Y. Konisi.
23 Hypersthene, phenocryst, pitchstone, top of rhyolite lava, Holmanes, Reydarfjordur, eastern Iceland (Carmichael, 1963). Anal. I. S. E. Carmichael.
24 Hypersthene, phenocryst, augite-bearing hypersthene dacite (obsidian), Kaziya, Hakone Volcano, Kanagawa Prefecture, Japan (Kuno, 1954). Anal. K. Nagashima.
25 Ferrohypersthene, rhyolite pumice, Taupo, New Zealand (Ewart, 1967). Anal. J. A. Ritchie (includes P_2O_5 0·30).

Table 3. Orthopyroxene Analyses (igneous rocks)—*continued*

	26		27		28		29		30	
SiO_2	50·06		49·43		50·00		49·0		44·52	
TiO_2	0·32		0·17		0·49		0·24		1·39	
Al_2O_3	1·84		0·38		0·47		0·49		4·76	
Cr_2O_3	—		—		—		—		—	
Fe_2O_3	2·06		0·04		0·56		—		1·26	
FeO	29·39		34·91		33·83		37·2		38·66	
MnO	0·19		1·19		0·84		1·58		0·28	
NiO	—		—		—		—			
MgO	13·63		12·96		11·51		10·1		6·59	
CaO	1·43		0·71		1·74		1·35		1·40	
Na_2O	—		0·02 (Na₂O + K₂O)		0·20		—		0·39	
K_2O	—				0·05		—		0·19	
H_2O^+	0·69		—		0·49		—		0·41	
H_2O^-	0·17		—		0·19		—		—	
Total	99·78		99·81		100·37		99·96		99·85	
α	1·715		—		1·721		—		1·752	
β	1·728		—		1·735		—		1·759	
γ	1·731		1·7385		1·738		—		1·765	
$2V_\alpha$	51°		53°		54°		—		78°	
D	3·60		—		—		—		—	
Si	1·966	2·00	1·976	1·99	1·998	2·00	1·984	2·00	1·838	2·00
Al	0·034		0·018		0·002		0·016		0·162	
Al	0·050		0·000		0·020		0·007		0·070	
Ti	0·009		0·005		0·031		0·007		0·043	
Cr	—		—		—		—		—	
Fe^{3+}	0·060		0·001		0·017		—		0·038	
Mg	0·798		0·772		0·686		0·609		0·405	
Ni	—	1·95	—	2·02	—	2·00	—	2·00	—	2·01
Fe^{2+}	0·965		1·167		1·129		1·260		1·335	
Mn	0·006		0·040		0·029		0·054		0·010	
Ca	0·060		0·030		0·074		0·059		0·062	
Na	—		0·001		0·014		—		0·032	
K	—		—		0·002		—		0·010	
Mg	42·4		39·2		35·5		30·7		22·5	
ΣFe	54·4		59·3		60·7		66·3		74·1	
Ca	3·2		1·5		3·8		3·0		3·4	
mg*	43·6		39·0		36·9		31·7		22·6	

26 Ferrohypersthene, contaminated norite, Craig Wood, Glen Buchat, Aberdeenshire (Henry, 1935). Anal. N. F. M. Henry.

27 Ferrohypersthene, ferrohypersthene–hornblende gabbro, Osaka, Japan (Kuno, 1954). Anal. Y. Konisi.

28 Ferrohypersthene, gabbro, Guadalupe igneous complex, California (Best, 1963). Anal. M. G. Best.

29 Ferrohypersthene, porphyritic rhyolite obsidian, Deadmans Creek, Inyo Crater, California (Carmichael, 1967a). Anal. I. S. E. Carmichael (probe analysis).

30 Eulite, granite, Rubideaux Mountain, Riverside, California (Larsen and Draisin, 1950). Anal. F. A. Gonyer.

Table 4. Orthopyroxene Analyses (metamorphic rocks)

	1		2		3		4		5	
SiO_2	58·07		53·95		55·12		53·37		53·17	
TiO_2	tr.		0·11		0·32		0·51		0·83	
Al_2O_3	0·91		4·24		1·25		1·86		2·46	
Cr_2O_3	—		0·05		0·16		0·00		0·06	
Fe_2O_3	0·76		2·24		1·26		1·61		0·91	
FeO	tr.		8·40		11·96		12·96		13·06	
MnO	0·016		0·25		0·20		0·43		0·28	
MgO	39·49		30·17		29·10		27·33		24·74	
CaO	0·10		0·57		0·40		1·85		4·19	
Na_2O	0·03		0·03		—		0·08		0·18	
K_2O	tr.		0·01		—		0·01		0·03	
H_2O^+	0·96 (H₂O⁺ + H₂O⁻)		—		—		—		—	
H_2O^-			0·00		—		0·03		0·00	
Total	100·48		100·02		99·77		100·04		99·91	
α	1·649		—		1·677		1·680		—	
β	—		1·683		1·688		—		1·694	
γ	1·657		—		1·691		1·692		1·699	
$2V_\alpha$	131·5°		84°		88°		69°		63°	
D	—		—		3·323–3·340		—		—	
	Numbers of ions on the basis of six O									
Si	1·962	2·00	1·893	2·00	1·963	2·00	1·922	2·00	1·925	2·00
Al	0·036		0·107		0·037		0·078		0·075	
Al	0·000	2·01	0·069	1·99	0·015	1·98	0·000	2·01	0·031	1·99
Ti	0·000		0·003		0·009		0·014		0·023	
Cr	0·000		0·002		0·004		0·000		0·002	
Fe^{3+}	0·019		0·058		0·034		0·044		0·024	
Mg	1·988		1·578		1·544		1·467		1·335	
Fe^{2+}	0·000		0·247		0·356		0·390		0·395	
Mn	0·001		0·007		0·006		0·013		0·008	
Ca	0 004		0 021		0·015		0·071		0·162	
Na	0·002		0·002		—		0·006		0·012	
K	0·000		0·001		—		0·001		0·002	
Mg	98·8		82·9		79·2		74·4		69·6	
ΣFe	1·0		16·0		20·0		22·0		21·9	
Ca	0·2		1·1		0·8		3·6		8·5	
mg*	99·0		82·5		79·6		76·6		75·8	

* $mg = 100\,Mg/(Mg+Fe^{2+}+Fe^{3+}+Mn)$.

1 Enstatite, (mean of two analyses) talc deposits, Mul'vodzha, Pamirs (Dobrokhotova *et al.*, 1967). (Includes P_2O_5 0·10, SO_3 0·04) (*a* 18·180, *b* 8·780, *c* 5·177 kX).

2 Bronzite, olivine–hypersthene pyroxenite, Scourie, Sutherland (Muir and Tilley, 1958). Anal. J. H. Scoon.

3 Bronzite, plagioclase-hypersthene granulite, Pahaoja, Sotajoki, Lapland (Eskola, 1952). Anal. E. Nordensvan.

4 Bronzite, olivine-bearing pyroxene granulite, Eilean Carrach, Ardnamurchan, Scotland (Muir and Tilley, 1958). Anal. J. H. Scoon.

5 Bronzite, metamorphosed picrite basalt, ejected block, Kilauea (Muir and Tilley, 1957). Anal. J. H. Scoon.

Table 4. Orthopyroxene Analyses (metamorphic rocks) – *continued*

	6	7	8	9	10
SiO_2	51·03	47·30	53·20	46·91	48·48
TiO_2	0·09	0·63	0·13	0·51	0·52
Al_2O_3	5·83	10·81	1·15	8·26	7·21
Cr_2O_3	—	—	—	—	—
Fe_2O_3	0·21	7·80	0·00	3·02	1·97
FeO	16·26	9·20	21·64	19·88	20·62
MnO	0·12	0·03	0·78	0·20	0·49
MgO	25·18	23·60	22·50	20·02	19·97
CaO	0·48	0·28	0·82	0·34	0·46
Na_2O	—	—	—	0·10	0·02
K_2O	—	—	—	0·06	0·00
H_2O^+	—	—	—	0·30	—
H_2O^-	—	—	—	0·03	0·07
Total	99·20	'99·71'	100·22	99·63	99·81
α	1·686	1·703	1·693	1·705	—
β	—	—	1·704	—	—
γ	1·698	1·712	1·709	1·716	1·719
$2V_\alpha$	60°–62°	—	58°	—	55°
D	3·39	—	—	3·42–3·51	3·41

Numbers of ions on the basis of six O

	6		7		8		9		10	
Si	1·861	2·00	1·703	2·00	1·973	2·00	1·762	2·00	1·811	2·00
Al	0·139		0·297		0·027		0·238		0·189	
Al	0·112	2·01	0·162	1·95	0·023	2·00	0·128	2·00	0·128	1·99
Ti	0·002		0·017		0·003		0·014		0·016	
Cr	—		—		—		—		—	
Fe^{3+}	0·006		0·211		0·000		0·085		0·055	
Mg	1·369		1·266		1·243		1·121		1·112	
Fe^{2+}	0·496		0·277		0·671		0·625		0·644	
Mn	0·004		0·001		0·024		0·006		0·018	
Ca	0·019		0·011		0·032		0·014		0·018	
Na	—		—		—		0·007		0·001	
K	—		—		—		0·003		0·000	
Mg	72·3		71·7		63·9		60·8		60·2	
ΣFe	26·7		27·7		34·5		38·5		38·8	
Ca	1·0		0·6		1·6		0·7		1·0	
mg*	73·0		72·1		64·1		61·0		60·8	

6 Bronzite, pyroxene granulite, Wieselburg, Austria (Scharbert, 1963). Anal. H. G. Scharbert.

7 Bronzite, pyrope–sapphirine rock, Anabar Massif, U.S.S.R. (Lutts and Kopaneva, 1968).

8 Hypersthene, amphibolite, Hôkizawa, Kanagawa Prefecture, Japan (Kuno, 1947a). Anal. K. Tada and M. Huzimoto.

9 Hypersthene, plagioclase–hypersthene–garnet granulite, Kevuavdshi, Kevujoki, Lapland (Eskola, 1952). Anal. A. Huhma.

10 Hypersthene, hypersthene–spinel–plagioclase hornfels, Belhelvie, Aberdeenshire (Howie, 1964). (*a* 18·257, *b* 8·825, *c* 5·193 Å.)

11		12		13		14		15		
50·34		48·19		49·44		50·08		48·85		SiO_2
0·13		0·42		0·15		0·64		0·24		TiO_2
3·14		2·14		2·21		1·23		13·91		Al_2O_3
—		—		—		—		—		Cr_2O_3
2·97		1·86		1·93		2·34		4·77		Fe_2O_3
22·53		27·45		28·06		27·85		18·22		FeO
0·70		0·42		1·03		0·85		0·86		MnO
19·52		17·26		16·61		15·78		11·51		MgO
0·60		1·67		0·23		1·44		0·42		CaO
0·05		0·13		0·10		0·05		—		Na_2O
0·03		0·18		tr.		0·02		—		K_2O
—		—		0·04		—		—		H_2O^+
0·03		—		0·29		0·00		—		H_2O^-
100·04		99·72		100·09		100·28		98·78		Total
1·701		—		1·707		1·707		—		α
—		—		1·720		—		1·719		β
1·717		1·728		1·724		1·722		—		γ
59°		57°		52°		50°		—		$2V_\alpha$
—		—		3·60		—		—		D
1·899	2·00	1·877	1·98	1·918	2·00	1·937	1·99	1·814	2·00	Si
0·101		0·098		0·082		0·056		0·186		Al
0·039		—		0·018		—		0·423		Al
0·004		0·012		0·004		0·019		0·007		Ti
—		—		—		—		—		Cr
0·084		0·056		0·056		0·068		0·133		Fe^{3+}
1·098	1·99	1·002	2·06	0·960	2·00	0·910	1·99	0·627	1·80	Mg
0·711		0·894		0·911		0·901		0·566		Fe^{2+}
0·022		0·014		0·034		0·028		0·027		Mn
0·024		0·070		0·010		0·060		0·017		Ca
0·004		0·005		0·007		0·004		—		Na
0·002		0·005		0·000		0·001		—		K
57·3		49·2		49·6		46·9		45·8		Mg
41·5		47·4		49·9		50·0		53·0		ΣFe
1·2		3·4		0·5		3·1		1·2		Ca
57·3		49·0		49·0		47·7		46·3		mg*

11 Hypersthene, hypersthene–diopside–plagioclase gneiss, Scourie, Sutherland (Muir and Tilley, 1958). Anal. J. H. Scoon.

12 Hypersthene, hypersthene granulite, Mont Tremblant Park, Quebec (Katz, 1970). Anal. W. B. Katz (*a* 18·2783, *b* 8·8954 Å).

13 Ferrohypersthene, intermediate rock, Ambagamudam Pothai, Tinnevelly, India (Howie, 1955). Anal. R. A. Howie.

14 Ferrohypersthene, hypersthene–diopside–plagioclase hornfels, Aavold Quarry, Oslo district, Norway (Muir and Tilley, 1958). Anal. J. H. Scoon.

15 Ferrohypersthene, hypersthene–cordierite–biotite granulite, Satnur, Mysore State, India (Devaraju and Sadashivaiah, 1971). Anal. G. V. Subbarayadu.

Table 4. Orthopyroxene Analyses (metamorphic rocks) – *continued*

	16	17	18	19	20
SiO_2	49·92	49·50	48·10	47·23	48·29
TiO_2	0·26	0·11	0·31	1·02	0·40
Al_2O_3	0·18	2·01	4·18	2·47	2·83
Cr_2O_3	—	—	—	—	
Fe_2O_3	1·60	1·31	1·34	1·60	1·23
FeO	30·70	32·60	31·46	34·03	33·67
MnO	0·56	0·59	0·66	0·89	0·67
MgO	15·16	13·74	12·92	11·14	10·77
CaO	1·10	0·16	0·31	1·57	2·29
Na_2O	0·05	0·17	0·27	0·05	0·03
K_2O	0·03	0·07	0·10	0·12	0·04
H_2O^+	0·08	0·01	0·00	—	0·03
H_2O^-	0·05	0·02	—	0·06	0·04
Total	99·76	100·29	99·65	100·18	100·29
α	1·717	1·720	—	1·725	—
β	—	1·732	—	1·736	—
γ	1·733	1·735	1·741	1·743	1·745
$2V_\alpha$	52°	56°	60°	58°	57°
D	3·60	—	—	—	3·68
	Numbers of ions on the basis of six O				
Si	1·966 } 1·97	1·943 } 2·00	1·894 } 2·00	1·892 } 2·00	1·921 } 2·00
Al	0·008	0·057	0·106	0·108	0·079
Al	— } 2·03	0·035 } 1·99	0·080 } 1·98	0·008 } 2·00	0·052 } 1·99
Ti	0·008	0·003	0·009	0·031	0·012
Cr	—	—	—	—	—
Fe^{3+}	0·047	0·039	0·040	0·048	0·039
Mg	0·890	0·804	0·758	0·665	0·642
Fe^{2+}	1·011	1·071	1·036	1·140	1·115
Mn	0·019	0·020	0·022	0·030	0·023
Ca	0·046	0·007	0·013	0·067	0·098
Na	0·004	0·012	0·020	0·004	0·002
K	0·002	0·003	0·005	0·006	0·002
Mg	44·2	41·8	40·6	34·6	33·9
ΣFe	53·5	57·8	58·7	61·9	60·9
Ca	2·3	0·4	0·7	3·5	5·2
mg*	45·2	41·5	40·8	35·3	35·3

16 Ferrohypersthene, ferrohypersthene–hornblende-bearing diopside gneiss, Lake Juurikka, Finland (Savolahti, 1966). Anal. A. Heikkinen (includes P_2O_5 0·07).

17 Ferrohypersthene, hypersthene 'diorite', Pallavaram, India (Howie, 1955). Anal. R. A. Howie.

18 Ferrohypersthene, cordierite-bearing xenolithic rock, Haddo House and Arnage district, Aberdeenshire (Gribble, 1968). Anal. C. D. Gribble.

19 Ferrohypersthene, intermediate member of the charnockite series Nambran Paramba, Tinnevelly, Madras (Howie, 1955).

20 Ferrohypersthene, two pyroxene granulite, Hitterö, Norway (Howie, 1963). Anal. R. A. Howie.

21	22	23	24	25	
48·66	48·21	45·15	47·55	47·44	**SiO_2**
0·31	0·17	<0·10	0·07	0·48	**TiO_2**
0·70	1·37	0·85	1·90	0·92	**Al_2O_3**
—	—	—	—	—	**Cr_2O_3**
1·51	1·46	6·00	0·41	0·32	**Fe_2O_3**
29·61	36·90	33·12	39·37	40·75	**FeO**
8·03	0·68	1·26	0·89	1·64	**MnO**
11·16	10·45	7·87	8·68	7·63	**MgO**
0·47	0·43	2·63	1·23	1·10	**CaO**
—	0·00	0·11	—	0·21	**Na_2O**
—	0·00	0·08	—	0·02	**K_2O**
—	—	2·30	0·05	—	**H_2O^+**
—	0·09	0·40	0·09	0·00	**H_2O^-**
100·45	99·76	99·77	100·24	100·51	**Total**
—	—	1·731–1·733	1·736	—	**α**
—	—	1·744–1·749	1·747	—	**β**
1·747	—	1·752–1·755	1·752	1·761	**γ**
—	—	57°–60°	63°	74°	**$2V_\alpha$**
3·72	3·68	—	3·78	3·761	**D**
1·954 ⎫ 1·99	1·952 ⎫ 2·00	1·942 ⎫ 2·00	1·941 ⎫ 1·99	1·952 ⎫ 2·00	**Si**
0·033 ⎭	0·048 ⎭	0·043 ⎭	0·059 ⎭	0·044 ⎭	**Al**
0·000	0·018	0·000	0·032	0·000	**Al**
0·009	0·005	0·003	0·002	0·015	**Ti**
—	—	—	—	—	**Cr**
0·046	0·045	0·193	0·012	0·010	**Fe^{3+}**
0·668 ⎬ 2·01	0·630 ⎬ 1·99	0·505 ⎬ 1·96	0·528 ⎬ 2·00	0·467 ⎬ 2·01	**Mg**
0·994	1·249	1·079	1·344	1·401	**Fe^{2+}**
0·273	0·024	0·046	0·031	0·057	**Mn**
0·020	0·018	0·121	0·054	0·047	**Ca**
—	—	0·009	—	0·017	**Na**
—	—	0·005	—	0·000	**K**
33·4	32·5	26·0	27·2	23·6	**Mg**
65·6	66·6	67·8	70·0	74·0	**ΣFe**
1·0	0·9	6·2	2·8	2·4	**Ca**
33·7	32·3	27·7	27·0	24·1	**mg***

21 Ferrohypersthene, metamorphosed iron formation, North Dangin, Western Australia (Davidson and Mathison, 1973). Anal. L. R. Davidson.

22 Ferrohypersthene, charnockite, British Guiana (Howie, 1963). Anal. R. A. Howie.

23 Eulite, charnockite, Vichan pluton, U.S.S.R. (Shemyakin *et al.*, 1967).

24 Eulite, eulysitic rock, Madial, Sudan (Howie, 1963). Anal. R. A. Howie.

25 Eulite, granulite, Broken Hill district, New South Wales (Binns, 1962). Anal. R. A. Binns.

Table 4. Orthopyroxene Analyses (metamorphic rocks) – *continued*

	26		27		28		29		30	
SiO_2	48·70		46·91		47·97		46·80		46·36	
TiO_2	0·10		0·23		0·03		0·12		0·16	
Al_2O_3	1·04		1·21		0·13		0·01		0·29	
Cr_2O_3	—		—		—		—		—	
Fe_2O_3	0·16		3·30		—		2·00		0·20	
FeO	42·35		40·45		43·88		39·63		44·93	
MnO	0·06		1·20		0·75		4·15		1·16	
MgO	6·88		6·80		6·26		5·18		5·09	
CaO	0·85		0·45		0·26		1·31		1·64	
Na_2O	0·01		—		0·01		0·07		—	
K_2O	0·00		—		0·03		0·03		—	
H_2O^+	0·02		—		0·59		0·35		0·03	
H_2O^-	0·08		—		0·10		0·18		0·07	
Total	100·25		100·55		100·01		100·07		99·93	
α	—		—		1·750		1·746		1·751	
β	—		1·751		1·761		—		1·760	
γ	1·769		—		1·770		1·764		1·769	
$2V_\alpha$	—		72°–74°		—		81·2°		83°	
D	3·78		—		3·85		—		3·84	
	Numbers of ions on the basis of six O									
Si	1·999	2·00	1·935	1·99	2·010	2·01	1·979	1·98	1·965	1·98
Al	0·001		0·059		—		0·000		0·014	
Al	0·050		0·000		0·006		0·000		0·000	
Ti	0·003		0·007		0·001		0·004		0·005	
Cr	—		—		—		—		—	
Fe^{3+}	0·005		0·102		—		0·064		0·006	
Mg	0·421	1·97	0·418	1·98	0·391	1·98	0·326	2·01	0·321	2·04
Fe^{2+}	1·454		1·395		1·538		1·401		1·593	
Mn	0·002		0·042		0·027		0·149		0·041	
Ca	0·038		0·020		0·012		0·059		0·074	
Na	0·001		—		0·001		0·006		—	
K	0·000		—		0·002		0·002		—	
Mg	22·0		21·1		19·9		16·3		16·1	
ΣFe	76·1		77·9		79·5		80·7		80·2	
Ca	1·9		1·0		0·6		3·0		3·7	
mg*	22·0		21·4		20·0		16·8		16·4	

26 Eulite, garnet–orthopyroxene–quartz rock, Pond Inlet, NE. Baffin Island (Howie, 1963). Anal. R. A. Howie.

27 Eulite, pyroxene–quartz–magnetite rock, Satnur-Halgura area, Mysore State, India (Devaraju and Sadashivaiah, 1966).

28 Eulite, quartz–eulite–grunerite–siderite rock, Wabush Iron Formation, south-western Labrador (Klein, 1966). (*a* 18·385, *b* 9·020, *c* 5·228 Å).

29 Eulite, garnet–orthopyroxene gneiss, Bear Mountain, Popolopan Lake Quadrangle, New York (Dodd, 1963). Anal. M. Chiba (includes P_2O_5 0·24).

30 Eulite, eulysite, Mansjö Mountain, Sweden (Henry, 1935). Anal. N. F. M. Henry.

31	32	33	34	35	
46·65	46·58	46·56	45·95	44·43	SiO_2
0·10	0·38	0·03	0·10	0·12	TiO_2
2·10	0·97	0·23	0·90	2·96	Al_2O_3
–	—	—	—	—	Cr_2O_3
0·57	1·94	0·20	0·31	0·70	Fe_2O_3
44·02	42·66	48·10	41·65	44·91	FeO
0·55	0·06	0·15	5·02	1·20	MnO
4·90	3·98	3·70	3·49	3·38	MgO
0·81	3·06	0·77	1·43	1·69	CaO
0·01	0·01	0·04 (Na₂O + K₂O)	—	0·07	Na_2O
0·01	0·00			0·05	K_2O
0·08	—	—	0·65	—	H_2O^+
0·03	—	—	0·09	—	H_2O^-
99·83	99·64	99·78	99·59	99·51	**Total**
—	1·753	1·7545	1·755	1·750	α
—	1·763	1·7645	1·763	1·758	β
—	1·774	1·7745	1·773	1·774	γ
—	89°	96°	97°	92°	$2V_\alpha$
—	3·93	—	3·88	3·61	D
1·954 } 2·00	1·960 } 2·00	1·988 } 2·00	1·972 } 2·00	1·896 } 2·00	Si
0·046	0·040	0·012	0·028	0·104	Al
0·058 } 1·99	0·008 } 1·97	0·000 } 2·00	0·018 } 2·00	0·045 } 2·02	Al
0·003	0·012	0·001	0·003	0·004	Ti
—	—	—	—	—	Cr
0·018	0·061	0·006	0·010	0·023	Fe^{3+}
0·306	0·250	0·235	0·223	0·215	Mg
1·542	1·501	1·718	1·495	1·603	Fe^{2+}
0·019	0·002	0·005	0·183	0·042	Mn
0·037	0·138	0·035	0·066	0·077	Ca
0·001	0·001	0·002	—	0·006	Na
0·001	0·000	—	—	0·003	K
15·9	12·8	11·8	11·3	11·0	Mg
82·2	80·1	86·4	85·4	85·1	ΣFe
1·9	7·1	1·8	3·3	3·9	Ca
16·2	13·8	12·0	11·7	11·4	mg*

31 Eulite, charnockitic adamellite, Natal (Howie, 1958). Anal. R. A. Howie.
32 Eulite, eulysite, Mariupol Iron Deposit (Val'ter, 1969).
33 Orthoferrosilite, eulysite, Wang-chang-tzu, Je-ho-shen, south-west Manchuria (Kuno, 1954). Anal. Y. Konisi.
34 Orthoferrosilite, thermally metamorphosed iron-rich rock, Yu hsi kou district, Manchuria (Tsuru and Henry, 1937). Anal. N. F. M. Henry.
35 Orthoferrosilite, bauchite (fayalite-bearing quartz monzonite), Bauchi, Nigeria ('Oyawoye and Makanjuola, 1972).

Table 5. Orthopyroxene Analyses (nodules in kimberlite, ultrabasic and basic rocks)

	1	2	3	4
SiO_2	58·48	51·77	54·13	55·04
TiO_2	tr.	0·05	0·12	tr.
Al_2O_3	0·88	4·00	1·62	3·24
Cr_2O_3	0·25	0·27	0·21	0·36
Fe_2O_3	0·72	1·68	0·85	1·30
FeO	3·93	4·58	5·70	5·85
MnO	0·02	0·14	0·14	0·10
NiO	—	0·12	0·08	0·14
MgO	34·71	33·52	34·55	33·30
CaO	0·50	1·59	0·57	0·39
Na_2O	0·23	0·18	0·16	0·14
K_2O	0·08	0·03	0·16	tr.
H_2O^+	—	2·05	1·17	0·00
H_2O^-	0·21	—	0·73	0·15
Total	100·01	99·98	100·19	100·01
α	1·658	—	1·672	1·667
β	1·662	1·664	—	1·671
γ	1·669	—	1·684	1·673
$2V_\alpha$	74°	—	—	84°
D	—	—	—	3·284
	Numbers of ions on the basis of six O			
Si	2·000 } 2·00	1·840 } 2·00	1·912 } 1·98	1·909 } 2·00
Al	0·000	0·160	0·067	0·091
Al	0·035 } 1·98	0·008 } 2·06	0·000 } 2·06	0·041 } 2·01
Ti	0·000	0·001	0·003	0·000
Cr	0·007	0·008	0·006	0·010
Fe^{3+}	0·019	0·045	0·023	0·034
Mg	1·769	1·776	1·819	1·722
Ni	—	0·003	0·002	0·004
Fe^{2+}	0·112	0·136	0·168	0·170
Mn	0·001	0·004	0·004	0·003
Ca	0·018	0·061	0·022	0·014
Na	0·015	0·012	0·011	0·009
K	0·003	0·001	0·007	0·000
Mg	92·2	87·8	89·3	88·6
ΣFe	6·9	9·2	9·6	10·7
Ca	0·9	3·0	1·1	0·7
mg*	93·1	90·6	90·3	89·3

1 Enstatite, lherzolite nodule in kimberlite, Maliba Matso, Basutoland, (Nixon *et al.*, 1963). Anal. M. H. Kerr.
2 Enstatite, lherzolite nodule, Green Knobs, Navajo County, Arizona (O'Hara and Mercy, 1966).
3 Enstatite, garnet lherzolite, southern Bohemia (Sobolev, 1964). Anal. F. L. Teleshova.
4 Enstatite, olivine nodule in basalt, Calton Hill, Derbyshire (Hamad, 1963). Anal. S. el D. Hamad (Co 80, Mo 8, V 40 p.p.m.). (*a* 18·255, *b* 8·821, *c* 5·192Å).

5		6		7		8		
55·81		52·35		53·30		51·67		SiO_2
0·04		0·23		0·36		0·50		TiO_2
1·54		7·72		4·94		2·11		Al_2O_3
0·23		0·37		0·20		—		Cr_2O_3
—		0·66		1·59		1·72		Fe_2O_3
7·43		7·23		9·21		12·65		FeO
0·06		0·15		0·15		0·30		MnO
—		0·11		—		—		NiO
34·38		29·00		29·56		29·33		MgO
0·52		1·92		0·81		1·42		CaO
—		0·18		0·15		—		Na_2O
—		0·00		0·03		—		K_2O
—		—		0·25		0·40		H_2O^+
0·03		—		0·10		—		H_2O^-
100·04		99·92		100·68		100·10		Total
—		1·677		—		1·684		α
—		1·683		—		—		β
—		1·687		—		1·695		γ
—		—		—		90°		$2V_\alpha$
3·27		—		—		—		D
1·939	2·00	1·831	2·00	1·872	2·00	1·869	1·96	Si
0·061		0·169		0·128		0·090		Al
0·002		0·149		0·077		0·000		Al
0·001		0·006		0·009		0·014		Ti
0·006		0·010		0·005		—		Cr
		0 017		0·042		0·047		Fe^{3+}
1·779		1·511		1·547		1·582		Mg
—	2·03	0·003	2·00	—	1·99	—	2·09	Ni
0·216		0·211		0·270		0·383		Fe^{2+}
0·002		0·004		0·004		0·009		Mn
0·019		0·072		0·030		0·055		Ca
—		0·012		0·010		—		Na
—		—		0·001		—		K
88·2		83·3		81·7		76·2		Mg
10·8		12·8		16·7		21·7		ΣFe
1·0		3·9		1·6		2·7		Ca
89·1		86·7		83·0		78·3		mg*

5 Enstatite, garnet lherzolite xenolith in ankaramite, Lashaine Volcano, northern Tanzania (Dawson *et al.*, 1970). Anal. D. G. Powell.

6 Bronzite, metacryst, mafic hawaiite, Walcha, New South Wales, Australia. (Binns *et al.*, 1970). Anal. G. I. Z. Kalocsai.

7 Bronzite, olivine eclogite nodule, Sale Lake, Hawaii (Kuno, 1969). Anal. H. Haramura (includes P_2O_5 0·03).

8 Bronzite, eclogitized schist in kimberlite pipe, Siberia (Lutts, 1965). Anal. A. F. Al'ferova.

Table 6 Orthopyroxene Analyses (meteorites)

	1.	2.	3.	Numbers of ions on basis of six O	1.		2.		3.	
SiO_2	59·92	54·30	53·23	Si	2·004	2·00	1·948	2·00	1·946	1·98
TiO_2	—	0·23	0·27	Al	—		0·052		0·037	
Al_2O_3	0·00	3·10	0·87	Al	—	1·99	0·079	1·97	—	2·03
Cr_2O_3	—	0·74	0·84	Ti	—		0·006		0·007	
Fe_2O_3	0·00	—	—	Cr	—		0·021		0·024	
FeO	0·38	9·07	17·17	Mg	1·969		1·528		1·412	
MnO	—	0·40	0·60	Fe^{2+}	0·011		0·272		0·524	
MgO	39·51	28·58	25·92	Mn	—		0·012		0·019	
CaO	0·32	1·34	1·16	Ca	0·011		0·052		0·045	
H_2O^+	—	1·67	—							
H_2O^-	—		—	Mg	98·9		82·0		70·6	
Total	100·13	99·43	'100·06'	ΣFe	0·5		15·2		27·1	
				Ca	0·6		2·8		2·3	
				mg	99·4		84·3		72·2	
α	1·653	1·673	1·686							
β	1·656	—	—							
γ	1·660	1·683	1·698							
$2V_\alpha$	125·5°	83°	—							
D	3·209	—	—							

1. Enstatite, Shallowater, Texas (Foshag, 1940). Anal. W. F. Foshag.
2. Bronzite, Kesen chondrite, Japan (Miyashiro, 1962). Anal. H. Haramura. (*a* 18·27, *b* 8·854, *c* 5·17 Å; powder data.)
3. Hypersthene, achondrite enclave in mesosiderite, Mount Padbury, Western Australia (McCall, 1966). Anal. H. B. Wiik.

substantial agreement. The numbers of Ca atoms in the orthopyroxenes of the pyroxene andesites and dacites of the Hakone volcano (Kuno, 1954) are 0·057 to 0·104 and 0·045 to 0·053 respectively. Compilations of orthopyroxene analyses with Si, Ti, Al, Cr, Fe, Mn, Ni, V, Mg, Ca, Na and K, by the microprobe X-ray emission technique include forty-eight orthopyroxenes by Howie and Smith (1966), and a statistical survey of the composition of 250 orthopyroxenes by Dobretsov (1968). The latter recalculated the analyses on the basis of 6000 oxygens, determined $[Al]^{IV}$ as 2000-Si, and $[Al]^{vi}$ as $Al_{total} - [Al]^{iv}$, and calculated the coefficients:

$$f = (Fe^{3+} + Ti + Cr + Fe^{2+} + Mn)/(Mg + Fe^{3+} + Ti + Cr + Fe^{2+} + Mn)$$
$$f' = (Fe^{2+} + Mn)/(Mg + Fe^{2+} + Mn)$$
$$K_{ox} = Fe^{3+}/(Fe^{2+} + Fe^{3+})$$

The survey showed that the Fe/(Mg + Fe) ratio and the contents of Fe^{2+}, Mn, Mg, Fe^{3+} and Cr are in general determined by the composition of the host rocks, and that the $[Al]^{iv}$, $[Al]^{vi}$, Ca and Na contents are strongly influenced by the temperature and pressure at the time of their formation. Dobretsov's compilation also includes fourteen aluminium-rich orthopyroxenes and associated minerals from metamorphic rocks. The variation in chemical composition between igneous and metamorphic orthopyroxenes has been evaluated by Bhattacharyya (1971). A plot of $(MgO + FeO + Fe_2O_3)$ against Al_2O_3 shows that the orthopyroxenes of the two major parageneses fall, with very few exceptions, in separate fields. The demarcation line intersects the ordinate at 44·3 per cent $(MgO + FeO + Fe_2O_3)$, the metamorphic field lying above and the igneous field below the demarcation divide given by the equation:

$$(MgO + FeO + Fe_2O_3) + 0{\cdot}775\ Al_2O_3 = 44{\cdot}304$$

Eleven analyses of charnockitic orthopyroxenes from the Madras area, India, are given by Howie (1955), and analyses of eighteen orthopyroxenes from a charnockite series, Alto Atentejo, south-east Portugal, are reported by Canilho (1974). Some trace element data have also been presented by Howie, one of the most notable features of which is the occurrence of gallium (5 to 30 p.p.m.) in greater abundance in the orthopyroxene than in the coexisting clinopyroxene; the Ga:Al ratio is highest in the more iron-rich members of the series. Chromium (15 to 300 p.p.m.) and nickel (30 to 650 p.p.m.) are both more abundant in the more magnesium-rich minerals, while the maximum content of cobalt (100 p.p.m.) is present in the orthopyroxenes ($\simeq$ Fs 50) of intermediate composition. Molybdenum was not detected in the bronzites but is present in variable but small amounts (3 to 10 p.p.m.) in the more iron-rich minerals, while vanadium (200 p.p.m.) and the $V:Fe^{3+}$ ratio rises to a maximum in the intermediate hypersthenes. Scandium (10 to 80 p.p.m.), lead (20 p.p.m.) and barium (20 to 50 p.p.m.) occur only in the more iron-rich orthopyroxenes. Ti, Zr, Cr, V, Zn, Cu, Ni, Co, Pb, Ba, Sr and Rb have been determined for seven orthopyroxenes and six clinopyroxenes from mafic granulites in the Strangways Range, central Australia (Woodford and Wilson, 1976). Trace element contents of some orthopyroxenes are also presented by DeVore (1955). The rare earth contents of the enstatite in the peridotite, Lizard, Cornwall, have been determined by Frey (1969).

The release of argon (radiogenic ^{40}Ar in excess of that which may be accounted for by the decay of ^{40}K in the mineral since crystallization) from two bronzites of the Stillwater complex, Montana, has been measured by Schwartzman and Giletti (1977). The Ar release occurred by two mechanisms one having an activation energy greater than 72 kcal/mole Ar between 600° and 1000°C, the second above 1000°C, a smaller Ar loss also occurring below 600°C. The computed ambient partial pressure of Ar in the melt at the time of crystallization of the bronzites is 0·1 atm. The amount of Ar in the melt is estimated to have been (3 to 80) $\times 10^{-6}$ scc/Ar/g melt.

Polymorphism. An account of the structural aspects of the polymorphism present in the $MgSiO_3$–$FeSiO_3$ series is given in the structure section (p. 29), and the following description is restricted to the temperature and pressure relationships of the various phase transformations. There are inconsistencies in the data some of which are due to the different origin and composition of the starting materials, heat treatment such as duration of runs and rates of cooling, and to pressure conditions, particularly the presence or absence of shear, used in the experimental procedures.

Much of the earlier work in the stability relationships and polymorphism of $MgSiO_3$ has been critically reviewed by Smith (1969a). Investigations since that date have clarified some of the previous uncertainties, but a full understanding of the polymorphic transitions has not as yet been achieved.

There are four polymorphs of $MgSiO_3$, protoenstatite, high-temperature clinoenstatite, orthoenstatite (enstatite) and low-temperature clinoenstatite.

Protoenstatite is the stable polymorph at temperatures above $\simeq$ 1000°C, and probably so remains to the incongruent melting temperature at about 1560°C. On quenching, protoenstatite inverts to a mixture of orthoenstatite and low-temperature clinoenstatite. The high-temperature modification of clinoenstatite likewise cannot be quenched. It occurs above $\simeq$ 980°C but it is uncertain whether high-temperature clinoenstatite has a stability field of its own.

Although enstatite is isostructural with bronzite, hypersthene and eulite, for which there is general agreement that these members of the $MgSiO_3$–$FeSiO_3$ series are the stable forms at low temperatures, some doubts have been expressed concerning the low-temperature stability of the enstatite composition. Its predominance, relative to low-temperature clinoenstatite, in natural rocks indicates that it is a form stable at low temperatures. The alternative view put forward by Grover (1972), following the synthesis of low-temperature clinoenstatite below 566°C, suggests that orthoenstatite is possibly not the stable polymorph at temperatures below 560°C.

Low-temperature clinoenstatite is produced on quenching protoenstatite, has been synthesized using a $MgCl_2$ flux, and occurs in kink bands in enstatite that has undergone shearing.

The orthorhombic polymorph cannot be synthesized at 1 atm except under hydrothermal conditions, e.g. fluxing with LiF or LiOH, or at high pressure. The temperature of the orthoenstatite–low-temperature clinoenstatite transformation was given as $\simeq$ 630°C by Boyd and England (1965), and as $\simeq$ 600°C by Schwab and Schwerin (1975). The transition boundary between the two phases at pressures between 5 and 40 kbar, $T = 630° + (26°C/kbar)P$ (Boyd and England) was obtained using a piston-cylinder solid-medium method, and a small shearing stress component was probably present during the experiment.

Pure synthetic enstatite, as shown by its refractive indices and cell volume is less dense than synthetic low-temperature clinoenstatite, α 1·647, β 1·649, γ 1·657, $\frac{1}{2}V = 416{\cdot}97$ Å^3 and α 1·650, β 1·653, γ 1·660, $\frac{1}{2}V = 415{\cdot}47$ Å^3 respectively (Ernst and Schwab, 1970). Densities, derived from the Gladstone–Dale relationship $(n-1)/d = k$ (constant) where $n = 3\sqrt{\alpha\beta\gamma}$, are 3·208 and 3·210 (Stephenson *et al.*, 1966). Thus the field of low-temperature clinoenstatite extends to somewhat higher temperatures at higher pressures (Fig. 18).

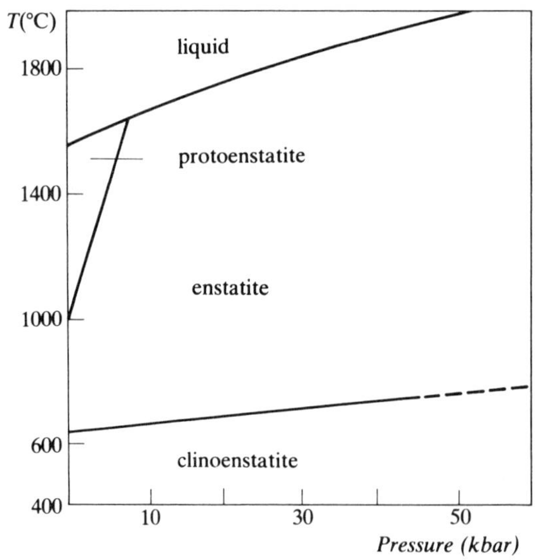

Fig. 18. Phase relations of $MgSiO_3$ polymorphs (after Boyd *et al.*, 1964; Boyd and England, 1965).

There is a similar, but much more rapid, expansion of the stability field of the orthorhombic polymorph, as pressure and temperature increase, and orthoenstatite is stable at the liquidus at $\simeq$ 1500°C and 7 kbar.

The transformation of orthoenstatite to low-temperature clinoenstatite under static confining pressure has been described by a number of workers including Rayleigh (1965), Riecker and Rooney (1967), Munoz (1968), Raleigh *et al.* (1971), Green and Radcliffe (1972) and Coe and Müller (1973). The product resulting from the deformation of enstatite commonly consists of two phases. Thus enstatite deformed at 800°C and 5 kbar confining pressure and a strain rate of $10^{-4}\,s^{-1}$ was found to consist of orthoenstatite and untwinned clinoenstatite (Müller, 1974).

The conversion of enstatite to low-temperature clinoenstatite, and the development of well-defined kink bands, due to translation on (100) planes parallel to the [001] direction, during the experimental deformation of an enstatite pyroxenite at temperatures between 500° and 800°C and a confining pressure of 5 kbar, have been reported by Turner *et al.* (1960). The formation of lamellae, 100–1000 Å in thickness, of untwinned clinoenstatite, in bronzite ($\simeq$ 14 mol. per cent $FeSiO_3$), joined along (100) planes and having common *y* and *z* crystallographic axes, resulting from uniaxial compression at 800°C and 5 kbar confining pressure has been described by Coe and Müller (1973). The effect of hydrostatic pressure on the ortho–clinoenstatite inversion is very small, and Lindsley and Munoz (1969b) report that no reaction was observed in enstatite in a hydrostatic cell at 20 kbar. The same authors have also described the transformation under stress of ortho- to clinoferrosilite analogous to that of the magnesium end-member of the series.

Coe and Kirby (1975) have shown that the production of clinoenstatite by deforming single crystals of orthoenstatite at 800°C and pressures of 5 and 15 kbar, results from a coherent transformation (a*, *b* and *c* crystallographic directions remaining unchanged) and involves shear on (100) parallel to [001]. Reversal of the transformation was achieved at 1 atm by annealing at 1100°C for 5 hours; at confining pressures of 5 and 15 kbar the reversal takes place at 960°C.

The kinetics of the transformation of low-temperature clinoenstatite to orthoenstatite has been determined for two orthopyroxenes, $En_{92\cdot8}Fs_{7\cdot0}Wo_{0\cdot2}$ and $En_{85\cdot4}Fs_{13\cdot9}Wo_{0\cdot6}$, by monitoring changes in electrical conductivity as a function of temperature under partial oxygen pressures corresponding to the quartz–fayalite–magnetite system at a total pressure of 1 atm (Boland, 1974; Boland *et al.*, 1974). Both minerals contain lamellae, parallel to (100), of clinoenstatite enriched in Fe and Ca relative to the orthopyroxene. The apparent activation energy of the exsolution of the clinoenstatite lamellae in the orthoenstatite, at temperatures between 800°–1000°C, is $\simeq$ 16 kcal/mole. The exsolution is considered to result from a homogenization process involving the breaking of Fe–, Mg– and Ca–O bonds, and the diffusion of Fe and Ca from the (100) lamellae followed by the transformation of the clinoenstatite to orthoenstatite (but see Raleigh *et al.*, 1971).

The thermodynamic effect of shear stress on the orthoenstatite–low-temperature clinoenstatite transition has been evaluated by Coe (1970). On the assumption that the entropy change is 1 cal/°C mole the transition temperature, depending on whether the angle of shear is 13·3° or 18·3° (for a discussion of the geometry of kink band boundaries between ortho- and clinoenstatite see Starkey, 1968), is raised by 177° or 248°C/kbar shear stress.

The protoenstatite–orthoenstatite inversion has been investigated by a number of workers including Atlas (1952), Boyd *et al.* (1964), Kushiro *et al.* (1968), Anastasiou and Seifert (1972), Smyth (1974a) and Chen and Presnall (1975). The effect of different heating and cooling rates on the proto–orthoenstatite transition, using enstatite consisting of orthoenstatite with up to 40 per cent low-temperature clinoenstatite from the Norton County meteorite, has been studied by Smyth (1974a). No reaction was observed on heating below 650°C; at temperatures above 975°C protoenstatite formed rapidly from the clinoenstatite, and above 1000°C formed more slowly from orthoenstatite. On cooling to 700°C the resulting protoenstatite showed no tendency to invert to either of the lower temperature phases and it was possible to maintain this phase metastably for periods of up to 24 hours. Below 700°C rapid inversion occurred, higher cooling rates favouring the formation of clinoenstatite, slower cooling rates favouring orthoenstatite. The proto–orthoenstatite inversion was not observed to take place independently of the proto–clinoenstatite transition, and it is possible that the inversion does not occur without clinoenstatite as an intermediary phase. The ortho–protoenstatite inversion is time and temperature dependent, is affected by the ratio of orthoenstatite to low-temperature clinoenstatite originally present and is relatively slow. It is an order–disorder type transition in contrast to the proto–clinoenstatite reaction which is a reversible, oriented, diffusionless and rapid inversion of martensitic type.

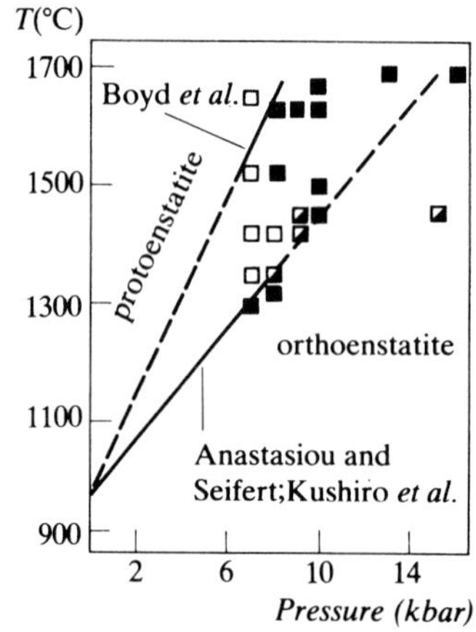

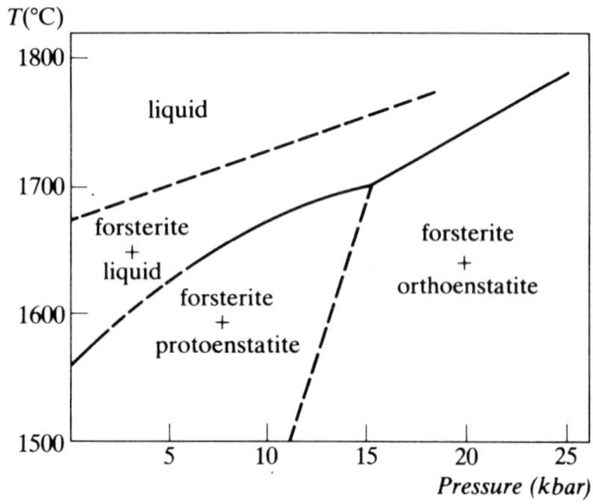

Fig. 19. Subsolidus data for the composition 34·93 per cent MgO, 65·07 per cent SiO_2 (see Fig. 61) (after Chen and Presnall, 1975). □ proto + clinoestatite; ◪ proto + clino + orthoenstatite; ■ orthoenstatite. Quartz is present in addition to phases indicated.

Fig. 20. Isopleth for the composition 43·56 per cent MgO, 56·44 per cent SiO_2 (see Fig. 61) (after Chen and Presnall, 1975).

The data of Chen and Presnall (1975) are compared with the results of earlier work on the protoenstatite–orthoenstatite transformation, and illustrated in Fig. 19. The later work shows that above 1450°C and 7 kbar orthoenstatite formed in the protoenstatite field as determined by Anastasiou and Seifert (1972) and Kushiro *et al.* (1968), whereas above 1555°C the data are comparable with the curve drawn by Boyd *et al.* (1964). Some of these discrepancies may be related to the observation that during quenching at lower pressures protoenstatite inverts partially or completely to clinoenstatite, while at higher pressures (and temperatures) protoenstatite inverts directly to orthoenstatite. This change in the character of the inversion occurs between 1300°C ($<$ 7 kbar) and 1450°C ($>$ 10 kbar). The solidus relations for the composition 43·56 weight per cent MgO, 56·44 per cent SiO_2 (composition A of Fig. 61, p. 97) are shown in Fig. 20. The solidus consists of two univariant curves, one at pressures 0 to 15·7 kbar (1700°C) involving forsterite, protoenstatite and liquid, and the other at pressures $>$ 15·7 kbar for forsterite, orthoenstatite and liquid. The liquidus curve approaches the solidus curve as pressure increases, and shows that the liquid composition in equilibrium with forsterite and enstatite shifts increasingly towards forsterite and would coincide with the composition of the initial mixture at an extrapolated pressure of 27 kbar.

The major difference between the polymorphism of $FeSiO_3$ and that of $MgSiO_3$ is the absence of a structural type corresponding to protoenstatite. Three polymorphs of $FeSiO_3$ were synthesized by Lindsley *et al.* (1964), orthoferrosilite, clino-

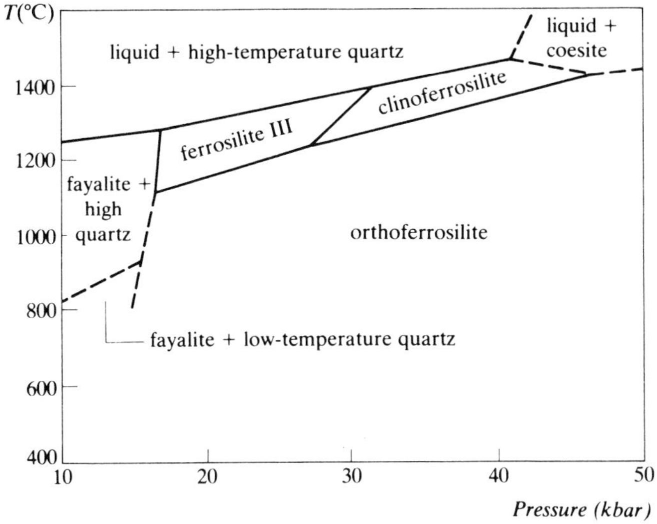

Fig. 21. Stability relations of ferrosilite (after Lindsley *et al.*, 1964).

ferrosilite, and a form, ferrosilite III (Fig. 21). The latter (not a pyroxene) is monoclinic but is optically distinctive (α 1·763, β 1·766, γ 1·785, $2V_\gamma$ 39°, $\gamma{:}z$ 48°, optic plane $\parallel$(010), compared with clinoferrosilite, α 1·764, β 1·767, γ 1·792, $2V_\gamma$ 23°, $\gamma{:}z$ 31°, optic plane $\perp$(010)). Subsequently it has been shown that clinoferrosilite occurs in both low-($P2_1/c$) and high-($C2/c$) temperature modifications. The transformation takes place rapidly, between 700° and 800°C for the composition $Mg_{0{\cdot}3}Fe_{0{\cdot}7}SiO_3$, and is reversible. The equation for the ortho–clinoferrosilite transition curve, P(kbars) $- -9+0{\cdot}067T$(°C), given by Akimoto *et al.* (1965) is considerably steeper than is shown in Fig. 21. The standard free energy for the formation of clinoferrosilite from orthoferrosilite, derived from the P–T equation and volume change data (the molar volumes in cm^3/mole for ortho- and clinoferrosilite respectively are 33·13 and 30·04) is $\Delta G = -40+0{\cdot}097T$ (cal/mole) where T ranges 900°–1500°K.

The sequence of the transformations, low-temperature clinoenstatite–enstatite–high-temperature clinoenstatite–protoenstatite does not change in solid solutions with diopside containing less than 8 mol. per cent $CaMgSi_2O_6$ (Schwab and Schwerin, 1975). With larger amounts of $CaMgSi_2O_6$ in solid solution, pigeonite develops over a narrow transition range (see pigeonite section, p. 162). The temperature of the enstatite–high-temperature clinoenstatite inversion is not appreciably affected by solid solution with diopside, but is stated to increase for the high-temperature clinoenstatite–protoenstatite transformation from 1250° for $MgSiO_3$ to 1370°C at the composition $En_{92}Di_8$.

The average linear thermal expansion coefficients for enstatite, low-temperature clinoenstatite and protoenstatite, in the temperature range 300°–700°C, are 120×10^{-7}, 135×10^{-7} and 98×10^{-7}°C^{-1} respectively (Sarver and Hummel, 1962).

Orthorhombic and monoclinic polymorphs of cobalt metagermanate, $CoGeO_3$, transition temperature 137°C, have been synthesized by solid state reaction of CoO and GeO_2 (Tauber and Kohn, 1965). The equilibrium phase boundary between pyroxmangite and $MnSiO_3$ (clinopyroxene structure) has been determined by Akimoto and Syono (1972). The boundary curve is represented approximately as P(kbars) $= 19 + 0{\cdot}057T$(°C) (see pyroxmangite section, p. 600).

Exsolution lamellae. Many orthopyroxenes of plutonic rocks display a characteristic development of lamellae, parallel to the (100) plane, of a calcium-rich monoclinic pyroxene; such orthopyroxenes have been termed *orthopyroxenes of the Bushveld type* (Hess and Phillips, 1940). The lamellae either form parallel-sided continuous sheets which resemble fine straight-ruled lines when the (010) plane of the crystal is observed between crossed polarizers or occur as rows of flattened blebs, which may be in optical continuity with adjacent crystals of clinopyroxene. Bushveld-type orthopyroxenes, $Mg_{67}Fe_{31}Ca_2$, have been described from the noritic marginal gabbro that forms the border of the Greenhills ultramafic complex, New Zealand (Mossman, 1971). The clinopyroxene lamellae, $Ca_{42}Mg_{47}Fe_{11}$, are 2 to 10 μm in width, and the y and z crystallographic axes of both host and lamellae are coincident. In the rocks of the Stillwater and Bushveld complexes lamellae of clinopyroxene parallel to the (100) plane occur in orthopyroxenes ranging in composition from Fs_{12} to Fs_{30}. In these two intrusions the more iron-rich orthopyroxenes, Fs_{30} to Fs_{44} in the Stillwater, and Fs_{30} to Fs_{55} in the Bushveld, are inverted pigeonites, and contain coarse plates of augite parallel to a plane which represents (001) of the original clinopyroxene. Hess (1960) has named these orthopyroxenes as *orthopyroxenes of the Stillwater type*. Such orthopyroxenes are common in plutonic rocks of basic composition but are rare in equivalent hypabyssal rocks. The calcium content of orthopyroxenes of the Stillwater type ($\simeq$ 4·0 weight per cent CaO) is approximately three times greater than that in orthopyroxenes of the Bushveld type. Orthopyroxenes with lamellae parallel to (100) occur in some high grade metamorphic rocks (e.g. Parras, 1958); the presence of fine lamellae of a monoclinic pyroxene in orthopyroxenes of volcanic rocks has also been described, and Kuno (1938) has observed orthopyroxene phenocrysts, in a lava, which show lamellae in the cores but not in the outer zones. Although in the latter case the $Mg:Fe^{2+}$ ratio of the cores and the margins may vary it is unlikely that there is any significant difference in the Ca content of the two parts of the phenocrysts, and the lamellae-free outer zones are probably due to more rapid cooling that prevented exsolution of the calcium-rich component. In the slowly cooled plutonic orthopyroxenes a higher degree of ordering is possible, and in such crystals unmixing of the calcium-rich phase generally occurs. Planes containing the minimal energy interface between two phases are favourable sites for unmixing (Buerger, 1948). Such an interface is one in which the two planes have a common structure, i.e. in the ortho- and clinopyroxenes the (100) plane. Champness and Lorimer (1973) have described orthopyroxenes with exsolution lamellae from the Stillwater complex in which the interfaces between host and lamellae consist of a regular network of dislocations. Nucleation occurred initially at grain boundaries and individual dislocations, and subsequent growth by the development of ledges along (100) interfaces (see also Kohlstedt and van der Sande, 1976).

In general, enstatites and bronzites of ultrabasic rocks contain less Ca than do the hypersthenes of basic igneous rocks, and the smaller number of lamellae present in the orthopyroxenes of ultrabasic rocks is probably a consequence of this difference in composition. The absence of lamellae in the orthopyroxenes of some metamorphic rocks may possibly be due to their crystallization at temperatures below that at which Ca-rich clinopyroxene enters in appreciable amounts into solid solution with orthopyroxene.

Orthopyroxenes showing fine exsolution lamellae parallel to (100) occur in the ultramafic rocks of Horoman, Hidaka and Horokanai, Japan (Yamaguchi, 1973). The compositions of two of the orthopyroxenes and their lamellae are $Mg_{86}Fe_{11}Ca_3$ and $Ca_{47}Mg_{48}Fe_5$ and $Mg_{77}Fe_{20}Ca_3$ and $Ca_{47}Mg_{46}Fe_7$ respectively. The lamellae show varying degrees of lattice distortion, and crystallized initially as twinned clinpyroxene with apparent orthorhombic symmetry. Subsequent growth leading to the development of broader lamellae was accompanied by a decrease in the degree of lattice distortion and the disappearance of twinning in the exsolved clinopyroxene.

Calcic orthopyroxene, $En_{76.1}Fs_{19.4}Wo_{4.5}$, and calcium-poor orthopyroxene, $En_{82}Fs_{16.8}Wo_{1.2}$, are associated with augite, $Wo_{38.2}En_{50.5}Fs_{11.3}$, in the Shaw (L-7) chondrite (Dodd *et al.*, 1975). The Ca-poor phase does not contain optically visible or submicroscopical lamellae of exsolved augite, and is considered to have formed at a temperature $\geq 1200°C$, and crystallized initially as protobronzite and subsequently inverted to bronzite during very slow cooling. The more calcic orthopyroxene is possibly a retrograde phase, formed by reaction of augite and Ca-poor orthopyroxene at a temperature below 120°C.

The formation of clinopyroxene lamellae in orthopyroxenes and the relationship of orthopyroxene to pigeonite and augite are discussed further in the sections on pigeonite (p. 176) and augite (p. 342); see also Frontispiece, Figs. A–C.

The early more magnesium-rich members of the orthopyroxene series in general crystallize below the appropriate orthorhombic–monoclinic pyroxene inversion temperature. As crystallization proceeds, orthopyroxenes richer in the ferrosilite molecule continue to form until the cooling curve of the magma intersects the inversion curve, when the initial Ca-poor pyroxene phase crystallizes as a monoclinic pigeonite (see Fig. 89, p. 173). In many rocks this has occurred when the $Mg:Fe^{2+}$ ratio was approximately 70:30 (McDougall, 1962; Sinitsȳn, 1965; Best, 1963; Best and Mercy, 1967), though examples with ratios between 85:15 and 60:40 have been recorded (Poldervaart and Hess, 1951: Philpotts, 1966). With further crystallization and iron-enrichment of the Ca-poor pyroxene, pigeonite continues to crystallize until the two-pyroxene boundary (see p. 334) is reached at a $Mg:Fe^{2+}$ ratio of approximately 45:55.

The early orthopyroxenes crystallize from basic magmas in equilibrium with a member of the diopside–hedenbergite series, and in many volcanic rocks the two pyroxenes are intergrown. In general, the monoclinic pyroxene has crystallized around the orthopyroxene and the junction between the two minerals indicates that there was no mutual reaction. Orthopyroxene phenocrysts in some basalts, however, are rimmed by a corona of augite and olivine, and in others the orthopyroxene is completely replaced by these two minerals. Such features are common in the olivine–augite basalts of the Hakone volcano, and here Kuno (1950) has ascribed the instability of the orthopyroxenes to their xenocrystal character.

Elongate inclusions of rutile in the (010) plane of a bronzite, from an ultramafic intrusion of the Giles complex, central Australia, have been described by Moore (1968). The inclusions are elongated parallel to the (601), ($\bar{6}$01) and (001) crystallographic planes in the pyroxene, and are considered to have formed by exsolution rather than epitaxial growth. A green spinel (15 weight per cent Cr_2O_3) is also exsolved from some of the orthopyroxenes but very few contain both rutile and spinel exsolution products. The breakdown at high pressure of orthopyroxene to spinel and stishovite, demonstrated by Ringwood and Major (1966a), led Moore to suggest that the exsolution of the rutile and spinel may represent the first stage of this transformation. Thin blades of a Cr–Al non-silicate, probably a spinel, have also been observed (White, 1966), in the enstatite of lherzolite inclusions in basalt. Here the blades are oriented in two directions, parallel to {100} and {010}, and the blade edges and their elongation are parallel to [001]. Similar intergrowths of enstatite and chromite, due to re-equilibration of an original garnet ($\pm$ pyroxene) assemblage under lower pressure conditions, have been described from the lherzolite xenoliths in a carbonatitic member of the lamprophyric dyke swarm at Mocraki River, New Zealand (Wallace, 1975). The compositions of the orthopyroxene within and outside the symplectite are almost the same; both contain chromium (0·20 and 0·26 weight per cent Cr_2O_3 in the symplectite and non-symplectite mineral respectively), and the symplectite reaction appears to have gone to completion.

The exsolution of a yellow-green spinel in the hypersthene of the gabbroic xenoliths in the basalts and trachybasalts of Gough Island has been described by Le Maitre (1965). The spinel occurs in two orientations with (111) parallel to (100), and either [100] or [1$\bar{1}$0] parallel to the *b* axis of the hypersthene.

The conditions relating to the formation of pyroxene–spinel symplectites from garnet–pyroxene assemblages has been investigated experimentally at pressures between 7·5 and 40 kbar and at temperatures ranging from 850°–1400°C. The univariant curve for the reaction, T (°C) = 44·4 P(kbars) + 534, has been determined by Tazaki *et al.* (1972), using mixtures consisting of orthopyroxene, $Mg_{89\cdot4}Fe_{9\cdot2}Ca_{1\cdot4}$, diopside, $Ca_{45\cdot1}Mg_{50\cdot7}Fe_{4\cdot2}$, and spinel in the ratio 5:2:3, from the pyroxene–spinel symplectites of the Horoman peridotite intrusion, Hokkaido, Japan.

Intergrowths of lamellar ilmenite in enstatite occur in the Monastery kimberlite pipe, Orange Free State (Boyd and Dawson, 1972), and have also been described from other South African kimberlite pipes as well as from Mir, Yakutia, and Stockdale, Kansas.

Orthopyroxene and ilmenite intergrowths from the lower zone of the Skaergaard intrusion and from a lunar breccia have been described by Hasleton and Nash (1975). The former is considered to result from the subsolidus oxidation of titanomagnetite included in the orthopyroxene, and the lunar example due to eutectic crystallization (see discussion of pyroxene–ilmenite intergrowths, p. 269).

Enstatite occurs in symplectic intergrowths with magnetite as an intermediate oxidation product during the alteration of basaltic olivines (Haggerty and Baker, 1967), according to the reaction:

$$\underset{\text{olivine}}{3(Fe_2SiO_4 + Mg_2SiO_4)} + O_2 = \underset{\text{enstatite}}{6MgSiO_3} + \underset{\text{magnetite}}{2Fe_3O_4}$$

Graphic intergrowths of hypersthene and magnetite in a monzonite–shonkinite porphyry resulting from the replacement of olivine have also been described by Yur'ev (1966).

Unusual exsolution features, the development of 'hourglass' shapes by clinopyroxene lamellae (Fig. 22), and the exsolution of spinel and rutile are displayed by the

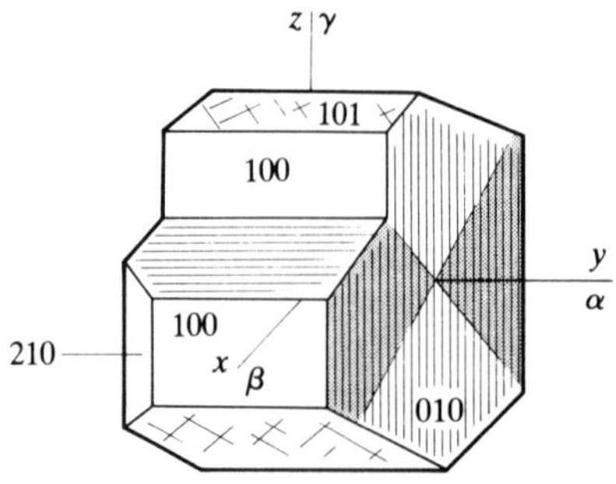

Fig. 22. Diagram showing the form of 'hourglass' exsolution in orthopyroxene (after Moore, 1971).

orthopyroxenes of the Gosse Pile layered ultramafics, central Australia (Moore, 1971). The orthopyroxenes have a relatively low CaO content and Moore has suggested that the fine lamellae parallel to (100) of the orthopyroxene may not be augitic in composition but more likely are the result of the distribution of calcium between phases involved in the polymorphic inversion of Ca-poor pyroxenes. The fine lamellae in parts of the crystals are accompanied by elongate and irregular blebs, and display a type of hourglass structure. The exsolved spinel (picotite Al > Cr, Fe:Mg 1:1) occurs as elongate plates parallel to *z*, and flattened on (100); the rutile needles are oriented parallel to the (601), ($\bar{6}$01) and (001) crystallographic planes of the host, and are contained within the (010) plane. Relict orthopyroxene augen, enclosed in a finer grained matrix of olivine, enstatite, diopside and spinel has been described from the peridotite at Tinaquillo, Venezuela (Green, 1963), and deformation fractures in the enstatite are occupied by a spinel derived from the breakdown of the aluminous enstatite to a less aluminium-rich variety.

Large orthopyroxene megacrysts, with Al_2O_3 contents ranging from 3·9 to 9·4 weight per cent, occurring in anorthositic rocks, have been described by Emslie (1975). Some of the orthopyroxenes contain regular parallel lamellae of plagioclase in the (100) planes of the pyroxene, and in some of the thicker lamellae the plagioclase has recrystallized and has a granular habit. The lamellae are more calcic in composition than the main plagioclase of the anorthosites, and commonly show strong zoning with the central part of the lamellae richer in Na than the margin. Bulk analysis of two large cleavage fragments of the orthopyroxenes show 8·38 and 7·28 per cent Al_2O_3, compared with microprobe analysis of the pyroxene adjacent to the lamellae of 2·74 and 3·57 per cent respectively. The conclusion, based on the textural evidence, that the pyroxenes were originally homogeneous aluminous orthopyroxenes, and have subsequently exsolved plagioclase is further supported by the variation in compositon from the central to marginal areas of the feldspar lamellae. In clinopyroxenes an increase in the pressure of formation leads initially to increasing solid solution of the $CaAl_2SiO_6$ relative to the jadeite component, and with a further pressure increase to greater solid solution of $NaAlSi_2O_6$ (Kushiro, 1969). Extrapolation of this relationship to orthopyroxenes provides an explanation of the zoning in the exsolved plagioclase lamellae, as in the progressive decompression, consequent on the upward movement of the high-pressure megacrysts, exsolution of a jadeite-rich pyroxene would be followed by increasing amounts of the $CaAl_2SiO_6$ component.

Another example of hypersthene megacrysts containing plagioclase lamellae in the (100) plane of the orthopyroxene has been reported from the andesine anorthosite, Tikkoatokhakh, Labrador (Morse, 1975). The plagioclase lamellae, amounting to between 2 and 10 per cent of the pyroxene volume, range in composition from An_{43} to An_{93}, and as those described by Emslie, are zoned from more Na-rich cores to more Ca-rich rims, and are invariably accompanied by grains or lamellar aggregates of magnetite. A similar exsolution origin is indicated by the regularity of the lamellar spacing, and their uniform thickness and orientation in the host pyroxene (Frontispiece, Fig. D), due to coupled reactions of the type

$$\underset{\text{Ca-Tsch}}{2CaAl_2SiO_6} + 2SiO_2 = \underset{\text{anorthite}}{2CaAl_2Si_2O_8}$$

$$\underset{\text{jadeite}}{NaAlSi_2O_6} + SiO_2 = \underset{\text{albite}}{NaAlSi_3O_8}$$

$$\underset{\text{ferrosilite}}{3FeSiO_3} + \tfrac{1}{2}O_2 = \underset{\text{magnetite (to plagioclase)}}{Fe_3O_4 + 3SiO_2}$$

the oxygen possibly being derived from a ferri-pyroxene component of the original orthopyroxene.

Overgrowths of augite on orthopyroxene are well displayed in the picrite dykes of the Lochinver area, Scotland (Tarney, 1969). In these rocks the orthopyroxene crystallized earlier than the augite, and formed favourable sites for the nucleation of the clinopyroxene. The overgrowths are oriented with the (100) plane and *y* and *z* axes of the host and overgrowth coincident. In small lens-shaped bodies formed during the late stage of the picrite crystallization, a second generation of a more iron-rich orthopyroxene is present, some of which displays an epitaxic relationship with the augite, and more rarely occurs as overgrowths on the clinopyroxene. The calcium (2·53–2·95 CaO) and aluminium (2·50–2·61 Al_2O_3) contents of the orthopyroxene are relatively high giving rise to an increase in *c*, so reducing the misfit on (100) between the two structures.

Lamellae, 2–10 μm in thickness, parallel to (100), of an edenitic amphibole in the orthopyroxene, $Mg_{92.3}Fe_{7.2}Ca_{0.5}$, of a peridotite xenolith in lava of the Ataq volcano, southern Arabia, have been described by Desnoyers (1975).

Ortho–clinopyroxene tie-lines. The tie-lines joining the compositions, expressed in terms of their Wo, En and Fs contents, of many coexisting ortho- and clinopyroxene pairs, as first noted by Hess (1941), for igneous pyroxenes, intersect the $CaSiO_3$–$MgSiO_3$ join at approximately Wo_{70}–Wo_{80}, and show that in the partition of Mg and Fe^{2+} there is a relatively higher concentration of Fe^{2+} in the orthorhombic member of the pair. The possible use of this relationship to estimate the composition of one pyroxene, knowing that of the other, either from the chemical analysis or the optical properties, was suggested by Hess, Muir and Tilley (1958), have also noted that, in using this tie-line relationship to estimate chemical composition of coexisting pyroxenes, caution is particularly necessary when the estimate is made solely on the basis of the optical properties. The discrepancies in the compositional data obtained by this method are mainly due to the effects of relatively large contents of minor constituents on the optical properties of the clinopyroxenes.

The orthopyroxene phenocrysts in some volcanic rocks are enriched in magnesium relative to the clinopyroxenes, and Muir and Tilley (1958) have suggested that this relationship may indicate non-equilibrium between the two pyroxenes. The same authors have observed that in metamorphic rocks carrying two pyroxenes the tie-line trends generally show no significant departure from the normal relationship of plutonic assemblages. Howie (1955) has also shown that the tie-line convergence of coexisting pyroxene pairs in charnockites and pyroxene granulites does not depart from the usual relationship, indicating that the associated ortho- and clinopyroxenes are equilibrium assemblages. Wilson (1960), however, from his investigations of the metamorphic rocks of the Musgrave Ranges, central Australia, has shown that the pyroxene tie-lines of some of the rocks of this area do not conform strictly to the normal orthopyroxene–clinopyroxene compositional relationship, and has suggested that these anomalies may be due to the different metamorphic histories of the rocks concerned (see also O'Hara, 1960; Vorma, 1975). The partition of Mg and Fe between ortho- and clinopyroxene is dependent on temperature and pressure, and their distribution between coexisting mineral pairs is better discussed in terms of the distribution coefficient (see p. 64).

O'Hara and Mercy (1963) have plotted the ferrosilite content of Ca-poor pyroxenes against the fayalitic content of the coexisting olivines in 160 natural orthopyroxene–olivine pairs, including pyroxene–olivine pairs from layered peridotite nodules in kimberlite, nodules in basalts, and peridotite and garnet lenses in gneiss, in which the Fe/(Fe + Mg) ratio of neither mineral exceeds 0·4. Comparison of these data with that of Bowen and Schairer (1935) on the synthetic system MgO–FeO–SiO_2, shows considerable discrepancies. The largest discrepancies are shown by orthopyroxene–olivine pairs formed during rapid cooling, and are probably related to the higher pressure, lower temperature of formation of the natural assemblages. Moreover, as Muir and Tilley have noted, some assemblages may not be in equilibrium due to the rapid crystallization subsequent to the formation of olivine, zoning or crystal settling.

Coronas. Coronas, consisting of an inner zone of orthopyroxene around olivine and rimmed by a symplectic intergrowth of orthopyroxene, clinopyroxene and spinel are common in some ultramafic and gabbroic rocks, for example Stjernöy, Norway (Öosterom, 1963). In other ultrabasic and basic rocks the outer rim of the corona consists of clinopyroxene and garnet. The formation of the latter mineral from olivine–plagioclase assemblages occurs at higher pressures than pyroxene–spinel in compositional environments with high $Mg/(Mg + Fe^{2+})$ ratios, and the absence of garnet in the corona assemblages at Stjernöy may be attributable to the highly magnesium character of these rocks. Thus in the experimental study of the gabbro → eclogite transition, Green and Ringwood (1967b) showed that an intermediate assemblage of aluminous pyroxene + plagioclase + spinel occurs between the lower pressure olivine-bearing and higher pressure garnet-bearing assemblage particularly in the more magnesium-rich basalts. The orthopyroxene of the inner rim of the coronas in the Sulitjelma troctolite has the composition $En_{82}Fe_{16\cdot7}Wo_{0\cdot3}$, and the pyroxene is considered (Mason, R., 1967) to have formed by the diffusion of the appropriate ions in solution in an aqueous medium across the olivine–plagioclase interface and not as the product of a late magmatic reaction.

Orthopyroxene forming the inner part of the coronas around olivine is a common feature of the metagabbros of the Adirondacks (Whitney and McLelland, 1973). In the southern area of the region the orthopyroxene shell is succeeded by a partial or complete symplectic rim of clinopyroxene and spinel while in the north-east area of the region the outer shell consists of garnet. The origin of the olivine–pyroxene–spinel coronas may be illustrated (Kushiro and Yoder, 1966) by the reaction:

$$\underset{\text{olivine}}{(6-4x)(\mathrm{Mg,Fe})_2\mathrm{SiO}_4} + \underset{\text{anorthite}}{3\mathrm{CaAl}_2\mathrm{Si}_2\mathrm{O}_8} =$$

$$\underbrace{(6-4x)(\mathrm{Mg,Fe})\mathrm{SiO}_3 + 2x(\mathrm{Mg,Fe})\mathrm{Al}_2\mathrm{SiO}_6}_{\text{Al-orthopyroxene}} +$$

$$+ \underbrace{(3-2x)\mathrm{Ca(Mg,Fe)Si}_2\mathrm{O}_6 + 2x\mathrm{CaAl}_2\mathrm{SiO}_6}_{\text{Al-clinopyroxene}} + \underset{\text{spinel}}{(3-4x)(\mathrm{Mg,Fe})\mathrm{Al}_2\mathrm{O}_4}$$

where $0 < x < 0\cdot75$.

The garnet-bearing corona reaction is given by the following reaction:

$$\underbrace{(12+4y+2x)(\mathrm{Mg,Fe})\mathrm{SiO}_3 + 2x(\mathrm{Mg,Fe})\mathrm{Al}_2\mathrm{SiO}_6}_{\text{Al-orthopyroxene}} +$$

$$+ \underbrace{x\mathrm{Ca(Mg,Fe)Si}_2\mathrm{O}_6 + (3+y)(\mathrm{Mg})\mathrm{Al}_2\mathrm{SiO}_6}_{\text{Al-clinopyroxene}} +$$

$$+ \underset{\text{plagioclase}}{(3+y)\mathrm{CaAl}_2\mathrm{Si}_2\mathrm{O}_8} = \underset{\text{garnet}}{(3+y+x)(\mathrm{Mg,Fe})_5\mathrm{CaAl}_4\mathrm{Si}_6\mathrm{O}_{24}}$$

where $x \simeq 0\cdot1$, and y the number of formula units of anorthite per two albite. As in the case of Stjernöy rocks, these reactions are compatible with the data of the experimental anorthite–forsterite system (Kushiro and Yoder, 1966), basalt composition relationships (Green and Ringwood, 1967a) and the olivine–plagioclase system (Green and Hibberson, 1970). The indicated pressures and temperatures of the corona reactions are of the order 8 kbar and 800°C for the garnet, and slightly lower pressures and temperatures for the pyroxene–spinel coronites.

Coronas, consisting of an inner zone of orthopyroxene and an outer clinopyroxene–spinel symplectite, around olivine where the latter was originally in contact with plagioclase, occur in the Ewarara and Kalka layered intrusions of the Giles complex, central Australia (Goode and Moore, 1975). The coronas are independent of the cumulus or intercumulus nature of the olivine and plagioclase. In the metagabbros of the southern Adirondacks and Adirondack Highlands the reaction giving rise to the replacement of plagioclase by a vermicular intergrowth of clinopyroxene and spinel is given by Whitney and McLelland (1973):

$$(4+2y-4x)(Mg,Fe^{2+})+3Ca^{2+}+yCaAl_2Si_2O_8+2NaAlSi_3O_8 =$$
$$= (3+y-2x)Ca(Mg,Fe)Si_2O_6+2xCaAl_2SiO_6+(1+y-4x)(Mg,Fe)Al_2O_4+$$
$$+2x(Mg,Fe)SiO_3+4xAl^{3+}+2Na^{+}$$

where $x \simeq 0{\cdot}1$ and y the number of formula units of anorthite per two albite.

The chemical and physical conditions of corona formation, with particular reference to the orthopyroxene, spinel (garnet) coronas in the olivine gabbros and troctolites of the Hadlington gabbroic complex, Ontario, Canada, have been investigated by Grieve and Gittins (1975). The orthopyroxenes vary in composition from $En_{75{\cdot}1}$ to $En_{56{\cdot}5}$ (maximum 1 mol. per cent Wo); the Mg/(Mg+Fe+Mn) ratio varies sympathetically with, and is always greater than, that of the associated olivine. The coronas are of two types, one with an inner rim of orthopyroxene, the z axis of which is normal to the olivine margin, and an outer rim of amphibole with inclusions of spinel; the other in which the inner rim of orthopyroxene is surrounded by amphibole containing minor amounts of spinel, and an outer, generally incomplete, rim of garnet.

The temperature of the corona formation, 850°C, estimated on the basis of the Mg–Fe partitioning between coexisting ortho- and clinopyroxene, on the assumption that equilibrium was established between the newly formed corona orthopyroxene and the pre-existing magmatic clinopyroxene, indicates that the corona reaction took place at subsolidus temperatures. Quantitative calculations, based on the chemical composition of the corona minerals and of the olivine and plagioclase do not provide a satisfactory chemical balance. This, however, is considerably improved if the observed orthopyroxene–amphibole (spinel) contact is (*a*) not considered to define precisely the original olivine–plagioclase contact, and (*b*) the present composition of the olivine is unrepresentative of its original composition. Thus Grieve and Gittins conclude that the subsolidus diffusion reaction between olivine and plagioclase involves a migration of the reaction boundary towards olivine, and an accompanying change towards a more Fe-rich composition of the orthosilicate.

Hypersthene–spinel coronas in the troctolitic cumulates, Wichita Mountains, Oklahoma, have been described by Hiss and Hunter (1966). The exsolution of spinel showing the (111) plane parallel to (100) and either [100] or $[1\bar{1}0]$ parallel to the y axis of the orthopyroxene, (Fs_{25}–Fs_{30}) of gabbro xenoliths during reheating of the basalt and trachybasalt flows and dykes has been described by Le Maitre (1965).

The development of orthopyroxene-bearing coronas due to the following reactions:

biotite + quartz → orthopyroxene (1)

garnet + quartz → orthopyroxene + plagioclase (2)

garnet + diopside-augite + quartz → orthopyroxene + pargasite + plagioclase (3)

garnet → orthopyroxene + spinel + quartz (4)

titano-magnetic oxides + quartz → orthopyroxene solid solutions (5)

has been described in the rocks of the granulite terrains from the Massif Central, France, and In'Ouzzal, Algeria, by Leyreloup *et al.* (1975). The corona-forming reactions are considered to have developed during the passage from an initial high pressure–temperature to a low pressure granulite facies phase resulting from a relatively rapid uplift of the lower crust and the consequent retention of a high geothermal gradient.

Other descriptions of orthopyroxene-bearing coronas between olivine and plagioclase include those of the garnet-bearing metaperidotites in the garnet–amphibolite complex of the Ötztal Alps, Austria (Miller, 1974), the mafic cumulates of the Seiland petrographic province, northern Norway (Gardner and Robins, 1974), the gabbroic rocks of Thessaloniki, Greece (Sapountzis, 1975), and pyroxene granulite xenoliths in the basalts of Nunivak Island, Alaska (Francis, 1976).

Orthopyroxene–ilmenite–plagioclase coronas around olivine in Apollo 14 breccia have been described by Cameron and Fisher (1975).

Cation distribution. Investigations by Mössbauer spectroscopy and X-ray diffraction of orthopyroxenes indicate that ordering occurs and the Fe^{2+} ions favour the more distorted, $M2$, sites in the structure (see p. 26). Assuming an ideal solution model for each site the distribution coefficient is given by

$$K = X_2(1-X_1)/X_1(1-X_2)$$

where X_1 and X_2 are the site occupancy factors, i.e. number of Fe atoms per six oxygens at $M1$ and $M2$ respectively. Typical values of K are those found by Ghose and Hafner (1967) using the Mössbauer technique. Metamorphic and volcanic orthopyroxenes had $K = 12$ and 48 respectively, while two metamorphic pyroxenes heated to 1000°C yielded a distribution isotherm with $K \simeq 7$. The volcanic orthopyroxenes isotherm lies close to this curve (Fig. 23) and it is thus likely that they were quenched from a temperature of about 1000°C.

The distribution and fractionation ratios of Mg and Fe^{2+}, as well as the distribution of Ti, Cr, Ni, Co and Mn, between coexisting orthopyroxenes and olivines and coexisting ortho- and clinopyroxenes, were investigated by DeVore (1955) who explained the ionic distribution, in terms of the different polarizing power of the constituent cations, in relation to the differences in the polarizabilities of the anions within the individual crystal phases and between different crystal phases. DeVore (1957) noted that although the orthopyroxene has twice the number of six-co-ordinated positions compared with that of the diopside structure, the latter contains more Fe^{3+}, Ti and $[Al]^6$ than the coexisting orthopyroxene. Another early investigation of cation distribution between ortho- and clinopyroxenes was presented by Carstens (1958) and showed Mn contents of 0·27, 0·19 and 0·28 per cent for the orthopyroxenes, compared with 0·14, 0·05 and 0·14 per cent for the clinopyroxenes, in a hypersthene gabbro and a banded inclusion. The vanadium

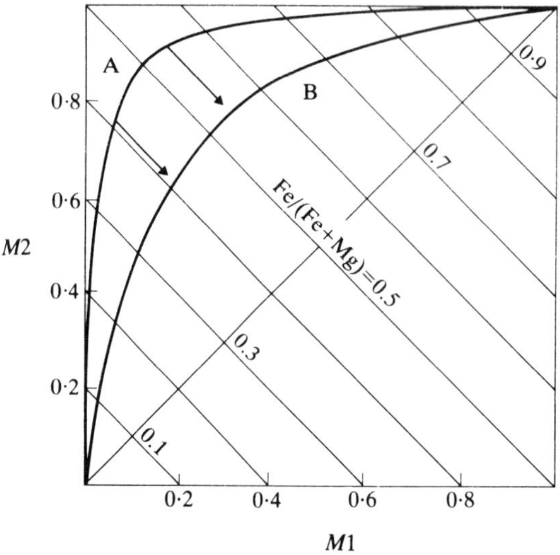

Fig. 23. Fe–Mg order–disorder in metamorphic, heated metamorphic and volcanic orthopyroxenes. Curves A and B show estimated Fe–Mg distribution isotherms at *M*1 and *M*2 sites for *K* = 48 and *K* = 7 at 1000°C. Arrows, heat treated metamorphic orthopyroxenes (after Ghose and Hafner, 1967).

contents of these pyroxene pairs also display a contrasted distribution with 0·02, 0·04, 0·03 and 0·04, 0·08, 0·06 atom % for the ortho- and clinopyroxenes respectively.

The thermodynamic basis for the distribution of Mg^{2+} and Fe^{2+} ions in coexisting orthopyroxenes and clinopyroxenes was considered by Kretz (1961a) and Bartholomé (1962), following the discovery by Mueller (1960, 1961), that for a number of coexisting pyroxene pairs, a plot of $X_{Mg}^{Opx} = Mg/(Mg+Fe^{2+})$ in orthopyroxene against $X_{Mg}^{Cpx} \pm Mg/(Mg v Fe^{2+})$ in clinopyroxene gives a curve of the type to be expected if both minerals are ideal or near ideal mixtures. The distribution coefficient with reference to the Mg–Fe exchange may be expressed:

$$K_D^{Mg\text{–}Fe} = \frac{X_{MgSiO_3}}{1-X_{MgSiO_3}} \cdot \frac{1-X_{CaMgSi_2O_6}}{X_{CaMgSi_2O_6}}$$

where X_{MgSiO_3} and $X_{CaMgSi_2O_6}$ are mole fractions and the exchange equilibrium is:

$$\underset{\text{orthopyroxene}}{FeSiO_3} + \underset{\text{calcic pyroxene}}{CaMgSi_2O_6} \rightleftharpoons \underset{\text{orthopyroxene}}{MgSiO_3} + \underset{\text{calcic pyroxene}}{CaFeSi_2O_6}$$

Kretz (1963) has presented data pertaining to the distribution coefficients for coexisting ortho- and clinopyroxenes from twenty-five high grade metamorphic and fifteen igneous rocks, and showed that K_D varied between 0·51 and 0·65 for the pyroxene pairs of the first group and from 0·65 to 0·86 for the second group. Subsequently Sen and Rege (1966) showed that the distribution coefficients of pyroxene pairs from the pyroxene–hornblende and pyroxene granulites, Saltora, eastern India, range from 0·56 to 0·61.

A more recent study of the distribution of Mg and Fe^{2+} between coexisting orthopyroxenes or pigeonites and Ca-pyroxenes in igneous and metamorphic rocks has been made by Saxena (1968a). Figure 24 shows two symmetric curves, which are solutions for K_D (= 0·54 and 0·73) at the pressure–temperature conditions of the granulite facies and igneous rocks respectively.

The magnesium–iron distribution coefficients in the ortho–clinopyroxene pairs in the basic granulites from Broken Hill, New South Wales, have been determined by Binns (1962), who shows that the K_D values (as defined above) decrease from $\simeq$ 0·50 for the more magnesium-rich to $\simeq$ 0·60 for the iron-rich pyroxenes.

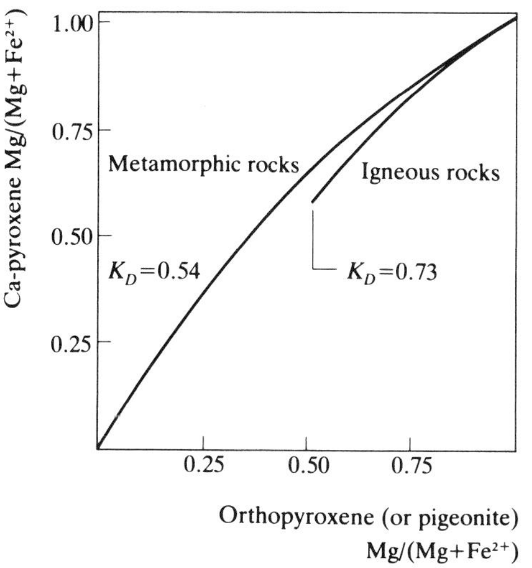

Fig. 24. Distribution of Mg and Fe^{2+} between coexisting orthopyroxene (or pigeonite) and Ca-pyroxene (after Saxena, 1968a).

Equilibrium temperatures for the Broken Hill granulites of New South Wales (Binns, 1962, 1965), and other granulites, based on the Wood-and-Banno temperatures (see p. 93) are estimated as 800° ± 50°C (Hewins, 1975).

A subsequent investigation of the distribution coefficients for ortho–clinopyroxene pairs from twelve basic granulites from the Quairading district, Western Australia, gave values ranging from ≃ 0·53 for the magnesium pyroxenes (orthopyroxene Mg ≃ 78) to ≃ 0·59 for the more iron-rich varieties (Mg ≃ 38). This distribution pattern led Davidson (1968) to suggest that one or both of the pyroxene structures do not conform completely with the assumed ideal Mg–Fe^{2+} distribution of the pyroxene equilibrium exchange, and the same author has discussed this non-ideal behaviour in terms of the geometry of the pyroxene cation sites, the relative bond energies of the sites and the structural ordering.

The marked dependence on chemical composition of the Fe–Mg distribution in the coexisting pyroxenes of the Quairading granulites, shown by Davidson's data, has been used by Froese and Gordon (1974) to derive the activity coefficients (γ) and the equilibrium constant (K) in terms of mole fractions (X), for regular solution.

$$K = \left(\frac{X^{Opx}_{FeSiO_3} X^{Cpx}_{CaMgSi_2O_6}}{X^{Opx}_{MgSiO_3} X^{Cpx}_{CaFeSi_2O_6}}\right)\left(\frac{\gamma^{Opx}_{FeSiO_3} \gamma^{Cpx}_{CaMgSi_2O_6}}{\gamma^{Opx}_{MgSiO_3} \gamma^{Cpx}_{CaFeSi_2O_6}}\right)$$

for the exchange reaction by solution of the equation:

$$\ln K = \ln K_D = \alpha^{Opx}(1 - 2X^{Opx}_{FeSiO_3}) - \alpha^{Cpx}(1 - 2X^{Cpx}_{CaFeSi_2O_6})$$

from which $\ln K$=0·412, K_D=1·51, $\alpha^{Opx}_{FeSiO_3}$=0.447, $\alpha^{Cpx}_{CaFeSi_2O_6}$=0·625, $\gamma^{Opx}_{MgSiO_3}$=1.56, $\gamma^{Cpx}_{CaMgSi_2O_6}$ ^{f} 1.87 (γ at infinite dilution).

Wilson (1976a) has presented data for six pairs of coexisting ortho- and clinopyroxenes with a wide range of Fe content (ΣFeO 26·05–37·29 and 10·28–15·94 weight per cent respectively) from two mafic granulite localities in the Frazer Range, Western Australia. The pyroxene compositions show a systematic variation and a good correlation of Fe (or Mg) with Al, Mn and Na, such that the Al content of the pyroxenes can be predicted provided the Mg or ΣFe content is known. As the element distribution, particularly of Al and Na, may be differently affected by both pressure and temperature, comparisons of pyroxenes from one region to another is facilitated by normalizing Al (or other elements) to an appropriate Mg value. In a subsequent paper, Wilson (1976b) has shown that the Al contents of the pyroxenes from the higher pressure compared with those from the lower pressure locality are less dependent on the Al content of the host rock, and are more effective discrimnants of the metamorphic conditions than K_D of the pyroxenes pairs (see also Woodford and Wilson, 1976).

The Fe–Mg distribution coefficients (1·40 to 1·90) of coexisting ortho- and clinopyroxenes in the amphibolites of the New Jersey Highlands (Maxey and Vogel, 1974), show a contrasted trend compared with those of the granulites at Broken Hill (Binns, 1962) and Quairading, Australia (Davidson, 1968). K_D is systematically related to bulk rock compositions and these, rather than crystallization temperatures appear to be the dominant influence on the distribution coefficients.

Values of $K_D^{\text{Mg-Fe}}$ for chromian orthopyroxenes ($Fe_{10-17\cdot4}$)–chrome diopsides ($Ca_{50\cdot3}Mg_{46\cdot5}Fe_{3\cdot2}$–$Ca_{49\cdot8}Mg_{44\cdot9}Fe_{5\cdot3}$), orthopyroxene–phlogopite, and orthopyroxene–chromite pairs from the ultrabasic rocks of Kondapalle and Gangineni, Andhra Pradesh, India are given by Mall (1973).

Numerous other studies of the Mg–Fe^{2+} distribution between coexisting ortho- and clinopyroxene have been reported. These include the igneous and metamorphic pyroxenes from the anorthosite–mangerite rocks of the Morin series, southern Quebec (Philpotts, 1966), containing an orthopyroxene with a Mg:Fe^{2+} ratio of 60:40 and differing from most Ca-poor pyroxenes with a comparable iron content in having crystallized as a primary rhombic pyroxene. Naldrett and Kullerud (1967) have presented analyses of thirty-three ortho–clinopyroxene pairs from the xenolithic norite, norite, footwall breccia matrix and footwall gneiss of the Strathcona Mine, Sudbury, and noted that there is a systematic increase in the value of K_D with increasing Fe:Mg ratio of the individual pairs. $K_D^{\text{Mg/Fe}}$ values for some ortho- and clinopyroxene pairs in granulite facies and intermediate gneisses are given by Giguère (1972), and for the amphibolite interlayers in the gneiss complex, north-west Adirondack Mountains, by Engel *et al.* (1964). The form of the function describing the partitioning of cations between coexisting single- and multi-site phases can be deduced, as shown by Grover and Orville (1969), from the cation exchange free energies between pairs of sites and conversely, the cation exchange free energies between pairs of sites can, in some cases, be deduced from the form of the partition curve. Although clinopyroxene and orthopyroxene both have two structurally distinct cation sites, the system may be considered on the basis of coexisting single- and multi-site phases if the Ca ($M2$) site in the clinopyroxene is completely filled with Ca, i.e. it is assumed that the $M2$ site does not participate in the Fe–Mg exchange. The analysis may be further simplified by assuming that the $M1$ site in clinopyroxene is energetically equivalent to the $M1$ site in orthopyroxene. With these assumptions it can be shown that ΔG_T°, the molar free energy of exchange for the intercrystalline reaction

$$2\text{Mg}^{M1}_{\text{Cpx}} + (\text{Fe}^{M1} + \text{Fe}^{M2})_{\text{Opx}} \rightleftharpoons 2\text{Fe}^{M1}_{\text{Cpx}} + (\text{Mg}^{M1} + \text{Mg}^{M2})_{\text{Opx}}$$

is equal to $-\Delta G_E^\circ$ for the intracrystalline reaction

$$(\text{Fe})^{M1}_{\text{Opx}} + (\text{Mg})^{M2}_{\text{Opx}} \rightleftharpoons (\text{Fe})^{M2}_{\text{Opx}} + (\text{Fe})^{M1}_{\text{Opx}}$$

The theoretical fit with the above model for Fe/Mg distribution between coexisting ortho- and clinopyroxene (Ca-rich) was found with a value of $\Delta G_E^\circ = -1538$ cal/mole and is shown in Fig. 25. For the full theoretical treatment and explicit solution for ΔG_E° see Grover and Orville (1969).

The partition of Mg and Fe^{2+} in coexisting pyroxenes may be approached by

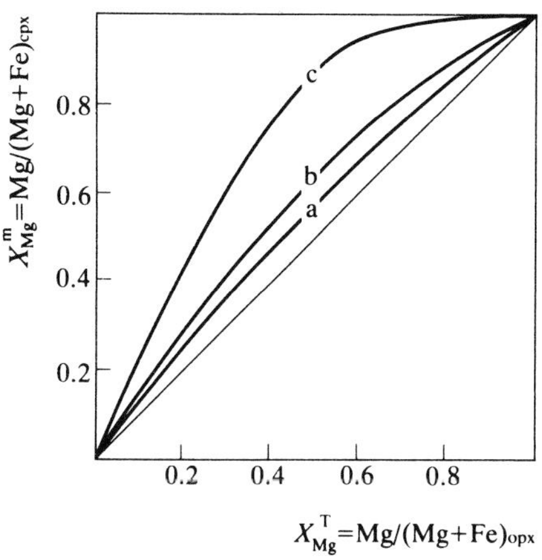

Fig. 25. Distribution of Mg between coexisting clinopyroxenes and orthopyroxenes (after Grover and Orville, 1969). Curves (a) and (b) theoretical fits for igneous and metamorphic data computed according to an ideal one-site–two-site model in which ΔG°_T ($= \Delta G^\circ_E = 1538$ cal mol^{-1}) constant over temperature range 673°K (curve b) and 1373°K (curve a). Curve (c) derived from data given by Ghose and Hafner (1967). X^{m}_{Mg} is the mol. fraction of Mg in site m, and X^{T}_{Mg} is the average mol. fraction of Mg in both sites.

considering the intercrystalline partitions separately rather than through the exchange reaction:

$$MgSiO_3^{Cpx} + FeSiO_3^{Opx} \rightleftharpoons FeSiO_3^{Cpx} + MgSiO_3^{Opx}$$

In both phases, as noted earlier, Mg and Fe^{2+} are accommodated in different proportions in two non-equivalent sites $M1$ and $M2$. The equilibrium constant for the Mg–Fe partitioning may thus be expressed:

$$K_{MgSiO_3} = a^{Opx}_{MgSiO_3}/a^{Ca\text{-}px}_{MgSiO_3}$$

$$K_{FeSiO_3} = a^{Opx}_{FeSiO_3}/a^{Ca\text{-}px}_{FeSiO_3}$$

$$K_{MgSiO_3/FeSiO_3} = a^{Ca\text{-}px}_{FeSiO_3} \cdot a^{Opx}_{MgSiO_3}/a^{Ca\text{-}px}_{MgSiO_3} \cdot a^{Opx}_{FeSiO_3}$$

and

$$K_{MgSiO_3 \cdot FeSiO_3} = a^{Opx}_{MgSiO_3} \cdot a^{Opx}_{FeSiO_3}/a^{Ca\text{-}px}_{MgSiO_3} \cdot a^{Ca\text{-}px}_{FeSiO_3}$$

Fleet (1974a) has illustrated this approach with reference to Ca-poor and Ca-rich pyroxene pairs from the Skaergaard and Bushveld intrusions, and from charnockites. The mole fractions X_{MgSiO_3} and X_{FeSiO_3}, X_{MgSiO_3}/X_{FeSiO_3} and $X_{MgSiO_3} \cdot X_{FeSiO_3}$ are shown in Fig. 26 and are plots respectively of the functions K'_{MgSiO_3}, K'_{FeSiO_3}, $K'_{MgSiO_3}/K'_{FeSiO_3}$ and $K'_{MgSiO_3 \cdot FeSiO_3}$. The $K'_{MgSiO_3/FeSiO_3}$ distributions are linear, and the form of the curves of K'_{MgSiO_3} and K'_{FeSiO_3} are complementary although the latter are more widely spaced, consequential on the location of Fe^{2+} in most of the Ca-rich pyroxene $M2$ sites not occupied by (Ca + Na). This illustrates the trend of Fe enrichment and Ca depletion in the pyroxene $M2$ sites in igneous rock suites and emphasizes that the differences in K_D reported for igneous and metamorphic pyroxenes is largely a reflection of this trend rather than the temperature dependence of the intracrystalline partition of the Ca-poor pyroxene.

In a subsequent paper, Fleet (1974b) presented data on the partition of Si, Al, Ti, Fe^{3+}, Mg, Fe^{2+}, Mn, Ca and Na between coexisting Ca-poor and Ca-rich pyroxenes for a variety of igneous and metamorphic rocks. Partition trends for Ti, Mg, Fe^{3+}, Mn and Na are shown to be characteristic of different petrological groups. The partition of Mg, Fe^{2+} and Mn also correlate with inferred cooling rates, and the partition coefficients of pyroxenes from more quickly cooled rocks show the closest approach to unity.

A study of the distribution of calcium ions between coexisting calcic pyroxenes

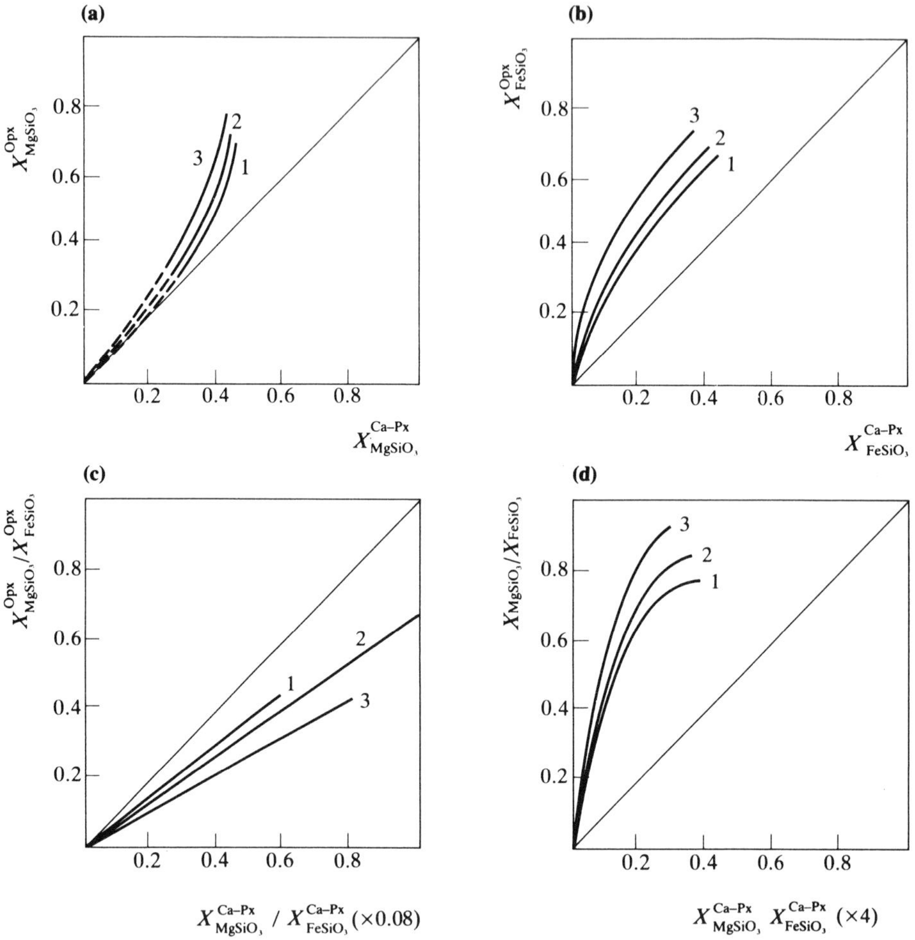

Fig. 26. Coexisting Ca-poor and Ca-rich pyroxenes: (a) mole fractions $MgSiO_3$, K'_{MgSiO_3} distributions; (b) mole fractions $FeSiO_3$, K'_{FeSiO_3} distributions; (c) quotients of mole fractions of $MgSiO_3$ and $FeSiO_3$, $K'_{MgSiO_3/FeSiO_3}$ distributions; (d) products of mole fractions of $MgSiO_3$ and $FeSiO_3$, $K'_{MgSiO_3/FeSiO_3}$ distributions (after Fleet, 1974a). Curves 1, 2, Skaergaard and Bushveld intrusions, curve 3, charnockites.

and orthopyroxenes in thirteen igneous and forty-one high grade metamorphic rocks has been presented by Saxena (1968a). A plot of the variation in the calcium content of the pyroxene pairs is shown in Fig. 27(a) in which ln $K_D^{Mg\text{-}Fe}$ is used as a measure of crystallization temperature. The concentration of Ca decreases in the Ca-pyroxenes and increases in the orthopyroxenes as ln K_D (and T) increase, in general agreement with the experimentally determined solvus on the $CaMgSi_2O_6$–$MgSiO_3$ join (see p. 84). The data show some discrepancies in relation to the experimentally determined solvus; that on the calcium-rich limb is probably due to variations in the concentrations of other octahedral ions. The mole fractions, $Mg/(Mg+Fe^{2+})$, of the pyroxene pairs fall into two distinct fields corresponding to the metamorphic and igneous parageneses, and are dependent on crystallization temperature rather than bulk chemical composition. In both parageneses the Mg/Fe^{2+} ratio is enriched at higher temperature (Fig. 27(b)).

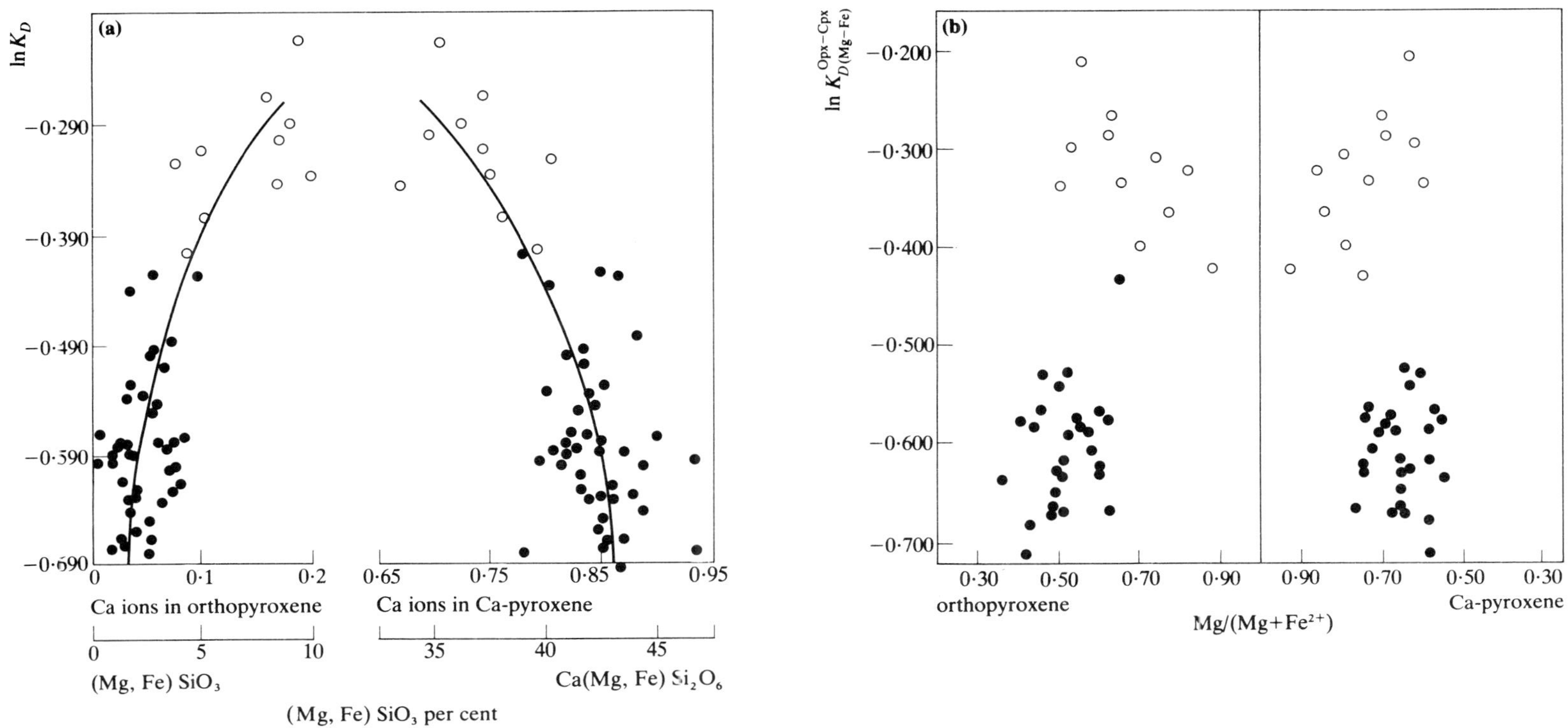

Fig. 27. **(a) Variation in the concentrations of calcium in Ca-poor and Ca-rich pyroxenes. Values of ln $K_{Mg-Fe}^{opx-cpx}$ have been used as a measure of temperature (and pressure) of their crystallization (after Saxena, 1968a). (b) Variation of mole fraction Mg/(Mg + Fe^{2+}) in Ca-poor and Ca-rich pyroxenes with temperature (= ln K_D) (after Saxena, 1968a). Solid circles, pyroxenes from metamorphic rocks. Open circles, pyroxenes from igneous rocks.**

An analytical method has been used by Mueller (1966) to examine the stability relations of the orthopyroxene, Ca-rich pyroxene, calcite, quartz, graphite assemblages from the iron formations of Quebec (Kranck, 1961). The pertinent reactions may be expressed:

$$\underset{\text{orthopyroxene}}{MgSiO_3} + \underset{\text{calcite}}{CaCO_3} + \underset{\text{quartz}}{SiO_2} \rightleftharpoons \underset{\text{Ca-pyroxene}}{CaMgSi_2O_6} + \underset{\text{fluid}}{CO_2} \quad (1)$$

$$\underset{\text{orthopyroxene}}{MgSiO_3} + \underset{\text{Ca-pyroxene}}{CaFeSi_2O_6} \rightleftharpoons \underset{\text{orthopyroxene}}{FeSiO_3} + \underset{\text{Ca-pyroxene}}{CaMgSi_2O_6} \quad (2)$$

On the assumption that an ideal solution model holds for both pyroxenes, the equations corresponding to reactions (1) and (2) are:

$$K_{(1)} \exp\left(\frac{-P\Delta V_1}{RT}\right) = X_{Mg}^{Cpx}/X_{Mg}^{Opx} \,.\, P^*_{CO_2} \quad (3)$$

$$K_{(2)} \exp\left(\frac{-P\Delta V_2}{RT}\right) = X_{Mg}^{Cpx}(1-X_{Mg}^{Opx})/X_{Mg}^{Opx}(1-X_{Mg}^{Cpx}) \quad (4)$$

in which $P^*_{CO_2}$ is the fugacity of CO_2, ΔV_1 and ΔV_2 are the differences in volume between the solid reactants and the products of reaction. For reaction (1) ΔV_1 has a large negative value, $\simeq -25\,cm^3$ at conditions of STP, and the reaction moves to the right by the total pressure operating. For reaction (2) ΔV_2 is small, $\simeq -1{\cdot}8\,cm^3$, and the effect of pressure can be neglected. Expressing $K_{(2)}\exp(-P\Delta V_2/RT)$ as $K'_{(2)}$, and eliminating X_{Mg}^{Cpx} from equations (3) and (4) gives:

$$P_{CO_2} = \frac{K_1}{K_2}\exp\left(\frac{-P\Delta V_1}{RT}\right)[(K'_2-1)X_{Mg}^{Opx}+1]$$

from which the compositional variation of the coexisting pyroxenes may be evaluated, either in relation to the effect of pressure at constant CO_2 fugacity and temperature, or in terms of the effect of CO_2 fugacity at constant pressure and temperature. The latter case is illustrated in Fig. 28, and shows that the iron content increases with decreasing CO_2 fugacity. The figure may also be used to illustrate the relation between mineral composition, the stability range of the clino- and orthopyroxenes and CO_2 fugacity at constant temperature and total pressure.

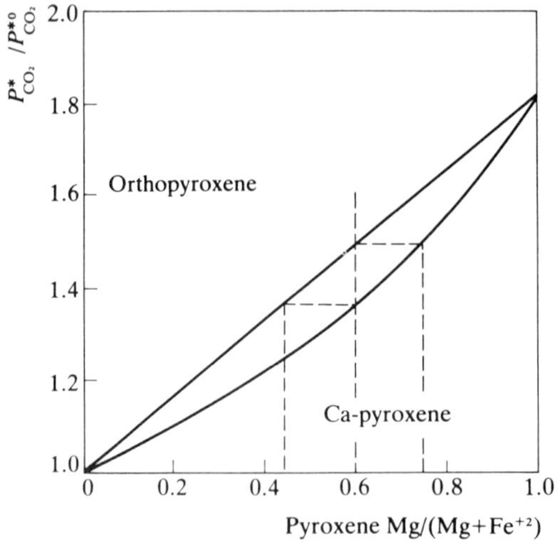

Fig. 28. Relative activity or fugacity ratio of CO_2 as a function of the orthopyroxene and clinopyroxene composition from equations (3) and (4) (see text) (after Mueller, 1966).

Consider the system orthopyroxene–calcite–quartz at 900°K with $X_{Mg}^{Opx} = 0{\cdot}6$ and $P_{CO_2}^{*}/P_{CO_2}^{*\circ}$ (fugacity ratio of CO_2) = 1·6. When the latter value falls to about 1·5 the reaction $MgSiO_3 + CaCO_3 + SiO_2 \rightleftharpoons CaMgSi_2O_6 + CO_2$ moves to the right and Ca-pyroxene, $X_{Mg}^{Cpx} = 0{\cdot}75$, is initially produced in accordance with reaction (2). Further reduction of $P_{CO_2}^{*}$ is accompanied by a further movement to the right until the composition of the orthopyroxene is $X_{Mg}^{Opx} = 0{\cdot}45$, when all the orthopyroxene is consumed. With still lower values of $P_{CO_2}^{*}$ the system consists of Ca-pyroxene, $X_{Mg}^{Cpx} = 0{\cdot}6$, with some calcite and quartz. Mueller has also considered the assemblage Ca-pyroxene, orthopyroxene, olivine and calcite, characteristic of rocks having a lower silica content than considered earlier, and has shown that the reaction:

$$3MgSiO_3 + CaCO_3 \rightleftharpoons CaMgSi_2O_6 + Mg_2SiO_4 + CO_2$$

is strongly displaced to the right under most metamorphic conditions, and that contrary to the case of reaction (1), $P_{CO_2}^{*}$ decreases with increasing values of X_{Mg}^{Opx}.

The Mg/Fe^{2+} partition and the distribution of Ni, Zn, Co and Mn in coexisting ortho- and clinopyroxenes in lenses in eclogitic rocks has been studied by Matsui *et al.* (1966). A plot of ionic radii against $K_{M/Fe}^{Opx-Cpx}$, where M is the divalent ion of comparable size to Mg and Fe^{2+}, led the authors to conclude that the partition is essentially controlled by the radii of the cations, in which the smaller ions enter preferentially into the clinopyroxene structure.

The distribution of transition metal ions between coexisting orthopyroxenes and clinopyroxenes in metamorphic rocks has been investigated by Schwarcz (1967). The partition ratio $K^{Opx-Cpx} = (R_E)_{Opx}/(R_E)_{Cpx}$, where R_E is the weight ratio of the minor or trace element to Fe^{2+}, has been evaluated for (1) the pyroxenes of amphibolite interlayers in a gneiss complex, north-west Adirondack Mountains (Engel *et al.*, 1964), and for (2) the charnockite series of Madras, India (Howie, 1955), from which the following sequences were obtained:

$$K_{Cu} > K_{Fe} > K_{Mn} > K_{Co} > K_{Ni} \quad (1)$$

$$K_{Fe} \simeq K_{Mn} > K_{Co} > K_{Ni} \quad (2)$$

$$K_{Fe^{3+}} > K_{Cr}$$

Partition coefficients for V, Sc, Mn and Co for the basaltic andesite–dacite sequences of the younger volcanic islands of Tonga increase with increasing iron enrichment (Ewart *et al.*, 1973).

The distribution of manganese in 115 coexisting Ca-poor and Ca-rich pyroxenes has been investigated by Lindh (1974). The distribution is regular indicating a close approach to equilibrium. Manganese is equally distributed between the pyroxenes when the mole fraction Mn/(Mg + Fe + Mn) in the orthopyroxene is less than 0·008. With more Mn-rich pyroxene pairs there is a small enrichment of Mn in the orthopyroxene; the average distribution coefficient is 1·4.

The distribution coefficient, $K_D^{Opx\text{-}ol}$, for an orthopyroxene, $Mg/(Mg + \Sigma Fe) = 13{\cdot}8$ (Table 4, anal. 32) – fayalite, $\simeq Fa_{96}$, pair in eulysite, from the Mariupol iron formation, has been determined by Val'ter (1969).

The partition of Mg and Fe between coexisting orthopyroxene and spinel in the granulite facies hornblende-bearing alpine peridotites of Klamath Mountain Province, northern California, and Finero, northern Italy, have been evaluated by Medaris (1975). The regression equation for the orthopyroxene–spinel pairs of the latter mass is

$$\ln (X_{Mg}/X_{Fe})^{Opx}(X_{Fe}/X_{Mg})^{Sp} = 0{\cdot}894 + 2{\cdot}310\,Y_{Cr}^{Sp}$$

where Y_{Cr}^{Sp} is the fraction of Cr^{3+} in spinel.

The partition of Ni between orthopyroxene and olivine, in the prehistoric Makaopuhi Lava Lake, Hawaii, has been determined by Evans (1969), and shows that the poikilitic orthopyroxene probably crystallized in exchange equilibrium with the phenocrystal olivine at about 1040°C. The nickel partition coefficients of the pairs orthopyroxene–olivine, orthopyroxene–augite and orthopyroxene–hornblende of the Parikkala gabbro, south-east Finland, have been presented by Häkli (1968), and between orthopyroxene and olivine in the metamorphosed layered 'intrusion' of the Fiskenæsset complex, west Greenland, by Windley and Smith (1974).

The distribution coefficient of Ni between the orthopyroxene and the coexisting sulphide phase in the peridotites and gabbros of the Kotalahti area is 0·00758 (Häkli, 1963).

In a study of the phase equilibria in the charnockites of the Varberg district, Sweden, Saxena (1968b) has determined the distribution of Fe between orthopyroxene and biotite, orthopyroxene and hornblende, as well as ortho- and clinopyroxene.

The compositional relationship with respect to magnesium and iron in orthopyroxene–hornblende pairs in basic granulite has been investigated by Sen (1973). No significant departure from ideality is indicated by the Mg–Fe distribution, and Sen noted that increasing X_{Mg} may be related to higher temperature of equilibrium. Saxena (1968c) has also calculated the distribution coefficient, $K_{DFe}^{Gar\text{–}Opx} = 2{\cdot}5$, for orthopyroxene–garnet pairs in rocks of the granulite facies, and an empirical study of the Fe^{2+}/Mg/Mn distribution in orthopyroxene–garnet pairs from the Quairading iron formation by Davidson and Mathison (1973), demonstrated that the Fe^{2+}/Mg distribution is independent of the Mn content.

Consideration by Schwarcz (1967) of the distribution of transition elements between orthopyroxene–olivine pairs in dunite and olivine-rich inclusions in basalts (Ross *et al.*, 1954) showed that for $D^{Ol\text{–}Opx}$

$$\underset{(5\cdot2)}{D_{Ni}} > \underset{(2\cdot8)}{D_{Co}} > \underset{(1\cdot4)}{D_{Mn}} > \underset{(1\cdot2)}{D_{Fe^{2+}}}$$

where D_E = concentrations of E in olivines/concentrations of E in clinopyroxenes indicating that for cations of the same charge, ionic radius has a smaller effect than crystal field stabilization energy on the order of distribution coefficient.

Other distribution studies include Peters' (1968) investigation of Mg, Fe, Al, Ca and Na in coexisting clinopyroxenes, orthopyroxenes and olivines of the Totalp serpentinite, Switzerland, and the Malenco serpentinite, northern Italy.

The distribution of Fe and Mg in orthopyroxene–augite pairs, synthesized at 800°–810°C and $P_{fluid} = P_{total} = 15$ kbar, has been determined by Lindsley *et al.* (1974). The distribution coefficients, $K_D = (X_{Mg}^{Opx}.X_{Fe}^{Aug})/(X_{Fe}^{Opx}.X_{Mg}^{Aug})$, have values of $\simeq 0{\cdot}690$ and show a small increase with increasing Fe/(Fe+Mg) ratios.

The partitioning of Fe^{2+} and Mg between synthetic orthopyroxene and olivine at 900°C, a total fluid pressure of 0·5 kbar, and oxygen fugacities defined by the fayalite–magnetite–quartz buffer has been investigated by Medaris (1969). The partition equation is:

$$\log\left(\frac{X_{Fe}}{X_{Mg}}\right)_{Ol} = 0{\cdot}1650 + 1{\cdot}1128\log\left(\frac{X_{Fe}}{X_{Mg}}\right)_{Opx}$$

and the partitioning of Fe^{2+} and Mg between the orthopyroxene and olivine is not appreciably dependent on temperature between 900° and 1300°C. Evaluation of the data on Fe^{2+}–Mg^{2+} exchange equilibria between the *M*1 and *M*2 sites in orthopyroxenes, and between the orthopyroxene and olivine phases, by Blander (1972) showed that the free energy change ΔG_T°, for the reaction:

$$Fe_2SiO_4 + 2MgSiO_3 \rightleftharpoons Mg_2SiO_4 + 2FeSiO_3$$

is 1·65 kcal/mole, from which it may be deduced that the standard free energy of formation of ferrosilite from wüstite and β-cristabolite at 900°C is – 1·8 kcal/mole. The partition of Fe^{2+} and Mg between coexisting orthopyroxene and olivine is according to Wood (1975), however, not suitable for use as a geothermometer.

Saxena and Nehru (1975) considered the ternary system $Mg_2Si_2O_6$–$FeMgSi_2O_6$–$CaMgSi_2O_6$ and estimated crystallization temperatures assuming ideal mixing for Fe–Mg and Fe–Ca and non-ideal mixing for Ca–Mg and binary *M*1 and ternary *M*2 sites. Saxena (1968a) has also discussed the distribution of Mn between coexisting clino- and orthopyroxenes, and shown that in general Ca-pyroxenes have approximately half the Mn content of the orthopyroxenes. The variation of the concentration of Fe^{3+} with distribution coefficient for sodium, $K_{D(Na)}$, between coexisting diopside-salites and orthopyroxenes, as well as the variation of the ratio Fe^{2+}/Mg with $\Sigma(Al + Fe^{3+} + Ti)$ ions in the *M*1 site, and $\Sigma(Ca + Na + K)$, have been studied by Saxena (1967).

The distribution of Mg and Fe in pyroxenes, in equilibrium with olivine, magnesioferrite and vapour, in the system MgO–FeO–Fe_2O_3–SiO_2 at subsolidus temperatures has been shown by Speidel and Osborn (1967) to be a function of temperature and oxygen fugacity, and these authors have determined the iron oxide contents of the three coexisting phases for a number of oxide mixtures at temperatures between 1080° and 1350°C and $\log f_{O_2}$ between −0·7 and −10·0. Applied to basic and ultrabasic rocks these compositional trends show that the Mg : Fe ratios in the three phases may differ significantly at temperatures lower than those at which crystallization occurred and, in the case of orthopyroxene, may involve an increase of the Mg : Fe ratio on cooling.

The distribution of Ti and Al between ortho- and clinopyroxene, garnet and ilmenite/rutile has been determined experimentally at temperatures and pressures of 1050°C and 25 kbar and 1100°C and 40 kbar respectively (Akella and Boyd, 1972). TiO_2 is low (0·2–0·5 weight per cent) and the atomic proportion of Al is more than twice that of Ti in the orthopyroxene.

The partition of Mg and Fe^{2+} between coexisting $(Mg,Fe)SiO_3$ pyroxene and $(Mg,Fe)_2SiO_4$ (spinel), at pressures of 80 and 90 kbar and temperatures of 840° and 1050°C has been investigated by Nishizawa and Akimoto (1973).

The distribution of nickel between orthopyroxene and liquid in a system with aluminous enstatite, forsterite and hydrous silicate liquid (12–17 weight per cent H_2O) at 20 kbar and 1025° ($\simeq$ 275 p.p.m.) and 1075°C ($\simeq$ 325 p.p.m.) has been determined by Mysen (1976b).

Experimental and Synthetic Systems

The incongruent melting of enstatite at 1 atm, first demonstrated by Bowen and Anderson (1914), has subsequently been shown to extend to 5 kbar under anhydrous conditions (Boyd *et al.*, 1964; see Taylor, 1973 and Chen and Presnall, 1975). Under water-saturated conditions enstatite melts incongruently at pressures above 10 to at least 60 kbar (Kushiro *et al.*, 1968; Kushiro, 1972a). In the system $MgSiO_3$–H_2O the incongruent melting products are dependent on the water content, thus in the relatively H_2O-rich region (> 11 weight per cent at 10 kbar)

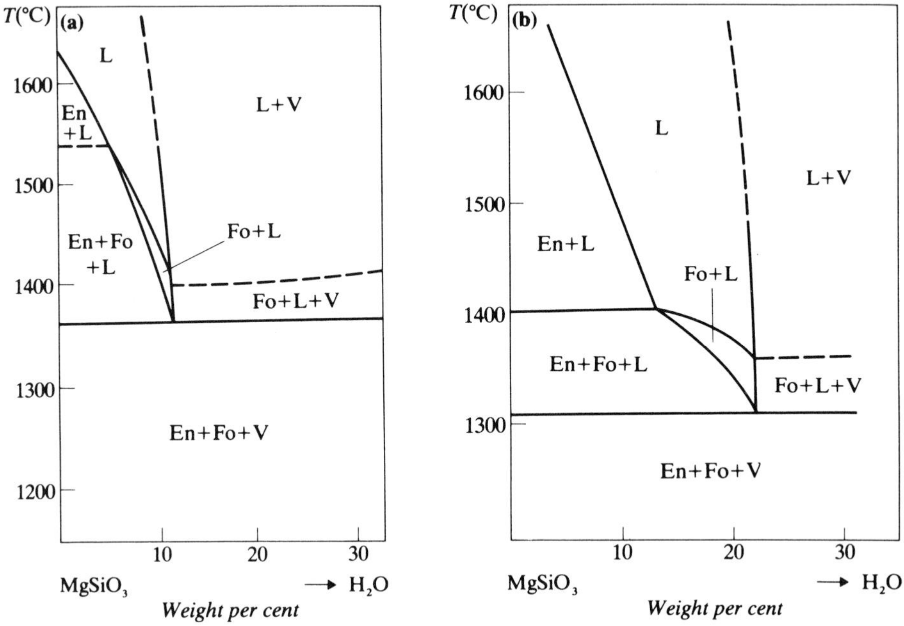

Fig. 29. (a) Phase equilibrium relations on the join $MgSiO_3$–H_2O at 10 kbar (after Kushiro *et al.*, 1968). (b) Phase equilibrium relations on the join $MgSiO_3$–H_2O at 20 kbar (after Eggler, 1973).

enstatite melts to forsterite + liquid + vapour, and in the relatively water-poor region, depending on the precise content of water and temperature, to enstatite + forsterite + liquid, forsterite + liquid, liquid, and enstatite + liquid (Fig. 29a). The phase equilibrium relations for the join $MgSiO_3$–H_2O have also been determined at 20 kbar (Eggler, 1973); at this pressure the melting temperature of the enstatite–forsterite–vapour assemblage is 1305°C (Fig. 29b) some 60°C below the melting temperature at 10 kbar. The phase relations on the join $5MgSiO_3.4CO_2$–H_2O at 20 kbar pressure have also been determined by Eggler (1973), and are illustrated in Fig. 30 (see also Eggler, 1975, 1976). The diagrams show that the solidus temperature varies continuously with changing vapour composition, that the incongruent melting temperature of enstatite is 1400°C, the same as in the system Mg_2SiO_4–SiO_2–H_2O, and that the subsolidus assemblage changes from enstatite + forsterite + vapour to enstatite + vapour when the vapour contains more than 38 mol. per cent CO_2.

The study of the system MgO–FeO–SiO_2 (Bowen and Schairer, 1935) showed that there is a complete solid solution series extending from $MgSiO_3$ to a maximum of 56 mol. per cent $FeSiO_3$ at solidus temperatures. The subsolidus phase equilibrium in the ternary system under low oxygen partial pressure has been investigated by Kitayama and Katsura (1968), and an isothermal section at 1204°C is shown in Fig. 31 at which temperature the solid solution of $FeSiO_3$ in the pyroxene (point A) is 48 mol. per cent. A determination of the activity–composition relations in the solid solutions, and the free formation of the end-members, by Nafziger and Muan (1967) showed the pyroxene solid solution to be essentially ideal at temperatures between 1200° and 1250°C.

The activity–composition relations in the $MgSiO_3$–$FeSiO_3$ solid solution series, and the standard free energy change in the formation of ferrosilite (Table 7) have

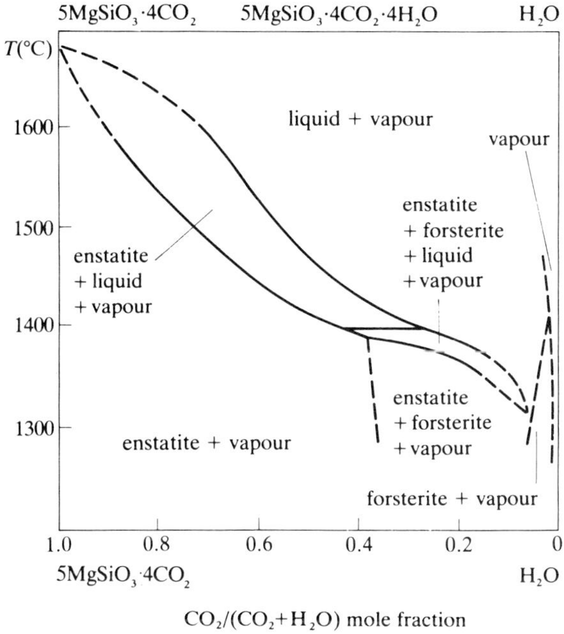

Fig. 30. Phase relations on the join $5MgSiO_3.4CO_2$–H_2O at 20 kbar (after Eggler, 1973).

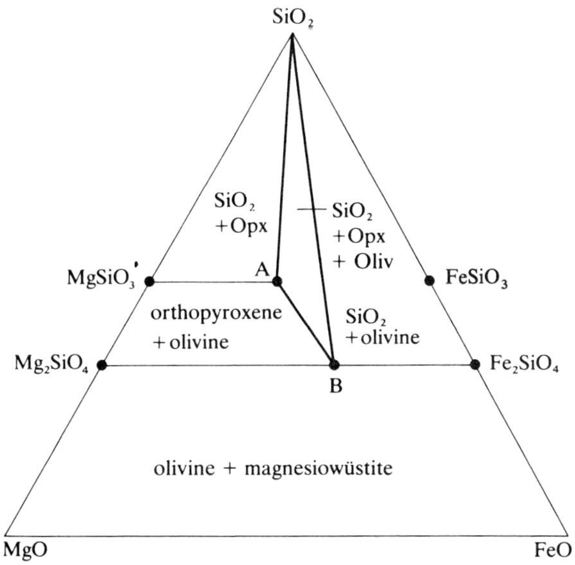

Fig. 31. Isothermal section for the system MgO–FeO–SiO_2 at 1204°C (after Kitayama and Katsura, 1968).

Table 7 Standard free energy of formation of ferrosilite from $Fe_2SiO_4+O_2$ (A) and from wüstite and SiO_2 (B)

T(°C)	(A) ΔG° (kcal/mole)	(B) ΔG° (kcal/mole)
1154	−42·5	−1·8
1204	−41·7	−1·5
1250	−41·3	−1·8

been determined also by Kitayama and Katsura (1968) and Kitayama (1970). The activity of $FeSiO_3$ in the pyroxene solid solution is represented by:

$$^{a}(FeSiO_3) = (P_{O_2}/P^*_{O_2})^{\frac{1}{2}}$$

where P_{O_2} is the oxygen partial pressures of the decomposition of the solid solution with varying compositions and $P^*_{O_2}$ is the oxygen partial pressure corresponding to the reaction of pure ferrosilite. The corresponding chemical reaction is given by the equation:

$$(FeSiO_3)_{ss} = Fe + SiO_2 + \tfrac{1}{2}O_2$$

The determination, by a quenching method in an atmosphere of a gas mixture of CO_2 and H_2 at 1154°, 1204° and 1250°C, of the activity–composition relations in $MgSiO_3$–$FeSiO_3$ solid solutions, ranging from ≃ 0·10 to 0·50 of the ferrosilite components (Fig. 32), by Kitayama is at variance with the conclusion of Nafziger and Muan, and showed a degree of non-ideality in the solutions at these temperatures, with increasing departure from ideality as the temperature decreases (see also p.65).

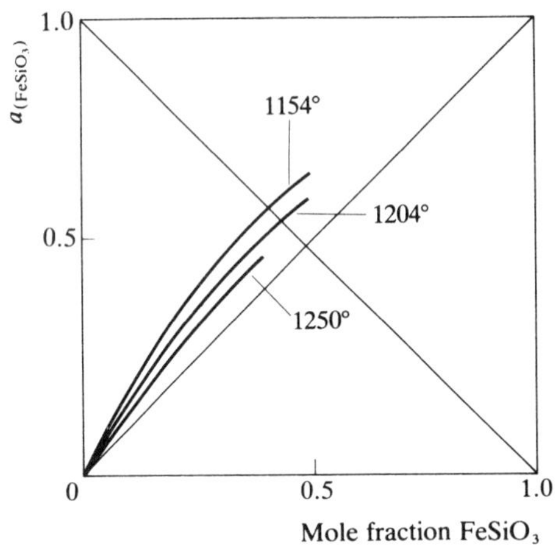

Fig. 32. Activity–composition relations in $MgSiO_3$–$FeSiO_3$ solid solutions at 1154°, 1204° and 1250°C (after Kitayama, 1970).

The observed distributions of Fe and Mg between the *M*1 and *M*2 sites in orthopyroxene, however, have been shown by Navrotsky (1971) to be consistent with the activity composition relations in the solid solutions determined by Nafziger and Muan. The cation distribution is substantially non-random and gives rise to a small negative excess free energy of mixing which may be partially or totally compensated by a small positive lattice term not closely related to the degree of order.

The oxidation of synthetic (Mg,Fe)SiO_3 pyroxenes has been studied thermogravimetrically at the constant oxygen pressure of air, the weight change being measured during oxidation with linear increasing temperature to 1350°C (Angeletos, 1975). Thus for each composition

$$Mg_{1-x}Fe_xSiO_3 \quad \text{or} \quad (1-x)MgO\left(\frac{Fe+Fe_2O_3}{3}\right)SiO_2$$

after complete oxidation:

$$xFe^{2+} - xe \rightleftharpoons xFe^{3+} \quad \text{or} \quad 2xFeO + \tfrac{1}{2}xO_2 \rightleftharpoons xFe_2O_3$$

Results obtained from compositions $x = 0{\cdot}0$, $0{\cdot}1$, $0{\cdot}3$, $0{\cdot}6$, $0{\cdot}7$, $0{\cdot}8$, $0{\cdot}9$ and $1{\cdot}0$ enabled Angeletos to construct the determination curve of oxidation (Fig. 33).

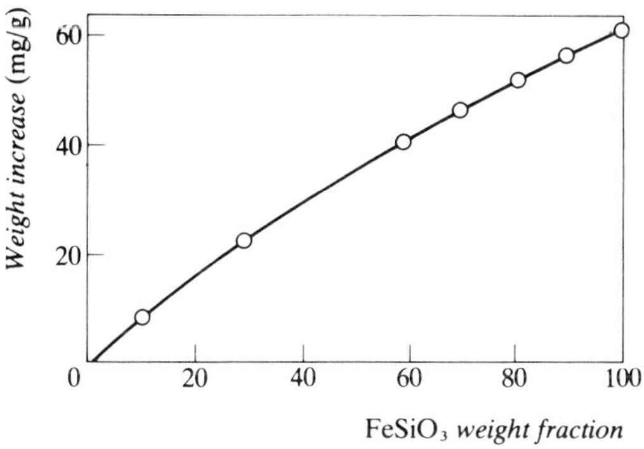

Fig. 33. Determinative curve of oxidation of (Mg,Fe)SiO_3 pyroxenes (after Angeletos, 1975). Open circles, theoretical weight increase.

If the composition of the sample, but not the ionic state of iron, is known the Fe^{3+}/Fe^{2+} ratio after complete oxidation can be obtained from the determinative curve as the ratio, theoretical weight increase minus measured weight increase/ measured weight increase.

Magnesium-rich orthopyroxenes are stable over a wide pressure range, but ferrosilite is stable only at high pressures, thus the $MgSiO_3$–$FeSiO_3$ series is a potentially useful geobarometer. Orthopyroxenes containing as much as 57 per cent of the $FeSiO_3$ component are stable at 1250°C and 1 atm (p. 78), while orthoferrosilite has been shown (Lindsley, 1965) to be stable at pressures above 17 kbar at the same temperature. An investigation of orthopyroxene stability at 1 atm, using stoichiometric mixtures of composition Fs_{50-70}, has been made by Wood and Strens (1971), the results of which are shown in Fig. 34(a). From the data presented in the figure, augmented by experimental work on the pressure–temperature field of orthoferrosilite, and data on the distribution of iron between orthopyroxene and olivine, the same authors have calculated a series of temperature–composition sections for pressures between 1 atm and 15 kbar at intervals of 2·5 kbar. Figure 34(b) shows that the shift of the boundary of the pyroxene field at constant temperature is approximately linear (2·6 per cent $FeSiO_3$/kbar) at a temperature of 1100°C. At constant pressure the Fs component decreases with increasing temperature, the rate varying between approximately 1 and 2 per cent $FeSiO_3$/100°C. Provided the temperature of crystallization can be estimated to ± 50°C, and the composition determined to ± 1 per cent $FeSiO_3$, an approximate pressure ($\pm 0{\cdot}5$ kbar) of crystallization can be obtained for those orthopyroxenes in which the replacement of (Mg,Fe) in $M1$ sites by Ca or Mn ions, or of (Mg,Fe) in $M2$ sites by Al is small ($\simeq < 2{\cdot}5$ mol. per cent). The possible use of the orthopyroxene geobarometer for the determination of palaeogravity is discussed by Wood and Strens (1971).

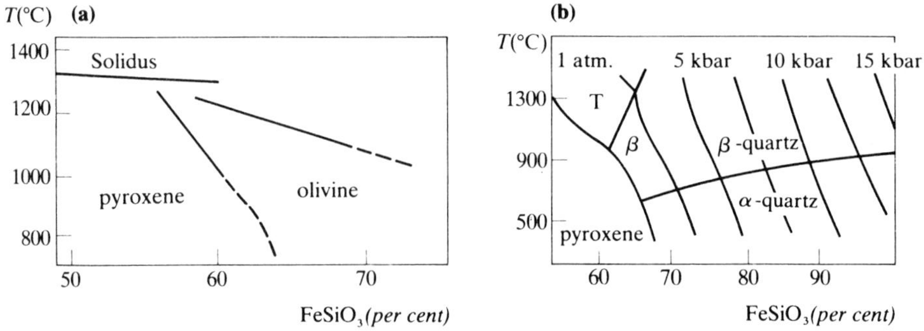

Fig. 34. (a) Temperature–composition diagram of part of the $MgSiO_3$–$FeSiO_3$ system at 1 atm showing boundary of pyroxene and olivine fields. Solidus from Nafziger and Muan, 1967 (after Wood and Strens, 1971). (b) Temperature–composition sections of part of the $MgSiO_3$–$FeSiO_3$ system at pressures of 1 atm to 15 kb, showing the shift of the boundary of the pyroxene field to higher $FeSiO_3$ contents with increasing pressure. The low (α), high (β) quartz and tridymite (T) fields are also shown (after Wood and Strens, 1971).

Ferrosilite. The rarity of orthoferrosilite compositions in natural assemblages is indicative of the limited stability field, at relatively low pressures, of the iron-rich members of the orthopyroxene series, a conclusion that is in accord with the early work by Bowen and Schairer (1935) on the MgO–FeO–SiO_2 system at 1 atm. The composition $FeSiO_3$ crystallizes at low pressures to the decomposition products fayalite and quartz as shown by the phase stability fields at low pressures in the Mg_2SiO_4–Fe_2SiO_4–SiO_2 system (Fig. 35). Ferrosilite is, however, stabilized, relative to fayalite and quartz at high pressures, and has been synthesized at temperatures between 1150° and 1400°C and 18 and 45 kbar respectively (Lindsley *et al.*, 1964). Ferrosilite has also been synthesized by Akimoto *et al.* (1964, 1965) under pressure–temperature conditions ranging from 12 to 73 kbar and 620° to 1270°C. The boundary curve for the reaction:

$$\tfrac{1}{2}(Fe_2SiO_4)+\tfrac{1}{2}SiO_2 = FeSiO_3 \quad \text{(orthoferrosilite)}$$

is shown in Fig. 36 and the equation for the curve is:

$$P_{(\text{kbars})} = 2{\cdot}7+0{\cdot}014T\,(°\text{C})$$

The standard free energy of formation of orthoferrosilite, assuming that the volume change of the reaction is relatively insensitive to pressure and temperature, may be obtained from the equation $\Delta G° = -36+0{\cdot}45T$ (cal/mole). The pressure–temperature stability curve for $FeSiO_3$ (Smith, 1971) is shown in Fig. 37.

The effect on the stability relations of the fayalite–quartz–orthopyroxene assemblage of adding $MgSiO_3$ to the Fe_2SiO_4–SiO_2 system has been investigated by Olsen and Mueller (1966). To evaluate the effect of the addition of MgO to the pure iron silicate system, the exchange reaction:

$$\tfrac{1}{2}Mg_2SiO_4+FeSiO_3 = \tfrac{1}{2}Fe_2SiO_4+MgSiO_3$$

may be used in combination with the reaction $Fe_2SiO_4+SiO_2 = 2FeSiO_3$, by which the general equilibrium equations may be expressed:

$$K_a\phi_a^{-2}\exp\left(\frac{-P\Delta V_a}{RT}\right) = \left(\frac{X_{Fe}^{Px}}{X_{Fe}^{Ol}}\right)^2 \qquad (1)$$

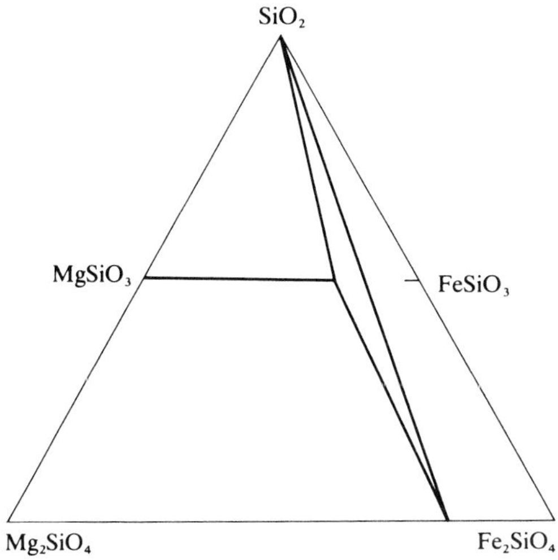

Fig. 35. Phase stability fields in the system Mg_2SiO_4–Fe_2SiO_4–SiO_2 at low pressure (after Smith, 1971).

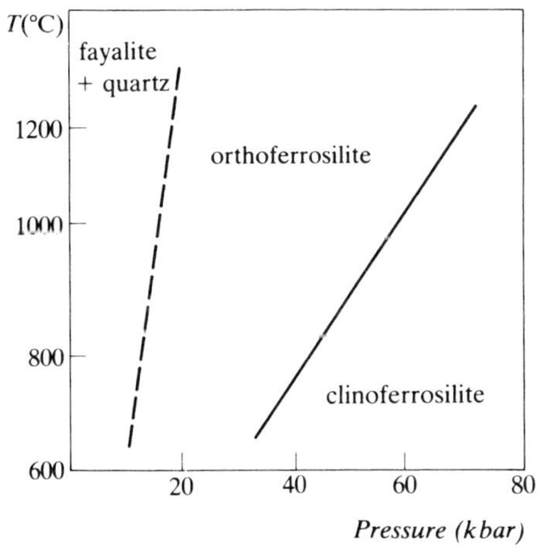

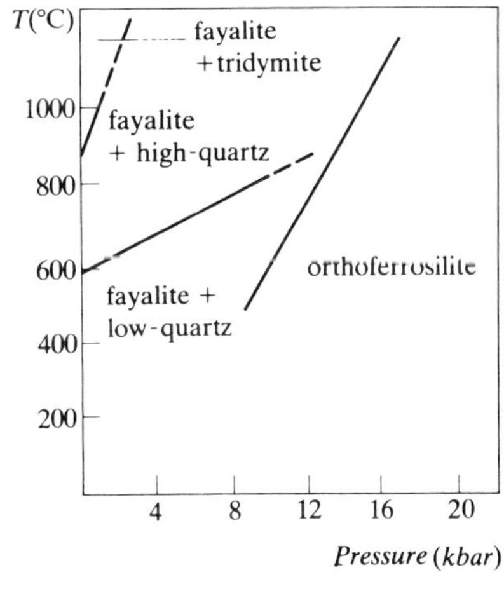

Fig. 36. Stability relations of $FeSiO_3$ (after Akimoto *et al.*, 1965).

Fig. 37. Pressure–temperature stability curve for orthoferrosilite (after Smith, 1971).

$$K_b \phi_b^{-1} \exp\left(\frac{-P\Delta V_b}{RT}\right) = \frac{(1-X_{Fe}^{Px})X_{Fe}^{Ol}}{(1-X_{Fe}^{Ol})X_{Fe}^{Px}} \qquad (2)$$

where X_{Fe} represents the atomic fraction $Fe^{2+}/(Fe^{2+}+Mg)$ in the pyroxene and olivine, P the pressure in atm, ΔV the volume change for the reaction, and the ϕ functions are defined as:

$$\phi_a = (f_{Fe}^{Px}/f_{Fe}^{Ol}) \quad \text{and} \quad \phi_b = (f_{Mg}^{Px}f_{Fe}^{Ol}/f_{Mg}^{Ol}f_{Fe}^{Px})$$

where f refers to the activity coefficients of the iron and magnesium end-members. The results of this evaluation, on the assumption that the pyroxenes and olivines behave as ideal solutions, are shown in Fig. 38, and indicate that the equilibrium pressure is reduced by $\simeq 2$ kbar/0·1 decrement in $Fe^{2+}/(Fe^{2+}+Mg)$. Olsen and Mueller have also considered the effect of the possible deviation from ideality of the pyroxene solutions. The 900°K isothermal section for the ideal and non-ideal solution model for pyroxene solid solutions is given in Fig. 39. From both figures it may be observed that non-ideality of the pyroxene solution greatly increases the effect of the decrement. The variations in the compositions of orthopyroxene and olivine with temperature at pressure less than 1 kbar, and with pressure at a temperature of 900°C (Smith, 1971) are shown in Fig. 40. The Fe/(Fe+Mg) ratios of orthopyroxene at 5 and 10 kbar are 76 and 88.

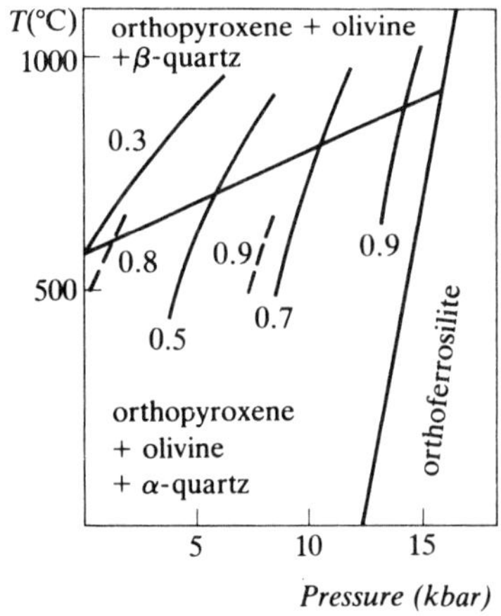

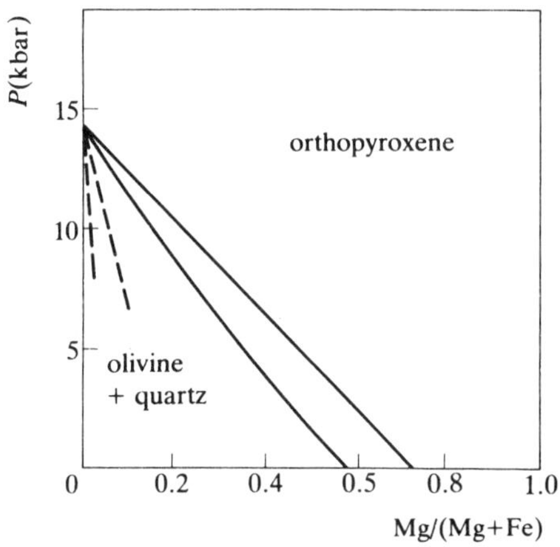

Fig. 38. Theoretical effect of adding $MgSiO_3$ to the system Fe_2SiO_4–SiO_2. Solid lines, pyroxene values of $Fe^{2+}/(Fe^{2+}+Mg)$ ideal solution model for both orthopyroxene and olivine. Broken lines, small non-ideality for pyroxene solutions (after Olsen and Mueller, 1966).

Fig. 39. Isothermal section of Fig. 38 for 900°K. Solid lines, ideal solution model. Broken lines, small non-ideality for pyroxene solutions (after Olsen and Mueller, 1966).

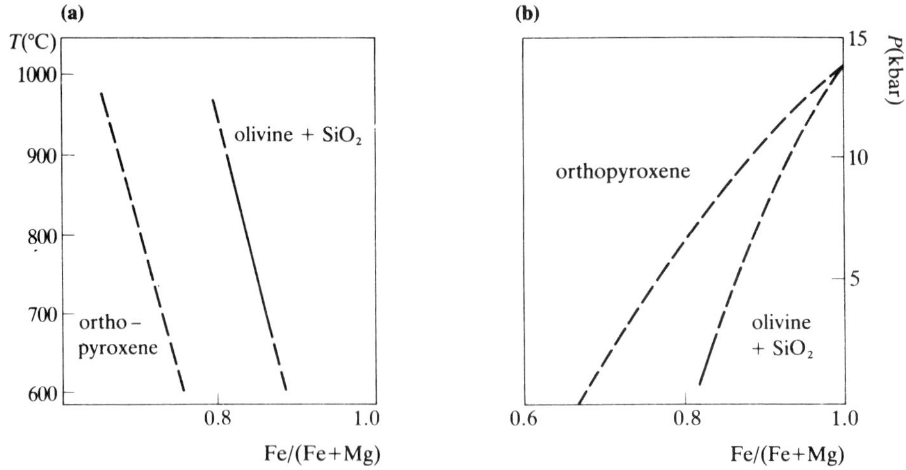

Fig. 40. (a) Variation of orthopyroxene and olivine compositions with temperature at pressures of less than 1 kbar. (b) Variation of composition of orthopyroxene and olivine with pressure at a temperature of 900°C. Broken lines are projections of the boundaries of the three phase field (after Smith, 1971).

The stability of ferrosilite may be considered in relation to its equilibrium coexistence, as a member of the $MgSiO_3$–$FeSiO_3$ solid solution series with tridymite, metallic iron and a gas phase, according to the equation:

$$FeSiO_3 \text{ in solid solution} = Fe_{(s)} + SiO_{2(s)} + \tfrac{1}{2}O_{2(g)} \quad (1)$$

for which reaction the free energy change may be expressed:

$$\Delta G = -2{\cdot}303RT\log_{10}\frac{(P_{O_2})^{\frac{1}{2}}}{N_{FeSiO_3}\,\gamma_{FeSiO_3}} \quad (2)$$

where N_{FeSiO_3} is the molar fraction of $FeSiO_3$ in solid solution, $\gamma FeSiO_3$ the activity coefficient of $FeSiO_3$ in solid solution, and P_{O_2} the oxygen pressure of the gas phase in equilibrium with the three solid phases. The α-function for the solid solution is given by the equation:

$$\log\gamma FeSiO_3 = \alpha(1 - N_{FeSiO_3})^2$$

and equation (2) can be expressed:

$$\log N_{FeSiO_3} = \tfrac{1}{2}\log P_{O_2} = -\alpha(1 - N_{FeSiO_3})^2 + \frac{\Delta G}{2{\cdot}303RT} \quad (3)$$

Solutions of this equation for $(Mg,Fe)SiO_3$ solid solutions of composition 20–60 mol. per cent $FeSiO_3$, in equilibrium with tridymite and metallic iron, have been determined as functions of oxygen pressure (10^{-11} to 10^{-13} atm) and temperatures of 1250° and 1300°C for which ΔG is $\simeq 1$ and $\simeq 2$ kcal respectively (Muan *et al.*, 1964).

The pressure–temperature–oxygen fugacity relations in the system Fe–O–SiO_2 have been considered by Lindsley *et al.* (1968), and in the system Fe–O–MgO–SiO_2 by Speidel and Nafziger (1968). Two invariant assemblages occur in the Fe–O–SiO_2 system, ferrosilite + fayalite + silica + iron + liquid at 17·5 kbar, 1280°C and $10^{-10{\cdot}85}$ atm f_{O_2}, and ferrosilite + fayalite + silica + magnetite + liquid at 17 kbar total pressure, 1205°C and $10^{-7{\cdot}25}$ atm f_{O_2}. Data are also given for the reaction, ferrosilite + magnetite = silica polymorph + liquid at pressures between 13 and 20 kbar and temperatures of 1205° to 1250°C, as well as temperature–pressure and temperature–f_{O_2} projections of part of the system. The phase relations in the $MgSiO_3$–$FeSiO_3$ system at pressures between 100 and 200 kbar at 1000°C, determined by Ringwood and Major (1968), are illustrated in Fig. 41.

At very high pressures, ferrosilite is transformed to spinel and stishovite. The pressure of this transformation is given by Ringwood and Major (1966a) as approximately 120 kbar at a temperature of about 800°C. Ringwood and Major (1966c) have synthesized orthopyroxenes, in the compositional range $(Mg_{0{\cdot}75}Fe_{0{\cdot}25})SiO_3$ to $FeSiO_3$ at temperatures of about 900°C and pressures up to 200 kbar, above which the pyroxene break down to a spinel (ringwoodite) and stishovite:

$$\underset{\text{pyroxene}}{2(Mg,Fe)SiO_3} \rightarrow \underset{\text{spinel}}{(Mg,Fe)_2SiO_4} + \underset{\text{stishovite}}{SiO_2}$$

Ahrens and Gaffney (1971) undertook a series of thermochemical calculations to estimate the temperatures and pressures necessary to produce the above transformations in more magnesium-rich pyroxenes, and also investigated the reaction:

$$2(Mg,Fe)SiO_3 \rightarrow 2(MgO \text{ periclase, FeO wüstite}) + SiO_2 \text{ (stishovite)}$$

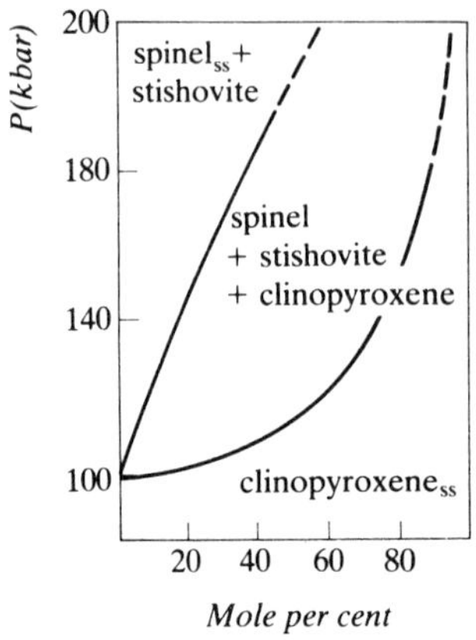

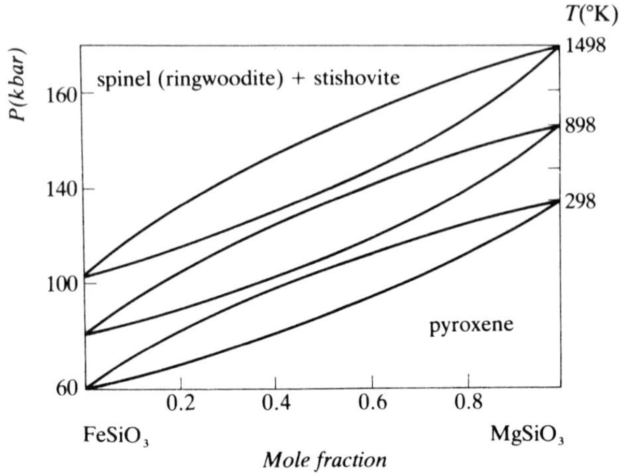

Fig. 41. Phase relations in the system $FeSiO_3$–$MgSiO_3$ at pressures between 100 and 200 kbar at approximately 1000°C (after Ringwood and Major, 1968).

Fig. 42. Theoretical phase diagram for the breakdown of $MgSiO_3$–$FeSiO_3$ pyroxenes to $(Mg,Fe)_2SiO_4$ (spinel-ringwoodite) + SiO_2 (stishovite) at 298°, 898° and 1498°K (after Ahrens and Gaffney, 1971).

The transformation pressure, P_i^T, at temperature T, for a given end-member may be calculated from:

$$\Delta G_i^{T,p} = \Delta H_i^T - T\Delta S_i^T = -\int_0^{P_i^T} (V_i^2 - V_i^1)\,dp$$

where i is the end-member component (1 or 2) of a solid solution series, superscripts 1 and 2 refer to the high-pressure and low-pressure phase respectively, and ΔH_i^T and ΔS_i^T are the differences in enthalpy and entropy at zero pressure and at temperature T. The theoretical phase diagram for the breakdown of the $MgSiO_3$–$FeSiO_3$ pyroxenes to $(Mg,Fe)_2SiO_4$ (ringwoodite) + SiO_2 (stishovite) is shown in Fig. 42.

Static high-pressure and high-temperature phase equilibrium data for iron-rich members of the $MgSiO_3$–$FeSiO_3$ pyroxenes have also been presented by Akimoto and Syono (1970).

The subsolidus relations for the join hedenbergite–ferrosilite under hydrothermal conditions at 20 kbar (Lindsley and Munoz, 1969b), show a two-pyroxene field in which the coexisting pyroxenes are orthoferrosilite solid solution, $Fs_{95}Wo_5$, and a clinopyroxene, the composition of which varies from approximately $Fs_{60}Wo_{40}$ at 800°C to $Fs_{92}Wo_8$ at 950°C (Fig. 43). The pyroxene equilibria of a number of synthetic pyroxenes on part of the join $Fs_{85}En_{15}$–wollastonite at 15 kbar have been examined by Smith (1972). The extrapolation, of these and the data of Lindsley and Munoz, to lower pressures and their application to the magmatic crystallization of pyroxenes is considered in the section on pigeonite (see also diopside section, p. 230).

The nickel pyroxene, $NiSiO_3$, is unstable at pressures of 1 atm, and like $FeSiO_3$ is probably stable only at high pressures (Campbell and Roeder, 1968), ΔG for the reaction $\frac{1}{2}Co_2SiO_4 + \frac{1}{2}SiO_2 \rightleftharpoons CoSiO_3$, and the ortho–clino inversions of the cobalt metasilicate are similar to those of the iron analogue. The manganese metagermanate $MnGeO_3$ (*a* 19·29, *b* 9·25, *c* 5·48 Å) has been synthesized, melts incongruently at 1290°C and is isostructural with enstatite (Tauber *et al.*, 1963).

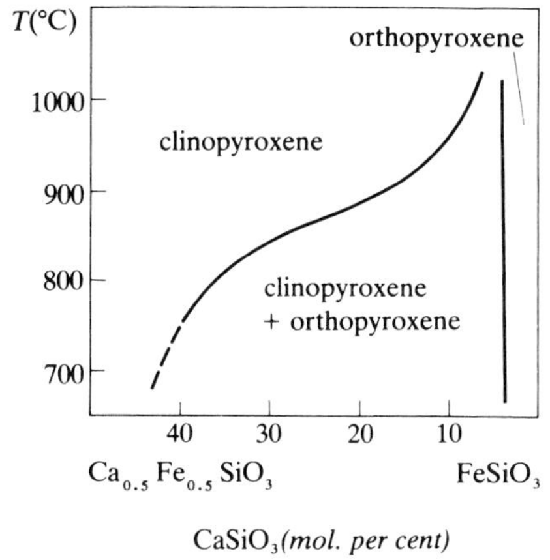

Fig. 43. Subsolidus phase relations for the join ferrosilite–hedenbergite at 20 kbar (after Lindsley and Munoz, 1969b).

The cobalt orthopyroxene, $CoSiO_3$, shows a similar high pressure transformation to $FeSiO_3$ at comparable temperature and pressure ($\simeq$ 900°C at 125 kbar). The iron germanate, $FeGeO_3$, breaks down to spinel and rutile at lower pressure and temperature (10 kbar and 700°C):

$$\underset{\text{pyroxene}}{2FeGeO_3} \rightarrow \underset{\text{spinel}}{Fe_2GeO_4} + \underset{\text{rutile}}{GeO_2}$$

Similar transformations take place with other metasilicates and metagermanates (see Table 8) according to the equation:

$$\underset{\text{pyroxene}}{2ABO_3} \rightarrow \underset{\text{spinel}}{A_2BO_4} + \underset{\text{rutile}}{BO_2}$$

where A = Fe, Mg, Co, Ni; B = Si, Ge.

Table 8 High-pressure transformations in pyroxenes (after Ringwood and Major, 1966c)

	$\simeq$ ***P*(kbar)**	$\simeq$ ***T*(°C)**	**Transformation products**
$MnGeO_3$	28	700	$MnGeO_3$ (ilmenite structure)
$MgGeO_3$	25	700	$MgGeO_3$ (ilmenite structure)
$FeGeO_3$	10	700	Fe_2GeO_4 (spinel) + GeO_2 (rutile)
$CoGeO_3$	10	700	Co_2GeO_4 (spinel) + GeO_2 (rutile)
$FeSiO_3$	125	900	Fe_2SiO_4 (spinel) + SiO_2 (rutile)
$CoSiO_3$	125	900	Co_2SiO_4 (spinel) + SiO_2 (rutile)

$MgGeO_3$ and $MnGeO_3$ transform directly to the ilmenite structure (Ringwood and Major, 1966c). Solid solutions in the compositional range $Mg(Ge_{0\cdot9}Si_{0\cdot1})O_3$–$Mg(Ge_{0\cdot5}Si_{0\cdot5})O_3$ break down to $Mg(Fe,Si)O_2$ (spinel) + $(Ge,Si)O_2$ (ilmenite).

Clinoenstatite–clinoferrosilite. During the investigation of the system MgO–FeO–SiO_3 at 1 atm, Bowen and Schairer (1935) synthesized a series of Mg–Fe clinopyroxenes varying in composition from Fs_0 to Fs_{70}, and by inverting a natural orthopyroxene from Tunaberg demonstrated that the series extends to at least Fs_{87}. The temperature of the ortho–clino inversion decreases from 1140°C (Fs_0) to 955°C at the composition Fs_{87}.

Iron-rich clinopyroxene, $\simeq Mg_{1\cdot5}(Fe^{2+},Mn)_{98\cdot5}$, together with silica and water, has been produced by the transformation of fibrous grunerite $(Fe_{6\cdot69}Mn_{0\cdot14}Ca_{0\cdot11}Mg_{0\cdot10})(Si_{7\cdot97}Al_{0\cdot02})O_{22}(OH)_2$ at 775°C and 0·5 kbar argon pressure (Ghose and Weidner, 1971). The transformation of the amphibole to clinopyroxene involves the movement of Fe^{2+} cations from the donor to acceptor regions and the release of protons. The donor regions are destroyed in the process, and in addition release oxygen and part of the protons to form water, the residue remaining as silica. The reactions in the acceptor and donor regions may be expressed:

$$7[Fe_7Si_8O_{24}H_2+Fe^{2+}] \rightarrow 7[4(Fe_2Si_2O_6)+2H^+]$$
$$Fe_7Si_8O_{24}H_2+14H^+ \rightarrow 7Fe^{2+}+8SiO_2+8H_2O$$

The *c* cell dimension of the manganese-rich clinoferrosilite, 5·30 Å, is intermediate between those of grunerite, 5·34 Å, and pure clinoferrosilite, 5·23 Å (Burnham, 1966), probably as a result, during the atomic rearrangement of a relaxation effect due to oxygen and silicon atoms not attaining the positions required by the pure clinoferrosilite structure.

Clinoenstatite breaks down to forsterite and stishovite at 115 kbar and $\simeq$ 650°C (Sclar *et al.*, 1964). ΔV for the reaction is $-5{\cdot}3\,cm^3/mole$, and the approximate equation for the equilibrium boundary of the reaction is $P(kbars) = (97{\cdot}5 \pm 5) + 0{\cdot}027T(°C)$. The high pressure and temperature phase transformations of $FeSiO_3$ have been investigated at pressures of 80–280 kbar in the temperature range 1200–1600°C, by Liu (1976). The transformations in order of increasing pressure are

$$\underset{\text{clinoferrosilite}}{2FeSiO_3} \rightarrow \underset{\text{spinel}}{Fe_2SiO_4} + \underset{\text{stishovite}}{SiO_2} \rightarrow \underset{\text{wüstite}}{2FeO} + \underset{\text{stishovite}}{2SiO_2}$$

$MgSiO_3$–$CaMgSi_2O_6$ system. The subsolidus and liquidus phase relations in the system $MgSiO_3$–$CaMgSi_2O_6$ at 1 atm, determined by Boyd and Schairer (1964) and revised by Kushiro (1972d) are shown in Fig. 44. Two calcium-poor pyroxenes are

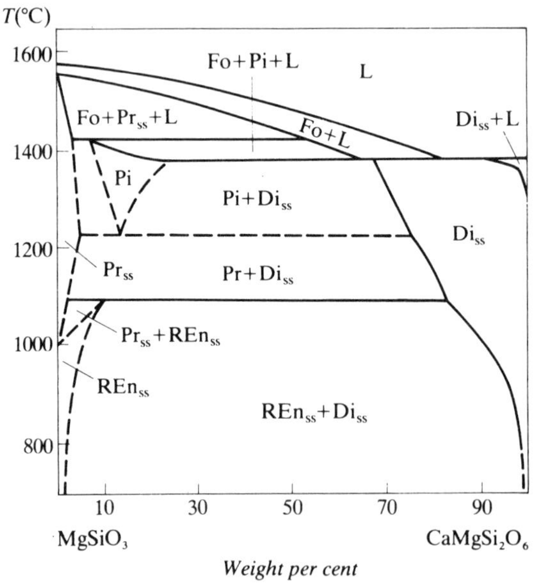

Fig. 44. Equilibrium relations along the join $MgSiO_3$–$CaMgSi_2O_6$ at 1 atm. Fo forsterite, Di_{ss} diopside solid solution, Pi iron-free pigeonite, Pr_{ss} protoenstatite solid solution, REn_{ss} orthoenstatite solid solution, L liquid (after Kushiro, 1972d).

found to exist in the enstatite-rich part of the $MgSiO_3$–$CaMgSi_2O_6$ join, one a low-calcium member, the other relatively rich in the diopside component. The former is an enstatite and the latter monoclinic iron-free pigeonite. The enstatite, formed by inversion from the high-temperature protoenstatite solid solution contains a maximum of 10 weight per cent $CaMgSi_2O_6$ at a temperature of 1100°C, the amount of solid solution decreasing to approximately 2 per cent at 800°C.

The effect of high pressure on the system $MgSiO_3$–$CaMgSi_2O_6$ has been investigated by Davis and Boyd (1966). At a pressure of 30 kbar the stability field of protoenstatite is eliminated and enstatite was found to be the stable polymorph at liquidus temperatures (see Fig. 18, p. 52). Orthoenstatite melts congruently and the liquidus curve descends smoothly to the melting point of diopside at 1715°C. The join is binary with a small melting interval and shows a peritectic relationship, at temperatures between 1750° and 1775°C, in the $MgSiO_3$-rich part of the system. A two-pyroxene field intersects the solidus at compositions between En_{92} and En_{79}, and the two phases are in equilibrium with a liquid of composition En_{65}. The diopside solvus is sharply inflected towards diopside at $\simeq$ 1500°C, and at temperatures below 1400°C the diopside solvus at 30 kbar is close to the boundary of the two pyroxene fields at atmospheric pressure. The melting temperature of both end-members and of intermediate compositions is increased by pressure, by 280°C for enstatite and 325°C for diopside. A later investigation of the $MgSiO_3$–$CaMgSi_2O_6$ join at a pressure of 20 kbar under anhydrous conditions has been undertaken by Kushiro (1969), and the equilibrium diagram near solidus and liquidus temperatures is shown in Fig. 45. Under these high-pressure conditions forsterite does not appear on the enstatite–diopside join and all the phases have compositions on the join. The iron-free pigeonite field present at 1 atm persists at this pressure but has a more restricted compositional range.

The phase reaction relations at water pressures in the range 1–10 kbar for the more magnesian part of the $MgSiO_3$–$CaMgSi_2O_6$ compositional region show that there is an invariant point involving enstatite$_{ss}$, protenstatite$_{ss}$, pigeonite$_{ss}$ and diopside$_{ss}$ at $P \simeq 1{\cdot}5$ kbar, $T \simeq 1140$°C (Warner, 1975). At pressures lower than

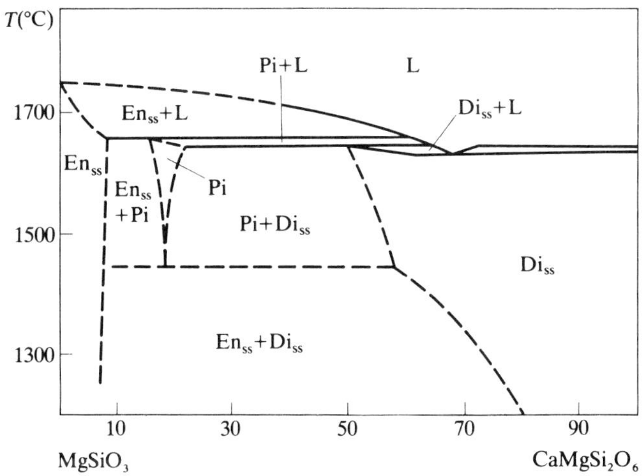

Fig. 45. Equilibrium relations of the diopside–enstatite system at 20 kbar under anhydrous conditions. En$_{ss}$ enstatite solid solution, Di$_{ss}$ diopside solid solution, Pi pigeonitic clinopyroxene, L liquid (after Kushiro, 1969).

that of the invariant point enstatite$_{ss}$ breaks down at high temperature to protoenstatite$_{ss}$ + diopside$_{ss}$, an assemblage that on further increase in temperature reacts to form pigeonite$_{ss}$ (Fig. 46(a)). At pressures above that of the invariant point the protoenstatite$_{ss}$ + diopside$_{ss}$ assemblage is not stable, and pigeonite forms from reaction between enstatite and diopside (Fig. 46(b)). A tentative phase diagram for the system $Mg_2Si_2O_6$–$CaMgSi_2O_6$ at approximately 5 kbar is given by Mori and Green (1976).

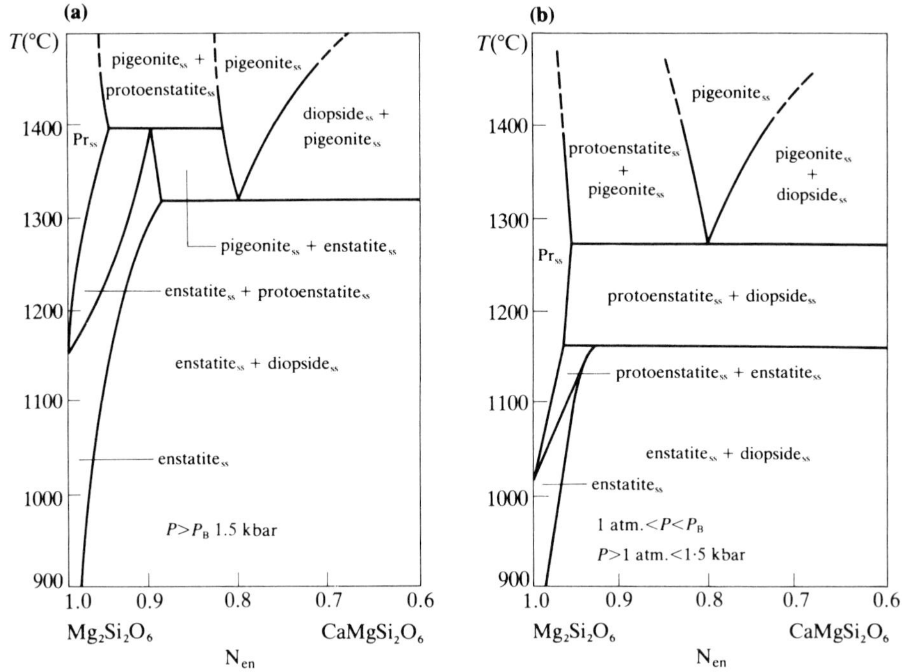

Fig. 46. Temperature–composition sections for the Ca-poor region of the join $Mg_2Si_2O_6$–$CaMgSi_2O_6$. (a) Inferred subsolidus pyroxene phase relations at pressures > 1 atm < 1·5 kbar. (b) At pressure > 2 kbar (after Warner, 1975).

A recent determination of the solvi bounding the enstatite–diopside two-phase region at 2, 5 and 10 kbar over the temperature range 900° to 1300°C, has been made by Warner and Luth (1974). Their calculated 30 kbar solvus is in good agreement with that determined experimentally by Davis and Boyd (1966). The maximum solubility of enstatite in diopside solid solutions as a function of pressure and temperature, together with the *P–T* curves outlining the lower stability limit of pigeonite on the enstatite–diopside join, and the reaction enstatite solid solution = diopside solid solution + protoenstatite solid solution, are shown in Fig. 47. The intersection of these curves at A is an invariant point in the binary system. The experimentally determined solvus for the join $Mg_{0.4}Fe_{0.6}SiO_3$–$Ca_{0.5}Mg_{0.2}Fe_{0.3}SiO_3$ at a pressure of 20 kbar shows a maximum $CaSiO_3$ content of 7 mol. per cent for a range of temperature between 800° and 950°C.

The equilibrium relations along the join $En_{47}Di_{53}$–H_2O at 20 kbar pressure have been determined by Kushiro (1972a). The phase relations in the water-present region are simple and essentially independent of the amount of water present, in contrast to the vapour-absent region in which enstatite solid solution and diopside

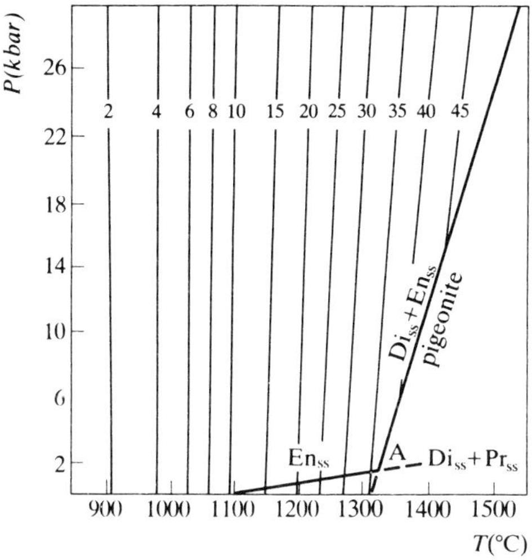

Fig. 47. Maximum solubility of $Mg_2Si_2O_6$ in diopside solid solution (mol. per cent) as a function of pressure and temperature. The lower stability limit of pigeonite on the $Mg_2Si_2O_6$–$CaMgSi_2O_6$ join, and the lower pressure breakdown of En_{ss} to Di_{ss} and Pr_{ss} are also shown (after Warner and Luth, 1974). A, invariant point in binary system.

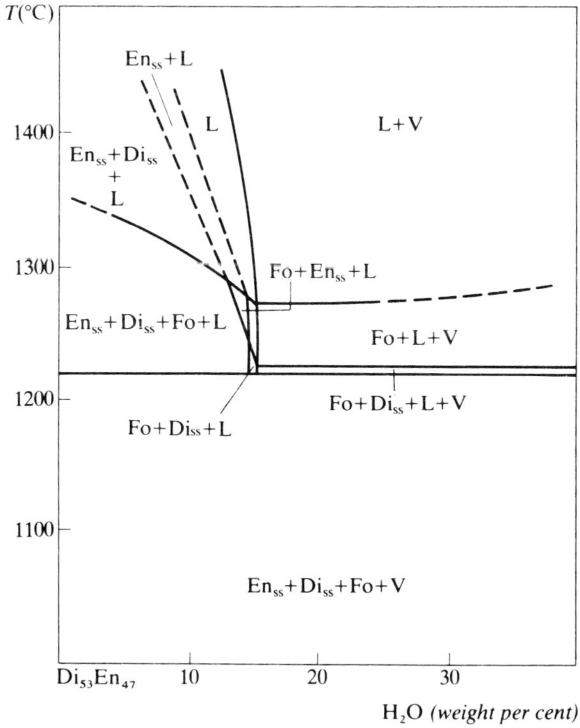

Fig. 48 Equilibrium relations along the join $Di_{53}En_{47}$–H_2O at 20 kbar (after Kushiro, 1972a).

solid solution and not forsterite are the stable phases at the higher temperatures (Fig. 48).

An investigation of the $Mg_2Si_2O_6$–$CaMgSi_2O_6$ solvus using mixtures of synthetic enstatite and diopside, reacted with small water contents at 30 kbar in the temperature range 1000° to 1700°C, has been presented by Nehru and Wyllie (1974). Most of the runs were made with the composition $En_{50}Di_{50}$ (weight per cent) and 5 per cent H_2O. The tentative phase fields intersected by the join $En_{50}Di_{50}$–H_2O are shown in Fig. 49(a). A large two-pyroxene field plus water-undersaturated liquid occur above the solidus, and the liquid coexists with diopside above 1500°C. The compositions of the coexisting pyroxenes at temperatures between 1000° and 1500°C fitted to second-order regression curves, the data between 1000° and 1300°C give the following expressions:

Mole fraction of $Mg_2Si_2O_6$ in enstatite solid solution =

$$-2{\cdot}86\times10^{-8}T^2-5{\cdot}49\times10^{-5}T+1{\cdot}0585$$

Mole fraction of $CaMgSi_2O_6$ in diopside solid solution =

$$1{\cdot}05\times10^{-7}T^2-6{\cdot}04\times10^{-4}T+1{\cdot}3950$$

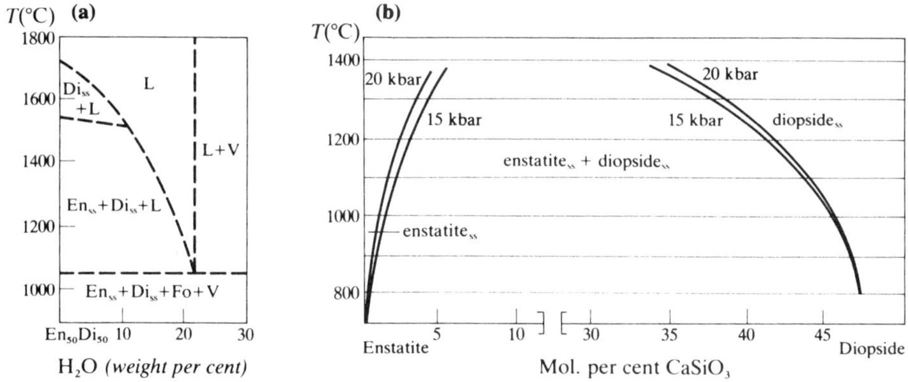

Fig. 49. (a) Tentative phase fields on the join $Di_{50}En_{50}$–H_2O at 30 kbar (after Nehru and Wyllie, 1974). (b) Coexisting enstatite and diopside solid solutions at pressures of 15 and 20 kbar (after Lindsley and Dixon, 1976).

A critical review of earlier experimental work, and further data relating to the composition of coexisting enstatite and diopside solid solutions, is given by Lindsley and Dixon (1976). The results of their dissolution and exsolution data, obtained in the temperature range 850° to 1400°C at 15 kbar and 900° to 1400°C at 20 kbar (Fig. 49(b)), show that there is a small pressure effect on the enstatite and diopside solvi, the effect increasing with increasing temperature.

The enstatite–diopside solvus is commonly used to estimate crystallization temperatures of natural rocks, and equilibration temperatures and pressures based on the amount of Ca in coexisting pyroxene pairs have been used by Mercier and Carter (1975) to construct continental and oceanic geotherms. Although the parameter Ca/(Ca + Mg − ΣFe), as well as Mg/(Mg + ΣFe) and $[Al]^6/[Al]^4$, increase isobarically with increasing temperature, and decrease isothermally with increasing pressure (Mysen and Boettcher, 1975), the ratios show a positive correlation with the starting material (Fig. 50), and their use for geothermometry and geobarometry is thus limited. The variation of the Ca/(Ca + Mg) ratio, as a function of temperature and pressure, in orthopyroxene in equilibrium with clinopyroxene, has been determined experimentally by Hensen (1973) using natural coexisting pyroxene pairs as starting material. This study shows that the ratio increases with temperature and decreases with pressure, and that orthopyroxenes with lower Ca contents at high pressure have the highest Mg/(Mg + Fe) ratios (Fig. 51 (see also Nehru, 1976, and Fig. 52)). Until recently the possible effects of solid solution of forsterite or quartz with enstatite and diopside in applying the pyroxene

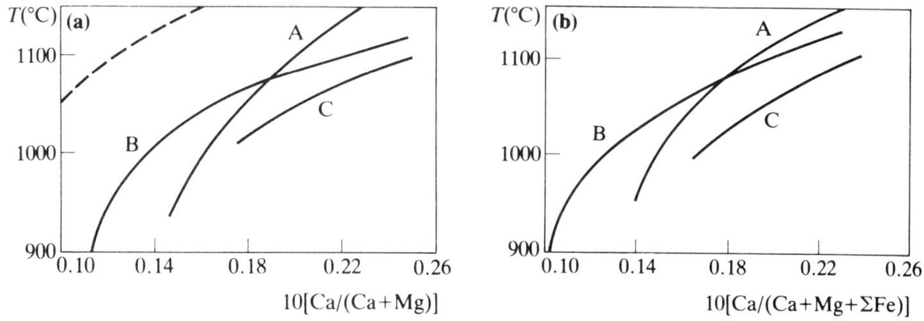

Fig. 50. (a) Variation of Ca/(Ca + Mg) ratio with temperature for orthopyroxene coexisting with clinopyroxene (after Mysen and Boettcher, 1975).
A, spinel lherzolite (1) + H_2O at P_{total} = 15 kbar
B, spinel lherzolite (2) + H_2O at P_{total} = 15 kbar
C, spinel lherzolite (2) + H_2O at P_{total} = 7·5 kbar
Broken line (Boyd, 1973).
(b) Variation of Ca/(Ca + Mg + ΣFe) ratio with temperature for orthopyroxene coexisting with clinopyroxene (after Mysen and Boettcher, 1975).

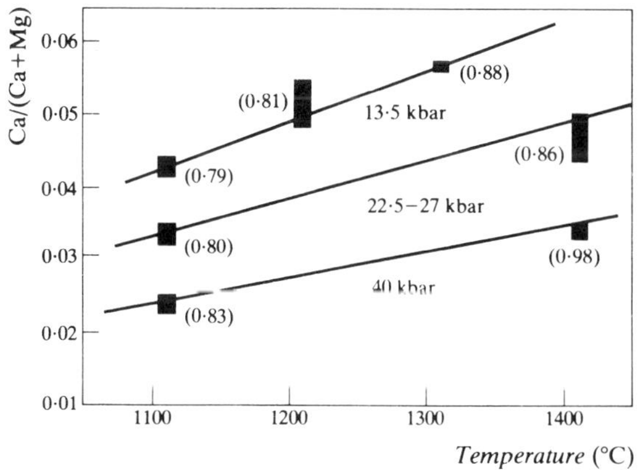

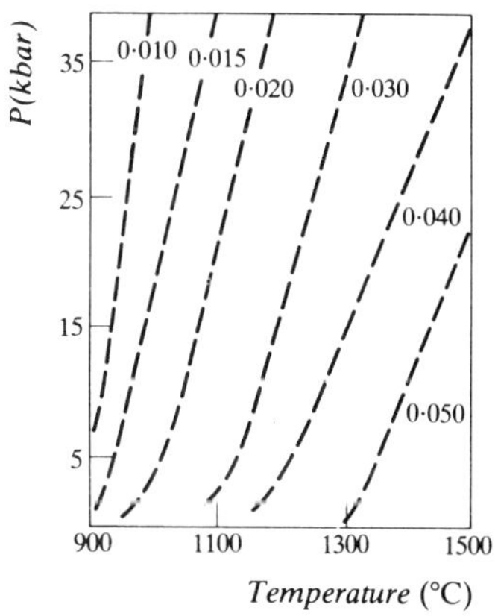

Fig. 51. Variation of Ca/(Ca + Mg) ratio in orthopyroxene in equilibrium with clinopyroxene as a function of temperature and pressure (after Hensen, 1973). Mg/(Mg + Fe^{2+}) ratio of the orthopyroxene indicated in brackets.

Fig. 52. Pressure–temperature grid showing the variation of Ca/(Ca + Mg) for enstatite solid solution coexisting with diopside solid solution in the iron-free $Mg_2Si_2O_6$–$CaMgSi_2O_6$ system (after Nehru, 1976).

solvus to estimate equilibration temperature have been ignored. The uncertainty due to these factors has now been examined by Howells and O'Hara (1975) and Mori and Green (1976). The extent of the forsterite and quartz solid solution at 30 kbar and 1500°C has been determined experimentally by the latter authors, and is shown to be small. Thus although the effect of the solid solution is to widen the pyroxene solvus when the pyroxenes are saturated with forsterite rather than with quartz the difference of the diopside limb is approximately 3 mol. per cent $CaMgSi_2O_6$ at the above temperature and pressure, and shows that silica activity does not appreciably affect ($\simeq$ 20°C) the pyroxene solvus as a geothermometer.

As the Ca content and the Fe–Mg distribution in pyroxene pairs vary with temperature the mixing of Ca with Fe and Mg is generally non-ideal, and so affects the simple correlation of Fe–Mg distribution with temperature. By combining compositional data and theoretical results on solution models, Saxena (1976) has suggested that such an approach may lead to greater precision in the use of coexisting pyroxene compositions in geothermometry. His tentative model involves solving the equation:

$$RT \ln X^{Cpx}_{Mg\text{-}M1} X^{Cpx}_{Mg\text{-}M2} \gamma^{Cpx}_{Mg\text{-}M2} = RT \ln \alpha_{Mg\text{-}opx} - 500$$

where R is the gas constant, T temperature in °K, $X^{Cpx}_{Mg\text{-}M1}$ the site occupancy of Mg on the site $M1$ or $M2$ in clinopyroxene, γ the $M2$ site activity coefficient, $\alpha_{Mg\text{-}opx}$ the activity of Mg in orthopyroxene, thus providing a relative temperature scale to evaluate models of equilibrium and differentiation in metamorphic and igneous rocks.

Estimates of crystallization temperatures and pressures derived from cation distribution between coexisting pyroxenes are based on the assumption that complex natural and relatively simple synthetic systems can be correlated directly. Equilibration of the cation distribution is also assumed. In many cases these assumptions appear to be justified, but some investigations indicate that the assumptions are not always valid. Thus large apparent variations in pressure within single spinel- and garnet-bearing ultramafic xenoliths in basalts, derived from the pyroxene temperature–pressure grid have been noted by Wilshire and Jackson (1975). The same authors have also noted apparent pressures that are not consistent with the size of some Alpine peridotite massifs, as well as inclusions from kimberlite that yield very high pressures and temperature yet display sheared and mylonitic textures that are unlikely to have survived their apparent temperature of equilibrium. Pike (1976) has reported significant variations in composition in Cr-rich orthopyroxene and endiopside pairs from the spinel–peridotite xenoliths in the basanitoid and pyroclastic flows of Black Rock Summit, Nevada. The compositional differences lead to large apparent gradients in pressure and temperature over a distance of a few millimetres, but are more probably due to metasomatic reactions, and an unusual overall composition of the source area.

Mysen and Boettcher (1975) have shown that the contents of chromium in the ortho- and clinopyroxenes increase with temperature and are apparently independent of the total pressure. Subsequently Mysen (1976a) suggested that, because of the similarity between octahedrally co-ordinated Al and Cr, the partitioning of $[Al]^6$ and Cr between ortho- and clinopyroxene is potentially useful as a geothermometer. Assuming that the activity coefficients of $[Al]^6$ and Cr are close to unity, the distribution coefficient for the exchange reaction

$$CaAlAlSiO_6 + MgCrAlSiO_6 \rightleftharpoons CaCrAlSiO_6 + MgAlAlSiO_6$$

can be expressed:

$$K_D = ([Al]^6 Cr)^{Opx} / ([Al]^6 Cr)^{Cpx}$$

Using data for natural peridotite nodules at pressures and temperatures corresponding to supersolvus conditions of a hydrous upper mantle, Mysen calibrated $\ln K_D$ v. $1/T$ over the compositional range and a pressure of 7·5–30 kbar and derived the equation of the straight line through the data points of Fig. 53

$$1/T = (0{\cdot}26 \pm 0{\cdot}01) \ln K_D + (0{\cdot}67 \pm 0{\cdot}01)$$

where T is the absolute temperature and the uncertainty in $1/T$ is $\simeq$ 2 per cent.

The solubility of Al_2O_3 in orthorhombic enstatite by the substitution AlAl $\rightleftharpoons$ MgSi has been investigated experimentally at 1, 3 and 5 kbar water pressure (Anastasiou

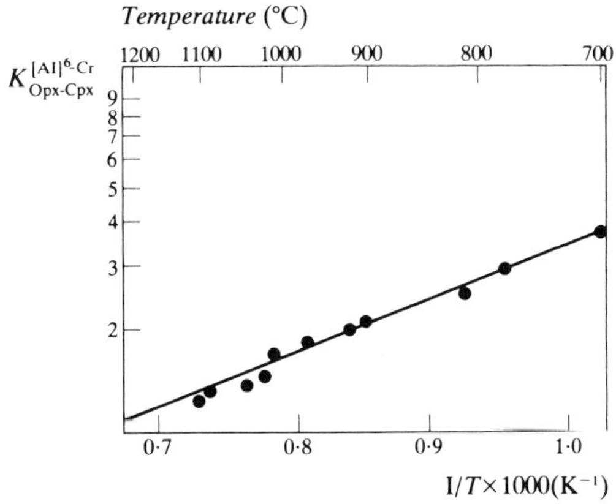

Fig. 53. $[Al]^6$-Cr geothermometry of coexisting ortho- and clinopyroxenes. $K_D = ([Al]^6/Cr)^{Opx}/([Al]^6/Cr)^{Cpx}$. Pressure range 7–30 kbar (after Mysen, 1976a).

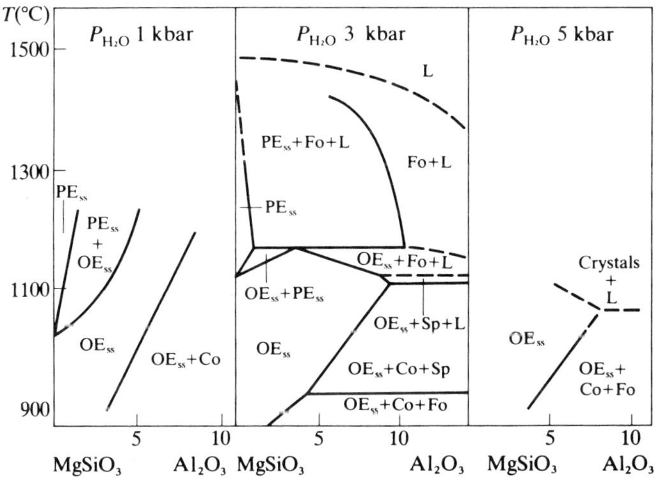

Fig. 54. Isobaric temperature–composition relations of the pseudobinary join $MgSiO_3$–Al_2O_3 at P_{H_2O} 1, 3 and 5 kbar. Melting relation of $MgSiO_3$ at 3 kbar (Kushiro *et al.*, 1968). PE_{ss} aluminous protoenstatite, OE_{ss} aluminous orthoenstatite, Co cordierite, Fo forsterite, Sp spinel, L liquid (after Anastasiou and Seifert, 1972).

and Seifert, 1972), and is illustrated in the isobaric temperature–composition diagrams of Fig. 54. At 1 kbar the content of Al_2O_3 increases from 3 per cent, at 900°C to 9 per cent at 1200°C. The 3 kbar temperature–composition section is more complex due to the incoming of the reaction forsterite + cordierite = enstatite + spinel at about 920°C; the solubility of Al_2O_3 in the orthoenstatite increases more rapidly with increasing temperature at the higher pressure and reaches a maximum of approximately 9 per cent at 1100°C, at which temperature the orthoenstatite solvus intersects the solidus. At 5 kbar the solubility is still markedly temperature dependent, and the solvus rises from 4 per cent Al_2O_3 at 900°C to 6·8 weight per cent at 1000°C.

Earlier Boyd and England (1964) and Boyd (1970) had shown that the solid solution of Al_2O_3 in enstatite in equilibrium with pyrope is both pressure and temperature dependent. Thus at a pressure of 30 kbar the alumina content of the pyroxene increases from 5 per cent at 1100°C to 16 per cent at 1650°C (Fig. 55). As

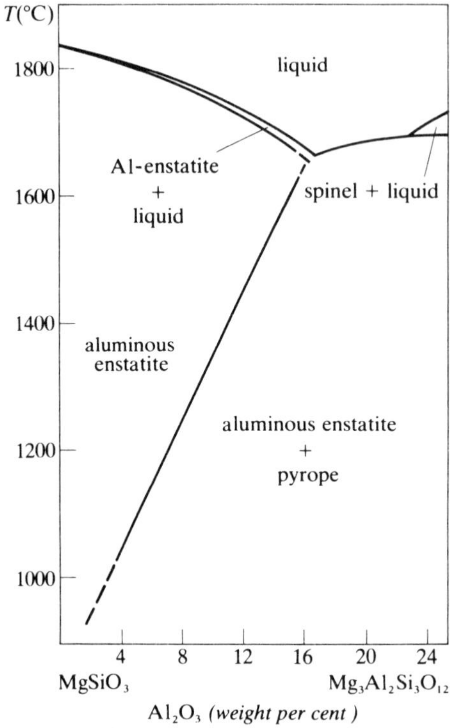

Fig. 55. The enstatite–pyrope system at 30 kbar (after Boyd and England, 1964).

the density of pyrope is considerably greater than that of enstatite the reaction aluminous enstatite → less aluminous enstatite + pyrope proceeds with a decrease in volume. The equilibrium pressure dependence is illustrated in Fig. 56 showing the 1100° and 1600°C isotherms for the enstatite solvus in the system $MgSiO_3$–$Mg_3Al_2Si_3O_{12}$. At the lower temperature the solubility of Al_2O_3 in the enstatite decreases from 5 per cent at 30 kbar to less than 1 per cent at 60 kbar; at the higher temperature the corresponding decrease is from 15 to 2 per cent. Similar temperature–pressure dependence solubility relationships have been shown by

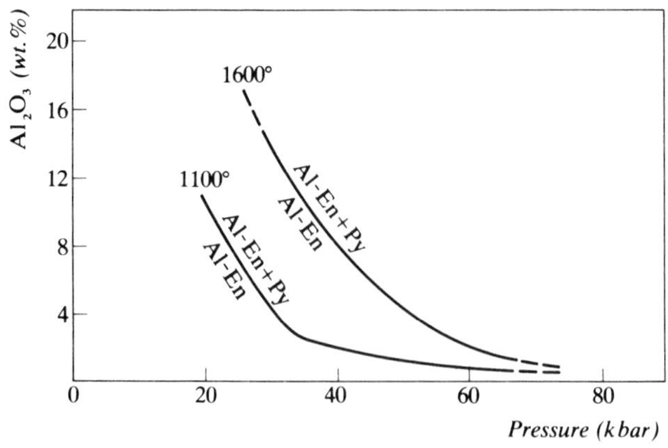

Fig. 56. Isotherms for the solvus Al_2O_3 in $MgSiO_3$ in the system $MgSiO_3$–$Mg_3Al_2Si_3O_{12}$ at 1100° and 1600°C (after Boyd and England, 1964; Boyd, 1970).

MacGregor and Ringwood (1964) also to apply to aluminous iron-bearing orthopyroxenes.

Experimental data on model natural systems (Green and Hibberson, 1970) demonstrated that the solubility of Al_2O_3 in enstatite in equilibrium with pyrope is reduced in systems containing Ca^{2+} and Fe^{2+}, and further experimental data on the solubility of alumina in orthopyroxenes, crystallized in the $MgSiO_3$–$FeSiO_3$–Al_2O_3 and the $MgSiO_3$–$CaSiO_3$–$FeSiO_3$–Al_2O_3 systems, have been presented by Wood (1974).

The reaction between aluminous orthopyroxene and pyrope may be expressed:

$$\underset{\text{orthopyroxene}_{\text{ss}}}{Mg_2Si_2O_6 + MgAl_2SiO_6} \rightleftharpoons \underset{\text{garnet}}{Mg_3Al_2Si_3O_{12}} \qquad (1)$$

The experimental data relating to the reaction in simple and complex systems may be represented by the following equation in which phase compositions are related to the equilibration pressure and temperature

$$P = 1 + \frac{RT}{\Delta V_r}\left[\ln\left(\frac{(X_{Mg}^{M1})_{Opx}\{X_{Mg\,Opx}^{M2}\}^2(X_{Al}^{M1})_{Opx}}{\{X_{Mg}^{Gt}\}_1^3}\right) - \frac{\Delta G^\circ(1,T)}{RT}\right] \qquad (2)$$

where $(X_{Mg}^{M1})_{Opx}$ and $(X_{Mg}^{M2})_{Opx}$ are the fractions of the $M1$ and $M2$ sites occupied by Mg atoms, $(X_{Mg}^{Gt})_1$ the fraction of the garnet sites occupied by Mg atoms, and the volume change of reaction (1) ΔV_r is obtained from the 1 bar/298°K volumes of pyrope and aluminous enstatites (Wood, 1974; Wood and Banno, 1973). In the $MgSiO_3$–Al_2O_3 system, $(X_{Mg}^{M2})_{Opx}$ and $(X_{Mg}^{Gt})_1$ are both equal to 1·0 and the equation above reduces to

$$P = 1 + \frac{RT}{\Delta V_r}\left[\ln\left((X_{Al}^{M1})_{Opx}\cdot(X_{Mg}^{M1})_{Opx}\right) - \frac{\Delta G^\circ(1,T)}{RT}\right] \qquad (3)$$

Using Macgregor's (1974) data for X_{Al}^{M1} ($=2X_{Al_2O_3}^{Opx}$) and X_{Mg}^{M1} the values ΔH° and ΔS° of -7012 cal/mole and $-3{\cdot}89$ Gbar/mole are obtained from which isopleths of alumina contents of $MgSiO_3$–Al_2O_3 orthopyroxene in equilibrium with pyrope have been calculated (Fig. 57). Wood has also applied the solution model, using equation (2), to more complex orthopyroxenes and has presented a T–X section for the composition $(Mg_{0{\cdot}5}Fe_{0{\cdot}5})SiO_3$–$Al_2O_3$ at pressures between 10 and 30 kbar.

The phase relations, as liquidus temperatures and P_{H_2O} between 5 and 30 kbar, of orthopyroxene in association with clinopyroxene, olivine and garnet, have been determined for a garnet websterite nodule from the Honolulu volcanic series, Oahu, Hawaii (Mysen and Boettcher, 1976).

The reaction:

$$\underset{\text{pyrope}}{Mg_3Al_2Si_3O_{12}} + (1-x)\underset{\text{forsterite}}{Mg_2SiO_4} \rightleftharpoons \qquad (1)$$

$$(4-x)\underset{\text{Al-enstatite}}{MgSiO_3.xAl_2O_3} + (1-x)\underset{\text{spinel}}{MgAl_2O_4}$$

is of petrological interest in that it defines the boundary between low (spinel-bearing) and high (garnet-bearing) pressure peridotite assemblages. Within each stability field the Al_2O_3 content of the orthopyroxene coexisting with an aluminous phase may be used to provide more precise indications of the pressure–temperature environment during the formation of the assemblages, as illustrated by the reactions:

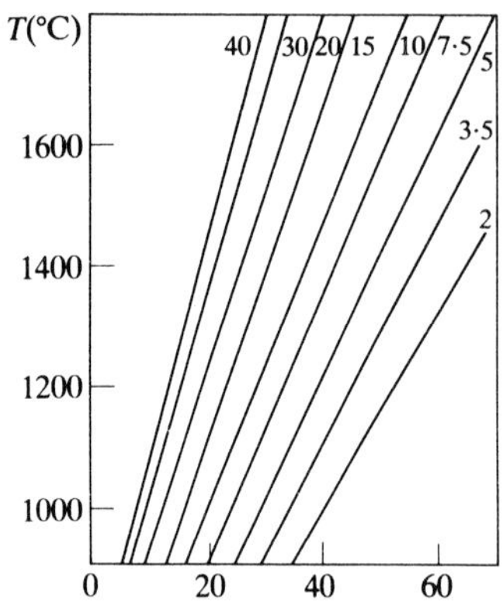

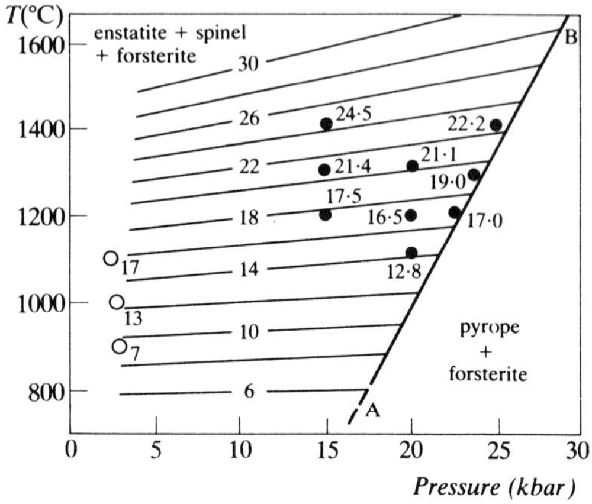

Fig. 57. Calculated composition of aluminous $MgSiO_3$–Al_2O_3 orthopyroxenes in equilibrium with pyrope. Isopleths refer to percentages of the $M1$ positions occupied by Al: $X_{Al}^{M1} = 2X\ Al_2O_3$ (after Wood, 1974).

Fig. 58. Solubility of alumina in orthopyroxene as a function of pressure and temperature (after Fujii, 1976). Numbered lines 6–30, calculated isopleths; percentages of Al in the orthopyroxene $M1$ site. Solid circles, Fujii's experimental data. Open circles, data of Anastasiou and Seifert (1972). A–B, univariant line delimiting the spinel and garnet stability regions in the system MgO–Al_2O_3–SiO_2, i.e. the reaction boundary for $Mg_3Al_2Si_3O_{12} + (1+x)Mg_2SiO_4 \rightleftharpoons (4-x)MgSiO_3.xAl_2O_3 + (1+x)MgAl_2O_4$.

$$\underset{\text{spinel}}{xMgAl_2O_4} + \underset{\text{enstatite}}{(1+x)MgSiO_3} \rightleftharpoons \underset{\text{Al-enstatite}}{MgSiO_3.xAl_2O_3} + \underset{\text{forsterite}}{xMg_2SiO_4} \qquad (2)$$

$$\underset{\text{pyrope}}{xMg_3Al_2Si_3O_{12}} + \underset{\text{enstatite}}{3(1-x)MgSiO_3} \rightleftharpoons \underset{\text{Al-enstatite}}{3MgSiO_3.xAl_2O_3} \qquad (3)$$

The solubility of alumina in orthopyroxenes coexisting with spinel and olivine may be expressed:

$$\underset{\text{En in Opx}_{ss}}{Mg_2Si_2O_6} + \underset{\text{Sp in sp}_{ss}}{MgAl_2O_4} = \underset{\text{Mg-tsch in Opx}_{ss}}{MgAl_2SiO_6} + \underset{\text{Fo in Ol}_{ss}}{Mg_2SiO_4} \qquad (1)$$

for which the condition for equilibrium in the system MgO–Al_2O_3–SiO_2 is:

$$(\Delta G^\circ)_{P,T} = -RT \ln \frac{a_{MgAl_2SiO_6}}{a_{Mg_2Si_2O_6}} \qquad (2)$$

The activities of the orthopyroxene components, enstatite and Mg-tschermakite, on the basis of the two-site model of Wood and Banno (1973) can be expressed:

$$a_{MgAl_2SiO_6}^{Opx} = (X_{Mg}^{M2})(X_{Al}^{M1}) \cdot \gamma_{MgAl_2SiO_6} \qquad (3)$$

$$a_{Mg_2Si_2O_6}^{Opx} = (X_{Mg}^{M2})(X_{Mg}^{M1}) \cdot \gamma_{Mg_2Si_2O_6} \qquad (4)$$

(γ the activity coefficient of the components in orthopyroxene, X the mole fraction of cations in $M1$ and $M2$ sites). Substituting the activity expressions into equation (2), Fujii and Takahashi (1976) and Fujii (1976) have calculated the alumina solubility in orthopyroxene as a function of pressure and temperature (Fig. 58). Also shown in the figure are compositions of aluminous enstatites coexisting with spinel and forsterite expressed as the percentage of Al in the $M1$ site ($X_{Al}^{M1} \times 100$; $X_{Al}^{M1} = 2XAl_2O_3$) determined using a mixture of glass and crystals of the composition $(MgSiO_3)_4$ and $MgAl_2O_4$ (mol. per cent). The solubility of alumina in orthopyroxene in the spinel stability region is not pressure sensitive, particularly below 1000°C but, as indicated by the earlier experimental data of Anastasiou and Seifert (1972), can be used as a geothermometer.

Experimental data on the subsolidus phase equilibria of ortho- and clinopyroxenes, olivine and spinel in the $CaO–MgO–Al_2O_3–SiO_2$ and natural ultrabasic rock systems at 16 kbar and 1200°C have been presented by Mori (1977). The work is based on the alumina contents in ortho- and clinopyroxene, compositions of coexisting pyroxene pairs in relation to the pyroxene solvus as well as quasi-thermodynamic modelling. Equilibration temperatures derived from the reactions:

$$\underset{\text{orthopyroxene}}{Mg_2Si_2O_6} = \underset{\text{clinopyroxene}}{Mg_2Si_2O_6} \tag{1}$$

$$\underset{\text{clinopyroxene}}{CaMgSi_2O_6} + \underset{\text{spinel}}{MgAl_2O_4} = \underset{\text{clinopyroxene}}{CaAl_2SiO_6} + \underset{\text{olivine}}{Mg_2SiO_4} \tag{2}$$

$$\underset{\text{orthopyroxene}}{Mg_2Si_2O_6} + \underset{\text{spinel}}{MgAl_2O_4} = \underset{\text{orthopyroxene}}{MgAl_2SiO_6} + \underset{\text{olivine}}{Mg_2SiO_4} \tag{3}$$

have been applied to the geothermometry of spinel lherzolite nodules in basalts and to intrusive lherzolites. Equilibration temperatures for the nodules are found to vary between 1000° to 1300°C. Considerable disequilibrium is indicated for the intrusive lherzolites, probably the result of retrogressive metamorphism subsequent to their emplacement.

Free energy values, estimated by combining analytical data for lherzolite xenoliths with interpretations of experimental data, have been used by Nicholls (1977) to calculate mineral compositions and modes for two-pyroxene–olivine–spinel assemblages. $[Al]^6$ contents in orthopyroxenes for lherzolite and pyrolite compositions have been calculated for pressures between 10 and 40 kbar and temperatures of 1200° to 1350°C, and illustrate that the bulk composition of the assemblage has a considerable influence on the solubility of alumina in orthopyroxenes.

The use of the alumina content of enstatite as a geobarometer, for spinel and plagioclase lherzolites, has also been considered by Akella (1976), Obata (1976) and Presnall (1976). The latter has determined the univariant curve for spinel and plagioclase lherzolite with reference to the system $CaO–MgO–Al_2O_3–SiO_2$ (Fig. 59). The univariant line of the reaction orthopyroxene + spinel = forsterite + pyrope in the system $MgO–Al_2O_3–SiO_2$ (Fig. 60), and for the reaction anorthite + forsterite = orthopyroxene + clinopyroxene + spinel in the system $CaO–MgO–Al_2O_3–SiO_2$ have been calculated by Obata.

The stability of alumina in orthopyroxene coexisting with spinel, and its possible use as a geobarometer with reference to Alpine-type peridotites has been examined by Stroh (1976). The pressure formula is based on equilibrium between aluminous orthopyroxene, olivine and spinel according to the reaction:

$$\underset{\text{Al-orthopyroxene}_{ss}}{MgAl_2SiO_6} + \underset{\text{olivine}}{Mg_2SiO_4} = \underset{\text{orthopyroxene}_{ss}}{Mg_2Si_2O_6} + \underset{\text{spinel}}{MgAl_2O_4}$$

Under isothermal conditions, increasing pressure lowers the Al content of the orthopyroxene and favours the association orthopyroxene–spinel. For natural peridotite compositions the presence of Fe^{2+}, due to its preferential partition into the spinel phase, leads to lower pressures ($\geq$ 15 kbar) than those calculated from the formula:

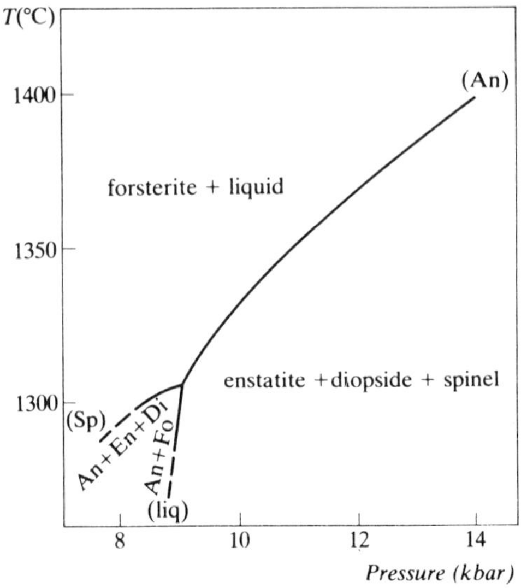

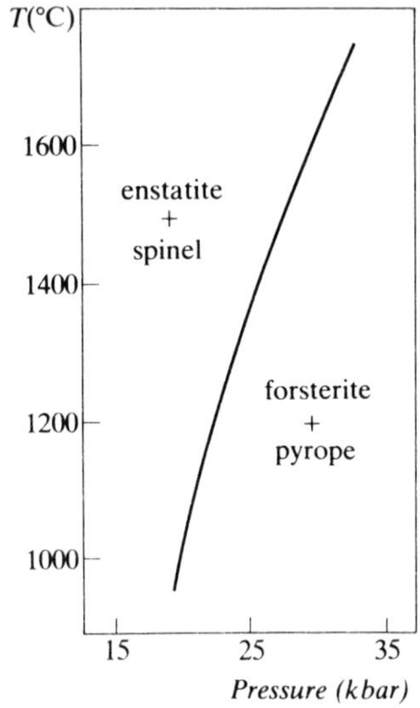

Fig. 59. Univariant solidus curve for plagioclase and spinel lherzolite in the system CaO–MgO–Al_2O_3–SiO_2 (after Presnall, 1976). Mineral names in brackets label the univariant curves according to the absent phase.

Fig. 60. Calculated univariant line of the reaction orthopyroxene + spinel ⇌ forsterite + pyrope in the system MgO–Al_2O_3–SiO_2 (after Presnall, 1976).

$$P = 1 + \left\{(9{\cdot}116 - 5{\cdot}065T_0) - RT_0 \ln\left(\frac{(X_{\text{Mg}})_{\text{Sp}}(X_{\text{Al}})^2_{\text{Sp}}(X^{M1}_{\text{Mg}})_{\text{Opx}}}{(X_{\text{Mg}})^2_{\text{Ol}}(X^{M1}_{\text{Al}})_{\text{Opx}}}\right)\right\}/\Delta V_r$$

consistent with an origin in the upper oceanic mantle and subsequent emplacement in continental crust.

Enthalpy, entropy and change in volume values for equilibrium reactions pertinent to the enstatite–diopside join, and the Al-solubility of pyroxenes in equilibrium with an Al-rich phase, spinel or garnet, have been derived from the available experimental data by Mercier (1976). These values, in conjunction with partition coefficients obtained from natural assemblages, have been used by Mercier to derive a set of equations from which pressure and temperature can be obtained for a single phase in equilibrium with a second pyroxene and with either spinel or garnet.

The effect of titanium on the solubility of Al in orthopyroxene coexisting with garnet has been examined by Akella and Boyd (1973). Compared with a titanium-free environment the solubility is lower for similar conditions of pressure and temperature, and also decreases as the $Mg/(Mg+Fe^{2+})$ ratio of the bulk composition becomes smaller.

The system Mg_2SiO_4–SiO_2, first determined at 1 atm by Bowen and Anderson (1914), has recently been investigated at 1 atm, 12 and 25 kbar by Chen and Presnall (1975). Temperature–composition sections for the system at these pressures are shown in Fig. 61. The section at 1 atm is in complete agreement with the earlier study. The incongruent melting temperature of enstatite (to forsterite + liquid) is 1559°C and the composition of the liquid at the peritectic point is 60·6 per cent SiO_2. Chen and Presnall found that this reaction disappears at about 1·3 kbar when

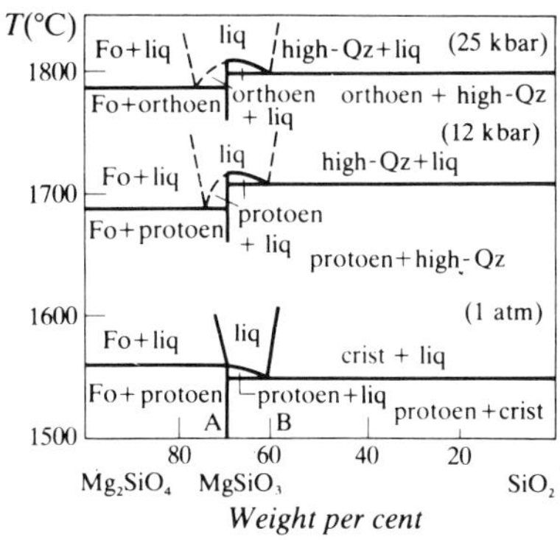

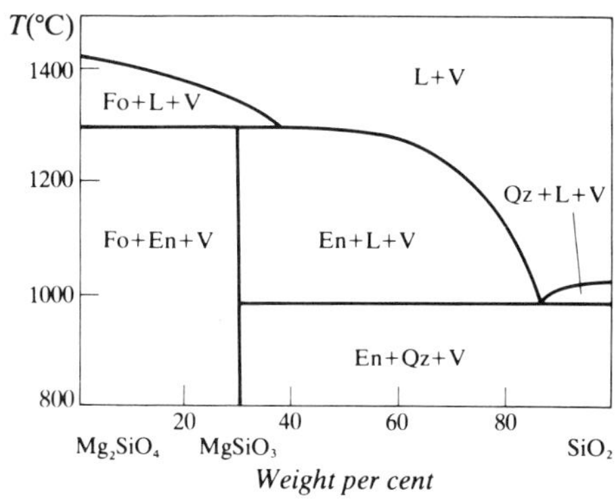

Fig. 61. Temperature–composition sections for the system Mg_2SiO_4–SiO_2 at 1 atm, 12 kbar and 25 kbar pressure (after Chen and Presnall, 1975).

Fig. 62. Projection of the system Mg_2SiO_4–SiO_2–H_2O at P_{H_2O} = 20 kbar (after Kushiro and Yoder, 1969). Fo, forsterite; En, orthoenstatite; Qz, quartz; L, liquid; V, vapour.

$MgSiO_3$ melts congruently and forms a eutectic with forsterite. The absence of the reaction relationship cannot, however, be inferred to occur in natural magmas as the presence of iron causes the relation to disappear; aluminium has the opposite effect (Taylor, 1973). The enstatite–quartz eutectic is not affected by pressure, while the enstatite–forsterite eutectic moves towards the orthosilicate composition as the pressure increases.

The addition of water to the Mg_2SiO_4–SiO_2 system has marked effects on the phase relations. The projection of the ternary system at P_{H_2O} = 20 kbar (Kushiro and Yoder, 1969) on to the join forsterite–silica from the H_2O apex is shown in Fig. 62. The melting temperature of quartz is greatly reduced relative to enstatite, and in consequence the enstatite primary phase field is enlarged. In addition, the forsterite liquidus extends beyond the metasilicate composition to about $Fo_{61}Qz_{39}$ when forsterite reacts with liquid to form enstatite in the presence of vapour. Thus fractional crystallization of a melt between Mg_2SiO_4 and $MgSiO_3$ in composition would give rise to the development of a silica-oversaturated liquid provided forsterite is removed before the reaction point (see also Taylor, 1973; Warner, 1973, p. 239).

The phase boundaries between the assemblages aluminous enstatite–cordierite–quartz, aluminous enstatite–sillimanite–quartz and pyrope–quartz, synthesized from pyrope+quartz composition at pressures between 10 and 20 kbar and temperatures between 1000° and 1400°C (Fig. 63), have been determined by Hensen and Essene (1971). The exchange reactions, in the form such as to produce one mole aluminous enstatite, are:

$$(1-2n)MgSiO_3 + \tfrac{n}{2}Mg_2Al_4Si_5O_{18} \rightleftharpoons (1-n)MgSiO_3.nAl_2O_3 + \tfrac{3}{2}nSiO_2 \quad (1)$$
enstatite — cordierite — aluminous enstatite — quartz

$$(1-n)MgSiO_3 + nAl_2SiO_5 \rightleftharpoons (1-n)MgSiO_3.nAl_2O_3 + nSiO_2 \quad (2)$$
enstatite — sillimanite — aluminous enstatite — quartz

$$(1-4n)MgSiO_3 + nMg_3Al_2Si_3O_{12} \rightleftharpoons (1-n)MgSiO_3.nAl_2O_3 \quad (3)$$
enstatite — pyrope — aluminous enstatite

A subsequent experimental study by Hensen (1972) of the reactions:

$$\text{sapphirine}_{ss} + \text{quartz} \rightleftharpoons \text{enstatite}_{ss} + \text{sillimanite} \quad (1)$$

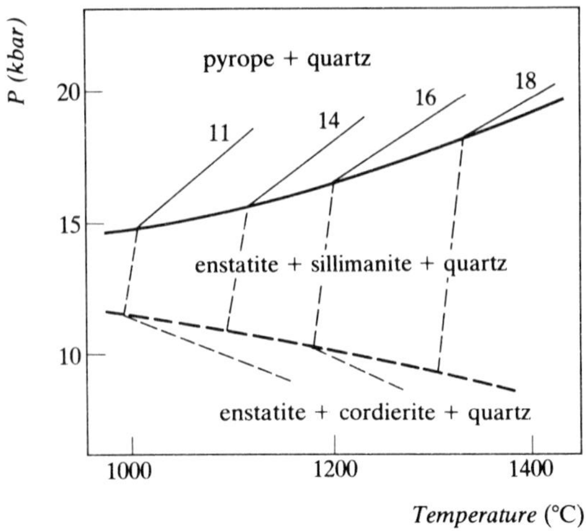

Fig. 63. Pressure–temperature relationships, for pyrope + quartz composition, of assemblages, enstatite + cordierite + quartz, enstatite + sillimanite + quartz and pyrope + quartz. Thin broken lines show possible location of contours of mol. per cent Al_2O_3 in enstatite (after Hensen and Essene, 1971).

$$\text{enstatite}_{ss} + \text{sapphirine}_{ss} + \text{sillimanite} \rightleftharpoons \text{pyrope} \quad (2)$$

$$\text{enstatite}_{ss} + \text{sillimanite} \rightleftharpoons \text{pyrope} + \text{quartz} \quad (3)$$

has shown that the reactions occur sequentially with increasing pressure in the temperature range 1100°–1400°C, and that the average slope of the reaction boundaries is 15 ± 5 bar/°C. These data, and other data relating to reactions involving enstatite in the system MgO–Al_2O_3–SiO_2, together with the effect of pressure and temperature on the solubility of Al_2O_3 in the orthopyroxene, have been collated by Hensen to indicate the topological relations of the Al_2O_3 isopleths (at 10 and 12 weight per cent) and the univariant phase boundaries. The effects of the addition of CaO to the MgO–Al_2O_3–SiO_2 system, particularly in relation to the relative stabilities of garnet granulite and eclogite, have also been discussed by Hensen.

Newton *et al.* (1974) have investigated the reactions:

$$\underset{\text{enstatite}}{2MgSiO_4} + \underset{\text{sillimanite}}{2Al_2SiO_5} = \underset{\text{sapphirine}}{Mg_2Al_4SiO_{10}} + \underset{\text{quartz}}{3SiO_2} \quad (1)$$

$$\underset{\text{cordierite}}{Mg_2Al_4Si_5O_{18}} = \underset{\text{enstatite}}{2MgSiO_3} + \underset{\text{sillimanite}}{2Al_2SiO_5} + \underset{\text{quartz}}{SiO_2} \quad (2)$$

$$\underset{\text{enstatite}}{8MgSiO_3} + \underset{\text{sillimanite}}{8Al_2SiO_5} = \underset{\text{sapphirine}}{Mg_2Al_4SiO_{10}} + \underset{\text{cordierite}}{3Mg_2Al_4Si_5O_{18}} \quad (3)$$

Values of dP/dT for reactions 1 and 3 are 47 bar/°C and $\simeq 10$ bar/°C respectively, and the data indicate that, provided the conditions are dry, pressures not greater than between 6 and 8 kbar are required for the development of the enstatite–sillimanite–quartz and enstatite–sapphirine–quartz assemblages.

Chatterjee and Schreyer (1972) determined the univariant pressure–temperature curve of the reaction enstatite solid solution + sillimanite $\rightleftharpoons$ sapphirine solid solution + quartz in the pressure range 12 to 20 kbar (Fig. 64). They used the mixture $MgO.Al_2O_3.2SiO_2$, i.e. the composition in the system MgO–Al_2O_3–SiO_2 at the intersection of the tie-lines enstatite–sillimanite and sapphirine–quartz

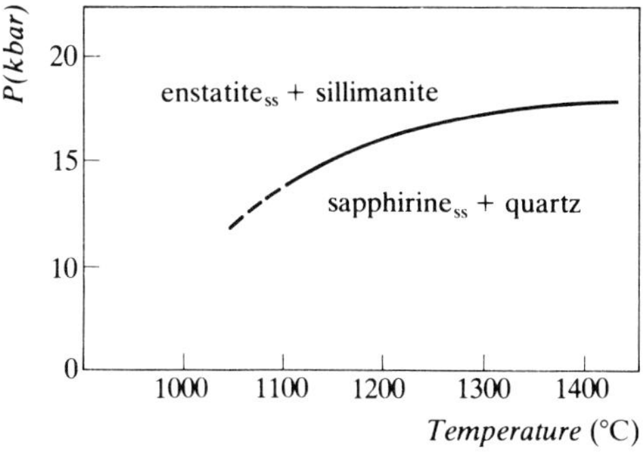

Fig. 64. Univariant curve for the reaction enstatite$_{ss}$ + sillimanite ⇌ sapphirine$_{ss}$ + quartz (after Chatterjee and Schreyer, 1972).

according to the theoretical equation

$$2MgSiO_3 + 2Al_2SiO_5 \rightleftharpoons Mg_2Al_4SiO_{10} + 3SiO_2$$

Enstatite with up to 18·9 weight per cent Al_2O_3 occurs as a liquidus phase, at 1420°C in association with sapphirine and quartz at 1420°C and 15 kbar, in the system MgO–Al_2O_3–SiO_2.

Aluminous enstatite crystallizes from the 1:1 composition (molecular ratio) anorthite + forsterite at pressures between 7·5 and 10·5 kbar at 1000°C (Kushiro and Yoder, 1966). The stability field of the orthopyroxene solid solution-bearing assemblage widens to between 9 and 17 kbar at 1335° and 1400°C, at which pressures and temperatures melting begins. Curve A (Fig. 65) represents the following reaction in which $0 < x < 1$ (x being estimated to be $> 0{\cdot}3$).

$$\underset{\text{anorthite}}{2CaAl_2Si_2O_8} + \underset{\text{forsterite}}{2Mg_2SiO_4} = \underset{\text{aluminous enstatite}}{2MgSiO_3.xMgAl_2SiO_6} +$$

$$\underset{\text{aluminous diopside}}{CaMgSi_2O_6.xCaAl_2SiO_6} + \underset{\text{anorthite}}{(1-x)CaAl_2Si_2O_8} + \underset{\text{spinel}}{(1-x)MgAl_2O_4}$$

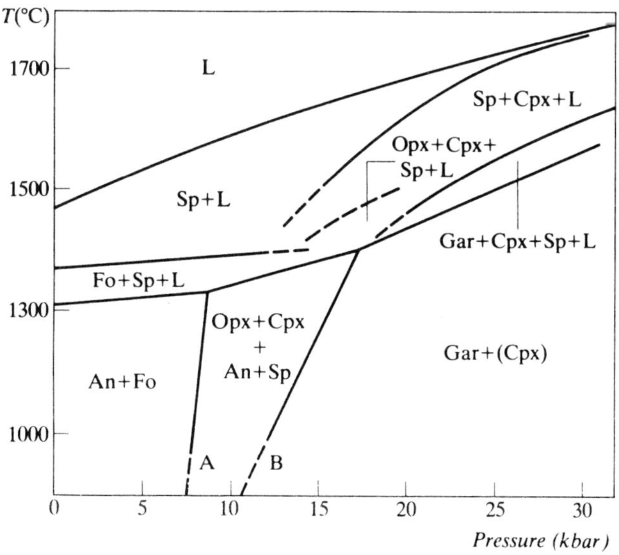

Fig. 65. Pressure–temperature plane for 1:1 anorthite–forsterite composition (after Kushiro and Yoder, 1966).

At higher pressures enstatite is unstable and the enstatite–diopside–anorthite–spinel assemblage is replaced by garnet + clinopyroxene (the pyroxene is probably unstable and decreases in amount with increasing duration of the experiment). The reaction on the univariant curve B may be represented:

$$\underset{\text{aluminous diopside}}{CaMgSi_2O_6.xCaAl_2SiO_6} + \underset{\text{aluminous enstatite}}{2MgSiO_3.xMgAl_2SiO_6} + \underset{\text{anorthite}}{(1-x)CaAl_2Si_2O_8} + \underset{\text{spinel}}{(1-x)MgAl_2O_4} = \underset{\text{garnet}}{2CaMg_2Al_2Si_3O_{12}}$$

where x is estimated to be between 0·4 and 0·5. Similar relationships are found for the 1 : 2 anorthite–forsterite composition, except that forsterite and not anorthite is present in the intermediate pressure field as indicated by the reaction:

$$\underset{\text{anorthite}}{2CaAl_2Si_2O_8} + \underset{\text{forsterite}}{4Mg_2SiO_4} = \underset{\text{aluminous diopside}}{(2-x)CaMgSi_2O_6.xCaAl_2SiO_6} + \underset{\text{aluminous enstatite}}{(4-2x)MgSiO_3.xMgAl_2SiO_6} + \underset{\text{spinel}}{(2-2x)MgAl_2O_4} + \underset{\text{forsterite}}{2xMg_2SiO_4}$$

where $x \simeq 0{\cdot}46$.

The investigation of forsterite–anorthite relationships has been extended by Green and Hibberson (1970) to compositions closer to those of natural rock association, e.g. to olivine–labradorite. This work indicates that a somewhat higher pressure is required for the beginning of the magnesium olivine–plagioclase reaction:

$$\underset{\text{labradorite}}{3CaAl_2Si_2O_8.2NaAlSi_3O_8} + \underset{\text{olivine}}{8Mg_2SiO_4} \rightleftharpoons \underset{\text{enstatite}}{10MgSiO_3} + \underset{\text{omphacite}}{3CaMgSi_2O_6.2NaAlSi_2O_6} + \underset{\text{spinel}}{3MgAl_2O_4}$$

The reaction occurs at about 11 kbar compared with the anorthite–forsterite assemblage where anorthite is absent at pressures greater than 8·5 kbar at a temperature of 1200°C. Investigation of the composition 2 enstatite + anorthite shows that the two phases are stable to pressures of approximately 13 and 15 kbar at 1150° and 1400°C respectively, and are replaced at higher pressures by ortho- and clinopyroxene solid solutions and quartz:

$$\underset{\text{anorthite}}{2CaAl_2Si_2O_8} + \underset{\text{enstatite}}{4MgSiO_3} = \underset{\text{aluminous enstatite}}{2MgSiO_3.MgAl_2SiO_6} + \underset{\text{aluminous diopside}}{CaMgSi_2O_6.CaAl_2SiO_6} + \underset{\text{quartz}}{2SiO_2}$$

In mixes of anorthite, enstatite and diopside with composition 2:2:1 (molecular proportions) at a temperature of 1200°C, the stable mineral assemblage at a pressure of 13·5 kbar is orthopyroxene, anorthite, and clinopyroxene solid solution (Green, 1969). At higher pressures the amount of orthopyroxene decreases, and at 18 kbar the stable assemblage consists of clinopyroxene solid solution, garnet and quartz, and results from a reaction of the form:

$$\underset{\text{enstatite}}{16MgSiO_3} + \underset{\text{anorthite}}{16CaAl_2Si_2O_8} + \underset{\text{diopside}}{8CaMgSi_2O_6} \rightleftharpoons \underset{\text{garnet solid solution}}{4Mg_3Al_2Si_3O_{12} + Ca_3Al_2Si_3O_{12}} + \underset{\text{pyroxene solid solution}}{10CaMgSi_2O_6 + 11CaAl_2SiO_6 + 2MgSiO_3} + 16SiO_2$$

Two isobaric invariant points in the ternary system $CaO–MgO–Al_2O_3–SiO_2$ at 10 kbar have been located by Kushiro (1972b), orthopyroxene + clinopyroxene + forsterite + spinel + liquid at 1350°C, and orthopyroxene + clinopyroxene + anorthite + spinel + liquid at 1345°C. The two invariant points, I and II, projected onto the plane $Mg_2SiO_4–CaAl_2Si_2O_8–SiO_2$ from diopside composition, are shown in Fig. 66. Data are also given by Kushiro for the system $Mg_2SiO_4–CaMgSi_2O_6–CaAl_2Si_2O_8–NaAlSi_3O_8$ at 10, 20 and 25 kbar, as well as for natural garnet and spinel lherzolites at 10 and 20 kbar, and are discussed in relation to the formation of alkali and tholeiitic basalt liquids by the partial melting of spinel lherzolite.

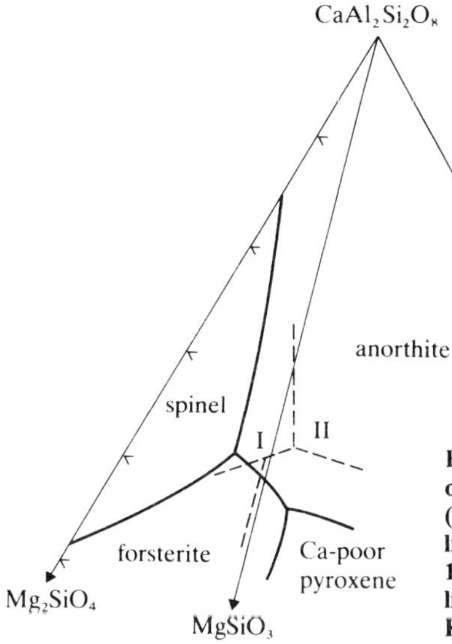

Fig. 66. Projection of the isobaric (10 kbar) invariant points, orthopyroxene + clinopyroxene + forsterite + spinel + liquid (I), and orthopyroxene + clinopyroxene + anorthite + spinel + liquid (II), and the liquidus boundaries (broken lines) at 10 kbar onto the plane $Mg_2SiO_4–CaAl_2Si_2O_8–SiO_2$. The liquidus boundaries at 1 atm are shown by solid lines (after Kushiro, 1972b).

In the system $CaO–MgO–Al_2O_3–SiO_2–Na_2O–H_2O$ at a pressure of 5 kbar the temperature of invariant equilibrium between orthopyroxene, forsterite, Ca-rich clinopyroxene, amphibole, plagioclase, liquid and vapour occurs at 960°C, and a similar invariant assemblage, in which clinopyroxene is replaced by spinel, exists at 950° (Cawthorn, 1976).

The effect of adding CO_2 to the system $CaO–MgO–SiO_2$ has been investigated by Wyllie and Huang (1976). They have shown that increase in pressure on peridotite composition leads to the reaction forsterite + clinopyroxene + $CO_2 \rightleftharpoons$ orthopyroxene + carbonate (Ca:Mg:70:30). The reaction curves pass through 15 kbar and 960°C with a slope of 45 bar/°C, and terminate at approximately 25 kbar–1200°C when melting begins.

The crystallization of orthopyroxene and its relationships with other phases have been studied experimentally for a number of natural rock compositions, including olivine tholeiite and olivine basalt by Green and Ringwood (1967b), olivine tholeiite, nepheline basanite and peridotite by Ito and Kennedy (1967, 1968), and the olivine-rich lavas of Nuanetsi from the lower part of the Karroo basaltic sequence (Cox and Jamieson, 1974). The subsolidus phase relations, at pressures up to 15 kbar, for a two-pyroxene granulite xenolith (Fig. 67) in the Delegate basaltic breccia pipes, Australia, have been determined by Irving (1974b).

Protoenstatite occupies a relatively small compositional field in the $MgO–Al_2O_3–SiO_2$ system at 1 atm pressure (Schreyer and Schairer, 1961), and is present at two ternary eutectics, 1364°C with forsterite and cordierite, and 1355°C with cordierite and tridymite (Fig. 68). At higher pressures enstatite is the stable magnesium metasilicate phase at liquidus temperatures, and occupies a larger primary field of crystallization as in the system forsterite–diopside–silica–water (Fig. 69), in which the field of enstatite solid solution occupies a large part of the SiO_2-rich compositions (Kushiro, 1969).

The association of enstatite and nepheline has generally been regarded as incompatible, and the occurrence of enstatite-bearing peridotite nodules in alkali

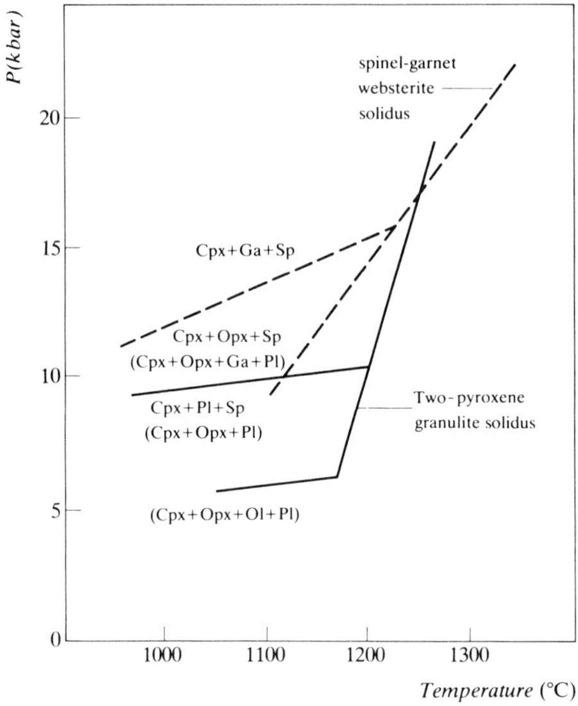

Fig. 67. Experimentally determined subolidus phase relationship for spinel–garnet websterite and two-pyroxene granulite (after Irving, 1974b). Websterite assemblages unbracketed, granulite assemblages in brackets. Cpx, clinopyroxene; Opx, orthopyroxene; Ga, garnet; Sp, spinel; Ol, olivine; Pl, plagioclase.

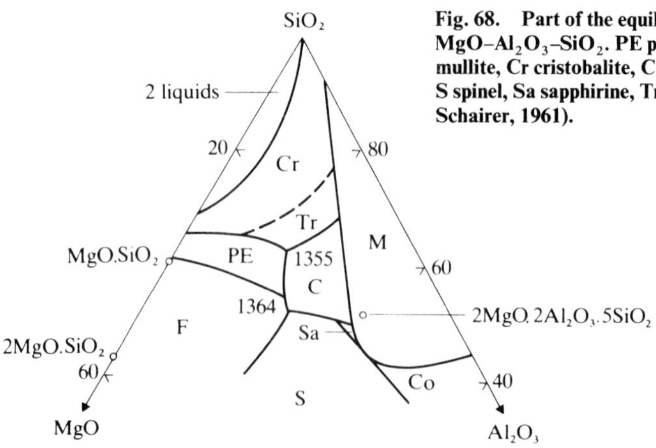

Fig. 68. Part of the equilibrium diagram of the system $MgO–Al_2O_3–SiO_2$. PE protoenstatite, F forsterite, M mullite, Cr cristobalite, C cordierite, Co corundum, S spinel, Sa sapphirine, Tr tridymite (after Schreyer and Schairer, 1961).

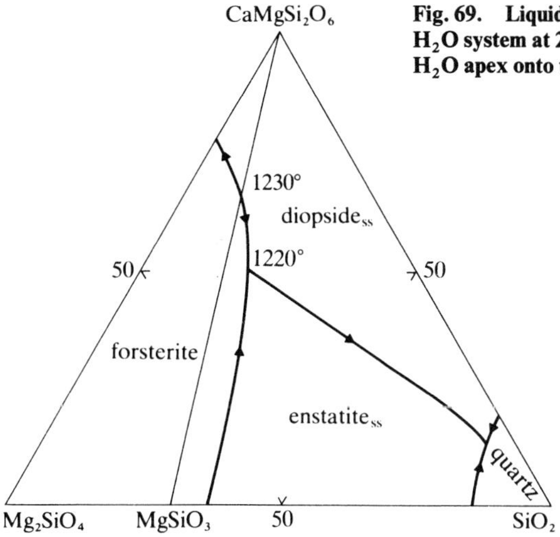

Fig. 69. Liquidus diagram of the diopside–forsterite–silica–H_2O system at 20 kbar. Boundaries are projected from the H_2O apex onto the plane Di–Fo–SiO_2 (after Kushiro, 1969).

basalts has usually been interpreted either as a non-equilibrium relationship, resulting from their occurrence as accidental xenoliths, or as an equilibrium association at high pressure. Enstatite + nepheline is isochemical with forsterite + albite:

$$4MgSiO_3 + NaAlSiO_4 = 2Mg_2SiO_4 + NaAlSi_3O_8$$

From a glass of composition nepheline$_{62}$, forsterite$_{18}$, silica$_{20}$, Kushiro (1965a) has crystallized forsterite + nepheline + albite at 10 kbar and 1000°C and enstatite + nepheline + albite at pressures between 15 and 20 kbar. Enstatite has also been shown to be a stable phase in the subsystem, forsterite–nepheline–silica–H_2O of the system MgO–CaO–Na_2O–Al_2O_3–SiO_2–H_2O (Kushiro, 1972a).

In the system MgO–SiO_2–H_2O (Bowen and Tuttle, 1949), enstatite crystallizes at temperatures and water vapour pressures above 700°C and 5000 lb/in^2 respectively, and is stable in the presence of water vapour below approximately 900°C; the characteristic reactions are:

$$\underset{\text{talc}}{(OH)_2Mg_3Si_4O_{10}} + \underset{\text{forsterite}}{Mg_2SiO_4} \rightleftharpoons \underset{\text{enstatite}}{5MgSiO_3} + H_2O \quad (1)$$

$$\underset{\text{talc}}{(OH)_2Mg_3Si_4O_{10}} \rightleftharpoons \underset{\text{enstatite}}{3MgSiO_3} + SiO_2 + H_2O \quad (2)$$

The enthalpy and the P_{H_2O}–T equilibrium curve for the second reaction have been given by Leonidov and Khitarov (1967), see also Kitahara *et al.*, 1966 (Fig. 23). The equation of the equilibrium curve is given by Skippen (1971) as:

$$\log_{10} K = \log_{10} fH_2O = -7422/T + 10{\cdot}54$$

and with an additional term for pressure adjustment:

$$\log_{10} K = -7422/T + 10{\cdot}54 + 0{\cdot}104\frac{(P-2000)}{T}$$

The transformation of talc to enstatite in an electron beam has been described by Akizuki (1967).

Enstatite crystallizes from a glass of appropriate composition at temperatures above 660°C, in the $MgO–Al_2O_3–SiO_2–H_2O$ system (Yoder, 1952), and in addition to the above two reactions, forms in equilibrium with serpentine and talc at 15000 lb/in^2 water vapour pressure.

$$\underset{\text{serpentine}}{(OH)_4Mg_3Si_2O_5} + \underset{\text{talc}}{(OH)_2Mg_3Si_4O_{10}} \rightleftharpoons \underset{\text{enstatite}}{6MgSiO_3} + 3H_2O \qquad (3)$$

The reaction forsterite + talc = enstatite + water, at pressures up to 30 kbar and 900°C has been investigated by Kitahara *et al.* (1966). At pressures above 5 kbar the *P–T* curve for the reaction has a negative slope (Fig. 70), and values of dP/dT for

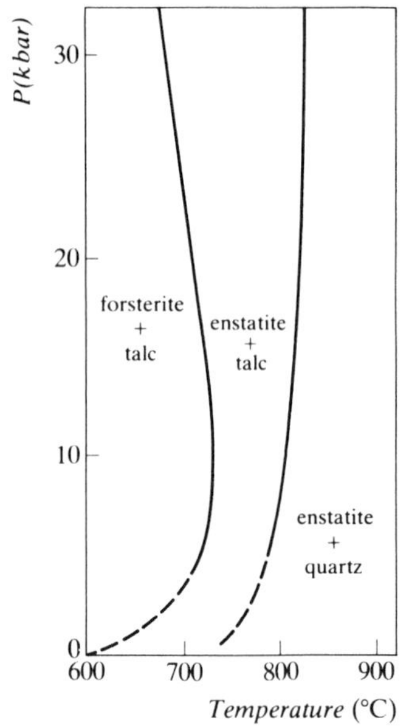

Fig. 70. Phase equilibria in the system $MgO–SiO_2–H_2O$ (after Kitahara *et al.*, 1966).

Table 9 Calculated value of dP/dT for the reaction talc + forsterite = 5 enstatite + water

Temperature °C	Pressure kb	ΔS cal/deg	ΔV cm^3	$\Delta S/\Delta V = dP/dT$ kbar/deg
700	5	18·7	5·05	0·155
735	10	17·4	0·16	4·52
713	20	16·2	−2·98	−0·228
690	30	15·9	−4·24	−0·152

some points, calculated from thermodynamic data (Table 9) are in good agreement with the experimental data (see also Kitahara and Kennedy, 1967). Enstatite may also form from anthophyllite according to the equation

$$Mg_7Si_8O_{22}(OH)_2 = 7MgSiO_3 + SiO_2 + H_2O$$

and Fyfe (1962) has shown that enstatite and quartz can be converted to anthophyllite at temperatures between 670° and 800°C at a pressure of 2 kbar. The equilibrium curve for the reaction

$$\underset{\text{tremolite}}{Ca_2Mg_5Si_8O_{22}(OH)_2} = \underset{\text{enstatite}}{3MgSiO_3} + \underset{\text{diopside}}{2CaMgSi_2O_6} + \underset{\text{quartz}}{SiO_2} + H_2O$$

$$\log_{10} K = \log_{10} fH_2O = \frac{-6977}{T} + 9{\cdot}332 + \frac{0{\cdot}125(P-2000)}{T}$$

where T is in degrees kelvin; f and P in bars, have been determined for a total pressure of 2 kbar by Skippen (1971). Akella and Winkler (1966) have investigated experimentally the reaction

chlorite + quartz ⇌ hypersthene + cordierite + H_2O

for a system composition with an atomic ratio $Fe^{2+} + Fe^{3+}/(Fe^{2+} + Fe^{3+} + Mg)$ of approximately 0·44, and shown that the lower stability limit of the hypersthene–cordierite pair is 663°C at 0·5 kbar, 703°C at 1 kbar, and 755°C at 2 kbar H_2O pressure (Fig. 71).

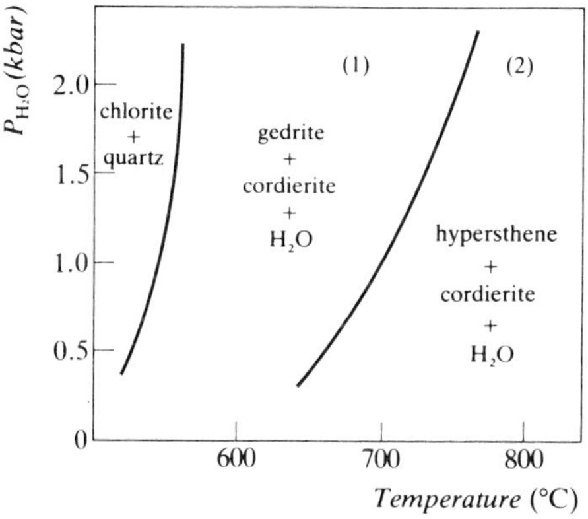

Fig. 71. Equilibrium curves for the reactions chlorite + quartz ⇌ gedrite + cordierite + H_2O and chlorite + quartz ⇌ hypersthene + cordierite + H_2O for $(Fe^{2+} + Fe^{3+})/(Fe^{2+} + Fe^{3+} + Mg)$ ratio of 0·4 (after Akella and Winkler, 1966). (1) *P–T* field growth of gedrite + cordierite; breakdown of chlorite in presence of quartz. (2) *P–T* field growth of hypersthene + cordierite, decrease of gedrite and quartz.

Reactions involving orthopyroxene and anthophyllite in the presence of fluids with a high CO_2-content in the system $MgO–SiO_2–H_2O–CO_2$ have been examined by Johannes (1969) who has presented equilibrium curves for the following reactions:

$$\underset{\text{enstatite}}{8MgSiO_3} + H_2O + CO_2 \rightleftharpoons \underset{\text{anthophyllite}}{Mg_7Si_8O_{22}(OH)_2} + \underset{\text{magnesite}}{MgCO_3} \quad (1)$$

$$2Mg_2SiO_4 + 2CO_2 \rightleftharpoons 2MgSiO_3 + 2MgCO_3 \quad (2)$$

$$2MgSiO_3 + 2CO_2 \rightleftharpoons 2MgCO_3 + 2SiO_2 \quad (3)$$

at a fluid pressure of 2 kbar.

The reactions:

hornblende + anthophyllite ⇌ enstatite + hornblende + anorthite + H_2O
hornblende ⇌ enstatite + diopside + anorthite + H_2O

produced on heating chlorite, talc, tremolite and quartz at water pressures between 0·5 and 2 kbar, have been described by Choudhuri and Winkler (1967). The equilibrium temperatures vary between 690°C at 0·5 kbar and 770°C at 2 kbar water pressure.

The formation of enstatite, together with forsterite and spinel, by the breakdown of chlorite at pressures above 3·5 kbar have been reported by Fawcett and Yoder (1966):

$$\underset{\text{clinochlore}}{5Mg_5Al_2Si_3O_{10}(OH)_8} = \underset{\text{enstatite}}{10MgSiO_3} + \underset{\text{forsterite}}{5Mg_2SiO_4} + \underset{\text{spinel}}{5MgAl_2O_4} + 20H_2O$$

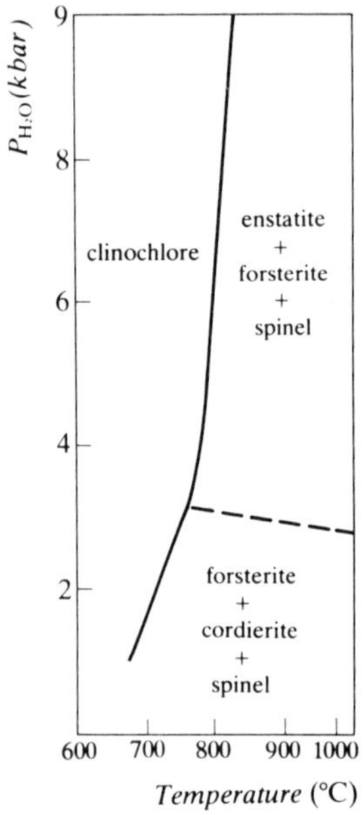

Fig. 72. Breakdown curve for the reaction clinochlore ⇌ enstatite + forsterite + spinel (after Fawcett and Yoder, 1966).

The breakdown curve (Fig. 72) is defined by the points 768°, 787° and 830°C at 3·5, 5 and 10 kbar respectively.

The join $MgSiO_3$–$KMg_3AlSi_3O_{10}(OH)_2$–H_2O has been studied by Modreski (1972). A 1:1 mixture of enstatite and phlogopite begins to melt to forsterite + liquid at 1085°C and 10 kbar, and 1265°C and 30 kbar, and enstatite + phlogopite + water vapour to forsterite + liquid at 1060°C and 10 kbar, and 1170°C and 30 kbar. At pressures up to 25 kbar, the alumina content of the enstatite coexisting with phlogopite ± vapour, and with phlogopite + forsterite + liquid ± vapour is small (0·09 to 0·67 weight per cent). Enstatites crystallized at subsolidus temperatures and in the pressure range 30 to 35 kbar have higher alumina contents, up to 1·34 per cent in 'dry', and 1·25 per cent in 'wet' runs. In the system orthorhombic $KAlSiO_4$–Mg_2SiO_4–SiO_2–H_2O, enstatite coexists with potassium feldspar, quartz and a water-undersaturated liquid, and the univariant reaction:

phlogopite + quartz = enstatite + potassium feldspar + liquid

occurs at 2 kbar (Luth, 1967).

Protoenstatite is stable with tridymite, forsterite, diopside, anorthite and cordierite at atmospheric pressure and subsolidus temperatures in the system CaO–MgO–Al_2O_3–SiO_2. The solid solution of CaO and Al_2O_3, in the temperature range 1220° to 1390°C varies between 0·25 and 4·10, and 0·17 and 1·62 weight per cent, respectively (Biggar and Clarke, 1972). The range of solid solution of FeO, CaO and Al_2O_3 in protoenstatites and orthopyroxenes in the system CaO–MgO–Al_2O_3–SiO_2–Fe–O at temperatures between 1200° and 1350°C at 1 atm have been determined by Clarke and Biggar (1972). The same authors have also presented the phase relations in parts of the system $Ca_3Al_2Si_{15}O_{36}$–MgO–$FeO_{1\cdot5}$ at 1200°, 1250°, 1300° and 1350°C, and the phase relations in the join MgO–$Ca_3Al_2Si_{15}O_{36}$. In the first system the pyroxene phase at 1250°C is dominantly orthopyroxene, at 1350°C protopyroxene.

The effect of certain oxides on the liquidus boundary between protoenstatite and forsterite at 1 atm has been examined by Kushiro (1973). The movement of the liquidus boundaries in the system $MgO–SiO_2–X$, where X is K_2O, Na_2O, CaO, FeO, Al_2O_3, Cr_2O_3 and TiO_2 are shown in Fig. 73(a). Solution of oxides of the monovalent elements causes the boundary to shift towards more silica-rich compositions, and solution of oxides of polyvalent elements towards more silica-poor compositions; the oxides of the divalent elements, Fe and Ca, have a smaller and intermediate effect. These changes can be related to the tendency of the monovalent elements to break the Si–O chains, and so prevent the polymerization of the $(SiO_4)^{4-}$ tetrahedra in the melt, thus favouring the crystallization of the orthosilicate, forsterite, with the consequent expansion of the liquidus field of the olivine relative to protoenstatite. By contrast the polyvalent elements tend to polymerize the Si–O tetrahedra and favour the crystallization of the pyroxene. The liquidus boundary between protoenstatite and the SiO_2 polymorphs in the system $MgO–SiO_2–X$ at 1 atm is similarly affected, except that the effect of the monovalent oxides is more pronounced, and a change of curvature of the boundary occurs with the cristobalite ⇌ tridymite transition (Fig. 73(b)).

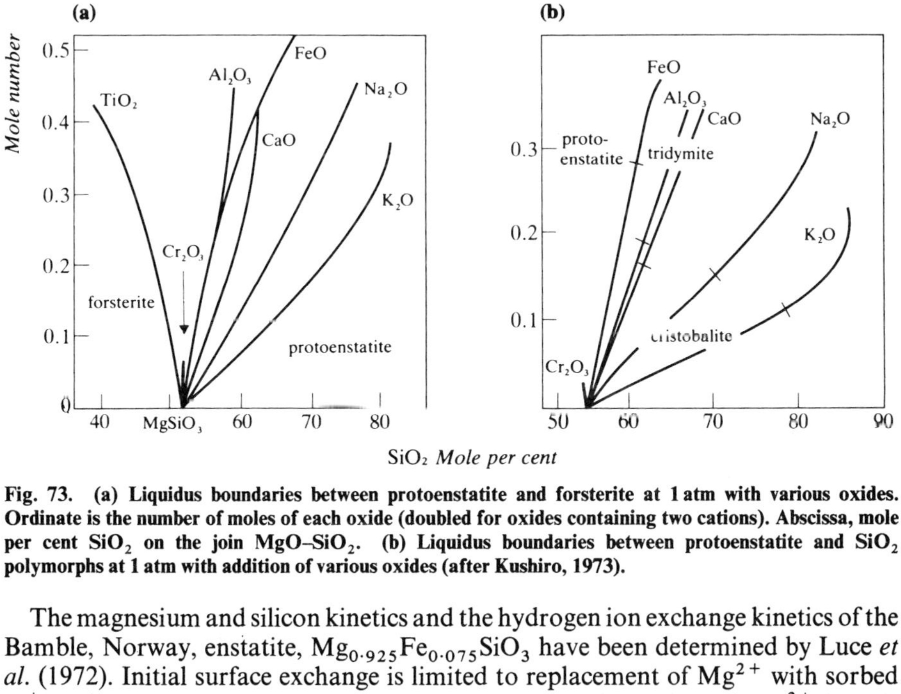

Fig. 73. (a) Liquidus boundaries between protoenstatite and forsterite at 1 atm with various oxides. Ordinate is the number of moles of each oxide (doubled for oxides containing two cations). Abscissa, mole per cent SiO_2 on the join $MgO–SiO_2$. (b) Liquidus boundaries between protoenstatite and SiO_2 polymorphs at 1 atm with addition of various oxides (after Kushiro, 1973).

The magnesium and silicon kinetics and the hydrogen ion exchange kinetics of the Bamble, Norway, enstatite, $Mg_{0\cdot925}Fe_{0\cdot075}SiO_3$ have been determined by Luce *et al.* (1972). Initial surface exchange is limited to replacement of Mg^{2+} with sorbed H^+ and is complete in a few minutes, and the total number of Mg^{2+} ions is dependent on the pH of the solution but is restricted to a layer thinner than one unit cell. Over longer periods the extraction of magnesium and silicon give parabolic extraction curves, at pH values of 3·2, 5·0 and 9·6, showing the relationships:

$$Q_{Mg} = K_{Mg}t^{\frac{1}{2}} + Q_{Mg_0}$$
$$Q_{Si} = K_{Si}t^{\frac{1}{2}} + Q_{Si_0}$$

where Q is the number of moles/cm^2 extracted, K is the parabolic rate constant for extraction and is determined from the slope of the straight line plot of Q against $t^{\frac{1}{2}}$, and Q_0 is the number of moles/cm^2 rapidly exchanged between the surface of the mineral and the solution. The parabolic kinetics relationship indicates that the diffusion controls the dissolution kinetics, and because diffusion within the solution is too rapid, that transport in the enstatite is the rate controlling factor. In more strongly acid solution, pH $\simeq$ 1·65, there is a transition from parabolic to linear kinetics when the relationship can be expressed

$$Q_{Mg} = K_{Mg}t + Q_{Mg_0}$$

where K is the linear experimental rate constant.

The heats of solution of three orthopyroxenes in 20·1 per cent HF at 93·7°C have been determined by Sahama and Torgeson (1949). These data indicated that no heat of mixing occurred and that the $Mg \rightleftharpoons Fe^{2+}$ replacement in the orthopyroxene structure is ideal (but see p. 24) and may be represented by the equation:

$$\Delta H = -57010 - 58{\cdot}7\, X_{Mg}$$

in which X_{Mg} is the molar percentage of $MgSiO_3$ in the pyroxene. The extrapolated heats of solution for the pure end-members are ΔH_{MgSiO_3} −62880 cal/mole; ΔH_{FeSiO_3} −57010 cal/mole. The heat of solution of an aluminium-rich orthopyroxene (Al_2O_3 6·39 per cent) is −62560±90 cal/mole, a value which deviates from the heat of solution curve of the pure $MgSiO_3$–$FeSiO_3$ series. This deviation may be due to the substitution of Al for tetravalent Si and divalent (Mg,Fe) ions, resulting in individual oxygens not being equally balanced and leading to a consequent weakening of the structure. The standard enthalpy of formation of enstatite at 298°K (−8·12±0·21 kcal/mole), calculated from data obtained at 965° and 1173°K, is given by Shearer and Kleppa (1973), and the Gibbs free energy of formation (−1457·27±4 kJ mol^{-1}) by Zen and Chernosky (1976).

Campbell and Roeder (1968) have investigated the stability of pyroxene in the Ni–Mg–SiO–O system at temperatures between 1300° and 1500°C at controlled oxygen fugacities.

Alteration. Orthopyroxenes, especially the more magnesium-rich varieties, are sometimes altered to serpentine, and where alteration is complete the pseudomorphs show a characteristic bronze-like metallic lustre or schiller and are known as bastite. Orthopyroxene, in association with olivine, is also involved in the general serpentinization reaction (Hostetler, *et al.*, 1966):

$$\text{olivine} + \text{orthopyroxene} + H_2O + O_2 \rightarrow \text{serpentine} + \text{brucite} + \text{magnetite} + MgO$$

Alteration to a pale-green amphibole, usually referred to as uralite, is not uncommon. The alteration of orthopyroxene to garnet during the retrograde metamorphism of the charnockitic rocks of Uganda has been described by Groves (1935). In some of these rocks the formation of the garnet was preceded by the development of a fibrous serpentine, while in others complex reaction rims, consisting of successive zones of hornblende and quartz, biotite and iron ore, and garnet and quartz, are developed between the orthopyroxene and plagioclase (see also Mason, R., 1967).

Optical and Physical Properties

The relationship between the optical properties and the chemical composition of the orthopyroxenes has been studied by a number of investigators (Walls, 1935; Henry, 1935; Hess and Phillips, 1940; Burri, 1941; Taneda, 1947; Poldervaart, 1947, 1950;

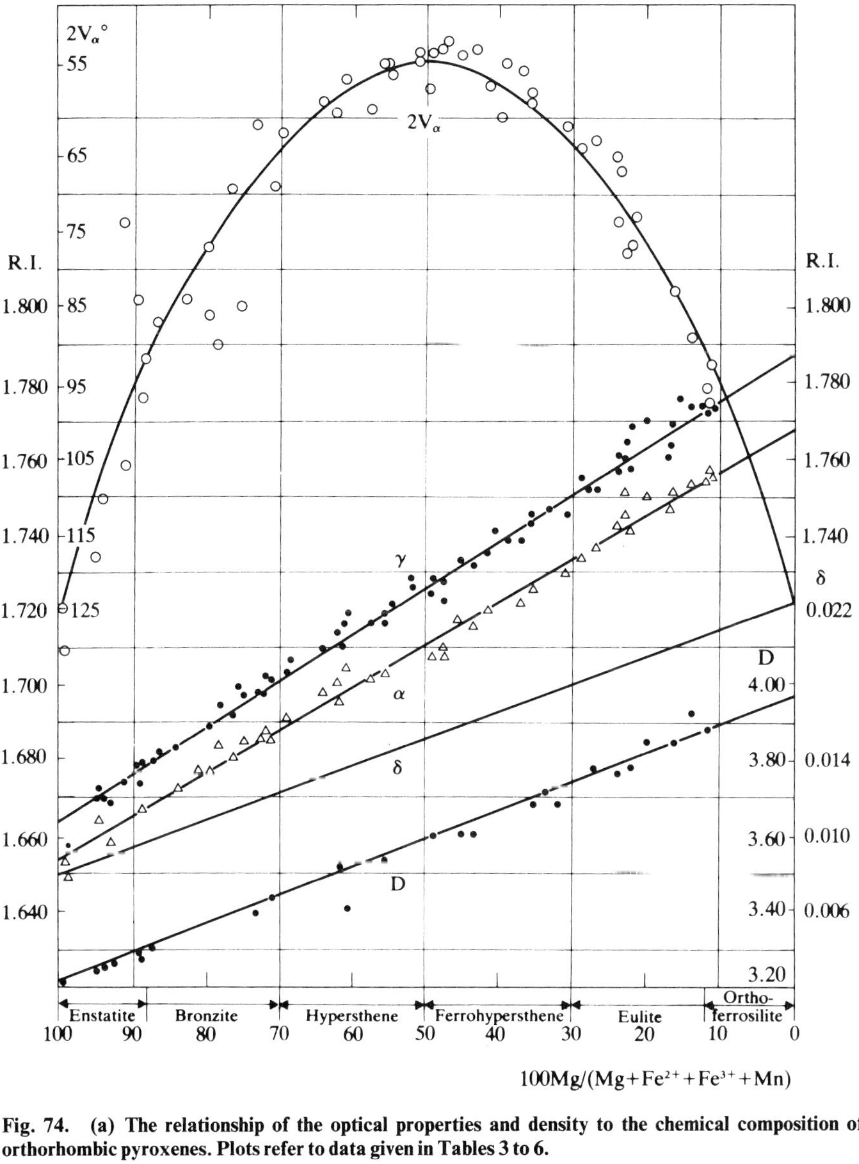

Fig. 74. (a) The relationship of the optical properties and density to the chemical composition of the orthorhombic pyroxenes. Plots refer to data given in Tables 3 to 6.

Ramberg and DeVore, 1951; Hess, 1952; Clavan *et al.*, 1954; Hori, 1956; Henriques, 1958; Dobretsov, 1959; Winchell and Leake, 1965; Leake, 1968; Shemyakin, 1968; Kosoy and Shemyakin, 1971), and a precise correlation between the optical properties and the replacement of Mg by Fe^{2+} has been established. Figure 74(a) shows the variation in refractive indices, optic axial angle and density with the atomic ratio Mg:(Mg + Fe^{2+} + Fe^{3+} + Mn). The refractive indices vary linearly with the content of (Fe^{2+} + Fe^{3+} + Mn) and the γ refractive index increases approximately 0·001 25 for each atomic per cent (Fe^{2+} + Fe^{3+} + Mn). The relation between mg = Mg/(Mg + Fe^{2+} + Fe^{3+} + Mn) and refractive indices computed by

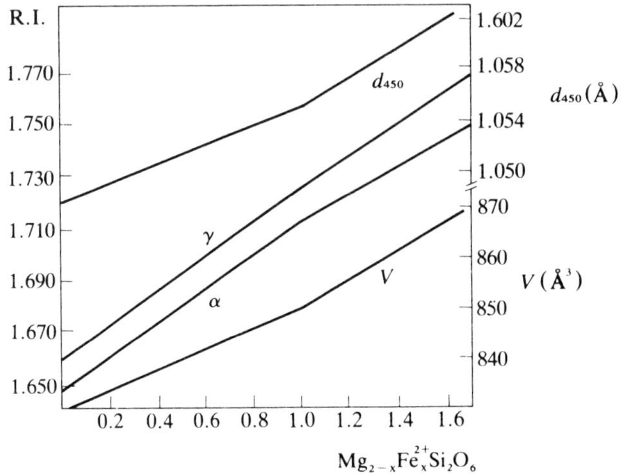

Fig. 74. (b) The relationship between refractive indices, γ and α, unit cell volume, and interplanar spacing d_{450} and the Fe:Mg ratio in orthopyroxenes ($Mg_{2-x}Fe_x^{2+}Si_2O_6$).

Leake (1968) for 240 orthopyroxenes is mg $= 14{\cdot}082 - 7{\cdot}870\,\gamma$, standard error[1] 0·05 mg, and mg $= 15{\cdot}315 - 8{\cdot}661\,\alpha$, standard error 0·07 mg.

Using much of the data on which the variation of the γ index shown in Fig. 74(a) is based, together with additional data obtained on orthopyroxenes of metamorphic rocks from the Adirondacks, Hudson Highlands and the Cortlandt complex, New York, and the Belchertown complex, Massachusetts, Jaffe *et al.* (1975) have presented a curve showing the variation of the γ index of refraction and the ratio 100 $(Fe^{2+}+Fe^{3+}+Mn)/(Mg+Fe^{2+}+Fe^{3+}+Mn)$ the equation of which

$$\gamma = 1{\cdot}6626 + 0{\cdot}1297[(Fe^{2+}+Fe^{3+}+Mn)/(Mg+Fe^{2+}+Fe^{3+}+Mn)]$$

gives compositional variations not greater than 1 mol. per cent for γ values less than 1·710, and between 1·5 and 3 for higher γ values, compared with those obtained using Fig. 74(a).

Shemyakin (1968) has investigated the relation between composition and γ and α refractive indices based on thirty-five chemical analyses of orthopyroxenes from various rock types. Graphs, using the method of least squares, have been constructed for the ferruginous coefficient, $F = 100\,(FeO + F_2O_3)/(FeO + Fe_2O_3 + MgO)$ mol. per cent $= 820\,\gamma - 1{\cdot}366 = 862\,\alpha - 1{\cdot}426$; $\gamma = 0{\cdot}00119F + 1{\cdot}667$, $\alpha = 0{\cdot}00116F + 1{\cdot}655$, and for the ferruginous coefficient $fm = 100\,FeO/(FeO + MgO)$ mol. per cent $= 840\,\gamma - 1403 = 885\alpha - 1467$, $\gamma = 0{\cdot}00119fm + 1{\cdot}670$, $\alpha = 0{\cdot}00113fm + 1{\cdot}658$. The ferruginous coefficient gives results comparable with those obtained from Fig. 74(a); iron–magnesium ratios obtained using the ferroginous coefficient give approximately 3 per cent lower iron values.

In a later paper, Kosoy and Shemyakin (1971), based on data for sixty-one orthopyroxenes from deep-seated metamorphic and igneous rocks, found that γ increases by $0{\cdot}065 \pm 0{\cdot}002$ and α by $0{\cdot}060 \pm 0{\cdot}002$ per (Fe + Mn) atom on the basis of six oxygens. Isomorphus substitution $MgSi \rightleftharpoons AlAl$ results in γ increasing by $0{\cdot}050 \pm 0{\cdot}014$ and α by $0{\cdot}060 \pm 0{\cdot}016$ per one AlAl atom; the substitution $MgSi \rightleftharpoons Fe^{3+}Al$ has a comparable effect, that of $Mg \rightleftharpoons Ca$ results in only small

[1] Standard error $= \sqrt{\Sigma(\mathrm{mg} - \mathrm{mg_c})^2/(n-1)}$, mg chemically determined mg; $\mathrm{mg_c}$ calculated mg from the refractive index; n the number of samples.

changes in refractive indices. The data also show that the increase in refractive indices is not linear throughout the isomorphus series, but that a small change occurs in slope of the curve at Mg:Fe = 1:1 (Fig. 74(b)). A comparable change in the variation of the unit cell volume and the interplanar distance d_{450} also takes place at $En_{50}Fs_{50}$.

The optical properties of six iron-rich orthopyroxenes from the metamorphic iron formation of the Mount Reed area, Quebec, have been presented by Moore *et al.* (1969).

The influence of Mn on the refractive indices does not differ significantly from that of Fe^{2+}; the γ indices of orthopyroxenes 34 and 35 (Table 4), with highly contrasted Mn but comparable (Fe^{2+} + Mn) contents, differ only by 0·0015. The effect of the Al content on the refractive indices of the orthopyroxenes has been investigated by Hess (1952) and Kuno (1954), and both concluded that a content of between 0·140 and 0·070 Al atoms per formula unit increases the γ index by 0·005 for the composition Fs_0 and that the effect decreases to zero for the extrapolated value for Fs_{100}. The refractive indices of enstatite, Fs_0, hypersthene, Fs_{50}, ferrosilite, Fs_{100}, for both pure compositions and those containing aluminium and aluminium + calcium (Table 10) have been calculated by Winchell and Leake (1965),

Table 10 Refractive indices, and densities of enstatite, hypersthene and orthoferrosilite (after Winchell and Leake, 1965)

	Enstatite			**Hypersthene**			**Orthoferrosilite**		
	$Mg_2Si_2O_6$	+ Al[a]	+ Al,Ca[b]	$MgFeSi_2O_6$	+ Al[a]	+ Al,Ca[b]	$Fe_2Si_2O_6$	+ Al[a]	+ Al,Ca[b]
α	1·6518 ± 0·0049	1·6505	1·6506	1·7115	1·7102	1·7085	1·7725	1·7682	1·7665
β	1·6562 ± 0·0031	1·6596	1·6607	1·7268	1·7302	1·7292	1·7840	1·7846	1·7836
γ	1·6621 ± 0·0017	1·6662	1·6671	1·7269	1·7310	1·7299	1·7926	1·7935	1·7924
D	3·216 ± 0·011	3·224	3·232	3·602	3·609	3·606	3·988	3·976	3·972

[a] 0 10 Al.
[b] 0·10 Al and 0·03 Ca.

and are in general agreement with the earlier conclusions of Hess and Kuno. The birefringence of the members of the orthopyroxene series also varies linearly with chemical composition. The line in Fig. 74(a) showing this relationship is taken from Hess (1952). Because most crushed fragments of the mineral lie on a {210} cleavage plane, the γ refractive index can be measured with the greatest ease and accuracy, and this is the most accurate optical method of determining the En:Fs ratio of an orthopyroxene. The α and β indices can be most readily measured in grains extracted from a thin section, showing a centred bisectrix figure.

Hori (1956) investigated the effect of the constituent cations on the refractive indices of the orthopyroxenes of a number of plutonic, volcanic and metamorphic rocks and found no evidence to doubt that there is a linear correlation between their refractive indices and chemical composition. Hori calculated the effect on the refractive indices of a number of ion substitutions and showed that α is decreased and β and γ increased by the substitution of Si by Al; Al in octahedral co-ordination causes an increase in γ and a decrease in α and β; α, β and γ are increased by the substitution of Mg by Ti and Fe^{3+}. Fe^{2+} and Mn have similar refractivities to each other, and Na and K cause the refractive indices to decrease and the birefringence to increase. A refinement of Hori's calculations has been made by Henriques (1958) who obtained a more precise correlation between the observed and calculated refractive indices.

A method of estimating the ratio of orthopyroxene and clinopyroxene in a rock, based on the construction of the frequency distribution of the extinction angle, $\gamma':z$ for random orientations of pyroxenes has been proposed by Takehita (1963).

Optic axial angle. The optic axial angles of the orthopyroxene series show a continuous and symmetrical variation with the replacement of Mg by Fe^{2+}; thus $2V_\gamma$ for enstatite, Fs_0, is 55°, the same as the extrapolated value for the pure iron end-member, orthoferrosilite. There are two changes in the optic sign; compositions between Fs_0 and Fs_{12} and between Fs_{88} and Fs_{100} are optically positive and compositions between Fs_{12} and Fs_{88} are optically negative. The minimum value of $2V_\alpha$ at the intermediate composition Fs_{50} was given by Deer *et al.* (1963) as 48·5° for the orthopyroxenes of plutonic rocks, and the curve for volcanic orthopyroxenes, with compositions in the range Fs_{65} to Fs_{35}, drawn at somewhat higher values with a minimum of 54° at Fs_{50}. As the result of a more recent survey of over 200 orthopyroxenes, Leake (1968) gave the minimum value of $2V_\alpha$ as 52·5 and concluded that volcanic orthopyroxenes show no consistent difference of 2V compared with plutonic and metamorphic orthopyroxenes. The standard error of determination of mg from 2V is 0·08 mg, and is greater than that using the γ refractive indices from which mg can be estimated to within about 0·05 mg providing the pyroxene does not have an unusual content of minor constituents. A linear dependence of 2V (and cell parameters) with Fe–Mg ordering in ferrohypersthene annealed at 500°, 700° and 900°C, however, has been demonstrated by Tarasov and Nikitina (1974; Tarasov *et al.*, 1975, see Fig. 75).

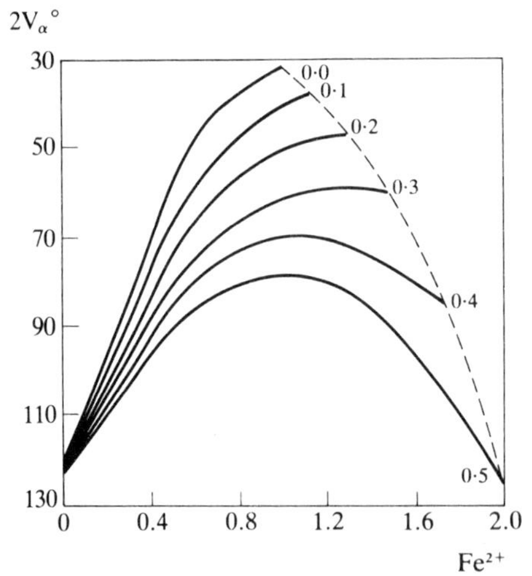

Fig. 75. Relationship of $2V_\alpha$ and chemical composition of orthopyroxenes for different degrees of Fe^{2+} ordering (after Tarasov *et al.*, 1975).

In the deformed and strained, but not recrystallized, anorthosites and gabbros of the Adirondacks and Labrador, many orthopyroxenes have abnormally large optic axial angles, departing by as much as 15° from those of unstrained crystals of the same composition. Unusually large optic axial angles in orthopyroxene have also been reported by Savolahti (1966). A volcanic orthopyroxene with $2V_\alpha$ as low as 35° has been recorded by Lewis (1960), but this feature is probably due to oxidation and heating. Variable optic axial angles in orthopyroxenes in the same thin section from rocks of the West Uusimaa complex, Finland have been reported by Parras (1958).

The dispersion of forty orthopyroxenes, varying in composition from enstatite to orthoferrosilite, has been investigated by Kuno (1941) who confirmed the earlier observation by Henry (1935) that the character of the dispersion of the optic axes changes at two compositions in the series. The first change in the dispersion, measured over α (from $r < v$ to $r > v$), takes place at approximately Fs_{15} (γ 1·682) and the second (from $r > v$ to $r < v$) at Fs_{50} (γ 1·726), thus at approximately 2V 90° and $2V_\alpha$ 50° there is no dispersion. The second change in dispersion may be of diagnostic value in distinguishing minerals in the two compositional ranges Fs_{15} to Fs_{50} and Fs_{50} to Fs_{80}.

Pleochroism. Many orthopyroxenes in the bronzite–hypersthene compositional range are pleochroic, and until the more iron-rich members of the series were investigated the strength of the pleochroism was considered to be related to the increasing replacement of Mg by Fe^{2+}. Many iron-rich orthopyroxenes, however, are only weakly pleochroic and moreover variable intensity of pleochroism has been observed in orthopyroxenes in a single rock section. In the Varberg area, Sweden, the orthopyroxenes in the basic granulites are more strongly pleochroic than the orthopyroxenes in the rocks of intermediate composition (Quensel, 1951), whereas in the Lapland granulites the strongly pleochroic orthopyroxenes occur in the more acid rocks (Eskola, 1952), and in the norites of the Caribou Lake complex pleochroism is less pronounced in the orthopyroxenes with higher contents of iron (Friedman, 1957). No obvious relationship between pleochroism and chemical composition has been established, although Kuno (1954) suggested a possible relationship with the content of titanium (see also Hess, 1960). Exceptions to this relationship are common in orthopyroxenes of igneous and metamorphic rocks and are considered by Kuno to be possibly associated with the exsolution of titanium together with calcium or iron. Howie (1955) has shown that this explanation is not acceptable for the orthopyroxenes of the Madras charnockites as the most pleochroic bronzites have a very low content of titanium, and exsolution lamellae and schiller inclusions are present. Howie suggested that the pleochroism may be related to the presence of such oriented inclusions. He has also noted that the pleochroic tints appear to be related to the Fe^{2+} content, though the strength of pleochroism is not; in the Madras charnockites, bronzite is brownish red in grains, and has a salmon-pink α absorption colour, while the ferrohypersthenes are greenish black in grains and show a pinkish yellow α absorption colour. The strongest pleochroism is shown by orthopyroxenes with a relatively high alumina content (> 4 weight per cent Al_2O_3), and has led to the suggestion that the strength of the pleochroism can be correlated with the alumina content and the consequent contraction of the cell parameters (Howie, 1965).

Parras (1958) has reported that a close relationship exists between the pleochroism and the presence of diopsidic lamellae in the orthopyroxenes of the West Uusimaa complex, Finland. In these orthopyroxenes the intensity of the pleochroism is proportional to the abundance of lamellae. Parras tentatively suggested that this relationship is due to the exsolution of diopsidic pyroxene and ferric iron, the latter possibly in the form of oxide, being concentrated on the surfaces of the growing lamellae. The intensity of the pleochroism is thus considered to depend on the amount of exsolution of both components and on their common orientation. More recently the conditions necessary for optical pleochroism in orthopyroxenes were suggested by Burns (1966) to be ordering of Fe^{2+} ions with preferential occupation of the *M*2 positions, the presence of small, high valence cations (Al^{3+}, Fe^{3+}, Ti^{4+}) in *M*1 positions, replacement of Si by Al, and an Fe content exceeding about 15 mol. per cent ferrosilite.

Density. The density of orthopyroxene varies linearly with composition. The density of the Shallowater enstatite (Table 6, anal. 1) has been calculated by Hess (1952):

$$D = \frac{M.n}{V.N_0} = 3{\cdot}2104 \text{ g/cm}^3 \text{ at } 23^\circ\text{C}$$

where M is the relative molecular mass (= 201·057), n the number of molecules in the unit cell (= 8), V the unit cell volume (= 18·230 × 8·815 × 5·177 Å = 831·931 Å^3 at 23°C), and N_0 is Avogadro's number (= 6·023 × 10^{23}). The density measured by flotation in Clerici solution is 3·209 ± 0·003 g/cm^3 at 27°C. The density of synthetic clinoenstatite is 3·19 g/cm^3 (Spencer and Coleman, 1969).

The extrapolated density of the pure $FeSiO_3$ end-member was given by Hess as 3·96 g/cm^3. Winchell and Leake (1965) give the density of ferrosilite as 3·988 g/cm^3 and this value has been used in Fig. 74(a). Howie (1963) has determined the specific gravity of thirty-five orthopyroxenes from metamorphic rocks ranging in composition from $Fs_{16{\cdot}4}$ to $Fs_{83{\cdot}8}$, and shown that there is a reasonable fit with the linear variation shown in Fig. 74(a).

Zoning. Zoned orthopyroxenes have not commonly been reported in plutonic and metamorphic rocks, but zoning is universal in volcanic orthopyroxenes. In the majority of zoned orthopyroxenes the cores are more magnesium-rich than the margins, e.g. the orthopyroxene phenocrysts in many Japanese andesites show marked zoning with $2V_\gamma$ in the range 87° to 98° and 106° to 113° in the cores and margins respectively. In these rocks the orthopyroxene in the groundmass is also zoned ($2V_\gamma$ 107° to 111°), and in both phenocrysts and the groundmass the transition between core and margin is rapid (Kuno, 1947b). Reversed zoning in orthopyroxenes of volcanic rocks, however, is not uncommon, and has been reported in the orthopyroxenes of the dacites of the Garibaldi area, British Columbia (Mathews, 1957). An unusual zonal arrangement in an orthopyroxene in andesite has been noted by Kuno (1947b); in this rock the mineral has a more iron-rich margin but the core shows reversed zoning. The compositions of individual grains of orthopyroxene, particularly in volcanic rocks, commonly show considerable variation, and Kuno (1950) has reported a range of Fs_{32} to Fs_{48} in the hypersthene of the pigeonite–quartz diorite of Hakone.

Orthopyroxenes of volcanic rocks in the compositional range Fs_{15} to Fs_{35} generally extinguish normally but in some dolerite orthopyroxenes of similar composition the extinction is uneven.

Six-rayed star enstatite, dark brown in colour, has been reported by Eppler (1967); the asterism is caused by fine needles probably of rutile. A short account of star enstatite and other varieties of enstatite is given by Eppler (1971). The presence of liquid CO_2 inclusions in an orthopyroxene from olivine-bearing nodules has been reported by Roedder (1965).

Euhedral phenocrysts of bronzite, ranging in composition from $Mg_{78}Fe_{17}Ca_5$ to $Mg_{72}Fe_{23}Ca_5$, in the chilled margins of a quartz-diabase dyke, Connecticut, have been described by Philpotts and Gray (1974). The crystals are prismatic in habit with the forms {100} and {010} equally developed, and the prism {110} less prominent. The crystals are doubly terminated by {011} or {$\bar{1}$11} prisms, and the morphology generally provides evidence of their previous monoclinic history. Weak oscillatory and sector zoning is present, the three dimensional symmetry of which is $2/m$, identical with the external morphology, and the crystals originally must have been monoclinic. Twin-like lamellae parallel to (100) and (001) are also present. The lamellae do not appear to be due either to twinning or exsolution and their origin has not been resolved.

Lamellar structure. The fine lamellar structure described in the chemistry section is not considered by all investigators to be due, in every example, to exsolution of a clinopyroxene phase in the orthorhombic host, and this structure has been ascribed by some authors to translation gliding, and by others to twinning. The fine lamellar structure is present in some orthopyroxenes with small contents of calcium; Tilley (1936) for example, has described a ferrohypersthene with 0·83 per cent CaO in which the structure is well developed. An enstatite containing as little as 0·32 per cent CaO (Shallowater meteorite, Table 6, anal. 1) displays undulatory and other extinction anomalies which Foshag (1940) considered could be due to twinning.

The lamellar structure is best observed under crossed polarizers in sections lying in the [100] zone, and appears to be due to very fine regular sheets, generally between 0·001 and 0·002 mm in thickness, oriented parallel to the (100) plane of the main crystal. The refractive indices and optic axial angles of the alternating sets of lamellae are reported by some workers to be identical, and the mineral appears to be completely homogeneous; the lamellae sometimes have a bent, pinched-out appearance. In suitably oriented sections, and when the main part of the crystal is in extinction, the fine lamellae are observed because their extinction position is different from that of the major part of the crystal. According to Scholtz (1936) the lamellae have common α directions, but the β and γ directions of the two sets are inclined to each other usually at angles of between 5° and 11°. From an X-ray and optical study of this lamellar structure, Henry (1942) showed that in a Stillwater bronzite the x and z axes of the lamellae are inclined at angles up to 15° to the main crystal in the (010) plane of the latter; in some examples the displacement is not strictly in the (010) plane and the y axes of the two sets of lamellae are slightly inclined to each other. In view of the range in orientation Henry concluded that the structure is not one of regular twinning, but he considered it reasonable to ascribe it to translation on (100) in the direction [100] accompanied by bending about [010].

As a result of an X-ray single-crystal investigation of some Stillwater bronzites, Hess (1960) corroborated Henry's observations but considered that the disorientation is entirely in the orthopyroxene host, and that the lamellae differ in chemical composition from the matrix and have the optical properties of a diopside (Fig. 76). A subsequent examination of this bronzite, using EMMA, a combined electron microscope-microanalyser (Lorimer and Champness, 1973) showed that the lamellae consist of clinopyroxene up to 0·5 μm in thickness. Direct determination of the Ca content as approximately 24 weight per cent CaO and Fe/(Fe + Mg) less than 10 per cent is in accord with the assignment as diopside by Hess.

A detailed description of the fine lamellar structure in the orthopyroxenes of the Insizwa intrusion, East Griqualand, has been given by Bruynzeel (1957). In some of these orthopyroxenes the lamellae extend continuously across strongly zoned crystals and terminate abruptly at the crystal margins without diminution in thickness, while in others the lamellae are irregularly distributed, some areas of the crystal being comparatively free from lamellae, so that under crossed polarizers the overall extinction is patchy. In sections parallel to (010) the extinction positions of the two sets of lamellae are inclined respectively at between 3° and 6° and minus 3° and 6° to the composition face (100). The presence of some crystals showing lamellar cores and non-lamellar margins, as well as the reverse relationship, led Bruynzeel to conclude that this lamellar structure in orthopyroxenes is due to multiple twinning about the z axis with (100) as the composition plane.

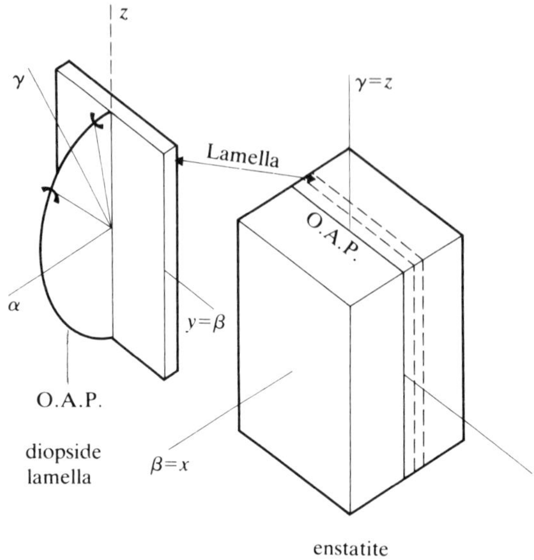

Fig. 76. The optical and crystallographic directions for enstatite and included diopside lamellae (after Hess, 1960).

The presence of kinking, due to deformation at high temperature and pressure, in natural orthopyroxene crystals has been described by Turner *et al.* (1960). Other examples include the orthopyroxene of the ultramafic xenoliths in the lavas of the Kerguelen Archipelago (Talbot *et al.*, 1963), and the bronzite in a highly deformed and partially metamorphosed gabbro of the Giles complex, Western Australia (Trommsdorff and Wenk, 1968). The glide plane T = {100} and the glide direction t = [001], and the kinked and unkinked domains have [010] = α in common (Fig. 77). The kinked and bent domains may show significant inclined extinction (γ:[001] = 32°) and high birefringence (0·018) and probably consist of clinoenstatite. Shock effects in the gas-rich enstatite achondrite, Khor Temiki, including

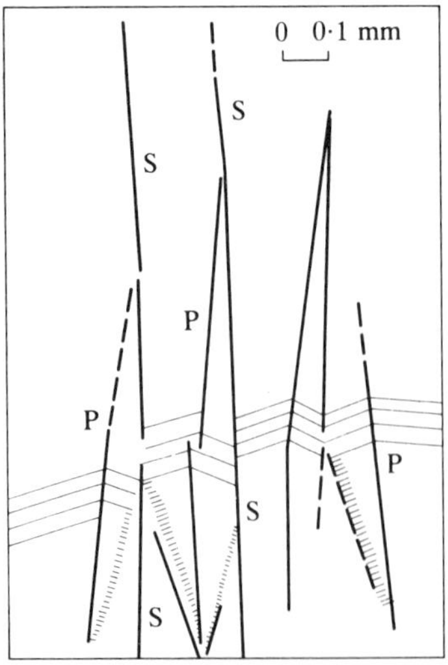

Fig. 77. Conjugate kinks in enstatite section normal to α = [001]. Glide system T = {100}, t = [001], kink axis = [010]. P and S, primary and secondary kink boundaries respectively. Stippled area partially inverted to clinoenstatite. Fine lines running approximately east-west are {100} lamellae (after Trommsdorff and Wenk, 1968).

the effects of unshielded solar radiation, an *in situ* lithology, due to intense impact shock, have been studied by Ashworth and Barber (1975).

Although the monoclinic polymorph is extremely rare in naturally deformed rocks the experimental deformation of orthorhombic enstatite, in the temperature range 200–1000°C, leads to the formation of clinoenstatite. Raleigh *et al.* (1971) have suggested that the transformation of ortho- to clinoenstatite and slip in orthopyroxene are competing rate-controlled processes of deformation in which slip has the higher activation energy, and have shown that, at strain rate of 10^{-2} and 10^{-1}, slip occurs in orthoenstatite without transformation to the monoclinic polymorph at temperatures above 1300° and about 1000°C respectively. Raleigh *et al.* consider that, above a temperature of between 450° and 650°C, clinoenstatite is not formed as a result of the natural deformation of orthoenstatite unless the strain rates are abnormally high (see also Carter, 1971), and that the rarity of its occurrence in kinked orthoenstatite is not the result of inversion back to orthoenstatite which may be kinetically difficult.

A model for the accommodation of a lamella of clinopyroxene in an orthopyroxene matrix, such that the (100) and [010] of the two structures are parallel, has been proposed by McLaren and Etheridge (1976). The model is based on the recognition that the orthopyroxene structure consists of alternate slabs of clinopyroxene structure, 0·5 *a* wide, joined along (100) and related to each other by *b*/2 glide on (100). Observation by transmission electron microscopy of refraction contrast fringes in naturally deformed orthopyroxene is consistent with the model.

The mechanical properties of natural orthopyroxenes, $Fs_{0·5}$ and $Fs_{13·3}$ and synthetic enstatite, Fs_0, have been investigated by Riecker and Rooney (1967). The natural orthopyroxenes have a shear strength of 14·9 kbar at 50 kbar normal pressure and 270°C. The shear strength decreases with increasing temperature and is 5·5 kbar at 40 kbar normal pressure and 920°C. In all cases clinoenstatite was obtained from orthorhombic starting material sheared for longer than one second.

An evaluation of the thermodynamic effect of shear stress on the ortho–clino inversion in enstatite has been undertaken by Coe (1970).

The enthalpies of solution, measured in a melt of $2PbO.B_2O_3$ at 970±2°K, for synthetic enstatite and aluminous enstatite $(MgSiO_3)_{0·9}(Al_2O_3)_{0·1}$, and the enstatites (< 0·1 weight per cent FeO) from the Cumberland Falls and Mt. Egerton chondrites, have been determined by Charlu *et al.* (1975). The values for the synthetic and meteoritic enstatites do not differ significantly and are respectively 8·84, 8·77 and 8·72 kcal/mole. The enthalpy of solution of the aluminous enstatite is lower (7·74 kcal/mole) than that for pure $MgSiO_3$.

The thermal expansion of pure $MgSiO_3$ is given by Sarver and Hummel (1962) as $0·120 \times 10^{-4}$°C^{-1}. This value is comparable with the isotropic thermal expansion coefficient, $\alpha_v/3 = 0·158 \times 10^{-4}$ °C^{-1} for a bronzite, $Mg_{0·8}Fe_{0·2}SiO_3$, measured by Frisillo and Buljan (1972), using an X-ray diffractometer technique at temperatures between 25° and 1000°C in air. The approximate linear rates of change of the lattice parameters with temperature are, $a(Å) = 18·225 + (0·300 \pm 0·003) \times 10^{-3}T$, $b(Å) = 8·863 + (0·129 \pm 0·006) \times 10^{-3}T$ and $c(Å) = 5·205 + (0·875 \pm 0·006) \times 10^{-3}T$(°C).

The electrical conductivity of three Mg-rich orthopyroxenes (36·88, 34·68 and 32·87 weight per cent MgO) has been measured at temperatures between 400° and 1300°C at oxygen fugacities from 10^{-7} to 10^{-15} atm (Duba *et al.*, 1973). The conductivity is largely dependent on the oxygen fugacity, but is also influenced by the presence of cleavage cracks and exsolution lamellae (see also Kobayashi and Maruyama, 1974). The activity energy, $E = 1·0 \pm 0·1$ eV has been obtained by

Duba *et al.* (1976) from measurements of the electrical conductivity, $\sigma = \sigma_0 \exp(-E/KT)$ where σ is a constant, K Boltzmann's constant, between 600° and 1200°K; the same authors have also shown that the conductivity is slightly anisotropic. Thermal conductivities of 10·50 and 9·94 mcal/cm s°C for enstatite, Fs_2, and bronzite, Fs_{22}, respectively are reported by Horai (1971). Schatz and Simmons (1972) have determined the total thermal conductivity, lattice plus radiative, for a bronzite, Fs_{18}, and shown that the conductivity varies from 0·0100 to 0·0120 cal/cm s °C at 600°K, to 0·0068 to 0·0071 cal/cm s °C at 1600°K.

The single-crystal elastic constants, as a function of pressure and temperature of the bronzite used by Frisillo and Buljan in their thermal expansion experiment, have been determined by Frisillo and Barsch (1972). The linear compressibility of the *c* axis decreases more rapidly than the linear compressibilities of the *a* and *b* axes as the pressure is increased, in general agreement with the earlier determination of the elastic constants of a gem quality bronzite (density 3·35 g/cm³) by Kumazawa (1969). Shock wave data for the Bamble, Norway, orthopyroxene, $Mg_{0\cdot86}Fe_{0\cdot14}SiO_3$, at shock pressures between 57 and 483 kbar have been presented by Ahrens and Gaffney (1971). The average first shock velocity, 7·78 km/s, is consistent with the longitudinal elastic velocity along (001), of 7·853 km/s, given by Kumazawa, and the value of 7·865 ± 0·015 km/s obtained by Ahrens and Gaffney.

Clinoenstatite–clinoferrosilite. The optical properties of the synthetic (Mg,Fe) clinopyroxenes are shown in Fig. 78. All members of the clinoenstatite–clinoferrosilite series are optically positive and have the optic axial plane perpendicular to (010). Different values of the optic axial angle have been given by various authors, thus Bowen and Schairer (1935) stated that the angle varies between 20° and 25° for the whole series, while Atlas (1952) gives the angle of 53·5° for pure clinoenstatite, a value comparable with the earlier data of Allen and White (1909) and Bowen (1914). The optic axial angle decreases with increasing replacement of magnesium by iron and is 23° for clinoferrosilite (Table 11). The optical properties of a manganoan clinoferrosilite, $Fe_{0\cdot95}Mn_{0\cdot05}SiO_3$, are α 1·763, β 1·763, γ 1·794, $\gamma{:}z$ 24·5, 2V very small (Bowen, 1935). Multiple twinning parallel to (100) occurs in the phenocrysts of clinoenstatite (Fig. 79) of the porphyritic volcanic rock from Cape Vogel (Dallwitz *et al.*, 1966). Such twinning, usually attributed to inversion from protopyroxene, is almost universally present in members of the series, but the presence of an untwinned clinobronzite, in the kinked bronzite crystals of the deformed and partially metamorphosed gabbro of the Giles complex, Western Australia, has been reported by Trommsdorff and Wenk (1968). Cleavages or partings parallel to (0*kl*), directions not previously reported for pyroxenes, making an angle of about 64° with (010), are present in the clinoenstatite of Cape Vogel (Dallwitz *et al.*, 1966). Needles of clinoferrosilite in lithophysae in obsidian from Lake Naivasha, Kenya, show a morphological relationship to magnetite crystals to which they are commonly attached such that $(001)_{pyx} \parallel (\bar{1}\bar{1}3)_{mag}$, and $[010]_{pyr} \parallel [110]_{mag}$ (Bown, 1965).

Distinguishing Features

Many orthopyroxenes can be distinguished from clinopyroxenes by their characteristic pale-pink to green pleochroism. In the absence of pleochroism they are distinguished from clinopyroxenes by their lower birefringence and their straight

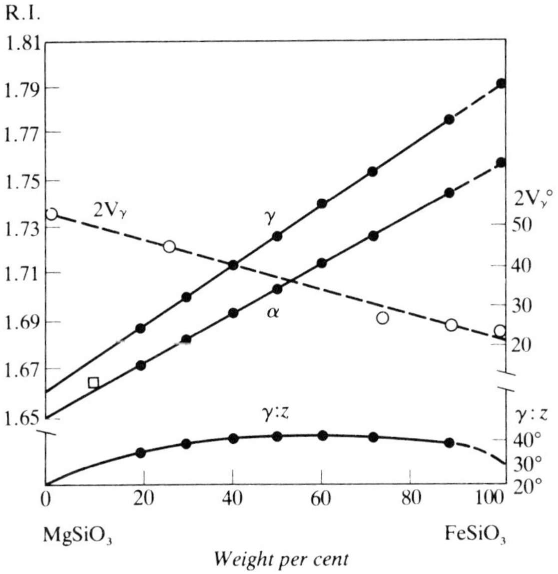

Fig. 78. Refractive indices and extinction angles of the synthetic clinoenstatite–clinoferrosilite series (after Bowen and Schairer, 1935). Open circles, from Lindsley *et al.* (1964). Open square, Cape Vogel clinoenstatite (Dallwitz *et al.*, 1966).

Table 11 Optical properties of some natural and synthetic $MgSiO_3$–$FeSiO_3$ clinopyroxenes

Composition	α	β	γ	δ	$2V_\gamma$	$\gamma:z$	Reference
En_{100} (S)	1·651	1·654	1·660	0·009	53·5°	22°	Atlas, 1952
En_{100} (S)	1·651	—	1·660	0·009	25°	22°	Bowen and Schairer, 1935
En_{100} (S)	1·650	1·652	1·658	0·008	—	—	Stephenson *et al.*, 1966
En_{100} (S)	1·650	1·653	1·660	0·010	—	—	Schwab, 1967
En_{91-89} (N)	1·662	—	—	—	25–56°	—	Dallwitz *et al.*, 1966
$En_{73\cdot5}$ (N)	—	—	—	—	44°	32°	Turner *et al.*, 1960
En_{30} (S)	1·725	—	1·752	0·027	25°	42°	Bowen and Schairer, 1935
En_{13} (S)	1·743	—	1·777	0·034	25°	36°	Bowen and Schairer, 1935
En_0 (S)	1·764	1·767	1·792	0·028	23°	31°	Lindsley *et al.*, 1964

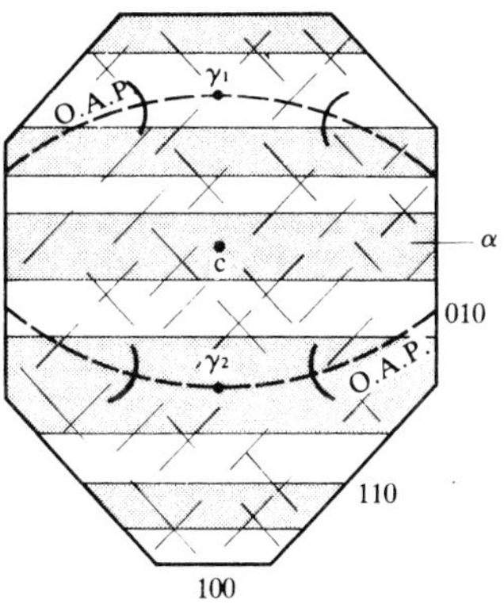

Fig. 79. Optical orientation of clinoenstatite. γ_1 and γ_2 acute bisectrices in alternating sets of twin lamellae (after Dallwitz *et al.*, 1966).

extinction in all [001] zone sections. Furthermore, bronzite, hypersthene, ferrohypersthene and eulite are all optically negative. Orthopyroxene is distinguished from sillimanite by the presence of (210) and by the absence of (010) cleavage, as well as by the smaller optic axial angle of sillimanite; it is distinguished from andalusite by the positive sign of the magnesium-rich, and the greater birefringence of the more iron-rich, orthopyroxenes.

Paragenesis

Igneous rocks

Ultrabasic rocks. Magnesium-rich orthopyroxenes, normal compositional range Fs_6 to $Fs_{13 \cdot 5}$, are important constituents of many ultrabasic and ultramafic rocks, in which they are commonly associated with olivine, diopside and spinel. They occur in the spinel peridotites of the axial regions of large fold belts, the nodules in alkaline olivine basalts and intrusions along major fault zones, and in the garnet peridotites of high grade metamorphic fold belts and nodules in kimberlite pipes. The bulk compositions of the orthopyroxenes in the Dun Mountain, Red Hills and Red Mountain ultramafic intrusions, South Island, New Zealand (Challis, 1965) vary between Fs_7 and $Fs_{11 \cdot 5}$. The enstatite in the harzburgite is zoned, and has an euhedral core, Fs_6, and anhedral margin, Fs_{14}. The orthopyroxenes are of the 'Bushveld type', relatively rich in calcium (CaO 0·90–1·32 weight per cent), and contain exsolution lamellae of clinopyroxene parallel to (100). In the layered ultrabasic rocks of Dawros, Eire (Rothstein, 1958), orthopyroxene (Table 3, anals. 3, 6, 8) is the sole pyroxene in the harzburgite, but occurs in association with clinopyroxene in the lherzolites. An enstatite pyroxenite, consisting almost entirely of orthopyroxene (Table 3, anal. 2), has been described from the websterite zone of the Webster–Addie ultrabasic mass (Miller, 1953).

The orthopyroxenes of the Pyrenees lherzolites have been described by Rio (1968) and Monchoux and Besson (1969), those of the ultramafic rocks of Mahrirrakhi, Greece, by Menzies (1973), and of a griquaite, associated with serpentinite, at Nihov, western Moravia, by Paděra and Procházha (1970). In the Kalskaret area, south Norway the orthopyroxenes, $Fs_{6 \cdot 6}$–$Fs_{7 \cdot 8}$, of the garnet-free peridotites are less iron-rich than those, $Fs_{9 \cdot 7}$–$Fs_{16 \cdot 6}$, in the garnet-bearing peridotites (Carswell, 1968). The orthopyroxene porphyroblasts in the two-pyroxene peridotite mylonite and amphibole enstatite peridotite mylonite of St. Paul Rocks, Atlantic (Tilley and Long, 1967), contain 5·3 weight per cent Al_2O_3, comparable with the orthopyroxene ($\simeq$ 5·5–6·6 Al_2O_3), of the high-pressure core of the Lizard peridotite (Green, 1964). The orthopyroxenes of the outer envelope of that intrusion contain less aluminium ($\simeq$ 2·1–3·3 Al_2O_3 weight per cent) and form part of an assemblage that crystallized either at a lower temperature or lower temperature and pressure.

Analyses of nineteen orthopyroxenes (Mg 83·3–78·2, Fe 13·0–15·6, Ca 2·9–6·5) from the olivine orthopyroxenite, orthopyroxenite and websterite cyclic units in the Gosse Pile layered ultramafic sequence within the Giles complex of central Australia, are given by Moore (1971, 1973) and Goode and Moore (1975). Compared with orthopyroxenes from two-pyroxene or pyroxene–plagioclase assemblages in other layered igneous bodies, those of the Gosse Pile intrusion have high Al_2O_3 and Cr_2O_3 contents, the latter varying between 0·32–0·95 weight per cent.

The orientation of enstatite in deformed peridotite has been examined by Darot and Boudier (1975). The lineation displayed by the orthopyroxene, in which the length/thickness ratio may be as high as sixty, corresponds with the long axis of the strain ellipsoid.

Orthopyroxene is present in many of the rocks of the Caribou Lake complex (Friedman, 1957), and the composition of the pyroxene changes systematically with that of the host rock, varying between Fs_{12} and Fs_{15}, in the picrite, to Fs_{18} in the pyroxenite, and Fs_{23} to Fs_{40} in the norite. In the picrite and olivine norite of this complex much of the orthopyroxene occurs as rims round the olivine, and the prisms of the orthopyroxene are aligned with their z axes normal to the olivine margins. The compositions of the orthopyroxenes and plagioclase feldspars show a systematic relationship, and both the ortho- and clinopyroxenes exhibit progressive iron enrichment.

Microlites of orthopyroxene, the relatively high calcium contents of which decrease from 4·7 to 3·9 per cent CaO, towards the less quickly cooled areas of the marginal facies of a picritic intrusion, Zanaga, Brazzaville (Congo), have been reported by Pouclet and Bizouard (1974).

The presence of clinoenstatite and clinobronzite as exsolution phases in the diopside of the ultramafic rocks of the Horoman intrusion, Hokkaido, Japan has been reported by Yamaguchi and Tomita (1970).

Nodules in basic rocks. Orthopyroxene is one of the main constituents of the ultramafic xenoliths of alkali basalts. The minerals, particularly in the lherzolite nodules, are commonly deformed, and a full description and discussion of the textures and fabrics displayed by orthopyroxenes (and associated clinopyroxenes and olivines) is given by Mercier and Nicolas (1975).

Numerous accounts of orthopyroxene in the nodules have been presented. Kuno and Aoki (1970) have described orthopyroxene, $Mg_{91\cdot6}Fe_{8\cdot4}Ca_{0\cdot0}$ and $Mg_{89\cdot2}Fe_{9\cdot4}Ca_{1\cdot4}$, from the lherzolite and websterite nodules from the Itinome-gata, and the Dreiser Weiher crater, and also from the lherzolite in the Hualalai flow, Hawaii. The orthopyroxene, $Mg_{89\cdot3}Fe_{9\cdot5}Ca_{1\cdot2}$, with 4·5 per cent Al_2O_3 and 0·25 per cent Cr_2O_3, of the spinel harzburgite inclusions in the basanites of the alkaline volcanics, Monaro, south-west Australia, have been described by Kesson (1973). Megacrysts, with compositional range $Mg_{85\cdot5-87\cdot1}Fe_{11\cdot2-10\cdot1}Ca_{2\cdot1-2\cdot9}$, in the diopside-bearing orthopyroxenite xenoliths in trachybasalt, Glen Innes area, north-eastern New South Wales, are reported by Wilkinson (1973). Other Australian examples include the wehrlite, lherzolite, websterite, harzburgite and orthopyroxenite xenoliths in the lavas of the New Basalt province of Victoria (Irving, 1974a; see also Ellis, 1976). Irving has also described the orthopyroxene in the garnet and spinel–garnet websterites and the two-pyroxene granulite xenoliths from the Delegate basalt pipes, Australia (Irving, 1974b). The contrasted composition of the orthopyroxene of the two nodule types, 8 and 22 per cent ($FeO+Fe_2O_3$), in the websterite and two-pyroxene granulite respectively, are related to the former having originated as cumulates in local pockets of alkali basaltic magma in the upper mantle, whereas the two-pyroxene granulites are accidental crustal xenoliths. An experimental investigation of the phase relations of the nodules indicates that the websterite assemblage formed in the pressure and temperature range 13–17 kbar and 1050°–110°C, in contrast to the equilibration of the two-pyroxene granulite at 6–10 kbar and about 1100°C. Two suites of inclusions, ultramafic and basic granulites, the orthopyroxenes of which show extensive compositional ranges $Mg_{82\cdot1-69\cdot1}Fe_{16\cdot8-29\cdot7}Ca_{1\cdot0-1\cdot6}$ and $Mg_{80\cdot6-57\cdot3}Fe_{18\cdot5-41\cdot4}$

$Ca_{0\cdot9-1\cdot3}$ respectively have been described from an analcitite sill, Barraba, New South Wales (Wilkinson, 1975).

Orthopyroxene-bearing nodules in which the orthopyroxene and spinel have resulted from a subsolidus reaction between olivine and plagioclase occur in the basic and ultrabasic dykes of the Seiland petrographic province, north Norway (Robins, 1975). A second type of nodule is also present. In these, orthopyroxene occurs as strongly deformed porphyroblasts that are relicts of a coarse-grained texture indicative of a non-accumulative relationship between the xenoliths and transporting magma. The original Al-rich composition of the orthopyroxene is demonstrated by the presence of exsolved spinel as well as clinopyroxene. The conversion of the original Al-rich to a deformed and less Al-rich orthopyroxene is considered to have been the first stage in the recrystallization of the xenolith assemblages along a decreasing *P–T* gradient. This was followed, at lower pressures, by a period of annealing that gave rise to an anorthite + olivine + chromite from an enstatite + diopside + spinel assemblage, a reaction comparable with that suggested by Green (1964) to account for different assemblages of the inner and outer parts of the Lizard peridotite, Cornwall. Subsequently a hydrous assemblage developed due to the formation of amphibole from the reaction of enstatite + diopside + anorthite + olivine + H_2O.

Small compositional variations in the orthopyroxenes of the peridotite nodules of the basalts of the Kassel area, West Germany (Gramse, 1970) are related to the type of basalt in which the nodules occur, $En_{89\cdot1-90\cdot4}Fs_{9\cdot5-8\cdot6}Wo_{0\cdot9-1\cdot8}$, and $En_{88\cdot5}Fs_{9\cdot8}Wo_{1\cdot7}$ for the orthopyroxene of the nodules in the alkali olivine basalts and nephelinites respectively. Enstatite, En_{91}, has been reported from the ultramafic inclusions in the lava of Teneguía volcano, Canary Islands (Munoz *et al.*, 1974).

Orthopyroxene, $Mg_{85\cdot6}Fe_{12\cdot4}Ca_{2\cdot0}$, constitutes some 60 per cent of the inclusions of graphite-bearing pyrope peridotite in the breccia of the Mir kimberlite pipe (Frantsesson and Lutts, 1970). Compared with the orthopyroxene in the associated pyrope–spinel peridotite nodules it has lower contents of aluminium (0·11 per cent Al_2O_3) and chromium (0·01 per cent Cr_2O_3) which may be related either to a greater depth of formation or lower temperatures of crystallization, and it is thus probable that the graphite-bearing nodules equilibrated under conditions intermediate between those of the diamond-bearing eclogite and the pyrope–spinel peridotite nodules. Orthopyroxene, En_{72}, with a high aluminium content (up to 8·5 weight per cent Al_2O_3), associated with sapphirine, sillimanite and garnet, occurs in the granulite xenoliths from the Stockdale kimberlite, Kansas (Meyer and Brookins, 1976).

The orthopyroxenes of the ultramafic nodules in the Obnazhennaya kimberlite pipe, north Yakutia, have been studied (Ukhanov and Mockalova, 1971) by a high-temperature emanation method (activation with an alcoholic solution of radiothorium). Changes in the emanation rates on heating were found to occur, and a number of peaks were observed between 550°–600°C and 760°–800°C. The absence of these peaks in reheated specimens suggests that irreversible structural changes, probably order–disorder transformations, took place at 550° to 600°C during the first heating. These changes are not present in the natural orthopyroxenes, and Ukhanov and Mockalova consider that the ultramafic nodules in the kimberlite melt were not heated above a temperature of between 500° and 600°C due to their rapid ascent in the conducting fissure.

Pressures and temperatures of equilibration of the Cr-rich (0·4–0·6 weight per cent Cr_2O_3) and Cr-poor (0·1–0·3 Cr_2O_3) groups of enstatite megacrysts in the Sloan kimberlite pipes, Colorado, based on the Ca and Ca/(Ca + Mg) molar contents of coexisting pyroxene pairs, and pressures calculated from orthopyroxene Al_2O_3 isopleths for garnet peridotites (see p. 94), show that although the majority equilibrated close to the shield geotherm, a number equilibrated at a temperature some 100°C higher. Comparable suites of endiopside (see p. 267) and garnet megacrysts are also present in the Sloan kimberlites indicating a possible relationship between the thermal disturbance and the intrusion of mantle diapirs (Eggler and McCallum, 1976).

The content of alumina in the orthopyroxenes of the lherzolite and eclogite nodules in the basaltic rocks of Hawaii ranges from 3 to 4 and 4·3 to 5·7 per cent respectively (Kuno, 1969). Analyses of orthopyroxenes, alumina contents varying between 2·1 and 4·6 weight per cent Al_2O_3, in the lherzolite nodules in the Itinome-gata crater, Oga Peninsula, Akitu Prefecture, Japan, are given by Kuno and Aoki (1970) and Aoki (1971), but values as high as 6·4 have been reported by Yamaguchi (1964). In the ultramafic nodules in the damtjernite of the Fen alkaline complex, Norway (Griffin, 1973), the orthopyroxene is an enstatite, $Fs_{9.6}$, and has a higher Al_2O_3 content, 4·2 to 5·0 weight per cent, than is found in the pyroxene of most lherzolite nodules, but close to the normal maximum found in orthopyroxenes synthesized in the course of experimental studies on pyrolite compositions. Orthopyroxenes of other ultrabasic and ultramafic xenoliths, however, contain considerably smaller amounts of aluminium, and Dawson *et al.* (1970) have reported orthopyroxenes, respectively with 1·54, 1·10, 1·24 and 1·67 per cent Al_2O_3, from garnet lherzolite, mica garnet lherzolite, lherzolite and spinel harzburgite xenoliths in the lavas of the Lashaine volcano, northern Tanzania; enstatite lamellae with a very low content of aluminium, in the diopside nodules of the Jagenfontein kimberlite, have been described by Borley and Suddaby (1975) (see diopside section, p. 266).

The effect of pressure on the paragenesis of ultrabasic rocks and on the reaction:

$$\underset{\text{orthopyroxene}}{2(Mg,Fe)SiO_3} + \underset{\text{diopside}}{Ca(Mg,Fe)Si_2O_6} + \underset{\text{chrome-spinel}}{(Mg,Fe)(Al,Cr)_2O_4} \rightleftharpoons \underset{\text{garnet}}{Ca(Mg,Fe)_2(Al,Cr)_2Si_3O_{12}} + \underset{\text{olivine}}{(Mg,Fe)_2SiO_4}$$

has been discussed by Mikhaylov and Rovsha (1965). The phase equilibrium data for the systems $Mg_2Si_2O_6$–$CaMgSi_2O_6$ and $MgSiO_3$–$Mg_3Al_2Si_3O_{12}$ (see chemistry section) have been used by Boyd (1973) to estimate equilibration conditions of the enstatite + diopside + garnet assemblages in the lherzolite nodules in the granular and sheared kimberlites of northern Lesotho.

Basic rocks. Magnesium-rich orthopyroxenes are present in the earlier formed rocks of many layered intrusions. In the Stillwater complex, Montana (Hess, 1941, 1960), the orthopyroxenes in the rocks of the lower horizons vary in composition between $En_{82}Fs_{14}Wo_4$ and $En_{73}Fs_{23}Wo_4$, while in the higher part of the intrusion the Ca-poor pyroxene, $En_{64}Fs_{27}Wo_9$, is an inverted pigeonite. The orthopyroxenes of the picrite of the marginal border group ($Mg_{77.6}Fe_{18.5}Ca_{3.9}$), and in the perpendicular-feldspar of the Skaergaard intrusion (Wager and Deer, 1939), are bronzitic in composition and have a relatively low calcium content, most of which is now present in fine-scale exsolution lamellae of augite parallel to (100). In contrast the calcium-poor pyroxenes, with bulk composition $Mg_{55.8}Fe_{35.2}Ca_{9.0}$, of the lowest part of the layered series are hypersthenes containing abundant blebs, or relatively thick oriented lamellae, of exsolved augite and have inverted from pigeonite. Iron enrichment of the inverted pigeonites

($Mg_{48\cdot2}Fe_{42\cdot9}Ca_{8\cdot9}$) continues throughout the middle zone, but at higher levels of this zone some of the pigeonite crystals are only partially inverted and in some cases have remained uninverted. Two hundred metres above the base of the upper zone the composition of the Ca-poor pyroxene is $Mg_{45\cdot5}Fe_{45\cdot9}Ca_{8\cdot6}$; but at higher levels, Ca-poor pyroxene did not precipitate from the strongly differentiated liquid, and the ferrodiorites in the upper part of the upper zone only contain a Ca-rich pyroxene (Brown *et al.*, 1957; Wager and Brown, 1968; see also sections on augite and pigeonite). The presence of clinoferrosilite in inverted β-wollastonite of the upper ferrodiorite of the intrusion has been described by Nwe (1975, see Fig. 138).

The change from orthopyroxene to pigeonite (total compositional range, $Fs_{13\cdot1}$ to $Fs_{39\cdot7}$) in the rocks of the lower layered series of the Jimberlana intrusion, Western Australia (Campbell and Borley, 1974) occurs at a somewhat lower ferrosilite content than in many layered intrusions. This feature, together with the relatively narrow miscibility gap between the calcium-rich and calcium-poor pyroxene, may be due to a higher temperature of crystallization than in other intrusions of comparable composition. Orthopyroxene, $Mg_{72}Fe_{27}Ca_1$, is the sole pyroxene in the anorthosites and norites of the differentiated anorthosite–mangerite intrusion, Bjerkrem-Sogndal, Norway (Duchesne, 1972), while in the more differentiated monzonorites the orthopyroxene, $Mg_{66}Fe_{32}Ca_2$, is associated with augite, $Ca_{46}Mg_{41}Fe_{13}$. Both pyroxenes became progressively enriched in iron as the differentiation continued, and in the mangerite the Ca-poor phase ($Mg_{19}Fe_{74}Ca_7$) crystallized initially with monoclinic symmetry, in association with ferroaugite, $Ca_{38}Mg_{15}Fe_{47}$.

Orthopyroxene, as a cumulus phase, is present throughout the layered sequence of the Bushveld complex (Atkins, 1969). In the basal series the earliest orthopyroxene is bronzite, $\simeq Mg_{87}Fe_{13}$, in the main zone an inverted pigeonite, $Mg_{70}Fe_{30}$ to $Mg_{60}Fe_{40}$, and in the upper part of the upper zone is an inverted ferropigeonite, $Mg_{29}Fe_{71}$. The iron enrichment is, however, not wholly progressive and for approximately 150 m above the base of the critical zone the iron content of the Ca-poor pyroxene decreases slightly before increasing in the last 225 m of the sequence. The general trend is further disrupted by a number of overlapping fluctuations in iron content of the orthopyroxene, changes that may be due either to separate intrusions of undifferentiated magma during the evolution of the sequence, or to resorption of the sinking crystals in the magma chamber (McDonald, 1967). The major and minor changes in the composition of the orthopyroxenes in the transition and critical zone of the eastern part of the complex have been discussed by Cameron (1970), and the orthopyroxene–pigeonite phase change in the main and upper zones has been examined by Gruenewaldt (1970). A more general account of the changing relationships and compositions of the Ca-poor and Ca-rich pyroxenes to crystallize during the differentiation of basic, and particularly tholeiitic magmas, is given in the section on augite.

The systematic variation of coexisting orthopyroxene and chromite in the bronzite–chromite cumulate sequence of the eastern Bushveld complex has been reported by Cameron (1975). The Mg/Fe ratio of the orthopyroxene varies sympathetically with the Cr/Fe and Mg/Fe ratios and the Al_2O_3 content of the chromite, and is associated with the modal percentage of chromite present in the cumulates. Although there are some indications that local equilibration occurred during postcumulus crystallization and at subsolidus temperatures, Cameron considers that the main variation took place during the cumulus phase of crystallization consequent on changes in oxygen fugacity due to the intermittent diffusion of hydrogen from the magma through the walls of the chamber, or from the diffusion of water inwards from the margin of the intrusion.

Orthopyroxene, except in the dunite, is present throughout the basic and ultrabasic sequence of the Great Dyke, Rhodesia (Lensch, 1971). The pyroxene occurs as a cumulus phase and shows increasing iron, aluminium and calcium contents, as well as ferric/ferrous iron ratios, with height in the layered sequence. In the Insch layered intrusion (Clarke and Wadsworth, 1970), orthopyroxene occurs in the lower, middle and upper zones and varies in composition from Fs_{18} in the cumulates of the lower, to Fs_{48}–Fs_{75} in the middle and upper zones. In the Noril'sk I basic and ultrabasic intrusion, orthopyroxene formed early in the crystallization sequence, was followed by pigeonite, subsequently inverted to bronzite and hypersthene, Fs_{26} to Fs_{56} in composition, and finally by orthopyroxene. The pyroxene crystallization relationship of other basic intrusions of the Siberian platform are also given by Vilenskiy (1966), and compared with those of the Skaergaard and Red Hill intrusions.

Orthopyroxene is a common ferromagnesian constituent of calc-alkaline rock series and occurs, for example, in many gabbros, norites, diorites, granodiorites and granites of central Ostrobothnic, western Finland (Wahl, 1964), the pyroxene becomes increasingly richer in iron as the acidity of the rock increases, and in the granites is a ferrohypersthene ($\simeq Fs_{60}$).

Orthopyroxenes, compositional range $Mg_{68\cdot8}Fe_{28\cdot0}Ca_{3\cdot2}$ to $Mg_{35\cdot4}Fe_{60\cdot8}Ca_{3\cdot8}$, are characteristic of the earlier gabbroic differentiates of the Guadalupe igneous complex, California (Best and Mercy, 1967). The orthopyroxenes display fine lamellae of augite, parallel to (100), and show no evidence of having inverted from pigeonite. The partition of Mg–Fe^{2+}, and of Ca, between the coexisting ortho- and clinopyroxenes of the complex differs from the usual pattern of tholeiitic rock suites, and is comparable with those of pyroxene pairs formed in high grade regional metamorphic environments. Many of the Guadalupe orthopyroxenes have a somewhat higher Fe^{3+} content than is general in the orthopyroxene of igneous rocks, and this feature, together with their formation directly as an orthorhombic phase, and their common association with hornblende, suggest that the Ca-poor pyroxene crystallized under conditions of relatively high water fugacities and lower temperatures than the majority of tholeiitic plutonic intrusions. Orthopyroxene occurs both as a cumulus and an intercumulus phase in the gabbro, as well as in the spinel–amphibole symplectites of the Sulitjelma mass, Norway (Mason, 1971). The decreasing calcium content of the pyroxene, expressed as the percentage of $CaSiO_3$, from 3·9 in the cumulus to 1·0–2·8 in the intercumulus phase, and to 0·2–0·4 in the orthopyroxene of the coronas, is attributed here to a decrease in solubility of Ca in the pyroxene with falling temperature.

Orthopyroxene is a common ferromagnesian constituent of anorthosites and in these rocks is usually a bronzite or hypersthene in composition. Thus in the Allard Lake anorthosite suite, Quebec (Hargraves, 1962), the orthopyroxene has an Mg:Fe ratio of 65:35. The aluminium content is higher than in most equivalent pyroxenes of layered intrusions and is probably related to crystallization at higher pressures. Two generations of Ca-poor pyroxenes, both close to Fs_{69} in composition occur in the ferrodiorite of the Western Redhills complex, Skye (Bell, 1966). The earlier ferrohypersthene displays exsolution lamellae and formed by inversion from pigeonite; the latter commonly rims the inverted pigeonite and precipitated directly as orthopyroxene.

Orthopyroxene is the essential ferromagnesian constituent of norite, and although in some it may have crystallized from an uncontaminated basic magma, in others (e.g. the cordierite and biotite norites of the Haddo House complex,

Aberdeenshire, Read, 1935), there is evidence of assimilation of metamorphosed aluminium-rich sediments. The result of such a reaction, as shown by Bowen (1928), would be to increase the amount of orthopyroxene crystallizing and to increase the anorthite content of the plagioclase at the expense of the calcium-rich pyroxene:

$$\underset{\text{augite}}{Ca(Mg,Fe)Si_2O_6} + Al_2SiO_5 \rightarrow \underset{\text{hypersthene}}{(Mg,Fe)SiO_3} + \underset{\text{anorthite}}{CaAl_2Si_2O_8}$$

Read's interpretation of the origin of the Haddo House norites has, however, been questioned by Gribble (1967, 1968) on the grounds of the absence of transitional rock types between the magmatic and country rocks. In Gribble's view orthopyroxene, plagioclase, cordierite and quartz are potential phases in the liquid derived by partial fusion of the aluminous xenoliths, and the orthopyroxene may result from the combination of the cordierite component with the potential clinopyroxene of the quartz-norite magma:

$$\underset{\text{cordierite}}{(Mg,Fe)_2Al_4Si_5O_{18}} + \underset{\text{diopside}}{2Ca(Mg,Fe)Si_2O_6} = \underset{\text{orthopyroxene}}{4(Mg,Fe)SiO_3} + \underset{\text{anorthite}}{2CaAl_2Si_2O_8} + \underset{\text{quartz}}{SiO_2}$$

Such an origin, by partial melting of the Dalradian country rocks, is in agreement with the suggestion of Chinner and Schairer (1962) that xenolithic complexes, such as those of Haddo House and Arnage, may not be the final products of gabbro–cordierite-norite contamination but due to the partial melting of country rock. In most norites the orthopyroxene is either bronzite or hypersthene. Thus in the Memesagamesing Lake complex, Ontario (Friedman, 1955), the composition of the orthopyroxene in the olivine norites is $Fs_{24\cdot5}$, while in the norites its composition varies between $Fs_{27\cdot5}$ and Fs_{36}. In the Morvan–Cabrach basic intrusion, Aberdeenshire (Allan, 1970), the composition of the orthopyroxene is $\simeq Fs_{50}$. More iron-rich orthopyroxenes, however, are not uncommon, and Henry (1935) has described a ferrohypersthene (Table 3, anal. 26) from a contaminated norite at Glen Buchat, Aberdeenshire.

Iron-rich orthopyroxenes occur in the adamellites and granodiorites associated with the Nain anorthosite massif, Labrador (Smith, 1974). The orthopyroxene in the adamellite displays two sets of exsolution lamellae and crystallized initially as pigeonite. The Fe/(Fe + Mg) ratio of orthopyroxene, including the (100) lamellae, is $Fe_{86}Mg_{14}$; it also contains between 6 and 7 mol. per cent of the $CaSiO_3$ component. The broad lamellae, interpreted as '(001)' lamellae in the original crystals of pigeonite, are ferroaugite containing 43–44 mol. per cent $CaSiO_3$. Some orthopyroxene is intergrown with olivine, quartz and ferroaugite, for which the Fe/(Fe + Mg) ratios are orthopyroxene 0·75–0·78, ferroaugite 0·67–0·69 and olivine 0·90–0·94, indicative of equilibration under similar conditions, and in accord with the textural evidence that the pyroxenes are subsolidus products of pigeonite. The breakdown of pyroxene to olivine and quartz is also supported by the compositions of the orthopyroxenes (Fig. 80) which lie within the area of the pyroxene quadrilateral in which Fe-rich pyroxenes are unstable in relation to the olivine–quartz assemblage at 1 atm and 925°C (Lindsley and Munoz, 1969a; Smith, 1972). As the area of the 'forbidden zone' contracts with increasing pressure towards the composition $FeSiO_3$, and Fe-rich, Ca-poor orthopyroxenes are stabilized relative to olivine + quartz by decreasing temperature, the breakdown of pigeonite, to an isochemical assemblage containing olivine and quartz, is interpreted by Smith as a response to a decrease in pressure during the subsequent uplift of the massif.

Orthopyroxenes (inverted pigeonites) as Fe-rich as Fe_{65}, associated with ferroaugite, $Ca_{34\cdot8}Mg_{26\cdot7}Fe_{38\cdot5}$, and olivine, $Mg_{11\cdot3}Fe_{88\cdot7}$, occur in the layered

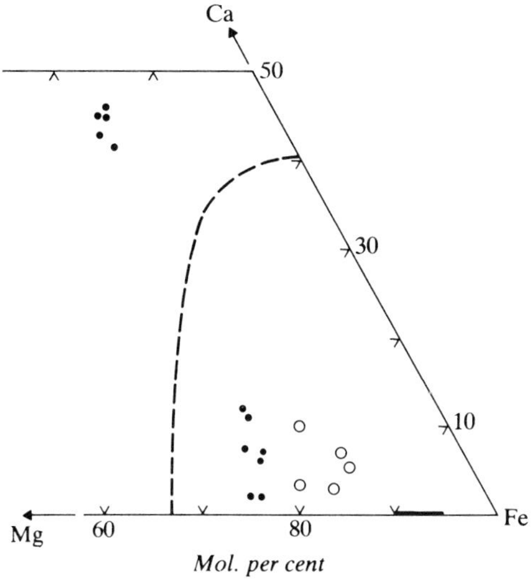

Fig. 80. Compositions of pyroxene and olivine in the Nain adamellite and granodiorite. Open circles, compositions of orthopyroxenes not showing breakdowns to olivine. Solid circles, compositions of orthopyroxene, ferroaugite and olivine (solid bar) in subsolidus reaction intergrowths. Broken line encloses area in which pyroxene is unstable at 1 atm and 925°C (after Smith, 1974).

mangerite intrusion, Raftsund, Lofoten-Vesterålen, north Norway (Griffin *et al.*, 1974). More Fe-rich orthopyroxene is present as rims, $Fe_{85\cdot4}$ in composition, between ferroaugite, $Ca_{41\cdot4}Mg_{13\cdot0}Fe_{45\cdot6}$, and olivine, $Mg_{3\cdot8}Fe_{96\cdot2}$.

Iron-rich orthopyroxenes, $Fe_{77-78}Mn_6Mg_{15}Ca_{1-2}$, containing two sets of Ca-rich clinopyroxene ($\simeq Ca_{44}Fe_{39}Mn_2Mg_{15}$ || (001) and (100), probably inverted from pigeonite, occur in an alkali-feldspar quartz syenite in the Precambrian gneisses south of Julianehaab, south Greenland (Frisch and Bridgwater, 1976). The exsolution is commonly restricted to the central parts of the grains while the marginal zones are homogeneous, presumably due to the overgrowth of eulite on the original, but now inverted, pigeonite.

Hypabyssal rocks. The bronzite phenocrysts in the devitrified glassy mesostasis of the chilled Lydenburg tholeiitic sill, Transvaal, have a high aluminium content (Al_2O_3 5·7 weight per cent), most of which is in octahedral co-ordination. The phenocrysts are commonly bent and oriented parallel to flow lines, and Frick (1970) has suggested that the pyroxenes crystallized intratellurically and, as indicated by their composition, at relatively high pressures. Orthopyroxenes in the Waracobra noritic gabbro sill, Guyana (Rust, 1963), are zoned from Fe_{21} to Fe_{46}. The cores display very fine exsolution lamellae parallel to (100), while the margins show a gradation from irregular exsolution blebs to two regularly oriented sets, one parallel to, and the other at an angle of 71° to the (100) plane of the host. In pigeonite the angle (100):(001) is 71·5°, and it is probable that the peripheral pyroxene crystallized with monoclinic symmetry and subsequently inverted to an orthorhombic pyroxene with exsolution of a Ca-rich clinopyroxene. Some 70 per cent of the central parts of the thick Ferrar dolerite sills, Antarctica, consist of orthopyroxene varying in composition from Fe_{17} to Fe_{37} (Gunn, 1963), while in the Tasmanian dolerites the composition of the orthopyroxene is Fe_{15} in the chilled rocks of the margins and Fs_{30} in the main part of the sills (McDougall, 1962). Iron-rich orthopyroxenes, eulites, have been described from the iron-rich dolerites of the new Amalfi sheet, Matatiele (Poldervaart, 1944), and from the iron-rich diabase of Beaver Bay (Muir, 1954). The formation of hypersthene at the junction of a rheomorphic vein and olivine dolerite has been described by Wilson (1952).

Volcanic rocks. Orthopyroxene phenocrysts and the orthopyroxene of ultramafic crystal cumulates derived from basaltic liquids are typically bronzitic in composition. The orthopyroxene of the olivine-tholeiite, Kolbeinsey islet, north of Iceland is, however, an enstatite ranging in composition (ten analyses) from $Mg_{88\cdot4}$–$Mg_{91\cdot7}$, $Fe_{4\cdot1}$–$Fe_{7\cdot8}$ and $Ca_{2\cdot5}$–$Ca_{5\cdot4}$ (Sigurdsson and Brown, 1970). The content of calcium is high for such a magnesium-rich composition, and by analogy with synthetic data the orthopyroxene is unlikely to have inverted from a protopyroxene. The magnesium-rich pyroxene in the porphyritic volcanic rock, Cape Vogel, Papua (Dallwitz *et al.*, 1966), is a polysynthetically twinned clinoenstatite with a low content of calcium (0·15–0·25 weight per cent CaO), and is associated with bronzite, $Mg_{82-85}Fe_{13-16}Ca_{1\cdot2-2\cdot0}$, in composition.

The occurrence of orthopyroxene, presumed to have crystallized at high pressure, has been described from the tholeiitic basalt of the Hirschberg diateme, south-west Mainz (Nicholls and Lorenz, 1973). The orthopyroxene phenocrysts, containing up to 5·5 weight per cent Al_2O_3, are partially or completely surrounded by a reaction rim of olivine and clinopyroxene. Melting experiments on the Hirschberg tholeiite indicate that the pyroxene is formed under a load pressure of between 6 and 10 kbar, temperatures between 1280° and 1080°C and 2–4 weight per cent water. The orthopyroxene phenocrysts are coarse and unzoned, and show considerable intergrain compositional variation, indicating rapid growth of the crystals and their failure to re-equilibrate with the derivative liquid. During a later period of slower cooling, clinopyroxene was also precipitated and subsequently, during the ascent of the partly crystallized magma, the cooling path reached the field of olivine + clinopyroxene resulting in the resorption and rimming of the orthopyroxene (Fig. 81).

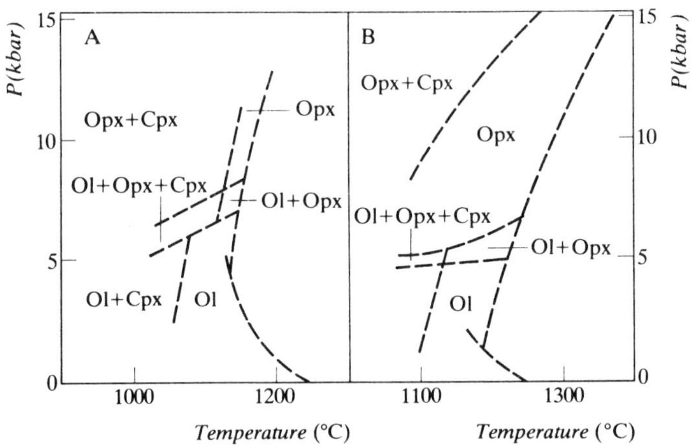

Fig. 81. Experimentally determined phase relations of the Hirschberg tholeiite. A, with 4 per cent H_2O added. B, with 2 per cent H_2O added (after Nicholls and Lorenz, 1973).

Comparable high-pressure megacrysts of orthopyroxene, in a high-alumina tholeiitic andesite from the Tweed shield volcano, New South Wales–Queensland border, have been described by Duggan and Wilkinson (1973). The orthopyroxene, $\simeq Mg_{71}Fe_{25}Ca_4$, contains 4·16 weight per cent Al_2O_3, individual crystals are partially or completely enclosed within the associated plagioclase megacrysts except in contact with the groundmass where they are mantled by aggregates of fine-grained low-pressure olivine and clinopyroxene. The orthopyroxene phenocrysts in the olivine-rich lavas of the lower part of the Karroo basaltic sequence, Nuanetsi province, have been described by Cox and Jamieson (1974). The orthopyroxenes vary in composition from En_{84} to En_{89}, have a moderate content of Al_2O_3 (1·4–1·7 weight per cent), and are jacketed by clinopyroxene. An experimental determination of the phase relation of the lavas at 7 and 10 kbar indicate that the orthopyroxene equilibration took place through a pressure range of approximately 6 to 12 kbar. Although present in some of the olivine andesites of the San Juan region, Colorado (Larsen *et al.*, 1936), orthopyroxene occurs only rarely in the olivine-bearing lavas, and is uncommon in rocks in which the SiO_2 content is less than 54 per cent. Orthopyroxene is a common constituent of many of the lavas with SiO_2 contents between 57 and 59 per cent, but when present in those with more than 65 per cent SiO_2 occurs only in very small amounts. Hypersthene, particularly as a groundmass constituent, is the characteristic ferromagnesian mineral in the Cainozoic calc-alkali volcanic series of Japan. The composition of the orthopyroxene varies from magnesian-rich hypersthene in the basalts and andesites to more iron-rich hypersthene or ferrohypersthene in the dacites and rhyolites (Kuno, 1959). A similar compositional range is shown by the orthopyroxene phenocrysts, $Mg_{71\cdot5}Fe_{23\cdot7}Ca_{4\cdot8}$ in the basaltic andesites to $Mg_{50\cdot2}Fe_{45\cdot2}Ca_{4\cdot6}$ in the dacites of the younger volcanic islands of Tonga (Ewart *et al.*, 1973). Two generations of hypersthene are present in some Hawaiian basalts in which it occurs both as phenocrysts often jacketed by clinopyroxene, and as more iron-rich and zoned micro-phenocrysts (Muir and Long, 1965). The relationship between the hypersthene and the bulk chemistry of the Hawaiian basalts has been examined by Tilley (1961), who concluded that its presence there is largely restricted to rocks high in silica and with a low CaO content.

Aluminium-rich orthopyroxene megacrysts occur in the alkaline lavas, hawaiites and analcite basanites of the Cainozoic volcanic province of north-eastern New South Wales (Binns *et al.*, 1970). The megacrysts are believed to be high-pressure cognate phases, and their high aluminium content (Al_2O_3 3·45 to 8·03 per cent) is comparable with the synthetic orthopyroxenes (4·6 to 10·6 per cent Al_2O_3) crystallized from equivalent basaltic compositions (Green and Ringwood, 1967a, b; Bultitude and Green, 1967). In the associated lherzolite xenoliths the coexisting ortho- and clinopyroxene show less mutual solid solution, indicating that they crystallized at lower temperatures than the megacryst pyroxenes. The orthopyroxenes in the basalts, dacites and rhyodacites of the Carboniferous volcanic series of New South Wales (Wilkinson, 1971), range in composition from $Mg_{63\cdot3}Fe_{31\cdot6}Ca_{5\cdot1}$ in the basalts, to $Mg_{54\cdot4}Fe_{43\cdot1}Ca_{2\cdot5}$ (Table 3, anal. 20) in the hypersthene–hornblende dacites. These pyroxenes have a low content of aluminium suggesting that they crystallized at relatively low pressures in contrast to the high-pressure megacrysts of some alkaline lavas (e.g. Frick, 1970; Binns *et al.*, 1970).

In the pigeonite andesite, Weiselberg, Germany, hypersthene and ferrohypersthenes ($Ca_4Mg_{54}Fe_{42}$–$Ca_4Mg_{47}Fe_{49}$) are associated with pigeonite and ferropigeonite ($Ca_8Mg_{48}Fe_{44}$–$Ca_9Mg_{35}Fe_{56}$), and with augite and ferroaugite $Ca_{38}Mg_{37}Fe_{25}$–$Ca_{35}Mg_{39}Fe_{36}$) (Nakamura and Kushiro, 1970a). The orthopyroxenes are zoned and the Fe/Mg ratio increases from the core outwards. The compositional and textural relations of the three phases show that the phenocryst assemblages vary from hypersthene–augite with relatively low Fe/Mg ratios, through hypersthene–augite–pigeonite to an augite–pigeonite association with relatively high Fe/Mg ratios, and that the replacement of hypersthene during the later crystallization sequence is the result of the reaction hypersthene + liquid $\rightleftharpoons$ pigeonite. The compositional relationships of the bronzite phenocryst and groundmass hypersthene in a tholeiitic andesite, Hakone volcano, Japan, with the associated pigeonite and augite have been described by Nakamura and Kushiro (1970a), and are discussed in the pigeonite section.

Orthopyroxene phenocrysts have been described from an Icelandic rhyolite pitchstone (Carmichael, 1963). The mineral, $\simeq Fe_{49}$, is unusually Mg-rich for such an acid rock, but the absence of appreciable zoning or other evidence of reaction indicate that the phenocrysts crystallized in equilibrium with the liquid. Orthopyroxene, together with magnetite, occurs in some lavas as reaction rims around olivine. Such rims have been attributed to either partial reaction of olivine with the liquid, or to oxidation of olivine without reaction with liquid. A peritectic point, at which olivine dissolves while pyroxene and spinel crystallize as heat is removed, occurs in the system $MgO–FeO–Fe_2O_3–SiO_2$, and Presnall (1966) has suggested that the reaction rims of orthopyroxene and magnetite around olivine phenocrysts may result from such a reaction. Orthopyroxene xenocrysts, mantled by olivine aggregates and invaded by the orthosilicate along planes parallel to (100) of the orthopyroxene, have been described from the alkaline olivine basalts and nephelinites of Hawaii. The possible origin of the mantling, due to the incongruent melting of orthopyroxene or to the fluxing action of a gas rich in water, alkalis and carbon dioxide has been discussed by White (1966).

Hypersthene and cordierite reaction rims, representing a low pressure–high temperature equilibrium assemblage, around garnet phenocrysts in the rhyodacite and granodiorite porphyries of central and north-eastern Victoria. and resulting from the instability of the garnet in the rhyodacite liquid under near-surface conditions, have been described by Green and Ringwood (1968). Orthopyroxene, $Mg_{62}Fe_{34}Ca_4$, rimmed by cummingtonite and more rarely by brown hornblende, has been reported from the quartz-bearing biotite–orthopyroxene–cummingtonite–clinopyroxene–hornblende gabbros of the En mass of the Tabito complex, Nakoso district, Abukuma plateau, Japan (Onuki and Kato, 1971).

Phenocrysts of clinoenstatite, 92 to 87 atomic per cent Mg, and less common phenocrysts of bronzite and a chrome spinel, together with acicular pyroxene microlites, occur in the glassy groundmass of a volcanic rock in the Cape Vogel area, Papua (Dallwitz *et al.*, 1966). The clinoenstatite phenocrysts show no exsolution lamellae but are characterized by multiple twin lamellae. Conspicuous curved cracks are visible in (010) sections and are presumably related to the process of inversion from protoenstatite (Nakamura, 1971). The majority of the clinoenstatite crystals have a narrow margin of a calcic iron-rich clinopyroxene, and may show the development of wedge-shaped and acicular outgrowths of the pyroxene on the terminal faces. Some pyroxene microlites are composite and consist of clinoenstatite, bronzite and the calcic iron-rich clinopyroxene. The magnesium content in the central area of the clinoenstatite phenocrysts is less than in the outer part except at the margin where the zoning is towards enrichment in Fe and Ca prior to an abrupt change to the rim of augite composition. The reverse zoning may be due either to change from lower to higher temperature or to increased oxidation in the magma during the course of the crystallization. The occurrence of clinoenstatite, inverted from protoenstatite, and orthopyroxene with somewhat higher contents of Fe and Ca is consistent with the magma temperature having passed through the protoenstatite$\rightleftharpoons$orthopyroxene inversion curve during crystallization of the phenocrysts, with the magma liquidus curve intersecting the pyroxene inversion curve at an atomic ratio $Mg/(Mg+Fe)$ of 87.

Needles of clinoferrosilite in the lithophysae of an obsidian, Lake Naivasha, Kenya, identified by Bowen (1935) on the basis of the optical properties, have subsequently been shown (Bown, 1965) to contain an appreciable amount of manganese and to have the composition $Fe_{0\cdot95}Mn_{0\cdot05}SiO_3$. The occurrence of clinoferrosilite in the lithophysae of other obsidians, was also reported by Bowen.

Metamorphic Rocks

Orthopyroxenes are the most characteristic and important ferromagnesian minerals in rocks of the granulite facies; in the charnockite series they occur either exclusively or in association with diopside–salite clinopyroxene, brown hornblende, biotite and garnet. In the ultrabasic member of the charnockite series of Madras (Howie, 1955), the orthopyroxene is bronzite (Fs_{24}), in the intermediate members it is hypersthene (Fs_{40} to Fs_{50}), and in the acid rocks ferrohypersthene (Fs_{58} to Fs_{61}). The equilibrium relations between coexisting orthopyroxene and garnet, and orthopyroxene and biotite, in the charnockites of intermediate composition from the type area, St. Thomas' Mount Pallavaram–Tambaran, have been investigated by Sen and Sahu (1970), and a further account of the mineral assemblages of the type area is given by Sen and Roy (1971). In the charnockites of the Kondapalli area, Andhra Pradesh, the orthopyroxenes range in composition from Fs_8 to Fs_{58}. Leelanandam (1967, 1968) has shown that the distribution $K_{\text{Fe–Mg}}^{\text{Opx–Cpx}}$ in the coexisting ortho- and clinopyroxenes is 1·8, a value closely comparable with the earlier data of Howie (1955), Howie and Subramanian (1957) and Subramanian (1962). The distribution coefficients of six orthopyroxenes (hypersthene and ferrohypersthene)–clinopyroxene pairs, from the pyroxene granulites and intermediate charnockites of the Amaravathi area, south India are given by Ramaswamy and Murty (1973). The orthopyroxenes in the leptynites of the Eastern Ghats, India, are described by Mukherjee and Rege (1972).

In the charnockites and associated rocks of the west Uusimaa complex, Finland, the compositions of the orthopyroxenes show no apparent relationship to the acidity of the host rocks. The composition of the orthopyroxene varies between Fs_{50} and Fs_{65}, and Parras (1958) has suggested that this may represent their compositional stability range in the pressure–temperature conditions prevailing during the formation of the complex. Here the orthopyroxene with the highest content of iron, Fs_{65}, occurs in rocks with a Mg:Fe ratio of 37:63; in those with a lower Mg:Fe ratio the additional iron is taken up, not by the orthopyroxene, but in the formation of biotite or magnetite. More iron-rich orthopyroxenes (Table 4, anal. 31) have been reported from a charnockitic adamellite, Natal (Howie, 1958). The orthopyroxenes, and associated clinopyroxenes, from the charnockites in the middle and upper Bug and Dniester region of the Ukraine, have been described by Kononova (1967b), orthopyroxene-bearing rocks of charnockitic affinities from the South Savanna-Kanuku complex, Guyana, by Singh (1966) and orthopyroxenes from the enderbites, charnockitic migmatites and metasomatic charnockites of the Baltic Shield by Shemyakin (1973).

Iron-rich eulite and orthoferrosilite (Table 4, anal. 35) are present in the bauchites (fayalite-bearing quartz monozonites) that occur, associated with charnockites, at a number of localities in the basement complex of Nigeria (Oyawoye and Makanjuola, 1972).

Orthopyroxenes occur in the ultrabasic granulites of Lapland (Eskola, 1952), and are the most common ferromagnesian minerals in rocks of noritic and dioritic composition. Orthopyroxene is also present in the quartz-rich and garnet-bearing granulites of this area; the aluminium-rich hypersthene (Table 4, anal. 9) occurs in a garnet-bearing aluminous granulite. An account of the pyroxenes of the granulite facies rock of Jotunheimen, Norway, is given by Battey and McRitchie (1975). Two orthopyroxene fractions, D 3·49 to 3·55 and mean refractive index 1·706, and D 3·55 to 3·58, *n* 1·715, which differ essentially only in their Mg and Fe^{2+} contents, were separated from the pyroxene granulite of Hartmannsdorf, Saxony (Phillipsborn, 1930). Orthopyroxene is an important constituent of the zoned ultrabasic and basic gneisses in the early Lewisian metamorphic complex, Scourie, Scotland, and here formed by reaction between ultrabasic intrusives and country rock under conditions of the granulite facies.

The occurrence of clinohypersthene lamellae in the calcium-rich pyroxenes of the high grade regionally metamorphosed basic rocks of the Willyama complex, and in the augite of the two-pyroxene granulites of Broken Hill, New South Wales, has been described by Binns (1965) and Binns *et al.* (1963).

An unusual hypersthene-bearing assemblage consisting of pyrope, sapphirine, sillimanite, cordierite and biotite, in association with the plagiogneiss of the Anabar massif, U.S.S.R. has been described by Lutts and Kopaneva (1968). The hypersthene (Table 4, anal. 7) is very rich in aluminium (10·81 weight per cent) in conformity with its occurrence in a granulite facies terrain. Other highly aluminous orthopyroxenes include hypersthene (9·0 weight per cent Al_2O_3) in the sapphirine-bearing granulites from Labwor, Uganda (Nixon *et al.*, 1973), and hypersthene (Al_2O_3 13·91, Table 4, anal. 15) in the cordierite and biotite granulites of Satnur, Mysore State, India (Devaraju and Sadashivaiah, 1971). Although orthopyroxenes formed under the physical conditions of the granulite facies characteristically have a high aluminium content, this feature is not invariably present; thus granulite facies orthopyroxenes (2–2·4 weight per cent Al_2O_3) have been reported from Mont Tremblant Park, Quebec (Katz, 1970). An aluminous bronzite (7·1 per cent Al_2O_3) occurs in a sapphirine–cordierite–phlogopite assemblage in an interbedded schist and quartzite succession of the Kheiz System, Namaqualand (Clifford *et al.*, 1975). The orthopyroxene encloses inclusions of cordierite and phlogopite, the latter occurring as optically continuous but physically discrete grains, and indicative of the formation of the bronzite in part from the mica. On textural grounds the sapphirine also appears to have crystallized later than the earlier-formed cordierite–phlogopite rock consequent on a reaction of the type:

$$4\ \text{phlogopite} + 2\ \text{cordierite} \rightarrow 3\ \text{bronzite} + 2\ \text{sapphirine} + \text{rutile} + \text{MgO}$$

The association of orthopyroxene with either sillimanite or kyanite is uncommon and is probably due to a combination of the rarity of highly magnesian argillaceous rocks, and the high pressure (> 10 kbar) required for the formation of stable enstatite–aluminium silicate assemblages. A rock from the pyroxene-granulite facies terrain in Rhodesia, consisting of enstatite, sillimanite, corundum, cordierite and quartz, and considered to have earlier consisted of enstatite, kyanite and quartz, has been described by Chinner and Sweatman (1968). The enstatite has a high content of aluminium and the maximum value (8·35 per cent Al_3O_3) was measured at the centre of a grain showing a decreasing Al_2O_3 content (to 5·56 per cent) at the margin. The cordierite is derived by a retrogressive reaction and the sillimanite is probably pseudomorphous after kyanite. Chinner and Sweatman consider that 8·35 weight per cent Al_2O_3 is probably the minimum aluminium content of orthopyroxene at the pressure necessary for the formation of the enstatite–kyanite assemblage.

Hypersthene–sillimanite–garnet gneisses, in which the hypersthene and sillimanite textural relationships indicate simultaneous crystallization, have been described from the Sutam River area of the Aldan shield, U.S.S.R. (Marakushev and Kudryavtsev, 1965). The two minerals occur in fine intergrowths and poikilitic inclusions of idiomorphic sillimanite are present in the pyroxene. The hypersthene has a high ferric/ferrous iron ratio and contains 9·35 per cent Al_2O_3, most of the aluminium being in tetrahedral co-ordination. Garnet–cordierite gneisses are present in the same region and represent a lower pressure assemblage. The formation of the higher pressure hypersthene–sillimanite gneiss association may be represented by the reaction:

$$\underset{\text{almandine}}{Mg_{1\cdot5}Fe_{1\cdot5}Al_2Si_3O_{12}} + 1{\cdot}14(\underset{\text{cordierite}}{Mg_{1\cdot66}Fe_{0\cdot34}Al_4Si_5O_{18}})$$

$$= 3{\cdot}96\underset{\text{Al-hypersthene}}{Mg_{0\cdot74}Fe_{0\cdot36}Al_{0\cdot2}Si_{0\cdot8}O_3} + 2{\cdot}576\underset{\text{sillimanite}}{Al_2SiO_5} + 2{\cdot}032\underset{\text{quartz}}{SiO_2}$$

Hypersthene, $Fe_{41\cdot5}$, containing 3·75 per cent Al_2O_3, associated with osumilite, cordierite, orthoclase, plagioclase, quartz, graphite and pyrrhotite, in a granulite from the contact aureole of the Precambrian anorthositic Nain complex, Labrador, has been described by Berg and Wheeler (1976).

The occurrence of ferrohypersthene, $Mg_{29}Fe_{70}Ca_1$, in a lava, with the remarkable assemblage andesine–titanomagnetite–ilmenite–ferrohypersthene–iron cordierite, from St. Helena, South Atlantic has been described by Ridley and Baker (1973). The ferrohypersthene has a relatively high Al-content (4·05 per cent Al_2O_3) that reflects the Al-rich composition of the lava which also shows strong iron enrichment and internal fractionation towards a granitic mesostasis.

Evidence of a hydration reaction involving the hypersthene in a granulite facies rock from the Johannsen's mine, southern Strangways Range, central Australia, has been presented by Vernon (1972). The hypersthene is partly surrounded by aluminous anthophyllite which occurs within the inferred original margins of the pyroxene, including the boundaries against quartz inclusions in the hypersthene. The alteration of the hypersthene is thus unlikely to have arisen from the reaction enstatite + silica + water = anthophyllite, and the reaction, based on the formulae derived from the mineral analyses, may be expressed:

$$\underset{\text{Al-hypersthene}}{Mg_{1\cdot2}Fe_{0\cdot6}Al_{0\cdot3}Si_{1\cdot9}O_{6\cdot0}} + 0{\cdot}5H_2O + 0{\cdot}1Al \rightleftharpoons$$

$$\underset{\text{Al-anthophyllite}}{Mg_{1\cdot1}Fe_{0\cdot5}Al_{0\cdot4}Si_{1\cdot9}O_{6\cdot0}(OH)_{0\cdot5}} + 0{\cdot}1Mg + 0{\cdot}1Fe^{2+} + 0{\cdot}5H$$

Orthopyroxenes occur in regionally metamorphosed rocks formed under somewhat less severe pressure–temperature conditions than those of the granulite facies, and in the metamorphosed gabbroic rocks of the Adirondacks (Buddington, 1939, 1952; Engel *et al.*, 1964), orthopyroxene is present in amphibolite gneiss, hornblende–hypersthene gneiss, and almandine-bearing ultrabasic pyroxene gneiss. In the latter, as well as in pegmatitic recrystallization veins, the orthopyroxene is either ferrohypersthene or eulite. Hypersthene associated with cordierite, almandine, biotite, potassium feldspar, plagioclase and quartz, also occurs in the pelitic gneisses, particularly those with a low alumina content. Waard's (1966) suggestion that, in this paragenesis, the typical empirical reactions involving hypersthene may be expressed:

$$\text{biotite} + \text{quartz} \rightleftharpoons$$

$$\rightleftharpoons \text{hypersthene} + \text{almandine} + \text{cordierite} + \text{potassium feldspar} + H_2O \quad (1)$$

$$\text{hornblende} + \text{almandine} + \text{quartz} \rightleftharpoons \text{hypersthene} + \text{plagioclase} + H_2O \quad (2)$$

has been critically discussed by Chesworth (1967).

The formation of hypersthene and plagioclase at the interface of pyroxene-bearing amphibolites and quartzo–feldspathic rocks of the hornblende–granulite facies terrain in the Grenville province have been described by Schrijver (1973). The pyroxene–plagioclase rims at the interface are ascribed to an isochemical reaction of the type:

$$\text{hydrous mafic minerals} + \text{quartz} \rightleftharpoons \text{pyroxene(s)} + \text{feldspare} + \text{Fe–Ti oxides} + H_2O$$

Enstatite occurs with forsterite, chlorite and magnetite, in an ultramafic schist of the amphibolite facies in the Lepontine Alps (Trommsdorff and Evans, 1969). The assemblage is considered to have formed, either during the metamorphism of a dry protolith, or by crystallization under conditions of high CO_2-activity, rather than by the simple progressive metamorphism of serpentinite (see also Trommsdorff and Evans, 1974). Hypersthene (Table 4, anal. 8), associated with cummingtonite and labradorite, in a rock belonging to the amphibole facies has been described by Kuno (1947a).

The stability relationships of the metamorphic assemblage hypersthene–sapphirine–gedrite have been described and discussed by Vogt (1947). In the pyroxene-hornfels facies, hypersthene may develop from hornblende or gedrite:

$$\underset{\text{gedrite}}{(Mg,Fe^{2+})_5Al_4Si_6O_{22}(OH)_2} + \underset{\text{quartz}}{2SiO_2} \rightarrow \underset{\text{hypersthene}}{3(Mg,Fe^{2+})SiO_3} + \underset{\text{cordierite}}{(Mg,Fe^{2+})_2Al_4Si_5O_{18}} + H_2O$$

or from the breakdown of biotite:

$$\underset{\text{biotite}}{K(Mg,Fe)_3AlSi_3O_{10}(OH)_2} + 3SiO_2 \rightarrow \underset{\text{hypersthene}}{3(Mg,Fe^{2+})SiO_3} + \underset{\text{orthoclase}}{KAlSi_3O_8} + H_2O$$

(Tilley, 1924; Loomis, 1966).

Orthopyroxene also occurs in medium grade thermally metamorphosed argillaceous rocks originally rich in chlorite and with a low calcium content:

$$\underset{\text{chlorite}}{(Mg,Fe^{2+})_4Al_4Si_2O_{10}(OH)_8} + \underset{\text{quartz}}{5SiO_2} \rightarrow \underset{\text{hypersthene}}{2(Mg,Fe^{2+})SiO_3} + \underset{\text{cordierite}}{(Mg,Fe^{2+})_2Al_4Si_5O_{18}} + 4H_2O$$

In such hornfelses, the orthopyroxene may also be associated with biotite and quartz, but its association with andalusite is excluded by the reaction:

$$\underset{\text{hypersthene}}{2(Mg,Fe^{2+})SiO_3} + \underset{\text{andalusite}}{2Al_2SiO_5} + SiO_2 \rightarrow \underset{\text{cordierite}}{(Mg,Fe^{2+})_2Al_4Si_5O_{18}}$$

or in an SiO_2 deficient environment by the reaction:

$$\underset{\text{hypersthene}}{5(Mg,Fe^{2+})SiO_3} + \underset{\text{andalusite}}{5Al_2SiO_5} \rightarrow \underset{\text{cordierite}}{2(Mg,Fe^{2+})_2Al_4Si_5O_{18}} + \underset{\text{spinel}}{(Mg,Fe^{2+})Al_2O_4}$$

Although experimental work has indicated that under high *P–T* conditions appreciable amounts of alumina can enter the orthopyroxene structure (e.g. Boyd and England, 1960), in natural minerals pressure is not necessarily the dominant factor. For example, a strongly pleochroic hypersthene with 7·21 per cent Al_2O_3 occurs (Table 4, anal. 10) in an Aberdeenshire hypersthene–spinel–plagioclase hornfels, a paragenesis in which pressure is not likely to have played a dominant role. The controlling factor here would appear to have been chemical, i.e. the availability of alumina in the environment of crystallization.

Orthopyroxenes are present in the antigorite–olivine–orthopyroxene–clinopyroxene and the olivine–orthopyroxene zones of the thermally metamorphosed Tari–Misaka ultramafic complex, Chugoku district, western Japan (Arai, 1975). The orthopyroxene, $Wo_{2\cdot9}En_{88\cdot9}Fs_{8\cdot2}$, of the first zone is a chromium-rich enstatite ($\simeq$ 0·90 weight per cent Cr_2O_3), and is associated with olivine, Fo_{85}–Fo_{92}, and chromian diopside. In the higher metamorphic zone the Mg/(Mg+Fe) ratio of the orthopyroxene varies between 0·90 and 0·95, the pyroxene is depleted in chromium and has a very low calcium content and is probably derived from the reaction:

$$\underset{\text{anthophyllite}}{(Mg,Fe)_7Si_8O_{22}(OH)_2} + \underset{\text{olivine}}{(Mg,Fe)_2SiO_4} \rightleftharpoons \underset{\text{orthopyroxene}}{9(Mg,Fe)SiO_3} + H_2O$$

Orthopyroxene is uncommon in the more calcareous hornfelses, due to the reactions (Mueller, 1966):

$$MgSiO_3 + CaCO_3 + SiO_2 \rightleftharpoons CaMgSi_2O_6 + CO_2 \qquad (1)$$

$$3MgSiO_3 + CaCO_3 \rightleftharpoons Mg_2SiO_4 + CaMgSi_2O_6 + CO_2 \qquad (2)$$

Eulite and orthoferrosilite, associated with fayalite, hedenbergite, grunerite or almandine-spessartine garnet, are present in some regionally metamorphosed iron-rich sediments, while in the eulysite at Mansjö (Eckermann, 1922), eulite, originally described as iron-anthophyllite, occurs with knebelite. Eulite, $Fs_{81\cdot9}$, has been described from the eulysite of the Lake Chudz'yavr region, Kola (Bondarenko and Dagelaĭskiĭ, 1961), and eulite, with $(Fe,Mn)SiO_3$ 84·6 per cent, occurs in the metamorphosed iron formation at Payne Bay, near Quebec (Moorhouse and Shepherd, 1963).

Large poikiloblasts of orthopyroxene, up to several centimetres across, enclosing olivine, Fa_{87}, plagioclase, An_{94}, and Fe–Ti oxides, $Mt_{62}Usp_{34}He_4$, are present in the Gunflint Iron Formation, an original ferruginous sediment subsequently metamorphosed at the time of the intrusion of the Duluth complex (Simmons *et al.*, 1974). The orthopyroxene ($\simeq Mg_{26}Fe_{71\cdot5}Ca_{2\cdot5}$) contains blebs of augite ($\simeq Ca_{43}Mg_{19}Fe_{38}$) and has inverted from a pigeonite ($\simeq Ca_{10\cdot5}Mg_{24\cdot5}Fe_{65}$). Orthopyroxenes, together with calcium-rich clinopyroxenes and fayalite, are present throughout the sequence of the Biwabik Iron Formation, Minnesota (Bonnichsen, 1969). The orthopyroxene, a significant amount of which has inverted from pigeonite, is typically a ferrohypersthene in composition[1]. In local pegmatitic segregations, orthopyroxene is associated with hornblende and quartz, and crystallized simultaneously with the amphibole at a temperature below that of the pigeonite–ferrohypersthene inversion. The Fe/(Mg+Fe+Ca) ratio of the orthopyroxenes varies from 0·307 to 0·770, and the possible relationship between the calcium-poor pyroxene, fayalite and quartz, coexisting with the clinopyroxene, is shown in Fig. 82.

Ferrohypersthene is the commonest ferromagnesian mineral in the Mount Reed area at the southern end of the Quebec–Labrador Iron Range (Kranck, 1961). The original sediments had a high carbonate and low water content, and the ferrohypersthene was probably derived by reaction between siderite and/or ferrodolomite and quartz: $FeCO_3 + SiO_2 \rightarrow FeSiO_3 + CO_2$

[1] Clinoeulite, $En_{19}Fs_{78}Wo_{3}$, containing exsolution lamellae of ferroaugite, $Wo_{40}En_{17}Fs_{43}$, probably derived from a ferropigeonite, $Wo_{13}En_{17}Ds_{70}$, in composition, is the main constituent of a eulysite in the centre of the Vredefort structure, South Africa (Schreyer *et al.*, 1978. *Contr. Min. Petr.*, **65**, 351–361).

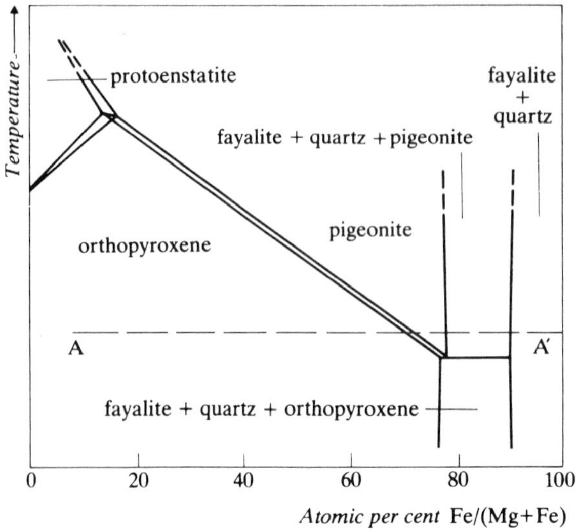

Fig. 82. Suggested equilibrium relations between orthopyroxene, protoenstatite, pigeonite and fayalite + quartz coexisting with Ca-pyroxene at relatively low pressures. Line A–A′ maximum metamorphic temperature (700°–750°C) of the Dunka River area, Biwabik Iron Formation (Bonnichsen, 1969).

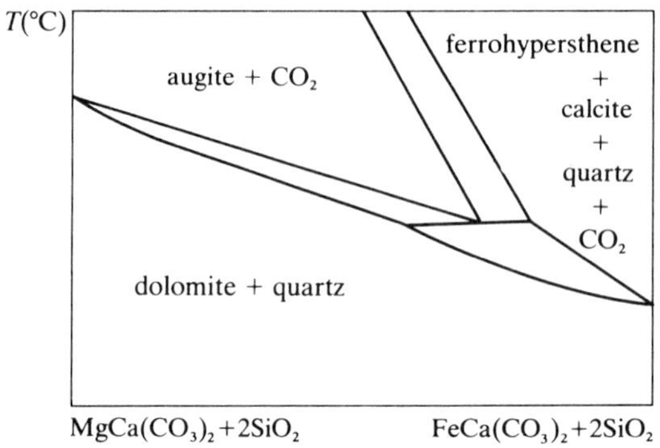

Fig. 83. Hypothetical phase relations by reaction between dolomite and quartz at different MgO/FeO ratios in a dry environment (after Kranck, 1961).

or by the reaction ferrodolomite + quartz → ferroaugite, the latter becoming unstable as the temperature increased and reacting with CO_2 to form ferrohypersthene and calcite (Fig. 83):

$$(Fe,Mg)Ca(CO_3)_2 + 2SiO_2 \rightarrow (Fe,Mg)CaSi_2O_6 + 2CO_2 \quad (1)$$

$$(Fe,Mg)CaSi_2O_6 + CO_2 \rightarrow (Fe,Mg)SiO_3 + CaCO_3 + SiO_2 \quad (2)$$

Analyses of thirty-eight orthopyroxenes (SiO_2, FeO, MnO, MgO and CaO) from the Gagnon region of the Wabush Iron Formation have been presented by Butler (1969). The majority of the minerals are ferrohypersthenes or eulites, although the full compositional range varies from 16·48–72·40 Fe, 0·62–15·85 Mn, 20·79–79·32 Mg, 0·45–2·97 Ca atomic per cent. Ortho- and clinopyroxene are the most abundant and ubiquitous phases present, and coexist with grunerite, actinolite, quartz, magnetite, hematite, calcite, ankerite, siderite, pyrite and pyrrhotite. The variations in the compositions of the orthopyroxenes, and the other silicates, in the subassemblages orthopyroxene + grunerite + quartz, orthopyroxene + clinopyroxene + quartz + calcite, and orthopyroxene + clinopyroxene + grunerite + quartz + calcite, are indicative of the influence of the μ_{H_2O} and μ_{CO_2} gradients present during the metamorphism, and are illustrated in Fig. 84 showing the variation in the composition of the orthopyroxene along two traverses across the strike of the iron formation.

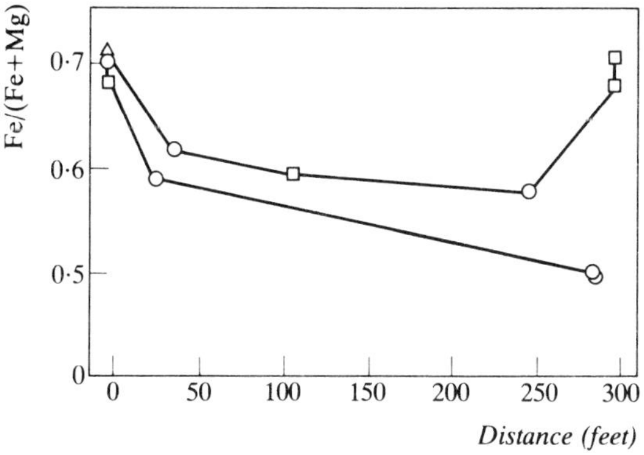

Fig. 84. Compositional change of orthopyroxene along two traverses of the Wabush Iron Formation (after Butler, 1969). (a) Δ assemblage quartz + orthopyroxene + grunerite which would respond to changes in μ_{H_2O}; (b) □ assemblage quartz + orthopyroxene + clinopyroxene + calcite and indicates changes in μ_{CO_2}; (c) ○ assemblage quartz + orthopyroxene + clinopyroxene + grunerite + calcite and indicates change in both μ_{H_2O} and μ_{CO_2}.

Orthoferrosilite (Table 4, anal. 34), in a thermally metamorphosed iron-rich rock from the Yu hsi kou district, Manchuria, has been described by Tsuru and Henry (1937). In the metamorphosed iron formations associated with the gneisses and granulites of the Quairading district, Western Australia (Davidson and Mathison, 1973), the orthopyroxenes, coexisting with spessartine-almandine garnet, have a high manganese content (MnO 7–8 weight per cent, Table 4, anal. 21). Other iron-rich orthopyroxenes ($En_{31\cdot1}Fs_{67\cdot5}Wo_{1\cdot4}$–$En_{19\cdot3}Fs_{79\cdot9}Wo_{0\cdot8}$) have been reported from the iron formation, magnetite-bearing quartzites, of the Tiris area, Mauritania (Cuney *et al.*, 1975), and from the iron formations of the Tobacco Root and Ruby Mountains and the Gravelly Range of south-western Montana (Immega and Klein, 1976). Here their compositional range is between Fs_{60} and Fs_{75}, and the orthopyroxenes are present in assemblages containing ferrosalite, almandine, quartz, magnetite, hematite, grunerite and/or hornblende.

The stability of orthopyroxene under conditions of retrograde metamorphism in a system open to H_2O and SiO_2 has been investigated by Himmelberg and Phinney (1967). In the granulite facies pyroxene–hornblende gneiss in the Granite Falls–Montevideo area, Minnesota, orthopyroxene is replaced by cummingtonite either as partial or complete mantles around the pyroxene, as well as by aggregates of cummingtonite blades with little or no relict orthopyroxene remaining. Where the replacement of the pyroxene is incomplete the cummingtonite replacement is generally restricted to hornblende–orthopyroxene contacts, the former acting as a nucleus for the crystallization of cummingtonite. The orthopyroxene is also replaced along fractures by a 'serpentine'-type mineral, and is commonly associated with cummingtonite mantles around the pyroxene. These relationships are illustrated in the chemical potential diagram of Fig. 85. Idealized reactions along the univariant curves are:

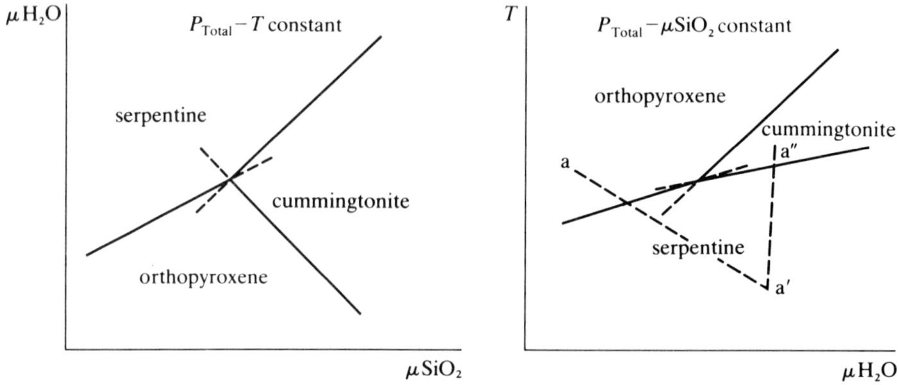

Fig. 85. Relative stabilities of orthopyroxene, cummingtonite and serpentine with respect to temperature and chemical potential of H_2O and SiO_2 (after Himmelberg and Phinney, 1967).

$$\underset{\text{orthopyroxene}}{7(Fe,Mg)SiO_3} + H_2O + SiO_2 \rightleftharpoons \underset{\text{cummingtonite}}{(Fe,Mg)_7Si_8O_{22}(OH)_2} \qquad \frac{\partial\mu_{H_2O}}{\partial\mu_{SiO_2}} = -1$$

$$\underset{\text{orthopyroxene}}{6(Fe,Mg)SiO_3} + 4H_2O \rightleftharpoons \underset{\text{serpentine}}{(Fe,Mg)_6Si_4O_{10}OH)_8} + 2SiO_2 \qquad \frac{\partial\mu_{H_2O}}{\partial\mu_{SiO_2}} = +\tfrac{1}{2}$$

$$\underset{\text{cummingtonite}}{6(Fe,Mg)_7Si_8O_{22}(OH)_2} + 22H_2 \rightleftharpoons \underset{\text{serpentine}}{7(Fe,Mg)_6Si_4O_{10}(OH)_8} + 20SiO_2 \qquad \frac{\partial\mu_{H_2O}}{\partial\mu_{SiO_2}} = +\frac{10}{11}$$

The textural relations indicate that the alteration sequence began initially with the replacement of orthopyroxene by serpentine, consequent on an increase in μ_{H_2O} and a decrease in temperature, line a–a′ in Fig. 85, and was followed by a replacement of both orthopyroxene and serpentine by cummingtonite due to increasing temperature (line a′–a″), or a higher level in the chemical potential of SiO_2.

Orthopyroxene and magnesite are the two essential constituents of the sagvandites (carbonate-orthopyroxenites) in which the orthopyroxene typically constitutes some 70 and magnesite 15 volume per cent. Other phases include olivine, talc, chlorite and phlogopite (Schreyer *et al.*, 1972; Ohnmacht, 1974). At Troms, northern Norway, the sagvandites occur as concordant lenses in high grade garnet–kyanite–biotite gneisses and amphibolites, and include orthopyroxene–magnesite–dolomite–tremolite, as well as orthopyroxene–magnesite rocks. In both subgroups the orthopyroxene is an enstatite or bronzite in composition. The orthopyroxene encloses olivine (Fa_{10}) relics, and is itself partially replaced by magnesite and sometimes by talc or talc + magnesite. The primary materials of the magnesite–orthopyroxenites and related rocks were probably ultrabasic, dunitic to saxonitic, in composition that have undergone CO_2–H_2O–SiO_2 metasomatism involving reactions of the type:

$$\underset{\text{olivine}}{(Mg,Fe)_2SiO_4} + CO_2 \rightarrow \underset{\text{orthopyroxene}}{(Mg,Fe)SiO_3} + \underset{\text{magnesite/breunnite}}{(Mg,Fe)CO_3}$$

$$\underset{\text{olivine}}{(Mg,Fe)_2SiO_4} + SiO_2 \rightarrow \underset{\text{orthopyroxene}}{2(Mg,Fe)SiO_3}$$

The possibility that the carbonate-orthopyroxenites are products of hydrothermal reactions, involving either SiO_2-undersaturated or saturated fluids, developed during the evolution of ultramafic bodies, and representing the end product of the series dunite → saxonite → orthopyroxenite → sagvandite, is discussed by Ohnmacht (1974). Sagvandites from western Norway, and the Lepontine area, Swiss Alps, have been described respectively by Kendel (1970) and Trommsdorff and Evans (1970). Veins of olivine-bearing sagvandite, in the hornblende–biotite gneiss of Berneray, South Harris, Outer Hebrides, have been reported by Livingstone (1965).

Stability conditions for the assemblage enstatite–magnesite in relation to the system $MgO–SiO_2–CO_2–H_2O$ (Johannes, 1969) are discussed by Schreyer *et al.* (1972) and a pressure–temperature plot of the stability range of the assemblage is given for $P_{\text{fluid}} = P_{\text{total}}$.

The metamorphic rocks of the Tabor massif, Czechoslovakia, of granite to syenite modal composition, show retrogressive metamorphism from the granulite to amphibolite facies (Jakeš, 1968). Reactions involving orthopyroxene and leading to the formation of the amphibolite facies assemblage include:

$$\underset{\text{orthopyroxene}}{3(Mg,Fe)SiO_3} + \underset{\text{K-feldspar}}{KAlSi_3O_8} + H_2O \rightleftharpoons \underset{\text{biotite}}{K(Mg,Fe)_3AlSi_3O_{10}(OH)_2} + \underset{\text{quartz}}{3SiO_2} \quad (1)$$

$$\underset{\text{orthopyroxene}}{3(Mg,Fe)SiO_3} + \underset{\text{diopside}}{Ca(Mg,Fe)Si_2O_6} + \underset{\text{plagioclase}}{NaAlSi_3O_8 + CaAl_2Si_2O_8} + H_2O \quad (2)$$

$$\rightleftharpoons \underset{\text{amphibole}}{NaCa_2(Mg,Fe)_4Al_3Si_6O_{22}(OH)_2} + \underset{\text{quartz}}{4SiO_2}$$

An orthopyroxene ($\simeq Fs_{20}$) in eclogite, in boundinaged and amphibolitized layers within serpentinized dunite of the Sunndal–Grubse ultramafic mass, Almklovdalen, Norway, has been described by Lappin (1974).

The formation of enstatite, constituting the inner part of a zoned sequence, enstatite, anthophyllite, tremolite, chlorite, in a vein cutting peridotite located in the basal gneiss region of southern Norway has been described by Carswell (1968) and Carswell *et al.* (1974). The vein sequence is considered to have developed as the result of a metasomatic reaction between a pore fluid phase, rich in alkali halides, and peridotite during a period of regional metamorphism at a temperature of about 700°C and a total pressure in excess of 6 kbar.

Meteorites. Orthopyroxenes are a common constituent of the chondrite, achondrite and stony-iron meteorite groups. Compared with terrestrial orthopyroxenes they are characterized by low contents of ferric iron and calcium. Enstatite or clinoenstatite is the essential phase in the enstatite chondrites and achondrites, and commonly approximates closely to the end-member composition (Fig. 86). In chondritic meteorites the smaller the amount of nickel–iron present, the greater is the content of the ferrosilite component in the orthopyroxene. In the iron meteorites the orthopyroxenes have relatively low iron contents, e.g. Fs_6 in the Woodbine, Illinois, meteorite, an octahedrite with $\simeq 80$ weight per cent Ni–Fe (Mason, 1967). The compositions of three meteoritic orthopyroxenes, ranging from $\simeq Mg_{99}$ to Mg_{72}, detailed in Table 6, do not represent the full extent of their chemical variation, and hypersthenes[1] containing up to 45 per cent of the ferrosilite component have been reported (see Fig. 105, p. 191).

Most chondrite enstatites have a low content of aluminium, whether or not the bulk composition of the meteorite is rich in alumina. The Al-rich orthoenstatite (7·5

[1] The nomenclature used by meteoriticists takes bronzite as 10–20 per cent $FeSiO_3$, and greater than 20 per cent $FeSiO_3$ as hypersthene.

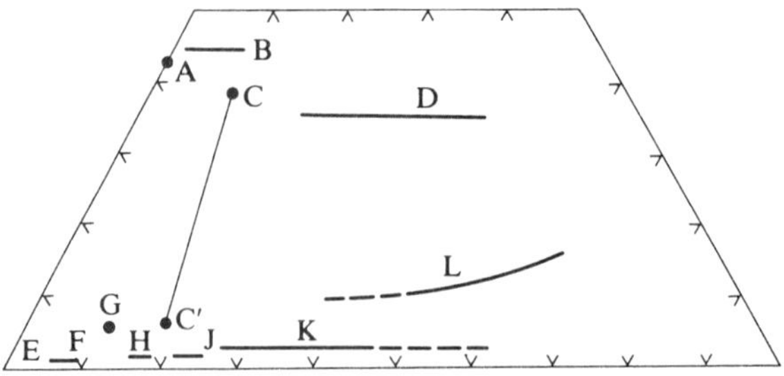

Fig. 86. Pyroxene compositions in meteorites (after Mason, 1968). A, diopside in enstatite achondrite. B, diopsides in chondrites and in silicate inclusions in irons. C–C′, coexisting orthopyroxene and augite in chondrite. D, augites in achondrites and mesosiderites. E, enstatite and clinoenstatite in enstatite chondrites and enstatite achondrites. F, enstatite in silicate inclusions in irons. G, clinoenstatite in ureilite. H, orthopyroxene in bronzite chondrites. J, orthopyroxene in hypersthene chondrite. K, orthopyroxene in achondrites and stony-iron mesosiderites. L, pigeonite in achondrites and mesosiderites.

weight per cent Al_2O_3) in the carbonaceous cordierite-bearing chondrite Attende, Chihuahua, Mexico (Fuchs, 1969) is an exception, and its association in this meteorite with cordierite, by analogy with terrestrial assemblages, indicates that crystallization occurred at relatively high pressure.

Many enstatites in achondrites display the effects of more intense deformation than is commonly present in terrestrial rocks, and the frequent occurrence of clinoenstatite, often intergrown with orthoenstatite, is usually attributed to its formation from the orthorhombic phase due to stress. The luminescence of enstatite from enstatite achondrites, due to excitation by X-rays, proton or electron bombardment is reported by Reid and Cohen (1967) to be related to their low content of iron and influenced by the amount of manganese present in the pyroxene.

References

Ahrens, T. J. and **Gaffney, E. S.**, 1971. Dynamic compression of enstatite. *J. Geophys. Res.*, **76**, 5504–5513.

Akella, J., 1976. Garnet pyroxene equilibria in the system $CaSiO_3$–$MgSiO_3$–Al_2O_3, and in a natural mineral mixture. *Amer. Min.*, **61**, 589–598.

Akella, J. and **Boyd, F. R.**, 1972. Partitioning of Ti and Al between pyroxenes, garnets and oxides. *Carnegie Inst. Washington, Ann. Rept. Dir. Geophys. Lab.*, 1971–72, 378–384.

Akella, J. and **Boyd, F. R.**, 1973. Effect of pressure on the composition of coexisting pyroxenes and garnet in the system $CaSiO_3$–$MgSiO_3$–$FeSiO_3$–$CaAlTi_2O_6$. *Carnegie Inst. Washington, Ann. Rept. Dir. Geophys. Lab.*, 1972–73, 523–526.

Akella, J. and **Winkler, H. G. F.**, 1966. Orthorhombic amphibole in some metamorphic reactions. *Contr. Min. Petr.*, **12**, 1–12.

Akimoto, S., **Fujisawa, H.** and **Katsura, T.**, 1964. Synthesis of $FeSiO_3$ pyroxene (ferrosilite) at high pressures. *Proc. Japan. Acad.*, **40**, 272–275.

Akimoto, S., **Katsura, T.**, **Syono, Y.**, **Fujisawa, H.** and **Komada, E.**, 1965. Polymorphic transition of pyroxenes $FeSiO_3$ and $CoSiO_3$ at high pressures and temperatures. *J. Geophys. Res.*, **70**, 5269–5278.

Akimoto, S., **Komada, E.** and **Kushiro, I.**, 1966. Preliminary experiments on the stability of natural pigeonite and enstatite. *Proc. Japan. Acad.*, **42**, 482–487.

Akimoto, S. and **Syono, Y.**, 1970. High-pressure decomposition in the system $FeSiO_3$–$MgSiO_3$. *Phys. Earth Planet. Interiors*, **3**, 186–188.

Akimoto, S. and **Syono, Y.**, 1972. High pressure transformations in $MnSiO_3$. *Amer. Min.*, **57**, 76–84.

Akizuki, M., 1967. Transformation of talc, chlorite and muscovite by electron bombardment. *J. Japanese Assoc. Min. Petr. and Econ. Geol.*, **58**, 161–169 (in Japanese, English summary).

Allan, W. C., 1970. The Morven–Cabrach basic intrusion. *Scottish J. Geol.*, **6**, 53–72.

Allen, E. T. and **White, W. P.**, 1909. Diopside and its relation to calcium and magnesium metasilicates, with optical study by F. E. Wright and E. S. Larsen. *Amer. J. Sci.*, ser. 4, **27**, 1–47.

Anastasiou, P. and **Seifert, F.**, 1972. Solid solubility of Al_2O_3 in enstatite at high temperatures and 1–5 kbar water pressure. *Contr. Min. Petr.*, **34**, 272–287.

Angeletos, S., 1975. Developments in oxidation of the $(Mg,Fe)SiO_3$ pyroxene series and use of thermogravimetry as a quantitative analytical method. *Chem. Geol.*, **15**, 145–153.

Aoki, K-I., 1968. Petrogenesis of ultrabasic and basic inclusions in alkali basalts, Iki Island, Japan. *Amer. Min.*, **53**, 241–256.

Aoki, K-I., 1971. Petrology of mafic inclusions from Itinome-gata, Japan. *Contr. Min. Petr.*, **30**, 314–331.

Arai, S., 1975. Contact metamorphosed dunite–harzburgite complex in the Chugoku district, western Japan. *Contr. Min. Petr.*, **52**, 1–16.

Ashworth, J. R. and **Barber, D. J.**, 1975. Electron petrography of shock effects in a gas-rich achondrite. *Contr. Min. Petr.*, **49**, 149–162.

Atkins, F. B., 1969. Pyroxenes of the Bushveld intrusion, South Africa. *J. Petr.*, **10**, 222–249.

Atlas, L., 1952. The polymorphism of $MgSiO_3$ and solid state equilibria in the system $MgSiO_3$–$CaMgSi_2O_6$. *J. Geol.*, **60**, 125–147.

Baker, P. E., 1968. Petrology of Mt. Misery Volcano, St. Kitts, West Indies. *Lithos*, **1**, 124–150.

Bancroft, G. M. and **Burns, G. R.**, 1967. Interpretation of the electronic spectra of iron in pyroxenes. *Amer. Min.*, **52**, 1275–1277.

Bancroft, G. M. and **Burns, G. R.**, 1968. Applications of the Mössbauer effect in mineralogy. *Min. Soc., I.M.A. vol (I.M.A. Papers and Proc. 5th Gen. Meeting, Cambridge, 1966)*, 36–42.

Bancroft, G. M., **Burns, G. R.** and **Howie, R. A.**, 1967. Determination of the cation distribution in the orthopyroxene series by the Mössbauer effect. *Nature*, **213**, 1221–1223.

Banno, S., 1964. Alumina content of orthopyroxene as a geologic barometer. *Japan. J. Geol. Geogr.*, **35**, 115–121.

Barth, T. F. W., 1951. Sub-solidus diagram of pyroxenes from common mafic magmas. *Norsk Geol. Tidsskr.*, **29**, 218–221.

Bartholomé, P., 1961. Coexisting pyroxenes in igneous and metamorphic rocks. *Geol. Mag.*, **98**, 346–348.

Bartholomé, P., 1962. Iron–magnesium ratio in associated pyroxenes and olivines. *Bull. Geol. Soc. Amer.*, Buddington Vol., 1–20.

Battey, M. H. and **McRitchie, W. D.**, 1975. The petrology of the pyroxene-granulite facies rocks of Jotunheimen, Norway. *Norsk. Geol. Tiddskr.*, **55**, 1–49.

Bauer, G. R., **Fodor, R. V.**, **Husler, J. W.** and **Keil, K.**, 1973. Contributions to the mineral chemistry of Hawaiian rocks. III. Composition and mineralogy of a new rhyodacite occurrence on Oahu, Hawaii. *Contr. Min. Petr.*, **40**, 183–194.

Bell, H. D., 1966. Granites and associated rocks of the eastern part of the Western Redhills complex, Isle of Skye. *Trans. Roy. Soc. Edin.*, **66**, 307–343.

Berg, H. J. and **Wheeler, E. P.**, 1976. Osumilite of deep seated origin in the contact-aureole of the anorthositic Nain complex, Labrador. *Amer. Min.*, **61**, 29–37.

Best, M. G., 1963. Petrology of the Guadalupe igneous complex, southwestern Sierra Nevada Foothills. *J. Petr.*, **4**, 223–259.

Best, M. G. and **Mercy, E. L. P.**, 1967. Composition and crystallization of mafic minerals in the Guadalupe igneous complex, California. *Amer. Min.*, **52**, 436–474.

Bhattacharyya, C., 1971. An evaluation of the chemical distinctions between igneous and metamorphic orthopyroxenes. *Amer. Min.*, **56**, 499–506.

Biggar, G. M. and **Clarke, D. B.**, 1972. Protoenstatite solid solution in the system CaO–MgO–Al_2O_3–SiO_2. *Lithos*, **5**, 125–130.

Binns, R. A., 1962. Metamorphic pyroxenes from the Broken Hill district, New South Wales. *Min. Mag.*, **33**, 320–338.

Binns, R. A., 1965. The mineralogy of metamorphosed basic rocks from the Willyama Complex, Broken Hill district, New South Wales. Part II. Pyroxenes, garnets, plagioclases and opaque oxides. *Min. Mag.*, **35**, 561–587.

Binns, R. A., **Duggan, M. B.** and **Wilkinson, J. F. G.**, 1970. High pressure megacrysts in alkaline lavas from north-eastern New South Wales, with chemical analyses by G. I. Z. Kalocsai. *Amer. J. Sci.*, **269**, 132–168.

Binns, R. A., **Long, J. V. P.** and **Reed, S. J. B.**, 1963. Some naturally occurring members of the clinoenstatite–clinoferrosilite mineral series. *Nature*, **198**, 777–778.

Blander, M., 1972. Thermodynamic properties of orthopyroxenes and clinopyroxenes based on the ideal two-site model. *Geochim. Cosmochim. Acta*, **36**, 787–799.

Boland, J. N., 1972. Electron petrography of exsolution in an enstatite-rich orthopyroxene. *Contr. Min. Petr.*, **37**, 229–234.

Boland, J. N., 1974. Lamellar structures in low-calcium orthopyroxenes. *Contr. Min. Petr.*, **47**, 215–222.

Boland, J. N., **Duba, A.** and **Eggleton, A.**, 1974. Kinetic and microstructural studies of magnesium-rich orthopyroxenes. *J. Geol.*, **82**, 507–514.

Bondarenko, L. P. and **Dagelaĭskiĭ, B. V.**, 1961. [Eulysites from the Lake Chudz'yavr region (Kola Peninsula).] *Zap. Vses. Min. Obshch.*, **90**, 408–424 (in Russian).

Bonnichsen, B., 1969. Metamorphic pyroxenes and amphiboles in the Biwabik Iron Formation, Dunka River area, Minnesota. *Min. Soc. America, Spec. Paper* **2**, 217–239.

Borchert, W. and **Kramer, V.**, 1969. Ein synthetisches Magnesium-Kupfer-Silikat mit Klinopyroxene-struktur $(Mg,Cu)_2[Si_2O_6]$. *Neues Jahrb. Min., Mh.*, 6–14.

Borg, I. and **Handin, J.**, 1966. Experimental deformation of crystalline rocks. *Tectonophys.*, **3**, 249–368.

Borg, I. and **Smith, D. K.**, 1969. Calculated X-ray powder patterns. *Mem. Geol. Soc. America*, **122**.

Borley, G. D. and **Suddaby, P.**, 1975. Stressed pyroxenite nodules from the Jagersfontein kimberlite. *Min. Mag.*, **40**, 6–12.

Bowen, N. L., 1914. The ternary system diopside–forsterite–silica. *Amer. J. Sci.*, 4th ser., **38**, 207–264.

Bowen, N. L., 1928. *Evolution of the Igneous Rocks*. Princeton Univ. Press.

Bowen, N. L., 1935. 'Ferrosilite' as a natural mineral. *Amer. J. Sci.*, 5th ser., **30**, 481–494.

Bowen, N. L. and **Andersen, O.**, 1914. The binary system $MgO–SiO_2$. *Amer. J. Sci.*, 4th ser., **37**, 487–500.

Bowen, N. L. and **Schairer, J. F.**, 1935. The system $MgO–FeO–SiO_2$. *Amer. J. Sci.*, 5th ser., **29**, 151–217.

Bowen, N. L. and **Tuttle, O. F.**, 1949. The system $MgO–SiO_2–H_2O$. *Bull. Geol. Soc. America*, **60**, 439–460.

Bown, M. G., 1965. Reinvestigation of clinoferrosilite from Lake Naivasha, Kenya. *Min. Mag.*, **34** (Tilley volume), 66–70.

Bown, M. G. and **Gay, P.**, 1959. The identification of oriented inclusions in pyroxene crystals. *Amer. Min.*, **44**, 592–602.

Bown, M. G. and **Gay, P.**, 1960. An X-ray study of exsolution phenomena in the Skaergaard pyroxenes. *Min. Mag.*, **32**, 379–388.

Boyd, F. R., 1970. Garnet peridotites and the system $CaSiO_3–MgSiO_3–Al_2O_3$. *Min. Soc. America Spec. Paper*, **3**, 63–75.

Boyd, F. R., 1973. The orthopyroxene geothermometer. *Geochim. Cosmochim. Acta*, **37**, 2533–2546.

Boyd, F. R. and **Dawson, J. B.**, 1972. Kimberlite garnets and pyroxene–ilmenite intergrowths. *Carnegie Inst. Washington, Ann. Rept. Dir. Geophys. Lab.*, 1971–72, 373–378.

Boyd, F. R. and **England, J. L.**, 1960. Minerals of the mantle. *Carnegie Inst. Washington, Ann. Rept. Dir. Geophys. Lab.*, 1959–60, 47–52.

Boyd, F. R. and **England, J. L.**, 1963. Some effects of pressure on phase relations in the system $MgO–Al_2O_3·SiO_2$. *Carnegie Inst. Washington, Ann. Rept. Dir. Geophys. Lab.*, 1962–63, 121–124.

Boyd, F. R. and **England, J. L.**, 1964. The system enstatite–pyrope. *Carnegie Inst. Washington, Ann. Rept. Dir. Geophys. Lab.*, 1963–64, 157–161.

Boyd, F. R. and **England, J. L.**, 1965. The rhombic enstatite–clinoenstatite inversion. *Carnegie Inst. Washington, Ann. Rept. Dir. Geophys. Lab.*, 1964–65, 117–120.

Boyd, F. R., **England, J. L.** and **Davis, T. C.**, 1964. Effects of pressure on the melting and polymorphism of enstatite, $MgSiO_3$. *J. Geophys. Res.*, **69**, 2101–2109.

Boyd, F. R. and **Schairer, J. F.**, 1964. The system $MgSiO_3–CaMgSi_2O_6$. *J. Petr.*, **5**, 275–309.

Brown, G. E., **Prewitt, C. T.**, **Papike, J. J.** and **Sueno, S.**, 1972. A comparison of the structures of low and high pigeonite. *J. Geophys. Res.*, **77**, 5778–5789.

Brown, G. M., 1960. The effects of ion substitution on the unit cell dimensions of the common clinopyroxenes. *Amer. Min.*, **45**, 15–38.

Brown, G. M., 1961. Co-existing pyroxenes in igneous assemblages: a re-valuation of the existing data on tie-line orientations. *Geol. Mag.*, **98**, 333–343.

Brown, G. M., 1967. Mineralogy of basaltic rocks. In: *Mineralogy of Basaltic Rocks*. I, *Basalt*. H. H. Hess and A. Poldervaart (eds.). Interscience, New York, 103–162.

Brown, G. M., 1968. Experimental studies on inversion relations in natural pigeonitic pyroxenes. *Carnegie Inst. Washington, Ann. Rept. Dir. Geophys. Lab.*, 1966–67, 347–353.

Brown, G. M., **Vincent, E. A.** and **Brown, P. E.**, 1957. Pyroxenes from the early and middle stages of fractionation of the Skaergaard intrusion, east Greenland. *Min. Mag.*, **31**, 511–543.

Brown, P. E., 1961. Co-existing pyroxenes. *Geol. Mag.*, **98**, 523–535.

Brown, W. L., 1971. Entropie de configuration, solution solide de structure cristalline. *Bull. Soc. franç. Min. Crist.*, **94**, 38–44.

Brown, W. L., **Morimoto, N.** and **Smith, J. V.**, 1961. A structural explanation of the polymorphism and transitions of $MgSiO_3$. *J. Geol.*, **69**, 609–616.

Brown, W. L. and **Smith, J. V.**, 1963. High-temperature X-ray studies on the polymorphism of $MgSiO_3$. *Z. Krist.*, **118**, 186–212.

Bruynzeel, D., 1957. A petrographic study of the Waterfall Gorge profile at Insizwa. *Anal. Univ. Stellenbosch*, Shand Mem. Vol., 484.

Buddington, A. F., 1939. Adirondack igneous rocks and their metamorphism. *Mem. Geol. Soc. America*, **7**.

Buddington, A. F., 1952. Chemical petrology of some metamorphosed Adirondack gabbroic, syenitic and quartz syenitic rocks. *Amer. J. Sci.* (Bowen Vol.), 37–84.

Buerger, M. J., 1948. The role of temperature in mineralogy. *Amer. Min.*, **33**, 101–121.

Bultitude, R. J. and **Green, D. H.**, 1967. Experimental study at high pressures on the origin of olivine nephelinite and olivine melilite nephelinite magmas. *Earth Planet. Sci. Letters*, **3**, 325–337.

Burnham, C. W., 1966. The crystal structures of the ferrosilite ($FeSiO_3$) polymorphs. *Abstr. I.M.A. 5th Gen. Meeting, Cambridge.*

Burnham, C. W., 1967. Ferrosilite. *Carnegie Inst. Washington, Ann. Rept. Dir. Geophys. Lab.*, 1965–66, 285–290.

Burnham, C. W., 1971. The crystal structure of pyroxferroite from Mare Tranquillitatis. *Proc. Second Lunar Sci. Conf. (Suppl. to Geochim. Cosmochim. Acta* **35**), **1**, 47–57.

Burnham, C. W., Ohashir, Y., Hafner, S. S. and **Virgo, D.**, 1971. Cation distribution and atomic thermal vibrations in an iron-rich orthopyroxene. *Amer. Min.*, **56**, 850–876.

Burns, R. G., 1966. Origin of optical pleochroism in orthopyroxenes. *Min. Mag.*, **35**, 715–719.

Burri, C., 1941. Zur optischen Bestimmung der orthorhombischen Pyroxene. *Schweiz. Min. Petr. Mitt.*, **21**, 177–182.

Buseck, P. R. and **Iijima, S.**, 1975. High resolution electron microscopy of enstatite. II: geological application. *Amer. Min.*, **60**, 771–784.

Butler, P., *Jr.*, 1969. Mineral compositions and equilibrium in the metamorphosed iron formation of the Gagnon Region, Quebec, Canada. *J. Petr.*, **10**, 56–101.

Byström, A., 1943. Röntgenuntersuchung des Systems $MgO–Al_2O_3–SiO_2$. *Ber. Dtsch. Keram. Ges.*, **24**, 2–12.

Cameron, E. N., 1970. Compositions of certain coexisting phases in the eastern part of the Bushveld complex. *Geol. Soc. South Africa, Spec. Publ.* **1**, 46–58.

Cameron, E. N., 1975. Postcumulus and subsolidus equilibration of chromite and coexisting silicates in the eastern Bushveld complex. *Geochim. Cosmochim. Acta*, **39**, 1021–1033.

Cameron, K. L. and **Fisher, G. W.**, 1975. Olivine-matrix reactions in thermally metamorphosed Apollo 14 breccias. *Earth Planet. Sci. Letters*, **24**, 197–207.

Cameron, M., Sueno, S., Prewitt, C. T. and **Papike, J. J.**, 1973. High temperature crystal chemistry of acmite, diopside, hedenbergite, jadeite, spodumene and ureyite. *Amer. Min.*, **58**, 594–618.

Campbell, F. E. and **Roeder, P.**, 1968. The stability of olivine and pyroxene in the Ni–Mg–Si–O system. *Amer. Min.*, **53**, 257–268.

Campbell, I. H. and **Borley, G. D.**, 1974. The geochemistry of pyroxenes from the Lower Layered Series of the Jimberlana intrusion, western Australia. *Contr. Min. Petr.*, **47**, 281–298.

Campbell, I. H. and **Nolan, J.**, 1974. Factors affecting the stability field of Ca-poor pyroxene and the origin of the Ca-poor minimum in Ca-rich pyroxenes from tholeiitic intrusions. *Contr. Min. Petr.*, **48**, 205–219.

Canilho, M. H., 1974. Ortopiroxenas das rochas charnoquiticas de Campo Maior (Alto Alentejo). *Bol. Mus. Lab. Min. Geol. Fac. Ciencias Lisboa*, **14**, 41–47 (M.A. 76–1915).

Carmichael, I. S. E., 1963. The occurrence of magnesian pyroxenes and magnetite in porphyritic acid glasses. *Min. Mag.*, **33**, 394–403.

Carmichael, I. S. E., 1967a. The iron–titanium oxides of salic volcanic rocks and their associated ferromagnesian silicates. *Contr. Min. Petr.*, **18**, 36–64.

Carmichael, I. S. E., 1967b. The mineralogy of Thingmuli, a tertiary volcano in eastern Iceland. *Amer. Min.*, **52**, 1815–1841.

Carstens, H., 1958. Note on the distribution of some minor elements in coexisting ortho- and clinopyroxene. *Norsk. Geol. Tiddskr.*, **38**, 257–260.

Carswell, D. A., 1968. Picritic magmas – residual dunite relationships in garnet peridotite at Kalskaret near Tafjord, south Norway. *Contr. Min. Petr.*, **19**, 97–124.

Carswell, D. A., Curtis, C. D. and **Kanaris-Sitoriou, R.**, 1974. Vein metasomatism in peridotite at Kalskaret, near Tafjord, south Norway. *J. Petr.*, **15**, 383–402.

Carter, N. L., 1971. Static deformation of silica and silicates. *J. Geophys. Res.*, **76**, 5514–5540.

Cawthorn, R. G., 1976. Melting relations in part of the system $CaO–MgO–Al_2O_3–SiO_2–Na_2O–H_2O$ under 5 kbar pressure. *J. Petr.*, **17**, 44–72.

Challis, G. S., 1965. The origin of New Zealand ultramafic intrusions. *J. Petr.*, **6**, 322–364.

Champness, P. E. and **Copley, P. A.**, 1976. The transformation of pigeonite to orthopyroxene. In *Electron Microscopy in Mineralogy*, 228–233. H. R. Wenk (Ed.), Berlin, Heidelberg, New York (Springer-Verlag).

Champness, P. E., Dunham, A. C., Gibb, F. G. F., Giles, H. N., MacKenzie, W. S., Stumpfl, E. F. and **Zussman, J.**, 1971. Mineralogy and petrology of some Apollo 12 lunar samples. *Proc. 2nd Lunar Sci. Conf. Geochim. Cosmochim. Acta, Suppl. 2*, **1**, 359–376.

Champness, P. E. and **Lorimer, G. W.**, 1973. Precipitation (exsolution) in an orthopyroxene. *J. Mat. Sci.*, **8**, 467–474.

Champness, P. E. and **Lorimer, G. W.,** 1974. A direct lattice-resolution study of precipitation (exsolution) in orthopyroxene. *Phil. Mag.*, **30**, 357–366.

Champness, P. E. and **Lorimer, G. W.,** 1976. Exsolution in silicates. In *Electron Microscopy in Mineralogy*, 174–204. H. R. Wenk (ed.), Berlin, Heidelberg, New York (Springer-Verlag).

Charlu, T. V., Newton, R. C. and **Kleppa, O. J.,** 1975. Enthalpies of formation at 970 K of compounds in the system $MgO–Al_2O_3–SiO_2$ from high temperature solution calorimetry. *Geochim. Cosmochim. Acta*, **39**, 1487–1497.

Chatterjee, N. D. and **Schreyer, W.,** 1972. The reaction $enstatite_{ss}$ + sillimanite $\rightleftharpoons$ $sapphirine_{ss}$ + quartz in the system $MgO–Al_2O_3–SiO_2$. *Contr. Min. Petr.*, **36**, 49–62.

Chayes, F., 1968. On locating field boundaries in simple phase diagrams by means of discriminant functions. *Amer. Min.*, **53**, 359–371.

Chen, C-H. and **Presnall, D. C.,** 1975. The system $Mg_2SiO_4–SiO_2$ at pressures up to 25 kilobars. *Amer. Min.*, **60**, 398–406.

Chesworth, W., 1967. The biotite–cordierite–almandite subfacies of the hornblende–granulite facies: a discussion. *Can. Min.*, **9**, 263–268.

Chinner, G. A. and **Schairer, J. R.,** 1962. The join $Ca_3Al_2Si_3O_{12}–Mg_3Al_2Si_3O_{12}$ and its bearing on the system $CaO–MgO–Al_2O_3–SiO_2$ at atmospheric pressure. *Amer. J. Sci.*, **260**, 611–634.

Chinner, G. A. and **Sweatman, T. R.,** 1968. A former association of enstatite and kyanite. *Min. Mag.*, **36**, 1052–1060.

Choudhuri, A and **Winkler, H. G. F.,** 1967. Anthophyllit und Hornblende in einigen metamorphen Reaktionen. *Contr. Min. Peti.*, **14**, 293, 315.

Clarke, D. B. and **Biggar, G. M.,** 1972. Calcium-poor pyroxenes in the system $CaO–MgO–Al_2O_3–SiO_2–Fe–O_2$. *Lithos*, **5**, 293–216.

Clarke, P. D. and **Wadsworth, W. J.,** 1970. The Insch layered intrusion. *Scottish J. Geol.*, **6**, 7–25.

Clavan, W., McNabb, W. M. and **Watson, E. H.,** 1954. Some hypersthenes from southeastern Pennsylvania and Delaware. *Amer. Min.*, **39**, 566–580.

Clifford, T. N., Stumpfl, E. F. and **McIver, J. R.,** 1975. A sapphirine–cordierite–bronzite–phlogopite paragenesis from Namaqualand, South Africa. *Min. Mag.*, **40**, 347–356.

Coe, R. S., 1970. The thermodynamic effect of shear stress on the ortho–clino inversion in enstatite, and other coherent phase transitions characterized by finite simple shear. *Contr. Min. Petr.*, **26**, 247–264.

Coe, R. S. and **Kirby, S. H.,** 1975. The orthoenstatite to clinoenstatite transformation by shearing and reversion by annealing: mechanisms and potential applications. *Contr. Min. Petr.*, **52**, 29–55.

Coe, R. S. and **Müller, W. F.,** 1973. Crystallographic orientation of clinoenstatite produced by deformation of orthoenstatite. *Science*, **180**, 64–66.

Collins, L. G. and **Hagner, A. F.,** 1965. Refractive index field diagrams and mineral relationships for coexisting ferromagnesian silicates. *Amer. Min.*, **50**, 1381–1389.

Cox, K. G. and **Jamieson, B. G.,** 1974. The olivine-rich lavas of Nuanetsi: a study of polybaric magmatic evolution. *J. Petr.*, **51**, 269–301.

Cuney, M., Bronner, G. and **Barbey, P.,** 1975. Les paragenèses catazonales des quartzites à magnétite de la province ferrifère du Tiris (Précambrien de la Dorsale Requibat, Mauretanie). *Pétrologie*, **1**, 103–120.

Dallwitz, W. B., Green, D. H. and **Thompson, J. E.,** 1966. Clinoenstatite in a volcanic rock from the Cape Vogel area, Papua. *J. Petr.*, **7**, 375–403.

Darot, M. and **Boudier, R.,** 1975. Mineral lineations in deformed peridotites: kinematic meaning. *Pétrologie*, **1**, 225–236.

Davidson, L. R., 1968. Variation in ferrous iron–magnesium distribution coefficients of metamorphic pyroxenes from Quairading, Western Australia. *Contr. Min. Petr.*, **19**, 239–259.

Davidson, L. R. and **Mathison, C. I.,** 1973. Manganiferous orthopyroxenes and garnets from metamorphosed iron formations of the Quairading district, Western Australia. *Neues Jahrb. Min., Mh.*, 47–57.

Davis, B. T. C. and **Boyd, F. R.,** 1966. The join $Mg_2Si_2O_6–CaMgSi_2O_6$ at thirty kilobars pressure and its application to pyroxenes from kimberlites. *J. Geophys. Res.*, **71**, 3567–3576.

Dawson, J. B., Powell, D. G. and **Reid, A. M.,** 1970. Ultrabasic xenoliths and lava from the Lashaine volcano, northern Tanzania. *J. Petr.*, **11**, 519–548.

Deer, W. A., Howie, R. A. and **Zussman, J.,** 1963. *Rock-Forming Minerals: Vol.* 2 *Chain Silicates* (First Edition).London (Longman).

Desborough, G. A. and **Rose, H. H.,** *Jr.*, 1968. X-ray and chemical analysis of orthopyroxenes from the lower part of the Bushveld complex, South Africa. *Prof. Paper U.S. Geol. Surv.*, **600-B**, 1–5.

Desnoyers, C., 1975. Exsolutions d'amphibole, de grenat et de spinelle dans les pyroxènes des roches ultrabasiques: péridotite et pyroxénolites. *Bull. Soc. franç. Min. Crist.*, **98**, 65–77.

Devaraju, T. C. and **Sadashivaiah, M. S.,** 1966. Pyroxene–quartz–magnetite rocks from Satnur–Halguru area, Mysore State. *J. Geol. Soc. India*, **7**, 70–85.

Devaraju, T. C. and **Sadashivaiah, M. S.**, 1971. Some orthopyroxene-bearing rocks constituting an integral part of high-grade metapelites of Satnur–Halaguru area, Mysore State. *J. Geol. Soc. India*, **12**, 1–13.

DeVore, G. W., 1955. Crystal growth and the distribution of elements. *J. Geol.*, **63**, 471–494.

DeVore, G. W., 1957. The association of strongly polarising cations with weakly polarising cations as a major influence in element distribution, mineral composition and crystal growth. *J. Geol.*, **65**, 178–195.

De Waard, D., 1967. The occurrence of garnet in the granulite-facies terrain of the Adirondack Highlands and elsewhere; an amplification and a reply. *J. Petr.*, **8**, 210–232.

Dobretsov, N. L., 1959. [On mutual relations between the principal ions of the rhombic pyroxenes and their influence on the optical properties of the mineral.] *Mem. All-Union Min. Soc.*, **88**, 672–685 (in Russian).

Dobretsov, N. L., 1968. Paragenetic types and compositions of metamorphic pyroxenes. *J. Petr.*, **9**, 358–377.

Dobrokhotova, E. S., **Romanovich, I. F.** and **Sidorenko, G. A.**, 1967. [Enstatite with a low iron content from the Mul'vodzha deposits.] *Zap. Vses. Min. Obshch.*, **96**, 340–343 (in Russian).

Dodd, R. T., 1963. Garnet–pyroxene gneisses at Bear Mountain, New York. *Amer. Min.*, **48**, 811–820.

Dodd, R. T., **Grover, J. T.** and **Brown, G. E.**, 1975. Pyroxenes in the Shaw (L-7) chondrite. *Geochim. Cosmochim. Acta*, **39**, 1585–1594.

Dostal, J. and **Capedri, S.**, 1975. Partition coefficients of uranium for some rock-forming minerals. *Chem. Geol.*, **15**, 285–294.

Dowty, E. and **Lindsley, D. H.**, 1971. Mössbauer spectroscopy of synthetic Ca–Fe pyroxenes. *Carnegie Inst. Washington, Ann. Rept. Dir. Geophys. Lab.*, 1969–70, 190–193.

Duba, A., **Boland, J. N.** and **Ringwood, A. E.**, 1973. The electrical conductivity of pyroxene. *J. Geol.*, **81**, 727–735.

Duba, A., **Piwinskii, A. J.**, **Heard, H. C.** and **Schock, R. N.**, 1976. The electrical conductivity of forsterite, enstatite and albite. *In: The Physics and Chemistry of Minerals and Rocks.* R. G. J. Strens (ed.). London and New York (Wiley–Interscience), 249–260.

Duchesne, J. C., 1972. Pyroxènes et olivines dans le massif de Bjerkrem–Sogndal (Norvège méridionale). Contribution à l'étude de la séries anorthosite–mangérite. *Rept. Intern. Geol. Congress, 24th Session*, Sect. **2**, 320–328.

Duggan, M. B. and **Wilkinson, J. F. G.**, 1973. Tholeiitic andesite of high pressure origin from the Tweed shield volcano, northeastern New South Wales. *Contr. Min. Petr.*, **39**, 267–276.

Duke, M. B. and **Silver, L. T.**, 1967. Petrology of eucrites, howardites and mesosiderites. *Geochim. Cosmochim. Acta*, **31**, 1637–1665.

Dundon, R. W. and **Hafner, S. S.**, 1971. Cation disorder in shocked orthopyroxene. *Science*, **174**, 581–583.

Dundon, R. W. and **Lindsley, D. H.**, 1968. Mössbauer study of synthetic Ca–Fe clinopyroxenes. *Carnegie Inst. Washington, Ann. Rept. Dir. Geophys. Lab.*, 1966–67, 366–369.

Eckermann, H. von, 1922. The rocks and contact minerals of the Mansjö Mountain. *Geol. För. Förh. Stockholm*, **44**, 205–410.

Eggler, D. H., 1973. Role of CO_2 in melting processes in the mantle. *Carnegie Inst. Washington, Ann. Rept. Dir. Geophys. Lab.*, 1972–73, 457–467.

Eggler, D. H., 1975. CO_2 as a volatile component of the mantle: the system Mg_2SiO_4–SiO_2–H_2O–CO_2. *Phys. Chem. Earth*, **9**, 869–881.

Eggler, D. H., 1976. Composition of the partial melt of carbonated peridotite in the system CaO–MgO–SiO_2–CO_2. *Carnegie Inst. Washington, Ann. Rept. Dir. Geophys. Lab.*, 1975–76, 623–626.

Eggler, D. H. and **McCallum, M. E.**, 1976. A geotherm from megacrysts in the Sloan kimberlite pipes, Colorado. *Carnegie Inst. Washington, Ann. Rept. Dir. Geophys. Lab.*, 1975–76, 538–541.

Ellis, D. J., 1976. High pressure cognate inclusions in the Newer volcanics of Victoria. *Contr. Min. Petr.*, **58**, 149–180.

Emslie, R. F., 1975. Pyroxene megacrysts from anorthositic rocks: new clues to the sources and evolution of the parent magmas. *Can. Min.*, **13**, 138–145.

Engel, A. E. J., **Engel, C. G.** and **Havens, R. G.**, 1964. Mineralogy of amphibolite interlayers in the gneiss complex, north-west Adirondack Mountains, New York. *J. Geol.*, **72**, 131–156.

Eppler, W. F., 1967. Star diopside and star enstatite. *J. Gemmology*, **10**, 185–188.

Eppler, W. F., 1971. Some rare materials. *J. Gemmology*, **12**, 256–262.

Ernst, Th. and **Schorer, G.**, 1969. Die Pyroxene des 'Maintrapps', einer Gruppe tholeiitischer Basalte des Vogelsberges. *Neues Jahrb. Min., Mh*, 108–130.

Ernst, Th. and **Schwab, R.**, 1970. Stability and structural relations of (Mg,Fe) metasilicates. *Phys. Earth Planet. Interiors*, **3**, 451–455.

Eskola, P., 1952. On the granulites of Lapland. *Amer. J. Sci.* (Bowen Vol.), 133–172.

Evans, B. J., Ghose, S. and **Hafner, S.**, 1967. Hyperfine splitting of Fe^{57} and Mg–Fe order–disorder in orthopyroxenes ($MgSiO_3$–$FeSiO_3$ solid solution). *J. Geol.*, **75**, 306–322.

Evans, B. W., 1969. The nickel partition geothermometer applied to the prehistoric Makaopuhi Lava Lake, Hawaii. *Geochim. Cosmochim. Acta*, **33**, 409–411.

Ewart, A., 1967. Pyroxene and magnetite phenocrysts from the Taupo quaternary rhyolitic pumice deposits, New Zealand. *Min. Mag.*, **36**, 180–194.

Ewart, A., Bryan, W. B. and **Gill, J. B.**, 1973. Mineralogy and geochemistry of the younger volcanic islands of Tonga, S.W. Pacific, *J. Petr.*, **14**, 429–465.

Fawcett, J. J. and **Yoder, H. S.**, 1966. Phase relationships of chlorites in the system MgO–Al_2O_3–SiO_2–H_2O. *Amer. Min.*, **51**, 352–380.

Ferguson, J., 1969. Compositional variation in minerals from mafic rocks of the Bushveld complex. *Trans. Geol. Soc. South Africa*, **72**, 61–78.

Flaschen, S. S. and **Osborn, E. F.**, 1957. Studies of the system iron oxide–silica–water at low oxygen partial pressures. *Econ. Geol.*, **52**, 923–943.

Fleet, M. E., 1974a. Partition of Mg and Fe^{2+} in coexisting pyroxemes. *Contr. Min. Petr.*, **44**, 251–257.

Fleet, M. E., 1974b. Partition of major and minor elements and equilibrium in coexisting pyroxenes. *Contr. Min. Petr.*, **44**, 259–274.

Foshag, W. F., 1940. The Shallowater meteorite: a new aubrite. *Amer. Min.*, **25**, 779–786.

Foster, W. R., 1951. High-temperature X-ray diffraction study of the polymorphism of $MgSiO_3$. *J. Amer. Ceram. Soc.*, **34**, 255–259.

Francis, D. M., 1976. Corona-bearing pyroxene granulite xenoliths and the lower crust beneath Nunivak Island, Alaska. *Can. Min.*, **14**, 291–298.

Frantsesson, Ye E. V. and **Lutts, B. G.**, 1970. A find of graphite-bearing pyrope peridotite in the Mir kimberlite pipe. *Dokl. Acad. Sci. U.S.S.R., Earth Sci. Sect.*, **191**, 159–161.

Frey, F. A., 1969. Rare earth abundances in a high temperature peridotite intrusion. *Geochim. Cosmochim. Acta*, **33**, 1429–1447.

Frick, C., 1970. The petrology and chemical composition of a tholeiite sill on Schaapkrall 68JT Lydenburg district, Transvaal, South Africa. *Min. Mag.*, **37**, 909–915.

Friedman, G. M., 1955. Petrology of the Memesagamesing Lake norite mass, Ontario, Canada. *Amer. J. Sci.*, **253**, 590–608.

Friedman, G. M., 1957. Structure and petrology of the Caribou Lake intrusive body, Ontario, Canada. *Bull. Geol. Soc. America*, **68**, 1531–1564.

Frisch, T. and **Bridgwater, D.**, 1976. Iron- and manganese-rich minor intrusions emplaced under late-orogenic conditions in the Proterozoic of south Greenland. *Contr. Min. Petr.*, **57**, 25–48.

Frisillo, A. L. and **Barsch, G. R.**, 1972. Measurement of single-crystal elastic constants of bronzite as a function of pressure and temperature. *J. Geophys. Res.*, **77**, 6360–6384.

Frisillo, A. L. and **Buljan, S. T.**, 1972. Linear thermal expansion coefficients of orthopyroxene to 1000°C. *J. Geophys. Res.*, **77**, 7115–7117.

Froese, E. and **Gordon, T. M.**, 1974. Activity coefficients of coexisting pyroxenes. *Amer. Min.*, **59**, 204–205.

Frost, B. R., 1975. Contact metamorphism of serpentinite, chlorite blackwall and rodingite at Paddy-Go-Easy Pass, central Cascades, Washington. *J. Petr.*, **16**, 272–313.

Fuchs, L. H., 1969. Occurrence of cordierite and aluminous orthoenstatite in the Allende meteorite. *Amer. Min.*, **54**, 1645–1653.

Fujii, T., 1976. Solubility of Al_2O_3 in enstatite coexisting with forsterite and spinel. *Carnegie Inst. Washington, Ann. Rept. Dir. Geophys. Lab.*, 1975–76, 566–571.

Fujii, T. and **Takahashi, E.**, 1976. On the solubility of alumina in orthopyroxene coexisting with olivine and spinel in the system MgO–Al_2O_3–SiO_3. *Min. J. Japan*, **8**, 122–128.

Fyfe, W. S., 1962. On the relative stability of talc, anthophyllite and enstatite. *Amer. J. Sci.*, **260**, 460–466.

Gardner, P. M. and **Robins, B.**, 1974. The olivine–plagioclase reaction: geological evidence from the Seiland petrographic province, northern Norway. *Contr. Min. Petr.*, **44**, 149–156.

Ghose, S., 1962. The nature of Mg^{2+} and Fe^{2+} distribution in some ferromagnesian silicate minerals. *Amer. Min.*, **47**, 388–394.

Ghose, S., 1965. Mg–Fe^{2+} order in an orthopyroxene, $Mg_{0.93}Fe_{1.07}Si_2O_6$. *Z. Krist.*, **122**, 81–99.

Ghose, S. and **Hafner, S.**, 1967. Mg^{2+}–Fe^{2+} distribution in metamorphic and volcanic orthopyroxenes. *Z. Krist.*, **125**, 157–162.

Ghose, S., Okamura, F. P., Wan, C. and **Ohashi, H.**, 1974. Site preference of transition metal ions in pyroxenes and olivines. *Trans. Amer. Geophys. Union*, **55**, 467.

Ghose, S. and **Weidner, J. R.**, 1971. Oriented transformation of grunerite to clinoferrosilite at 775°C and 500 bar argon pressure. *Contr. Min. Petr.*, **30**, 64–71.

Giguère, J. F., 1972. Coexisting pyroxenes in some granulite-facies gneisses from Somerset Island. *Can. Min.*, **11**, 548–551.

Ginsburg, I. V., 1968. An attempt to rationalize the classification of natural pyroxenes of space group *C2/c*. *I.M.A. (Papers and Proc. Fifth Gen. Meeting, Cambridge, 1966)*, 212–221.

Goode, A. D. T. and **Moore, A. C.,** 1975. High pressure crystallization of the Ewarara, Kalka and Gosse Pile intrusions, Giles complex, central Australia. *Contr. Min. Petr.*, **51**, 77–97.

Gossner, B. and **Mussgnug, F.,** 1929. Über Enstatit und sein Verhaltnis zur Pyroxen und Amphibol gruppe. *Z. Krist.*, **70**, 234–248.

Gramse, M., 1970. Quantitative Undersuchungen mit der Elektron-Mikrosonde an Pyroxenen aus Basalten und Peridotite–Einschlüssen. *Contr. Min. Petr.*, **29**, 43–73.

Green, D. H., 1963. Alumina content of enstatite in a Venezuelan high-temperature peridotite. *Bull. Geol. Soc. America*, **74**, 1397–1401.

Green, D. H., 1964. The petrogenesis of the high-temperature peridotite in the Lizard area, Cornwall. *J. Petr.*, **5**, 135–188.

Green, D. H. and **Hibberson, W.,** 1970. The instability of plagioclase in peridotite at high pressure. *Lithos*, **3**, 209–221.

Green, D. H. and **Ringwood, A. E.,** 1967a. The genesis of basaltic magmas. *Contr. Min. Petr.*, **15**, 104–190.

Green, D. H. and **Ringwood, A. E.,** 1967b. An experimental investigation of the gabbro to eclogite transformation. *Geochim. Cosmochim. Acta*, **31**, 767–833.

Green, H. W. and **Radcliffe, S. V.,** 1972. Deformation processes in the upper mantle. In: *Flow and fracture of rocks. Amer. Geophys. Union, Monograph*, **16**, 139–156.

Green, T. H., 1967. An experimental investigation of sub-solidus assemblages formed at high pressure in high alumina basalt, kyanite eclogite and grosspydite compositions. *Contr. Min. Petr.*, **16**, 84–114.

Green, T. H., 1969. The diopside–kyanite join at high pressures and temperatures. *Lithos*, **2**, 334–341.

Green, T. H. and **Ringwood, A. E.,** 1968. Origin of garnet phenocrysts in calc-alkaline rocks. *Contr. Min. Petr.*, **18**, 163–174.

Gribble, C. D., 1967. The basic intrusive rocks of Caledonian age of the Haddo House and Arnage districts, Aberdeenshire. *Scottish J. Geol.*, **3**, 125–136.

Gribble, C. D., 1968. The cordierite bearing rocks of the Haddo House and Arnage districts, Aberdeenshire. *Contr. Min. Petr.*, **17**, 315–330.

Grieve, R. A. F. and **Gittins, J.,** 1975. Composition and formation of coronas in the Hadlington gabbro, Ontario, Canada. *Can. J. Earth Sci.*, **12**, 289–299.

Griffin, W. L., 1973. Lherzolite nodules from the Fen alkaline complex, Norway. *Contr. Min. Petr.*, **38**, 135–146.

Griffin, W. L., Heier, K. S., Taylor, P. N. and **Weigand, P. W.,** 1974. General geology, age and chemistry of the Raftsund mangerite intrusion, Lofoten-Vesterålen. *Norges Geol. Unders.*, **312**, 1–30.

Grover, J. E., Lindsley, D. H. and **Turnock, A. C.,** 1972. Ca–Fe–Mg pyroxenes: subsolidus phase relations in iron-rich portions of the pyroxene quadrilateral. *Geol. Soc. America, Abstracts with Programs*, 521–522.

Grover, J. E., 1972. The stability of low-clinoenstatite in the system $Mg_2Si_2O_6$–$CaMgSi_2O_6$. *Trans. Amer. Geophys. Union*, **53**, 539 (abstract).

Grover, J. E., 1974. On calculating activity coefficients and other excess functions from the intracrystalline exchange properties of a double-site phase. *Geochim. Cosmochim. Acta*, **38**, 1527–1548.

Grover, J. E. and **Orville, P. M.,** 1969. The partitioning of cations between coexisting single- and multi-site phases with application to the assemblages: orthopyroxene–clinopyroxene and orthopyroxene–olivine. *Geochim. Cosmochim. Acta*, **33**, 205–226.

Groves, A. W., 1935. The charnockite series of Uganda, British East Africa. *Quart. J. Geol. Soc.*, **91**, 150–207.

Gruenewaldt, G. von, 1970. On the phase-change orthopyroxene–pigeonite and the resulting textures in the Main and Upper Zones of the Bushveld complex in the eastern Transvaal. *Geol. Soc. South Africa, Spec. Publ.* **1**, 67–73.

Gunn, B. M., 1963. Layered intrusions in the Ferrar dolerites, Antarctica. *Min. Soc. Amer., Spec. Paper* **I**, 124–133.

Hafner, S. S. and **Virgo, D.,** 1969. Cooling history of orthopyroxenes. *Science*, **165**, 285–287.

Haggerty, S. E. and **Baker, I.,** 1967. The alteration of olivine in basaltic and associated lavas. Part I: High temperature alteration. *Contr. Min. Petr.*, **16**, 233–257.

Häkli, A., 1963. Distribution of nickel between the silicate and sulphide phases in some basic intrusions in Finland. *Bull. Comm. géol. Finlande*, no. 209, 1–54.

Häkli, A., 1968. An attempt to apply the Makaopuhi nickel fractionation data to the temperature determination of a basic intrusive. *Geochim. Cosmochim. Acta*, **32**, 449–460.

Hamad, S. el D., 1963. The chemistry and mineralogy of the olivine nodules of Calton Hill, Derbyshire. *Min. Mag.*, **33**, 483–497.

Hargraves, R. B., 1962. Petrology of the Allard Lake anorthosite suite. *Geol. Soc. America* (Buddington vol.), 163–190.

Hasleton, J. D. and **Nash, W. P.**, 1975. Ilmenite–orthopyroxene intergrowths from the moon and the Skaergaard intrusion. *Earth Planet. Sci. Letters*, **26**, 287–291.

Henriques, A., 1958. The influence of cations on the optical properties of orthopyroxenes. *Arkiv. Min. Geol.*, **2**, 385–340.

Henry, N. F. M., 1935. Some data on the iron-rich hypersthenes. *Min. Mag.*, **24**, 221–226.

Henry, N. F. M., 1938. A review of the data of Mg–Fe-clinopyroxenes. *Min. Mag.*, **25**, 23–29.

Henry, N. F. M., 1942. Lamellar structure in orthopyroxenes. *Min. Mag.*, **26**, 179–189.

Hensen, B. J., 1972. Phase relations involving pyrope, enstatite$_{ss}$ and sapphirine$_{ss}$ in the system MgO–Al_2O_3–SiO_2. *Carnegie Inst. Washington, Ann. Rept. Dir. Geophys. Lab.*, 1971–72, 421–426.

Hensen, B. J., 1973. Pyroxenes and garnets as geothermometers and barometers. *Carnegie Inst. Washington, Ann. Rept. Dir. Geophys. Lab.*, 1972–73, 527–534.

Hensen, B. J. and **Essene, E. J.**, 1971. The stability of pyrope–quartz in the system MgO–Al_2O_3–SiO_2. *Contr. Min. Petr.*, **30**, 72–83.

Hess, H. H., 1941. Pyroxenes of common mafic magmas, part 2. *Amer. Min.*, **26**, 573–594.

Hess, H. H., 1952. Orthopyroxenes of the Bushveld type, ion substitution and changes in unit cell dimensions. *Amer. J. Sci.* (Bowen Vol.), 173–188.

Hess, H. H., 1960. Stillwater igneous complex, Montana. *Mem. Geol. Soc. America*, **80**.

Hess, H. H. and **Phillips, A. H.**, 1938. Orthopyroxenes of the Bushveld type. *Amer. Min.*, **23**, 450–456.

Hess, H. H. and **Phillips, A. H.**, 1940. Optical properties and chemical composition of magnesian orthopyroxenes. *Amer. Min.*, **25**, 271–285.

Hewins, R. H., 1975. Pyroxene geothermometry of some granulite facies rocks. *Contr. Min. Petr.*, **50**, 205–209.

Hietanen, A., 1971. Distribution of elements in biotite–hornblende pairs and in an orthopyroxene–clinopyroxene pair from zoned plutons, northern Sierra Nevada, California. *Contr. Min. Petr.*, **30**, 161–176.

Himmelberg, G. R. and **Jackson, E. D.**, 1967. X-ray determination curve for some orthopyroxenes of composition Mg_{48-85} from the Stillwater complex, Montana. *Prof. Paper U.S. Geol. Surv.*, **575B**, 101–102.

Himmelberg, G. R. and **Phinney, W. D.**, 1967. Granulite-facies metamorphism, Granite Falls-Montevideo area, Minnesota. *J. Petr.*, **8**, 325–348.

Hiss, W. L. and **Hunter, H. E.**, 1966. Primary orthopyroxene–spinel intergrowths in Cambrian cumulates, Wichita Mountains, Oklahoma. *Oklahoma Geol. Notes*, **26**, 231–235.

Horai, K-I., 1971. The thermal conductivity of rock-forming minerals. *J. Geophys. Res.*, **76**, 1278–1308.

Hori, F., 1956. Effects of constituent cations on the refractive indices of orthopyroxenes. *Min. J. Japan*, **1**, 359–371.

Hostetler, P. B., **Colman, R. G.**, **Mumpton, F. A.** and **Evans, B. W.**, 1966. Brucite in Alpine serpentinites. *Amer. Min.*, **51**, 75–98.

Howells, S. and **O'Hara, M. J.**, 1975. Palaeogeotherms and the diopside–enstatite solvus. *Nature*, **254**, 406–408.

Howie, R. A., 1955. The geochemistry of the charnockite series of Madras, India. *Trans. Roy. Soc. Edin.*, **62**, 725–768.

Howie, R. A., 1958. African charnockites and related rocks. *Bull. Serv. Géol. Congo Belge*, **8**, 1–16.

Howie, R. A., 1963. Cell parameters of orthopyroxenes. *Min. Soc. America, Spec. Paper* **1**, 213–222.

Howie, R. A., 1964. Some orthopyroxenes from Scottish metamorphic rocks. *Min. Mag.*, **33**, 903–911.

Howie, R. A., 1965. The pyroxenes of metamorphic rocks. In: *Controls of Metamorphism*, 319–326. London (Oliver and Boyd).

Howie, R. A. and **Smith, J. V.**, 1966. X-ray emission microanalysis of rock-forming minerals. V. Orthopyroxenes. *J. Geol.*, **74**, 443–462.

Howie, R. A. and **Subramanian, A. P.**, 1957. The petrogenesis of garnet in charnockite, enderbite and related granulites. *Min. Mag.*, **31**, 565–586.

Iijima, S. and **Buseck, P. R.**, 1975. High resolution electron microscopy of enstatite I: twinning, polymorphism and polytypism. *Amer. Min.*, **60**, 758–770.

Immega, I. P. and **Klein, C.**, 1976. Mineralogy and petrology of some metamorphic Precambrian iron-formations in southwestern Montana. *Amer. Min.*, **61**, 1117–1144.

Irving, A. J., 1974a. Pyroxene-rich ultramafic xenoliths in the newer basalts of Victoria, Australia. *Neues Jahrb. Min., Abhdl.*, **120**, 147–167.

Irving, A. J., 1974b. Geochemical and high pressure experimental studies of garnet pyroxenite and pyroxene granulite xenoliths from the Delegate basaltic pipes, Australia. *J. Petr.*, **15**, 1–40.

Ito, T., 1935. On the symmetry of the rhombic pyroxenes. *Z. Krist*, **90**, 151–162.

Ito, T., 1950. *X-ray Studies on Polymorphism*. Maruzen Co., Tokyo.

Ito, T. and **Kennedy, G. C.**, 1967. Melting and phase relations in a natural peridotite to 40 kilobars. *Amer. J. Sci.*, **265**, 519–538.

Ito, T. and **Kennedy, G. C.**, 1968. Melting and phase relations in the plane tholeiite-lherzolite-nepheline basanite to 40 kilobars with geological implications. *Contr. Min. Petr.*, **19**, 177–211.

Jaffe, H. W., **Robinson, P.**, **Tracy, R. J.** and **Ross, M.**, 1975. Orientation of pigeonite exsolution lamellae in metamorphic augite: correlation with composition and calculated optimal phase boundaries. *Amer. Min.*, **60**, 9–28.

Jakeš, P., 1968. Ferromagnesian minerals from the rocks of Tábor massif, Czechoslovakia. *Neues Jahrb. Min., Mh.*, 193–208.

Johannes, W., 1969. An experimental investigation of the system $MgO–SiO_2–H_2O–CO_2$. *Amer. J. Sci.*, **267**, 1083–1104.

Kano, H., 1971. [On the magma eruption of Akita Komgatake volcano in 1970.] *J. Geol. Soc. Japan*, **77**, 45–51 (in Japanese with English tables) (M.A. 72–1532).

Katz, M., 1970. Notes on the mineralogy and coexisting pyroxenes from the granulites of Mont Tremblant Park, Quebec. Can. Min., **10**, 247–251.

Kendel, F., 1970. CO_2-Metasomatose in Peridotiten des Aaheimer Gebietes, (W-Norwegen). *Fortschr. Min.*, **48**, 13–14.

Kesson, S. F., 1973. The primary geochemistry of the Monaro alkaline volcanics, southeastern Australia – evidence for upper mantle heterogeneity. *Contr. Min. Petr.*, **42**, 92–108.

Kirby, S. H., 1976. The role of crystal defects in shear-induced transformation of orthoenstatite to clinoenstatite. In: *Electron Microscopy in Mineralogy*, 465–472. H. R. Wenk (ed.). Berlin, Heidelberg, New York (Springer-Verlag).

Kirby, S. H. and **Coe, S.**, 1974. The role of crystal defects in the enstatite inversion. *Trans. Amer. Geophys. Union*, **55**, 419.

Kitahara, J., 1958. Chromian enstatite from Hirose mine, Tari district, Tottori Prefecture. *J. Min. Soc. Japan*, **3**, 539–542 (in Japanese, English summary).

Kitahara, S. and **Kennedy, G. C.**, 1967. The calculated equilibrium curves for some reactions in the system $MgO–SiO_2–H_2O$ at pressures up to 30 kilobars. *Amer. J. Sci.*, **265**, 211–217.

Kitahara, S., **Takeneuchi, S.** and **Kennedy, G. C.**, 1966. Phase relations in the system $MgO–SiO_2–H_2O$ at high temperatures and pressures. *Amer. J. Sci.*, **264**, 223–233.

Kitayama, K., 1970. Activity measurements in orthosilicate and metasilicate solid solutions: II $MgSiO_3–FeSiO_3$ at 1154°, 1204° and 1250°C. *Bull. Chem. Soc. Japan*, **43**, 1390–1393.

Kitayama, K. and **Katsura, T.**, 1968. Activity measurements in orthosilicate and metasilicate solid solutions. I $Mg_2SiO_4–Fe_2SiO_4$ and $MgSiO_3–FeSiO_3$ at 1204°C. *Bull. Chem. Soc. Japan*, **41**, 1146–1151.

Klein, C., 1966. Mineralogy and petrology of the metamorphosed Wabush iron formation, Southwestern Labrador. *J. Petr.*, **7**, 246–305.

Knutson, J. and **Green, T. H.**, 1975. Experimental duplication of a high-pressure megacryst/cumulate assemblage in a near-saturated hawaiite. *Contr. Min. Petr.*, **52**, 121–132.

Kobayashi, Y. and **Maruyama, H.**, 1974. Electrical conductivity of orthopyroxene and garnet single crystals at high temperature. *Sci. Rept. Kanazawa Univ.*, **19**, 79–85 (M.A. 76–3053).

Kohlstedt, D. L. and **van der Sande, D. L.**, 1976. On the detailed structure of ledges in an augite-enstatite interface. In: *Electron Microscopy in Mineralogy*, 234–237. H. R. Wenk (ed.). Berlin, Heidelberg, New York (Springer-Verlag).

Koltermann, M., 1965. [Conditions for the formation of $MgSiO_3$ modifications during reactions between MgO and SiO_2.] *Ber. dtsch. keram. Ges.*, **42**, 6–10.

Kononova, M. M., 1967a. [The composition of lamellar intergrowths in orthorhombic pyroxenes from charnockite rocks.] *Zap. Vses. Min. Obshch.*, **96**, 104–107 (in Russian).

Kononova, M. M., 1967b. [Pyroxenes from the charnockites of the Bug and Dniester areas.] *Akad. Nauk Ukr. RSR. Dopov.*, ser. B, No. 7, 609–611 (Russian with English summary).

Kosoi, A., **Malkova, L. A.** and **Frank-Kamenetskii, V. A.**, 1974. Crystal-chemical characteristics of rhombic pyroxenes. *Soviet Physics – Crystallography*, **19**, 171–174.

Kosoy, A. L. and **Shemyakin, V. M.**, 1971. Determination of the chemical composition of orthopyroxenes. *Dokl. Acad. Sci., U.S.S.R., Earth Sci. Sect.*, **201**, 182–185.

Kranck, S. H., 1961. A study of phase equilibria in a metamorphic iron formation. *J. Petr.*, **2**, 137–184.

Kretz, R., 1961a. Co-existing pyroxenes. *Geol. Mag.*, **98**, 344–345.

Kretz, R., 1961b. Some applications of thermodynamics to co-existing minerals of variable compositions. Examples: Orthopyroxenes–clinopyroxenes and orthopyroxene–garnet. *J. Geol.*, **69**, 361–387.

Kretz, R., 1963. Distribution of magnesium and iron between orthopyroxene and calcic pyroxene in natural mineral assemblages. *J. Geol.*, **71**, 773–785.

Kronert, W. and **Schwiete, H. E.,** 1966. [Research on the systems MgO–SiO_2 and MgO–Al_2O_3–SiO_2.] *Keram. Z.*, **18**, 323–324 (in German). *Brit. Ceram. Abstr.*, abstr. 3236.

Krstanovic, I., 1975. X-ray diffraction studies of orthoenstatite. *Acta Cryst.* **A 31**, *Suppl.* **3**, 73 (Abstr. – *10th Intern. Congress Cryst.*).

Kumazawa, M., 1969. The elastic constants of single-crystal orthopyroxene. *J. Geophys. Res.*, **74**, 5973–5980.

Kuno, H., 1938. Hypersthene from Odawara-mati, Japan. *Proc. Imp. Acad., Tokyo*, **14**, 218–220.

Kuno, H., 1941. Dispersion of optic axes in the orthorhombic pyroxene series. *Proc. Imp. Acad., Tokyo*, **17**, 204–209.

Kuno, H., 1947a. Hypersthene in a rock of amphibolite facies from Tanzawa Mountainland, Kanagawa Prefecture. Japan. *Proc. Jap. Acad.*, **23**, 114–116.

Kuno, H., 1947b. Two orthopyroxenes from the so-called bronzite–andesite of Japan. *Proc. Jap. Acad.*, **23**, 117–120.

Kuno, H., 1950. Petrology of Hakone volcano and the adjacent areas, Japan. *Bull. Geol. Soc. Amer.*, **61**, 957–1020.

Kuno, H., 1954. Study of orthopyroxenes from volcanic rocks. *Amer. Min.*, **39**, 30–46.

Kuno, H., 1959. Origin of Cenozoic petrographic provinces of Japan and surrounding areas. *Bull. Volcan.*, ser. II, **20**, 37–76.

Kuno, H., 1966. Review of pyroxene relations in terrestrial rocks in the light of recent experimental works. *Min. J. Japan*, **5**, 21–43.

Kuno, H., 1969. Mafic and ultramafic nodules in basaltic rocks of Hawaii. *Mem. Geol. Soc. Amer.*, **115** (Poldervaart vol.), 189–234.

Kuno, H. and **Aoki, K.,** 1970. Chemistry of ultramafic nodules and their bearing on the origin of basaltic magmas. *Phys. Earth Planet. Interiors*, **3**, 273–301.

Kuno, H. and **Hess, H. H.,** 1953. Unit cell dimensions of clinoenstatite and pigeonite in relation to other common clinopyroxenes. *Amer. J. Sci.*, **251**, 741–752.

Kuno, H. and **Nagashima, K.,** 1952. Chemical compositions of hypersthene and pigeonite in equilibrium in magma. *Amer. Min.*, **37**, 1000–1006.

Kushiro, I., 1964. The system diopside–forsterite–enstatite at 20 kilobars. *Carnegie Inst. Washington, Ann. Rept. Dir. Geophys. Lab.*, 1963–64, 101–108.

Kushiro, I., 1965a. Coexistence of nepheline and enstatite at high pressures. *Carnegie Inst. Washington, Ann. Rept. Dir. Geophys. Lab.*, 1964–65, 109–112.

Kushiro, I., 1965b. The liquidus relations in the systems forsterite–$CaAl_2SiO_6$–silica and forsterite–nepheline–silica at high pressure. *Carnegie Inst. Washington, Ann. Rept. Dir. Geophys. Lab.*, 1964–65, 103–109.

Kushiro, I., 1969. The system diopside–silica with and without water at high pressures. *Amer. J. Sci.*, **267A**, 269–294.

Kushiro, I., 1972a. Effect of water on the composition of magmas formed at high pressures. *J. Petr.*, **13**, 311–334.

Kushiro, I., 1972b. Partial melting of synthetic and natural peridotites at high pressure. *Carnegie Inst. Washington, Ann. Rept. Dir. Geophys. Lab.*, 1971–72, 357–362.

Kushiro, I., 1972c. Determination of the liquidus relations in the synthetic silicate systems with electron probe analysis: the system forsterite–diopside–silica at 1 atmosphere. *Amer. Min.*, **57**, 1260–1271.

Kushiro, I., 1972d. New method of determining liquidus boundaries with confirmation of incongruent melting of diopside and existence of iron-free pigeonite at 1 atm. *Carnegie Inst. Washington, Ann. Rept. Dir. Geophys. Lab.*, 1971–72, 603–607.

Kushiro, I., 1973. Regularities in the shift of liquidus boundaries in silicate systems and their significance in magma genesis. *Carnegie Inst. Washington, Ann. Rept. Dir. Geophys. Lab.*, 1972–73, 497–502.

Kushiro, I. and **Yoder, H. S.,** 1964. Experimental studies on the basalt–eclogite transformation. *Carnegie Inst. Washington, Ann. Rept. Dir. Geophys. Lab.*, 1963–64, 108–114.

Kushiro, I. and **Yoder, H. S.,** 1965. The reactions between forsterite and anorthite at high pressures. *Carnegie Inst. Washington, Ann. Rept. Dir. Geophys. Lab.*, 1964–65, 89–94.

Kushiro, I. and **Yoder, H. S.,** 1966. Anorthite–forsterite and anorthite–enstatite reactions and their bearing on the basalt eclogite transformation. *J. Petr.*, **7**, 337–362.

Kushiro, I. and **Yoder, H. S.,** 1969. Melting of forsterite and enstatite at high pressures under hydrous conditions. *Carnegie Inst. Washington, Ann. Rept. Dir. Geophys. Lab.*, 1967–68, 153–161.

Kushiro, I., Yoder, H. S. and **Nishikawa, M.,** 1968. Effect of water on the melting of enstatite. *Bull. Geol. Soc. America*, **79**, 1685–1692.

Lappin, M. A., 1974. Eclogites from the Sunndal–Grubse ultramafic mass, Almklovdalen, Norway and the T–P history of the Almklovdalen masses. *J. Petr.*, **15**, 567–601.

Larsen, E. S. and **Draisin, W. M.,** 1950. Composition of the minerals in the rocks of the southern California Batholith. *Rept. Intern. Geol. Congr. (18th Session)*, Pt. II, 66–79.

Larsen, E. S., Irving, J., Gonyer, F. A. and **Larsen, E. S.**, 3rd., 1936. Petrologic results of a study of the minerals from the tertiary volcanic rocks of the San Juan region, Colorado. *Amer. Min.*, **21**, 694–700.

Leake, B. E., 1968. Optical properties and composition in the orthopyroxene series. *Min. Mag.*, **36**, 745–747.

Leelanandam, C., 1967. Chemical study of pyroxenes from the charnockitic rocks of Kondapalli (Andhra Pradesh) India, with emphasis on the distribution of elements in coexisting pyroxenes. *Min. Mag.*, **36**, 153–179.

Leelanandam, C., 1968. Paired pyroxenes from Kondapalli. *J. Indian Geosci. Assoc.*, **8**, 89–92.

LeFèvre, C., 1969. Remarques sur la valeue du paramètre b de la maile des clinopyroxènes. *Bull. Soc. franç. Min. Crist.*, **92**, 95–98.

Le Maitre, R. W., 1965. The significance of the gabbroic xenoliths from Gough Island, South Atlantic. *Min. Mag.*, **34** (Tilley vol.), 303–317.

Leonidov, V. Ya. and **Khitarov, N. I.**, 1967. New data on the thermodynamic properties of talc. *Geochem. Intern.*, **4**, 944–949.

Lensch, G., 1971. Geochemistry of some rocks and minerals from the Great Dyke (Southern Rhodesia). *Neues Jahrb. Min., Mh.*, 366–370.

Lewis, J. F., 1960. The occurrence of orthopyroxene with low optic axial angle. *Amer. Min.*, **45**, 1125–1129.

Leyreloup, A., Lasnier, B. and **Marchand, J.**, 1975. Retrograde corona-forming reactions in high pressure granulite facies rocks. *Pétrologie*, **1**, 43–55.

Lindemann, W., 1961. Beitrag zur Enstatitstruktur (Verfeinerung der Parameterwerte). *Neues Jahrb. Min., Mh.*, **10**, 226–233.

Lindemann, W. and **Wögerbauer, R.**, 1974. Gitterkonstanten und Raumgruppe für Protoenstatit ($MgSiO_3$). *Naturwiss*, **61**, 500.

Lindh, A., 1974. Manganese distribution between coexisting pyroxenes. *Neues Jahrb. Min., Mh*, 335–345.

Lindsley, D. H., 1965. Ferrosilite. *Carnegie Inst. Washington, Ann. Rept. Dir. Geophys. Lab.*, 1964–65, 148–150.

Lindsley, D. H. and **Dixon, S. A.**, 1976. Diopside–enstatite equilibria at 850° to 1400°C, 5 to 35 kb. *Amer. J. Sci.*, **276**, 1285–1301.

Lindsley, D. H., King, H. E. and **Turnock, A. C.**, 1974. Compositions of synthetic augite and hypersthene coexisting at 810°C: application to pyroxenes from lunar highlands rocks. *Geophys. Res. Letters*, **1**, 134–136.

Lindsley, D. H., MacGregor, I. C. and **Davis, B. T. C.**, 1964. Synthesis and stability of ferrosilite. *Carnegie Inst. Washington, Ann. Rept. Dir. Geophys. Lab.*, 1963–64, 174–176.

Lindsley, D. H. and **Munoz, J. L.**, 1968. Subsolidus relationships in part of the hedenbergite–ferrosilite join at low pressures. *Carnegie Inst. Washington, Ann. Rept. Dir. Geophys. Lab.*, 1966–67, 363–366.

Lindsley, D. H. and **Munoz, J. L.**, 1969a. Ortho–clino inversion in ferrosilite. *Carnegie Inst. Washington, Ann. Rept. Dir. Geophys. Lab.*, 1967–68, 86–88.

Lindsley, D. H. and **Munoz, J. L.**, 1969b. Subsolidus relations along the join hedenbergite–ferrosilite. *Amer. J. Sci.*, **267-A** (Schairer vol.), 295–324.

Lindsley, D. H., Speidel, D. H. and **Nafziger, R. H.**, 1968. P–T–fo_2 relations in the system Fe–O–SiO_2. *Amer. J. Sci.*, **266**, 342–360.

Liu, L-G., 1976. The high pressure phases of $FeSiO_3$ with implications for Fe_2SiO_4 and FeO. *Earth Planet. Sci. Letters*, **33**, 101–106.

Livingstone, A., 1965. An olivine-bearing sagvandite from Berneray, Outer Hebrides. *Geol. Mag.*, **102**, 227–230.

Lokka, L., 1943. Beitrage zur Kenntnis des Chemismus der finnischen Minerale. *Bull. Comm. géol. Finlande*, no. 129, 1–68.

Loomis, A. A., 1966. Contact metamorphic reactions and processes in the Mt. Tallac roof remnant, Sierra Nevada, California. *J. Petr.*, **7**, 221–245.

Lorimer, G. W. and **Champness, P. E.**, 1973. Combined electron microscopy and analysis of an orthopyroxene. *Amer. Min.*, **58**, 243–248.

Luce, R. W., Bartlett, R. W. and **Parks, G. A.**, 1972. Dissolution kinetics of magnesium silicates. *Geochim. Cosmochim. Acta*, **36**, 35–40.

Luth, W. C., 1967. Studies in the system $KAlSiO_4$–Mg_2SiO_4–SiO_2–H_2O: I Inferred phase relations and petrologic applications. *J. Petr.*, **8**, 372–416.

Lutts, B. G., 1965. Eclogitization reactions in rocks formed at great depths. *Geochim. Intern.*, **2**, 1093–1104.

Lutts, B. G. and **Kopaneva, L. H.**, 1968. A pyrope–sapphirine rock from the Anabar massif and its conditions of metamorphism. *Dokl. Acad. Sci., U.S.S.R., Earth Sci. Sect.*, **179**, 161–163.

MacGregor, I. D., 1965. Stability fields of spinel and garnet peridotites in the synthetic systems MgO–CaO–Al_2O_3–SiO_2. *Carnegie Inst. Washington, Ann. Rept. Dir. Geophys. Lab.*, 1964–65, 126–134.

MacGregor, I. D., 1974. The system MgO–Al_2O_3–SiO_2: solubility of Al_2O_3 in enstatite for spinel and garnet peridotite compositions. *Amer. Min.*, **59**, 110–119.

MacGregor, I. D. and **Ringwood, A. E.**, 1964. The natural system enstatite–pyrope. *Carnegie Inst. Washington, Ann. Rept. Dir. Geophys. Lab.*, 1963–64, 161–163.

Mall, A. P., 1973. Distribution of elements in coexisting ferromagnesian minerals from ultrabasics of Kondapalle and Gangineni, Andhra Pradesh, India. *Neues Jahrb. Min., Mh.*, 323–336.

Marakushev, A. A. and **Kudryavtsev, V. A.**, 1965. Hypersthene–sillimanite paragenesis and its petrological implication. *Dokl. Acad. Sci., U.S.S.R., Earth Sci. Sect.*, **164**, 145–148.

Maske, S., 1957. The diorites of Yzerfontein, Darling, Cape Province. *Annal. Univ. Stellenbosch*, Shand Mem. vol., 23–53.

Mason, B., 1966. The enstatite chondrites. *Geochim. Cosmochim. Acta*, **30**, 23–40.

Mason, B., 1967. The Woodbine meteorite, with notes on silicates in iron meteorites. *Min. Mag.*, **36**, 120–126.

Mason, B., 1968. Pyroxenes in meteorites. *Lithos*, **1**, 1–11.

Mason, R., 1967. Electron-probe microanalysis of coronas in troctolite from Sulitjelma, Norway. *Min. Mag.*, **36**, 504–514.

Mason, R., 1971. The chemistry and structure of the Sulitjelma gabbro. *Norges Geol. Unders.*, **29**, 108–142.

Mathews, W. H., 1957. Petrology of Quaternary volcanics of the Mount Garilbaldi map-area, southwestern British Columbia. *Amer. J. Sci.*, **255**, 400–415.

Matsui, Y., **Banno, S.** and **Harnes, I.**, 1966. Distribution of some elements among minerals of Norwegian eclogites. *Norsk. Geol. Tidsskr.*, **46**, 364–368.

Matsui, Y., **Syono, Y.**, **Akimoto, S.** and **Kitayama, K.**, 1968. Unit cell dimensions of some synthetic orthopyroxene group solid solutions. *Geochem. J., Japan*, **2**, 61–70.

Matsumoto, M., 1974. Possible structure types derived from *Pbca*-orthopyroxenes. *Min. J. Japan*, **7**, 374–383.

Maxey, L. R. and **Vogel, T. A.**, 1974. Compositional dependence of the coexisting pyroxene iron–magnesium distribution coefficient. *Contr. Min. Petr.*, **43**, 295–306.

McCall, G. J. H., 1966. The petrology of the Mount Padbury mesosiderite and its achondrite enclaves. *Min. Mag.*, **35**, 1029–1060.

McDonald, J. A., 1967. Evolution of part of the lower Critical Zone Farm Ruighoek, western Bushveld. *J. Petr.*, **8**, 165–209.

McDougall, I., 1962. Differentiation of the Tasmanian dolerites Red Hill dolerite–granophyre association. *Bull. Geol. Soc. America*, **73**, 279–315.

McLaren, A. C. and **Etheridge, M. A.**, 1976. A transmission electron microscope study of naturally deformed orthopyroxene. I. Slip mechanisms. *Contr. Min. Petr.*, **57**, 163–177.

Medaris, L. C., *Jr.*, 1969. Partitioning of Fe^{++} and Mg^{++} between coexisting synthetic olivine and orthopyroxene. *Amer. J. Sci.*, **267**, 945–968.

Medaris, L. C., *Jr.*, 1975. Coexisting spinel and silicates in alpine peridotites of the granulite facies. *Geochim. Cosmochim. Acta*, **39**, 947–958.

Menzies, M., 1973. Mineralogy and partial melt textures within an ultramafic–mafic body, Greece. *Contr. Min. Petr.*, **42**, 273–285.

Mercier, J-C., 1976. Single pyroxene geothermometry and geobarometry. *Amer. Min.*, **61**, 603–615.

Mercier, J-C. and **Carter, N. L.**, 1975. Pyroxene geotherms. *J. Geophys. Res.*, **80**, 3349–3362.

Mercier, J-C. and **Nicolas, A.**, 1975. Textures and fabrics of upper-mantle peridotites as illustrated by xenoliths from basalts. *J. Petr.*, **16**, 454–487.

Mercy, E. and **O'Hara, M. J.**, 1965. Olivines and orthopyroxenes from garnetiferous peridotites and related rocks. *Norsk. Geol. Tiddskr.*, **45**, 457–461.

Mercy, E. and **O'Hara, M. J.**, 1967. Distribution of Mn, Cr, Ti and Ni in coexisting minerals of ultramafic rocks. *Geochim. Cosmochim. Acta*, **31**, 2331–2341.

Meyer, H. O. R. and **Brookins, D. G.**, 1976. Sapphirine, sillimanite and garnet in granulite xenoliths from Stockdale kimberlite, Kansas. *Amer. Min.*, **61**, 1194–1202.

Mikhaylov, N. P. and **Rovsha, V. S.**, 1965. Influence of pressure on the paragenesis of ultrabasic rocks. *Dokl. Acad. Sci. U.S.S.R., Earth Sci. Sect.*, **160**, 151–153.

Miller, C., 1974. Reaction rims between olivine and plagioclase in metaperidotites, Ötztal Alps, Austria. *Contr. Min. Petr.*, **43**, 333–342.

Miller, R., 1953. The Webster–Addie ultramafic ring, Jackson county, North Carolina, and secondary alteration of its chromite. *Amer. Min.*, **38**, 1134–1147.

Mitra, S., 1976. Mössbauer study of orthopyroxenes from Sukinda, Orissa, India. *Neues Jahrb. Min., Mh.*, 169–173.

Miyamoto, M., **Takeda, H.** and **Takano, Y.**, 1975. Crystallographic studies of a bronzite in the Johnstown achondrite. *Fortschr. Min.*, **52**, 389–397.

Miyashiro, A., 1962. The Kesen, Japan, chondrite. *Japan. J. Geol. Geogr.*, **30**, 73–77.

Modreski, P. J., 1972. The melting of phlogopite in the presence of enstatite, aluminous enstatite, diopside, spinel, corundum, and pyrope. *Carnegie Inst. Washington, Ann. Rept. Dir. Geophys. Lab.*, 1971–72, 392–396.

Monchoux, P. and **Besson, M.**, 1969. Sur les compositions chimiques des minéraux des lherzolites pyrénéennes et leur signification génétique. *Bull. Soc. franç. Min. Crist.*, **92**, 289–298.

Moore, A. C., 1968. Rutile exsolution in orthopyroxene. *Contr. Min. Petr.*, **17**, 233–236.

Moore, A. C., 1971. The mineralogy of the Gosse Pile ultramafic intrusion, central Australia. II. Pyroxenes. *J. Geol. Soc. Australia*, **18**, 243–258.

Moore, A. C., 1973. Studies of igneous and tectonic textures and layering in the rocks of the Gosse Pile intrusion, central Australia. *J. Petr.*, **14**, 49–79.

Moore, J. M., *Jr*., **Kranck, S. H.** and **Chao, G. Y.**, 1969. Optical and X-ray data for iron-rich orthopyroxenes from northern Quebec. *Can. Min.*, **10**, 101–104.

Moorhouse, W. W. and **Shepherd, N.**, 1963. Hypersthene and cummingtonite from Payne Bay, New Quebec. *Can. Min.*, **7**, 527–532.

Mori, T., 1977. Geothermometry of spinel lherzolites. *Contr. Min. Petr.*, **59**, 261–280.

Mori, T. and **Green, D. H.**, 1976. Subsolidus equilibria between pyroxenes in the $CaO–MgO–SiO_2$ system at high pressures and temperatures. *Amer. Min.*, **61**, 616–625.

Morimoto, N., 1959. The structural relations among three polymorphs of $MgSiO_3$ – enstatite, protoenstatite, and clinoenstatite. *Carnegie Inst. Washington, Ann. Rept. Dir. Geophys. Lab.*, 1958–59, 197–198.

Morimoto, N., **Appleman, D. E.** and **Evans, H. T.**, 1960. The crystal structures of clinoenstatite and pigeonite. *Z. Krist.*, **114**, 120–147.

Morimoto, N. and **Koto, K.**, 1969. The crystal structure of orthoenstatite. *Z. Krist.*, **129**, 65–83.

Morimoto, N., **Nakajima, Y.**, **Syono, Y.**, **Akimoto, S.** and **Matsui, Y.**, 1975. Crystal structures of pyroxene-type $ZnSiO_3$ and $ZnMgSi_2O_6$. *Acta Cryst.*, **B, 31**, 1041–1049.

Morimoto, N. and **Tokonami, M.**, 1969. Domain structure of pigeonite and clinoenstatite. *Amer. Min.*, **54**, 725–740.

Morse, S. A., 1975. Plagioclase lamellae in hypersthene, Tokkoatokhakh Bay, Labrador. *Earth Planet. Sci. Letters*, **26**, 331–336.

Mossman, D. J., 1971. Composition and orientation of clinopyroxene lamellae in Bushveld-type orthopyroxene from the Greenhills ultramafic complex, Bluff Peninsula, New Zealand. *Min. Mag.*, **38**, 160–164.

Muan, A., **Nafziger, R. H.** and **Roedder, P. L.**, 1964. A method of determining the instability of ferrosilite. *Nature*, **202**, 688–689.

Mueller, R. F., 1960. Compositional characteristics and equilibrium relations in mineral assemblages of a metamorphosed iron formation. *Amer. J. Sci.*, **258**, 449–497.

Mueller, R. F., 1961. Analysis of relations among Mg,Fe and Mn in certain metamorphic minerals. *Geochim. Cosmochim. Acta.*, **25**, 267–296.

Mueller, R. F., 1966. Stability relations of the pyroxenes and olivine in certain high grade metamorphic rocks. *J. Petr.*, **7**, 363–374.

Mueller, R. F., 1970. Two-step mechanism for order-disorder kinetics in silicates. *Amer. Min.*, **55**, 1210–1218.

Muir, I. D., 1954. Crystallization of pyroxenes in an iron-rich diabase from Minnesota. *Min. Mag.*, **30**, 376–388.

Muir, I. D. and **Long, J. V. P.**, 1965. Pyroxene relations in two Hawaiian hypersthene-bearing basalts. *Min. Mag.*, **34**, 358–369.

Muir, I. D. and **Tilley, C. E.**, 1957. Contributions to the petrology of Hawaiian basalts. I. The picrite basalts of Kilauea. *Amer. J. Sci.*, **255**, 241–253.

Muir, I. D. and **Tilley, C. E.**, 1958. The compositions of coexisting pyroxenes in metamorphic assemblages. *Geol. Mag.*, **95**, 403–408.

Mukherjee, A. and **Rege, S. M.**, 1972. Facies transition and growth of hypersthene in some high grade metamorphic rocks from the Eastern Ghats, India. *Neues Jahrb. Min., Mh.*, 116–132.

Müller, W. F., 1974. One-dimensional lattice imaging of a deformation-induced lamellar intergrowth of orthoenstatite and clinoenstatite [$(Mg,Fe)SiO_3$]. *Neues Jahrb. Min., Mh.*, 83–88.

Munoz, J. L., 1968. Effect of shearing on enstatite polymorphism. *Carnegie Inst. Washington, Ann. Rept. Dir. Geophys. Lab.*, 1966–67, 369–370.

Munoz, J. L., **Segredo, J.** and **Afonso, A.**, 1974. Mafic and ultramafic inclusions in the eruption of Teneguía volcano (La Palma, Canary Islands). *Estud. Geol. (Inst. 'Lucas Mallada')*, Teneguía vol., 65–74.

Mysen, B. O., 1976a. Experimental determination of some geochemical parameters relating to conditions of equilibration of peridotite in the upper mantle. *Amer. Min.*, **61**, 677–683.

Mysen, B. O., 1976b. Nickel partitioning between upper mantle crystals and partial melts as a function of pressure, temperature and nickel concentration. *Carnegie Inst. Washington, Ann. Rept. Dir. Geophys. Lab.*, 1975–76, 662–668.

Mysen, B. O. and **Boettcher, A. L.**, 1975. Melting of a hydrous mantle: II. Geochemistry of crystals and liquids by anatexis of mantle peridotite at high pressures and high temperatures as a function of controlled activities of water, hydrogen and carbon dioxide. *J. Petr.*, **16**, 549–593.

Mysen, B. O. and **Boettcher, A. L.**, 1976. Melting of a hydrous mantle: III. Phase relations of garnet–websterite + H_2O at high pressures and temperatures. *J. Petr.*, **17**, 1–14.

Nafziger, R. H. and **Muan, A.**, 1967. Equilibrium phase compositions and thermodynamic properties of olivines and pyroxenes in the system MgO–'FeO'–SiO_2. *Amer. Min.*, **52**, 1364–1385.

Nagasaki, H., 1966. A layered ultrabasic complex at Horoman, Hokkaido, Japan. *J. Fac. Sci. Univ. Tokyo*, Sect. II, **16**, pt. 2, 313–346.

Nakamura, Y., 1971. Equilibrium relations in Mg-rich part of the pyroxene quadrilateral. *Min. J. Japan*, **6**, 264–276.

Nakamura, Y. and **Kushiro, I.**, 1970a. Equilibrium relations of hypersthene, pigeonite and augite in crystallizing magmas; microprobe study of a pigeonite andesite from Weiselberg, Germany. *Amer. Min.*, **55**, 1999–2015.

Nakamura, Y. and **Kushiro, I.**, 1970b. Compositional relations of coexisting orthopyroxene, pigeonite and augite in a tholeiitic andesite from Hakone volcano. *Contr. Min. Petr.*, **26**, 265–275.

Naldrett, A. J. and **Kullerud, G.**, 1967. A study of the Strathcona Mine and its bearing on the origin of the nickel–copper ores of the Sudbury district, Ontario. *J. Petr.*, **8**, 453–531.

Navrotsky, A., 1971. The intracrystalline cation distribution and the thermodynamics of solid solution formation in the system $FeSiO_3$–$MgSiO_3$. *Amer. Min.*, **56**, 201–211.

Nehru, C. E., 1976. Pressure dependence of the enstatite limb of the enstatite–diopside solvus. *Amer. Min.*, **61**, 578–581.

Nehru, C. E. and **Wyllie, P. J.**, 1974. Electron microprobe measurement of pyroxenes coexisting with H_2O-undersaturated liquid in the join $CaMgSi_2O_6$–Mg_2SiO_6–H_2O at 30 kilobars, with applications to geothermometry. *Contr. Min. Petr.*, **48**, 221–228.

Newton, R. C., **Charlu, T. V.** and **Kleppa, O. J.**, 1974. A calorimetric investigation of the stability of anhydrous magnesium cordierite with application to granulite facies metamorphism. *Contr. Min. Petr.*, **44**, 295–311.

Nishizawa, O. and **Akimoto, S.**, 1973. Partition of magnesium and iron between olivine and spinel, and between pyroxene and spinel. *Contr. Min. Petr.*, **41**, 217–230.

Nicholls, J., 1977. The calculation of mineral compositions and modes of olivine–two pyroxene–spinel assemblages. Problems and possibilities. *Contr. Min. Petr.*, **60**, 119–142.

Nicholls, I. A. and **Lorenz, V.**, 1973. Origin and crystallization history of Permian tholeiites from the Saar–Nahe trough, S.W. Germany. *Contr. Min. Petr.*, **40**, 327–344.

Nicolas, A., 1966. Etude pétrochimique des Roches vertes et de leurs minéraux entre Dora Maira et Grand Paradis (Alpes piémontaises). *Nantes, Fac. Sci.*, 1–299.

Nixon, P. H., **von Knorring, O.** and **Rooke, J. M.**, 1963. Kimberlite and associated inclusions of Basutoland: A mineralogical and geochemical study. *Amer. Min.*, **48**, 1090–1132.

Nixon, P. H., **Reedman, A. J.** and **Burns, L. K.**, 1973. Sapphirine-bearing granulites from Labwar, Uganda. *Min. Mag.*, **39**, 420–428.

Norton, D. A. and **Clavan, W. S.**, 1959. Optical mineralogy, chemistry and X-ray crystallography of ten pyroxenes. *Amer. Min.*, **44**, 844–874.

Nwe, Y. Y., 1975. Chemistry, subsolidus relations and electron petrography of pyroxenes from the late ferrodiorites of the Skaergaard intrusion, east Greenland. *Contr. Min. Petr.*, **53**, 37–54.

Obata, M., 1976. The solubility of Al_2O_3 in orthopyroxenes in spinel and plagioclase peridotites and spinel pyroxenite. *Amer. Min.*, **61**, 804–815.

O'Hara, M. J., 1960. Co-existing pyroxenes in metamorphic rocks. *Geol. Mag.*, **97**, 498–503.

O'Hara, M. J., 1963. Distribution of iron between coexisting olivines and Ca-poor pyroxenes in peridotites, gabbros and other magnesium environments. *Amer. J. Sci.*, **261**, 32–46.

O'Hara, M. J. and **Mercy, E. P. L.**, 1963. Petrology and petrogenesis of some garnetiferous peridotites. *Trans. Roy. Soc. Edin.*, **65**, 251–314.

O'Hara, M. J. and **Mercy, E. P. L.**, 1966. Eclogite, peridotite and garnet from the Navajo County, Arizona and New Mexico. *Amer. Min.*, **51**, 336–352.

O'Hara, M. J. and **Yoder, H. S.**, 1967. Formation and fractionation of basic magmas at high pressures. *Scot. J. Geol.*, **3**, 67–117.

Ohashi, Y. and **Finger, L. W.**, 1973. A possible high–low temperature transition in orthopyroxenes and orthoamphiboles. *Carnegie Inst. Washington, Ann. Rept. Dir. Geophys. Lab.*, 1972–73, 544–547.

Ohnmacht, W., 1974. Petrogenesis of carbonate-orthopyroxenites (sagvandites) and related rocks from Troms, northern Norway. *J. Petr.*, **15**, 303–324.

Olsen, E. and **Mueller, R. F.**, 1966. Stability of orthopyroxenes with respect to pressure, temperature and composition. *J. Geol.*, **74**, 620–625.

O'Nions, R. K. and **Smith, D. G.**, 1973. Bonding in silicates; an assessment of bonding in orthopyroxene. *Geochim. Cosmochim. Acta*, **37**, 249–257.

Ono, A., 1971. Synthesis of Ca-poor clinopyroxenes. *J. Japan. Assoc. Min. Petr. and Econ. Geol.*, **65**, 211–220 (in Japanese) (M.A. 23–101).

Onuki, H. and **Kato, Y.**, 1971. Some gabbroic rocks of the Tabito plutonic complex in the Abukuma plateau. *Sci. Repts., Tohoku Univ.*, Ser. III, **11**, 113–123.

Öosterom, M. G., 1963. The ultramafites and layered gabbro sequences in the granulite facies rocks on Stjernöy (Finnmark, Norway). *Leidse Geol. Med.*, **28**, 177–296.

Ota, R., 1952. Petrographic study of the Akagi volcano lava. *Rept. Geol. Surv. Japan*, No. 151, 33–40.

Ott, W. D., 1971. Stabile Phasen im System Diopside–Enstatit–Eisenoxide bei 1 atm Druck. *Fortschr. Min.*, **49**, 36–37.

Oyawoye, M. O. and **Makanjuola, A. A.**, 1972. Bauchite: a fayalite-bearing quartz monzonite. *Intern. Geol. Congr., 24th Session*, Sect. **2**, 251–266.

Paděra, K. and **Procházka, J.**, 1970. Eclogite (griquaite) with orthorhombic pyroxene from Nihov, western Moravia. *Acta Univ. Carolinae. Geol.*, 171–186.

Pampuch, R. and **Ptak, W.**, 1969. Infra-red spectra and structure of 1:1 layer lattice silicates. Pt. I. The vibrations of the tetrahedral layer. *Polska Akad. Nauk, Prace Min.*, **15**, 1–55 (in English with Polish and Russian summaries).

Papike, J. J., **Prewitt, C. T.**, **Sueno, S.** and **Cameron, M.**, 1973. Pyroxenes: comparisons of real and ideal structural topologies. *Z. Krist.*, **69**, 254–273.

Parkes, G. A. and **Arltar, S.**, 1968. Magnetic moment of Fe^{2+} in paramagnetic minerals. *Amer. Min.*, **53**, 406–415.

Parras, K., 1958. On the charnockites in the light of a highly metamorphic rock complex in south-western Finland. *Bull. Comm. géol. Finlande*, No. 181, 1–137.

Perrotta, A. J. and **Stephenson, D. A.**, 1965. Clinoenstatite, high–low inversion. *Science*, **148**, 1090–1091.

Peters, Tj., 1968. Distribution of Mg, Fe, Al, Ca and Na in coexisting olivine, orthopyroxene and clinopyroxene in the Totalp serpentinite (Davos, Switzerland) and in the Alpine metamorphosed Malenco serpentine (N. Italy). *Contr. Min. Petr.*, **18**, 65–75.

Phillipsborn, H., 1930. Zur chemische-analytischen Erfassung der isomorphen Variation gesteinbildender Minerale. Die Mineralkomponenten des Pyroxen-granulits von Hartmannsdorf. *Chemie der Erde*, **5**, 233–253.

Philpotts, A. R., 1966. Origin of the anorthosite mangerite rocks in southern Quebec. *J. Petr.*, 7, 1–64.

Philpotts, A. R. and **Gray, N. H.**, 1974. Inverted clinobronzite in eastern Connecticut diabase. *Amer. Min.*, **59**, 374–377.

Pike, J. E., 1976. Pressures and temperatures calculated from chromium-rich pyroxene compositions of megacrysts and peridotite xenoliths, Black Rock Summit, Nevada. *Amer. Min.*, **61**, 725–731.

Poldervaart, A., 1944. The petrology of the Elephant's Head dyke and the New Amalfi sheet. *Trans. Roy. Soc. South Africa*, **30**, 85–119.

Poldervaart, A., 1947. The relationship of orthopyroxenes to pigeonite. *Min. Mag.*, **28**, 164–172.

Poldervaart, A., 1950. Correlation of physical properties and chemical composition in the plagioclase, olivine and orthopyroxene series. *Amer. Min.*, **35**, 1067–1079.

Poldervaart, A. and **Hess, H. H.**, 1951. Pyroxenes in the crystallization of basaltic magma. *J. Geol.*, **59**, 472–489.

Pollack, S. S., 1966. Disordered orthopyroxene in meteorites. *Amer. Min.*, **51**, 1722–1726.

Pollack, S. S., 1968. Disordered pyroxene in chondrites. *Geochim. Cosmochim. Acta*, **32**, 1209–1217.

Pollack, S. S. and **DeCarli, P. S.**, 1969. Enstatite: disorder produced by a megabar shock event. *Science*, **165**, 591–592.

Pollack, S. S. and **Ruble, W. D.**, 1964. X-ray identification of ordered and disordered orthoenstatite. *Amer. Min.*, **49**, 983–992.

Pouclet, A. and **Bizouard, H.**, 1974. Les trois phases pyroxéniques de la picrite de Zanaga (Congo–Brazzaville). *Bull. Soc. franç. Min. Crist.*, **97**, 470–474.

Presnall, D. C., 1966. The join forsterite–diopside–iron oxide and its bearing on the crystallization of basaltic and ultramafic magmas. *Amer. J. Sci.*, **264**, 753–809.

Presnall, D. C., 1976. Alumina content of enstatite as a geobarometer for plagioclase and spinel lherzolites. *Amer. Min.*, **61**, 582–588.

Prewitt, C. T., **Brown, G. E.** and **Papike, J. J.**, 1971. Apollo 12 clinopyroxenes: high temperature X-ray diffraction studies. *Proc. 2nd Lunar Sci. Conf. Geochim. Cosmochim. Acta, Suppl.* **2**, **1**, 59–68.

Quensel, P., 1951. The charnockite series of the Varberg district on the south-western coast of Sweden. *Arkiv. Min. Geol.*, **1**, 227–332.

Rabbitt, J. C., 1948. A study of the anthophyllite series. *Amer. Min.*, **33**, 263–323.

Rajagopalan, C., 1946–47. Studies in charnockites from St. Thomas' Mount, Madras, pts. I and II. *Proc. Indian Acad. Sci.*, **24**, 315–331; **26**, 237–260.

Raleigh, C. B., 1965. Glide mechanisms in experimentally deformed minerals. *Science*, **150**, 739–741.

Raleigh, C. B., Kirby, S. H., Carter, N. L. and **Ave Lallemant, H. G.,** 1971. Slip and the clinoenstatite transformation as competing rate processes in enstatite. *J. Geophys. Res.*, **76**, 4011–4022.

Ramaswamy, A. and **Murty, M. S.,** 1973. Ortho- and clino-pyroxenes from the charnockite series of Amaravathi, Andhra Pradesh, S. India. *Min. Mag.*, **39**, 74–77.

Ramberg, H., 1952. Chemical bonds and the distribution of cations in silicates. *J. Geol.*, **60**, 331–335.

Ramberg, H. and **DeVore, G.,** 1951. The distribution of Fe^{++} and Mg^{++} in coexisting olivine and pyroxenes. *J. Geol.*, **59**, 193–210.

Rao, B. R. and **Rao, L. R.,** 1937. On 'bidalotite', a new orthorhombic pyroxene derived from cordierite. *Proc. Indian Acad. Sci.*, **5**, 290–296.

Ray, S. and **Sen, S. K.,** 1970. Partitioning of major exchangeable cations among orthopyroxene, calcic pyroxene and hornblende in basic granulites from Madras. *Neues. Jahrb. Min., Abhdl.*, **114**, 61–88.

Read, H. H., 1935. The gabbros and associated xenolithic complexes of the Haddo House district, Aberdeenshire. *Quart. J. Geol. Soc.*, **91**, 591–638.

Reesman, A. L. and **Keller, W. D.,** 1965. Calculation of apparent standard free energies of formation of six rock-forming silicate minerals from solubility data. *Amer. Min.*, **50**, 1729–1739.

Reid, A. M. and **Cohen, A. J.,** 1967. Some characteristics of enstatite from enstatite achondrites. *Geochim. Cosmochim. Acta*, **31**, 661–672.

Reid, A. M., Williams, R. F. and **Takeda, H.,** 1974. Coexisting bronzite and clinobronzite and the thermal evolution of the Steinbach meteorite. *Earth Planet. Sci. Letters*, **22**, 67–74.

Ridley, W. I. and **Baker, I.,** 1973. The petrochemistry of a unique cordierite-bearing lava from St. Helena Island, South Atlantic. *Amer. Min.*, **58**, 813–818.

Riecker, R. E. and **Rooney, T. P.,** 1967. Deformation and polymorphism of enstatite under shear stress. *Bull. Geol. Soc. America*, **78**, 1045–1053.

Ringwood, A. E. and **Major, A.,** 1966a. High-pressure transformation of $FeSiO_3$ pyroxene to spinel plus stishovite. *Earth Planet. Sci. Letters*, **1**, 135–136.

Ringwood, A. E. and **Mayor, A.,** 1966b. High-pressure transformation of $CoSiO_3$ pyroxene and some geochemical implications. *Earth Planet. Sci. Letters*, **1**, 209–210.

Ringwood, A. E. and **Major, A.,** 1966c. High-pressure transformation in pyroxenes. *Earth Planet. Sci. Letters*, **1**, 351–571.

Ringwood, A. E. and **Major, A.,** 1968. High-pressure transformation in pyroxenes II. *Earth Planet. Sci. Letters*, **5**, 76–78.

Ringwood, A. E. and **Seabrook, M.,** 1962. High-pressure transition of $MgGeO_3$ from pyroxene to corundum structure. *J. Geophys. Res.*, **67**, 1690–1691.

Rio, M., 1968. Quelques précisions sur la composition minéralogique de la lherzolite de Moun Caou (Basses-Pyrénées). *Bull. Soc. franç. Min. Crist.*, **91**, 298–299.

Robins, B., 1975. Ultramafic nodules from Seiland, northern Norway. *Lithos*, **8**, 15–27.

Roedder, E., 1965. Liquid CO_2 inclusions in olivine-bearing nodules and phenocrysts from basalts. *Amer. Min.*, **50**, 1746–1782.

Ross, C. S., Foster, M. D. and **Myers, C. D.,** 1954. Origin of dunites and of olivine-rich inclusions in basaltic rocks. *Amer. Min.*, **39**, 693–737.

Rothstein, A. T. V., 1958. Pyroxenes from the Dawros peridotite and some comments on their nature. *Geol. Mag.*, **95**, 456–562.

Runciman, W. A., Sengupta, D. and **Marshall, M.,** 1973. The polarized spectra of iron in silicates. I. Enstatite. *Amer. Min.*, **58**, 444–450.

Rust, B. R., 1963. Some investigations of the Roraima Formation intrusives in British Guiana. *Geol. Mag.*, **100**, 24–32.

Sadanaga, R. and **Okamura, F.,** 1971. On the high-clino phase of enstatite. *Min. J. Japan*, **6**, 365–374.

Sadanaga, R., Okamura, E. and **Takeda, H.,** 1969. X-ray study of the phase transformations of enstatite. *Min. J. Japan*, **6**, 110–130.

Sahama, Th. G. and **Torgeson, D. R.,** 1949. Some examples of application of thermochemistry to petrology. *J. Geol.*, **57**, 255–262.

van der Sande, J. B. and **Kohlstedt, D.,** 1974. A high-resolution electron microscopy study of exsolution in enstatite. *Phil. Mag.*, **29**, 1041–1049.

Sapountzis, E. S., 1975. Coronas from the Thessaloniki gabbros (north Greece). *Contr. Min. Petr.*, **51**, 197–203.

Sarver, J. F. and **Hummel, F. A.,** 1962. Stability relations of magnesium metasilicate polymorphs. *J. Amer. Ceram. Soc.*, **45**, 152–156.

Savolahti, A., 1966. On rocks containing garnet, hypersthene, cordierite and gedrite in the Kiurevesi region, Finland. Part I: Juurikkajarvi. *Bull. Comm. géol. Finlande*, No. 222, 342–386.

Saxén, M., 1925. Om mangan-jarnmalmfyndigheten i Vittinki. *Fennia*, **45**, 18–40.

Saxena, S. K., 1967. Intracrystalline chemical variations in calcic pyroxenes and biotites. *Neues Jahrb. Min. Abhdb.*, **107**, 299–316.

Saxena, S. K., 1968a. Crystal-chemical aspects of distribution of elements among certain coexisting rock-forming silicates. *Neues Jahrb., Abhdl.*, **108**, 292–323.

Saxena, S. K., 1968b. Distribution of elements between coexisting minerals and the nature of solid solution in garnet. *Amer. Min.*, **53**, 994–1014.

Saxena, S. K., 1968c. Chemical study of phase equilibria of charnockites, Varberg, Sweden. *Amer. Min.*, **53**, 1674–1695.

Saxena, S. K., 1971. Mg^{2+}–Fe^{2+} order–disorder in orthopyroxene and the Mg^{2+}–Fe^{2+} distribution between coexisting minerals. *Lithos*, **4**, 345–354.

Saxena, S. K., 1972. Retrieval of thermodynamic data from a study of intercrystalline and intra-crystalline ion-exchange equilibrium. *Amer. Min.*, **57**, 1782–1800.

Saxena, S. K., 1976. Two-pyroxene geothermometer: a model with an approximate solution. *Amer. Min.*, **61**, 643–652.

Saxena, S. K. and **Ghose, S.**, 1970. Order–disorder and the activity–composition relation in a binary crystalline solution. I. Metamorphic orthopyroxene. *Amer. Min.*, **55**, 1219–1225.

Saxena, S. K. and **Ghose, S.**, 1971. Mg^{2+}–Fe^{2+} order–disorder and the thermodynamics of the orthopyroxene crystalline solution. *Amer. Min.*, **56**, 532–559.

Saxena, S. K. and **Nehru, C. E.**, 1975. Enstatite–diopside solvus and geothermometry. *Contr. Min. Petr.*, **49**, 259–267.

Schairer, J. F. and **Boyd, F. R.**, 1957. Pyroxenes, the join $MgSiO_3$–$CaMgSi_2O_6$. *Carnegie Inst. Washington, Ann. Rept. Dir. Geophys. Lab.*, 1956–57, 223.

Schairer, J. F. and **Yoder, H. S.**, 1962. Pyroxenes: the system diopside–enstatite–silica. *Carnegie Inst. Washington, Ann. Rept. Dir. Geophys. Lab.*, 1961–62, 75–82.

Scharbert, H. G., 1963. Die Granulite der südlichen Böhmischen Masse. *Geol. Rdsch.*, **52**, 112–123.

Schatz, J. F. and **Simmons, G.**, 1972. Thermal conductivity of Earth materials at high temperatures. *J. Geophys. Res.*, **77**, 6966–6983.

Schmidt, E. R., 1952. The structure and composition of the Merensky Reef and associated rocks on the Rustenburg platinum mine. *Trans. Geol. Soc. South Africa*, **55**, 233–279.

Scholtz, D. L., 1936. The magnetic nickelferous ore deposits of East Griqualand and Pondoland. *Trans. Geol. Soc. South Africa*, **39**, 81–210.

Schorer, G., 1970. Die Pyroxene tertiárer Vulkanite des Vogelsberges. *Chemie der Erde*, **29**, 69–138.

Schreyer, W. and **Schairer, J. F.**, 1961. Compositions and structural states of anhydrous Mg-cordierites: a re-investigation of the central part of the system MgO–Al_2O_3·SiO_2. *J. Petr.*, **2**, 324–406.

Schreyer, W., **Ohnmacht, W.** and **Mannchen, J.**, 1972. Carbonate-orthopyroxenites (sagvandites) from Troms, northern Norway. *Lithos*, **5**, 345–364.

Schrijver, K., 1973. Bimetasomatic plagioclase–pyroxene reaction zones in granulite facies. *Neues Jahrb. Min., Abhdl.*, **119**, 1–19.

Schwab, R. G., 1967. Die Bedeutung und die experimentelle Beherrschung des Sauerstoffpartialdruckes bei der Synthese und Untersuchung Fe^{2+}-haltiger Silikate. *Neues Jahrb. Min., Mh.*, **7**, 244–254.

Schwab, R. G., 1968. Das system $Mg_2Si_2O_6$–$MnNiSi_2O_6$. *Neues Jahrb. Min., Mh.*, 337–350.

Schwab, R. G. and **Schwerin, M.**, 1975. Polymorphie und Entmischungsreaktionen im System Enstatit ($MgSiO_3$)–Diopsid ($CaMgSi_2O_6$). *Neues Jahrb. Min., Abhdl.*, **124**, 223–245.

Schwarcz, H. P., 1967. The effect of crystal field stabilization on the distribution of transition metals between metamorphic minerals. *Geochim. Cosmochim. Acta*, **31**, 503–517.

Schwartzman, D. W. and **Giletti, B. J.**, 1977. Argon diffusion and absorption studies of pyroxenes from the Stillwater complex, Montana. *Contr. Min. Petr.*, **60**, 143–160.

Sclar, C. B., **Carrison, L. C.** and **Schwartz, C. M.**, 1964. High-pressure reaction of clinoenstatite to forsterite plus stishovite. *J. Geophys. Res.*, **69**, 325–330.

Semenov, I. V., 1970. Effect of chemical composition on the unit-cell parameters of pyroxenes as a function of the energy of the elements. *Dokl. Acad. Sci., U.S.S.R., Earth Sci., Sect.*, **192**, 128–131.

Sen, S. K., 1973. Compositional relations among hornblende and pyroxenes in basic granulites and an application to the origin of garnets. *Contr. Min. Petr.*, **38**, 299–306.

Sen, S. K. and **Rege, S. M.**, 1966. Distribution of magnesium and iron between metamorphic pyroxenes from Saltora, West Bengal, India. *Min. Mag.*, **35**, 759–762.

Sen, S. K. and **Roy, S.**, 1971. Hornblende–pyroxene granulites versus pyroxene granulites: a study from the type charnockite area. *Neues Jahrb. Min., Abhdl.*, **115**, 291–314.

Sen, S. K. and **Sahu, J. R.**, 1970. Phase relations in three charnockites from Pallavaram–Tambaram. *Contr. Min. Petr.*, **27**, 239–243.

Shearer, J. A. and **Kleppa, O. J.**, 1973. The enthalpies of formation, of $MgAl_2O_4$, $MgSiO_3$, Mg_2SiO_4 and Al_2SiO_5 by oxide melt calorimetry. *J. Inorg. Nucl. Chem.*, **35**, 1073–1078.

Shemyakin, V. M., 1968. [Relation between the composition and the refractive index of rhombic pyroxenes.] *Zap. Vses. Min. Obshch.*, **97**, 41–48 (in Russian).
Shemyakin, V. M., 1973. [Rhombic pyroxenes from charnockites of the Baltic Shield.] *Zap. Vses. Min. Obshch.*, **102**, 635–641 (in Russian).
Shemyakin, V. M., Afanas'yeva, L. L. and **Terent'yeva, M. V.**, 1967. A hydroxyl-bearing rhombic pyroxene from charnockites of northern Karelia. *Dokl. Acad. Sci., U.S.S.R., Earth Sci. Sect.*, **175**, 132–134.
Sigurdsson, H. and **Brown, G. M.**, 1970. An unusual enstatite–forsterite basalt from Kolbeinsey Island, North of Iceland. *J. Petr.*, **11**, 205–220.
Simmons, E. C., Lindsley, D. H. and **Papike, J. J.**, 1974. Phase relations and crystallization sequence in a contact-metamorphosed rock from the Gunflint Iron Formation, Minnesota. *J. Petr.*, **15**, 539–566.
Singh, S., 1966. Orthopyroxene-bearing rocks of charnockitic affinities in the South Savanna–Kanuku complex of British Guiana. *J. Petr.*, **6**, 171–192.
Sinitsȳn, A. V., 1965. [Pyroxenes in differentiated dolerite intrusion.] *Zap. Vses. Min. Obshch.*, **94**, 583–592 (in Russian).
Skinner, B. M. and **Boyd, F. R.**, 1964. Aluminous enstatites. *Carnegie Inst. Washington, Ann. Rept. Dir. Geophys. Lab.*, 1963–64, 163–165.
Skippen, G. B., 1971. Experimental data for reactions in siliceous marbles. *J. Geol.*, **79**, 457–481.
Smith, D., 1971. Stability of the assemblage iron-rich orthopyroxene–olivine–quartz. *Amer. J. Sci.*, **271**, 370–383.
Smith, D., 1972. Stability of iron-rich pyroxene in the system $CaSiO_3$–$FeSiO_3$–$MgSiO_3$. *Amer. Min.*, **57**, 1413–1428.
Smith, D., 1974. Pyroxene–olivine–quartz assemblages in rocks associated with the Nain anorthosite massif, Labrador. *J. Petr.*, **15**, 58–78.
Smith, J. V., 1959. The crystal structure of proto-enstatite $MgSiO_3$. *Acta Cryst.*, **12**, 515–519.
Smith, J. V., 1969a. Crystal structure and stability of the $MgSiO_3$ polymorphs; physical properties and phase relations of Mg,Fe pyroxenes. *Min. Soc. Amer., Spec. Paper* **2**, 3–29.
Smith, J. V., 1969b. Magnesium pyroxenes at high temperature: inversion in clinoenstatite. *Nature*, **222**, 256–257.
Smith, J. V., Stephenson, D. A., Howie, R. A. and **Hey, M. H.**, 1969. Relations between cell dimensions, chemical composition and site preference of orthopyroxene. *Min. Mag.*, **37**, 90–114.
Smyth, J. R., 1971. Protoenstatite: a crystal structure refinement at 1100°C. *Z. Krist.*, **134**, 262–274.
Smyth, J. R., 1973. An orthopyroxene structure up to 850°C. *Amer. Min.*, **58**, 636–648.
Smyth, J. R., 1974a. Experimental study on the polymorphism of enstatite. *Amer. Min.*, **59**, 345–352.
Smyth, J. R., 1974b. Low orthopyroxene from a lunar deep crustal rock: a new pyroxene polymorph of space group $P2_1ca$. *Geophys. Res. Letters*, **1**, 27–30.
Smyth, J. R. and **Burnham, C. W.**, 1972. The crystal structures of high and low clinohypersthene. *Earth Planet. Sci. Letters*, **14**, 183–189.
Snellenburg, J. W., 1975. Computer simulation of the distribution of octahedral cations in orthopyroxenes. *Amer. Min.*, **60**, 441–447.
Sobolev, N. V., 1964. Rhombic pyroxenes from garnet peridotite and eclogite. *Dokl. Acad. Sci. U.S.S.R., Earth Sci. Sect.*, **154**, 110–111.
Speidel, D. H. and **Nafziger, R. H.**, 1968. P–T–FO_2 relations in the system Fe–O–MgO–SiO_2. *Amer. J. Sci.*, **266**, 361–379.
Speidel, D. H. and **Osborn, E. F.**, 1967. Element distribution among coexisting phases in the system MgO–FeO–Fe_2O_3–SiO_2 as a function of temperature and oxygen fugacity. *Amer. Min.*, **52**, 1139–1152.
Spencer, D. R. F. and **Coleman, D. S.**, 1969. Densities of silicates in the CaO–MgO–SiO_2 system. *Trans. Brit. Ceram. Soc.*, **68**, 125–127.
Starkey, J., 1968. The geometry of kink bands in crystals – a simple model. *Contr. Min. Petr.*, **19**, 133–141.
Steele, I. M., 1975. Mineralogy of lunar norite 78235; second lunar occurrence of $P2_1ca$ pyroxene from Apollo 17 soils. *Amer. Min.*, **60**, 1086–1091.
Stephenson, D. A., Sclar, C. B. and **Smith, J. V.**, 1966. Unit cell volumes of synthetic orthoenstatite and low clinoenstatite. *Min. Mag.*, **35**, 838–846.
Stroh, J. M., 1976. Solubility of alumina in orthopyroxene plus spinel as a geobarometer in complex systems. Applications to spinel-bearing Alpine-type peridotites. *Contr. Min. Petr.*, **54**, 173–188.
Subramanian, A. P., 1962. Pyroxenes and garnets from charnockites and associated granulites. *Bull. Geol. Soc. America* (Buddington vol.), 21–36.
Sueno, S., Cameron, M. and **Prewitt, C. T.**, 1976. Orthoferrosilite: high-temperature crystal chemistry. *Amer. Min.*, **61**, 38–53.

Sundius, N., 1932. Über den sogenannten Eisenanthophyllit der Eulysite. *Årsbok Sveriges Geol. Undersök*, **26**, No. 2, 1–8.

Takane, K., 1932. Crystal structure of bronzite from Chichijima in the Bonin Islands. *Proc. Imp. Acad., Tokyo*, **8**, 308–311.

Takeda, H., 1972. Crystallographic studies of coexisting aluminian orthopyroxene and augite of high-pressure origin. *J. Geophys. Res.*, **77**, 5798–5811.

Takeda, H. and **Ridley, W. I.**, 1972. Crystallography and chemical trends of orthopyroxene–pigeonite from rock 14310 and coarse fine 12033. *Proc. 3rd Lunar Sci. Conf., Geochim. Cosmochim. Acta, Suppl.* **3, 1**, 423–430.

Takehita, H., 1957. On extinction angles c ∧ Z′ of monoclinic and rhombic pyroxenes in random sections. *J. Japan. Assoc. Min. Petr. and Econ. Geol.*, **41**, 181–184.

Takehita, H., 1963. A procedure for estimation of the mineral ratio of orthopyroxene to sum of both clinopyroxene and orthopyroxene in the groundmass of calc-alkali volcanic rocks. *J. Japan. Assoc. Min. Petr. and Econ. Geol.*, **50**, 142–150 (in Japanese with English summary). (M.A. **19**-83.)

Talbot, J. L., Hobbs, B. E., Wilshire, H. G. and **Sweatman, T. R.**, 1963. Xenoliths and xenocrysts from lavas of the Kerguelan Archipelago. *Amer. Min.*, **48**, 159–179.

Taneda, S., 1947. Variations in chemical composition and optic properties in rhombic pyroxenes. *Mem. Fac. Sci. Kyusyu Univ., Ser. Geol.*, **3**, 14–21.

Tarasov, V. I. and **Nikitina, L. P.**, 1974. [The optic axial angle and unit cell parameters as criteria of order–disorder measurement of Fe^{2+}–Mg in the structure of orthorhombic pyroxenes.] *Zap. Vses. Min. Obshch.*, **103**, 268–271 (in Russian).

Tarasov, V. I., Nikitina, L. P. and **Ekimov, S. N.**, 1975. [Crystal optic and X-ray methods for determining cation ordering in structures of rhombic pyroxenes.] *Zap. Vses. Min. Obshch.*, **104**, 748–750 (in Russian).

Tarney, J., 1969. Epitaxic relations between coexisting pyroxenes. *Min. Mag.*, **37**, 115–122.

Tauber, A. and **Kohn, J. A.**, 1965. Orthopyroxene and clinopyroxene polymorphs of $CoGeO_3$. *Amer. Min.*, **50**, 13–21.

Tauber, A., Kohn, J. A. Whinfrey, C. G. and **Babbage, W. D.**, 1963. The occurrence of an enstatite phase in the sybsystem GeO_2–$MnGeO_3$. *Amer. Min.*, **48**, 555–564.

Taylor, H. C. J., 1973. Melting relations in the system MgO–Al_2O_3–SiO_2 at 15 kbar. *Bull. Geol. Soc. America*, **84**, 1335–1348.

Tazaki, K., Ito, E. and **Komatsu, M.**, 1972. Experimental study in a pyroxene–spinel symplectite at high pressures and temperatures. *J. Geol. Soc. Japan*, **78**, 347–354.

Thompson, J. B., 1970. Geometrical possibilities for amphibole structures: model biopyriboles. *Amer. Min.*, **55**, 292–293.

Tilley, C. E., 1924. Contact-metamorphism in the Comrie area of the Perthshire Highlands. *Quart. J. Geol. Soc.*, **80**, 21–71.

Tilley, C. E., 1936. Eulysites and related rock types from Loch Duich, Ross-shire. *Min. Mag.*, **24**, 331–342.

Tilley, C. E., 1961. The occurrence of hypersthene in Hawaiian basalts. *Geol. Mag.*, **98**, 257–260.

Tilley, C. E., 1966. A note on the dunite (peridotite) mylonites of St. Paul's Rocks (Atlantic). *Geol. Mag.*, **143**, 120–123.

Tilley, C. E. and **Long, J. V. P.**, 1967. The porphyroclast minerals of the peridotite–mylonites of St. Paul's Rocks (Atlantic). *Geol. Mag.*, **104**, 46–48.

Trommsdorff, V. and **Evans, B. W.**, 1969. The stable association enstatite–forsterite–chlorite in amphibole facies ultramafics of the Lepontine Alps. *Schweiz. Min. Petr. Mitt.*, **49**, 325–332.

Trommsdorff, V. and **Evans, B. W.**, 1970. Stabile Paragenesen mit Enstatit in der zentralalpinen Amphibolitfazies. *Fortschr. Min.*, **37**, 73–87.

Trommsdorff, V. and **Evans, B. W.**, 1974. Alpine metamorphism of peridotitic rocks. *Schweiz. Min. Petr. Mitt.*, **54**, 333–352.

Trommsdorff, V. and **Wenk, H-R.**, 1968. Terrestrial metamorphic clinoenstatite in kink bands of bronzite crystals. *Contr. Min. Petr.*, **19**, 158–168.

Tsuru, K. and **Henry, N. F. M.**, 1937. An iron-rich optically positive hypersthene from Manchuria. *Min. Mag.*, **24**, 527–528.

Turner, F. J., Heard, H. C. and **Griggs, D. T.**, 1960. Experimental deformation of enstatite and accompanying inversion to clinoenstatite. *Rept. 21st Int. Geol. Congr., Copenhagen*, **18**, 399–408.

Turnock, A. C., Lindsley, D. H. and **Grover, J. E.**, 1973. Synthesis and unit cell parameters of Ca–Mg–Fe pyroxenes. *Amer. Min.*, **58**, 50–59.

Ukhanov, A. V. and **Mockalova, Yu. Z.**, 1971. High-temperature emanation study of heat-induced transformations in enstatite from a kimberlite pipe. *Dokl. Acad. Sci. U.S.S.R., Earth Sci. Sect.*, **198**, 157–158.

Val'ter, A. A., 1969. Distribution of magnesium and iron between coexisting iron-rich solid solutions of olivine and orthopyroxene in eulysite from the Mariupol' iron deposit. *Dokl. Acad. Sci. U.S.S.R. Earth Sci. Sect.*, **187**, 106–109.

Vernon, R. H., 1972. Reactions involving hydration of cordierite and hypersthene. *Contr. Min. Petr.*, **35**, 125–137.

Vilenskiy, A. M., 1966. Phase equilibria and evolution in the chemical nature of the pyroxenes of intrusive traps of the Siberian platform. *Geochem. Intern.*, **3**, 1275–1306.

Virgo, D. and **Hafner, S. S.**, 1969. Fe^{2+},Mg order–disorder in heated orthopyroxenes. *Min. Soc. Amer. Spec. Paper* **2**, 67–81.

Virgo, D. and **Hafner, S. S.**, 1970. Fe^{2+},Mg order–disorder in natural orthopyroxenes. *Amer. Min.*, **55**, 201–223.

Vogt, T., 1947. Mineral assemblages with sapphirine and kornerupine. *Bull. Comm. géol. Finlande*, no. 140, 15–24.

Vorma, A., 1975. A contribution to the mineralogy of iron–magnesium silicates, especially pyroxenes, contained in certain noritic rocks in central Finland. *Bull. Geol. Soc. Finland*, **277**, 25pp.

Waard, D. de., 1966. The biotite–cordierite–almandite subfacies of the hornblende-granulite facies. *Can. Min.*, **8**, 481–492.

Wager, L. R. and **Brown, G. M.**, 1968. *Layered Igneous Rocks*, 42–48. Edinburgh (Oliver and Boyd).

Wager, L. R. and **Deer, W. A.**, 1939. Geological investigations in east Greenland. Part III. The petrology of the Skaergaard intrusion, Kangerdlugssuaq, east Greenland. *Medd. om Grønland*, **105**, no. 4.

Wager, L. R. and **Vincent, E. A.**, 1962. Ferrodiorite from the Isle of Skye. *Min. Mag.*, **33**, 26–36.

Wahl, W., 1964. The hypersthene granites and unakites of central Finland. *Bull. Comm. géol. Finlande*, no. 212, 83–100.

Walker, F., 1940. Differentiation of the Palisade diabase, New Jersey. *Bull. Geol. Soc. America*, **51**, 1059–1105.

Walker, F. and **Poldervaart, A.**, 1949. Karroo dolerites of the Union of South Africa. *Bull. Geol. Soc. America*, **60**, 591–701.

Wallace, R. C., 1975. Mineralogy and petrology of xenoliths in a diatreme from South Westland, New Zealand. *Contr. Min. Petr.*, **49**, 191–199.

Walls, R., 1935. A critical review of the data for a revision of the enstatite–hypersthene series. *Min. Mag.*, **24**, 165–172.

Warner, R. D., 1973. Liquidus relations in the system $CaO–MgO–SiO_2–H_2O$ at 10 kbar P_{H_2O} and their petrologic significance. *Amer. J. Sci.*, **273**, 925–946.

Warner, R. D., 1975. New experimental data for the system $CaO–MgO–SiO_2–H_2O$ and a synthesis of inferred phase relations. *Geochim. Cosmochim. Acta*, **39**, 1413–1421.

Warner, R. D. and **Luth, W. C.**, 1974. The diopside–orthoenstatite two-phase region in the system $CaMgSi_2O_6–Mg_2Si_2O_6$. *Amer. Min.*, **59**, 98–109.

Warren, B. E. and **Modell, D. I.**, 1930. Structure of enstatite $MgSiO_3$. *Z. Krist.*, **75**, 1–14.

Washington, H. S., 1932. The use of 'ferrosilite' as a name for the normative molecule $MgSiO_3$. *Tschermak Min. Petr. Mitt.*, **43**, 63–66.

White, W. B., 1966. Ultramafic inclusions in basaltic rocks from Hawaii. *Contr. Min. Petr.*, **12**, 245–314.

White, W. B. and **Keester, K. L.**, 1966. Optical absorption spectra of iron in the rock-forming silicates. *Amer. Min.*, **51**, 774–791.

Whitney, P. R. and **McLelland, J. M.**, 1973. Origin of coronas in metagabbros of the Adirondack Mts., N.Y. *Contr. Min. Petr.*, **39**, 81–98.

Wilkinson, J. F. G., 1971. The petrology of some vitrophyric calc-alkaline volcanics from the Carboniferous of New South Wales. *J. Petr.*, **12**, 587–620.

Wilkinson, J. F. G., 1973. Pyroxene xenoliths from an alkali trachybasalt in the Glen Innes area, northeastern New South Wales. *Contr. Min. Petr.*, **42**, 15–32.

Wilkinson, J. F. G., 1975. An Al-spinel ultramafic–mafic inclusion suite and high-pressure megacrysts in an analcimite and their bearing on basaltic magma fractionation at elevated pressures. *Contr. Min. Petr.*, **53**, 71–104.

Wilkinson, J. F. G. and **Binns, R. A.**, 1969. Hawaiite of high pressure origin from north-eastern New South Wales. *Nature*, **222**, 553–555.

Wilshire, H. G. and **Jackson, E. D.**, 1975. Problems in determining mantle geotherms from pyroxene compositions of ultramafic rocks. *J. Geol.*, **83**, 313–329.

Wilson, A. F., 1952. Occurrence of metasomatic hypersthene and its petrogenetic significance. *Amer. Min.*, **37**, 633–636.

Wilson, A. F., 1960. Co-existing pyroxenes: some causes of variation and anomalies in the optically derived compositional tie-lines, with particular reference to charnockitic rocks. *Geol. Mag.*, **97**, 1–17.

Wilson, A. F., 1976a. A normalizing procedure for comparing chemical variations in pyroxenes in mafic granulites. *Contr. Min. Petr.*, **55**, 131–138.

Wilson, A. F., 1976b. Aluminium in coexisting pyroxenes as a sensitive indicator of changes in metamorphic grade within the mafic granulite terrain of the Frazer Range, Western Australia. *Contr. Min. Petr.*, **56**, 255–277.

Winchell, H. and **Leake, B. E.**, 1965. Regressions of refractive indices, density and lattice constants on the composition of orthopyroxenes. *Amer. Min.*, **50**, 294.

Windley, B. F. and **Smith, J. V.**, 1974. The Fiskenaesset complex, west Greenland. Part II. General mineral chemistry from Qeqertarssuatsiaq. *Medd. om Grønland*, **196**, no. 4, 1–54.

Wood, B. J., 1974. The solubility of alumina in orthopyroxene co-existing with garnet. *Contr. Min. Petr.*, **46**, 1–15.

Wood, B. J., 1975. The application of thermodynamics to some subsolidus equilibria involving solid solutions. *Fortschr. Min.*, **52** (Spec. vol. *Papers and Proc. 9th Gen. Meeting I.M.A., Berlin (West) – Regensberg*), 21–45.

Wood, B. J. and **Banno, S.**, 1973. Garnet–orthopyroxene and orthopyroxene–clinopyroxene relationships in simple and complex systems. *Contr. Min. Petr.*, **42**, 109–124.

Wood, B. J. and **Strens, R. G. J.**, 1971. The orthopyroxene geobarometer. *Earth Planet. Sci. Letters*, **11**, 1–6.

Wood, B. J. and **Strens, R. G. J.**, 1972. Calculation of crystal field splittings in distorted co-ordination polyhedra: spectra and thermodynamic properties of minerals. *Min. Mag.*, **38**, 909–917.

Woodford, P. J. and **Wilson, A. F.**, 1976. Chemistry of coexisting pyroxenes, hornblendes and plagioclases in mafic granulites, Strangways Range, central Australia. *Neues Jahrb. Min., Abhdl.*, **128**, 1–40.

Wyckoff, R. W. G., **Merwin, H. E.** and **Washington, H. S.**, 1925. X-ray diffraction measurements upon the pyroxenes. *Amer. J. Sci.*, ser. 4, **10**, 383–397.

Wyllie, P. J. and **Huang, W-L.**, 1976. Carbonation and melting reactions in the system CaO–MgO–SiO_2–CO_2 at mantle pressures with geophysical and petrological applications. *Contr. Min. Petr.*, **54**, 79–107.

Yamaguchi, M., 1964. Petrogenic significance of ultrabasic inclusions in basaltic rocks from Southwest Japan. *Mem. Fac. Sci., Kyushu Univ., Sect. D., Geology*, **15**, 163–219.

Yamaguchi, M., 1973. Study on the exsolution phenomena of pyroxenes. *J. Fac. Sci., Hokkaido Univ.*, ser. 4, **16**, 133–165.

Yamaguchi, M. and **Tomita, K.**, 1970. Clinoenstatite as an exsolution phase in diopside. *Mem. Fac. Sci. Koyoto Univ.*, **37**, 174–180.

Yin, L., **Ghose, S.** and **Adler, I.**, 1971. Core binding energy difference between bridging and non-bridging oxygen atoms in a silicate chain. *Science*, **173**, 633–635.

Yoder, H. S., 1950. The jadeite problem, Part II. *Amer. J. Sci.*, **248**, 312–334.

Yoder, H. S., 1952. The MgO–Al_2O_3–SiO_2–H_2O system and the related metamorphic facies. *Amer. J. Sci.* (Bowen vol.), 569–627.

Yur'ev, L. D., 1966. [Symplectic intergrowth of magnetite with pyroxene.] *Zap. Vses. Min. Obshch.*, **95**, 283–241 (in Russian).

Zen, E-an and **Chernosky, J. V.**, 1976. Correlated free energy values of anthophyllite, brucite, clinochrysotile, enstatite, forsterite, quartz and talc. *Amer. Min.*, **61**, 1156–1166.

Zwaan, P. C., 1954. On the determination of pyroxenes by X-ray powder diagrams. *Leidse Geol. Mededel.*, **19**, 167–276.

Pigeonite $(Mg,Fe^{2+},Ca)(Mg,Fe^{2+})[Si_2O_6]$

Monoclinic (+)

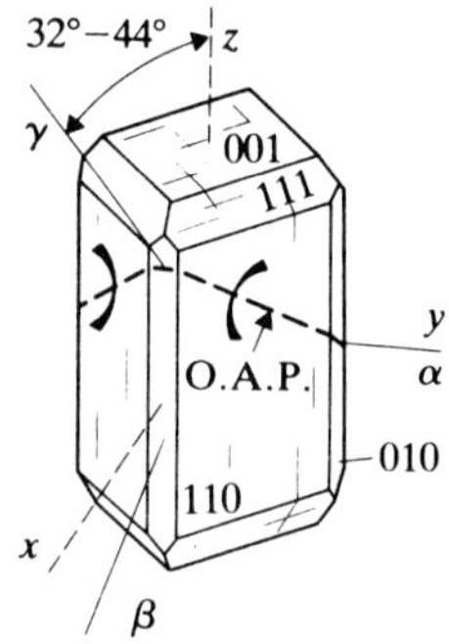

α	1·682–1·732
β	1·684–1·732
γ	1·705–1·757
δ	0·023–0·029
$2V_\gamma$	0°–30°
$\gamma{:}z$	32°–44°
$\alpha = y$	O.A.P. $\perp$ (010);
more rarely $\beta = y$,	O.A.P. (010)
Dispersion:	$r \gtrless v$, moderate.
D	3·17–3·46
H	6
Cleavage:	{110} good; {100}, {010}, {001} partings; (110):(1$\bar{1}$0) $\simeq$ 87°
Twinning:	{100} or {001}, simple or lamellar, common.
Colour:	Brown, greenish brown black; colourless, pale brownish green, pale yellow-green in thin section.
Pleochroism:	Often absent but may be weak to moderate. α colourless, pale green, yellowish green or smoky brown. β pale brown, pale brownish green, brownish pink or smoky brown. γ colourless, pale green or pale yellow.
Unit cell:	Most within ranges *a* 9·67–9·73, *b* 8·90–8·97, *c* 5·22–5·25 Å, β 108·3–108·7° Z = 4. Space group $P2_1/c$.

Insoluble in HCl.

Pigeonites are calcium-poor monoclinic pyroxenes containing between 5 and 15 per cent of the $CaSiO_3$ component. This low content of Ca compared with augites is accompanied by some distortion of the pyroxene chain configuration, and the space group of pigeonite differs from that of augite. Pigeonites are distinguished optically from other pyroxenes by low optic axial angles which invariably are less than 30°, and in most examples lower than 25°; the optic axial plane may be either perpendicular to or parallel with (010). Pigeonite occurs mainly in quickly chilled lavas and minor intrusions, in which it is commonly the most abundant pyroxene. Pigeonite has sometimes been regarded as a metastable phase but such a view is not substantiated by its occurrence as phenocrysts in some volcanic rocks; moreover it has commonly formed under conditions of plutonic crystallization. In such an environment, however, the pigeonite inverts, at subsolidus temperatures to an orthorhombic pyroxene; the excess of the diopside–hedenbergite component is exsolved and occurs as augite lamellae parallel to original (001) plane of the pigeonite. It is uncommon in metamorphic rocks, and in this environment only iron-rich compositions occur, probably because metamorphic temperatures are not sufficiently high for the formation of the more magnesium-rich members of the series. A comprehensive review of pigeonite and its relationships to other low-calcium pyroxenes is given by Brown (1972). Pigeonite was named by A. N. Winchell after the locality Pigeon Point, Minnesota.

Structure

The first structure determination for a pigeonite was that by Morimoto *et al.* (1960) from two-dimensional projections, and subsequently three-dimensional refinements of the structures of several pigeonites have been carried out. Since the calcium content of pigeonites is low the structures have similar features to those of the clinoenstatite–clinoferrosilite series (see p. 30). The average size of cation in *M*2 is small compared with diopside in which *M*2 is filled by Ca. This results in two structurally distinct Si–O chains; one chain (A) is more extended than the other (B) but not as fully extended as the (A) chain in jadeite. The O3–O3–O3 angles are 170° and 149° for the A and B chain respectively as compared with 175° for the single chain in jadeite. The tetrahedra of A chains are *S* rotated while those of B chains are *O* rotated (see Thompson, 1970). The space group of pigeonite is $P2_1/c$ as compared with $C2/c$ in diopside. As in the calcium-free pyroxenes, the *M*1 and *M*2 cations are in general positions as compared with special positions on the diad axis in the structure of diopside.

The X-ray reflections used in the structure refinements of pigeonites can be divided into two classes, those with $(h+k)$ even and those with $(h+k)$ odd. The latter reflections, often referred to as (b) reflections, are attributable entirely to the deviations of the structure from conformity with the $C2/c$ space group (non-equivalence of Si–O chains etc.). The (b) reflections also show different degrees of diffuseness for different pigeonite specimens. It should be noted that the structure refinements of pigeonites have treated the different classes of reflections in different ways. Some have ignored the (b) reflections entirely, some have used all of the reflections and others have carried out refinements based upon the (b) reflections only.

Morimoto and Güven (1970), ignoring the somewhat diffuse (b) reflections, determined the structure of a pigeonite $Mg_{0\cdot39}Fe_{0\cdot52}Ca_{0\cdot09}SiO_3$ from Mull, with *a* 9·706, *b* 8·950, *c* 5·246 Å, *β* 108·59°. They showed that the *M*1 site has almost regular octahedral co-ordination and is occupied by 0·73 Mg and 0·27 Fe, while the *M*2 cations are in rather irregular seven-fold co-ordination (eight-fold in diopside) and the site occupation is 0·05 Mg, 0·77 Fe and 0·18 Ca. They described the structure as intermediate between those of clinoferrosilite and diopside. Unusually large temperature factors determined for this structure were attributed to positional rather than thermal disorder. It was suggested that the calcium atoms would locally have the diopside-type co-ordination and that the net result of the structure determination would be a space average between this and the calcium-free structure.

The domain structure described for clinoenstatite (p. 33) exists also for pigeonites. At high temperatures the $P2_1/c$ structure transforms to $C2/c$ in which all Si–O chains are equivalent. In cooling through the transition temperature equivalent chains become either (A)- or (B)-type and hence out-of-step domains can occur. It is possible that mistakes or twins can occur and give rise to this feature (see Iijima and Buseck, 1975). The domain sizes appear to vary from 100 to 500 Å for pigeonites from volcanic rocks but are greater than 1000 Å for those from slowly cooled intrusive rocks. The diffuseness of the (b) X-ray reflections is attributed to the domain structure, with smaller domains producing more diffuse reflections. Correlations between shapes and sizes of domains and diffuseness of (b) reflections have been demonstrated by Ghose *et al.* (1972) and by Hamil *et al.* (1975). Domain structures have also been observed directly by electron microscopy (Fig. 87), e.g. by Christie *et al.* (1971), Champness *et al.* (1971), and for the pigeonite from Mull, the observed domain sizes agreed with those estimated from X-ray data (Champness and Lorimer, 1972). Although domain sizes appear to be a function of cooling rate, they may also vary with composition since the temperature of transition from the *P* to *C* structure is composition dependent. When the domain structure is produced it is thought that calcium atoms tend to concentrate at domain boundaries (see Ohashi and Finger, 1973). Brown *et al.* (1972) observed that heating a pigeonite above the inversion temperature followed by rapid quenching serves to sharpen the (b) reflections, i.e. enlarges the domains. Williams and Takeda (1972), however, showed that a synthetic pigeonite cooled rapidly from the melt nevertheless exhibited sharp (b) reflections.

Other determinations of the 'average' structure of pigeonite similar to that by Morimoto and Güven (1970) have been by Takeda (1972), Brown *et al.* (1971) and Brown and Wechsler (1973). All three were lunar pigeonites.

The structure of another lunar pigeonite was determined by Clark *et al.* (1971). This pigeonite has *a* 9·678, *b* 8·905, *c* 5·227 Å, *β* 108·71°, and its chemical formula is $Ca_{0·163}Fe_{0·718}Mg_{1·05}Ti_{0·026}Al_{0·043}(Al_{0·017}Si_{1·983})O_6$. From the broadness of its (b) reflections the domain thickness was estimated as about 100 Å. The results of the structure determination were very similar to those for the Mull pigeonite, but refinement using only the (b) reflections gave more reasonable temperature factors. For the *M*2 cations six *M*–O distances are less than 2·557 Å and the next smallest is 3·021 Å. The *M*2 co-ordination is therefore better described as six-fold rather than seven-fold. Ignoring the seventh oxygen, only one edge of the *M*2 polyhedron is shared with an Si–O tetrahedron. This is similar to the situation found in enstatite and in clinoferrosilite. The structure is therefore described as essentially one of a Ca-free clinopyroxene $Mg_{0·54}Fe_{0·46}SiO_3$ existing in the domains, the domain boundaries being Ca-rich regions with a diopside-like structure. The structure within the domains, therefore, is not itself a statistical mean structure like that which emerges from refining all data or only '$(h+k)$ even' data. Similar results were obtained for a lunar pigeonite by Ohashi and Finger (1973).

High Pigeonite

Like the Ca-free pyroxenes, the pigeonites also exhibit polymorphism with $P2_1/c$ at low and $C2/c$ at high temperatures (Morimoto and Tokonami, 1969a; Smyth, 1969; Prewitt *et al.* 1970). The transformation is displacive and rapid and cannot be quenched. The crystal structure of high pigeonite has been determined by Brown *et al.* (1972) for a crystal $Wo_9En_{39}Fs_{52}$ from Mull. It has previously been suggested (Morimoto and Tokonami, 1969a), that in the high-temperature structure the Si–O chains were only apparently equivalent and were really a space average of the two kinds shown in $P2_1/c$. The structure determination showed this not to be so, and that the chains are truly equivalent. In comparison with the low-temperature structure the *M*–O polyhedra are seen to expand more than do the Si–O tetrahedra, and the Si–O (B) chains become more straight. In the high pigeonite the O3–O3–O3 angle is 173·4° for both chains, whereas in the low pigeonite it is 173·9 for (A) and 151° for (B). The *M*2–O co-ordination changes in detail but remains six-fold.

Prewitt *et al.* (1971), by studies on lunar pigeonites, showed that the low–high transition temperature depended greatly upon Fe content, being about 1000°C for a magnesium-rich pigeonite and 200°C for one with composition $Fs_{85}Wo_{15}$. They also noted a hysteresis effect in that once a specimen had been heated it transformed subsequently at a lower temperature. Expansion coefficients were also determined and it was noted that *c* of pigeonites always expands to about 5·33 Å at the transition temperature regardless of the value of *c* at room temperature.

The precise relationship between the *P–C* inversion process and exsolution needs further investigation.

Yang (1973) showed that an iron-free pigeonite with $Ca_{0\cdot05}$ to $Ca_{0\cdot09}$ can exist at high temperatures. The range of Ca content decreases with decreasing temperature disappearing at 1276°C and $Ca_{0\cdot07}$. Between 1276° and 1432°C, protonestatite coexists with this iron-free pigeonite (see p. 179). Further polymorphic transformations of pigeonite from *C*2/*c* to *P*2$_1$/*c* and *Pbca* pyroxene at about 1360° and 1410°C respectively, are reported by Schwab and Schwerin (1975).

Fe–Mg Distribution

As with the Ca-free pyroxenes, crystal structure refinements and Mössbauer spectroscopy show that Fe atoms are preferentially sited in *M*2 rather than *M*1. The extent to which *M*2 is preferred can be expressed by the exchange energy difference ΔG_E. Some results for pigeonites compared with some for orthopyroxenes are given in Table 12 (from Smyth, 1974). Whether ΔG_E is always greater for Ca-rich than for

Table 12 Distribution coefficients and Mg–Fe exchange energies of low-Ca pyroxenes (Smyth, 1974)

Pyroxene	Composition			T(°C)	k[a]	ΔG_E kcal/mole	Reference
	En	Fs	Wo				
Clinopyroxenes							
Lunar 12021, 150	59	32	9	555	0·08		Hafner *et al.* (1971)
				800	0·15	4·1	Hafner *et al.* (1971)
				1000	0·18	4·3	Hafner *et al.* (1971)
Lunar 14053, 47	58	30	12	1000	0·16	4·7	Schürmann and Hafner (1972)
Mull	39	52	9	960	0·14	4·9	Brown *et al.* (1972)
B1-9. Heated opx.	31	67	1·5	700	0·11	4·2	Smyth (1974)
				760	0·12	4·3	Smyth (1974)
				825	0·16	4·0	Smyth (1974)
Orthopyroxenes							
B1-9	31	67	1·5	500	0·08	4·0	Smyth (1973)
				700	0·14	3·8	Smyth (1973)
				850	0·18	3·8	Smyth (1973)
Lunar 14310, 116	70	24	6	1000	0·21	4·0	Schürmann and Hafner (1972)
3209-40	42	56	2	600	0·12	3·7	Virgo and Hafner (1969)
				700	0·14	3·8	Virgo and Hafner (1969)
				800	0·18	3·6	Virgo and Hafner (1969)
				1000	0·22	3·9	Virgo and Hafner (1969)

[a] $k = (Fe/Mg)_{M1}/(Fe/Mg)_{M2}$.

Ca-poor pyroxenes is unclear at present. While for orthopyroxenes, Mössbauer and X-ray results have been in fairly good accord, there are some discrepancies for Ca-bearing pyroxenes. Virgo and Hafner (1969) concluded from Mössbauer studies that ΔG_E is higher in augite than in orthopyroxene, and similarly in augite compared with pigeonite. For nearly Ca-free pyroxene they gave ΔG_E as approximately

3·6 kcal/mole, and two or three times this for a calcium-rich pyroxene. Hafner *et al.* (1971), from Mössbauer studies of pigeonites from lunar basalts found Mg almost entirely in *M*1. In this instance they concluded that there is a two-stage ordering process: short range above 580°–600°C and long-range ordering below this temperature. They deduced that at high temperatures (approximately 1000°C) the (Fe,Mg) distribution cannot be quenched in, but at lower temperatures (approximately 500°C) it can. Brown *et al.* (1972), in their X-ray study of the crystal structures of high and low pigeonites, obtained a distribution constant of K_D 0·045 at room temperature and 0·134 at 960°C, which gave ΔG_E approximately equal to 4·9 kcal/mole at 960°C. This, in conformity with Mössbauer results, is higher than ΔG_E (1000°C) of approximately 3·66 kcal/mole for a metamorphic orthopyroxene (Virgo and Hafner, 1969), and considerably less than ΔG_E (1000°C) approximately equal to 8 kcal/mole for a lunar augite (Hafner and Virgo, 1970). However, crystal structure determinations of coexisting lunar pigeonite and augite by Takeda (1972) showed that the augite is less ordered than the pigeonite, and a similar result was found for augite and aluminous orthopyroxene coexisting in an alkali olivine basalt. He suggests that the unusual *M*2 co-ordination in subcalcic augites may be too unfavourable for Fe^{2+} to gain much crystal field stabilization energy (see p. 295).

There has been considerable discussion of the difficulty of obtaining correct Fe/Mg distributions from Mössbauer spectroscopy on Ca-bearing pyroxenes, with a tendency to overestimate Fe in *M*2, particularly in the Fe-rich pyroxenes. Williams *et al.* (1971) attributed this to the existence of domains in the *C*2/*c* pyroxene but these have not been directly observed. Dowty *et al.* (1972) suggest rather that the correct interpretation of the Mössbauer spectrum must take into account next nearest neighbour influences, and also that the effects of exsolution are perhaps not fully understood. Dowty and Lindsley (1973) obtained Mössbauer spectra at liquid nitrogen temperatures for pyroxenes on the hedenbergite–ferrosilite join synthesized at high temperatures and pressures. They confirmed that different next nearest neighbour configurations influenced the peaks for *M*1 which overlap those for *M*2, and that the resulting complexity makes interpretation of Mössbauer spectra of high Ca pyroxenes difficult even at liquid nitrogen temperatures. They also confirmed that exsolution can cause difficulties.

There are other difficulties concerned with the estimation of Fe,Mg distribution in pigeonites. As noted earlier, some structure determinations give only average structures, and the meaning of Fe,Mg distribution in such circumstances is not clear. Mössbauer results, even if correct, could appear to be wrong if calcium which is really in diopside-like domain boundaries is assumed to be additional to Fe and Mg within the domains. As noted by Ghose *et al.* (1972) some Mössbauer results can clearly give only 'bulk order' because of exsolution and also varying composition within pyroxene grains. Brown *et al.* (1972) in comparing their results with those for the Mull pigeonite (Hafner *et al.*, 1971) suggest that ΔG_E for clinopyroxene increases with Fe content.

Assuming that the Fe,Mg distribution can be accurately determined, the deduction of thermal history for pigeonite is still not straightforward. As with Ca-free pyroxenes, there may be indication of high order and slow cooling from a specimen which has in fact been quenched from a high temperature and subsequently re-heated. The presence or absence of exsolution lamellae, and their coarseness or fineness if present, can be a useful indication of cooling rate, and so also can the pigeonite domain structure. A combination of X-ray, electron microscope and Mössbauer studies may be needed to unravel a complex thermal history (see for example, Ghose *et al.*, 1972).

Exsolution

The way in which exsolution lamellae of augite can occur in pigeonite has been studied by Morimoto and Tokonami (1969b), who calculated the strain energies as 0·02 kcal/mole and 0·04 kcal/mole for '(001)' and '(100)' lamellae respectively. The compositions for host and lamellae were $En_{48}Fs_{46}Wo_6$ and $En_{36}Fs_{21}Wo_{43}$, for the

Fig. 87. Electron micrograph showing a modulated structure (001) consistent with spinodal decomposition in a lunar pigeonite (Champness and Lorimer, 1976).

sample studied (from the Moore County meteorite). The misfit in cell parameters as between augite and pigeonite, and the consequent ideal exsolution plane is discussed more fully under augite (p. 343).

Electron microscope studies of exsolution in pyroxenes have been reviewed by Champness and Lorimer (1976). An example of exsolution by spinodal decomposition in a basaltic pigeonite is illustrated in Fig. 87. By contrast, clinopyroxenes from plutonic rocks tend to show microstructures which are consistent with a heterogeneous nucleation and growth exsolution process.

Chemistry

Analyses of twelve pigeonites and four inverted pigeonites, together with the numbers of ions on the basis of six oxygens, are given in Table 13. The major variation in pigeonite compositions is due to the replacement of Mg by Fe^{2+}, and the range of composition is approximately from Mg:Fe 70:30 to 30:70. The CaO content of the pigeonite analyses shown in Table 13 varies between 3·14 and 7·05 per cent, corresponding respectively with 0·13 and 0·22 Ca atoms per mole. With smaller amounts of calcium there is a gradation to the structurally similar subcalcic pigeonite (clinoenstatite–clinoferrosilite), but such compositions are restricted to products of exsolution from augites and ferroaugites. The development of this subcalcic phase is generally considered to result from re-equilibration, subsequent to the initial exsolution, the augite becoming richer in calcium and the exsolved pigeonite less calcic. Clinopyroxenes containing between 15 and 30 per cent of the $CaSiO_3$ component are generally regarded as subcalcic augites. They are probably metastable at subsolidus temperatures, and may crystallize either as calcium-rich pigeonite or subcalcic augite with space group $P2_1/c$ or $C2/c$ respectively.

The content of Al_2O_3 does not usually exceed 2 per cent and in the majority of pigeonites Al is mainly in tetrahedral co-ordination. In a substantial proportion of pigeonites the content of Al, together with Si, is insufficient to bring the Z group up to the ideal two atoms per mole, and in some minerals it is necessary to include Fe^{3+}, as well as Ti, with the Z group atoms. The amount of manganese present in most pigeonites is relatively high even in the more magnesium-rich varieties, thus the average content of MnO is 0·42 weight per cent in the eight pigeonites of Table 13. The ferropigeonites, although showing no consistent trend of increasing manganese with enrichment in iron, generally contain more manganese (the average content of MnO in the five ferropigeonites is 1·08 weight per cent). More manganese-rich pigeonites have been described, and Bonnichsen (1969) has reported ferropigeonite with 13·0 per cent MnO ($Ca_{0\cdot20}Mg_{0\cdot16}Fe_{1\cdot10}Mn_{0\cdot47}Si_{2\cdot04}O_6$) from the Biwabik Iron Formation, Minnesota.

The polymorphism of iron-free pigeonite at high temperature, using high-temperature X-ray diffractometry and thermochemical methods, has been investigated by Schwab and Schwerin (1975). The temperatures of the polymorphic, non-quenchable, transformations and the space group of the various phases for the composition $Ca_{0\cdot05}Mg_{0\cdot95}SiO_3$ ($CaMgSi_2O_6$ 10 mol. per cent), are:

$$\underset{(P2_1/c)}{\text{pigeonite}} \overset{1050^\circ}{\rightleftharpoons} \underset{(C2/c)}{\text{high pigeonite}} \overset{1360^\circ}{\rightleftharpoons} \underset{(P2_1/c)}{\text{protopigeonite}} \overset{1410^\circ}{\rightleftharpoons} \underset{Pbca}{\text{orthopyroxene}}$$

The solid solution of $CaMgSi_2O_6$ in Fe-free pigeonite increases with temperature

Table 13. Pigeonite Analyses

	1		2		3		4		5	
SiO_2	52·84		50·40		50·35		50·61		51·47	
TiO_2	0·22		0·55		0·55		0·65		0·29	
Al_2O_3	0·44		1·99		2·23		1·68		1·56	
Fe_2O_3	1·06		0·13		1·14		0·84		1·42	
FeO	16·89		21·30		21·12		21·35		21·72	
MnO	0·56		—		0·38		0·39		0·52	
MgO	23·51		18·28		20·03		18·71		21·68	
CaO	4·06		6·43		4·50		5·52		1·45	
Na_2O	0·19		1·33		—		0·11		0·07	
K_2O	0·00		0·02		—		0·00		0·03	
H_2O^+	—		—		—		0·10		—	
H_2O^-	0·22		—		—		—		0·02	
Total	99·99		100·43		100·30		99·96		10·23	
α	—		—				1·696		—	
β	—		—		—		1·698		—	
γ	—		—		1·714		1·721		—	
γ:z	—		—		—		41°		—	
$2V_\gamma$	—		—		127°[f]		25°		—	
D	—		—		—		3·383		—	
	Numbers of ions on the basis of six O									
Si	1·955	2·00[a]	1·902	2·00[b]	1·895	2·00[c]	1·919	2·00[d]	1·927	2·00[e]
Al	0·018		0·088		0·099		0·076		0·067	
Al	0·000		0·000		0·000		—		0·000	
Ti	0·006		0 010		0·016		0·018		0·009	
Fe^{3+}	0·003		0·000		0·028		0·019		0·034	
Mg	1·296		1·031		1·123		1·057		1·210	
Fe^{2+}	0·523	2·02	0·674	2·07	0·665	2·03	0·677	2 02	0·679	2·01
Mn	0·017		—		0·012		0·012		0·016	
Ca	0·161		0·261		0·181		0·224		0·058	
Na	0·014		0·097		—		0·008		0·004	
K	0·000		0·001		—		—		0·000	
Mg	63·9		52·3		55·8		53·0		60·4	
*ΣFe	28·1		34·4		35·2		35·8		36·7	
Ca	8·0		13·3		9·0		11·2		2·9	

* $\Sigma Fe = Fe^{2+} + Fe^{3+} + Mn$.
[a] Includes 0·026 Fe^{3+}.
[b] Includes 0·004 Fe^{3+}, 0·006 Ti^{4+}.
[c] Includes 0·005 Fe^{3+}.
[d] Includes 0·005 Fe^{3+}.
[e] Includes 0·006 Fe^{3+}.
[f] These values related to the orthopyroxene to which the original pigeonite has inverted.

1 Magnesian pigeonite, hypersthene–olivine andesite, Hakone, Japan (Kuno and Nagashima, 1952). Anal. K. Nagashima.
2 Magnesian pigeonite, augite–pigeonite–hypersthene andesite, Hakone, Japan (Kuno, 1955). Anal. T. Sameshima.
3 Inverted pigeonite, olivine gabbro, Skaergaard, Kangerdlugssuaq, east Greenland (Brown, 1957). Anal. P. E. Brown.
4 Pigeonite, fine-grained dolerite, Hindubagh, Pakistan (Bilgrami, 1964). Anal. S. A. Bilgrami.
5 Inverted pigeonite, gabbro, Bushveld complex (Atkins, 1969). Anal. F. B. Atkins (includes Cr_2O_3 < 0·01; trace element data for Cr, Ni, Co, Cu, V, Zr, Sc, Sr, Ba).

Table 13. Pigeonite Analyses – *continued*

	6	7	8	9	10
SiO_2	50·56	51·53	51·13	49·37	50·5
TiO_2	0·58	0·51	0·50	0·62	0·4
Al_2O_3	1·41	1·64	1·76	1·55	1·3
Fe_2O_3	0·12	0·18	4·27	1·83	0·7
FeO	23·17	23·35	20·17	26·44	27·0
MnO	0·54	0·49	0·47	0·37	0·2
MgO	16·10	17·27	16·51	15·54	15·5
CaO	7·05	4·47	4·50	4·60	4·1
Na_2O	0·26	0·10	0·28	—	0·2
K_2O	0·23	0·00	0·06	—	0·1
H_2O^+	—	0·32	0·00	—	—
H_2O^-	0·07	0·14	0·00	—	—
Total	100·09	100·00	99·65	100·32	100·0
α	—	1·6980	1·695	1·7085	—
β	—	1·6988	1·719	1·7095	—
γ	—	1·7228	1·722	1·736	1·727
$\gamma{:}z$	—	40°–41°	—	40°	—
$2V_\gamma$	—	12°–23° (aver. 18°)	22°–25°	21°	134°[f]
D	—	—	3·173	—	—

Numbers of ions on the basis of six O

	6		7		8		9		10	
Si	1·940	2·00	1·963	2·00	1·943	2·00	1·911	2·00[a]	1·95	2·00
Al	0·060		0·037		0·057		0·070		0·05	
Al	0·004		0·036		0·022		—		0·01	
Ti	0·017		0·015		0·014		0·018		0·01	
Fe^{3+}	0·003		0·005		0·122		0·033		0·02	
Mg	0·921		0·981		0·935		0·896		0·89	
Fe^{2+}	0·734	2·02	0·744	1·99	0·640	1·96	0·856	2·01	0·87	1·99
Mn	0·017		0·016		0·015		0·012		0·01	
Ca	0·290		0·182		0·183		0·191		0·17	
Na	0·019		0·007		0·021		—		0·01	
K	0·010		—		0·003		—		0·00	
Mg	46·7		50·9		49·3		44·6		45·5	
*ΣFe	38·6		39·7		41·0		45·9		45·9	
Ca	14·7		9·4		9·7		9·5		8·6	

[a] Includes 0·019 Fe^{3+}.

f These values related to the orthopyroxene to which the original pigeonite has inverted.

6 Pigeonite, andesite, Hakone Volcano, Japan (Kuno, 1955). Anal. K. Nagashima.

7 Pigeonite, coarse diabase pegmatite, Goose Creek, Virginia (Hess, 1949). Anal. L. C. Peck (analysis corrected for augite impurity: original sample $\frac{5}{8}$ pigeonite, $\frac{3}{8}$ augite).

8 Pigeonite, andesite, Mt. Malyy, Sinyak, Transcarpathia (Ginzburg *et al.*, 1964). Anal. R. L. Teleshova, L. S. Abramova and D. A. Pchelintsev (*a* 9·720, *b* 8·964, *c* 5·255, *β* 109°24′).

9 Ferropigeonite, meteorite, Moore County, North Carolina (Hess and Henderson, 1949). Anal. E. P. Henderson.

10 Inverted ferropigeonite, ferrodiorite, Skaergaard, Kangerdlugssuaq, east Greenland (Brown, 1957). Anal. E. A. Vincent.

11	12	13	14	15	16	
51·24	49·30	48·9	49·72	48·36	49·17	SiO_2
0·50	0·60	0·5	0·85	0·39	0·68	TiO_2
1·07	0·68	1·5	0·90	0·41	1·42	Al_2O_3
0·05	3·83	2·1	1·72	1·36	1·30	Fe_2O_3
26·85	23·17	28·7	27·77	32·86	32·93	FeO
0·55	2·44	0·6	0·98	0·84	0·59	MnO
14·85	15·38	13·8	12·69	11·34	8·87	MgO
4·31	3·14	3·6	3·80	4·58	5·25	CaO
0·02	—	0·2	0·23	tr.	0·30	Na_2O
0·02	—	0·0	0·12	tr.	0·21	K_2O
0·36	—	—	1·27	0·15	0·18	H_2O^+
0·18	—	0·0	0·08	tr.	0·06	H_2O^-
100·00	98·54	99·9	100·13	100·33	'99·96'	Total
1·7055	1·713	—	1·7137	1·715–1·732	1·710	α
1·7066	1·713 (range 0·003)	—	1·7137	1·715–1·732	1·712	β
1·7325	—	—	1·7417	1·740–1·757	1·742	γ
42°	—	—	39°–42°	40°	32°	$\gamma{:}z$
14°–29° (aver. 22°)	11° (range 20°)	—	0°–12°	16°–5°	14°	$2V_\gamma$
—	—	—	3·44	—	—	D
1·985 } 2·00	1·936 } 2·00[a]	1·923 } 2·00[b]	1·968 } 2·00	1·940 } 2·00[c]	1·958 } 2·00	Si
0·015	0·032	0·069	0·032	0·019	0·042	Al
0·034	0·000	0·000	0·010	0·000	0·025	Al
0·015	0·018	0·012	0·025	0·012	0·020	Ti
0·001	0·081	0·052	0·051	0·000	0·039	Fe^{3+}
0·857	0·900	0·810	0·749	0·678	0·526	Mg
0·870 } 1·97	0·761 } 1·97	0·945 } 2·01	0·919 } 1·97	1·102 } 2·02	1·096 } 1·98	Fe^{2+}
0·018	0·081	0·021	0·033	0·028	0·020	Mn
0·171	0·132	0·151	0·161	0·197	0·224	Ca
0·001	—	0·014	0·018	—	0·023	Na
0·001	—	0·002	0·006	—	0·010	K
44·5	45·3	40·7	39·2	33	27·6	Mg
46·2	48·1	51·7	52·4	57	60·6	*ΣFe
9·3	6·6	7·6	8·4	10	11·8	Ca

[a] Includes 0·032 Fe^{3+}.
[b] Includes 0·009 Fe^{3+}.
[c] Includes 0·041 Fe^{3+}.

11 Ferropigeonite, diabase sill, south of Lambertville, New Jersey (Hess, 1949). Anal. L. C. Peck (analysis corrected for 17 per cent augite).
12 Ferropigeonite, remelted quartz diorite, ejected block in pumice, Hakone Volcano (Kuno, 1955). Anal. T. Katsura.
13 Inverted ferropigeonite, ferrogabbro, Bushveld complex (Atkins, 1969). Anal. F. B. Atkins.
14 Ferropigeonite, andesite, Mull, Scotland (Hallimond, 1914). Anal. E. G. Radley (includes (Co,Ni)O nil, Li_2O trace).
15 Ferropigeonite, phenocryst, andesite, Asio, Japan (Kuno, 1969). Anal. H. Haramura (includes P_2O_5 0·04).
16 Ferropigeonite, dolerite sill, Kola Peninsula (Sinitsÿn, 1965). Anal. E. F. Tumina, A. N. Filonova and A. P. Arkhangel'skaya.

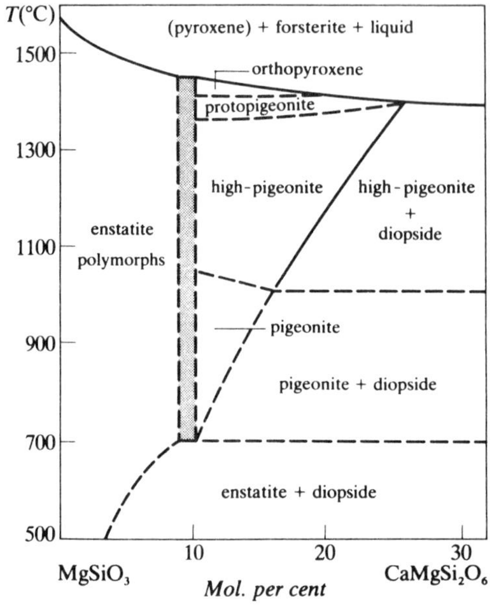

Fig. 88. $MgSiO_3$-rich part of the pseudobinary $MgSiO_3$–$CaMgSi_2O_6$ system (after Schwab and Schwerin, 1975). Stippled area, transition range between pigeonite and enstatite polymorphs.

and reaches a maximum of 26 mol. per cent in high pigeonite at $\simeq$ 1390°C, and then decreases to a maximum of 19 mol. per cent in the high-temperature orthopyroxene phase (Fig. 88).

Pigeonite containing less than 20 mol. per cent $FeSiO_3$ has not been reported from igneous rocks. An assemblage consisting of a low-iron pigeonite, low-Ca augite, olivine and plagioclase has, however, been described from the Allende and Vigarano meteorites (Yang, 1973). In most saturated basic magmas crystallization of the monoclinic calcium-poor phase does not occur until the Mg:Fe ratio of the Ca-poor pyroxene is $\simeq Mg_{70}Fe_{30}$. Subsequently in a plutonic environment, increasingly more iron-rich pigeonites crystallize together with augite or ferroaugite as cumulus phases. The composition of the pigeonite at the two-pyroxene boundary (see p. 334), at which stage in the fractionation the Ca-poor clinopyroxene is replaced by fayalitic olivine, varies for different intrusions. In the Skaergaard and Bushveld intrusions the most Fe-rich cumulus pigeonites (now inverted) crystallized with compositions $Ca_{10}Mg_{33}Fe_{57}$ and $Ca_9Mg_{28}Fe_{63}$ respectively. In the strongly differentiated mangerite of the layered intrusion of Bjerkrem-Sogndal, Norway (Duchesne, 1972) the composition of the inverted ferropigeonite is $Ca_8Mg_{19}Fe_{73}$.

The compositional gap between the microphenocrysts of pigeonite (4·06 weight per cent CaO, Table 13, anal. 1) and orthopyroxene (2·67 per cent CaO) in the groundmass of the hypersthene–olivine andesite of the Hakone volcano, Japan was considered by Kuno and Nagashima (1952) to represent the compositional change across the monoclinic$\rightleftharpoons$orthorhombic inversion interval. This relatively large compositional difference is in contrast to the inversion curve determined by Bowen and Schairer (1935). The latter, however, related to the pure $(Mg,Fe)SiO_3$ series, and the effect of the addition of Ca appears to result in increasing the magnitude of the inversion interval. A compositional gap, comparable with that between the pigeonite and orthopyroxene of the Hakone andesite, has been reported for Fe-free pigeonite and magnesium orthopyroxene in the forsterite–diopside–silica system (see p. 239).

The possible relationship between the crystallization curve of Ca-poor pyroxenes

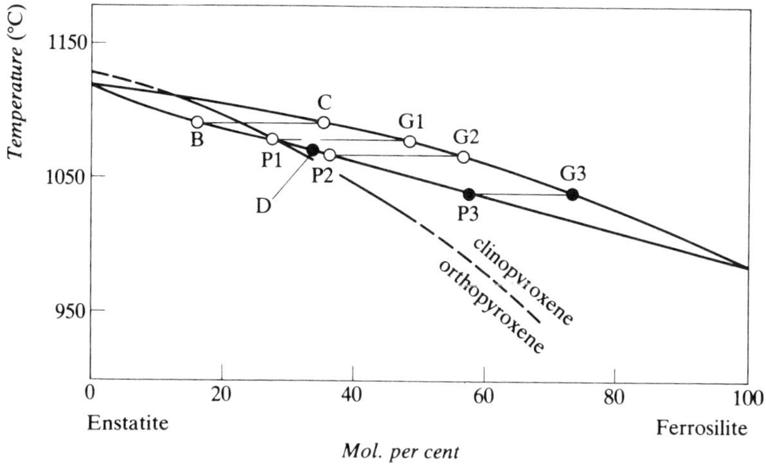

Fig. 89. Possible relationship between the crystallization curves of calcium-poor pyroxenes and the orthorhombic–monoclinic inversion curve. D. pigeonite of chilled marginal gabbro, B orthopyroxene of gabbro picrite, Skaergaard intrusion, crystallized at temperature C: P_1G_1–P_3G_3, coexisting phenocryst and groundmass pyroxenes of Japanese lavas (after Brown, 1957).

and the monoclinic ⇌ orthorhombic inversion has been illustrated by Brown (1957). In Fig. 89 the orthopyroxene of the gabbro picrite, and the inverted pigeonite of the Skaergaard chilled marginal gabbro, and three pigeonite phenocrysts of some Japanese lavas, fall on a smooth solidus curve; the groundmass pyroxenes associated with the phenocrysts form a smooth liquidus (see also Vilenskiy, 1966). The crystallization of Ca-poor clinopyroxene in place of orthopyroxene occurs when the crystallization curve of the calcium-poor pyroxene intersects the monoclinic ⇌ orthorhombic inversion curve, and variations in ratio at which this occurs (commonly at a Mg/Fe ratio of about 70:30) may be correlated with the different temperatures of fractionating liquids in individual intrusions.

During the fractionation of basic liquids the crystallizing pigeonites have higher Fe/Mg ratios than the associated Ca-rich pyroxenes. Thus in the hortonolite ferrogabbro of the Beaver Bay intrusion, Minnesota (Nakamura and Konda, 1974), ferropigeonite, $Ca_{10.4}Mg_{40.3}Fe_{49.3}$ (range, including exsolved augite lamellae, $Ca_{11}Mg_{41}Fe_{48}$–$Ca_{8}Mg_{36}Fe_{56}$), occurs mantling augite $Ca_{36.5}Mg_{33.9}Fe_{29.6}$ in composition, and is a product of an early stage of crystallization of the intercumulus liquid, but the two phases are considered to have been close to that of equilibrium at the time of crystallization. The composition of the ferropigeonite, exclusive of the exsolved augite lamellae, is $Ca_{5.3}Mg_{40.9}Fe_{53.8}$. The presence of uninverted ferropigeonite in these ferrogabbros is atypical and probably related to the faster cooling rate of this intrusion compared with those of larger layered masses. The pigeonite–orthopyroxene compositional relationships in the ferrogabbros of the Beaver Bay intrusion are unusual in that the rhombic pyroxenes ($Ca_{3.0}Mg_{39.2}Fe_{57.8}$–$Ca_{3.5}Mg_{32.7}Fe_{63.8}$ have higher Fe/Mg ratios than the associated pigeonite. The ferrohypersthene which does not contain exsolution lamellae, however, crystallized directly from late intercumulus liquids, after the pigeonite, and it is likely that the two phases do not represent an equilibrium association.

The main factors, silica activity, temperature, pressure and oxygen fugacity, affecting the crystallization relationships of Ca-poor and Ca-rich pyroxenes under plutonic conditions are discussed more fully in the augite section (p. 335, see also Campbell and Nolan, 1974).

The compositions of the calcium-poor pigeonite and the calcium-rich augite which crystallize under plutonic conditions indicate that there is a field of immiscibility between the two pyroxenes. Similar evidence is, however, less evident from volcanic rocks in which zoned pyroxenes are common, and Kuno (1950, 1955) suggested that a complete range of composition between augite and pigeonite may occur, under certain conditions, in the rapidly crystallized clinopyroxenes of volcanic rocks. Much of the support for the hypothesis of continuous variation has been based on the evidence of optical axial angles of zoned crystals in quickly chilled rocks, but there are few records of pyroxenes showing continuous zoning between augite and pigeonite. Poldervaart and Hess (1951) did not accept complete miscibility between pigeonite and augite in volcanic rocks, and suggested that the amount of solid solution between diopside–hedenbergite and clinoenstatite–ferrosilite is a function more of the temperature than the speed of crystallization, the latter factor merely preserving but not controlling the amount of solid solution at the temperature of crystallization. It is now clear that complete miscibility between pigeonite and augite does not occur in terrestrial rocks, as shown by the coexistence of Ca-poor and Ca-rich clinopyroxenes in both volcanic as well as plutonic rocks.

The thermal relations of Ca-poor pyroxenes at or close to atmospheric pressures have been discussed by Brown (1972), and his conclusions are illustrated in Fig. 90. In particular the probability that high-temperature pigeonite, space group *C*2/*c*, may crystallize at liquidus temperatures and later undergo transformation to a

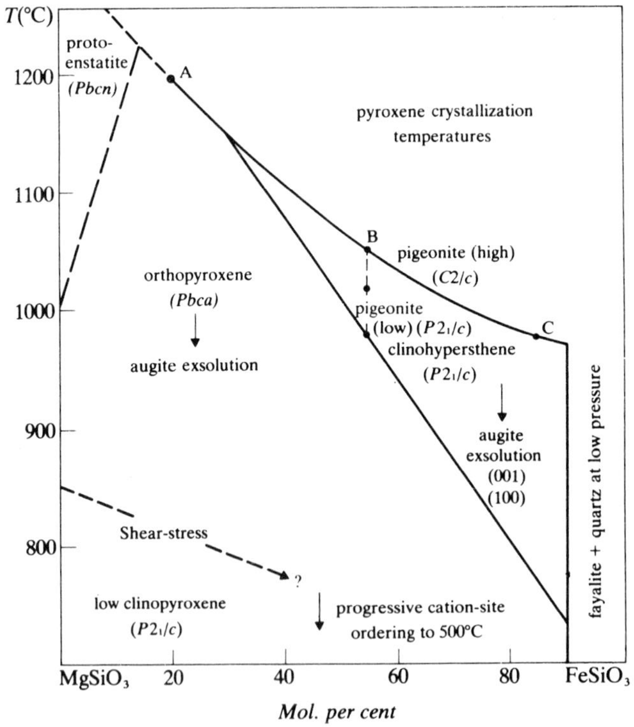

Fig. 90. Hypothetical thermal relations at 1 atm among Ca-poor pyroxenes (after Brown, 1972). Data for bronzite (A) and pigeonite (B) including temperatures of notable exsolution and of inversion from Brown (1968). Crystallization temperature for ferropigeonite (C) assumed from data on ferroaugite (Lindsley and Munoz, 1969). The relations show the small cooling range, especially for Mg-rich composition, within which *C*2/*c* pigeonite could crystallize.

$P2_1/c$ structure accompanied by exsolution of augite is indicated, evidence of which is provided by the diffuseness of the characteristic ($h+k$ odd) pigeonite reflections (p. 163) reported for some pigeonite crystals. Thus pigeonite with a Fe/(Mg + Fe) ratio of $\simeq$ 0·55 would crystallize with space group $C2/c$, at approximately 1050°C, undergo transformation to the $P2_1/c$ structure with exsolution of some augite at 1020°C before finally inverting to orthopyroxene at 980°C. At lower temperatures, to about 500°C, further (100) exsolution from orthopyroxene and cation migration takes place (Boyd and Brown, 1969). The figure also shows the low-temperature clinoenstatite field found by Boyd and England (1965) during their investigation of the rhombic enstatite–clinoenstatite inversion. This has subsequently been attributed to shearing stress developed during the experimental procedure, and the absence of this phase in most metamorphic rocks can be ascribed to the persistence of temperatures greater than 600°C after shear stress had ceased.

In the $CaSiO_3$–$MgSiO_3$–$FeSiO_3$ system, and during the fractionation of basic magmas, the lower stability limit of pigeonite decreases as the Fe/Mg ratios of the pyroxenes increase. This series of liquid compositions along which pigeonite can coexist with augite and orthopyroxene may be represented by a univariant line, the so-called 'pigeonite eutectoid reaction line' of Ishii (1975).

In the groundmass of the basalt of Mihara-yama, Ō-shima island, and in the andesite of Akita-kamagatake, north-eastern Japan, pigeonites are in equilibrium with orthopyroxene and augite. The Fe/(Mg + Fe) ratios of the pigeonites of the two lavas are 0·30 and 0·37. The highest recorded temperatures during the eruption of the basalt and andesite are 1125° and 1090°C respectively corresponding with the equilibrium temperatures of the two three-pyroxene assemblages. These data, in conjunction with the lower stability limit of iron-free pigeonite determined by Yang (1973), and the experimental inversion temperature of a natural 'inverted pigeonite', $Ca_{7\cdot6}Mg_{40\cdot7}Fe_{51\cdot7}$, to pigeonite (Brown, 1968), have been used by Ishii (1975) to define the 'pigeonite eutectoid reaction line' as $T = 1270 - 480X_{Fe} \pm 20$ (T in °C, X_{Fe} the atomic ratio Fe/(Mg + Fe)). This is shown in Fig. 91 from which the formation

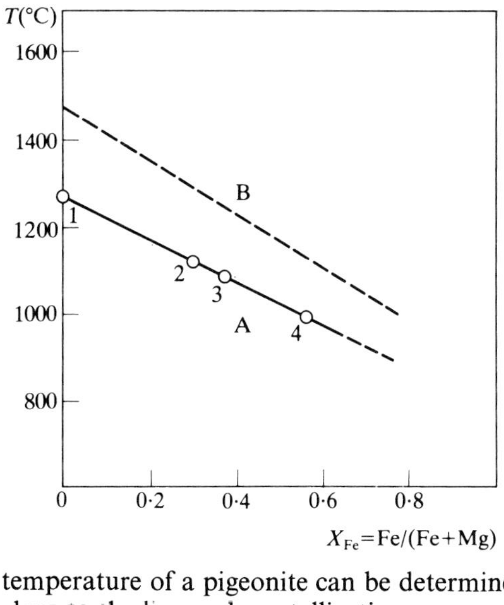

Fig. 91. The lower stability limit of pigeonite, or 'pigeonite eutectoid reaction line', A, at or near 1 atm, B at 20 kbar (after Ishii, 1975). (1) lower stability limit of iron-free pigeonite at 1 atm (Yang, 1973). (2) pigeonite, basalt, Mihara-yama, O-shima island. (3) pigeonite, andesite, Akita-kamagatake, north-eastern Japan. (4) pigeonite, $Ca_{7\cdot6}Mg_{40\cdot7}Fe_{51\cdot7}$ (Brown, 1968).

temperature of a pigeonite can be determined provided its composition lies on or close to the line and crystallization occurred at or near 1 atm. The lower stability limit of iron-free pigeonite increases linearly with pressure from 1270° at 1 atm to

1480°C at 20 kbar (Kushiro and Yoder, 1970), and of the natural inverted pigeonite from 1000° at 1 atm to 1140°C at 20 kbar (Brown, 1968). Thus the lower stability limit of pigeonite is raised by ≃ 10° and 7°C/kbar for Fe/(Mg+Fe) ratios 0·0 and 0·56 respectively, and can be represented by the equation:

$$T = 1270 - 480X_{Fe} + (10 - 5X_{Fe})P(\text{kbars})$$

A detailed investigation of the pyroxene relationships, and particularly their cooling history as revealed by the zonal patterns, in the lunar maria pyroxene-phyric basalts, has been presented by Dowty *et al.* (1974). These rocks are characterized by large zoned pyroxenes in a groundmass of iron-rich pyroxenes, plagioclase and ilmenite. A brief account of the zoned pyroxene in these basalts is given on p. 183 and in the augite section (p. 362). Typically the {110} sectors of the phenocrysts show a low-calcium magnesium pigeonite core grading outwards to a somewhat more Ca- and Fe-rich composition, succeeded by magnesian augite with zones to Fe-rich augite, and commonly one or more oscillations back to pigeonite. Different trends may be shown by {010} and {100} sectors which often display continuous zoning to more Fe-rich pigeonite. The groundmass pyroxenes are similar in composition to the more Fe-rich parts of the phenocrysts. From a detailed consideration of the porphyritic texture and zonal arrangements Dowty *et al.* concluded that the cooling history of the pyroxenes is related to a single-stage rapid cooling under supercooled conditions rather than a two-stage cooling in which the phenocrysts crystallized at depth and the groundmass on extrusion.

Pyroxenes, believed to have crystallized originally with compositions intermediate between pigeonite and augite, have been described from a number of lunar rocks. Thus intergrowths of pigeonite and augite derived by partial intragranular recrystallization of a primary subcalcic augite have been reported by Ross *et al.* (1970), and submicroscopic unmixed strongly zoned pyroxene, the two phases of the lighter fraction having compositions $Ca_7Mg_{65}Fe_{28}$ and $Ca_{44}Mg_{53}Fe_3$ and the denser fraction $Ca_9Mg_{26}Fe_{65}$ and $Ca_{37}Mg_{24}Fe_{39}$ respectively have been described by Bailey *et al.* (1970). A similar example of unmixing of an originally homogeneous pyroxene, in which subordinate (100), as well as (001) sectors are distributed in subgrains, 2–10 μm in diameter, with one lamellae orientation usually predominating in individual subgrains, has been reported by Radcliffe *et al.* (1970).

Exsolution

In many pigeonites some 10 per cent of the total divalent ions are calcium, an amount approximately three times in excess of the number of Ca ions that the orthopyroxene structure can accommodate. With slow cooling, however, the pigeonite exsolves some of the excess Ca ions as calcium-rich augite (Fig. 92) along planes parallel to '(001)'.[1] At the monoclinic⇌orthorhombic inversion temperature further exsolution accompanies the pigeonite inversion to orthopyroxene which retains the exsolved lamellae on the relict (001) plane of the original pigeonite. With intermediate rates of cooling, e.g. hypabyssal conditions, the exsolution of the calcium-rich phase is often incomplete before the inversion temperature is reached. Below the inversion temperature the calcium-rich phase is rapidly exsolved from the orthopyroxene, and commonly gives rise to a 'graphic intergrowth' between the orthopyroxene and exsolved augite (Walker and Poldervaart, 1949). The more

[1] Plane of exsolution is only approximately (001); see augite section.

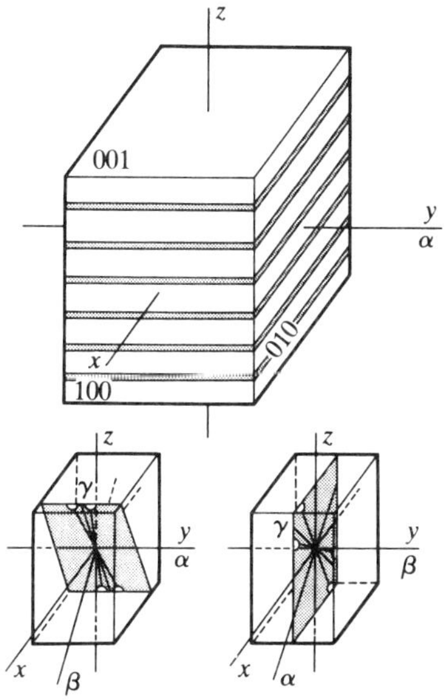

Fig. 92. Augite lamellae parallel to (001) in pigeonite (after Poldervaart and Hess, 1951). See also Frontispiece Figs. B and C.

ferriferous pigeonites generally do not show the same degree of exsolution, but fine augite lamellae parallel to (001) have been described (Brown, 1957). In volcanic and other quickly quenched rocks the pigeonite does not display optically visible exsolution lamellae, and such crystals have been considered to be chemically homogeneous. In some instances, however, this is not substantiated by electron microscopy. Thus optically homogeneous phenocrysts have been shown to contain lamellae 40 Å in width at the margins, and up to 3200 Å wide in the central parts of the Whin sill (Dunham *et al.*, 1972).

The (100) composition plane of twinned original pigeonite crystals persists in the orthopyroxene after inversion, in which the exsolved Ca-rich augitic lamellae mark the original (001) plane of the pigeonite. The earlier view (Poldervaart and Hess, 1951) that the *y* and *z* crystallographic axes of the orthopyroxene retain the same orientation as the original Ca-poor monoclinic phases has been shown to be incorrect, and it has since been demonstrated that in most inverted pigeonites the orientation of the orthopyroxene is random with respect to the original phase (Brown, 1972; Champness and Copley, 1976).

Pigeonite phenocrysts in which exsolution is absent in the central area, but showing fine scale coherent composition modulations on (001) and (100) in the more marginal areas and surrounded by an epitaxial rim of augite, occur in lunar rock 12052 (Champness and Lorimer, 1971). The augite rim has a coarse exsolution texture, with pigeonite lamellae 500 Å in width, parallel to (001) of the host matrix. A second and much finer exsolution texture on (001) and (100) is present in both matrix and unmixed pigeonite; the latter texture probably developing during rapid cooling at the surface, and the former during the period before the lava was extruded. The mechanism by which the fine scale composition modulations developed is consistent with the breakdown of a single-phase pyroxene to Ca-poor pigeonite and Ca-rich augite by spinodal decomposition, and is illustrated in Fig. 93. Thus a Ca-poor pyroxene of composition X_1 would form zoned pigeonite and augite at the invariant temperature. On further cooling below the spinodal temperature pigeonite formed during the later stages of solidification would exsolve

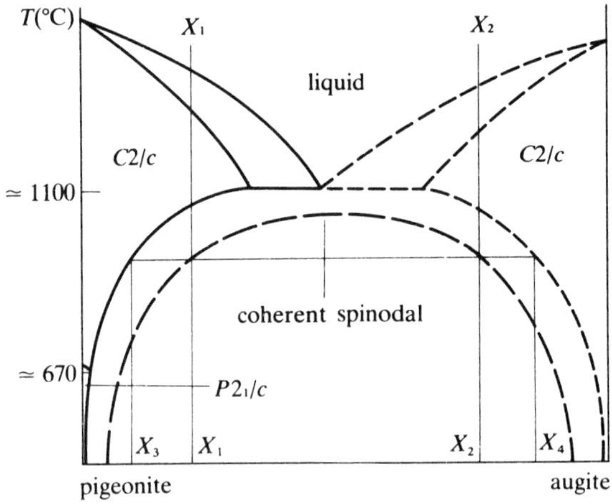

Fig. 93. Schematic section across the pyroxene quadrilateral parallel to a pigeonite–augite tie-line. Augite side shown as broken line because a peritectic, rather than a eutectic may be present for some compositions (after Champness and Lorimer, 1971). Temperature at which exsolution begins, ≃ 1100°C from Ross *et al.* (1970), temperature of $P2_1/c \rightarrow C2/c$ transition, ≃ 670°C from Prewitt *et al.* (1970).

Ca-rich augite and Ca-poor pigeonite of compositions X_3 and X_4 respectively. At a still lower temperature, the early-formed, and the exsolved, pigeonite undergoes a polymorphic transition from *C2/c* to a $P2_1/c$ structure, leading to the formation of antiphase domain boundaries between the different lattices. A comparable sequence would take place, except for the absence of a polymorphic transition, on cooling a Ca-rich pyroxene of composition X_2. The presence of antiphase domains in such an early crystallized pyroxene, and in the exsolved pigeonites, confirms that pigeonites of all compositions have *C2/c* symmetry at high temperatures.

Chemically and crystallographically complex pigeonite, $Ca_{6\cdot6}Mg_{59\cdot9}Fe_{33\cdot5}$, with optically visible augite lamellae parallel to (001) and submicroscopic augite lamellae parallel to (001), as well as chromite exsolution lamellae parallel to and perpendicular to (100), have been described from Luna 20 soil (Ghose *et al.*, 1973). Subsequent to the primary exsolution of augite and chromite, the resulting subcalcic pigeonite further exsolved (001) augite, and the consequent second generation subcalcic pigeonite later partially inverted to orthopyroxene with the same composition. Exsolution of second generation subcalcic pigeonite from the (001) augite has also occurred, and the exsolution sequence and inversion may be shown diagrammatically:

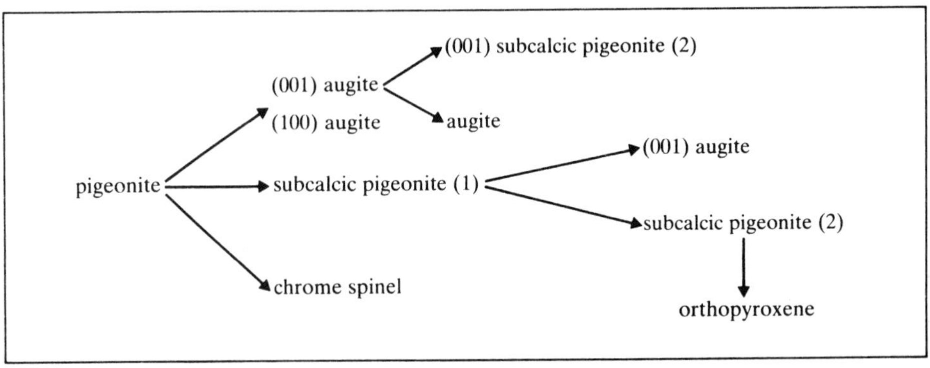

The two host-exsolution pairs, pigeonite 1–augite$_{(001)}$ and pigeonite 2–augite$_{(100)}$, in pigeonite crystals ($Wo_{6-14}En_{50-54}Fs_{26-34}$) from a lunar basalt have been studied by Nakazawa and Hafner (1977). The change of the plane of intergrowth is interpreted in terms of two-dimensional misfit between pigeonite host and exsolved augite lattices. Calculated misfit ratios of the *a* and *c* axes are shown to be temperature dependent, and to cross over at a 'critical temperature', the *a* axis giving the better fit above, and the *c* axis the better fit below this temperature.

Exsolution lamellae $\simeq$ 200 Å in width, oriented along (001) or (100) and other planes, in pigeonite, $Wo_9En_{60}Fs_{31}$, from Apollo 12 specimen 12021·150, have been homogenized by heating at 1125°C *in vacuo* for eight days (Fernández-Morán *et al.*, 1971).

Experimental and synthetic systems

Reports of the synthesis of pigeonite include that from a dry melt (Güssvegen, 1964), an incomplete solid solution series between $MgSiO_3$ and $CaSiO_3$ in the temperature range 800°–1050°C (Borchert and Kramer, 1969), and from starting materials with the composition $Wo_9En_{27\cdot3}Fs_{63\cdot7}$ with borax at 1000° and 1060°C (Ono, 1971). A pigeonitic pyroxene [α 1·745, γ 1·785, $\gamma:z$ 36°, $2V_\gamma$ 20–25°, optic axial plane (010)], with the probable composition $Wo_8En_{10}Fs_{82}$, from a copper furnace slag, was reported by Bowen (1933).

Iron-free pigeonite was synthesized at 20 kbar by Kushiro (1969) during an investigation of the $MgSiO_3$–$CaMgSi_2O_6$ system (see p. 4 and Fig. 45). Subsequently Kushiro and Yoder (1970), using a starting mixture of $En_{80}Di_{20}$ determined the stability field of iron-free pigeonite, $\simeq Ca_{0\cdot1}Mg_{0\cdot9}SiO_3$, at 2, 5, 10, 15 and 17·5 kbar. The CaO content of the iron-free pigeonite in equilibrium with diopside in the system enstatite–diopside (Fig. 94) decreases with decreasing temperature, the maximum content occurring at 1386°C when the composition is $Ca_{0\cdot09}Mg_{0\cdot91}SiO_3$, and the minimum at 1276°C at the composition $Ca_{0\cdot07}Mg_{0\cdot93}SiO_3$, when the pigeonite solid solution coexists with both diopside and protoenstatite solid solutions (Yang, 1973). The pigeonite, as do the other pyroxene

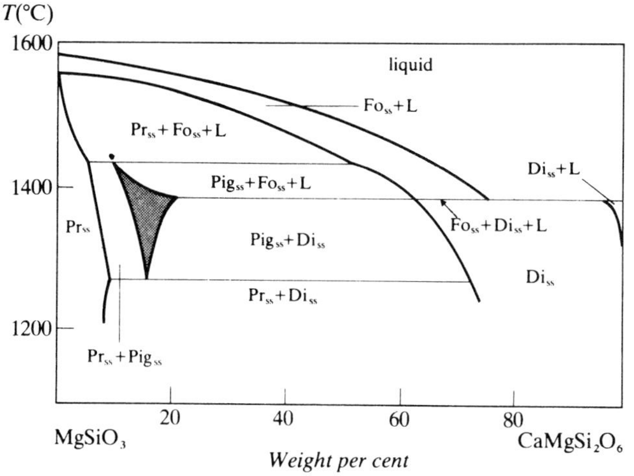

Fig. 94. **Phase relations in the upper part of the subsolidus region of the enstatite–diopside system (after Yang, 1973). Fe-free pigeonite solid solution field shaded; Fo_{ss} forsterite solid solution; Pr_{ss} protoenstatite solid solution; Di_{ss} diopside solid solution; Pig_{ss} iron-free pigeonite solid solution; L liquid.**

phases in Fig. 94, contains about 0·5 per cent Al_2O_3. The pigeonite field in the system forsterite–diopside–silica (see p.238), Fig. 124) at 1 atm has been determined by Kushiro (1972), and shows that the composition of the pigeonite lies in the join enstatite–diopside between $En_{94}Di_6$ and $En_{77}Di_{23}$. A parallel investigation of the same system by Yang and Foster (1972) indicated that the iron-free pigeonite has a CaO content of 2·59 weight per cent at 1420° and 4·92 per cent at 1387°C, and showed that a compositional gap occurs between the iron-free pigeonite and the orthorhombic phase, the composition of which varies from 0·54 weight per cent CaO at 1470° to 1·35 per cent at 1435°C (Fig. 95). Other investigations relating to the stability of pigeonite in the $MgSiO_3$–$CaMgSi_2O_6$ system include those of Warner and Luth (1974) at P_{H_2O} 2–10 kbar (see p. 86 and Fig. 47), and at 5 kbar by Mori and Green (1975).

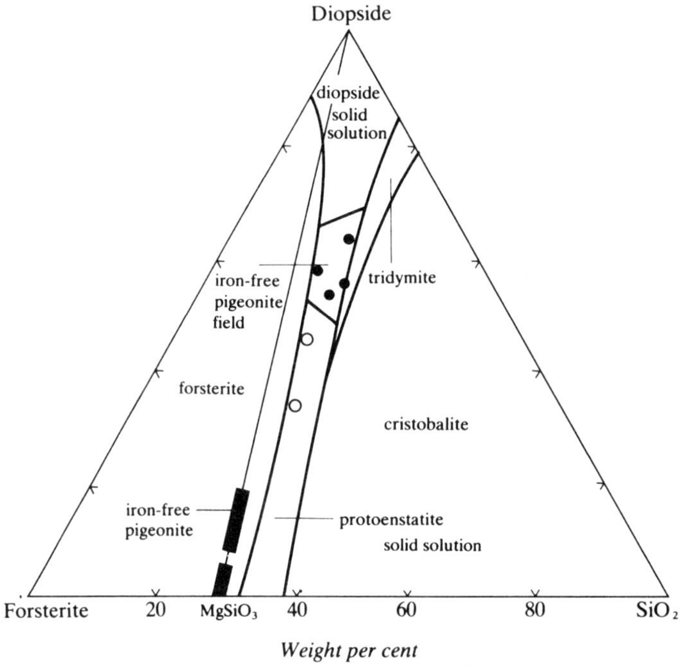

Fig. 95. Iron-free pigeonite field and compositional range of pigeonite and orthorhombic phases plotted on the system forsterite–diopside–silica (after Yang and Foster, 1972). Solid circles: compositions from which pigeonite crystallizes near liquidus. Open circles: compositions from which orthopyroxene crystallizes near liquidus.

The phase relations on the join $Mg_{0\cdot2}Fe_{0\cdot3}Ca_{0\cdot5}SiO_3$–$Mg_{0\cdot4}Fe_{0\cdot6}SiO_3$ at 20 kbar have been determined using synthetic clinopyroxenes containing 5, 10, 20 and 30 weight per cent $CaSiO_3$ as starting material. Compositions with 5 and 10 per cent $CaSiO_3$ react to form orthopyroxene and a two-phase mixture of augite and orthopyroxene respectively. The 20 per cent $CaSiO_3$ composition reacts to augite and orthopyroxene at 810°C, and augite alone at 950°C, while the more augitic composition with 30 per cent of the wollastonite component undergoes only minor recrystallization and an increase in grain size. The effect of temperature at 20 kbar on the amount of $CaSiO_3$ in solid solution in orthopyroxene, as shown by the steepness of the low-calcium limb of the $MgSiO_3$–$CaMgSi_2O_6$ solvus, is small in contrast to the marked increase in the stability of augitic compositions at the

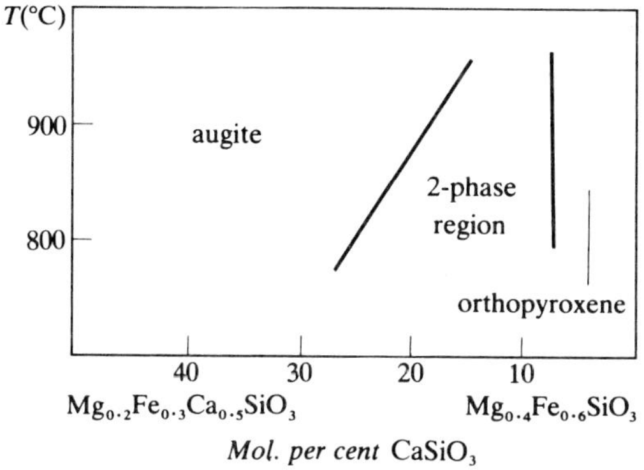

Fig. 96. Temperature–composition section across the pyroxene quadrilateral at pressure of 20 kbar (after Turnock, 1970).

expense of the two-phase region as the temperature increases (Fig. 96). This enlargement of the range of augite solid solution suggests that a hypersolvus low-calcium augite, rather than pigeonite, may be the stable phase at the pressure–temperature–composition represented in Fig. 96 (Turnock, 1970).

The stability field of iron-free pigeonite in the system diopside–enstatite–anorthite–silica at 1 atm has been investigated by Yang (1973). The primary volume of the pigeonite is not contiguous with that of anorthite at any temperature in the compositional tetrahedron at this pressure. A limited amount of iron–magnesium substitution in the pigeonite may, however, lower the temperature of the stability field sufficiently for low-iron pigeonite and anorthite to coexist in equilibrium, and low-iron pigeonite, low-Ca clinopyroxene, olivine, plagioclase assemblages have been reported from the Allende and Vigarano meteorites.

The relationships between subsolidus exsolution and inversion in low-calcium pyroxenes at pressures between 2 and 20 kbar, and at 1 atm pressure in the presence of andesitic liquid has been investigated (Brown, 1968). The study, using natural inverted pigeonites, $Ca_{7.6}Mg_{40.7}Fe_{51.7}$ and $Ca_{8.9}Mg_{48.2}Fe_{42.9}$, from the Bushveld and Skaergaard intrusions respectively, indicates that the pigeonites crystallized at about 1050°C, began to exsolve augite at 1020°C and inverted to orthopyroxene at about 990°C. The slope of the *P–T* inversion curve, as shown by the pressure data, is positive but only slightly pressure sensitive.

Optical and Physical Properties

Pigeonites are characterized optically by low values of the optic axial angle which is always less than 30° and generally below 25°. The optic axial plane may be either parallel with or perpendicular to (010); the majority of pigeonites display the latter orientation. The major replacement in pigeonites, i.e. the substitution of Mg by Fe^{2+}, is accompanied by an increase in the refractive indices, and Hess (1949) has presented a diagram showing the variation of the optical properties with percentage of Fe atoms. Although more data are now available these are considered insufficient to warrant the production of a set of curves in addition to those of Fig. 168 (see p. 353): this figure also has the advantage of giving an estimate of the Ca content, provided that both the optic axial angle and β refractive index are measured.

It is difficult to locate precisely the α, β and γ directions, particularly in those pigeonites which are practically uniaxial. A procedure for the determination of the optic axial angle and extinction angle of a pigeonite has been given by Turner

(1940). Twinning is common in pigeonite and may be either simple or lamellar, the composition plane is {100}; the {001} parting is frequently well developed. Interpenetration twins with (101) as the twin plane have been described by Prasad and Naidu (1968), and multiple twinning, with twin axis $\perp(\bar{1}\bar{2}2)$, in a lunar pigeonite, has been reported by Johnston and Gibb (1973).

Pleochroism is usually absent or weak but is occasionally moderate in pale greens and greenish browns. The manganese-rich ferropigeonite (MnO 13·0 per cent) from the Biwabik Iron Formation, Minnesota (Bonnichsen, 1969) is yellowish tan in hand specimen, very pale yellow in thin section, is non-pleochroic, and has an unusually low birefringence, 0·012. The iron-rich pigeonite from Mull is markedly pleochroic with α and β smoky brown, γ pale yellow. There does not, however, appear to be any correlation between the strength of the pleochroism and the content of iron.

Zoning is a common feature in pigeonites; e.g. from the Mount Wellington dolerite sill, Edwards (1942) has described pigeonite in which in the central part of the crystal $2V_\gamma$ is 7° and the optic axial plane perpendicular to (010) while in the marginal zone $2V_\gamma$ is 5° and the optic axial plane is parallel to (010). Similarly from the Mount Arthur dolerite complex, Poldervaart (1946) has recorded a magnesium-rich pigeonite the core of which has a $2V_\gamma$ of 12° and the optic axial plane perpendicular to (010), and the margin a $2V_\gamma$ of 8° and the optic axial plane parallel to (010).

The optical properties of iron-free pigeonite synthesized at 1 atm (Yang and Foster, 1972) are α 1·648, β 1·649, γ 1·663, $\gamma{:}z$ 25°, $2V_\gamma$ 26°, $\beta = y$, O.A.P. $\|$(010).

The clinopyroxenes of lunar rocks display great variety and complexity of zoning. A number of examples are briefly described here and in the augite section (p. 363). Pigeonite cores zoned by subcalcic augite and augite and margined by a ferrohedenbergitic pyroxene in the microgabbros of Apollo 12 basalts have been described by Klein *et al.* (1971). Large phenocrysts with pigeonite cores (16·9 per cent FeO) showing a progressive change to subcalcic augite with increasing iron content (19·8–23·4 per cent FeO) in the later stages of core growth, followed by successive zones of augite $Wo_{33}En_{45}Fs_{22}$–$Wo_{27}En_{45}Fs_{28}$, subcalcic augite ($\simeq Wo_{15}En_{50}Fs_{35}$) and ferroaugite also occur in Apollo 12 basalts (Dence *et al.*, 1971).

Phenocrysts with zoned cores, $Ca_9En_{65}Fs_{26}$–$Ca_{17}Mg_{55}Fs_{28}$, mantled by subcalcic augite grading outwards from $Ca_{27}Mg_{50}Fs_{23}$ to $Ca_{31}Mg_{45}Fs_{24}$ and rimmed by ferroaugite have been described by Gay *et al.* (1971). In {110} sectors the boundary between pigeonite core and augite margin is sharp; {100} sectors in contrast may display a continuous compositional variation, the variable sector chemistry possibly resulting from metastable crystallization at high growth rates in a supercooled magma of low viscosity. The surface of the pigeonite at the core–rim boundary is rounded probably due to resorption, and may represent a peritectic type reaction relationship. Part of the cores show curved fractures similar to those observed in synthetic clinoenstatite derived from inverted protoenstatite, and displayed by the clinoenstatite phenocrysts of the Cape Vogel tholeiite described by Dallwitz *et al.* (1966). Clinopyroxene phenocrysts displaying sector growth with distinct compositional trends on (110), (100), (010) and parallel to z, with erratic oscillatory effects superimposed on the normal sector zoning have been reported by Boyd and Smith (1971). The occurrence of sector zoning in Apollo 11 pyroxenes has also been described by Hollister and Hargraves (1970). The sector zoning is ascribed to rapid growth such that the initial compositional variatons on the surfaces of crystallographically different growth faces are preserved, and the continuous zoning within sectors explained by the reduction of magnesium in the melt consequent on the crystallization of abundant pyroxene.

Large pigeonite phenocrysts in Apollo 15 Mare basalts (Kushiro, 1973) show a smooth compositional variation from the low Ca ($Ca_5Mg_{69}Fe_{26}$) core to a high Ca ($Ca_{12}Mg_{58}Fe_{30}$) rim. An overgrowth of subcalcic augite surrounds the pigeonite and is succeeded by ferropigeonite and subcalcic ferroaugite (Fig. 97). The structural characterization and site occupancy of Fe^{2+} and Mg in the $M1$ and $M2$ pyroxene sites of the pigeonites are discussed by Virgo (1973).

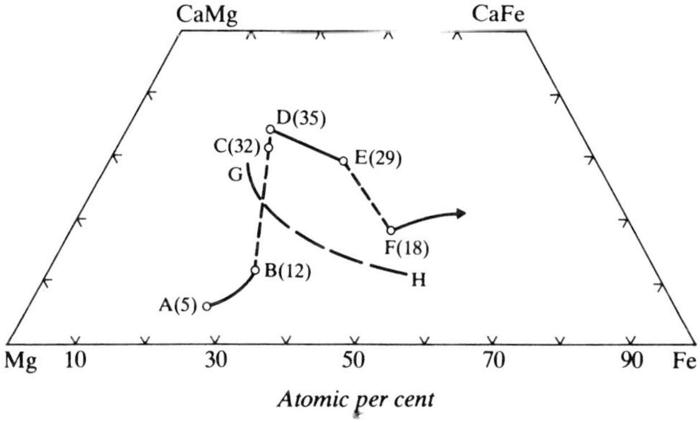

Fig. 97. Variation of Ca–Mg–Fe of clinopyroxenes from Apollo 15 Mare basalts (after Kushiro, 1973). Solid lines, continuous zoning in a single crystal. Broken lines, discontinuous zoning. Trend A → E variation along one direction of a single composite grain. Long dashed line G–H, continuous variation in single crystal from Apollo 11 rock (Ross *et al.*, 1970). Figures in brackets, atomic per cent Ca.

Progressive reversed zoning in lunar pigeonites of Apollo 14 basalt (Johnston and Gibb, 1973), and marginal magnesium-rich pigeonite zones (Agrell *et al.*, 1970) in Apollo 11 basalt have been reported.

Distinguishing Features

Pigeonite is unlikely to be confused with minerals other than members of the pyroxene group. The essential distinguishing features of pigeonites are the low values of the optic axial angle which vary between 0° and 30°; the optic axial plane may be parallel with or perpendicular to (010). Some members of the clinoenstatite–clinoferrosilite series.have optic axial angles close to 30° but these minerals are very rare in igneous rocks. Pigeonite is distinguished from orthorhombic pyroxenes by the lower birefringence and straight extinction in all [001] zone sections of the latter minerals. In small grains pigeonite may be confused with olivine but the higher birefringence of the latter is usually sufficiently diagnostic. In plutonic rocks inverted pigeonites are identified by the presence of abundant augite lamellae commonly located along the (001) planes of the original pigeonite and, in twinned crystals, displaying a herring-bone pattern.

Paragenesis

Pigeonite is a particularly characteristic constituent of andesites and dacites. It is less common in basalts as magmatic temperatures of such Mg-rich liquids are in general higher than the orthopyroxene–pigeonite inversion for compositions in which the $FeSiO_3$ component is less than about 30 per cent. As a mineral of tholeiitic rocks, pigeonite, particularly as microphenocrysts and smaller grains, often as coronas around olivine and orthopyroxene, is a typical constituent of the volcanic phase of the tholeiitic rock series, and is present in the lavas of the Deccan, Columbia River region, Parana Basin, the Hawaiian primitive shields, the Izu-Hakone and Hachijo-jima regions, Japan, and parts of the Thulean igneous province. The common presence of pigeonite and ferropigeonite among the fractionation products of these tholeiitic liquids led to their designation (Kuno, 1959) as the pigeonitic rock series. In some areas (e.g. Izu-Hakone) the series is associated with plutonic and hypabyssal rocks containing inverted pigeonite that occurs commonly in tholeiitic gabbros and dolerites.

Pigeonite phenocrysts have been reported from the andesite of Minami-Aizu (Kuno and Inoue, 1949), and in the andesites of the Hakone volcano, Japan (Kuno, 1950; Ito and Morimoto, 1956). In the latter, pigeonite, $Wo_{16}En_{46}Fs_{39}$, is intergrown with augite, and the two phases are products of unmixing, both minerals having identical *b* parameters (8·940 Å) and oriented with *x* and *y* directions coincident. Ferropigeonite occurs in the ferrohypersthene andesite and ferrohortonolite andesite dykes of Asio, Japan (Kuno, 1969), and has been described from drusy cavities in dacite (Kuno, 1950).

In the early and middle stages of the evolutionary sequence displayed by members of the Hachijō-jima volcanic group, Japan (Isshiki, 1963), the crystallization of pyroxene began with augite and orthopyroxene. Both phases continued to crystallize together until the En:Fs ratio of the latter was close to unity when pigeonite replaced orthopyroxene as the Ca-poor phase. Thereafter ferroaugite and ferropigeonite crystallized together, and were followed in the late stage of fractionation, by the crystallization of ferropigeonite, $Wo_{12}En_{32}Fs_{56}$, as the sole pyroxene phase. An unusual occurrence of ferropigeonite in trachyandesite and alkali rhyolite members of the alkali volcano series of the Higashimatsuura and Ikutsuki-jima district, Japan, has been reported by Ishibashi (1971).

The Mull andesite, from which the pigeonite of Table 13 (anal. 14) was originally described (Hallimond, 1914), has been re-examined by Virgo and Ross (1973), and a new analysis gives a smaller Wo component than previously assumed ($Wo_{7\cdot2}En_{43\cdot4}Fs_{49\cdot4}$ compared with $Wo_{8\cdot6}En_{39\cdot8}Fs_{51\cdot6}$). The recent investigation has also shown that two pyroxene phases are present in other samples of the andesite, an unzoned pigeonite with a low calcium content, and orthopyroxene the zoning of which ($Wo_2En_{52}Fs_{46}$–$Wo_2En_{46}Fs_{52}$) indicates progressive crystallization from a fractionating liquid (Fig. 98(a)). The pigeonite ($Wo_4En_{42}Fs_{54}$)–orthopyroxene ($Wo_2En_{45}Fs_{53}$) assemblage is considered to represent a narrow two-phase field close to the En–Fs join (Fig. 99), and is not at variance with the experimental results obtained on heating an orthopyroxene ($Wo_{1\cdot5}En_{42\cdot4}Fs_{56\cdot3}$) at 1038°C for 17 hours which was found to contain 2 per cent pigeonite epitaxially oriented on (100) of the host.

Iron-rich pigeonite, average composition $Ca_8Mg_{40}Fe_{52}$, ferroaugite and orthopyroxene phenocrysts are present in the pitchstones, felsites and granophyres of Rhum and Ardnamurchan (Emeleus *et al.*, 1971). The pigeonites are variable in composition due in part to marginal zoning which, in extreme cases, gives rise to the composition $Ca_{8\cdot3}Mg_{23\cdot1}Fe_{68\cdot6}$. The most Mg-rich pigeonites occur in the cores of composite grains that are mantled by augite, $Ca_{37\cdot8}Mg_{36\cdot5}Fe_{25\cdot7}$; other composite grains consist of hypersthene cores, $Ca_{4\cdot2}Mg_{45\cdot5}Fe_{50\cdot5}$, surrounded by ferropigeonite $Ca_{7\cdot4}Mg_{40\cdot5}Fe_{52\cdot1}$. The composite grains are interpreted as a sequential record of the course of crystallization rather than equilibrium features, and further evidence of the disequilibrium associations are indicated by the presence of two pigeonite compositional groups in the Ben Hiant pitchstone. The formation of pigeonite unusually rich in iron, and the absence of a ferrohedenbergite–fayalite–quartz assemblage, is considered to result from crystallization in a magmatic environment of low f_{O_2} and high silica activity, the latter indicated by the coprecipitation of high-temperature quartz with the Fe-rich portion of the pyroxene.

Pigeonite, $\simeq Wo_{13}En_{57}Fs_{30}$, with orthopyroxene and augite, or augite alone, occurs in the tholeiitic basalts of the Vogelsberg area, Germany (Ernst and Schorer, 1969). Orthopyroxene was the first pyroxene phase to crystallize and was followd by pigeonite and augite. In those basalts in which orthopyroxene is not present the pigeonite occurs mainly as cores surrounded, with a marked discontinuity, by augite.

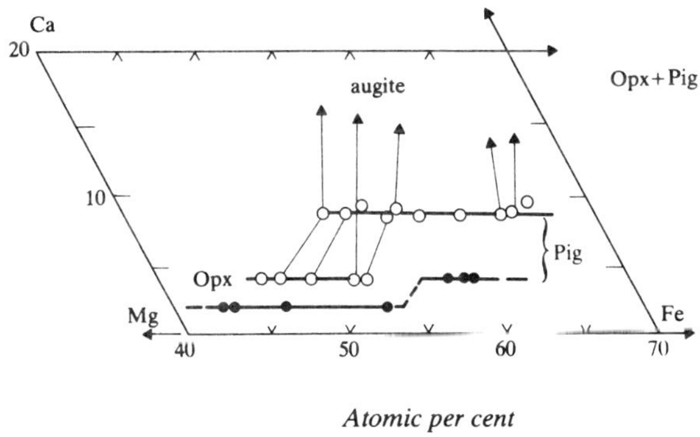

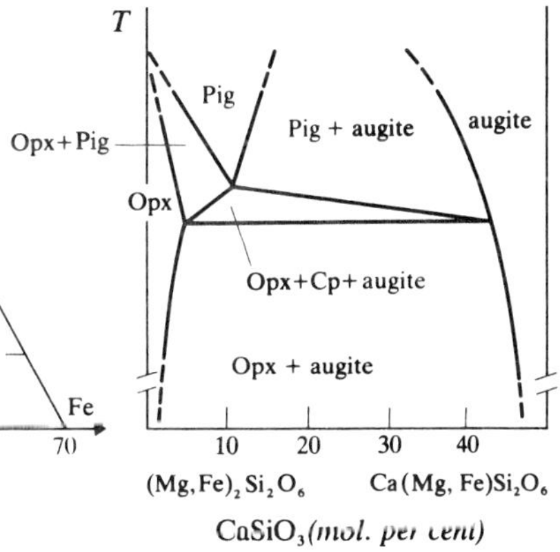

Fig. 98. Variation of Ca–Mg–Fe of pigeonites and orthopyroxenes of the Mull and Weiselberg andesites. Solid circles, Mull andesite; open circles, Weiselberg andesite (after Virgo and Ross, 1973).

Fig. 99. Schematic subsolidus relationships of a pseudobinary join across the pyroxene quadrilateral (after Virgo and Ross, 1973).

The compositional relations of augite, hypersthene and pigeonite phenocrysts and ferroaugite and ferropigeonite of the groundmass in the andesite from Weiselberg, Germany, have been examined in detail by Nakamura and Kushiro (1970b). During the process of fractional crystallization the phenocryst assemblage changed from augite–hypersthene, through augite–hypersthene–pigeonite, to augite–pigeonite, coincident with an increasing Fe/Mg ratio. Ferroaugite and ferropigeonite occur in the groundmass and represent the final products of the pyroxene crystallization. Pigeonite and orthopyroxene phenocrysts both display normal zoning, and the Fe/Mg ratio of the groundmass pigeonites is a little higher than at the margins of the phenocrysts, indicating that the physical conditions changed gradually between the crystallization of the phenocrysts and the groundmass. The compositions of contiguous crystals of orthopyroxene and pigeonite, orthopyroxene and augite, and pigeonite and augite plotted on the pyroxene quadrilateral (Fig. 100) show that a number of three-phase triangles can be drawn for the orthopyroxene–pigeonite–augite assemblage. As a three-phase triangle relates to one set of pressure–temperature conditions, the compositional relationships of the pyroxenes in the Weiselberg andesite indicate that equilibration conditions changed during crystallization. In the present case pressure would remain relatively constant, and changing temperature would be the main cause for the shift of the three-phase triangle (Fig. 101). The bulk compositions and the Fe/Mg ratios of the pyroxene phases are similar to those of the Mull andesite. The three-phase assemblage, orthopyroxene (Wo_4)–pigeonite ($Wo_{8\cdot5}$)–augite ($Wo_{25\cdot4}$) present in the Weiselberg andesite may represent a unique phase equilibrium, developed at a lower temperature than the two-phase assemblage of the Mull andesite, due to quenching across the narrow inversion interval between orthopyroxene and low Ca-pigeonite at a somewhat higher temperature.

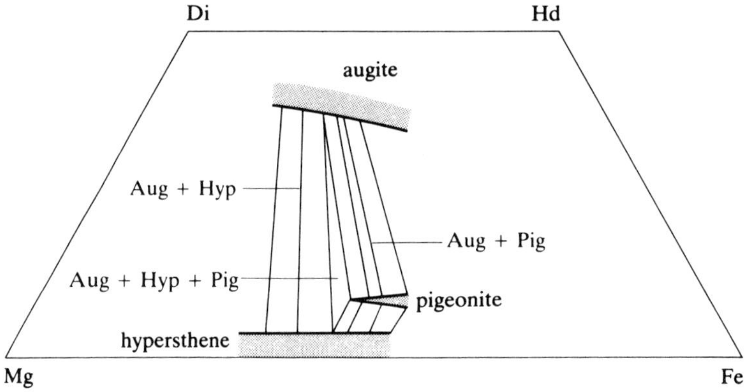

Fig. 100. Schematic isothermal section at a subsolidus temperature, showing the compositional relations of orthopyroxene, pigeonite and augite phenocrysts, and ferropigeonite and ferroaugite in the groundmass of the Weiselberg pigeonite andesite (after Nakamura and Kushiro, 1970b).

In the Mount Wellington sill, Tasmania (Edwards, 1942), pigeonite is absent in the lower parts of the sheet, and the calcium-poor pyroxene in the earlier formed rocks is an orthopyroxene. Pigeonite first occurs 150 m below the top and has an estimated composition of $Wo_5En_{65}Fs_{30}$. At successively higher levels the iron content of the pigeonite increases, and in the uppermost part of the sill its estimated composition is $Wo_{10}En_{20}Fs_{70}$. The occurrence of orthopyroxene in the lower parts and the presence of pigeonite in the upper parts of some of the dolerite sheets of the Karroo has been described by Walker and Poldervaart (1941a,b, 1949). In these sheets, as in the Mount Wellington sill, the orthopyroxene–pigeonite sequence is doubtless consequent on the ortho–clinopyroxene inversion curve falling below the temperature of the magma. Two pigeonite generations have been described by Poldervaart (1944) from the pigeonite-bearing dolerites in the Elephant's Head dyke and New Amalfi sheet, South Africa. The early pigeonite is colourless and magnesium-rich and forms cores to augite; the late pigeonite is brown and iron-rich and occurs as rims around augite.

Pigeonite is present in the lower and inverted pigeonite in the upper part of the quartz diabase horizon of the Nipissing diabase sill (Hriskevich, 1968). Both pigeonite and inverted pigeonite decrease progressively, and orthopyroxene increases in the overlying hypersthene diabase zone, a relationship that is reversed in the higher levels of the sill in which the amount of pigeonite and inverted pigeonite increases and is accompanied by a decrease in orthopyroxene.

Inverted pigeonites occur in the olivine gabbros, middle gabbros, and lower ferrodiorites of the Skaergaard intrusion; inverted pigeonite is present also in the chilled marginal gabbro. During the formation of the early accumulative gabbro picrite the ortho–clinopyroxene inversion curve was above the magmatic temperature and in consequence the calcium-poor pyroxene crystallized as the orthorhombic phase (see Fig. 89). In the lower olivine gabbros the composition of the pigeonite is $Ca_{9\cdot0}Mg_{55\cdot8}Fe_{35\cdot2}$ (Table 13, anal. 3) compared with the associated augite, $Ca_{37\cdot7}Mg_{41\cdot4}Fe_{20\cdot9}$. During fractionation the pigeonite became richer in iron and in conformity with its normal relationship with augite, is richer in iron than the calcium-rich phase. The composition of the most iron-rich pigeonite analysed (Table 13, anal. 10), from a ferrodiorite, is $Ca_{8\cdot6}Mg_{45\cdot5}Fe_{45\cdot9}$ and the mineral is associated with augite, $Ca_{34\cdot6}Fe_{37\cdot0}Fe_{28\cdot4}$ in composition. Pigeonite occurs in small amounts in the more highly fractionated ferrodiorites but Brown (1957) considered it unlikely that the iron content of the final pigeonite would contain more than Fe_{55}, and that at approximately this composition pigeonite ceased to crystallize as a cumulus phase from the Skaergaard magma, subsequent cooling giving rise to the formation of a single, ferroaugite, phase. Ferropigeonites, richer in the $FeSiO_3$ component, however, occur as products of intercumulus crystallization (Nwe, 1975; see also p. 333 and Fig. 150).

In the Bushveld complex the change from orthopyroxene to pigeonite occurs at a composition between Fe_{30} and Fe_{35} (Atkins, 1969). Iron enrichment of the inverted cumulus pigeonite–ferropigeonite reaches a maximum of about Fe_{60}, at the fractionation stage represented on the Ca-rich pyroxene trend by a marked change in composition towards more calcic ferroaugite, and probably represents the limit of coprecipitation of Ca-poor and Ca-rich pyroxenes in this intrusion. More Fe-rich Ca-poor pyroxenes (the most Fe-rich ferropigeonite has an estimated composition Fe_{71}) are present in small quantities in the more highly fractionated cumulate. These

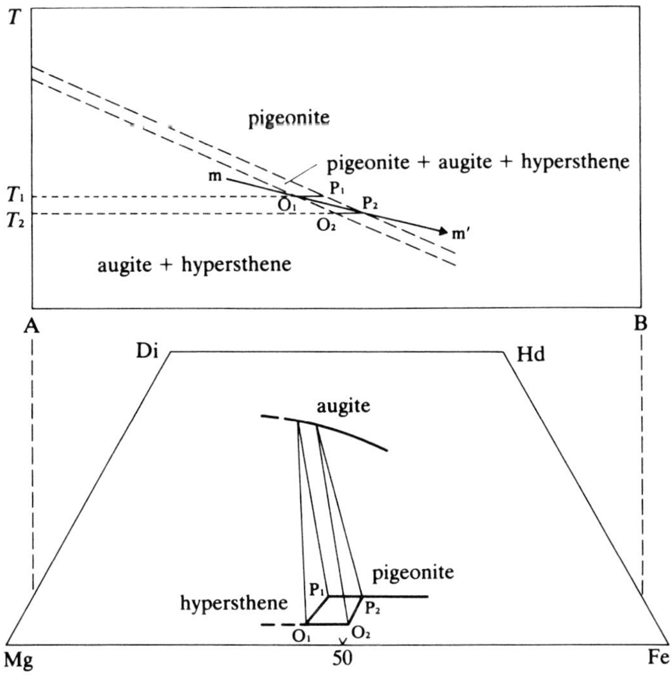

Fig. 101. **Schematic diagram illustrating the crystallization of pyroxenes in the Weiselberg andesite. Two three-phase triangles based on analyses of pyroxenes (after Nakamura and Kushiro, 1970b). m–m′ temperature change during crystallization of the andesite. Intervals O_1–P_1 and O_2–P_2 are not the 'inversion interval', but the maximum widths along the join A–B of three-phase triangle, pigeonite–hypersthene–augite at temperatures T_1 and T_2 respectively.**

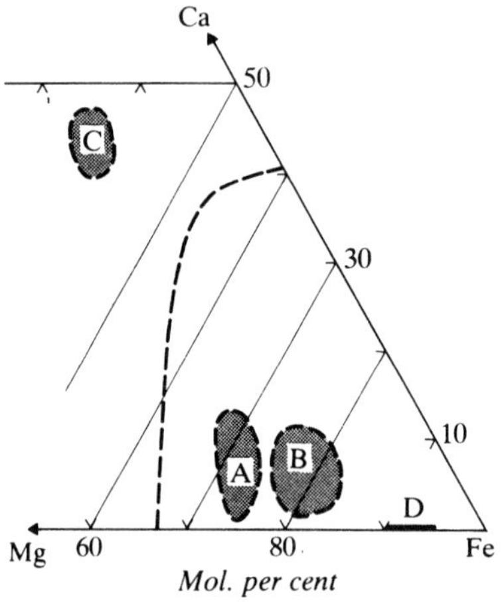

Fig. 102. Pyroxene compositions in the adamellite and granodiorite of the Nain anorthosite massif (after Smith, 1974). Compositional field (A), Ca-poor pyroxenes now represented by intergrowths of orthopyroxene, ferroaugite (field C), fayalite and quartz. Compositional field (B), orthopyroxenes, (D) olivines. Broken curve encloses compositional field in which pyroxene is not stable at 925°C and 1 atm.

occur, however, adjacent to cumulus olivine and are likely to have formed by reaction of the orthosilicate with the liquid. The pigeonites–ferropigeonites of the Bushveld are somewhat less rich in calcium than those of equivalent Mg/Fe ratio in the Skaergaard intrusion and, as the Ca-rich pyroxenes are correspondingly slightly richer in calcium, it is likely that the miscibility gap was greater during the crystallization of the Bushveld compared with those of the Skaergaard liquids. The extension of the two-pyroxene field closer towards the $FeSiO_3$ corner of the pyroxene trapezium, and the larger miscibility gap, can be explained on the assumption that the pyroxene solidus was depressed, both limbs of the solvus being truncated at a somewhat lower temperature for pyroxene pairs of equivalent composition than during the crystallization of the Skaergaard magma (see also augite section, p. 332 and Fig. 150).

Ca-poor pyroxenes, very rich in iron, occur in the adamellites and monzonites associated with the Nain anorthosite massif, Labrador (Smith, 1974). The rocks have not been affected by any major metamorphic event, and the phase assemblages probably reflect the conditions at the time emplacement took place. Two types of Fe-rich, Ca-poor pyroxene are present in the adamellites: an orthopyroxene, $Ca_6Mg_{12}Fe_{82}$, inverted from pigeonite and containing exsolution of a Ca-rich pyroxene, and the breakdown products of a Ca-poor pyroxene and now consisting of intergrowths of less Fe-rich orthopyroxene, $Ca_7Mg_{21}Fe_{72}$, ferroaugite, fayalite and quartz (Fig. 102). The composition of the original pigeonite lies within the 'forbidden zone' of the pyroxene quadrilateral, the size of which is both pressure and temperature dependent. With increasing pressure the width of the zone contracts towards ferrosilite composition, and the stability field of Fe-rich pyroxene expands in comparison with that for 1 atm at 925°C. In the Ca-free synthetic system (see p. 232), orthopyroxene is stabilized relative to olivine + quartz by increasing pressure as well as by decreasing temperature, and the breakdown of these high pressure–temperature pigeonites to isochemical associations containing fayalite and quartz is considered to be consequential on the decrease in pressure during the uplift of the massif, closer to the surface. Pyroxenes with calcium contents intermediate between augite (Wo_{35-45}) and pigeonite (Wo_{10}) occur in some of the monzonites. Many of the crystals are zoned with core compositions in the range 13 to 31 mol. per cent $CaSiO_3$, and rims close to $Ca_{40}Mg_{15}Fe_{45}$. Other monzonites contain crystals consisting of alternating broad lamellae, in approximately equal proportions, of ferroaugite, $\simeq Ca_{43}Mg_{17}Fe_{40}$, and orthopyroxene, $\simeq Ca_3Mg_{19}$

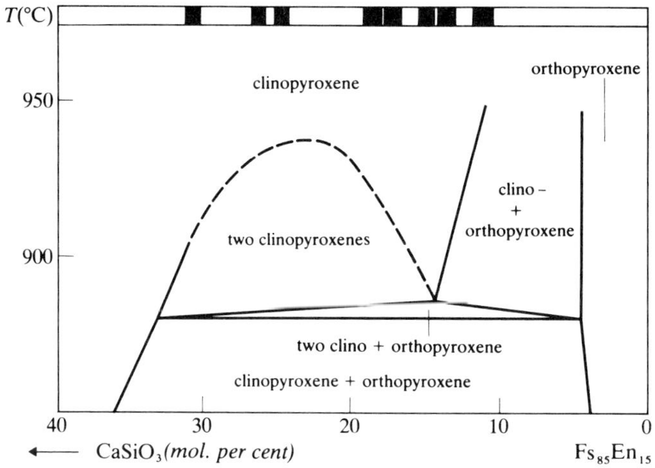

Fig. 103. Molecular per cent $CaSiO_3$ (■) in presumed primary pyroxenes in monzonites of the Nain anorthosite massif shown together with phase relations on part of the join $Fs_{85}En_{15}$–Wo_{100} at 15 kbar (after Smith, 1974).

Fe_{78}, and are considered to have formed from pyroxenes that crystallized initially with a calcium content between Wo_{20} and Wo_{30} that subsequently unmixed to ferroaugite and ferropigeonite, the latter finally inverting to orthopyroxene. Intergrowths of ferroaugite, fayalite and quartz are also present in the monzonites but, unlike those in the adamellites, pigeonite and orthopyroxene do not occur in the intergrowths, and reaction of a Ca-poor pyroxene with residual melt rather than simple breakdown is more likely in this case. If this is the correct interpretation then a more calcic pyroxene should form during and after the reaction, and the occurrence of grains with Ca-poor cores and ferroaugite rims is consistent with such a pyroxene-melt reaction. The formation of such an intermediate Ca-pyroxene would require that crystallization occurred initially at temperatures above the crest of the augite–pigeonite miscibility gap (Fig. 103).

Pigeonites and orthopyroxenes occur in parts of the Biwabik Iron Formation and developed during the prograde metamorphism (pyroxene hornfels facies) of the original sediments by the Duluth intrusion (Bonnichsen, 1969). Much of the pigeonite has subsequently inverted to orthopyroxene during cooling. The texture of the inversion product is unusual not only in that the orthopyroxene displays a random crystallographic orientation with regard to the pre-existing pigeonite, but also shows that several pigeonite grains became encompassed by individual orthopyroxene crystals during the transformation. The resulting ferrohypersthene grains are considerably larger than other grains with which they coexist, indicating that growth began initially at widely spread points. The microstructure is considered by Champness and Copley (1976) to result from a diffusionless transformation with nucleation taking place preferentially at grain boundaries, and subsequent growth occurring by propagation of an incoherent interface disregarding pre-existing grain boundaries. The Ca-poor pyroxenes occur in association with Ca- and Fe-rich clinopyroxenes, fayalitic olivine and quartz (Fig. 104). The content of CaO in the uninverted ferropigeonites is approximately 3·2 weight per cent, that of the ferrohypersthene generally varying between 0·5 and 1·4 per cent; the inverted pigeonites contain more than 71 per cent Fe/(Mg + Fe) compared with 69 per cent of the pre-inversion phase. The monoclinic ⇌ orthorhombic inversion temperature for compositions close to the fayalite + quartz stability field is estimated (Bonnichsen, 1969) on the basis of the $^{18}O/^{16}O$ ratios of quartz and magnetite, to be between 700° and 750°C.

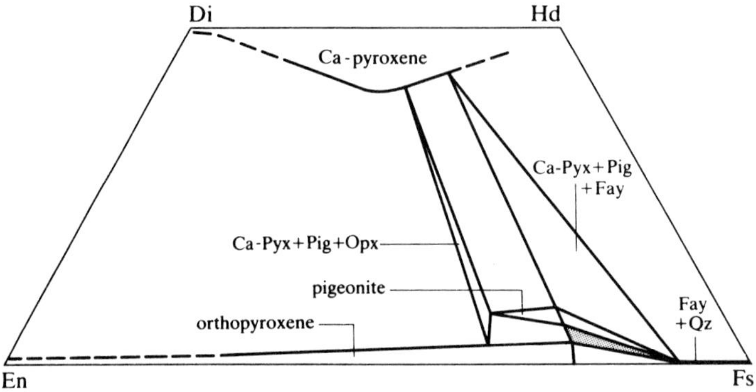

Fig. 104. Possible equilibrium phase relations for coexisting pyroxenes and fayalite and quartz at 700°–750°C and other conditions attained during metamorphism of the Biwabik Iron Formation (after Bonnichsen, 1969). Stippled field, pigeonite + orthopyroxene + fayalite + quartz.

A manganese-rich pigeonite (SiO_2 49·0, MgO 2·5, FeO 31·5, MnO 13·0, CaO 4·5) occurs in a quartz-rich facies from the lower part of the formation. This specimen has not inverted to an orthorhombic structure, and it is possible that the high content of Mn may have depressed the inversion temperature, so lowering the diffusion rate and thus, due to the sluggishness of the reaction, preserving the monoclinic phase. The relatively large content of manganese (average of twelve specimens, 2·0 weight per cent MnO) of many of the Ca-poor pyroxenes of the Biwabik Iron Formation may account for the apparently lower temperatures of pigeonite stability in these rocks compared with that for comparable Fe-rich compositions derived by extrapolation of the 'pigeonite eutectoid reaction line' (Ishii, 1975) for magmatic pyroxenes.

An assemblage, pigeonite ($Wo_{10}En_{24}Fs_{66}$), olivine (Fa_{87}), augite ($Wo_{40}En_{20}Fs_{40}$), plagioclase (An_{94}), titanomagnetite and quartz, similar to those of the Biwabik Iron Formation, has been described from the metamorphosed ferruginous sediment of the Gunflint Iron Formation, Minnesota(Simmons *et al.*, 1974). The major component of the rock is an inverted pigeonite, and the present orthopyroxene phase contains thin ($\leqslant$ 1 μm) lamellae of augite parallel to (100), and tabular lamellae and blebs exsolved on (001) of the original pigeonite. The exsolved material on (001) has in some cases coalesced to form blebs up to 150 μm in width, consisting of an intergrowth of augite and a Ca-poor pyroxene, possibly pigeonite. Textures of this type are interpreted by Bonnichsen (1969) as a result of the difficulty with which orthopyroxene nucleated, possibly after supercooling had taken place before the pigeonite inverted. This conclusion is supported by the observation that the large orthopyroxene crystals in the Gunflint Formation enclose a number of randomly oriented sets of (001) augite lamellae resulting from the rapid growth of the orthopyroxene across the grain boundaries of the original pigeonite grains. Additional augite was exsolved during the inversion process, and in some cases merged into the larger blebs. After exsolution of augite on the pigeonite (001) plane ceased the composition of the lamellae ranged from $Wo_{41-45}En_{20-23}Fs_{35-39}$. With the subsequent inversion to orthopyroxene, augite then exsolved along the (100) plane of the orthopyroxene host, the final compositions of which are $\simeq Wo_{45}$

$En_{20}Fs_{35}$ and $Wo_{2-4}En_{21-28}Fs_{70-75}$ respectively. Nucleation of orthopyroxene within the primary augite did not occur and the uniform exsolution lamellae on (001), 1–2 μm in thickness, in the Ca-rich pyroxene consist of a low-calcium pigeonite that exsolved metastably in the orthopyroxene stability field.

Coronas consisting of an inner zone of pigeonite and ilmenite and an outer monomineralic zone of pigeonite around olivine in metamorphosed Apollo 14 breccias have been described by Cameron and Fisher (1975). The composition of the pigeonite is related to that of the olivine, thus compositions $Wo_{7\cdot7}En_{55\cdot9}Fs_{36\cdot4}$ and $Wo_{8\cdot5}En_{52\cdot0}Fs_{39\cdot5}$ are associated with olivine $Fo_{44\cdot0}$ and $Fa_{48\cdot5}$ respectively. The formation of the corona assemblage is discussed in terms of the potential gradients of FeO and TiO_2 developed between olivine and matrix during the thermal annealing.

Large ($\simeq 2 \times 10$ mm) prismatic phenocrysts of pigeonite showing a progressive compositional variation from a low-Ca core, $Ca_5Mg_{69}Fe_{26}$, to a relatively high-Ca margin, $Ca_{12}Mg_{58}Fe_{30}$, and overgrown 'discontinuously' by subcalcic augite occur in a lunar pigeonite porphyry. The groundmass has a variolitic texture indicating rapid crystallization. Part of the high-calcic rims, as well as the low-calcic phenocrysts probably began to crystallize below the lunar surface, and the final crystallization of the subcalcic augite under surface conditions (Kushiro, 1973).

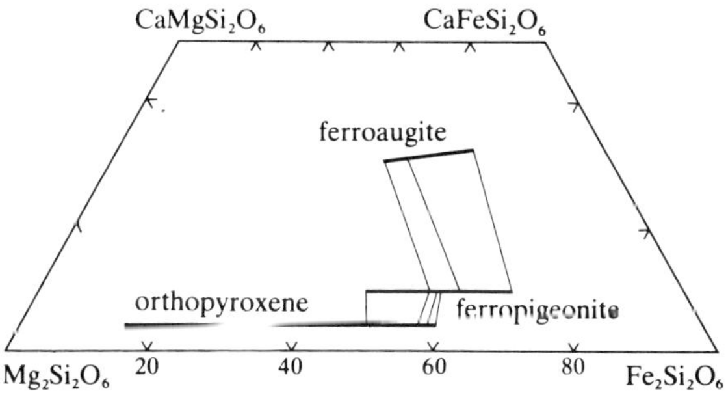

Fig. 105. Compositions (deduced from optical measurements) of pyroxenes in eucrites, a howardite and a mesosiderite showing tie-lines between coexisting ferropigeonites and ferroaugite. Tie-lines between pigeonites and orthopyroxenes indicate coexisting pyroxenes resulting from the partial inversion of pigeonite → orthopyroxene (after Duke and Silver, 1967).

Pigeonite is not uncommon in meteorites and a wide range of compositions including relatively iron-rich varieties are represented, e.g. pigeonite, $Wo_{12}En_{45}Fs_{43}$, in achondrite enclaves in the Mount Padbury meteorite, Western Australia (McCall, 1966) and ferropigeonite (Table 13, anal. 9) in the Moore County meteorite (Hess and Henderson, 1949). Ferropigeonites (Fig. 105) many of which contain exsolution lamellae of ferroaugite, and show partial or complete inversion to orthopyroxene are particularly common constituents of eucrites (Duke and Silver, 1967).

References

Agrell, S. O., Scoon, J. H., Muir, I. D., Long, J. V. P., McConnell, J. D. C. and **Peckett, A.,** 1970. Observations on the chemistry, mineralogy and petrology of some Apollo 11 lunar samples. *Proc. Apollo 11 Lunar Sci. Conf. (Suppl. to Geochim. Cosmochim Acta* **34**), **1**, 93–128.

Atkins, F. B., 1969. Pyroxenes of the Bushveld intrusion, South Africa. *J. Petr.*, **10**, 222–249.

Bailey, J. C., Champness, P. E., Dunham, A. C., Esson, J., Fyfe, W. S., Mackenzie, W. S., Stumpfl, E. F. and **Zussman, J.,** 1970. Mineralogy and petrology of Apollo 11 lunar samples. *Proc. Apollo 11 Lunar Sci. Conf. (Suppl. to Geochim. Cosmochim Acta* **34***)*, **1**, 169–194.

Bilgrami, S. A., 1964. Mineralogy and petrology of the central part of the Hindubagh igneous complex, Hindubagh mining district, Zhob valley, West Pakistan. *Rec. Geol. Surv. Pakistan*, **10**, pt. 2-C, 1–28.

Binns, R. A., 1965. The mineralogy of metamorphosed basic rocks from the Willyama Complex, Broken Hill district, New South Wales. Part II. Pyroxenes, garnet, plagioclase and opaque oxides. *Min. Mag.*, **35**, 561–587.

Binns, R. A., Long, J. V. P. and **Reed, S. J. B.,** 1963. Some naturally occurring members of the clinoenstatite–clinoferrosilite mineral series. *Nature*, **198**, 777–778.

Bonnichsen, B., 1969. Metamorphic pyroxenes and amphiboles in the Biwabik Iron Formation, Dunka River Area, Minnesota. *Min. Soc. America, Spec. Paper* **2**, 217–239.

Borchert, W. and **Kramer, V.,** 1969. Ein synthetisches Magnesium-Kupfer-Silikat mit Klinopyroxene-struktur, $(Mg,Cu)_2[Si_2O_6]$. *Neues Jahrb. Min., Mh*, 6–14.

Bowen, N. L., 1933. Crystals of iron-rich pyroxene from a slag. *J. Washington Acad. Sci.*, **23**, 83–87.

Bowen, N. L. and **Schairer, J. F.,** 1935. The system $MgO–FeO–SiO_2$. *Amer. J. Sci.*, ser. 5, **29**, 151–217.

Bown, M. G. and **Gay, P.,** 1957. Observations on pigeonite. *Acta Cryst.*, **10**, 440–441.

Boyd, F. R. and **Brown, G. M.,** 1969. Electron-probe study of pyroxene exsolution. *Min. Soc. America, Spec. Paper* **2**, 211–216.

Boyd, F. R. and **England, J. L.,** 1965. The rhombic enstatite–clinoenstatite inversion. *Carnegie Inst. Washington, Ann. Rept. Dir. Geophys. Lab.*, 1964–65, 117–120.

Boyd, F. R. and **Smith, D.,** 1971. Compositional zoning in pyroxenes from lunar rock 12021, Oceanus Procellarum. *J. Petr.*, **12**, 439–465.

Brown, G. E., Prewitt, C. T. and **Papike, J. J.,** 1971. Structural studies of intergrown lunar pyroxenes (abstract). *Eos. Trans. A.G.U.*, **52**, 270.

Brown, G. E., Prewitt, C. T., Papike, J. J. and **Sueno, S.,** 1972. A comparison of the structures of low and high pigeonite. *J. Geophys. Res.*, **77**, 5778–5789.

Brown, G. E. and **Wechsler, B. A.,** 1973. Crystallography of pigeonites from basaltic vitrophyre 15597. *Proc. 4th Lunar Sci. Conf., Geochim. Cosmochim. Acta*, Suppl. **4**, **1**, 887–900.

Brown, G. M., 1957. Pyroxenes from the early and middle stages of fractionation of the Skaergaard intrusion, east Greenland. *Min. Mag.*, **31**, 511–543.

Brown, G. M., 1968. Experimental studies on inversion relations in natural pigeonitic pyroxenes. *Carnegie Inst. Washington, Ann. Rept. Dir. Geophys. Lab.* 1966–67, 347–353.

Brown, G. M., 1972. Pigeonitic pyroxenes: A review. *Mem. Geol. Soc. America*, **132**, 523–534.

Cameron, K. L. and **Fisher, G. W.,** 1975. Olivine-matrix reactions in thermally metamorphosed Apollo 14 breccias. *Earth Planet. Sci. Letters*, **24**, 197–207.

Campbell, I. H. and **Nolan, J.,** 1974. Factors affecting the stability field of Ca-poor pyroxene and the origin of the Ca-poor minimum in Ca-rich pyroxenes from tholeiitic intrusions. *Contr. Min. Petr.*, **48**, 205–219.

Champness, P. E. and **Copley, P. A.,** 1976. The transformation of pigeonite to orthopyroxene. In: *Electron Microscopy in Mineralogy*. Berlin, Heidelberg and New York (Springer-Verlag), pp. 228–233.

Champness, P. E., Dunham, A. C., Gibbs, F. G. F., Giles, H. N., Mackenzie, W. S., Stumpfl, E. F. and **Zussman, J.,** 1971. Mineralogy and petrology of some Apollo 12 lunar samples. *Proc. 2nd Lunar Sci. Conf., Geochim. Cosmochim. Acta*, Suppl. **2**, **1**, 359–376 (M.I.T. Press).

Champness, P. E. and **Lorimer, G. W.,** 1971. An electron microscopic study of a lunar pyroxene. *Contr. Min. Petr.*, **33**, 171–183.

Champness, P. E. and **Lorimer, G. W.,** 1972. Electron microscopic studies of some lunar and terrestrial pyroxenes. *Proc. 5th Intern. Materials Symp.*, Berkeley, 1245–1255.

Champness, P. E. and **Lorimer, G. W.,** 1976. Exsolution in silicates. In *Electron Microscopy in Mineralogy*, H. R. Wenk (ed.). Berlin, Heidelberg and New York (Springer-Verlag), pp. 174–204.

Christie, J. M., Lally, J. S., Heuer, A. H., Fisher, R. M., Griggs, D. T. and **Radcliffe, S. V.,** 1971. Comparative electron petrography of Apollo 11, Apollo 12 and terrestrial rocks. *Proc. 2nd Lunar Sci. Conf., Geochim. Cosmochim. Acta*, Suppl. **3**, **1**, 69–89 (M.I.T. Press).

Clark, J. R., Ross, M. and **Appleman, D. E.,** 1971. Crystal chemistry of a lunar pigeonite. *Amer. Min.*, **56**, 888–908.

Dallwitz, W. G., Green, D. H. and **Thompson, J. E.,** 1966. Clinoenstatite in a volcanic rock from the Cape Vogel Area, Papua. *J. Petr.*, **7**, 375–403.

Dence, M. R., Douglas, J. A. V., Plant, A. G. and **Traill, R. J.,** 1971. Mineralogy and petrology of some Apollo 12 samples. *Proc. Second Lunar Sci. Conf., Geochim. Cosmochim. Acta*, Suppl. **2**, **1**, 285–299.

Dowty, E. and **Lindsley, D. H.**, 1973. Mössbauer spectra of synthetic hedenbergite–ferrosilite pyroxenes. *Amer. Min.*, **58**, 850–868.
Dowty, E., Keil, K. and **Prinz, M.**, 1974. Lunar pyroxene-phyric basalts: crystallization under supercooled conditions. *J. Petr.*, **15**, 419–453.
Dowty, E., Ross, M. and **Cuttita, F.**, 1972. Fe^{2+}–Mg site distribution in Apollo 12021 pyroxenes: evidence for bias in Mössbauer measurements and relation of ordering to exsolution. *Proc. 3rd Lunar Sci. Conf., Geochim. Cosmochim. Acta*, Suppl. **3, 1**, 481–492.
Duchesne, J. C., 1972. Pyroxènes et olivines dans le massif de Bjerkrem-Sogndal (Norvège méridionale). Contribution à l'étude de la série anorthosite-mangérite. *Intern. Geol. Congr. (Montreal) 24th session*, Pt. 2, 320–328.
Duke, M. B. and **Silver, L. T.**, 1967. Petrology of eucrites, howardites and mesosiderites. *Geochim. Cosmochim Acta*, **31**, 1637–1665.
Dunham, A. C., Copley, P. A. and **Strasser-King, V. H.**, 1972. Submicroscopic exsolution lamellae in pyroxenes in the Whin sill, northern England. *Contr. Min. Petr.*, **37**, 211–220.
Edwards, A. B., 1942. Differentiation of the dolerites of Tasmania I & II. *J. Geol.*, **50**, 451, 579–610.
Emeleus, C. H., Dunham, A. C. and **Thompson, R. N.**, 1971. Iron-rich pigeonite from acid rocks in the Tertiary igneous province of Scotland. *Amer. Min.*, **56**, 940–951.
Ernst, Th. and **Schorer, G.**, 1969. Die Pyroxene des 'Maintrapps', einer Gruppe tholeiitischer Basalte des Vogelsberges. *Neues Jahrb. Min., Mh.*, 108–130.
Evans, B. W. and **Moore, J. G.**, 1968. Mineralogy as a function of depth in the prehistoric Makaopuhi tholeiitic lava lake, Hawaii. *Contr. Min. Petr.*, **17**, 85–115.
Fernández-Morán, H., Ohtsuki, M. and **Hibino, A.**, 1971. Correlated electron microscopy and diffraction of lunar clinopyroxenes from Apollo 12 samples. *Proc. Second Lunar Sci. Conf., Geochim. Cosmochim. Acta*, Suppl. **2, 1**, 109–116.
Gay, P., Bown, M. G., Muir, I. D., Bancroft, G. M. and **Williams, P. G. L.**, 1971. Mineralogical and petrographic investigation of some Apollo 12 samples. *Proc. 2nd Lunar Sci. Conf., Geochim. Cosmochim. Acta*, Suppl. **2, 1**, 377–392.
Ghose, S., McCallum, I. S. and **Tidy, E.**, 1973. Lunar 20 pyroxenes: exsolution and phase transformation as indicators of petrologic history. *Geochim. Cosmochim. Acta*, **37**, 831–839.
Ghose, S., Ng, G. and **Walker, L. S.**, 1972. Clinopyroxenes from Apollo 12 and 14; exsolution, domain structure and cation order. *Proc. 3rd Lunar Sci. Conf., Geochim. Cosmochim. Acta*, Suppl. **3, 1**, 507–531 (M.I.T. Press).
Ghose, S. and **Weider, J. R.**, 1971. Oriented transformation of grunerite to clinoferrosilite at 775°C and 500 bars argon pressure. *Contr. Min. Petr.*, **30**, 64–71.
Ginzburg, I. V., Maleeyev, Ye. F., Sidorenko, G. A. and **Teleshova, R. L.**, 1964. Another find of pigeonite in the U.S.S.R. *Dokl. Acad. Sci., U.S.S.R. Earth Sci. Sect.*, **159**, 99–102.
Güssvegen, H., 1964. Synthese von Pigeoniten und deren Beziehung zu den Mg–Fe–Pyroxenen in Gesteinen. *Beitrage Min. Petr.*, **11**, 1–14.
Güven, N., 1969. Nature of the co-ordination polyhedra around *M*2 cations in pigeonite. *Contr. Min. Petr.*, **24**, 268–274.
Hafner, S. S. and **Virgo, D.**, 1970. Temperature dependent cation distributions in lunar and terrestrial pyroxenes. *Proc. Apollo 11 Lunar Sci. Conf., Geochim. Cosmochim. Acta*, Suppl. **1, 3**, 2183–2198.
Hafner, S. S., Virgo, D. and **Warburton, D.**, 1971. Cation distributions and cooling history of clinopyroxenes from Oceanus Procellarum. *Proc. 2nd Lunar Sci. Conf., Geochim. Cosmochim. Acta*, Suppl. **2, 1**, 91–108.
Hallimond, A. F., 1914. Optically uniaxial augite from Mull. *Min. Mag.*, **17**, 97–99.
Hamil, M. M., Ghose, S. and **Sparks, R. A.**, 1975. Antiphase domains in a lunar pigeonite: determination of the average shape, size and orientation from a measurement of three-dimensional intensity profiles of diffuse ($h+k$ = odd) reflections. *Acta Cryst.*, **A31**, 126–130.
Hess, H. H., 1941. Pyroxenes of common mafic magmas. Part I, Part II. *Amer. Min.*, **26**, 515–535, 573–594.
Hess, H. H., 1949. Chemical composition and optical properties of common clinopyroxenes, Part I. *Amer. Min.*, **34**, 621–666.
Hess, H. H. and **Henderson, E. P.**, 1949. The Moore County meteorite; a further study with comment on its primordial environment. *Amer. Min.*, **34**, 494–507.
Hollister, L. S. and **Hargreaves, R. B.**, 1970. Compositional zoning and its significance in pyroxenes from two coarse grained Apollo 11 samples. *Proc. Apollo 11 Lunar Sci. Conf., Geochim. Cosmochim. Acta*, Suppl. **1, 1**, 541–550.
Hriskevich, M. E., 1968. Petrology of the Nipissing diabase sill of the Cobalt area, Ontario, Canada. *Bull Geol. Soc. America*, **79**, 1387–1403.
Iijima, S. and **Buseck, P. R.**, 1975. High resolution electron microscopy of enstatite I: twinning, polymorphism and polytypism. *Amer. Min.*, **60**, 758–770.

Ishibashi, K., 1971. [Petrochemical study of basaltic rocks from Higashimatsuura and Ikutsuki-jima district northern Kyushu, Japan.] (In Japanese, English abstract.) *Sci. Rept. Dept. Geol., Kyushu Univ.*, **10**, 177–221.

Ishii, T., 1975. The relations between temperature and composition of pigeonite in some lavas and their application to geothermometry. *Min. J. Japan*, **8**, 48–57.

Isshiki, N., 1963. Petrography of Hachijō-jima volcanic group, Seven Izu Islands, Japan. *J. Fac. Sci. Univ. Tokyo*, Sec. 2, **15**, 91–134.

Ito, T. and **Morimoto, N.**, 1956. Pseudotwins in certain pyroxenes. *Jubilee publication in commemoration of the sixtieth birthday of Professor Jun Suzuki.* M. J. A. Sapporo, Japan, 337–341 (In Japanese, English abstract.)

Johnston, R. and **Gibb, F. G. F.**, 1973. Multiple-twinned and reverse-zoned pigeonite in Apollo 14 basalt 14310. *Min. Mag.*, **39**, 248–251.

Klein, C., *Jr.*, **Drake, J. C.** and **Frondel, C.**, 1971. Mineralogical, petrological and chemical features of four Apollo 12 lunar microgabbros. *Proc. 2nd Lunar Sci. Conf., Geochim. Cosmochim. Acta*, Suppl. **2**, **1**, 265–284.

Konda, T., 1970. Pyroxenes from the Beaver Bay gabbro complex of Minnesota. *Contr. Min. Petr.*, **29**, 338–344.

Kuno, H., 1936. Petrological notes on pyroxene andesites from Hakone volcano, with special reference to some types with pigeonite phenocrysts. *Japan. J. Geol. Geogr.*, **13**, 107–140.

Kuno, H., 1940. Pigeonite in the groundmass of some andesites from Hakone volcano. *J. Geol. Soc. Japan*, **47**, 347–351.

Kuno, H., 1947. Occurrence of porphyritic pigeonite in 'Weiselbergite' from Weiselberg, Germany. *Proc. Jap. Acad.*, **23**, 111.

Kuno, H., 1950. Petrology of Hakone volcano and the adjacent areas, Japan. *Bull. Geol. Soc. America*, **61**, 947–1019.

Kuno, H., 1955. Ion substitution in the diopside–ferropigeonite series of clinopyroxenes. *Amer. Min.*, **40**, 70–93.

Kuno, H., 1959. Origin of Cenozoic petrographic provinces of Japan and surrounding area. *Bull. Volcanol.*, ser. 2, **20**, 37–76.

Kuno, H., 1966. Review of pyroxene relations in terrestrial rocks in the light of recent experimental works. *Min. J. Japan*, **5**, 21–43.

Kuno, H., 1969. Pigeonite-bearing andesite and associated dacite from Asio, Japan. *Amer. J. Sci.*, **367-A** (Schairer vol.), 257–268.

Kuno, H. and **Inoue, T.**, 1949. On porphyritic pigeonite in andesite from Okubo-yama, Minami-Aizu, Hukusima Prefecture. *Proc. Japan. Acad.*, **25**, 128–132.

Kuno, H. and **Nagashima, K.**, 1952. Chemical compositions of hypersthene and pigeonite in equilibrium in magma. *Amer. Min.*, **37**, 1000–1006.

Kushiro, I., 1969. Synthesis and stability of iron-free pigeonite in the system $MgSiO_3$–$CaMgSi_2O_6$ at high pressures. *Carnegie Inst. Washington, Ann. Rept. Dir. Geophys. Lab.*, 1967–68, 80–83.

Kushiro, I., 1972. Determination of liquidus relations in synthetic silicate systems with electron probe analysis: the system forsterite–diopside–silica at 1 atmosphere. *Amer. Min.*, **57**, 1260–1271.

Kushiro, I., 1973. Crystallization of pyroxenes in Apollo 15 Mare basalts. *Carnegie Inst. Washington, Ann. Rept. Dir. Geophys. Lab.*, 1972–73, 647–650.

Kushiro, I. and **Yoder, H. S.**, 1970. Stability field of iron-free pigeonite in the system $MgSiO_3$–$CaMgSi_2O_6$. *Carnegie Inst. Washington, Ann. Rept. Dir. Geophys. Lab.*, 1968–69, 226–229.

Lally, J. S., Heuer, A. H., Nord, G. L. and **Christie, J. M.**, 1975. Subsolidus reactions in lunar pyroxenes: an electron petrographic study. *Contr. Min. Petr.*, **51**, 263–282.

Lindsley, D. H. and **Munoz, J. L.**, 1969. Ortho–clino inversion in ferrosilite. *Carnegie Inst. Washington, Ann. Rept. Dir. Geophys. Lab.*, 1967–68, 86–88.

Maske, S., 1966. The petrography of the Ingeli Mountain Range. *Ann. Univ. Stellenbosch*, **41**, Ser. A, no. 1, 1–109.

McCall, G. J. H., 1966. The petrology of the Mount Padbury mesosiderite and its achondrite enclaves. *Min. Mag.*, **35**, 1029–1060.

Mori, T. and **Green, D. H.**, 1975. Pyroxenes in the system $Mg_2Si_2O_6$–$CaMgSi_2O_6$ at high pressure. *Earth Planet. Sci. Letters*, **26**, 277–286.

Morimoto, N., Appleman, D. E. and **Evans, H. T.**, 1960. The crystal structures of clinoenstatite and pigeonite. *Z. Krist.*, **114**, 120–147.

Morimoto, N. and **Güven, N.**, 1970. Refinement of the crystal structure of pigeonite. *Amer. Min.*, **55**, 1195–1209.

Morimoto, N. and **Ito, T.**, 1968. Pseudo-twin of augite and pigeonite. *Bull. Geol. Soc. America*, **69**, 1616 (abstract).

Morimoto, N. and **Tokonami, M.**, 1969a. Domain structure of pigeonite and clinoenstatite. *Amer. Min.*,

54, 725–740.
Morimoto, N. and **Tokonami, M.**, 1969b. Oriented exsolution of augite in pigeonite. *Amer. Min.*, **54**, 1101–1117.
Muir, I. D., 1954. Crystallization of pyroxenes in an iron-rich diabase from Minnesota. *Min. Mag.*, **30**, 376–388.
Nakamura, Y. and **Konda, T.**, 1974. Compositional relations of pyroxenes in a ferrogabbro from Beaver Bay intrusion, Minnesota. *Lithos*, **7**, 7–14.
Nakamura, Y. and **Kushiro, I.**, 1970a. Compositional relations of coexisting orthopyroxene, pigeonite and augite in a tholeiitic andesite from Hakone volcano. *Contr. Min. Petr.*, **26**, 265–275.
Nakamura, Y. and **Kushiro, I.**, 1970b. Equilibrium relations of hypersthene pigeonite and augite in crystallizing magmas; microprobe study of a pigeonite andesite from Weiselberg, Germany. *Amer. Min.*, **55**, 1999–2015.
Nakazawa, H. and **Hafner, S. S.**, 1977. Orientation relations of augite exsolution lamellae in pigeonite hosts. *Amer. Min.*, **62**, 79–88.
Nwe, Y. Y., 1975. Two different pyroxene crystallization trends in the trough bands of the Skaergaard intrusion, east Greenland. *Contr. Min. Petr.*, **49**, 285–300.
Ohashi, Y. and **Finger, L. W.**, 1973. A lunar pigeonite: crystal structure of primitive-cell domains (abstract). *Amer. Min.*, **58**, 1106.
Ono, A., 1971. [Synthesis of Ca-poor clinopyroxene.] *J. Japanese Assoc. Min. Petr. and Econ. Geol.*, **65**, 211–220 (in Japanese with English Summary).
Poldervaart, A., 1944. The petrology of the Elephant's Head dike and the New Amalfi sheet (Matatiele). *Trans. Roy. Soc. South Africa*, **30**, 85–119.
Poldervaart, A., 1946. The petrology of the Mount Arthur dolerite complex, East Griqualand. *Trans. Roy. Soc. South Africa*, **31**, 84–110.
Poldervaart, A., 1947. The relationship of orthopyroxene to pigeonite. *Min. Mag.*, **28**, 164–172.
Poldervaart, A. and **Hess, H. H.**, 1951. Pyroxenes in the crystallization of basaltic magmas. *J. Geol.*, **59**, 472–489.
Prasad, E. A. V. and **Naidu, M. G. C.**, 1968. On interpenetration twinning in pigeonite. *Current Sci.*, **37**, 75–76.
Prewitt, C. T., **Brown, G. E.** and **Papike, J. J.**, 1971. Apollo 12 clinopyroxenes: high temperature X-ray diffraction studies. *Proc. 2nd Lunar Sci. Conf., Geochim. Cosmochim. Acta*, Suppl. **2**, **1**, 59–68.
Prewitt, C. T., **Papike, J. J.** and **Ross, M.**, 1970. Cummingtonite: a reversible non-quenchable transition from $P2_1/m$ to $C2/m$ symmetry. *Earth Planet. Sci. Lett.*, **8**, 448–450.
Radcliffe, S. V., **Heuer, A. H.**, **Fisher, R. M.**, **Christie, J. M.** and **Griggs, D. T.**, 1970. High voltage (800 kV) electron petrography of type B rock from Apollo 11. *Proc. Apollo 11 Lunar Sci. Conf., Geochim. Cosmochim. Acta*, Suppl. **34**, **1**, 731–748.
Ross, M., **Bence, A. E.**, **Dwornik, E. J.**, **Clark, J. R.** and **Papike, J. J.**, 1970. Mineralogy of the lunar clinopyroxenes, augite and pigeonite. *Proc. Apollo 11 Lunar Sci. Conf., Geochim. Cosmochim. Acta*, Suppl. **34**, **1**, 839–848.
Sathe, R. V. and **Oka, S. S.**, 1972. High calcic pigeonite from pegmatitic segregations in a dolerite dyke from Huli, Mysore State, India. *Min. Mag.*, **38**, 975–976.
Schürmann, K. and **Hafner, S. S.**, 1972. Distinct sub-solidus cooling histories of Apollo 14 basalts. *Proc. 3rd Lunar Sci. Conf., Geochim. Cosmochim. Acta*, Suppl. **3**, **1**, 493–506.
Schwab, R. G. and **Schwerin, M.**, 1975. Polymorphie und Entmischungsreaktionen der Pyroxene im system Enstatit ($MgSiO_3$)–Diopsid ($CaMgSi_2O_6$). *Neues Jahrb. Min., Abhdl.*, 124, 223–245.
Simmons, E. C., **Lindsley, D. H.** and **Papike, J. J.**, 1974. Phase relations and crystallization sequence in a contact-metamorphosed rock from the Gunflint Iron Formation, Minnesota. *J. Petr.*, **15**, 539–565.
Sinitsÿn, A. V., 1965. [Pyroxenes in differentiated dolerite intrusion.] *Zap. Vses. Min. Obshch.*, **94**, 583–592 (in Russian).
Smith, D., 1974. Pyroxene–olivine–quartz assemblages in rocks associated with the Nain anorthosite massif, Labrador. *J. Petr.*, **15**, 58–78.
Smyth, J. R., 1969. Orthopyroxene–high-low-clinopyroxene inversions. *Earth Planet. Sci. Letters*, **6**, 406–407.
Smyth, J. R., 1973. An orthopyroxene structure up to 850°C. *Amer. Min.*, **58**, 636–648.
Smyth, J. R., 1974. The high temperature crystal chemistry of clinohypersthene. *Amer. Min.*, **59**, 1069–1082.
Takeda, H., 1972. Structural studies on rim augite and core pigeonite from lunar rock 12052. *Earth Planet. Sci. Letters*, **15**, 65–71.
Thompson, J. B., *Jr.*, 1970. Geometrical possibilities for amphibole structures: model biopyriboles. *Amer. Min.*, **55**, 292–293.
Trommsdorff, V. and **Wenk, H-R.**, 1968. Terrestrial metamorphic clinoenstatite in kink bands of bronzite crystals. *Contr. Min. Petr.*, **19**, 158–168.

Turner, F. J., 1940. Note on determination of optic axial angle and extinction angle in pigeonite. *Amer. Min.*, **25**, 831–825.

Turner, F. J., Heard, H. and **Griggs, D. J.,** 1960. Experimental deformation of enstatite and accompanying inversion to clinoenstatite. *Intern. Geol. Congr. Rept., 21, Copenhagen*, Pt. **18**, 399–408.

Turnock, A. C., 1970. A pyroxene solvus section. *Can. Min.*, **10**, 744–747.

Vilenskiy, A. M., 1966. Phase equilibria and evolution in the chemical nature of the pyroxenes of intrusive traps of the Siberian platform. *Geochem. Intern.*, **3**, 1275–1306.

Virgo, D., 1973. Crystallization and subsolidus cooling history of Apollo-15 basalts 15076 and 15476. *Carnegie Inst. Washington, Ann. Rept. Dir. Geophys. Lab.*, 1972–73, 650–656.

Virgo, D. and **Hafner, S. S.,** 1969. Fe^{2+},Mg order-disorder in heated orthopyroxenes. *Min. Soc. America, Spec. Paper* **2**, 67–81.

Virgo, D. and **Ross, M.,** 1973. Pyroxenes from Mull andesites. *Carnegie Inst. Washington, Ann. Rept. Dir. Geophys. Lab.*, 1972–73, 535–540.

Walker, F. and **Poldervaart, A.,** 1941a. The Karroo dolerites of the Calvinia District. *Trans. Geol. Soc. South Africa*, **44**, 127–150.

Walker, F. and **Poldervaart, A.,** 1941b. The Hangnest dolerite sill, South Africa. *Geol. Mag.*, **78**, 429–450.

Walker, F. and **Poldervaart, A.,** 1949. Karroo dolerites of the Union of South Africa. *Bull. Geol. Soc. America*, **60**, 591–705.

Warner, R. D. and **Luth, W. C.,** 1974. The diopside–orthoenstatite two-phase region in the system $CaMgSi_2O_6$–$Mg_2Si_2O_6$. *Amer. Min.*, **59**, 98–109.

Wheeler, E. P., 1965. Fayalitic olivine in northern Newfoundland–Labrador. *Can. Min.*, **8**, 339–346.

Williams P. G. L., Bancroft, G. M., Bown, M. G. and **Turnock, A. C.,** 1971. Anomalous Mössbauer spectra of *C2/c* clinopyroxenes. *Nature (Phys. Sci.)*, **230**, 149–151.

Williams, R. J. and **Takeda, H.,** 1972. Pigeonites and augites precipitate from synthetic lunar melts. *Trans. Amer. Geophys. Union*, **53**, 551.

Yang, H-Y., 1973. Crystallization of iron-free pigeonite in the system anorthite–diopside–enstatite–silica at atmospheric pressure. *Amer. J. Sci.*, **273**, 488–497.

Yang, H-Y. and **Foster, W. R.,** 1972. Stability of iron-free pigeonite at atmospheric pressure. *Amer. Min.*, **57**, 1232–1241.

Calcium Pyroxenes

Diopside–Hedenbergite $Ca(Mg,Fe^{2+})Si_2O_6$
Augite $(Ca,Mg,Fe^{2+},Al)_2(Si,Al)_2O_6$
Fassaite $Ca(Mg,Fe^{2+},Fe^{3+},Al)(Si,Al)_2O_6$
Johannsenite $CaMnSi_2O_6$

Diopside $CaMg[Si_2O_6]$
Hedenbergite $CaFe^{2+}[Si_2O_6]$

Monoclinic (+)

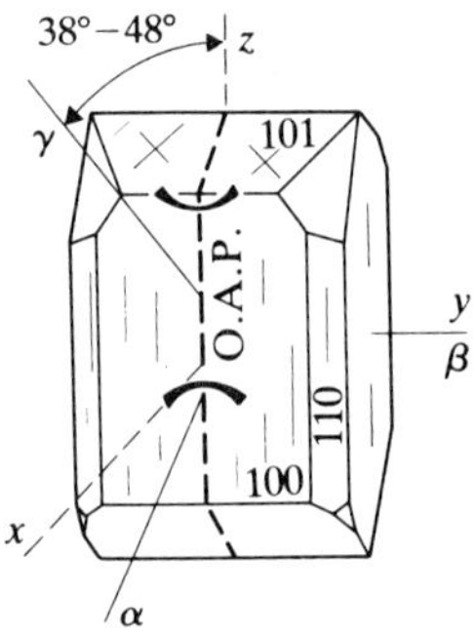

	Diopside[1]	Hedenbergite[1]
α	(1·664)[2] 1·664–1·695	1·1771–1·726(1·732)[2]
β	(1·6715) 1·672–1·701	1·723–1·730
γ	(1·694) 1·694–1.721	1·741–1·751 (1·755)
δ	(0·030) 0·024–0·031	0·025–0·034 (0·024)
$2V_\gamma$	(59·3°) 49°–64°	52°–64°
$\gamma:z$	(38°30′) 38°–46°	47°–48°
	$\beta = y$. O.A.P. (010)	$\beta = y$. O.A.P. (010)
Dispersion:	$r > v$, weak to moderate	$r > v$, strong
D	3·22–3·38	3·50–3·56
H	$5\frac{1}{2}$–$6\frac{1}{2}$	6

Cleavage: {110} good; {100}, {010} partings (110):(1$\bar{1}$0) ≃ 87°

Twinning: {100}, {010}, simple and multiple, common.

Colour:	White, pale green, dark green; diopside colourless in thin section, more Fe-rich members colourless to pale green.	Brownish green, dark green, black, pale green, yellow-green, brownish green in thin section.

Pleochroism: Diopside non-pleochroic; more Fe-rich members and hedenbergite may show weak pleochroism:

	α pale green.	pale green, bluish green.
	β pale greenish brown.	green, light bluish green.
	γ pale brownish green.	green, yellow-green.
Unit cell:[2]	a (Å) 9·752	9·844
	b (Å) 8·926	9·028
	c (Å) 5·248	5·246
	β° 105·83°	104·80°
	Z = 4. Space group $C2/c$.	

Insoluble in HCl.

[1] Intermediate members of the series have properties continuous between those of diopside and hedenbergite.

[2] Values for synthetic end-members.

The diopside–hedenbergite minerals form a complete solid solution series between $CaMgSi_2O_6$ and $CaFe^{2+}Si_2O_6$. The division between intermediate compositions is arbitrary as is that between the members of the series and endiopside, augite, ferroaugite and ferrohedenbergite (see Fig. 1, p. 3). Relatively pure diopsides occur in some thermally metamorphosed siliceous dolomites and in skarns; the more Fe-rich members of this series are found in rocks of similar paragenesis. Diopside, and particularly chrome diopside, is the common pyroxene as is endiopside, in nodules and xenocrysts in kimberlites and alkali olivine basalts. Salite occurs in calcium-rich sediments and basic igneous rocks which have been regionally metamorphosed in the environment of the higher grades of the amphibolite facies, and also occurs in some granulites and charnockites. Diopside–salite is the typical pyroxene in rocks that have formed from alkali olivine basalt magma, especially during its crystallization in a hypabyssal environment. Hedenbergite and manganoan hedenbergite occur in limestone skarns, in thermally metamorphosed iron-rich sediments, in quartz syenites and in granophyres. Ferrohedenbergites have also been reported from quartz syenites and granophyres as well as from ferrodiorites.

The name schefferite has been used to describe the manganese-rich salites occurring at Pajsberg and Långban, Sweden, and at Franklin, New Jersey, and jeffersonite to describe the manganese- and zinc-rich ferrosalite at Franklin. The name diopside is from Latin *di*, two, and the Greek *opsis*, appearance, in allusion to its variable habit; hedenbergite is named after M. A. Ludwig Hedenberg, the Swedish chemist who discovered and described it; salite (sahlite) is a locality name, from Sala in Västmanland, Sweden. The term diallage has been used to describe diopside containing appreciable amounts of trivalent ions, but is more often used to describe both diopsides and augites with well developed and closely spaced {100} partings which are commonly occupied by magnetite or ilmenite.

Structure

The first of the pyroxene structures to be determined was that of diopside (Warren and Bragg, 1928), and this work established the essential features of all pyroxene structures (see p. 5). Clark *et al.* (1969) refined the crystal structure of a very nearly pure diopside from the Gouverneur talc district, New York State, U.S.A. It has *a* 9·746, *b* 8·899, *c* 5·251 Å, *β* 105·63°, Z = 4, space group *C*2/*c*, and calculated density 3·278 g/cm³. In diopside the pyroxene chains are linked laterally by Ca and Mg ions in *M*2 and *M*1 sites respectively as shown in Fig. 6, p. 8. Pairs of chains come back to back without displacement in the *y* direction but they are staggered in the *z* direction so that the monoclinic cell results. The Ca ions are in irregular eight-fold co-ordination, having four short Ca–O distances with a mean of 2·36 Å and four longer distances with a mean of 2·64 Å. The mean bridging Si–O distance is 1·676 Å and the mean non-bridging Si–O distance is 1·594 Å. In the diopside structure the edges of four tetrahedra are shared with the large *M*2 polyhedra. The shared edges have a reduced tetrahedral angle compared with unshared edges.

In diopside, as in other clinopyroxenes with space group *C*2/*c*, all Si–O chains are equivalent. The chain is somewhat kinked (with *O* rotations – see Thompson, 1970) relative to the almost straight chain in jadeite (angle O3–O3–O3 is 166·38° compared with 174·7° for jadeite).

The structure of hedenbergite (*a* 9·844, *b* 9·028, *c* 5·246, *β* 104·80°, *C*2/*c*) is similar to that of diopside (Veblen and Burnham, 1969). The latter authors also determined the structure of an intermediate member (ferrosalite) of the Mg,Fe diopside–hedenbergite series, and showed it to be isostructural with diopside. Its cell parameters are *a* 9·788, *b* 8·969, *c* 5·253 Å, *β* 105·63°. The extent of kinking in the Si–O chain is similar: O3–O3–O3 is 164·5° for hedenbergite, and 165·9° for ferrosalite.

The thermal expansion of diopside has been studied by Deganello (1973) and crystal structure determinations were carried out at various temperatures by Cameron *et al.* (1973) and by Finger and Ohashi (1976); mean thermal expansion coefficients were given for cell parameters, cell volumes and the sizes of cation polyhedra. The cell edges *a*, *b* and *c* increase linearly with temperature whereas for *β* the increase is non-linear; principal strain components were also calculated. The Si–O bond lengths and chain configurations are very little affected by heating. The maximum expansion is parallel to *y*, and the minimum expansion is parallel to the shortest Ca–O bond rather than to the tetrahedral chain. Similar results were found for hedenbergite by Cameron *et al.* (1973).

The Mössbauer spectra of five *C*2/*c* clinopyroxenes lying on or close to the diopside–hedenbergite join are described by Bancroft *et al.* (1971). As expected, since the *M*2 site is normally fully occupied by Ca, the Fe^{2+} ions reside entirely in *M*1. For one specimen, however, there was evidence of a small amount of Fe^{2+} in *M*2.

The variation of cell parameters with composition for synthetic members of the diopside–hedenbergite series was studied by Rutstein and Yund (1969). The cell volume and *a* and *b* vary linearly over the range whereas *β* does not; *c* does not vary significantly. Thus the solid solution series seems to be ideal, a result to be expected since only *M*1 is available for magnesium and iron substitution. Synthetic diopside has *a* 9·752, *b* 8·926, *c* 5·248 Å, *β* 105·83°. Similar values were reported by Clark *et al.* (1962) and Sakata (1957). Coleman (1962) synthesized diopsides with small amounts of various substituents and deduced relationships between the cell parameters and composition.

The variations in cell parameters with Ti substitution in synthetic Ca-pyroxenes were reported by Onuma *et al.* (1968). For the substitution $MgSi \rightleftharpoons Ti^{4+}Al$ (i.e. Ti in *M*1) changes are linear with coefficients $a+0{\cdot}00009$, $b-0{\cdot}0022$, $c+0{\cdot}0025$ Å, $\beta+0{\cdot}024°$, $V+0{\cdot}09$ Å^3 for each weight per cent $CaTiAl_2O_6$ up to 11 per cent. The changes with MgSi = TiAl substitution were found to be not very different when Fe^{3+} was present (up to 10 weight per cent Fe_2O_3). For the alternative substitution where Ti replaces Si, the coefficients are $a+0{\cdot}0042$, $b+0{\cdot}0008$, $c+0{\cdot}0039$ Å, $\beta+0{\cdot}03$, $V+0{\cdot}49$ Å^3 for each weight per cent of $CaMgTi_2O_6$ up to 10 per cent. The effects of Ti substitution on the cell parameters of synthetic diopsides were also studied by Schröpfer (1968, 1971). The substitution of MgSi by $2\,Fe^{3+}$ (Fe^{3+} in T and *M*1) was shown by Huckenholz *et al.* (1969) to produce increases in *a*, *c* and *V*, a decrease in *b* and no change in *β*.

They suggested that the substitution CaSi = $2Fe^{3+}$ could also occur (i.e. Fe^{3+} in T and *M*2). Cell parameters for clinopyroxenes along the join $CaMgSi_2O_6$–$CaFe^{3+}AlSiO_6$ are given by Hijikata (1968), and for some aluminous natural clinopyroxenes by Lewis (1967). The ordering of Fe^{3+} and Al between octahedral and tetrahedral sites has been reported by Ohashi and Hariya (1973) to increase with increasing pressure.

The cell parameters of clinopyroxenes of the Di, Hd, En, Fs quadrilateral are discussed further in the section on augite (p. 299).

Nolan and Edgar (1963) demonstrated linear relationships between composition and *d* values of the planes 220, 310, $13\bar{1}$ and 150 which are usable determinatively for clinopyroxenes known to be close to the diopside–hedenbergite join (e.g. those from calc-silicate skarns). For the system diopside–hedenbergite–acmite, in which there is complete solid solution, the variations in cell parameters have been presented by Nolan (1969); see p. 486.

Chemistry

The diopside–salite–ferrosalite–hedenbergite minerals (Tables 14–19) form a complete solid solution series between $CaMgSi_2O_6$ and $CaFe^{2+}Si_2O_6$. Most of the minerals of this series, however, contain other ions, although some diopside from metamorphosed siliceous limestones and dolomites, and salites from skarns, approximate to the composition $Ca(Mg,Fe)Si_2O_6$, and nearly pure diopside, $Ca_{49{\cdot}0}Mg_{49{\cdot}2}Fe_{1{\cdot}8}$, in the serpentinite within the contact aureole of the Mount Stuart Batholith has been reported by Frost (1975). In both the igneous and metamorphic members of the series the replacement of Si by Al usually does not exceed 10 per cent and in the majority of the minerals the Al_2O_3 content varies between 1 and 3 per cent. Ferrian diopsides, from biotite pyroxenites and biotite peridotites of the Gardar alkaline province, south Greenland, containing trivalent cations ($Al_2O_3+Fe_2O_3$ between 6 and 14·5 per cent) and TiO_2 contents varying from 1 to 3·27 per cent have been described by Upton and Thomas (1973).

Many diopsides and magnesium-rich salites in basic and ultrabasic rocks contain appreciable amounts of chromium, and the average content of Cr_2O_3 in the chrome-diopsides detailed in Table 15 is 1·09 per cent. The incorporation of chromium in diopside may be considered as arising from the $NaCrSi_2O_6$ and $CaCrAlSiO_6$ type substitutions. Although nickel diopside ($CaNiSi_2O_6$) has been synthesized (Gjessing, 1941), natural diopsides containing appreciable percentages of nickel are unknown, due most probably to the small quantities of Ni in the magmas from which the pyroxene crystallized. The majority of diopside–hedenbergites have a small content of titanium; the average content of TiO_2 in eighty-four of the analyses detailed in Tables 14–19 is 0·54 per cent; those having a titania content greater than 0·5 per cent are mainly confined to salitic minerals particularly of igneous rocks.

The manganese content of the magnesium-rich members of the series is generally low, but the amount increases in the more iron-rich minerals. Thus the average MnO content in diopsides, chrome-diopsides and endiopsides is 0·13 per cent, in salites 0·24 per cent, in ferrosalites 0·35 per cent, in the hedenbergites 3·60 per cent (anal. 26, Table 17 and anal. 10, Table 18 are omitted from these averages), and in the ferrohedenbergites 0·70 per cent (Table 20). The replacement of Mg and Fe^{2+} by Mn is, however, much greater in some minerals of the diopside–hedenbergite series; thus manganese-rich diopsides and salites (see Table 43) from skarn rocks at Altin Topkan, Central Asia, have been described by Zharikov and Vlasova (1955), and it is most probable that diopside, hedenbergite and johannsenite form a complete isomorphous group. A study by Vinokurov (1966) of the Mn^{2+} electron paramagnetic resonance spectra in a number of diopsides has shown that Mn^{2+} may replace Ca, as well as Mg ions. Zn also may substitute for Mg and Fe^{2+}, and Palache (1937) has described two zinc-bearing manganese-rich pyroxenes from Franklin, New Jersey, to which he gave the names zinc schefferite and jeffersonite.

Abundance levels of K, Rb, Sr and Ba in the Ca-rich pyroxenes of ultrabasic and eclogite nodules in kimberlite are given by Griffin and Murthy (1969). Values of Cr, Ni and V for twenty-two pyroxenes close to diopside–hedenbergite compositions have been given by Smith (1966). Cr, V, Ni, Co, Li, Sc, Rb, Sr and Ga have been

Table 14. Diopside Analyses

	1		2		3		4		5	
SiO_2	54·66		51·10		54·61		54·02		50·74	
TiO_2	—		0·00		tr.		tr.		0·11	
Al_2O_3	0·07		5·60		1·87		0·51		3·67	
Cr_2O_3	—		—		—		—		—	
Fe_2O_3	0·68		0·00		1·22		0·89		0·96	
FeO	0·07		0·94		0·00		1·00		0·79	
MnO	0·02		—		0·00		0·13		0·08	
NiO	—		—		—		—		—	
MgO	18·78		15·63		18·42		16·73		16·92	
CaO	25·85		26·35		23·14		26·37		25·78	
Na_2O	—		tr.		—		0·16		—	
K_2O	—		tr.		—		tr.		—	
H_2O^+	0·22		0·30		—		} 0·39		0·54	
H_2O^-	—		0·25		0·61				0·11	
Total	100·35		100·17		99·87		100·43		99·94	
α	1·664		—		1·669		1·669		1·669	
β	1·673		—		—		1·675		—	
γ	1·695		—		1·700		1·698		1·695	
$\gamma:z$	38°		—		38°–40°		45°		—	
$2V_\gamma$	60°		—		58°		59°		—	
D	3·259		—		—		3·29		—	
	Numbers of ions on the basis of six O									
Si	1·976	1·98	1·864	2·00	1·973	2·00	1·971	1·99	1·873	2·00
Al	0·002		0·136		0·027		0·022		0·127	
Al	—		0·105		0·053		0·000		0·032	
Ti	—		—		—		0·000		0·004	
Cr	—		—		—		—		—	
Fe^{3+}	0·018		—		0·033		0·007		0·026	
Mg	1·012		0·850		0·992		0·915		0·931	
Ni	—	2·03	—	2·01	—	1·97	—	2·00	—	2·02
Fe^{2+}	0·002		0·029		—		0·031		0·024	
Mn	0·001		—		—		0·004		0·002	
Ca	1·001		1·029		0·894		1·031		1·005	
Na	—		—		—		0·011		—	
K	—		—		—		0·000		—	
Mg	49·8		44·6		51·7		46·0		46·8	
*ΣFe	1·0		1·5		1·7		2·1		2·6	
Ca	49·2		53·9		46·6		51·9		50·6	

*ΣFe = Fe^{2+} + Fe^{3+} + Mn.

1 Diopside, limestone, Juva, Finland (Juurinen and Hytönen, 1952). Anal. A. Juurinen.
2 Diopside, altered zone at gabbro–serpentinite contact, Jordanów, Lower Silesia, Poland (Heflik, 1967).
3 Diopside, veinlet in serpentinized peridotite, Karachai district, northern Caucasus (Serdyuchenko, 1937).
4 Diopside, vein in serpentinite, Alpe della Rossa, Alpes Piémontaises (Nicolas, 1966) (includes CO_2 0·23).
5 Diopside, contact skarn, Kilbride, Skye (Tilley, 1951). Anal. J. H. Scoon (includes CO_2 0·24).

6	7	8	9	10	
53·65	49·81	52·34	54·3	53·60	SiO_2
0·04	0·14	0·18	0·62	0·07	TiO_2
1·89	6·42	2·72	0·19	0·00	Al_2O_3
—	—	—	0·03	—	Cr_2O_3
0·58	1·25	1·31	—	1·12	Fe_2O_3
1·53	1·03	1·77	3·3	1·33	FeO
0·09	0·06	0·24	0·09	1·33	MnO
—	—	—	0·05	—	NiO
16·23	15·22	16·15	16·9	16·38	MgO
25·30	25·92	24·23	23·9	25·24	CaO
0·19	—	0·57	0·12	0·04	Na_2O
tr.	—	0·08	—	0·01	K_2O
0·47	0·13	0·66	—	0·07	H_2O^+
0·07	0·06	0·00	—	—	H_2O^-
100·04	100·19	100·25	99·5	99·70	**Total**
1·665	1·680	1·674	—	1·6794	α
1·672	—	1·683	—	1·6855	β
1·695	1·704	1·699	—	1·7099	γ
39°	—	40°	—	40°	$\gamma:z$
57°	—	60°	—	56°	$2V_\gamma$
3·28	—	3·25	—	3·29	**D**

6		7		8		9		10		
1·960	2·00	1·824	2·00	1·920	2·00	1·991	2·00	1·988	1·99	Si
0·040		0·176		0·080		0·008		—		Al
0·042		0·101		0·038		—		—		Al
0·001		0·004		0·005		0·017		0·002		Ti
—		—		—		0·001		—		Cr
0·016		0·034		0·036		—		0·031		Fe^{3+}
0·884		0·832		0·883		0·923		0·905		Mg
—	2·00	—	2·01	—	2·02	0·002	1·99	—	2·01	Ni
0·047		0·030		0·054		0·101		0·041		Fe^{2+}
0·003		0·002		0·007		0·003		0·042		Mn
0·991		1·011		0·952		0·939		0·984		Ca
0·014		—		0·040		0·008		0·004		Na
—		—		0·003		—		0·000		K
45·5		43·6		45·7		47·0		45·2		Mg
3·4		3·4		5·0		5·3		5·7		*ΣFe
51·1		53·0		49·3		47·7		49·1		Ca

6 Diopside, diopside–zoisite–amphibole–calcite skarn. Lochan an Sgòr Gaoithe, Glen Urquhart (Francis, 1958). Anal. D. I. Bothwell and K. C. Chaperlin (spectrographic determinations: Ga 5, Cr 45, V 67, Li 8, Ni 10, Co 3, Zr 37, Sr 112, Ba 10 p.p.m.).

7 Diopside, contact skarn, Kilbride, Skye (Tilley, 1951). Anal. J. H. Scoon (includes CO_2 0·15).

8 Diopside, calc silicate gneiss, Rodil, South Harris (Davidson, 1943). Anal. W. H. Herdsman.

9 Diopside, wyomingite, Leucite Hills, Wyoming (Carmichael, 1967). Anal. I. S. E. Carmichael (microprobe analysis).

10 Diopside, carbonate rock, Oka, Quebec (Gold, 1966). Anal. C. O. Ingamells (includes Sr 0·067, Ba 0·019, P_2O_5 0·38, F 0·06 less O = F 0·02).

Table 14. Diopside Analyses – *continued*

	11	12	13	14	15
SiO_2	53·20	50·30	53·78	51·48	50·67
TiO_2	0·12	0·41	0·08	0·91	0·31
Al_2O_3	0·75	4·90	1·29	7·99	2·56
Cr_2O_3	—	—	—	—	0·32
Fe_2O_3	0·76	2·09	0·80	0·84	2·35
FeO	2·92	1·89	3·89	3·23	3·02
MnO	0·21	0·08	0·19	0·06	0·13
NiO	—	—	—	—	—
MgO	17·32	14·93	15·08	12·22	16·51
CaO	24·41	25·20	24·21	20·14	23·46
Na_2O	0·07	0·14	0·50	1·84	0·16
K_2O	tr.	0·00	0·06	0·49	0·02
H_2O^+	0·05	—	0·06	1·00	0·55
H_2O^-	0·06	—	0·00	—	—
Total	99·87	99·94	99·96	100·20	100·06
α	1·680	1·684	—	1·676	—
β	1·693	—	1·6826	—	—
γ	1·707	1·708	—	1·706	—
$\gamma:z$	39°	36°–40°	—	—	—
$2V_\gamma$	62°	—	—	56°	—
D	3·266	3·10	3·33	—	3·26

Numbers of ions on the basis of six O

	11	Σ	12	Σ	13	Σ	14	Σ	15	Σ
Si	1·955	1·99	1·848	2·00	1·977	2·00	1·883	2·00	1·878	1·99
Al	0·032		0·152		0·023		0·117		0·112	
Al	0·000		0·060		0·033		0·227		0·000	
Ti	0·003		0·011		0·002		0·025		0·009	
Cr	—		—		—		—		0·009	
Fe^{3+}	0·021		0·058		0·022		0·023		0·066	
Mg	0·948		0·818		0·826		0·666		0·912	
Ni	—	2·03	—	2·01	—	2·00	—	1·98	—	2·04
Fe^{2+}	0·088		0·058		0·120		0·099		0·094	
Mn	0·007		0·002		0·006		0·002		0·004	
Ca	0·959		0·992		0·954		0·789		0·931	
Na	0·003		0·010		0·036		0·130		0·011	
K	0·000		0·000		0·003		0·023		0·001	
Mg	46·9		42·4		42·8		42·2		45·4	
*ΣFe	5·7		6·1		7·7		7·8		8·2	
Ca	47·4		51·5		49·5		50·0		46·4	

11 Diopside, rodingite dyke, Pastoki, Hindubagh, Pakistan (Bilgrami and Howie, 1960). Anal. R. A. Howie.

12 Diopside, diopside–fassaite rock, metasomatized dolerite, Vilyuy, Yakutia (Ginsburg, 1969). Anal. O. G. Unanova.

13 Diopside, scapolite–pyroxene skarn, Huddersfield Township, Pontiac County, Quebec (Shaw *et al.*, 1963b). Anal. C. O. Ingamells (includes F 0·02, Cl 0·01 less O = F, Cl 0·01, BaO tr., Li_2O tr.).

14 Diopside, plagioclase–hypersthene–diopside–garnet nodule, Aeromagnitnaya kimberlite pipe, Siberia (Lutts, 1965). Anal. A. F. Al'ferova.

15 Diopside, olovine pyroxenite, Baranchinsk massif, Urals (Borisenko, 1967a). Anal. IMGRE Analytical Laboratories (Sc_2O_3 0·015, V_2O_3 0·027, Co 0·002, Ni 0·008, Ga 0·0006, Li 0·0003).

16		17		18		19		20		21		
53·84		52·67		51·99		49·90		49·23		48·90		SiO_2
0·09		0·14		0·65		0·51		2·25		0·37		TiO_2
2·00		1·42		2·49		5·31		2·01		3·98		Al_2O_3
—		—		—		0·28		—		0·03		Cr_2O_3
0·12		1·23		0·83		2·09		2·87		3·76		Fe_2O_3
4·90		4·34		5·09		4·01		3·53		2·73		FeO
0·17		0·13		0·11		0·08		0·09		0·14		MnO
—		—		—		–		—		—		NiO
15·51		15·59		15·49		15·68		14·69		15·17		MgO
23·07		23·81		22·42		21·62		24·23		23·11		CaO
0·70		0·45		0·44		0·35		0·46		0·42		Na_2O
0·01		0·00		0·10		0·04		0·15		tr.		K_2O
—		0·20		—		—		0·38		0·57		H_2O^+
—		—		—		—		0·00				H_2O^-
100·41		99·98		99·61		99·87		99·89		'99·72'		**Total**
—		1·6791		1·678		—		1·695		1·684		α
—		1·6843		1·683		—		1·701		1·692		β
—		1·707		1·705		—		1·721		1·714		γ
—		38°		—		—		46°		38°		$\gamma{:}z$
—		50·5°		57°		—		60°		55°		$2V_\gamma$
—		3·303		—		—		—		3·35		D
1·966	2·00	1·946	2·00	1·922	2·00	1·837	2·00	1·846	1·94	1·835	2·00	Si
0·034		0·054		0·078		0·163		0·089		0·165		Al
0·052		0·008		0·030		0·067		—		0·011		Al
0·002		0·004		0·018		0·014		0·064		0·010		Ti
—		—		—		0·008				0·001		Cr
0·004		0·034		0·023		0·058		0·081		0·106		Fe^{3+}
0·846		0·859		0·859		0·860		0·821		0·849		Mg
—	2·01	—	2·02	—	2·01	—	2·01	—	2·09	—	2·03	Ni
0·150		0·134		0·156		0·123		0·111		0·086		Fe^{2+}
0·005		0·004		0·003		0·002		0·003		0·004		Mn
0·903		0·943		0·887		0·853		0·972		0·929		Ca
0·050		0·032		0·031		0·025		0·033		0·031		Na
0·000		0·000		0·005		0·002		0·007		0·000		K
44·3		43·5		44·6		45·4		41·3		43·0		Mg
8·3		8·7		9·4		9·6		9·8		9·9		*ΣFe
47·4		47·8		46·0		45·0		48·9		47·1		Ca

16 Diopside, pyroxene amphibolite, Nero Hill, central Tanzania (Haslam and Walker, 1971). Anal. H. W. Haslam.
17 Diopside, fine-grained gabbro xenolith in granite gneiss, Upper Bridge, Crum Creek Reservoir, Pennsylvania (Norton and Clavan, 1959). Anal. W. S. Clavan.
18 Diopside, hornblende-bearing biotite–olivine pyroxenite, Ichinohe, Kitakami, northern Japan (Onuki and Tiba, 1964). Anal. H. Onuki.
19 Diopside, megacrysts in alkali basalt scoria, Itinome-gata crater, north-eastern Japan (Aoki, 1971). Anal. K. Aoki.
20 Diopside, potassic ankaratrite, Nyamunuka crater, Katwe-Kilorongo field, south-western Uganda (Sahama, 1952). Anal. H. B. Wiik.
21 Diopside, gabbro pyroxenite, Kachkanar, Urals (Shteinberg and Tominyh, 1967).

Table 15. Chrome-diopside Analyses

	1	2	3	4	5
SiO_2	50·97	53·04	51·56	50·70	54·07
TiO_2	0·49	0·45	0·25	0·28	0·21
Al_2O_3	4·86	6·22	2·76	4·90	2·08
Cr_2O_3	1·47	1·20	1·90	1·15	0·98
Fe_2O_3	0·92	0·72	1·22	1·37	0·56
FeO	1·41	1·08	1·46	1·61	2·53
MnO	0·10	0·05	0·11	0·06	0·09
MgO	17·89	14·56	17·18	15·90	17·39
CaO	21·28	18·90	21·52	22·26	22·12
Na_2O	0·83	2·29	1·56	2·60	0·41
K_2O	tr.	0·07	0·09	tr.	0·00
H_2O^+	0·23	1·20	—	—	0·04
H_2O^-	0·07	—	—	—	0·06
Total	100·52	99·78	99·61	100·83	100·58
α	1·678–1·684	1·678	—	—	1·6742
β	1·687–1·693	1·690	—	—	1·6800
γ	1·697–1·703	1·708	—	—	1·7019
$\gamma:z$	38°	41°	—	—	40·5°
$2V_\gamma$	58–54°	64°	—	—	55·5°
D	3·309	—	—	—	—

Numbers of ions on the basis of six O

	1		2		3		4		5	
Si	1·847	2·00	1·927	2·00	1·898	2·00	1·845	2·00	1·953	2·00
Al	0·153		0·073		0·102		0·155		0·047	
Al	0·055	2·03	0·193	1·98	0·011	2·07	0·055	2·10	0·041	1·99[a]
Ti	0·013		0·012		0·006		0·008		0·006	
Cr	0·042		0·034		0·057		0·033		0·028	
Fe^{3+}	0·025		0·020		0·035		0·038		0·015	
Mg	0·965		0·788		0·950		0·863		0·936	
Fe^{2+}	0·043		0·033		0·044		0·049		0·076	
Mn	0·003		0·002		0·002		0·002		0·003	
Ca	0·826		0·736		0·849		0·868		0·856	
Na	0·058		0·161		0·110		0·183		0·029	
K	0·000		0·003		0·004		0·000		0·000	
Mg	51·8		49·9		50·5		47·4		49·6	
*ΣFe	3·8		3·5		4·3		4·9		5·0	
Ca	44·4		46·6		45·2		47·7		45·4	

$*\Sigma = Fe^{2+} + Fe^{3+} + Mn$.
[a] Includes Ni 0·001.

1 Chrome-diopside, peridotite, Horoman, Hokkaido, Japan (Yamaguchi, 1961), Anal. T. Katsura.
2 Chrome-diopside, enstatite–forsterite–chrome-diopside–garnet–spinel nodule, Slyudyanka kimberlite pipe, Siberia (Lutts, 1965). Anal. A. F. Al'ferova.
3 Chrome-diopside, garnet lherzolite, block in carbonatite tuff and glassy scoria, ankaramitic glassy scoria, Lashaine volcano, northern Tanzania (Dawson *et al.*, 1970). Anal. D. G. Powell.
4 Chrome-diopside, saxonite, Beni-Bouchera, Morocco (Kornprobst, 1966). Anal. P. Blot.
5 Chrome-diopside, bronzitite, Bushveld complex (Hess, 1949). Anal. R. B. Ellestad (includes NiO 0·04).

6	7	8	9	10	11	
51·72	53·42	52·26	50·14	52·67	54·09	SiO_2
0·61	0·12	0·73	0·75	0·17	0·28	TiO_2
6·69	0·68	1·94	4·85	3·83	1·57	Al_2O_3
0·72	0·56	0·54	0·95	0·50	2·03	Cr_2O_3
1·80	1·36	1·08	1·70	1·61	0·74	Fe_2O_3
1·08	2·01	2·29	1·51	2·46	1·47	FeO
0·09	0·07	0·07	0·09	0·10	0·09	MnO
15·22	15·99	16·73	16·49	15·06	16·96	MgO
20·23	25·75	22·80	21·83	22·19	21·10	CaO
1·78	0·12	0·87	0·78	1·23	1·37	Na_2O
0·00	0·06	0·18	—	0·02	0·15	K_2O
—	0·16	—	0·00	—	0·22	H_2O^+
—	0·01	—	0·73	—	0·08	H_2O^-
100·01	'100·34'	99·49	99·90	99·89	100·64	**Total**
—	1·672–1·676	—	—	1·678	—	α
—	1·679–1·683	—	1·689	1·685	> 1·66	β
—	1·699–1·703	—	—	1·705	—	γ
—	39°	—	—	—	50°	γ:z
—	56°	—	—	56°	70°–75°	$2V_\gamma$
—	—	—	—	—	—	D
1·867 } 2·00	1·957 } 1·99	1·922 } 2·00	1·817 } 2·00	1·923 } 2·00	1·961 } 2·00	Si
0·133	0·030	0·078	0·183	0·077	0·039	Al
0·152	—	0·006	0 024	0·088	0·029	Al
0·016	0·003	0·020	0·020	0·005	0·008	Ti
0·020	0·016	0·016	0·027	0·014	0·058	Cr
0·049	0·037	0·030	0·047	0·044	0·020	Fe^{3+}
0·819 } 2·00[a]	0·873 } 2·01	0·917 } 2·05	0·890 } 1·96[a]	0·819 } 2·01[b]	0·917 } 2·01[b]	Mg
0·033	0·062	0·070	0·046	0·075	0·045	Fe^{2+}
0·003	0·002	0·002	0·003	0·003	0·003	Mn
0·783	1·011	0·917	0·847	0·868	0·820	Ca
0·125	0·008	0·062	0·055	0·088	0·096	Na
0·000	0·002	0·008	0·000	0·001	0·007	K
48·6	44·0	47·4	48·6	45·3	50·8	Mg
5·0	5·1	5·2	5·2	6·7	3·8	*ΣFe
46·4	50·9	47·4	46·2	48·0	45·4	Ca

[a] Includes Ni 0·002.
[b] Includes Ni 0·001.

6 Chrome-diopside, lherzolite nodule in hawaiite, Kyogle, New South Wales (Wilkinson and Binns, 1969). Anal. G. I. Z. Kalocsai (includes NiO 0·07).

7 Chrome-diopside, basalt, Kozōri, Sano, Japan (Kuno, 1957). Anal. H. Haramura (includes P_2O_5 0·03; NiO 0·016 and V_2O_5 0·004 determined as Ni and V respectively are not included in the total).

8 Chrome-diopside, jumillite, Jumilla, Murcia, Spain (Borley, 1967). Anal. G. D. Borley.

9 Chrome-diopside, plagioclase peridotite, Horoman, Hokkaido, Japan (Nagasaki, 1966). Anal. H. Haramura (includes NiO 0·08; K_2O and P_2O_5 reported < 0·05).

10 Chrome-diopside, websterite, Glenelg, Inverness-shire, Scotland (Mercy and O'Hara, 1965). Anal. L. P. Mercy (includes NiO 0·046, P_2O_5 0·004).

11 Chrome-diopside, kimberlite, Dutoitspan, South Africa (Holmes, 1937). Anal. L. S. Theobald (includes NiO 0·03, ZrO 0·12, V_2O_3 0·07, Sr 0·01, CO_2 0·26, Ce tr., S tr.).

Table 16. Endiopside Analyses

	1	2	3	4	5
SiO_2	53·88	54·41	49·35	52·40	51·78
TiO_2	0·14	0·00	0·34	—	0·38
Al_2O_3	0·68	2·21	4·92	2·46	5·14
Cr_2O_3	0·87	0·48	1·06	—	0·95
Fe_2O_3	0·16	0·00	0·77	0·20	1·75
FeO	2·11	2·39	2·19	3·67	2·14
MnO	0·07	0·09	0·07	0·11	0·12
NiO	0·03	—	0·05	—	0·04
MgO	18·84	18·36	19·86	24·43	16·04
CaO	22·38	21·59	19·75	15·01	20·32
Na_2O	0·78	0·21	1·23	1·37	1·06
K_2O	0·04	0·30	0·00	0·29	0·08
H_2O^+	0·14	0·35	0·17	—	—
H_2O^-	—	0·00	0·08	—	—
Total	100·15	100·39	99·84	100·03	99·84
α	—	—	1·676	—	—
β	—	—	1·689	—	—
γ	—	—	—	—	—
$\gamma:z$	—	—	—	—	—
$2V_\gamma$	—	56°	58°	—	—
D	3·303	—	3·306	—	3·322

Numbers of ions on the basis of six O

	1	Σ	2	Σ	3	Σ	4	Σ	5	Σ
Si	1·958	1·99	1·950	2·00	1·795	2·00	1·832	1·93	1·883	2·00
Al	0·029		0·050		0·205		0·101		0·117	
Al	—		0·043		0·006		—		0·023	
Ti	0·004		—		0·009		—		0·010	
Cr	0·025		0·014		0·030		—		0·027	
Fe^{3+}	0·004		—		0·021		0·005		0·048	
Mg	1·020		0·981		1·083		1·272		0·869	
Ni	0·001	2·05	—	1·97	0·001	2·07	—	2·06	0·001	1·92[a]
Fe^{2+}	0·064		0·072		0·066		0·107		0·065	
Mn	0·002		0·003		0·002		0·003		0·004	
Ca	0·871		0·829		0·770		0·562		0·790	
Na	0·055		0·015		0·086		0·093		0·075	
K	0·002		0·014		0·000		0·013		0·003	
Mg	52·0		52·0		55·8		65·2		48·9	
*ΣFe	3·6		4·0		4·6		5·9		6·6	
Ca	44·4		44·0		39·6		28·9		44·5	

*$\Sigma Fe = Fe^{2+} + Fe^{3+} + Mn$.

[a] Includes V^{+5} 0·001.

1 Endiopside, dunite, Twin Sisters, northern Whatcom County, Washington (Ross *et al.*, 1954). Anal. M. D. Foster (includes V_2O_5 0·02; spectrographic determination Co 0·004, Sr 0·003).

2 Endiopside, harzburgite of layered sequence, Red Hills, South Island, New Zealand (Challis, 1965). Anal. G. A. Challis.

3 Endiopside, olivine-rich nodule, Calton Hill, Derbyshire (Hamad, 1963). Anal. S. el D. Hamad (*a* 9·743, *b* 8·897, *c* 5·184 Å, β 106°56′).

4 Endiopside, peridotite inclusion in basalt, La Palmas, Canary Islands (Muñoz and Sagredo, 1974) (includes P_2O_5 0·09).

5 Endiopside, olivine-rich inclusion in basalt, Camargo, Chihuahua, Mexico (Ross *et al.*, 1954). Anal. M. D. Foster (includes V_2O_5 0·04; spectrographic determination Co 0·005).

6		7		8		9		10		
54·61		53·30		52·55		51·65		51·43		SiO_2
0·23		0·07		0·24		0·31		0·27		TiO_2
1·30		2·65		2·98		3·89		3·00		Al_2O_3
0·92		—		1·18		0·42		1·09		Cr_2O_3
1·14		1·54		1·17		1·77		1·37		Fe_2O_3
3·02		2·94		3·98		3·84		4·23		FeO
0·10		0·10		0·12		0·12		0·14		MnO
—		—		0·05		—		—		NiO
20·88		17·52		18·21		17·28		19·44		MgO
16·20		21·63		19·23		20·88		17·86		CaO
1·28		0·51		0·32		0·26		0·28		Na_2O
0·12		—		0·00		0·01		0·02		K_2O
0·41		—		0·14		0·42		0·70		H_2O+
0·07		—		0·08		—		0·10		H_2O-
100·28		100·26		100·25		100·85		99·93		Total
1·670		—		1·6805		—		1·6810		α
1·677		—		1·6852		—		1·6853		β
1·693		—		1·7075		—		1·7057		γ
38·5°		43°		40·5° (aver.)		—		40·5°		$\gamma:z$
56·5°		—		49° (aver.)		—		48·75°		$2V_\gamma$
3·269		—		3·299		—		—		D
1·966	2·00	1·933	2·00	1·912	2·00	1·880	2·00	1·888	2·00	Si
0·034		0·067		0·088		0·120		0·112		Al
0·021		0·046		0·039		0·047		0·018		Al
0·006		0·002		0·007		0·008		0·007		Ti
0·026		—		0·034		0·012		0·032		Cr
0·031		0·042		0·032		0·048		0·038		Fe^{3+}
1·121		0·947		0·987		0·938		1·063		Mg
—	2·02	–	2·01	0·002	2·00	—	2·01	—	2·02	Ni
0·091		0·089		0·121		0·117		0·130		Fe^{2+}
0·003		0·003		0·004		0·004		0·004		Mn
0·625		0·840		0·748		0·814		0·702		Ca
0·089		0·036		0·023		0·018		0·020		Na
0·006		—		0·000		0·000		0·001		K
59·9		49·3		52·2		48·8		54·9		Mg
6·7		7·0		8·2		8·8		8·9		$*\Sigma Fe$
33·4		43·7		39·5		42·4		36·2		Ca

6 Endiopside, lherzolite nodule, Thaba Putsoa kimberlite pipe (Nixon *et al.*, 1963). Anal. M. A. Kerr.

7 Endiopside, core of phenocryst in porphyritic olivine diabase. Ile Ronda breccia, Montreal (Clark *et al.*, 1967). Anal. H. Soutar.

8 Endiopside, pegmatitic gabbro, below chromite horizon, Stillwater complex, Montana (Hess, 1949). Anal. R. B. Ellestad.

9 Endiopside, calc-alkaline basalt, Ryozen, Fukushima, Japan (Konda, 1970).

10 Endiopside, pegmatitic olivine plagioclase rock, Mount View Lake, Stillwater Valley, Montana (Hess, 1949). Anal. R. B. Ellestad.

Table 17. Salite Analyses

	1	2	3	4	5
SiO_2	47·15	46·20	49·74	49·86	53·80
TiO_2	1·50	1·72	0·72	0·41	0·02
Al_2O_3	7·20	7·15	5·02	5·48	0·47
Fe_2O_3	3·51	3·68	1·99	2·42	0·93
FeO	2·96	2·98	4·70	4·23	5·63
MnO	0·11	0·11	0·13	0·15	0·35
MgO	14·71	14·25	14·42	15·02	14·15
CaO	22·10	22·80	23·51	22·34	24·46
Na_2O	0·58	0·71	0·15	0·00	0·37
K_2O	0·07	0·20	0·10	0·00	0·004
H_2O^+	—	—	0·01	0·20	0·00
H_2O^-	—	—	0·05	0·11	0·04
Total	100·25	100·26	100·54	100·22	100·22
α	—	—	1·694	1·689	1·674
β	1·695	1·697	1·708	1·696	1·681
γ	—	—	1·718	1·715	1·703
$\gamma:z$	—	—	44°	43°	—
$2V_\gamma$	58°	60°	60°	59°	—
D	—	—	3·38	—	—
	Numbers of ions on the basis of six O				
Si	1·741 } 2·00	1·718 } 2·00	1·827 } 2·00	1·839 } 2·00	1·990 } 2·00
Al	0·259	0·282	0·173	0·161	0·010
Al	0·054	0·031	0·045	0·077	0·010
Ti	0·042	0·048	0·022	0·011	0·001
Fe^{3+}	0·098	0·103	0·058	0·067	0·026
Mg	0·810	0·790	0·802	0·825	0·780
Fe^{2+}	0·091 } 2·03[a]	0·093 } 2·05[b]	0·144 } 2·02	0·131 } 2·00	0·174 } 2·00
Mn	0·003	0·003	0·004	0·005	0·011
Ca	0·875	0·908	0·934	0·881	0·970
Na	0·042	0·051	0·010	—	0·027
K	0·003	0·009	0·004	—	0·000
Mg	43·2	41·6	41·3	43·2	39·8
*ΣFe	10·2	10·5	10·6	10·6	10·8
Ca	46·6	47·9	48·1	46·2	49·4

*ΣFe = $Fe^{2+} + Fe^{3+}$ + Mn.

[a] Includes Cr 0·011.

[b] Includes Cr 0·014.

1 Chromian salite, alkali olivine basalt, Hocheifel, West Germany (Huckenholz, 1965a). Includes Cr_2O_3 0·36.
2 Chromian salite, phenocryst core, picritic basalt (ankaramite), Hocheifel, West Germany (Huckenholz, 1966). Includes Cr_2O_3 0·46.
3 Salite, pyroxenite dyke, Pammal, Madras, India (Howie, 1955). Anal. R. A. Howie.
4 Salite, from hypersthene-bearing olivine–augite basalt, isolated crystals in tuff of the Taga volcano, Wadaki, north Izu, Japan (Kuno and Sawatari, 1934). Anal. S. Tanaka.
5 Salite, salite-actinolite skarn, Collins Creek Cabin, Clearwater County, Idaho (Hietanen, 1971). Anal. C. O. Ingamells and S. T. Neill (trace element data).

6		7		8		9		10		
52·96		49·20		48·73		51·20		51·43		**SiO_2**
0·10		1·83		0·81		0·27		0·28		**TiO_2**
1·21		3·33		5·98		2·71		2·63		**Al_2O_3**
0·49		1·85		2·47		2·76		1·70		**Fe_2O_3**
6·00		5·30		4·96		4·94		5·92		**FeO**
0·56		0·18		0·15		0·05		0·19		**MnO**
14·02		15·03		14·43		15·43		14·83		**MgO**
24·29		23·33		22·49		22·62		22·18		**CaO**
0·36		0·40		0·26		0·24		0·44		**Na_2O**
0·02		tr.		0·00		0·12		0·00		**K_2O**
0·23		—		—		—		0·31		**H_2O^+**
0·20		0·03		—		0·06		—		**H_2O^-**
100·44		100·48		100·33		100·40		99·95		**Total**
1·677		1·697		1·690		1·690 (mean)		1·6862		**α**
1·686		1·707		1·697		1·696 (mean)		1·6908		**β**
1·706		1·722		1·716		1·715 (mean)		1·714		**γ**
41°		41°		43°		—		39·5°		**γ:z**
58°		48°		57°		59° (3° range)		48°		**$2V_\gamma$**
3·34		—		3·36		—		3·332		**D**
1·967	2·00	1·828	1·97	1·803	2·00	1·891	2·00	1·913	2·00	**Si**
0·033		0·146		0·197		0·109		0·087		**Al**
0·020		0·000		0·063		0·009		0·029		**Al**
0·003		0·051		0·022		0·007		0·009		**Ti**
0·014		0·052		0·069		0·077		0·048		**Fe^{3+}**
0·776		0·832		0·796		0·849		0·822		**Mg**
0·186	2·01	0·165	2·06	0·153	2·02[a]	0·153	2·01	0·184	2·02	**Fe^{2+}**
0·018		0·006		0·005		0·001		0·006		**Mn**
0·967		0·929		0·891		0·895		0·884		**Ca**
0·026		0·028		0·019		0·017		0·032		**Na**
0·001		0·000		0·000		0·006		0·000		**K**
39·6		42·0		41·8		43·0		42·3		**Mg**
11·1		11·2		11·4		11·7		12·2		***ΣFe**
49·3		46·8		46·8		45·3		45·5		**Ca**

[a] Includes Cr 0·001.

6 Salite, salite–hornblende–zoisite–calcite–prehnite skarn, Sgòr Gaoithe, Glen Urquhart (Francis, 1958). Anal. G. H. Francis (spectrographic determinations Ga 2, Cr 500, V 100, Li 10, Ni 500, Co 50, Sc 40, Zr 50, Y 10, Sr 5 p.p.m.).

7 Salite, picrite, Garbh Eilean, Shiant Isles, Scotland (Murray, 1954). Anal. R. J. Murray.

8 Salite, plagioclase–amphibole–magnetite cumulate, erupted block, St. Vincent (Lewis, 1967). Anal. J. F. Lewis (includes Cr_2O_5 0·05).

9 Salite, olivine–augite basalt, Taga Volcano, Tyōzyahagara, north Izu, Japan (Kuno, 1955). Anal. K. Nagashima.

10 Salite, coarse-grained hornblende eucrite, West Chester, Pennsylvania (Norton and Clavan, 1959). Anal. W. S. Clavan (includes Cr_2O_3 0·04).

Table 17. Salite Analyses – *continued*

	11	12	13	14	15
SiO_2	47·86	49·31	46·95	48·93	42·70
TiO_2	1·67	1·26	2·12	0·60	2·60
Al_2O_3	6·17	4·60	7·33	5·55	9·70
Fe_2O_3	2·44	2·38	2·96	5·02	5·10
FeO	5·07	5·57	4·60	3·29	3·40
MnO	0·14	—	0·10	0·09	0·20
MgO	13·87	14·02	12·95	14·26	13·90
CaO	21·54	22·71	21·42	21·93	21·30
Na_2O	0·72	—	0·69	0·00	0·60
K_2O	0·13	—	tr.	0·00	0·15
H_2O^+	0·11	—	0·47	0·09	—
H_2O^-	0·03	—	0·16	0·17	—
Total	99·77	99·85	99·75	99·93	99·65
α	1·693	1·698	1·695	1·698	—
β	1·698	—	1·699	—	1·714
γ	1·720	1·730	1·720	1·719	—
$\gamma:z$	45°	43°	—	45°–48°	—
$2V_\gamma$	48°	—	46°–54°	55°–64°	50°
D	3·35	—	—	—	—
	Numbers of ions on the basis of six O				
Si	1·786 } 2·00	1·835 } 2·00	1·758 } 2·00	1·814 } 2·00	1·610 } 2·00
Al	0·214	0·165	0·242	0·186	0·390
Al	0·057	0·037	0·082	0·057	0·040
Ti	0·047	0·035	0·060	0·017	0·074
Fe^{3+}	0·068	0·067	0·083	0·140	0·144
Mg	0·771	0·778	0·723	0·788	0·780
Fe^{2+}	0·158 } 2·02	0·173 } 2·00	0·144 } 2·00	0·102 } 1·98	0·107 } 2·07
Mn	0·004	—	0·003	0·003	0·010
Ca	0·861	0·906	0·859	0·871	0·859
Na	0·052	—	0·050	0·000	0·044
K	0·006	—	0·000	0·000	0·007
Mg	41·4	40·4	39·9	41·4	41·1
*ΣFe	12·3	12·5	12·7	12·9	13·6
Ca	46·3	47·1	47·4	45·7	45·3

11 Salite, melanocratic camptonite, Skaergaard, Kangerdlugssuaq, east Greenland (Vincent, 1953). Anal. E. A. Vincent (includes Cr_2O_3 0·02).

12 Salite, pyroxenite, chalk–dolerite contact, Scawt Hill, Co. Antrim (Tilley and Harwood, 1931). Anal. Fresenius Chemical Laboratories, Wiesbaden.

13 Salite, olivine theralite, Nundle, New South Wales (Wilkinson, 1966). Anal. N. Chiba.

14 Salite, andesite-basalt, Zvare, Georgia, U.S.S.R. (Gvakhariya, 1968). Anal. V. I. Kobiashvili.

15 Salite, analcite alkali trachyte, Selberg, Hockeifel region, West Germany (Gutberlet and Huckenholz, 1965). Anal. H. Grünhagen.

16	17	18	19	20	
46·93	49·57	51·55	45·86	49·71	SiO_2
3·04	2·05	0·31	2·34	0·34	TiO_2
6·61	3·82	2·44	8·30	2·85	Al_2O_3
1·53	2·00	2·03	2·08	2·76	Fe_2O_3
6·64	6·59	7·28	7·03	7·45	FeO
0·10	0·13	0·24	0·17	0·50	MnO
12·32	13·75	13·65	12·65	13·35	MgO
22·35	21·44	21·73	20·23	22·36	CaO
0·65	0·69	0·52	0·68	0·60	Na_2O
0·04	0·08	0·00	0·11	0·04	K_2O
0·11	0·10	0·26	0·54	0·16	H_2O^+
tr.	0·02	—	0·02	0·13	H_2O^-
100·32	100·33	100·03	100·26	100·25	Total
1·699	1·690	1·6911	—	1·684–1·689	α
1·704	1·697	1·6951	·714	1·689–1·695	β
1·725	1·715	1·7167	—	1·711–1·715	γ
—	42°	39·5°	48°	43·5°	$\gamma:z$
—	52°–54°	46°–75°	55°	55°	$2V_\gamma$
3·42	3·34	3·358	—	—	D

16		17		18		19		20		
1·755	2·00	1·846	2·00	1·926	2·00	1·725	2·00	1·874	2·00	Si
0·245		0·154		0·074		0·275		0·126		Al
0·047	2·02	0·014	2·01[a]	0·031	2·00	0·093	2·02	0·000	2·04	Al
0·086		0·057		0·009		0·066		0·010		Ti
0·043		0·056		0·056		0·058		0·078		Fe^{3+}
0·686		0·763		0·760		0·709		0·750		Mg
0·208		0·205		0·227		0·221		0·235		Fe^{2+}
0·003		0·004		0·008		0·005		0·016		Mn
0·895		0·855		0·870		0·815		0·903		Ca
0·047		0·050		0·038		0·050		0·044		Na
0·002		0·003		0·000		0·006		0·002		K
37·4		40·5		39·6		39·2		37·8		Mg
13·8		14·1		15·1		15·7		16·6		*ΣFe
48·8		45·4		45·3		45·1		45·6		Ca

[a] Includes Ni 0·001.

16 Salite, magnesium-rich fraction, teschenite, 20 feet above lower contact, Black Jack Sill, Gunnedah, New South Wales (Wilkinson, 1957). Anal. J. F. G. Wilkinson.

17 Salite, teschenite, Red Hill, Brocken Range, eastern Wellington, New Zealand (Hutton, 1943). Anal. F. T. Seelye (includes SrO 0·006, V_2O_5 0·045, NiO 0·04).

18 Salite, fine-grained gabbro, Wilmington, Delaware (Norton and Clavan, 1959). Anal. W. S. Clavan (includes Cr_2O_3 0·01, NiO 0·01).

19 Salite, olivine basalt, Old Pallas, Co. Limerick, Eire (Ashby, 1946). Anal. Geochemical Laboratories (includes P_2O_5 0·05, CO_2 0·20).

20 Salite, inner zone of corona to quartzite xenolith in granodiorite, Kentallen, Scotland (Muir, 1953). Anal. I. D. Muir.

Table 17. Salite Analyses – *continued*

	21	22	23	24	25	26
SiO_2	50·19	48·48	50·02	50·66	50·75	51·70
TiO_2	0·20	2·15	0·41	0·27	0·38	—
Al_2O_3	2·73	4·08	1·98	2·16	1·33	0·36
Fe_2O_3	2·98	1·46	2·67	2·18	4·41	0·37
FeO	7·54	10·00	9·72	10·88	10·69	—
MnO	0·40	0·19	0·83	0·45	0·36	7·43
MgO	12·38	10·90	9·84	10·53	9·53	12·57
CaO	23·58	21·66	23·37	23·02	21·82	23·68
Na_2O	0·45	0·67	0·12	—	0·94	0·12
K_2O	0·00	0·08	0·08	—	0·03	—
H_2O^+	0·00	0·17	1·58	—	0·02	0·65
H_2O^-	0·02	0·02	0·04	—	0·03	—
Total	100·47	99·86	100·66	100·15	100·34	100·19
α	1·6915	1·707	1·693	1·704	1·696–1·700	1·676
β	1·6980	1·711	—	1·711	—	1·683
γ	1·7185	1·732	1·716	1·729	1·722–1·725	1·705
γ:z	45°	—	—	—	48°–54°	—
$2V_\gamma$	59°	—	59°	60°	60°–77°	60°
D	3·373	3·43	—	—	—	3·39

Numbers of ions on the basis of six O

	21		22		23		24		25		26	
Si	1·886	2·00	1·838	2·00	1·926	2·00	1·926	2·00	1·937	2·00	1·971	1·99
Al	0·114		0·162		0·074		0·074		0·060		0·016	
Al	0·007		0·020		0·016		0·023		0·000		—	
Ti	0·006		0·061		0·012		0·008		0·011		—	
Fe^{3+}	0·084		0·041		0·077		0·062		0·126		0·010	
Mg	0·693		0·615		0·564		0·597		0·542		0·714	
Fe^{2+}	0·237	2·02	0·317	1·99	0·313	1·99	0·346	1·99	0·341	2·00[a]	—	2·03[b]
Mn	0·013		0·006		0·027		0·014		0·012		0·240	
Ca	0·950		0·879		0·964		0·938		0·892		0·967	
Na	0·032		0·049		0·009		—		0·070		0·008	
K	0·000		0·004		0·003		—		0·001		—	
Mg	35·1		33·1		29·0		30·5		28·3		37·0	
*ΣFe	16·9		19·7		21·4		21·6		25·1		12·9	
Ca	48·0		47·2		49·6		47·9		46·6		50·1	

[a] Includes Sr 0·001.
[b] Includes Zn 0·093.

21 Salite, skarn, magnetite deposit, Clifton mine, Adirondacks (Hess, 1949). Anal. N. Davidson.
22 Salite, iron-rich fraction, teschenite, 500 feet above lower contact, Black Jack Sill, Gunnedah, New South Wales (Wilkinson, 1957). Anal. J. F. G. Wilkinson.
23 Salite, amphibolite, Domeki, Hurudono-mura, Gosaisyo-Takanuki district, Japan (Miyashiro, 1958). Anal. H. Haramura.
24 Salite, skarn in hypersthene gneiss, Amerdloq Fjord, west Greenland (Pauly, 1948). Anal. A. H. Nielson.
25 Salite, pyroxene-rich xenolith in diorite, Glenmore River, Suardalan, Glenelg–Ratagain, Scotland (Nicholls, 1951). Anal. G. D. Nicholls (includes Sr 0·05, BaO 0·005).
26 Zincian salite (zinc schefferite), skarn, Franklin, New Jersey (Palache, 1937). Anal. W. F. Hillebrand (includes ZnO 3·31).

Table 18. Ferrosalite Analyses

	1	2	3	4	5
SiO_2	51·00	47·91	45·90	52·07	50·34
TiO_2	0·00	0·31	1·40	0·12	0·19
Al_2O_3	0·00	5·21	5·90	0·10	1·47
Fe_2O_3	2·38	3·27	8·10	—	0·64
FeO	14·22	12·44	7·60	16·76	16·24
MnO	0·30	0·41	0·35	0·30	0·32
MgO	8·80	7·86	8·00	8·72	6·24
CaO	22·42	22·04	20·00	21·30	24·08
Na_2O	0·26	0·35	1·35	0·01	0·47
K_2O	0·00	0·02	0·04	0·03	0·12
H_2O^+	0·38	0·18	0·25	0·53	0·05
H_2O^-	—	0·08	0·10	0·13	0·02
Total	100·10	100·11	100·62	100·07	100·18
α	1·700	1·720	1·725	1·701	1·706
β	—	1·728	1·734	1·708	1·712
γ	1·725	1·744	1·750	1·736	1·731
$\gamma:z$	43°	42°–48°	—	49°	—
$2V_\gamma$	60°	60°–63°	72°	—	—
D	3·46	—	—	3·48	—
	Numbers of ions on the basis of six O				
Si	1·984 } 1·98	1·846 } 2·00	1·779 } 2·00	2·021 } 2·02	1·962 } 2·00
Al	—	0·154	0·221	0·000	0·038
Al	—	0·082	0·048	0·005	0·029
Ti	—	0·009	0·041	0·004	0·005
Fe^{3+}	0·070	0·097	0·236	0·000	0·019
Mg	0·510	0·451	0·462	0·504	0·362
Fe^{2+}	0·463 } 2·01	0·402 } 1·99	0·246 } 1·98	0·544 } 1·95	0·529 } 2·00
Mn	0·010	0·014	0·011	0·010	0·011
Ca	0·934	0·909	0·830	0·886	1·006
Na	0·020	0·028	0·101	0·001	0·036
K	0·000	0·000	0·002	0·001	0·006
Mg	25·7	24·1	25·9	25·9	18·8
*ΣFe	27·3	27·4	27·6	28·5	29·0
Ca	47·0	48·5	46·5	45·6	52·2

*ΣFe = $Fe^{2+} + Fe^{3+}$ + Mn.

1 Ferrosalite, skarn, Fabian mine, Herräng, Sweden (Magnusson, 1940). Anal. G. Assarsson (includes CO_2 0·16, F 0·16, Ce 0·02).
2 Ferrosalite, metasomatic, granite–calc-granulite contact, Garividi, Andhra Pradesh, India (Rao and Rao, 1971). Anal. K. S. R. Rao (*a* 9·750, *b* 8·904, *c* 5·279 Å, β 105°59′) (includes Cr_2O_3 0·03).
3 Sodian ferri-ferrosalite, absarokite, Grabels, Monpellier, France (Babkine *et al.*, 1968). Anal. Patureau (includes org. residue 1·55, P_2O_5 0·08).
4 Ferrosalite, quartz–grunerite–ferrosalite rocks, Lower Wabush Iron Formation, Labrador (Klein, 1966). Anal. J. Ito (*a* 9·786, *b* 8·974, *c* 5·248, β 105·61°).
5 Ferrosalite, calc granulite, Garividi, Andhra Pradesh, India (Mukherjee and Rege, 1972).

Table 18. Ferrosalite Analyses – *continued*

	6	7	8	9	10	11
SiO_2	50·18	45·80	50·71	50·69	48·79	49·03
TiO_2	0·13	0·31	0·07	0·08	0·09	—
Al_2O_3	0·00	5·11	1·06	1·49	1·32	0·86
Fe_2O_3	0·70	5·87	0·53	3·64	0·00	4·22
FeO	17·33	12·44	18·57	14·51	12·36	3·95
MnO	0·28	0·41	0·18	0·56	9·53	7·91
MgO	6·87	6·86	5·70	6·77	5·56	5·81
CaO	23·28	22·54	22·86	20·71	21·35	19·88
Na_2O	—	0·35	0·16	1·16	—	—
K_2O	—	0·02	0·02	0·09	—	—
H_2O^+	0·09	0·17	0·08	0·09	0·84	0·70
H_2O^-	0·10	0·09	0·04	0·14	0·24	0·60
Total	99·70	99·97	99·98	99·94	100·08	100·10
α	1·709	1·7210	1·708	—	1·710	1·713
β	—	1·7290	1·714	1·7218	1·720	1·722
γ	1·729	1·7463	1·736	—	1·736	1·745
$\gamma:z$	44°	51°	43°	—	—	55°
$2V_\gamma$	56°	67·5°	59·75°	—	60°	74°
D	—	3·44	3·413	3·40	3·46	3·55
	Numbers of ions on the basis of six O					
Si	1·991 } 1·99	1·794 } 2·00	1·991 } 2·00	1·971 } 2·00	1·957 } 2·00	1·976 } 2·00
Al	—	0·206	0·009	0·029	0·043	0·023
Al	—	0·030	0·040	0·039	0·019	0·017
Ti	0·004	0·009	0·002	0·002	0·003	—
Fe^{3+}	0·021	0·173	0·016	0·107	—	0·127
Mg	0·406	0·400	0·333	0·392	0·332	0·347
Fe^{2+}	0·575 } 2·00	0·407 } 2·01	0·610 } 1·98	0·472 } 1·98	0·415 } 2·01	0·133 } 1·96[a]
Mn	0·009	0·013	0·006	0·018	0·324	0·269
Ca	0·989	0·946	0·962	0·863	0·917	0·855
Na	—	0·026	0·012	0·087	—	—
K	—	0·001	0·001	0·004	—	—
Mg	20·3	20·6	17·3	20·1	16·7	20·0
*ΣFe	30·4	30·6	32·8	35·7	37·2	30·6
Ca	49·3	48·8	49·9	44·2	46·1	49·4

[a] Includes Zn 0·211.

6 Ferrosalite, calcite–wollastonite–meionite–quartz gneiss, Kirrelä-Korppila, Vihti, south-western Finland (Parras, 1958). Anal. P. Ojanperä (includes CO_2 0·74).

7 Ferrosalite, skarn, St. Laurence County, New York (Hess, 1949). Anal. L. C. Peck.

8 Ferrosalite, marble in anorthosite, Pockamoonshine quarry, Ausable quadrangle, Adirondacks (Hess, 1949). Anal. L. C. Peck.

9 Ferrosalite, pyroxene granodiorite, Grand Calumet Township, Quebec (Shaw *et al.*, 1963b). Anal. C. O. Ingamells (includes F 0·02, less O = F 0·01).

10 Manganoan ferrosalite, Treburland maganese mine, Altarnun, Cornwall (Tilley, 1946). Anal. H. C. G. Vincent.

11 Zincian ferrosalite (jeffersonite), skarn, Franklin, New Jersey (Palache, 1937). Anal. G. Steiger (includes ZnO 7·14).

Table 19. Hedenbergite Analyses

	1	2	3	4	5
SiO_2	48·00	48·03	48·34	44·76	51·82
TiO_2	—	0·11	0·08	—	tr.
Al_2O_3	0·63	0·45	0·30	1·70	0·47
Fe_2O_3	3·32	1·04	1·50	tr.	0·18
FeO	22·25	19·96	22·94	19·27	14·47
MnO	0·81	3·68	3·70	6·22	10·50
MgO	2·12	2·73	1·06	3·25	0·94
CaO	20·35	22·56	21·30	22·20	19·78
Na_2O	0·34	0·35	0·14	0·14	0·07
K_2O	0·18	0·05	0·03	—	0·03
H_2O^+	—	0·85	0·46	0·17	1·97
H_2O^-	1·72	0·15	—	—	0·08
Total	99·72	100·10	99·85	'98·71'	'100·29'
α	—	1·711	1·7225	1·716	—
β	—	1·729	1·7300	1·723	—
γ	—	1·745	1·7505	1·741	—
$\gamma{:}z$	42°12′	46°	47·5°	47°	—
$2V_\gamma$	52°	64°	62·5°	—	—
D	—	3·53	3·535	—	—

Numbers of ions on the basis of six O

	1		2		3		4		5	
Si	1·979	2·00	1·967	1·99	1·988	2·00	1·883	1·97	2·069	2·07
Al	0·021		0·022		0·012		0·084		0·000	
Al	0·009		0·000		0·006		—		0·024	
Ti	—		0·003		0·002		—		0·000	
Fe^{3+}	0·103		0·032		0·046		—		0·005	
Mg	0·130		0·166		0·065		0·204		0·058	
Fe^{2+}	0·767	1·97	0·683	2·03[a]	0·789	1·99	0·678	2·10	0·488	1·80
Mn	0·028		0·128		0·129		0·222		0·359	
Ca	0·899		0·990		0·939		0·999		0·856	
Na	0·027		0·028		0·011		—		0·005	
K	0·010		0·002		0·001		—		0·001	
Mg	6·7		8·3		3·3		9·7		3·3	
*ΣFe	46·6		42·2		49·0		42·8		48·2	
Ca	46·7		49·5		47·7		47·5		48·5	

* $\Sigma Fe = Fe^{2+} + Fe^{3+} + Mn$.
[a] Includes Zn 0·001.

1 Hedenbergite, skarn, Tignitoio iron-ore deposit, Elba (Federico and Fornaseri, 1953).
2 Manganhedenbergite, skarn, Kamioka mine, Gifu Prefecture, Japan (Shimazaki, 1967). Includes P_2O_5 0·11, ZnO 0·03.
3 Hedenbergite, Herault, California (Wyckoff *et al.*, 1925). Anal. E. S. Shepherd (optical data from Hess, 1949).
4 Manganoan hedenbergite, skarn, western Karamazar, U.S.S.R. (Zharikov and Podlessky, 1955).
5 Manganoan hedenbergite, bustamite–clinopyroxene skarn, Nakatatsu mine, Fukui Prefecture, Japan (Tokunaga, 1965). Includes P_2O_5 0·04; X-ray powder data.

Table 20. Ferrohedenbergite Analyses

	1	2	3	4	5
SiO_2	47·22	49·54	48·00	46·56	46·98
TiO_2	0·30	0·13	0·31	0·60	0·22
Al_2O_3	2·02	1·13	1·26	1·42	0·48
Fe_2O_3	4·90	2·08	2·97	2·01	4·23
FeO	22·52	25·08	25·59	28·05	28·08
MnO	0·02	0·52	0·59	1·24	0·10
MgO	2·97	2·92	1·40	1·82	2·20
CaO	19·52	18·21	18·16	17·96	17·38
Na_2O	0·35	0·43	1·17	0·45	0·10
K_2O	0·02	0·05	0·34	0·14	0·15
H_2O^+	—	—	0·20	0·17	—
H_2O^-	—	—	—	0·04	—
Total	99·84	100·09	99·99	100·46	99·92
α	1·724	—	1·738	1·735	1·726
β	1·731	1·728	1·746	1·740	—
γ	1·750	—	1·765	1·765	1·754
$\gamma:z$	—	—	—	47°	43°
$2V_\gamma$	57°	54°	55°	56°	59°
D	3·64	—	3·639	—	—

Numbers of ions on the basis of six O

	1		2		3		4		5	
Si	1·907	2·00	1·986	2·00	1·961	2·00	1·916	1·98	1·936	1·96
Al	0·093		0·014		0·039		0·068		0·023	
Al	0·003	1·98	0·039	1·95	0·022	2·01	0·000	2·04	0·000	2·03
Ti	0·009		0·002		0·010		0·018		0·007	
Fe^{3+}	0·149		0·063		0·091		0·062		0·131	
Mg	0·179		0·174		0·085		0·111		0·135	
Fe^{2+}	0·761		0·841		0·874		0·966		0·968	
Mn	0·001		0·017		0·020		0·043		0·003	
Ca	0·845		0·781		0·795		0·792		0·767	
Na	0·027		0·034		0·093		0·036		0·008	
K	0·001		0·002		0·018		0·008		0·008	
Mg	9·2		9·3		4·6		5·6		6·7	
*ΣFe	47·1		48·6		52·8		54·3		55·0	
Ca	43·7		42·1		42·6		40·1		38·3	

*$\Sigma Fe = Fe^{2+} + Fe^{3+} + Mn$.

1 Ferrohedenbergite, eulysite, Mariupol' iron deposit (Val'ter, 1969). Anal. A. A. Shvakova.
2 Ferrohedenbergite, melanogranophyre, Skaergaard, east Greenland (Brown and Vincent, 1963). Anal. E. A. Vincent.
3 Ferrohedenbergite, microsyenite, Carnarvon Range, Queensland, Australia. (Bryan, 1969). Anal. L. J. Sutherland.
4 Ferrohedenbergite, granophyre, Meall Dearg, Skye (Anwar, 1955). Anal. Y. M. Anwar.
5 Ferrohedenbergite, eulysite, Lake Chudzyavr region, Kola Peninsula (Bondarenko and Dagelaiskii, 1961). Anal. V. D. Bugrova.

6	7	8	9	10	
46·69	46·90	46·59	48·24	48·69	**SiO_2**
0·24	0·10	1·50	0·64	0·86	**TiO_2**
1·45	2·20	0·95	1·37	2·04	**Al_2O_3**
5·11	1·66	1·82	1·44	5·60	**Fe_2O_3**
25·82	29·10	28·93	27·32	21·96	**FeO**
0·86	0·67	0·53	0·98	1·25	**MnO**
2·12	0·58	1·04	1·42	0·89	**MgO**
16·74	18·87	18·08	16·93	16·36	**CaO**
0·68	0·32	0·25	0·67	2·86	**Na_2O**
0·28	0·04	0·17	0·16	0·09	**K_2O**
—	—	0·31	0·29	—	**H_2O^+**
—	—	—	0·09	—	**H_2O^-**
99·99	100·44	100·17	99·64	100·60	**Total**
—	—	1·735–1·736	—	—	α
—	1·737	1·742–1·744	—	1·747	β
—	—	1·764–1·765	—	—	γ
—	—	—	—	55°	$\gamma{:}z$
—	56°	57·5°–54°	—	66°	$2V_\gamma$
—	—	—	—	—	**D**

6		7		8		9		10		
1·915	1·99	1·928	2·00	1·926	1·97	1·964	2·00	1·950	2·00	**Si**
0·072		0·072		0·047		0·036		0·050		**Al**
—		0·034		—		0·030		0·046		**Al**
0·005		0·002		0·047		0·019		0·026		**Ti**
0·144		0·052		0·055		0·044		0·168		**Fe^{3+}**
0·128		0·037		0·065		0·086		0·053		**Mg**
0·885	2·00	1·000	2·01	1·002	2·02	0·931	1·94	0·737	2·00	**Fe^{2+}**
0·032		0·022		0·017		0·034		0·043		**Mn**
0·735		0·832		0·803		0·739		0·703		**Ca**
0·054		0·025		0·020		0·053		0·221		**Na**
0·015		0·002		0·010		0·008		0·005		**K**
6·7		1·9		3·3		4·5		3·1		**Mg**
55·1		55·3		55·3		55·4		55·7		***ΣFe**
38·2		42·8		41·4		40·1		41·2		**Ca**

6 Ferrohedenbergite, epigranite, Loch Ainort, Skye (Bell, 1966). Anal. J. D. Bell (calculated to 100·0 per cent on water-free basis).

7 Ferrohedenbergite, transitional granophyre, Skaergaard, east Greenland (Brown and Vincent, 1963). Anal. E. A. Vincent.

8 Ferrohedenbergite, intergrowth of purple and green varieties, granophyre, Red Hill, Tasmania (McDougall, 1961). Anal. I. McDougall (optics for green (left) and purple variety (right)).

9 Ferrohedenbergite, quartz syenite, Berkshire Valley, New Jersey Highlands (Young, 1972). Anal. A. G. Loomis (includes P_2O_5 0·03, CO_2 0·06).

10 Sodian ferrohedenbergite, obsidian, Pantelleria (Carmichael, 1962). Anal. I. S. E. Carmichael.

Table 20. Ferrohedenbergite Analyses – *continued*

	11		12		13		14	
SiO_2	47·00		46·76		46·71		48·28	
TiO_2	1·48		0·69		0·95		0·19	
Al_2O_3	0·74		1·14		0·93		0·79	
Fe_2O_3	1·90		2·34		0·59		1·25	
FeO	29·01		29·21		31·48		30·78	
MnO	0·47		0·80		0·26		0·93	
MgO	0·94		0·15		0·14		0·36	
CaO	17·96		18·82		18·75		17·67	
Na_2O	0·20		0·23		0·26		0·12	
K_2O	0·15		0·02		0·03		0·08	
H_2O^+	0·10		—		—		—	
H_2O^-	0·06		—		—		—	
Total	100·01		100·16		100·10		100·45	
α	1·738		—		—		—	
β	1·745		—		1·742		1·743	
γ	1·766		—		—		—	
$\gamma:z$	44°		—		—		—	
$2V_\gamma$	59°		—		56°		55°	
D	—		—		—		—	
	Numbers of ions on the basis of six O							
Si	1·941	1·98	1·947	2·00	1·938	1·99	1·985	2·00
Al	0·036		0·052		0·047		0·015	
Al	0·000	2·00	0·000	2·01	—	2·01	0·025	1·97
Ti	0·046		0·022		0·031		0·005	
Fe^{3+}	0·059		0·072		0·017		0·037	
Mg	0·058		0·010		0·007		0·022	
Fe^{2+}	1·002		1·020		1·093		1·059	
Mn	0·016		0·027		0·010		0·032	
Ca	0·795		0·842		0·833		0·778	
Na	0·016		0·015		0·020		0·010	
K	0·008		0·000		0·002		0·005	
Mg	3·0		0·5		0·4		1·2	
*ΣFe	55·8		56·8		57·1		58·5	
Ca	41·2		42·7		42·5		40·3	

11 Ferrohedenbergite, differentiated dolerite sill, Kola Peninsula (Sinitsyn, 1965).

12 Ferrohedenbergite, ferro-syenodiorite, Upper Zone C, 7320 m, eastern Bushveld (Atkins, 1969). Anal. F. B. Atkins (trace element data).

13 Ferrohedenbergite, fayalite ferrodiorite, Skaergaard, east Greenland (Brown and Vincent, 1963). Anal. E. A. Vincent.

14 Ferrohedenbergite, inverted from ferriferous β-wollastonite, fayalite ferrodiorite, Skaergaard, east Greenland (Brown and Vincent, 1963). Anal. E. A. Vincent.

determined in a number of diopsides, containing between 9·4 and 13·7 mol. per cent $FeSiO_3$, from the ultrabasic rocks of the Gussevogorsk, Baranchinsk and Nizhe-Tagilsk massifs. The diopside–salites in the dunites and olivine pyroxenites, compared with those of the hornblendites and pyroxenites, are characterized by higher Sc, Ti, V, Ga and Sr and lower Cr and V contents (Borisenko, 1967a). The Y, La, Sr, Ba, Ga, Sn, Be, V, Cu and Zr contents of hedenbergites from the pulaskites, foyaites and naujaite of the Ilímaussaq intrusion, south-west Greenland, are given by Larsen (1976). The rare earth contents of the diopside in the high-temperature peridotite of the Lizard, Cornwall, and the distribution coefficients for La and Yb of coexisting diopside–olivine and diopside–orthopyroxene pairs have been presented by Frey (1969). The data are interpreted as evidence that the peridotite equilibrated with a basic magma either as a deep-seated accumulate, or as a residue after partial melting.

Cation distribution

The equilibrium distribution of Mg^{2+} and Fe^{2+} in *M*1 and *M*2 sites as a function of temperature has been determined for a Ca-rich clinopyroxene, $Wo_{27\cdot7}En_{64\cdot3}Fs_{8\cdot0}$, from the Thuba Putsoa kimberlite pipe, Lesotho, and for a megacryst $Wo_{38\cdot4}En_{53\cdot4}Fs_{8\cdot2}$ in the Kakanui nepheline breccia, New Zealand (McCallister *et al.*, 1976). The distribution coefficient:

$$K_D = \frac{[Fe^{2+}(M1)/Mg(M1)]}{[Fe^{2+}(M2)/Mg(M2)]}$$

calculated from the site occupancy, obtained by a complete crystal structure refinement, has been used to construct calibration curves for $\ln K_D$ v. $1/T\,^{\circ}K$. The Thuba Putsoa pyroxene has a low K_D value (0·068), implying a high degree of ordering, and corresponds with a temperature of 530° ± 50°C, that of the Kakanui mineral is 0·289, indicative of considerable disorder, and related to a temperature of 1367° ± 120°C. Both temperatures are in fair agreement with those deduced from the Mg–Fe distribution of the pyroxene and associated garnet megacrysts, and with the experimentally determined values of K_D.

Matsui *et al.* (1972) have investigated by Mössbauer spectroscopy the distribution of Fe^{2+} in *M*1 and *M*2 sites in members of two synthetic systems, $CaMgSi_2O_6$–$CaFeSi_2O_6$ and $Ca_{0\cdot8}Mg_{1\cdot2}Si_2O_6$–$Ca_{0\cdot8}Fe_{1\cdot2}Si_2O_6$, and in a number of natural pyroxenes.

The partitioning of Mg and Fe^{2+} and of other cation pairs between coexisting clino- and orthopyroxenes in igneous and metamorphic rocks has been investigated by many authors and an account of some of their findings is included in the section on orthopyroxenes (pp. 63–68, see also Nehru and Wyllie, 1974). A number of investigations relating to the cation distribution between members of the diopside–hedenbergite series and phases other than orthopyroxene include the distribution of Mg and Fe in coexisting chrome-diopside and phlogopite in the ultrabasic rocks of Andhra Pradesh (Mall, 1973), diopside and phlogopite in the silicate-bearing marble lenses in the metamorphic terrain of the Lafoten–Vesterålen region, north Norway (Glassley, 1975), and ferrosalites and garnets from the calc-granulites of the Eastern Ghats, India (Mukherjee and Rege, 1972). The Fe/Mg partition in diopside–orthopyroxene, diopside–olivine and diopside–garnet pairs of the Mt. Higasi-Akaisi area, Japan, has been investigated by Mori and Banno (1973); equilibration temperatures, 550°–600°C, estimated from the three sets of data, indicated that the formation temperature was lower than those of similar associations from other occurrences. The distribution coefficient for Fe/Mg and Mn/Mg between coexisting diopsides and garnets of the Norwegian garnet lherzolites is discussed by Carswell (1974), and of calcic pyroxenes and biotites by Saxena (1967).

In the hedenbergite from the eulysite of the Mariupol iron deposits Fe occurs in both the $M2$ and $M1$ sites. This atypical distribution of Fe in both of the octahedral sites in hedenbergite is regarded as the consequence of the high pressure, probably combined with the high temperature and the large amount of iron present in the system (Zverev *et al.*, 1971).

The interdependence of $K_D{}^{\text{Cpx-hbl}}$ and $K_D{}^{\text{Opx-hbl}}$, and the effect of the tetrahedral aluminium content of hornblende on the Mg–Fe distribution between calcic pyroxene–hornblende, as well as orthopyroxene–hornblende pairs in basic granulites, has been examined by Sen (1973).

The distribution of iron and magnesium between coexisting clinopyroxene and garnet may be considered on a one cation exchange basis:

$$MgCaSi_2O_6 + \tfrac{1}{3}(Fe_3Al_2Si_3O_{12}) \rightleftharpoons FeCaSi_2O_6 + \tfrac{1}{3}(Mg_3Al_2Si_3O_{12})$$

for which the distribution coefficient may be expressed:

$$K_{D\text{Fe}}^{\text{Cpx-Gar}} = \frac{X_{\text{Fe}}^{\text{Cpx}}(1 - X_{\text{Fe}}^{\text{Gar}})}{(1 - X_{\text{Fe}}^{\text{Cpx}})X_{\text{Fe}}^{\text{Gar}}}$$

A statistical study of the Fe–Mg distribution between clinopyroxene and garnet (Saxena, 1969), has shown that small contents of aluminium in the pyroxene do not lead to any significant departure from the apparent ideal binary solution of Mg and Fe in both minerals.

The influence of temperature on the partition coefficient for exchange equilibria between clino–orthopyroxene pairs and clinopyroxene–garnet pairs, using a spinel lherzolite as the starting mixture, has been demonstrated by Hensen (1973). The insignificant decrease in K_D at 1100°C from 0·43 at 22·5 kbar to 0·41 at 40·5 kbar indicates that the pressure effect is much smaller than the theoretical prediction by Banno (1970). The K_D–temperature plot for clinopyroxene–garnet pairs (temperatures estimated from the Di(En) solvus, Boyd and Nixon, 1973) shows a linear relationship between 950° and 1420°C suggesting that K_D clinopyroxene–garnet is a potential geothermometer, and can be applied as a first approximation to high-temperature, high-pressure ultrabasic rocks.

An experimental study of the influence of temperature and pressure on the partitioning of Cr and Al between clinopyroxenes and spinel in the system CaO–MgO–Al_2O_3–Cr_2O_3–SiO_2 has been presented by Dickey and Yoder (1972). In the latter system the compositions of the diopside solid solution, crystallized from bulk compositions midway between the $MgAl_2O_4$–$MgCr_2O_4$ join and the points $Di_{95}(CaCrAlSiO_6)_5$ and $Di_{90}(CaCrAlSiO_6)_{10}$ on the $CaMgSi_2O_6$–$CaCrAlSiO_6$ join, have been determined at pressures and temperatures between 5 and 20 kbar, and 900° and 1200°C respectively. The range of their molecular components was shown to be $CaMgSi_2O_6$ 65–96, $CaCrAlSiO_6$ 2·9–6·5, $CaAl_2SiO_6$ 1·0–20·9 and $MgSiO_3$ 0·0–7·6, and the distribution coefficient $K_D = (\text{Al/Cr})_{\text{Cpx}}/(\text{Al/Cr})_{\text{Sp}}$ greater than unity in conformity with the lower Cr/Al ratios of diopsides coexisting with spinel in natural assemblages.

The exchange reaction

$$CaMgSi_2O_6 + \tfrac{1}{3}KFe_3^{2+}(AlSi_3O_{10})(OH)_2 \rightleftharpoons CaFeSi_2O_6 + \tfrac{1}{3}KMg_3(AlSi_3O_{10})(OH)_2$$

using 1 M MgI_2 with an Ag–AgI internal buffer to facilitate exchange at 500°–680°C at 1 kbar has been studied by Gunter. $K_D = (X_{Fe}^{Cpx}/X_{Mg}^{Cpx})/(X_{Mg}^{Bi}/X_{Fe}^{Bi})$, is between 0–64 and 1·00 at 600°C, and between 0·61 and 0·97 at 500°C indicating the small temperature dependence of the distribution coefficient.

An investigation of the partitioning of K, Rb, Cs, Sr and Ba between diopside phenocrysts and matrix from an ankaramite and an andesitic basalt (Hart and Brooks, 1974) gave partition coefficients in the range 0·001–0·004 for K, Rb, Cs and Ba, values in good agreement with those for clinopyroxene and liquid determined experimentally in the system diopside–albite–anorthite–water at 15–30 kbar and 1100°–1200°C (Shimizu, 1974).

The distribution of Ti^{4+}, V^{3+}, Cr^{3+}, Mn^{2+}, Fe^{2+}, Co^{2+}, Ni^{2+} and Cu^{2+} between diopside crystals and liquid in the system $CaMgSi_2O_6$–$Na_2Si_2O_6$–H_2O at 1 kbar pressure has been determined by Seward (1971). With the exception of iron and manganese the elements are enriched in the diopside relative to the coexisting liquid. Thermodynamic data calculated for the exchange reactions involving Ti^{4+}, V^{3+}, Cr^{2+}, Co^{2+} and Ni^{2+}, e.g.

$$(CaMgSi_2O_6)_s + (Ni^{2+})_{liq} \rightleftharpoons (CaNiSi_2O_6)_s + (Mg^{2+})_{liq}$$

indicate the dominant role of the ligand field effect in the distribution of the transition element in the crystal–liquid equilibria. The octahedral site preference energies for the various exchange reactions involving Ti^{4+}, V^{3+}, Cr^{3+}, Co^{2+} and Ni^{2+} at 1198°K are given in Table 21.

Table 21 Octahedral site preference energies with $\Delta H°$ for the exchange reactions (after Seward, 1971)

Ion	Octahedral site preference energies (kcal/mole)	$\Delta H°$ (kcal/mole)
Cr^{3+}	37·7	−9·96
Ni^{2+}	20·6	−7·54
V^{3+}	12·8	−7·01
Co^{2+}	7·4	−4·57
Ti	0	−2·55

The partition coefficients of ten lanthanides and barium between diopside and liquid in the synthetic system diopside–enstatite–silica–H_2O at 20 kbar have been measured by Masuda and Kushiro (1970).

The distribution of Sr and REE elements between diopside and silicate liquid has been investigated by Grutzeck *et al.* (1974).

An experimental study of the partitioning of gadolinium in the system diopside–aqueous vapour showed a maximum diffusion rate of $D = 2 \times 10^{-15}\ cm^2\ s^{-1}$ at 800°C and 1 kbar (Zielinski and Frey, 1974).

Experimental and synthetic systems

The melting point of synthetic diopside was first determined by Day and Sosman (1910); a more precise determination by Adams (1914) gave the value of 1391·2° ± 1·5°C; the presently accepted value is 1391·5°C (Biggar, 1972). The melting relations of pure diopside at 1 atm have been studied more recently by Kushiro (1973a). The earlier view that diopside melted congruently has been shown to be in error, and at temperatures above at least 1375°C diopside solid solution and a liquid coexist (Fig. 106). At higher temperatures the amount of the liquid increases and the composition of the diopside solid solution is less diopside-rich than $Di_{98}En_2$ (weight per cent). At temperatures close to the liquidus the solid solution contains a small amount, less

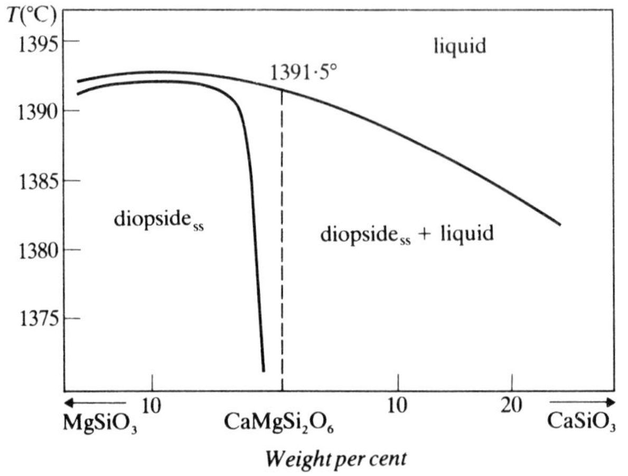

Fig. 106. Diopside solid solution liquidus and solidus (after Kushiro, 1973a).

than 5 weight per cent, of a forsterite component, and the coexisting liquid carries an excess of silica over the compositions of the join $MgSiO_3$–$CaSiO_3$. Diopside with up to 4 per cent excess silica has been synthesized at atmospheric pressure (Shinno, 1968).

The effect of pressure on the melting temperature of diopside has been investigated by Yoder (1952), Boyd and England (1963), and by Williams and Kennedy (1969). The melting point at 50 kbar, $\simeq$ 2000°C, determined by the latter authors is substantially higher than the earlier value by Boyd and England, and the melting curves given by the two pairs of authors show some discrepancy at pressure greater than 30 kbar. Yoder (1952) has calculated the volume change of diopside on melting from the Clapeyron equation:

$$\frac{dT}{dP} = \frac{T\Delta V}{\Delta H}$$

Substituting the values of ΔH 102 cal/g, dT/dP 0·01297°/bar, and T 1664·5°K gives a volume change of 0·033 cm^3/g. A different value, 0·0566 cm^3/g was obtained by Dane (1941) who measured the density of a synthetic diopside liquid above the melting point, and used the calculated specific volume of diopside and the measured specific volume of the liquid at the melting point to compute the volume change. Williams and Kennedy have plotted the melting temperature of diopside against the volume of melting solid using the relationship:

$$t = t_0\left[1 + C, \frac{\Delta V}{V_0}\right]$$

where t is the melting point, $\Delta V/V_0$ is the compression of the solid phase at the same pressure (C), and t_0 and V_0 the melting temperature and specific volume of the solid at 1 bar. The data of Boyd and England (1963) on the effect of pressure on the melting of diopside have been examined by Chayes (1968), using the discriminant function method, and the derived equation $T = 649{\cdot}0 + 1{\cdot}95P$ is in good agreement with the form of the phase diagram.

The enthalpy of melting of synthetic diopside, using high pressure differential thermal analysis (differential scanning calorimetry) has been measured by Rosenhauer (1976). The enthalpy value is 85·5 cal/g at 1 atm; the value increases under dry

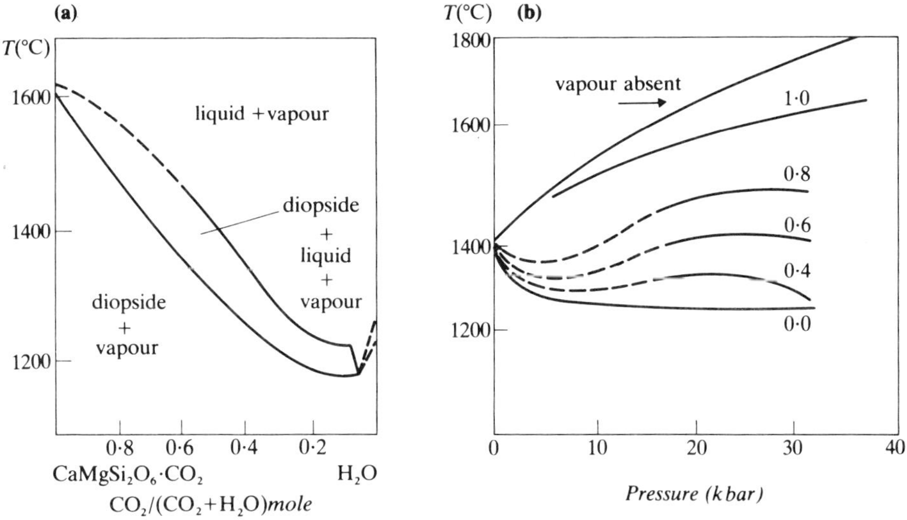

Fig. 107. (a) Phase relations on the join $CaMgSi_2O_6.CO_2$–H_2O at 30 kbar pressure (after Rosenhauer and Eggler, 1975). Abscissa indicates molar $CO_2/(CO_2+H_2O)$ content of the volatile portion of the bulk composition. (b) Melting of diopside in the presence of CO_2 and H_2O vapour, and in the absence of vapour (after Rosenhauer and Eggler, 1975). Numbers indicate $CO_2/(CO_2+H_2O)$ of vapour.

melting with increasing pressure to 95 cal/g at 5 kbar, whereas under water-saturated conditions the enthalpy decreases to 70 cal/g at 1 kbar and then remains constant at this level up to 4 kbar. A considerably smaller estimate of the effect of pressure under water-saturated conditions, 25–40 cal/g between 1 and 2 kbar, is given by Khitarov and Kadik (1973).

The melting curve for diopside with excess CO_2, and the effect of CO_2 and H_2O in combination, at high pressures has been determined by Eggler (1973) and Rosenhauer and Eggler (1975). The phase relations in the join $CaMgSi_2O_6.CO_2$–H_2O at 30 kbar are shown in Fig. 107(a) and the melting of diopside in the presence of CO_2 and H_2O vapour in Fig. 107(b). The standard free energy of a diopside, $(Ca_{0\cdot45}Mg_{0\cdot46}Fe_{0\cdot047}Al_{0\cdot03})_2(Si_{0\cdot92}Al_{0\cdot02})_2O_6$, -725 kcal/mole, calculated from data obtained from its solubility in water, has been determined by Reesman and Keller (1965); the melting entropy of diopside is 77 kJ/°K (Ferrier, 1971).

Nickel diopside, $CaNiSi_2O_6$, has been synthesized at high pressures, and the melting curve determined between 10 and 40 kbar (Higgins and Gilbert, 1973). On the assumption that the melting point at 1 atm is 1340°C the equation of the curve is $T(°C) = 1340+12\cdot81P(\text{kbars})-0\cdot113P^2$. Compared with diopside, the nickel analogue melts about 65°C lower at 10 kbar and approximately 120°C lower at 40 kbar. In the presence of water the melting temperature at 10 kbar is 1250°C and is comparable with that of diopside with excess water, but is approximately 200°C lower than in the absence of water. Nickel diopside appears to melt incongruently at 1 atm, but the evidence for incongruent melting at higher pressures is not conclusive.

Hedenbergite has been synthesized using various buffers by Gustafson (1974). The isobaric 2 kbar fluid pressure, $\log f_{O_2}$–T diagram for hedenbergite bulk composition is shown in Fig. 108. The f_{O_2}–T stability range of hedenbergite is extensive, thus at $P_f = 2$ kbar it is stable below $f_{O_2} = 10^{-13}$ bar at 800°C, and below $f_{O_2} = 10^{-28}$ bar at 400°C. At higher f_{O_2} values the pyroxene reacts with oxygen to form the anhydrous oxidized assemblage andradite + magnetite + quartz:

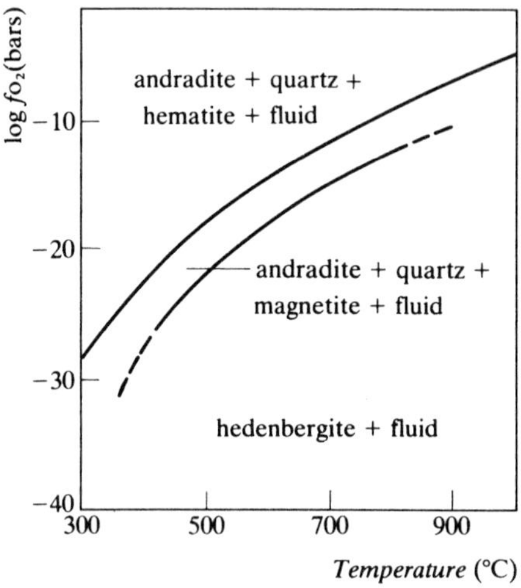

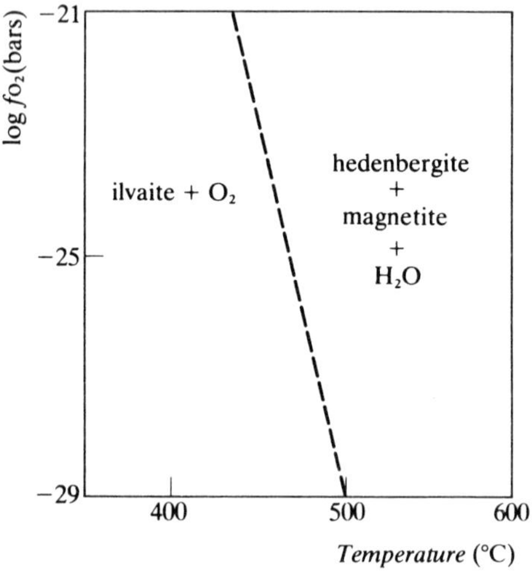

Fig. 108. Logf_{O_2}–T diagram for hedenbergite bulk composition + excess H_2O at P_{fluid} = 2 kbar (after Gustafson, 1974).

Fig. 109. Logf_{O_2}–T diagram for the reaction ilvaite + $O_2 \rightleftharpoons$ hedenbergite + magnetite + H_2O at P_{fluid} = 2 kbar (after Gustafson, 1974).

$$9CaFeSi_2O_6 + 2O_2 \rightleftharpoons 3Ca_3Fe_2^{3+}Si_3O_{10} + Fe_3O_4 + 9SiO_2$$

Using a starting mixture $CaO.2FeO.\frac{1}{2}Fe_2O_3.2H_2O$ + excess H_2O, and 50–50 synthetic hedenbergite + magnetite and ilvaite, Gustafson has determined the log f_{O_2}–T curve for the reaction:

$$6CaFe_2^{2+}Fe^{3+}Si_2O_8(OH) + \tfrac{1}{2}O_2 \rightleftharpoons 6CaFeSi_2O_6 + 4Fe_3O_4 + 3H_2O$$

at 2 kbar (Fig. 109).

Hedenbergite has also been synthesized from a gel heated at temperatures between 400° and 600°C at a pressure of 1 kbar. Mössbauer spectra taken at intervals during the synthesis showed that the reaction was complete earlier than was indicated by X-ray diffraction techniques. The time difference is considered to be due to the interval required for crystal growth from rapidly formed domains of short-range order (Kinrade *et al.*, 1975).

The equilibrium diagram of the join $CaMgSi_2O_6$–$MgSiO_3$ at 1 atm, first presented by Boyd and Schairer (1964), has been revised by Kushiro (1972b), and the liquidus boundaries determined directly by microprobe analysis. The latter investigation has confirmed the incongruent melting of diopside, the presence of silica deficiency in diopside solid solution, and the existence of an iron-free pigeonite on the diopside–enstatite join (see Fig. 44, and enstatite section). The melting temperatures of diopside and diopside–enstatite solid solutions are greatly increased by pressure, thus for diopside the temperature rises from 1390°C at 1 atm to 1715°C at 30 kbar. Above 1385°C, as a result of the increase in solvus temperatures, the extent of the solid solution of $MgSiO_3$ in diopside increases with pressure (Fig. 110). Below this temperature, pressure has little effect on the composition of the diopside solid solution in equilibrium with orthopyroxene. The diopside solvus at 1 atm and 30 kbar is shown in Fig. 111; apart from the inflection of the boundary of the two-pyroxene field at 1 atm, due to the enstatite $\rightleftharpoons$ protoenstatite inversion, the two curves are similar and the maximum compositional difference between them is

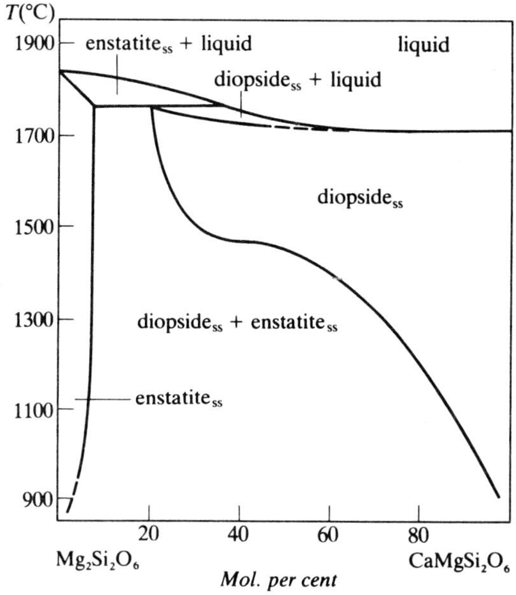

Fig. 110. Phase relations along the join $Mg_2Si_2O_6$–$CaMgSi_2O_6$ at 30 kbar (after Davis and Boyd, 1966).

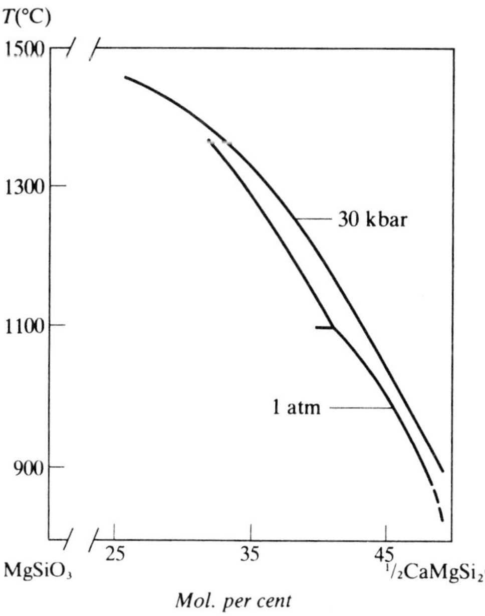

Fig. 111. Projection of part of the diopside solvus in the system $CaMgSi_2O_6$–$Mg_2Si_2O_6$ at 30 kbar and 1 atm. The inflection in curve at 1 atm is due to the enstatite ⇌ protoenstatite inversion. Protoenstatite is unstable at high pressure and the solvus at 30 kbar is a smooth curve (after Davis and Boyd, 1966).

approximately 2 mol. per cent $CaSiO_3$. The phase relations of Ca-rich, in equilibrium with Mg-rich, pyroxenes at temperatures up to 1400°C may thus provide a useful geothermeter that is relatively independent of pressure. Consequent on an investigation of the subsolidus and vapour-saturated phase relations for the composition region $CaMgSi_2O_6$–Mg_2SiO_4–SiO_2–H_2O at P_{H_2O} between 1 and 10 kbar, Warner (1975) has presented two temperature–composition sections for the Ca-poor region of the join $CaMgSi_2O_6$–$Mg_2Si_2O_6$ (see orthopyroxene section, Fig. 46(a), (b)).

A preliminary study of the melting relations of synthetic pyroxenes along the join diopside–hedenbergite at 1 atm has been presented by Turnock (1962). There is a complete series of monoclinic pyroxenes between the two end-members at subsolidus temperatures, although since the earlier work by Bowen *et al.* (1933) on the system CaO–FeO–SiO_2 it has been known that hedenbergite is not a stable phase at liquidus temperatures (Fig. 112). Compositions between Di_{100} and approximately $Di_{60}Hd_{40}$ form an essentially binary system (Fig. 113). At temperatures above the solidus, solid solutions richer in the hedenbergite component are converted to a $wollastonite_{ss}$.

The temperature–composition relationships between hedenbergite and wollastonite solid solutions in the temperature range 600° to 955°C, at a P_{H_2O} of 1 kbar and f_{O_2} defined by the fayalite–quartz–iron buffer are shown in Fig. 248(a). The marked inflection of the solvus between 800° and 815°C, and compositions $(Ca_{0.88}Fe_{0.12})SiO_3$ and $(Ca_{0.80}Fe_{0.20})SiO_3$, may be due to the presence of a second pyroxenoid phase (see Fig. 248(b), p. 554, showing the equilibrium relations between hedenbergite and wollastonite and bustamite solid solutions, Rutstein, 1971).

The phase relations along the join hedenbergite–ferrosilite at 1 atm were investigated by Bowen *et al.* (1933). The results showed that $FeSiO_3$-rich pyroxenes are unstable at low pressure; hedenbergite solid solutions invert to ferriferous wollastonite solid solutions at subsolidus temperature. All compositions between $CaFeSi_2O_6$ and $FeSiO_3$ melt incongruently. More recently studies of the

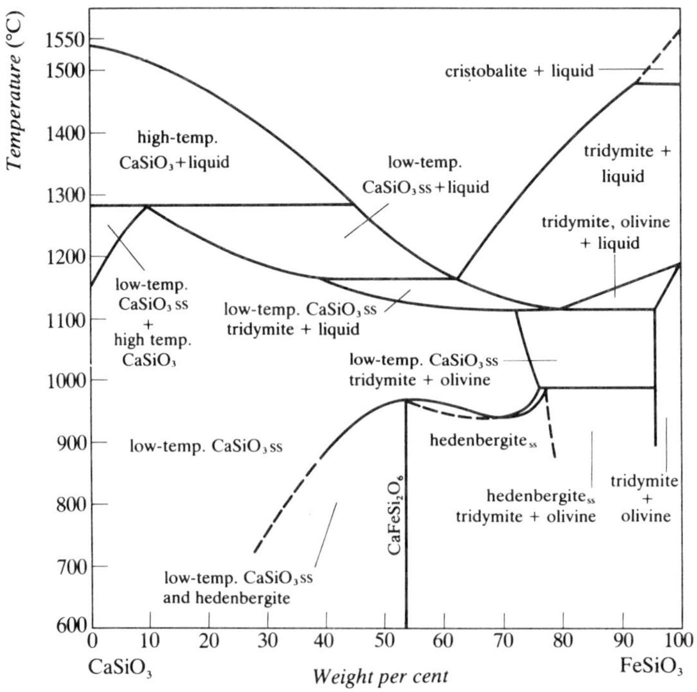

Fig. 112. Phase diagram of mixtures with a metasilicate ratio in the system CaO–FeO–SiO_2. Heavy curves refer to binary equilibrium and light curves to ternary equilibrium (after Bowen *et al.*, 1933).

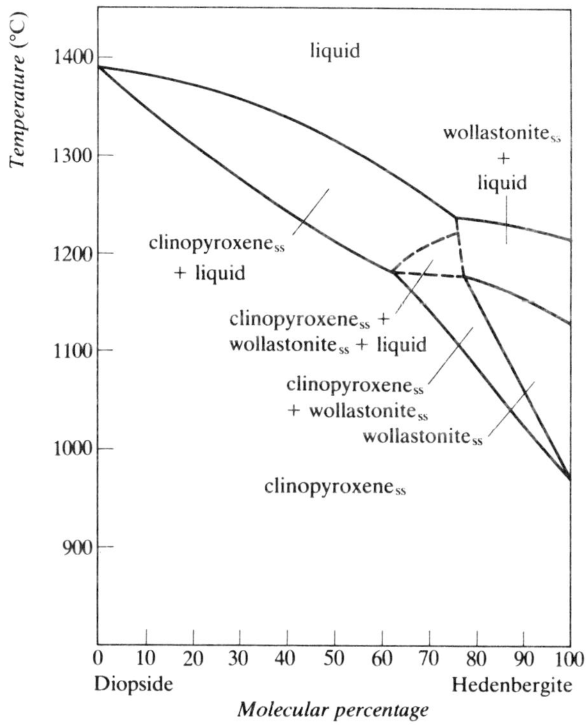

Fig. 113. Phase diagram for the join diopside–hedenbergite (after Schairer and Yoder, 1962).

hedenbergite–ferrosilite join at pressures up to 2·8 kbar have been undertaken by Lindsley (1967), and by Lindsley and Munoz (1968, 1969).

Hedenbergite and wollastonite solid solutions, in equilibrium with fayalite+ quartz or tridymite are controlled by the reactions:

$FeSiO_3$-rich hedenbergite$_{ss}$ $\rightleftharpoons$ $FeSiO_3$-poor hedenbergite$_{ss}$ + fayalite + SiO_2

$FeSiO_3$-rich wollastonite$_{ss}$ $\rightleftharpoons$ $FeSiO_3$-poor wollastonite$_{ss}$ + fayalite + SiO_2

Redox reactions are not involved, and the following reactions are independent of oxygen fugacity in the range of f_{O_2} bounded by the breakdown of fayalite:

$$Fe_2SiO_4 \rightleftharpoons 2Fe + SiO_2 + O_2$$
$$3Fe_2SiO_4 + O_2 \rightleftharpoons 2Fe_3O_4 + 3SiO_2$$

and correspond with fayalite–iron–quartz and fayalite–magnetite–quartz buffers, under which conditions the individual phases are stable.

The revised subsolidus relationships for the hedenbergite–ferrosilite join, based on the more recent studies by Lindsley and Munoz (1969) at 2 kbar pressure and below, combined with the melting relationships along the join determined by Bowen *et al.* (1933), are shown in Fig. 114. Curve A marks the limit of stable solid solution of $FeSiO_3$ in hedenbergite at pressures below 2 kbar. At higher pressures the position of the curve shifts towards the $FeSiO_3$ composition with a consequent increase in the size of the field of hedenbergite solid solution, and at 20 kbar and temperature 1000°C there is complete solid solution between hedenbergite and clinoferrosilite.

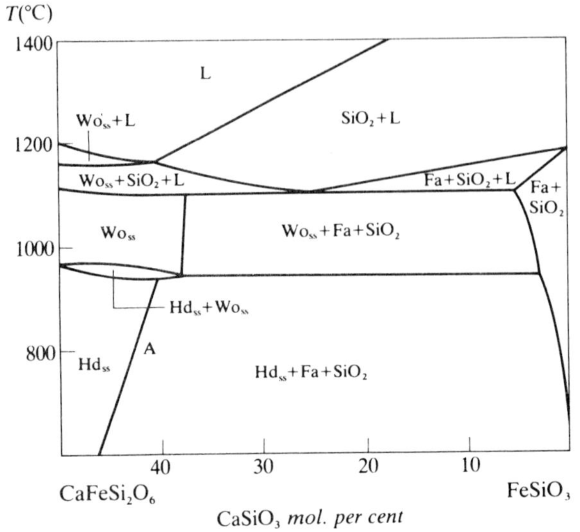

Fig. 114. **The join hedenbergite–ferrosilite at and below 2 kbar. Subsolidus from Lindsley and Munoz (1969). Above solidus from Bowen *et al.*, 1933. Hd$_{ss}$, clinopyroxene solid solutions close to $Ca_{0.5}Fe_{0.5}SiO_3$ and extending towards $FeSiO_3$; Fa, olivine close to Fe_2SiO_4 in composition; Wo$_{ss}$, iron-wollastonite solid solutions; SiO_2, tridymite or quartz; L, liquid.**

Within the pyroxene quadrilateral, Di–Hd–En–Fs, the low-pressure instability field, for which the term 'forbidden zone' is commonly used, extends from the $FeSiO_3$ apex to $Ca_{0.4}Fe_{0.6}SiO_3$ along the ferrosilite–hedenbergite join, and to approximately $Mg_{0.32}Fe_{0.68}SiO_3$ along the ferrosilite–enstatite join (Fig. 115). The extent, at subsolidus temperatures, of the miscibility gap within the pyroxene quadrilateral, using a number of compositions on the join $Fs_{85}En_{15}$–wollastonite, have been studied by Smith (1971, 1972). The minimum pressures for the stability of iron-rich pyroxenes at 925°C are shown in Fig. 115. Thus on the $Fs_{85}En_{15}$–$CaSiO_3$ join a minimum pressure of 3·5 kbar at 925°C is required to stabilize a composition less calcium-rich and more iron-rich than Wo_{15}. On the $MgSiO_3$–$FeSiO_3$ join the

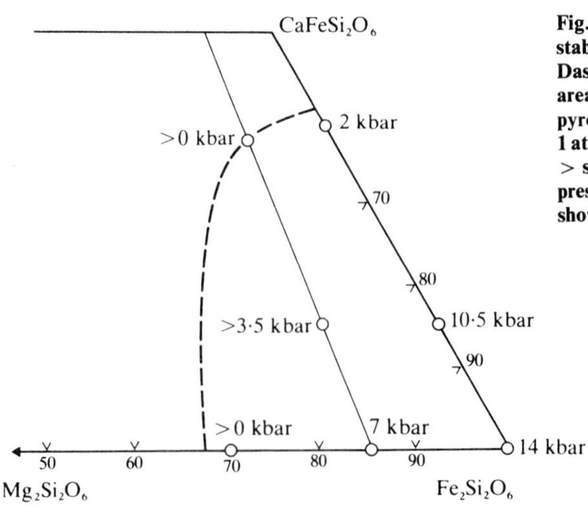

Fig. 115. **Minimum pressure to stabilize iron-rich pyroxenes at 925°C. Dashed line outlines the approximate area of the 'forbidden zone' within which pyroxene is not stable at 925°C and 1 atm pressure (after Smith, 1972). The > symbols indicate that the minimum pressures may be greater than the values shown.**

forbidden zone expands and minimum pressures increase with increasing temperature, in contrast to the contraction of the zone along the $CaFeSi_2O_6$–$FeSiO_3$ join with increasing temperature. In both cases the change in the limiting compositions for pyroxene stability is 3–4 mol. per cent $FeSiO_3$/100°C, and the temperature effect upon minimum pressures is probably small. The effect of temperature on the minimum pressures has not been investigated for compositions within the quadrilateral, but it is probably small as the effects of temperature on the 'forbidden zone' at the boundary joins are not large and are of opposite sense.

Ernst (1966) has synthesized a number of hedenbergite–ferrosilite solid solutions, of bulk composition Hd_{100}, $Hd_{80}Fs_{20}$ and $Hd_{60}Fs_{40}$, under oxygen fugacities controlled by the magnetite–wüstite buffer at a fluid pressure of 2 kbar. At pressures and temperatures between 0·5 and 3 kbar and 437° and 543°C hedenbergite pyroxene, fayalite, quartz and liquid, is the stable assemblage for the ferrotremolite composition, $Ca_2Fe_5^{2+}Si_8O_{22}(OH)_2$.

The solid solution of $CaAl_2SiO_6$ (Ca-Tschermak's molecule, CaTs) in synthetic diopside has been investigated by Zvetkov (1945) and Sakata (1957), and the limits of solid solution along the joins $CaMgSi_2O_6$–Al_2O_3, $CaMgSi_2O_6$–$CaAl_2SiO_6$ and $CaMgSi_2O_6$–$MgAl_2SiO_6$ were determined by Segnit (1953). More recently the phase relations for compositions on the join diopside–$CaAl_2SiO_6$ at 1 atm have been examined by de Neufville and Schairer (1962). The maximum substitution of AlAl for (CaMg)Si for compositions on the join $CaMgSi_2O_6$–$CaAl_2SiO_6$ is approximately 45 mol. per cent.

Additional data on the relationships on the join $CaMgSi_2O_6$–$CaAl_2SiO_6$, and some revision of the phase equilibrium diagram at 1 atm have been presented by Schairer and Yoder (1970). Crystallization of the composition $Di_{65}CaTs_{35}$ gives rise to liquids that follow the surface Di_{55} + An + liquid, and these phases are joined by melilite and spinel at 1238°C. During crystallization the composition $Di_{45}CaTs_{55}$ follows the univariant line, An + Di_{ss} + Sp + liquid to the quaternary univariant point An + Di_{ss} + Sp + melilite + liquid at 1238°C (Fig. 116(a)).

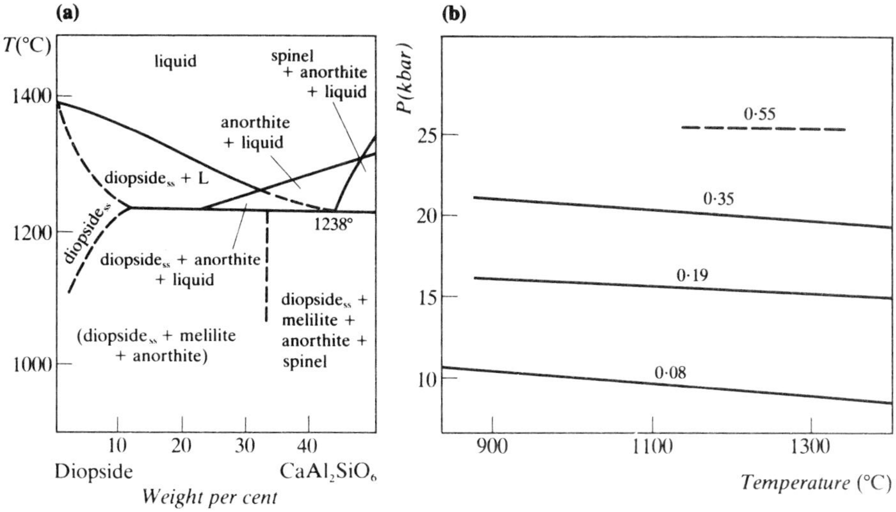

Fig. 116. (a) Phase equilibrium diagram for part of the $CaMgSi_2O_6$–$CaAl_2SiO_6$ join (after Shairer and Yoder, 1970). (b) Mole fraction of $CaAl_2SiO_6$ component in $CaMgSi_2O_6$–$CaAl_2SiO_6$ solid solutions coexisting with anorthite and quartz at different temperatures and pressures (after Wood, 1976a). 0·55, 0·35, 0·19 and 0·08 represent approximate isopleths.

Table 22 Crystallization products of $CaMgSi_2O_6$–$CaAl_2SiO_6$ glasses (after Toropov and Khotimchenko, 1967)

Composition of glasses						Products
$CaMgSi_2O_6$	$CaAl_2SiO_6$	CaO	MgO	Al_2O_3	SiO_2	
100	0	25·89	18·62	—	55·49	diopside$_{ss}$
90	10	25·87	16·75	4·67	52·71	
80	20	25·85	14·89	9·35	49·90	
70	30	25·83	13·03	14·02	47·10	diopside$_{ss}$ + anorthite + melilite
60	40	25·82	11·17	18·70	44·31	
50	50	25·81	9·31	23·37	41·52	An + Mel + forsterite
40	60	25·79	7·45	28·04	38·73	An + Mel + Fo + spinel

The crystallization products of glasses of Di–CaTs composition, varying from Di_{100} to Di_{40}, have been examined by Toropov and Khotimchenko (1967). Their investigation showed that the single phase region of solid solutions is restricted to 20 per cent $CaAl_2SiO_6$ (Table 22), and that the probable limit of this component in diopside is 40 per cent (= 18·7 weight per cent Al_2O_3). Toropov and Khotimchenko also investigated the crystallization, at various temperatures, of glasses with the stoichiometric composition of diopside. At 800°C a solid of Mg-rich diopside and åkermanite crystallize simultaneously; with greater duration of heating the concentration of Mg and Ca changes to that of pure diopside while the amount of åkermanite crystallizing remains unchanged. At temperatures of 900°C and above diopside is the only product of crystallization (see Table 23, p. 251).

The phase relations for the compositions (mol. per cent) $(CaMgSi_2O_6)_{50}$ $(CaAl_2SiO_6)_{50}$ and $(CaMgSi_2O_6)_{25}(CaAl_2SiO_6)_{75}$ have been investigated at temperatures between 880° and 1093°C and pressures between 3 and 13 kbar (Hijikata,

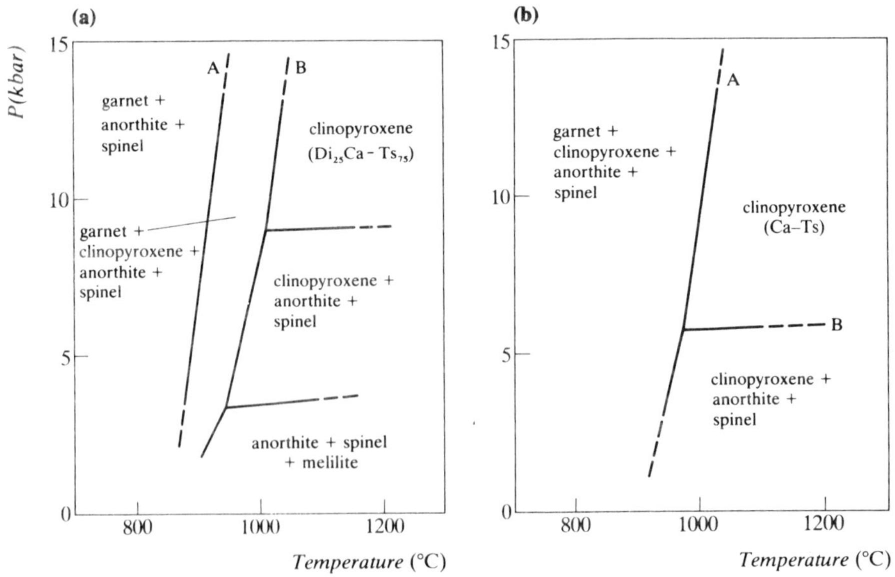

Fig. 117. (a) Pressure–temperature plane for the composition $(CaMgSi_2O_6)_{25}(CaAl_2SiO_6)_{75}$. (b) Pressure–temperature plane for the composition $(CaMgSi_2O_6)_{50}(CaAl_2SiO_6)_{50}$ (after Hijikata, 1973).

1973). In the higher CaTs composition (Fig. 117(a)) the assemblage anorthite–melilite–spinel is stable between 1 atm and 2·5 kbar and temperatures greater than $\simeq$ 950°C. In the presure range 2·5 to 9 kbar the assemblage consists of anorthite–spinel–clinopyroxene$_{ss}$. The latter phase is not $Di_{25}CaTs_{75}$; this composition is stable only at pressures and temperatures above 9 kbar and 1000°C respectively. At lower temperatures the stable assemblage is garnet$_{ss}$ + clinopyroxene$_{ss}$ + anorthite + spinel. In this phase field between curves A and B, the relative amount of clinopyroxene$_{ss}$ to anorthite and spinel decreases, and the amount of garnet$_{ss}$ increases with decreasing temperature until clinopyroxene$_{ss}$ disappears completely and the stable assemblage is garnet$_{ss}$ + anorthite + spinel. Similar phase relationships are shown by the composition $Di_{50}CaTs_{50}$. The lower pressure limit of the clinopyroxene$_{ss}$ + anorthite + spinel assemblage has not been determined, but it is known, from the earlier work of de Neufville and Schairer, that the stable low-pressure assemblage for this composition is anorthite + melilite$_{ss}$ + forsterite. Clinopyroxene$_{ss}$ with the composition $Di_{50}CaTs_{50}$ is restricted to the high-pressure portion of the composition plane bounded by the curves A and B (Fig. 117(b)).

The compositions of these high alumina clinopyroxenes indicated the possible presence of excess silica in the structure, as earlier foreshadowed by Kushiro's (1969c) investigation of clinopyroxene solid solutions formed by reactions between diopside and plagioclase at high pressure. The presence of excess silica, probably present as a $Ca_{0\cdot5}AlSi_2O_6$ component, would lead to a cation total of less than four (on the basis of six oxygens), as previously indicated by Cawthorn and Collerson (1974) who found that natural clinopyroxenes show decreasing cation totals as the amounts of octahedrally co-ordinated aluminium in the pyroxenes increases. Confirmation of these conclusions has been obtained experimentally by Wood (1976b) who has shown that there is a correlation between the aluminium content of clinopyroxenes coexisting with quartz and their stoichiometry, similar in magnitude to that observed by Cawthorn and Collerson for natural clinopyroxenes.

The relationship between the activity and mole fraction of the Ca-Tschermak's component in clinopyroxene solid solutions has been investigated by Wood (1976a). Mole fractions of $CaAl_2SiO_6$ were determined experimentally (Fig. 116(b)) in clinopyroxene, anorthite, quartz assemblages by crystallizing compositions in the systems $CaMgSi_2O_6$–$CaAl_2SiO_6$–SiO_2, $CaFeSi_2O_6$–$CaAl_2SiO_6$–SiO_2 and $Ca(Mg_{0\cdot6}Fe_{0\cdot4})Si_2O_6$–$CaAl_2SiO_6$–$SiO_2$ at pressures of 10–25 kbar and temperatures between 900° and 1300°C.

The activity of the $CaAl_2SiO_6$ component, determined by the reaction $CaAl_2Si_2O_8 \rightleftharpoons CaAl_2SiO_6 + SiO_2$ was calculated from Hariya and Kennedy's (1968) data for the reverse reaction, and the high-temperature entropy and 298°K/1 bar data of Robie and Waldbaum (1968), shows that the activity is a simple function of the clinopyroxene composition such that:

$CaMgSi_2O_6$–$CaAl_2SiO_6$	$a^{Cpx}_{CaAl_2SiO_6} \geqslant X^{Cpx}_{CaAl_2SiO_6}$
$CaMg_{0\cdot6}Fe_{0\cdot4}Si_2O_6$–$CaAl_2SiO_6$	$a^{Cpx}_{CaAl_2SiO_4} = X^{Cpx}_{CaAl_2SiO_6}$
$CaFeSi_2O_6$–$CaAl_2SiO_6$	$a^{Cpx}_{CaAl_2SiO_4} \leqslant X^{Cpx}_{CaAl_2SiO_4}$

The kinetics of linear crystal growth in the system $CaMgSi_2O_6$–$CaAl_2SiO_6$ have been investigated by Kirkpatrick (1974).

The equilibrium diagram of part of the join diopside–silica (Fig. 118) illustrates the non-binary character of the system as shown by the three-phase assemblage diopside solid solution + tridymite + liquid. The liquidus minimum is a piercing point and is located at a silica content of between 15 and 16 weight per cent at a

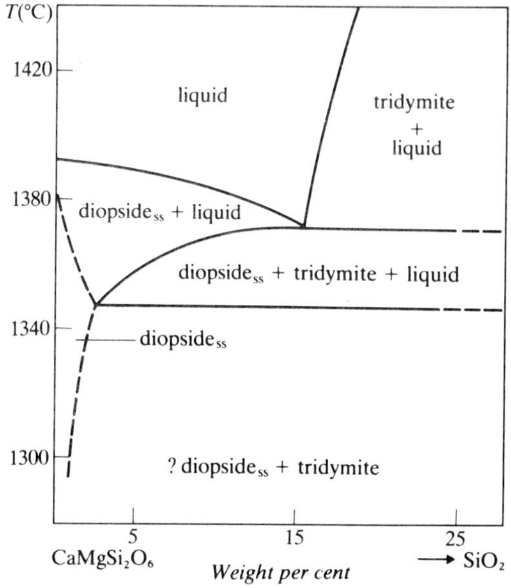

Fig. 118. Equilibrium diagram of the diopside-rich portion of the system diopside–silica (after Schairer and Kushiro, 1964).

temperature of 1371°C. In the diopside-rich part of the system the temperature at which diopside is joined by tridymite decreases as the content of diopside in the mixture increases (Schairer and Kushiro, 1964).

The liquidus relations in the system $CaMgSi_2O_6$–$CaAl_2SiO_6$–SiO_2 at 1 atm and 20 kbar (Fig. 119a,b) have been presented by Clark *et al.* (1962). The join diopside–anorthite is not binary due to the solid solubility of Al_2O_3 in the pyroxene. At 20 kbar the temperature of the diopside–silica eutectic is more than 200°C higher than at 1 atm although the eutectic composition remains essentially unchanged. The relationships in the subsystem diopside–anorthite are considerably modified at high

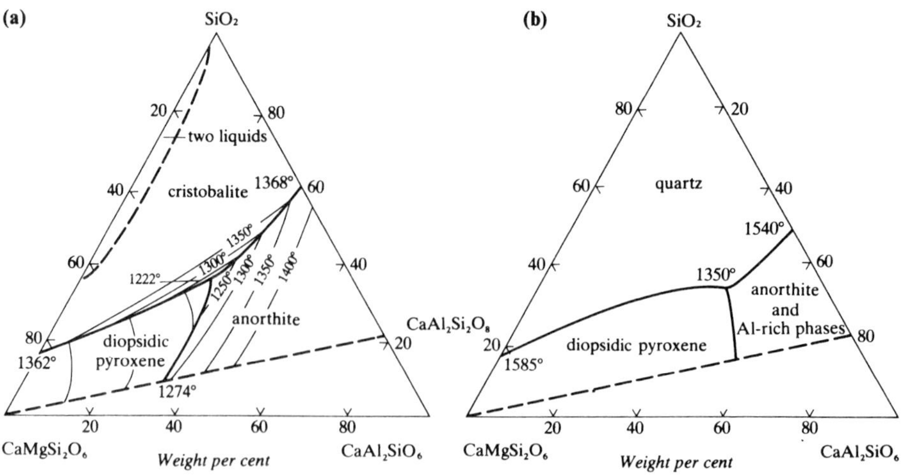

Fig. 119. (a) Equilibrium diagram for the system diopside–Ca-Tschermak's molecule–SiO_2 at 1 atm (after Clark *et al.*, 1962). (b) Equilibrium diagram for the system diopside–Ca-Tschermak's molecule–SiO_2 at 20 kbar. At compositions near anorthite, corundum and 'β-alumina' appear at high temperatures but precise relationship between these phases is unknown (after Clark *et al.*, 1962).

pressure due to the incongruent melting of anorthite, and the greatly increased amount of Al_2O_3 in the pyroxene, the composition of which lies on the $CaMgSi_2O_6$–$CaAl_2SiO_6$ join.

The stability field of $CaAl_2SiO_6$ (Hays, 1966, 1967) is defined by the reactions:

$$\underset{\text{anorthite}}{CaAl_2Si_2O_8} + \underset{\text{gehlenite}}{Ca_2Al_2SiO_7} + \underset{\text{corundum}}{Al_2O_3} = 3\,CaAl_2SiO_6 \quad (1)$$

$$3\,CaAl_2SiO_6 = \underset{\text{grossular}}{Ca_3Al_2Si_3O_{12}} + \underset{\text{corundum}}{2\,Al_2O_3} \quad (2)$$

which may be expressed by the linear equations:

$$P(\text{bars}) = 12\,500 + 9{\cdot}9\,(T°\text{C} - 1250) \quad (1)$$

$$P(\text{bars}) = 17\,500 + 63{\cdot}8\,(T°\text{C} - 1250) \quad (2)$$

The pure Ca-Tschermak's molecule is not stable below 1160°C (Fig. 120), and has a relatively restricted pressure stability field, thus accounting for the absence of the $CaAl_2SiO_6$ pyroxene in natural environments, as well as indicating that diopside solid solutions rich in this component have probably crystallized at high temperatures and moderate, rather than high, pressures.

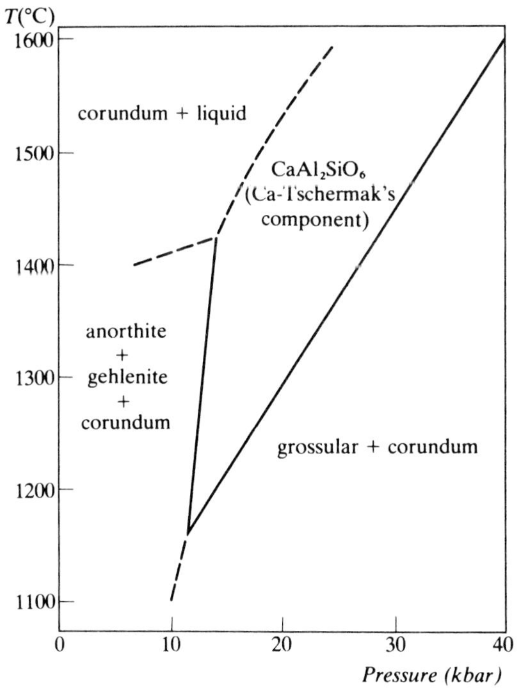

Fig. 120. The stability field of Ca-Tschermak's molecule, $CaAl_2SiO_6$ (after Hays, 1967).

A hexagonal polymorph of $CaAl_2SiO_6$, grown from a stoichiometric, dry melt of the same composition, in the temperature range 950° to 1050°C at ambient atmosphere has been reported by Kirkpatrick and Steele (1973).

Compositions along the joins $CaMgSi_2O_6$–$CaFe^{3+}_2SiO_6$ (ferri-Tschermak's molecule) and $CaMgSi_2O_6$–$Ca_3Fe^{3+}_2Si_3O_{10}$ (andradite), parts of the quaternary system CaO–MgO–Fe_2O_3–SiO_2, have been examined at 1 atm by Huckenholz *et al.* (1969). The maximum stable solid solution of the ferri-Tschermak's molecule in

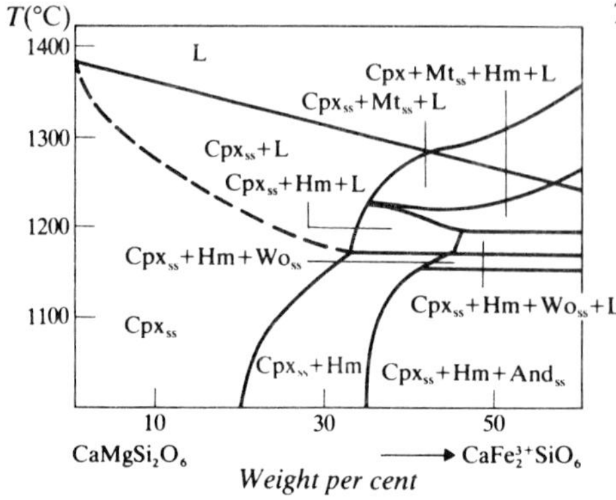

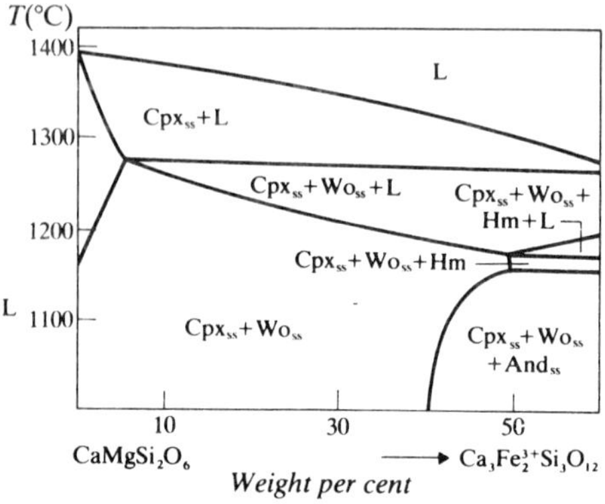

Fig. 121. Phase relations on the join diopside–ferri-Tschermak's molecule at 1 atm. Cpx_{ss} clinopyroxene solid solution, Wo_{ss} wollastonite solid solution, Hm hematite, Mt magnetite solid solution, And andradite solid solution, L liquid (after Huckenholz *et al.*, 1969).

Fig. 122. Phase relations on the join diopside–andradite at 1 atm. Cpx_{ss} clinopyroxene solid solution, Wo_{ss} wollastonite solid solution, Hm hematite, And_{ss} andradite solid solution, L liquid (after Huckenholz *et al.*, 1969).

diopside is approximately 33 weight per cent at 1175°C (Fig. 121). The clinopyroxene solid solutions at solidus temperatures, in compositions between 20 and 33 per cent $CaFe_2^{3+}SiO_6$, at 1000° to 1175°C break down on cooling to a less Fe^{3+}-rich clinopyroxene + hematite. The clinopyroxene solid solution + hematite assemblage is replaced by $clinopyroxene_{ss}$ + hematite + andradite solid solution in compositions containing more than 35 and 44 weight per cent $CaFe_2^{3+}SiO_6$ at 1000° and 1175°C respectively. The primary clinopyroxene solid solution coexisting with liquid is about $Di_{94}FeTs_6$ at 1300°C and $Di_{76}FeTs_{24}$ at 1200°C. In diopside-rich compositions along the diopside–$Ca_3Fe_2^{3+}Si_3O_{10}$ join (Fig. 122), $clinopyroxene_{ss}$ and $wollastonite_{ss}$ are the stable phases, but in compositions between And_{40} and And_{49} the $clinopyroxene_{ss}$–$wollastonite_{ss}$ assemblage is replaced at lower temperatures by $clinopyroxene_{ss}$ + $wollastonite_{ss}$ + $andradite_{ss}$.

The phase equilibrium in air in the system $CaMgSi_2O_6$–$CaFe^{3+}AlSiO_6$ has been investigated by Hijikata and Onuma (1969). The system is binary between pure diopside and $Di_{53{\cdot}5}$ and the composition of the liquidus minimum, 1268° ± 5°C. With greater amounts of the $CaFe^{3+}AlSiO_6$ component, magnetite solid solution is the liquidus phase. Solid solution between the end-members of the system is complete at subsolidus temperatures; compositions with more than 67 per cent $CaFe^{3+}AlSiO_6$ form stable clinopyroxene between 1100° and 1293°C (Hijikata, 1968).

The synthesis of highly silica-deficient pyroxenes from a mix of $CaCO_3$, SiO_2, Fe_2O_3 and Al_2O_3 has been reported by Dyson and Juckes (1972). The pyroxenes show marked pleochroism from yellow-green to orange, and have a birefringence of between 0·025–0·030. The refractive indices range 1·9–1·97, the higher values relating to the more iron-rich specimens. Electron probe analyses of these pyroxenes give the approximate formula $Ca_{1{\cdot}0}(Ca_{0{\cdot}14}Fe^{3+}_{0{\cdot}86})(Fe^{3+}_{0{\cdot}32}Al_{0{\cdot}60}Si_{1{\cdot}08})O_6$. The cell parameters are *a* 9·840, *b* 8·825, *c* 5·398 Å, *β* 105°40′, closely resembling those of diopside.

Diopside–jadeite forms a solid solution series at 30 and 40 kbar pressure (Bell and Davis, 1969). The melting interval is small, with a maximum of about 40°C (see Fig. 202, p. 000). A temperature–composition section at 30 kbar showing tentative subsolidus relationships, and an investigation of the solvus in the system diopside–jadeite has been presented by Bell and Davis (1965, 1967). Hydrothermal syntheses of $Di_{70}Jd_{30}$, $Di_{60}Jd_{40}$ and $Di_{50}Jd_{50}$ at temperatures between 450° and 650°C and pressures between 1000 and 5000 atm P_{H_2O}, have been made by Wikström (1970). The jadeite content of the clinopyroxene solid solutions decreases significantly with increasing temperature and decreasing pressure, and is accompanied by a smaller increase in the proportion of the $CaAl_2SiO_6$ molecule.

Nolan and Edgar (1963) synthesized a series of diopside–acmite solid solutions, with compositions at 10 per cent intervals, between Di_{100} and Ac_{100} at 750°C and 1 kbar water vapour pressure. The complete solid solution at low temperature between the end-members in the diopside–acmite system was later confirmed by Yagi (1966). At higher temperatures the field of pyroxene solid solution is truncated by the hematite + liquid and hematite + pyroxene$_{ss}$ + liquid fields, and the system is thus binary only for diopside-rich compositions, $Di_{100}Ac_0$–$Di_{40}Ac_{60}$ (see Fig. 230, p. 499).

Nolan (1966) has also studied the phase equilibrium within the pyroxene compositional planes contained in the $CaMgSi_2O_6$–$NaFe^{3+}Si_2O_6$–$NaAlSi_3O_8$–$NaAlSiO_4$–H_2O system at 1 kbar water vapour pressure. The introduction of small amounts of diopside composition into the $CaMgSi_2O_6$-free system has a marked effect on the pyroxene phase volume. Thus successive increments of $CaMgSi_2O_6$ in the pyroxene solid solution lead to a shift in the position of the boundary curve separating the phase volumes of pyroxene and feldspar towards the albite–nepheline join (see Fig. 233, p. 501).

The solid solution of the pyroxene ureyite ($NaCrSi_2O_6$) in diopside and the phase equilibrium along the join $CaMgSi_2O_6$–$NaCrSi_2O_6$ at 1 atm has been investigated by Ikeda and Yagi (1972) and at 20 kbar by Vredevoogd and Forbes (1975); details are given in the section on ureyite.

The presence of a single phase area of diopside solid solution in the $CaMgSi_2O_6$–$CaCrCrSiO_6$ system has been demonstrated by Ikeda and Yagi (1977). The maximum solubility of $CaCrCrSiO_6$ (Cr-Tschermak's component) in diopside is 6·7 weight per cent at 940°C; a diopside$_{ss}$ + uvarovite$_{ss}$ subsolidus field is present between 6·7 and 55 per cent Cr–Ts, and with greater amounts of this component the subsolidus assemblage consists of uvarovite$_{ss}$ and eskolaite. The diopside$_{ss}$ crystallized in the diopside-rich portion of the system, from Di_{100} to Di_{70}Cr–Ts_{30}, is blue in colour (see p. 256).

The phase equilibrium relationships in the diopside–forsterite system at 1 atm (Kushiro and Schairer, 1963) and at 20 kbar (Kushiro, 1964) are shown in Fig. 123(a). At 1 atm the piercing point (A) exists at $Di_{89}Fo_{11}$ and a temperature of 1385°C; at 20 kbar the composition of the piercing point (B) is $Di_{77}Fo_{23}$ at a temperature of 1635°C. The addition of CO_2 to the diopside–forsterite system at 30 kbar results in a marked change of the phase relationships (Fig. 123(b)) due to the appearance of enstatite at liquidus and subsolidus temperatures (Eggler, 1974, 1975). The solidus assemblages of such a peridotitic composition containing small amounts of CO_2 consist of diopside, enstatite, forsterite, dolomite and liquid. Eggler (1976) has also investigated two other joins, Ca_2SiO_4–Mg_2SiO_4–CO_2 and $CaMgSi_2O_6$–$CaMgO_2$–CO_2, in the CaO–MgO–SiO_2–CO_2 system.

A number of investigations (e.g. Schairer and Yoder, 1962) of the system diopside–forsterite–silica at 1 atm have been made since it was first studied by Bowen (1914). The most recent, by Kushiro (1972a, b), shows that the location of the liquidus boundaries between the pyroxenes and the silica minerals is unchanged, but those between forsterite solid solution and the pyroxenes are located somewhat more to the silica side of the system (Fig. 124) than shown in the earlier phase diagrams. The joins diopside–forsterite and diopside–silica are neither binary nor

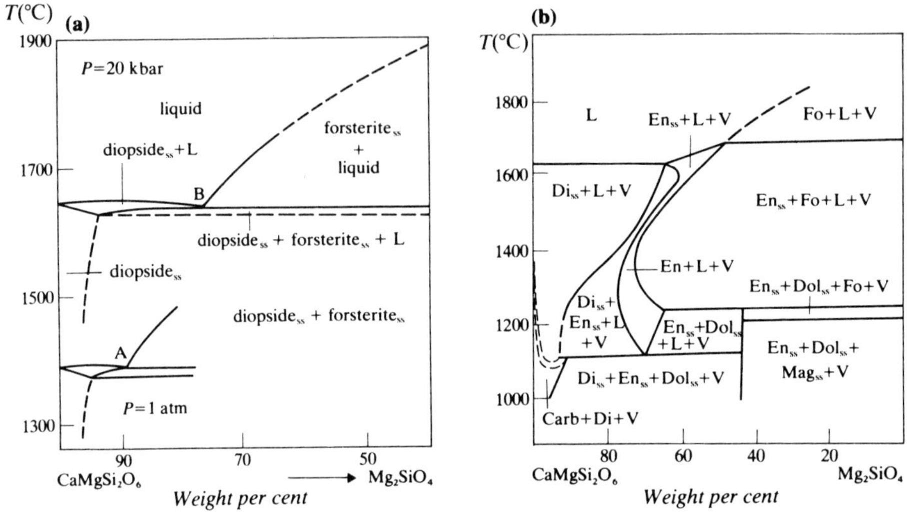

Fig. 123. (a) Equilibrium diagram of part of the system diopside–forsterite at 1 atm and 20 kbar (after Kushiro, 1964). (b) Phase relations on the join $CaMgSi_2O_6$–Mg_2SiO_4 in the presence of excess CO_2 (greater than ≃ 20 weight per cent) (after Eggler, 1975).

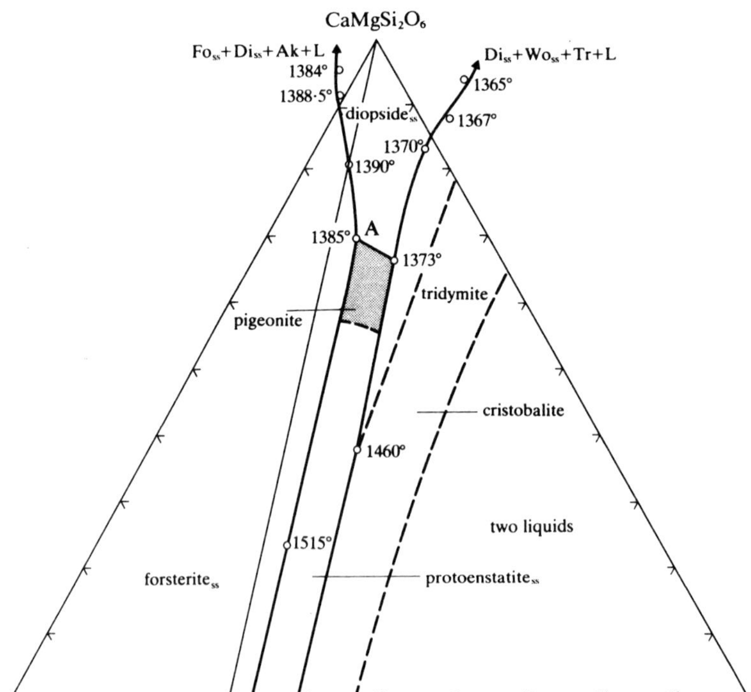

Fig. 124. Liquidus boundaries in the system $CaMgSi_2O_6$–Mg_2SiO_4–SiO_2 at 1 atm (after Kushiro, 1972b). Ak åkermanite, Wo$_{ss}$ wollastonite solid solution, Tr tridymite, A invariant point Fo$_{ss}$ + Di$_{ss}$ + Pig + Liquid.

thermal barriers. Liquids formed on the first join, at temperatures (1388·5° and 1384°C) a little above the solidus, and coexisting with diopside and forsterite solid solutions, have compositions outside the system, and contain an åkermanite component. On the diopside–silica join, liquids formed just above the solidus (1367° and 1365°C) likewise lie outside the system and have a wollastonite component. At the invariant point A, forsterite solid solution + $diopside_{ss}$ + pigeonite + liquid are in equilibrium, and forsterite reacts with liquid:

$$\text{forsterite} + \text{liquid} \rightleftharpoons \text{diopside}_{ss} + \text{pigeonite}$$

The compositions of the diopside solid solutions, and of the coexisting liquids, do not lie on the diopside–enstatite join, those of the solid phase lying between the joins diopside–enstatite and diopside–forsterite, and those of the liquid to the silica side of the pyroxene join. The range in composition of $diopside_{ss}$ at 1390° and approximately 1350°C is shown in Fig. 125(a).

The system diopside–forsterite–silica at 20 kbar under hydrous conditions (water pressure equal to load pressure) is shown in Fig. 69, p. 103, in which the boundaries between primary phase volumes are projected from the H_2O apex on to the diopside–forsterite–quartz plane in the diopside–forsterite–quartz–H_2O system (Kushiro, 1969b). Compared with the relationships at 1 atm the primary phase

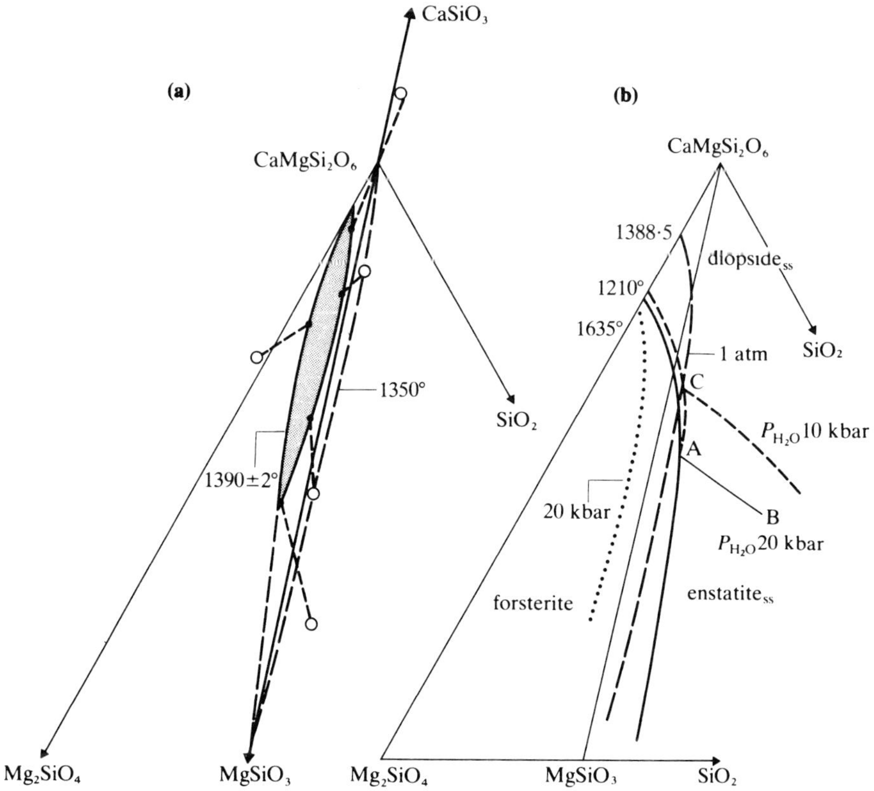

Fig. 125. **(a) Diopside solid solution at 1390° ± 2° (stippled area) and approximately 1350°C (area enclosed by long dashed lines). The lines between coexisting diopside solid solution (solid circles) and liquids (open circles) at 1390° ± 2°C (after Kushiro, 1972b). (b) Pyroxene–forsterite boundary at 1 bar, 20 kbar anhydrous conditions and 20 kbar P_{H_2O} (after Kushiro, 1969b) A–B liquidus boundary between diopside and diopside–enstatite solid solutions.**

fields at 20 kbar of both the diopside and enstatite solid solutions are much enlarged in the direction of the SiO_2 apex, due to the greater depressive effect of high water pressure on the melting temperature of quartz. In consequence the liquidus boundary between the pyroxenes and forsterite is located more towards SiO_2-rich compositions than at 1 atm, in contrast to the position of the boundary, closer to the $CaMgSi_2O_6$–Mg_2SiO_4 join at 20 kbar under anhydrous conditions (Fig. 125(b)).

Diopside is one of the ternary components in the system CaO–MgO–SiO_2 (Ferguson and Merwin, 1919). The equilibrium diagram (Fig. 126) is based on the

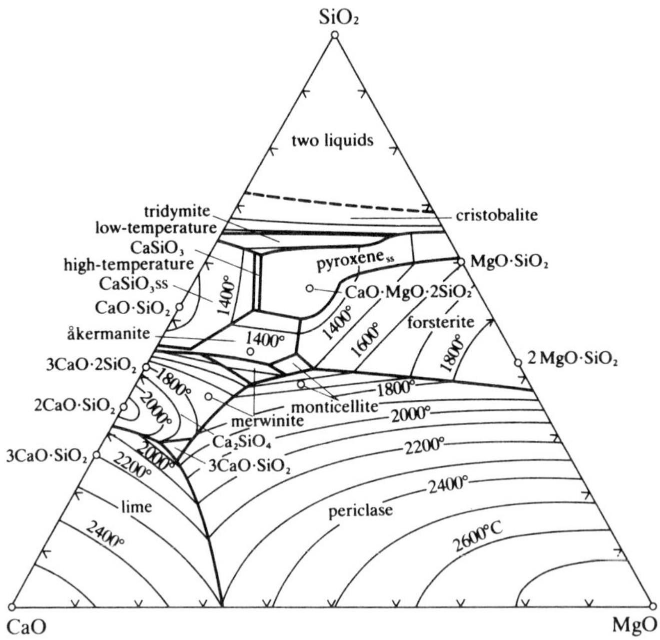

Fig. 126. Phase diagram of the system CaO–MgO–SiO_2 (after Ricker and Osborn, 1954).

earlier work of Ferguson and Merwin, together with additional data by Schairer and Bowen (1942), Schairer and Osborn (1950) and Ricker and Osborn (1954). The join diopside–iron oxide–silica, and its relationship to the join diopside–forsterite–silica, has been examined by Nafziger (1970), and the phase relationship determined at oxygen fugacities defined by a constant room temperature and a CO_2/H_2 ratio of 50 at 1 atm pressure.

The vapour-saturated liquidus relations for the join $CaMgSi_2O_6$–Mg_2SiO_4–SiO_2–H_2O have been determined by Warner (1973). The univariant equilibrium involving Di_{ss}, En_{ss}, Fo_{ss}, L and vapour at $1230° \pm 20°C$ (point C in Fig. 125(b)) is significantly more $CaMgSi_2O_6$-rich than at 20 kbar water pressure, and shows that there is a general trend away from diopside composition with increasing P_{H_2O}. The phase reaction at C is deduced to be:

$$\text{diopside}_{ss} + \text{enstatite}_{ss} + \text{vapour} \rightleftharpoons \text{forsterite}_{ss} + \text{liquid}$$

The diopside_{ss} contains about 20 weight per cent dissolved $MgSiO_3$, and the enstatite_{ss} approximately 8 weight per cent $CaMgSi_2O_6$. Compared with the anhydrous system the join $CaMgSi_2O_6$–$MgSiO_3$–H_2O is not a thermal barrier to fractional crystallization of silica-undersaturated and vapour-undersaturated liquid, and quartz normative liquids may thus be derived from initially silica-undersaturated compositions either by fractional crystallization or by partial melting.

In the system $CaMgSi_2O_6$–Mg_2SiO_4–$CaAl_2Si_2O_8$ (Osborn and Tait, 1952) the sequence of equilibrium crystallization is simple, and the liquid changes in composition with the crystallization of the appropriate phases until the ternary eutectic is reached at 1270°C when diopside, forsterite and anorthite crystallize together until the liquid is exhausted. The diopside in equilibrium with anorthite, forsterite and liquid contains some aluminium and the ternary eutectic is not strictly a ternary invariant point. The composition of the pseudoternary eutectic is diopside 49, forsterite 7·5 and anorthite 43·5 weight per cent.

The join diopside–albite was first studied by Bowen (1915), and later by Schairer and Yoder (1960) as part of their investigation of the system diopside–nepheline–silica. The latter authors found that the liquidus minimum, between diopside solid solution and albite-rich plagioclase occurs at $Di_{10}Ac_{90}$ weight per cent. A more recent examination of the system at 1150°, 1250° and 1350°C, in the pressure range 13 to 32 kbar has been made by Kushiro (1965). Diopside solid solutions containing a jadeite component, are formed by reactions between the pyroxene and albite:

$$CaMgSi_2O_6 + nNaAlSi_3O_8 = CaMgSi_2O_6.nNaAlSi_2O_6 + nSiO_2$$

and at relatively high pressures the pyroxene is omphacitic in composition (see omphacite section).

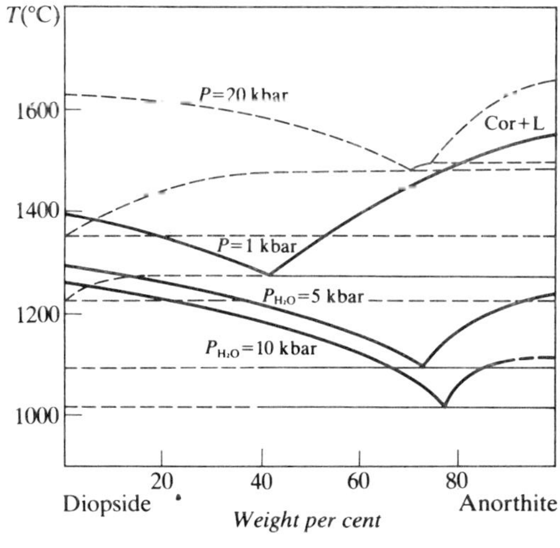

Fig. 127. Diopside–anorthite system at 1 bar, 20 kbar, P_{H_2O} 5 bar and P_{H_2O} 10 kbar (after Yoder, 1965). Crystalline assemblage at all pressures probably includes a form of SiO_2 in addition to diopside solid solution and anorthite. Cor corundum.

Liquidus temperatures for the diopside–anorthite system at various pressures, with and without water (Fig. 127), have been presented by Yoder (1965). The effect of pressure, as well as water under pressure, results in large changes in the initial coprecipitation point of the diopside solid solution and anorthite. At 1 atm, diopside contains about 3 weight per cent of the $CaAl_2SiO_6$ component, the amount increasing with increasing pressure (Kushiro, 1965) due to the reaction:

$$CaMgSi_2O_6 + nCaAl_2Si_2O_8 = CaMgSi_2O_6.nCaAl_2SiO_6 + nSiO_2$$

and the diopside solid solutions become fassaitic in composition (p. 407).

Six compositions on the join diopside–anorthite, $Di_{80}An_{20}$, $Di_{70}An_{30}$, $Di_{60}An_{40}$, $Di_{50}An_{50}$, $Di_{40}An_{60}$ and $Di_{35}An_{65}$ (weight per cent), at temperatures of 1150° and 1350°C and pressures between 8 and 37 kbar, have been studied by Kushiro (1969c). The extent of the solid solution of Ca-Tschermak's molecule in diopside varies with the composition of the initial material and with pressure and temperature (Fig. 192, p. 408). The Di_{ss} + quartz field, in relation to the Di_{ss} + anorthite + quartz field, expands with increasing pressure at both 1150° and 1350°C and reaches $Di_{60}An_{40}$ and $Di_{40}An_{60}$ at the lower and higher temperature respectively. Above 18 kbar at 1150°C the Di_{ss} + anorthite + quartz assemblage is replaced by the garnet-bearing assemblages, Di_{ss} + Gr_{ss} + quartz, and Di_{ss} + Gr_{ss} + anorthite + quartz, probably due to the reaction:

$$CaMgSi_2O_6 + CaAl_2Si_2O_8 = Ca_2MgAl_2Si_3O_{12} + SiO_2$$

although it is likely that the diopside solid solution contains some excess $MgSiO_3$ over diopside, and that the garnet is more grossular-rich than is indicated above.

The relationships in the system diopside–albite–anorthite have been calculated by Barron (1972a, b) and the graphically constructed liquidus surfaces and fractionation paths are comparable with those originally presented by Bowen (1928).

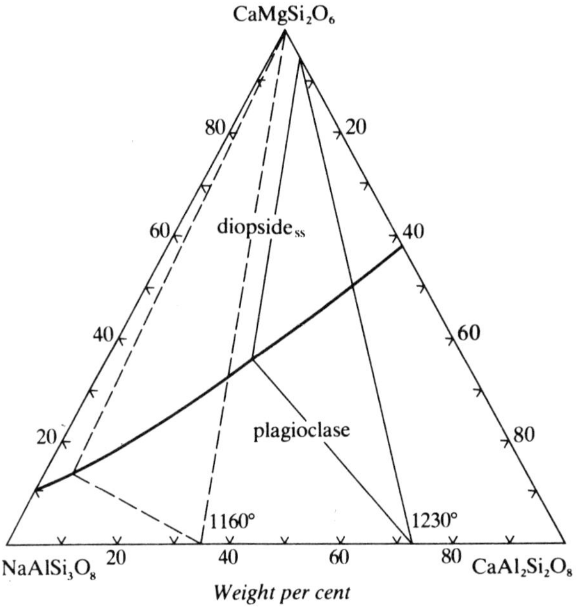

Fig. 128. Liquidus boundary of the system diopside–albite–anorthite at 1 atm (after Kushiro, 1973b). Diopside solid solutions are projected on to the diopside–anorthite join from SiO_2. Three phase triangles at 1230° and 1160°C show compositions of plagioclase at different temperatures.

Kushiro (1973b) has re-examined the diopside–albite–anorthite system at 1 atm (Fig. 128). The Al_2O_3 content of the diopside solid solution crystallizing along the pyroxene–plagioclase liquidus boundary varies from 2·4 (= 5·1 per cent $CaAl_2SiO_6$) at higher, to 0·37 weight per cent (= 0·8 $CaAl_2SiO_6$) at lower temperatures. The diopside solid solutions also contain an excess of Mg relative to Ca, thus for example the composition of a pyroxene with 2·41 per cent Al_2O_3 is $Na_{0 \cdot 02}Ca_{0 \cdot 89}Mg_{0 \cdot 94}Al_{0 \cdot 10}Si_{2 \cdot 00}O_6$. The effects of the changes in slope occurring on the liquidus and solidus paths in the system diopside–albite–anorthite, in relation to the relative amounts of ultrabasic, basic, intermediate and acid rocks, have been discussed by Wyllie (1963).

Earlier, Kushiro (1969c) had studied the reaction of diopside and plagioclase, using the compositions $Di_{70}An_{15}Ab_{15}$, $Di_{50}An_{25}Ab_{25}$ and $Di_{30}An_{35}Ab_{35}$ at temperatures of 1150° and 1350°C and pressures between 15 and 31 kbar. The diopside solid solution is presumed to contain both the Ca-Tschermak's and jadeite component. The phase assemblage and general disposition of the field boundaries at 1150°C are similar to those in the diopside–anorthite system, except that the $diopside_{ss}$–plagioclase–quartz assemblage persists to higher pressures due to the presence of the albite component in the plagioclase (isothermal and isobaric projections at 15, 20, 28 and 35 kbar at 1150°C are given in Fig. 207 (p. 443). General equations for the isotherms and equilibrium crystallization path in the diopside–albite–anorthite system, calculated from Bowen's (1915) data, have been presented by Mueller (1964).

The system diopside–nepheline was first studied by Bowen (1922) and later investigated by Schairer *et al.* (1962). The system is not binary and includes stability fields for $diopside_{ss}$ + liquid, $diopside_{ss}$ + olivine + liquid, $diopside_{ss}$ + olivine + melilite + liquid, $diopside_{ss}$ + olivine + melilite + $nepheline_{ss}$ + L, the successive assemblages appearing as the temperature decreases. The composition of the residual liquids in the diopside–nepheline system, and their significance in relation to the genesis of olivine melilitite, melilite nephelinite, melilitite and related rocks, possibly by reactions of the type:

3 diopside + 2 nepheline = $(\text{åkermanite} + \text{soda melilite})_{ss}$ + forsterite + albite

have been discussed by Yoder and Kushiro (1972).

The phase relations for the $NaAlSi_3O_8$-rich part of the system diopside–nepheline–albite–H_2O at 1 kbar water vapour pressure, the compositions of which approximate to rocks of nepheline syenite affinities, have been given by Edgar (1964). The phase equilibrium diagram for the join diopside–nepheline–anorthite at 1 atm is given by Schairer *et al.* (1968) and the joins diopside–nepheline–sanidine and diopside–albite–leucite, as part of the diopside–nepheline–kalsilite–silica system, have been examined by Sood *et al.* (1970). A later investigation of the system diopside–nepheline–sanidine has been presented by Platt and Edgar (1972), who also discuss the phase relationships in the system with reference to the genesis of melilite and olivine-bearing alkaline rocks. The phase relations in the pseudobinary system $(Di_{38}Ak_3Ne_{59})_{100-x}$–$Lc_x$ at temperatures between 1075° and 1175°C have been determined by Gupta, Venkateswara *et al.* (1973).

The join diopside–FeO, at a total pressure of 1 atm and an oxygen fugacity of $10^{-0.68}$, has been studied by Presnall (1966), and the same author has determined the liquidus surface of the join diopside–forsterite–FeO at oxygen fugacities varying between $10^{-0.68}$ and 10^{-6} bars. Under these conditions the phase field of diopside is small, and the minimum liquidus temperature remains nearly constant at 1300°C at the composition $Di_{79}FeO_{21}$ (weight per cent).

A study of the diopside–kyanite join at high pressures and temperature has been made by Green (1969). The composition 2 diopside + enstatite + 2 anorthite (molecular proportions), equivalent to 3 diopside + 2 kyanite, crystallized at 1200°C and 18 kbar, reacts to form $garnet_{ss}$ + $diopside_{ss}$ + quartz, while at 27 kbar kyanite is also produced.

The diopside–åkermanite system has been studied by Ferguson and Merwin (1919) and by Schairer *et al.* (1967), diopside–åkermanite–anorthite by Wys and Forster (1958, see also Wys, 1972); diopside–åkermanite–nepheline by Onuma and Yagi (1967), diopside–åkermanite leucite by Gupta (1972) and diopside–nepheline–leucite by Gupta and Lidiak (1973). The pressure–temperature relation of åkermanite–CO_2, using the diopside + calcite (1 : 1 mole) composition has been presented by Yoder (1975). Investigations of other systems in which diopside and a feldspathoid are components include diopside–leucite–SiO_2 (Schairer and Bowen, 1938) and diopside–nepheline–SiO_2 (Schairer and Yoder, 1960) in both of which diopside has a large field of crystallization extending respectively towards the leucite–silica and nepheline–silica joins.

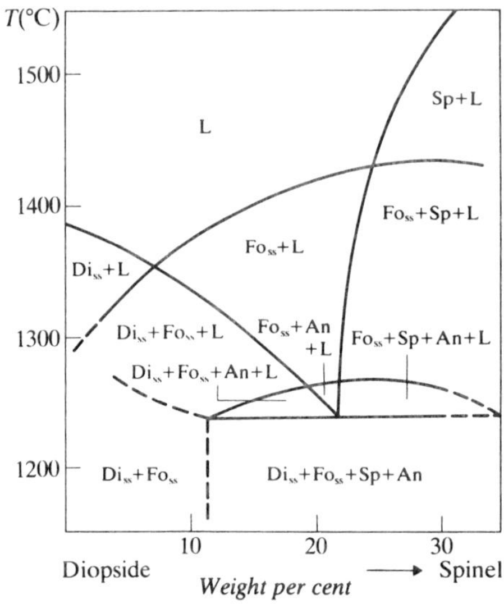

Fig. 129. Phase equilibrium for part of the $CaMgSi_2O_6$–$MgAl_2O_4$ join (after Schairer and Yoder, 1970). Di$_{ss}$, diopside solid solution; Fo, forsterite solid solution with monticellite; An, anorthite; Sp, spinel; L, liquid.

The phase relations at 1 atm along the join diopside–spinel were first investigated by Schairer and Kushiro (1965) and later revised (Fig. 129) by Schairer and Yoder (1970). Diopside–spinel compositions are isochemical with the assemblages anorthite + forsterite + melilite, and forsterite + garnet:

$$\underset{\text{diopside}}{7CaMgSi_2O_6} + \underset{\text{spinel}}{4MgAl_2O_4} = \underset{\text{anorthite}}{3CaAl_2Si_2O_8} + \underset{\text{forsterite}}{5Mg_2SiO_4} + \underset{\text{melilite}}{Ca_2MgSi_2O_7.Ca_2Al_2SiO_7},$$

$$\underset{\text{diopside}}{2CaMgSi_2O_6} + \underset{\text{spinel}}{MgAl_2O_4} = \underset{\text{garnet}}{Ca_2MgAl_2Si_3O_{12}} + \underset{\text{forsterite}}{Mg_2SiO_4}$$

and the diopside solid solution + forsterite assemblage can be illustrated by the equation:

$$\underset{\text{diopside}}{xCaMgSi_2O_6} + \underset{\text{spinel}}{MgAl_2O_4} = \underset{\text{diopside}_{ss}}{(x-1)CaMgSi_2O_6.CaAl_2SiO_6} + \underset{\text{forsterite}}{Mg_2SiO_4}$$

Schairer and Yoder found that melilite is not a stable phase in some of the subsolidus assemblages, but the conversion of the assemblage forsterite + anorthite + melilite to diopside + spinel at temperatures between 1200° and 1250°C has been reported by O'Hara and Biggar (1969).

Investigations relating to the phase relationships in the garnet peridotites include the diopside–pyrope join at 40 kbar (Davis, 1964) and the diopside–forsterite–pyrope system at 1 atm and 40 kbar by Davis and Schairer (1965). The pressure–

temperature relationships of the aluminous diopside–forsterite–spinel, and aluminous diopside–forsterite–garnet–spinel assemblages have been determined by MacGregor (1965a, b). Within the aluminous diopside + spinel + forsterite field the solid solution of Al_2O_3 in the pyroxene results in reactions of the types:

$$2MgSiO_2 + MgAl_2O_4 \rightleftharpoons MgAl_2SiO_6 + Mg_2SiO_4 \quad (1)$$

$$CaMgSi_2O_6 + MgAl_2O_4 \rightleftharpoons CaAl_2SiO_6 + Mg_2SiO_4 \quad (2)$$

Within the forsterite–pyroxene–spinel stability field the amount of solid solution of Al_2O_3 in diopside decreases with increasing pressure at constant temperature (Fig. 130).

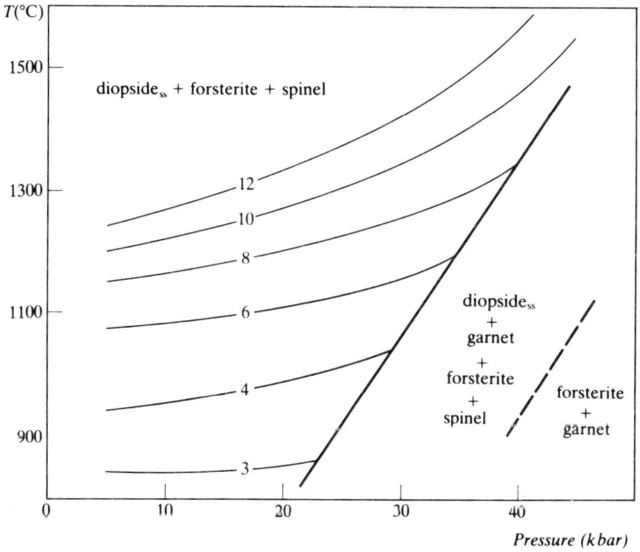

Fig. 130. Temperature–pressure boundary between assemblages diopside solid solution + forsterite + spinel, and diopside solid solution + forsterite + spinel + garnet. Numbered curves weight per cent Al_2O_3 isopleths for diopside solid solution (after MacGregor, 1965a).

The phase relations in the system $CaSiO_3$–$MgSiO_3$–Al_2O_3 have been determined at a temperature of 1200°C and pressure of 30 kbar (Boyd, 1970). The system contains four unvariant assemblages (three-phase fields). Two predominantly above the garnet join, namely diopside$_{ss}$–Mg-garnet–corundum$_{ss}$, and diopside$_{ss}$–Ca-garnet–wollastonite$_{ss}$, and diopside$_{ss}$–orthopyroxene$_{ss}$–Mg-garnet and diopside–Ca-garnet–wollastonite below the join (Fig. 131). The solid solution of $CaAl_2SiO_6$ in $CaMgSi_2O_6$ extends to 56 weight per cent $CaAl_2SiO_6$ and interrupts the solid solution, in the interval 23–51 weight per cent pyrope, between the Ca- and Mg-garnet. Pyroxenes along the $CaMgSi_2O_6$–$CaAl_2SiO_6$ join also dissolve $MgSiO_3$, the solubility of which decreases as the alumina content of the diopside$_{ss}$ increases.

The reaction $Di_{50}Hd_{50}$–anorthite → garnet + quartz at temperatures and pressure in the range 1000°–1500°C and 10–40 kbar respectively has been investigated by Akella and Kennedy (1971). The lower pressure assemblage, pyroxene–anorthite, reacts at pressures between 12 and 20 kbar and temperatures of about 1000°–1300°C to form pyroxene + anorthite + garnet + quartz, and at pressure > 20 kbar to pyroxene + garnet + quartz (Fig. 132).

The reactions involving the crystallization of aluminium-rich clinopyroxenes for

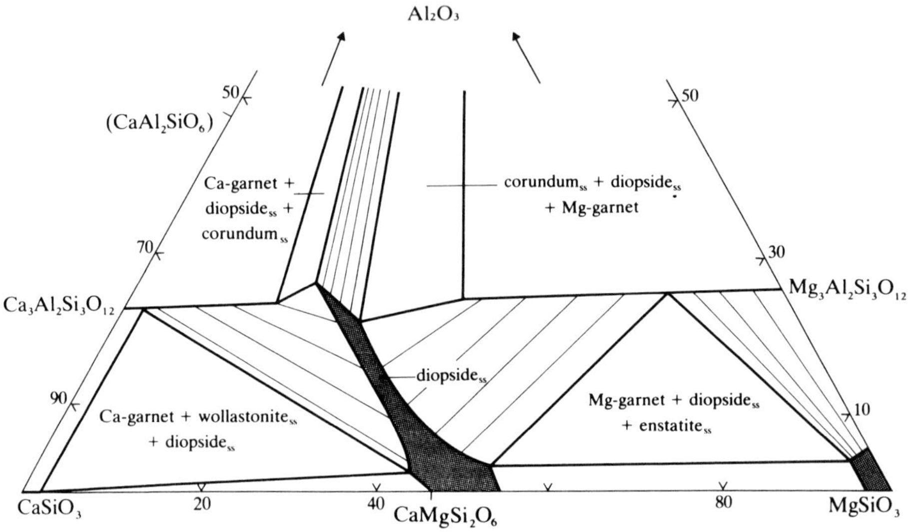

Fig. 131. Phase relations in the system $CaSiO_3$–$MgSiO_3$–Al_2O_3 (weight per cent) at 1200°C and 30 kbar (after Boyd, 1970).

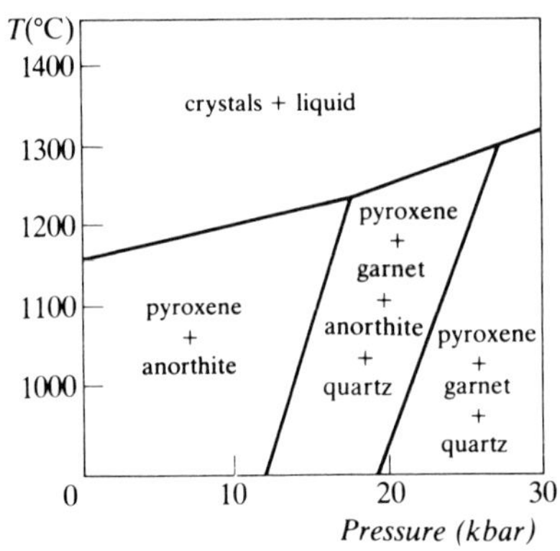

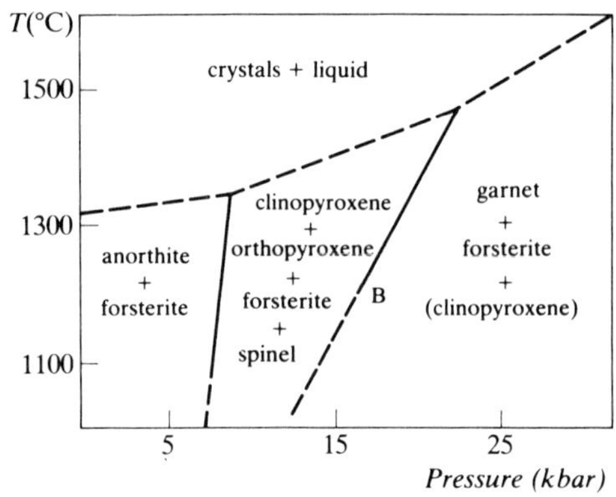

Fig. 132. Pressure–temperature plane for the reaction (diopside$_{50}$hedenbergite$_{50}$)–anorthite ⇌ pyroxene + garnet + quartz (after Akella and Kennedy, 1971).

Fig. 133. Pressure–temperature plane for the 1:2 composition (molecular ratio) anorthite:forsterite (after Kushiro and Yoder, 1966).

the compositions 1:1 (molecular ratio) anorthite–forsterite and 1:2 anorthite–enstatite at pressures between 7 and 28 kbar (Kushiro and Yoder, 1966) have been discussed in the section on orthopyroxene (see Fig. 65, p. 99). The stable assemblage at intermediate pressure for the composition 1:2 anorthite–forsterite (Fig. 133), can be derived from the reaction:

$$\underset{\text{anorthite}}{2CaAl_2Si_2O_6} + \underset{\text{forsterite}}{4Mg_2SiO_4} = \underset{\text{diopside solid solution}}{(2-x)CaMgSi_2O_6.xCaAl_2SiO_6} +$$

$$\underset{\text{aluminous enstatite}}{(4-2x)MgSiO_3.xMgAl_2SiO_6} + \underset{\text{spinel}}{(2-2x)MgAl_2O_4} + \underset{\text{forsterite}}{2Mg_2SiO_4}$$

where $x \equiv 0{\cdot}46$.

In the high-pressure field both pyroxenes are unstable and the reaction at curve B can be expressed:

$$\underset{\text{diopside solid solution}}{(2-x)CaMgSi_2O_6.xCaAl_2SiO_6}+\underset{\text{aluminous enstatite}}{(4-2x)MgSiO_3.xMgAl_2SiO_6}+$$

$$\underset{\text{spinel}}{(2-2x)MgAl_2O_4}=\underset{\text{garnet}}{2CaMg_2Al_2Si_3O_{12}}+\underset{\text{forsterite}}{(2-2x)Mg_2SiO_4}$$

The boundary curve between the intermediate–high-pressure assemblage at 1400°C is at 21 kbar compared with 17 kbar for the 1 : 1 anorthite–forsterite composition.

Mysen and Boettcher (1975) have investigated the system peridotite–H_2O–CO_2. Using four mantle peridotites in the presence of H_2O- and (H_2O+CO_2)-bearing vapours at controlled hydrogen fugacities, and pressures and temperatures between 7·5–25 kbar and 820°–1120°C respectively, they have shown that the Mg/(Mg+ ΣFe) ratios of the pyroxenes ($Wo_{39-48}En_{46-57}Fs_{3-10}$) increase, while the $[Al]^6$/ $[Al]^4$ ratios, the overall contents of Al_2O_3 and the Na/(Na+Ca) ratios decrease with increasing temperature; chromium appears to be pressure independent.

Using mixtures of 20:50:30 diopside, $Ca_{45\cdot1}Mg_{50\cdot7}Fe_{4\cdot2}$, enstatite, $Mg_{89\cdot4}Fe_{9\cdot2}Ca_{1\cdot4}$, and spinel (90 weight per cent $Mg_{0\cdot8}Fe_{0\cdot2}Al_2O_4$) from the symplectites of the Horoman layered peridotite intrusion, Hokkaido, Japan, Tazaki *et al.* (1972) determined the univariant curve, in the pressure and temperature ranges 7·5–40 kbar and 850°–1400°C respectively, for the reaction Ca pyroxenes + (Mg,Fe) pyroxenes + spinel = garnet + pyroxenes + (olivine) as:

$$T(°C) = 44{\cdot}4P(\text{kbars}) + 534$$

in good agreement with curve B for the composition 1 : 2 anorthite–forsterite (Fig. 133).

Diopside solid solutions have been crystallized during an investigation of the melting interval and phase equilibria, at pressures up to 40 kbar, of a garnet peridotite nodule from the kimberlite pipe, Dutoitspan Mine, South Africa (Ito and Kennedy, 1967). The assemblage diopside solid solution, orthopyroxene solid solution, olivine and chrome spinel is stable to 23 kbar. At higher pressures the stable assemblage consists of diopside and orthopyroxene solid solutions, olivine and garnet. A comparable study of the phase assemblages to crystallize from an anhydrous tholeiite basalt glass, in the pressure range 1 atm to 40 kbar, has been reported by Cohen *et al.* (1967). High pressure studies of a garnet-lherzolite inclusion from the Bultfontein kimberlite, and a spinel-lherzolite inclusion in the tuff of Salt Lake Crater, Hawaii, as well as synthetic peridotites (see Fig. 66, p. 101) have been described by Kushiro (1972c).

The effect of the degree of melting at 20 and 35 kbar on the coexisting phases in two peridotites has been studied by Mysen and Kushiro (1976). The clinopyroxene crystallizing in the melting interval 1350°–1375°C in a peridotite nodule (higher Al_2O_3, CaO and Na_2O) from a nephelinite tuff is diopsidic in composition, whereas that in the melting interval 1450°–1500°C in a peridotite nodule (higher MgO) from kimberlite is a pigeonite, consistent with the lower stability limit of iron-free pigeonite on the diopside–enstatite join at 1450°C, 20 kbar (see Fig. 45, p. 85). In both peridotite compositions the pyroxene stability interval is greater at the higher pressure (see also Arndt, 1976).

Investigations of the subsolidus fields of crystallization of synthetic high-alumina basalt, anorthite-enriched high-alumina basalts, kyanite eclogite, grosspydite and gabbroic anorthosite at pressures up to 30 kbar (Green, 1967) show that, in the pressure range 5–12 kbar, clinopyroxene increases in amount relative to plagioclase, and becomes more aluminous due to reactions of the type:

$$\underset{\text{diopside}}{x\text{Ca(Mg,Fe)Si}_2\text{O}_6} + \underset{\text{anorthite}}{\text{CaAl}_2\text{Si}_2\text{O}_8} \rightleftharpoons \underset{\text{clinopyroxene}_{\text{ss}}}{x\text{Ca(Mg,Fe)Si}_2\text{O}_6.\text{CaAl}_2\text{SiO}_6} + \text{SiO}_2$$

At pressure above 12–14 kbar the amount of clinopyroxene decreases as the result of reactions leading to the formation of garnet:

$$\underset{\text{clinopyroxene}_{\text{ss}}}{x\text{Ca(Mg,Fe)Si}_2\text{O}_6.\text{CaAl}_2\text{SiO}_6} + \underset{\text{orthopyroxene}}{2\text{(Mg,Fe)SiO}_3} \rightleftharpoons$$

$$\underset{\text{garnet}_{\text{ss}}}{\text{Ca(Mg,Fe)}_2\text{Al}_2\text{Si}_3\text{O}_{12}} + \underset{\text{diopside}}{x\text{Ca(Mg,Fe)Si}_2\text{O}_6}$$

The albite component of the plagioclase persists to some 22 kbar, at which pressure its incorporation as the jadeitic component of the clinopyroxene solid solution begins (see omphacite section, p.441). At higher pressures the jadeite content of the clinopyroxene increases, and is associated with decreasing solubility of $CaAl_2SiO_6$ in the pyroxene solid solution, and a greater grossular-enrichment in the garnet.

Other experimental studies of the pressure–temperature relationship of natural rock compositions, in which clinopyroxene solid solutions are important liquidus and subsolidus phases, include Thompson's (1972) investigation of an olivine tholeiite and an andesine-normative lava, both relatively iron-rich, from the Snake River Plains, Idaho.

The relationships along the join $CaMgSi_2O_6$–$CaTiAl_2O_6$ at 1 atm pressure (Yagi and Onuma, 1967) are comparable with those along the join $CaMgSi_2O_6$–$CaAl_2SiO_6$ at the same pressure. A series of binary solid solutions occur between diopside and Di_{89}–$(CaTiAl_2O_6)_{11}$, the latter composition representing the maximum solubility of $CaTiAl_2O_6$ ($\simeq$ 4 weight per cent TiO_2) in diopside. Compositions richer in the $CaTiAl_2O_6$ component crystallize to diopside solid solution and perovskite, $CaTiO_3$. The phase equilibrium diagrams for $(CaMgSi_2O_6)_{90,80,70}$·$(Ca_2MgSi_2O_7)_{10,20,30}$–$CaTiAl_2O_6$ have been determined by Onuma and Yagi (1971) who discuss the diopside$_{ss}$ in this join in relation to natural titanopyroxenes, particularly of melilite-bearing alkalic rocks. Thus the solubility of $CaTiAl_2O_6$ in diopside decreases with pressure, and at still higher pressures the metastable composition $Di_{47\cdot5}(CaTiAl_2O_6)_{52\cdot5}$ crystallizes as garnet, perovskite and corundum.

$$\underset{\text{pyroxene}_{\text{ss}}}{3\text{CaMgSi}_2\text{O}_6.3\text{CaTiAl}_2\text{O}_6} \rightarrow \underset{\text{garnet}}{2\text{(Ca,Mg)}_3\text{Al}_2\text{Si}_3\text{O}_{12}} + \underset{\text{perovskite}}{3\text{CaTiO}_3} + \underset{\text{corundum}}{\text{Al}_2\text{O}_3}$$

Crystallization of Ti- and Al-rich diopsides at liquidus temperature between 1235° and 1242°C at 1 atm, have been reported by Yang (1975). The primary crystals, containing between 7·5 and 10·3 weight per cent TiO_2 and 16·3 to 19·7 weight per cent Al_2O_3, coexist in equilibrium with diopside$_{ss}$, spinel, perovskite, anorthite and melilite in the subliquidus and subsolidus regions of the system $CaMgSi_2O_6$–$CaAl_2SiO_6$–$CaTiAl_2O_6$.

Gupta *et al.* (1973) have investigated the effect of silica on the solubility of the $CaTiAl_2O_6$ molecule in diopsidic pyroxenes, and have determined the equilibrium phase relations in the system $CaMgSi_2O_6$–$CaTiAl_2O_6$–SiO_2. Increasing silica concentration is accompanied by a decrease in the solubility of both aluminium and titanium in the pyroxene, and is also the main factor controlling the stability of the coexisting titanium-bearing phases, perovskite, sphene and rutile. The exsolution of perovskite from homogeneous titaniferous pyroxene has been demonstrated experimentally by Yagi and Onuma (1969).

The formation of diopside from the reaction of tremolite, calcite and quartz at 1 kbar total pressure in the temperature range 300°–600°C, has been investigated by Metz and Winkler (1964). The temperature at which diopside is formed is dependent on the mole fraction (X_{CO_2}) of CO_2 in the equilibrium composition of the gas phase. With $X_{CO_2} \simeq 0$ the formation of the pyroxene occurs at 350° ± 20°C but the temperature increases rapidly with increasing CO_2 content of the gas phase, and is 500°C at $X_{CO_2} = 0{\cdot}12$. As the mole fraction of CO_2 increases further the equilibrium curve rises less steeply, and at $X_{CO_2} = 0{\cdot}75$ is essentially parallel to the abscissa, at 540° ± 10°C. The effect of low pressures of CO_2 on the crystallization temperature of diopside has also been examined by Kalinin (1967a). Thus in a series of hydrothermal syntheses at 500 atm, diopside was observed to form in the temperature range 350°–380°C and tremolite between 400° and 450°C. Kalinin has also reported the formation of diopside and talc from tremolite composition below 420°–430°C at total pressures of 200 to 600 atm and low partial pressure of CO_2. Such low CO_2 concentrations are, however, unlikely to occur during the progressive metamorphism of siliceous dolomites as H_2O enters the tremolite structure at lower temperatures and thus releases CO_2, and the formation of diopside in such rocks occurs almost exclusively from the reaction of tremolite, calcite and quartz.

A more detailed investigation of the equilibrium conditions of the following diopside-forming metamorphic reactions has been made by Metz (1970).

$$\underset{\text{tremolite}}{Ca_2Mg_5Si_8O_{22}(OH)_2} + \underset{\text{calcite}}{3CaCO_3} + \underset{\text{quartz}}{2SiO_2} \rightleftharpoons \underset{\text{diopside}}{5CaMgSi_2O_6} + 3CO_2 + H_2O \qquad (1)$$

$$\underset{\text{tremolite}}{Ca_2Mg_5Si_8O_{22}(OH)_2} + \underset{\text{calcite}}{3CaCO_3} \rightleftharpoons \underset{\text{diopside}}{4CaMgSi_2O_6} + \underset{\text{dolomite}}{CaMg(CO_3)_2} + CO_2 + H_2O \qquad (2)$$

$$\underset{\text{dolomite}}{CaMg(CO_3)_2} + \underset{\text{quartz}}{2SiO_2} \rightleftharpoons \underset{\text{diopside}}{CaMgSi_2O_6} + 2CO_2 \qquad (3)$$

The equilibrium data for total fluid pressures of 0·5, 1·0, 3·0 and 5 kbar for reaction (1) are shown in the temperature–X_{CO_2} diagram of Fig. 134. The equilibrium temperatures for $X_{CO_2} > 0{\cdot}75$ and fluid pressures of 0·5 and 1 kbar, calculated from equilibrium constants derived from experimental data at smaller values of X_{CO_2} and fugacities of CO_2 and H_2O, show that a decrease in the equilibrium temperature occurs when the X_{CO_2} value is greater than 0·9.

An experimental investigation, using a solid phase buffer technique, of reactions involving diopside in the system CaO–MgO–SiO_2–C–O–H has been presented by Skippen (1971). The reactions have been studied at total pressures of between 0·5 and 3 kbar, and are shown below, together with the equations of the equilibrium curves:

$$\underset{\text{talc}}{Mg_3Si_4O_{10}(OH)_2} + \underset{\text{calcite}}{3CaCO_3} + \underset{\text{quartz}}{2SiO_2} \rightleftharpoons \underset{\text{diopside}}{3CaMgSi_2O_6} + \underset{\text{fluid}}{3CO_2 + H_2O} \qquad (1)$$

$$\log_{10}K = \log f^3_{CO_2} \cdot f_{H_2O} = -12\,930/T + 29{\cdot}16 + [0{\cdot}494(P - 2\,000)]/T$$

where T is in degrees Kelvin, f and P in bars.

$$\underset{\text{tremolite}}{Ca_2Mg_5Si_8O_{22}(OH)_2} \rightleftharpoons \underset{\text{diopside}}{2CaMgSi_2O_6} + \underset{\text{enstatite}}{3MgSiO_3} + \underset{\text{quartz}}{SiO_2} + H_2O \qquad (2)$$

$$\log_{10}K = \log f_{H_2O} = -6\,966/T + 9{\cdot}33 + [0{\cdot}125(P - 2\,000)]/T$$

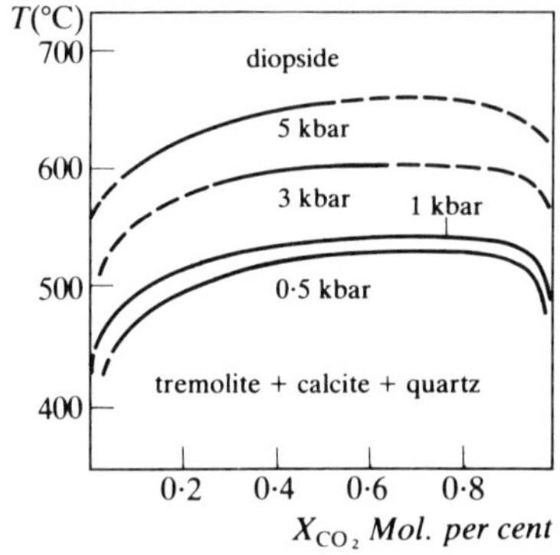

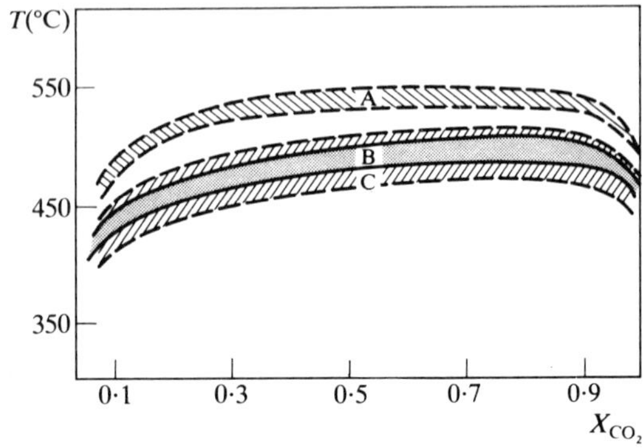

Fig. 134. Temperature–X_{CO_2} relationships at 0·5, 1, 3 and 5 kbar for the reaction tremolite + 3 calcite + 2 quartz = 5 diopside + 3CO_2 + H_2O (after Metz, 1970).

Fig. 135. Temperature–X_{CO_2} relations at 1 kbar for the reaction tremolite + 3 calcite + 2 quartz = 5 diopside + 3CO_2 + H_2O showing the uncertainty limits of the data of (A) Metz[1] (1970), (B) Slaughter *et al.* (1975) and (C) Skippen (1974).

$$\underset{\text{talc}}{4Mg_3Si_4O_{10}(OH)_2} + \underset{\text{dolomite}}{5CaMg(CO_3)_2} \rightleftharpoons \underset{\text{diopside}}{5CaMgSi_2O_6} + \underset{\text{forsterite}}{6Mg_2SiO_4} + \underset{\text{fluid}}{10CO_2 + 4H_2O} \quad (3)$$

$$\log_{10}K = \log f^{10}_{CO_2} \cdot f^{4}_{H_2O} = -75\,400/T + 133{\cdot}7 + [1{\cdot}438(P - 2\,000)]/T$$

In a subsequent investigation, Skippen (1974) showed that diopside occurs as a component in a number of stable assemblages in the system CaO–MgO–SiO_2–CO_2–H_2O, viz. diopside–tremolite–quartz–calcite–dolomite, diopside–tremolite–forsterite–calcite–dolomite, diopside–enstatite–tremolite–quartz–dolomite, and diopside–enstatite–forsterite–tremolite–dolomite. At pressure above approximately 3 kbar the formation of diopside occurs after forsterite in fluids with a high CO_2 content, but at lower pressures diopside will generally develop before forsterite.

During an experimental and thermodynamic study of the equilibria in the system CaO–MgO–SiO_2–H_2O–CO_2, Slaughter *et al.* (1975) examined the same reactions as those investigated by Metz (1970), the equilibrium conditions of which are:

$P_f = 1$ kbar, 494° ± 10°C, X_{CO_2} 0·50 (1)
$P_f = 1$ kbar, 486° ± 15°C, X_{CO_2} 0·90
$P_f = 5$ kbar, 637° ± 5°C, X_{CO_2} 0·85

$P_f = 1$ kbar, 442° ± 15°C, X_{CO_2} 0·95 (2)
$P_f = 2$ kbar, 515° ± 15°C, X_{CO_2} 0·95

$P_f = 5$ kbar, 649° ± 5°C, $X_{CO_2} > 0{\cdot}82$ (3)

These data, and those of other reactions not involving diopside, together with the free energy and activity data for CO_2 and H_2O, are used by Slaughter *et al.* to calculate phase equilibria in the CaO–MgO–SiO_2–H_2O–CO_2 system, the results of which are comparable with Skippen's (1971) experimental data but have lower uncertainties (Fig. 135).

The sequence of low-pressure decarbonation reactions occurring in the metamorphism of siliceous dolomites (Skippen, 1974) has been extrapolated to mantle pressures by Wyllie and Huang (1976). This evaluation of the effect of CO_2 on

[1] Using natural tremolite containing considerable amounts of fluorine. Equilibrium temperatures will be some 10°C lower for pure synthetic (OH)-tremolite.

model mantle peridotite assemblages demonstrates that in the peridotite–CO_2 system, peridotite can be carbonated to yield diopside + orthopyroxene + forsterite + calcic dolomite at higher pressures (lower temperatures) and diopside + orthopyroxene + forsterite + vapour at higher temperatures (lower pressures).

Alteration. The hydrothermal treatment of hedenbergite at 1 kbar and temperatures to 500°C gives little evidence of the occurrence of alteration. In acid solution, in the presence of $4MgCO_3.Mg(OH)_2$, however, alteration to olivine, magnetite, siderite, calcite and serpentine has been reported by Chao (1974). The alteration of hedenbergite and natural and synthetic diopside has been investigated experimentally in solutions at various pH values, in the temperature and pressure ranges 300°–800°C and 200–2000 atm respectively (Chao and Tsao, 1975). The topotaxial alteration by weathering of hedenbergite in a skarn to nontronite has been reported by Eggleton (1975). All the Ca, most of the Mg, and some Si is removed, the Fe is oxidized, permitting the Si–Fe tetrahedral–octahedral chains to move || {010}. The chains coalesce to form a dioctahedral Fe^{3+} talc layer in crystallographic continuity with the pyroxene. Single crystals, ≃ 0·01–0·1 mm, of hedenbergite are transformed to single crystals of nontronite.

Andraditization of hedenbergite is not an uncommon feature in skarns, and the experimental synthesis of andradite from hedenbergite has been described by Kalinin (1967b). Synthetic hedenbergite was heated with $CaCl_2$, $CaCO_3$, CaO, Na_2CO_3 and other mineralizers, at 550°C and a pressure of 1000 atm and gave rise to variable amounts of andradite, in some cases as garnet pseudomorphs, due to the reactions of the type:

$$2CaFeSi_2O_6 + CaCl_2 + 2H_2O + Fe \rightarrow Ca_3Fe_2Si_3O_{12} + FeCl_2 + SiO_2 + 2H_2 \quad (1)$$

$$2CaFeSi_2O_6 + CaCO_3 + H_2O \rightarrow Ca_3Fe_2Si_3O_{12} + SiO_2 + CO_2 + H_2 \quad (2)$$

The alteration of a mangan hedenbergite, to amphibole, stilpnomelane and hisingerite, in a skarn of the Upper mine, Tetyukhe, U.S.S.R., has been described by Mozgova (1959).

Optical and Physical Properties

The refractive indices of synthetic diopside are α 1·664, β 1·6715, γ 1·694 (Allen and White, 1909). The indices depend, however, on the mode and temperature of the synthesis (Table 23), thus diopside crystallized from glass of composition

Table 23 Refractive indices of diopside$_{ss}$, diopside and åkermanite crystallized from $CaMgSi_2O_6$ glass (after Toropov and Khotimchenko, 1967)

Thermal regime		Diopside		Åkermanite	
Temp (°C)	Time (h)	α	γ	ε	ω
800	2	1·661	1·684	1·632	1·636
800	4	1·662	1·687	1·633	1·638
800	22	1·664	1·691	1·632	1·638
800	38	1·664	1·691	1·632	1·638
800	100	1·664	1·692	1·632	1·638
900	15	1·664	1·694	—	—
1350	2	1·664	1·694	—	—

$CaMgSi_2O_6$ at 800°C has α 1·661, and at 1350°C α 1·664 (Toropov and Khotimchenko, 1967). A number of natural diopsides have closely comparable indices, and Merriam and Laudermilk (1936) have described two diopsides containing Al_2O_3 0·12 and 1·28, Fe_2O_3 0·24 and 0·30, FeO 0·00 and 0·93, the refractive indices of which are respectively α 1·663, β 1·671, γ 1·693 and α 1·666, β 1·675, γ 1·695. Closely similar values were also obtained by Juurinen and Hytönen (1952) on an exceptionally pure diopside (Table 14, anal. 1) from Finland. The replacement of Mg by Fe^{2+} in the series diopside–hedenbergite is accompanied by an increase in the refractive indices. The indices of the Herault hedenbergite (Table 19, anal. 3) are α 1·7225, β 1·7300, γ 1·7505, and the indices of synthetic hedenbergite are α 1·730, β 1·754 (Myer and Lindsley, 1969). The α and β indices of the diopside–hedenbergites detailed in Tables 14–19, plotted against the (Fe^{2+} + Fe^{3+} + Mn) ions per mole, are shown in Fig. 136. It is evident from this figure that the indices of many minerals deviate not only from the assumed linear variation between the synthetic end-members but also from the slightly convex upward curves drawn by Hess (1949). Previously published data for the variation of α and γ

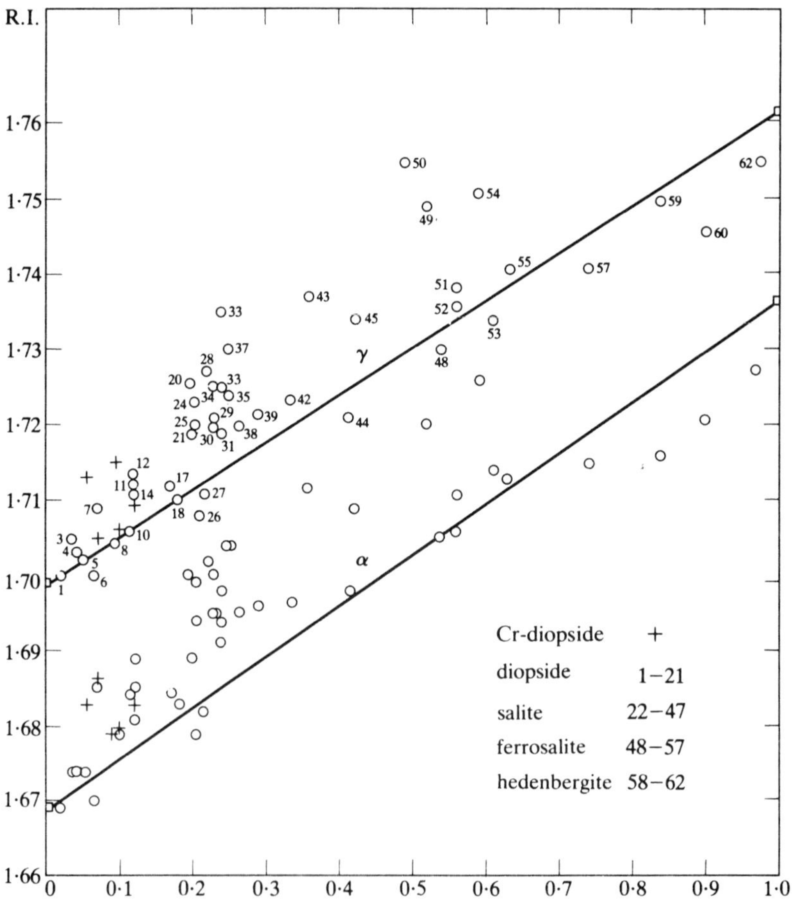

Fig. 136. The relationship between the refractive indices and composition in the diopside–hedenbergite minerals.

refractive indices for the diopside–hedenbergite series have been recalculated in terms of the indices of refraction measured on (100) cleavage flakes (Parker, 1961). The variation is sensibly linear, α 1·669 to 1·720, β 1·691 to 1·750. The major discrepancies in the correlation of the indices and composition are due to unusually high contents of Al and Fe^{3+} (e.g. Table 14, anal. 7). Zvetkov (1945) has shown that compared with pure diopside the α and γ indices of a diopside containing 40 mol. per cent $CaAl_2SiO_6$ are increased by 0·020 (see also Hijikata, p. 254). The effect of ferric iron on the refractive indices of diopside has been investigated by Segnit (1953) who demonstrated that the indices are increased by 0·003 for each 1·0 per cent Fe_2O_3. Comparable effects are shown by natural minerals, thus the indices of the ferrosalite (Table 18, anal. 7), which contains approximately 20 mol. per cent. $Ca(Al,Fe^{3+})_2SiO_6$, are about 0·010 higher than ferrosalites with the same $Mg:Fe^{2+}$ ratios and normal amounts of (Al,Fe^{3+}).

The optic axial angles of the members of the diopside–hedenbergite series do not show any major variation. The optic axial angle of synthetic diopside is 59·3°, and in many natural diopsides is between 56° and 60°; the $2V_\gamma$ of the Herault hedenbergite is 62·5°. Smaller optic axial angles are recorded in Tables 14–19 for minerals of normal compositions but the high optic angles of the synthetic series with large amounts of ferric iron are not shown by natural samples. Moderately high optic axial angles are shown, however, by diopsides with higher Fe^{3+} or with unusually high Cr (Table 17, anal. 15; Table 15, anal. 11).

The extinction angle $\gamma{:}z$ increases from 38° to 48°, and the density from around 3·22 to 3·555 g/cm³ between diopside and hedenbergite, but the variation of these properties in natural minerals is too erratic to be of any great diagnostic value.

The refractive indices of endiopsides (Table 16) are similar to those of diopsides having comparable $Mg:Fe^{2+}$ ratios; those of chrome-diopsides (Table 15) are generally somewhat higher, although in some examples this is attributable, at least in part, to a higher alumina content.

The refractive indices of synthetic ferrohedenbergites were determined by Bowen *et al.* (1933), and showed that both the α and γ indices increase with increasing tenor of iron. More precise measurements of the refractive indices of clinopyroxenes along the join hedenbergite–ferrosilite, synthesized at pressures of 20 kbar and above, have been made by Myer and Lindsley (1969). The α and γ curves intersect between $Fs_{95}Wo_5$ and $Fs_{100}Wo_0$ and are most probably coincident at $Fs_{98}Wo_2$, at which composition the pyroxene is pseudouniaxial. The indices of refraction of synthetic hedenbergite have also been determined by Gustafson (1974). The refractive indices of the ferrohedenbergite from the differentiated dolerite sill, Kola Peninsula (Table 20, anal. 11) are in good agreement with those of synthetic material of comparable composition (Table 24).

The relationship between the α-refractive indices, optic axial angle, and extinction angle $\alpha{:}z$ and chemical composition in the diopside–hedenbergite–acmite system (see Fig. 237) has been presented by Kostyuk (1965), see also Perchuk (1966). The α and γ refractive indices of synthetic pyroxenes on the join diopside–hedenbergite, and the α index for compositions on the joins diopside–acmite and hedenbergite–acmite have been measured by Nolan (1969). The same author has also constructed a diagram showing the relationship between chemical composition and the α-refractive index and b unit cell dimensions for synthetic and natural pyroxenes in the system diopside–hedenbergite–acmite (see Fig. 226).

The relation between the optical properties and composition of the manganese-rich pyroxenes of the diopside–hedenbergite–johannsenite series has been investigated by Zharikov and Vlasova (1955) and some of the relevant data given in

Table 24 Refractive indices of synthetic and natural ferrohedenbergites

Composition		Refractive indices		Reference
$CaSiO_3$	$FeSiO_3$	α	γ	
50	50	1·730	1·754	Myer and Lindsley (1969)
50	50	1·731	1·755	Gustafson (1974)
48·1	51·9	1·724	1·750	Val'ter (1969)
44·6	55·4	1·738	1·765	Bryan (1969)
42·5	57·5	1·738	1·765	Sinitsȳn (1965)
40	60	1·736	1·762	Myer and Lindsley (1969)
35	65	1·739	1·765	Myer and Lindsley (1969)
30	70	1·743	1·769	Myer and Lindsley (1969)
25	75	1·746	1·772	Myer and Lindsley (1969)
20	80	1·750	1·776	Myer and Lindsley (1969)
10	90	1·756	1·784	Myer and Lindsley (1969)

Table 43. The variation in the refractive indices and optic axial angles in the hedenbergite–johannsenite series is shown in Fig. 195, p. 419, and the variation in the optical properties of the diopside–hedenbergite–johannsenite pyroxenes in Fig. 196, p. 420.

Table 25 Optical properties of some natural and synthetic diopside–hedenbergite containing Mn, Zn, Ni, Al and Cr.

Composition mol. per cent	α	β	γ	$2V_γ$	γ:z	Reference
Diopside	1·644	1·6715	1·694	—	—	Allen and White (1909)
$Diopside_{56}(CaMnSiO_6)_{44}$	1·700	1·715	1·735	—	—	Hess (1949)
$CaMnSi_2O_6$	1·710	1·719	1·728	70°	48°	Schaller (1938)
Schefferite(MnO 9·69)	1·690	1·699	1·721	60°	43°	Palache 1937)
Zinc schefferite(MnO 7·4, ZnO 3·31)	1·676	1·683	1·705	60°	—	Palache (1937)
Jeffersonite(MnO 7·91, ZnO 7·14)	1·713	1.722	1·745	74°	55°	Palache (1937)
$CaNiS_2O_6$	1·7362	1·7513	1·7742	79°	—	Higgins and Gilbert (1973)
$Diopside_{75}(CaAl_2SiO_6)_{25}$	1·676	—	1·702	—	—	Hijikata (1973)
$Diopside_{50}(CaAl_2SiO_6)_{50}$	1·682	—	1·714	—	—	Hijikata (1973)
$Diopside_{25}(CaAl_2SiO_6)_{75}$	1·692	—	1·721	—	—	Hijikata (1973)
$CaAl_2SiO_6$	1·710	—	1·735	—	—	Hijikata (1973)
$CaAl_2SiO_6$	1·709	1·714	1·730	59°	—	Hays (1966)
$Diopside_{90}(CaTiAl_2O_6)_{10}$	1·683	1·695	1·716	—	—	Yagi and Onuma (1967)
Hedenbergite	1·730	—	1·754	—	—	Myer and Lindsley (1969)
Chrome–diopside(Cr_2O_3 1·47)	1·681	1·690	1·700	—	—	Yamaguchi (1961)

The α and γ indices of clinopyroxene solid solutions along the join $CaMgSi_2O_6$–$CaAl_2SiO_6$, containing 25, 50, 75 and 100 mol. per cent $CaAl_2SiO_6$, are given by Hijikata (1973); the values for $CaAl_2SiO_6$ being 1·710 and 1·735±0·005 respectively. The optical properties of a synthetic pyroxene close to the CaTs composition, $Ca_{1\cdot04}Al_{2\cdot07}Si_{0\cdot93}O_6$, are given by Hays (1966) as α 1·709, β1·714, γ 1·730, $2V_γ$ 59° (calculated) and the density 3·43 g/cm³. The refractive indices of synthetic clinopyroxene solid solutions between $CaMgSi_2O_6$–$CaFe^{3+}AlSiO_6$ have been measured by Hijikata and Onuma (1969) who give the α and γ indices of

$Di_{90}(CaFe^{3+}AlSiO_6)_{10}$ as 1·684 and 1·710 respectively, and $Di_0(CaFe^{3+}$-$AlSiO_6)_{100}$ as 1·855 and 1·873. The refractive indices of diopsides containing up to 11 weight per cent $CaTiAl_2O_6$ (3·75 weight per cent TiO_2) in solid solution (Yagi and Onuma, 1967) are detailed in Table 26. The increase in the indices is 0·006 for each per cent TiO_2.

Table 26 Refractive indices of Ti-bearing pyroxene solid solutions (Yagi and Onuma, 1967)

Composition (weight per cent)		Refractive indices (± 0·003)		
$CaMgSi_2O_6$	$CaTiAl_2O_6$	α	β	γ
97·5	2·5	1·669	1·677	1·700
95	5	1·673	1·683	1·705
92·5	7·5	1·678	1·690	1·711
90	10	1·683	1·695	1·716
89	11	1·684	1·697	1·718

The properties of a zinc-bearing manganoan salite and of a jeffersonite are given in Tables 17 and 18 respectively. The optical properties of synthetic $CaNiSi_2O_6$ are α 1·7362, β 1·7513, γ 1·7742, $2V_\gamma$ 79° and dispersion $r > v$ (Higgins and Gilbert, 1973).

Hijikata (1968) has calculated the densities of the pyroxenes, at 10 mol. per cent intervals, along the join $CaMgSi_2O_6$–$CaFe^{3+}AlSiO_6$, and gives 3·274 for pure diopside and 3·692 g/cm³ for the $CaFe^{3+}AlSiO_6$ end-member.

Rare Fe^{3+}-rich diopsides, pleochroic in yellows and yellow-green (maximum absorption parallel to α, minimum parallel to γ with strong dispersion), anomalous (inky-blue) interference, high $2V_\gamma$ (→ 67°) and extinction angle $\gamma:z$ (→ 80°), occur in biotite pyroxenites and biotite peridotites of the Gardar provinces, south Greenland (Upton and Thomas, 1973).

Gem quality chrome diopside is present in the chromiferous skarns of Outokumpu, Finland (Vuorelainen, 1963) and has been reported from Kwale, Kenya (Schmetzer and Medenbach, 1974). The pyroxene, α 1·668, β 1·678, γ 1·700, $2V_\gamma \simeq 60°$, D 3·286, is pleochroic from emerald to deep emerald green and twinned on (100). Blackish brown star diopside from Nammakal, southern India, α 1·67–1·68, D 3·33, has been described by Eppler (1967). The asterism is caused by the intersection of acicular inclusions at angles of 73° or 107°. Black star-diopside has also been reported by Martin (1967). The inclusions, the nature of which have not been determined, are oriented in two directions intersecting at 105°. A turquoise blue fluorescent diopside (SiO_2 55·2, TiO_2 0·03, Al_2O_3 0·1, FeO 0·08, CaO 25·7, MgO 18·8), from a calc-silicate rock at the contact of the Bergell granite and the metamorphic rocks of the Margna Nappe has been described by Wenk and Maurizio (1970). The blue diopside is associated with a colourless variety that apart from a difference in vanadium content (0·11 and 0·03 per cent V_2O_3 respectively) is almost identical in composition.

Deep blue diopside, colourless in thin section, has been described from a xenolith of carbonate rock included in an olivine gabbro-norite of the North Baikal region. The carbonate rock contains vein-like bodies consisting of blue, greenish blue and grey diopside. A similar blue diopside occurs in the diopside–wollastonite and diopside–garnet skarns at the contact of dolomitized limestone and the granites of the Andet intrusion, Kuznetsk Ala-Tau, Siberia (Gurulev *et al.*, 1965). The chemical compositions of the two blue varieties approximate closely to pure diopside (SiO_2 53·40, 55·09, Al_2O_3 0·75, 0·57, FeO 0·04, 0·07, MgO 19·42, 18·78, CaO 26·29, 25·35,

Na_2O 0·03, 0·22), and chromophores such as Cu and Co, although present in trace amounts, are not appreciably different in their content in the associated green and pale coloured varieties. Electron microscopy shows that the blue diopside contains many inclusions, a fraction of a micron in size, the green diopside a much smaller number, and the absence of inclusions in the pale-coloured variety. The composition of the inclusions is unknown but the diopsides contain 0·01 and 0·02 P_2O_5 respectively, and Gurulev *et al.* have suggested that the colour may be due to phorphorus-bearing mineral crystallites.

Diopside solid solutions containing a $CaCrCrSiO_6$ component (see p. 237), blue in colour and showing weak pleochroism from γ' bluish purple to α' reddish purple, have been synthesized by Ikeda and Yagi (1977). The optical spectra of the solid solutions are interpreted by Ikeda and Yagi by means of crystal field theory from which they conclude that the blue colour of these diopsides is probably attributable to tetrahedrally co-ordinated Cr^{3+} ions in low spin state.

Rose-coloured, radial, stellar aggregates, and spherules of diopside have been reported from vein rocks of the Gusevogorsk pyroxenite massif (Volchenko, 1971). The diopside has α 1·673, β 1·680, γ 1·699, $2V_\gamma$ 60°, γ:z 42° and composition SiO_2 50·10, TiO_2 0·10, Al_2O_3 4·36, FeO 5·74, MnO 0·13, MgO 14·41, CaO 20·75, Na_2O 0·89, H_2O+ 2·90. The absorption spectrum displays a maximum close to 520 nm due to Mn^{3+} ions. The absorption spectra have been recorded, between 290 and 1·7°K, for a chrome diopside (Cr_2O_3 0·30 weight per cent, α 1·668, β 1·675, γ 1·688), and at room temperature showed broad bands with peaks at 20020 and 15150 cm^{-1}. The spectra at low temperatures showed a narrow band at 14570 and 15860 cm^{-1} (Grum-Grzhimailo *et al.*, 1967). The infrared absorption spectra of pyroxenes of the diopside–jadeite series have been measured by Kuznetsova and Moskaleva (1968) and of three chrome-diopsides by Boksha *et al.*, 1974.

The optical spectra and crystal field parameters of synthetic $CaNiSi_2O_6$, $CaMg_{1-x}Ni_xSi_2O_6$ and $CaMg_{1-x}Co_xSi_2O_6$ have been measured by White *et al.* (1971).

Sector zoning in members of the diopside–hedenbergite series has been reported by a number of authors. A detailed investigation by Hollister and Gancarz (1971) of the compositional sector zoning in the diopside–salites from the Narce area, Italy, shows that the relative order of enrichment in Al,Ti,Fe^{3+},Na in the four crystallographic sectors (see augite section) is (100) > (110) > (010) > ($\bar{1}$11). Thus in a crystal showing strong sector zoning, the (100) sector has approximately 5 and 4 per cent more $CaAl_2SiO_6$ and $CaFe^{3+}AlSiO_6$ respectively, and 1 per cent more $NaTiAlSiO_6$ and $CaTiAl_2O_6$ than the ($\bar{1}$11) sector. Reversed Mg–Fe zoning in the salitic augites of inclusions and microphenocrysts in the Rogue Nublo basalts, Gran Canaria Island, has been described by Frisch and Schmincke (1969, 1971). The pyroxenes in the inclusions have patchy brown-coloured cores enclosed by a greenish variety richer in Al, Ti and Na and containing less Si and Mg. The microphenocrysts are similarly zoned, with brown inner areas, $Ca_{48}Mg_{35}Fe_{17}$, and greenish or greenish brown margins $Ca_{44}Mg_{45}Fe_{11}$ in composition.

Well developed sectoral zoning, in which the principal differences between alternate sectors are in the Al and Ti contents (5·98 and 9·54 weight per cent Al_2O_3 and 2·89 and 4·82 per cent TiO_2 in ($\bar{1}$11) and (100) sectors respectively), is shown by the salite phenocrysts and groundmass crystals in the leucite nephelinite of Mt. Nyiragongo volcano, Zaire (Sahama, 1976).

Concentric and sector zoning is commonly displayed by the phenocrystal augitic pyroxenes of a mildly alkaline basaltic group of the Oslo area (Weigand, 1975). In some the zonation is marked by distinct core and rim compositions. Typically the zoning is from more diopsidic augite centres to more salitic margins (e.g. $Wo_{45\cdot8}En_{47\cdot4}Fs_{6\cdot8}$ to $Wo_{49\cdot1}En_{36\cdot7}Fs_{14\cdot2}$). Aluminium and titanium show a much greater variation, and in the above example the contents of Al_2O_3 and TiO_2 are 2·7 and 0·6 per cent and 10·3 and 3·1 per cent in the core and margin respectively.

Lamellar twinning, parallel to (100), and to a smaller extent parallel to (001), is not uncommon in diopsides from deformed basic and metamorphic rocks. An experimental study of the twin-glide elements of diopside, under confining pressures of 5 kbar and temperatures ranging from 200° to 1000°C, and constant strain rates between $5 \times 10^{-2}\,s^{-1}$ and $5 \times 10^{-5}\,s^{-1}$, has been made by Rayleigh (1965) and Rayleigh and Talbert (1967). Mechanical twinning on (100) is the principal mechanism of the plastic deformation, with the twin-glide direction t = [001]. The twins on (100) are narrow and closely spaced and indistinguishable from natural polysynthetic twins in diopside. Twinning on (001) was also produced in the experimentally deformed material but, compared with the (100) twins, the basal twins are few in number, form broader lamellae and are discontinuous. Growth twins have the same symmetry elements as those of mechanical origin, but are broad, and individual crystals consist only of one or two twinned units. The clinopyroxene in the harzburgite of the Isle of Euboea, Greece (Capedri, 1974) shows mechanical twinning, with the composition plane (100) and [100] as the twin axis. Where in contact with orthopyroxene displaying kinking, the monoclinic pyroxene shows secondary lamellar twins, parallel or nearly parallel, to the kink band boundary of the orthopyroxene. As this boundary is parallel to the maximum resolved shear stress, the composition plane of the twinned clinopyroxene must also be parallel to the shear stress. Polysynthetic deformation twinning on (001), in single crystals of diopside and in pyroxenite, at shock pressures of 50 to 390 kbar, has been described by Hornemann and Müller (1971). Small suboptical domains twinned on {100} have been observed in diopside deformed at pressure of 10 kbar and temperatures between 500° and 650°C (Wenk, 1970).

Diopsides, $\simeq Ca_{45}Mg_{50}Fe_5$, containing lamellae of orthopyroxene, from the ultrabasic rocks of the Hidaka metamorphic belt, Hokkaido, Japan, have been studied by X-ray precession photography (Yamaguchi and Tomita, 1970). Some samples, displaying optical homogeneity, show the presence of a single, clinoenstatite ($Wo_1En_{89}Fs_{10}$) exsolution phase, others in which broad lamellae can be observed microscopically are composed of either orthoenstatite and clinoenstatite, or solely of orthoenstatite. The submicroscopic clinoenstatite may represent a metastable phase to lattice distortion in the early stage of the exsolution process; at a later stage the distortion was released and the clinoenstatite converted to the orthoenstatite stable phase.

Distinguishing Features

The members of the diopside–hedenbergite series cannot always be distinguished from clinopyroxenes of augite and ferroaugite composition (see p. 294), but in general the diopside–hedenbergite minerals have higher optic axial angles than those of augite and ferroaugite with comparable refractive indices. Members of the diopside–hedenbergite series may be distinguished from the orthopyroxenes by their higher birefringence. Moreover many orthopyroxenes display a characteristic pleochroism, and except for the magnesium-rich orthopyroxenes their optic sign is negative; the extinction is straight in all sections of the [001] zone in orthopyroxenes. Diopside is distinguished from wollastonite by the presence of only two cleavages and by higher birefringence and refractive indices. Ferrosalite and hedenbergite are distinguished from rhodonite by the higher birefringence, lower optic axial angle, and by the two cleavages of the clinopyroxenes.

Paragenesis

Igneous rocks

Ultrabasic rocks. The clinopyroxene of ultrabasic and ultramafic rocks is generally diopsidic in composition and commonly has a moderate to high content of chromium. Thus chrome diopsides have been described from the peridotitic rocks of Coasta lui Rusu, southern Carpathians (Pavelescu, 1968), from the lherzolite of Louvie-Juzon, Basses-Pyrénées (Rio, 1968), and from the harzburgite and peridotite of the layered ultramafic bodies of Dun Mountain, Red Hills and Red Mountain, South Island, New Zealand (Challis, 1965); chrome endiopsides are also present in the rocks of the latter area.

Diopside–enstatite–pyrope–olivine–chrome spinel peridotites have been described from the Borovnik and Dragonia, West Moravia, ultrabasic intrusives. This association, spanning the olivine–pyroxene–spinel and olivine–pyroxene–pyrope assemblages is attributed to a pressure–temperature regime in which the reaction:

$$\underset{\text{orthopyroxene}}{2(Mg,Fe)SiO_3} + \underset{\text{Ca-pyroxene}}{Ca(Mg,Fe)Si_2O_6} + \underset{\text{chrome spinel}}{(Mg,Fe)(Al,Cr)_2O_4} \rightleftharpoons \underset{\text{garnet}}{Ca(Mg,Fe)_2(Al,Cr)_2Si_3O_{12}} + \underset{\text{olivine}}{(Mg,Fe)_2SiO_4}$$

approximates to an equilibrium state. Other diopside-bearing ultrabasic rocks include the peridotite and olivine pyroxenites of the Duke Island ultramafic complex, Alaska (Irvine, 1974), and those of the ultramafic–mafic belt, Makrirrakhi, Othis Mountains, Greece (Menzies, 1973).

The diopsides, $Ca_{50}Mg_{47\cdot7}$–$Ca_{45\cdot7}Mg_{49\cdot2}$, of the garnet peridotite from Kalskaret, south Norway have a higher sodium content than the typical diopside of ultrabasic rocks, but an even higher value (Na_2O 2·60 per cent) has been reported (Kornprobst, 1966) for a chromian diopside from the Beni-Bouchera ultrabasic massif, Morocco. In the peridotite intrusion of the Lizard, Cornwall (Green, 1964), the diopsides of the primary core have a higher alumina content (7·09 and 6·80 weight per cent Al_2O_3) than the pyroxenes (3·8 and 4·5 per cent Al_2O_3) of the recrystallized marginal shell. The bulk chemical compositions of the core and margin of the intrusion are comparable, and the contrasted alumina contents of the two diopsides are attributed by Green to the pressure differences operating during the crystallization of the primary core and at the time the outer shell recrystallized (see p. 120). Similar alumina contents to those of the core diopsides are shown by the diopside porphyroblasts ($Ca_{49\cdot3}Mg_{46\cdot6}Fe_{4\cdot1}$) and the clinopyroxene ($Ca_{48\cdot3}Mg_{47\cdot8}Fe_{3\cdot9}$) intergrowths with enstatite, in the peridotite mylonites of St. Paul's Rocks, Atlantic (Tilley and Long, 1967). Diopside-bearing inclusions in serpentinite have been reported from Kanto Mountains, central Japan (Seki and Kuriyagawa, 1962). The inclusions display a zonal arrangement, with the inner zone consisting of diopside, grossular and chlorite, followed by a diopside–chlorite zone, and a chlorite rim adjacent to the enclosing serpentinite. Diopside also occurs in some more alkaline ultrabasic rocks, e.g. the olivine–melilite–nephelinite of d'Essey-la-Côte, Meurthe-et-Moselle, France (Velde and Thiebaut, 1973).

In the lower ultrabasic, gabbro and upper ultrabasic zones of the layered alkali ultrabasic–gabbro ring complex, South Island, New Zealand (Grapes, 1975), endiopside, $Ca_{45}Mg_{48}Fe_7$–$Ca_{36}Mg_{55}Fe_9$, occurs as a cumulus phase. The minerals have high contents of chromium (0·72–1·87 per cent Cr_2O_3), are rimmed by pre-depositional overgrowths of titanaugite, and in some cases display oscillatory zoning between endiopside and titanaugite. The Cr-diopsides and endiopsides of the Salt Lick Creek intrusion, East Kimberley, Western Australia, show no significant Mg–Fe variation, and the main chemical change is restricted to their Cr contents, from 1·1 weight per cent Cr_2O_3 in the pyroxenes of the lowest olivine–plagioclase cumulates of the basal zone to 0·47 Cr_2O_3 in those of the plagioclase–orthopyroxene cumulates of the main zone (Wilkinson *et al.*, 1975). Endiopside ($\simeq Ca_{41}Mg_{50}Fe_9$) phenocrysts in the olivine-rich basaltic rocks (picrites) from the lower part of the Karroo basalt sequence, Nuanetsi province, have been described by Cox and Jamieson (1974). The occurrence of diopside in garnet pyroxenite and corundum–garnet amphibolite blocks from a slumped breccia deposit of Miocene age, Sabah, Malaysia, is described by Morgan (1974).

Basic rocks. Although many clinopyroxenes of early crystallization from basaltic magmas have an augitic composition, diopside and salite are not uncommon constituents of basic rocks and particularly those of alkali olivine basalt parentage. Thus among the rocks of the Hakone volcano, diopside is present in the less differentiated lavas (Kuno and Sawatari, 1934; Kuno, 1950). In the olivine basalts and olivine trachybasalts of the Lower Carboniferous of the Old Pallas area, Co. Limerick (Ashby, 1946), the phenocrystal pyroxenes vary in composition from diopside in the less differentiated lavas to salite (Table 17, anal. 19) in the more differentiated rocks. Similarly the clinopyroxenes of the Bridget Cove volcanics, Juneau area, Alaska (Irvine, 1973) range in composition from $Ca_{47}Mg_{48}Fe_5$ to $Ca_{47}Mg_{38}Fe_{15}$. High-aluminium diopside–salites (5–8 weight per cent Al_2O_3, 1·3–2·4 per cent TiO_2) are present in the nephelinites, basanites and alkali basalts of the Tertiary Monaro volcanics, south-eastern Australia (Kesson, 1973).

Diopside is also an essential constituent of the minettes of the Carboniferous-Permian dyke suite, north of Loch Sunart, Argyllshire (Gallagher, 1963). Phenocrysts of ferridiopside, in camptonite have been described by Velde and Touron (1970). The phenocryst cores are pale brown in colour and are surrounded by a brown margin; the zoning is related to a decrease in Si and Mg and an increase in Ti (3·77 per cent TiO_2) and Al (6·71 per cent Al_2O_3) in the composition of the brown margins of the phenocrysts. Pyroxenes of salitic composition are particularly characteristic in hypabyssal rocks derived from alkali basalt magmas, e.g. a magnesium-rich salite (Table 17, anal. 7 and Table 27), associated with magnesium-rich olivine and plagioclase, An_{80}, present in the picrite and picro-dolerite of the Garbh Eilean sill, Shiant Isles (Murray, 1954), and the pyroxenes (Table 17, anals. 16 and 22) in the differentiated rocks of the Black Jack teschenite sill, New South Wales (Wilkinson, 1957) (see also Hutton, 1943). Salite is also the characteristic pyroxene of the alkaline diabase of the Prospect intrusion (Wilshire, 1967), and the analcite olivine theralite (Table 17, anal. 13) and analcite tinguaite of the Square Top intrusion, Nundle, New South Wales (Wilkinson, 1966).

Salite and calcium-rich augite are the characteristic pyroxenes of the rocks of the olivine basalt–hawaiite–mugearite–trachyte series of the Hockeifel area, West Germany (Huckenholz, 1965a, b). The salite phenocrysts of the ankaramite

Table 27 Al_2O_3, Fe_2O_3 **and** TiO_2 **contents of salites from rocks of alkali basalt parentage**

Rock type	Locality	Al_2O_3	Fe_2O_3	TiO_2	Reference
Picrite-picrodolerite	Shiant Isles	2·6–3·9	1·5–2·7	1·2–1·8	Murray, 1954
Teschenite	Black Jack	3·6–6·6	1·5–1·8	2·2–3·2	Wilkinson, 1957
Analcite theralite-analcite tinguaite	Square Top	2·1–7·3	3·0–7·3	0·6–2·2	Wilkinson, 1966
Alkaline diabase	Prospect	4·2–5·1	3·6–5·1	1·9–2·5	Wilshire, 1967
Alkaline olivine basalt	Hocheifel	6·4–7·5	2·5–3·5	1·1–1·5	Huckenholz, 1965a
Ankaramite	Hocheifel	6·3–9·3	2·1–4·3	1·7–3·6	Huckenholz, 1966
Analcite trachyte	Hocheifel	9·7	5·1	2·6	Gutberlet and Huckenholz, 1965
Alkali basalt	Tenerife	4·7–8·2	—	1·0–4·1	Scott, 1976

have a chromium-rich core, $Ca_{47\cdot9}Mg_{41\cdot6}Fe_{10\cdot5}$ (Table 17, anal. 2) rimmed by a strongly zoned titansalite (TiO_2 2·42 per cent) of composition $Ca_{45}Mg_{42}Fe_{12}$. The latter occurs as microphenocrysts and the groundmass pyroxene also is a titansalite, $Ca_{45}Mg_{37}Fe_{18}$ (Huckenholz, 1966). The pyroxenes show only limited iron-enrichment in the more differentiated members of the series, e.g. the composition of the salite in the analcite alkali trachyte is $Ca_{45\cdot3}Mg_{41\cdot1}Fe_{13\cdot6}$ (Gutberlet and Huchenholz, 1965), but has a very high content of alumina and ferric iron (Table 17, anal. 15). Calcium-rich salites ($Ca_{49\cdot9}Mg_{36\cdot9}Fe_{13\cdot2}$ and $Ca_{55\cdot0}Mg_{32\cdot4}Fe_{12\cdot6}$), containing 7·37 and 10·70 weight per cent Al_2O_3, and 3·34 and 3·38 per cent Fe_2O_3 respectively occur in the Forez region, central France (Hernandez, 1973).)

Members of the diopside–hedenbergite series are present in a shoshonite from Stromboli (Girod, 1975). The pyroxenes, associated with plagioclase and alkali feldspars, olivine, phlogopite, analcite and titanomagnetite, occur in large phenocrysts, zoned from $Wo_{49\cdot2}En_{45\cdot1}Fs_{5\cdot7}$ to $Wo_{49\cdot0}En_{44\cdot5}Fs_{6\cdot5}$, microphenocrysts, zoned from $Wo_{47\cdot8}En_{42\cdot7}Fs_{9\cdot5}$ to $Wo_{47\cdot4}En_{40\cdot8}Fs_{11\cdot8}$, and as microlites $Wo_{47}En_{36}Fs_{17}$ in composition. The ferric/ferrous ratios of the large phenocrysts pyroxene vary from 61 to 58 for core and rim respectively, from 40 to 25 in the small phenocrysts and is 13 in the microlites.

Calcium-rich pyroxenes, diopside–salite and salite, are not, however, restricted to alkali basalt parageneses, and their presence as products of subalkaline magmas has been reported by a number of authors. These include salite from the basalts and andesites of the Sheveluch volcano, Kamchatka (Naboko and Shavrova, 1954), salite (Table 17, anal. 10) from eucrites and gabbros, West Chester area, Pennsylvania (Norton and Clavan, 1959), and from hornblende gabbro inclusions (Yamazaki *et al.*, 1966). More recently, aluminium-rich salites (Table 17, anal. 8) have been described from the ejected plutonic blocks of the Soufrière volcano, St. Vincent (Lewis, 1967, 1973), and from the pillow lavas and sills of the Karawanken Mountains, south-east Austria (Loeschke, 1973). Endiopside, $Ca_{42\cdot3}Mg_{49\cdot1}Fe_{8\cdot5}$, also occurs in some basalts of calc alkali affinities, e.g. the phenocrystal pyroxene (Table 16, anal. 9) of the Nakaizuka basalt, Ryozen, Fukushima, Japan (Konda, 1970).

Although less common in plutonic rocks, diopside and salite are not untypical of coarse grained rocks of alkali basalt parentage, e.g. the Lilloise layered intrusion, east Greenland (Brown, 1973). Like those of the hypabyssal rocks the clinopyroxenes characteristically have high Al_2O_3 and Fe_2O_3 contents. Diopside (Table 14, anal. 18) is a constituent of the early differentiates of the alkali plutonic complex, Ichinohe, Kitakami Mountainland, northern Japan (Onuki and Tiba, 1964).

Ferrohedenbergite occurs as a product of both primary crystallization, and as the result of inversion from a ferrowollastonite solid solution, in the ferrodiorites of upper zone C of the Skaergaard intrusion (Wager and Deer, 1939; Brown and Vincent, 1963; Wager and Brown, 1968). The pyroxene from the top of the zone is a brown-coloured ferrohedenbergite solid solution, $Ca_{42 \cdot 5}Mg_{0 \cdot 4}Fe_{57 \cdot 1}$ in composition. At slightly lower levels in the zone the ferrohedenbergites are green in colour and display a mosaic texture, and are rimmed by an intergrowth with a brown variety. On the basis of this textural evidence the green ferrohedenbergite mosaics are considered to have been derived by inversion from an iron-rich wollastonite solid solution. The pressure–temperature relations of this green hedenbergite solid solution, at pressures between 0 to 5 kbar, have been investigated by Lindsley *et al.* (1969). These relationships are illustrated in Fig. 137 and show

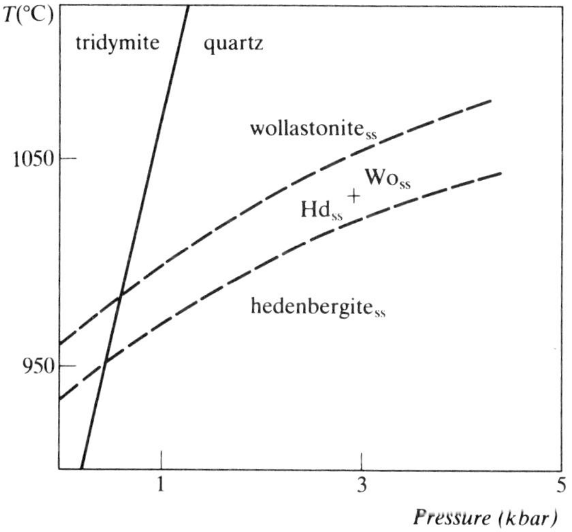

Fig. 137. Hedenbergite$_{ss}$–wollastonite$_{ss}$ inversion for clinopyroxenes of upper zone C, Skaergaard intrusion (after Lindsley *et al.*, 1969).

that the transition from Wo_{ss} to Hd_{ss} probably occurred over a temperature interval of 950° to 980°C at a pressure of approximately 0·6 kbar (the pressure value is based on the relationship of quartz and tridymite in the rocks of the upper zone and upper border zone).

The transition of $Wo_{ss} \rightarrow Hd_{ss}$ may occur in three ways, (*1*) $Wo_{ss} \rightarrow Hd_{ss}$ + fayalite + SiO_2, (*2*) $Wo_{ss} \rightarrow Hd_{ss}$ + metastable clinoferrosilite, (*3*) inversion of $Wo_{ss} \rightarrow$ metastable Hd_{ss} of the same composition. The first reaction is not substantiated, as earlier predicted (Bown and Gay, 1960), by a later detailed investigation of the inverted wollastonite solid solutions in the ferrodiorites of the Skaergaard upper zone C (Nwe and Copley, 1975). The latter data indicate that a Wo_{ss} with a solidus composition of Wo_{39} may invert by either reaction (*2*) or (*3*), and that both reactions are represented in different crystals within a single large mosaic-patterned grain. The compositions of the inverted Wo_{ss} fall into two groups, $Wo_{45 \cdot 1}En_{1 \cdot 1}Fs_{53 \cdot 5}$–$Wo_{43 \cdot 1}En_{1 \cdot 1}Fs_{55 \cdot 8}$ and $Wo_{40 \cdot 8}En_{1 \cdot 1}Fs_{58 \cdot 1}$–$Wo_{39 \cdot 2}En_{1 \cdot 1}Fs_{59 \cdot 7}$. Thus a Wo_{ss} with a solidus composition of Wo_{39} (Fig. 138) would on cooling react to a Hd_{ss} of composition $Wo_{42 \cdot 6}$ (the compositions of the Wo_{ss} and primary Hd_{ss} coexisting in a ferrodiorite of upper zone C), and close to the corresponding compositions, respectively Wo_{38} and Wo_{40} obtained, in the synthetic Mg-free system at pressures of 2 kbar and below, by Lindsley and Munoz (1969). With further cooling (curve 2) the final subsolidus compositions would be Wo_{46}, the maximum content of the wollastonite component, determined by probe analysis, in

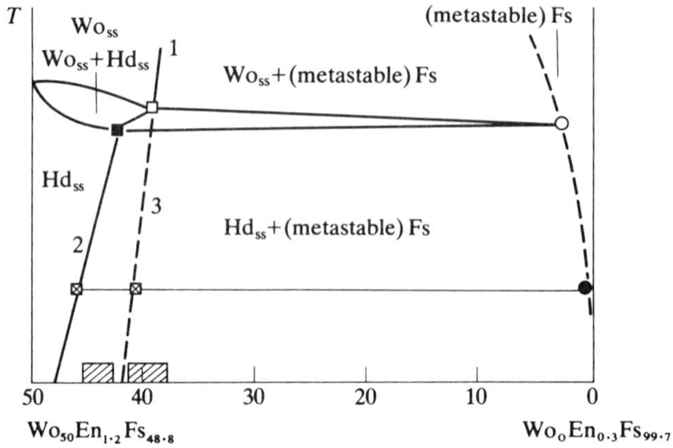

Fig. 138. Schematic pseudobinary *T–X* diagram illustrating the cooling history of inverted wollastonite$_{ss}$ in upper ferrodiorites of the Skaergaard intrusion (after Nwe and Copley, 1975). □, solidus composition of Wo_{ss}: ■, subsolidus composition of Hd_{ss}; ⊠ final composition of inverted Wo_{ss}; ○ ●, initial and final composition of coexisting exsolution lamellae; ▨, bimodal ranges of composition of inverted Wo_{ss} in different crystals within a large mosaic patterned grain. Curve 2, subsolidus cooling path; curve 3 metastable subsolidus cooling path.

a Hd_{ss} containing an exsolved Ca-poor phase. On the assumption that the exsolved phase is pure ferrosilite this would require the exsolution of some 15 per cent, a value that is not consistent with the electron microscope data.

The reaction $Wo_{ss} \rightarrow Hd_{ss}$ + metastable clinoferrosilite, as the sole operative factor, also does not account for the bimodal distribution of the analyses. Thus those grains approximating to the composition Wo_{41} probably inverted to metastable Hd_{ss}, and followed a cooling path below the transition interval illustrated by curve 3, a conclusion implying that certain grain boundaries were more favourably oriented than others for nucleation of clinoferrosilite lamellae.

In the ferrosyenodiorites of the Bushveld complex (Atkins, 1969), ferrohedenbergite (Table 20, anal. 12) crystallized directly as a primary phase. On both general grounds, and on the basis of the Mg–Fe distribution in coexisting Ca-rich and Ca-poor pyroxenes, the intrusion is considered to have solidified at a higher pressure than the Skaergaard magma. The absence of inverted wollastonite solid solution in the late differentiates of the Bushveld and its occurrence in those of the Skaergaard is illustrated in Fig. 139. The effect of increasing pressure on the temperature of the Wo_{ss}–Cpx inversion is greater than its effect on that of magmatic

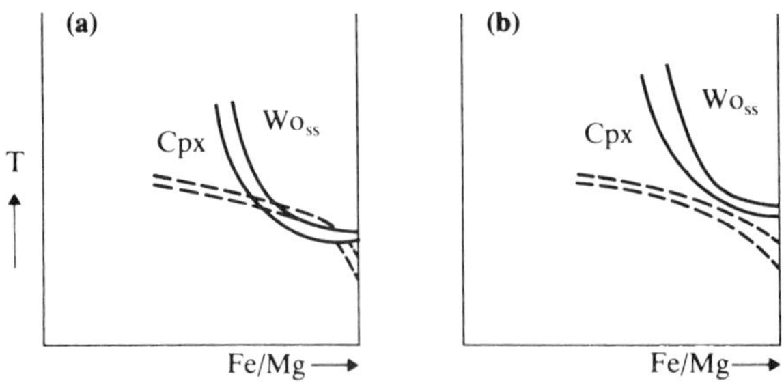

Fig. 139. Hypothetical diagram showing relationship between clinopyroxene–wollastonite solid solution interval and the crystallization curves for (a) Skaergaard (b) Bushveld magmas. Full lines, approximate boundaries of clinopyroxene–wollastonite$_{ss}$ inversion intervals. Broken lines, crystallization curves (after Lindsley *et al.*, 1969).

crystallization, so that during the fractionation of the Bushveld liquid the crystallization curve did not intersect the wollastonite solid solution field.

Alkaline and acid rocks. Diopside also occurs in some more strongly alkaline rocks. Thus a chrome-diopside (Table 15, anal. 8), associated with forsterite, titaniferous phlogopite, katophorite, sanidine and leucite, has been described from the jumillites of the Murcia and Almeria provinces, Spain (Borley, 1967). Diopside (Table 14, anal. 9) is present in the potassium-rich lavas and plugs of the Leucite Hills, Wyoming (Carmichael, 1967), and there occurs as a groundmass constituent with leucite, magnophorite and apatite in wyomingite, with sanidine, leucite and magnophorite in orendite, and as phenocrysts and in the groundmass of madupite. The pyroxene is a relatively pure diopside and the composition is comparable with the diopside of a carbonate rock in the carbonatite and alkaline complex at Oka, Quebec (Gold, 1966).

Chrome-diopside, $\simeq Ca_{49}Mg_{46 \cdot 5}Fe_{4 \cdot 5}$, megacrysts, and salite phenocrysts, $\simeq Ca_{48 \cdot 5}Mg_{28 \cdot 5}Fe_{23}$, occur in a camptonite dyke, Wiedemanns Fjord, east Greenland (Brooks and Rucklidge, 1973). Both pyroxenes contain significant amounts of the $CaAl_2SiO_6$ component and, in contrast to the uniform core compositions, display strongly zoned margins involving enrichment in titanium, and in the case of the salite reversed zoning from $Ca_{48}Mg_{29}Fe_{23}$ to $Ca_{51}Mg_{36}Fe_{13}$. The diopside (and kaersutite) megacrysts are considered to have been formed as an early cumulate phase at depth, while the salite phenocrysts, although precipitated under similar pressure conditions, are probably derived from a more differentiated liquid, or as the consequence of an increase in oxygen fugacity of the magma.

Homogenization at atmospheric pressure of the primary multiphase, gas + glass + crystalline phases including leucite, melt inclusions in the diopside of a wyomingite of Leucite Hills (Sobolev *et al.*, 1975) shows that the pyroxene crystallized at temperatures between 1270° and 1220°C.

Salite, rich in aluminium (5·95–7·63 weight per cent Al_2O_3), as phenocrysts, microphenocrysts and groundmass constituents is the characteristic ferromagnesian constituent of the leucite tephrites of Vesuvius and the phonolitic tephrites of Monte Somma (Rahman, 1975). The pyroxenes have high contents of aluminium (5·95–7·63 weight per cent Al_2O_3) and show oscillatory and sector zoning. Salite is also the common microphenocrystal phase in the leucite-bearing lavas in the northern part of the Roman volcanic region (Cundari, 1975). Zoning in the pyroxenes is relatively small but shows a general Fe-enrichment trend parallel to the diopside–hedenbergite join, as well as contrasting trends in the distribution of Ti relative to Al between core and rim. The variation in the content of $CaAl_2SiO_6$, within and between pyroxenes in the same lava, differs by a factor of two, and is considered to be related to crystal–liquid equilibria rather than pressure changes in the volcanic region. An unusual occurrence of diopside, associated with Ti-rich phlogopite, armalcolite and sanidine in lamproite dykes and plugs, Jordan, Montana, has been reported by Velde (1975). Salites rich in Al (maximum 13·6 weight per cent Al_2O_3) and Ti (5·9 per cent TiO_2) are reported from the leucite–rhönite basanites of the Sillon Houiller region, central France (Magonthier and Velde, 1976, see also Magonthier, 1975).

Zoned phenocrysts of ferrosalite in a micro-kakortokite dyke, Ilímaussaq, south Greenland, have been described by Larsen (1974). The ferrosalite core is rimmed continuously from Na-rich ferrosalite ($Di_{48 \cdot 2}Hd_{46 \cdot 4}Ac_{5 \cdot 4}$) through Na-rich hedenbergite ($Di_{9 \cdot 6}Hd_{76 \cdot 5}Ac_{13 \cdot 9}$) to almost pure acmite ($Hd_{5 \cdot 4}Ac_{94 \cdot 6}$), a trend analogous to that displayed by the pyroxenes of the Ilimaussaq intrusion (see Fig. 238b).

Hedenbergite occurs in the pulaskite, heterogeneous foyaite, sodalite foyaite and naujaite, and ferrosalite, $Ca_{47}Mg_{26\cdot5}Fe_{26\cdot5}$–$Ca_{48\cdot5}Mg_{10\cdot5}Fe_{41\cdot0}$, in the augite syenite of the Ilímaussaq intrusion, south Greenland (Larsen, 1976). The hedenbergite, $\simeq Ca_{47}Mg_2Fe_{51}$, of the pulaskite and heterogeneous foyaite occurs as partially resorbed cores surrounded by successive rims of aegirine-augite, alkali amphibole and aegirine. In the sodalite foyaite and naujaite the hedenbergite cores show greater resorption and are completely rimmed by amphibole. Sodium-rich hedenbergites have been reported from a wide range of alkaline rocks, e.g. from shonkinite (Larsen, 1941), and from alkali granite (Tilley, 1949).

Hedenbergite also occurs as a constituent of some acid igneous rocks, and is not uncommon in quartz-bearing syenites, in which it is usually associated with fayalite, e.g. Pikes Peak batholith, Colorado (Barker *et al.*, 1975). A hedenbergite–fayalite granite occurs in the Vesturhorn intrusion, south-east Iceland (Roobol, 1974), and hedenbergite–fayalite granites and porphyries have been described from the ring complexes of northern Nigeria (Jacobson *et al.*, 1958); in the rocks of this area the hedenbergite is commonly rimmed by arfvedsonite.

Ferrohedenbergite occurs in a number of acid rocks and oversaturated syenites and has been described from the granite ring dykes (Table 20, anal. 6) of the Western Redhills intrusive complex, Skye (Bell, 1966), and the granophyre (anal. 4) of the Meall Dearg, Skye (Anwar, 1955). The ferrohedenbergite (Table 20, anal. 8) in the granophyre of Red Hill, southern Tasmania (McDougall, 1961, 1962) consists of irregular intergrowths of a pale purple and pale green pyroxene in which both varieties display the same optical and crystallographic orientation. Ferrohedenbergite has been reported from a quartz-syenite sheet, Berkshire Valley, New Jersey Highlands (Young, 1972), a microsyenite (anal. 3), Carnarvon Range, Queensland, Australia (Bryan, 1969), and from a differentiated sill, Kola Peninsula (Sinitsÿn, 1965).

An unusual intergrowth of lamellar ferrohedenbergite ($Ca_{36}Mg_8Fe_{56}$, estimated composition based on optical properties) with ferrohortonolite and quartz, occurs in the adamellite associated with the Nain anorthosite mass, northern Newfoundland, Labrador, and is interpreted by Wheeler (1960) to have resulted from the partial breakdown of the pyroxene to olivine + quartz at lower temperatures.

The variable composition of salite in rims around partly fused rhyolite inclusions in basanite, in which the $CaAl_2SiO_6$ (1·34–6·46 mol. per cent) and $CaTiAl_2O_6$ (2·72–5·46 mol. per cent) components of the pyroxene increase in the less contaminated areas of the rims is interpreted by Maury and Bizouard (1974) as a consequence of decreasing Si activity away from the acid inclusion.

Nodules in basic rocks. Clinopyroxenes, ranging in composition from diopside to endiopside and calcic augite, occur in ultrabasic nodules in alkali olivine basalts, and many of the earlier analyses of these minerals have been collated by Ross *et al.* (1954).

White (1966) and Kuno (1969) have made comprehensive studies of the ultramafic inclusions in the basaltic rocks of Hawaii. The inclusions are predominantly of two types: (*1*) lherzolite nodules mainly confined to strongly undersaturated hosts, olivine nephelinite, nepheline basanite and ankaratrite; (*2*) dunite, wehrlite, feldspathic peridotite and pyroxenite nodules occurring preferentially in moderately undersaturated hosts, alkaline olivine basalts, hawaiite and ankaramite. The clinopyroxenes in both types of nodules are diopsides, or calcic augites but each characteristically displays distinctive chemical features. Thus the contents of Al_2O_3 and Cr_2O_3 (5·55 and 1·0 weight per cent respectively, average of four specimens) of the diopsides in the lherzolite nodules are considerably greater than those of the diopsides of the second group (Al_2O_3 2·9, Cr_2O_3 0·5 per cent, average of four).

Extensive exsolution of orthopyroxene and garnet is displayed by the clinopyroxenes of the garnet pyroxenite and garnet websterite xenoliths in the Salt Lake Crater tuff, Oahu (Beeson and Jackson, 1970). The reconstructed compositions of the clinopyroxenes (i.e. the observed compositions plus those of the exsolved phases) range from $Ca_{29\cdot3-33\cdot6}$, $Mg_{58\cdot4-52\cdot6}$ and $Fe_{12\cdot2-15\cdot0}$ compared with the present compositions of the clinopyroxene hosts, $Ca_{41\cdot4-44\cdot7}$, $Mg_{45\cdot1-46\cdot1}$ and $Fe_{9\cdot6-12\cdot5}$. Clinopyroxene is also present as exsolved lamellae in the orthopyroxene of the garnet websterite, and is similar in composition to the clinopyroxene hosts from which garnet and orthopyroxene has exsolved. The reconstructed compositions, by analogy with the synthetic system $CaMgSi_2O_6$–$MgSiO_3$, indicate that the original clinopyroxenes equilibrated at temperatures between 1300° to 1400°C, in contrast to the subsequent re-equilibration between 1000° and 1100°C indicated by the compositions of the clinopyroxene hosts and the clinopyroxene lamellae exsolved from orthopyroxene (see Fig. 140).

The pyroxenes in the ultramafic nodules in the alkali olivine basalts, Victoria, Australia, have been described by Irving (1974), the diopsides and endiopsides of the lherzolite inclusions of the Victorian basanites by Frey and Green (1974), and diopside and salite in the Al-spinel pyroxenite, Cr-spinel lherzolite and wehrlite xenoliths in an analcitite sill, Barraba, New South Wales by Wilkinson (1975a). The pyroxenes of the inclusions in the basaltic lavas and dykes of the Auckland province, New Zealand (Rodgers and Brothers, 1969) are more variable, ranging from iron-rich diopsides to magnesium salites and calcic augites, with moderate contents of Al_2O_3 (3·6 to 4·2 weight per cent) and relatively rich in chromium (Cr_2O_3 0·6–1·1 per cent), in contrast to those of the ultrabasic and basic inclusions in the alkali basalts and allied rocks of Iki Island, Japan (Aoki, 1964, 1968). In these inclusions the pyroxenes are diopsidic salites and calcic augites (Ca 44·2 to 45·3 atomic per cent) with a high content of alumina (range 3·94 to 8·19 and averaging 7·21 for nine samples), and a low chromium content (0·0 to 0·18 and averaging 0·09 for eight samples). This contrast in the chemistry of the clinopyroxenes with that of many in the lherzolite and websterite inclusions in alkali basalt may possibly be related to their derivation from the lower crust at a depth of some 20 to 30 km, rather than from the upper mantle. Diopside and diopsidic augite in the lherzolite and websterite nodules from the Itinome-gata. Oga peninsula, Japan, and the Dreisen Weiher, Eifel district, West Germany, craters have been described by Kuno and Aoki (1970), and from the lherzolitic inclusions in the undersaturated basalts of the Massif Central, France, by Hutchison *et al.* (1975). Aoki (1971) has described diopside–salites (Al_2O_3 3·27–6·70 and averaging 5·08 weight per cent for eleven samples, see anal. 19, Table 14) from the spinel-bearing olivine hornblendite and gabbro xenoliths in the lapilli tuff of alkali basalt, and as megacrysts in the basalt scoria of the Itinome-gata crater north-eastern Japan. Diopside occurs in the clinopyroxenite and spinel pyroxenite nodules of the Ladinian volcanic rocks, Predazzo–Monzoni complex, northern Italy (Ferrara *et al.*, 1974). The diopsides of the clinopyroxenites have low contents of Al (1·26–2·36 weight per cent Al_2O_3) in contrast to those of the associated augite megacrysts (13·8 per cent Al_2O_3). The nodules are interpreted as cognate xenoliths derived by crystal settling from an alkali-basalt magma at 15–25 km in the crust, whereas the megacrysts crystallized from the same magma at the crust–mantle boundary or in the upper mantle.

Chrome-diopsides (Table 15, anal. 3) comparable in composition with the clinopyroxene of the garnet peridotite nodules in kimberlite diatremes, occur in the garnet lherzolites and wehrlite blocks in the carbonatitic and ankaramitic glassy scoria of the Lashine volcano, northern Tanzania (Dawson *et al.*, 1970), in the xenoliths and xenocrysts of the olivine basalts of northern Kyushu, Japan (Ishibashi, 1970), in the ultramafic inclusions and as megacrysts in a nephelinite sill, Nandewar Mountains, north-eastern New South Wales (Wilkinson, 1975b), and in the peridotite xenoliths in the kimberlite dyke and sheet intrusions of south-west Greenland (Emeleus and Andrews, 1974).

Aggregates of a diopsidic pyroxene, commonly consisting of independent and irregularly shaped grains, and with outlines reminiscent of sections through garnet porphyroblasts, from the alkali olivine basalt of the Gimi River area, Nigeria, have been described by Wright (1972). Disequilibrium of the pyroxene is indicated by the presence of rims of titanaugite around the margins of the aggregates, similar to that in the groundmass of the basalt, and Wright has suggested that they may have developed by inversion from an original high-pressure garnet.

The diopside salites in the spinel lherzolite nodules in the damkjernite of the Fen alkaline complex, Norway, contain relatively large amounts of Al_2O_3 (5·87–6·80 weight per cent), but somewhat smaller contents of Cr_2O_3 (0·57–0·83 per cent) than is common in the clinopyroxenes of many lherzolite nodules in basic rocks, and indicate a crystallization pressure between 10–15 kbar in the temperature range 1200°–1250°C (Griffin, 1973; see also Griffin and Taylor, 1975). Other chrome diopsides include those of the lherzolite nodules in the hawaiite, Kyogle, New South Wales (Wilkinson and Binns, 1969), the garnet pyroxenite nodules in the alkali basalt breccias, Atakor, Hoggar, Sahara (Girod, 1967), and the diopside xenoliths in a camptonite dyke, associated with the Oktyabr'skiy alkalic stock, south-east Ukraine (Bayrakov, 1965).

Chrome-diopside nodules containing partially or wholly serpentinized exsolution lamellae of enstatite, Fe_{10}, and minute crystals of pyrope-almandine aligned along the enstatite lamellae have been described from the Jagersfontein kimberlite (Borley and Suddaby, 1975). The diopside crystals are highly sheared, and in kink bands the lamellae have been destroyed and the pyroxene recrystallized. On the textural evidence the enstatite exsolution probably occurred before the deformation, while the garnet exsolution is syntectonic in origin.

Zoned salites with titaniferous cores (2·5 per cent TiO_2) and outer zone of brownish titansalite (3·08 per cent TiO_2) have been described from the clinopyroxene–amphibole kaersutite inclusions of the Rogue Nublo alkalic basalts (Frisch and Schmincke, 1969), and endiopside, diopside and salite occur respectively in the peridotite, gabbro and amphibole-bearing xenoliths of the Teneguia volcano, La Palma, Canary Islands (Munoz *et al.*, 1974).

Endiopside (Table 16, anal. 3) has been described from the olivine-rich nodules in the basalt of Calton Hill, Derbyshire (Hamad, 1963). Al-rich endiopsides, containing between 10 and 14 weight per cent of the $CaAl_2SiO_6$ component, are present in amphibole-bearing cumulate inclusions in basanite within the western Grand Canyon (Best, 1975), and as phenocrysts consisting of a green core (Table 16, anal. 7), bordered by a brownish violet rim, in the porphyritic olivine diabase matrix of the Ile Ronde breccia, Montreal (Clark *et al.*, 1967).

The fabric and textural characteristics of diopside in lherzolite xenoliths in basalts from Western Europe and Hawaii have been studied by Mercier and Nicolas (1975).

Nodules in kimberlite. The calcium-rich pyroxenes occurring in the nodules of kimberlites have compositions approximating to $CaMgSi_2O_6$–$MgSiO_3$ solid solutions. Other cations are generally present in the range $\Sigma(Al+Fe^{3+}+Cr) = 0{\cdot}1–0{\cdot}2$, $\Sigma(Na+K) = 0{\cdot}05–0{\cdot}1$, and Fe/(Ca+Mg+Fe) ratios are between 0·02 and 0·04. The diopsides display a bimodal chemical character and form two groups, one calcium-rich embracing about 95 per cent of the analyses examined by Boyd (1970) and showing variable but restricted solid solution with enstatite and ferrosilite. The numerically smaller group consists of endiopsides and shows extensive solid solution with enstatite. The average content of Al_2O_3 of the calcium-rich and subcalcic pyroxenes is 2·1 weight per cent (range 1·0 to 2·5). The chromium content is high, up to about 3 per cent and averaging 1·3 per cent Cr_2O_3.

Two compositionally, Cr-rich and Cr-poor, endiopside megacryst groups are present in the Sloan kimberlite pipes. Colorado (Eggler and McCallum, 1976). The Cr-rich pyroxenes (1·0–2·3 weight per cent Cr_2O_3) have higher calcium content and Mg/(Mg+Fe) ratios (92–93) in contrast to the Cr-poor endiopsides (0·4–0·8 weight per cent Cr_2O_3, and Mg/(Mg+Fe) between 87 and 90). Both groups are associated with orthopyroxene (see p. 122) and garnet megacrysts that show comparable compositional characteristics.

The solvus curves limiting the solid solution of enstatite in diopside at 1 atm and 30 kbar indicate that the composition of diopside in equilibrium with enstatite in the temperature range 1000°–1400°C is relatively insensitive to pressure. The diopside–enstatite solvus temperatures at 30 kbar are shown in Fig. 140, and indicate that the calcium-rich diopsides equilibrated at temperatures between 1000°–1100°C, and the endiopsides around 1350°C. The lack of diopside–enstatite solid solutions with compositions intermediate between the two groups is difficult to explain in the context of the data at pressures up to 30 kbar, and the occurrence of diamond among the primary phases in kimberlites may indicate that the two compositional groups equilibrated at higher pressures (> 50 kbar) than those so far studied experimentally. A possible explanation may be related to the presence of a narrow pigeonite field in the $CaMgSi_2O_6$–$MgSiO_3$ system at 20 kbar (Kushiro, 1969d). The pigeonite field, however, is restricted to compositions that are considerably more magnesium-rich than the kimberlite subcalcic diopsides (see Figs. 44, 45 (p. 85)). In addition, the lower temperature limit of the pigeonite field is higher (1450°C) than that indicated for equilibration of the subcalcic diopsides, and an explanation of the compositional gap, based on the relations in the $CaMgSi_2O_6$–$MgSiO_3$ system at 20 kbar requires a shift of the pigeonite field at higher pressures towards more diopsidic compositions.

A detailed study of the diopsides in the nodules of the Thuba Putsoa kimberlite pipe (Dawson, 1962; Nixon *et al.*, 1963), by Boyd and Nixon (1972) has shown that there is a wide variation in the pyroxene composition that is related to the type of nodule in which the diopside occurs. Based on their study of the Thuba Putsoa diopsides, and the analyses of some forty-six diopsides from twenty African kimberlite pipes, Boyd and Nixon have demonstrated the presence of three compositional fields, namely one containing the common chrome diopsides of the garnet lherzolite nodules, one consisting of diopsides that have crystallized with ilmenite, sometimes in lamellar intergrowths, and a third group composed of the diopsides from tectonically deformed lherzolite nodules in discrete nodules and from griquaites (Fig. 140(a)). The diopsides of the first and third groups show a marked contrast in the ratio of CaO/(CaO+MgO+FeO), and are further distinguished by the high chromium and low titanium contents of the calcium-rich variety, and the low chromium and higher titanium in the subcalcic pyroxenes. Diopsides of the sheared and granular lherzolite nodules of the Premier kimberlite pipe, Pretoria, South Africa are described by Danchin and Boyd (1976).

Pyroxene–ilmenite intergrowths have been observed in South African, Yakutia, the Stockdale, Kansas, and Iron Mountain, Wyoming kimberlite pipes. The pyroxene–ilmenite intergrowths occurring as inclusions in kimberlite of the Monastery mine, Orange Free State, have been examined by Ringwood and

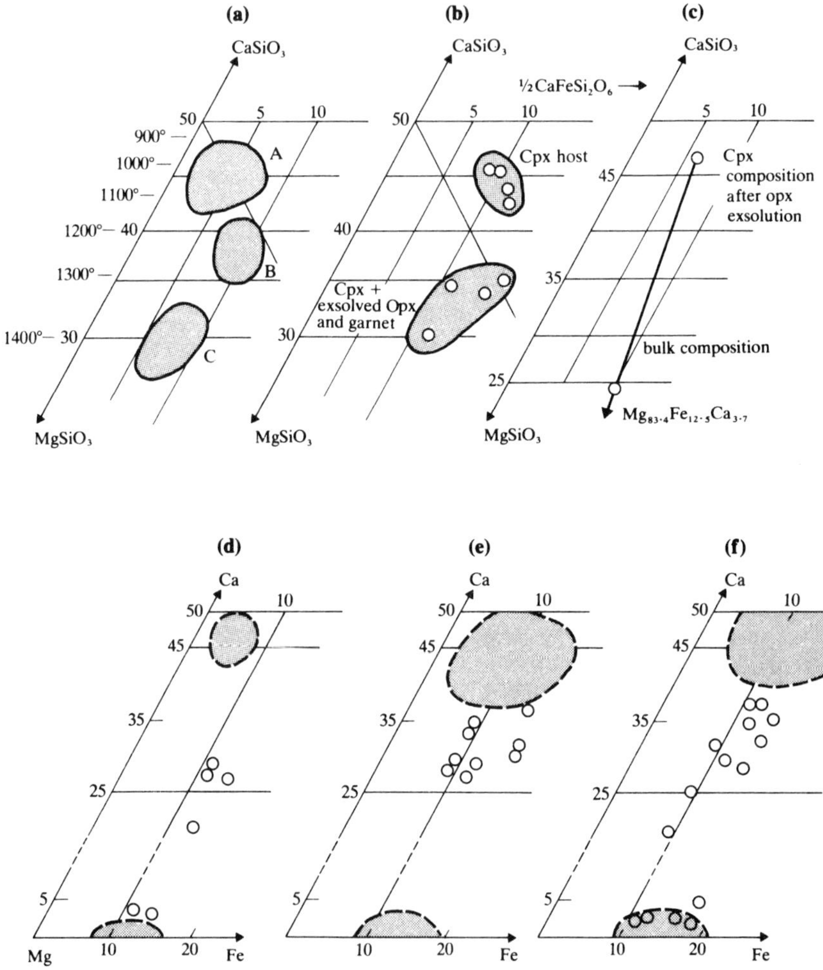

Fig. 140. (a) Composition fields of diopsides and endiopsides from African kimberlites plotted in a portion of the pyroxene quadrilateral (after Boyd and Nixon, 1972). Temperatures are for the Di(En) solvus in the system $Mg_2Si_2O_6$–$CaMgSi_2O_6$ at 30 kbar. A, field of common chrome-diopsides from garnet lherzolite nodules. B, diopside–endiopsides that have crystallized with ilmenite commonly in lamellar intergrowth. C, endiopsides of tectonically deformed lherzolite nodules, discrete nodules and griquaites. (b) Bulk and reconstructed compositions of diopsides of garnet pyroxenite xenoliths in the nepheline basalt tuff, Salt Lake Crater, Oahu, Hawaiian Islands (after Beeson and Jackson, 1970). (c) Compositions of subcalcic clinopyroxene, $Ca_{24.7}Mg_{65.1}Fe_{10.2}$, and exsolved diopside, $Ca_{47}Mg_{47.2}Fe_{5.8}$, and orthopyroxene, $Ca_{3.7}Mg_{83.4}Fe_{12.5}$. Subcalcic clinopyroxene megacryst in alkali trachybasalt, Glen Innes area, New South Wales (after Wilkinson, 1973). (d) Compositions of pyroxenes in pyroxenites, Ariège, France. (e) Compositions of pyroxenes in pyroxenites of other lherzolite complexes. (f) Compositions of pyroxenes in spinel lherzolite xenoliths in alkali basalts (after Conquéré, 1977). Open circles, primary pyroxenes. Stippled areas recrystallized pyroxenites.

Lovering (1970). The composition of the pyroxene, $(Na_{0\cdot12}Ca_{0\cdot63})(Mg_{0\cdot97}Fe^{2+}_{0\cdot16}Fe^{3+}_{0\cdot05}Ti_{0\cdot01}Al_{0\cdot07})(Si_{1\cdot96}Al_{0\cdot04})O_6$ lies just outside the endiopside field. The intergrowth fused to a glass and subsequently recrystallized at temperatures of 900°, 1200° and 1500°C at a pressure of 35 kbar, and at 1000°C and pressures of 60 and 90 kbar formed a pyroxene–ilmenite intergrowth essentially the same as the starting material. At a pressure of 103 kbar and temperature of 1000°C a small amount of garnet developed; at pressures greater than 110 kbar the recrystallized product contained between 85 and 90 per cent garnet. These data thus provide strong evidence that the pyroxene–ilmenite intergrowths crystallized as a single-phase high-pressure garnet (see also Wyatt, 1977). A comparable pyroxene $(Ca_{35\cdot0}Mg_{54\cdot8}Fe_{10\cdot2})$–ilmenite intergrowth from the Uintjes Berg pipe has been described by Boyd and Dawson (1972, see also Boyd and Nixon (1975)).

Endiopside–ilmenite intergrowths, the compositional range of which compares closely with that of the associated Cr-poor pyroxene megacrysts, have been described from the kimberlites of the Iron Mountain kimberlite district, Wyoming (Smith *et al.*, 1976).

Mechanisms to account for the intergrowths in addition to the eutectoid breakdown of Ti-rich garnet suggested by Ringwood and Lovering (1970), include exsolution of ilmenite from a clinopyroxene–ilmenite solid solution (Dawson *et al.*, 1970), and eutectic crystallization. The latter explanation is supported by the experimental crystallization of a melt of a natural clinopyroxene–ilmenite intergrowth, $Cpx_{71}Ilm_{29}$, from the Monastery mine, cooled from 1570°C ($\simeq$ 50°C above the liquidus) to 1206°C ($\simeq$ 200°C below the solidus) at 20°C/h at 38 kbar. The textural relationships of the crystallized melt, the composition of the clinopyroxene and ilmenite phases, and their crystallographic orientation are consistent with a eutectic origin (Wyatt *et al.*, 1975).

The transformation of hypersthene + plagioclase to diopside + garnet in nodules from the Aeromagnitnaya kimberlite pipe has been reported by Lutts (1965). The diopside, unusually rich in sodium (Table 14, anal. 14) occurs as reaction rims around the hypersthene where the orthopyroxene is in contact with plagioclase. The assemblage chrome-diopside–enstatite–spinel–forsterite–garnet has been observed in nodules of the Slyudyanka pipe, and Lutts has suggested that the assemblage may be due to the reaction:

enstatite + spinel + diopside = pyrope + olivine + chrome-diopside

Diopside is present in the sheared lherzolite and granular peridotite nodules of the diamondiferous Udachnaya and non-diamondiferous Obnazhennaya kimberlite pipes, Yakutia (Boyd *et al.*, 1976). Those of the latter have high sodium contents (2·6–3·9 weight per cent Na_2O). Diopside megacrysts in the Obnazhennaya pipe contain garnet lamellae oriented parallel to (110), (010) and (130), as well as small exsolved lamellae and blebs of orthopyroxene.

Inclusions of chrome-diopside occur in diamond of the Udachnaya pipe (Sobolev *et al.*, 1970). Microprobe analyses of two of the diopsides, together with the analysis of a pigeonite-type pyroxene that occurs as small rounded segregations within an isolated inclusion of a chrome diopside, are given in Table 28. The presence of the pigeonite segregation indicates that the equilibration temperature could not have been less than 1100°C even at a pressure of 1 atm.

Aluminium- and sodium-rich endiopside, $\simeq Ca_{37\cdot5}Mg_{55\cdot5}Fe_7$, (6–7 per cent Al_2O_3 and $\simeq$ 1·5 Na_2O) have been reported in lherzolite xenoliths, in a carbonatitic diatreme member of the lamprophyric dyke swarm, Moeraki River, New Zealand (Wallace, 1975).

Table 28 Microprobe analyses of chrome-diopside (and pyroxene segregation) inclusions in diamonds Udachnaya kimberlite pipe, Yakutia (Sobolev *et al.*, 1970)

					Number of ions		
	1.	**2.**	**3.**		**1.**	**2.**	**3.**
SiO_2	55·4	54·1	62·0				
TiO_2	0·07	0·10	—	Si	1·999	1·979	2·09
Al_2O_3	1·75	1·50	0·7	Al	0·001	0·021	—
Cr_2O_3	1·65	1·62	0·7	Al	0·072	0·045	0·03
FeO	1·36	1·11	1·6	Ti	0·002	0·002	—
MnO	0·03	0·03	—	Cr	0·048	0·048	0·02
MgO	16·6	16·8	27·0	Mg	0·893	0·917	1·36
CaO	21·4	22·1	9·3	Fe	0·041	0·033	0·04
Na_2O	1·40	1·31	0·4	Mn	—	—	—
K_2O	0·15	0·14	—	Na	0·100	0·092	0·02
				Ca	0·828	0·866	0·34
				K	0·007	0·007	—

Metamorphic rocks

Diopside, salite and ferrosalite are typical minerals of many metamorphic rocks. They are particularly characteristic of thermally metamorphosed calcium-rich sediments, but are also formed under the *P–T* conditions of regional metamorphism, and are common constituents of calcium- and magnesium-rich schists of both igneous and sedimentary parentage. Although diopside is a typical mineral of the pyroxene hornfels facies, in carbonate-rich environments it may form at lower temperatures, e.g. the occurrence of ferrosalite in layers of garnet–pyroxene gneiss in rocks of the hornblende granulite subfacies at Bear Mountain, New York, described by Dodd (1963). The pyroxene $Ca_{47·6}Mg_{15·1}Fe_{37·3}$ (MnO 0·76 per cent) and manganese-rich garnet assemblage was probably derived from an originally chert–manganoan siderite sediment. During the thermal metamorphism of siliceous dolomites diopside occurs relatively early in the mineralogical sequence of increasing decarbonation, and in such an environment is commonly preceded by tremolite or forsterite (Bowen, 1940):

$$\underset{\text{tremolite}}{Ca_2Mg_5Si_8O_{22}(OH)_2} + 3CaCO_3 + 2SiO_2 \rightarrow \underset{\text{diopside}}{5CaMgSi_2O_6} + 3CO_2 + H_2O$$

The P_{CO_2}–T equilibrium curve for the reaction:

$$\underset{\text{dolomite}}{CaMg(CO_3)_2} + \underset{\text{quartz}}{2SiO_2} \rightleftharpoons \underset{\text{diopside}}{CaMgSi_2O_6} + 2CO_2$$

in terms of thermodynamic quantities has been calculated by Weeks (1956a) who has shown, however, that diopside forms at a lower temperature than forsterite:

$$\underset{\text{dolomite}}{2CaMg(CO_3)_2} + \underset{\text{quartz}}{SiO_2} \rightleftharpoons \underset{\text{calcite}}{2CaCO_3} + \underset{\text{forsterite}}{Mg_2SiO_4} + 2CO_2$$

This sequence is at variance with that suggested by Bowen (1940) on the basis of a phase rule study of the progressive decarbonation of siliceous dolomites during contact metamorphism. The calculated equilibrium curves (Fig. 141) are in agreement, however, with the evidence from the regional metamorphism of dolomites (e.g. Laitakari, 1921; Eskola, 1922; Eckermann, 1950). The formation of forsterite before diopside in contact metamorphism is attributed by Weeks to the metastable persistence, as a consequence of rapid heating, of the dolomite–quartz

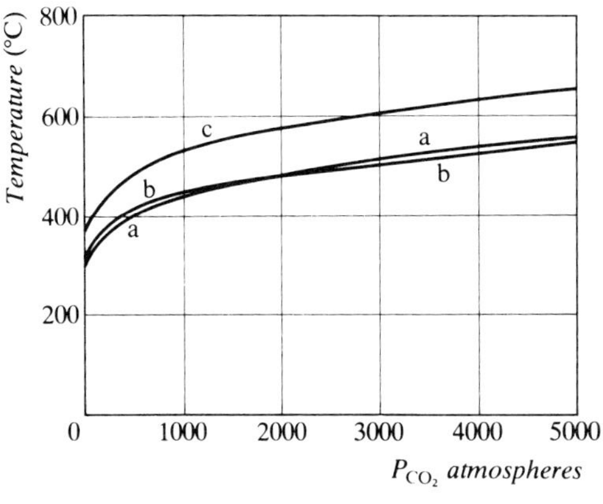

Fig. 141. Equilibrium *P–T* curves for reactions (a) $CaMg(CO_3)_2 + 2SiO_2 \rightarrow CaMgSi_2O_6 + 2CO_2$; (b) $CaMg(CO_3)_2 + SiO_2 \rightarrow CaCO_3 + MgSiO_3 + CO_2$; (c) $2CaMg(CO_3)_2 + SiO_2 \rightarrow 2CaCO_3 + Mg_2SiO_4 + 2CO_2$ during metamorphism of siliceous dolomites (after Weeks, 1956a).

assemblage above the reaction curve: dolomite + silica $\rightleftharpoons$ diopside + CO_2. With such a rise in temperature both of the above reactions would be thermodynamically possible at the same time. Under these conditions, which minerals of the pair diopside–forsterite will be the first to crystallize will depend on the relative degree of ordering of the Mg and Ca ions in the two structures. Thus forsterite, in which there is only one divalent ion, and in which all the Mg positions are equivalent, may be expected to form in preference to diopside. In rocks remaining at a high temperature and containing excess SiO_2 diopside will form subsequently from the reaction:

$$\underset{\text{forsterite}}{Mg_2SiO_4} + \underset{\text{calcite}}{2CaCO_3} + 3SiO_2 \rightarrow \underset{\text{diopside}}{2CaMgSi_2O_6} + 2CO_2$$

The consequent forsterite–diopside, or forsterite–calcite assemblages will persist unless the temperature rises sufficiently for the reaction:

$$\underset{\text{diopside}}{CaMgSi_2O_6} + \underset{\text{forsterite}}{Mg_2SiO_4} + \underset{\text{calcite}}{2CaCO_3} \rightarrow \underset{\text{monticellite}}{3CaMgSiO_4} + 2CO_2$$

to take place.

Bowen (1940) and Weeks (1956a) both considered the decarbonation reactions in siliceous dolomites in the absence of water. Those occurrences in which such hydrous phases as talc and tremolite are present imply, however, that water was present in the pore fluid during the decarbonation process. A reassessment of the significance, the order of some possible reactions in siliceous limestones and dolomites, and the effects of the partial pressure of water, particularly for $P_{H_2O} = P_{CO_2}$, on the reactions, has been discussed by Turner (1967) (see also chemistry section, p. 249).

Diopside, associated with brucite, serpentine and olivine, is present in the lower grade assemblages associated with the regional and contact metamorphism of peridotic rocks of the pennine region between the Rhaetic and Wallis Alps (Trommsdorff and Evans, 1974). Pyroxene is not present in the intermediate grades due to the reaction:

$$\underset{\text{serpentine}}{5Mg_3Si_2O_5(OH)_4} + \underset{\text{diopside}}{2CaMgSi_2O_6} = \underset{\text{tremolite}}{Ca_2Mg_5Si_8O_{22}(OH)_2} + \underset{\text{forsterite}}{6Mg_2SiO_4} + 9H_2O$$

but occurs with enstatite consequent on the instability of tremolite under the *P–T* conditions of the higher grade.

The formation of diopside as a result of the metamorphism of serpentinite by a later intrusion of gabbro, and from the leucocratic altered zones in the basic and ultrabasic Sobótka (Lower Silesia) Massif, has been described by Heflik (1967). In the latter occurrence the diopside, commonly associated with zoisite and quartz, formed during the pegmatite or pneumatolytic stage of the crystallization of the gabbro.

The formation of diopside in reaction zones between interbedded carbonate and pelitic beds of the calc-mica schists, Gassetts, Vermont, due to reactions of the type:

$$\underset{\text{phlogopite}}{KMg_3AlSi_3O_{10}(OH)_2} + \underset{\text{calcite}}{3CaCO_3} + \underset{\text{quartz}}{6SiO_2} = \underset{\text{diopside}}{3CaMgSi_2O_6} + \underset{\text{microcline}}{KAlSi_3O_8} + H_2O + 3CO_2$$

is discussed in terms of Fe–Mg exchange and X_{H_2O}/X_{CO_2} gradients by Thompson (1975a, b). The association of diopside and clinozoisite in the calc-silicate zones between interlayered marble and pelite in a roof-pendant of the granitic intrusive, Lake Willoughby, northern Vermont, is described by Thompson (1975a, b).

The phase relationships in the silicate-bearing calcite–dolomite marble pods and lenses in the high grade metamorphic terrain of the Lofoten–Vesteralen region, northern Norway, have been discussed by Glassley (1975). The carbonate bodies, enclosed in gneiss and mantled by a metasomatically altered zone of pyroxenite and a banded assemblage of clinopyroxene + phlogopite ± amphibole, contain both diopside–dolomite and forsterite + calcite associations. The different assemblages show a regional distribution and are discussed by Glassley in relation to variations in the fugacities of CO_2, H_2O and HF.

Salite and ferrosalite occur commonly in regionally metamorphosed calcium-rich sediments and basic igneous rocks belonging to the higher grades of the amphibolite facies, and are particularly characteristic in calc-silicate rocks in which the ratio of $CaO:Al_2O_3$ is $\simeq 1{\cdot}0$ (Winchester, 1974). Typical reactions include:

$$\underset{\text{hornblende}}{Ca_2(Mg,Fe^{2+})_3Al_4Si_6O_{22}(OH)_2} + \underset{\text{calcite}}{3CaCO_3} + \underset{\text{quartz}}{4SiO_2} = \underset{\text{salite}}{3Ca(Mg,Fe^{2+})Si_2O_6} + \underset{\text{anorthite}}{2CaAl_2Si_2O_8} + 3CO_2 + H_2O \quad (1)$$

$$\underset{\text{hornblende}}{2Ca_2(Mg,Fe^{2+})_3Al_4Si_6O_{22}(OH)_2} + \underset{\text{grossular}}{Ca_3Al_2Si_3O_{12}} + \underset{\text{quartz}}{2SiO_2} = \underset{\text{salite}}{3Ca(Mg,Fe^{2+})Si_2O_6} + \underset{\text{anorthite}}{4CaAl_2Si_2O_8} + \underset{\text{almandine}}{(Mg,Fe^{2+})_3Al_2Si_3O_{12}} + 2H_2O \quad (2)$$

The association of diopside–actinolite with plagioclase and quartz, or plagioclase, scapolite and quartz, occurs in the amphibolite facies dolomitic sandstones of the Precambrian Wallace formation, north-west of the Idaho batholith. Layers consisting wholly of salite or actinolite are present in the pyroxene–actinolite gneisses, and in some areas salite has crystallized around lenses of calcite, and in turn is surrounded by a shell of actinolite (Hietanen, 1971). The Fe/Mg ratios of the pyroxene (Table 17, anal. 5) and actinolite are respectively 0·22 and 0·21 and the main difference in composition of the pyroxene and amphibole, apart from the presence of hydroxyl in the latter, is the higher calcium content of the salite. The association of calcite with both the pyroxene and actinolite layers indicates that the controlling factor leading to the formation of the one rather than the other mineral, was not related to either high P_{CO_2} or P_{H_2O}, but was probably due to the amount of CaO available, and Hietanen concluded that pyroxene crystallized in those layers in which $[Ca/(Mg+Fe)] > 1$; salite and actinolite when $[Ca/(Mg+Fe)] < 1 > 0{\cdot}4$,

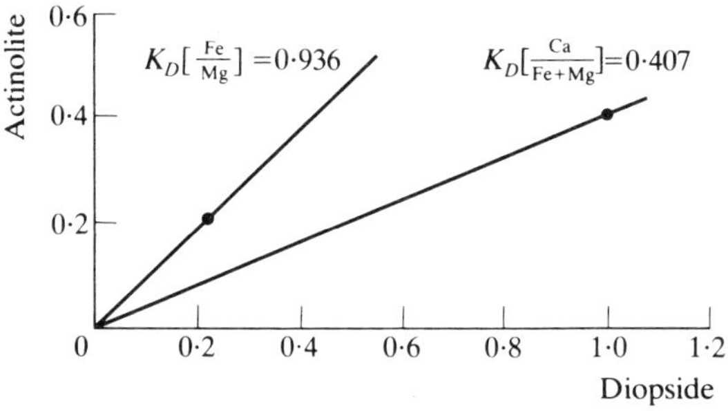

Fig. 142. Distribution coefficients for [Fe/Mg] and [Ca/(Fe + Mg)] in coexisting diopside and actinolite in metamorphosed dolomitic sandstone (after Hietanen, 1971).

and actinolite when the ratio was < 0·4 (Fig. 142). Ferrosalite (Table 18, anal. 5), associated with wollastonite, scapolite, garnet, potassium feldspar, quartz, sphene and magnetite, occurs in calc-silicate bands within the khondalites and leptynites of the granulite facies of the Eastern Ghats, India (Mukherjee and Rege, 1972).

In the regionally metamorphosed rocks of the amphibolite facies of the central Abukuma Plateau, Japan (Shidô, 1958; Miyashiro, 1958), the $(Mg+Fe^{2+})$:Ca ratio of the salites (e.g. Table 17, anal. 23) and ferrosalites increases with increasing grade of metamorphism in conformity with the greater solubility at higher temperatures of $(Mg+Fe^{2+})$ in diopside–hedenbergite. The clinopyroxenes of amphibolites of basic igneous origin are commonly somewhat richer in sodium. Thus the salites of the amphibolites associated with the Alpine-type ultramafic rocks of the Dinaride ophiolite zone, Yugoslavia (Pamić *et al.*, 1973) have a jadeite–acmite content of between 10 and 20 per cent ($Di_{62}Hd_{26}Jd_{12}$–$Di_{56}Hd_{24}Jd_{20}$).

Although in many regionally metamorphosed rocks of higher grade the clinopyroxene has an augitic composition, salite (Table 17, anal. 3) occurs in some charnockitic rocks from Madras (Howie, 1955) and Sri Lanka (Jayawardena and Carswell, 1976). Salites and ferrosalites, ranging in composition from 35 to 75 mol. per cent of the hedenbergite component, are common constituents of the charnockitic rocks of the West Uusimaa complex, south-western Finland (Parras, 1958). The development of salite, $Ca_{0.87}(Mg_{0.71}Fe_{0.33}Ti_{0.01}Al_{0.07})(Si_{1.93}Al_{0.06})O_6$, and plagioclase at the interface of pyroxene-bearing amphibolites and quartzo-feldspathic rocks in the hornblende-granulite facies of the Grenville province has been described by Schrijver (1973). Diopside-rich lenses, layers and intercalations, formed during the progressive metamorphism of siliceous dolomites in the Precambrian granulites of the Brazilian basement are reported by Sighinolfi and Fujimori (1974).

In some calc-silicate rocks, hedenbergite is a product of thermal metamorphism of iron-rich sediments, where its formation is probably due to the instability of ferroactinolite with rising temperature:

$$\underset{\text{ferroactinolite}}{Ca_2Fe_5^{2+}Si_8O_{22}(OH)_2} + 3CaCO_3 + 2SiO_2 \rightarrow \underset{\text{hedenbergite}}{5CaFeSi_2O_6} + 3CO_2 + H_2O$$

The association of hedenbergite with grunerite has been described from a number of localities and its formation here may be due to the reaction:

$$\underset{\text{ferroactinolite}}{7Ca_2Fe_5Si_8O_{22}(OH)_2} \rightarrow \underset{\text{hedenbergite}}{14CaFeSi_2O_6} + \underset{\text{grunerite}}{3Fe_7Si_8O_{22}(OH)_2} + 4SiO_2 + 4H_2O$$

Magnetite–hedenbergite–grunerite quartzites and hedenbergite eulysites have been described by Miles (1946) in the regionally metamorphosed sedimentary banded ironstones (jasper bars) of Western Australia and the occurrence of manganhedenbergite associated with the manganese-rich grunerite, dannemorite, from the Broken Hill orebodies by Stillwell (1959). Hedenbergite, associated with eulite, ferroactinolite, andradite and pyroxmangite has been described from the metamorphosed iron-rich sediments in the Satnur-Halguru area, Mysore State (Devaraju and Sadashivaiah, 1966). Hedenbergite (γ 1·755 $\simeq$ 95 per cent $CaFeSi_2O_6$), associated with grunerite and fayalite, also occurs in the metamorphosed iron-rich formation, Raahe, Finland (Hytönen, 1968), and manganoan hedenbergite in the metamorphosed (kyanite to sillimanite zone) quartz–carbonate iron formation of the Labrador Trough due to the reaction

$$\underset{\text{ferrodolomite}}{Ca(Fe,Mg)(CO_3)_2} + 2SiO_2 = \underset{\text{hedenbergitic pyroxene}}{Ca(Fe,Mg)Si_2O_6} + 2CO_2$$

has been described by Klein (1973).

Ferrohedenbergite is uncommon in metamorphic rocks but has been described from the eulysite of the Mariupol' iron deposit (Val'ter, 1969) and from the Lake Chudzyavr region, Kola Peninsula (Bondarenko and Dagelaĭskiĭ, 1961).

Metasomatic Rocks and Skarns

Diopside, salite and ferrosalite are characteristic minerals in many skarn rocks. Thus diopside-bearing skarns, in which the diopside (Table 14, anal. 7) is associated with forsterite, chondrodite or clinohumite, monticellite and magnetite have been described by Tilley (1951) from the dolomite contact skarns of the Broadford area, Skye. During the assimilation of limestone and chert by the olivine dolerite magma of the plug at Carneal, Co. Antrim (Sabine and Young, 1975), salite was the earlier phase to crystallize in the resulting contaminated rock in which the pyroxene replaced the olivine of the original dolerite magma. Symplectite intergrowths of diopside and scapolite have been reported in the contact assemblages between pegmatites and limestone at Mansjö Mountain (Eckermann, 1922), and of diopside and plagioclase in the contact rocks of Pargas (Laitakari, 1921). A massive development of diopside in the metasomatized volcanic and intrusive rocks of the Kustanai region, Kazakstan, has been reported by Zavaritsky and Kirova (1973). In the calc-silicate skarns in limestone at Lough Anure, Co. Donegal (Pitcher, 1950), diopside is commonly associated with wollastonite, grossular, vesuvianite and tremolite and displays an antipathetic relationship with the latter mineral. Salite (e.g. Table 17, anal. 24) is also an extremely widespread and abundant constituent of the skarns in the gneisses of the Holsteinborg district, west Greenland (Pauly, 1948).

Pyroxenes of the diopside–hedenbergite series, ranging in composition from $Ca_{1\cdot03}(Mg_{0\cdot88}Fe_{0\cdot10}Mn_{0\cdot01})(Si_{1\cdot96}Al_{0\cdot05})O_6$, to nearly pure hedenbergite $Ca_{1\cdot01}(Mg_{0\cdot02}Fe_{0\cdot94}Mn_{0\cdot07})(Si_{1\cdot96}Al_{0\cdot02})O_6$, occur in the skarns at the junction of the carbonate-bearing rocks of the Mount Morrison pendant in the Sierra Nevada batholith (Morgan, 1975). The more diopside-rich members are associated with andradite and the more iron-rich occur in hedenbergite–grossular assemblages. Some of the skarns have been affected by a retrograde reaction and single crystals of hedenbergite are commonly replaced by a symplectic intergrowth consisting of amphibole, calcite and quartz:

$$\underset{\text{hedenbergite}}{5CaFeSi_2O_6} + 3CO_2 + H_2O \rightarrow \underset{\text{ferroactinolite}}{Ca_2Fe_5Si_8O_{22}(OH)_2} + \underset{\text{calcite}}{3CaCO_3} + \underset{\text{quartz}}{2SiO_2}$$

Diopside-bearing 'pegmatites' are common in many limestone areas which have been intruded by granites. The formation of diopside in this paragenesis is considered by Eskola (1922) to be due to the assimilation of calcium within the pegmatite fissures and not to crystallization from granitic parent magmas, as such diopside-bearing pegmatites are only developed where they are in contact with the limestone. Ferrosalite (Hd_{64}), together with wollastonite and andradite, in pegmatites contaminated by skarn material at Carlingford, Co. Lough, Eire, has been recorded by Osborne (1932).

Irregular transgressive veins of ferrosalite ($Di_{35}Hd_{65}$) with zoisite, quartz and calcite, and of diopside and grossular, as well as the extensive development of diopside along the original bedding planes of metamorphosed sediments and lavas, occur in the amphibolite schists near Southern Cross, Western Australia. At this locality the formation of diopside is considered by Wilson (1953) to be the result of calcium metasomatism, and the calcium-rich emanations responsible for the diopsidization are believed to have been derived from the associated basic lavas during the alteration of the latter by alkali metasomatism. The formation from metasomatic fluids, of diopside and salite, associated with hornblende, clinozoisite, prehnite, xonotlite, biotite, muscovite, quartz and calcite, in reaction skarns between calcareous and aluminous rocks, has been described by Francis (1958).

Skeletal inclusions of diopside (α 1·671, γ 1·697, $\simeq Di_{91}$), associated with olivine, Fa_{10}, occur in the platiniferous dunite zone of the dunite–pyroxenites of the Urals (Yefimov and Ivanova, 1964). The passage outwards to the pyroxenite zone is marked by an increase in iron content, and the mineral develops a phenocrystal habit. The continuous variation in the composition of the diopside–hedenbergite solid solution, to Di_{85}–Di_{80} in the pyroxenite, is considered to result from a process of diffusive metasomatism. The rodingitic metasomatic replacement of gabbro, by a green chrome diopside (Cr_2O_3 1·15 per cent), white diopside, zoisite and hydrogrossular, at Rockville, Maryland, has been reported by Larrabee (1970). Bilgrami and Howie (1960) have described diopside (Table 14, anal. 11) from a rodingite dyke, Hindubagh, Pakistan, and it has been reported, associated with prehnite in veins in the rodingites from the fracture zones of the equatorial Mid-Atlantic Ridge (Honnorez and Kirst, 1975). The occurrence of metasomatic ferroan salite in narrow zones bordering intrusive fingers of microcline granite in the calc granulites, and in the calc granulite xenoliths, in granite have been reported by Rao and Rao (1971). Diopside (Table 14, anal. 12) occurs with fassaite (see Table 39, anal. 7) in a pyroxene-rich rock formed by the metasomatism of dolerite, Vilyuy, Yakutia (Ginzburg, 1969). The two pyroxenes do not display either intergrowths or overgrowths and are considered to be cogenetic.

Diopside is one of the principal minerals of the skarns in the Grenville province (Shaw *et al.*, 1963a, b). The predominant assemblage is diopside (Table 14, anal. 13)–scapolite–sphene formed under upper amphibolite facies metamorphism, but diopside also occurs in the pink calcite and fluorite, as well as the pyroxenite, skarns. More iron-rich members of the diopside–hedenbergite series occur in the Grenville amphibolites, gneisses and granodiorite (e.g. ferrosalite, Table 18, anal. 9). In the Lower Wabush Iron Formation, Labrador, diopside is associated with grunerite, actinolite and magnetite, and ferrosalite (Table 18, anal. 4) occurs in the quartz–grunerite–ferrosalite assemblage (Klein, 1966).

Bed-like ridges and irregular masses of diopside ($Wo_{48\cdot7}$–$Wo_{50\cdot3}$, $En_{48\cdot4}$–$En_{43\cdot8}$, $Fs_{1\cdot8}$–$Fs_{7\cdot5}$), surrounded successively by zones of tremolite and calcite, enclosed within the siliceous dolomite marbles of the Tweed area, Ontario, have been described by Cermignani and Anderson (1973). Diopside was the first calc-silicate to form during the metamorphism and was derived from the reaction of the dolomitic host rock and the enclosed silica-rich beds:

$CaMg(CO_3)_2 + 2SiO_2 = CaMgSi_2O_6 + 2CO_2$

The introduction of water into the system at a later stage resulted in the crystallization of tremolite, some partially replacing the pyroxene, and calcite due to the reactions:

$$5CaMgSi_2O_6 + 3CO_2 + H_2O = Ca_2Mg_5Si_8O_{22}(OH)_2 + 3CaCO_3 + 2SiO_2 \quad (1)$$

$$4CaMgSi_2O_6 + CaMg(CO_3)_2 + CO_2 + H_2O = Ca_2Mg_5Si_8O_{22}(OH)_2 + 3CaCO_3 \quad (2)$$

The occurrence of acicular chrome diopside, associated with uvarovite and calcite, in vug-like cavities in a diopside pyroxenite dyke, Normandie, Quebec, has been described by Grubb (1965). The diopside displays a strong longitudinal colour zoning, from off-white (Cr_2O_3 nil), adjacent to the cavity walls, to a greenish colour (Cr_2O_3 0·92 per cent) at the centre of the cavity. Salite ($Di_{74-55}Hd_{26-45}$) is a common constituent of the skarn of the Åsgruvan mine, Norberg, central Sweden (Sarap, 1957), and a ferrosalite–spinel skarn zone has been described at the junction of the nepheline syenite of the Synnyr pluton with the adjacent marble (Andreyev, 1965). Sodium-rich diopsides have been reported from the manganese formation, India (Roy, 1966, 1971; Roy and Mitra, 1964), and manganoan ferrosalites (MnO 3·76 and 8·39) from skarn zones at Dognecea, Romania (Vlad and Vasiliu, 1969).

A manganoan ferrosalite (Table 18, anal. 10) associated with bustamite has been described by Tilley (1946) from the Treburland manganese mine, Cornwall. This paragenesis is of considerable interest in view of the relationship between hedenbergite and the low-temperature wollastonite solid solution series, and between johannsenite and bustamite. Material of the bulk composition of the pyroxene and bustamite could exist at higher temperatures as a single (Fe,Mn)-wollastonite phase or, at a lower temperature, as a single clinopyroxene phase. The hedenbergite $\rightleftharpoons$ wollastonite inversion temperature is 965°C, that of johannsenite $\rightleftharpoons$ bustamite is approximately 830°C (see p. 416); thus at Treburland the metamorphism probably occurred at an intermediate temperature at which both manganoan ferrosalite and bustamite crystallized in association.

Hedenbergite occurs commonly in limestone skarns (Federico and Fornaseri, 1953; Zharikov and Podlessky, 1955; Ito, 1962), and in this paragenesis is often associated with Pb, Zn and Ca metasomatism. Manganoan hedenbergite (MnO 11·47 per cent) occurs in the skarn zones at Dognecea, Banat, Romania, and formed by the pyrometasomatic action of iron- and manganese-rich fluids with carbonate rocks (Vlad and Vasiliu, 1969). The reaction of later and lower temperature hydrothermal fluids subsequently led to the alteration of the hedenbergite to ilvaite. The alteration of manganoan hedenbergite to ilvaite has also been described by Machairas and Blais (1966). Manganoan hedenbergite is found in a pyrometasomatic deposit, Paymaster Canyon, Tonopah, Nevada (Gulbrandsen and Gielow, 1960), from the Perran lode breccia, Cornwall (Henley, 1971), and has been described with bustamite in the skarn at the Nakatatsu mine, Fukui Prefecture, Japan (Tokunaga, 1965). Hedenbergite (Table 19, anal. 2), associated with epidote, occurs in the graphite-bearing skarn at the marble–altered gneiss contact of the Hida complex, Gifu Prefecture (Shimazaki, 1967) and is associated with andradite, magnetite and quartz in layers in the marble of Hartenstein, Austria (Scharbert, 1966).

References

Adams, L. H., 1914. Calibration table for copper-constantan and platinum-platinrhodium thermoelements. *J. Amer. Chem. Soc.*, **36**, 65.

Akella, J. and **Kennedy, G. C.,** 1971. Studies on anorthite + $diopside_{ss}$–$hedenbergite_{ss}$ at high pressures and temperatures. *Amer. J. Sci.*, **270**, 155–165.

Aleksandrov, K. S., Ryzhova, T. V. and **Belikov, B. P.,** 1964. The elastic properties of pyroxenes. *Soviet Physics: Crystallography*, **8**, 589–591.

Allen, E. T. and **White, W. P.,** 1909. Diopside and its relations to calcium and magnesium metasilicates: with optical study by F. E. Wright and E. S. Larsen. *Amer. J. Sci.*, ser. 4, **27**, 1–47.

Allen, V. T. and **Fahey, J. J.,** 1953. Rhodonite, johannsenite and ferroan johannsenite at Vanadium, New Mexico. *Amer. Min.*, **38**, 883–890.

Andreyev, G. V., 1965. Magnesian skarn at a dolomite-nepheline syenite contact. *Dokl. Acad. Sci., U.S.S.R., Earth Sci. Sect.*, **161**, 159–162.

Anwar, Y. M., 1955. A clinopyroxene from the granophyre of Meall Dearg, Skye. *Geol. Mag.*, **92**, 367–373.

Aoki, K-I., 1964. Clinopyroxenes from alkaline rocks of Japan. *Amer. Min.*, **49**, 1199–1223.

Aoki, K-J., 1968. Petrogenesis of ultrabasic and basic inclusions in alkali basalts, Iki Island, Japan. *Amer. Min.*, **53**, 241–256.

Aoki, K-I., 1971. Petrology of mafic inclusions from Itinome-gata, Japan. *Contr. Min. Petr.*, **30**, 314–331.

Arndt, N. T., 1976. Melting relations of ultramafic lavas (komatiites) at 1 atm and high pressure. *Carnegie Inst. Washington, Ann. Rept. Dir. Geophys. Lab.*, 1975–76, 555–562.

Ashby, D. F., 1946. Pyroxenes from the Lower Carboniferous basalts of the Old Pallas area, Co. Limerick. *Min. Mag.*, **27**, 195–197.

Atkins, F. B., 1969. Pyroxenes of the Bushveld intrusion, South Africa. *J. Petr.*, **10**, 222–249.

Babkine, J., Conquéré, F. and **Vilminot, J-C.,** 1968. Les caractères particuliers de volcanisme au nord Montpellier: l'absarokite du Pouget; la ferrisalite sodique de Grabels. *Bull. Soc. franç. Min. Crist.*, **91**, 141–150.

Bancroft, G. M., Williams, P. G. L. and **Burns, R. G.,** 1971. Mössbauer spectra of minerals along the diopside hedenbergite tie line. *Amer. Min.*, **56**, 1617–1625.

Banno, S., 1970. Classification of eclogites in terms of physical conditions of their origin. *Phys. Earth Planet. Interiors*, **3**, 405–421.

Barker, F., Wones, D. R., Sharp, W. N. and **Desborough, G. A.,** 1975. The Pikes Peak batholith, Colorado Front Range and a model for the origin of the gabbro–anorthosite–syenite–potassic granite suite. *Precambrian Res.*, **2**, 97–160.

Barron, L. M., 1972a. Thermodynamic multi-component silicate equilibrium phase calculations. *Amer. Min.*, **57**, 809–823.

Barron, L. M., 1972b. Cooling paths calculated in some ternary systems. *Proc. 24th Intern. Geol. Congr., Montreal*, Sect. 2, 241–250.

Bayrakov, V. V., 1965. Occurrence of eclogitic rock xenoliths in a lamprophyre dike of the Oktyabr'skiy alkalic stock. *Dokl. Acad. Sci. U.S.S.R., Earth Sci. Sect.*, **156**, 137–140.

Beeson, M. H. and **Jackson, E. D.,** 1970. Origin of the garnet pyroxenite xenoliths at Salt Lake Crater, Oahu. *Min. Soc. America, Spec. Paper* **3**, 95–112.

Bell, J. D., 1966. Granites and associated rocks of the eastern part of the Western Redhills complex, Isle of Skye. *Trans. Roy. Soc. Edin.*, **66**, 307–343.

Bell, P. M. and **Davis, B. T. C.,** 1965. Temperature-composition section for jadeite–diopside. *Carnegie Inst., Washington, Ann. Rept. Dir. Geophys. Lab.*, 1964–65, 120–123.

Bell, P. M. and **Davis, B. T. C.,** 1967. Investigation of a solvus in the system jadeite–diopside. *Carnegie Inst., Washington, Ann. Rept. Dir. Geophys. Lab.*, 1965–66, 239–241.

Bell, P. M. and **Davis, B. T. C.,** 1969. Melting relations in the system jadeite–diopside at 30 and 40 kilobars. *Amer. J. Sci.*, **267-A** (Schairer vol.), 17–32.

Best, M. G., 1975. Amphibole-bearing cumulate inclusions, Grand Canyon, Arizona and their bearing on silica-undersaturated hydrous magmas in the upper mantle. *J. Petr.*, **16**, 212–236.

Biggar, G. M., 1972. Diopside, lithium metasilicate, and the 1968 temperature scale. *Min. Mag.*, **38**, 768–770.

Bilgrami, S. A. and **Howie, R. A.,** 1960. The mineralogy and petrology of a rodingite dike, Hindubagh, Pakistan. *Amer. Min.*, **45**, 791–801.

Boksha, O. N., Varina, T. M. and **Kostyukova, I. G.,** 1974. Structure of the absorption spectrum of chromediopsides. *Soviet Physics – Crystallography*, **19**, 241–242.

Bondarenko, L. P. and **Dagelaĭskiĭ, B. V.,** 1961. Eulysites from the Lake Chudz'yavr region (Kola peninsula). *Zap. Vses. Min. Obshch. (Mem. All-Union Min. Soc.)*, **90**, 408–424 (in Russian).

Borisenko, L. F., 1967a. Trace elements in pyroxenes and amphiboles from ultramafic rocks of the Urals. *Min. Mag.*, **36**, 403–410.

Borisenko, L. F., 1967b. Correlation among iron, vanadium and titanium in pyroxenes. *Geochem. International*, **4**, 263–269.
Borley, G. D., 1967. Potash-rich volcanic rocks from southern Spain. *Min. Mag.*, **36**, 364–379.
Borley, G. D. and **Suddaby, P.**, 1975. Stressed pyroxenite nodules from the Jagersfontein kimberlite. *Min. Mag.*, **40**, 6–12.
Bowen, N. L., 1914. The ternary system: diopside–forsterite–silica. *Amer. J. Sci.*, 4th ser., **38**, 207–264.
Bowen, N. L., 1915. The crystallization of haplobasaltic, haplodioritic and related magmas. *Amer. J. Sci.*, 4th ser., **40**, 161–185.
Bowen, N. L., 1922. Genetic features of alnoitic rocks from Isle Cadieux, Quebec. *Amer. J. Sci.*, 5th ser., **3**, 1–34.
Bowen, N. L., 1928. *The evolution of the igneous rocks*. Princeton Univ. Press.
Bowen, N. L., 1940. Progressive metamorphism of siliceous dolomite. *J. Geol.*, **48**, 225–274.
Bowen, N. L., **Schairer, J. F.** and **Posnjak, E.**, 1933. The system $CaO–FeO–SiO_2$. *Amer. J. Sci.*, 5th ser., **26**, 193–284.
Bown, M. G. and **Gay, P.**, 1957. Observations on pigeonite. *Acta Cryst.*, **10**, 440–441.
Bown, M. G. and **Gay, P.**, 1959. The identification of oriented inclusions in pyroxene crystals. *Amer. Min.*, **44**, 592–602.
Bown, M. G. and **Gay, P.**, 1960. An X-ray study of exsolution phenomena in the Skaergaard pyroxenes. *Min. Mag.*, **32**, 379–388.
Boyd, F. R., 1970. Garnet peridotites and the system $CaSiO_3–MgSiO_3–Al_2O_3$. *Min. Soc. America, Spec. Paper* **3**, 63–75.
Boyd, F. R. and **Dawson, J. B.**, 1972. Kimberlite garnets and pyroxene–ilmenite intergrowths. *Carnegie Inst. Washington, Ann. Rept. Dir. Geophys. Lab.*, 1971–72, 373–378.
Boyd, F. R. and **England, J. L.**, 1958. Melting of diopside under high pressure. *Carnegie Inst. Washington, Ann. Rept. Dir. Geophys. Lab.*, 1957–58, 173–174.
Boyd, F. R. and **England, J. L.**, 1963. Effect of pressure on the melting of diopside, $CaMgSi_2O_6$ and albite, $NaAlSi_3O_8$, in the range up to 50 kilobars. *J. Geophys. Res.*, **68**, 311–323.
Boyd, F. R., **Fujii, T.** and **Danchin, R. V.**, 1976. A noninflected geotherm for the Udachnaya kimberlite pipe, U.S.S.R. *Carnegie Inst. Washington, Ann. Rept. Dir. Geophys. Lab.*, 1975–76, 523–531.
Boyd, F. R. and **Nixon, P. H.**, 1972. Ultramafic nodules from the Thuba Putsoa kimberlite pipe. *Carnegie Inst. Washington, Ann. Rept. Dir. Geophys. Lab.*, 1971–72, 362–373.
Boyd, F. R. and **Nixon, P. H.**, 1973. Structure of the upper mantle beneath Lesotho. *Carnegie Inst. Washington, Ann. Rept. Dir. Geophys. Lab.*, 1972–73, 431–445.
Boyd, F. R. and **Nixon, P. H.**, 1975. Origins of the ultramafic nodules from some kimberlites of northern Lesotho and the Monastery Mine, South Africa. *Phys. Chem. Earth*, **9**, 431–454.
Boyd, F. R. and **Schairer, J. F.**, 1957. The join $MgSiO_3–CaMgSi_2O_6$. *Carnegie Inst. Washington, Ann. Rept. Dir. Geophys. Lab.*, 1956–57, 223–225.
Boyd, F. R. and **Schairer, J. F.**, 1964. The system $MgSiO_3–CaMgSi_2O_6$. *J. Petr.*, **5**, 545–560.
Bragg, W. L., 1937. *Atomic Structure of Minerals*. Cornell Univ. Press.
Brooks, C. K. and **Rucklidge, J. C.**, 1973. A Tertiary lamprophyre dike with high pressure xenoliths and megacrysts from Wiedemanns Fjord, east Greenland. *Contr. Min. Petr.*, **42**, 197–212.
Brown, G. M., 1960. The effect of ion substitution on the unit cell dimensions of the common clinopyroxenes. *Amer. Min.*, **45**, 15–38.
Brown, G. M. and **Vincent, E. A.**, 1963. Pyroxenes from the late stages of fractionation of the Skaergaard intrusion, east Greenland. *J. Petr.*, **4**, 175–197.
Brown, P. E., 1973. A layered plutonic complex of alkali basalt parentage: the Lilloise intrusion, east Greenland. *J. Geol. Soc.*, **129**, 405–418.
Bryan, W. B., 1969. An olivine gabbro–microsyenite intrusion from the Carnarvon Range, Queensland, Australia. *Carnegie Inst. Washington, Ann. Rept. Dir. Geophys. Lab.*, 1967–68, 247–250.
Bunch, T. E. and **Olsen, E.**, 1974. Restudy of pyroxene–pyroxene equilibration temperatures for ordinary chondritic meteorites. *Contr. Min. Petr.*, **43**, 83–90.
Cameron, M., **Sueno, S.**, **Prewitt, C. T.** and **Papike, J. J.**, 1973. High temperature crystal chemistry of acmite, diopside, hedenbergite, jadeite, spodumene and ureyite. *Amer. Min.*, **58**, 594–618.
Capedri, S., 1974. Genesis and evolution of a typical alpine-type peridotite mass under deep seated conditions (central Euboea, Greece). *Bull. Soc. Geol. Italy*, **93**, 81–114.
Carmichael, I. S. E., 1962. Pantelleritic liquids and their phenocrysts. *Min. Mag.*, **33**, 86–113.
Carmichael, I. S. E., 1967. The mineralogy and petrology of the volcanic rocks from the Leucite Hills, Wyoming. *Contr. Min. Petr.*, **15**, 24–66.
Carswell, D. A., 1968. Picritic magmas – residual dunite relationships in garnet peridotite at Kalskaret nr. Tufjord, South Norway. *Contr. Min. Petr.*, **19**, 97–124.
Carswell, D. A., 1974. Comparative equilibration temperatures and pressures of garnet lherzolites in Norwegian gneisses and in kimberlite. *Lithos*, **7**, 113–121.

Cawthorn, R. G. and **Collerson, K. D.,** 1974. The recalculation of pyroxene end-member parameters and the estimation of ferrous and ferric iron content from electron microprobe analyses. *Amer. Min.*, **59**, 1203–1208.

Cermignani, C. and **Anderson, G. M.,** 1973. Origin of a diopside–tremolite assemblage near Tweed, Ontario. *Can. J. Earth Sci.*, **10**, 84–90.

Challis, G. A., 1965. The origin of New Zealand ultramafic intrusions. *J. Petr.*, **6**, 322–364.

Chao, P., 1974. [Experimental studies of hedenbergite alteration.] *Geochimica*, 196–202 (Chinese with English abstract) (M.A. 75-2150).

Chao, P. and **Tsao, R-L.,** 1975. [Experimental study of the alteration of hedenbergite, diopside and andradite.] *Geochimica*, 63–74 (Chinese with English abstract) (M.A. 76-503).

Chayes, F., 1968. On locating field boundaries in simple phase diagrams by means of discriminant functions. *Amer. Min.*, **53**, 359–371.

Christie, J. M., Lally, J. S., Heuer, A. H., Fisher, R. M., Griggs, D. T. and **Radcliffe, S. V.,** 1971. Comparative electron petrography of Apollo 11, Apollo 12 and terrestrial rocks. *Proc. Second Lunar Sci. Conf. (Suppl. to Geochim. Cosmochim. Acta* **35**), **1**, 69–89.

Clark, J. R., Appleman, D. E. and **Papike, J. J.,** 1969. Crystal-chemical characterization of clinopyroxenes based on eight new structure refinements. *Min. Soc. America, Spec. Paper*, **2**, 31–50.

Clark, S. P., Schairer, J. F. and **de Neufville, J.,** 1962. Phase relations in the system $CaMgSi_2O_6$–$CaAl_2SiO_6$–H_2O at low and high pressure. *Carnegie Inst. Washington, Ann. Rept. Dir. Geophys. Lab.*, 1961–62, 59–68.

Clark, T. H., Kranck, E. H. and **Philpotts, A. R.,** 1967. Ile Ronde breccia – Montreal. *Can. J. Earth Sci.*, **4**, 507–514.

Cohen, L. H., Ito, K. and **Kennedy, G. C.,** 1967. Melting and phase relations in an anhydrous basalt to 40 kilobars. *Amer. J. Sci.*, **265**, 475–518.

Coleman, L. C., 1962. Effect of ionic substitution in the unit-cell dimensions of synthetic diopside. *Geol. Soc. America* (Buddington vol.), 429–446.

Conquéré, F., 1977. Pétrologie des pyroxénites litées dans les complexes ultra-mafiques de l'Ariège (France) et autres gisements de lherzolite à spinelle. I. Compositions minéralogique et chimique évolution des conditions d'équilibre des pyroxénites. *Bull. Soc. franç. Min. Crist.*, **100**, 42–80.

Cox, K. G. and **Jamieson, B. G.,** 1974. The olivine-rich lavas of Nuanetsi: a study of polybaric magmatic evolution. *J. Petr.*, **15**, 269–301.

Cundari, A., 1975. Mineral chemistry and petrogenetic aspects of the Vico lavas, Roman volcanic region, Italy *Contr. Min. Petr.*, **53**, 129–144.

Czamanske, G. K. and **Wones, D. R.,** 1973. Oxidation during magmatic differentiation, Finnmarka complex, Oslo area, Norway: Part 2, The mafic silicates. *J. Petr.*, **14**, 349–380.

Danchin, R. V. and **Boyd, F. R.,** 1976. Ultramafic nodules from the Premier kimberlite pipe, South Africa. *Carnegie Inst. Washington, Ann. Rept. Dir. Geophys. Lab.*, 1975–76, 531–538.

Dane, E. B., 1941. Densities of molten rocks and minerals. *Amer. J. Sci.*, **239**, 809–821.

Davidson, C. F., 1943. The Archaean rocks of the Rodil district, South Harris, Outer Hebrides. *Trans. Roy. Soc. Edin.*, **61**, 71–112.

Davis, B. T. C., 1964. The system diopside–forsterite–pyrope at 40 kilobars. *Carnegie Inst. Washington, Ann. Rept. Dir. Geophys. Lab.*, 1963–64, 165–171.

Davis, B. T. C. and **Boyd, F. R.,** 1966. The join $Mg_2Si_2O_6$–$CaMgSi_2O_6$ at 30 kilobars pressure and its applications to pyroxenes from kimberlites. *J. Geophys. Res.*, **71**, 3567–3576.

Davis, B. T. C. and **Schairer, J. F.,** 1965. Melting relations in the join diopside–forsterite–pyrope at 40 kilobars and at one atmosphere. *Carnegie Inst. Washington, Ann. Rept. Dir. Geophys. Lab.*, 1964–65, 123–126.

Dawson, J. B., 1962. Basutoland kimberlites. *Bull. Geol. Soc. America*, **73**, 545–559.

Dawson, J. B., Powell, D. G. and **Reid, A. M.,** 1970. Ultrabasic xenoliths and lava from the Lashaine volcano, northern Tanzania. *J. Petr.*, **11**, 519–548.

Day, A. L. and **Sosman, R. B.,** 1910. The nitrogen thermometer from zinc to palladium. *Amer. J. Sci.*, 4th ser., **39**, 93–161.

Deganello, S., 1973. The thermal expansion of diopside. *Z. Krist.*, **137**, 127–131.

Deganello, S., 1974. Atomic vibrations and thermal expansion of some silicates at high temperatures. *Z. Krist.*, **139**, 297–316.

Devaraju, T. C. and **Sadashivaiah, M. S.,** 1966. Pyroxene–quartz–magnetite rocks from Satnur-Halguru area, Mysore State. *J. Geol. Soc. India*, **7**, 70–85.

Dickey, J. S. and **Yoder, H.,** 1972. Partitioning of chromium and aluminium between clinopyroxene and spinel. *Carnegie Inst. Washington, Ann. Rept. Dir. Geophys. Lab.*, 1971–72, 384–392.

Dodd, R. T., 1963. Garnet–pyroxene gneisses at Bear Mountain, New York. *Amer. Min.*, **48**, 811–820

Dowty, E. and **Lindsley, D. H.,** 1973. Mössbauer spectra of synthetic hedenbergite–ferrosilite pyroxene. *Amer. Min.*, **58**, 850–868.

Dunham, K. C. and **Peacock, M. A.**, 1936. Xenoliths in the Organ batholith, New Mexico. *Amer. Min.*, **21**, 312–320.

Dyson, C. J. and **Juckes, L. M.**, 1972. A silica-deficient pyroxene in iron-ore sinters. *Min. Mag.*, **38**, 872–877.

Eckermann, H. von, 1922. The rocks and contact minerals of the Mansjö Mountain. *Geol. För. Förh., Stockholm*, **44**, 205.

Eckermann, H. von, 1950. A comparison between the Fennoscandian limestone contact minerals and those of the Alnö alkaline rocks, associated with carbonates. *Min. Mag.*, **29**, 304–312.

Edgar, A. D., 1964. Phase equilibrium relations in the system $CaMgSi_2O_6$ (diopside)–$NaAlSiO_4$ (nepheline)–$NaAlSi_3O_8$ (albite)–H_2O at 1000 kg/cm^2 water vapour pressure. *Amer. Min.*, **49**, 573–585.

Eggler, D. H., 1973. Role of CO_2 in melting processes in the mantle. *Carnegie Inst. Washington, Ann. Rept. Dir. Geophys. Lab.*, 1972–73, 457–467.

Eggler, D. H., 1974. Volatiles in ultrabasic and derivative rock systems. Effect of CO_2on the melting of peridotite. *Carnegie Inst. Washington, Ann. Rept. Dir. Geophys. Lab.*, 1973–74, 215–224.

Eggler, D. H., 1975. Peridotite–carbonate relations in the system CaO–MgO–SiO_2–CO_2. *Carnegie Inst. Washington, Ann. Rept. Dir. Geophys. Lab.*, 1974–75, 468–474.

Eggler, D. H., 1976. Composition of the partial melt of carbonated peridotite in the system CaO–MgO–SiO_2–CO_2. *Carnegie Inst. Washington, Ann. Rept. Dir. Geophys. Lab.*, 1975–76, 623–626.

Eggler, D. H. and **McCallum, M. E.**, 1976. A geotherm from megacrysts in the Sloan kimberlite pipes, Colorado. *Carnegie Inst. Washington, Ann. Rept. Dir. Geophys. Lab.*, 1975–76, 538–541.

Eggleton, R. A., 1975. Nontronite topotaxial after hedenbergite. *Amer. Min.*, **60**, 1063–1068.

Emeleus, C. H. and **Andrews, J. R.**, 1974. Mineralogy and petrology of kimberlite dyke and sheet inclusions and included peridotite xenoliths from south-west Greenland. *Phys. Chem. Earth.* **9**, 179–197.

Eppler, W. F., 1967. Star diopside and star enstatite. *J. Gemmology*, **14**, 185–188.

Ernst, W. G., 1966. Synthesis and stability relations of ferrotremolite. *Amer. J. Sci.*, **264**, 37–65.

Eskola, P., 1922. On contact phenomena between gneiss and limestone in western Massachusetts. *J. Geol.*, **30**, 265.

Farquhar, O. C., 1960. Occurrences and origin of the hourglass structure. *Rept. 21st Intern. Geol. Congr., Norden*, Part 21, 194–200.

Favorskaia, M. A., **Volchanskaia, I. K.** and **Nissenbaum, P. N.**, 1963. Some new data of pyroxene features from Upper Cenozoic effusives of Kamehatka. *Akad. Nauk. S.S.S.R. Lab. Vulkanol. Vulkanizm Kamchatki*, 111–118 (in Russian).

Federico, M. and **Fornaseri, M.**, 1953. Fenomeni di trasformazione dei pirosseni del giacimenti ferriferi dell'isola d'Elba. *Periodico Min. Roma*, **22**, 107–127.

Ferguson, J. B. and **Merwin, H. E.**, 1919. The ternary system CaO–MgO–SiO_2. *Amer. J. Sci.*, 4th ser., **48**, 81–123.

Ferrara, G., **Lucchini, F.**, **Morten, L.**, **Rita, F.**, **Rossi, P. L.** and **Simboli, G.**, 1974. Clinopyroxenite inclusions in the Triassic volcanic rocks from Latemar, Predazzo, north Italy. *Rend. Soc. Italiana Min. Petr.*, **30**, 141–163.

Ferrier, A., 1971. An experimental study of the enthalpy of crystallization of diopside and synthetic anorthite. *Rev. Inst. Hautes Temp. et Refract.*, **8**, 31 (in French). *British Ceram. Abstr.*, abstr. 3323/71, 1971.

Finger, L. W. and **Ohashi, Y.**, 1976. The thermal expansion of diopside to 800°C and a refinement of the crystal structure at 700°C. *Amer. Min.*, **61**, 303–310.

Francis, G. H., 1958. Petrological studies in Glen Urquhart, Inverness-shire. *Bull. Brit. Mus. (Min.)*, **1**, 123.

Frantsesson, E. V., 1970. *The petrology of the kimberlites.* Transl. from the Russian by D. A. Brown, Canberra (Australian Nat. Univ.), 194pp.

Frey, F. A., 1969. Rare earth abundances in a high temperature peridotite intrusion. *Geochim. Cosmochim. Acta*, **33**, 1429–1447.

Frey, F. A. and **Green, D. H.**, 1974. The mineralogy, geochemistry and origin of lherzolite inclusions in Victorian basanites. *Geochim. Cosmochim. Acta*, **38**, 1023–1059.

Friedman, G. M., 1957. Structure and petrology of the Caribou Lake intrusive body, Ontario, Canada. *Bull. Geol. Soc. America*, **68**, 1531–1564.

Frisch, T. and **Schmincke, H-U.**, 1969. Petrology of clinopyroxene–amphibole inclusions from the Rogue Nublo volcanics, Gran Canaria, Canary Islands. *Bull. Vulc.*, **33**, 1073–1088.

Frisch, T. and **Schmincke, H-U.**, 1971. A comment on 'formation of the hour-glass structure in augite' by D. F. Strong. *Min. Mag.*, **38**, 251–252.

Frost, B. R., 1975. Contact metamorphism of serpentinite, chloritic Blackwall and rodingite at Paddy-Go-Easy Pass, Central Cascades, Washington. *J. Petr.*, **16**, 272–313.

Gallagher, M. J., 1963. Lamprophyre dykes from Argyll. *Min. Mag.*, **33**, 415–430.

Ghose, S., **McCallum, I. S.** and **Tidy, E.**, 1973. Luna 20 pyroxenes: exsolution and phase transformation as indicators of petrologic history. *Geochim. Cosmochim. Acta*, **37**, 831–839.

Ghose, S. and **Schindler, P.**, 1969. Determination of the distribution of trace amounts of Mn^{2+} in diopsides by electron paramagnetic resonance. *Min. Soc. America, Spec. Paper*, **2**, 51–58.

Ginzburg, I. V., 1969. Immiscibility of the natural pyroxenes diopside and fassaite and the criterion for it. *Dokl. Acad. Sci. U.S.S.R., Earth Sci. Sect.*, **186**, 106–109.

Girod, M., 1967. Données pétrographiques sur des pyroxénolites à grenat en enclaves dans des basaltes du Hoggar (Sahara central). *Bull. Soc. franç. Min. Crist.*, **90**, 202–213.

Girod, M., 1975. Données pétrologiques sur une shoshonite de Stromboli (Îles Eoliennes). *Petrologie*, **1**, 189–196.

Gjessing, L., 1941. Contribution à l'étude des metasilicates. *Norsk. Geol. Tidsskr.*, **20**, 265–267.

Glassley, W. E., 1975. High grade regional metamorphism of some carbonate bodies: significance of the orthopyroxene isograd. *Amer. J. Sci.*, **275**, 1133–1163.

Gold, D. P., 1966. The minerals of the Oka carbonatite and alkaline complex, Oka, Quebec. Min. Soc. India, I.M.A. vol. (*I.M.A. Papers and Proc. 4th Gen. Meeting, New Delhi, 1964*), 109–125.

Gragnani, R., 1972. Le vulcaniti melilititiche di Cupaello (Rieti). *Rend. Soc. Italiana Min. Petr.*, **28**, 165–189.

Grapes, R. H., 1975. Petrology of the Blue Mountain complex, Marlborough, New Zealand. *J. Petr.*, **16**, 371–428.

Green, D. A., 1964. The petrogenesis of the high-temperature peridotite intrusion in the Lizard area, Cornwall. *J. Petr.*, **5**, 135–188.

Green, T. H., 1967. An experimental investigation of sub-solidus assemblages formed at high pressure in high alumina basalt, kyanite eclogite and grosspydite compositions. *Contr. Min. Petr.*, **16**, 84–114.

Green, T. H., 1969. The diopside–kyanite join at high pressures and temperatures. *Lithos*, **2**, 334–341.

Griffin, W. L., 1973. Lherzolite nodules from the Fen alkaline complex, Norway. *Contr. Min. Petr.*, **38**, 135–146.

Griffin, W. L. and **Murthy, V. R.**, 1969. Distribution of K, Rb, Sr and Ba in some minerals relevant to basalt genesis. *Geochim. Cosmochim. Acta*, **33**, 1389–1414.

Griffin, W. L. and **Taylor, P. N.**, 1975. The Fen damkjernite: Petrology of a 'central-complex kimberlite'. *Phys. Chem. Earth*, **9**, 163–177.

Grover, J. E. and **Orville, P. M.**, 1969. The partitioning of cations between coexisting single- and multi-site phases with application to the assemblages: orthopyroxene–clinopyroxene and orthopyroxene–olivine. *Geochim. Cosmochim. Acta*, **33**, 205–226.

Grubb, P. L. C., 1965. An unusual occurrence of diopside and uvarovite near Thetford, Quebec. *Can. Min.*, **8**, 241–248.

Grum-Grzhimailo, S. V., **Bokshaw, O. N.** and **Varina, T. M.**, 1967. The absorption spectrum of chrome-diopside. *Soviet Physics – Crystallography*, **12**, 935.

Grutzeck, M., **Kridelbaugh, S.** and **Weill, D.**, 1974. The distribution of Sr and REE between diopside and silicate liquid. *Geophys. Res. Letters*, **1**, 272–275.

Gulbrandsen, R. A. and **Gielow, D. G.**, 1960. Mineral assemblage of a pyrometasomatic deposit near Tonopah, Nevada. *Prof. Paper U.S. Geol. Surv.*, **400B**, 20–21.

Gunter, A. E., 1974. An experimental study of iron-magnesium exchange between biotite and clinopyroxene. *Can. Min.*, **12**, 258–261.

Gupta, A. K., 1972. The system forsterite–diopside–akermanite–leucite and its significance in the origin of potassium-rich mafic and ultramafic volcanic rocks. *Amer. Min.*, **57**, 1242–1259.

Gupta, A. K. and **Lidiak, E. G.**, 1973. The system diopside–nepheline–leucite. *Contr. Min. Petr.*, **41**, 231–239.

Gupta, A. K., **Onuma, K.**, **Yagi, K.** and **Lidiak, E. G.**, 1973. Effect of silica concentration on the diopside pyroxenes in the system diopside–$CaTiAl_2O_6$–SiO_2. *Contr. Min. Petr.*, **41**, 333–344.

Gupta, A. K., **Venkateswaran, G. P.**, **Lidiak, E. G.** and **Edgar, A. D.**, 1973. The system diopside–nepheline–akermanite–leucite and its bearing on the genesis of alkali-rich mafic and ultramafic volcanic rocks. *J. Geol.*, **81**, 209–218.

Gurulev, S. A., **Kostyuk, V. P.**, **Manuylova, M. M.** and **Rafiyenko, N. I.**, 1965. Discovery of blue diopside in Siberia. *Dokl. Acad. Sci. U.S.S.R., Earth Sci. Sect.*, **163**, 100–103.

Gustafson, W. I., 1974. The stability of andradite, hedenbergite and related minerals in the system Ca–Fe–Si–O–H. *J. Petr.*, **15**, 455–496.

Gutberlet, H. G. and **Huckenholz, H. G.**, 1965. Mineralbestand und Chemismus der Tertiären Analcimalkalitrachyte vom Selberg und der Grader Seife Bei Quiddelback in der Hockeifel. *Neues Jahrb. Min., Mh.*, 10–19.

Gvakhariya, G. B., 1968. Some aspects of the chemical compositions of non-alkaline monoclinic pyroxenes from effusive rocks. Min. Soc., I.M.A. Vol. (*I.M.A. Papers and Proc. 5th Gen. Meeting,*

Cambridge 1966), 319–322.
Hafner, S. S. and **Huckenholz, H. G.**, 1971. Mössbauer spectrum of synthetic ferri-diopside. *Nature (Phys. Sci.)*, **233**, 9–11.
Hamad, S. el D., 1963. The chemistry and mineralogy of the olivine nodules of Calton Hill, Derbyshire. *Min. Mag.*, **33**, 483–497.
Hargraves, R. B., 1962. Petrology of the Allard Lake anorthosite suite. *Geol. Soc. America* (Buddington vol.), 163.
Hariya, Y. and **Kennedy, G. C.**, 1968. Equilibrium study of anorthite under high pressure and high temperature. *Amer. J. Sci.*, **59**, 549–557.
Hart, S. R. and **Brooks, C.**, 1974. Clinopyroxene–matrix partitioning of K, Rb, Cs, Sr and Ba. *Geochim. Cosmochim. Acta*, **38**, 1799–1806.
Haslam, H. W. and **Walker, B. G.**, 1971. A metamorphosed pyroxenite at Nero Hill, central Tanzania. *Min. Mag.*, **38**, 58–63.
Hays, J. F., 1966. Stability and properties of the synthetic pyroxene $CaAl_2SiO_6$. *Amer. Min.*, **51**, 1524–1529.
Hays, J. F., 1967. Lime–alumina–silica. *Carnegie Inst. Washington, Ann. Rept. Dir. Geophys. Lab.*, 1965–66, 234–236.
Heflik, W., 1967. Mineralogical and petrographic investigation of the leucocratic altered zone of Jordanów (Lower Silesia) (in Polish, English summary). *Polska Akad. Nauk, Prace Min.*, **10**, 123pp.
Henley, S., 1971. Hedenbergite and sphalerite from the Perran iron Lode, Cornwall. *Proc. Ussher Soc.*, **2**, 329–334.
Henson, B. J., 1973. Pyroxenes and garnets as geothermometers and barometers. *Carnegie Inst. Washington, Ann. Rept. Dir. Geophys. Lab.*, 1972–73, 527–534.
Hess, H. H., 1949. Chemical composition and optical properties of common clinopyroxenes, Part I. *Amer. Min.*, **34**, 621–666.
Hernandez, J., 1973. Le volcanisme tertiaire des monts du Forez (Massif central français): basanites à analcime, à leucite et néphélinites à mélilite. *Bull. Soc. franç. Min. Crist.*, **96**, 303–312.
Hietanen, A., 1971. Diopside and actinolite from skarn, Clearwater County, Idaho. *Amer. Min.*, **56**, 234–239.
Higgins, B. B. and **Gilbert, M. C.**, 1973. High pressure stability of nickel diopside. *Amer. J. Sci.*, **273-A** (Cooper vol.), 511–521.
Hijikata, K., 1968. Unit-cell dimensions of the clinopyroxenes along the join $CaMgSi_2O_6$–$CaFe^{3+}AlSiO_6$. *J. Fac. Sci., Hokkaido Univ., Ser. IV, Geol. Min.*, **4**, 149–157.
Hijikata, K., 1973. Phase relations in the system $CaMgSi_2O_6$–$CaAl_2SiO_6$ at high pressures and temperatures. *J. Fac. Sci., Hokkaido Univ., Ser. 4*, **16**, 167–178.
Hijikata, K. and **Onuma, K.**, 1969. Phase equilibria of the system $CaMgSi_2O_6$–$CaFe^{3+}AlSiO_6$ in air. *J. Japan Assoc. Min. Petr. Econ. Geol.*, **62**, 209–217.
Hollister, L. S. and **Gancarz, A. J.**, 1971. Compositional sector-zoning in clinopyroxene from the Narce area, Italy. *Amer. Min.*, **56**, 959–979.
Holmes, A., 1937. A contribution to the petrology of kimberlite and its inclusions. *Trans. Geol. Soc. South Africa*, **39**, 379–427.
Honnorez, J. and **Kirst, P.**, 1975. Petrology of rodingites from the Equatorial Mid-Atlantic fracture zones and their geotectonic significance. *Contr. Min. Petr.*, **49**, 233–257.
Hornemann, U. and **Müller, W. F.**, 1971. Shock-induced deformation twins in clinopyroxene. *Neues Jahrb. Min., Mh.*, 247–256.
Howie, R. A., 1955. The geochemistry of the charnockite series of Madras, India. *Trans. Roy. Soc. Edin.*, **62**, 725–768.
Huckenholz, H. G., 1965a. Der petrogenetische Werdegang der Klinopyroxene in den tertiären Vulkaniten der Hocheifel. I. Die Klinopyroxene der Alkaliolivinbasalt–Trachyte–Assoziation. *Beitr. Min. Petr.*, **11**, 138–195.
Huckenholz, H. G., 1965b. Der petrogenetische Werdegang der Klinopyroxene in den tertiären Vulkaniten der Hocheifel. II. Die Klinopyroxene der Basanitoide. *Beitr. Min. Petr.*, **11**, 415–448.
Huckenholz, H. G., 1966. Der petrogenetische Werdegang der Klinopyroxene in den tertiären Vulkaniten der Hocheifel. III. Die Klinopyroxene der Pikritbasalte (Ankaramite). *Contr. Min. Petr.*, **12**, 73–95.
Huckenholz, H. G., **Schairer, J. F.** and **Yoder, H. S.**, *Jr.*, 1969. Synthesis and stability of ferri-diopside. *Min. Soc. America, Spec. Paper*, **2**, 163–177.
Hutchison, R., **Chambers, A. L.**, **Paul, D. K.** and **Harris, P. G.**, 1975. Chemical variation among French ultramafic xenoliths – evidence for a heterogeneous upper mantle. *Min. Mag.*, **40**, 153–170.
Hutton, C. O., 1943. The igneous rocks of the Brocken range-Ngahape area, eastern Wellington. *Trans. Roy. Soc. New Zealand*, **72**, 353–370.
Hytönen, K., 1968. A preliminary report on an iron-rich formation near Raahe in the Gulf of Bothnia,

Finland. *Bull. Geol. Soc. Finland*, **40**, 135–144.

Ikeda, K. and **Yagi, K.**, 1972. Synthesis of kosmochlor and phase equilibria in the join $CaMgSi_2O_6$–$NaCrSi_2O_6$. *Contr. Min. Petr.*, **36**, 63–72.

Ikeda, K. and **Yagi, K.**, 1977. Experimental study on the phase equilibria in the join $CaMgSi_2O_6$–$CaCrCrSiO_6$ with special references to blue diopside. *Contr. Min. Petr.*, **61**, 91–106.

Irvine, T. N., 1973. Bridget Cove volcanics, Juneau Area, Alaska: Possible parental magma of Alaskan-type ultramafic complexes. *Carnegie Inst. Washington, Ann. Rept. Dir. Geophys. Lab.*, 1972–73, 478–491.

Irvine, T. N., 1974. Petrology of the Duke Island ultramafic complex, southeastern Alaska. *Mem. Geol. Soc. America*, **138**, 240pp.

Irving, A. J., 1974. Pyroxene-rich ultramafic xenoliths in the newer basalts of Victoria, Australia. *Neues Jahrb. Min., Abhdl.*, **120**, 147–167.

Ishibashi, K., 1970. Petrochemical study of basic and ultrabasic inclusions in basaltic rocks from northern Kyushu, Japan. *Mem. Fac. Sci., Kyushu Univ., Ser. D. Geol.*, **20**, 85–146.

Ito, K., 1962. Zoned skarn of the Fujigatani mine, Yamaguchi Prefecture. *Jap. J. Geol. Geogr.*, **33**, 169–190.

Ito, K. and **Kennedy, G. C.**, 1967. Melting and phase relations in a natural peridotite to 40 kilobars. *Amer. J. Sci.*, **265**, 519–538.

Jacobson, R. R. E., **Macleod, W. N.** and **Black, R.**, 1958. Ring-complexes in the younger granite province of northern Nigeria. *Mem. Geol. Soc. London*, no. 1.

Jayawardena, D. E. de S. and **Carswell, D. A.**, 1976. The geochemistry of 'charnockites' and their constituent ferromagnesian minerals from the Precambrian of south-east Sri Lanka (Ceylon). *Min. Mag.*, **40**, 541–554.

Jong, W. F. de, 1959. *General Crystallography*. A brief compendium. San Francisco (Freeman and Co.).

Juurinen, A. and **Hytönen, K.**, 1952. Diopside from Juva, Finland. *Bull. Comm. géol. Finlande*, No. 157, 145–146.

Kalinin, D. V., 1967a. Lower temperature boundaries for the formation of tremolite, diopside and wollastonite under hydrothermal conditions: experimental data. *Geol. Geofiz. Akad. Nauk SSSR, Sib. Otd.*, no. 1, 123–126 (in Russian) (see also English transl. in *Geochem. Intern.*, **4**, 836–839).

Kalinin, D. V., 1967b. Chemical mode of replacement of hedenbergite by andradite (experimental data). *Dokl. Acad. Sci. U.S.S.R., Earth Sci. Sect.*, **173**, 112–114.

Kent, D. and **Webster, R.**, 1973. Star-diopside and labradorite as paramagnetic minerals. *J. Gemmology*, **13**, 308–311.

Kesson, S. F., 1973. The primary geochemistry of the Monaro alkaline volcanics, southeastern Australia evidence for upper mantle heterogeneity. *Cont. Min. Petr.*, **42**, 92–108.

Khitarov, N. I. and **Kadik, A. A.**, 1973. Water and carbon dioxide in magmatic melts and peculiarities of the melting process. *Contr. Min. Petr.*, **41**, 205–215.

Kinrade, J., **Skippen, G. D.** and **Wiles, D. R.**, 1975. A Mössbauer observation of hedenbergite synthesis. *Geochim. Cosmochim. Acta*, **39**, 1325–1327.

Kirkpatrick, R. J., 1974. Kinetics of crystal growth in the system $CaMgSi_2O_6$–$CaAl_2SiO_6$. *Amer. J. Sci.*, **274**, 215–242.

Kirkpatrick, R. J. and **Steele, I. M.**, 1973. Hexagonal $CaAl_2SiO_6$: a new synthetic phase. *Amer. Min.*, **58**, 945–946.

Klein, C., 1966. Mineralogy and petrology of the metamorphosed Wabush Iron Formation, Southwestern Labrador. *J. Petr.*, **7**, 246–305.

Klein, C., 1973. Changes in mineral assemblages with metamorphism of some banded Precambrian iron-formations. *Econ. Geol.*, **68**, 1075–1088.

Konda, T., 1970. [Endiopside in the calc-alkaline basalt in the Ryozen area.] (Japanese with English summary.) *J. Geol. Soc. Japan.*, **76**, 7–12.

Kornprobst, J., 1966. A propos des péridotites du massif des Beni-Bouchera (Rif septentional, Maroc). *Bull. Soc. franç. Min. Crist.*, **89**, 399–505.

Kostyuk, V. P., 1965. Additional data on the diopside–hedenbergite–aegirine diagram. *Dokl. Acad. Sci. U.S.S.R., Earth Sci. Sect.*, **156**, 110–113.

Kozu, S. and **Ueda, J.**, 1933. Thermal expansion of diopside. *Proc. Imp. Acad. Japan*, **9**, 317–319.

Kresten, P. and **Persson, L.**, 1975. Discrete diopside in alnöite from Alnö Island. *Lithos*, **8**, 187–192.

Krivenko, A. P. and **Guletskaya, E. S.**, 1968. Composition of pyroxene from the gabbro–syenite association in Bol'shoy Taskyl pluton, Kuzetskiy Alatau. *Dokl. Acad. Sci. U.S.S.R., Earth Sci. Sect.*, **180**, 149–152.

Kuno, H., 1950. Petrology of Hakone volcano and the adjacent areas, Japan. *Bull. Geol. Soc. America*, **61**, 957–1014.

Kuno, H., 1955. Ion substitution in the diopside–ferropigeonite series of clinopyroxenes. *Amer. Min.*, **40**, 70–93.

Kuno, H., 1957. Chromian diopside from Sano, Yamanasi Prefecture. *J. Geol. Soc. Japan*, **63**, 523–526.

Kuno, H., 1969. Mafic and ultramafic nodules in basaltic rocks of Hawaii. *Mem. Geol. Soc. America*, **115**, 189–234.

Kuno, H. and **Aoki, K-I.**, 1970. Chemistry of ultramafic nodules and their bearing on the origin of basaltic magmas. *Phys. Earth Planet. Interiors*, **3**, 273–301.

Kuno, H. and **Hess, H. H.**, 1953. Unit cell dimensions of clinoenstatite and pigeonite in relation to other common clinopyroxenes. *Amer. J. Sci.*, **251**, 741–752.

Kuno, H. and **Sawatari, M.**, 1934. On the augites from Wadaki, Izu, and from Yoneyama, Etigo, Japan. *Jap. J. Geol. Geogr.*, **11**, 327–343.

Kushiro, I., 1964. The system diopside–forsterite–enstatite at 20 kilobars. *Carnegie Inst. Washington, Ann. Rept. Dir. Geophys. Lab.*, 1963–64, 101–108.

Kushiro, I., 1965. Clinopyroxene solid solutions at high pressure. The join diopside–albite. *Carnegie Inst. Washington, Ann. Rept. Dir. Geophys. Lab.*, 1964–65, 112–117.

Kushiro, I., 1969a. Stability of omphacite in the presence of excess silica. *Carnegie Inst. Washington, Ann. Rept. Dir. Geophys. Lab.*, 1967–68, 98–100.

Kushiro, I., 1969b. The system forsterite–diopside–silica with and without water at high pressure. *Amer. J. Sci.*, **267-A**, 269–294.

Kushiro, I., 1969c. Clinopyroxene solid solutions formed by reactions between diopside and plagioclase at high pressures. *Min. Soc. America, Spec. Paper*, **2**, 179–191.

Kushiro, I., 1969d. Synthesis and stability of iron-free pigeonite in the system $MgSiO_3$–$CaMgSi_2O_6$ at high pressures. *Carnegie Inst. Washington, Ann. Rept. Dir. Geophys. Lab.*, 1967–68, 80–83.

Kushiro, I., 1972a. New method of determining liquidus boundaries with confirmation of incongruent melting of diopside and existence of iron-free pigeonite at 1 atm. *Carnegie Inst. Washington, Ann. Rept. Dir. Geophys. Lab.*, 1971–72, 603–607.

Kushiro, I., 1972b. Determination of liquidus relations in synthetic silicate systems with electron probe analysis: the system forsterite–diopside–silica at 1 atmosphere. *Amer. Min.*, **57**, 1260–1271.

Kushiro, I., 1972c. Partial melting of synthetic and natural peridotites at high pressure. *Carnegie Inst. Washington, Ann. Rept. Dir. Geophys. Lab.*, 1971–72, 357–362.

Kushiro, I., 1973a. Incongruent melting of pure diopside. *Carnegie Inst. Washington, Ann. Rept. Dir. Geophys. Lab.*, 1972–73, 708–710.

Kushiro, I., 1973b. The system diopside–anorthite–albite: determination of compositions of coexisting phases. *Carnegie Inst. Washington, Ann. Rept. Dir. Geophys. Lab.*, 1972–73, 502–507.

Kushiro, I. and **Schairer, J. F.**, 1963. New data on the system diopside–forsterite–silica. *Carnegie Inst. Washington, Ann. Rept. Dir. Geophys. Lab.*, 1962–63, 95–103.

Kushiro, I. and **Schairer, J. F.**, 1964. The join diopside–akermanite. *Carnegie Inst. Washington, Ann. Rept. Dir. Geophys. Lab.*, 1963–64, 132–133.

Kushiro, I. and **Thompson, R. H.**, 1972. Origin of some abyssal tholeiites from the Mid-Atlantic Ridge. *Carnegie Inst. Washington, Ann. Rept. Dir. Geophys. Lab.*, 1971–72, 403–406.

Kushiro, I. and **Yoder, H. S.**, 1966. Anorthite–forsterite and anorthite–enstatite reactions and their bearing on the basalt eclogite transformation. *J. Petr.*, **7**, 337–362.

Kuznetsova, L. G. and **Moskaleva, V. N.**, 1968. [Infrared absorption spectra of pyroxenes of the diopside–jadeite isomorphous series.] *Zap. Vses. Min. Obshch.*, **97**, 715–718 (in Russian).

Laitakari, A., 1921. Über die Petrographie und Mineralogie der Kalksteinlagerstätten von Parainen (Pargas). *Bull. Comm. géol. Finlande*, no. 54.

Larrabee, D. M., 1970. Serpentine and rodingite in the Hunting Hill quarry Montgomery County, Maryland. *Bull. U.S. Geol. Surv.*, **1283**, 1–34.

Larsen, E. S., 1941. Alkalic rocks of Iron Hill, Gunnison County, Colorado. *Prof. Paper U.S. Geol. Surv.*, **197-A**.

Larsen, L. M., 1976. Clinopyroxenes and coexisting mafic minerals from the alkaline Ilímaussaq intrusion, south Greenland. *J. Petr.*, **17**, 258–290.

Lewis, J. F., 1967. Unit cell dimensions of some aluminous natural clinopyroxenes. *Amer. Min.*, **52**, 42–54.

Lewis, J. F., 1973. Mineralogy of the ejected plutonic blocks of the Soufrière volcano, St. Vincent: olivine, pyroxene, amphibole and magnetite paragenesis. *Contr. Min. Petr.*, **38**, 197–220.

Lindsley, D. H., 1967. The join hedenbergite–ferrosilite at high pressures and temperatures. *Carnegie Inst. Washington, Ann. Rept. Dir. Geophys. Lab.*, 1965–66, 231–234.

Lindsley, D. H., **Brown, G. M.** and **Muir, I. D.**, 1969. Conditions of the ferrowollastonite–ferrohedenbergite inversion in the Skaergaard intrusion, east Greenland. *Min. Soc. America, Spec. Paper*, **2**, 193–201.

Lindsley, D. H. and **Munoz, J. L.**, 1968. Subsolidus relations along the join hedenbergite–ferrosilite. join at low pressures. *Carnegie Inst. Washington, Ann. Rept. Dir. Geophys. Lab.*, 1966–67, 363–366.

Lindsley, D. H. and **Munoz, J. L.**, 1969. Subsolidus relations along the join hedenbergite–ferrosilite.

Amer. J. Sci., **267-A**, 295–324.
Loeschke, J., 1973. Petrochemistry of Paleozoic spilites of the eastern Alps (Austria). *Geol. Mag.*, **110**, 19–28.
Lutts, B. G., 1965. Eclogitization reactions in rocks formed at great depths. *Geochem. Intern.*, **2**, 1093–1104.
MacGregor, I. D., 1965a. Stability fields of spinel and garnet peridotites in the synthetic systems MgO–CaO–Al_2O_3–SiO_2. *Carnegie Inst. Washington, Ann. Rept. Dir. Geophys. Lab.*, 1964–65, 126–134.
MacGregor, I. D., 1965b. Aluminous diopsides in the three-phase assemblage diopside solid solution and forsterite and spinel. *Carnegie Inst. Washington, Ann. Rept. Dir. Geophys. Lab.*, 1964–65, 134–135.
Machairas, G. and **Blais, R.**, 1966. La transformation de l'hédenbergite manganésifère en ilvaite dans les sulfures de cuivre et de zinc de la région de Noranda (Québec). *Bull. Soc. franç. Min. Crist.*, **89**, 372–376.
McCallister, R. H., **Finger, L. W.** and **Ohashi, Y.**, 1976. Intracrystalline Fe^{2+}–Mg equilibria in three natural Ca-rich clinopyroxenes. *Amer. Min.*, **61**, 671–676.
McCallister, R. H. and **Yund, R. A.**, 1975. Kinetics and microstructure of pyroxene exsolution. *Carnegie Inst. Washington, Ann. Rept. Dir. Geophys. Lab.*, 1974–75, 433–436.
McDougall, I., 1961. Optical and chemical studies of pyroxenes in a differentiated Tasmanian dolerite. *Amer. Min.*, **46**, 661–687.
McDougall, I., 1962. Differentiation of the Tasmanian dolerites: Red Hill dolerite–granophyre association. *Bull. Geol. Soc. America*, **73**, 279–315.
Magonthier, M-C., 1975. Les basanites à leucite tertiaires de la partie nord du Sillon Houiller (d'Herment à Pontaumur, Puy-de-Dôme). Comparaison avec deux autres provinces à caractère potassique du Massif central français (Sioule et Sillon Houiller Sud). *Bull. Soc. franç. Min. Crist.*, **98**, 245–253.
Magonthier, M. C. and **Velde, D.**, 1976. Mineralogy and petrology of some Tertiary leucite–rhönite basanites from central France. *Min. Mag.*, **40**, 817–826.
Magnusson, N. H., 1940. [The Herräng field and its iron ores.] *Årsbok Sveriges Geol. Undersökning*, **34**, nol 1, 1–72 (in Swedish, English summary).
Mall, A. P., 1973. Distribution of elements in coexisting ferromagnesian minerals from ultrabasics of Kondapalle and Gangineni, Andhra Pradesh, India. *Neues Jahrb. Min., Mh.*, 323–336.
Martin, B. F., 1967. The characteristics of black star-diopside. *J. Gemmology*, **10**, 235–241.
Mason, B., 1968. Pyroxenes in meteorites. *Lithos*, **1**, 1–11.
Masuda, A. and **Kushiro, I.**, 1970. Experimental determination of partition coefficients of ten rare earth elements and barium between clinopyroxene and liquid in the synthetic silicate system at 20 kilobar pressure. *Contr. Min. Petr.*, **26**, 42–49.
Matsui, Y., **Syono, Y.** and **Maeda, Y.**, 1972. Mössbauer spectra of synthetic and natural calcium-rich clinopyroxenes. *Min. J. Japan*, **7**, 88–107.
Maury, R. C. and **Bizouard, H.**, 1974. Clinopyroxènes des contacts acides-basiques. *Bull. Soc. franç. Min. Crist.*, **97**, 465–469.
Melchoir Larsen, L. and **Steenfelt, A.**, 1974. Alkali loss and retention in an iron-rich peralkaline phonolite dyke from the Gardar province, south Greenland. *Lithos*, **7**, 81–90.
Menzies, M., 1973. Mineralogy and partial melt textures within an ultramafic body, Greece. *Contr. Min. Petr.*, **42**, 273–285.
Mercier, J-C. C. and **Nicolas, A.**, 1975. Textures and fabrics of upper mantle peridotites as illustrated by xenoliths from basalts. *J. Petr.*, **16**, 454–487.
Mercy, E. L. P. and **O'Hara, M. J.**, 1965. Websterite from Glenelg, Inverness-shire. *Scott. J. Geol.*, **1**, 282–284.
Mercy, E. L. P. and **O'Hara, M. J.**, 1967. Distribution of Mn, Cr, Ti and Ni in co-existing minerals of ultramafic rocks. *Geochim. Cosmochim. Acta*, **31**, 2331–2341.
Merriam, R. and **Laudermilk, J. D.**, 1936. Two diopsides from southern California. *Amer. Min.*, **21**, 715–718.
Metz, P., 1970. Experimentalle Untersuchung der Metamorphose von kieselig dolomitischen Sedimenten. II. Die Bildungsbedingungen des Diopsids. *Contr. Min. Petr.*, **28**, 221–250.
Metz, P. and **Winkler, H. G. F.**, 1964. Experimental investigation of the formation of diopside from tremolite, calcite and quartz. *Geochemistry Intern.*, **2**, 388–389.
Mikhaylov, N. P. and **Rovsha, V. S.**, 1965. Influence of pressure on the paragenesis of ultrabasic rocks. *Dokl. Acad. Sci., U.S.S.R., Earth Sci. Sect.*, **160**, 151–153.
Miles, K. R., 1946. Metamorphism of the jasper bars of Western Australia. *Quart. J. Geol. Soc.*, **102**, 115–156.
Miyashiro, A., 1958. Regional metamorphism of the Gosaisyo-Takanuki district in the Central Abukuma Plateau. *J. Fac. Sci. Univ. Tokyo*, **11**, 219–272.
Morgan, B. A., 1974. Chemistry and mineralogy of garnet pyroxenites from Sabah, Malaysia. *Contr. Min. Petr.*, **48**, 301–314.

Morgan, B. A., 1975. Mineralogy and origin of skarns in the Mount Morrison pendant, Sierra Nevada, California. *Amer. J. Sci.*, **275**, 119–142.

Mori, T. and **Banno, S.,** 1973. Petrology of peridotite and garnet clinopyroxenite of the Mt. Higasi-Akaisi mass, Sikoku, Japan – subsolidus relation of anhydrous phases. *Contr. Min. Petr.*, **41**, 301–323.

Morimoto, N., 1956. The existence of monoclinic pyroxenes with space group C^5_{2h}–$P2_1/c$. *Proc. Jap. Acad.*, **32**, 750.

Morimoto, N., Appleman, D. E. and **Evans, H. T.,** 1959. The structural relations between diopside, clinoenstatite and pigeonite. *Carnegie Inst. Washington, Ann. Rept. Dir. Geophys. Lab.*, 1958–59, 192–197.

Morimoto, N. and **Ito, T.,** 1958. Pseudo-twin angle of augite and pigeonite. *Bull. Geol. Soc. America*, **69**, 1616.

Mozgova, N. N., 1959. [Hypogenic alteration of hedenbergite in the skarn-polymetallic deposit of the Upper mine (Tetyukhe).] *Mat. Geol. Ore-Deposits, Petr. Min., Geochem., Acad. Sci. U.S.S.R.*, 279–293. (In Russian) [M.A. **15**-337].

Mueller, R. F., 1964. Theory of the equilibria between complex silicate melts and crystalline solutions. *Amer. J. Sci.*, **262**, 643–652.

Mueller, R. F., 1966. Stability relations of the pyroxenes and olivine in certain high grade metamorphic rocks. *J. Petr.*, **7**, 363–374.

Muir, I. D., 1951. The clinopyroxenes of the Skaergaard intrusion, eastern Greenland. *Min. Mag.*, **29**, 690–714.

Muir, I. D., 1953. Quartzite xenoliths from the Ballachulish granodiorite. *Geol. Mag.*, **90**, 409–427.

Mukherjee, A. and **Rege, S. M.,** 1972. Stability of wollastonite in the granulite facies: some evidences from the Eastern Ghats, India. *Neues Jahrb. Min., Abhdl.*, **118**, 23–42.

Muñoz, M. and **Sagredo, J.,** 1974. Clinopyroxenes as geobarometric indicators in mafic and ultramafic rocks from Canary Islands. *Contr. Min. Petr.*, **44**, 139–147.

Muñoz, M., Sagredo, J. and **Afonso, A.,** 1974. Mafic and ultramafic inclusions in the eruption of Teneguia volcano (La Palma, Canary Islands). *Estud. Geol. (Inst. 'Lucas Mallada'), Teneguia vol.*, 65–74.

Murray, R. M., 1954. The clinopyroxenes of the Garbh Eilean sill, Shiant Isles. *Geol. Mag.*, **91**, 17–31.

Myer, G. H. and **Lindsley, D. H.,** 1969. Optical properties of synthetic clinopyroxenes on the join hedenbergite–ferrosilite. *Carnegie Inst. Washington, Ann. Rept. Dir. Geophys. Lab.*, 1967–68, 92–94.

Mysen, B. O. and **Boettcher, A. L.,** 1975. Melting of a hydrous mantle: II. Geochemistry of crystals and liquids formed by anatexis of mantle peridotite at high pressures and high temperatures as a function of controlled activities of water, hydrogen and carbon dioxide. *J. Petr.*, **16**, 549–593.

Mysen, B. O. and **Kushiro, I.,** 1976. Compositional variation of coexisting phases with degree of melting of peridotite under upper mantle conditions. *Carnegie Inst. Washington, Ann. Rept. Dir. Geophys. Lab.*, 1975–76, 546–555.

Naboko, S. I. and **Shavrova, N. N.,** 1954. [On pyroxenes in lavas of modern and recent eruptions of certain Kamchatka volcanos] *Bull. Volc. Station, Acad. Sci. U.S.S.R.*, No. 23, 47–50 (in Russian). [M.A. **13**-392.]

Nafziger, R. H., 1970. The join diopside–iron oxide–silica and its relation to the join diopside–forsterite–iron oxide–silica. *Amer. Min.*, **55**, 2042–2052.

Nagasaki, H., 1966. A layered ultrabasic complex at Horoman, Hokkaido, Japan. *J. Fac. Sci. Univ. Tokyo*, Sect. II, **16**, Pt. 2, 313–346.

Nehru, C. E. and **Wyllie, P. J.,** 1974. Electron microprobe measurements of pyroxenes coexisting with H_2O-undersaturated liquid in the join $CaMgSi_2O_6$–$Mg_2Si_2O_6$–H_2O at 30 kilobars, with applications to geothermometry. *Contr. Min. Petr.*, **48**, 221–228.

de Neufville, J. and **Schairer, J. F.,** 1962. Pyroxenes: The join diopside–Ca Tschermak's molecule at atmospheric pressure. *Carnegie Inst. Washington, Ann. Rept. Dir. Geophys. Lab.*, 1961–62, 56–59.

Nicolas, A., 1966. Etude pétrochimique des roches vertes et de leurs minéraux entre Dora Maira et Grand Paradis (Alpes piémontaises). *Nantes, Fac. Sci.*, 299 pp.

Nicholls, G. D., 1951. An unusual pyroxene-rich xenolith in the diorite of the Glenelg-Ratagain igneous complex. *Geol. Mag.*, **88**, 284–295.

Nixon, P. H., von Knorring, O. and **Rooke, J. M.,** 1963. Kimberlite and associated inclusions of Basutoland; a mineralogical and geochemical study. *Amer. Min.*, **48**, 1090–1132.

Nolan, J., 1966. Melting relations in the system $NaAlSi_3O_8$–$NaAlSiO_4$–$NaFeSi_2O_6$–$CaMgSi_2O_6$–H_2O and their bearing on the genesis of alkaline-undersaturated rocks. *Quart. J. Geol. Soc.*, **122**, 119–157.

Nolan, J., 1969. Physical properties of synthetic and natural pyroxenes in the system diopside–hedenbergite–acmite. *Min. Mag.*, **37**, 216–229.

Nolan, J. and **Edgar, A. D.,** 1963. An X-ray investigation of synthetic pyroxenes in the system acmite–diopside–water at 1000 kg/cm^2 water-vapour pressure. *Min. Mag.*, **33**, 624–634.

Norton, D. A. and **Clavan, W. S.**, 1959. Optical mineralogy, chemistry and X-ray crystallography of ten pyroxenes. *Amer. Min.*, **44**, 844–874.

Nwe, Y. Y. and **Copley, P.**, 1975. Chemistry, subsolidus relations and electron petrography of pyroxenes from the late ferrodiorites of the Skaergaard intrusion, east Greenland. *Contr. Min. Petr.*, **53**, 37–54.

O'Hara, M. J. and **Biggar, G. M.**, 1969. Diopside + spinel equilibria, anorthite and forsterite reaction relationships in silica-poor liquids in the system $CaO–MgO–Al_2O_3–SiO_2$ at atmospheric pressure and their bearing on the genesis of melilitites and nephelinites. *Amer. J. Sci.*, 267-A (Schairer vol.), 364–390.

Ohashi, H. and **Hariya, Y.**, 1973. Order–disorder of ferric iron and aluminium in the system $CaMgSi_2O_6–CaFeAlSiO_6$ at high pressure. *J. Japan. Assoc. Min. Petr. Econ. Geol.*, **68**, 230–233.

Ohashi, Y. and **Finger, L. W.**, 1973. Thermal vibration ellipsoids and equipotential surfaces at the cation sites of olivine and clinopyroxenes. *Carnegie Inst. Washington, Ann. Rept. Dir. Geophys. Lab.*, 1972–73, 547–551.

Omori, K., 1971. Analysis of the infrared absorption spectrum of diopside. *Amer. Min.*, **56**, 1607–1616.

Onuki, H. and **Tiba, T.**, 1964. Petrochemistry of the Ichinole alkali province complex, Kitakami Mountainland, northern Japan. *Sci. Rept. Tohoku Univ.*, ser. 3, **9**, 124–154.

Onuma, K., **Hijikata, K.** and **Yagi, K.**, 1968. Unit cell dimensions of synthetic titan-bearing clinopyroxene. *J. Fac. Sci. Hokkaido Univ.*, ser IV, **14**, 111–121.

Onuma, K. and **Yagi, K.**, 1967. The system diopside–akermanite–nepheline. *Amer. Min.*, **54**, 227–243.

Onuma, K. and **Yagi, K.**, 1971. The join $CaMgSi_2O_6–Ca_2MgSi_2O_6–CaTiAl_2O_6$ in the system $CaO–MgO–TiO_2–SiO_2$ and its bearing on the titanpyroxenes. *Min. Mag.*, **38**, 471–480.

Osborn, E. F., 1942. The system $CaSiO_3$–diopside–anorthite. *Amer. J. Sci.*, **240**, 751–788.

Osborn, E. F. and **Tait, D. B.**, 1952. The system diopside–forsterite–anorthite. *Amer. J. Sci.* (Bowen vol.), 413–433.

Osborne, C. D., 1932. The metamorphosed limestones and associated contaminated igneous rocks of the Carlingford district, Co. Lough. *Geol. Mag.*, **69**, 209–233.

Palache, C., 1937. The minerals of Franklin and Sterling Hill, Sussex County, New Jersey. *Prof. Paper U.S. Geol. Surv.*, **180**.

Pamić J., **Šćavničar, S.** and **Medjimorec, S.**, 1973. Mineral assemblages of amphibolites associated with Alpine-type ultramafics in the Dinaride ophiolite zone (Yugoslavia). *J. Petr.*, **14**, 133–157.

Parker, R. B., 1961. Rapid determination of the approximate composition of amphiboles and pyroxenes. *Amer. Min.*, **46**, 892–900.

Parkes, G. A. and **Arltar, S.**, 1968. Magnetic moment of Fe^{2+} in paramagnetic minerals. *Amer. Min.*, **53**, 406–415.

Parras, K., 1958. On the charnockites in the light of a highly metamorphic rock complex in south-western Finland. *Bull. Comm. géol. Finlande*, No. 181, 1–138.

Pauly, H., 1948. Calcite and skarn minerals in the gneisses of the Holsteinborg district, west Greenland. *Medd. Dansk. Geol. För.*, **11**, 328–350.

Pavelescu, L., 1968. Contribution to the study of chromiferous diopsides and other minerals of the peridotitic rocks from Coasta lui Rusu (southern Carpathians). *Rev. Roum. Geol. Geophys. Geogr., ser. geol.*, **12**, 47–53 [M.A. 69-1439].

Perchuk, L. L., 1966. [Composition property diagram for the aegirine–hedenbergite–diopside system and possible limited miscibility in the subsolidus of the system.] *Zap. Vses. Min. Obshch.*, **95**, 619–626 (in Russian).

Peters, Tj., 1968. Distribution of Mg, Fe, Al, Ca and Na in coexisting olivine, orthopyroxene and clinopyroxene in the Totalp Serpentinite (Davos, Switzerland) and in the Alpine metamorphosed Malenco serpentine (N. Italy). *Contr. Min. Petr.*, **18**, 65–75.

Pitcher, W. S., 1950. Calc-silicate skarn veins in the limestone of Lough Anure, Co. Donegal. *Min. Mag.*, **29**, 126–141.

Platt, R. G. and **Edgar, A. D.**, 1972. The system nepheline–diopside–sanidine and its significance to the genesis of melilite- and olivine-bearing alkaline rocks. *J. Geol.*, **80**, 224–236.

Presnall, D. C., 1966. The join forsterite–diopside–iron oxide and its bearing on the crystallization of basaltic and ultramafic magmas. *Amer. J. Sci.*, **264**, 753–809.

Prewitt, C. T., **Brown, G. E.** and **Papike, J. J.**, 1971. Apollo 12 clinopyroxenes: high temperature X-ray diffraction studies. *Proc. Second Lunar Sci. Conf. (Suppl. to Geochim. Cosmochim. Acta* **35***)*, **1**, 56–68.

Rahman, S., 1975. Some aluminous clinopyroxenes from Vesuvius and Monte Somma. *Min. Mag.*, **40**, 43–52.

Rao, A. T. and **Rao, K. S. R.**, 1971. Ferroan sahlite from the Eastern Ghats, Andhra Pradesh, India. *Min. Mag.*, **38**, 377–379.

Rayleigh, C. B., 1965. Glide mechanisms in experimentally deformed minerals. *Science*, **150**, 739–741.

Rayleigh, C. B. and **Talbert, J. L.**, 1967. Mechanical twinning in naturally and experimentally deformed diopside. *Amer. J. Sci.*, **265**, 151–165.

Reesman, A. L. and **Keller, W. D.**, 1965. Calculation of apparent standard free energies of formation of six rock-forming silicate minerals from solubility data. *Amer. Min.*, **50**, 1729–1739.

Ricker, R. W. and **Osborn, E. F.**, 1954. Additional phase equilibrium data for the system CaO–MgO–SiO_2. *J. Amer. Ceram. Soc.*, **37**, 133.

Ringwood, A. E. and **Lovering, J. F.**, 1970. Significance of pyroxene–ilmenite intergrowths among kimberlite xenoliths. *Earth, Planet. Sci. Letters*, **7**, 371–375.

Rio, M., 1968. Quelques précisions sur la composition minéralogique de la lherzolite de Moun Caou (Basses-Pyrénées). *Bull. Soc. franç. Min. Crist.*, **91**, 298–299.

Robie, R. A. and **Waldbaum, D. R.**, 1968. Thermodynamic properties of minerals and related substances at 298·15°K (25·0°C) and one atmosphere (1·013 bars) pressure and at higher temperatures. *U.S. Geol. Surv. Bull.*, **1259**, 256 pp.

Rodgers, K. A. and **Brothers, R. N.**, 1969. Olivine, pyroxene, feldspar and spinel in ultramafic nodules from Auckland, New Zealand. *Min. Mag.*, **37**, 375–390.

Roobol, M. J., 1974. The geology of the Vesturhorn intrusion, S.E. Iceland. *Geol. Mag.*, **111**, 273–286.

Rosenhauer, M., 1976. Effect of pressure on the melting enthalpy of diopside under dry and H_2O-saturated conditions. *Carnegie Inst. Washington, Ann. Rept. Dir. Geophys. Lab.*, 1975–76, 648–651.

Rosenhauer, M. and **Eggler, D. H.**, 1975. Solution of H_2O and CO_2 in diopside melt. *Carnegie Inst. Washington, Ann. Rept. Dir. Geophys. Lab.*, 1974–75, 474–479.

Ross, C. S., Foster, M. D. and **Myers, A. T.**, 1954. Origin of dunites and of olivine-rich inclusions in basaltic rocks. *Amer. Min.*, **39**, 693–737.

Roy, S., 1966. *Syngenetic manganese formations of India.* Jadarpar Univ., Calcutta.

Roy, S., 1971. Studies on manganese-bearing silicate minerals from metamorphosed manganese formations of India. II. Blanfordite, manganoan diopside and brown manganiferous pyroxene. *Min. Mag.*, **38**, 32–41.

Roy, S. and **Mitra, F. N.**, 1964. Mineralogy and genesis of the gondites associated with metamorphic manganese ore bodies of Madhya Pradesh and Mahreshtra, India. *Proc. Nat. Inst. Sci. India*, **30**, 395–438.

Rutstein, M. S., 1971. Re-examination of the wollastonite–hedenbergite ($CaSiO_3$–$CaFeSi_2O_6$) equilibria. *Amer. Min.*, **56**, 2040–2052.

Rutstein, M. S. and **Yund, R. A.**, 1969. Unit-cell parameters of synthetic diopside hedenbergite solid solutions. *Amer. Min.*, **54**, 238–245.

Ryall, W. T. and **Threadgold, I. M.**, 1966. Evidence for $[(SiO_3)_5]_\infty$ type chains in inesite as shown by X-ray and infrared absorption studies. *Amer. Min.*, **51**, 754–761.

Sabine, P. A. and **Young, B. R.**, 1975. Metamorphic processes at high temperature and low pressure: the petrogenesis of the metasomatized and assimilated rocks of Carneal, Co. Antrim. *Phil. Trans. Roy. Soc., A*, **280**, 225–269.

Sahama, Th. G., 1952. Leucite, potash nepheline and clinopyroxene from volcanic lavas from southwestern Uganda and adjoining Belgian Congo. *Amer. J. Sci.*, Bowen vol., 457–470.

Sahama, Th. G., 1976. Composition of clinopyroxene and melilite in the Nyiragongo rocks. *Carnegie Inst. Washington, Ann. Rept. Dir. Geophys. Lab.*, 1975–76, 585–591.

Sakata, Y., 1957. Unit cell dimensions of synthetic aluminian diopsides. *Jap. J. Geol. Geogr.*, **28**, 161–168.

Sarap, H., 1957. Studien an den Skarnmineralien der Asgrube im Eisenerzfeld von Norberg, Mittelschweden. *Geol. För. Förh. Stockholm*, **79**, 542–571.

Saxena, S. K., 1967. Intracrystalline chemical variations in certain calcic pyroxenes and biotites. *Neues Jahrb. Min. Abhdl.*, **107**, 299–316.

Saxena, S. K., 1968. Crystal-chemical aspects of distribution of elements among certain coexisting rock-forming silicates. *Neues Jahrb., Abhdl.*, **108**, 292–323.

Saxena, S. K., 1969. Silicate solid solutions and geothermometry. 4. Statistical study of chemical data on garnets and clinopyroxene. *Contr. Min. Petr.*, **23**, 140–156.

Schairer, J. F. and **Bowen, N. L.**, 1938. The system leucite–diopside–silica. *Amer. J. Sci.*, 5th ser., **35A**, 289–309.

Schairer, J. F. and **Bowen, N. L.**, 1942. The binary system $CaSiO_3$–diopside and the relations between $CaSiO_3$ and åkermanite. *Amer. J. Sci.*, **240**, 725–742.

Schairer, J. F. and **Kushiro, I.**, 1964. The join diopside–silica. *Carnegie Inst. Washington, Ann. Rept. Dir. Geophys. Lab.*, 1963–64, 130–132.

Schairer, J. F. and **Kushiro, I.**, 1965. The join diopside–spinel at atmospheric pressure and the significance of the diopside–spinel assemblage. *Carnegie Inst. Washington, Ann. Rept. Dir. Geophys. Lab.*, 1964–65, 100–103.

Schairer, J. F. and **Osborn, E. F.**, 1950. The system CaO–MgO–FeO–SiO_2: I. Preliminary data on the join $CaSiO_3$–MgO–FeO. *J. Amer. Ceram. Soc.*, **33**, 160–167.

Schairer, J. F., **Tilley, C. E.** and **Brown, G. M.**, 1968. The join nepheline–diopside–anorthite and its relation to alkali basalt fractionation. *Carnegie Inst. Washington, Ann. Rept. Dir. Geophys. Lab.*, 1966–67, 467–471.

Schairer, J. F., **Yagi, K.** and **Yoder, H. S.**, 1962. The system nepheline–diopside. *Carnegie Inst. Washington, Ann. Rept. Dir. Geophys. Lab.*, 1961–62, 96–98.

Schairer, J. F. and **Yoder, H. S.**, 1960. The nature of residual liquids from crystallization, with data on the system nepheline–diopside–silica. *Amer. J. Sci.*, **258A**, 273–283.

Schairer, J. F. and **Yoder, H. S.**, 1962. Pyroxenes: The system diopside–enstatite–silica. *Carnegie Inst. Washington, Ann. Rept. Dir. Geophys. Lab.*, 1961–62, 75–82.

Schairer, J. F. and **Yoder, H. S.**, 1970. Critical planes and flow sheet for a portion of the system CaO–MgO–Al_2O_3–SiO_2 having petrological implications. *Carnegie Inst. Washington, Ann. Rept. Dir. Geophys. Lab.*, 1968–69, 202–214.

Schairer, J. F., **Yoder, H. S.** and **Tilley, C. E.**, 1967. The high temperature behaviour of synthetic melilites in the join gehlenite–soda melilite–akermanite. *Carnegie Inst. Washington, Ann. Rept. Dir. Geophys. Lab.*, 1965–66, 217–226.

Schaller, W. T., 1938. Johannsenite, a new manganese pyroxene. *Amer. Min.*, **23**, 575–582.

Scharbert, H. G., 1966. Andraditführende Einschaltungen in Marmor von Hartenstein (Kl. Kremstal, N.O.). *Neues Jahrb. Min., Mh.*, 221–223.

Schiavinato, G., 1953. Sulla johannsenite dei giacimenti a silicati manganesiferi del Monte Civillina presso Recoaro (Vicenza). *Rend. Soc. Min. Ital.*, **9**, 210–218 (M.A. **12**-259).

Schmetzer, K. and **Medenbach, O.**, 1974. Chrom-Diopsid aus Kenya. *Zeit. Deutschen Gemmologischen Gessell.*, **23**, 178–179 (M.A. 75-1090).

Schrijver, K., 1973. Bimetasomatic plagioclase–pyroxene reaction zones in granulite facies. *Neues Jahrb. Min. Abhdl.*, **119**, 1–19.

Schröpfer, L., 1968. Über den Einbau von Titan in Diopsid. *Neues Jahrb. Min., Mh.*, 441–453.

Schröpfer, L., 1971. Über den gekoppelten Ersatz von $Mg^{2+}Si^{4+}$ durch $Ti^{4+}Al_2^{3+}$ im Diopsid. *Neues Jahrb. Min., Abhdl.*, **116**, 20–40.

Schwab, R. G. and **Schwerin, M.**, 1975. Polymorphie und Entmischungsreaktionen im System Enstatit ($MgSiO_3$)–Diopsid ($CaMgSi_2O_6$). *Neues Jahrb. Min., Abhdl.*, **124**, 223–245.

Schwarcz, H. P., 1967. The effect of crystal field stabilization on the distribution of transition metals between metamorphic minerals. *Geochim. Cosmochim. Acta*, **31**, 503–517.

Scott, P. W., 1976. Crystallization trends of pyroxenes from the alkaline volcanic rocks of Tenerife, Canary Islands. *Min. Mag.*, **40**, 805–816.

Segnit, E. R., 1953. Some data on synthetic aluminous and other pyroxenes. *Min. Mag.*, **30**, 218–226.

Seki, Y. and **Kuriyagawa, S.**, 1962. Mafic and leucocratic rocks associated with serpentinite of Kanasaki, Kantô Mountains, Central Japan. *Jap. J. Geol. Geogr.*, **33**, 15–32.

Semenov, I. V., 1970. Effect of chemical composition on the unit-cell parameters of pyroxenes as a function of the energy of the elements. *Dokl. Acad. Sci. U.S.S.R., Earth Sci., Sect.*, **192**, 128–131.

Sen, S. K., 1973. Compositional relations among hornblende and pyroxenes in basic granulites and an application to the origin of garnets. *Contr. Min. Petr.*, **38**, 299–306.

Serdyuchenko, D. P., 1937. Diopside from the Ekhresku range in the Karachai. *Mem. Soc. Russe Min.*, ser. 2, **66**, 474–478 (M.A. **7**-181).

Seward, T. M., 1971. The distribution of transitional elements in the system $CaMgSi_2O_6$–$Na_2Si_2O_5$–H_2O at 1000 bars pressure. *Chem. Geol.*, **7**, 73–95.

Shaw, D. M., **Moxham, R. L.**, **Filby, R. H.** and **Lapkowsky, W. W.**, 1963a. The petrology and geochemistry of some Grenville skarns. Part I. Geology and petrography. *Can. Min.*, **7**, 420–442.

Shaw, D. M., **Moxham, R. L.**, **Filby, R. H.** and **Lapkowsky, W. W.**, 1963b. The petrology and geochemistry of some Grenville skarns. Part II. Geochemistry. *Can. Min.*, **7**, 576–616.

Shidô, F., 1958. Plutonic and metamorphic rocks of the Nakoso and Iritōno districts in the central Abukuma Plateau. *J. Fac. Sci. Univ. Tokyo*, **11**, 131–217.

Shimazaki, H., 1967. Hedenbergite and epidote from a graphite-bearing skarn of the Kamioka mine, Gifu Prefecture. *J. Geol. Soc. Japan*, **73**, 259.

Shimizu, N., 1974. An experimental study of the partitioning of K, Rb, Cs, Sr and Ba between clinopyroxene and liquid at high pressures. *Geochim. Cosmochim. Acta*, **38**, 1789–1798.

Shinno, I., 1968. [On the 'hybrid solid solution' of diopside in the system $(CaMg)SiO_3$–SiO_2.] *Rept. Earth Sci. Dept. Coll. Gen. Edu. Kyushu Univ.*, **15**, 7–17 (in Japanese, English summary).

Shinno, I., 1970. A consideration on the crystallization process of diopside. *J. Japanese Assoc. Min. Petr. Econ. Geol.*, **63**, 146–159.

Shteinberg, D. C. and **Tominyh, V. G.**, 1967. [Clinopyroxenes of pyroxenites from Kachkanar (Urals).] *Zap. Vses. Min. Obshch.*, **96**, 133–140 (in Russian).

Sighinolfi, G. P. and **Fujimori, S.**, 1974. Petrology and chemistry of diopsidic rocks in granulite terrains from the Brazilian basement. *Atti. Soc. Tosc. Sci. Natur. Mem., Ser. A*, **81**, 103–120.
Sinistÿn, A. V., 1965. [Pyroxenes in differentiated dolerite intrusion.] *Zap. Vses. Min. Obschch.*, **94**, 583–592 (in Russian).
Skippen, G. B., 1971. Experimental data for reactions in siliceous marbles. *J. Geol.*, **79**, 457–481.
Skippen, G. B., 1974. An experimental model for low pressure metamorphism of siliceous dolomitic marble. *Amer. J. Sci.*, **274**, 487–500.
Slaughter, J., **Kerrick, D. M.** and **Wall, V. J.**, 1975. Experimental and thermodynamic study of equilibria in the system $CaO–MgO–SiO_2–H_2O–CO_2$. *Amer. J. Sci.*, **275**, 143–162.
Smith, C. B., **McCallum, M. E.** and **Eggler, D. H.**, 1976. Clinopyroxene–ilmenite intergrowths from the Iron Mountain kimberlite district, Wyoming. *Carnegie Inst. Washington, Ann. Rept. Dir. Geophys. Lab.*, 1975–76, 542–546.
Smith, D., 1971. Iron-rich pyroxenes. *Carnegie Inst. Washington, Ann. Rept. Dir. Geophys. Lab.*, 1969–70, 285–290.
Smith, D., 1972. Stability of iron-rich pyroxene in the system $CaSiO_3–FeSiO_3–MgSiO_3$. *Amer. Min.*, **57**, 1413–1428.
Smith, J. V., 1966. X-ray emission microanalysis of rock-forming minerals. VI. Clinopyroxenes near the diopside–hedenbergite join. *J. Geol.*, **74**, 463–477.
Sobolev, N. V., **Bartoshinskiy, Z. V.**, **Yefimova, E. S.**, **Larrent'yev, Yu. G.** and **Pospelova, L. N.**, 1970. Olivine–garnet–chrome diopside assemblage from Yakutian diamond. *Dokl. Acad. Sci. U.S.S.R., Earth Sci. Sect.*, **192**, 134–137.
Sobolev, V. S., **Bazarova, T. J.** and **Yagi, K.**, 1975. Crystallization temperature of wyomingite from Leucite Hills. *Contr. Min. Petr.*, **49**, 301–308.
Sood, M. K., **Platt, R. G.** and **Edgar, A. D.**, 1970. Phase relations in portions of the system diopside–nepheline–kalsilite–silica and their importance in the genesis of alkaline rocks. *Can. Min.*, **10**, 380–394.
Spencer, A. C. and **Page, S.**, 1935. Geology of the Santa Rita Mining area, New Mexico. *Bull. U.S. Geol. Surv.*, **859**.
Stillwell, F. L., 1959. Petrology of the Broken Hill Lode and its bearing on ore genesis. *Proc. Austral. Inst. Mining Metall.*, **190**, 1–84.
Tazaki, K., **Ito, E.** and **Komatsu, M.**, 1972. Experimental study of pyroxene–spinel symplectite at high pressures and temperatures. *J. Geol. Soc. Japan*, **76**, 347–354.
Thompson, A. B., 1975a. Calc-silicate diffusion zones between marble and pelitic schist. *J. Petr.*, **16**, 314–346.
Thompson, A. B., 1975b. Mineral reactions in a calc-mica schist from Gassetts, Vermont, U.S.A. *Contr. Min. Petr.*, **53**, 105–127.
Thompson, J. B., 1970. Geometrical possibilities for amphibole structures: model biopyriboles. *Amer. Min.*, **55**, 292–293.
Thompson, R. N., 1972. Melting behaviour of two Snake River lavas at pressures up to 35 kbar. *Carnegie Inst. Washington, Ann. Rept. Dir. Geophys. Lab.*, 1971–72, 406–410.
Tilley, C. E., 1920. The metamorphism of the pre-Cambrian dolomites of southern Eyre Peninsula, South Australia. *Geol. Mag.*, **57**, 449–462.
Tilley, C. E., 1946. Bustamite from Treburland manganese mine, Cornwall, and its paragenesis. *Min. Mag.*, **27**, 236–241.
Tilley, C. E., 1949. An alkali facies of granite at granite–dolomite contacts in Skye. *Geol. Mag.*, **86**, 81–93.
Tilley, C. E., 1951. The zoned contact-skarns of the Broadford area, Skye: a study of boron-fluorine metasomatism in dolomites. *Min. Mag.*, **29**, 622–666.
Tilley, C. E. and **Harwood, H. F.**, 1931. The dolerite–chalk contact of Scawt Hill, Co. Antrim; the production of basic alkalic-rocks by the assimilation of limestone by basaltic magma. *Min. Mag.*, **22**, 439–468.
Tilley, C. E. and **Long, J. V. P.**, 1967. The porphyroclast minerals of the peridotite-mylonites of St. Paul's Rocks (Atlantic). *Geol. Mag.*, **104**, 46–48.
Tokunaga, M., 1965. On the zoned skarn including bustamite, ferroan johannsenite, and manganoan hedenbergite from Nakatatsu mine, Fukui Prefecture, Japan. *Tokyo Univ., Sci. Rept.*, **9**, 67–87.
Toropov, N. A. and **Khotimchenko, V. C.**, 1967. Isomorphous substitution in aluminous pyroxenes. *Geochem. Intern.*, **4**, 831–835.
Trommsdorff, V. and **Evans, B. W.**, 1974. Alpine metamorphism of peridotitic rocks. *Schweiz. Min. Petr. Mitt.*, **54**, 333–352.
Turner, F. J., 1967. Thermodynamic appraisal of steps in progressive metamorphism of siliceous dolomitic limestones. *Neues Jahrb. Min., Mh.*, 1–22.
Turnock, A. C., 1962. Synthetic pyroxenes phase relationships on the join diopside–hedenbergite. *J. Geophys. Res.*, **67**, 3605–3610.

Turnock, A. C., Lindsley, D. H. and **Grover, J. E.**, 1973. Synthesis and unit cell parameters of Ca–Mg–Fe pyroxenes. *Amer. Min.*, **58**, 50–59.
Upton, B. G. J. and **Thomas, J. E.**, 1973. Precambrian potassic ultramafic rocks: south Greenland. *J. Petr.*, **14**, 509–534.
Vaidya, S. N., Bailey, S., Pasternack, T. and **Kennedy, G. C.**, 1973. Compressibility of fifteen minerals to 45 kilobars. *J. Geophys. Res.*, **78**, 6893–6898.
Val'ter, A. A., 1969. Distribution of magnesium and iron between coexisting iron-rich solid solutions of olivine and orthopyroxene in eulysite from the Mariupol' iron deposit. *Dokl. Acad. Sci. U.S.S.R. Earth Sci. Sect.*, **187**, 106–109.
Val'ter, A. A., Gorogotskaya, L. I., Zverev, N. D. and **Romanov, V. P.**, 1970. Two types of distribution of iron atoms in pyroxenes similar to hedenbergite, determined by Mössbauer spectroscopy. *Dokl. Acad. Sci., U.S.S.R.*, **192**, 104–107.
Veblen, D. R. and **Burnham, C. W.**, 1969. The crystal structures of hedenbergite and ferrosilite (abstract). *Can. Min.*, **10**, 147.
Velde, D., 1975. Armalcolite–Ti-phlogopite–diopside–analcite-bearing lamproites from Smoky Butte, Garfield County, Montana. *Amer. Min.*, **60**, 566–573.
Velde, D. and **Thiebaut, J.**, 1973. Quelques précisions sur la constitution minéralogique de la néphélinite à olivine et mélilite d'Essey-la-Côte (Meurthe-et-Moselle). *Bull. Soc. franç. Min. Crist.*, **96**, 298–302.
Velde, D. and **Touron, J.**, 1970. La camptonite de San Feliú de Buxallan (Province de Gérone, Espagne). *Bull. Soc. franç. Min. Crist.*, **93**, 482–487.
Vilenskiy, A. M., 1966. Phase equilibria and evolution in the chemical nature of the pyroxenes of intrusive traps of the Siberian platform. *Geochem. Intern.*, **3**, 1275–1306.
Vincent, E. A., 1953. Hornblende–lamprophyre dykes of basaltic parentage from the Skaergaard area, East Greenland. *Quart. J. Geol. Soc.*, **109**, 21–50.
Vinokurov, V. M., 1966. Electron paramagnetic resonance data on isomorphism of manganese and iron ions in certain minerals. *Geochem. Intern.*, **3**, 996–1002.
Vlad, S.-N. and **Vasiliu, C.**, 1969. Some chemical characteristics of the pyroxenic skarns from Dognecea, Banat, Romania. *Norsk Geol. Tidsskr.*, **49**, 361–366.
Volchenko, Yu. A., 1971. [Find of rose-coloured diopside in the Urals.] *Zap. Vses. Min. Obschch.*, **100**, 341–344. (In Russian) (M.A. 72-3232).
Vredevoogd, J. J. and **Forbes, W. C.**, 1975. The system diopside–ureyite at 20 kbars. *Contr. Min. Petr.*, **52**, 147–156.
Vuorelainen, Y., 1963. Notes on the gem variety of the Outokumpu chrome-diopside. *J. Gemmology*, **9**, 42–43.
Wager, L. R. and **Brown, G. M.**, 1968. *Layered Igneous Rocks*. Edinburgh (Oliver and Boyd).
Wager, L. R. and **Deer, W. A.**, 1939. Geological investigations in east Greenland. Part III. The petrology of the Skaergaard intrusion, Kangerdlugssuaq, east Greenland. *Medd. om Grønland*, **105**, no. 4, 1–352.
Wallace, R. C., 1975. Mineralogy and petrology of xenoliths in a diatreme from South Westland, New Zealand. *Contr. Min. Petr.*, **49**, 191–199.
Warner, R. D., 1973. Liquidus relations in the systems $CaO–MgO–SiO_2–H_2O$ at 10 kbar P_{H_2O} and their petrologic significance. *Amer. J. Sci.*, **273**, 925–946.
Warner, R. D., 1975. New experimental data for the system $CaO–MgO–SiO_2–H_2O$ and a synthesis of inferred phase relations. *Geochim. Cosmochim. Acta*, **39**, 1413–1421.
Warren, B. E. and **Bragg, W. L.**, 1928. The structure of diopside $CaMg(SiO_3)_2$. *Z. Krist.*, **69**, 168–193.
Weeks, W. F., 1956a. A thermochemical study of equilibrium relations during metamorphism of siliceous carbonate rocks. *J. Geol.*, **64**, 245–270.
Weeks, W. F., 1956b. Heats of formation of metamorphic minerals in the systems $CaO–MgO–SiO_2–H_2O$ and their petrological significance. *J. Geol.*, **64**, 456–472.
Weigand, P. W., 1975. Geochemistry of the Oslo basaltic rocks. *Norske Videnskaps-Akad. I. Mat.-Naturv. Klasse Skr.* Ser., **34**, 5–38.
Wenk, H.-R., 1970. Submicroscopical twinning in lunar and experimentally deformed pyroxenes. *Contr. Min. Petr.*, **26**, 315–323.
Wenk, H.-R. and **Maurizio, R.**, 1970. Geological observations in the Bergell area (SE Alps). II. Contact minerals from Mt. Sissone–Cima di Vazzeda (a mineralogical note). *Schweiz. Min. Petr. Mitt.*, **50**, 349–354.
Wheeler, E. P., 1960. Anorthosite–adamellite complex of Nain, Labrador. *Bull. Geol. Soc. America*, **71**, 1755–1762.
Wheeler, E. P., 1965. Fayalitic olivine in northern Newfoundland–Labrador. *Can. Min.*, **8**, 339–346.
White, W. B., 1966. Ultramafic inclusions in basaltic rocks from Hawaii. *Contr. Min. Petr.*, **12**, 245–314.
White, W. B. and **Keester, K. L.**, 1966. Optical absorption spectra of iron in the rock-forming silicates. *Amer. Min.*, **51**, 774–791.

White, W. B., McCarthy, G. J. and **Scheetz, B. E.**, 1971. Optical spectra of chromium, nickel and cobalt-containing pyroxenes. *Amer. Min.*, **56**, 72–89.

Whitney, P. R. and **McLelland, J. M.**, 1973. Origin of coronas in metagabbros of the Adirondack Mts., N.Y. *Contr. Min. Petr.*, **39**, 81–98.

Wikström, A., 1970. Hydrothermal experiments in the system jadeite–diopside. *Norsk Geol. Tidsskr.*, **50**, 1–14.

Wilkinson, J. F. G., 1956. Clinopyroxenes of alkali-basalt magma. *Amer. Min.*, **41**, 724–743.

Wilkinson, J. F. G., 1957. The clinopyroxenes of a differentiated teschenite sill near Gunnedah, New South Wales. *Geol. Mag.*, **94**, 123–134.

Wilkinson, J. F. G., 1966. Clinopyroxenes from the Square Top intrusion, Nundle, New South Wales. *Min. Mag.*, **35**, 1061–1070.

Wilkinson, J. F. G., 1973. Pyroxenite xenoliths from an alkali trachybasalt in the Glen Innes Area, northeastern New South Wales. *Contr. Min. Petr.*, **42**, 15–32.

Wilkinson, J. F. G., 1975a. An Al-spinel ultramafic–mafic inclusion suite and high pressure megacrysts in an analcimite and their bearing on basaltic magma fractionation at elevated pressures. *Contr. Min. Petr.*, **53**, 71–104.

Wilkinson, J. F. G., 1975b. Ultramafic inclusions and high pressure megacrysts from a nephelinite sill, Nandewar Mountains, north-eastern New South Wales, and their bearing on the origins of certain ultramafic inclusions in alkaline volcanic rocks. *Contr. Min. Petr.*, **51**, 235–262.

Wilkinson, J. F. G. and **Binns, R. A.**, 1969. Hawaiite of high pressure origin from northeastern New South Wales. *Nature*, **222**, 553–555.

Wilkinson, J. F. G., Duggan, M. B., Herbert, H. K. and **Kalocsai, G. I. Z.**, 1975. The Salt Lick Creek layered intrusion, east Kimberley region, Western Australia. *Contr. Min. Petr.*, **50**, 1–23.

Williams, D. W. and **Kennedy, G. C.**, 1969. Melting curve of diopside to 50 kilobars. *J. Geophys. Res.*, **74**, 4359–4366.

Wilshire, H. G., 1967. The Prospect alkaline diabase-picrite intrusion, New South Wales, Australia. *J. Petr.*, **8**, 97–163.

Wilson, A. F., 1953. Diopsidization and hornblendization – important metasomatic phenomena in the basic schists near Southern Cross, Western Australia. *J. Roy. Soc. W. Australia*, **37**, 97–103.

Winchester, J. A., 1974. The zonal pattern of regional metamorphism in the Scottish Caledonides. *J. Geol. Soc.*, **130**, 509–524.

Windley, B. F. and **Smith, J. V.**, 1974. The Fiskenaesset complex, west Greenland. Part II. General mineral chemistry from Qeqertarssuatsiaq. *Medd. om Grønland*, **196**, no. 4, 1–54.

Wolfe, C. W., 1955. Crystallography of jadeite crystals from near Cloverdale California. *Amer. Min.*, **40**, 248–260.

Wood, B. J., 1976a. Mixing properties of tschermakitic clinopyroxenes. *Amer. Min.*, **61**, 599–602.

Wood, B. J., 1976b. On the stoichiometry of clinopyroxenes in the system $CaO–MgO–Al_2O_3–SiO_2$. *Carnegie Inst. Washington, Ann. Rept. Dir. Geophys. Lab., 1975–76*, 741–742.

Wright, J. B., 1972. Natural garnet–pyroxene transformation in Cenozoic alkali basalt from Nigeria? *Min. Mag.*, **38**, 579–582.

Wyatt, B. A., 1977. The melting and crystallization behaviour of a natural clinopyroxene–ilmenite intergrowth. *Contr. Min. Petr.*, **61**, 1–10.

Wyatt, B. A., McCallister, R. H., Boyd, F. R. and **Ohashi, Y.**, 1975. An experimentally produced clinopyroxene–ilmenite intergrowth. *Carnegie Inst. Washington, Ann. Rept. Dir. Geophys. Lab.*, 1974–75, 536–539.

Wyckoff, R. W. G., Merwin, H. E. and **Washington, H. S.**, 1925. X-ray diffraction measurements upon the pyroxenes. *Amer. J. Sci.*, 4th ser., **10**, 389–397.

Wyllie, P. J., 1963. Effects of the changes in slope occurring in liquidus and solidus paths in the system diopside–anorthite–albite. *Min. Soc. America Spec. Paper*, **1**, 204–212.

Wyllie, P. J. and **Huang, W-L.**, 1976. Carbonation and melting reactions in the system $CaO–MgO–SiO_2–CO_2$ at mantle pressures with geophysical and petrological applications. *Contr. Min. Petr.*, **54**, 79–107.

Wys, E. C. De, 1960. Silicate melts with indications of ino structures. *Min. Mag.*, **32**, 640–643.

Wys, E. C. De, 1972. Corrections to the system anorthite–åkermanite–diopside by means of Gibb's free energy analyses of the systems anorthite–åkermanite and anorthite–diopside. *Min. Mag.*, **38**, 632–634.

Wys, E. C. De and **Foster, W. R.**, 1958. The system diopside–anorthite–åkermanite. *Min. Mag.*, **31**, 736–743.

Yagi, K., 1966. The system acmite–diopside and its bearing on the stability relations of natural pyroxenes of the acmite–hedenbergite–diopside series. *Amer. Min.*, **51**, 976–1000.

Yagi, K. and **Onuma, K.**, 1967. The join $CaMgSi_2O_6–CaTiAl_2O_6$ and its bearing on the titanaugites. *Journ. Fac. Sci. Hokkaido Univ.*, Ser. 4, *Geol. Min.*, **13**, 463–483.

Yagi, K. and **Onuma, K.**, 1969. An experimental study on the role of titanium in alkali basalts in light of the system diopside–akermanite–nepheline–$CaTiAl_2O_6$. *Amer. J. Sci.* (Schairer vol.), 509–549.
Yamaguchi, M., 1961. Chrome-diopside in the Horoman and Higashi-Akaishi peridotites, Japan. *Mem. Fac. Sci., Kyushu Univ., ser. D., Geol.*, **10**, 233–245.
Yamaguchi, Y., 1973. Study on the exsolution phenomena of pyroxenes. *J. Fac. Sci., Hokkaido Univ.*, Ser. 4, **16**, 133–165.
Yamaguchi, Y. and **Tomita, K.**, 1970. Clinoenstatite as an exsolution phase in diopside. *Mem. Fac. Sci., Kyoto Univ., Ser. Geol. Min.*, **37**, 173–180.
Yamazaki, T., **Onuki, H.** and **Tiba, T.**, 1966. Significance of hornblende gabbroic inclusions in calc-alkaline rocks. *J. Japanese Assoc. Min. Petr. Econ. Geol.*, **55**, 87–103.
Yang, H-Y., 1975. Al- and Ti-rich clinopyroxenes in the system $CaMgSi_2O_6$–$CaAl_2SiO_6$–$CaTiAl_2O_6$. *Proc. Geol. Soc. China [Formosa]*, **18**, 48–58.
Yefimov, A. A. and **Ivanova, L. I.**, 1964. Metasomatic zoning in platiniferous dunite and pyroxenite of the Urals. *Dokl. Acad. Sci. U.S.S.R., Earth Sci. Sect.*, 1964, **151**, 157–160.
Yoder, H. S., 1950. The jadeite problem, Part II. *Amer. J. Sci.*, **248**, 312–334.
Yoder, H. S., 1952. Change of melting point of diopside with pressure. *J. Geol.*, **60**, 364–374.
Yoder, H. S., 1954. The system diopside–anorthite–water. *Carnegie Inst. Washington, Ann. Rept. Dir. Geophys. Lab.*, 1953–54, 106–107.
Yoder, H. S., 1965. Diopside–anorthite–water at five and ten kilobars and its bearing on explosive volcanism. *Carnegie Inst. Washington, Ann. Rept. Dir. Geophys. Lab.*, 1964–65, 82–89.
Yoder, H. S., 1975. Relationship of melilite-bearing rocks to kimberlite: A preliminary report on the system akermanite–CO_2. *Phys. Chem. Earth*, **9**, 883–894.
Yoder, H. S. and **Kushiro, I.**, 1972. Composition of residual liquids in the nepheline–diopside system. *Carnegie Inst. Washington, Ann. Rept. Dir. Geophys. Lab.*, 1971–72, 413–416.
Young, D. A., 1972. A quartz syenite intrusion in the New Jersey Highlands. *J. Petr.*, **13**, 511–528.
Zavaritsky, V. A. and **Kirova, T. V.**, 1973. [The succession of metamorphic processes during the formation of skarn iron ore deposits of the Kustanai region.] *Zap. Vses. Min. Obschch.*, **102**, 600–611 (in Russian).
Zharikov, V. A. and **Podlessky, K. V.**, 1955. On the behaviour of pyroxene as a mineral of variable composition in infiltration skarn zones. *Dokl. Acad. Sci. U.S.S.R.*, **105**, 1096–1099.
Zharikov, V. A. and **Vlasova, D. K.**, 1955. The diagram: composition–properties for pyroxenes of the isomorphous series, diopside–hedenbergite–johannsenite. *Dokl. Acad. Sci. U.S.S.R.*, **105**, 814–817.
Zielinski, R. A. and **Frey, F. A.**, 1974. An experimental study of the partitioning of a rare earth element (Gd) in the system diopside–aqueous vapour. *Geochim. Cosmochim. Acta*, **38**, 545–565.
Zverev, N. D., **Valter, A. A.**, **Romanov, V. P.** and **Gorogotskaia, L. I.**, 1971. Character of Fe^{2+} ion distribution in pyroxenes from eulysite. *Lithos*, **4**, 17–21.
Zvetkov, A. I., 1945. Synthesis of alumina pyroxenes and dependence of their optics on composition. *Mém. Soc. Russe Min.*, Ser. 2, **74**, 215–222 (M.A. **11**-92).

Augite $(Ca,Mg,Fe^{2+},Fe^{3+},Ti,Al)_2[(Si,Al)_2O_6]$

Monoclinic (+)

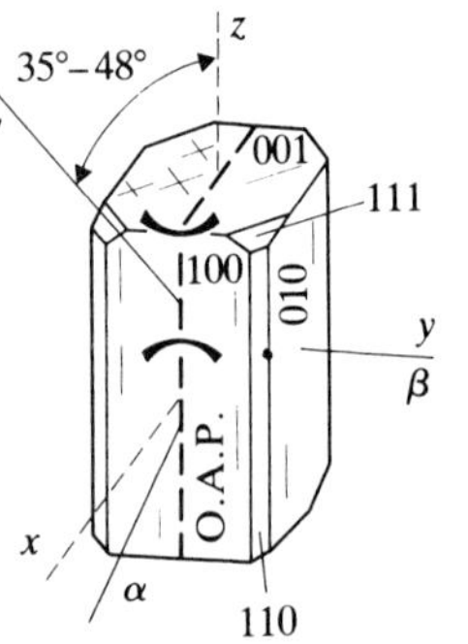

α	1·671–1·735
β	1·672–1·741
γ	1·703–1·774
δ	0·018–0·033
$2V_\gamma$	25°–61°[1]
$\gamma:z$	35°–48°[1]
$\beta = y$	O.A.P. (010)
Dispersion:	$r > v$ weak to moderate, strong in titanaugite.
D	3·19–3·56
H	$5\frac{1}{2}$–6
Cleavage:	{110} good; {100}, {010} parting. (110):(1$\bar{1}$0) ≃ 87°
Twinning:	{100} simple, multiple; common: {001} multiple.
Colour:	Pale brown, brown, purplish brown, green, black; colourless, pale brown, pale purplish brown, pale greenish brown or pale green in thin section.
Pleochroism:	More strongly coloured varieties show weak to moderate pleochroism; titanaugite moderate to strong. α pale greenish, pale brownish, green, greenish yellow. β pale brown, pale yellow-green, violet. γ pale green, greyish green, violet.
Unit cell:	$a \simeq 9{\cdot}8$, $b \simeq 9{\cdot}0$, $c \simeq 5{\cdot}25$ Å, $\beta \simeq 105°$ Z = 4. Space group $C2/c$.

Insoluble in HCl.

Augites occur in a wide variety of igneous rocks, and are particularly characteristic constituents of gabbros, dolerites and basalts; they occur also, but less frequently, in ultrabasic and intermediate rocks. Augites are present in some granulites and charnockites and other high grade metamorphic rocks, but are far less common in these than in igneous rocks. The ferroaugites are largely restricted to the highly differentiated ferrogabbros and iron-rich dolerites and their pegmatites, and to syenites, but have also been reported in acid volcanic glasses and in some metamorphosed iron formations. Subcalcic augite and subcalcic ferroaugite are the characteristic clinopyroxene phases of quickly chilled basic magmas and occur almost exclusively in basalts and andesites. Titanaugites are the typical pyroxenes of basic alkaline rocks, e.g. teschenite, essexite and nepheline dolerite.

The clinopyroxenes common in gabbros, dolerites and basalts have been the subject of varying nomenclature, particularly those with optic axial angles in the range of 20° to 40°. Suggestions for a standardized nomenclature have been put forward by Hess (1941), Benson (1944), Poldervaart (1947) and Poldervaart and

[1] Values up to $2V_\gamma$ 88° and $\gamma:z$ as high as 62° have been reported for sodian augites.

Hess (1951). The nomenclature here adopted is that of Poldervaart and Hess, and the division between subcalcic augite (and subcalcic ferroaugite) and pigeonite is taken at 15 mol. per cent $CaSiO_3$ (Fig. 1, p. 3). This chemical division corresponds with an optic axial angle of approximately 30°. The division of the main augite compositional field into augite, ferroaugite, subcalcic augite and subcalcic ferroaugite is an arbitrary one, and there is a continuous variation in the chemistry, as well as in the optical properties, of these minerals.

The name augite, is from the Greek *auge*, lustre, from the appearance of the cleavage plane, and was used by Werner in 1792 to describe the dark crystals found in many basaltic rocks.

Structure

The crystal structure of an augite (from Kakanui, New Zealand) has been determined by Clark *et al.* (1969). Its composition is $Ca_{0\cdot61}Na_{0\cdot09}Mg_{0\cdot90}Fe^{2+}_{0\cdot11}Fe^{3+}_{0\cdot10}Ti_{0\cdot02}Al_{0\cdot16}(Al_{0\cdot17}Si_{1\cdot83})O_6$, and cell parameters *a* 9·699, *b* 8·844, *c* 5·272 Å, β 106·97°, *V* 432·5 Å^3, with space group *C*2/*c*, Z = 4 and $D_{calc.}$ 3·31 g/cm^3.

The structure is very similar to that of diopside. All (Si,Al)–O chains are equivalent and they are more kinked, with O3–O3–O3 = 165·8° compared with Na (174·7° for jadeite) and Li pyroxenes. The mean non-bridging T–O distance is 1·612 Å and the mean bridging T–O distance is 1·668 Å (T = tetrahedral cations).

Whereas the *M*1 site contains no Ca and *M*2 no Al, the Mg,Fe cations show considerable disorder ($k = (Fe/Mg)_{M1}/(Fe/Mg)_{M2} = 0{\cdot}25$, compared with *k* approximately 0·1 for Ca-poor pyroxenes). This (Mg,Fe) distribution was confirmed by Mössbauer spectroscopy (Hafner and Virgo, 1970).

The *M*1 co-ordination is very nearly a regular octahedron (mean *M*1–O = 2·054 Å) while *M*2 is in eight-fold co-ordination with four short distances (mean 2·297 Å) two at 2·563 Å and two at 2·760 Å. The mixed population of *M* sites causes little essential difference between this structure and that of diopside.

The structure of an aluminous augite of high-pressure origin from Takasima, Japan, was determined by Takeda (1972). Its composition was given as $K_{0\cdot001}Na_{0\cdot064}Ca_{0\cdot753}Mg_{0\cdot792}Mn_{0\cdot006}Fe^{2+}_{0\cdot177}Ti_{0\cdot035}Fe^{3+}_{0\cdot086}Al_{0\cdot116}(Al_{0\cdot269}Si_{1\cdot731})O_6$, and its cell parameters as *a* 9·707, *b* 8·858, *c* 5·274 Å, β 106·52°, with space group *C*2/*c* and $D_{calc.}$ 3·40g/cm^3. Comparison with a coexisting aluminous orthopyroxene shows that the Mg,Fe ordering is less in the augite (= 0·16) than in the orthopyroxene (*k* = 0·08), but the difference is not large in relation to the margins of error. Details of the structure are very similar to those described above for the augite from Kakanui.

Takeda (1972) also determined the crystal structure of an aluminian augite from lunar rock 12052. Its composition is $Ca_{0\cdot607}Mg_{0\cdot764}Fe^{2+}_{0\cdot453}Al_{0\cdot059}Ti^{4+}_{0\cdot081}Cr_{0\cdot036}(Al_{0\cdot214}Si_{1\cdot786})O_6$, and its cell parameters *a* 9·726, *b* 8·909, *c* 5·268 Å, β 106·8°, with space group *C*2/*c* and $D_{calc.}$ 3·45 g/cm^3. In this augite there is a considerable amount of Fe and Mg in the *M*2 site and the distribution of (Fe,Mg) between *M*1 and *M*2 shows some ordering (*k* = 0·15). The extent of ordering is less than that for a coexisting pigeonite (*k* = 0·11), see p. 167, but again the difference is slight.

In the *M*2–O polyhedra, the four shortest distances (*M*2–O1, *M*2–O2 with mean 2·251 Å) are shorter than in diopside (mean 2·357 Å) but the *M*2–O3 distance is larger (mean 2·718 compared with 2·639 Å in diopside). Takeda suggests that this is a consequence of very rapid cooling without exsolution, which is also responsible for large anistropic temperature factors. Thus this structure of a subcalcic augite is thought to be a metastable one quenched from a high temperature. The (Fe,Mg) distribution for the augite and the coexisting pigeonite might be taken to indicate slow cooling, but this is in conflict with other evidence (lack of exsolution, diffuse (b) reflections from the pigeonite). It probably relates to continued subsolidus equilibration after an initial rapid cooling phase is over.

Takeda usefully compares the room temperature structures of three augites with that of diopside, with particular attention to the *M*2–O distances as a function of Ca content. The *M*2 cations are shown to approach closer to the two O1 and two O2 and becomes more distant from the four O3 atoms. At about 40 per cent Ca in *M*2, the *M*–O3 distances would be so large as to leave *M*2 effectively in four-fold co-ordination, an unstable configuration, which may explain the presence of the miscibility gap between Ca-rich and Ca-poor pyroxenes.

Takeda noted also that whereas the *M*2 thermal ellipsoid for diopside has its longest axis in the *xz* plane, with decreasing Ca, elongation along *y* develops. The former can be regarded as a true thermal vibration ellipsoid whereas the latter is taken to indicate positional disorder due to the mixed character of the *M*2 cations.

Dowty and Lindsley (1973) synthesized pyroxenes at 950°C and 20 kbar across the series $CaFeSi_2O_6$–$Fe_2Si_2O_6$. Those which were more iron-rich than about Fs_{80} had space group $P2_1/c$ while others were $C2/c$, but there was no definite evidence of an augite–pigeonite solvus. It was suggested that the less calcic pyroxene may have formed with $C2/c$ symmetry and inverted to $P2_1/c$ on cooling, but there was no X-ray evidence for a domain structure. Mössbauer spectra obtained at liquid nitrogen temperatures were studied by Dowty and Lindsley. The change in quadrupole splitting for Fe in the *M*2 site as the Ca content increases is related to the way the Fe co-ordination changes (even at low Ca concentrations) from the six-fold of ferrosilite towards the 4+4 co-ordination described by Takeda for subcalcic augites. It is also thought to be indicative of the structural differences responsible for the miscibility gap for the more magnesian clinopyroxenes.

The effect of the substitution of Ca by Fe on the crystal structure of a clinopyroxene was studied by Ohashi *et al.* (1975) who determined the structures of four synthetic clinopyroxenes on the hedenbergite–clinoferrosilite join with compositions $Fs_{65}Wo_{35}$, $Fs_{75}Wo_{25}$, $Fs_{80}Wo_{20}$, $Fs_{85}Wo_{15}$. The first three had space group $C2/c$ while the fourth was $P2_1/c$. Comparison was also made with the structure of clinoferrosilite Fs_{100} as determined by Burnham (1967). The *M*1–O co-ordination octahedron remains virtually unchanged across the series. With increasing Fe the *M*2–O polyhedron decreases in volume, and *M*2–O1 and *M*2–O2 distances decrease. All four *M*2–O3 distances, however, increase from Fs_{50} to Fs_{80}. Beyond this, and with the change to $P2_1/c$, two *M*2–O3 distances increase markedly, while two decrease, leaving *M*2 in six-fold co-ordination (Fig. 143). The intermediate structures are considered to be true structures rather than space averages of the extreme hedenbergite and clinoferrosilite structures.

The configuration of the Si–O chains through the series are illustrated by Fig. 144. With increasing Fe content the chain kinking (*O* rotation) becomes more marked. Beyond the transition to $P2_1/c$ the B chain becomes still more kinked while A straightens to 180° and then kinks in the opposite direction (*S* rotation).

Augites not uncommonly show exsolution lamellae of orthopyroxene on (100) and of pigeonite on planes which have been described as (001) and (100). Robinson *et al.* (1971) and Jaffe *et al.* (1975) noted that the pigeonite lamellae are parallel to an irrational plane of the host augite. The exact orientations vary with Fe/Mg ratio of the host (and lamellae) and are in agreement with the optimal phase boundary as

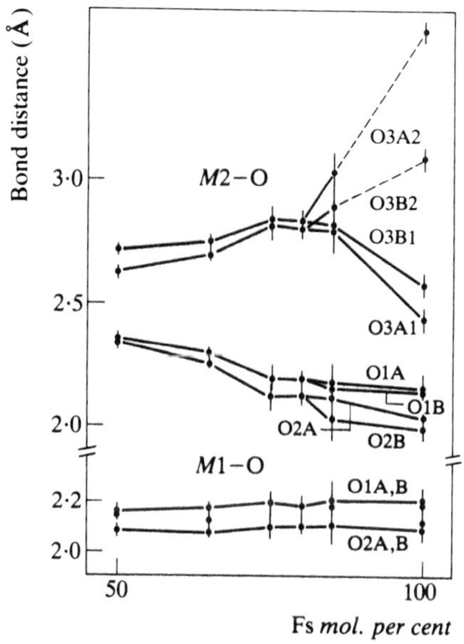

Fig. 143. Variation of *M*1–O and *M*2–O distances with Fe content in Ca–Fe clinopyroxenes (Ohashi *et al.*, 1975). Error bars represent ±1 standard deviation. Data for hedenbergite and clinoferrosilite are after Cameron *et al.* (1973) and Burnham (1967) respectively.

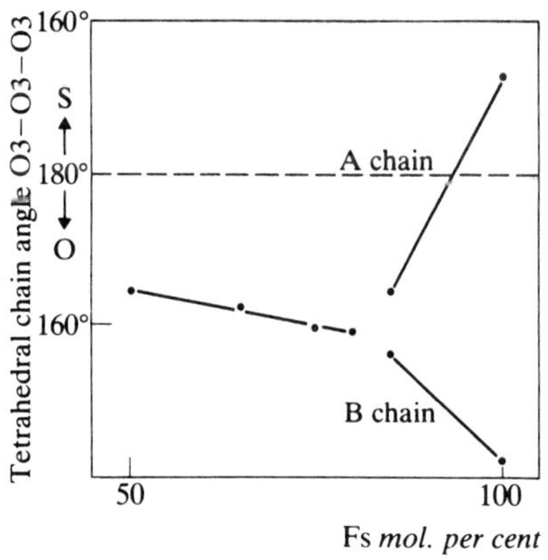

Fig. 144. Variation of tetrahedral chain angle O3–O3–O3 with Fe content in Ca–Fe clinopyroxenes (Ohashi *et al.*, 1975). Angles for hedenbergite and clinoferrosilite were computed from data by Cameron *et al.* (1973) and Burnham (1967) respectively.

calculated from the respective cell parameters. The deviations from (001) and (100) are appreciable (up to about 20°) and increase with decreasing Fe/Mg ratio. This trend correlates well with the greater misfit in cell parameters between corresponding augites and pigeonites for the more magnesian members. Although the composition plane between lamellae and host departs from rationality in this manner, the two lattices are almost parallel, deviating by less than 10′.

Fine-scale exsolution of pigeonite from augite can be studied by electron microscopy. Figure 145, for example, shows (001) and (100) pigeonite lamellae in augite from a lunar basalt (Nord *et al.*, 1976).

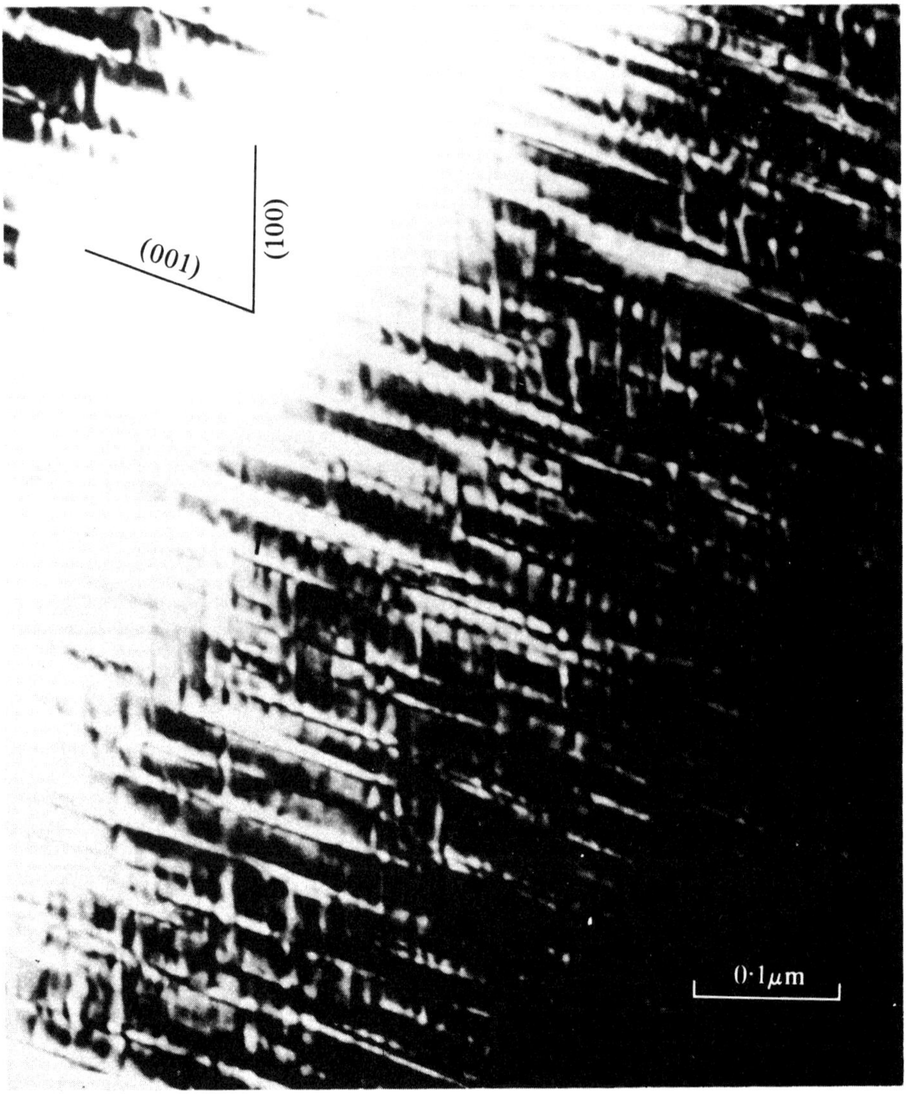

Fig. 145. Electron micrograph showing (001) and (100) fine pigeonite lamellae in augite from a lunar basalt (Nord *et al.*, 1976).

The cell parameters and optical properties of finely exsolved phases are not necessarily those intrinsic to their chemical composition since they may be structurally distorted (see for example Yamaguchi, 1973). The structural distortion may tend to favour the formation of metastable forms of exsolved pyroxene. Yamaguchi also noted that Ca-poor clinopyroxenes, more Fe-rich than Fs_{30}, tend to share (001) with the host phase while those with less iron tend to share (100), and associates this with lattice parameter requirements rather than the pigeonite–orthopyroxene inversion. In view of these observations Yamaguchi questions the usefulness of pyroxene exsolution as a geothermometer. For further discussion of exsolution see chemistry section, p. 342.

Cell Parameters of Ca–Fe–Mg clinopyroxenes

Relationships between cell parameters and composition for clinopyroxenes in the Ca–Fe–Mg field were given by Kuno and Hess (1953) and Kuno (1955), and modified versions were given by Brown (1960) and by Viswanathan (1966). Curves of *a* sin *β* and *b* plotted on the compositional trapezium (Fig. 146) are useful even though they do not take specific account of substituents other than Ca, Mg and Fe, e.g. Na, Al, Fe^{3+}, etc. Coleman (1962) determined the cell parameters for diopsides synthesized with small amounts of various substituents and thereby deduced relationships between cell parameters and composition. Winchell and Tilling (1960) and Winchell and Leake (1965) set up regression equations for the variation of cell parameters with composition for a large number of clinopyroxenes. These can be used to predict cell parameters from composition, but not composition from cell parameters.

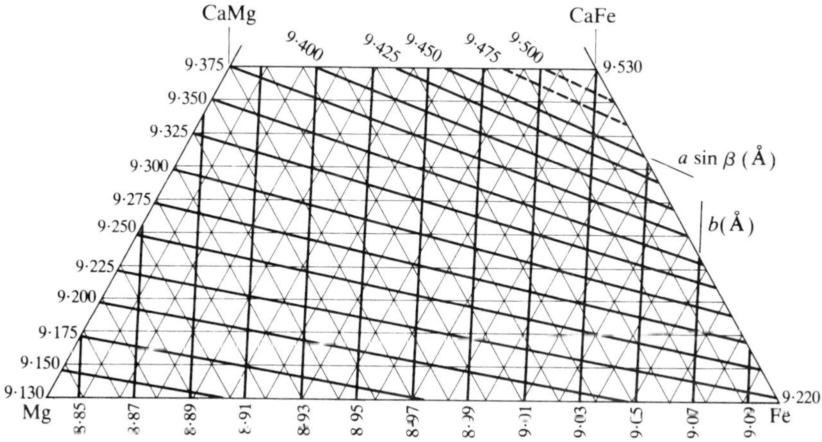

Fig. 146. Diagram showing isodimensional lines for *b* (after Brown, 1960) and *a* sin *β* (after Viswanathan, 1966).

As in the case of the calcium-poor pyroxenes, the way in which Mg and Fe are distributed between *M*1 and *M*2 sites will influence cell parameters, except for those which have *M*2 filled or nearly filled by Ca. Thus different determinative curves would be expected to be appropriate for different conditions of crystallization i.e. volcanic versus metamorphic clinopyroxenes. Accurate analyses are better obtained by electron probe methods if available.

It would be convenient if finely intergrown augite–pigeonite pairs could have their two components determined by use of a *β*–*b* plot, but here there is a further complication that the structures are probably strained and give rise to distorted cell parameters (the scale of intergrowth may in this instance rule out electron-probe methods).

The latter problem is discussed by Papike *et al.* (1971) who use a plot of *β* against *b* on the Di–Hd–En–Fs trapezium which gives consistent results for the suite of lunar clinopyroxenes examined by them. The *β* angle appears to be a good indicator of Ca content, and for an augite–pigeonite pair, Δ*β*, is seen to be a measure of the miscibility gap. For augites, *β* decreases with increasing iron content, but for pigeonites *β* stays constant. A qualitative correlation between *β* and Ca content had previously been noted by Whittaker (1960).

For synthetic pyroxenes on or near the diopside–hedenbergite join, Rutstein and

Yund (1969) showed that *a* and *b* vary linearly over the range of composition. Turnock *et al.* (1973) measured cell parameters for synthetic pyroxenes with compositions representing the whole of the Di–Hd–En–Fs quadrilateral and calculated trend surfaces and graphical plots for *a*, *b*, *c*, *β*, *V* and molar volume. They emphasize that these are not strictly comparable with results for natural specimens.

Several attempts have been made to use various features of X-ray powder patterns instead of measured cell parameters for the determination of pyroxenes (e.g. Zwaan, 1954). The use of powder patterns alone, and indexing by analogy over a wide range of composition may be rather unreliable. Ginsburg and Sidorenko (1964) found distinctive sections of powder patterns for distinguishing not only Ca-rich, Na-rich, and (Fe,Mg) pyroxenes, but also subdivisions within these groups, e.g. seven divisions within diopside–hedenbergite. They also reported two subgroups within the orthopyroxenes, as indicated by optical properties and later confirmed by accurate cell parameter measurements. As in the orthopyroxenes, it is possible that cation ordering processes can give rise to some at least of the other subgroups reported.

Successful methods for determinative use over limited ranges of compositions have been those by Boyd and Schairer (1957) using 2θ $(31\bar{1})$, Kushiro and Schairer (1963) using 2θ (311)–2θ (310), and Davis (1963) using 2θ (220).

Calculated X-ray powder patterns for diopside, pigeonite and clinoenstatite are given by Borg and Smith (1969).

Chemistry

The chemical composition, together with the numbers of ions on the basis of six oxygens, for some augites, ferroaugites, subcalcic augites and ferroaugites, titanaugites, ferrian augites and sodian augites are given in Tables 29–38. In the great majority of these analyses there is sufficient Al to compensate for the deficiency of Si in the tetrahedrally co-ordinated Z group. A small number of the analyses, however, still show a residual deficiency after all the Al has been allocated to the four-fold positions in the structure. In the previous edition, Ti as first suggested by Barth (1931b), was added to (Si,Al) to complete the ideal two Z atoms per mole, the presumed replacement $Si \rightleftharpoons Ti$ being based on the valency of Ti (see also Shkodzinskiy, 1968). Other investigators have questioned the validity of this assumption and Hartman (1969) has pointed out that the Ahren's radius for Ti^{4+} is slightly greater than the radius of Fe^{3+}. Furthermore titanium minerals with tetrahedrally co-ordinated Ti are unknown whereas Fe^{3+} occurs in four-fold positions in synthetic iron biotite and iron feldspar. In consequence the substitutions $[Si]^4+[Fe]^6 \rightarrow [Fe]^4+[Ti]^6$ or $[Si]^4+[Al]^6 \rightarrow [Al]^4+[Ti]^6$ are preferred to $[Si]^4 \rightarrow [Ti]^4$. Following the nomenclature of Yagi and Onuma (1967), augites containing between 1 and 2 per cent TiO_2 are described as titaniferous, and those with more than 2 per cent as titanaugite.

The majority of augites of igneous and metamorphic rocks contain between 2·5 and 4·0 per cent Al_2O_3, and most ferroaugites have a smaller content, the Al_2O_3 generally varying between 1·5 and 3·0 per cent. Augites in the nodules in basalts and volcanic breccias (Table 31), however, are normally much richer in Al, and the average content of Al_2O_3 in the eight analyses of the table is 7·3 weight per cent. Most titanaugites also have high contents of Al, those of Table 36 ranging from 2·6 to 14·3 with an average value of 7·5 weight per cent.

Table 29. Augite Analyses (plutonic rocks)

	1	2	3	4	5
SiO_2	52·92	49·49	51·20	52·70	50·14
TiO_2	0·50	0·50	0·91	0·34	0·00
Al_2O_3	2·80	6·69	3·01	1·84	0·95
Fe_2O_3	0·85	2·37	2·46	2·12	0·91
FeO	5·57	5·01	5·32	5·42	7·45
MnO	0·15	0·11	tr.	0·16	0·08
MgO	16·40	18·70	15·48	15·15	17·10
CaO	19·97	15·86	20·73	21·58	21·26
Na_2O	0·35	0·74	0·34	0·49	0·24
K_2O	0·01	0·04	—	0·01	0·07
H_2O^+	0·10	0·12	0·51	—	0·76
H_2O^-	0·07	0·08	—	—	—
Total	100·67	100·24	99·96	99·96	99·62
α	1·6818	—	1·680	1·684	1·6751
β	1·6865	—	1·686	1·691	1·6820
γ	1·7085	1·683–1·688	1·705	1·714	1·7039
$\gamma:z$	40·75°	—	46°	—	35°
$2V_\gamma$	49°	52°	61°	—	58°20′
D	—	—	—	—	3·236
	Numbers of ions on the basis of six O				
Si	1·929 ⎤ 2·00	1·801 ⎤ 2·00	1·897 ⎤ 2·00	1·947 ⎤ 2·00	1·906 ⎤ 1·97[d]
Al	0·071 ⎦	0·199 ⎦	0·103 ⎦	0·053 ⎦	0·042 ⎦
Al	0·049	0·087	0·029	0·027	0·000
Ti	0·014	0·013	0·025	0·009	0·000
Fe^{3+}	0·024	0·066	0·068	0·059	0·000
Mg	0·891	1·022	0·855	0·834	0·969
Fe^{2+}	0·170 — 1·99[a]	0·151 — 2·03[b]	0·165 — 1·99	0·167 — 1·99[c]	0·237 — 2·10
Mn	0·005	0·002	0·000	0·005	0·003
Ca	0·780	0·620	0·823	0·854	0·866
Na	0·024	0·048	0·024	0·035	0·018
K	0·000	0·001	—	—	0·003
Mg	47·6	54·9	44·7	43·2	46·1
[e]ΣFe	10·7	11·8	12·2	12·5	12·7
Ca	41·7	33·3	43·1	44·3	41·2

[a] Includes Cr 0·026, Ni 0·003.
[b] Includes Cr 0·013, Ni 0·002.
[c] Includes Cr 0·003.
[d] Includes Fe^{3+} 0·026.
[e] $\Sigma Fe = Fe^{2+} + Fe^{3+} + Mn$.

1 Chromium augite, gabbro, Rustenberg platinum mine, Bushveld complex (Hess, 1949). Anal. R. B. Ellestad (includes Cr_2O_3 0·88, NiO 0·10).
2 Chromian augite, chrome peridotite, Takashima, northern Kyushu, Japan (Ishibashi, 1970). Anal. K. Ishibashi (includes Cr_2O_3 0·45, NiO 0·08).
3 Augite, peridotite, Garabal Hill, Loch Lomond, Scotland (Nockolds, 1941). Anal. S. R. Nockolds.
4 Augite, pyroxene diorite, Feather River area, northern Sierra Nevada (Hietanen, 1971). Anal. S. T. Neil (includes Cr_2O_3 0·10, V_2O_5 0·05).
5 Fluorine-bearing augite, mansjöite, mansjöite-fels facies of pegmatite dyke, Mansjö Mountain, Sweden (Eckermann, 1922). Anal. N. Sahlbom (includes F 0·63, P_2O_5 0·03).

Table 29. Augite Analyses (plutonic rocks) – *continued*

	6	7	8	9	10
SiO_2	49·69	51·83	50·93	50·73	47·51
TiO_2	1·05	0·49	1·02	1·45	1·25
Al_2O_3	3·95	3·07	3·34	3·61	5·11
Fe_2O_3	1·75	1·38	1·72	1·24	3·57
FeO	6·25	7·21	6·70	7·70	6·53
MnO	0·17	0·17	0·31	0·26	0·14
MgO	15·35	16·00	14·72	14·48	14·51
CaO	20·91	19·21	20·61	19·46	20·99
Na_2O	0·49	0·27	0·68	0·49	0·67
K_2O	0·01	0·02	0·04	0·10	0·07
H_2O^+	0·03	0·47	0·02	0·49	0·02
H_2O^-	—	0·11	0·01	0·10	0·00
Total	100·40	100·23	100·12	'100·14'	100·37
α	1·692	1·6832	1·689	—	1·693
β	1·705	1·6870	1·694	—	1·697
γ	1·719	1·7100	1·716	—	1·719
$\gamma:z$	43°	42·1°	—	—	42°
$2V_\gamma$	51°	53°	47·5°	50°	46·5°
D	3·33	—	3·34	—	3·37

Numbers of ions on the basis of six O

	6		7		8		9		10	
Si	1·841	2·00	1·917	2·00	1·888	2·00	1·873	2·00	1·778	2·00
Al	0·159		0·083		0·112		0·127		0·222	
Al	0·016	2·01[a]	0·051	2·00	0·034	2·01	0·030	1·95	0·004	2·05
Ti	0·029		0·014		0·028		0·040		0·035	
Fe^{3+}	0·048		0·038		0·048		0·034		0·100	
Mg	0·830		0·882		0·813		0·797		0·809	
Fe^{2+}	0·194		0·223		0·208		0·238		0·204	
Mn	0·005		0·005		0·010		0·008		0·004	
Ca	0·830		0·762		0·819		0·770		0·841	
Na	0·035		0·019		0·049		0·033		0·048	
K	0·000		0·001		0·002		0·002		0·003	
Mg	44·1		46·1		42·8		43·2		41·3	
*ΣFe	12·8		13·9		14·0		15·1		15·7	
Ca	43·1		40·0		43·2		41·7		43·0	

[a] Includes Cr 0·022.

6 Augite, plagioclase-augite-olivine cumulate, Kap Edvard Holm, east Greenland (Deer and Abbott, 1965). Anal. J. H. Scoon (includes Cr_2O_3 0·75, *a* 9·747, *b* 8·922, *c* 5·234 Å, β 73°43′; trace element data).
7 Augite, gabbro, Stillwater complex, Montana (Hess, 1949). Anal. R. B. Ellestad.
8 Augite, ferrigabbro, Somerset Dam, Queensland, Australia (Mathison, 1967). Anal. C. I. Mathison (includes Cr_2O_3 0·02; trace element data for V, Ni, Co).
9 Augite, gabbro, Upper Layered Series, Kap Edvard Holm, east Greenland (Elsdon, 1971). Anal. R. Elsdon.
10 Augite, olivine gabbro, Okonjeje, South-West Africa (Simpson, 1954). Anal. E. S. W. Simpson.

11		12		13		14		15		
50·69		50·89		51·76		51·86		51·67		SiO_2
0·50		1·10		0·55		0·55		0·28		TiO_2
3·60		3·06		1·47		2·33		1·38		Al_2O_3
2·91		1·35		2·35		1·60		0·08		Fe_2O_3
6·88		8·53		8·76		9·45		11·63		FeO
0·30		0·28		0·23		0·24		0·32		MnO
13·34		14·24		13·48		14·50		12·78		MgO
21·19		20·48		20·30		18·92		21·42		CaO
0·49		0·39		—		0·23		0·31		Na_2O
0·02		0·01		—		0·00		0·06		K_2O
0·12		—		0·53		0·37		—		H_2O^+
0·01		—		0·16		0·09		0·00		H_2O^-
100·05		100·33		99·59		100·17		99·93		**Total**
—		1·693		1·684		1·6870		—		α
—		1·705		1·690		1·6921		—		β
—		1·720		1·713		1·7149		—		γ
—		42°		42°		42°		—		γ:z
—		48°		58°		51°		—		$2V_\gamma$
3·369		3·38		—		—		—		D
1·891	2·00	1·893	2·00	1·954	2·00	1·936	2·00	1·954	2·00	Si
0·109		0·107		0·046		0·064		0·046		Al
0·049		0·027		0·019		0·038		0·016		Al
0·014		0·031		0·016		0·015		0·008		Ti
0·082		0·038		0·067		0·045		0·002		Fe^{3+}
0·742		0·789		0·758		0·807		0·720		Mg
0·215	1·99	0·265	2·00	0·277	1·97	0·295	1·98[a]	0·368	2·02	Fe^{2+}
0·009		0·009		0·007		0·008		0·010		Mn
0·847		0·816		0·821		0·757		0·868		Ca
0·035		0·028		—		0·016		0·023		Na
0·001		0·000		—		0·000		0·003		K
39·2		41·1		39·3		42·2		36·6		Mg
16·1		16·3		18·2		18·2		19·3		*ΣFe
44·7		42·6		42·5		39·6		44·1		Ca

[a] Includes Ni 0·001.

11 Augite, anorthosite, Allard Lake, Quebec (Hargraves, 1962). Anal. C. O. Ingamells (*a* 9·754, *b* 8·916, *c* 5·264 Å, β 105°58′).

12 Augite, plagioclase–augite–olivine cumulate, Kap Edvard Holm, east Greenland (Deer and Abbott, 1965). Anal. J. H. Scoon (trace element data: *a* 9·751, *b* 8·936, *c* 5·233 Å, β 73°49′).

13 Augite, pegmatoid gabbro, Salmenkylä, Kangasniemi, Finland (Savolahti and Kujansuu, 1966). Anal. S. Turkka (X-ray diffraction data).

14 Augite, gabbro, Stillwater complex, Montana (Hess, 1949). Anal. R. B. Ellestad (includes Cr_2O_3 0·01, NiO 0·02).

15 Augite, outer quartz diorite, Ben Nevis, Scotland (Haslam, 1968). Anal. H. W. Haslam (trace element data).

Table 29. Augite Analyses (plutonic rocks) – *continued*

	16	17	18	19
SiO_2	50·59	50·66	54·58	50·58
TiO_2	0·45	1·30	0·43	0·61
Al_2O_3	2·59	2·45	2·11	2·20
Fe_2O_3	1·78	1·33	1·18	1·57
FeO	9·90	11·24	12·44	15·53
MnO	0·48	0·29	0·52	0·28
MgO	12·58	14·25	15·23	12·60
CaO	20·68	18·01	11·02	16·40
Na_2O	0·62	0·36	0·44	0·24
K_2O	0·08	0·08	0·12	0·03
H_2O^+	0·29	—	1·35	—
H_2O^-	—	—	0·20	—
Total	100·07	99·97	100·34	100·04
α	1·696	—	1·676	—
β	1·701	1·697	—	1·704
γ	1·723	—	1·689	—
γ:z	43°–45°	—	—	—
$2V_\gamma$	53·5°	42°	—	40°–42°
D	3·40	—	3·19	—

Numbers of ions on the basis of six O

	16		17		18		19	
Si	1·912	2·00	1·903	2·00	2·036	2·04	1·926	2·00
Al	0·088		0·097		0·000		0·074	
Al	0·025	2·02	0·011	2·00	0·093	1·89[a]	0·025	1·99
Ti	0·014		0·037		0·012		0·017	
Fe^{3+}	0·050		0·037		0·033		0·044	
Mg	0·709		0·798		0·847		0·715	
Fe^{2+}	0·313		0·352		0·388		0·495	
Mn	0·016		0·009		0·016		0·009	
Ca	0·839		0·725		0·449		0·669	
Na	0·045		0·026		0·032		0·018	
K	0·005		0·004		0·006		0·001	
Mg	36·8		41·5		49·0		37·0	
*ΣFe	19·7		20·8		25·0		28·4	
Ca	43·5		37·7		26·0		34·6	

[a] Includes Cr 0·003, Ni 0·001, Zn 0·011.

16 Augite, monzonite, Tilba, Mount Dromedary, New South Wales, Australia (Boesen, 1964). Anal. R. S. Boesen (includes Cr_2O_3 0·02, P_2O_5 0·01).

17 Augite, olivine gabbro, Skaergaard, Kangerdlugssuaq, east Greenland (Brown, 1957). Anal. P. E. Brown.

18 Calcium-poor augite, 'Tábor syenite', central Bohemia (Jakeš, 1968). Anal. F. Chaluš (includes Cr_2O_3 0·09, Ni 0·014, ZnO 0·40, P_2O_5 0·22: equivalent amount of CaO, 0·28, subtracted for apatite in calculation of numbers of cations; birefringence probably too small).

19 Augite, ferrogabbro, Skaergaard, Kangerdlugssuaq, east Greenland (Brown, 1957). Anal. E. A. Vincent.

Table 30. Augite Analyses (hypabyssal and volcanic rocks)

	1	2	3	4	5
SiO_2	47·99	50·09	49·25	48·50	46·76
TiO_2	1·10	0·88	1·52	1·07	2·96
Al_2O_3	6·37	4·38	5·26	4·50	6·26
Fe_2O_3	2·73	2·16	2·27	1·36	1·45
FeO	3·44	6·08	6·40	8·18	8·92
MnO	0·12	0·21	—	0·30	0·16
EgO	15·17	14·39	15·08	15·06	12·07
CaO	21·30	21·41	19·14	19·32	20·34
Na_2O	0·65	0·29	0·78	0·32	0·43
K_2O	0·07	0·02	0·05	0·17	0·28
H_2O^+	—	—	0·04	1·70	—
H_2O^-	—	0·04		0·15	0·89
Total	99·55	99·95	99·79	100·64	100·52
α	—	1·690	—	1·673	$1{\cdot}694^b$
β	—	—	1·700	1·678	—
γ	1·693	1·717	—	1·705	$1{\cdot}719^b$
$\gamma:z$	—	45°	—	42°	36°–43°
$2V_\gamma$	57°	48·5°	52°	48°–49°	38°–53°
D	—	—	—	—	—

Numbers of ions on the basis of six O

	1		2		3		4		5	
Si	1·779	2·00	1·858	2·00	1·827	2·00	1·833	2·00	1·768	2·00
Al	0·221		0·142		0·173		0·167		0·232	
Al	0·058		0·049		0·057		0·033		0·047	
Ti	0·031		0·024		0·042		0·030		0·084	
Fe^{3+}	0·076		0·060		0·063		0·039		0·041	
Mg	0·838		0·801		0·834		0·848		0·680	
Fe^{2+}	0·106	$2{\cdot}03^c$	0·188	2·00	0·199	2·01	0·259	2·03	0·282	2·01
Mn	0·003		0·007		—		0·009		0·005	
Ca	0·846		0·851		0·761		0·782		0·824	
Na	0·047		0·021		0·056		0·024		0·031	
K	0·003		0·000		0·002		0·008		0·014	
Mg	44·0		42·0		44·9		43·8		37·1	
$^a\Sigma Fe$	11·8		13·4		14·1		15·8		17·9	
Ca	44·2		44·6		41·0		40·4		45·0	

[a] $\Sigma Fe = Fe^{2+} + Fe^{3+} + Mn$.
[b] On (110).
[c] Includes Cr 0·018.

1 Chromian augite, hawaiite, Hocheifel, Germany (Huckenholz, 1965). Anal. H. G. Huckenholz (includes Cr_2O_3 0·61).
2 Augite, picrite dolerite, Nosappu Cape, Nemuro Peninsula, Japan (Yagi, 1969). Anal. K. Onuki.
3 Augite, trachybasalt dyke, Gough Island, South Atlantic (Le Maitre, 1965). Anal. R. W. Le Maitre.
4 Augite, andesine dolerite, Squilver Hill, Shropshire (Blyth, 1948). Anal. Geochemical Laboratories (includes Cr_2O_3 0·01).
5 Augite, dolerite, Morotu Cape, Sakhalin, U.S.S.R. (Yagi, 1953). Anal. K. Yagi.

Table 30. Augite Analyses (hypabyssal and volcanic rocks) – *continued*

	6	7	8	9	10
SiO_2	52·41	50·29	50·92	50·74	49·38
TiO_2	0·40	1·33	1·18	0·95	1·25
Al_2O_3	1·48	2·82	2·90	2·98	2·28
Fe_2O_3	1·92	1·76	0·47	2·37	2·24
FeO	9·17	9·75	11·11	10·04	11·62
MnO	0·33	0·26	0·33	0·17	0·29
MgO	13·84	12·49	15·63	14·24	13·86
CaO	20·02	21·14	17·28	17·88	17·97
Na_2O	0·41	0·44	0·12	} 0·67	0·50
K_2O	0·00	tr.	0·12		0·05
H_2O^+	0·00	—	0·07	0·17	0·23
H_2O^-	—	—	0·00	0·03	0·02
Total	99·98	100·28	100·13	100·24	'100·00'
α	1·694	1·700	—	1·695	1·699
β	1·699	1·707	1·696–1·700	1·701	1·704
γ	1·718	1·724	—	1·719	1·729
γ:z	—	44°	42°–44°	41°	39°
$2V_\gamma$	53°	50°–53°	42°–50°	55°	43°
D	—	—	3·375–3·391	3·405	3·38
	Numbers of ions on the basis of six O				
Si	1·968 } 2·00	1·890 } 2·00	1·900 } 2·00	1·899 } 2·00	1·880 } 1·98
Al	0·032	0·110	0·100	0·101	0·102
Al	0·033 } 2·01	0·015 } 2·00	0·027 } 2·00	0·021 } 1·98	0·000 } 2·04
Ti	0·011	0·037	0·033	0·027	0·036
Fe^{3+}	0·054	0·050	0·013	0·066	0·064
Mg	0·775	0·700	0·869	0·794	0·787
Fe^{2+}	0·288	0·306	0·347	0·314	0·370
Mn	0·011	0·008	0·010	0·005	0·009
Ca	0·806	0·851	0·691	0·717	0·733
Na	0·030	0·032	0·008	0·024	0·037
K	0·000	0·000	0·006	0·016	0·002
Mg	40·1	36·6	45·1	41·9	40·1
[a]ΣFe	18·2	19·0	19·0	20·3	22·6
Ca	41·7	44·4	35·9	37·8	37·3

6 Augite, vitrophyric hypersthene–augite dacite, Mt. Hudson, Gosforth area, New South Wales, Australia (Wilkinson, 1971). Anal. G. I. Z. Kalocsai.

7 Augite, crinanite, Garbh Eilean, Shiant Isles, Scotland (Murray, 1954). Anal. R. J. Murray.

8 Augite, tholeiite, Kinkell, Stirlingshire (Walker *et al.*, 1952). Anal. H. C. G. Vincent (trace element data).

9 Augite, andesite, Summitville Quadrangle, Colorado (Larsen *et al.*, 1936). Anal. F. A. Gonyer.

10 Augite, olivine diabase, Ausio, Padasjoki, Finland (Savolahti, 1964). Anal. A. Heikkinen (includes P_2O_5 0·19; trace element and X-ray powder data).

11	12	13	
51·37	51·76	49·77	SiO_2
0·79	0·75	0·70	TiO_2
1·56	1·44	3·19	Al_2O_3
1·62	1·63	0·95	Fe_2O_3
12·82	13·03	16·14	FeO
0·79	0·92	0·24	MnO
12·28	11·14	14·49	MgO
17·86	18·44	13·77	CaO
0·69	0·81	0·30	Na_2O
0·04	0·18	0·04	K_2O
—	0·00	0·34	H_2O^+
—	0·00	—	H_2O^-
99·82	100·10	99·93	**Total**
—	—	—	α
1·705	—	—	β
—	—	—	γ
—	—	—	γ:z
50°	—	—	$2V_\gamma$
—	—	—	D
1·950 ⎤ 2·00	1·966 ⎤ 2·00	1·894 ⎤ 2·00	Si
0·050 ⎦	0·034 ⎦	0·106 ⎦	Al
0·020 ⎤	0·030 ⎤	0·037 ⎤	Al
0·022	0·021	0·020	Ti
0·046	0·047	0·027	Fe^{3+}
0·695	0·631	0·822	Mg
0·407 2·00	0·414 1·99	0·513 2·01	Fe^{2+}
0·025	0·030	0·008	Mn
0·727	0·750	0·561	Ca
0·051	0·060	0·022	Na
0·002 ⎦	0·009 ⎦	0·002 ⎦	K
36·6	33·7	42·7	Mg
25·1	26·2	28·1	[a]ΣFe
38·3	40·1	29·2	Ca

11 Augite, phenocryst, rhyolite, Holmanes, Reydarfjordur, eastern Iceland (Carmichael, 1963). Anal. I. S. E. Carmichael.
12 Augite, trachyte, Cantina la Croce, Pantelleria (Chayes and Zies, 1964). Anal. E. G. Zies.
13 Augite, spilite, Bhoiwada, Bombay Island, India (Vallance, 1974). Anal. T. G. Vallance.

Table 31. Augite Analyses (nodules in basalts and volcanic breccias)

	1	2	3	4
SiO_2	52·43	48·75	48·11	48·72
TiO_2	0·37	1·94	1·14	2·28
Al_2O_3	4·16	6·54	7·26	8·20
Fe_2O_3	2·66	2·58	3·13	3·60
Cr_2O_3	1·06	0·25	—	—
FeO	3·53	4·21	4·86	4·98
MnO	0·12	—	0·11	—
NiO	0·08	—	—	—
MgO	18·19	14·45	14·04	13·70
CaO	16·56	19·78	20·46	16·91
Na_2O	0·96	1·14	0·66	1·80
K_2O	0·07	0·01	0·04	0·09
H_2O^+	0·09	—	0·33	—
H_2O^-	0·11	0·16	0·10	—
Total	100·32	99·81	100·24	100·28
α	1·6814	1·672	—	1·686
β	1·6861	—	1·700	1·694
γ	1·7047	1·702	—	1·715
$\gamma:z$	38°59′	—	48°	35°
$2V_\gamma$	55°10′	60°	51°–53°	66°
D	3·33	—	—	—
	Numbers of ions on the basis of six O			
Si	1·895 } 2·00	1·801 } 2·00	1·782 } 2·00	1·783 } 2·00
Al	0·105	0·199	0·218	0·217
Al	0·072	0·085	0·099	0·137
Ti	0·010	0·054	0·032	0·063
Fe^{3+}	0·072	0·072	0·087	0·099
Cr	0·030	0·003	—	—
Mg	0·980	0·795	0·775	0·747
Ni	0·002 } 1·99	— } 2·00	— } 2·01	— } 1·99
Fe^{2+}	0·107	0·129	0·150	0·152
Mn	0·004	—	0·003	—
Ca	0·641	0·782	0·812	0·663
Na	0·067	0·081	0·047	0·128
K	0·003	—	0·002	0·003
Mg	45·9	43·1	42·7	35·0
[a]ΣFe	15·3	16·3	17·0	18·9
Ca	38·8	40·6	40·3	46·1

[a] $\Sigma Fe = Fe^{2+} + Fe^{3+} + Mn$.

1 Chromian augite, olivine-pyroxene nodule in melilite basalt, Hofgeismar, Kassel, Germany (Ernst, 1936). Includes V_2O_5 0·01, BaO 0·03.

2 Augite, garnet pyroxenite enclave in alkali basalt breccia, Hoggar, Sahara (Girod, 1967). Anal. P. Blot and A. Nétillard (augite contains garnet exsolution lamellae).

3 Augite, pyroxene nodule in trachybasalt, Numaza, Gonoura-machi, Higashi-Matsuura district, Japan (Aoki, 1959). Anal. K. Aoki.

4 Augite, igneous breccia, d'Eglazines, Rozier-Peyreleau, France (Brousse and Berger, 1965). Anal. R. Coquet.

5	6	7	8	
48·12	49·91	47·43	50·09	SiO_2
1·95	1·24	1·47	1·15	TiO_2
8·76	8·19	9·88	5·34	Al_2O_3
2·20	2·63	3·66	4·23	Fe_2O_3
—	—	0·02	—	Cr_2O_3
6·30	6·27	5·84	6·13	FeO
0·16	0·14	0·11	0·15	MnO
0·05	—	—	—	NiO
14·21	12·97	13·04	10·49	MgO
16·68	16·97	17·10	19·17	CaO
1·55	1·95	1·77	2·78	Na_2O
0·05	0·03	0·00	0·04	K_2O
0·06	—	—	0·07	H_2O^+
0·02	—	—	0·09	H_2O^-
100·16	100·35	100·32	99·73	**Total**
1·700	1·695	—	1·705	α
1·705	1·701	—	1·711	β
1·725	—	—	1·729	γ
46°	—	—	50°	$\gamma:z$
47°	57°	—	63°	$2V_\gamma$
3·38	—	—	3·38	D

5		6		7		8		
1·770	2·00	1·827	2·00	1·744	2·00	1·875	2·00	Si
0 230		0·173		0·256		0·125		Al
0·149	2·01	0·180	2·00	0·173	2·01	0·111	2·02	Al
0·054		0·034		0·040		0·032		Ti
0·061		0·072		0·101		0·119		Fe^{3+}
—		—		0·001		—		Cr
0·778		0·708		0·714		0·585		Mg
0·002		—		—		—		Ni
0·194		0·192		0·179		0·192		Fe^{2+}
0·005		0·004		0·004		0·005		Mn
0·657		0·666		0·674		0·769		Ca
0·110		0·138		0·128		0·201		Na
0·002		0·001		0·000		0·002		K
54·3		44·7		42·4		45·0		Mg
10·1		11·3		13·1		15·1		[a]ΣFe
35·6		44·0		44·5		39·9		Ca

5 Sodian augite, megacryst in basanitoid, Armidale, New South Wales, Australia (Binns, 1969). Anal. G. I. Z. Kalocsai (includes P_2O_5 0·05).
6 Sodian augite, eclogite inclusion in breccia, Kakanui, New Zealand (White *et al.*, 1972). Anal. B. W. Chappell and Z. L. Wasik (includes P_2O_5 0·05; rare earth data).
7 Sodian augite, crystal, basanitic ejecta, Grand Canyon, Arizona (Best, 1970). Anal. K. Aoki.
8 Sodian augite, xenocryst in pipe breccia, Okonjeje. South-West Africa (Simpson, 1954). Anal. E. S. W. Simpson (trace element data).

Table 32. Augite Analyses (metamorphic rocks)

	1	2	3	4	5
SiO_2	51·96	51·99	51·95	50·27	49·60
TiO_2	0·80	0·54	0·41	1·06	0·53
Al_2O_3	2·18	3·27	3·15	2·39	2·41
Fe_2O_3	1·43	0·73	1·89	2·05	1·66
FeO	8·85	9·82	9·62	10·24	13·57
MnO	0·21	0·28	0·29	0·58	0·49
MgO	14·49	13·88	13·63	12·68	9·97
CaO	20·25	18·40	19·16	20·54	21·13
Na_2O	—	0·64	0·46	0·46	0·43
K_2O	—	0·06	0·06	0·17	0·07
H_2O^+	—	0·77	—	—	0·42
H_2O^-	—	0·12	0·03	—	0·26
Total	100·28	100·50	100·65	100·44	100·54
α	1·693	—	—	—	1·694
β	—	—	—	—	1·702
γ	—	—	—	1·725	1·727
$\gamma:z$	—	—	—	—	—
$2V_\gamma$	60°	—	—	56°	55·5°–56°
D	3·484–3·510	—	—	—	—

Numbers of ions on basis of six O

	1		2		3		4		5	
Si	1·929	2·00	1·908	2·00	1·926	2·00	1·896	2·00	1·909	2·00
Al	0·071		0·092		0·074		0·104		0·091	
Al	0·025		0·049		0·064		0·000		0·018	
Ti	0·022		0·015		0·011		0·029		0·015	
Fe^{3+}	0·039		0·020		0·053		0·059		0·048	
Mg	0·802		0·759		0·753		0·713		0·572	
Fe^{2+}	0·275	1·98[b]	0·301	1·93	0·298	1·99	0·324	1·99	0·437	2·01
Mn	0·007		0·009		0·009		0·018		0·016	
Ca	0·806		0·724		0·761		0·829		0·872	
Na	—		0·046		0·033		0·016		0·032	
K	—		0·003		0·003		0·005		0·003	
Mg	41·6		41·9		40·2		36·7		29·4	
*ΣFe	16·6		18·2		19·2		20·6		25·8	
Ca	41·8		39·9		40·6		42·7		44·8	

[a] $\Sigma Fe = Fe^{2+} + Fe^{3+} + Mn$.
[b] Includes Cr 0·003.

1 Augite, hypersthene granulite, Härkäselkä, Lapland (Eskola, 1952). Anal. E. Nordensvan (includes Cr_2O_3 0·11).
2 Augite, pyroxene hornfels, Red Hills, Wairau Valley, New Zealand (Challis, 1965). Anal. G. A. Challis.
3 Augite, charnockite, Kondapalli, India (Leelanandam, 1967a). Anal. C. Leelanandam (trace element data including Cr 125, Ni 125, Co 45, V 260 p.p.m.).
4 Augite, hypersthene granulite, Mont Tremblant Park, Quebec (Katz, 1970). Anal. M. B. Katz.
5 Augite, amphibolite, Sekimoto-mura, Ibaragi Prefecture, Japan (Shidô, 1958). Anal. H. Haramura.

6		7		8		9		
50·09		50·54		49·80		46·13		SiO_2
0·30		0·25		0·16		0·41		TiO_2
2·86		2·74		2·50		5·28		Al_2O_3
1·32		2·42		2·95		0·98		Fe_2O_3
14·09		12·80		16·60		16·96		FeO
0·24		0·15		0·28		6·16		MnO
10·51		9·65		8·60		8·02		MgO
20·11		20·16		18·66		15·46		CaO
0·39		0·75		0·50		—		Na_2O
0·18		0·07		0·05		—		K_2O
0·01		0·17		—		0·43		H_2O^+
0·09		0·04		0·05		0·48		H_2O^-
100·19		99·74		100·15		100·31		Total
1·695		—		—		—		α
1·700		—		—		—		β
1·720		—		1·730		—		γ
43°		—		—		42°		$\gamma{:}z$
50°		—		55°		56·5°		$2V_\gamma$
3·52		—		3·45		—		D
1·918	2·00	1·936	2·00	1·927	2·00	1·830	2·00	Si
0·082		0·064		0·073		0·170		Al
0·046		0·060		0·041		0·077		Al
0·009		0·007		0·005		0·012		Ti
0·037		0·069		0·086		0·029		Fe^{3+}
0·604		0·551		0·496		0·474		Mg
0·450	2·01	0·410	1·99	0·537	1·99	0·563	2·02	Fe^{2+}
0·008		0·005		0·009		0·207		Mn
0·822		0·827		0·774		0·657		Ca
0·028		0·056		0·038		—		Na
0·009		0·003		0·002		—		K
31·4		29·6		26·2		24·6		Mg
25·8		26·0		32·9		41·4		$^a\Sigma$Fe
42·8		44·4		40·9		34·0		Ca

6 Augite, hypersthene granulite, Pallavaram, Madras, India (Howie, 1955). Anal. R. A. Howie (trace element data including Cr 75, Ni 50, V 400, Co 65 p.p.m.).

7 Augite, garnet–augite–oligoclase granulite, Elizabethtown, New York (Buddington, 1952). Anal. L. C. Peck.

8 Ferroaugite, two-pyroxene granulite, Hitterö, south-west Norway (Howie, 1964). Anal. R. A. Howie.

9 Manganoan ferroaugite, manganese formation, Forno, Novara, Italy (Bertolani, 1967).

Table 33. Ferroaugite Analyses (plutonic rocks)

	1	2	3	4	5
SiO_2	51·53	48·42	49·73	48·86	49·39
TiO_2	0·19	0·65	0·77	0·93	0·35
Al_2O_3	1·50	1·89	1·39	1·51	1·59
Fe_2O_3	2·72	2·47	1·50	2·26	2·71
FeO	13·13	16·82	19·28	18·71	19·03
MnO	0·82	0·68	0·41	0·38	0·52
MgO	8·92	6·78	9·40	9·48	11·13
CaO	20·17	20·47	17·75	17·58	15·52
Na_2O	0·67	0·61	0·24	0·24	0·21
K_2O	0·00	0·05	0·02	0·00	0·04
H_2O^+	0·36	0·75	—	0·25	—
H_2O^-	0·06	0·24	—	0·06	—
Total	100·07	100·01	100·49	100·26	100·49
α	1·6995	—	—	1·704	—
β	1·7063	1·715	1·714	1·711	—
γ	1·7280	—	—	1·733	—
$\gamma:z$	42·5°	—	—	—	—
$2V_\gamma$	48·5°	58°	46·5°	50°	—
D	—	—	—	—	—

Numbers of ions on the basis of six O

	1		2		3		4		5	
Si	1·978	2·00	1·921	2·00	1·931	1·99	1·910	1·98	1·913	1·99
Al	0·022		0·079		0·063		0·070		0·072	
Al	0·046	1·97	0·009	2·00	0·000	2·01	0·000	2·02	0·000	2·03
Ti	0·005		0·019		0·022		0·027		0·010	
Fe^{3+}	0·078		0·074		0·044		0·066		0·079	
Mg	0·510		0·401		0·544		0·552		0·642	
Fe^{2+}	0·422		0·558		0·626		0·612		0·616	
Mn	0·027		0·023		0·013		0·012		0·016	
Ca	0·830		0·870		0·738		0·736		0·644	
Na	0·050		0·047		0·018		0·018		0·016	
K	0·000		0·003		0·001		0·000		0·002	
Mg	27·3		20·8		27·7		27·9		32·1	
$^a\Sigma Fe$	28·2		34·0		34·8		34·9		35·6	
Ca	44·5		45·2		37·5		37·2		32·3	

$^a\Sigma Fe = Fe^{2+} + Fe^{3+} + Mn$.

1 Ferroaugite, ilmenite-magnetite band in syenite, Kalurah, Adirondacks (Hess, 1949). Anal. L. C. Peck.
2 Ferroaugite, pyroxene syenite, gabbro-syenite pluton, Bol'shoy Taskyl, U.S.S.R. (Krivenko and Guletskaya, 1968). Anal. E. S. Guletskaya (includes P_2O_5 0·18).
3 Ferroaugite, ferrogabbro, Skaergaard, Kangerdlugssuaq, east Greenland (Brown, 1957). Anal. E. A. Vincent.
4 Ferroaugite, iron-rich diabase, Beaver Bay, Minnesota (Muir, 1954). Anal. I. D. Muir.
5 Ferroaugite, ferrodiorite, upper zone C, 7130 m, eastern Bushveld complex (Atkins, 1969). Anal. F. B. Atkins (trace element data).

6	7	8	9	10	
50·33	44·22	46·61	49·57	47·67	SiO_2
0·28	2·27	1·18	0·78	0·85	TiO_2
2·32	6·46	3·47	1·36	1·64	Al_2O_3
1·88	1·48	0·90	1·38	1·19	Fe_2O_3
18·23	19·36	20·18	21·92	22·71	FeO
0·83	0·73	1·11	0·45	0·51	MnO
6·92	6·74	7·27	10·03	5·67	MgO
18·39	17·79	17·24	14·32	19·24	CaO
0·61	0·47	1·04	tr.	0·29	Na_2O
0·07	0·08	0·27	tr.	0·07	K_2O
0·16	0·41	0·42	0·44	0·71	H_2O^+
0·09	—	0·04	0·04	—	H_2O^-
100·11	100·01	99·73	100·29	100·55	Total
—	1·706	1·710	1·712	—	α
—	1·716	1·716	—	—	β
—	—	1·736	1·733	—	γ
—	43°	45°	44°	—	$\gamma{:}z$
—	58°	52°	43°	—	$2V_\gamma$
—	—	3·49	3·47	—	D

6	Σ	7	Σ	8	Σ	9	Σ	10	Σ	
1·962	2·00	1·753	2·00	1·859	2·00	1·941	2·00	1·911	1·99	Si
0·038		0·247		0·141		0·059		0·077		Al
0·068		0·055		0·021		0·003		0·000		Al
0·008		0·068		0·035		0·023		0·025		Ti
0·056		0·044		0·026		0·040		0·036		Fe^{3+}
0·402		0·398		0·432		0·585		0·339		Mg
0·595	1·97	0·642	2·03	0·673	2·06	0·718	1·99	0·759	2·01	Fe^{2+}
0·027		0·025		0·037		0·015		0·017		Mn
0·768		0·756		0·737		0·601		0·825		Ca
0·046		0·036		0·080		0·000		0·010		Na
0·003		0·004		0·014		0·000		0·003		K
21·8		21·4		22·8		29·9		17·2		Mg
36·7		38·1		38·6		39·4		41·0		$^{a}\Sigma$Fe
41·5		40·5		38·6		30·7		41·8		Ca

6 Ferroaugite, ferroaugite syenite gneiss, Moody Lake, Adirondacks (Buddington 1952). Anal. L. C. Peck.
7 Ferroaugite, ferroaugite granophyre, Kap Edvard Holm, east Greenland (Deer and Abbott, 1965). Anal. W. A. Deer and J. Johnson (trace element data).
8 Ferroaugite, syenite, Okonjeje, South-West Africa (Simpson, 1954). Anal. E. S. W. Simpson.
9 Ferroaugite, dolerite pegmatite, Mount Arthur, East Griqualand (Poldervaart, 1947). Anal. W. H. Herdsman.
10 Ferroaugite, fayalite ferrogabbro, Beaver Bay, Minnesota (Konda, 1970). Anal. J. Konda.

Table 33. Ferroaugite Analyses (plutonic rocks) – *continued*

	11	12	13	14	15
SiO_2	46·73	48·13	47·87	46·06	47·59
TiO_2	0·46	0·46	0·70	1·58	0·66
Al_2O_3	2·91	1·51	1·32	4·06	3·22
Fe_2O_3	2·74	2·48	1·24	1·21	2·24
FeO	23·35	23·27	24·66	24·02	23·80
MnO	0·82	0·66	0·49	0·42	0·56
MgO	3·92	4·56	3·96	4·51	4·24
CaO	16·85	18·46	19·07	17·37	17·05
Na_2O	0·78	0·83	0·30	0·53	0·34
K_2O	0·45	0·30	0·07	0·21	0·04
H_2O^+	0·46 (H₂O⁺ + H₂O⁻)	0·10	0·31	0·21	—
H_2O^-		—	0·16	0·12	—
Total	99·47	100·76	100·15	100·30	99·74
α	—	1·718	—	1·735	—
β	1·717	1·723	—	1·741	—
γ	—	1·748	—	1·761	—
γ:z	46°	44°–46°	—	40°	
$2V_\gamma$	57°	49°–59°	—	51°	—
D	—	—	—	—	—
	Numbers of ions on the basis of six O				
Si	1·901 (Si+Al = 2·00)	1·925 (2·00)	1·938 (2·00)	1·849 (2·00)	1·908 (2·00)
Al	0·099	0·072	0·062	0·151	0·092
Al	0·040 (Al–K = 2·01)	0·000 (2·03)	0·000 (2·00)	0·041 (2·01)	0·059 (1·97)
Ti	0·014	0·014	0·021	0·048	0·019
Fe^{3+}	0·083	0·074	0·037	0·036	0·067
Mg	0·239	0·272	0·239	0·270	0·255
Fe^{2+}	0·791	0·778	0·834	0·806	0·795
Mn	0·028	0·022	0·017	0·014	0·019
Ca	0·735	0·791	0·827	0·747	0·732
Na	0·061	0·064	0·023	0·040	0·026
K	0·023	0·016	0·003	0·010	0·002
Mg	13·6	14·0	12·2	14·4	13·6
[a]ΣFe	44·8	45·1	45·5	45·7	47·3
Ca	41·6	40·9	42·3	39·9	39·1

11 Ferroaugite, syenite, Sivamalai, India (Bose, 1968). Anal. M. K. Bose (X-ray data).

12 Ferroaugite, ferroaugite-ferrohortonolite orthopyre vein, Portrush, Northern Ireland (Murray, 1954). Anal. R. J. Murray.

13 Ferroaugite, fayalite ferrogabbro, Beaver Bay, Minnesota (Konda, 1970). Anal. T. Konda.

14 Ferroaugite, ferrohortonolite ferrogabbro, Skaergaard, Kangerdlugssuaq, east Greenland (Muir, 1951). Anal. I. D. Muir.

15 Ferroaugite, upper zone, Insch layered intrusion, Scotland (Clarke and Wadsworth, 1970). Anal. E. L. P. Mercy.

16		17		18		19		
47·58		48·18		49·68		49·44		SiO_2
0·37		0·70		0·60		0·51		TiO_2
1·16		1·06		1·86		1·58		Al_2O_3
2·60		1·46		2·29		—		Fe_2O_3
24·21		26·08		25·60		27·06		FeO
0·59		0·53		0·51		0·51		MnO
3·34		3·52		7·14		7·11		MgO
18·80		18·90		12·72		11·29		CaO
0·47		0·23		0·22		0·44		Na_2O
0·21		0·04		0·12		0·06		K_2O
0·34		—		—		—		H_2O^+
		—		—		0·16		H_2O^-
99·67		100·70		100·80		'100·25'		Total
1·730		—		—		—		α
1·736		—		1·725		—		β
1·755		—		—		—		γ
45°		—		—		—		$\gamma{:}z$
60°		—		39·5°		—		$2V_\gamma$
—		—		3·56		—		D
1·940	2·00	1·941	1·99	1·950	2·00	1·964	2·00	Si
0·056		0·051		0·050		0·036		Al
0·000	2·01	0·000	2·01	0·036	1·95	0·038	1·97	Al
0·011		0·021		0·018		0·015		Ti
0·080		0·044		0·067		0·063		Fe^{3+}
0·203		0·211		0·415		0·421		Mg
0·826		0·879		0·844		0·899		Fe^{2+}
0·020		0·019		0·016		0·017		Mn
0·821		0·816		0·535		0·481		Ca
0·038		0·017		0·016		0·034		Na
0·010		0·002		0·006		0·003		K
10·4		10·7		22·1		22·4		Mg
47·5		47·8		49·4		52·0		[a]ΣFe
42·1		41·5		28·5		25·6		Ca

16 Ferroaugite, syenite, Pilot Range, Percy quadrangle, New Hampshire (Chapman and Williams, 1935). Anal. F. A. Gonyer.

17 Ferroaugite, ferrogabbro, Skaergaard, Kangerdlugssuaq, east Greenland (Brown, 1960). Anal. E. A. Vincent.

18 Calcium-poor ferroaugite, granophyre, Great Lake dolerite sheet, Tasmania (McDougall, 1964). Anal. A. J. Easton includes P_2O_5 0·06, Cr_2O_3 < 0·01).

19 Calcium-poor ferroaugite, quartz mangerite, Grenville Township (Philpotts, 1966). Anal. A. R. Philpotts (trace element data).

Table 34. Ferroaugite Analyses (hypabyssal and volcanic rocks)

	1	2	3	4
SiO_2	49·78	48·78	47·80	49·27
TiO_2	0·75	0·65	1·04	0·61
Al_2O_3	0·98	2·25	4·65	3·00
Fe_2O_3	2·01	1·76	1·75	1·24
FeO	20·71	19·69	22·42	23·16
MnO	0·42	1·42	0·42	1·02
MgO	10·02	5·70	9·66	9·05
CaO	14·31	18·98	11·31	12·39
Na_2O	0·23	0·62	0·51	0·34
K_2O	0·19	0·05	0·28	0·09
H_2O^+	0·30	—	—	—
H_2O^-	0·30	—	—	—
Total	100·00	99·90	99·84	100·17
α	—	—	—	—
β	1·730	1·716–1·723	1·711	1·712–1·718
γ	1·754	—	—	—
$\gamma:z$	46°	—	—	—
$2V_\gamma$	51°	54°–57°	10°–46°	52°–24°
D	—	—	—	—

Numbers of ions on the basis of six O

	1		2		3		4	
Si	1·952	2·00	1·930	2·00	1·867	2·00	1·926	2·00
Al	0·045		0·070		0·133		0·074	
Al	0·000	1·99	0·035	1·99	0·081	2·00	0·054	1·98
Ti	0·022		0·019		0·031		0·018	
Fe^{3+}	0·059		0·052		0·051		0·036	
Mg	0·586		0·338		0·562		0·530	
Fe^{2+}	0·679		0·649		0·732		0·754	
Mn	0·014		0·047		0·014		0·034	
Ca	0·601		0·805		0·473		0·519	
Na	0·017		0·047		0·039		0·026	
K	0·010		0·002		0·014		0·005	
Mg	30·2		17·8		30·7		28·3	
[a]ΣFe	38·8		39·6		43·5		44·0	
Ca	31·0		42·6		25·8		27·7	

[a] $\Sigma Fe = Fe^{2+} + Fe^{3+} + Mn$.

1 Ferroaugite, dolerite sill, Kola Peninsula, U.S.S.R. (Sinitsyn, 1965). Anal. G. T. Tumina, A. N. Filonova and A. P. Archangel'skaya.
2 Ferroaugite, fayalite-ferroaugite trachyte, Matsushima Island, Japan (Aoki, 1964). Anal. K. Aoki.
3 Calcium-poor ferroaugite, dolerite dyke, Scourie, Sutherland, Scotland (O'Hara, 1961). Anal. M. J. O'Hara (trace element data given: pyroxene probably zoned from ferropigeonite core to ferroaugite margin).
4 Calcium-poor ferroaugite, hortonolite trachyte, Kakarashima Island, Japan (Aoki, 1964). Anal. K. Aoki.

Table 35. Subcalcic Augite Analyses

	1		2		3		4	
SiO_2	52·39		50·66		49·68		50·15	
TiO_2	0·31		0·52		0·56		2·40	
Al_2O_3	6·38		2·80		0·78		1·95	
Fe_2O_3	0·64		2·45		3·29		2·15	
FeO	5·28		13·64		18·15		17·22	
MnO	0·18		0·62		0·59		0·65	
MgO	21·60		17·24		16·19		13·08	
CaO	11·42		11·10		9·90		11·02	
Na_2O	0·82		0·44		0·65		0·38	
K_2O	0·00		0·16		0·15		0·12	
H_2O^+	—		} 0·75		0·10		0·60	
H_2O^-	—				0·00		0·40	
Total	99·74		100·39		100·04		100·12	
α	—		1·692		1·709		—	
β	—		—		1·711		—	
					(range 0·042)			
γ	—		1·721		1·738 (aver.)		—	
γ:z	—		—		—		45°	
$2V_\gamma$	—		27°		28°		38°	
					(range 30°)			
D	—		—		—		—	
	Numbers of ions on the basis of six O							
Si	1·873	2·00	1·901	2·00	1·905	2·00[d]	1·923	2·00
Al	0·127		0·099		0·034		0·077	
Al	0·142		0·025		0·000		0·011	
Ti	0·008		0·015		0·016		0·069	
Fe^{3+}	0·017		0·069		0·033		0·062	
Mg	1·151		0·964		0·925		0·747	
Fe^{2+}	0·158	2·00[b]	0·427	2·01	0·582	2·04	0·552	1·95
Mn	0·005		0·020		0·019		0·021	
Ca	0·437		0·446		0·407		0·453	
Na	0·057		0·032		0·048		0·028	
K	0·000		0·008		0·008		0·006	
Mg	65·1		50·0		45·6		40·7	
[a]ΣFe	10·2		26·8		34·3		34·6	
Ca	24·7		23·2		20·1		24·7	

[a] $\Sigma Fe = Fe^{2+} + Fe^{3+} + Mn$.
[b] Includes Cr 0·018, Ni 0·003.
[d] Includes Fe^{3+} 0·061.

1 Subcalcic augite, megacryst in alkali trachybasalt, Glen Innes area, north-eastern New South Wales (Wilkinson, 1973) (probe analysis, includes Cr_2O_3 0·62, NiO 0·10).
2 Subcalcic augite, andesine basalt, Kamchatka (Favorskaia *et al.*, 1963). Anal. P. N. Nissenbaum (includes SrO 0·01).
3 Subcalcic augite, hypersthene–augite basalt, Ō-sima Island, Japan (Kuno, 1955). Anal. J. Ossaka.
4 Subcalcic augite, olivine dolerite, Portrush, Co. Antrim, N. Ireland (Harris, 1937).

Table 35. Subcalcic Augite Analyses – *continued*

	5	6	7	8
SiO_2	49·72	49·98	49·46	48·90
TiO_2	0·73	0·27	0·57	0·12
Al_2O_3	0·59	0·04	1·79	3·86
Fe_2O_3	3·74	1·64	1·65	4·65
FeO	18·12	23·22	25·51	25·35
MnO	0·78	0·27	0·81	0·51
MgO	16·44	12·73	10·94	6·87
CaO	9·56	11·11	8·57	7·96
Na_2O	0·42	0·29	0·23	0·58
K_2O	0·07	0·16	0·05	0·20
H_2O^+	0·17	—	—	0·57
H_2O^-	0·00	0·12	0·20	0·35
Total	100·34	99·99	99·78	99·92
α	1·711 (av.)	1·719 (av.)	—	—
β	1·713	1·722	1·704 –1·720	—
γ	(range 0·043)	(range 0·068)		
	1·739 (av.)	1·745 (av.)	—	—
$\gamma:z$	—	—	—	—
$2V_\gamma$	25°	29°	0°–47°	—
D	(range 41°)	(range 47°)		
	—	—	—	—
	Numbers of ions on the basis of six O			
Si	1·902 ⎤ 2·00[f]	1·956 ⎤ 2·00[e]	1·945 ⎤ 2·00	1·941 ⎤ 2·00
Al	0·027 ⎦	0·002 ⎦	0·055 ⎦	0·059 ⎦
Al	0·000 ⎤	0·000 ⎤	0·028 ⎤	0·121 ⎤
Ti	0·021	0·008	0·017	0·004
Fe^{3+}	0·037	0·006	0·049	0·139
Mg	0·937	0·742	0·641	0·406
Fe^{2+}	0·580 – 2·03	0·760 – 2·02[c]	0·839 – 1·98	0·842 – 1·92
Mn	0·025	0·009	0·027	0·017
Ca	0·392	0·466	0·361	0·338
Na	0·031	0·022	0·018	0·046
K	0·003 ⎦	0·008 ⎦	0·003 ⎦	0·010 ⎦
Mg	45·9	36·6	33·5	23·3
[a]ΣFe	34·9	40·4	47·7	57·3
Ca	19·2	23·0	18·8	19·4

[c] Includes Sr 0·003.
[e] Includes Fe^{3+} 0·042.
[f] Includes Fe^{3+} 0·071.

5 Subcalcic augite, hypersthene basalt, Ō-sima Island, Japan (Kuno, 1955). Anal. J. Ossaka.
6 Subcalcic ferroaugite, segregation vein in hypersthene–olivine basalt, Ō-sima Island, Japan (Kuno, 1955). Anal. K. Nagashima (includes SrO 0·16; analysis corrected for 0·5 per cent ilmenite impurity).
7 Subcalcic ferroaugite, dolerite, Otaki, Yamagata Prefecture, Japan (Aoki, 1962). Anal. K. Aoki.
8 Subcalcic ferroaugite, 'ferropigeonite' andesite, Ōkubo-yama, Minami-Aizu, Hukusima Prefecture, Japan (Kuno and Inoue, 1949). Anal. T. Inoue.

Table 36. Titanaugite Analyses

	1	2	3	4	5	6
SiO_2	44·16	44·71	45·30	47·58	43·96	46·20
TiO_2	2·91	2·92	4·33	3·01	5·50	2·96
Al_2O_3	9·56	7·85	6·72	2·60	7·35	8·58
Fe_2O_3	4·86	4·46	3·98	4·35	4·42	2·68
FeO	3·72	4·23	5·15	6·12	4·96	5·47
MnO	0·09	0·10	0·19	0·13	0·19	0·18
MgO	12·30	11·74	11·20	13·85	10·62	11·66
CaO	21·09	22·37	22·24	21·76	21·91	20·10
Na_2O	0·49	0·90	0·78	0·68	1·06	0·91
K_2O	0·12	0·09	0·18	0·14	0·13	0·20
H_2O^+	0·42	0·26	0·15	—	0·04	0·80
H_2O^-	0·20	0·09	0·05	—	0·12	—
Total	99·92	'100.34'	100·27	100·22	100·26	99·91
α	—	1·721	1·719	—	1·725	—
β	—	1·725	1·722	—	1·730	—
γ	—	1·746	1·742	—	1·750	1·712
$\gamma:z$	—	50°	—	—	—	—
$2V_\gamma$	48°	51°	42°	—	39·6°	37°
D	—	3·401	3·414	—	3·421	—

Numbers of ions on the basis of six O

	1	Σ	2	Σ	3	Σ	4	Σ	5	Σ	6	Σ
Si	1·659	2·00	1·690	2·00	1·706	2·00	1·795	2·00[b]	1·660	1·99	1·733	2·00
Al	0·341		0·310		0·294		0·121		0·327		0·267	
Al	0·082		0·040		0·004		0·000		0·000		0·113	
Ti	0·082		0·083		0·123		0·081		0·156		0·084	
Fe^{3+}	0·137		0·127		0·113		0·040		0·126		0·076	
Mg	0·689		0·661		0·629		0·777		0·598		0·653	
Fe^{2+}	0·117	2·00	0·134	2·02	0·162	2·00	0·193	2·03	0·156	2·01	0·172	1·99[c]
Mn	0·003		0·003		0·006		0·004		0·006		0·007	
Ca	0·849		0·906		0·897		0·879		0·885		0·807	
Na	0·036		0·066		0·057		0·050		0·078		0·066	
K	0·006		0·004		0·008		0·004		0·006		0·005	
Mg	38·4		36·1		34·8		39·3		33·8		37·9	
[a]ΣFe	14·3		14·4		15·6		16·2		16·2		18·0	
Ca	47·3		49·5		49·6		44·5		50·0		44·1	

[a] $\Sigma Fe = Fe^{2+} + Fe^{3+} + Mn$.
[b] Includes Fe^{3+} 0·084.
[c] Includes Cr 0·005.

1 Titanaugite, basalt, Świeta Anna mountain, Poland (Chodyniecka, 1967). Anal. L. Chodyniecka (X-ray powder data).
2 Titanaugite, monzonitic teschenite, Marklowice, Poland (Wawryk, 1935). Includes P_2O_5 0·12.
3 Titanaugite, high density fraction, essexite porphyry, Stöffel, Germany (Holzner, 1934). Anal. J. Holzner.
4 Titanaugite, bulk silicate phase with exsolved ilmenite, sagenitic mafic olivine gabbro, gabbro zone, ultrabasic-gabbroic complex, Blue Mountain, New Zealand (Grapes, 1975). Anal. R. Grapes (includes Cr_2O_3 trace).
5 Titanaugite, high density fraction, nepheline dolerite, Meiches, Vogelberg, Germany (Holzner, 1934). Anal. J. Holzner.
6 Titanaugite, ankaramite, Hocheifel (Huckenholz, 1966). Anal. H. G. Huckenholz (includes Cr_2O_3 0·17).

Table 36. Titanaugite Analyses – *continued*

	7	8	9	10	11	12
SiO_2	47·25	37·52	40·28	44·76	47·11	39·9
TiO_2	3·45	5·72	3·85	3·69	3·75	5·2
Al_2O_3	5·36	14·29	10·30	5·83	3·00	9·5
Fe_2O_3	2·04	4·43	5·35	5·37	3·84	3·9
FeO	9·06	7·12	7·92	8·22	12·20	12·9
MnO	0·18	0·14	—	0·58	—	0·25
MgO	10·85	6·72	7·78	9·59	16·65	4·6
CaO	21·44	24·06	23·57	20·49	13·54	22·7
Na_2O	0·65	0·09	0·36	1·34	0·22	0·55
K_2O	0·00	tr.	tr.	0·11	0·03	—
H_2O^+	0·18	0·00	0·19	—	—	0·20
H_2O^-	0·00	—	—	—	—	—
Total	100·46	100·20	99·60	99·98	100·34	99·70
α	—	1·741	1·74	—	1·695	1·734
β	—	1·741	—	—	1·701	1·746
γ	—	1·762	1·76	—	1·728	1·774
$\gamma:z$	—	32°	—	—	39°	53°
$2V_\gamma$	—	& 0°	—	—	46°	—
D	—	3·43	3·391	—	—	—

Numbers of ions on the basis of six O

	7		8		9		10		11		12	
Si	1·783	2·00	1·443	2·00	1·565	2·00	1·723	2·00[b]	1·775	2·00[c]	1·581	2·00
Al	0·217		0·557		0·435		0·264		0·133		0·419	
Al	0·021	1·99	0·091	2·00[a]	0·037	2·02	0·000	2·01	0·000	2·01	0·024	2·01
Ti	0·098		0·165		0·112		0·106		0·106		0·154	
Fe^{3+}	0·058		0·128		0·156		0·143		0·017		0·116	
Mg	0·610		0·385		0·450		0·553		0·935		0·271	
Fe^{2+}	0·286		0·229		0·257		0·264		0·384		0·427	
Mn	0·006		0·005		—		0·019		—		0·008	
Ca	0·867		0·992		0·981		0·824		0·547		0·963	
Na	0·048		0·006		0·027		0·100		0·016		0·042	
K	0·000		0·000		0·000		0·005		0·001		—	
Mg	33·4		22·1		24·4		30·4		47·3		15·2	
[a]ΣFe	19·2		20·9		22·4		24·2		25·0		30·9	
Ca	47·4		57·0		53·2		45·4		27·7		53·9	

[a] Includes Cr 0·003.
[b] Includes Fe^{3+} 0·013.
[c] Includes Fe^{3+} 0·092.

7 Titanaugite, olivine basalt, Fishnish Peninsula, Mull (Tilley and Muir, 1962). Anal. J. H. Scoon.
8 Uniaxial titanaugite, hornfels xenolith, Haddo norite, Schivas, Aberdeenshire (Dixon and Kennedy, 1933). Anal. B. E. Dixon (includes Cr_2O_3 0·11).
9 Titanaugite, melilite–nepheline dolerite, Scawt Hill, Co. Antrim (Tilley and Harwood, 1931). Anal. Fresenius Chemical Laboratories.
10 Sodian titanaugite, nepheline-syenite, Vallee de Papenoo, Tahiti-nui (McBirney and Aoki, 1968).
11 Titanaugite, basalt, Hiva Oa, Marquesas Islands (Barth, 1931b). Anal. T. F. W. Barth.
12 Titanaugite, melilite–titanaugite rock, Carneal plug, Ballyvallagh Bridge, Co. Antrim (Sabine, 1975). Anal. G. A. Sargeant.

Table 37. Ferrian Augite Analyses

	1	2	3	4	5
SiO_2	47·44	46·47	45·81	47·01	41·75
TiO_2	1·48	—	1·77	0·86	0·50
Al_2O_3	3·41	7·21	6·60	3·22	10·55
Fe_2O_3	8·66	6·92	6·57	7·83	13·05
FeO	1·88	4·08	4·72	5·03	0·72
MnO	0·15	0·10	0·09	—	—
MgO	14·12	10·45	12·40	13·10	13·71
CaO	22·37	23·34	21·44	22·14	19·25
Na_2O	0·06	0·08	0·33	0·24 (Na_2O + K_2O)	—
K_2O	0·01	0·14	0·42		—
H_2O^+	0·94	0·43	0·13	0·74	—
H_2O^-	0·16	—	0·06	—	—
Total	100·68	'99·89'	100·34	100·17	99·53
α	1·695	1·7028	—	1·692	
β	1·700	1·7082	—	—	—
γ	1·713	1·7308	—	1·711	—
$\gamma:z$	43·5°	44°3′	—	—	53°
$2V_\gamma$	60·5°	52°36′	—	—	—
D	—	3·335	—	3·4625	3·48

Numbers of ions on basis of six O

	1		2		3		4		5	
Si	1·783	2·00[c]	1·763	2·00	1·724	2·00	1·791	2·00[d]	1·572	2·00
Al	0·150		0·237		0·276		0·145		0·428	
Al	0·000	1·98	0·085	1·98[b]	0·017	2·01	0·000	2·01	0·040	1·99
Ti	0·042		—		0·050		0·025		0·014	
Fe^{3+}	0·177		0·197		0·186		0·160		0·370	
Mg	0·791		0·591		0·695		0·744		0·769	
Fe^{2+}	0·059		0·129		0·148		0·160		0·023	
Mn	0·005		0·000		0·003		—		—	
Ca	0·900		0·949		0·865		0·904		0·777	
Na	0·004		0·006		0·024		0·009		—	
K	0·000		0·007		0·020		0·006		—	
Mg	39·6		31·7		36·6		36·6		39·7	
[a]ΣFe	15·4		17·5		17·8		18·9		20·3	
Ca	45·0		50·8		45·6		44·5		40·0	

[a] $\Sigma Fe = Fe^{2+} + Fe^{3+} + Mn$.
[b] Includes Cr 0·004, Ni 0·003, Sb 0·001, Pb 0·002, Cu 0·001, Co 0·002.
[c] Includes Fe^{3+} 0·067.
[d] Includes Fe^{3+} 0·064.

1 Ferrian augite, monchiquite, Khibina, Kola Peninsula (Lupanova, 1934). Anal. O. N. Kobylina.
2 Ferrian augite, crystal, lava surface, Vesuvius (Alfani, 1934). Includes Cr_2O_3 0·16, NiO 0·11, Sb_2O_3 0·05, As_2O_3 0·04, PbO 0·18, CuO 0·05, CoO 0·08.
3 Ferrian augite, crystal, Vesuvius (Müller, 1936). Anal. E. Narici.
4 Ferrian augite, black augite tuff, Pirazora, Azerbaidjan (Kashkai, 1944).
5 Ferrian augite, crystal, Cape Tourmente, St. Joachim, Montmorency Co., Quebec (Putman, 1942).

Table 38. Sodian Augite and Ferroaugite Analyses

	1	2	3	4
SiO_2	51·7	48·15	38·40	51·92
TiO_2	0·2	1·35	4·30	0·64
Al_2O_3	6·9	4·07	12·95	1·34
Fe_2O_3	3·4	5·66	7·35	7·15
FeO	6·0	5·40	5·05	6·36
MnO	0·2	0·24	—	0·28
MgO	10·8	11·61	13·00	10·31
CaO	18·8	20·02	15·90	19·73
Na_2O	2·7	2·30	2·00	2·28
K_2O	0·05	0·34	0·55	0·15
H_2O^+	—	0·59	—	0·05
H_2O^-	—	0·08	—	0·00
Total	'100·6'	99·81	99·50	100·21
α	—	—	—	1·697–1·700
β	—	1·713	—	—
γ	—	—	—	1·724–1·734
$\gamma:z$	—	—	—	29°–22°
$2V_\gamma$	—	55°	—	—
D	—	—	—	3·415
	Numbers of ions on the basis of six O			
Si	1·895 } 2·00	1·830 } 2·00	1·464 } 2·00	1·965 } 2·00
Al	0·105	0·170	0·536	0·035
Al	0·193 }	0·013 }	0·046 }	0·024 }
Ti	0·005	0·039	0·123	0·018
Fe^{3+}	0·093	0·162	0·211	0·200
Mg	0·590	0·662	0·739	0·571
Fe^{2+}	0·183 } 2·00	0·171 } 2·05	0·161 } 2·10	0·197 } 1·98
Mn	0·006	0·007	—	0·009
Ca	0·738	0·814	0·649	0·785
Na	0·192	0·169	0·148	0·164
K	0·000	0·016	0·027	0·007
Mg	36·7	36·4	42·0	32·4
[a]ΣFe	17·5	18·7	21·1	23·0
Ca	45·8	44·9	36·9	44·6

[a] $\Sigma Fe = Fe^{2+} + Fe^{3+} + Mn$.

1 Sodian augite, clinopyroxene–plagioclase symplectite, eclogite, Hareidland Norway (Mysen, 1972). Anal. B. O. Mysen (microprobe anal., Fe_2O_3 and FeO calculated from structural formula).

2 Sodian augite, biotite melteigite, Itapirapuã, São Paulo, Brazil (Gomes *et al.*, 1970). Anal. S. L. Moro (trace element data including V 230, Zr 760, Sr 960 p.p.m.).

3 Ferri-titano-sodian augite, ordanchite, Luscade, Mont Dore, France (Brousse, 1961). Anal. R. Brousse.

4 Sodian augite, cancrinite ijolite, Iivaara, Kuusamo, Finland (Lehijärvi, 1960). Anal. P. Ojanperä (zoned dark green at centre to lighter green marginally).

5		6		7		8		
50·62		48·30		52·66		48·85		**SiO_2**
0·39		1·42		0·57		0·66		**TiO_2**
2·41		3·39		1·00		3·35		**Al_2O_3**
6·11		3·82		9·06		8·58		**Fe_2O_3**
7·30		11·69		5·60		10·52		**FeO**
0·96		0·17		0·28		0·28		**MnO**
9·95		10·95		9·90		5·88		**MgO**
18·75		18·28		16·83		18·32		**CaO**
2·24		1·62		4·15		2·77		**Na_2O**
0·19		0·29		0·06		0·60		**K_2O**
—		—		0·06		—		**H_2O^+**
0·58		0·55		0·02		0·18		**H_2O^-**
99·50		100·48		100·19		100·25		**Total**
—		1·698[b]		1·708		1·720		α
—		—		1·720		1·732		β
—		1·721[b]		1·741		1·752		γ
—		42°		18°[c]		62°		$\gamma:z$
—		54°–56°		88°		74°		**$2V_\gamma$**
—		—		—		3·501		**D**
1·933	2·00	1·850	2·00	1·975	2·00	1·890	2·00	**Si**
0·067		0·150		0·025		0·110		**Al**
0·041		0·003		0·019		0·043		**Al**
0·011		0·041		0·016		0·019		**Ti**
0·175		0·110		0·256		0·250		**Fe^{3+}**
0·570		0·625		0·554		0·339		**Mg**
0·232	2·00	0·374	2·04	0·176	2·01	0·340	2·02[d]	**Fe^{2+}**
0·031		0·005		0·009		0·009		**Mn**
0·767		0·750		0·676		0·759		**Ca**
0·165		0·120		0·302		0·222		**Na**
0·009		0·014		0·003		0·030		**K**
32·1		33·5		33·1		20		**Mg**
24·7		26·2		26·4		35·3		**[a]ΣFe**
43·2		40·3		40·5		44·7		**Ca**

[b] 0 On (110).
[c] $\alpha:z$.
[d] Includes Ba 0·003, Sr 0·001.

5 Sodian augite, syenite veins in alkali dolerite, Nosappu Cape, Nemuro Peninsula, Japan (Yagi, 1969). Anal. K. Onuki.

6 Sodian augite, monzonite, Tyaki, Morotu district, Sakhalin, U.S.S.R. (Yagi, 1953). Anal. K. Yagi.

7 Sodian-ferri augite, melteigite, Iivaara, Kuusamo, Finland (Lehijärvi, 1960). Anal. P. Ojanperä (a 9·71, b 8·96, c 5·74 Å, β 105·3°, V 439·7 Å^3).

8 Sodian ferroaugite, feldspar–nepheline pyroxenite (shonkinite), Iron Hill, Colorado (Larsen, 1942). Anal. F. A. Gonyer (includes BaO 0·22, Sr 0·04).

A statistical study of the substitution of Si by Al in clinopyroxenes has shown that the replacement is less in the pyroxenes of oversaturated tholeiitic compared with those of undersaturated alkali basalt magmas (Kushiro, 1960). The data also indicated that the number of tetrahedral sites occupied by Al decreases with fractionation in tholeiitic and related rocks and increases in the clinopyroxenes of alkalic rocks.

Common augites contain relatively small amounts of titanium and ferroaugites have a somewhat lower average Ti content. There is, however, no apparent relationship between the amount of Ti and the replacement of Mg by Fe^{2+} in clinopyroxenes; the subcalcic augites and ferroaugites of volcanic rocks have a low Ti content. Many titanaugites contain between 3 and 5 weight per cent TiO_2 but higher values are not uncommon, and a titanaugite with 8·97 weight per cent TiO_2 has been described by Lebedev and Lebedev (1934).

The contents of Ti in twenty-three pyroxenes from the Chernogorsk intrusion of the Noril'sk region, the weakly differentiated Padunsk and Kayerkansk intrusions and the strongly fractionated products, including ferrogabbros and granophyres, of the Alamadzhakh intrusion, Vilny River, Siberian platform, reported by Vilenskiy (1966) show a direct relationship with the amount of Ti in the individual intrusions, as well as a general tendency for the pyroxenes in the later products of differentiation to be richer in titanium (e.g. Chernogorsk, mean value TiO_2 0·55, with pyroxene range 0·59–2·32, Kayerkansk, 0·78 and 1·57–3·27 Padunsk 0·85 and 1·32–4·03 respectively).

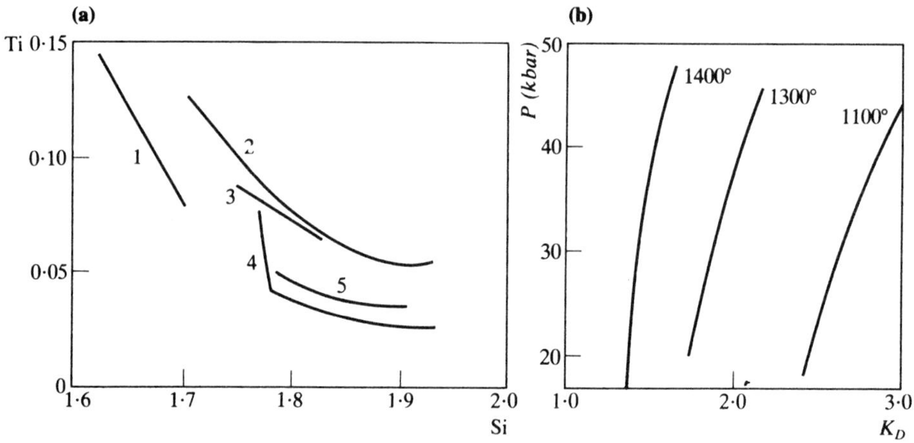

Fig. 147. (a) Atomic proportions of Si and Ti in titanaugites in the middle and late stages of fractional crystallization of alkali suites (after Yagi and Onuma, 1967). (1) Nyiragongo (Sahama and Meyer, 1958), (2) Takakusayama (Tiba, 1966), (3) Black Jack (Wilkinson, 1957), (4) Morotu (Yagi, 1953), (5) Shiants (Murray, 1954). (b) K_D (garnet–clinopyroxene) as a function of pressure at 1100°, 1300° and 1400°C (after Wood, 1976b).

Variation in the titanium content of pyroxenes with magmatic differentiation (Fig. 147(a)) has been noted by Yagi and Onuma (1967). Thus in the Morotu alkalic rocks (Yagi, 1953) the early diopsidic augite has a low content of TiO_2; this increases in the titanaugites to 2·9 per cent and then decreases through the sodian titanaugite (2·7 per cent) and aegirine-augites (1–2 per cent) to 0·6 per cent in the aegirines. A similar trend occurs in the pyroxenes of the Garbh Eilean sill, Shiant Isles, and Aoki (1964) and Kushiro (1964) have noted comparable trends in the pyroxenes from the alkalic rocks of the Iki Islands and the Atumi dolerites respectively. In contrast the titanium content of the clinopyroxenes of the alkalic rocks of the Hocheifel volcanic district, Germany (Huckenholz, 1965), continues to increase with fractionation. The latter trend conforms with the phase relationships in the system $CaMgSi_2O_6$–$CaTiAl_2O_6$ in which the later-formed pyroxene solid solutions are enriched in the $CaTiAl_2O_6$ component. The cause of the first trend, initial increase followed by a decreasing content of titanium with fractionation, is not known but may be related to the crystallization of titanium-iron oxides.

The content of ferric iron in augites rarely exceeds 3 weight per cent Fe_2O_3, and a similar value is normal in ferroaugites, thus the ratio of ferric to ferrous iron is smaller in the more iron-rich varieties. A number of augites from volcanic rocks, however, contain considerably greater amounts of Fe^{3+} and have $Fe^{3+}:Fe^{2+}$ ratios greater than unity, and are described as ferrian augite (Table 37). High Fe^{3+} also occurs in the sodian augites that are essentially restricted to alkaline rocks, and which form a continuous series with the aegirine-augites. Segnit (1953) has recalculated, as percentages of $CaSiO_3$, $(Mg,Fe^{2+},Mn)SiO_3$ and $(Al,Fe^{3+})_2O_3.TiO_2$, seventy-six of the pyroxene analyses detailed in Niggli's (1943) tables, and showed that with few exceptions their compositions lie within the experimentally determined solid solution limits of Al_2O_3, TiO_2 and Fe_2O_3 in synthetic diopside.

Chromium is commonly present in significant amounts (0·1 to 1·0 weight per cent Cr_2O_3) in magnesium-rich augites but occurs in only trace amounts in those containing more than approximately 15 per cent of the $FeSiO_3$ component. The nickel content in the magnesium-rich augites rarely exceeds 0·1 weight per cent NiO, and is commonly not determined in augite analyses. It has been investigated in the pyroxenes of the Kutlyn platiniferous belt of the Urals (Yefimov and Ivanova, 1963), and in the pyroxenes of the Noril'sk gabbro (Genkin *et al.*, 1963). In the latter the Ni content of the pyroxenes varies between 0·01 and 0·05 weight per cent. The content of manganese varies sympathetically with that of iron; thus in Table 29, ten minerals with less than 15 per cent Fe have a mean value of MnO of 0·16, for nine in which Fe varies between 16 and 28 per cent the mean value is 0·32, while for the nineteen ferroaugites of Table 33 the value is 0·61 weight per cent. The MnO content of seven ferroaugites from British and Icelandic Tertiary volcanic rocks ranges from 0·72 to 1·28 weight per cent (Carmichael, 1960). The entry of Cr^{3+}, Ni^{2+} and Mn^{2+} ions into the pyroxene structure has been discussed, in relation to their crystal field stabilization energy, by Campbell and Borley (1974). Alkalis, largely Na_2O, usually amount to between 0·3 and 0·7 weight per cent but, in addition to the sodian augites, higher contents are generally present in titanaugites and, more particularly, in the augites of the nodules in volcanic rocks; the mean value for the eight examples in Table 31 is 1·58 weight per cent Na_2O.

The content of vanadium in the diopsidic augites of the pyroxenites of Vasil'yevsk Mountains and Nizhniy Tagil' and Gusevogorsk massifs, Urals, varies between 0·024 to 0·56 weight per cent V_2O_5, and displays a positive correlation with the amount of iron present in the pyroxene (Borisenko, 1967). The distribution of Zr and Hf in the clinopyroxenes of the ultrabasic alkaline rocks of the Kola Peninsula has been presented by Kukharenko *et al.* (1960), and that of the Be contents (8·1 to 20 p.p.m.) of five pyroxenes in the nepheline-syenite of north-east Tuva, Siberia, by Popolitov *et al.* (1967). A detailed investigation of the boron content of thirty-six pyroxenes of the magnesian skarns (0·0014–0·009, average 0·0031 weight per cent B), twenty from the non-magnesian skarns (0·0018–0·01, average 0·0049), and fifteen pyroxenes from the plagioclase–pyroxene rocks (0·000–0·006, average 0·0036) of the Precambrian gneisses and associated rocks of Northern Kazakhstan has been presented by Malinko (1967). The boron content of sixteen clinopyroxenes from a number of igneous (0·007–0·016 weight per cent B) and metamorphic rocks of the Middle Urals have been determined by Lisitsyn and Khitrov (1962). There is a close correlation between the boron and aluminium contents of the pyroxenes, confirming the earlier theoretical discussion of the diadochy between B and Al by Barsukov (1960). The phosphorus content of nine pyroxenes has been determined by Koritnig (1965). The mean value is 89 p.p.m. with a range from 18 to 630 p.p.m. and the phosphorus is considered to replace Si in tetrahedral sites. The contents of the rare earth elements in the clinopyroxenes of the garnet pyroxenites and lherzolite xenoliths of Salt Lake Crater, Hawaii, are given by Reid and Frey (1971), and show a small enrichment of La to Eu and a higher degree of depletion of Gd to Lu in the pyroxenes of the pyroxenite. The $^{18}O/^{16}O$ ratios of clinopyroxenes from two gabbros, two pyroxenites and an amphibolite have been determined by Taylor and Epstein (1962). The variation in the content of the minor constituents in a series of pyroxenes from the layered gabbros and ferrodiorites of the Skaergaard intrusion have been investigated by Wager and Mitchell (1951). The earlier formed magnesium-rich clinopyroxenes of this intrusion contain significant amounts of Cr, V, Ni, Co, Ca, Sc and Zr, while in the iron-rich pyroxenes which crystallized later from the strongly fractionated magma, Cr, V, Ni and Zr are virtually absent, and the minerals are enriched in Li, Y, La and Sr. Ag, As, Ba, Co, Cr, Cu, Ga, Mn, Ni, Sb, Sc, Zn and *REE* data for four augites of the upper zone of the Skaergaard intrusion have been presented by Paster *et al.* (1974). Other trace element data for the pyroxenes of a single complex or series of related rocks include those of the Okonjeje complex (Simpson, 1954), Caledonian plutonic rocks, Scotland (Nockolds and Mitchell, 1948), anorthosite–mangerite rocks of southern Quebec (Philpotts, 1966), and charnockites (Howie, 1955; Leelanandam, 1967a).

Minor and trace element data, including the lanthanides (e.g. Sm $\simeq$ 2 p.p.m., Ho 0·5 p.p.m.) for augite megacrysts in the volcanic breccia, Kakanui, New Zealand, are given by Mason and Allen, 1973).

Liquid CO_2 inclusions in the clinopyroxenes of olivine-bearing nodules and phenocrysts in basalts have been described by Roedder (1965b), and the presence of inclusions containing a gas and a solid phase or phases in augites of the gabbro-diabase of the Noril'sk Mass have been reported by Bulgakova (1969). The multiphase variety is considered to have originated from small amounts of trapped melt, and the gas phase is accompanied by three or four solid phases including pyroxene, plagioclase, oxides and possibly olivine. Homogenization temperatures vary with the composition of the host rock, 1210°–1230°C for olivine-free gabbro-diabase, 1255°–1265°C for olivine gabbro-diabase, and 1265°–1270°C for picritic gabbro-diabase.

Element distribution. References to the distribution of Mg–Fe between coexisting Ca-rich and Ca-poor pyroxenes, as well as their partition between pyroxenes and other associated minerals, are common in the literature. Brief reference to some examples is given here and additional data are detailed in the sections on orthopyroxenes and the diopside–hedenbergite series. The compositional relationships between coexisting augites and orthopyroxenes show that in general the calcium-rich pyroxenes have higher contents of Fe^{3+}, Ti, and both four- and six-coordinated Al, as well as higher $Mg:(Mg+Fe^{2+})$ ratios. An early study by Carstens (1958) of the distribution of some minor elements in coexisting clino- and orthopyroxenes showed that the augite member of the pair contained more Ti, Cr, V, and less Mn and Cu than the associated orthopyroxene, and previously Harry

(1950) had drawn attention to the relatively small amounts of aluminium in tetrahedral positions in the clinopyroxene structure compared with the aluminium in similar positions in hornblendes and biotites of plutonic rocks.

Studies of Mg–Fe distribution between augite and orthopyroxene pairs in metamorphic assemblages include those for the Precambrian pyroxene–hornblende and pyroxene granulites of Saltora, eastern India (Sen and Rege, 1966), the mafic granulites of the Frazer Range, Western Australia (Wilson, 1976), and augite–orthopyroxene pairs in hypersthene granulites (Katz, 1970). Data on pyroxene pairs from charnockites are given by Naidu and Rao (1967) and Leelanandam (1967b). The former have also reported on the distribution of Mn in pyroxene and clinopyroxene–hornblende pairs, and the latter on the partition of Ti, ($[Al]^6 + Fe^{3+} + Ti$) and the ratio $Ti/(Ti + Fe^{2+} + Fe^{3+} + Mg + Mn)$, as well as the distribution of Cr, Ni, Co and V in the coexisting pyroxene pairs.

The compositional (Fe/Mg ratios, $[Al]^4$, and Ca contents of clinopyroxene) and thermal effects on the distribution of Mg between clino- and orthopyroxenes, based on a statistical study of 117 coexisting pairs from non-alkaline rocks, have been evaluated by Lindh (1975). The same author (Lindh, 1974) has also studied the Mn distribution between 115 coexisting clino- and orthopyroxene pairs. The distribution of Fe, Mg, Mn, Ni, Sc, Al, Ti and V in coexisting Ca-rich and Ca-poor pyroxene phenocrysts in the basaltic andesite–dacite series of the younger volcanic islands of Tonga, S.W. Pacific, is given by Ewart *et al.* (1973), of Mg, Fe, Al, Ca and Na between augite and orthopyroxene in a partly serpentinized Alpine-type ultrabasic mass, Davos, Switzerland, and in the Malenca serpentinite, northern Italy, by Peters (1968), and of B, Be, Ga, Cr, Ni, Li, Mo, Co, Cu, V, Zr, Sc, Y, Zn, Sr, Pb and Ba, in the pyroxenes of the skarns, pyroxenites and hybrid rocks of the Grenville province, Canada, by Moxham (1960).

The distribution preference of Ni^{2+} and Co^{2+} between the pyroxene *M*1 and *M*2 sites, based on a study of sixteen augites from the Duluth complex, has been examined by Dasgupta (1972). Based on a statistical approach he concluded that both cations are more likely to occupy *M*2 rather than the *M*1 sites predicted on the basis of crystal field splitting parameters (see Burns, 1972).

DeVore (1955, 1957) has investigated the distribution of cations between coexisting clinopyroxenes and hornblendes, and between clinopyroxenes and biotites on the basis of the ionization potentials of the cations and the polarizabilities of the anions. Distribution ratios of Fe^{2+} and Mg in relation to the fractionation of Al in some coexisting pyroxenes and hornblendes are also given by DeVore, as well as the fractionation ratios of Ti, Zr, Cr, Ga, Sc, Y, Cu, Ni, Zn, Co, Mn and Pb in associated pairs of these minerals.

The magnesium–iron compositional variance of Ca-rich pyroxenes, orthopyroxenes and hornblendes in hornblende pyroxene granulites has been examined by Sen (1970). The generalized metamorphic reaction, hornblende + quartz → Ca-rich pyroxene + orthopyroxene + plagioclase + water, of this assemblage may be expressed:

$$\underset{\text{hornblende}}{Na_{0\cdot5}Ca_{2\cdot0}(Mg_{x_1}Fe_{1-x_1})_4Al_{2\cdot5}Si_{6\cdot5}O_{22}(OH)_2} + 2SiO_2 \rightleftharpoons$$

$$\underset{\text{orthopyroxene}}{3(Mg_{x_2}Fe_{1-x_2})SiO_3} + \underset{\text{Ca-pyroxene}}{Ca\left(\frac{Mg_{2x_2}}{1+x_2} + \frac{Fe_{1-x_2}}{1+x_2}\right)Si_2O_6} +$$

$$+ \underset{\text{plagioclase}}{CaAl_2Si_2O_8 + 0\cdot5NaAlSi_3O_8} + H_2O$$

where x_1 and x_2 are the atomic concentrations of Mg in hornblende and orthopyroxene respectively. Chemical data for twenty-two Ca-pyroxene–orthopyroxene–hornblende triads of basic granulites and the partition coefficients for the clinopyroxene–orthopyroxene, clinopyroxene–hornblende and ortho-pyroxene–hornblende pairs have been presented by Sen.

Zn and Mn distribution coefficients of augite–hornblende pairs in the ejecta of the Laacher-See area are given by Jasmund and Seck (1964), the partition of Ni between the augites and olivines of the Parikkala intrusive, Finland, by Häkli (1968), and of Mg–Fe in a pyroxene pair from a pyroxene diorite by Hietanen (1971). The distribution of Ti between silicates and oxides in igneous rocks has been discussed by Verhoogen (1962).

An ideal distribution pattern is displayed by the sodian augite–biotite pair in the metamorphosed syenite, Glen Dessary, Scotland (Richardson, 1968). The augite–hornblende pair, however, show irregular distribution coefficients indicating a disequilibrium association that was frozen during the metamorphic reactions to an equilibrium biotite–hornblende-bearing assemblage.

The iron–magnesium exchange reaction between calcium-rich clinopyroxene and olivine:

$$2CaMgSi_2O_6 + Fe_2SiO_4 = 2CaFeSi_2O_6 + Mg_2SiO_4$$

has been formulated as a geothermometer, using mixing parameters calculated from groundmass clinopyroxene–olivine pairs in lavas for which there are groundmass iron-titanium oxide temperatures (Powell and Powell, 1974). There is a pressure dependence of approximately 5°C/kbar, and an additional uncertainty due to the likelihood that the minerals may re-equilibrate to lower temperature depending on the rate of cooling (but see Wood, 1976a).

Magnesium–iron distribution between coexisting clinopyroxene and garnet in rocks of the granulite and glaucophane schist facies and in eclogites has been investigated by Saxena (1968), and the distribution of Mg, Fe, Mn, Ni, Co and Zn between coexisting clinopyroxene, orthopyroxene and garnet in a Norwegian eclogite has been examined by Matsui *et al.* (1966). The cation partition between the two pyroxenes of the latter is controlled by ionic size, and is independent of their chemical properties. The cation partition between the clinopyroxene–garnet pair is more complex and is controlled by two factors, the preference of the larger ions for, and the exclusion of the less electronegative ions from, the garnet. A thermodynamic approach to the partition of Mg–Fe^{2+} between coexisting clinopyroxenes and garnets, as illustrated by the exchange reaction:

$$CaMgSi_2O_6 + FeAl_{\frac{2}{3}}SiO_4 \rightleftharpoons FeCaSi_2O_6 + MgAl_{\frac{2}{3}}SiO_4$$

has been investigated by Oka and Matsumoto (1974). Their study showed that variation of the distribution coefficient:

$$K' = (X_{Alm}/X_{Py})/(X_{Hd}/X_{Di})$$

can be partially explained by the compositional dependence of K' under isophysical conditions, and the non-ideality of both the pyroxene and garnet solid solutions. The pressure dependence of K_D for clinopyroxene–garnet pairs in the temperature range 1100° to 1400°C and pressures between 20 and 45 kbar (Fig. 147(b)) has been determined experimentally by Wood (1976(b)).

Distribution coefficients for Mg and Ca in clinopyroxene–garnet and clino-pyroxene–amphibole pairs, and for Na in clinopyroxene–hornblende pairs in basic granulite facies metagabbro and clinopyroxene–garnet–amphibole gneiss of the

Mellid area, Galicia, Spain, determined by Hubregtse (1973) demonstrate that approximate chemical equilibrium, particularly of the distribution of Mg with respect to Fe and Mn was attained in both metagabbro and gneiss. The partition of K, Rb, Sr and Ba between clinopyroxene and garnet in inclusions in kimberlite and inclusions in alkali basalt has been investigated by Griffin and Murthy (1969). The distribution coefficient K_{Rb} ranges from 1·6 to 0·85 for the nodules in the alkali basalts and from 0·5 to 0·6 in the nodules in kimberlite.

The relationship between partition coefficients and ionic radii for major and minor elements in augite phenocrysts and groundmass of some volcanic rocks shows that the maximum value for the partition coefficient is correlated with the optimum cation ionic radius for a particular lattice site (Jensen, 1973). The curves, obtained by plotting partition coefficients and ionic radii, provide information relating to the site or sites occupied by a given element, the valency state of the element, and the proportion of different valency states present.

Partition coefficients of uranium between augite phenocrysts and groundmass of the acid volcanic rocks of Tuscany and Elba have been determined by Dostal and Capedri (1975), and the partition of thorium and uranium between augite and groundmass of two basalts, two andesites and a dacite pumice has been determined by Nagasawa and Wakita (1968). The partition mechanism is interpreted on the assumption that the ion-exchange equilibria may be expressed:

$$(\mathrm{Ca})_s + (\mathrm{Th})_l \overset{k_{\mathrm{Th}}}{\rightleftharpoons} (\mathrm{Th})_s + (\mathrm{Ca})_l \qquad (1)$$

$$(\mathrm{Ca})_s + (\mathrm{U})_l \overset{k_{\mathrm{U}}}{\rightleftharpoons} (\mathrm{U})_s + (\mathrm{Ca})_l \qquad (2)$$

where s and l denote solid and liquid phases respectively.

Considering the groundmass as representative of the liquid phase and augite as the solid phase under chemical equilibrium then

$$(\mathrm{Th/U})_{\mathrm{augite}}/(\mathrm{Th/U})_{\mathrm{groundmass}} = (\mathrm{Th/U})_s/(\mathrm{Th/U})_l$$

The data obtained by Nagasawa and Wakita using this equation show greater uniformity than those derived from either $U_{augite}/U_{groundmass}$ or $Th_{augite}/Th_{groundmass}$. The data also show a smooth variation in relation to the SiO_2 content of the groundmass, the value $(Th/U)_{augite}/(Th/U)_{groundmass}$ decreasing with increasing SiO_2, and thus indicating a possible temperature effect on the partition coefficient.

The partitioning of thorium and uranium at 20 kbar, 1340°C, between clinopyroxene and silicate melt (composition $Di_{50}Ab_{25}An_{25}$ with 15–25 weight per cent $Ca_3(PO_4)_2$) has been determined by Benjamin *et al.* (1976); $K_{Th}^{Cpx\text{-}liq} = 0{\cdot}042 \pm 0{\cdot}006$, $K_{U}^{Cpx\text{-}liq} = 0{\cdot}012 \pm 0{\cdot}002$.

The temperature and oxygen fugacity effects on the distribution coefficients of Sr and Eu between augite and coexisting liquid have been determined at temperatures between 1140° and 1190°C and oxygen fugacities of 10^{-8} to 10^{-14} atm (Sun *et al.*, 1974). K_{Sr}^{Aug} was found to be highly dependent on temperature and independent of oxygen fugacity; K_{Eu}^{Aug} not strongly temperature dependent but highly dependent on oxygen fugacity, and the molar distribution coefficients given by the equations:

$$\log K_{Sr}^{Aug} = 18\,020/T - 13{\cdot}10$$

$$\log K_{Eu}^{Aug} = 6\,580/T + 0{\cdot}04 \log f_{O_2} - 4{\cdot}37$$

The distribution coefficients are related to bulk composition, in this investigation an oceanic ridge basalt, and are not applicable to substantially different compositions.

Data for the rare earths in augite phenocryst and groundmass of a Tongan basaltic andesite are given by Ewart (1976).

The distribution of iron and magnesium between calcic pyroxene ($Wo_{43\cdot8}En_{53\cdot5}Fs_{2\cdot7}$–$Wo_{41\cdot1}En_{39\cdot5}Fs_{19\cdot4}$) and olivine ($Fo_{95\cdot4}Fa_{3\cdot9}La_{0\cdot7}$–$Fo_{61\cdot0}Fa_{38\cdot5}La_{0\cdot7}$), and between calcic pyroxene and mafic silicate liquid, has been determined at temperatures between 1125° and 1250°C and 1 bar total pressure under anhydrous conditions (Duke, 1976). The distribution coefficient is not affected by temperature and the partition of Mg and Fe between pyroxene, olivine and liquid may be described by the equations:

$$\log\frac{(X_{Fe}^{Ol})}{(X_{Mg}^{Ol})} = 0{\cdot}198 + 1{\cdot}30\log\frac{(X_{Fe}^{Cpx})}{(X_{Mg}^{Cpx})}$$

$$\log\frac{(X_{Fe}^{Cpx})}{(X_{Mg}^{Cpx})} = -0{\cdot}564 + 0{\cdot}755\log\frac{(X_{Fe}^{liq.})}{(X_{Mg}^{liq.})}$$

Duke also studied the distribution of Ti^{4+}, V^{3+}, Cr^{3+}, Mn^{2+}, Co^{2+} and Ni^{2+} between calcic pyroxene, olivine and silicate liquid by introducing 0·5 weight per cent of one of the transition element oxides into the melt. The observed relative enrichment pattern of titanium is liquid > pyroxene ≫ olivine, and of chromium, pyroxene ≫ olivine > liquid; vanadium is concentrated in pyroxene and liquid but partitioning between the two phases is very variable. The magnitude of the enrichment of Mn, Co and Ni in olivine relative to pyroxene, in conformity with crystal field theory as predicted by Burns (1970), is

$$D_{NiO}^{Ol/Cpx} > D_{CoO}^{Ol/Cpx} > D_{MnO}^{Ol/Cpx}$$

The theory of trace element partition coefficients, particularly between crystal and liquid phases, and the effects of temperature, pressure and phase composition have been reviewed by McIntire (1963). The partition between crystal and liquid phase is defined by the equation:

$$D = \left(\frac{Tr}{Cr}\right)_s \Big/ \left(\frac{Tr}{Cr}\right)_l$$

where $(Tr/Cr)_s$ and $(Tr/Cr)_l$ are the ratios of trace element to carrier (macrocomponent) in the solid phase and in the liquid phase respectively.

The experimentally determined distribution coefficients between augite and orthopyroxene, $K_D = (X_{Mg}^{Opx}.X_{Fe}^{Aug})(X_{Fe}^{Opx}.X_{Mg}^{Aug})$, synthesized at 810°C and a fluid pressure of 15 kbar have been found to be compatible with a constant value of 0·690, but probably increase with increasing bulk Fe/(Fe + Mg) (Lindsley *et al.*, 1974).

The partitioning of Ti and Al between clinopyroxene, orthopyroxene, garnet and ilmenite has been determined experimentally for the composition $(Wo_{25}En_{45}Fs_{30})_{90}(CaAl_2TiO_6)_{10}$ at 1050°C and 25 kbar, and 1100°C and 40 kbar respectively (Akella and Boyd, 1972). The lower pressure assemblage consists of augite, hypersthene, garnet, metallic iron and liquid; at the higher pressure ilmenite is replaced by rutile. The content of titanium (TiO_2 0·9–1·5 weight per cent) in the augites of the ilmenite-bearing assemblage is higher than in the augites (TiO_2 0·4–0·5 per cent) associated with rutile. The atomic proportions of Al in the augites of both assemblages is greater than twice that of Ti, and there is thus a small Tschermak's component in these synthetic pyroxenes.

Crystallization of pyroxenes in igneous rocks. The compositional variation of pyroxenes associated with the fractional crystallization of basaltic magmas is related to the two main series, tholeiite and alkali olivine basalt. Thus the Ca-rich pyroxenes of the tholeiitic series show strong iron enrichment coupled with an initial decrease, followed at a later stage by an increase in Ca content. The tholeiitic series is also characterized until a last stage of fractionation by a two-pyroxene assemblage. In contrast the single Ca-rich pyroxene of the alkalic series is richer in calcium, shows more limited iron enrichment, and the fractionation trend is approximately parallel to the diopside–hedenbergite join.

The crystallization of pyroxenes from basaltic magmas has been considered by many workers, e.g. Asklund (1923), Barth (1931a, 1936), Tsuboi (1932), Kennedy (1933), Kuno (1936, 1950, 1955), Wager and Deer (1939), Hess (1941), Edwards (1942), Benson (1944), Poldervaart and Hess (1951), Muir (1951), Murray (1954), Brown (1957), Wilkinson (1957), Carmichael (1960), Carmichael *et al.* (1970), Brown and Vincent (1963), Atkins (1969) and Walker *et al.* (1973); a review of some of the earlier results and hypotheses was given by Hess (1941).

Many augites, and particularly those which have crystallized from basaltic magmas can as a first approximation be considered to belong to the diopside–hedenbergite–clinoenstatite–clinoferrosilite system, the relationships of the four boundary pairs of which are known either from experimental investigations or from natural minerals. Thus in the $MgSiO_3$–$FeSiO_3$ series there is complete solid solution between Fs_0 and Fs_{88} at solidus temperatures, with the inversion of the metastable clinopyroxenes to orthopyroxenes at subsolidus temperatures (Bowen and Schairer, 1935; see also p. 83). In the $CaMgSi_2O_6$–$Mg_2Si_2O_6$ system the solvus intersects the solidus in the composition range 9 to 50 weight per cent $CaMgSi_2O_6$, so that even at high temperatures solid solution is not complete between $CaMgSi_2O_6$ and $Mg_2Si_2O_6$ (Kushiro, 1969). The experimentally determined range of solid solution in the ferrohedenbergite series is from $Wo_{50}Fs_{50}$ to $Wo_{20}Fs_{80}$ (Bowen *et al.*, 1933; see also p. 228); a complete solid solution series between diopside and hedenbergite occurs in natural minerals although in the experimental system a wollastonite solid solution occurs as the liquidus phase for compositions more iron rich than $Di_{25}Hd_{75}$.

The miscibility gap in the synthetic $CaMgSi_2O_6$–$Mg_2Si_2O_6$ system, as shown by natural pyroxene pairs, extends into the more iron-rich augite and subcalcic augite fields, but the compositional limits of the gap within the pyroxene quadrilateral have not been determined experimentally. Consideration of natural compositions shows that the Ca-rich members of pyroxene pairs may extend to $\simeq Ca_{37}Mg_{16}Fe_{47}$ and of the coexisting Ca-poor phases to $\simeq Ca_8Mg_{18}Fe_{74}$. Pyroxenes with compositions intermediate between augite and pigeonite also occur, particularly in volcanic rocks; thus clinopyroxenes zoned from subcalcic augite to pigeonite have been described by Kuno (1950, 1955), who has presented evidence to show that under conditions of rapid cooling at high temperatures, compositions ranging from salite through augite and subcalcic augite to pigeonite occur in the volcanic rocks of the Izu-Hakone province, Japan, and clinopyroxenes continuously zoned from augite to subcalcic augite have been reported from some quickly chilled dolerites (e.g. Benson, 1944). It is probable, however, that some of these optically homogeneous pyroxenes are unmixed on a submicroscopic scale. The compositions of the more calcium-rich clinopyroxenes which have crystallized from basaltic magmas under conditions of slow cooling show a more restricted range of solid solution, and in such clinopyroxenes of plutonic rocks the usual limit of $(Mg,Fe)SiO_3$ initially in solid solution in $Ca(Mg,Fe)Si_2O_6$ is normally less than $\simeq 40$ mol. per cent. With slow cooling much of the $(Mg,Fe)SiO_3$ in solid solution at the temperature of crystallization is exsolved either as pigeonite or orthopyroxene.

Two pyroxene phases are commonly formed during the earlier stages of the fractional crystallization of basic magmas, and the association of a calcium-rich and a calcium-poor pyroxene is particularly characteristic of rocks of tholeiitic affinities. During the initial period of crystallization the composition of the calcium-rich phase is close to $Ca_{45}Mg_{45}Fe_{10}$, and the associated calcium-poor phase is an orthopyroxene $\simeq Ca_4Mg_{77}Fe_{19}$ in composition. With fractionation and enrichment of the residual liquids in iron, a monoclinic calcium-poor phase, pigeonite, crystallizes in place of the orthopyroxene, and in many rocks this change occurs when the Mg:Fe ratio of the calcium-poor phase is about 70:30 (the range of Mg:Fe ratios at which this change from an orthorhombic to a monoclinic calcium-poor phase has been observed is given by Hess (1941), as 85:15 to 60:40).

In the Skaergaard intrusion the transition from the orthopyroxene to the pigeonite stability field occurs at an En:Fs ratio of 72:28. At this stage in the fractionation Ca-poor pyroxene with exsolved augite occurs with a Ca-poor pyroxene in which the exsolved phase is absent, the latter is considered to have crystallized at a slightly lower temperature as primary orthopyroxene (Nwe, 1976).

Further cooling of the magma proceeds with the cotectic crystallization of Ca-rich and Ca-poor pyroxene, pigeonite, phases, both of which become more iron-rich as fractionation takes place, but in many intrusions the sequence of crystallization has not continued beyond magmatic compositions from which the two pyroxenes crystallize cotectically. With stronger fractionation, and further iron enrichment of the residual liquids, however, the precipitation of the Ca-poor pyroxene ceases, and pyroxene crystallization is represented by a single Ca-rich ferroaugite. In these intrusions, e.g. Bushveld and Skaergaard (Figs. 148, 149), in which extreme fractionation has occurred the latest pyroxenes to crystallize are ferrohedenbergitic in composition, and in the case of the Skaergaard intrusion the pyroxenes are inverted β-wollastonite solid solutions and the Ca-rich pyroxene trend extends from $\simeq Wo_{45}En_{45}Fs_{10}$ to $Wo_{45}En_0Fs_{55}$ (see also pigeonite section, p. 187). The pyroxene trends represent the compositions of the phases at solidus temperatures. The subsolidus trends of the Skaergaard pyroxenes, i.e. the compositions of the augite–ferroaugite sequence after exsolution of a Ca-poor phase, and of the orthopyroxene–pigeonite compositions subsequent to exsolution of a Ca-rich phase have been determined by Nwe (1976), and are shown in Fig. 149.

An additional factor in the crystallization of the Skaergaard pyroxene has recently been described by Nwe (1975). The pyroxenes of the trough bands show two distinct trends, a cumulus trend following that of the main layered series, and an intercumulus trend (Fig. 150). The pyroxenes of the latter vary widely in

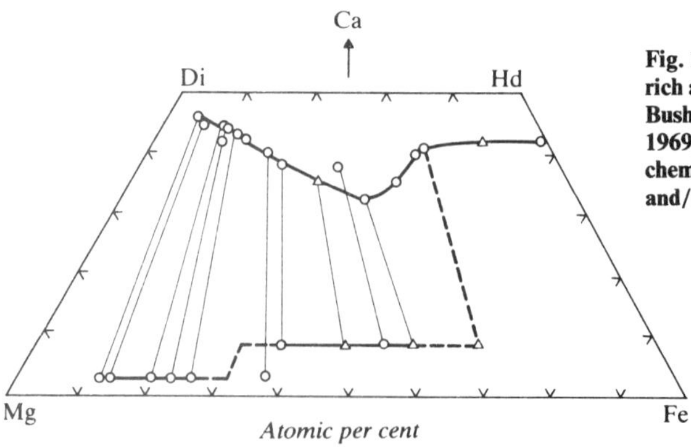

Fig. 148. Crystallization trends of Ca-rich and Ca-poor pyroxenes of the Bushveld intrusions (after Atkins, 1969). Circles, compositions from chemical analysis, triangles from optical and/or X-ray data.

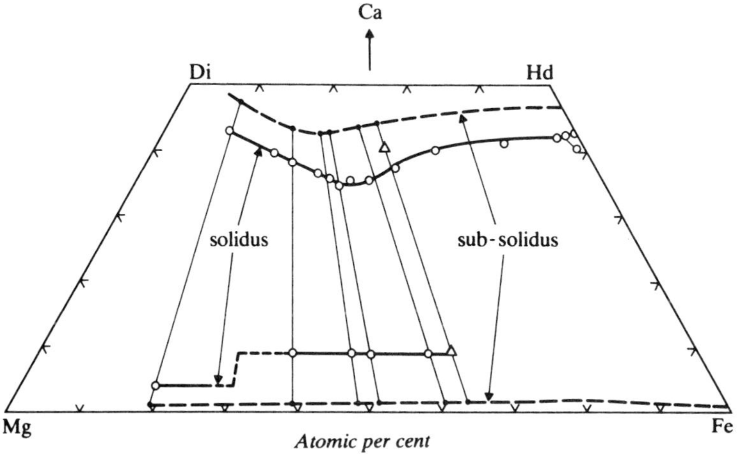

Fig. 149. Crystallization trends of the Ca-rich and Ca-poor pyroxenes of the Skaergaard intrusions (after Brown and Vincent, 1963; subsolidus trend after Nwe, 1976).
○ pyroxene compositions on the solidus trend.
△ composition of pyroxene pair which include the most ferriferous Ca-poor pyroxene on the Skaergaard main trend.
• postulated composition at which the Ca-poor pyroxene started to react with the liquid.

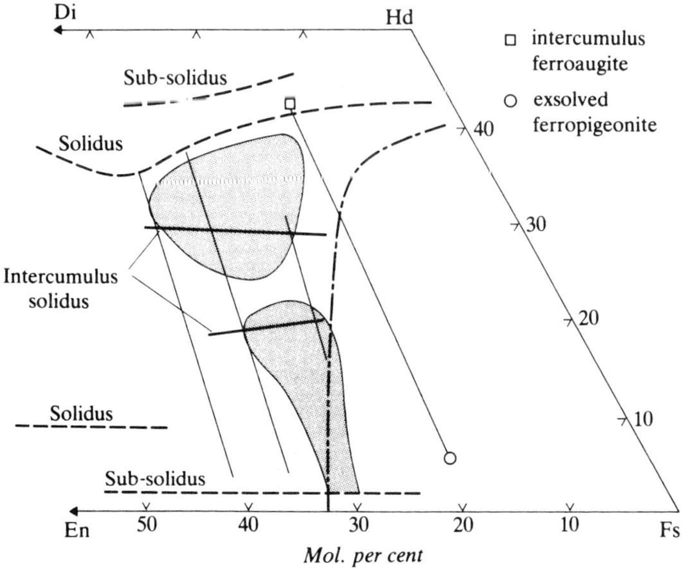

Fig. 150. Pyroxene crystallization trends in the trough bands of the Skaergaard intrusion (after Nwe, 1975). Dot-dash curve, boundary of the 'forbidden zone' at 925°C and 1 atm (Smith, 1972). Stippled areas, compositional fields of intercumulus phases or exsolved material in intercumulus phases.

composition from ferroaugites comparable with the cumulus phase, through calcic ferroaugites (e.g. $Wo_{17\cdot9}En_{27\cdot2}Fs_{54\cdot9}$) to $Wo_{3\cdot5}En_{30\cdot8}Fs_{65\cdot7}$ (Nwe, 1975). This feature might be related to metastable crystallization with regard to the pyroxene solvus, or the metastable extension of the solvus. Such an explanation implies a delay in nucleation and undercooling of the intercumulus liquid resulting in a 'quench' trend. To provide these conditions the movement of a supercooled current over the trough band cumulates subsequent to their deposition would be necessary. As the cumulus pyroxenes are unzoned there is no evidence to support the supposition that the intercumulus pyroxenes are the product of rapid crystallization, and the difference between the cumulus and intercumulus trends is more likely to be related to the isolation of individual intercumulus cells both from each other and from the supernatant liquid, at least during the later stages of crystallization. The narrowing of the miscibility gap shown by the pyroxene compositions implies a contraction of the solvus for the intercumulus liquid due to different crystallization conditions, e.g. f_{H_2O}, from those prevailing during the formation of the cumulus pyroxenes. A second characteristic of the intercumulus is the greater iron enrichment resulting from crystallization at lower temperatures, shown by the pyroxenes compared with those of the cumulus trend at the same level in the intrusion, and their extension to more iron-rich parts of the quadrilateral.

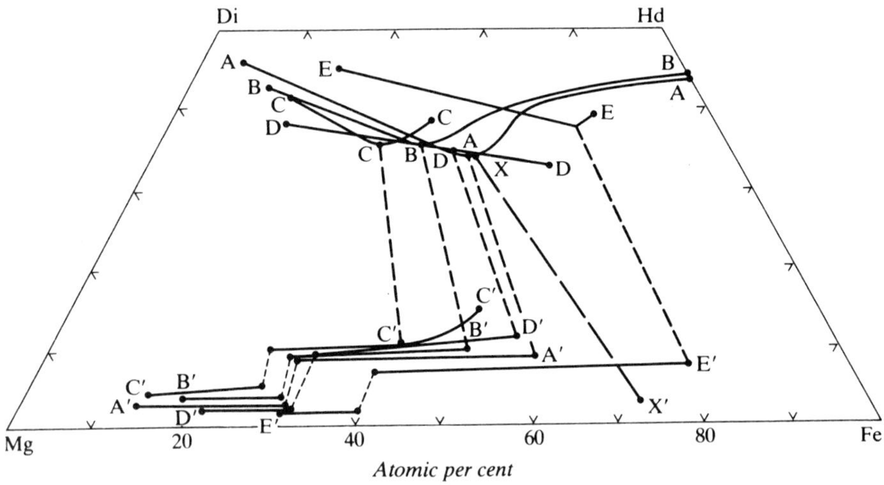

Fig. 151. Ca-rich and Ca-poor pyroxene fractionation trends of the Bushveld (AA′), Skaergaard (BB′), Jimberlana (CC′), Dufek (DD′) and Bjerkrem-Sogndal (EE′) intrusions. Short broken lines, boundaries of the two-pyroxene field. XX′ boundary of the two-pyroxene field in thePalisades sill.

The limit to the compositional field in which two pyroxene phases crystallize together has been termed the two-pyroxene boundary (Hess, 1941). Although this term was first introduced by Tsuboi (1932) for the boundary separating the field of monoclinic and orthorhombic pyroxenes in the system $CaSiO_3$–$MgSiO_3$–$FeSiO_3$ its use in this sense has now been abandoned. The limit of the two-pyroxene field is delineated by the join between ferroaugite with minimum Wo content, the 'pyroxene minimum', and the coexisting Ca-poor pyroxene, at which stage in the fractionation process the composition of the ferroaugite generally coincides with a change in the trend towards more Ca-rich ferroaugite. The limit of the two-pyroxene field, and of the miscibility gap in slowly cooled liquids varies in individual intrusions (Fig. 151) and is probably related to the effects of different concentrations of Al, Fe^{3+} and Na on the phase equilibria, as well as variations in the vapour pressure of H_2O and in silica activity. The origin of the 'pyroxene minimum' has been considered, among others, by Yoder *et al.* (1963), Lindsley and Munoz (1969), Carmichael *et al.* (1970), Williams *et al.* (1971), Oyawoye and Makanjuola (1972), Walker *et al.* (1973) and Nwe (1976).

Hess (1941) suggested that the earlier cotectic relationship between Ca-rich and Ca-poor phases might be replaced, at high iron concentrations, by a solid solution series with a minimum, such that the compositions of the augite and pigeonite phases converge, the one becoming less Ca-rich and the other more Ca-rich. The evidence of the mineralogy of ferrogabbros does not support this suggestion, and in general ferroaugites, the compositions of which lie in the single pyroxene field, become progressively more Ca-rich.

A reaction between Ca-poor pyroxene and melt to form a fayalitic olivine is indicated by the phase relationships, at high iron concentrations, in both the CaO–FeO–SiO_2 and MgO–FeO–SiO_2 systems. In those intrusions in which a Ca-poor phase ceases to crystallize its cessation is coincident with the reappearance of olivine, and the absence of more iron-rich orthopyroxene or pigeonite may be due to the reaction:

$$\underset{\text{(in Ca-poor pyroxene)}}{2FeSiO_3} = \underset{\text{(in olivine)}}{2FeSi_{\frac{1}{2}}O_2} + \underset{\text{(in melt)}}{SiO_2}.$$

Thus the termination of crystallization of the Ca-poor phase may be related to a decrease in the silica activity of the melt.

The relative influence of silica activity, pressure and oxygen fugacity on the stability field of Ca-poor pyroxenes has been investigated by Campbell and Nolan (1974). For equilibrium coexistence of Ca-poor pyroxene and olivine the following conditions must be satisfied:

$$\underset{\text{(in olivine)}}{2MgSi_{\frac{1}{2}}O_2} + \underset{\text{(in liquid)}}{SiO_2} = \underset{\text{(in Ca-poor pyroxene)}}{2MgSiO_3} \tag{1}$$

The silica activity may be calculated (Carmichael *et al.*, 1970) from

$$a_{SiO_2} = \frac{1}{K_1}\frac{(a_{MgSiO_3})^2}{(a_{MgSi_{\frac{1}{2}}O_2})^2} \tag{1a}$$

Provided the solid solutions are ideal:

$$a^{Pyx}_{MgSiO_3} = X^{Pyx}_{MgSiO_3} = \frac{En}{100} \quad \text{and} \quad a^{Oliv}_{MgSi_{\frac{1}{2}}O_2} = X^{Oliv}_{MgSi_{\frac{1}{2}}O_2} = \frac{Fo}{100}$$

thus:

$$a_{SiO_2} = \frac{1}{K_1}\frac{(En)^2}{(Fo)^2} \tag{1b}$$

Similarly from the equation:

$$2FeSiO_3 = 2FeSi_{\frac{1}{2}}O_2 + SiO_2 \tag{2}$$

$$a_{SiO_2} = \frac{1}{K_2}\frac{(Fs)^2}{(Fa)^2}$$ and since a_{SiO_2} is the same for (1b) and (2):

$$K_3 = \frac{(En)^2\,(Fa)^2}{(Fo)^2\,(Fs)^2} = \frac{(En)^2}{(1-En)^2}\,\frac{(1-Fo)^2}{(Fo)^2} \tag{3}$$

where $K_3 = K_1/K_2$.

The absence of Ca-poor pyroxenes in the latter stages of fractionation of the Skaergaard and Bushveld intrusions is accompanied by the reappearance of iron-rich olivine but without primary quartz (i.e. not as a cumulus phase). Because it is the $FeSiO_3$ component that leads to the instability of the Ca-poor pyroxene the effect of the SiO_2 activity on the above reaction can be illustrated from the equilibrium condition:

$$K = \frac{a_{Fe_2SiO_4} \cdot a_{SiO_2}}{a_{Fe_2Si_2O_6}}$$

showing that decreasing activity of SiO_2 results in either a decrease in the activity of $Fe_2Si_2O_6$ or an increase in the activity of Fe_2SiO_4 or in the activity of both (Lindsley and Munoz, 1969).

The maximum stability of the $FeSiO_3$-rich pyroxene will occur when the liquid is saturated with respect to SiO_2 and quartz is a primary phase, i.e. when $a_{SiO_2} = 1$. This corresponds to the intersection of the crystallization trend of the Ca-poor pyroxene with the boundary of the 'forbidden zone' (see Fig. 150). When, however, the silica activity is less than unity the Ca-poor pyroxene is less stable and the reaction $2FeSiO_3 + 2FeSi_{\frac{1}{2}}O_2 = SiO_2$ will occur prior to the Ca-poor pyroxene trend reaching the 'forbidden zone'. This does not imply that a_{SiO_2} decreases during fractionation, but only that the silica activity will be insufficient to stabilize Ca-poor pyroxenes with high Fe/Mg ratios. Thus variations in silica activity values may account for the failure of the two-pyroxene field to reach the 'forbidden zone', as well as to effect its termination in individual intrusions, such as the Bushveld and Skaergaard, at different compositional limits of the coexisting Ca-rich and Ca-poor pyroxenes.

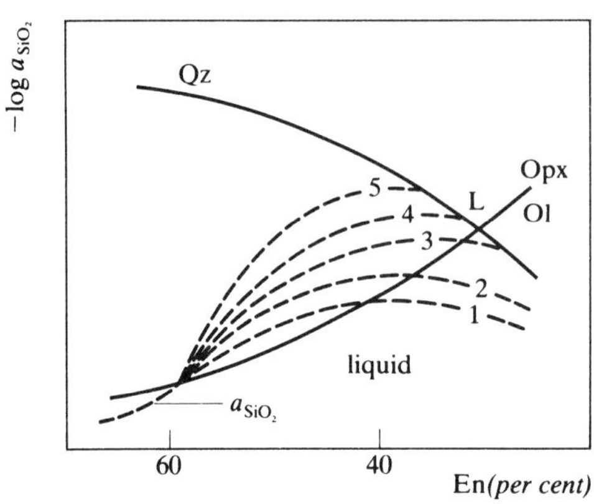

Fig. 152. Diagrammatic representation showing the influence of different rates of increase in a_{SiO_2} of the melt during the later stages of crystallization. L, limiting point; Qz, quartz; Opx, orthopyroxene, Ol, olivine (after Campbell and Nolan, 1974).

This is illustrated in Fig. 152; Fe-rich olivine crystallizes before quartz from liquids one to three, whereas for liquids four and five the rate of increase of a_{SiO_2} is such that the a_{SiO_2} liquid curve lies above the orthopyroxene:olivine curve until it intersects the quartz curve, with consequent crystallization of quartz prior to the precipitation of olivine. At L the Ca-poor pyroxene is unstable and breaks down to Fe-rich olivine and quartz, and this point, described by Campbell and Nolan (1974) as the limiting point, represents the lower En limit of the stability field of Ca-poor pyroxene. The position of the limiting point is also dependent on the temperature and pressure of magmatic crystallization. The temperature effect is illustrated by considering three gradients, 1300°–914°, 1175°–789° and 1050°–664°C, at a pressure of 0·6 kbar (Fig. 153) showing that the En value of the limiting point increases as the crystallization temperature is reduced, and results in an extension of the a_{SiO_2} range of the stability field of Ca-poor pyroxenes with Mg contents greater than En_{50}.

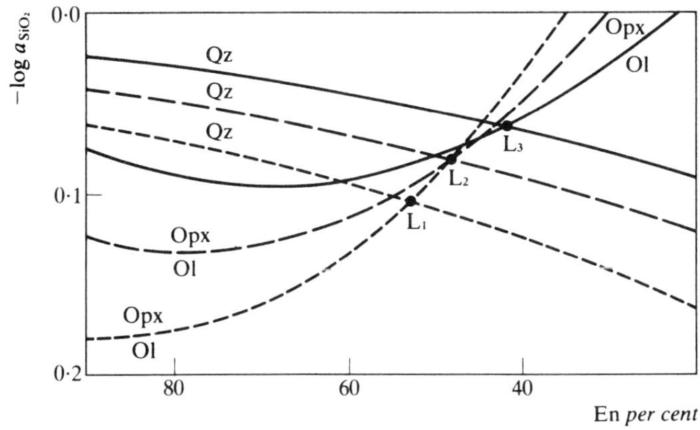

Fig. 153. Orthopyroxene, olivine and quartz curves for three sets of temperature gradients (after Campbell and Nolan, 1974). Full line 1300°–914°C; long broken line 1175°–789°C; short broken line 1050°–664°C at 0·6 kbar pressure. L_1, L_2, L_3 limiting points for low, moderate and high gradients. Note that lowering the crystallization temperature reduces the Fe/Mg ratio of the limiting point and increases the range of a_{SiO_2} at which Ca-poor pyroxene is stable.

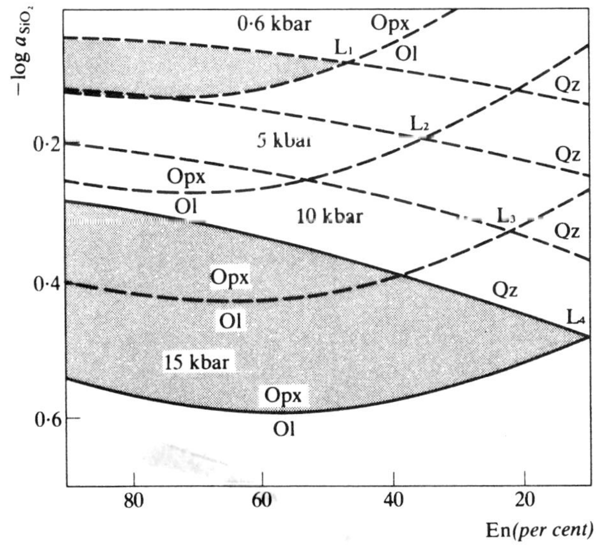

Fig. 154. Silica activity curves of orthopyroxene, olivine and quartz at 0·6, 5, 10 and 15 kbar for the temperature gradient 1175°–789°C (after Campbell and Nolan, 1974). L_1–L_4 limiting points for pressures of 0·6, 5, 10 and 15 kbar. Shaded areas show the a_{SiO_2} at which Ca-poor pyroxene is stable at pressures of 0·6 and 15 kbar.

Increasing pressure of crystallization reduces the En value of the limiting point, and increases the range of a_{SiO_2} at which Ca-poor pyroxene is stable (Fig. 154). The replacement of Ca-poor pyroxene by Fe-rich olivine will, however, only occur if the a_{FeO} of the magma is in excess of the level required by the equation:

$$FeSiO_3 + FeO = 2FeSi_{\frac{1}{2}}O_2$$
(in pyroxene)(in liquid)(in olivine)

Decreasing oxygen fugacity increases the a_{FeO} by reducing the Fe_2O_3:FeO ratio, and low f_{O_2} can thus reduce the stability field of Ca-poor pyroxene.

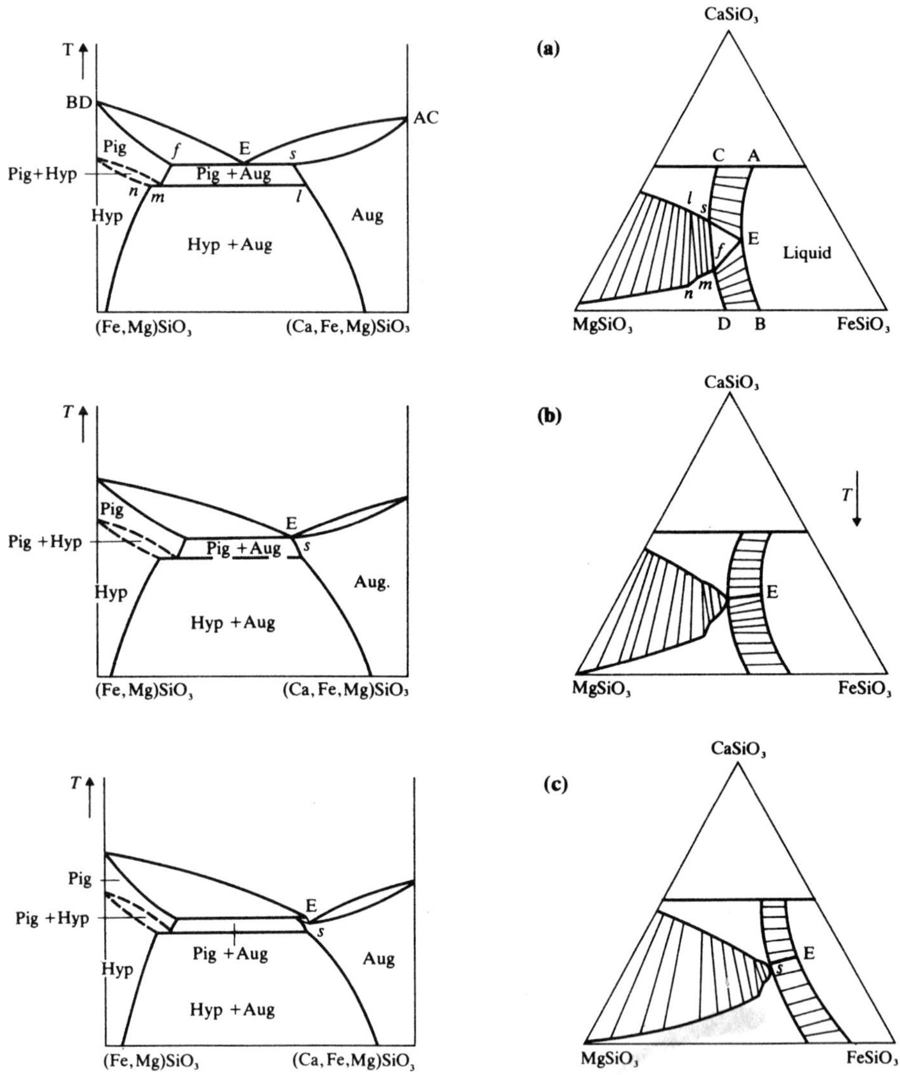

Fig. 155. Hypothetical equilibrium diagrams for pyroxenes (after Muir, 1954). The diagrams on the left represent the phase relations in a simplified manner as three pseudobinary sections in the system $MgSiO_3$–$FeSiO_3$–$CaMgSi_2O_6$–$CaFeSi_2O_6$ at increasing Fe:Mg ratios. Diagrams on the right show hypothetical phase equilibria for a series of isothermal planes at successively lower temperatures and at different stages of iron-enrichment.

(a) Two-pyroxene field. The diagram on the right represents an isothermal plane at the temperature of the line *f*E*s* of the left-hand diagram. From a liquid of composition E on the pyroxene cotectic curve, two pyroxenes *f* (pigeonite) and *s* (augite) should separate. On the right-hand diagram the triangle *lmn* represents the pigeonite–hypersthene inversion interval. Comparable points in the two diagrams are designated by the same letter.

(b) Limit of two-pyroxene field. The cotectic curve at E has migrated to the calcium-rich side of the intersection with the solvus. From liquid E, a single pyroxene *s* of similar calcium content will separate. Right-hand diagram represents the isothermal plane of temperature E*s* on the left-hand diagram.

(c) One-pyroxene field. The cotectic curve at E has become detached from the solvus, and ferroaugite *s* alone should separate from the liquid. Right-hand diagram represents the isothermal plane of temperature E*s* on left-hand diagram.

In some intrusions a Ca-poor pyroxene continues to crystallize after the reappearance of olivine and after the inflection in the ferroaugite trend. Thus the formation of Fe-rich orthopyroxene (in the range Fs_{60}–Fs_{80}) subsequent to the cessation of pigeonite has been reported in some tholeiitic intrusions (e.g. New Amalfi sheet, Beaver Bay, Palisade sill and the Noril'sk complex). Such examples usually show evidence that the orthopyroxene did not crystallize at the same time as the ferroaugite, but probably formed at a later magmatic or deuteric stage and thus should not be regarded as extending the compositional range of the two-pyroxene field (see also Fig. 151).

An alternative explanation of the cessation of crystallization of the Ca-poor pyroxene has been suggested by Muir (1954), and is illustrated by the series of hypothetical phase diagrams of Fig. 155. Here the absence of iron-rich pigeonite is shown to result from a gradual shift with falling temperature, of the liquidus minimum of the cotectic pair, ferroaugite–pigeonite, towards the calcium-rich side of its intersection with the solvus, i.e. from E(a) to E(b) (Fig. 155). If the liquidus minimum passes beyond the intersection with the solvus (to E(c), Fig. 155) then the calcium-poor phase would cease to crystallize as a separate phase. The migration of the liquidus to the Ca-rich end of the solvus, and its subsequent decoupling to become a liquidus minimum is supported by the most recent work on the Skaergaard pyroxenes. At the limit of the two-pyroxene field in this intrusion the boundaries between ferroaugite and ferropigeonite display a corroded relationship (Nwe and Copley, 1975), and two stages of pyroxene crystallization appear to be represented, firstly the cotectic crystallization of Ca-rich and Ca-poor pyroxene, secondly the peritectic reaction pigeonite + liquid → augite, the first stage by the earlier cumulus crystals and the latter by more fractionated later pyroxenes. The composition of the ferropigeonite at which the peritectic reaction occurred during the crystallization of the Skaergaard magma is $Wo_{10}En_{36.7}Fs_{53.3}$.

Kushiro (1960) and Le Bas (1962) have shown that the Al content of Ca-rich pyroxenes is inversely related to the SiO_2 content of the magmas in which they crystallized. The latter in particular has demonstrated the distinctive relationship between Al_2O_3 in the pyroxene and the SiO_2 content of the rocks of the non-alkaline, alkaline and peralkaline suites (Fig. 156). The relationship of the silica activity of the melt with the aluminium content of pyroxenes has been discussed by Carmichael *et al.* (1970). The reaction:

$$CaAl_2SiO_6 + SiO_2 = CaAl_2Si_2O_8$$

(Ca-Tschermak's molecule)

may be used in illustration of the role of Al_2O_3 in Ca-rich pyroxenes. The reaction recast to:

$$\frac{\overset{\text{(plagioclase)}}{aCaAl_2Si_2O_8}}{\underset{\text{(pyroxene)}}{aCaAl_2SiO_6}.\underset{\text{(melt)}}{aSiO_2}} = K$$

shows that a decrease in silica activity leads to an increase in the $CaAl_2SiO_6$ content in the pyroxene for a given plagioclase composition. An example of this reaction is provided by the marginal enrichment in Al_2O_3 of the zoned augites of basanites, the residual liquids of which are considered to have crystallized under conditions of reduced silica activity (Brown and Carmichael, 1969).

The above equation applies to the special case in which half the Al^{3+} enters

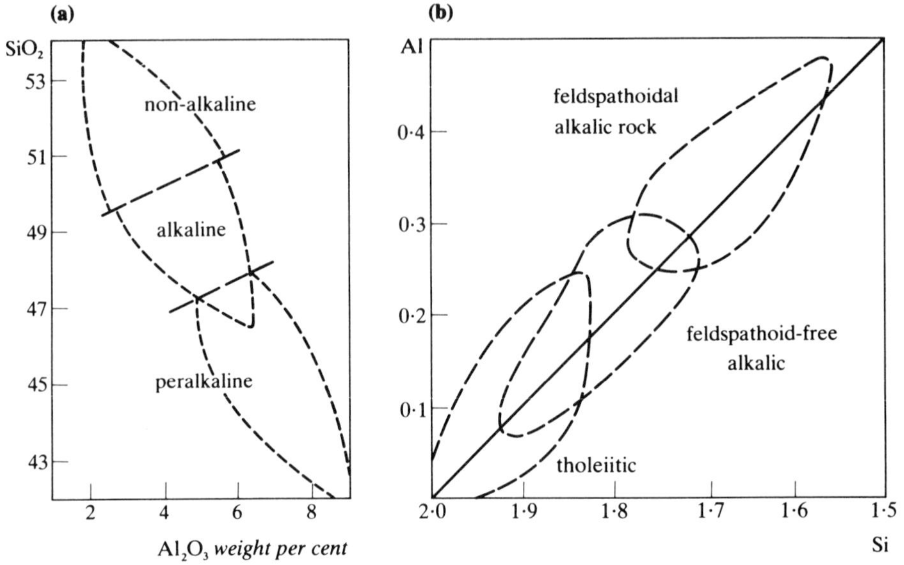

Fig. 156. (a) Relationship of SiO_2 and Al_2O_3 in groundmass pyroxenes of peralkaline, alkaline and tholeiitic, high alumina and calc-alkaline rocks (after Le Bas, 1962). (b) Atomic proportions, on the basis of six oxygen atoms, of Si and Al in clinopyroxenes from tholeiitic, feldspathoid-free and feldspathoidal alkalic rocks (after Kushiro, 1960).

octahedral sites and, as emphasized by Campbell and Borley (1974) the general equation takes the form:

$$\tfrac{1}{2}Y_2O_3+\tfrac{1}{2}Al_2O_3+Mg_2Si_2O_6=(MY)AlSiO_6+SiO_2+MO \qquad (1)$$

or

$$ZO_2+Al_2O_3+2Mg_2Si_2O_6=(M_3Z)Al_2Si_2O_{12}+2SiO_2+MO \qquad (2)$$

where $Y = Al^{3+}$, Cr^{3+} or Fe^{3+}, $Z = Ti^{4+}$, and $M = Fe^{2+}$, Ca^{2+} or Mg^{2+}. If K is the equilibrium constant for equation (1) at constant temperature and pressure then:

$$(MY)AlSiO_6 = K\cdot\frac{(Y_2O_3)^{\frac{1}{2}}(Al_2O_3)^{\frac{1}{2}}M_2Si_2O_6}{(SiO_2)(MO)}$$

from which it follows that the Al^{3+} content of the pyroxene increases with increasing $a_{Al_2O_3}$ and decreasing a_{SiO_2} in the melt and is, in addition, dependent on the nature and activity of the octahedrally co-ordinated cations providing the charge balance.

The formation of subcalcic augite and ferroaugite is essentially restricted to environments of rapid cooling at comparatively high temperatures. The stability of the subcalcic pyroxenes is considered (Kuno, 1955) to be due to the occupation of tetrahedral sites by Fe^{3+}, which lowers the temperature of the solvus for such pyroxenes. Thus under conditions of rapid cooling the temperature of the magma may lie above the solvus temperature with the consequent formation of subcalcic augite and ferroaugite as stable phases. Kuno (1955) has suggested that under these conditions subcalcic augites may crystallize with an Mg:Fe ratio as high as 65:35. For many slowly cooled basic magmas the Mg:Fe ratio at which the magmatic and solvus temperatures are coincident is approximately 35 : 65. Two pyroxene fractionation trends may be distinguished in the crystallization of tholeiitic basalts. One is the phenocrystal trend of intratelluric equilibrium crystallization leading to the formation of hedenbergitic compositions, and the 'quench' trend typical of the subcalcic augite sequence of the groundmass (Muir and Tilley, 1964). In a study of the Kilauean series, these authors have shown that those rocks containing Ca-rich pyroxene phenocrysts, the composition of which plotted in the pyroxene quadrilateral, display a characteristic distribution on the diopside–hedenbergite side of the 'cotectic' curve (approximately parallel to the enstatite–ferrosilite join at 25 per cent $CaSiO_3$), in contrast to those with orthopyroxene phenocrysts, which plot below the curve. The quench trend of the groundmass pyroxenes, consisting of subcalcic augites or ferroaugites, in which the principal substitution is $Ca \rightleftharpoons Fe^{2+}$ transgresses the 'cotectic' curve, and in the more acid differentiates continues to ferropigeonite compositions (see pigeonite section).

The effect of the rate of crystallization on the type of chemical variation in augite is illustrated by the pyroxenes in the chilled basal margin and the more slowly cooled interior of a flow of the Picture Gorge basalt (Smith and Lindsley, 1971). The composition of the augites from the more central parts of the flow display a normal plutonic trend from $\simeq Fe_{12}$ to Fe_{27} with only a small decrease in Ca. Those of the quickly chilled base of the flow show the characteristic $Ca \rightleftharpoons Fe$ substitution, Ca_{45} to Ca_{36}, with constant Mg, of the 'quench trend' and probably reflect a metastable crystal–liquid partition consequent on the rapid crystallization.

Two contrasted crystallization trends of augite in the tholeiitic dolerite, Semi, northern Japan, have been described by Yamakawa (1971). In one the augite composition varies from $Ca_{37}Mg_{41}Fe_{22}$ to $Ca_{35}Mg_{32}Fe_{33}$, with the Ca/(Ca + Mg + Fe) ratio remaining nearly constant; in the other augite, compositions range from $Ca_{36}Mg_{40}Fe_{24}$ to $Ca_{28}Mg_{35}Fe_{37}$. The augite occurs both as phenocrysts and as a groundmass constituent, as also does the associated pigeonite. On the evidence of the textural relations the augite of the first trend is considered to have crystallized in equilibrium with the pigeonite and the liquid, whereas the augite– and subcalcic augite–pigeonite pairs are interpreted as disequilibrium associations. The first trend is accounted for by the cotectic crystallization of Ca-rich and Ca-poor pyroxenes, and the second by metastable crystallization resulting from the undercooling of the liquid with respect to pigeonite, in accord with the quench trend of Muir and Tilley. An explanation of the contrasted trends is illustrated by the hypothetical pseudobinary $(Mg,Fe)SiO_3$–$Ca(Mg,Fe)Si_2O_6$ system of Fig. 157. Starting with liquid L_1, the composition of the augite changes from A_1 to A_2 as the temperature

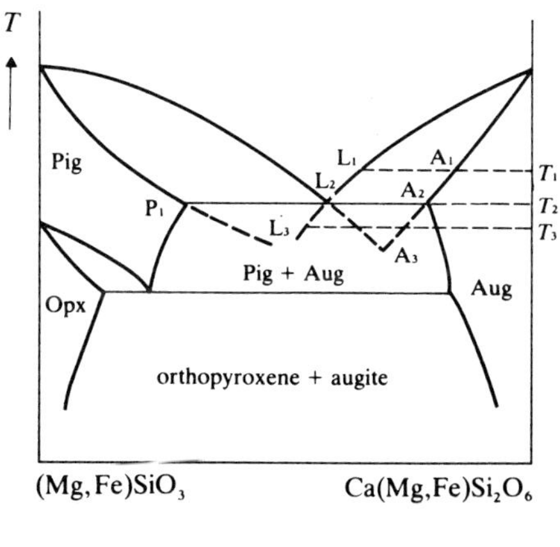

Fig. 157. Hypothetical pseudobinary diagram showing crystallization of the pyroxenes in the tholeiitic dolerite of the Semi sheet (after Yamakawa, 1971).

falls from T_1 to T_2, at which temperature crystallization of pigeonite of compositions P_1 begins, and under equilibrium conditions augite and pigeonite crystallize with a cotectic relation. If, however, pigeonite nucleation is delayed and the liquid in contact with augite A_2 is undercooled with respect to pigeonite, the liquid composition would change from L_2 to L_3, that of the augite changing metastably from A_2 to A_3.

Exsolution lamellae. The augites of many basic plutonic rocks, especially those of tholeiitic affinities, enclose exsolution lamellae of a calcium-poor pyroxene phase but, in some cases due to their narrow width, it is not always possible to distinguish optically whether the lamellae consist of orthopyroxene or pigeonite. Orthopyroxene lamellae are exsolved parallel to the (100) plane, and it has been generally accepted that the orientation of the pigeonite lamellae is parallel to the (001) plane of the augite host (but see below). Orthopyroxene exsolution lamellae may be distinguished from narrow mechanical twins in augite on (100) by the fact that exsolution lamellae have [001] and (100) in common with the augite host, and thus the γ optical direction in the lamellae is parallel to [001] of the host. Exsolution lamellae should have parallel extinction whereas in the twin lamellae, the extinction will in general be inclined. Moreover, in contrast to lamellar twins on (100), the lamellae of which extend to the grain boundary, exsolution lamellae commonly terminate within the crystal a short distance from the boundary.

The more magnesian-rich clinopyroxenes crystallize below the pigeonite–orthopyroxene inversion temperature, and the exsolution lamellae in such augites consist of orthopyroxene (Fig. 158). Augites more iron-rich than approximately $Wo_{41}En_{44}Fs_{15}$ commonly crystallize above the pigeonite–orthopyroxene inversion temperature and the exsolved phase crystallizes as pigeonite. In many such pyroxenes the exsolved pigeonite subsequently inverts to orthopyroxene on cooling through the monoclinic–orthorhombic inversion temperature, thus a first set of lamellae is approximately parallel to (001), and a second set of lamellae is parallel to the (100) plane of the augite host.

The usually accepted statement that exsolution lamellae in clinopyroxene are

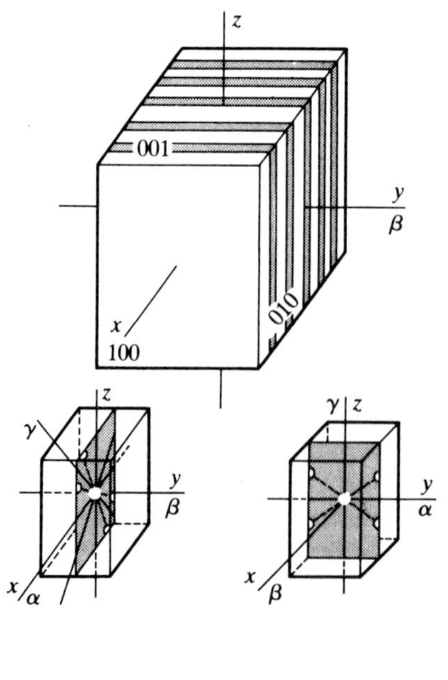

Fig. 158. Lamellae of orthopyroxene parallel to (100) in magnesium-rich augite host (after Poldervaart and Hess, 1951).

oriented parallel to (001) and (100) has been shown to be in error (Robinson *et al.*, 1971). Lamellar orientation (Fig. 159) is determined by the relationship between a and c dimensions of the lattice with the smaller β angle (augite) and those of the lattice with the larger β angle (pigeonite):

	'001' lamellae		'100' lamellae
$a_{augite} > a_{pigeonite}$	in β acute	$c_{augite} > c_{pigeonite}$	in β acute
$a_{augite} < a_{pigeonite}$	in β obtuse	$c_{augite} < c_{pigeonite}$	in β obtuse

Changes in lattice parameters may occur subsequent to lamellar nucleation; at constant composition as a result of thermal contraction, reduction in pressure or reversible polymorphic transitions, or with changing composition due to the exsolution or by element fractionation between matrix and lamellae. The lamellar angles are thus a potential means of elucidating the pressure–temperature regimes of the pyroxenes.

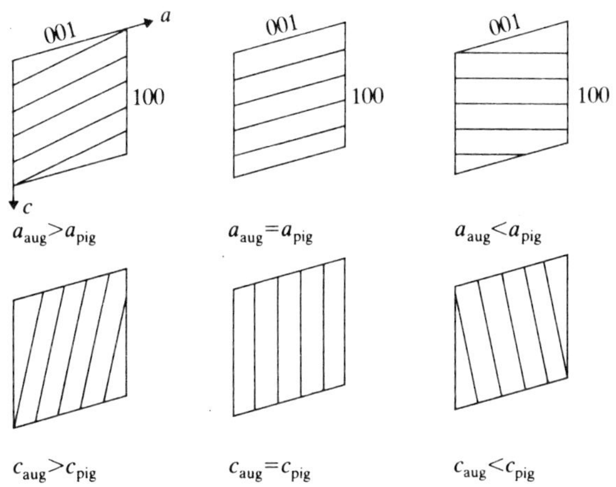

Fig. 159. Relation between relative *a* and *c* dimensions and orientation of exsolution lamellae in clinopyroxene (after Robinson *et al.*, 1971).

In augite–pigeonite pairs of igneous rocks crystallized at high temperature the miscibility gap is relatively small, differences in the a dimensions would also be small, and thus the exsolution angles could be fairly close to β. Under metamorphic conditions at lower temperatures the miscibility gap is generally larger and the exsolution angles correspondingly larger; this temperature dependent effect of the solvus on the angle of '001' exsolution lamellae has been described as the *solvus effect*. The contrast in exsolution angles in augite of the Duluth gabbro, and augite in a pyroxene granulite of the Hudson Highlands, New York, is illustrated in Fig. 160; a detailed discussion of their petrological significance is given by Robinson *et al.* (1971).

Three sets of optically visible exsolution are present in metamorphic augites from the Adirondacks, Hudson Highlands and the Cortlandt and Belcherton complexes (Jaffe *et al.*, 1975). X-ray single-crystal photographs show that the lamellae consist of orthopyroxene on (100), and of pigeonite lamellae, '001' and '100', oriented on irrational planes close to (001) and (100) respectively (Fig. 161). The augites coexist with orthopyroxenes and both the pigeonite and orthopyroxene lamellae were exsolved at a temperature below that of the pigeonite–orthopyroxene inversion curve. In comparison with igneous pyroxenes the exsolution lamellae are less

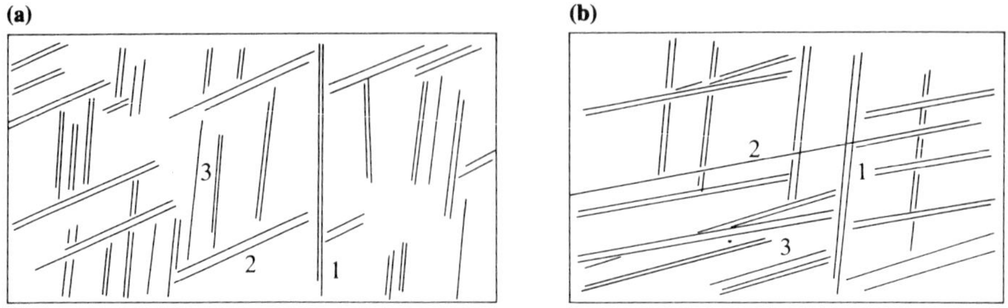

Fig. 160. (a) Augite, in pyroxene granulite, normal to (010) showing three sets of exsolution lamellae. (1) prominent continuous lamellae parallel to (100) of host; (2) lamellae at 116° to z crystallographic axis; (3) very fine lamellae lying in acute β angle at 6° to z crystallographic axis. (b) Augite, Duluth gabbro, normal to (010). (1) coarse lamellae containing oxide blebs and orthopyroxene parallel to (100) of the host; (2) pigeonite and oxide blebs at 108° to z; (3) pigeonite lying at 111° to 115° to the z crystallographic axis. Width of sections 0·16 mm (after Robinson *et al.*, 1971).

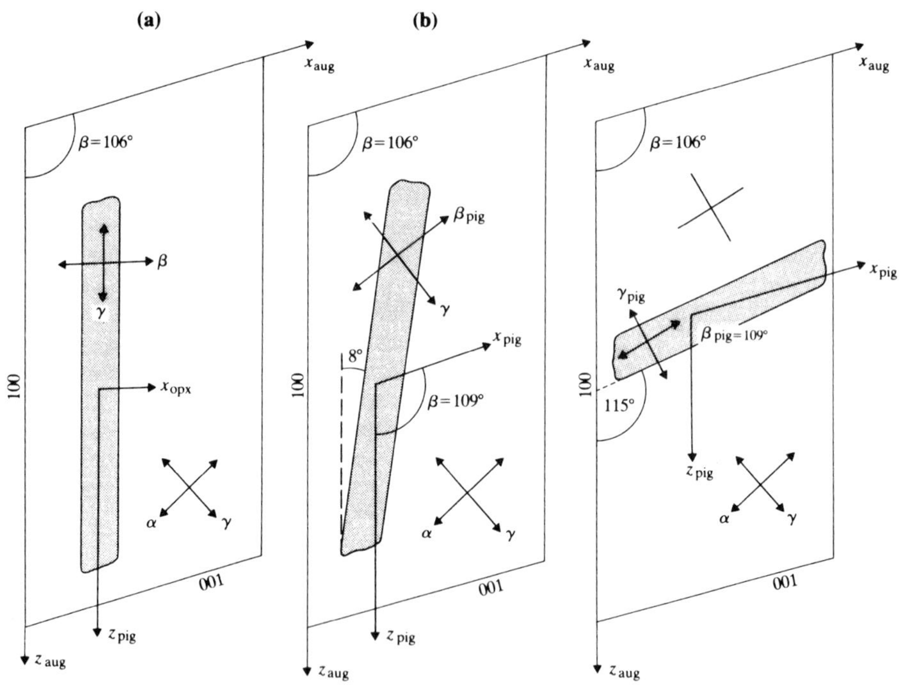

Fig. 161. (a) Augite with (100) orthopyroxene lamellae, z_{aug} parallel to z_{opx}, $x_{aug} \wedge x_{opx} = 16°$, $y_{aug} \wedge y_{opx}$ and $z_{aug} \wedge z_{opx} = 0°$, phase boundary of lamellae is parallel to augite (100). (b) Augite with '100' pigeonite lamellae, $x_{aug} \wedge x_{pig} \simeq 3°$, $y_{aug} \wedge y_{pig} = 0°$, $z_{aug} \wedge z_{pig} \simeq 0°$, phase boundary of lamellae is parallel to y but at 8° to z_{aug}. (c) Augite with '001' pigeonite lamellae, $x_{aug} \wedge x_{pig} \simeq 0°$, $y_{aug} \wedge y_{pig} = 0°$, $z_{aug} \wedge z_{pig} \simeq 3°$, phase boundary of lamellae is parallel to y but at an angle of 115° to z_{aug}. All sections are parallel to augite (010) (after Jaffe *et al.*, 1975).

numerous and finer in scale, indicating a more restricted initial solid solution at the lower temperature of the metamorphic recrystallization.

The angle between phase boundaries of the '001' and '100' pigeonite lamellae and the augite z axis varies with the iron content of the pyroxenes, the more magnesium-rich augites having larger '001' $\wedge$ '100', z angles. This relationship between chemical composition and lamellar orientation is due to the effect of composition on the lattice parameters. The angles approximate closely to those of optimum phase boundaries computed from the measured lattice parameters of the augite host and the pigeonite lamellae. Little change has occurred in the relative lattice parameters of lamellae and host subsequent to the nucleation of the lamellae under the conditions of the metamorphic regime unlike the change observed in some igneous pyroxenes.

A so-called striped ferroaugite (Table 33, anal. 11) from the Sivamalai syenite, Madras State, has been described by Bose (1968). This pyroxene consists of an intergrowth of two phases, the bulk composition of which is $Ca_{41\cdot6}Mg_{13\cdot6}Fe_{44\cdot8}$. The greater part of the crystal consists of a more Ca-rich host containing lamellae of a Ca-poor pyroxene. The calcium content of the exsolved phase is related to the thickness of individual lamellae, and the variable composition may represent different stages of the diffusion of calcium into the host structure. The exsolution lamellae are described as being oriented parallel to (001) and (100) planes of the Ca-rich host. The (001) are coarser than the (100) lamellae, and reflect the greater ease of Ca migration across the (001) plane at a relatively higher temperature, whereas the thinner (100) lamellae result from a more limited diffusion at lower temperatures.

Small areas of optical and larger areas of submicroscopic exsolution lamellae of pigeonite in an augite, $Ca_{0\cdot35}Mg_{0\cdot35}Fe_{0\cdot30}SiO_3$, from an inninmorite pitchstone have been described by Copley *et al.* (1974). The lattice planes of the two phases are considered to be continuous across the interface and the small difference between the *a* and *b* lattice parameters is probably such that the misfit is accommodated by elastic strain. The distribution of the pigeonite lamellae is consistent with heterogeneous nucleation at the interface boundaries, and growth of the lamellae by the migration of ledges along the (001) interface, and their thickening by the propagation of ledges across the (001) faces. This mechanism of lamellar thickening requires the lamellae to decrease in width as they grow from the original nucleation site into the matrix.

Submicroscopic unmixing in the augites of the Whin sill dolerite has been described by Dunham *et al.* (1972). The optically homogeneous augites contain pigeonite lamellae varying in width from 40 Å at the margins to 3200 Å at the centre of the sill. Unmixing of subcalcic ferroaugite to ferroaugite and ferropigeonite, the latter forming lamellae approximately 250 Å in width has been reported from a lunar basalt by Bailey *et al.* (1970). Other investigations (Radcliffe *et al.*, 1970) have revealed alternating exsolution lamellae, typically between 60 and 100 Å wide, of augite and pigeonite oriented parallel to (001) and less commonly (100).

Well developed pigeonite exsolution lamellae in augite, analogous to those observed in pyroxenes of layered intrusions, have been described from anorthositic clasts in Apollo 11 soils (McCallum *et al.*, 1975). The bulk compositions of the original pyroxene and exsolved augite and exsolution lamellae are $(Ca_{0\cdot85}Mg_{0\cdot68}Fe_{0\cdot45}Al_{0\cdot02})(Si_{1\cdot95}Al_{0\cdot05})O_6$, $(Ca_{0\cdot62}Mg_{0\cdot75}Fe_{0\cdot60}Al_{0\cdot03})(Si_{1\cdot97}Al_{0\cdot03})O_6$ and $(Ca_{0\cdot04}Mg_{0\cdot94}Fe_{0\cdot98}Al_{0\cdot01})(Si_{1\cdot98}Al_{0\cdot02})O_6$, respectively. The '001' pigeonite lamellae occur in three lattice orientations, two of which do not share (001) with the augite host but make angles ($c^*_{Aug} \wedge c^*_{Pig}$) of 1·89° and 1·60°; the third set show lattice coherence across the host–lamellae interface.

Needles of ilmenite, approximately parallel to the (100) and (010) planes of the pyroxene, are present in the titanaugites of the ultrabasic–gabbroic rocks of the Blue Mountain complex, New Zealand (Grapes, 1975). The pyroxenes (Table 36, anal. 4) with the exsolved ilmenite are richer in Fe^{2+} and Fe^{3+} than the titanaugites of other rocks of the complex, and the needles may be due to the breakdown of the iron-rich augite in the presence of titanium with the formation of ilmenite and Ti–Fe-poor augite. Augite of the gabbro xenoliths in the basalts of Gough Island, South Atlantic (Le Maitre, 1965), contain exsolved magnetite and spinel. The (111) and ($\bar{1}10$) planes of both oxide phases are parallel respectively with the (100) and (010) planes of the host. Exsolved ilmenite is also present in some augites in which the (0001) ilmenite plane is parallel with the augite (100), and either ($2\bar{1}\bar{1}0$) or ($\bar{2}110$) parallel with the *z* crystallographic axis of the host. Exsolution of spinel, in the (010) plane, and elongated either parallel or perpendicular to [001], in the augite of the layered intrusions of the Giles complex, central Australia have been described by Goode and Moore (1975).

The crystallographic relationship of spinel $(Mg_{0\cdot08}Fe_{1\cdot18})(Cr_{1\cdot16}Al_{0\cdot36}Ti_{0\cdot22})O_4$, exsolved from a lunar augite (McCallum *et al.*, 1975) is $[113]^*_{Sp} \parallel c^*_{Aug}$ and $[110]^*_{Sp} \parallel b^*_{Aug}$. There are also small amounts of ilmenite with the (0001) plane parallel to the (111) plane of the spinel.

Needles and blebs of titanomagnetite, with (111) and ($\bar{1}\bar{1}3$) parallel respectively to the (100) and (001) planes of the augite are a common feature of the Kap Edvard Holm upper layered series clinopyroxenes. The exsolution process (Elsdon, 1971) is considered to be related to an increase in the equilibrium ratio Mg:Fe as a function of df_0/dT, with the consequent oxidation of the ferrosilite, $6FeSiO_3 + O_2 \rightarrow 2Fe_3O_4 + 6SiO_2$, component of the augite.

Plagioclase lamellae, parallel to (100), with the same orientation as the exsolution lamellae of orthopyroxene and ilmenite, occur in augites of the Morin series of the Grenville Township (Philpotts, 1966). Although some features, such as the termination of the lamellae within the augite, are indicative of exsolution, others, including the uneven distribution of the lamellae within the pyroxene, their presence in relatively few pyroxene grains, and the variable orientation of the plagioclase in different lamellae, suggest that the lamellae are intergrowths of plagioclase in pyroxene.

The presence of amphibole lamellae in augite has been described from a number of localities, e.g. the Sierra Nevada batholith and the San Carlos peridotite, Arizona (Papike *et al.*, 1969), peridotite xenoliths in the lava of the Ataq volcano, southern Arabia (Desnoyers, 1975), and harzburgite, Harzburg, hypersthene gabbro, Belhelvie, Scotland, perpendicular feldspar rock of the Skaergaard marginal border group, and gabbro from the critical zone of the Bushveld complex (Smith, 1977). The estimated composition of the amphibole lamellae in the augite of the harzburgite is $Ca_{2\cdot09}(Mg_{4\cdot08}Fe^{2+}_{0\cdot47}Cr_{0\cdot17}Al_{0\cdot62})(Si_{6\cdot35}Al_{1\cdot65}O_{22})(OH)_2$. Possible origins of the amphibole lamellae have been put forward by Papike *et al.*: late alteration of the pyroxene, primary epitaxial intergrowth between the two phases, development of the pyroxene from original amphibole, and exsolution of amphibole from the pyroxene host. The lamellae described by Smith are less than 0·08 μm in thickness, lie in the (010) plane, and the crystallographic axes of the clinoamphibole (space group *I*2/*m*) are parallel to those of the pyroxene. As the formation of the lamellae occurred subsequently to the exsolution of orthopyroxene on (100), hypotheses 2 and 3 can thus be eliminated, and the unaltered condition of the augites invalidates the first suggestion. Papike *et al.* suggested that finite solid solution between pyroxene and clinoamphibole may occur at high pressures and temperatures. The morphological data of Smith are consistent with this view and an exsolution mechanism appears to be the most probable explanation of the formation of the amphibole lamellae.

A technique, for bringing out exsolution lamellae not detectable in thin section, has been described by Miller and Philpotts (1973) using polished thin sections placed over hydrofluoric acid for ten to fifteen minutes, washed in distilled water and dried. An etching method for use in observing submicron-size exsolution lamellae with the scanning electron microscope is described by Chapman and Meagher (1975).

Experimental

The limiting binary systems of the pyroxene quadrilateral, $CaMgSi_2O_6$–$CaFeSi_2O_6$–$Mg_2Si_2O_6$–$Fe_2Si_2O_6$ are well documented but data relating to compositions within the quadrilateral are still scanty. Some preliminary results (Fig. 162) have been given by Yoder *et al.* (1963, 1964), Roedder (1965a), and more recently by Huebner *et al.* (1972), Lindsley *et al.* (1974) and Ross *et al.* (1973).

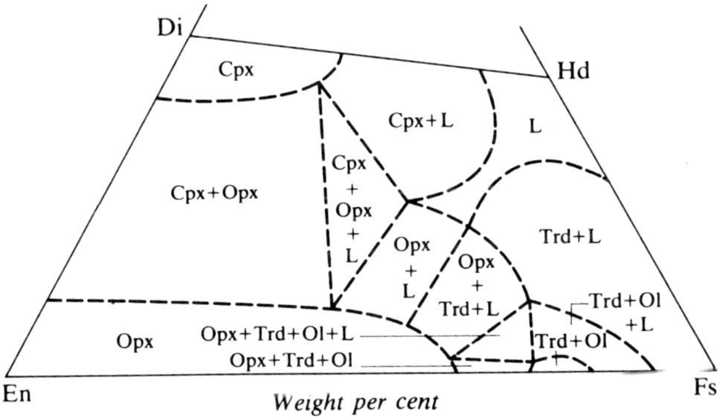

Fig. 162. Preliminary diagram in the pyroxene plane of the CaO–MgO–FeO–SiO_2 system at 1250°C (after Yoder *et al.*, 1963).

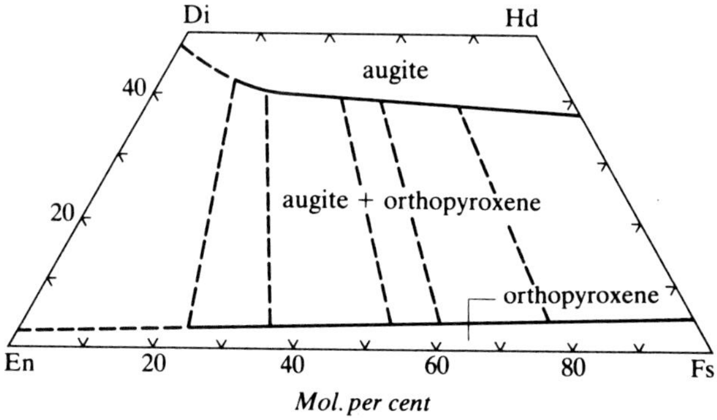

Fig. 163. Compositions of augite and orthopyroxene coexisting at 810°C at 15 kbar fluid pressure (after Lindsley *et al.*, 1974)

An experimental study to determine the composition of coexisting augites and orthopyroxenes in the temperature range 800°–810°C has been reported by Lindsley *et al.* (1974). The data, obtained at $P_{fluid} = P_{total} = 15$ kbar, show except for Fe-poor compositions, that the content of $CaSiO_3$ of the augites is not greatly affected by the Fe/(Fe + Mg) ratio, $CaSiO_3$ ranging from 43 mol. per cent at Fe/(Fe + Mg) = 0·18 to 38 mol. per cent at Fe/(Fe + Mg) = 0·75 (Fig. 163).

The subsolidus relationships along the join $Mg_{0\cdot2}Fe_{0\cdot3}Ca_{0\cdot5}SiO_3$–$Mg_{0\cdot4}Fe_{0\cdot6}SiO_3$ at 20 kbar (Turnock, 1970) are reported in the pigeonite section (p.181).

The augite–pigeonite miscibility gap has been determined experimentally under controlled oxygen pressures at 1 atm for Al,Ti-bearing pyroxenes from lunar basalt 12021 (Ross *et al.*, 1973). The augites contain a submicroscopic exsolved pigeonite phase and similarly the pigeonite contains exsolved augite. Homogenization of both phases to a single-phase pyroxene occurs just below or at the solidus temperature, and the data, illustrated in Fig. 164, show that the $CaSiO_3$ component of the coexisting augite and pigeonite is a function of the temperature. The solidus intersects the pyroxene solvus over the whole range of Fe/Mg compositions investigated (augite $\simeq Fe_{30}$–Fe_{85}, pigeonites Fe_{33}–Fe_{82}), the temperature decreasing with increasing Fe/(Mg+Fe). The miscibility gap is asymmetric towards pigeonite, the pigeonite solvus being steeper than the augite solvus for the more Mg-rich compositions. At the solidus the gap narrows with increasing Fe/(Fe+Mg), from $\simeq$ 20 per cent $CaSiO_3$ at 0·30 to $\simeq$ 10 per cent $CaSiO_3$ at Fe(Fe+Mg) = 0·85 (Fig. 165). The miscibility gap is smaller than is known for terrestrial rocks, possibly

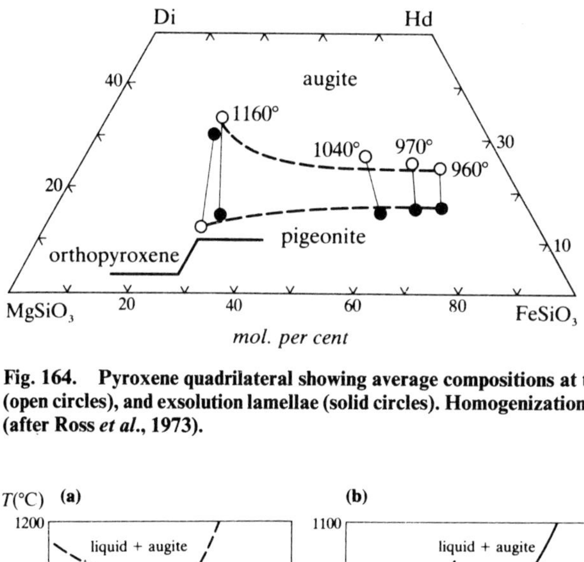

Fig. 164. Pyroxene quadrilateral showing average compositions at the solidus of host augite or pigeonite (open circles), and exsolution lamellae (solid circles). Homogenization temperatures are shown on the figure (after Ross *et al.*, 1973).

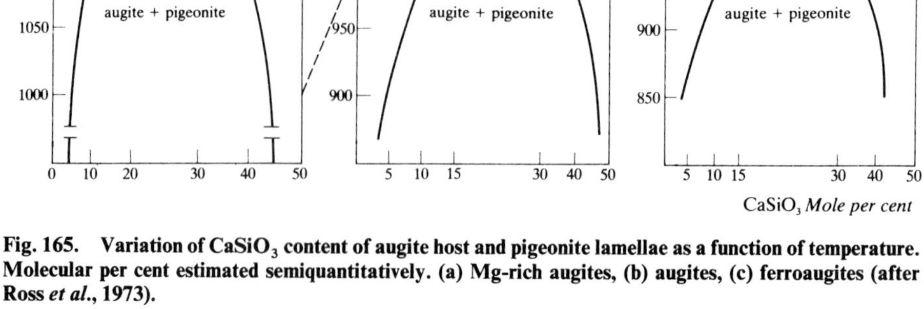

Fig. 165. Variation of $CaSiO_3$ content of augite host and pigeonite lamellae as a function of temperature. Molecular per cent estimated semiquantitatively. (a) Mg-rich augites, (b) augites, (c) ferroaugites (after Ross *et al.*, 1973).

due to higher f_{H_2O} and alkalis in terrestrial magmas. The melting temperatures observed in the investigation indicate that the cotectic crystallization of the two pyroxenes occurred over the temperature range 1175° ± 15° to 960° ± 15°C, in good agreement with temperatures obtained from whole-rock crystallization and melting studies.

The phase relations for the composition $CaAl_2SiO_6$, at temperatures between 900° and 1550°C and pressures varying from 9·5 to 25 kbar, have been determined by Hijikata and Yagi (1967). Clinopyroxene is the stable phase at temperatures above 1100°C and pressures greater than 11 kbar. At higher temperatures and pressures the pyroxene is replaced by grossular and corundum and the reaction is defined by the univariant curve $P(\text{kbars}) = 0{\cdot}028T(°\text{C}) - 17$. At temperatures greater than about 1400°C the clinopyroxene melts incongruently to corundum + liquid (1420°C at 13 kbar, 1450°C at 20 kbar), while anorthite–gehlenite–corundum is the stable assemblage at pressures less than 11 kbar.

In the pseudobinary system diopside–$CaTiAl_2O_6$ (Yagi and Onuma, 1967), the diopside solid solution is gradually enriched in TiO_2 as crystallization proceeds and this relationship is comparable with the variation of titanium-bearing augites in the early stages of the crystallization of alkali basalts, e.g. at Morotu and Iki (Yagi, 1953; Aoki, 1959, 1964). The solid solution of $CaTiAl_2O_6$ in diopside occurring at 1000°C and 1 atm pressure is greatly reduced at pressures between 10 and 25 kbar, and is at variance with Thompson's (1974) data based on pyroxenes crystallized from melts of a variety of natural basalts. In these more complex pyroxenes, Ti was found to increase with rising pressure at constant temperature showing that pyroxenes crystallized from basic magmas at depth may be expected to have comparatively low Ti contents due to their high temperature of formation. The observed difference between the pyroxenes of the simpler synthetic systems, and those consisting of a greater number of components formed from rock melts may be related to reactions such as:

$$CaTiAl_2O_6 + NaFe^{3+}Si_2O_6 \rightleftharpoons NaTiAlSiO_6 + CaFe^{3+}AlSiO_6$$

moving to the right as pressure increases.

During their study of the system $CaMgSi_2O_6$–$Ca_2MgSi_2O_7$–$CaTiAl_2O_6$, Onuma and Yagi (1971) showed that 11 weight per cent $CaTiAl_2O_6$, found to be incorporated in the diopside solid solution on the join $CaMgSi_2O_6$–$CaTiAl_2O_6$, may persist in the multicomponent system. Calculation of the constituent molecules of three natural titanaugites from undersaturated mafic alkalic rocks investigated by Onuma and Yagi show that the $CaTi(Al,Fe^{3+})_2O_6$ component may vary between 8·7 and 11·7 mol. per cent.

The equilibrium relationships of clinopyroxene with olivine, anorthite, tridymite and magnetite at 1 atm total pressure and oxygen partial pressures between 10^{-11} to $10^{-0{\cdot}7}$ atm in the system MgO–FeO–Fe_2O_3–$CaAl_2Si_2O_8$–SiO_2 have been investigated by Roedder and Osborn (1966).

The variations in solubility of both $CaAl_2SiO_6$ and $MgSiO_3$ in clinopyroxenes of the spinel lherzolite assemblage in the system CaO–MgO–Al_2O_3–SiO_2 with changes in temperature and pressure have been investigated experimentally by Herzberg and Chapman (1976). The $CaAl_2SiO_6$ content varies sympathetically with the $MgSiO_3$ content of the pyroxene and is largely temperature dependent. The solubility of both components decreases slightly with increasing pressure at constant temperature but, due to the parallel nature of the $CaAl_2SiO_6$ and $MgSiO_3$ isopleths of the clinopyroxenes in temperature–pressure space, equilibration pressures cannot be estimated (Fig. 166). The reaction relating the variation of the $CaAl_2SiO_6$

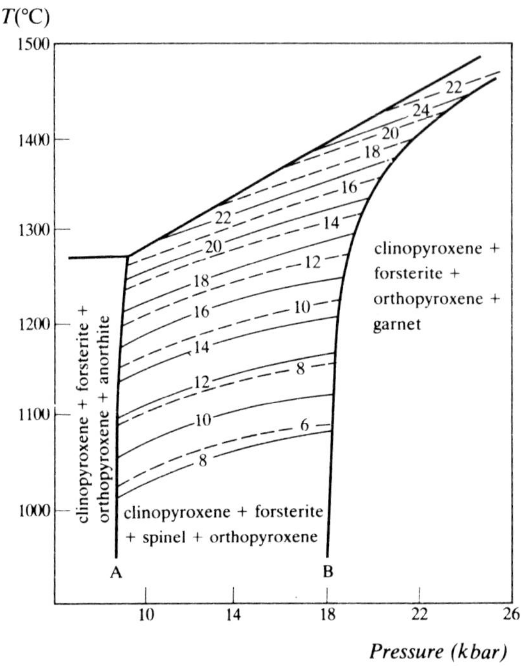

Fig. 166. Variations in the $CaAl_2SiO_6$ and $MgSiO_3$ contents of clinopyroxene with temperature and pressure in the spinel lherzolite assemblage of the system $CaO–MgO–Al_2O_3–SiO_2$ (after Herzberg and Chapman, 1976). Thin solid lines, contents of $CaAl_2SiO_6$ (weight per cent); thin broken lines, contents of $MgSiO_3$ (weight per cent). A. clinopyroxene–forsterite–anorthite–orthopyroxene–spinel univariant equilibrium. B. Clinopyroxene–forsterite–orthopyroxene–spinel–garnet equilibrium.

component of spinel-peridotite clinopyroxenes with pressure and temperature (MacGregor, 1965)

$$CaMgSi_2O_6 + MgAl_2O_4 \rightleftharpoons CaAl_2SiO_6 + Mg_2SiO_4$$

proceeds with a very small positive volume change ($\Delta V = +0{\cdot}96$ per cent), and is consistent with the small decrease in the $CaAl_2SiO_6$ component with increasing pressure at constant temperature. The experimental data, in conjunction with simple thermodynamic mixing models of the solid solution phase involved in the subsolidus exchange reactions, is used by Herzberg and Chapman to obtain two estimates of the temperature of equilibration of natural spinel lherzolites, the first based on the equilibrium exchange of $Mg_2Si_2O_6$ in clinopyroxene and orthopyroxene, and the second on the equilibrium exchange represented by the above reaction, both at 12 kbar. This method of estimating equilibration temperatures has been applied to spinel lherzolites from Alpine-type intrusions and as nodules in basalts. Temperatures estimated for the latter occurrence vary from the water-undersaturated peridotite solidus to the anhydrous solidus, indicating a close genetic relationship to the basalts or to those of the same fractionation series. Estimated temperatures for spinel lherzolites from Alpine-type intrusions which have not been affected by post intrusion metamorphism indicate solidus temperatures, followed by partial re-equilibration at subsolidus temperatures.

The phase relations of a plagioclase garnet pyroxenite layer of the Ronda peridotite massif, Spain, have been determined at pressures between 5 and 30 kbar, and temperatures of 950° to 1500°C (Obata and Dickey, 1976). The mineral assemblage of the original rock is stable over a broad *P–T* interval, and clinopyroxene, plagioclase and olivine all crystallize within 250°C of the liquidus at pressures up to 8 kbar. The clinopyroxene is exceptionally rich in aluminium (15·8 weight per cent Al_2O_3), and is the only liquidus phase at higher pressures. Three possible orgins of the layer are considered, partial fusion of peridotite at less than 8 kbar without subsequent fractional crystallization, partial fusion of dry peridotite followed by fractional crystallization of olivine from the liquid at a pressure less than 30 kbar, and fractional crystallization of a mafic or ultramafic magma at pressures between 23 and 30 kbar that gave rise to peridotite and garnet pyroxenite cumulates. Plagioclase lamellae in the clinopyroxenes occur in the mafic horizons of the massif and the feldspar may have formed as the result of subsolidus reactions from the jadeite component of the pyroxene during decomposition.

Melting experiments on mixtures of a Kilauea tholeiite at H_2O pressures less than total pressure (*P* 2–8 kbar) show that the augite is stable at 1100°C, the highest temperature used in the investigation (Holloway and Burnham, 1972). The compositions of the augites from eight runs at varying temperatures and pressures display no marked differences.

An investigation of the phase relations of the Picture Gorge tholeiite, the 1921 Kilauea olivine tholeiite, and the 1801 Hualalai alkali basalt at $P_{H_2O} = P_T = 5$ kbar, 680°–1000°C, and oxygen fugacities of quartz–fayalite–magnetite and hematite–magnetite buffers, showed that, in the temperature range 850° to 1000°C, augite forms part of an assemblage derived from the incongruent breakdown of hornblende, the stable phase at lower temperatures (Helz, 1973). The stability field of augite is restricted as oxygen fugacity increases, and the compositions of the pyroxenes become progressively more iron-rich as f_{O_2} decreases at a given temperature. The augites are richer in the $CaSiO_3$ component compared with those of anhydrous tholeiitic assemblages.

The stability fields of augitic pyroxenes in relation to the crystallization of plagioclase and olivine in tholeiites from the Mid-Atlantic Ridge have been determined experimentally at pressures and temperatures of 1–12 kbar and 1150°–1250°C respectively (Kushiro and Thompson, 1972).

Augites crystallized from a natural hawaiite in experimental runs with 2 per cent H_2O at 5 and 10 kbar, and 5 per cent H_2O as 10 kbar have $Mg/(Mg+Fe^{2+})$ ratios varying from 66·8 to 72·7 closely comparable with the range 67·2–78·9 for the natural megacryst and cumulate augite (Knutson and Green, 1975). The experimentally obtained augites and those of the hawaiite are both relatively rich in Al (3·5–6 weight per cent Al_2O_3) and indicate that the megacryst/cumulate assemblage of the hawaiite crystallized in the pressure range 6·5–8 kbar at temperatures between 1040°–1080°C, prior to rapid eruption to higher crustal levels.

Mixtures of Al-rich augite and spinel separated from a spinel clinopyroxenite xenolith in a tuff-filled volcanic vent have been investigated at 18 kbar and temperature between 1300° and 1500°C (Chapman, 1975). For compositions approximating to that of the xenolith the maximum amount of spinel entering the natural subsolidus augite at 1350°C is 6 weight per cent while at a temperature of 1290°C this composition intersects the spinel-augite solvus and the pyroxene would commence exsolving spinel.

Thompson (1974) has shown that augite or subcalcic augite is an important phase to crystallize in the melting intervals of alkali basalt, transitional basalt, olivine tholeiite, iron-rich tholeiitic andesite and augite leucitite at pressures and temperatures between 8 and 45 kbar and 1125° and 1475°C respectively. Except for a small pressure range in the case of the alkali basalt, in which the Ca-rich pyroxene is accompanied by a Ca-poor pyroxene, augite was found to crystallize either alone or in various associations with olivine, plagioclase or garnet. The $NaAlSi_2O_6$, $CaAl_2SiO_6$ and $Fe_2Si_2O_6$ components of augite crystallized near the liquidus, at the lowest and highest pressures, are shown in Fig. 167. The detailed variation is not linear, as indicated in the diagram, and fluctuations occur for each of the *P–T* sets of the different lava compositions. The liquidus curves for each specimen, except for a negative section at relatively low pressures in that part of the leucitite liquidus curve where leucite is the primary phase, have a positive slope on a *P–T* diagram, and the pyroxene compositions thus reflect the effects of both rising pressure and temperature. The jadeitic and Ca-Tschermak's components increase and the content of ferrosilite decreases (as also do the enstatite and wollastonite components except for the latter in the case of the tholeiitic andesite) with increasing pressure. Nevertheless the bulk composition of the lava is the predominating factor affecting the composition of the pyroxenes. Salitic augites and subcalcic augites, as products of melting experiments on olivine tholeiite and tholeiitic andesite of the Snake River Plain, Idaho, have been reported by Thompson (1975).

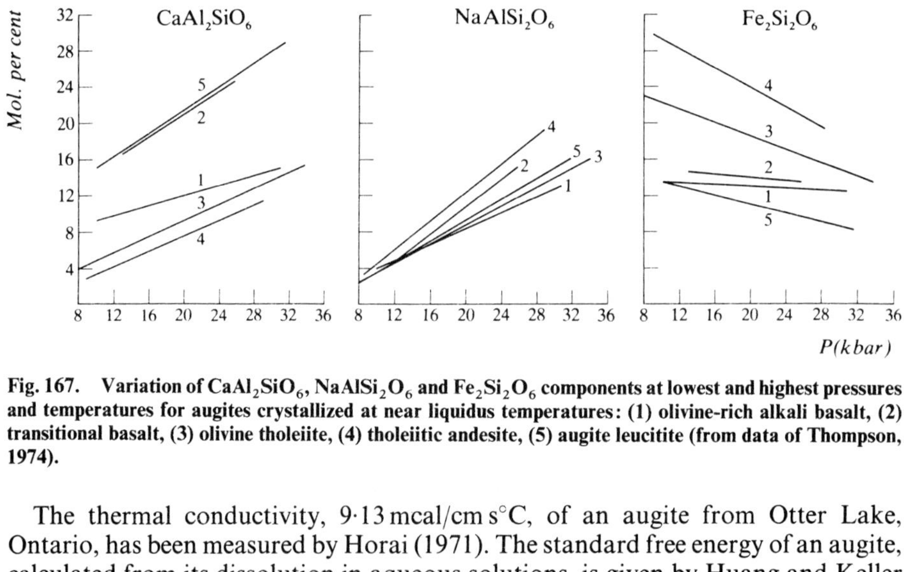

Fig. 167. Variation of $CaAl_2SiO_6$, $NaAlSi_2O_6$ and $Fe_2Si_2O_6$ components at lowest and highest pressures and temperatures for augites crystallized at near liquidus temperatures: (1) olivine-rich alkali basalt, (2) transitional basalt, (3) olivine tholeiite, (4) tholeiitic andesite, (5) augite leucitite (from data of Thompson, 1974).

The thermal conductivity, 9·13 mcal/cm s°C, of an augite from Otter Lake, Ontario, has been measured by Horai (1971). The standard free energy of an augite, calculated from its dissolution in aqueous solutions, is given by Huang and Keller (1972). The specific magnetic susceptibilities of a series of Ca-rich clinopyroxenes from the Skaergaard intrusion have been determined by Vernon (1961), and show a linear relationship with total Fe (as FeO) and MnO.

Alteration. Augite is frequently altered to a uralitic amphibole, which may form as a single crystal, but which occurs more commonly as an aggregate of small prismatic crystals. The alteration generally begins around the periphery or along the cleavages, and in the early stages the alteration may be accompanied by an irregular bleaching of the crystal resulting in the formation of patchy areas of colourless pyroxene flecked with small plates of amphibole. The bleached areas usually appear to have the same refractive indices and optic axial angle, and to be in optical continuity with the unbleached pyroxene. Chlorite is also a common alteration product of augite (see Fawcett, 1965); less common products include epidote and carbonates.

Optical and Physical Properties

The relationship between the optical properties and chemical composition of the clinopyroxenes of the diopside–hedenbergite–clinoenstatite–clinoferrosilite system has been investigated by a number of authors (Tomita, 1934; Winchell, 1935; Deer and Wager, 1938; Hess, 1949; Muir, 1951; Hori, 1954; Henriques, 1958). Winchell (1961, 1963), Benson (1944), Brown (1956, 1957), Norton and Clavan (1959) and Carmichael (1960) have presented curves relating the optical properties and composition of augites from the same intrusion or from related rocks for restricted parts of the system. Although the essential chemical variation of these pyroxenes can be expressed as atomic percentages of Ca, Mg and Fe^{2+}, other metal atoms are generally present, and the refractive index and optic axial angle curves (Fig. 168)

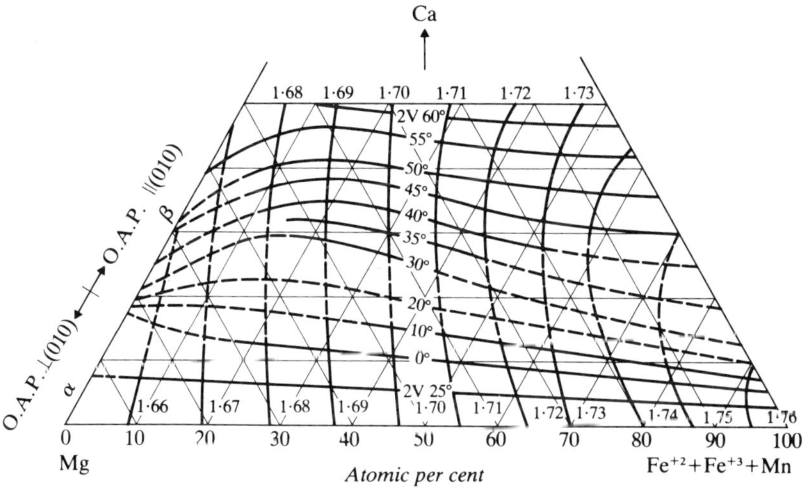

Fig. 168. Variation of the optical properties with chemical composition of clinopyroxenes included in the system $CaMgSi_2O_6$–$CaFeSi_2O_6$–$Mg_2Si_2O_6$–$Fe_2Si_2O_6$ (after Hess, 1949; Muir, 1951).

drawn by Hess (1949) and later slightly modified by Muir (1951) are based on the assumption that these other constituents, in terms of their oxides, are present in approximately the following amounts: Al_2O_3 3 per cent, Fe_2O_3 1·5 per cent, Na_2O 0·4 per cent, TiO_2 0·4 per cent, MnO 0·3 per cent but higher in the more iron-rich pyroxenes, Cr_2O_3 1·1 per cent in the magnesium-rich augites of igneous rocks but negligible in minerals more iron-rich than $(Ca,Mg)_{87}Fe_{13}$. In using the values of $2V_\gamma$ and the β refractive index to estimate the chemical composition it is necessary not only to consider the effects of minor constituents present in amounts differing appreciably from those quoted above, but also to take into account the effects of the exsolution of orthopyroxene or pigeonite. Moreover the effect of a given cation on the optical properties is dependent upon its position in the pyroxene structure (Hori, 1954). Thus Al and Fe^{3+} in tetrahedral co-ordination increase the optic axial angle and lower the β refractive index, whereas in octahedral co-ordination the same ions have the opposite effect. Because of all these factors compositions estimated from the optic axial angle and β index may differ by as much as 5 per cent of the Ca, Mg or Fe content from their true values. Brown (1957) has suggested that variable optic axial angles sometimes observed in a single crystal, and often regarded as zoning (i.e. variation mainly in the Fe:Mg ratio) are probably due to irregular Ca distribution which may result from the exsolution not being uniform throughout the grain.

In Fig. 168 the variation of the β index is plotted for pyroxenes in which the optic axial plane is parallel to (010), and the α index for those in which the optic axial plane is perpendicular to (010). As the β vibration direction is parallel to y for the former pyroxenes and the α vibration direction is parallel to y for the latter, and α and β are coincident at zero 2V, the α and β indices may be combined to construct a continuous set of smooth curves. Moreover the refractive index for the vibration along the symmetry axis is the index most easily and accurately measured for pyroxenes of either orientation.

Crushed fragments of the pyroxenes include flakes lying parallel to the prismatic cleavage $\{110\}$ and a smaller number of tablets parallel to the $\{100\}$ parting. The latter are easy to recognize by their low birefringence since an optic axis emerges from them at about 20° from the vertical, and if bounded by the prismatic cleavage or $\{010\}$ parting planes such fragments exhibit straight extinction. The orientation of these tablets can be checked by observing the interference figure which, if the fragment is parallel to the (100) plane, will show an off-centre optic axis figure. If on rotating the stage until the isogyre is east-west the brush divides the field exactly, then the optic axial plane is normal to the section and the β vibration direction north–south.

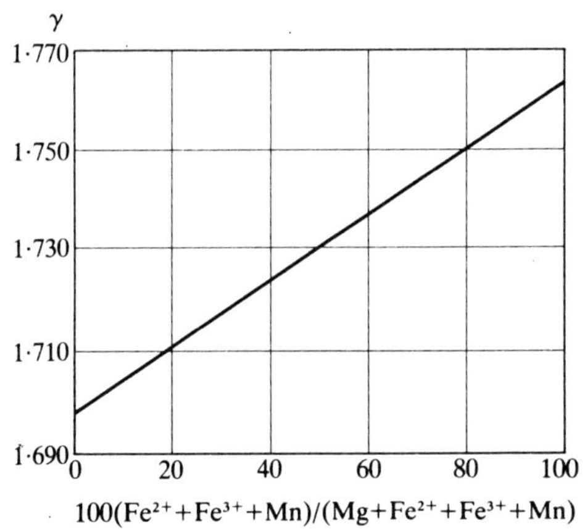

Fig. 169. Variation of γ index with ratio $100(Fe^{2+}+Fe^{3+}+Mn)/(Mg+Fe^{2+}+Fe^{3+}+Mn)$ (after Jaffe *et al.*, 1975).

Measurement of the γ index as a method of determining the iron/magnesium ratio has been shown (Jaffe *et al.*, 1975) to have a comparable accuracy with that of the β index. The variation of the γ index with $100(Fe^{2+}+Fe^{3+}+Mn)/(Mg+Fe^{2+}+Fe^{3+}+Mn)$ (Fig. 169) is given by the equation

$$\gamma = 1{\cdot}6977+0{\cdot}0669[(Fe^{2+}+Fe^{3+}+Mn)/(Mg+Fe^{2+}+Fe^{3+}+Mn)]$$

Regression equations for the optical properties α, β, γ and $2V_\gamma$ as a function of the chemical composition of clinopyroxenes have been calculated by Hori (1954) and Henriques (1958). Hori postulated equations of the type:

$$2V = a_{2V}+\Sigma a_{i2V} N_i$$

where a_{2V} is a constant, a_{i2V} represents the effects of the cation of type i on the optic axial angle, and N_i is the number of the cations in a definite volume. Better agreement between calculated and observed values is obtained (Henriques, 1958) from equations of the form:

$$2V = a_{2V}+\Sigma a_{i2V} N_i+\Sigma a'_{i2V} N_i^2$$

More recently Winchell (1961) calculated the regression coefficients of α, β, γ, $\gamma:z$ and density on fifteen compositional variables of the clinopyroxenes, (K,Na,Ca, Mn,Fe^{2+},Mg), (Mg,Fe^{2+},Mn,Ti,Fe^{3+},Al)(Fe^{3+},AlSi)$_2O_6$ by least squares from the data of 155 pyroxenes. Subsequently Winchell (1963) suggested that the pyroxene trapezoidal diagram is more precisely considered as a projection of one end and one side of a trigonal prism, as in the clinopyroxene formula the *X* and *Y* cations are normally Ca- and Mg-sized respectively, and correspondingly have different effects on the optical and physical properties; thus the composition of clinohypersthene should be expressed as $FeMgSi_2O_6$ and not $MgFeSi_2O_6$. In consequence small discontinuities are likely as the contours cross from the triangular, (Ca,Fe,Mg)$MgSi_2O_6$, to the parallelogram-shaped, (Ca,Fe)(Mg,Fe)Si_2O_6, field (Fig. 170). A linear relationship between the β refractive index and the β angle in the Skaergaard pyroxenes has been demonstrated by Layton (1965). The relationship between the β index and the *b* and *c* cell parameters, however, shows discontinuities, and a plot of the β index–*c* parameter indicates the occurrence of three distinct series that are coincident with changes related to the onset or cessation of crystallization of the associated phases.

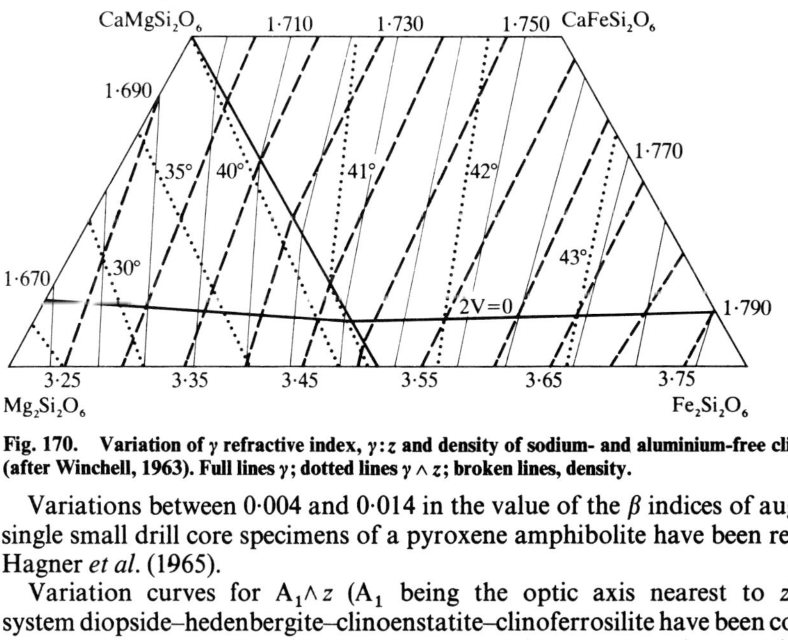

Fig. 170. Variation of γ refractive index, $\gamma:z$ and density of sodium- and aluminium-free clinopyroxenes (after Winchell, 1963). Full lines γ; dotted lines $\gamma \wedge z$; broken lines, density.

Variations between 0·004 and 0·014 in the value of the β indices of augites from single small drill core specimens of a pyroxene amphibolite have been reported by Hagner *et al.* (1965).

Variation curves for $A_1 \wedge z$ (A_1 being the optic axis nearest to *z*), for the system diopside–hedenbergite–clinoenstatite–clinoferrosilite have been constructed for pyroxenes with Fe/Mg < 1 (Ruegg, 1964). The curves trend across those of the optic axial angle, and allow the composition to be determined solely from universal-stage measurements. For more iron-rich pyroxenes the $A_1 \wedge z$ and 2V curves are nearly parallel and this procedure is not applicable.

The measurement of extinction angles in monoclinic pyroxenes has been discussed by Haff (1941), Turner (1942), Hess (1949) and Norton and Clavan (1959). Reasonable accuracy is not usually attained unless the measurements are made on twinned crystals. Correlation between extinction angles and chemical composition is not precise.

The more magnesium-rich augites are usually either non-pleochroic or only weakly so. Increasing tenor of iron, however, is usually accompanied by the development, predominantly in pale green colours, of perceptible pleochroism.

Titanium has a marked effect on the pleochroism and titanaugites are pleochroic in purplish tints. In some titanaugites the absorption indicatrix does not coincide with the optical indicatrix, and in such cases the pleochroism is more easily related to the directions of the *y* and *z* axes and the direction at right angles to the *yz* plane.

The polarized spectra of a (001) and of a cleavage section of titanaugite containing 6 and 2·8 atomic per cent total Fe and Ti respectively has been determined by Manning and Nickel (1969). The absorption bands are assigned to *d–d* electronic transitions in Fe^{2+}, Fe^{3+} and Ti^{3+}, and to $Fe^{2+} \rightarrow Fe^{3+}$ charge transfer; the pleochroism is considered to result mainly from the second process. The electronic transitions, except those in Fe^{2+}, are observed in the visible region and account for the colour of the titanaugite.

Augites generally show weak to moderate dispersion, but in titanaugite strong inclined dispersion of the optic axes is present. A detailed investigation of the dispersion in titanaugite (e.g. Table 36, anals, 3, 5) has been made by Holzner (1934). The refractive indices of the titanaugite from the Stöffel essexite porphyry are α 1·715, γ 1·738 (639 μm); α 1·726, γ 1·749 (539 μm).

Few data are available for augite relating the variation of density with chemical composition. In view of the difficulty associated with the accurate determination of density, optical methods of estimating chemical composition are considerably more reliable.

Augite is commonly twinned on {100}, either as simple or multiple twins, and less frequently shows multiple twinning on {001}. Interpenetration twins with twinning plane parallel to ($\bar{1}$22) and interpenetration or cruciform twins with (101) as twinning plane have been described by Gubbaiah and Varadarajan (1962). The secondary twinning, on {100}, of an augite in a dolerite dyke has been described by Prasad and Naidu (1971). The diagnosis of the secondary nature of the twinning is based on the lack of a regular relationship between the distribution of the twin lamellae and crystal morphology, and on the presence of irregular, bent and twisted lamellae. An unusual multiple twin aggregate, consisting of six individuals, twinned alternately on (100) and (110), has been reported by Maske (1966).

Zoning is a relatively common feature of augite and is particularly marked in the titaniferous pyroxenes and titanaugites of the more alkaline basic rocks. Thus the titanaugites (Table 36, anal. 10) of the alkaline rocks of Tahiti (McBirney and Aoki, 1968), characterized by high contents of Al, Fe^{3+}, Ti and Na, are zoned from a colourless core to a reddish brown or green rim. The colour zoning is accompanied by an increase in dispersion, $r > v$, and a decrease in optic axial angle, and is associated with enrichment in Ti towards the peripheral areas of the crystals. The greatest concentration of Na occurs in the outermost rims and leads to a reversal of the decreasing values of the optic axial angle, e.g. 2V 55° at the core, 45° close to the margin and > 66° at the edge of the crystal.

Oscillatory zoning between endiopside and titanaugite in the pyroxenes of the Blue Mountain complex, New Zealand, has been described by Grapes (1975). The contents of Al, Ti, Fe and Ca increase and those of Si and Mg decrease in each successive titaniferous zone from the core to the margin. An unusual type of zoning has been described in some ferroaugites in the Circle Creek rhyolite, Nevada (Coats, 1968). The majority of the crystals are unzoned but a small proportion show irregular cores with γ brownish green and $\gamma{:}z$ 39° in contrast to γ grey green and $\gamma{:}z$ 46° of the unzoned crystals. The core pyroxenes display exsolution lamellae, parallel to (001), that extend to the outer rim, and which, on the basis of their optical properties, probably consist of subcalcic ferroaugite $\simeq Ca_{20-30}Mg_{20-30}Fe_{50}$ in composition.

Three types of zoning are present in the clinopyroxenes of the ijolite and nepheline monzonite of the monzonitic complex, Mount Dromedary, New South Wales (Boesen, 1964). One is defined by inclusions of plagioclase, ilmenite-magnetite, apatite, biotite and hornblende alteration products, another by relatively broad zones showing small variations in colour, birefringence and optic orientation, and a third type consisting of an alternation of very narrow, closely spaced, light (containing less Al and Fe) and dark striae of more diopsidic composition, parallel to the crystal margins, resembling the fine-scale oscillatory zoning of plagioclase phenocrysts in andesite.

Diopside–hedenbergites, augites and titanaugites, particularly the latter, display sector zoning (hourglass zoning). This commonly takes the form of four triangular segments, the apices of which are directed to the mid-point of the grain. The crystal can thus be considered to consist of a number of pyramids, the apices of which are coincident at the centre of the crystal, and the crystal faces forming the bases of the pyramids. In Fig. 171, showing the forms $\{100\}$, $\{010\}$, $\{110\}$ and $\{\bar{1}11\}$, the sets of pyramids are designated the (100), (010), (110) and $(\bar{1}11)$ sectors. The opposite zone sectors are chemically and optically identical; adjacent sectors are distinguished by differences in colour, birefringence and dispersion. The chemical differences between sectors are probably produced on the surfaces of crystal faces growing simultaneously under the same conditions of composition, pressure and temperature. The mechanism of the growth is complex and probably involves a number of different factors. The regularity of sector zoning is variable and the growth surfaces may be irregular and the zoning patchy (Frontispiece, Fig. F).

In addition to the hourglass structure, titanaugites may also show concentric zoning, distinguished by changes in colour intensity and extinction angle. The concentric zoning is present throughout all sectors of the hourglass zones, may be normal or oscillatory, and appears to have been produced in the same way as the zoning in plagioclase, and thus represents time boundaries within the crystal. A review of the hourglass zoning in clinopyroxenes is given by Strong (1969) who distinguishes the 'swallow-tailed' or 'quench hourglass' form, common in augite of the fine-grained groundmass of many igneous rock types, from the 'hourglass' structure of the larger augite crystals of porphyritic and coarse-grained rocks.

Dowty (1976a, b) has developed a model, relating crystal structure and morphology, that illustrates the influence of internal structure on crystal growth, particularly in relation to the preferential nucleation of new layers on one rather than on other planes. The model, based on computing the fraction of the total bond energy on a growth layer attaching it to the substratum, and on predicting the atomic configuration on the surface of each face, applied to diopside (or augite) ranks the forms in the order $\{110\}$, $\{010\}$, $\{11\bar{1}\}$, $\{100\}$, the most important forms occurring in augite.

Euhedral phenocrysts of complexly zoned augite, consisting of a colourless core with an irregular corroded outline and showing no significant radial or sector zoning, followed by a brownish-green mantle containing many paler and darker oscillatory zones on which well developed sector zoning is superimposed, and finally surrounded by a thin discontinuous lemon yellow outer rim, comparable with the groundmass pyroxene, have been described from a Vesuvius lava (Kushiro and Thompson, 1972). The core–mantle and oscillatory zoning are optically sharp and reflect chemical discontinuities (core 0·05 and 0·005, paler zones of mantle 0·093 and 0·013, darker mantle zones 0·432 and 0·060 Al and Ti respectively on the basis of six oxygens). The resorption of the cores is presumably related to an abrupt upwards movement of the magma resulting in the temperature rising temporarily above the

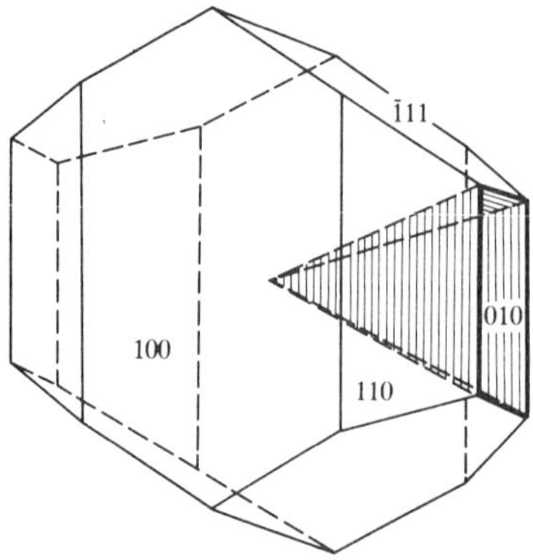

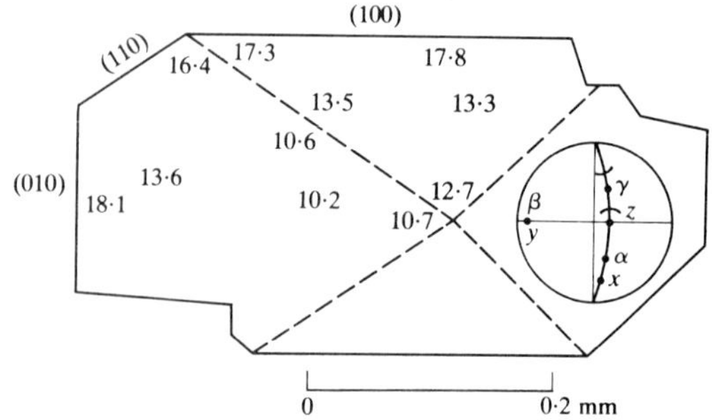

Fig. 171. Clinopyroxene showing forms {100}, {010}, {110} and [$\bar{1}$11}. Volume of (010) sector shown diagrammatically (after Hollister and Gancarz, 1971).

Fig. 172. Sector zoned augite. Section approximately normal to *z*, showing variation of total Fe as FeO (after Nakamura, 1973).

pyroxene liquidus at low pressures, prior to further crystallization during rapid cooling at higher crustal levels. The oscillatory and sector zoning formed prior to the eruption of the magma, and the eruption was marked by a large compositional break between the outermost part of the green augite phenocryst mantles and the discontinuous rim of the lemon yellow pyroxene that surrounds them and occurs in the groundmass of the lava. The strong euhedral form of the oscillatory zoning in the sector zoned green mantle of the Vesuvius pyroxene contrasts markedly with the suggestion, put forward to explain some forms of sector zoning, that 'hourglass' zoning is due to later infilling of pyramidal voids formed at the ends of rapidly growing prismatic crystals. The zoning here is more plausibly explained on the basis of a disequilibrium crystallization model in which precipitation at rapidly growing faces is controlled by varying diffusion rates in the liquid. Radially and sector zoned augites, in which the darker coloured zones are enriched in Al, Ti (and Fe), have also been described from the alkali basalt dykes of Fuerteventura, Canary Islands (Lopez Ruiz, 1970).

Well developed sector and oscillatory zoning is displayed by the augites (mean composition $Wo_{46}En_{40}Fs_{14}$) in the hawaiites and mugearites of Mt. Etna, Italy (Downes, 1974). The main chemical variation occurs between the ($\bar{1}$10) and the other sectors. The crystals are elongated parallel to the *z*-axis and indicate growth rates $\|z > \|y > \|x$, and the compositional variation can be accounted for by differences in growth parallel and perpendicular to the *z*-axis. The oscillatory zones occur on a scale of $< 20\,\mu m$, so that rates of diffusion of ions and polymeric complexes in the melt to the crystal/liquid interface, as well as growth rates were probably contributing factors during their formation. The principal compositional variations are shown by Ti and Al, and both cations vary by as much as 39 per cent, synchronous surfaces within each sector having approximately the same Ti and Al contents and the same Ti/Al ratio between sectors. The alternation of Al-rich and Al-poor zones is considered to relate to the tendency of the slower growing (100) or (010) faces to take more Al than the faster growing ($\bar{1}$11) face for which time for equilibration of the liquid with respect to Al distribution at the interface would be shorter. Such an explanation is consistent with the small width of the oscillations as the initial relatively high incorporation of Al would be followed by depletion as the growth rate increased before being succeeded by a temporary cessation of growth marked by a sharp boundary, and a return to higher Al values. Such a cyclic development might be expected with growth rates intermediate between those of very rapid cooling and those sufficiently slow to maintain equilibrium, together with supersaturation at the liquid/crystal interface immediately preceding the growth of each oscillation.

The mechanism of sector zoning in clinopyroxenes has been discussed by Hollister and Gancarz (1971), based on a study of the diopside–salites of the Narce area, Italy (see diopside section, p. 256). Four factors are considered to be involved in the development of the sector zoning; size and composition of ionic complexes added during crystal growth, the rate at which the material is added, rate of equilibration of the new blocks of material with the matrix at the surface of dislocation steps, and the rate of re-equilibration of surface layers with the matrix by ion exchange perpendicular to the crystal faces.

Sector zoning (Fig. 172) in the augites of the veins and pegmatoids of the tholeiitic dolerite of Tawhiroko, New Zealand, have been described by Nakamura and Coombs (1973). Relative to the {010} and {110}, the {100} sectors are enriched in Mg, Fe and Mn, and the Ca/(Ca + Mg + Fe) ratio is approximately 0·31 in {100} compared with 0·38 in the {110} and {010} sectors. The origin of the zoning is considered (Nakamura, 1973) in terms of partially formed structural sites, so-called *protosites*. In these, in contrast to the structural sites, the second and commonly the first co-ordination sphere of neighbours is incomplete, a feature that permits greater

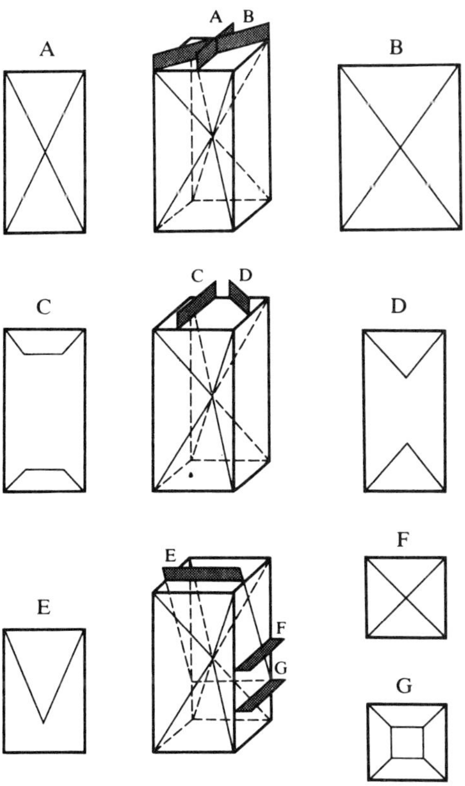

Fig. 173. Diagrammatic representation showing geometrical relation between growth sectors according to the nature of their cross-sections relative to idealized morphological model (after Leung, 1974).

flexibility in the accommodation of those atoms not ideally suited to the completed structural sites.

Leung (1974) has made a detailed study of the zoning in titanaugites of alkali-rich rocks. The crystals show one to four colourless and light brown, moderately pleochroic sectors in which {100}, {110} and {010} are the coloured sectors. Basal sections, generally displaying ultra-brown and ultra-blue interference colours, show light brown trapezoidal (prism) sectors surrounding a pseudo-octagonal colourless core, i.e. $(11\bar{1})$ sectors. The crystals consist of a series of growth layers, parallel to the crystal faces, and the geometric configuration of the sectors is closely related to the distance of the section from the centre of the crystal and the direction in which the crystal is viewed (Fig. 173). The optical properties vary considerably within a single crystal, e.g. $2V_\gamma$ ranging from 53° to 67° from core to rim, and extinction angles, $\gamma:z$, consistently 4°–8° smaller in the $\{11\bar{1}\}$ compared with the {100} sector. Chemical zoning is of two types, core to rim and sector zoning. Depletion of Al (up to 5 per cent Al_2O_3), enrichment in Fe and impoverishment in Mg (1–2 per cent MgO), and a decrease of 0·5–1·0 per cent TiO_2, commonly occurs from core to rim. The prism sectors, {100}, {010} and {010} contain more Al, Ti and Fe, and the $\{11\bar{1}\}$ sectors have higher contents of Si and Al, while the distribution pattern of Ca does not vary significantly in either the core to rim or sector zoning. The development of the latter is discussed by Leung with reference to the pyroxene structure and in particular to the distribution of the *M* sites. The sectoral fractionation of the elements is explained in terms of constitutional supercooling and diffusion during rapid disequilibrium crystallization.

The intratelluric and quench clinopyroxenes of the alkali basaltic rocks of the South Highlands, New South Wales (Wass, 1973), both display hourglass zoning, the more strongly coloured sectors of which are richer in Ti and Al and lower in Si than the paler coloured zones (Fig. 174). Many of the phenocrysts also have concentric zones which, except for Fe/Mg substitution amounting to some 0·5 per cent FeO, have identical chemical characteristics with the sector in which they are located. The crystals are bordered by a narrow, strongly pleochroic, rim, also rich in Ti and Al but distinguished by a higher Fe content which is chemically and optically comparable with the groundmass clinopyroxene.

An hour-glass structure in which the surface between the (010) and (100) sectors takes the form of a parabolic cylinder occurs in the titanaugite phenocrysts of a fourchite sill, Ste. Dorothée, Quebec. The zoning in relation to the growth kinetics of the controlling processes, e.g. rate of phase boundary reactions, surface nucleation and diffusion on adjacent faces of the crystal, is discussed by Gray (1971).

An unusual hourglass structure (Fig. 175) showing compositional and exsolution variations, in a common augite from a dolerite pegmatite, Doraville, Northern Ireland, has been described by Preston (1966b). The augite displays two distinct domains, one containing exsolution lamellae parallel to (001), the other incipient unmixing parallel to (100). The domains differ in composition, those showing exsolution being on average 2·85 atomic per cent richer in Ca, and 2·41 and 0·43 per cent poorer respectively in Fe and Mg, preferential incorporation of calcium occurring in the *x*, and of iron in the *y*, direction. Augites showing sector zoning, from the water quenched submarine basalt of the South Atlantic and West Atlantic margin have been reported by Bryan (1972).

Strong compositional zoning, averaging $Ca_{37}Mg_{24}Fe_{39}$ to $Ca_9Mg_{26}Fe_{65}$, in the pyroxenes of lunar basalts has been reported by Bailey *et al.* (1970). The clinopyroxenes of other lunar rocks may show even more extreme variation in chemistry with composition varying from augite and subcalcic augite to ferroaugite

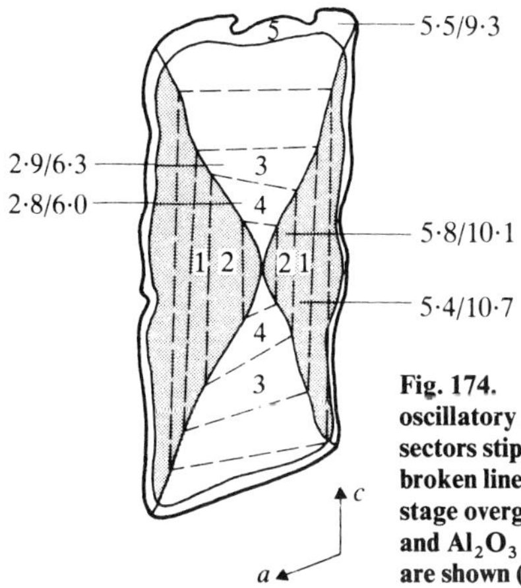

Fig. 174. **Hourglass and concentric normal and oscillatory zoning in titanaugite. Deeply coloured sectors stippled, concentric zoning indicated by broken lines. Outermost concentric rim is a late stage overgrowth. TiO_2 (first figure in each pair) and Al_2O_3 contents (weight per cent) of some zones are shown (after Wass, 1973).**

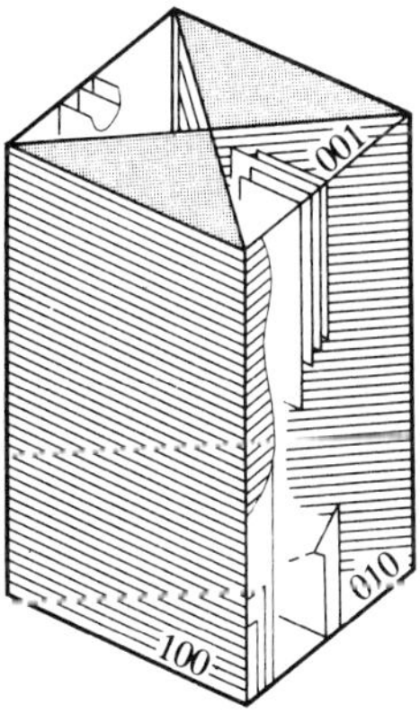

Fig. 175. **Ideal pattern of hourglass structure and exsolution features in common augite showing basal lamellae throughout (100) sectors, and partial development of lamellae in (010) sectors (after Preston, 1966b).**

and subcalcic ferroaugite, with single grains displaying different Ca/Mg and Mg/Fe ratios. The majority of the compositions lie within the range $Ca_{40}Mg_{43}Fe_{17}$ to $Ca_{25}Mg_5Fe_{70}$, most of which is usually present within systematically zoned crystals (Dence *et al.*, 1970). Essene *et al.* (1970) report single pyroxenes with compositional variation from subcalcic augite to ferrohedenbergite and in some examples to pyroxferroite $(Ca_{0\cdot15}Fe_{0\cdot85})SiO_3$. The pyroxenes show a systematic increase in Mn (0·15–0·75 weight per cent MnO) and decrease in Al (6·0–1·0 Al_2O_3). Ti (2·5–0·5 TiO_2) and Cr (0·6–< 0·1 Cr_2O_3) as the degree of iron enrichment increases. The Al-rich cores may be high-pressure xenocrysts brought up by the magma that subsequently formed nuclei for the low pressure and extreme local fractional crystallization. Compositional variations of the pyroxenes of other lunar rocks show a much greater replacement of Ca by Fe, changing initially from diopsidic augite to subcalcic augite, and later from subcalcic augite to subcalcic ferroaugite. The early replacement of Ca by Fe in *M*2 positions in these pyroxenes may be related to the depletion of Ca in the melt as a result of early and continuing crystallization of anorthite-rich plagioclase (Ross *et al.*, 1970).

Zoned augite–subcalcic ferroaugites (Fig. 176) are the most abundant constituents of the Apollo 11 lunar dolerites and microgabbros (Kushiro and Nakamura, 1970). The augites are associated with pigeonite, and the two pyroxenes

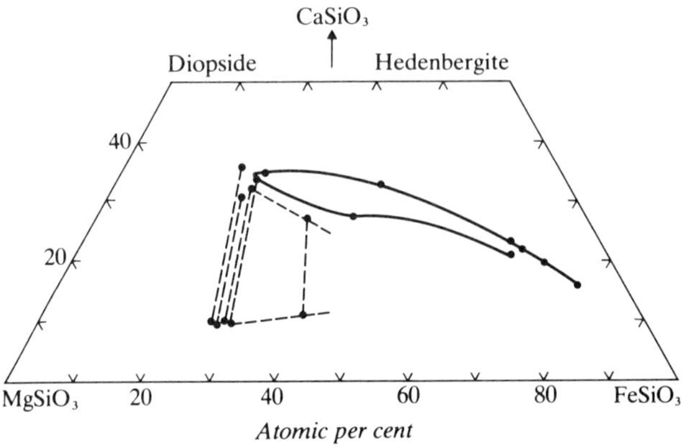

Fig. 176. Variation of Ca, Mg and Fe in pyroxene of lunar Apollo 11 microgabbro (after Kushiro and Nakamura, 1970). Solid lines, continuous zoning within a single crystal; long broken lines, tie-lines between augite and pigeonite; short broken lines, possible solvus between augite and pigeonite.

are commonly present in irregular intergrowths. The greater part of the augites has a high content of titanium (2–3 weight per cent TiO_2) and, as in the pyroxenes of the lunar basalts, the zoning is continuous, with Ti, Al and Cr decreasing and Mn increasing progressively from core to margin. The partition coefficients, K_D Mg–Fe, of the coexisting augites and pigeonites (average 0·96) are greater than those of the Ca-rich and Ca-poor pyroxenes of both the Skaergaard and Bushveld complexes, but comparable with those of the basaltic andesite of the Hakone volcano. The compositional break between augite and pigeonite is smaller than is usual in terrestrial plutonic rocks and this feature together with the presence of subcalcic augite, indicates that the miscibility gap of the lunar pyroxenes may represent a metastable solvus resulting from rapid crystallization.

Phenocrysts consisting of augite and pigeonite have been described from the porphyritic lunar basalts of *Oceanus Procellarum* (Bence, *et al.*, 1971). The phenocrysts have a central core, consisting of a granular aggregate of plagioclase, Fe-rich clinopyroxene and ilmenite, surrounded by a composite of augite and pigeonite in crystallographic continuity. The composite consists of an inner augite

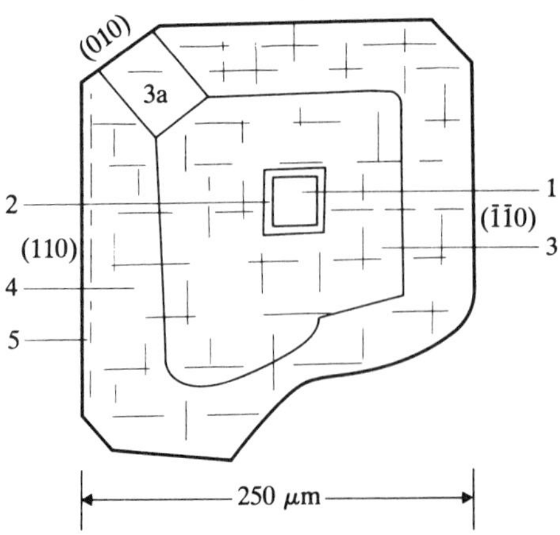

Fig. 177. Composite clinopyroxene phenocryst, viewed along the *z* axis, in lava from *Oceanus Procellarum* (after Bence *et al.*, 1971). (1) 'hollow' core; (2) 'inner' augite; (3) thick shell of pigeonite; (3a) pigeonite channelway parallel to *y*; (4) mantle of Al-rich augite; (5) discontinuous zone of Fe-rich and Al-poor augite.

shell, followed successively by pigeonite, a channelway parallel to *y* and optically continuous with the earlier pigeonite shell, a mantle of Al-rich augite, and finally a discontinuous zone of Fe-rich and Al-poor augite (Fig. 177).

Similar complex pyroxene zoning occurs in the Apollo 12 pyroxene-rich basalt, 12021. The phenocrysts consist of a core of pigeonite mantled epitaxially by augite, $Ca_{32}Mg_{46}Fe_{21}$, which is itself separated from a subcalcic ferroaugite, $Ca_{23}Mg_{12}Fe_{65}$, zone by a band of pigeonite more Fe-rich than the Ca-poor pyroxene of the core. The ferroaugite of the groundmass has a similar composition to that in the phenocrysts and here is intergrown with pyroxferroite. The zoning differs from one crystallographic plane to another, and distinct trends characterize growth on (100), (110) and (010), and parallel to the *c*-axis. Well defined but erratic oscillatory zoning is present in each sector, with compositional variations up to 10 mol. per cent $FeSiO_3$, superimposed on the normal zoning. Possible explanations of the cyclic crystallization of pigeonite and augite and the pyroxene zoning put forward by Boyd and Smith (1971) include a peritectic reaction between the more Ca-rich and Ca-poor pyroxenes, and turbulent flow of the lava.

Ca-poor augite and subcalcic augite is present in the Apollo 15 basalts, and is rimmed by ferropigeonite and subcalcic ferroaugite. Late-stage pyroxenes display diverse trends, some towards ferrosilite and the formation of pyroxferroite, some towards hedenbergite, both trends probably resulting from non-equilibrium and rapid crystallization (Kushiro, 1973). Microstructures in zoned lunar pyroxenes, due to compositional discontinuities that have occurred during continuous crystallization, exsolution lamellae of augite and pigeonite, and the development of 'denuded' zones (zones denuded of fine scale, late, exsolution adjacent to a set of coarse, high temperature, lamellae that is at least as wide as the exsolution itself), and anti-phase domains in pigeonite have been described by Lally *et al.* (1975), and are interpreted with the aid of a possible pseudobinary phase diagram at constant 2:1 En/Fs ratio (see Fig. 93, p.178).

Habit. Clinopyroxene usually crystallizes from basic magmas later than olivine, and in consequence skeletal forms are less commonly developed than with the orthosilicate. Pyroxenes with a skeletal habit, however, occur in some ultrabasic rocks, and have been described from rocks of the Archaean greenstone belt of Western Australia (Nesbitt, 1971). The augites vary in composition between Ca_{35-44}, Mg_{40-51}, Fe_{14-17}, and have a high content of aluminium (5·0–6·5 weight per cent Al_2O_3). The columnar growth of dendritic augite, $Ca_{39}Mg_{49}Fe_{12}$, up to 20 cm in length, occurs on the internal contact of a dolerite intrusion, Donegal (Preston, 1966a). The *z* crystallographic axis is oriented normal to the contact, and the main stem of the dendrite is inclined at approximately 30° to the *z* axis, initial growth probably following the trace of the (031) plane.

Feathery dendritic, spherulitic and acicular habits (spinifex texture), due to rapid cooling, are sometimes developed in the augite of basalts and dolerites. An account of these textures in blade megacrysts of the ultramafic lavas of Munro Township, Ontario is given by Fleet and MacRae (1975). In these megacrysts the acicular and skeletal augites, $\simeq Ca_{38}Mg_{45}Fe_{17}$, occupy a large part of the volume between crystals of clinochlore. The grain size of the analysed crystals varies from 1·0 × 0·12 mm to 0·5 × 0·05 mm, and with decreasing grain size the acicular augites successively develop fan and feathery spherulitic habit. A general discussion of the habit of clinopyroxenes in relation to crystal structure, and a detailed consideration of dendritic and other features associated with rapid crystallization is given by Fleet (1975).

An account of the epitaxial or parallel growth of augite and orthopyroxene has been described from the picritic dykes of Lochinver, Scotland (Tarney, 1969). In some dykes, augite occurs as oriented overgrowths on the (100) face of the orthopyroxene, such that the (100) planes, z and y axes of host and overgrowth are coincident. In other dykes, orthopyroxene forms similar overgrowths on augite. The latter has a relatively low Ca content (32–35 per cent Wo), and the high proportion of $(Mg,Fe)SiO_3$ in solid solution gives rise to a decrease in the c parameter so that it approaches more closely the c dimension of the orthopyroxene. The orthopyroxene is relatively rich in Ca (4·9–5·7 per cent Wo) thus increasing the c parameter of the Ca-poor pyroxene, the overall effect thus reducing the misfit between the two structures on (100).

Pyroxenes with a lath-shaped habit, and extending radially from the core into the inner shell of orbicules, have been described from the orbicular pyroxene-amphibole lamprophyre, Vestby, south-east Norway (Bryhni and Dons, 1975). The Ti content of the pyroxene increases from the core to the inner shell of the orbicules, and is accompanied by an overall compositional change from $Wo_{43.5}En_{46.5}Fs_7Ac_3$ to $Wo_{45.5}En_{40.5}Fs_{11}Ac_3$, the latter probably the result of recrystallization.

The elastic moduli of augite, as well as diopside, aegirine-augite and augite have been measured by Aleksandrov *et al.* (1964). The constants are greatest in the direction of the chains of silicon–oxygen tetrahedra, and are only slightly affected by the radii and valencies of the cations linking the chains.

Distinguishing Features

Because of the continuous chemical variation between diopside, hedenbergite, augite and ferroaugite it is not always possible to identify precisely these minerals on the basis of their optical properties. In general augite is distinguished from diopside by smaller birefringence, higher refractive indices and stronger dispersion, and the more (Mg,Fe)-rich augites and ferroaugites have smaller optic axial angles than salite and ferrosalite; ferroaugites have smaller optic axial angles and smaller extinction angles than hedenbergite. Titanaugites can usually be distinguished by their stronger colour and characteristic pleochroism with a violet colour in the γ vibration direction; sodian augites have a stronger green absorption colour, higher optic axial angles and extinction angles than augite with more normal contents of sodium. Augite is also distinguished from aegirine-augite and aegirine by the strong colour, higher birefringence, and higher $\gamma{:}z$ extinction angles and higher optic axial angles of the latter minerals. Magnesium-rich augites are difficult to distinguish from both omphacite and fassaite but these pyroxenes may sometimes be distinguished from augite by their smaller birefringence combined with their pale green colour. Augite is distinguished from spodumene and jadeite by higher extinction angle and refractive indices, from johannsenite by lower optic axial angle, and from the orthorhombic pyroxenes by higher birefringence, lower optic axial angle and by the negative optical character of many orthopyroxenes: moreover in the latter the extinction angle is straight in all sections of the [001] zone. Augite is distinguished from wollastonite by the presence of only two cleavages, higher refractive indices and birefringence; wollastonite has inclined extinction in all [001] zone sections. Rhodonite is similarly distinguished from augite except that there is an overlap with the refractive indices of the more iron-rich augites.

Paragenesis

Igneous rocks

Nodules and megacrysts. Although chrome-diopside is the characteristic pyroxene of lherzolite nodules in many kimberlites, augites, averaging $Wo_{37}En_{45}Fs_{18}$ in composition and often intergrown with ilmenite, have been reported from this paragenetic environment (Boyd and Nixon, 1972). Nodules of magnesium-rich augite, $Ca_{37\cdot6}Mg_{52\cdot5}Fe_{9\cdot9}$, and containing 1·28 weight per cent Na_2O, with intergrowths of ilmenite have been described from the Monastery Mine, South Africa, by Dawson and Reid (1970). The pyroxenes are distinguished from those of single nodules in other kimberlites by their very low chromium and relatively high sodium contents. The composition of the ilmenite lamellae also differs from that of the usual picroilmenite nodules in kimberlites, which together with the exsolution pattern is consistent with their development from a pyroxene with an ilmenite-type structure. The structural formula, $(Na_{0\cdot08}Ca_{0\cdot50}Mg_{1\cdot02}Fe^{2+}_{0\cdot35}Ti_{0\cdot10})(Si_{1\cdot48}Al_{0\cdot10}Fe^{3+}_{0\cdot14}Ti_{0\cdot28})O_6$, derived from the bulk composition of the nodules, resembles that of an augite very rich in titanium.

The Ca-rich pyroxenes of the garnet pyroxenite, spinel–garnet websterite and pyroxene granulite xenoliths in the Delegate basaltic breccia pipes, New South Wales (Irving, 1974) include both augite and endiopside. The augite of the websterite contains orthopyroxene lamellae, blebs of garnet and some spinel which on textural grounds are considered to have been exsolved from the clinopyroxene. The experimentally determined phase relations of the websterite composition support this conclusion, and demonstrate that the assemblage could have developed during isobaric cooling, at pressures in the range 13–17 kbar, from an accumulate formed in pockets of alkaline basaltic magma in the upper mantle. The two-pyroxene granulite equilibrated in a lower pressure range, 6–10 kbar, and is considered to represent accidental fragments of metamorphic rock from the deep crust.

Chromian augites (0·22–3·99 weight per cent Cr_2O_3) in ultrabasic nodules, including spinel lherzolites, websterites and clinopyroxenites, and as fragments of large augite crystals, in the alkali basalts of the Minusa and Transbaikalian regions, Siberia, are described by Kutolin and Frolova (1970).

Sodium-rich high aluminian augite, $(Na_{0\cdot23}Ca_{0\cdot60}Mg_{0\cdot75}Fe^{2+}_{0\cdot10}Al_{0\cdot25})(Si_{1\cdot91}Al_{0\cdot09})O_6$, occurs in a graphite-bearing pyrope peridotite inclusion from the Mir kimberlite pipe (Frantsesson and Lutts, 1970).

Aluminium-rich augites, containing exsolved garnet and orthopyroxene, occur in pyroxenite xenoliths in the nepheline basalt tuff of Salt Lake Crater, Oahu, Hawaiian Islands (Beeson and Jackson, 1970). The composition of the augite host varies, $Ca_{44\cdot9-41\cdot4}Mg_{45\cdot1-46\cdot1}Fe_{9\cdot6-12\cdot5}$, and the reconstructed compositions, i.e. prior to exsolution, range $Ca_{29\cdot3-34\cdot5}Mg_{58\cdot4-52\cdot6}Fe_{12\cdot2-15\cdot3}$. A later account, mainly restricted to the single-phase subcalcic clinopyroxenite xenoliths is given by Wilkinson (1976). These Al-rich subcalcic augites, $Ca_{33\cdot6-35\cdot1}Mg_{52\cdot8-54\cdot4}Fe_{12\cdot0-12\cdot2}$, have exsolved orthopyroxene $\simeq Mg_{81}Fe_{17}Ca_2$ ($\simeq$ 7 weight per cent Al_2O_3) either as broad lamellae $\parallel$ (100) or forming 'myrmekitic' intergrowths with the Ca-rich host. Other Al-rich augites have been described from the kaersutite peridotite inclusions in basanitic lava and ejecta, north-western Grand Canyon, Arizona (Best, 1970), the ultramafic xenoliths in basaltic rocks of the western United States (Wilshire and Shervais, 1975), the lavas and pyroclastic rocks of the northern Nigeria alkali-basaltic province (Frisch and Wright, 1971), as megacrysts in high-alumina tholeiitic andesite, Tweed Shield volcano (Duggan and Wilkinson, 1973) and in basanitoid, Armidale, New South Wales (Binns, 1969). Megacrysts of Al-rich (8·16–8·64 weight per cent Al_2O_3) subcalcic augite occur in the Elie Ness tuff, East Fife, Scotland. Two varieties are present, differing mainly in Ca/(Mg + Fe) ratio, both have low Cr_2O_3 and relatively high TiO_2 contents, and show affinities with augites crystallized from alkali basalts (Chapman, 1976). Compositions of twenty-five clinopyroxenes from the basic and ultrabasic inclusions, including Al- and Cr-rich pyroxenites, gabbros, dunite and peridotite, from the post Miocene alkali olivine basalts of south-western Japan are given by Ishibashi (1970).

Primary subcalcic augites of the layered pyroxenites in the lherzolites of Ariège, France, have been described by Conquéré (1977). Formed at an estimated temperature of 1600°C and pressures between 15 and 20 kbar, the pyroxenes subsequently recrystallized during a period of deformation at temperatures and pressures of 800° to 900°C and 12 to 15 kbar respectively. The recrystallization of the original subcalcic augite was accompanied by exsolution of orthopyroxene and garnet and a change to a diopsidic composition in the pyroxene host (see Fig. 140(d)). Following a review of pyroxenites associated with other spinel-bearing lherzolite complexes and as xenoliths in alkali basalt, Conquéré concluded that elsewhere these rocks have crystallized and recrystallized under *P–T* conditions similar to those of the lherzolites of the Ariège district, i.e. in the upper part of the upper mantle at depths of between 35 to 70 km.

The occurrence of Al-rich diopsidic augites (3·60–8·06 weight per cent Al_2O_3) in spinel gabbro xenoliths in a lava, Kerguelen Island, South Indian Ocean has been described by McBirney and Aoki (1973). The textural features of the gabbroic rocks provide evidence, in the form of embayed plagioclase surrounded by coronas of pyroxene and spinel, of the reaction plagioclase + olivine → pyroxene + spinel, at a depth estimated to be between 25 to 30 km. Subsequent reheating of the xenoliths at lower pressures has led to the reverse reaction with the formation of olivine and plagioclase from pyroxene and spinel. Mg-rich augites, $Ca_{43}Mg_{45}Fe_{12}$, have been reported in gabbro–quartz diorite inclusions in basalts of Okala, O-shima Island, Japan (Aoki and Kuno, 1973).

Titaniferous Al-rich augites occur both as megacrysts and as a constituent of olivine pyroxenite inclusions, in a nephelinite sill, Nandewar Mountains, New South Wales (Wilkinson, 1975). The megacrystal augites are similar in composition to the augite of the ultramafic inclusions and both megacrysts and inclusions are considered to be cognate in origin and derived at pressures between 15–20 kbar from olivine nephelinite. Spinel pyroxenite xenoliths in a tuff-filled volcanic vent Duncansby Ness, Caithness, Scotland, have been reported by Chapman (1975). The pyroxene is an Al-rich augite (9·19 weight per cent Al_2O_3) and encloses anhedral inclusions of aluminous pleonaste. The bulk composition of the pyroxenite, augite 95 per cent, spinel 5 per cent, suggests that the spinel could have been exsolved from a homogeneous pyroxene crystallized at a pressure $\geq$ 18 kbar in the temperature range 1450°–1350°C with subsequent exsolution of spinel at a temperature less than 1290°C.

A number of unusually calcium-poor, magnesium-rich pyroxenes, including subcalcic augites and endiopsides, occur in the pyroxenite xenoliths of the alkali trachybasalts, Glen Innes area, north-eastern New South Wales (Wilkinson, 1973). The subcalcic augite (Table 35, anal. 1) contains megascopic exsolution lamellae, up to 0·5 mm in width, of orthopyroxene which in some crystals show a graphic relation with the clinopyroxene host. The subcalcic augite has the bulk composition $Ca_{24\cdot7}Mg_{65\cdot1}Fe_{10\cdot2}$, and contains 11 mol. per cent of the $CaAl_2SiO_6$ component; after the exsolution of orthopyroxene, $Ca_{3.7}Mg_{83.8}Fe_{12.5}$, the composition of the Ca-rich phase is $Ca_{47.0}Mg_{47.2}Fe_{5.8}$ (see Fig. 140). The partition of the cations accompanying the process of unmixing shows that the Ca-rich phase is enriched in Ti, Al, Cr and Na, while Ni and Mn are preferentially concentrated in the orthopyroxene. Both phases are closely comparable in composition with the Ca-rich and Ca-poor pyroxenes of the matrix of the xenoliths. An apparent anomaly is displayed by the relatively high Ca content of the exsolved orthopyroxene, and may indicate that complete equilibration was not attained during the exsolution process. The normative compositions of the xenoliths contain both *ol* and *an*, and these potentially plagioclase-bearing pyroxenites presumably crystallized at temperatures and pressures sufficiently high to inhibit the formation of modal plagioclase, but lower than those necessary for the formation of garnet. From the experimental data of Green and Ringwood (1970), relating to pyroxene and garnet peridotites, the crystallization temperature of the xenoliths would have been > 1300°C. Comparison of the compositions of the coexisting subcalcic augite and orthopyroxene megacrysts with those of phases in equilibrium in the system $Mg_2Si_2O_6$–$CaMgSi_2O_6$ at 10–20 kbar indicates that crystallization occurred at temperatures between 1350° and 1450°C. From the compositions of the exsolved and host phases the subsequent unmixing of the subcalcic augite is inferred to have taken place at approximately 1000°C.

Plutonic rocks. The most complete range of clinopyroxene compositions for a single fractionated intrusion is shown by the Bushveld complex. The calcium-rich pyroxene of the lowest exposed rocks of the basal series is an intercumulus diopside, $Ca_{45}Mg_{50}Fe_5$, and here is associated with cumulus orthopyroxene, $Ca_3Mg_{85}Fe_{12}$ (Atkins, 1969). The trend of the clinopyroxene crystallization during the accumulation of the layered series from the basal and critical horizons to upper zone a and b, some 6500 m in thickness, is towards the Fe corner of the pyroxene trapezium with both Ca and Mg being progressively replaced in approximately equal proportions (see Fig. 148). At this level, when the augite composition is $Ca_{33}Mg_{31}Fe_{36}$ the trend of crystallization changes abruptly and the succeeding ferroaugites display a relatively small increase in Fe content, and the main compositional changes are enrichment in Ca and impoverishment in Mg. This trend continues to the composition $Ca_{40.5}Mg_{18.7}Fe_{40.8}$, and thereafter proceeds nearly parallel to the diopside–hedenbergite join, terminating close to the hedenbergite–ferrosilite join at the composition $Ca_{42.7}Mg_{0.5}Fe_{56.8}$. In contrast to some of the ferrohedenbergites in the upper ferrodiorites of the Skaergaard intrusion there are no indications that these iron-rich pyroxenes of the Bushveld intrusion are inverted ferriferous β-wollastonite solid solutions.

The trend of clinopyroxene crystallization associated with the fractionation of tholeiitic magma, although not as extensive as that of the Bushveld, is also well illustrated by the augite–ferroaugite–ferrohedenbergite sequence of the Skaergaard intrusion, east Greenland (Wager and Deer, 1939; Muir, 1951; Brown, 1957; Brown and Vincent, 1963; Nwe, 1975, 1976; Nwe and Copley, 1975). In the earliest crystal accumulates of this intrusion, represented by the gabbro picrite, the pyroxene crystallization began with the separation of an endiopside, $Ca_{42.4}Mg_{47.9}Fe_{9.7}$, and the clinopyroxene is associated with orthopyroxene, $Ca_{3.9}Mg_{77.6}Fe_{18.5}$ in composition. In the olivine gabbros of the layered series the clinopyroxenes which separated from the continuously fractionated and iron enriched liquids show increasing contents of Fe and decreasing amounts of Mg. A number of small reversals (maximum $\simeq$ 3 per cent Fs) occur and are attributed to local temperature differences within the magma chamber and periodic lateral movements of the convecting cell. This progressive change in the Mg:Fe ratio is accompanied by a decrease in the Ca content to a minimum, represented by an augite of composition $Ca_{34\cdot6}Mg_{37\cdot0}Fe_{28\cdot4}$ (Table 29, anal. 19), in the lower ferrogabbros. This pyroxene is associated with an inverted ferropigeonite $Ca_{8\cdot6}Mg_{45\cdot5}Fe_{45\cdot9}$ in composition. With greater fractionation of the Skaergaard magma the pyroxene crystallization passed from the two-pyroxene to the one-pyroxene field, and the earlier trend of decreasing Ca content was reversed, and the ferroaugite of the ferrodiorite of upper zone b has the composition $Ca_{40}Mg_{21}Fe_{39}$. In the latter stages of fractionation the iron enrichment continued with the Ca content showing only a small increase, the ferroaugite at the top level of upper zone b having the composition $Ca_{44\cdot5}Mg_{10\cdot7}Fe_{47\cdot8}$, a trend that is continued by the ferro-hedenbergites, and inverted β-wollastonite solid solutions (see p. 332) to $Ca_{42\cdot5}Mg_{0\cdot4}Fe_{57\cdot1}$ in the upper part of upper zone c (see Fig. 149).

In the layered gabbros and related rocks of the thick stratiform intrusion, Dufek, Antarctica (Himmelberg and Ford, 1976), the sequence of cumulus crystallization of the pyroxene phases shows a general similarity to those of the Bushveld and Skaergaard intrusions. The early crystallization of cumulus augite, $Ca_{36\cdot4}Mg_{48\cdot7}Fe_{14\cdot9}$, and bronzite, $Ca_{3\cdot5}Mg_{69\cdot1}Fe_{27\cdot4}$, is followed by augite–ferroaugite and pigeonite–ferropigeonite to compositions $Ca_{34\cdot7}Mg_{30\cdot6}Fe_{34\cdot7}$ and $Ca_{11\cdot4}Mg_{34\cdot0}Fe_{54\cdot6}$ respectively, and finally by ferroaugite, the most iron-rich composition of which is $Ca_{30\cdot0}Mg_{23\cdot5}Fe_{46\cdot5}$ (Fig. 178). In contrast to the Bushveld and Skaergaard

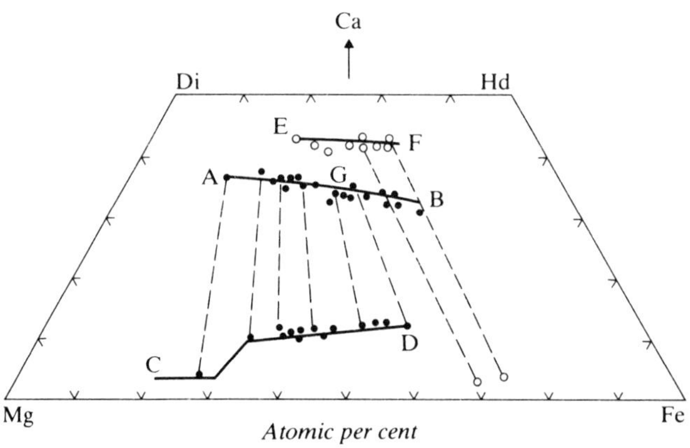

Fig. 178. Plot of Ca-rich (A–B) and Ca-poor (C–D), and late-stage (E–F) pyroxenes of the Dufek intrusion (after Himmelberg and Ford, 1976). Solid circles, cumulus pyroxenes. Open circles, late-stage pyroxenes. G–D, tie-line between coexisting Ca-rich and Ca-poor pyroxenes at the limit of the cumulus two-pyroxene field.

trends the early cumulus Ca-rich pyroxenes have smaller contents of calcium, and although there is a small progressive decrease in Ca with fractionation there is no apparent Ca-minimum at the limit of the two-pyroxene field. The composition of the Ca-rich pyroxene at the limit of the two-pyroxene field is similar to that of the Bushveld pyroxenes. Thus if the limit is related to silica activity in the melt according to the reaction (see p. 335):

$$\underset{\text{in pyroxene}}{(Fe_2Si_2O_6)_{ss}} = \underset{\text{in olivine}}{(Fe_2SiO_4)_{ss}} + \underset{\text{in melt}}{SiO_2}$$

the silica activity in the Dufek and Bushveld melts was presumably greater than in the Skaergaard melt. The Ca-poor pyroxenes are correspondingly more Ca-rich and the miscibility gap of the Dufek pyroxenes is smaller than in either of the other two intrusions. In addition, iron-enrichment in the cumulus Ca-rich pyroxenes is more restricted. As the most Fe-rich ferroaugites occur in the strongly fractionated granophyre derivative of the Dufek magma, the absence of more extreme iron compositions may be related to the occurrence of a higher oxygen fugacity compared with those present during the crystallization of the Bushveld and Skaergaard liquids.

Late stage Ca-rich and Ca-poor pyroxenes are also present in the gabbros of the Dufek intrusion. The augites are pale green in contrast to the brown colour of the cumulus phase, and occur as patches within and as margins to the cumulus pyroxene. The orthopyroxenes are associated with the pale green augite, hornblende and biotite and crystallized directly as ferrohypersthene. The late-stage Ca-rich augite-ferroaugites are richer in Ca and Si and poorer in Al, Ti and Mn than the cumulus augites. The ferrohypersthenes also have lower Al and Ti contents relative to the Ca-poor cumulus pyroxenes and extend beyond the primary trend to more Fe-rich, $Ca_{3.5}Mg_{24.5}Fe_{72.0}$, compositions (Fig. 178). The compositions of the late stage Ca-rich pyroxenes are comparable with those of the subsolidus Ca-rich pyroxene trend of the Skaergaard intrusion, and together with the primary formation of ferrohypersthene indicate lower crystallization temperatures, estimated by Himmelberg and Ford to be between 900° and 800°C.

The Ca-rich pyroxenes in the gabbros, ferrogabbros and iron-rich tholeiites of the Birds River gabbro complex, Dordrecht district, South Africa (Eales and Booth, 1974) display a similar trend to that of the Skaergaard intrusion. Pyroxene phenocrysts in the tholeiites are zoned from ferroaugite, $Ca_{40.5}Mg_{16}Fe_{43.5}$ to ferrohedenbergite, $Ca_{42}Mg_{6}Fe_{52}$. A mosaic structure is shown by some of the phenocrysts, and is probably due to the inversion of a β-wollastonite solid solution to a hedenbergitic pyroxene.

In some layered intrusions the clinopyroxenes are richer in Al and Cr than those of the Bushveld and Skaergaard intrusions. The clinopyroxenes of the Giles complex, central Australia (Goode and Moore, 1975) are a typical example, and their usually high contents of these cations are probably related to primary crystallization at high pressure, a conclusion supported by the development, resulting from solid state reactions, of clinopyroxene + orthopyroxene + spinel + Ca-poor plagioclase coronas between olivine and Ca-rich plagioclase in these rocks.

Clinopyroxene occurs as a cumulus phase and orthopyroxene as a late intercumulus phase in the early cumulates of the Muskox layered intrusion. At a later stage orthopyroxene formed before clinopyroxene. This reversal in crystallization order has been examined by Irvine (1970), and the different crystallization paths are considered to have developed by a combination of fractional crystallization, additions of fresh magma, and contamination of the magmas by sialic material from the roof of the intrusion.

Although the sequence of the layered rocks of the ultrabasic-basic Stillwater complex, Montana (Hess, 1960) is more extensive than the Skaergaard layered series the Ca-rich pyroxene shows only a restricted range of iron enrichment. In the ultrabasic zone the pyroxene, occurring as an intercumulus phase, is a chromium-rich endiopside, $Ca_{40}Mg_{52}Fe_{8}$. In the upper 4000 m the cumulus augite displays some irregular fluctuations in composition with an overall iron enrichment to $Ca_{40}Mg_{42}Fe_{18}$.

The clinopyroxenes of the tholeiitic series of the Okenjeje complex, South-West Africa (Simpson, 1954) show a remarkably small variation in calcium content and, compared with the associated olivines and orthopyroxenes, exhibit relatively little iron enrichment. Thus in the early tholeiitic gabbros of the complex the compositions of the clinopyroxene, olivine and orthopyroxene are respectively $Wo_{40}En_{43}Fs_{17}$, $Fo_{73}Fa_{27}$ and $En_{78}Fs_{22}$, while in the highly fractionated ferrogabbros the compositions of the three coexisting phases have compositions $Wo_{39}En_{29}Fs_{32}$, $Fo_{20}Fa_{80}$ and $En_{27}Fs_{73}$.

A series of pyroxenes ranging in composition from endiopside to ferroaugite in the rocks of a differentiated diabase at Harrisburg, Pennsylvania, have been described by Hotz (1953). In the upper chilled zone of the intrusion the pyroxene is an endiopside, augite or salite in the normal diabase, ferroaugite in the rocks transitional between diabase and granophyre, and in the granophyre it is close to ferrohedenbergite in composition.

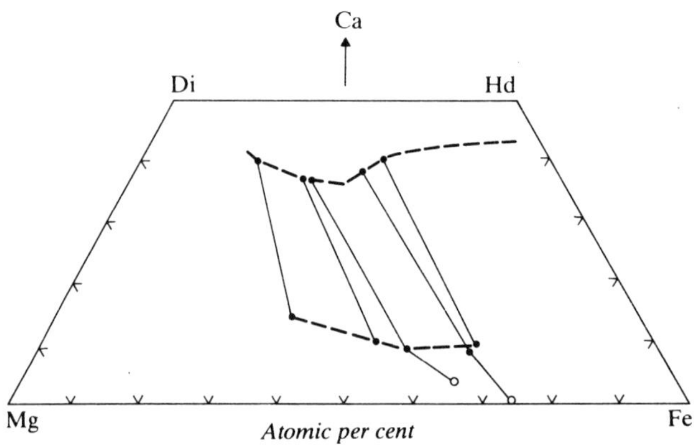

Fig. 179. Crystallization trend of Beaver Bay pyroxenes, showing tie-lines between coexisting cumulus augite–ferroaugite and intercumulus pigeonite and orthopyroxene (after Konda, 1970).

Other ferroaugites from highly differentiated basic rocks include the clinopyroxenes (Table 33, anals. 4, 10) from the Beaver Bay diabase (Muir, 1954; Konda, 1970), and from the iron-rich dolerite of the New Amalfi sheet (Poldervaart, 1944). The Ca-rich pyroxenes of the small layered intrusion of Beaver Bay range in composition from augite, $Ca_{38}Mg_{38}Fe_{24}$, to ferrohedenbergite, $Ca_{42}Mg_{10}Fe_{48}$ (Fig. 179). The associated Ca-poor pyroxenes include both pigeonite, $Ca_{11}Mg_{39}Fe_{50}$–$Ca_8Mg_{27}Fe_{65}$, and orthopyroxene, $Ca_4Mg_{32}Fe_{64}$–$Ca_0Mg_{25}Fe_{75}$. The ferrogabbro of the intrusion does not contain cumulus pigeonite, and the intercumulus iron-poor monoclinic pyroxene may have formed by reaction between olivine and residual liquid. Pigeonite crystallization terminated in the upper part of the ferrohortonolite ferrogabbro, the olivine of which has the composition Fa_{85}. Although the crystallization trend of the Ca-rich pyroxenes shows a similar pattern to that of the Bushveld and Skaergaard intrusion, the Beaver Bay minerals are somewhat richer in Ca and Ti. Exsolution lamellae in the ferroaugite are very thin, and the pigeonite has only partially inverted, presumably due to the more rapid cooling of this small intrusion.

In the higher levels of the Sudbury irruptive, Ontario, the compositions of the augite cores show a normal plutonic iron enrichment trend that is reversed in the pyroxenes of the border group. In the latter the augite mean core composition becomes less iron-rich in passing upwards from rocks that crystallized rapidly *in situ* to those displaying cumulate textures. The calcium content of the augite cores is relatively constant, but there is a marked enrichment in Ca, and a corresponding decrease in Fe, Cr, Ti and Al, in the composition of the rims. The rims do not coexist with orthopyroxene, but with late magnetite and biotite, and the Ca enrichment is considered to result from a movement of the augite composition, away from the solvus, into the one-pyroxene field, and the reverse Fe–Mg zoning to be due to a higher oxygen fugacity in the intercumulus liquid or, as a secondary result, consequent on the loss of orthopyroxene to the late liquid (Hewins, 1974).

Ca-rich and Ca-poor clinopyroxenes coexist throughout most of the lower layered series of the Jimberlana intrusion, Western Australia (Campbell and Borley, 1974). The Ca-rich pyroxenes show a restricted compositional range, $Wo_{41\cdot1}En_{51\cdot2}Fs_{7\cdot7}$ to $Wo_{36\cdot2}En_{40\cdot4}Fs_{23\cdot5}$, and represent the early and middle stages of crystallization only, due to injection of the upper layered series magma before fractionation of the lower series was completed. The augites display a relatively wide variation in the Wo component, the adcumulate and orthocumulate pyroxenes differing by between 2 and 3 per cent Wo, with the composition of all orthocumulate augites lying to the diopside–hedenbergite side of the average trend line. The change from orthopyroxene to pigeonite crystallization took place at the composition Fs_{27}, which together with the narrow miscibility gap suggests that the Jimberlana pyroxenes formed at a higher temperature than those of comparable compositions from other layered intrusions.

Iron enrichment in the augites, from $Mg_{80\cdot3}Fe_{19\cdot7}$ to $Mg_{55\cdot3}Fe_{44\cdot7}$, of the eucrites, gabbros and dolerites of Centre III igneous complex, Ardnamurchan, Scotland (Walsh, 1975), is similar to the early fractionation pattern of differentiated tholeiitic intrusions; the trend, however, is not continued by the augites of the intermediate rocks, tonalite and quartz monzonite, of the complex.

The augites of the Kap Edvard Holm intrusion, some 20 km south-west of the Skaergaard, display a different trend and considerably less iron enrichment. The intrusion (Deer and Abbott, 1965; Elsdon, 1971; Abbott and Deer, 1972) consists of three separate layered series, the augites of which, in addition to being more Ca-rich, have a higher content of aluminium and titanium than either the Skaergaard or Stillwater clinopyroxenes. The augites are, however, not as Ca-rich as pyroxenes which have crystallized from alkali olivine basalt magmas (e.g. in the Garbh Eilean picrodolerite and crinanite; Murray, 1954; the teschenite of the Black Jack sill; Wilkinson, 1956, 1957), and the pyroxene trend is characteristic of fractional crystallization of basic magmas the compositions of which are transitional between tholeiitic and alkali olivine basalts. The Ca contents of the Kap Edvard Holm pyroxenes show relatively little variation, and the compositional trend is essentially parallel to the diopside–hedenbergite join. The augites of the lower and middle series do not contain exsolved lamellae of orthopyroxene, those of the upper series, although optically homogeneous, as shown by X-ray oscillation photographs, enclose some exsolved pigeonite. Boyd and Schairer (1962) and Kushiro and Schairer (1963) have shown experimentally that augites crystallized at lower, compared with those formed at higher temperatures, are likely to contain less of the $(Mg,Fe)SiO_3$ component, and it is probable that the augites of Kap Edvard Holm crystallized at temperatures lower than the comparable pyroxenes of the Skaergaard intrusion due to the depression of the liquidus and solidus consequent on higher water pressure. This conclusion is supported by the presence only of submicroscopic pigeonite lamellae, for unless diffusion was greatly inhibited, the amount of the Ca-poor component in the original augites must have been small. In these circumstances the bulk composition would intersect the solvus, in the relevant $(Mg,Fe)SiO_3$–$Ca(Mg,Fe)Si_2O_6$ composition plane, at a temperature only a little above the minimum at which diffusion was effective at the cooling rate of the upper

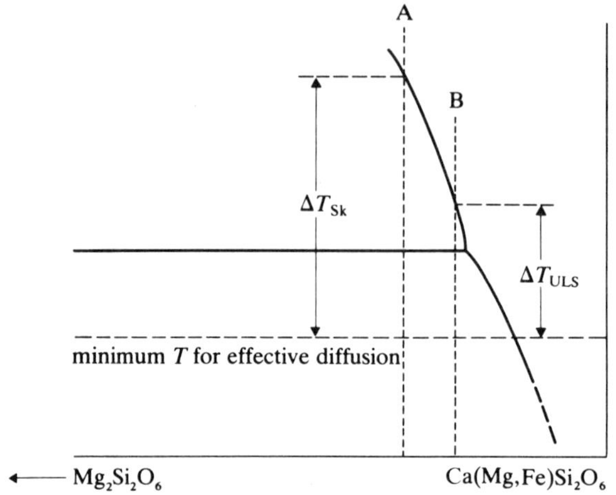

Fig. 180. Hypothetical diagram showing relationship between exsolution of Ca-poor pyroxenes from augites of the Skaergaard and Upper Layered series of the Kap Edvard Holm intrusion (after Elsdon, 1971). A and B are bulk compositions of Skaergaard and Kap Edvard Holm augites respectively.

layered series. The augites of the Skaergaard intrusion exsolved considerably greater amounts of Ca-poor pyroxene, are less rich in calcium, and it is thus likely that the temperature interval, between the intersection of the bulk composition with the solvus and the minimum temperature of effective diffusion, was greater in that intrusion (Fig. 180).

In some layered intrusions, augite shows little cryptic variation, e.g. in the Talnakh intrusion (Godlevskii and Polushkina, 1971) $Ca_{42}Mg_{48}Fe_{10}$ to $Ca_{39}Mg_{42}Fe_{19}$. In the Eulogie Park gabbro, eastern Queensland (Wilson and Mathison, 1968) the composition of the cumulus augite is $Ca_{43\cdot4}Mg_{43\cdot6}Fe_{13\cdot0}$ in the olivine gabbros and ferrigabbros, and as an intercumulus phase has the composition $Ca_{42\cdot8}Mg_{45\cdot0}Fe_{12\cdot2}$ in the troctolites and leucogabbros.

The sequence of pyroxene crystallization in the rocks of calc-alkaline affinities shows considerable variation but in general extreme compositions are less common than those displayed in highly fractionated tholeiitic series. The early formed clinopyroxenes of the anorthosite–mangerite differentiated sequence of the Bjerkrem-Sogndal intrusion, Norway (Duchesne, 1972) have the composition $Ca_{46}Mg_{41}Fe_{13}$ and are associated with orthopyroxene, $Ca_2Mg_{66}Fe_{32}$. Progressive enrichment in Fe occurred during the evolution of the intrusion and terminated with the crystallization of a ferroaugite, $Ca_{39}Mg_{14}Fe_{47}$ and pigeonite, now inverted, $Ca_7Mg_{19}Fe_{74}$ in composition (Fig. 181).

The principal variation in the composition of the augites of the Guadalupe calc-alkaline igneous complex, California (Best and Mercy, 1967) is the substitution of Fe^{2+} for Mg. The augites, $Ca_{43\cdot3}Mg_{44\cdot6}Fe_{12\cdot1}$–$Ca_{42\cdot8}Mg_{30\cdot0}Fe_{27\cdot2}$, are associated with orthopyroxene, $Ca_{3\cdot2}Mg_{68\cdot8}Fe_{28\cdot0}$–$Ca_{3\cdot0}Mg_{37\cdot3}Fe_{59\cdot7}$, throughout the sequence of gabbroic rocks. The Fe^{2+}–Mg and Ca partition between the coexisting pyroxene phases is similar to those of high-grade regional metamorphic rocks, and essentially the same as in other calc-alkaline plutonic suites. In contrast to the phase mineralogy of tholeiitic rocks, that of the Guadalupe complex includes abundant hydrous minerals even in the early gabbroic differentiates, a feature indicating high water fugacity which led not only to the early precipitation of hornblende, but also to the depression of the pyroxene liquidus to a temperature below the pigeonite–orthopyroxene inversion.

Aluminium-rich amphiboles are commonly the main ferromagnesian phases of intermediate plutonic rocks of the calc-alkaline series. In some intrusions and

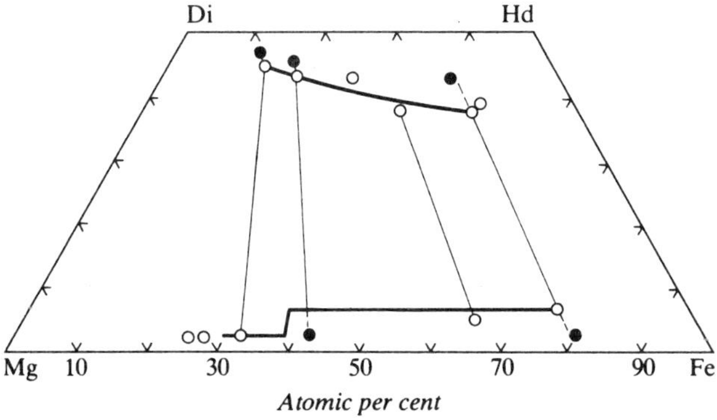

Fig. 181. Differentiation trends of Ca-rich and Ca-poor pyroxenes of the Bjerkrem–Sogndal anorthosite–mangerite intrusion (after Duchesne, 1972). Solid circles, compositions of host pyroxenes after exsolution.

especially those emplaced at relatively high crustal levels, as were the quartz diorites of the Ben Nevis complex, Scotland (Haslam, 1968), augite was the stable phase under the conditions of low pressure, and at near liquidus temperatures. With the continued cooling of the liquid, amphibole began to crystallize and the augite was replaced by an aluminium-poor hornblende.

Ca-rich, associated with Ca-poor pyroxenes, are the characteristic ferromagnesian minerals of the Morin series, southern Quebec (Philpotts, 1966; see also Emslie and Ermanovics, 1975). Many of these rocks, derived from a calc-alkaline parental magma underwent differentiation in a dry environment and gave rise to strong iron enrichment. Although there was intense deformation during the emplacement in the western Grenville area of the western zone of the province leading to extensive recrystallization and gneissic foliation, in other parts of the series only mild deformation occurred and some members of the anorthosite–mangerite series retain their original igneous textures. The pyroxenes of the more highly deformed rocks contain considerably larger amounts of Al than those in the less deformed varieties. The augite displays a wide composition range from salite, $Ca_{47\cdot0}Mg_{42\cdot1}Fe_{10\cdot9}$ to Ca-poor ferroaugite, $Ca_{25\cdot6}Mg_{22\cdot4}Fe_{52\cdot0}$ (Table 33, anal. 19), the latter coexisting with an inverted ferropigeonite, $\simeq Ca_9Mg_{23}Fe_{68}$. In the Belleau-Desaulnies area of Quebec province, orthopyroxene is the main ferromagnesian constituent of the anorthosites and norites, and the early clinopyroxene has a composition $Ca_{43\cdot4}Mg_{38\cdot5}Fe_{18\cdot1}$, and coexists with hypersthene, $Ca_{1\cdot5}Mg_{59\cdot4}Fe_{39\cdot1}$. The clinopyroxenes have higher Ca contents probably related to a lower temperature of crystallization leading to the intersection of the pyroxene solvus at more Ca-rich compositions.

Ferroaugite, intergrown with orthopyroxene, and associated with fayalite and quartz, an assemblage that on textural evidence is considered to have been derived from the breakdown of pigeonite (see also p. 189), occurs in some of the adamellites and granodiorites of the anorthosite massif, Nain, Labrador (Smith, 1974). In others, ferroaugite and olivine form discrete crystals, and the two phases, together with orthopyroxene and pigeonite, occur as primary minerals. Ferroaugite, as individual grains, and as rims, $Ca_{42}Mg_{14}Fe_{44}$, around cores with an average composition of $Ca_{26}Mg_{15}Fe_{59}$ and in crystals consisting of alternating broad lamellae of ferroaugite $\simeq Ca_{43}Mg_{17}Fe_{40}$ and orthopyroxene $\simeq Ca_2Mg_{19}Fe_{79}$ occur in the monzonite of the massif. The lamellae are present in nearly equal proportions, and it is likely that they have developed from a pyroxene with a calcium content in the range Wo_{20-30}, that subsequently unmixed to ferroaugite and pigeonite, the latter subsequently inverting to orthopyroxene. Intergrowths of ferroaugite, fayalite and quartz occur also in the monzonite and, in view of the absence of orthopyroxene, Smith has suggested that the intergrowths are due to a reaction of a Ca-poor pyroxene with residual melt, rather than from its simple breakdown. The occurrence of crystals with a Ca-poor core and ferroaugite rims is consistent with a pigeonite–melt reaction as Ca-rich pyroxene should crystallize both during and after the reaction.

In basic alkaline rock the characteristic pyroxene is usually titaniferous augite or titanaugite; sodium-poor pyroxenes more iron-rich than $Ca_{45}Mg_{35}Fe_{20}$ are rare, and the more strongly fractionated liquids usually precipitated acmitic pyroxene. Such a trend is shown by the pyroxenes in the alkaline rocks of the Morotu district, Sakhalin (Yagi, 1953; see also Neumann, 1976 and Fig. 238(b)). Here the earliest pyroxenes have a diopsidic composition; in many of the dolerites and in some of the monzonites of this area the pyroxene is an augite (Table 30, anal. 5) relatively enriched in titanium and commonly zoned by a more sodium-rich variety. In most of the monzonites, however, the pyroxene is a sodian augite (Table 38, anal. 6), and in the syenites it is an aegirine-augite or aegirine. A similar compositional trend is shown by the clinopyroxenes in the alkaline rocks of Iron Hill, Colorado (Larsen, 1942). With the exception of the syenites and nepheline-syenites, the pyroxenes of this complex have a low sodium and moderate iron content and show little chemical variation. The pyroxene (Table 38, anal. 8) of the shonkinite is a sodian augite, and in the syenites and nepheline syenites it is aegirine-augite.

Titanaugite occurs in the ilmenite gabbro, alkali gabbro and lamprophyre ring dykes of the Blue Mountain complex, New Zealand (Grapes, 1975). The alkali ultrabasic–gabbroic rocks of the complex contain both titanaugite and endiopside, and there is a complete transition from titaniferous endiopside to titaniferous augite ($CaTiAl_2O_6$ component varying between 2·1 to 5·5 per cent), and from titaniferous augite to titanaugite (3·3 to 8·1 per cent $CaTiAl_2O_6$). The titanaugites of the Tertiary sancyites, doreites and ordanchites of Mont Dore, France (Brousse, 1961) have high contents of Al (6·7–13·0 Al_2O_3), Fe^{3+} (4·5–9·8 Fe_2O_3), Na (1·05–2·0 Na_2O) and K (0·25–1·55 weight per cent K_2O). Zoning is a common feature of the pyroxenes in the doreites and ordanchites, cores of violet-coloured titanaugite being followed by a slightly pleochroic light brown zone and rimmed by aegirine-augite. Titanaugite occurs in the kaersutite-bearing xenoliths in the pyroclastic deposits and lava flows of Tristan da Cunha, South Atlantic (Le Maitre, 1969), and in biotite melteigite, Itapirapuã, São Paulo, Brazil (Gomes *et al.*, 1970).

Hypabyssal rocks. The clinopyroxenes in the picrites, and as phenocrysts in the chilled basalts of the Karroo dolerites range in composition from $Wo_{39}En_{50}Fs_{11}$ to $Wo_{41}En_{42}Fs_{17}$ (Walker and Poldervaart, 1949). The augites of the thick dolerite sheet of Ingeli Mountain display a greater compositional variation, from endiopside to iron-poor ferroaugite, $Wo_{39}En_{30}Fs_{31}$ (Maske, 1966). In many of the Karroo dolerites, however, the composition of the Ca-rich pyroxenes is essentially related to the replacement $Ca \rightleftharpoons Fe$, and the composition ranges from augite, $Wo_{39}En_{42}Fs_{19}$, to subcalcic augite, $Wo_{25}En_{41}Fs_{34}$. More iron-rich pyroxenes, $Wo_{25}En_{41}Fs_{34}$–$Wo_{30}En_{33}Fs_{37}$ occur in some of the marginal tholeiites and the dolerite pegmatites. In some basic magmas pyroxene crystallization began with the formation of an orthopyroxene, thus during the early fractionation of the sills and dykes of Tasmania (Edwards, 1942) bronzite formed prior to the magnesium-rich augite. During the early period of the fractionation both pyroxenes continued to crystallize and are richer in iron than the first formed pyroxene phases. Later crystallization, represented in the upper parts of the sills and the central portions of the dykes, led to the formation of pigeonite in place of orthopyroxene and, in many of these rocks, rims of pigeonite around augite, and rims of augite around pigeonite, as well as intimate intergrowths of the two pyroxenes, indicate that their crystallization was simultaneous. In contrast, the sequence of crystallization in some of the dolerites of north-western Otago, New Zealand (Benson, 1944), is diopside augite, augite and finally subcalcic augite. The formation of a single pyroxene phase in these rocks is considered by Benson to result from the rapid crystallization of the dolerite magma which here was injected in comparatively small volumes at shallow depths into recently deposited and often still plastic marine sediments.

The trend of pyroxene crystallization in the Red Hill dolerite, southern Tasmania (McDougall, 1961) is similar to that of the Skaergaard pyroxenes. Augite, except in the earlier stages, crystallized throughout the cooling history, initially with orthopyroxene and then together with pigeonite until the latter ceased to form at an Mg:Fe ratio of $\simeq$ 40:60. In the earlier and middle stages of the fractionation the Ca-content of the augite decreased from 40 to 31 weight per cent, and was followed in the later stages by an increasing content of Ca to a ferrohedenbergite composition $Ca_{41\cdot4}Mg_{3\cdot3}Fe_{55\cdot3}$. The pyroxenes of the Great Lake dolerite, Tasmania (McDougall, 1964), follow a broadly similar pattern, the augite becoming progressively enriched in Fe and impoverished in Ca as well as Mg, and are associated with orthopyroxene, Fs_{16} to $Fs_{27\cdot5}$ in the earlier, and pigeonite, $Ca_8Mg_{58}Fe_{34}$ to $Ca_{14\cdot5}Mg_{31\cdot5}Fe_{54}$, in the later stages of the differentiation.

The augite microphenocrysts of the chilled margin of the tholeiitic dolerite, Moeraki, New Zealand (Nakamura and Coombs, 1973), are around $Ca_{38}Mg_{50}Fe_{12}$ in composition and have low Ti/Al ratios ($< \frac{1}{4}$); the groundmass augites are more iron-rich, and have higher Ti/Al ratios ($\simeq \frac{1}{2}$). Some zoning is present, and the margins of the pyroxenes in later segregations are subcalcic ferroaugites $Ca_{17\cdot3}Mg_{33\cdot8}Fe_{48\cdot9}$. A subcalcic ferroaugite from the Scourie dolerite dyke has been described by O'Hara (1961). X-ray single-crystal examination has shown that it consists of an intergrowth of two clinopyroxenes, probably ferroaugite and ferropigeonite, in roughly equal proportions.

A series of pyroxenes ranging in composition from endiopside to ferroaugite in the rocks of a differentiated diabase at Harrisburg, Pennsylvania, have been described by Hotz (1953). In the upper chilled zone of the intrusion the pyroxene is an endiopside, in the normal diabase it is either augite or salite, in the rocks transitional between diabase and granophyre it is ferroaugite, and in the granophyre the pyroxene is close to ferrohedenbergite in composition. Two Ca-rich pyroxene series occur in the highly differentiated Palisade sill, New Jersey (Walker, 1940, Walker *et al.*, 1973). The first consists of an augite–mauve-brown ferroaugite sequence that follows the normal tholeiitic fractionation trend. The other, a pale green ferroaugite series first developed in the later middle stages of fractionation, is characterized by a higher Ca content and less Mg, Al and Ti, and in consequence the trend is located closer to the diopside–hedenbergite join. Thus the compositional change of the normal trend is from $Wo_{38}En_{51}Fs_{11}$ to $Wo_{38}En_{10}Fs_{52}$ and of the more Ca-rich series from $Wo_{44}En_{27}Fs_{29}$ to $Wo_{44}En_{10}Fs_{46}$ (Fig. 182). In addition to the presence of two series the Ca-rich pyroxenes of the Palisade sill differ, in the absence of a marked inflection at the Wo minimum composition, from those of either the Skaergaard or Bushveld intrusions. The coexisting pyroxene pairs may also be distinguished by the variation in the location of the tie-line intersections with the Wo–Di join in the pyroxene triangle. The pale green ferroaugite first developed in the middle fractionation stage, initially along lamellae in the augite, that

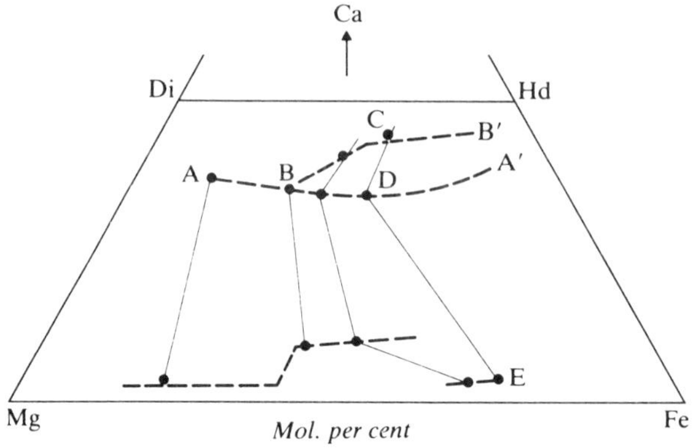

Fig. 182. Differentiation trends of pyroxenes of the Palisade sill, New Jersey (after Walker *et al.*, 1973). A–A′, augite–mauve-brown ferroaugite trend. B–B′, green ferroaugite trend. CDE, limit of two pyroxene field.

gradually increased in amount as the fractionation progressed. The pyroxenes have a common optical orientation, and the grain boundaries between them are often indistinct, features that together with the inclusion of hydrous phases and traces of the mauve ferroaugite in some of the pale green ferroaugite, indicate that the latter have been derived, at least in part, from the mauve ferroaugite of the normal trend. The pyroxenes of many dolerite sills, however, show relatively small compositional changes, e.g. in the large thick stratiform intrusion of Lower Tunguska (Naumov, 1965), in which the pyroxene constitutes between 23 and 39 per cent of the olivine dolerite and dolerite pegmatite; here the ferrosilite component varies between Fs_{23} and Fs_{31}. In the dykes and sheets of the Deep River Basin, North Carolina, augite constitutes 10, 32 and 40 per cent of the picritic, normal and micropegmatitic dolerites respectively. Small but significant variations in augite compositions of different members and within individual members of the Tertiary ultrabasic dykes of south-west Skye have been described by Gibb (1968).

The clinopyroxenes of basic alkaline rocks generally have a higher content of Al, Fe^{3+}, Ti and Ca than the clinopyroxenes of tholeiitic rocks. In addition to these chemical differences, the clinopyroxene is never associated with either pigeonite or orthopyroxene and is the sole pyroxene phase to crystallize from alkali magmas. In a review of clinopyroxene trends in alkali olivine basalt magmas, Wilkinson (1956) has listed the compositions (in atomic per cent Ca, Mg and Fe) and parageneses of fifty-six clinopyroxenes from basic alkaline and related rocks, including many representative of alkali olivine basalt magma. The pyroxenes of these rocks show a relatively small variation in their content of Ca, Mg and Fe, and the majority have a salite composition; the remainder lie in the augite field close to the boundary with salite.

This restricted compositional trend of clinopyroxenes of alkali basalt magmas is illustrated by the pyroxenes of the Garbh Eilean sill, Shiant Isles (Murray, 1954) and the Black Jack teschenite sill, Gunnedah, New South Wales (Wilkinson, 1957). The crystallization trends of the pyroxene cores in both these intrusions are sub-parallel to the diopside–hedenbergite join, and are limited to the compositional field Fs_9 to Fs_{29}, Wo_{43} to Wo_{50}. Thus in the Black Jack sill the pyroxenes are members of the diopside–hedenbergite series and, although the calcium content decreases slightly as fractionation proceeds, all the pyroxenes have a salite composition. The pyroxenes of earlier crystallization are richer in aluminium than those which formed later, and this compositional variation is considered by Wilkinson to result from the higher crystallization temperatures of the earlier pyroxenes. The titanium content is also lower in the later formed pyroxenes but is probably related to the composition and order of the mineral crystallization sequence, and not to the temperature of formation. The crystallization of the pyroxenes of the Black Jack sill probably extends beyond the iron enrichment shown by the analysed minerals, and an unzoned pyroxene in the teschenite from the roof of the sill has an optically determined composition of approximately $Ca_{46}Mg_{20}Fs_{34}$.

A recent examination of the clinopyroxenes of the Garbh Eilean sill (Gibb, 1973) has shown that zoning is present varying from barely detectable levels in the picrite to extreme zoning in the higher parts of the crinanite. Thus the pyroxene is zoned from $Ca_{45\cdot9}Mg_{39\cdot8}(Fe+Mn)_{14\cdot3}$ to $Ca_{45\cdot8}Mg_{35\cdot7}(Fe+Mn)_{18\cdot5}$ in the picrite, $Ca_{45\cdot0}Mg_{43\cdot7}(Fe+Mn)_{11\cdot3}$ to $Ca_{44\cdot9}Mg_{37\cdot6}(Fe+Mn)_{17\cdot5}$ in the lower picrodolerite, and reaches a maximum $Ca_{44\cdot0}Mg_{26\cdot8}(Fe+Mn)_{29\cdot2}$ to $Ca_{47\cdot2}Mg_{2\cdot3}(Fe+Mn)_{50\cdot5}$ at the 90 m level in the crinanite. The dominant substitution initially is Fe for Mg with Ca remaining constant or slightly decreasing, and then followed later by an increase in Ca. The crystallization trend, as defined by the zoning, is thus comparable in extent with that of large differentiated non-alkaline basic intrusions. The pyroxene trend, calcic augite → calcic ferroaugite → hedenbergite of this mildly alkaline basic magma, is, however, in contrast to the typical augite → ferroaugite → ferrohedenbergite trend of strongly differentiated tholeiitic intrusions, and the common salite → ferrosalite → aegirine sequence of alkaline basic suites.

Although most subcalcic augites and subcalcic ferroaugites have been described from lavas, they have also been reported from more slowly cooled intrusive rocks, e.g. the olivine dolerite, Portrush, Co. Antrim (Harris, 1937), and the Mount Arthur dolerite complex, East Griqualand (Poldervaart, 1946). At the latter locality the trend of crystallization towards subcalcic augite and subcalcic ferroaugite is shown by the zoned pyroxene ($Wo_{42}En_{41}Fs_{17}$ to $Wo_{32}En_{39}Fs_{29}$) of the olivine dolerite, and by the zoned pyroxene ($Wo_{39}En_{38}Fs_{23}$ to $Wo_{32}En_{35}Fs_{33}$) in tholeiite (see also Walker and Poldervaart, 1941a and b). The crystallization trend, ferroaugite, $Ca_{40}Mg_{16}Fe_{44}$, to ferrohedenbergite, $Ca_{42}Mg_{6}Fe_{52}$, of the Ca-rich clinopyroxene phenocrysts of the strongly differentiated shallow Karroo intrusion at Birds River, Dordrecht district, South Africa (Eales and Robey, 1976) is, however, similar to that displayed by the iron-rich clinopyroxenes of the Skaergaard intrusion.

Titanaugite is a relatively common constituent of alkaline dykes, e.g. in the Carboniferous–Permian monchiquites and camptonites of Argyll, Scotland (Gallagher, 1963), and is the principal constituent of the alkaline pyroxenite dykes of Seiland, northern Norway (Robins, 1974). The pyroxene occurs as large, commonly twinned, euhedral or subhedral crystals, and as small anhedral grains formed by the recrystallization of the larger crystals, in coronas along ilmenite–plagioclase grain boundaries and as symplectic intergrowths with ilmenite.

Highly aluminous (7·36–9·83 weight per cent Al_2O_3) titanaugites ($Ca_{49\cdot2-51\cdot8}Mg_{33\cdot1-36\cdot0}Fe_{14\cdot7-15\cdot1}$, and 3·27–3·96 per cent TiO_2) are the principal ferromagnesian constituents of the essexitic and theralitic gabbro veins and dykes in the basalts of Lanzarote and Fuerteventura, Canarian Archipelago (Santin, 1970).

Volcanic rocks. Augite is a usual constituent of basic lavas and occurs in members of both the tholeiitic and olivine alkali basalt suites. In the tholeiitic lavas the trend of the Ca-rich pyroxene crystallization commonly shows a marked contrast to the equilibrium trend of plutonic tholeiitic fractionation, those of the groundmass in particular are often subcalcic with compositions lying in the two-pyroxene field and are usually considered to be metastable phases resulting from rapid cooling.

The subcalcic nature of augite during the later stages of crystallization, as rims to phenocrysts or as a constituent of the groundmass, is well illustrated by the pyroxenes of the younger Tongan andesites and dacites (Ewart, 1976). The pyroxenes in these lavas display great compositional variability and span the augite–ferroaugite, subcalcic augite–ferroaugite and pigeonite–ferropigeonite fields, characteristic of the quench or metastable trend (Fig. 183). The groundmass

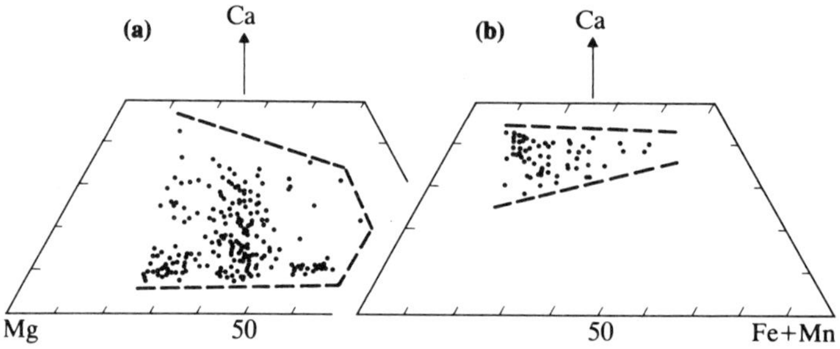

Fig. 183. (a) Compositional field (Ca,Mg,Fe + Mn atomic per cent) of groundmass pyroxenes of Tongan andesites and dacites. (b) Compositional field of groundmass pyroxenes of Niua Fo'ou olivine tholeiite (after Ewart, 1976).

augites of the associated olivine tholeiites of Niua Fo'ou Island, SW Pacific, in contrast, do not extend to the subcalcic field and the compositional change is closer to the pyroxene equilibrium trend. The random variation of ΣAl and ΣFe shown by the augites of the groundmass of the Tongan andesites and dacites is probably due to slow diffusion in the liquid relative to the pyroxene growth rate and the coprecipitation of plagioclase that led to irregular concentration gradients under the conditions of quench crystallization.

Comparable relationships are shown by the groundmass pyroxenes of the basalts and basaltic andesites of the Thingmuli volcano, eastern Iceland (Carmichael, 1967a). The augites have an outer zone of pigeonite, and there is a well defined miscibility gap between the Ca-rich and Ca-poor pyroxenes (Fig. 184). The presence of two coexisting pyroxene phases and the absence of subcalcic augite in some volcanic rock series indicate that the rate of cooling of the liquids is not the sole factor in controlling the trend of pyroxene crystallization. The Thingmuli tholeiites are relatively iron-rich and the pyroxene equilibrium assemblage in this instance may be related to the lower viscosity and temperature of the liquids from which they crystallized.

The first pyroxenes to crystallize in the volcanic rocks of the Izu-Hakone province, Japan (Kuno, 1955), have compositions close to diopside. As crystallization proceeded, the content of Ca decreased and that of Fe increased until the normal limit of replacement, i.e. a Ca:(Mg + Fe) ratio of 25:75, was reached. With further crystallization the composition of the magma became andesitic; the temperature of this magma was usually not sufficiently high to permit additional replacement of Ca by either Mg or Fe, with the result that in general two pyroxene phases crystallized, one with a higher $CaSiO_3$ content than 25 per cent and the other a pigeonite with $CaSiO_3$ content lower than 15 per cent. The rapid cooling of high-temperature magma, however, gave rise in certain cases to the crystallization of subcalcic augites in the groundmass of some of the basalts (e.g. Table 35, anals. 3 and 6). In the tholeiitic series of the Hachijō-Jima volcanic group (Isshiki, 1963), ranging from basalts and andesites to dacites, the clinopyroxenes cover a wide

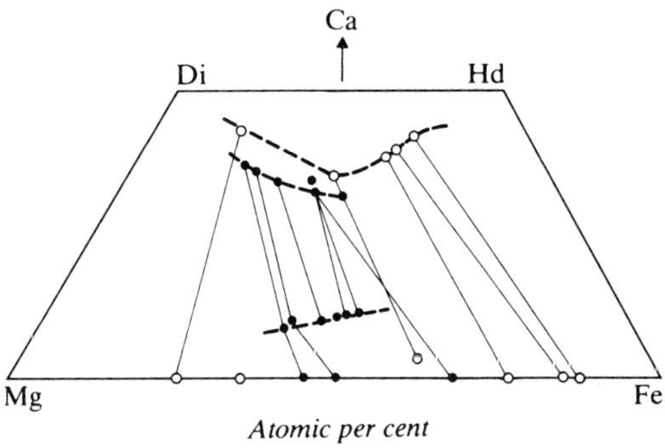

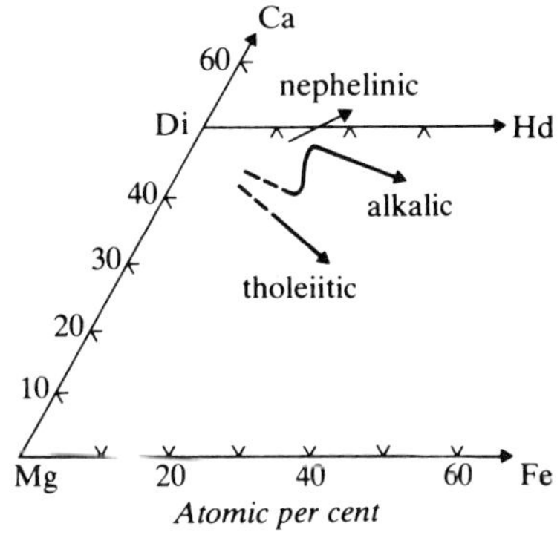

Fig. 184. Average compositions of the pyroxenes of the olivine tholeiite, basaltic andesite, islandite and pitchstone lavas of Thingmuli, Iceland (after Carmichael, 1967a). Open circles, phenocrysts; solid circles, groundmass constituents. Coexisting phenocrysts and coexisting groundmass associations joined by tie-lines. Olivine composition shown as open circles along Mg–Fe join.

Fig. 185. Differentiation trends of phenocrysts and groundmass pyroxenes of the tholeiitic, alkalic and nephelinic suites of Hakeakala and West Maui volcanoes, Hawaii (after Fodor *et al.*, 1975). Broken lines, phenocrysts; solid line, groundmass pyroxenes.

compositional spectrum. The augite of the olivine-eucrite is $Wo_{45}En_{37}Fs_{18}$ but with increasing differentiation of the parent magma the pyroxene phase became enriched in iron, and ferroaugite, $Wo_{36}En_{31}Fs_{33}$, occurs as phenocrysts and as included grains in the ferropigeonite of the andesites. More iron-rich compositions, $Wo_{41}En_{17}Fs_{42}$, occur in the quartz andesites and dacites, and a calcium-poor ferroaugite, $Wo_{30}En_{20}Fs_{50}$, has been described from the iron-rich basalt of the Nishi-yama volcano.

Electron probe analyses of coexisting phenocrysts of augite and orthopyroxene from a tholeiitic andesite of the Hakone volcano have been presented by Nakamura and Kushiro (1970a). Augite also occurs as inclusions and rims round the orthopyroxene phenocrysts and as rims round both the groundmass orthopyroxene and pigeonite. There is a marked break between the augite, $Ca_{33.1}Mg_{41.6}Fe_{25.3}$, and pigeonite, $Ca_{7.8}Mg_{61.4}Fe_{30.8}$, and the absence of transitional subcalcic augite composition indicates the lack of continuous solid solution between the two phases, at an Fe/(Fe+Mg) ratio of approximately 0·3, during the crystallization of the groundmass. Tie-lines between groundmass augite and pigeonite compositions differ considerably from those of the pyroxene pairs of the Skaergaard gabbros. The augite is richer in iron, possibly due to later crystallization from a more iron-rich liquid than that from which the two Ca-poor phases precipitated.

Augite microphenocrysts, zoned from $Ca_{39.4}Mg_{49.5}Fe_{11.1}$ to $Ca_{32.2}Mg_{54.6}Fe_{13.2}$ and showing oscillatory zoning have been described from the hypersthene basalt, Koolan Series, Oahu (Muir and Long, 1965). The groundmass pyroxene is a subcalcic augite some of which is intergrown with pigeonite. Strongly zoned augite also occurs in these basalts jacketing hypersthene phenocrysts forming parallel growths that began under intratelluric conditions and continued after the lava erupted.

The pyroxenes of the tholeiitic, alkalic and nepheline suites of the Hakeakala and West Maui volcanoes, Hawaii (Fodor *et al.*, 1975), show the characteristic augite, pigeonite and orthopyroxene association in the rocks of tholeiitic affinity, salite, augite and ferroaugite in the alkalic, and salite in the nepheline suites. The calcium contents of the Ca-rich pyroxenes range Wo_{30}–Wo_{40}, Wo_{38}–Wo_{48} and Wo_{47}–Wo_{51} in the tholeiitic, alkalic and nephelinic rocks respectively (Fig. 185). In the basalts of the alkalic suite the $CaSiO_3$ content of the Ca-rich pyroxenes increases from Wo_{38} to Wo_{45} to between Wo_{45} and Wo_{48} in the hawaiites and mugearites. The pyroxenes of the trachytes have intermediate contents of the $CaSiO_3$ component but are distinguished by their greater Fe content. The groundmass pyroxenes are invariably poorer in Ca and richer in Fe and Mn than the phenocrysts of the individual members of the tholeiitic, alkalic and nephelinic suites.

Evans and Moore (1968), as a result of their study of the augites of the prehistoric Makaopuhi lava lake, have shown that the pyroxene compositions are related to the depth at which they crystallized, and display a smooth gradation from a quench to an intratelluric trend. Zoning is most pronounced in the rapidly cooled rocks (Fe_{10}–Fe_{36}), and decreases with depth to Fe_{14}–Fe_{19} at 58 m below the lava surface.

In the rocks of the San Juan region, Colorado (Larsen *et al.*, 1936), the pyroxenes of the basalts have a higher Fe : Mg ratio than those of the latites and rhyolites. The decrease in iron content in the pyroxenes from basalt to rhyolite is due to the fact that the iron goes preferentially into the hornblende and biotite and into the iron oxide. The pyroxene within the individual rock type shows a normal zoning with an increased Fe : Mg ratio toward the margin.

Phenocrysts, up to 8 cm, of greenish black and black augites, in the basalt of the Paneake Range, Nevada, commonly display partial resorption. Others are zoned and consist of a clear colourless core and exterior zone separated by a narrow band containing many inclusions and probably representing disequilibrium of the early augites with the liquid. The smaller phenocrysts are not resorbed and both they and the outer zone of the larger crystals are most probably products of equilibrium crystallization.

Augites of the Keweenawan lavas of northern Minnesota, consisting of a series of tholeiitic olivine basalt, basaltic andesite, quartz latite and rhyolite, range in composition from $Ca_{44}Mg_{45}Fe_{11}$ to $Ca_{43}Mg_{12}Fe_{45}$ (Konda and Green, 1974). The composition trend is similar to that of the Ca-rich pyroxenes of the Skaergaard intrusion rather than the 'quench trend' of some volcanic rock series, and is interpreted as a result of the relatively slow cooling of the thick flows. Typical compositions of the ferroaugites of the pyroxene andesite and quartz latite are $Ca_{34.4}Mg_{32.2}Fe_{33.4}$ and $Ca_{40.1}Mg_{19.0}Fe_{40.9}$ respectively.

Magnesium-rich augite phenocrysts (Table 30, anal. 11) have been described from an acid pitchstone, Reydarfjordur, Iceland (Carmichael, 1963). The augites enclose small crystals of titanomagnetite, the iron ratios of which are higher than those of the pitchstones or their residual glasses. The crystallization of the relatively iron-poor augite from such an acid magma is believed to result from the impoverishment of the liquids in iron consequent on the early precipitation of the iron oxide. This suggestion is supported by the presence of sodic ferrohedenbergites in pantelleritic obsidians in which microphenocrysts of magnetite are absent (Carmichael, 1962). Magnesium-rich augite phenocrysts, $Ca_{39.2}Mg_{48.5}Fe_{12.3}$, have also been described from the rhyolitic pumice of the Taupo volcanic zone, New Zealand (Ewart, 1967), and titanaugite from trachyte (Chayes and Zies, 1964).

The compositions of ten augite phenocrysts from salic volcanic rocks, rhyolitic and phenolitic obsidians, kenyte and dacite examined by Carmichael (1967b) display a wide range of Fe/Mg ratios ($Ca_{43.5}Mg_{3.2}Fe_{53.3}$–$Ca_{45.2}Mg_{38.3}Fe_{16.5}$). The pyroxenes except for the more iron-rich varieties, show little zoning, and the trend of the tie-lines between augites and coexisting orthopyroxene, by comparison with the Skaergaard pyroxene data, are mutually in equilibrium. Precipitation temperatures, based on the equilibration of the associated iron oxides, show an absence of a systematic relationship with the compositions of the augites indicating that the initial chemistry of the pyroxenes is, in the main, controlled by the composition of the magmas.

Ophitic grains of a brownish augite surrounded by broad rims of a lemon-yellow variety, the latter also occurring in the groundmass, occur in the enstatite–forsterite basalt, Kolbeinsey Island, north of Iceland (Sigurdsson and Brown, 1970). The average composition, determined by electron probe, of the ophitic and groundmass pyroxenes is $Wo_{42}En_{45}Fs_{13}$. The yellow colour of the augite rims and the groundmass clinopyroxene, however, is indicative of a relatively high ferric iron content; the ferrosilite component given above may be greater than its real value, and it is possible that the pyroxenes have a ferri-diopside composition. Augites, some with high contents of aluminium (4–8 per cent Al_2O_3) and chromium (0·4–0·8 per cent Cr_2O_3) have been reported from submarine basalts (Bryan, 1972).

Augite and pigeonite are the main pyroxenes of the Akrotiri-Thira and main series lavas, basalts, basaltic andesites, andesites, dacites and rhyodacites, of the Santorini volcano, Cyclades (Nicholls, 1971). The phenocryst and groundmass trends (Fig. 186) are comparable with those of the Thingmuli series (Carmichael,

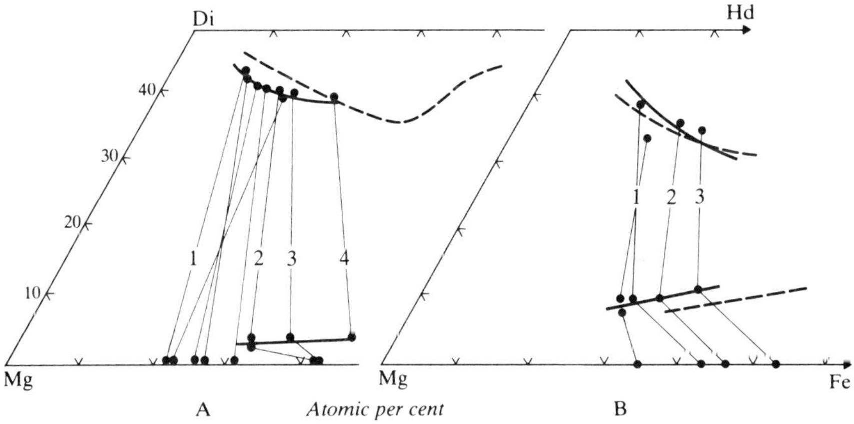

Fig. 186. Average composition of phenocrystal (A) and groundmass (B) pyroxenes and olivines in (1) basalts, (2) basaltic andesites, (3) andesites and (4) rhyodacites of the Santorini volcano (after Nicholls, 1971). Part of the pyroxene trends of the Thingmuli lavas are shown by broken lines.

1967a). The Wo and Fs components of the augite phenocrysts range from 46–39 and 8–24 per cent respectively, compared with values of 43–31 and 11–30 for the Ca-rich pyroxene of the groundmass. Subcalcic augite is not present in the Santorini lavas, and pigeonite, Fs_{29}–Fs_{38}, is restricted to the groundmass. Orthopyroxene occurs as phenocrysts, the Fs component increases with differentiation, from Fs_{29} to Fs_{44}, and thus shows a complete compositional overlap with the groundmass pigeonite. The possible relationship between the two Ca-poor pyroxenes is illustrated schematically in Fig. 187. On the assumption that the orthopyroxene phenocrysts represent intratelluric crystallization, the movement of basaltic and andesitic magmas to higher crustal levels could, on the reduction in pressure, lead the crystallizing pyroxene to move into the pigeonite stability field. In the case of the more highly differentiated rhyodacite liquid a similar decrease in total pressure, because of the lower initial temperature, would be insufficient to produce crystallization of a Ca-poor clinopyroxene. Variation in water pressure may also have an important influence on the crystallization of Ca-poor pyroxene. Thus initial high P_{H_2O} at depth would favour the formation of orthopyroxene, while decreasing

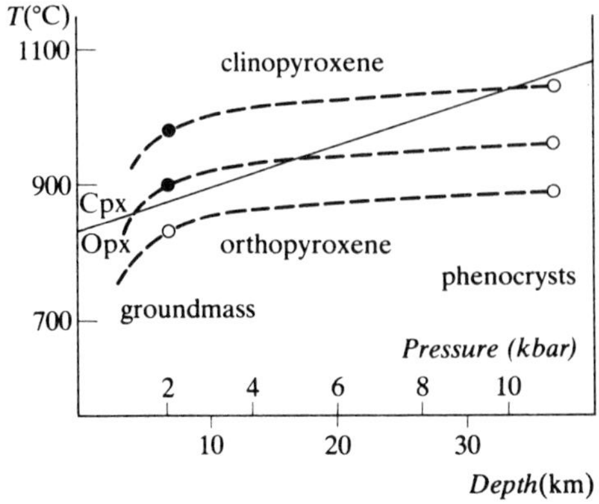

Fig. 187. Hypothetical pressure (depth)–temperature relationships during crystallization of phenocrystal and groundmass pyroxenes of the main series lavas, Santorini volcano (after Nicholls, 1971).

water pressure as the magma rose to the surface, with consequent rise in temperature, would lead to the crystallization of pigeonite.

The augites of alkali basalts are commonly richer in titanium than those of the tholeiitic basalts, and the presence of titaniferous augites and titanaugites is generally considered to be an indication of the alkalic nature of the host rock. Factors other than the initial magmatic composition may, however, affect the titanium content of pyroxenes. The influence of high magmatic water pressure on the nature of pyroxene has been discussed by Barberi *et al.* (1971). These authors have suggested that the alkaline series may be due to crystallization and differentiation under these conditions, the precipitation of pyroxene taking place before plagioclase with the consequence that the pyroxene is enriched in the $CaAl_2SiO_6$ component, and has a high titanium content due to the replacement $R^{2+} + 2Si \rightleftharpoons Ti^{4+} + 2Al$. At lower water pressures plagioclase crystallizes before pyroxene, leading to a smaller $CaAl_2SiO_6$ content and less titanium thus enabling iron enrichment to occur and the pyroxene trend to be more tholeiitic in character. A sympathetic relationship between titanium and aluminium is, however, not universal, and Gramse (1970) has described phenocrystal augites from the Nordhessen and Sudniedersachen basalts, the cores of which are richer in the $CaAl_2SiO_6$ component and have a smaller titanium content in the outer zones of the crystals as well as in the pyroxenes of the groundmass.

The Tertiary volcanics of the Volgelsberg area (Schorer, 1970) include members of both the tholeiitic and alkali basalt series. The Ca-rich pyroxenes of the former are augites, ranging in composition between $Ca_{40}Mg_{48}Fe_{12}$ and $Ca_{30}Mg_{35}Fe_{35}$, and are in equilibrium with pigeonite, $Ca_7Mg_{68}Fe_{25}$–$Ca_{13}Mg_{46}Fe_{41}$. The diopside-augites of the alkali series have a more varied composition and include representatives of megacrystal, phenocrystal and groundmass crystallization. All are rich in calcium ($Wo_{40\cdot4}$–$Wo_{48\cdot6}$), with up to 15 weight per cent in the high-pressure megacrysts; the phenocrystal and groundmass augites have a high titanium content. Analyses of the pyroxenes in spilites are rare, but present data indicate that augitic compositions coexist in equilibrium with chlorite and albite. Light and heavy fractions of the augites of a Deccan flow showing transition from black tholeiitic basalt to a green albite–chlorite spilite have been separated by Vallance (1974). The lighter fractions have an average composition $Ca_{34\cdot4}Mg_{47\cdot2}Fe_{18\cdot4}$. The darker fractions are more iron-rich and approach a subcalcic composition (Table 30, anal. 13) but, like the lighter fractions, there is little difference in the chemistry of the augites from the basaltic and spilitic parts of the flow.

The parent basaltic magma of the Erta'Ale volcanic series, Ethiopia, covering a differentiation series from basalt to alkaline and peralkaline rhyolite, is transitional between tholeiitic and alkali olivine basalts (Barberi *et al.*, 1971). In this series the zoning and the contrasted compositions between the augite microphenocrysts and groundmass crystals closely resemble those of a tholeiitic suite.

Ferrohedenbergite, $Wo_{42}En_8Fs_{50}$ (composition determined from X-ray single crystal and powder data) as phenocrysts in welded tuff, Devine Canyon, south-eastern Oregon, has been described by Greene (1973).

Augitic alkalic lava consisting mainly of titanaugite with subordinate amounts of titanomagnetite and olivine forming the cores of manganese nodules dredged from the ocean floor south of the Cook Islands, Pacific has been reported by Prokoptsev and Murdmaa (1970).

Metamorphic rocks

In most regionally metamorphosed rocks the composition of the clinopyroxene lies in or close to the salite field, e.g. in the rocks of the Wilmington complex, U.S.A. (Ward, 1959), the compositional range of the clinopyroxenes is from $Wo_{46}En_{42}Fs_{12}$ to $Wo_{44}En_{37}Fs_{19}$. In the regionally metamorphosed rocks of the Central Abukuma Plateau, Japan (Shidô, 1958; Miyashiro, 1958), the clinopyroxene in the rocks of the amphibolite facies is salitic, but in the higher grade rocks of the granulite and pyroxene-hornfels facies the pyroxenes are richer in ($Mg:Fe^{2+}$) and have augitic compositions. This relationship between the composition of the pyroxenes and the metamorphic environment of their crystallization may possibly be related to the increasing solubility of Mg and Fe^{2+} in diopside–hedenbergite at higher temperatures. The formation of augite during the reconstitution, at progressively higher temperatures and pressures, of the amphibolites of the Adirondack massif has been described by Engel *et al.* (1964). The pyroxenes are uniformly rich in calcium (43·7–45·3 atomic per cent Ca), and in some of the reconstructed amphibolites have a salite composition. The development of augite in pyroxene hornfels at the contact of an Alpine-type ultramafic intrusion with spilitic lavas has been reported, and the pyroxene geothermometry of the granulite facies assemblage of the Broken Hill area, New South Wales and other granulite terrains discussed by Hewins (1974). High aluminium augites (fassaitic augites) are present in clinopyroxene–orthopyroxene–plagioclase–scapolite, and clinopyroxene–garnet–plagioclase–scapolite assemblages of the granulite inclusions in breccia and basalt-filled pipes of eastern Australia (Lovering and White, 1964).

In charnockites, clinopyroxenes are normally restricted to the less acid members of the series. In the ultrabasic rocks of the Madras charnockite series (Howie, 1955), the clinopyroxene has a diopsidic composition, but although in the less basic rocks the clinopyroxene is an augite, and in the intermediate rock of Tinnevelly it is a ferroaugite, the clinopyroxenes of this series are all characterized by a high calcium content, and their compositions lie close to the diopside–hedenbergite join. The clinopyroxenes of the charnockitic rocks of the West Uusimaa complex, Finland (Parras, 1958), have even higher calcium contents and their compositions all lie within the salite–ferrosalite field.

Sodian augite (Table 38, anal. 1), formed by exsolution of the sodic component from omphacite, occurs in the clinopyroxene plagioclase symplectites of the Hareidland eclogite, Norway (Mysen, 1972), and an aluminium-rich sodian augite, containing 6·6 and 17·3 per cent of the $NaAlSi_2O_6$ and $CaAl_2SiO_6$ components respectively has been described.

Pyroxenes of the diopside–ferroaugite–hedenbergite series are the common calcium-bearing silicates in the metamorphic Iron Formation of northern Quebec (Kranck, 1961). The augite is typically associated with orthopyroxene and calcite, and in some assemblages with members of the cummingtonite–grunerite series. The formation of the augite involved, at least initially, a reaction between ferrodolomite and quartz:

$$Ca(Mg,Fe)(CO_3)_2 + 2SiO_2 \rightarrow Ca(Mg,Fe)Si_2O_6 + 2CO_2$$

and was possibly followed at higher temperatures by the instability of the ferroaugite in the presence of CO_2

$$Ca(Mg,Fe)Si_2O_6 + CO_2 \rightarrow (Mg,Fe)SiO_3 + CaCO_3 + SiO_2$$

Augite does not occur in the more extreme iron-rich rocks, in which only orthopyroxene and calcite are present, and ferroaugites containing 63 mol. per cent $CaFeSi_2O_6$ are the most iron-rich Ca-bearing pyroxenes in the Iron Formation (see also orthopyroxene section). More recently, Butler (1969) has described similar silicate–carbonate assemblages in which clino- and orthopyroxene and quartz are the dominant constituents, and which are associated with calcite and actinolite or grunerite. The augites and ferroaugites display a compositional range similar to that described by Kranck, as do the other silicates of the assemblages, and the regularity of the distribution of Fe, Mn and Ca between coexisting mineral pairs, as well as the distribution of Mn between coexisting silicates and carbonates indicate that equilibrium was generally attained. The compositional variations are discussed by Butler, in relation to changes in the chemical potential of CO_2 and H_2O (Fig. 188),

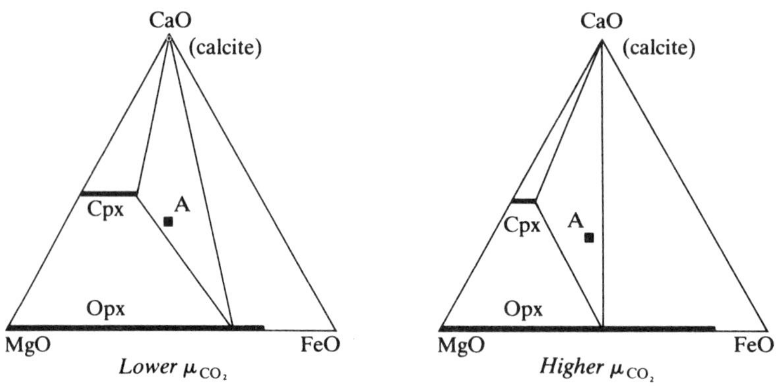

Fig. 188. Diagrammatic representation of the effects of variation in μ_{CO_2} at constant temperature and pressure for the assemblage clinopyroxene–orthopyroxene–quartz–calcite (after Butler, 1969). Compositions (mol. per cent) projected through SiO_2 and CO_2. For a given composition, A ■, the proportion of clinopyroxene decreases and the proportions of orthopyroxene and calcite increase with increasing μ_{CO_2}.

and are correlated with high μ_{CO_2} and μ_{H_2O} across, and low μ_{CO_2} and μ_{H_2O} along the strike of the formation. Augite is present in the metamorphosed ferruginous sediments of the Gunflint Iron Formation associated with the Duluth complex (Simmons *et al.*, 1974). The primary metamorphic augite, $\simeq Wo_{40}En_{20}Fs_{40}$, formed during the initial recrystallization of the sediment, and on subsequent cooling exsolved some 10 per cent pigeonite, $Wo_5En_{22}Fs_{73}$, with the composition of the augite host changing to $Wo_{45}En_{19}Fs_{36}$.

An unusual occurrence of subcalcic augite, reported to contain 0·40 per cent ZnO, has been described from the 'Tábor syenite', central Bohemian pluton, the rocks of which have been subjected to granulite and amphibolite facies metamorphism (Jakeš, 1968).

References

Abbott, D. and **Deer, W. A.,** 1972. Geological investigations in east Greenland. Part X. The gabbro cumulates of the Kap Edvard Holm lower layered series. *Medd. om Grønland*, **190**, 1–42.

Akella, J. and **Boyd, F. R.,** 1972. Partitioning of Ti and Al between pyroxenes, garnets and oxides. *Carnegie Inst. Washington, Ann. Rept. Dir. Geophys. Lab.*, 1971–72, 378–384.

Aleksandrov, K. S., Ryzhova, T. V. and **Belikov, B. P.,** 1964. The elastic properties of pyroxenes. *Soviet Physics: Crystallography*, **8**, 589–591.

Alfani, M., 1934. L'augite pneumatolitica raccolta sulle lave del fondo del cratere vesuviano del 1929. *Periodico Min., Roma*, **5**, 77–96 (M.A. **5**-440).

Aoki, K., 1959. Petrology of alkali rocks of the Iki islands and Higashi-Matsuura district, Japan. *Sci. Rept. Tohoku Univ.*, ser. 3, **6**, 261–310.

Aoki, K., 1962. [The clinopyroxenes of the Otaki dolerite sill] *J. Japanese Assoc. Min. Petr. Econ. Geol.*, **47**, 41–45 (in Japanese, English summary).

Aoki, K., 1964. Clinopyroxenes from alkaline rocks of Japan. *Amer. Min.*, **49**, 1199–1225.

Aoki, K., 1967. The role of titanium in alkali basalt magmas. *J. Japanese Assoc. Min. Petr. Econ. Geol.*, **57**, 188–199.

Aoki, K. and **Kuno, H.,** 1973. Gabbro–quartz diorite inclusions from Izu-Hakone region, Japan. *Bull. Volc.*, **36**, 164–173.

Ashby, D. F., 1946. Pyroxenes from the lower Carboniferous basalts of the Old Pallas area, Co. Limerick. *Min. Mag.*, **27**, 195–197.

Asklund, B., 1923. Petrological studies in the neighbourhood of Stavsjo at Kolmården. *Årsbok Sveriges Geol. Undersök*, **17**, 1–122.

Atkins, F. B., 1969. Pyroxenes of the Bushveld intrusion, South Africa. *J. Petr.*, **10**, 222–249.

Atlas, L., 1952. The polymorphism of $MgSiO_3$ and solid state equilibria in the system $MgSiO_3$–$CaMgSi_2O_6$. *J. Geol.*, **60**, 125–147.

Bailey, J. C., Champness, P. E., Dunham, A. C., Esson, J., Fyfe, W. S., Mackenzie, W. S., Stumpfl, E. F. and **Zussman, J.,** 1970. Mineralogy and petrology of Apollo 11 lunar samples. *Proc. Apollo 11 Lunar Sci. Conf. (Suppl. to Geochim. Cosmochim. Acta* **34***)*, **1**, 169–194.

Barberi, F., Bizouard, H. and **Varet, J.,** 1971. Nature of the clinopyroxene and iron enrichment in alkalic and transitional basaltic magmas. *Contr. Min. Petr.*, **33**, 93–107.

Barsukov, V. I., 1960. [On mineral-indicators of the boron content in skarns.] *Geokhimiya*, **5** (in Russian).

Barth, T. F. W., 1931a. Crystallization of pyroxenes from basalt. *Amer. Min.*, **16**, 195–208.

Barth, T. F. W., 1931b. Pyroxen von Hiva Oa, Marquesas-Inseln und die Formel titanhaltiger Augite. *Neues Jahrb. Min., Abt. A*, **64**, 217–224.

Barth, T. F. W., 1936. Crystallization process of basalt. *Amer. J. Sci.*, 5th ser., **31**, 321–351.

Barth, T. F. W., 1951. Subsolidus diagram of pyroxenes from common mafic magmas. *Norsk Geol. Tidsskr.*, **29**, 218–221.

Beeson, M. H. and **Jackson, E. D.,** 1970. Origin of the garnet pyroxenite xenoliths at Salt Lake Crater, Oahu. *Min. Soc. America, Spec. Paper* **3**, 95–112.

Bence, A. E., Papike, J. J. and **Lindsley, D. H.,** 1971. Crystallization histories of clinopyroxenes in two porphyritic rocks from *Oceanus Procellarum. Proc. Second Lunar Sci. Conf. (Suppl.* **2** *to Geochim. Cosmochim. Acta)*, **1**, 559–574.

Benjamin, T. M., Burnett, D. S. and **Seitz, M. G.,** 1976. Experimental plutonium, thorium and uranium partition coefficients with application to meteoritic assemblages. *Carnegie Inst. Washington, Ann. Rept. Dir. Geophys. Lab.*, 1975–76, 684–687.

Benson, W. N., 1944. The basic igneous rocks of eastern Otago and their tectonic environment. Part IV. The mid-Tertiary basalts, tholeiites and dolerites of north-eastern Otago. Section B: Petrology with special reference to the crystallization of pyroxene. *Trans. Roy. Soc. New Zealand*, **74**, 71–123.

Bertolani, M., 1967. Rocce manganesifere tra le granuliti della valle Strona (Novara). *Periodico Min., Roma*, **36**, 1011–1032.

Best, M. G., 1970. Kaersutite-peridotite inclusions and kindred megacrysts in basanitic lavas, Grand Canyon, Arizona. *Contr. Min. Petr.*, **27**, 25–44.

Best, M. G. and **Mercy, E. L. P.,** 1967. Composition and crystallization of mafic minerals in the Guadalupe igneous complex, California. *Amer. Min.*, **52**, 436–474.

Binns, R. A., 1969. High-pressure megacrysts in basanitic lavas near Armidale, New South Wales. *Amer. J. Sci.*, **267-A** (Schairer vol.), 33–49.
Blyth, F. G. H., 1948. Pyroxene from the Squilver dolerite, south Shropshire. *Min. Mag.*, **28**, 380–383.
Boesen, R. S., 1964. The clinopyroxenes of a monzonitic complex at Mount Dromedary, New South Wales. *Amer. Min.*, **49**, 1435–1457.
Borg, I. Y. and **Smith, D. K.**, 1969. Calculated X-ray powder patterns. *Mem. Geol. Soc. America*, **122**.
Borisenko, L. F., 1967. Correlation among iron, vanadium and titanium in pyroxenes. *Geochem. International*, **4**, 263–269.
Bose, M. K., 1968. Mineralogical study of striped pyroxene in syenitic rocks of Sivamalai, South India. *Amer. Min.*, **53**, 464–471.
Bowen, N. L. and **Schairer, J. F.**, 1935. The system MgO–FeO–SiO_2. *Amer. J. Sci.*, 5th ser., **29**, 151–217.
Bowen, N. L., **Schairer, J. F.** and **Posnjak, E.**, 1933. The system CaO–FeO–SiO_2. *Amer. J. Sci.*, 5th ser., **26**, 193–284.
Boyd, F. R. and **Brown, G. M.**, 1968. Electron-probe study of exsolution in pyroxenes. *Carnegie Inst. Washington, Ann. Rept. Dir. Geophys. Lab.*, 1966–67, 353–359.
Boyd, F. R. and **Nixon, P. H.**, 1972. Ultramafic nodules from the Thaba Putsoa kimberlite pipe. *Carnegie Inst. Washington, Ann. Rept. Dir. Geophys. Lab.*, 1971–72, 362–373.
Boyd, F. R. and **Schairer, J. F.**, 1957. The join $MgSiO_3$–$CaMgSi_2O_6$. *Carnegie Inst. Washington, Ann. Rept. Dir. Geophys. Lab.*, 1956–57, 223–225.
Boyd, F. R. and **Schairer, J. F.**, 1962. The system $MgSiO_3$–$CaMgSi_2O_6$. *Carnegie Inst. Washington, Ann. Rept. Dir. Geophys. Lab.*, 1961–62, 68–75.
Boyd, F. R. and **Smith, D.**, 1971. Compositional zoning in pyroxenes from Lunar Rock 12021 *Oceanus Procellarum. J. Petr.*, **12**, 439·465.
Brousse, R., 1961. Minéralogie et pétrographie des roches volcaniques du Massif du Mont-Dore (Auvergne). I. Les minéraux. II. Petrographie systematique. *Bull. Soc. franç. Min. Crist.*, **84**, 131–186.
Brousse, R. and **Berger, E.**, 1965. Grands cristaux d'augite dans une brèche volcanique intrusive. *Bull. Soc. franç. Min. Crist.*, **88**, 525–526.
Brown, F. H. and **Carmichael, I. S. E.**, 1969. The Quaternary volcanoes of the Lake Rudolph region. I. The basanite–tephrite series of the Korath Range. *Lithos*, **2**, 239–260.
Brown, G. M., 1956. The layered ultrabasic rocks of Rhum, Inner Hebrides. *Phil. Trans. Roy. Soc., ser. B*, **240**, 1–53.
Brown, G. M., 1957. Pyroxenes from the early and middle stages of fractionation of the Skaergaard intrusion, east Greenland. *Min. Mag.*, **31**, 511–543.
Brown, G. M., 1960. The effect of ion substitution on the unit cell dimensions of the common clinopyroxenes. *Amer. Min.*, **45**, 15–38.
Brown, G. M. and **Vincent, E. A.**, 1963. Pyroxenes from the late stages of fractionation of the Skaergaard intrusion, east Greenland. *J. Petr.*, **4**, 175–197.
Bryan, W. B., 1972. Mineralogical studies of submarine basalts. *Carnegie Inst. Washington, Ann. Rept. Dir. Geophys. Lab.*, 1971–72, 398–403.
Bryhni, I. and **Dons, J. A.**, 1975. Orbicular lamprophyre from Vestby, southern Norway. *Lithos*, **8**, 113–122.
Buddington, A. F., 1952. Chemical petrology of some metamorphosed Adirondack gabbroic, syenitic and quartz syenitic rocks. *Amer. J. Sci.*, Bowen vol., 37–84.
Bulgakova, Y. N., 1969. Investigations of inclusions of molten matter in rock-forming minerals of the Mount Chernaya intrusion, Noril'sk district. *Dokl. Acad. Sci. U.S.S.R., Earth Sci. Sect.*, **185**, 132–135.
Burnham, C. W., 1967. Ferrosilite. *Carnegie Inst. Washington, Ann. Rept. Dir. Geophys. Lab.*, 1965–66, 285–290.
Burnham, C. W., **Clark, J. R.**, **Papike, J. J.** and **Prewitt, C. T.**, 1967. A proposed crystallographic nomenclature for clinopyroxene structures. *Z. Krist.*, **125**, 109–119.
Burns, R. G., 1970. *Mineralogical Applications of Crystal Field Theory*. Cambridge University Press.
Burns, R. G., 1972. Site preferences of Ni^{2+} and Co^{2+} in clinopyroxene and olivine: limitations of the statistical approach. *Chem. Geol.*, **9**, 67–73.
Butler, P., *Jr.*, 1969. Mineral compositions and equilibrium in the metamorphosed iron formation of the Gagnon Region, Quebec, Canada. *J. Petr.*, **10**, 56–101.
Cameron, M., **Sueno, S.**, **Prewitt, C. T.** and **Papike, J. J.**, 1973. High temperature crystal chemistry of acmite, diopside, hedenbergite, jadeite, spodumene and ureyite. *Amer. Min.*, **58**, 594–618.
Campbell, I. H. and **Borley, G. D.**, 1974. The geochemistry of pyroxenes from the lower layered series of the Jimberlana intrusion, Western Australia. *Contra. Min. Petr.*, **47**, 281–297.
Campbell, I. H. and **Nolan, J.**, 1974. Factors effecting the stability field of Ca-poor pyroxene and origin of the Ca-poor minimum in Ca-rich pyroxenes from tholeiitic intrusions. *Contr. Min. Petr.*, **48**, 205–219.

Carmichael, I. S. E., 1960. The pyroxenes and olivines from some Tertiary acid glasses. *J. Petr.,* **1,** 309–336.

Carmichael, I. S. E., 1962. Pantelleritic liquids and their phenocrysts. *Min. Mag.,* **33,** 86–113.

Carmichael, I. S. E., 1963. The occurrence of magnesian pyroxenes and magnetite in porphyritic acid glasses. *Min. Mag.,* **33,** 394–403.

Carmichael, I. S. E., 1967a. The mineralogy of Thingmuli, a Tertiary volcano in eastern Iceland. *Amer. Min.,* **52,** 1815–1841.

Carmichael, I. S. E., 1967b. The iron-titanium oxides of salic volcanic rocks and their associated ferromagnesian silicates. *Contr. Min. Petr.,* **14,** 36–64.

Carmichael, I. S. E., Nicholls, J. and **Smith, A. L.,** 1970. Silica activity in igneous rocks. *Amer. Min.,* **55,** 246–263.

Carstens, H., 1958. Note on the distribution of some minor elements in coexisting ortho- and clino-pyroxenes. *Norsk Geol. Tidsskr.,* **38,** 257–260.

Challis, G. A., 1963. Layered xenoliths in a dyke, Awatere Valley, New Zealand. *Geol. Mag.,* **100,** 11–16.

Challis, G. A., 1965. High-temperature contact metamorphism at the Red Hills ultramafic intrusion–Wairau Valley, New Zealand. *J. Petr.,* **6,** 395–419.

Chapman, N. A., 1975. An experimental study of spinel clinopyroxenite xenoliths from the Duncansby Ness vent, Caithness, Scotland. *Contr. Min. Petr.,* **51,** 223–230.

Chapman, N. A., 1976. Inclusions and megacrysts from undersaturated tuffs and basanites, East Fife, Scotland. *J. Petr.,* **17,** 472–498.

Chapman, P. A. and **Meagher, E. P.,** 1975. A technique for observing exsolution lamellae in pyroxenes with the scanning electron microscope. *Amer. Min.,* **60,** 155–156.

Chapman, R. W. and **Williams, C. R.,** 1935. Evolution of the White Mountain magma series. *Amer. Min.,* **20,** 502–530.

Chayes, F. and **Zies, E. G.,** 1964. Notes on some Mediterranean comendite and pantellerite specimens. *Carnegie Inst. Washington, Ann. Rept. Dir. Geophys. Lab.,* 1963–64, 186–190.

Chodyniecka, L., 1967. [The basalt from the Swieta Anna mountain.] *Polska Akad. Nauk, Prace Min.,* **8,** 1–56 (Polish, with Russian and English summaries).

Clark, J. R., Appleman, D. E. and **Papike, J. J.,** 1969. Crystal-chemical characterization of clinopyroxenes based on eight new structure refinements. *Min. Soc. America, Spec. Paper* **2,** 31–50.

Clarke, P. D. and **Wadsworth, W. J.,** 1970. The Insch layered intrusion. *Scottish J. Geol.,* **6,** 7–25.

Coats, R. R., 1968. The Circle Creek rhyolite, a volcanic complex in northern Elko County, Nevada. *Mem. Geol. Soc. America,* **116,** 69–106.

Coleman, L. C., 1962. Effect of ionic substitution on the unit-cell dimensions of synthetic diopside. *Geol. Soc. America* (Buddington vol.), 429–446.

Collins, L. G. and **Hagner, A. F.,** 1965. Refractive index field diagrams and mineral relationships for coexisting ferromagnesian silicates. *Amer. Min.,* **50,** 1381–1409.

Conquéré, F., 1977. Pétrologie des pyroxénites litées dans les complexes metamafiques de l'Ariège (France) et autres gisements de lherzolite à spinelle. I. Compositions minéralogiques et chimiques, évolution des conditions d'équilibre des pyroxenites. *Bull. Soc. franç. Min. Crist.,* **100,** 42–79.

Copley, P. A., Champness, P. E. and **Lorimer, G. W.,** 1974. Electron petrography of exsolution textures in an iron-rich clinopyroxene. *J. Petr.,* **15,** 41–57.

Cundari, A., 1975. Mineral chemistry and petrogenetic aspects of the Vico lavas, Roman volcanic region, Italy. *Contr. Min. Petr.,* **53,** 129–144.

Dane, E. B., *Jr.,* 1941. Densities of molten rocks and minerals. *Amer. J. Sci.,* **239,** 809–818.

Dasgupta, H. C., 1972. Site preference of Ni^{2+} and Co^{2+} in clinopyroxene and olivine: a statistical study. *Chem. Geol.,* **9,** 57–65.

Davis, B. T. C., 1963. The system enstatite–diopside at 30 kilobars pressure. *Carnegie Inst. Washington, Ann. Rept. Dir. Geophys. Lab.,* 1962–63, 103–107.

Dawson, J. B. and **Reid, A. M.,** 1970. A pyroxene-ilmenite intergrowth from the Monastery mine, South Africa. *Contr. Min. Petr.,* **26,** 296–301.

Deer, W. A. and **Abbott, D.,** 1965. Clinopyroxenes of the gabbro cumulates of the Kap Edvard Holm complex, east Greenland. *Min. Mag.,* **34** (Tilley vol.), 177–193.

Deer, W. A. and **Wager, L. R.,** 1938. Two new pyroxenes included in the system clinoenstatite, clinoferrosilite, diopside and hedenbergite. *Min. Mag.,* **25,** 15–22.

Dence, M. R., Douglas, J. A. V., Plant, A. G. and **Traill, R. J.,** 1970. Petrology, mineralogy and deformation of Apollo 11 samples. *Proc. Apollo 11 Lunar Sci. Conf. (Suppl. to Geochim. Cosmochim. Acta* **34***),* **1,** 315–340.

Deriu, M., 1954. Biotite e augite di Monte Columbargui (Montiferro-Sardegna centro-occidentale). *Periodico Min., Roma,* **23,** 27–34.

Desnoyers, C., 1975. Exsolutions d'amphibole, de grenat et de spinelle dans les pyroxènes de roches ultrabasiques; péridotite et pyroxénolites. *Bull. Soc. franç. Min. Crist.,* **98,** 65–77.

DeVore, G., 1955. Crystal growth and the distribution of elements. *J. Geol.*, **63**, 471–494.

DeVore, G., 1957. The association of strongly polarizing cations with weakly polarizing cations as a major influence in element distribution mineral composition, and crystal growth. *J. Geol.*, **65**, 178–195.

Dixon, B. E. and **Kennedy, W. G.**, 1933. Optically uniaxial titanaugite from Aberdeenshire. *Z. Krist.*, **86**, 112–120.

Dostal, J. and **Capedri, S.**, 1975. Partition coefficients of uranium for some rock-forming minerals. *Chem. Geol.*, **15**, 285–294.

Downes, M. J., 1974. Sector and oscillatory zoning in calcic augites from Mt. Etna, Sicily. *Contr. Min. Petr.*, **47**, 187–196.

Dowty, E., 1976a. Crystal structure and crystal growth. I. The influence of internal structure on morphology. *Amer. Min.*, **61**, 448–459.

Dowty, E., 1976b. Crystal structure and crystal growth. II. Sector zoning in minerals. *Amer. Min.*, **61**, 460–469.

Dowty, E. and **Lindsley, D. H.**, 1973. Mössbauer spectra of synthetic hedenbergite–ferrosilite pyroxenes. *Amer. Min.*, **58**, 850–868.

Duchesne, J. C., 1972. Pyroxènes et olivines dans le massif de Bjerkrem–Sogndal (Norvège méridionale). Contribution à l'étude de la série anorthosite-mangérite. *Intern. Geol. Congr., 24th Session, Montreal*, Sect. **2**, 320–328.

Duggan, M. B. and **Wilkinson, J. F. G.**, 1973. Tholeiitic andesite of high-pressure origin from the Tweed Shield volcano, north-eastern New South Wales. *Contr. Min. Petr.*, **39**, 267–276.

Duke, J. M., 1976. Distribution of the period four transition elements among olivine, calcic clinopyroxene and mafic silicate liquid: experimental results. *J. Petr.*, **17**, 499–521.

Duke, M. B. and **Silver, L. T.**, 1967. Petrology of eucrites, howardites and mesosiderites. *Geochim. Cosmochim. Acta*, **31**, 1637–1665.

Dunham, A. C., **Copley, P. A.** and **Strasser-King, V. H.**, 1972. Submicroscopic exsolution lamellae in pyroxenes in the Whin sill, northern England. *Contr. Min. Petr.*, **37**, 211–220.

Eales, H. V. and **Booth, P. W. K.**, 1974. The Birds River gabbro complex, Dordrecht district. *Trans. Geol. Soc. South Africa*, **77**, 1–15.

Eales, H. V. and **Robey, J. van A.**, 1976, Differentiation of tholeiitic Karroo magma at Birds River, South Africa. *Contr. Min. Petr.*, **56**, 101–117.

Eckermann, H. von, 1922. The rocks and contact minerals of the Mansjö Mountain. *Geol. För. Förh. Stockholm*, **44**, 203–410.

Edwards, A. B., 1942. Diffe:entiation in the dolerites of Tasmania, I and II. *J. Geol.*, **50**, 451, 579–610.

Elsdon, R., 1971. Clinopyroxenes from the Upper Layered Series, Kap Edvard Holm, east Greenland. *Min. Mag.*, **38**, 49–57.

Emslie, R. F. and **Ermanovics, I. F.**, 1975. Major rock units of the Morin complex, south-western Quebec. *Paper Geol. Surv. Canada*, **74–48.**

Engel, A. E. J., **Engel, C. G.** and **Havens, R. G.**, 1964. Mineralogy of amphibolite interlayers in the gneiss complex, north-west Adirondack Mountains, New York. *J. Geol.*, **72**, 131–156.

Ernst, T., 1936. Der Melilith-Basalt des Westberges bei Hofgeismar, nördlich von Kassel, ein Assimilationsprodukt ultrabasischer Gesteine. *Chemie der Erde*, **10**, 631–666.

Ernst, T. and **Schorer, G.**, 1969. Die Pyroxene des 'Maintrapps', einer Gruppe tholeiitischer Basalte des Vogelsberges. *Neues Jahrb. Min., Mh.*, 108–130.

Eskola, P., 1952. On the granulites of Lapland. *Amer. J. Sci.* (Bowen vol.), 133–171.

Essene, E. J., **Ringwood, A. E.** and **Ware, N. G.**, 1970. Petrology of the lunar rocks from Apollo 11 landing site. *Proc. Apollo 11 Lunar Sci. Conf. (Suppl. to Geochim. Cosmochim. Acta* **34**), **1**, 385–397.

Evans, B. W. and **Moore, J. G.**, 1968. Mineralogy as a function of depth in the prehistoric Makaopuhi tholeiitic lava lake, Hawaii. *Contr. Min. Petr.*, **17**, 85–115.

Ewart, A., 1967. Pyroxene and magnetite phenocrysts from the Taupo quaternary rhyolitic pumice deposits, New Zealand. *Min. Mag.*, **36**, 180–194.

Ewart, A., 1976. A petrological study of the younger Tongan andesites and dacites, and the olivine tholeiites of Niua Fo'ou Island, SW Pacific. *Contr. Min. Petr.*, **58**, 1–22.

Ewart, A., **Bryan, W. B.** and **Gill, J. B.**, 1973. Mineralogy and geochemistry of the younger volcanic islands of Tonga, SW. Pacific. *J. Petr.*, **14**, 429–465.

Favorskaia, M. A., **Volchanskaia, I. K.** and **Nissenbaum, P. N.**, 1963. [Some new data of pyroxene features from Upper Cenozoic effusives of Kamchatka.] *Akad. Nauk. S.S.S.R. Lab. Vulkanol. Vulkanizm Kamchatki*, 111–118 (in Russian).

Fawcett, J. J., 1965. Alteration products of olivine and pyroxene in basalt lavas from the Isle of Mull. *Min. Mag.*, **35**, 55–68.

Fleet, M. E., 1975. Growth habits of clinopyroxene. *Can. Min.*, **13**, 336–341.

Fleet, M. E. and **MacRae, N. D.**, 1975. A spinifex rock from Munro Township, Ontario. *Can. J. Earth Sci.*, **12**, 928–939.

Fodor, R. V., Keil, K. and **Bunch, T. E.**, 1975. Contributions to the mineral chemistry of Hawaiian rocks. IV. Pyroxenes in rocks from Hakeakala and West Maui volcanoes, Maui, Hawaii. *Contr. Min. Petr.*, **50**, 173–195.
Frantsesson, Ye. V. and **Lutts, B. G.**, 1970. A find of graphite-bearing pyrope peridotite in the Mir kimberlite pipe. *Dokl. Acad. Sci. U.S.S.R., Earth Sci. Sect.*, **191**, 159–161.
Friedman, G. M., 1957. Structure and petrology of the Caribou Lake intrusive body, Ontario, Canada. *Bull. Geol. Soc. America*, **68**, 1531–1564.
Frisch, T. and **Wright, J. B.**, 1971. Chemical composition of high-pressure megacrysts from Nigerian Cenozoic lavas. *Neues Jahrb. Min., Mh.*, 289–304.
Gallagher, M. J., 1963. Lamprophyre dykes from Argyll. *Min. Mag.*, **33**, 415–430.
Genkin, A. D., Teleshova, R. L. and **Alekseyeva, O. A.**, 1963. Nickel content in the essential minerals of the mineralized gabbro-diabase of the Noril'sk deposit. *Geochemistry*, no. 11, 1086–1092.
Gibb, F. C. G., 1968. Flow differentiation in the xenolithic ultrabasic dykes of the Cuillins and the Strathaird Peninsula, Isle of Skye, Scotland. *J. Petr.*, **9**, 411–443.
Gibb, F. C. G., 1973. The zoned clinopyroxenes of the Shiant Isles sill, Scotland. *J. Petr.*, **14**, 203–230.
Ginzburg, I. V. and **Sidorenko, G. A.**, 1964. Use of X-ray powder patterns to determine some crystal chemical characteristics of the pyroxenes. *Geochemistry Intern.*, no. 3, 536–562.
Girod, M., 1967. Données pétrographiques sur des pyroxénolites à grenat en enclaves dans des basaltes du Hoggar (Sahara central). *Bull. Soc. franç. Min. Crist.*, **90**, 202–213.
Godlevskii, M. N. and **Polushkina, A. P.**, 1971. [Monoclinic pyroxenes in the Talnakh differentiates intrusion.] *Zap. Vses. Min. Obshch.*, **100**, 543–557 (in Russian)
Gomes, C. de B., Moro, S. L. and **Dutra, C. V.**, 1970. Pyroxenes from the alkaline rocks of Itapirapuã, São Paulo, Brazil. *Amer. Min.*, **55**, 224–230.
Goode, A. D. T. and **Moore, A. C.**, 1975. High pressure crystallization of the Ewarara, Kalka and Gosse Pile intrusions, Giles complex, central Australia. *Contr. Min. Petr.*, **51**, 77–97.
Gramse, M., 1970. Quantitative Untersuchungen mit der Elektronen-Mikrosonde an Pyroxenen aus Basalten und Peridotit-Einschlüssen. *Contr. Min. Petr.*, **29**, 43–73.
Grapes, R. H., 1975. Petrology of the Blue Mountain complex, Marlborough, New Zealand. *J. Petr.*, **16**, 371–428.
Gray, N. H., 1971. A parabolic hourglass structure in titanaugite. *Amer. Min.*, **56**, 952–958.
Green, D. H. and **Ringwood, A. E.**, 1970. Mineralogy of peridotite compositions under upper mantle conditions. *Phys. Earth Planet. Interiors*, **3**, 359–371.
Greene, R. C., 1973. Petrology of the welded tuft of Devine Canyon, south-eastern Oregon. *Prof. Paper U.S. Geol. Surv.*, **797**, 26 pp.
Gribble, C. D., 1967. The basic intrusive rocks of Caledonian age of the Haddo House and Arnage districts, Aberdeenshire. *Scott. J. Geol.*, **3**, 125–136.
Griffin, W. L. and **Murthy, V. R.**, 1969. Distribution of K, Rb, Sr and Ba in some minerals relevant to basalt genesis. *Geochim. Cosmochim. Acta*, **33**, 1389–1414.
Groome, D. R. and **Hall, A.**, 1974. The geochemistry of the Devonian lavas of the northern Lorne plateau, Scotland. *Min. Mag.*, **39**, 621–640.
Gubbaiah, K. G. and **Varadarajan, S.**, 1962. Unusual types of twins in clinopyroxene from a dolerite dyke, Bangalore. *Curr. Sci.*, **31**, 19–20.
Haff, J. C., 1941. Determination of extinction angles in augite and hornblende with the universal stage according to the method of Conrad Burri. *Amer. J. Sci.*, **239**, 489–492.
Hafner, S. S. and **Virgo, D.**, 1970. Temperature dependent cation distributions in lunar and terrestrial pyroxenes. *Proc. Apollo 11 Lunar Sci. Conf., Geochim. Cosmochim. Acta, Suppl.* **1, 3**, 2183–2198.
Hagner, A. F., Leung, S. S. and **Dennison, J. M.**, 1965. Optical and chemical variations in minerals from a single rock specimen. *Amer. Min.*, **50**, 341–355.
Häkli, T. A., 1968. An attempt to apply the Makaopuhi nickel fractionation data to the temperature determination of a basic intrusive. *Geochim. Cosmochim. Acta*, **32**, 449–460.
Hargraves, R. B., 1962. Petrology of the Allard Lake anorthosite suite, Quebec. *Geol. Soc. America* (Buddington vol.), 163–189.
Harris, N., 1937. A petrological study of the Portrush sill and its veins. *Proc. Roy. Irish Acad.*, **43**, sect. B, no. 9, 95–134.
Harry, W. T., 1950. Aluminium replacing silicon in some silicate lattices. *Min. Mag.*, **29**, 142–149.
Hartman, P., 1969. Can Ti^{4+} replace Si^{4+} in silicates? *Min. Mag.*, **37**, 366–369.
Haslam, H. W., 1968. The crystallization of intermediate and acid magmas at Ben Nevis, Scotland. *J. Petr.*, **9**, 84–104.
Helz, R. T., 1973. Phase relations of basalts in their melting range at $P_{H_2O} = 5$ kbar as a function of oxygen fugacity. *J. Petr.*, **14**, 249–302.
Henriques, A., 1958. The influence of cations on the optical properties of clinopyroxenes. Part I and Part II. *Arkiv Min. Geol.*, **2**, 341 and 381.

Hermes, O. D., 1964. A quantitative petrographic study of dolerite in the Deep River Basin, North Carolina. *Amer. Min.*, **49**, 1718–1729.

Herzberg, C. T. and **Chapman, N. A.**, 1976. Clinopyroxene geothermometry of spinel-lherzolite. *Amer. Min.*, **61**, 626–637.

Hess, H. H., 1941. Pyroxenes of common mafic magmas, Part II. *Amer. Min.*, **26**, 573–594.

Hess, H. H., 1949. Chemical composition and optical properties of common clinopyroxenes. *Amer. Min.*, **34**, 621–666.

Hess, H. H., 1952. Orthopyroxenes of the Bushveld type, ion substitutions and changes in unit cell dimensions. *Amer. J. Sci.* (Bowen vol.), 173–187.

Hess, H. H., 1960. Stillwater igneous complex, Montana. *Mem. Geol. Soc. America*, **80**, 230 pp.

Hewins, R. H., 1974. Pyroxene crystallization trends and contrasting augite zoning in the Sudbury nickel irruptive. *Amer. Min.*, **59**, 120–126.

Hietanen, A., 1971. Distribution of elements in biotite-hornblende pairs and in an orthopyroxene-clinopyroxene pair from zoned plutons, northern Sierra Nevada, California. *Contr. Min. Petr.*, **30**, 161–176.

Hijikata, K. and **Onuma, K.**, 1969. Phase equilibria of the system $CaMgSi_2O_6$–$CaFe^{3+}AlSiO_6$ in air. *J. Japanese Assoc. Min. Petr. Econ. Geol.*, **62**, 209–217.

Hijikata, K. and **Yagi, K.**, 1967. Phase relations of Ca-Tschermak's molecule at high pressures and temperatures. *J. Fac. Sci., Hokkaido Univ., ser. IV, Geol. Min.*, **13**, 407–417.

Himmelberg, G. R. and **Ford, A. B.**, 1976. Pyroxenes of the Dufek intrusion, Antarctica. *J. Petr.*, **17**, 219–243.

Hollister, L. S. and **Gancarz, A. J.**, 1971. Compositional sector-zoning in clinopyroxene from the Narce area, Italy. *Amer. Min.*, **56**, 959–979.

Holloway, J. R. and **Burnham, C. W.**, 1972. Melting relations of basalt with equilibrium water pressure less than total pressure. *J. Petr.*, **13**, 1–30.

Holzner, J., 1934. Beiträge zur Chemie und Optik sanduhrförmiger Titanaugite. *Z. Krist.*, **87**, 1–42.

Horai, K.-I., 1971. Thermal conductivity of rock-forming minerals. *J. Geophys. Res.*, **76**, 1278–1308.

Hori, F., 1954. Effects of constituent cations on the optical properties of clinopyroxenes. *Sci. Papers, Coll. Gen. Educ., Univ. Tokyo*, **IV**, 71–83.

Hotz, P. E., 1953. Petrology of granophyre in diabase near Dillsburg, Pennsylvania. *Bull. Geol. Soc. America*, **64**, 676–704.

Howie, R. A., 1955. The geochemistry of the charnockite series of Madras, India. *Trans. Roy. Soc. Edinburgh*, **62**, 725–768.

Howie, R. A., 1964. A pyroxene granulite from Hitterö, south-west Norway. *Adv. Frontiers Geol. Geophys., Hyderabad*, 297–307.

Huang, W. H. and **Keller, W. D.**, 1972. Standard free energies of formation calculated from dissolution data using specific mineral analyses. *Amer. Min.*, **57**, 1152–1162.

Hubregtse, J. J. M. W., 1973. Distribution of elements in some basic granulite-facies rocks. *Verh. Koninklijke Nederlandse Akad. Wetenschappen, Afd. Natuur., Eerste Reeks*, Veel 27, no. 1, 7–68.

Huckenholz, H. G., 1965. Der petrogenetische Werdegang der Klinopyroxene in den tertiären Vulkaniten der Hocheifel. I. Die Klinopyroxene der Alkaliolivinbasalt-Trachyt-Assoziation. *Beitr. Min. Petr.*, **11**, 138–195.

Huckenholz, H. G., 1966. Der petrogenetische Werdegang der Klinopyroxene in den tertiären Vulkaniten der Hocheifel. III. Die Klinopyroxene der Pikritbasalte (Ankaramite). *Contr. Min. Petr.*, **12**, 73–95.

Huckenholz, H. G., **Schairer, J. F.** and **Yoder, H. S.**, *Jr.*, 1969. Synthesis and stability of ferri-diopside. *Min. Soc. America, Spec. Paper* **2**, 163–177.

Huebner, J. S. and **Ross, M.**, 1972. Phase relations of lunar and terrestrial pyroxenes at one atm. *Lunar Sci. Inst. Contr.* no. 88, 410–412.

Huebner, J. S., **Ross, M.** and **Turnock, A. C.**, 1972. Phase diagrams for pyroxene composition (abstr.). *Geol. Soc. Amer., Abstr. Programs*, **4**, 547.

Irvine, T. N., 1970. Crystallization sequences in the Muskox intrusion, and other layered intrusions. I. Olivine–pyroxene–plagioclase relations. *Geol. Soc. South Africa, Spec. Publ.* **1**, 441–476.

Irving, A. J., 1974. Geochemical and high pressure experimental studies of garnet pyroxenite and pyroxene granulite xenoliths from the Delegate basaltic pipes, Australia. *J. Petr.*, **15**, 1–40.

Ishibashi, K., 1970. Petrochemical study of basic and ultrabasic inclusions in basaltic rocks from northern Kyushu, Japan. *Mem. Fac. Sci. Kyushu Univ., ser. D, Geol.*, **20**, 85–146.

Isshiki, N., 1963. Petrology of Hachijō-Jima volcano group, Seven Izu Islands, Japan. *J. Fac. Sci., Univ. Tokyo, sect. 2*, **15**, 91–134.

Jaffe, H. W., **Robinson, P.**, **Tracy, R. J.** and **Ross, M.**, 1975. Orientation of pigeonite exsolution lamellae in metamorphic augite: correlation with composition and calculated optimal phase boundaries. *Amer. Min.*, **60**, 9–28.

Jakeš, P., 1968. Ferromagnesian minerals from the rocks of Tábor massif, Czechoslovakia. *Neues Jahrb. Min., Mh.*, 193–208.
Jasmund, K. and **Seck, H. A.**, 1964. Geochemische Untersuchungen an Auswürflingen (Gleesiten) des Laacher-See-Gebietes. *Beitr. Min. Petr.*, **10**, 275–295.
Jensen, B. B., 1973. Patterns of trace element partitioning. *Geochim. Cosmochim. Acta*, **37**, 2227–2242.
Joropov, N. A., 1966. Structural alterations in aluminous pyroxenes. *Dokl. Acad. Sci. U.S.S.R., Earth Sci. Sect.*, **170**, 151–152.
Kashkai, M.-A., 1944. The augites from Talysh (Azerbaidjan). *Dokl. Acad. Sci. U.S.S.R.*, **43**, 351–353.
Katz, M., 1970. Notes on the mineralogy and coexisting pyroxenes from the granulites of Mont Tremblant Park, Quebec. *Can. Min.*, **10**, 247–251.
Kennedy, W. Q., 1933. Trends of differentiation in basaltic magmas. *Amer. J. Sci.*, 5th ser., **25**, 239–256.
Knutson, J. and **Green, T. H.**, 1975. Experimental duplication of a high-pressure megacryst/cumulate assemblage in a near-saturated hawaiite. *Contr. Min. Petr.*, **52**, 121–132.
Konda, T., 1970. Pyroxenes from the Beaver Bay gabbro complex of Minnesota. *Contr. Min. Petr.*, **29**, 338–344.
Konda, T. and **Green, J. C.**, 1974. Clinopyroxenes from the Keweenawan lavas of Minnesota. *Amer. Min.*, **59**, 1190–1197.
Koritnig, S., 1965. Geochemistry of phosphorus. I. The replacement of Si^{4+} by P^{5+} in rock-forming silicate minerals. *Geochim. Cosmochim. Acta*, **29**, 361–371.
Kranck, S. H., 1961. A study of phase equilibria in a metamorphic iron formation. *J. Petr.*, **2**, 137–184.
Krivenko, A. P. and **Guletskaya, E. S.**, 1968. Composition of pyroxene from the gabbro–syenite association in Bol'shoy Taskyl pluton, Kuzetskiy, Alatau. *Dokl. Acad. Sci. U.S.S.R., Earth Sci. Sect.*, **180**, 149–152.
Kudryashova, V. I., 1963. [Hydrothermal pyroxene from the traps of the Lower Tunguska river.] *Trudy Min. Muz. Akad. Nauk, S.S.S.R.*, no. 14, 238–242 (in Russian).
Kukharenko, A. A., **Vainshtein, E. E.** and **Shevaleevskii, I. D.**, 1960. Zirconium and hafnium ratios in rock-forming pyroxenes and in zirconium minerals of the Paleozoic complex of the ultrabasic and alkaline rocks of the Kola Peninsula. *Geochemistry*, no. 7, 730–739.
Kuno, H., 1936. On the crystallization of pyroxenes from rock-magmas, with special reference to the formation of pigeonite. *Japan. J. Geol. Geogr.*, **13**, 141–150.
Kuno, H., 1950. Petrology of Hakone volcano and adjacent areas, Japan. *Bull. Geol. Soc. America*, **61**, 957–1019.
Kuno, H., 1955. Ion substitution in the diopside–ferropigeonite series of clinopyroxenes. *Amer. Min.*, **40**, 70–93.
Kuno, H., 1964. Aluminian augite and bronzite in alkali olivine basalt from Taka-sima, North Tyusyu, Japan. *Adv. Frontiers Geol. Geophys.* (Krishnan vol.), Hyderabad (Indian Geophys. Union), 205–220.
Kuno, H. and **Hess, H. H.**, 1953. Unit cell dimensions of clinoenstatite and pigeonite in relation to the common pyroxenes. *Amer. J. Sci.*, **251**, 741–752.
Kuno, H. and **Inoue, T.**, 1949. On porphyritic pigeonite in andesite from Okubo-yama, Minami-Aizu, Hukusima Prefecture. *Proc. Japan. Acad.*, **25**, 128–132.
Kuno, H., **Yamasaki, K.**, **Iida, C.** and **Nagashima, K.**, 1957. Differentiation of Hawaiian magmas. *Japan. J. Geol. Geogr.*, **28**, 179–218.
Kushiro, I., 1960. Si–Al relations in clinopyroxenes from igneous rocks. *Amer. J. Sci.*, **258**, 548–554.
Kushiro, I., 1964. Petrology of the Atumi dolerite, Japan. *J. Fac. Sci. Univ. Tokyo*, sect. 2, **15**, 135–202.
Kushiro, J., 1969. Synthesis and stability of iron-free pigeonite in the system $MgSiO_3$–$CaMgSi_2O_6$ at high pressures. *Carnegie Inst. Washington, Ann. Rept. Dir. Geophys. Lab.*, 1967–68, 80–83.
Kushiro, I., 1973. Crystallization of pyroxenes in Apollo 15 Mare basalts. *Carnegie Inst. Washington, Ann. Rept. Dir. Geophys. Lab.*, 1972–73, 647–650.
Kushiro, I. and **Nakamura, Y.**, 1970. Petrology of some lunar crystalline rocks. *Proc. Apollo 11 Lunar Sci. Conf. (Suppl. to Geochim. Cosmochim. Acta,* **34***)*, **1**, 607–626.
Kushiro, I., **Nakamura, Y.** and **Haramura, H.**, 1970. Crystallization of some lunar mafic magmas and generation of rhyolite liquid. *Science*, **167**, 610–612.
Kushiro, I. and **Schairer, J. F.**, 1963. New data on the system diopside–forsterite–silica. *Carnegie Inst. Washington, Ann. Rept. Dir. Geophys. Lab.*, 1962–63, 95–103.
Kushiro, I. and **Thompson, R. N.**, 1972. Origin of some abyssal tholeiites from the Mid-Atlantic Ridge. *Carnegie Inst. Washington, Ann. Rept. Dir. Geophys. Lab.*, 1971–72, 403–406.
Kutolin, V. A. and **Frolova, V. M.**, 1970. Petrology of ultrabasic inclusions from basalts of Minusa and Transbaikalian regions (Siberia, U.S.S.R.). *Contr. Min. Petr.*, **29**, 163–179.
Lally, J. S., **Heuer, A. H.**, **Nord, G. L.**, *Jr.* and **Christie, J. M.**, 1975. Subsolidus reactions in lunar pyroxenes: an electron petrographic study. *Contr. Min. Petr.*, **51**, 262–281.

Larsen, E. S., 1942. Alkalic rocks of Iron Hill, Gunnison County, Colorado. *U.S. Geol. Surv., Prof. Paper*, **197A**, 1–64.
Larsen, E. S., Irving, J., Gonyer, F. A. and **Larsen, E. S.,** *3rd.*, 1936. Petrologic results of a study of the minerals from the Tertiary volcanic rocks of the San Juan region, Colorado. *Amer. Min.*, **21**, 679–701.
Layton, W., 1965. Some aspects of optical properties and cell dimensions of amphiboles and pyroxenes. *Neues Jahrb. Min., Mh.*, 369–379.
Le Bas, N. J., 1962. The role of aluminium in igneous clinopyroxenes with relation to their parentage. *Amer. J. Sci.*, **260**, 267–288.
Lebedev, P. I., 1936. Beitrag zur Petrographie und Mineralogie der basischen Paegmatite Volyniens. *Acad. Sci. U.S.S.R.* (Vernadsky vol.), **2**, 999–1012 (M.A. **7**-198).
Lebedev, P. I. and **Lebedev, A. P.**, 1934. The titanite-magnetite gabbro mass, Patyn (Western Siberia). *Dok. Acad. Sci. U.S.S.R.*, **3**, 294–297 (M.A. **6**-321).
Leelanandam, C., 1967a. Chemical study of pyroxenes from the charnockitic rocks of Kondapalli (Andhra Pradesh), India, with emphasis on the distribution of elements in coexisting pyroxenes. *Min. Mag.*, **36**, 153–179.
Leelanandam, C., 1967b. Paired pyroxenes from Kondapalli. *J. Indian Geosci. Assoc.*, **8**, 89–92.
Lefèvre, C., 1969. Remarques sur le valeur du parametre b de la maille des clinopyroxènes. *Bull. Soc. franç. Min. Crist.*, **92**, 95–98.
Lehijärvi, M., 1960. The alkaline district of Iivaara, Kuusamo, Finland. *Bull. Comm. géol. Finlande*, **185**, 1–62.
Le Maitre, R. W., 1965. significance of the gabbroic xenoliths from Gough Island, South Atlantic. *Min. Mag.* (Tilley vol.), **34**, 303–317.
Le Maitre, R. W., 1969. Kaersutite-bearing plutonic xenoliths from Tristan da Cunha, South Atlantic. *Min. Mag.*, **37**, 185–197.
Leung, I. S., 1974. Sector-zoned titanaugites: morphology, crystal chemistry and growth. *Amer. Min.*, **59**, 127–138.
Lewis, J. F., 1967. Unit cell dimensions of some aluminous natural clinopyroxenes. *Amer. Min.*, **54**, 42–54.
Lindh, A., 1974. Manganese distribution between coexisting pyroxenes. *Neues Jahrb. Min., Mh.*, 335–345.
Lindh, A., 1975. Coexisting pyroxenes–a multivariate statistical approach. *Lithos*, **8**, 151–161.
Lindsley, D. H., King, H. E. and **Turnock, A. C.**, 1974. Compositions of synthetic augite and hypersthene coexisting at 810°C: application to pyroxenes from lunar highlands rocks. *Geophys. Res. Letters*, **1**, 134–136.
Lindsley, D. H. and **Munoz, J. T.**, 1969. Subsolidus relations along the join hedenbergite–ferrosilite. *Amer. J. Sci.*, **267-A**, 295–324.
Lisitsyn, A. E. and **Khitrov, V. G.**, 1962. A microspectro-chemical study of the distribution of boron in minerals of some igneous and metamorphic rocks in the middle Urals. *Geochemistry*, no. 3, 293–305.
Lopez Ruiz, J., 1970. Sobre la génesis de las augites zonadas y con estructuras en reloj de arena. *Estud. Geol. (España)*, **26**, 237–244.
Lovering, J. F. and **White, A. J. R.**, 1964. The significance of primary scapolite in granulitic inclusions from deep-seated pipes. *J. Petr.*, **5**, 195–218.
Lupanova, N. P., 1934. Basaltic hornblende and augite from monchiquite of Hibina mountains. *Trav. Inst. Pétrogr. Acad. Sci. U.S.S.R.*, no. 6, 53–64 (M.A. **6**-420).
McBirney, A. R. and **Aoki, K.-I.**, 1968. Petrology of the Island of Tahiti. *Mem. Geol. Soc. America*, **116**, 524–556.
McBirney, A. R. and **Aoki, K.-I.**, 1973. Factors governing the stability of plagioclase at high pressures as shown by spinel-gabbro xenoliths from the Kerguelen Archipelago. *Amer. Min.*, **58**, 271–276.
McCallum, I. S., Okamura, F. P. and **Ghose, S.**, 1975. Mineralogy and petrology of sample 67075 and the origin of lunar anorthosites. *Earth Planet. Sci. Letters*, **26**, 36–53.
MacDonald, G. A., 1944. Pyroxenes in Hawaiian lavas. *Amer. J. Sci.*, **242**, 626–629.
McDougall, I., 1961. Optical and chemical studies of pyroxenes in a differentiated Tasmanian dolerite. *Amer. Min.*, **46**, 661–687.
McDougall, I., 1964. Differentiation of the Great Lake dolerite sheet, Tasmania. *J. Geol. Soc. Australia*, **11**, 107–132.
MacGregor, I. D., 1965. Aluminous diopsides in the three-part assemblage diopside solid solution + forsterite + spinel. *Carnegie Inst. Washington, Ann. Rept. Dir. Geophys. Lab.*, 1964–65, 134–135.
McIntire, W. L., 1963. Trace element partition coefficients – a review of theory and applications to geology. *Geochim. Cosmochim. Acta*, **27**, 1209–1264.
Malinko, S. V., 1967. On isomorphism of boron in pyroxenes. *Geochem. Intern.*, no. 4, 870–875.
Manning, P. G. and **Nickel, E. H.**, 1969. A spectral study of the origin of colour and pleochroism of a titanaugite from Kaiserstuhl and of a riebeckite from St. Peter's Dome, Colorado. *Can. Min.*, **10**, 71–83.

Maske, S., 1966. The petrography of the Ingeli Mountain Range. *Ann. Univ. Stellenbosch*, **41**, Ser. A, no. 1, 1–109.

Mason, B and **Allen, R. O.**, 1973. Minor and trace elements in augite, hornblende and pyrope megacrysts from Kakanui, New Zealand. *N.Z. J. Geol. Geophys.*, **16**, 935–947.

Mathison, C. I., 1967. The Somerset Dam layered basic intrusion, south-eastern Queensland. *J. Geol. Soc. Australia*, **14**, 57–86.

Matsui, Y., **Banno, S.** and **Hernes, I.**, 1966. Distribution of some elements among minerals of Norwegian eclogites. *Norsk Geol. Tidsskr.*, **46**, 364–368.

Miller, R. T. and **Philpotts, A. R.**, 1973. Etching of pyroxenes and their exsolution lamellae. *Amer. Min.*, **58**, 543–544.

Miyashiro, A., 1958. Regional metamorphism of the Gosaisyo-Takanuki district in the central Abukuma Plateau. *J. Fac. Sci., Univ. Tokyo*, sect. II, **11**, 219–272.

Morimoto, N., 1958. Pyroxenes. *Carnegie Inst. Washington, Ann. Rept. Dir. Geophys. Lab.*, 1957–58, 249–252.

Moxham, R. L., 1960. Minor element distribution in some metamorphic pyroxenes. *Can. Min.*, **6**, 522–545.

Muir, I. D., 1951. The clinopyroxenes of the Skaergaard intrusion, eastern Greenland. *Min. Mag.*, **29**, 690–714.

Muir, I. D., 1954. Crystallization of pyroxenes in an iron-rich diabase from Minnesota. *Min. Mag.*, **30**, 376–388.

Muir, I. D. and **Long, J. V. P.**, 1965. Pyroxene relations in two Hawaiian hypersthene-bearing basalts. *Min. Mag.* (Tilley vol.), **34**, 358–369.

Muir, I. D. and **Tilley, C. E.**, 1964. Iron enrichment and pyroxene fractionation in tholeiites. *Geol. Journ.*, **4**, 143–156.

Müller, K., 1936. Augit von Vesuv. *Zentr. Min., Abt. A*, 116–122.

Murray, R. J., 1954. The clinopyroxenes of the Garbh Eilean sill, Shiant Isles. *Geol. Mag.*, **91**, 17–31.

Mysen, B. O., 1972. Five clinopyroxenes in the Hareidland eclogite. *Contr. Min. Petr.*, **34**, 315–325.

Naboko, S. I. and **Shavrova, N. N.**, 1954. [On pyroxenes in lavas of modern and recent eruptions of certain Kamchatka volcanoes.] *Bull. Volcan. Station, Acad. Sci. U.S.S.R.*, no. 23, 47–50 (in Russian) (M.A. **13**-392).

Nagasawa, H. and **Wakita, H.**, 1968. Partition of uranium and thorium between augite and host lavas. *Geochim. Cosmochim. Acta*, **32**, 917–921.

Naidu, C. M. G. and **Rao, J. J.**, 1967. Coexisting pyroxenes from charnockitic rocks of Salur Biblili area, Srikakulam district, A.P. *Bull. Geochem. Soc. India*, **2**, 27–30.

Nakamura, Y., 1973. Origin of sector zoning in igneous clinopyroxenes. *Amer. Min.*, **58**, 986–990.

Nakamura, Y. and **Coombs, D. S.**, 1973. Clinopyroxenes in the Tawhiroko tholeiitic dolerite at Moeraki, north-eastern Otago, New Zealand. *Contr. Min. Petr.*, **42**, 213–228.

Nakamura, Y. and **Kushiro, I.**, 1970a. Compositional relations of coexisting orthopyroxene, pigeonite and augite in a tholeiitic andesite from Hakone volcano. *Contr. Min. Petr.*, **26**, 265–275.

Nakamura, Y. and **Kushiro, I.**, 1970b. Equilibrium relations of hypersthene pigeonite and augite in crystallizing magmas; microprobe study of a pigeonite andesite from Weiselberg, Germany. *Amer. Min.*, **55**, 1999–2015.

Naldrett, A. J. and **Kullerud, G.**, 1967. A study of the Strathcona Mine and its bearing on the origin of the nickel-copper ores of the Sudbury district, Ontario. *J. Petr.*, **8**, 453–531.

Naumov, V. A., 1965. A differentiated trap intrusion on the upper reaches of the Lower Tunguska. *Dokl. Acad. Sci. U.S.S.R., Earth Sci. Sect.*, **163**, 17–18.

Nesbitt, R. W., 1971. Skeletal crystal forms in the ultramafic rocks of the Yilgarn Block, Western Australia: evidence for an Archaean ultramafic liquid. *Geol. Soc. Australia, Spec. Publ.*, **3**, 331–347.

Nesterenko, G. V. and **Al'Mukhamedov, A. E.**, 1966. Titanium in the pyroxenes of differentiated traps. *Geochem. Intern.*, **3**, 767–774.

Neumann, E.-R., 1976. Compositional relations among pyroxenes, amphiboles, and other mafic phases in the Oslo region plutonic rocks. *Lithos*, **9**, 85–109.

Nicholls, I. A., 1971. Petrology of Santorini volcano, Cyclades, Greece. *J. Petr.*, **12**, 67–119.

Nicholls, I. A. and **Lorenz, V.**, 1973. Origin and crystallization history of Permian tholeiites from the Saar-Nahe trough, S.W. Germany. *Contr. Min. Petr.*, **40**, 327–344.

Niggli, P., 1943. Gesteinschemismus und mineralchemismus II. Die Pyroxene der magmatischen Erstarrung. *Schweiz. Min. Petr. Mitt.*, **23**, 538–607.

Nockolds, S. R., 1941. The Garabal Hill–Glen Fyne igneous complex. *Quart. J. Geol. Soc.*, **94**, 451–511.

Nockolds, S. R. and **Mitchell, R. L.**, 1948. The geochemistry of some Caledonian plutonic rocks: a study in the relationship between the major and trace elements of igneous rocks and their minerals. *Trans. Roy. Soc. Edinburgh*, **61**, 533–575.

Nord, G. L., *Jr.*, **Heuer, A. H.** and **Lally, J. S.**, 1976. Pigeonite exsolution from augite. In *Electron*

Microscopy in Mineralogy. H. R. Wenk (ed.). Berlin, Heidelberg, and New York (Springer-Verlag), 220–227.

Norton, D. A. and **Clavan, W. S.**, 1959. Optical mineralogy, chemistry and X-ray crystallography of ten pyroxenes. *Amer. Min.*, **44**, 844–874.

Nwe, Y. Y., 1975. Two different pyroxene crystallization trends in the trough bands of the Skaergaard intrusion, east Greenland. *Contr. Min. Petr.*, **49**, 285–300.

Nwe, Y. Y., 1976. Electron-probe studies of the earlier pyroxenes and olivines from the Skaergaard intrusion, east Greenland. *Contr. Min. Petr.*, **55**, 105–126.

Nwe, Y. Y. and **Copley, P. A.**, 1975. Chemistry, subsolidus relations and electron petrography of pyroxenes from the late ferrodiorites of the Skaergaard intrusion, east Greenland. *Contr. Min. Petr.*, **53**, 37–54.

Obata, M. and **Dickey, J. S.** *Jr.*, 1976. Phase relations of mafic layers in the Ronda peridotite. *Carnegie Inst. Washington, Ann. Rept. Dir. Geophys. Lab.*, 1975–76, 562–566.

O'Hara, M. J., 1961. Petrology of the Scourie dyke, Sutherland. *Min. Mag.*, **32**, 848–865.

Ohashi, Y. and **Burnham, C. W.**, 1972. Electrostatic and repulsive energies of the *M*1 and *M*2 cation sites in pyroxenes. *J. Geophys. Res.*, **77**, 5761–5766.

Ohashi, Y., **Burnham, C. W.** and **Finger, L. W.**, 1975. The effect of Ca–Fe substitution on the clinopyroxene crystal structure. *Amer. Min.*, **60**, 423–434.

Oka, Y. and **Matsumoto, T.**, 1974. Study on the compositional dependence of the apparent partition coefficient of iron and magnesium between coexisting garnet and clinopyroxene solid solution. *Contr. Min. Petr.*, **48**, 115–121.

Onuma, K., **Hijikata, K.** and **Yagi, K.**, 1968. Unit cell dimensions of synthetic titan-bearing clinopyroxenes. *J. Fac. Sci., Hokkaido Univ., ser. iv, Geol. Min.*, **14**, 111–121.

Onuma, K. and **Yagi, K.**, 1971. The join $CaMgSi_2O_6$–$Ca_2MgSi_2O_7$–$CaTiAl_2O_6$ in the system CaO–MgO–Al_2O_3–TiO_2–SiO_2 and its bearing on titanpyroxenes. *Min. Mag.*, **38**, 471–480.

Oyawoye, M. O. and **Makanjuola, A. A.**, 1972. Bauchite: a fayalite-bearing quartz monzonite. *Proc. 24th Intern. Geol. Congr., Montreal*, **2**, 251–266.

Papike, J. J., **Bence, A. E.**, **Browne, G. E.**, **Prewitt, C. T.** and **Wu, C. H.**, 1971. Apollo 12 clinopyroxenes: exsolution and epitaxy. *Earth Planet. Sci. Letters*, **10**, 307–315.

Papike, J. J., **Ross, M.** and **Clark, J. R.**, 1969. Crystal chemical characterization of clinoamphiboles based on five new structure determinations. *Min. Soc. America Spec. Paper* **2**, 117–136.

Parras, K., 1958. On the charnockites in the light of a highly metamorphic rock complex in south-western Finland. *Bull. Comm. géol. Finlande*, **181**, 1–45.

Paster, T. P., **Schauwecker, D. S.** and **Haskin, L. A.**, 1974. The behaviour of some trace elements during solidification of the Skaergaard layered series. *Geochim. Cosmochim. Acta*, **38**, 1549–1577.

Peacor, D. P., 1967. Refinement of the crystal structure of a pyroxene of formula M_IM_{II} $(Si_{1\cdot5}Al_{0\cdot5})O_6$. *Amer. Min.*, **52**, 31–41.

Peters, T., 1968. Distribution of Mg, Fe, Al, Ca and Na in coexisting olivine, orthopyroxene and clinopyroxene in the Totalp serpentinite (Davos, Switzerland) and in the Alpine metamorphosed Malenco serpentinite (N. Italy). *Contr. Min. Petr.*, **18**, 65–75.

Philpotts, A. R., 1966. Origin of the anorthosite-mangerite rocks in southern Quebec. *J. Petr.*, **7**, 1–64.

Poldervaart, A., 1944. The petrology of the Elephant's Head dyke and the New Amalfi sheet. *Trans. Roy. Soc. South Africa*, **30**, 85–119.

Poldervaart, A., 1946. The petrology of the Mount Arthur dolerite complex, East Griqualand. *Trans. Roy. Soc. South Africa*, **31**, 84–110.

Poldervaart, A., 1947. Subcalcic ferroaugite from Mount Arthur, East Griqualand. *Min. Mag.*, **28**, 159–163.

Poldervaart, A. and **Hess, H. H.**, 1951. Pyroxenes in the crystallization of basaltic magma. *J. Geol.*, **59**, 472–489.

Popolitov, E. I., **Petrov, L. L.** and **Kovalenko, V. I.**, 1967. Geochemistry of beryllium in the middle Paleozoic intrusives of north-eastern Tuva. *Geochem. Intern.*, **4**, 682–689.

Powell, M. and **Powell, R.**, 1974. An olivine–clinopyroxene geothermometer. *Contr. Min. Petr.*, **48**, 243–263.

Prasad, E. A. V. and **Naidu, M. G.**, 1971. On pyroxene twinning. *Norsk Geol. Tidsskr.*, **51**, 15–23.

Preston, J., 1966a. A columnar growth of dendritic pyroxene. *Geol. Mag.*, **103**, 548–557.

Preston, J., 1966b. An unusual hourglass structure in augite. *Amer. Min.*, **51**, 1227–1232.

Prokoptsev, N. and **Murdmaa, I. O.**, 1970, Alkalic augitite lava from the Pacific Ocean floor. *Dokl. Acad. Sci. U.S.S.R., Earth Sci. Sect.*, **191**, 231–234.

Putman, H. M., 1942. Analyse chimique d'une augite du cap Tourmente. *Nat. Canadien*, **69**, 161–163.

Radcliffe, S. V., **Heuer, A. H.**, **Fisher, R. M.**, **Christie, J. M.** and **Griggs, D. T.**, 1970. High voltage (800 kV) electron petrography of type B rock from Apollo 11. *Proc. Apollo 11 Lunar Sci. Conf. (Suppl. to Geochim. Cosmochim. Acta* **34**), **1**, 731–748.

Ray, S. and **Sen, S. K.**, 1970. Partitioning of major exchangeable cations among orthopyroxene, calcic pyroxene and hornblende in basic granulites from Madras. *Neues. Jahrb. Min., Abhdl.*, **114**, 61–88.

Reid, J. B., *Jr.* and **Frey, F. A.**, 1971. Rare earth distributions in lherzolite and garnet pyroxenite xenoliths and the constitution of the upper mantle. *J. Geophys. Res.*, **76**, 1184–1196.

Reznikov, A. P. and **Rodzyanko, N. G.**, 1967. [Pyroxenes of Tyrnyauz deposit (north Caucasus).] *Zap. Vses. Min. Obshch.*, **96**, 608–620 (in Russian).

Richardson, S. W., 1968. The petrology of the metamorphosed syenite in Glen Dessarry, Inverness-shire. *Quart. J. Geol. Soc.*, **124**, 9–52.

Robinson, P., **Jaffe, H. W.**, **Ross, M.** and **Klein, C.**, *Jr.*, 1971. Orientation of exsolution lamellae in clinopyroxenes and clinoamphiboles: consideration of optimal phase boundaries. *Amer. Min.*, **56**, 909–939.

Robins, B., 1974. Synorogenic alkaline pyroxenite dykes on Seiland northern Norway. *Norsk Geol. Tidsskr.*, **54**, 247–268.

Roedder, E., 1965a. A laboratory reconnaissance of the liquidus surface in the pyroxene system En–Di–Hd–Fs($MgSiO_3$–$CaMgSi_2O_6$–$CaFeSi_2O_6$–$FeSiO_3$). *Amer. Min.*, **50**, 696–703.

Roedder, E., 1965b. Liquid CO_2 inclusions in olivine-bearing nodules and phenocrysts from basalts. *Amer. Min.*, **50**, 1746–1782.

Roedder, P. L. and **Osborn, E. F.**, 1966. Experimental data for the system MgO–FeO–Fe_2O_3–$CaAl_2Si_2O_8$–SiO_2 and their petrological implications. *Amer. J. Sci.*, **264**, 428–480.

Ross, M., **Bence, A. E.**, **Dwornik, E. J.**, **Clark, J. R.** and **Papike, J. J.**, 1970. Mineralogy of the lunar clinopyroxenes, augite and pigeonite. *Proc. Apollo 11 Lunar Sci. Conf. (Suppl. to Geochim. Cosmochim. Acta* **34***)*, **1**, 839–848.

Ross, M., **Huebner, J. S.** and **Dowty, F.**, 1973. Delineation of the one atmosphere augite–pigeonite miscibility gap for pyroxene from lunar basalt 12021. *Amer. Min.*, **58**, 619–635.

Ruegg, N. R., 1964. Use of the angle $A_1 \wedge c$ in optical determination of the composition of augite. *Amer. Min.*, **49**, 599–606.

Rutstein, M. S. and **White, W. B.**, 1971. Vibrational spectra of high-calcium pyroxenes and pyroxenoids. *Amer. Min.*, **56**, 877–887.

Rutstein, M. S. and **Yund, R. A.**, 1969. Unit-cell parameters of synthetic diopside hedenbergite solid solutions. *Amer. Min.*, **54**, 238–245.

Ryall, N. R. and **Threadgold, I. M.**, 1966. Evidence for $[(SiO_3)_5]_\infty$ type chains in inesite as shown by X-ray and infrared absorption studies. *Amer. Min.*, **51**, 754–761.

Sabine, P. A., 1975. Metamorphic processes at high temperature and low pressure: the petrogenesis of the metasomatized and assimilated rocks of Carneal, Co. Antrim. *Phil. Trans. Roy. Soc. London, A*, **280**, 225–269.

Sahama, T. G. and **Meyer, A.**, 1958. Study of the volcano Nyiragongo. *Inst. Parc. Nat. Congo Belge, Expl. Parc National Albert*, Fasc. 2, 1–85.

Sakata, Y., 1957. Unit cell dimensions of synthetic aluminium diopsides. *Jap. J. Geol. Geogr.*, **28**, 161–168.

Santin, S. F., 1970 (for 1969). Pegmatitoides in the horizontal basalts (series I) of Lanzarote and Fuerteventura Islands. *Bull. Volc.*, **33**, 989–1007.

Sapountzis, E. S., 1975. Coronas from the Thessaloniki gabbros (north Greece). *Contr. Min. Petr.*, **51**, 197–203.

Savolahti, A., 1964. Olivine diabase dike of Ausio in Padasjoki, Finland. *Bull. Comm. géol. Finlande*, **215**, 99–112.

Savolahti, A. and **Kujansuu, R.**, 1966. Some features of the Salmenkylä gabbro in Kangasniemi commune, Finland. *Bull. Comm. géol. Finlande*, **222**, 109–115.

Saxena, S. K., 1968. Distribution of iron and magnesium between coexisting garnet and clinopyroxene in rocks of varying metamorphic grade. *Amer. Min.*, **53**, 2018–2024.

Schairer, J. R. and **Boyd, F. R.**, 1957. Pyroxenes: The join $MgSiO_3$–$CaMgSi_2O_6$. *Carnegie Inst. Washington, Ann. Rept. Dir. Geophys. Lab.*, 1956–57, 223–225.

Schorer, G., 1970. Die Pyroxene tertiärer Vulkanite des Vogelsberges. *Chem. der Erde*, **29**, 69–138.

Schwartzman, D. W. and **Giletti, B. J.**, 1977. Argon diffusion and absorption studies of pyroxenes from the Stillwater complex, Montana. *Contr. Min. Petr.*, **60**, 161–182.

Segnit, E. R., 1953. Some data on synthetic aluminous and other pyroxenes. *Min. Mag.*, **30**, 218–226.

Sen, S. K., 1970. Magnesium-iron compositional variance in hornblende pyroxene granulites. *Contr. Min. Petr.*, **29**, 76–88.

Sen, S. K. and **Rege, S. M.**, 1966. Distribution of magnesium and iron between metamorphic pyroxenes from Saltora, West Bengal, India. *Min. Mag.*, **35**, 759–762.

Shidô, F., 1958. Plutonic and metamorphic rocks of the Nakoso and Iritōno districts in the central Abukuma Plateau. *J. Fac. Sci. Univ. Tokyo*, sect. II, **11**, 131–217.

Shinno, I., 1971. Mössbauer spectra of pyroxenes in the system $MgSiO_3$–Fe_2O_3. *Sci. Rept., Dept. Geol., Kyushu Univ.*, **11** (Matsushita vol.), 365–370.

Shkodzinskiy, V. S., 1968. The position of titanium in clinopyroxene. *Dokl. Acad. Sci. U.S.S.R., Earth Sci. Sect.*, **182**, 142–145.

Sigurdsson, H. and **Brown, G. M.,** 1970. An unusual enstatite-forsterite basalt from Kolbeinsey Island, north of Iceland. *J. Petr.*, **11**, 205–220.

Simmons, E. C., Lindsley, D. H. and **Papike, J. J.,** 1974. Phase relations and crystallization sequence in a contact-metamorphosed rock from the Gunflint Iron Formation, Minnesota. *J. Petr.*, **15**, 539–565.

Simpson, E. S. W., 1954. The Okonjeje igneous complex, South-West Africa. *Trans. Geol. Soc. South Africa*, **57**, 126–172.

Sinitsȳn, A. V., 1965. [Pyroxenes in differentiated dolerite intrusion.] *Zap. Vses. Min. Obshch.*, **94**, 583–592 (in Russian).

Smith, D., 1972. Stability of iron-rich pyroxene in the system $CaSiO_3$–$FeSiO_3$–$MgSiO_3$. *Amer. Min.*, **57**, 1413–1428.

Smith, D., 1974. Pyroxene–olivine–quartz assemblages in rocks associated with the Nain anorthosite massif, Labrador. *J. Petr.*, **15**, 58–78.

Smith, D. and **Lindsley, D. H.,** 1971. Stable and metastable augite crystallization trends in a single basalt flow. *Amer. Min.*, **56**, 225–233.

Smith, P. P. K., 1977. An electron microscopic study of amphibole lamellae in augite. *Contr. Min. Petr.*, **59**, 317–322.

Strong, D. F., 1969. Formation of the hour-glass structure in augite. *Min. Mag.*, **37**, 472–479.

Subramaniam, A. P., 1962. Pyroxenes and garnets from charnockites and associated granulites. *Bull. Geol. Soc. America* (Buddington vol.), 21–36.

Sun, C.-O., Williams, R. J. and **Sun, S. S.,** 1974. Distribution coefficients of Eu and Sr for plagioclase–liquid and clinopyroxene–liquid equilibria in oceanic ridge basalt: an experimental study. *Geochim. Cosmochim. Acta*, **38**, 1415–1433.

Takashima, Y. and **Ohashi, S.,** 1968. The Mössbauer spectra of various natural minerals. *Bull. Chem. Soc. Japan*, **41**, 88–93.

Takeda, H., 1972. Crystallographic studies of coexisting aluminan orthopyroxene and augite of high-pressure origin. *J. Geophys. Res.*, **77**, 5798–5811.

Tarney, J., 1969. Epitaxic relations between coexisting pyroxenes. *Min. Mag.*, **37**, 115–122.

Taylor, H. P. and **Epstein, S.,** 1962. Relationship between O^{18}/O^{16} ratios in coexisting minerals of igneous and metamorphic rocks, Part I. Principles and experimental results. *Bull. Geol. Soc. America*, **73**, 461–480.

Thompson, R. N., 1972. Oscillatory and sector zoning in augites from a Vesuvian lava. *Carnegie Inst. Washington, Ann. Rept. Dir. Geophys. Lab.*, 1971–72, 463–470.

Thompson, R. N., 1974. Some high-pressure pyroxenes. *Min. Mag.*, **39**, 768–787.

Thompson, R. N., 1975. Primary basalts and magma genesis. *Contr. Min. Petr.*, **52**, 213–232.

Tiba, T., 1966. Petrology of the alkaline rocks of the Takakusayama district, Japan. *Sci. Rept. Tohoku Univ.*, Ser. 3, **9**, 541–610.

Tilley, C. E., 1952. Some trends of basaltic magma in limestone syntexis. *Amer. J. Sci.* (Bowen, vol.), 529–545.

Tilley, C. E. and **Harwood, H. F.,** 1931. The dolerite–chalk contact of Scawt Hill, Co. Antrim. The production of basic alkali-rocks by the assimilation of limestone by basaltic magma. *Min. Mag.*, **22**, 439–468.

Tilley, C. E. and **Muir, I. D.,** 1962. The Hebridean plateau magma type. *Trans. Edin. Geol. Soc.*, **19**, 208–215.

Tomita, J., 1934. Variations in optical properties, according to chemical composition in the pyroxenes of the clinoenstatite–clinohypersthene–diopside–hedenbergite system. *J. Shanghai Sci. Inst.*, sect. 2, **7**, 41–58.

Tsuboi, S., 1932. On the course of crystallization of pyroxenes from rock-magmas. *Japan. J. Geol. Geogr.*, **10**, 67–82.

Turner, F. J., 1942. Determination of extinction angles in monoclinic pyroxenes and amphiboles. *Amer. J. Sci.*, **240**, 571–583.

Turnock, A. C., 1970. A pyroxene solvus section. *Can. Min.*, **10**, 744–747.

Turnock, A. C., Lindsley, D. H. and **Grover, J. E.,** 1973. Synthesis and unit cell parameters of Ca–Mg–Fe pyroxenes. *Amer. Min.*, **58**, 50–59.

Vallance, T. G., 1974. Spilitic degradation of a tholeiitic basalt. *J. Petr.*, **15**, 79–96.

Velde, D. and **Tournon, J.,** 1970. La camptonite de San Feliú de Buxalleu (Province de Gérone, Espagne). *Bull. Soc. franç. Min. Crist.*, **93**, 482–487.

Verhoogen, J., 1962. Distribution of titanium between silicates and oxides in igneous rocks. *Amer. J. Sci.*, **260**, 211–220.

Vernon, R. H., 1961. Magnetic susceptibility as a measure of total iron plus manganese in some ferromagnesian silicate minerals. *Amer. Min.*, **46**, 1141–1153.

Vilenskiy, A. M., 1966. Phase equilibria and evolution in the chemical nature of the pyroxenes of intrusive traps of the Siberian platform. *Geochim. Intern.*, **3**, 1275–1306.
Viswanathan, K., 1966. Unit cell dimensions and ionic substitutions in common clinopyroxenes. *Amer. Min.*, **51**, 429–442.
Vitaliano, C. J. and **Harvey, R. D.,** 1965. Alkali basalt from Nye County, Nevada. *Amer. Min.*, **50**, 73–84.
Wager, L. R. and **Deer, W. A.,** 1939. Geological investigations in east Greenland. Part III. The petrology of the Skaergaard intrusion, Kangerdlugssuaq, east Greenland. *Medd. om Grønland*, **106**, no. 4, 1–352.
Wager, L. R. and **Mitchell, R. L.,** 1951. The distribution of trace elements during strong fractionation of basic magma – a further study of the Skaergaard intrusion, east Greenland. *Geochim. Cosmochim. Acta*, **1**, 129–208.
Walker, F., 1940. Differentiation of the Palisade diabase, New Jersey. *Bull. Geol. Soc. America*, **51**, 1059–1106.
Walker, F., 1943. Note on the pyroxenes of basalt magma. *Amer. J. Sci.*, **241**, 517–520.
Walker, F. and **Poldervaart, A.,** 1941a. The Karroo dolerites of the Calvinia district. *Trans. Geol. Soc. South Africa*, **44**, 127–148.
Walker, F. and **Poldervaart, A.,** 1941b. The Hangnest dolerite sill, South Africa. *Geol. Mag.*, **78**, 429–450.
Walker, F. and **Poldervaart, A.,** 1941c. The petrology of the dolerite sill of Downes Mountain, Calvinia. *Trans. Geol. Soc. South Africa*, **43**, 159–173.
Walker, F. and **Poldervaart, A.,** 1949. Karroo dolerites of the Union of South Africa. *Bull. Geol. Soc. America*, **60**, 591–701.
Walker, F., Vincent, H. C. G. and **Mitchell, R. L.,** 1952. The chemistry and mineralogy of the Kinkell tholeiite, Stirlingshire. *Min. Mag.*, **29**, 895–908.
Walker, K. R., Ware, N. G. and **Lovering, J. F.,** 1973. Compositional variations in the pyroxenes of the differentiated Palisades sill, New Jersey. *Bull. Geol. Soc. America*, **84**, 89–110.
Walsh, J. N., 1975. Clinopyroxenes and biotites from the Centre III igneous complex, Ardnamurchan, Argyllshire. *Min. Mag.*, **40**, 335–345.
Ward, R. F., 1959. Petrology and metamorphism of the Wilmington complex, Delaware, Pennsylvania, and Maryland. *Bull. Geol. Soc. America*, **70**, 1425–1458.
Wass, S. Y., 1973. The origin and petrogenetic significance of hour-glass zoning in titaniferous clinopyroxenes. *Min. Mag.*, **39**, 133–144.
Wawryk, W., 1935. Sur l'augite commune titanifère des teschénites en Pologne. *Arch. Min. Soc. Sci. Varsovie*, **11**, 175–181 (M.A. **6**-212).
White, A. J. R., Chappell, B. W. and **Jakeš, P.,** 1972. Coexisting clinopyroxene, garnet and amphibole from 'an eclogite', Kakanui, New Zealand. *Contr. Min. Petr.*, **34**, 185–191.
White, W. B. and **Keester, K. L.,** 1967. Selection rules and assignments for the spectra of ferrous iron in pyroxenes. *Amer. Min.*, **52**, 1508–1514.
Whittaker, E. J. W., 1960. Relationships between the crystal chemistry of pyroxenes and amphiboles. *Acta Cryst.*, **13**, 741–742.
Wilkening, L. L. and **Anders, E.,** 1975. Some studies of an unusual eucrite: Ibitira. *Geochim. Cosmochim. Acta*, **39**, 1205–1210.
Wilkinson, J. F. G., 1956. Clinopyroxenes of alkali-basalt magma. *Amer. Min.*, **41**, 724–743.
Wilkinson, J. F. G., 1957. The clinopyroxenes of a differentiated teschenite sill near Gunnedah, New South Wales. *Geol. Mag.*, **94**, 123–134.
Wilkinson, J. F. G., 1971. Petrology of some vitrophyric calc-alkaline volcanics from the Carboniferous of New South Wales. *J. Petr.*, **12**, 587–620.
Wilkinson, J. F. G., 1973. Pyroxenite xenoliths from an alkali trachybasalt in the Glen Innes area, north-eastern New South Wales. *Contr. Min. Petr.*, **42**, 15–32.
Wilkinson, J. F. G., 1975. Ultramafic inclusions and high pressure megacrysts from a nephelinite sill, Nandewar Mountains, north-eastern New South Wales, and their bearing on the origin of certain ultramafic inclusions in alkaline volcanic rocks. *Contr. Min. Petr.*, **51**, 235–262.
Wilkinson, J. F. G., 1976. Some subcalcic clinopyroxenes from Salt Lake Crater, Oahu and their petrogenetic significance. *Contr. Min. Petr.*, **58**, 181–202.
Williams, P. G. L., Bancroft, G. M. and **Bown, M. G.,** 1971. Anomalous Mössbauer spectra of *C2/c* clinopyroxenes. *Nature (Phys. Sci.)*, **230**, 149–151.
Wilshire, H. G. and **Shervais, J. W.,** 1975. Al-augite and Cr-diopside ultramafic xenoliths in basaltic rocks from western United States: structural and textural relationship. *Phys. Chem. Earth*, **9**, 257–272.
Wilson, A. F., 1976. Aluminium in coexisting pyroxenes as a sensitive indicator of changes in metamorphic grade within the mafic granulite terrain of the Frazer Range, Western Australia. *Contr. Min. Petr.*, **56**, 255–278.
Wilson, M. M. and **Mathison, C. I.,** 1968. The Eulogie Park gabbro, a layered basic intrusion from

eastern Queensland. *J. Geol. Soc. Australia*, **15**, 139–158.

Winchell, A. N., 1935. Further studies in the pyroxene group. *Amer. Min.*, **20**, 562–568.

Winchell, H., 1961. Regressions of physical properties in the compositions of clinopyroxenes. II. Optical properties and specific gravity. *Amer. J. Sci.*, **259**, 295–319.

Winchell, H., 1963. Regressions of physical properties on the composition of clinopyroxenes. Part III. The common soda-free alumina-free clinopyroxenes. *Amer. J. Sci.*, **261**, 168–185.

Winchell, H. and **Leake, B. E.**, 1965. Regressions of refractive indices, density and lattice constants on the composition of orthopyroxenes. *Amer. Min.*, **50**, 294 (abstract).

Winchell, H. and **Tilling, R. I.**, 1960. Regressions of physical properties on the compositions of clinopyroxenes. *Amer. J. Sci.*, **258**, 529–547.

Wood, B. J., 1976a. An olivine–clinopyroxene geothermometer. *Contr. Min. Petr.*, **56**, 297–303.

Wood, B. J., 1976b. The partitioning of iron and magnesium between garnet and clinopyroxene. *Carnegie Inst. Washington Ann. Report. Dir. Geophys. Lab.*, 1975–76, 571–574.

Yagi, K., 1953. Petrochemical studies of the alkalic rocks of the Morotu district, Sakhalin. *Bull. Geol. Soc. America*, **64**, 769–809.

Yagi, K., 1969. Petrology of the alkalic dolerites of the Nemuro Peninsula, Japan. *Mem. Geol. Soc. America*, **115** (Poldervaar vol.), 103–147.

Yagi, K. and **Onuma, K.**, 1967. The join $CaMgSi_2O_6$–$CaTiAl_2O_6$ and its bearing on the titanaugites. *J. Fac. Sci. Hokkaido Univ.*, ser. 4, **13**, 463–483.

Yagi, K. and **Onuma, K.**, 1969. An experimental study on the role of titanium in alkali basalts in the light of the system diopside–akermanite–nepheline–$CaTiAl_2O_6$. *Amer. J. Sci.*, **267A** (Schairer vol.), 509–549.

Yamaguchi, Y., 1973. Study of exsolution phenomena of pyroxenes. *J. Fac. Sci. Hokkaido Univ.*, ser. 4, *Geol. Min.*, **16**, 133–154.

Yamakawa, M., 1971. Two different crystallization trends of pyroxene in a tholeiitic dolerite, Semi, northern Japan. *Contr. Min. Petr.*, **33**, 232–238.

Yefimov, A. A. and **Ivanova, L. P.**, 1963. Behaviour of chromium, nickel and cobalt during the formation of the Kytlym platiniferous massif, Urals. *Geochemistry*, no. 11, 1076–1085.

Yoder, H. S., **Tilley, C. E.** and **Schairer, J. F.**, 1963. Pyroxene quadrilateral. *Carnegie Inst. Washington, Ann. Rept. Dir. Geophys. Lab.*, 1962–63, 84–95.

Yoder, H. S., **Tilley, C. E.** and **Schairer, J. F.**, 1964. Isothermal sections of pyroxene quadrilateral. *Carnegie Inst. Washington, Ann. Rept. Dir. Geophys. Lab.*, 1963–64, 121–129.

Zharikov, V. A. and **Podlessky, K. V.**, 1955. On the behaviour of pyroxene as a mineral of variable composition in infiltration skarn zones. *Dokl. Acad. Sci. U.S.S.R.*, **105**, 1096–1099.

Zwaan, P. C., 1954. On the determination of pyroxenes by X-ray powder diagrams. *Leidse Geol. Mededelingen*, **19**, 167–276.

Fassaite $Ca(Mg,Fe^{3+},Al)[(Si,Al)_2O_6]$

Monoclinic (+)

α	1·672–1·730
β	1·682–1·735
γ	1·702–1·750
δ	0·018–0·028
$2V_\gamma$	51°–64°
$\gamma:z$	33°–47°
O.A.P.	(010) $\beta = y$
Dispersion:	$r > v$ moderate to strong.
D	2·96–3·60
H	6
Cleavage:	{110} good, {100} parting; (110):(1$\bar{1}$0) ≃ 87°
Twinning:	{100} simple, lamellar.
Colour:	Pale to dark green, brown, crimson; colourless to pale green in thin section.
Pleochroism:	Feeble, α pale green, β pale yellow-green, γ pale green, yellow-grey.
Unit cell:	a 9·71–9·80, b 8·85–8·91, c 5·26–5·36Å, β 105°30′–106°. Z = 4. Space group $C2/c$.

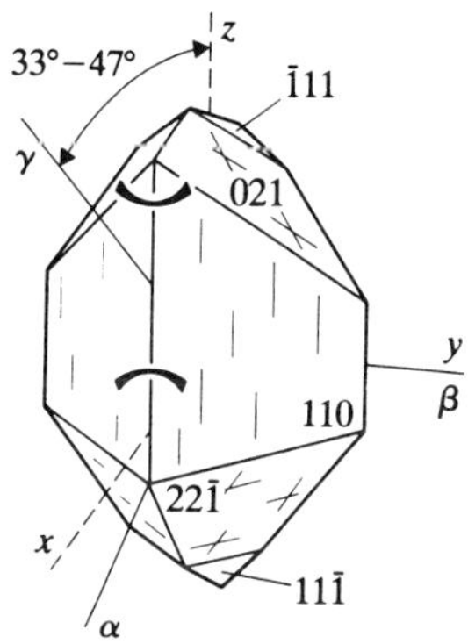

Insoluble in HCl.

The name fassaite was first introduced to describe a pyroxene from an augite syenite–limestone contact in the Fassa valley, Trentino, where it occurs in light to dark green crystals possessing a distinct habit with the zone {110} strongly developed. Subsequently the name has been more generally used to describe the aluminium-rich and sodium-poor pyroxenes commonly found in metamorphosed limestones and dolomites, but fassaite has also been reported from eclogitic inclusions in kimberlite, and in meteorites.

Structure

The structure of a fassaite (Table 39, anal. 3), in which approximately one-quarter of the tetrahedra are occupied by Al, was determined by Peacor, 1967. The unit cell has a 9·794, b 8·906, c 5·319 Å, β 105·90°, and space group $C2/c$. The crystal structure is essentially the same as that of diopside. Ordering of Ca into the $M2$ site is complete and it is co-ordinated by eight oxygens, in a similar fashion. Electrostatic charge unbalance at O2 and O3 is compensated by shorter M–O2 and longer M–O3 distances, and these are achieved by rotations of tetrahedra about the Si–O1 bond producing a somewhat kinked Si–O chain (Fig. 189). The angle O3–O3–O3 is approximately 166° in both fassaite and diopside as compared with 175° in jadeite, and 180° in the straight chain synthetic pyroxene $LiFe^{3+}Si_2O_6$. The average

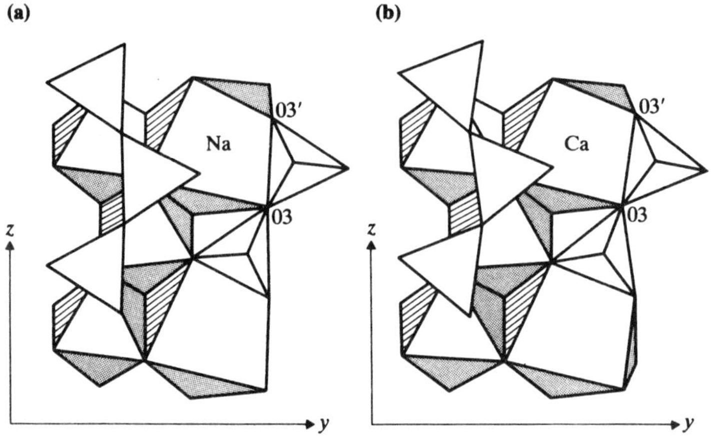

Fig. 189. Parts of the structures of (A) jadeite and (B) a fassaite, projected on (100). The larger relative rotation of tetrahedra in B is evident from the nearly equal Ca–O3 and Ca–O3′ bond lengths (after Peacor, 1967).

Si–O distance of 1·663 Å is consistent with one-quarter occupation of tetrahedra by Al, the remaining Al together with Ti, Fe^{3+}, Fe^{2+} and Mg being located in *M*1 sites.

The crystal structure of an extremely titanium-rich fassaite from the Allende meteorite was determined by Dowty and Clark (1973). Its formula was given as $Ca_{1\cdot00}Mg_{0\cdot39}Ti_{0\cdot48}Al_{0\cdot13}(Al_{0\cdot74}Si_{1\cdot26})O_6$, and its cell parameters *a* 9·80, *b* 8·85, *c* 5·36 Å, *β* 105°37′, Z = 4, space group *C*2/*c*. The structure is similar to that of the terrestrial fassaite and of diopside. Most of the Ti is present as Ti^{3+} and all of the Ti is in the *M*1 site which departs a little more from regularity than does the similar site in diopside. It was noted that the Allende pyroxene did not exhibit the normal {110} cleavage, and this was attributed to the effect of a high proportion of tetrahedral Al, and high Al and Ti in *M*1. Dowty and Clark (1973) ascribed the two infrared absorption bands at 21000 cm^{-1} and 16500 cm^{-1} respectively to Ti^{3+} crystal field transitions and to Ti^{3+}–Ti^{4+} charge transfer. Burns and Huggins (1973) ascribed both to crystal field transitions.

The crystal structure of the end-member $CaAlSiAlO_6$ (calcium Tschermak's molecule) was determined by Okamura *et al.* (1974). This was synthesized at 1300°C and 18 kbar and has cell parameters *a* 9·609, *b* 8·652, *c* 5·274 Å, *β* 106°, *V* 421·35 $Å^3$, Z = 4, space group *C*2/*c*(?). In this structure, in which the *M*1 octahedron is occupied entirely by Al, the average *M*1–O distance (1·947 Å) is longer than in spodumene (1·919 Å) or jadeite (1·928 Å), presumably because of the effect of Ca in *M*2 as compared with Na in the other two pyroxenes. The Ca–O eight-fold polyhedron is smaller and more regular than that in diopside, principally because of the close approach between Ca and the O3 atoms. This is consistent with the latter's charge balance consequential on Al substitution in the tetrahedral chain. The average T–O distance is appropriately larger (1·686 Å, compared with 1·634 Å in diopside) since half of the tetrahedra are occupied by Al.

Okamura *et al.* discussed the possibilities of Si,Al ordering. Assuming that there is only one type of chain, as in other Ca-rich pyroxenes, then the space groups *C*2, *P*2/*n*, $C\bar{1}$, and $P2_1/n$ are possible. They suggest that an ordered arrangement in one of these space groups may in fact occur but with a very fine scale 'out of phase' domain structure. This could result in the extra reflections demanded by these space groups being so weak and diffuse as to be unobservable. Although no extra reflections were in fact observed by them, and the space group assignment was *C*2/*c*, they reported that other workers had noted some extra reflections for other specimens of $CaAlSiAlO_6$. So far, however, Si/Al ordering has not been apparent for any fassaite or for the more highly aluminous end-member.

Variations in the cell parameters and X-ray powder patterns for members of the solid solution series $CaMgSi_2O_6$–$CaAlSiAlO_6$ have been investigated by several workers. In this series, equal numbers of Al atoms replace Mg and Si in the *M*2 and T sites of the diopside structure. The replacement of Si by Al in T increases *c* and the replacement of Mg by Al in *M*1 decreases *b* and *a*. For synthetic solid solutions up to 20 mol. per cent $CaAlSiAlO_6$, Sakata (1957) gave the following changes in cell parameters for each 0·1 Al per mole: *a*, −0·016 Å, *b*, −0·028 Å, *c*, +0·008 Å. Lewis (1967) noted a reduction in *b* and $a \sin\beta$ with increasing total Al for a suite of volcanic augites ('salites'), but in these most of the Al was substituting for Si and there were additional effects from Fe^{3+} and Ti.

Clark *et al.* (1962) showed that for the $CaMgSi_2O_6$–$CaAlSiAlO_6$ series the variations in cell volume and in cell parameters were non-linear and that the series was therefore not one of ideal solid solution. Okamura *et al.* (1974) discussed the likely changes in structure which are consistent with this as Al is progressively substituted into the T and *M*1 sites (see also Newton *et al.* (1977) and Fig. 190).

X-ray powder data for a synthetic approximation to the 'Ca-Tschermak' molecule, $Ca_{1\cdot04}Al_{1\cdot00}Si_{0\cdot93}Al_{1\cdot07}O_6$, are given by Hays (1966); cell parameters are *a* 9·619, *b* 8·659, *c* 5·278 Å, β 106·14°.

Coleman (1962) showed that substitution of the type $Fe^{3+}Al^{3+}$ for $Mg^{2+}Si^{4+}$ decreases *b* and increases *c*, and $Ti^{4+}Al_2^{3+}$ for $Mg^{2+}Si_2^{4+}$ decreases *b* and β and increases *a* and *c*. Hijikata (1968) determined the cell parameters of pyroxenes over the whole range of solid solution $CaMgSi_2O_6$–$CaFe^{3+}AlSiO_6$. He found that *a*, *c* and cell volume increased approximately linearly across the series, *a* from 9·745 to 9·783 Å, *c* from 5·250 to 5·372 Å, and *V* from 439·2 to 444·3 Å^3. The parameter *b* however, decreases from 8·925 to 8·787 Å with a slight change of slope at about Di_{70}. It appears that for *a* the substitution of Al for Si is more effective than Fe^{3+} for Mg. The angle β increases from 105·87° to 106·04° at Di_{70} and then decreases to 105·82° at Di_0.

The cell parameters of synthetic solid solutions ranging in composition between $CaFe_2^{3+}SiO_6$ and $CaAl_2SiO_6$ have been measured by Hijikata and Onuma (1969) and Huckenholz *et al.* (1974) and are shown in Fig. 191. The substitution of Fe^{3+} for Al expands the cell volume, the increase in the *c* and *b* parameters resulting from the replacement of Al by Fe^{3+} within the tetrahedral chains and the octahedra respectively. The maximum change is shown by the *a* parameter consequent on the involvement of two tetrahedra and one octahedron of the structure in this direction.

Chemistry

The main characteristics of the composition of fassaite are the high content of calcium (CaO $\simeq$ 25 per cent, equivalent to approximately one Ca ion per mole of fassaite, a variable but high aluminium content and a high ferric to ferrous iron ratio, and replacement of Si by Al generally between 0·25 and 0·50 per mole. In some fassaites the content of aluminium and ferric iron are both high (e.g. anal. 1, Table 39) and these minerals probably contain the maximum number of trivalent ions that

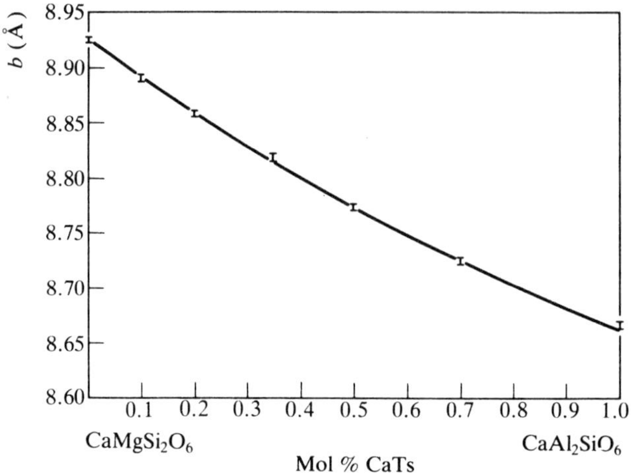

Fig. 190. The b cell dimensions of synthetic $CaMgSi_2O_6$–$CaAl_2SiO_6$ clinopyroxenes used in calorimetric work. Brackets show standard deviations ($\pm$ 1σ) about the means. Smoothed data of Clark *et al.* (1962) shown (after Newton *et al.*, 1977).

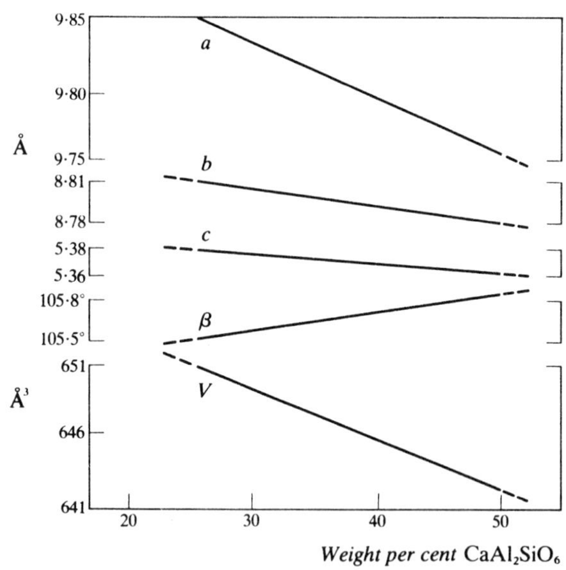

Fig. 191. Cell parameters of fassaite solid solutions along the join $CaFe^{3+}_2SiO_6$–$CaAl_2SiO_6$ (after Hijikata and Onuma, 1969; Huckenholz *et al.*, 1974).

has been found in a natural calcium-rich pyroxene. In general fassaite has less $Al+Fe^{3+}$ than omphacite, but the two minerals are distinguished chemically by the higher silicon and sodium and lower calcium content of the latter. Thus in fassaite the main replacement may be expressed $(Mg,Fe^{2+})Si \rightleftharpoons (Al,Fe^{3+})Al$. A number of the early analyses of fassaite are given by Tröger (1951). The trace element content of eighteen fassaites from phlogopitic, felspathic and scapolite pyroxenites from SE Madagascar are given by Majmundar (1971).

Table 39. Fassaite Analyses

	1		2		3		4	
SiO_2	41·36		41·78		39·89		48·67	
TiO_2	0·76		0·64		2·30		0·53	
Al_2O_3	15·75		15·43		14·94		13·89	
Fe_2O_3	6·10		5·45		5·62		1·38	
FeO	0·24		0·56		2·01		2·40	
MnO	0·03		0·10		0·21		0·05	
MgO	10·34		9·78		10·14		11·75	
CaO	25·27		25·33		24·33		19·37	
Na_2O	0·06		0·33		0·10		1·89	
K_2O	0·03		0·12		0·03		0·06	
H_2O^+	0·10		0·48		0·13		—	
H_2O^-	0·00		0·04		0·07		0·00	
Total	100·04		100·04		'99·87'		100·30	
α	1·712		1·730		1·725		1·677	
β	1·719		1·735		1·732		1·683	
γ	1·736		1·750		1·745		1·707	
$\gamma:z$	46°		—		33°–37°		—	
$2V_\gamma$	51°–58°		52°–58°		58°–64°		61°	
D	3·34		—		3·399		3·42	
	Numbers of ions on the basis of six O							
Si	1·537	2·00	1·561	2·00	1·506	2·00	1·751	2·00
Al	0·463		0·439		0·494		0·249	
Al	0·228		0·240		0·171		0·340	
Ti	0·021		0·018		0·065		0·014	
Fe^{3+}	0·170	1·00	0·153	0·98	0·159	1·04	0·037	1·10[a]
Mg	0·573		0·545		0·570		0·630	
Fe^{2+}	0·007		0·017		0·063		0·072	
Mn	0·001		0·003		0·007		0·002	
Na	0·004		0·024		0·007		0·132	
Ca	1·007	1·01	1·014	1·04	0·975	0·98	0·747	0·88
K	0·001		0·006		0·001		0·003	

[a] Includes Cr^{3+} 0·009.

1 Fassaite, tabular mass of spinel, fassaite, garnet and clintonite, Helena, Montana (Knopf and Lee, 1957). Anal. E. H. Oslund.

2 Fassaite, massive granular, spinel–pyroxene zone, Magnet Heights, Bushveld (Willemse and Bensch, 1964). Anal. A. Victor.

3 Fassaite, jacupirangite, Oka, Quebec (Peacor, 1967). Anal. H. Uik (includes P_2O_5 0·16; cell parameters *a* 9·794, *b* 8·906, *c* 5·319 Å, β 105·90°).

4 Fassaite, fassaite eclogite, Delegate, New South Wales, Australia (Lovering and White, 1969). Anal. Easton (includes Cr_2O_3 0·30, P_2O_5 0·01).

Table 39. Fassaite Analyses – *continued*

	5	6	7	8
SiO_2	48·02	47·04	46·00	44·99
TiO_2	0·46	0·28	0·25	0·36
Al_2O_3	13·39	12·43	9·50	8·08
Fe_2O_3	2·09	0·63	5·69	6·91
FeO	3·11	1·23	1·51	0·42
MnO	0·07	0·24	0·08	0·17
MgO	8·18	12·80	12·37	12·47
CaO	24·03	24·34	24·75	25·09
Na_2O	0·31	0·32	0·05	0·33
K_2O	0·06	0·19	0·00	0·36
H_2O^+	0·20	0·50	—	0·83
H_2O^-	0·07	0·06	—	0·34
Total	99·99	100·06	100·20	100·43
α	1·686	1·683	1·695	1·696
β	1·693	1·691	—	1·798
γ	1·714	1·710	1·726	1·712
$\gamma:z$	43°	43°–45°	42°–46°	46°–47°
$2V_\gamma$	56°	53°–56°	—	60°
D	3·33	—	3·26	3·34

Numbers of ions on the basis of six O

	5		6		7		8	
Si	1·760	2·00	1·721	2·00	1·704	2·00	1·696	2·00
Al	0·240		0·279		0·296		0·304	
Al	0·338	0·95	0·257	1·02	0·119	1·02	0·056	0·98
Ti	0·012		0·009		0·007		0·010	
Fe^{3+}	0·058		0·018		0·159		0·196	
Mg	0·447		0·697		0·683		0·701	
Fe^{2+}	0·095		0·037		0·047		0·013	
Mn	0·002		0·006		0·003		0·005	
Na	0·022	0·97	0·022	0·99	0·004	0·99	0·024	1·06
Ca	0·944		0·954		0·983		1·014	
K	0·002		0·009		0·004		0·018	

5 Fassaite, eclogite, Knockormal, Ayrshire (Bloxam and Allen, 1960). Anal. J. B. Allen.

6 Fassaite, calc-silicate skarn vein, Gondivalasa, Orissa, India (Rao and Rao, 1970). Anal. Rao and Vishnavardhana (cell parameters *a* 9·716, *b* 8·865, *c* 5·268 Å, β 105°57′).

7 Fassaite, fassaite–diopside metasomatized dolerite, Vilyuy, Yakutia (Ginzburg, 1969). Anal. O. G. Unanova.

8 Fassaite, metamorphic rock (conubianite) consisting of pyroxene, biotite and calcite, from moraine, Val di Solda, Venezia, Lombardy (Tomasi, 1940). Anal. L. Tomasi (includes Cr_2O_3 0·01, S 0·07).

9		10		11		12		
48·09		50·85		49·55		48·75		**SiO_2**
0·49		0·78		0·53		0·34		**TiO_2**
7·61		7·41		7·05		6·42		**Al_2O_3**
3·09		2·19		0·79		2·15		**Fe_2O_3**
0·42		4·02		1·72		0·29		**FeO**
0·03		0·13		—		—		**MnO**
15·11		15·13		15·57		16·15		**MgO**
25·34		17·37		24·08		25·94		**CaO**
0·00		1·84		0·12		—		**Na_2O**
0·00		0·01		0·00		—		**K_2O**
0·41		0·00		0·92		0·40		**H_2O^+**
0·00		0·00		0·08		0·05		**H_2O^-**
100·59		100·12		100·41		100·49		**Total**
1·686		—		1·680		—		**α**
1·694		—		1·689		—		**β**
1·712		—		1·706		—		**γ**
43°		—		43°		45°–47°		**$\gamma:z$**
—		—		—		57°–61°		**$2V_\gamma$**
3·319		—		3·300		—		**D**
1·760	2·00	1·843	2·00	1·815	2·00	1·785	2·00	**Si**
0·240		0·157		0·185		0·215		**Al**
0·088		0·161		0·119		0·062		**Al**
0·013		0·022		0·015		0·009		**Ti**
0·084	1·02	0·061	1·20[a]	0·022	1·06	0·059	1·02	**Fe^{3+}**
0·824		0·822		0·850		0·881		**Mg**
0·013		0·122		0·053		0·009		**Fe^{2+}**
0·001		0·004		—		—		**Mn**
—		0·130		0·008		—		**Na**
0·994	0·99	0·674	0·80	0·945	0·95	1·018	1·02	**Ca**
—		0·000		—		—		**K**

[a] Includes Cr^{3+} 0·011.

9 Fassaite, fassaite–spinel rock, Monzoni (Tilley, 1938). Anal. H. C. G. Vincent.

10 Fassaite, ejected nodule of hypersthene eclogite in tuff, Salt Lake Crater, Oahu (Yoder and Tilley, 1962). Anal. J. H. Scoon (includes Cr_2O_3 0·39).

11 Fassaite, fassaite–spinel rock, Adhekanwela, Ceylon (Tilley, 1938). Anal. H. C. G. Vincent.

12 Fassaite, contact metamorphosed dolomitic limestone, Monte Costone, Adamello (Hieke, 1945). Anal. M. Sesso.

Table 39. Fassaite Analyses – *continued*

	13	14	15	16
SiO_2	47·30	43·04	41·20	37·52
TiO_2	1·93	2·33	0·58	5·72
Al_2O_3	8·42	9·67	7·45	14·29
Fe_2O_3	6·21	5·95	23·00	4·43
FeO	5·64	5·65	—	7·12
MnO	0·52	0·31	0·40	0·14
MgO	8·10	9·78	3·60	6·72
CaO	20·16	21·90	23·60	24·06
Na_2O	1·20	1·34	0·68	0·09
K_2O	0·07	tr.	—	tr.
H_2O^+	—	—	—	0·00
H_2O^-	—	—	—	0·00
Total	99·62	100·21	100·51	100·20
α	—	—	1·724	1·741
β	—	—	—	—
γ	1·738	1·729	1·743	1·762
$\gamma:z$	—	—	45°	32°
$2V_\gamma$	77°	—	59°	0°
D	—	—	3·489	3·43

Numbers of ions on the basis of six O

	13		14		15		16	
Si	1·781	2·00	1·627	2·00	1·605	1·95	1·443	2·00
Al	0·219		0·373		0·342		0·557	
Al	0·154	1·03	0·056	1·03	0·000	0·91	0·091	1·01[a]
Ti	0·054		0·065		0·017		0·166	
Fe^{3+}	0·177		0·168		0·674		0·128	
Mg	0·454		0·551		0·209		0·385	
Fe^{2+}	0·177		0·179		0·000		0·229	
Mn	0·016		0·009		0·013		0·005	
Na	0·088	0·90	0·097	0·97	0·051	1·04	0·007	1·00
Ca	0·812		0·874		0·985		0·992	
K	0·002		0·000		—		0·000	

[a] Includes Cr^{3+} 0·003.

13 Fassaite, clinopyroxenite in nepheline basanite, Barsberg Hocheifel area, West Germany (Huckenholz, 1973). Anal. G. G. Huckenholz (includes P_2O_5 0·07).

14 Fassaite, alkali basalt, Burgkopf, Hocheifel area, West Germany (Huckenholz, 1973). Anal. G. Cammann (includes P_2O_5 0·24).

15 Ferrian fassaite, pyroxenite, Amboalengo, south-east Madagascar (Majmundar, 1971). Anal. H. H. Majmundar (total iron as Fe_2O_3; trace element data).

16 Titanium-rich fassaite ('uniaxial titanaugite'), sphene-rich plagioclase–diopside hornfels xenolith in norite, Schivas, Aberdeenshire (Dixon and Kennedy, 1933). Anal. B. E. Dixon (includes Cr_2O_3 0·11).

The fassaite from the Allende meteorite contains a higher content of titanium than has previously been reported in a natural pyroxene (Dowty and Clark, 1973). The composition, obtained by microprobe analysis, is SiO_2 32·8 (33·95), TiO_2 16·6 (17·35), Al_2O_3 19·3 (16·29), MgO 6·7 (7·80), and CaO 24·6 (23·86), figures in brackets from Mao and Bell (1974). On the assumption that the mineral is stoichiometric, and from spectral data, it is considered that the titanium is partitioned between Ti^{4+} and Ti^{3+} ($TiO_2 = 4{\cdot}7$ and $Ti_2O_3 = 10{\cdot}7$ per cent respectively) on the basis of which the structural formula is $Ca_{1{\cdot}0}Mg_{0{\cdot}39}Ti_{0{\cdot}48}Al_{0{\cdot}13}[Al_{0{\cdot}74}Si_{1{\cdot}26}O_6]$.

Garnet exsolution lamellae in a fassaitic pyroxene, containing 6·5 and 22·6 per cent respectively of $NaAlSi_2O_6$ and $CaAl_2SiO_6$, in the garnet pyroxenite nodules of the Roberts Victor mine kimberlite pipe have been described by Desnoyers (1975). Unlike the clinopyroxenes described by MacGregor and Carter (1970) from the same locality, the lamellae are oriented in five directions in the pyroxene, two parallel to the (110) and ($1\bar{1}0$) cleavages, the others parallel to (010), (130) and ($1\bar{3}0$).

Experimental

A series of $CaMgSi_2O_6$–$CaAl_2SiO_6$ solid solutions prepared by Zvetkov (1945) showed a maximum content of 40 per cent $CaAl_2SiO_6$ in diopside (18–19 weight per cent Al_2O_3), and a similar maximum was later reported by Sakata (1957) and by Toropov and Khotimchenko (1967). The solubility of Fe_2O_3 and TiO_2 in diopside was given as 10 and 6 per cent respectively by Segnit (1953). In the majority of fassaites and related pyroxenes, the total content of $Al_2O_3 + Fe_2O_3 + TiO_2$ does not exceed the experimentally determined solid solution limit of Al_2O_3. More extensive solid solution between $CaMgSi_2O_6$ and $CaAl_2SiO_6$, however, has been reported by Boyd (1970, see p. 245), and some fassaites contain more than 20 weight per cent $Al_2O_3 + Fe_2O_3 + TiO_2$ (Table 39, anals. 1, 2, 3, 15, 16). See also Newton *et al.* (1977).

Many fassaites are essentially diopside solid solutions containing a considerable proportion of the Ca-Tschermak component ($CaAl_2SiO_6$)

$$\underset{\text{diopside}}{CaMgSi_2O_6} + \underset{\text{anorthite}}{xCaAl_2Si_2O_8} = \underset{\text{fassaite}}{CaMgSi_2O_6.xCaAl_2SiO_6} + xSiO_2$$

The molar volume of the right-hand assemblage is less than the left-hand assemblage and the content of $CaAl_2SiO_6$ would therefore be expected to increase with pressure at a given temperature. At a pressure of 1 atm and temperature of 1135°C in the system diopside–anorthite, diopside contains approximately 3 weight per cent of $CaAl_2SiO_6$ (Hytönen and Schairer, 1961). Isothermal sections of diopside–anorthite compositions at subsolidus temperatures, 1350° and 1150°C in the pressure range 8–37 kbar (Fig. 192), have been presented by Kushiro (1965, 1969). Solid solution of diopside and $CaAl_2SiO_6$ is not complete under silica-saturated conditions. With increasing pressure at constant temperature the compositional range of the diopside solid solution + quartz increases at the expense of the diopside solid solution–anorthite–quartz assemblage, and the diopside solid solution becomes richer in the $CaAl_2SiO_6$ component; the maximum at 1150°C is 40 weight per cent at $\simeq$ 18 kbar, rising to $\simeq$ 60 per cent at $\simeq$ 30 kbar at 1350°C.

The relationships of the Fe^{3+} and Al-rich fassaites, $Ca(Al,Fe^{3+})(Al,Fe^{3+})SiO_6$, have been studied at atmospheric pressure by Huckenholz *et al.* (1974). $CaFe^{3+}AlSiO_6$, together with pseudowollastonite and hematite solid solution, is a liquidus phase within the plane $CaSiO_3$–Fe_2O_3–Al_2O_3 (Fig. 193) of the system CaO–Al_2O_3–Fe_2O_3–SiO_2. The maximum range of the fassaite solid solution

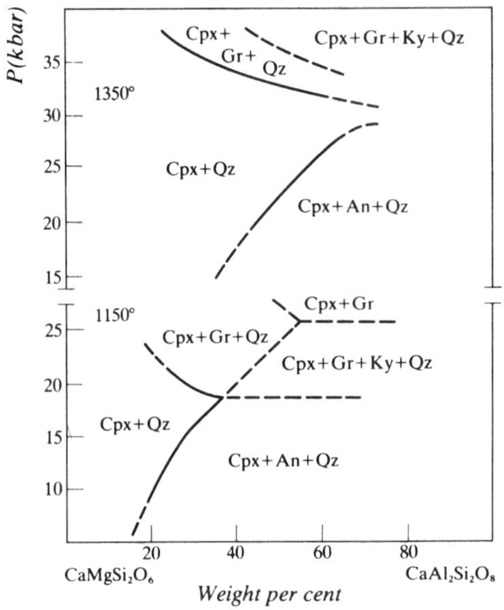

Fig. 192. Isothermal sections at 1150° and 1350°C of the join diopside–anorthite. Cpx, clinopyroxene solid solution; Gr, pyrope–grossular solid solution; An, anorthite; Ky, kyanite; Qz, quartz (after Kushiro, 1969).

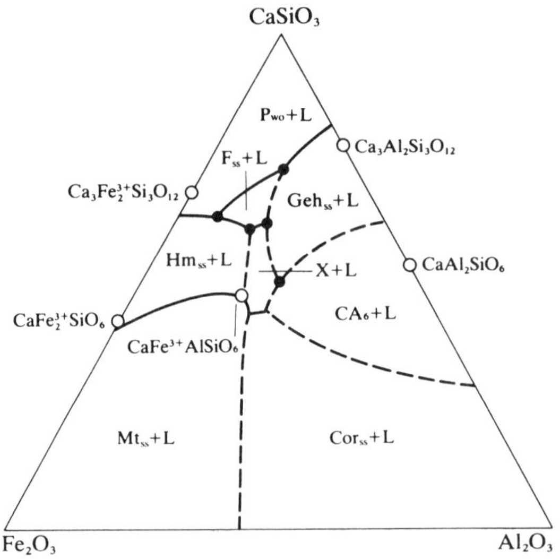

Fig. 193. Liquidus relationships of the join $CaSiO_3$–Al_2O_3–Fe_2O_3. Phases P_{wo}, F_{ss} and Hm_{ss} have compositions within the plane. Geh_{ss}, X, CA_6 and anorthite lie outside. Points with three solids + liquid, and curves with two solids and liquid are respectively piercing points of (isobaric) quaternary univariant curves and traces of divariant surfaces of the CaO–Al_2O_3–Fe_2O_3–SiO_2 system (after Huckenholz *et al.*, 1974). F_{ss}, $Ca(Al,Fe^{3+})(Al,Fe^{3+})SiO_6$; P_{wo}, parawollastonite; Geh_{ss}, gehlenite solid solution; Hm_{ss}, hematite solid solution; X, $2CaO.4(Al,Fe^{3+})_2O_3.SiO_2$; CA_6,$CaAl_{12}O_{10}$; Cor_{ss}, corundum solid solution; Mt_{ss} magnetite solid solution; L, liquid.

extends from $\simeq (CaFe_2^{3+}SiO_6)_{78}(CaAl_2SiO_6)_{22}$ to $(CaFe_2^{3+}SiO_6)_{48}(CaAl_2SiO_6)_{52}$ between 1178° and 1203°C. The composition $(CaFe_2^{3+}SiO_6)_{54}(CaAl_2SiO_6)_{46}$ melts incongruently at 1294°C to hematite solid solution $[(Al,Fe^{3+})_2O_3]$ + $[2CaO.4(Fe^{3+},Al)_2O_3.SiO_2]$ + anorthite$[CaAl_2Si_2O_8]$ + liquid (Fig. 194). Phase equilibria on the join $CaFe^{3+}AlSiO_6$–$CaMgSi_2O_6$ have been investigated by Hijikata and Onuma (1969) (see diopside section).

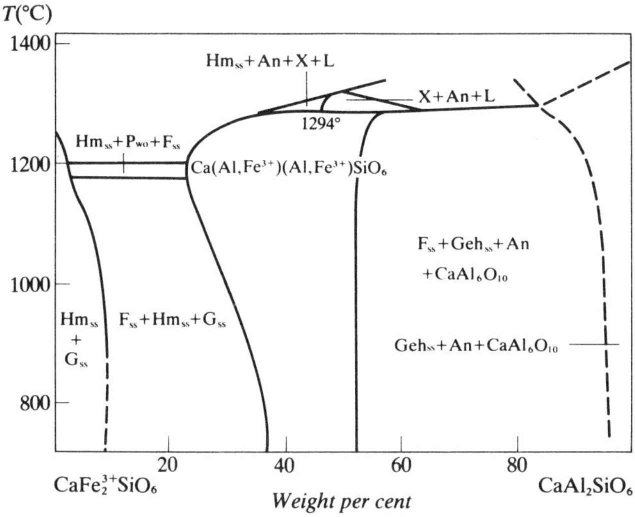

Fig. 194. **Temperature–composition relations on the join $CaFe_2^{3+}SiO_6$–$CaAl_2SiO_6$ at 1 atm. F_{ss}, $Ca(Al,Fe^{3+})(Al,Fe^{3+})SiO_6$; Hm_{ss}, $(Al,Fe^{3+})_2O_3$; G_{ss}, garnet solid solution; P_{wo}, parawollastonite; An, anorthite; Geh_{ss}, gehlenite, $Ca_2(Al,Fe^{3+})_2SiO_7$: X, $2CaO.4(Al,Fe^{3+})_2O_3.SiO_2$; L, liquid (after Huckenholz *et al.*, 1974).**

Optical and Physical Properties

There are not sufficient data to establish a precise relationship between the chemical composition and optical properties of fassaite. There is a general increase in the refractive indices with increasing tenor of Al and Fe^{3+}, and although a rough estimate of the total content of trivalent ions can be obtained from the refractive indices it is not possible to estimate the relative proportions of Al and Fe^{3+}. A diagram showing the relation between the γ index and total ferruginosity, fm $= 100Fe/(Fe+Mg)$, and aluminosity (al $= 100Al/(Al+Si+Fe+Mg)$, in fassaites and members of the diopside–hedenbergite series, based on the data for twenty-six clinopyroxenes from magnesian-rich skarns in the U.S.S.R., has been presented by Shabynin (1969). The refractive indices of the synthetic solid solutions series between pure diopside and a diopside containing 40 mol. per cent $CaAl_2SiO_6$ show a linear variation with composition (Zvetkov, 1945; Toropov and Khotimchenko, 1967), from pure diopside α 1·664, γ 1·694 and γ:z 38·5°, $2V_\gamma$ 59° to α 1·684, γ 1·714 and γ:z 35°, $2V_\gamma$ 62°. Segnit's (1953) data relating to the refractive indices of synthetic (Ca,Mg)-pyroxenes containing different amounts of Al_2O_3 in solid solution show singularly little variation (Table 40). These data, however, are not strictly comparable with those of Zvetkov, as the former pyroxenes contain variable proportions of $CaSiO_3$ and $MgSiO_3$. Segnit showed that the refractive indices of diopside increase by approximately 0·004 for each per cent of TiO_2; the γ index of diopside increases by approximately 0·0025 for each per cent of Fe_2O_3 and the α index increases by a slightly greater amount. The optic axial angles of synthetic pyroxenes containing either Al,Fe^{3+} or Ti are between 76° and 77°, but comparable values do not occur in natural fassaites. The refractive indices for a synthetic Ca-tschermakite, $Ca_{1.04}Al_{1.00}(Al_{1.07}Si_{0.93})O_6$, are α 1·709, β 1·714, γ 1·730 (Hays, 1966), and the density 3·431 g/cm^3.

Table 40 Optical properties of synthetic pyroxene solid solutions (after Segnit, 1953)

$CaSiO_3$	$MgSiO_3$	Al_2O_3	α	γ	δ
Aluminous					
48·3	41·7	10·0	1·668	1·693	0·025
53·5	37·0	9·5	1·669	1·694	0·025
45·6	46·9	7·5	1·669	1·695	0·026
58·5	34·5	7·0	1·670	1·695	0·025
35·0	60·0	5·0	1·664	1·682	0·018
40·0	55·0	5·0	1·667	1·690	0·023
Diopside	—	—	1·664	1·694	0·030
Ferriferous	Diopside	Fe_2O_3			
	98·0	2·0	1·670	1·702	0·032
	96·0	4·0	1·677	1·705	0·028
	94·0	6·0	1·684	1·710	'0·027'
	92·0	8·0	1·691	1·716	0·025
Titaniferous	Diopside	TiO_2			
	98·0	2·0	1·667	1·695	0·028
	96·0	4·0	1·677	1·704	0·027

Higher refractive indices, α 1·724, γ 1·743, than are shown above have been reported by Majmundar (1971) for an iron-rich fassaite (total iron as Fe_2O_3 = 23·0 per cent; Table 39, anal. 15), and for an aluminium and iron-rich fassaite α 1·730, β 1·735, γ 1·750 (Table 39, anal. 2). Considerably higher refractive indices (α 1·747, β 1·750, γ 1·762) are also reported for the very titanium-rich fassaite (TiO_2 = 16·6 per cent) from the Allende meteorite, by Dowty and Clark (1973). This fassaite with $\gamma{:}z = 58°$ and $2V_\gamma$ 64°, is pleochroic from dark green in (100) to red perpendicular to (100), and displays strong inclined dispersion. Data are also presented for the polarized optical absorption spectra and discussed in terms of the structure.

An investigation of the fassaite from the Angra dos Reis meteorite, containing ferrous iron (4·1 weight per cent) in addition to the possible presence of Ti^{3+}, has revealed that the maximum and minimum absorption directions in plane polarized light are not coincident with the crystallographic (except y) or optical (indicatrix) axes (Bell and Mao, 1976). The orientation of the optic axial plane of the pyroxene is parallel to (010), but the unique direction of minimum absorption, n, is normal to the z axis. There are five and not the usual three (α, β, γ) directions of absorption at 480 nm. The absorption maxima are not constant in intensity, the reason for which is as yet unknown.

The optic axial angle for the Bushveld fassaite (Table 39, anal. 2) is for λ 4861 Å = 53·5°, λ 5893 Å = 56·5°, and λ 6563 Å = 57° with one axis remaining in the same position for all wavelengths. This mineral shows anomalous violet-blue and brown interference colours, zonally arranged, with blue in the centre and brown towards the margins.

The following values of the dispersion for the fassaite from Lombardy (Table 39, anal. 8) are given by Tomasi (1940): $\gamma{:}z$ 46° to 47° (yellow), 44° to 44·5° (red), and $2V_\gamma$ 58°30′ (blue), 59°45′ (yellow), 61° (red). The titanium-rich fassaite (Table 39, anal. 16) from Schivas, described by Dixon and Kennedy (1933), occurs in black crystals which are highly pleochroic in thin section with $\alpha = \beta$ plum-coloured, γ light yellow; the mineral is uniaxial and $\gamma{:}z = 32°$. The dispersion of the bisectrices is very strong and sections approximately normal to the optic axis show abnormal

(Berlin blue) interference colours; sections other than those of the [010] zone do not extinguish in white light.

The colour of synthetic fassaite solid solutions is related to the amount of the ferri-Tschermak's molecule, $CaFe_2^{3+}SiO_6$, in the structure (Huckenholz *et al.*, 1974). Those members close to $CaFe^{3+}AlSiO_6$ in composition are slightly yellow becoming more brownish yellow as the amount of the $CaFe_2^{3+}SiO_6$ component increases. The colour change is accompanied by an increase in the birefringence.

Distinguishing Features

Fassaite is difficult to distinguish from diopsides containing moderate amounts of iron but has a lower birefringence, somewhat higher extinction angle and stronger dispersion; the latter property is the most distinctive diagonistic feature. Fassaite is difficult to distinguish from omphacite, but generally has a smaller optic axial angle, higher refractive indices, a larger extinction angle and stronger birefringence.

Paragenesis

Pyroxene–spinel–calcite assemblages are not uncommon in metamorphosed limestones, and Tilley (1938) has shown that the pyroxenes of such rocks are rich in aluminium and have a typical fassaite composition (Table 39, anals. 9 and 11). Fassaite occurs characteristically in parageneses devoid of quartz, and Tilley has suggested that fassaite is probably unstable in the presence of free silica. Knopf and Lee (1957) have described an aluminium- and ferric iron-rich fassaite (anal. 1) associated with spinel, vesuvianite, garnet, clintonite, calcite, diopside and chondrodite, in a rock, interpreted as a pyrometasomatically altered limestone septum, located between gabbro and granogabbro. At Mansjö (Eckermann, 1922), fassaite, associated with vesuvianite and garnet, occurs at pegmatite–limestone contacts, and in skarns situated between schist and limestone. Calcareous gneiss aggregates, consisting of fassaite, pargasite, phlogopite, scapolite and calcite, occur in the marble of Tiree, Inner Hebrides (Hallimond, 1947). Hieke (1945) has described fassaite (anal. 12), associated with garnet, epidote, tremolite and scapolite, at a hornblende diorite-dolomitic limestone contact. Its occurrence with spinel, forsterite and periclase in magnesian skarns, south-east Bulgaria, has been reported by Ivanova–Panayotova (1972).

Massive granular and fibrous fassaite, the latter having smaller Al_2O_3 and larger MgO and CaO contents, occurs in inclusions of original carbonate rocks in the gabbro and norite of the eastern part of the Bushveld complex. The massive fassaite (anal. 2) associated with calcite, occurs in the spinel-pyroxene zone; the assemblage is unstable and is rimmed, first by garnet and then by a symplectic intergrowth of garnet and/or vesuvianite. The reaction, massive fassaite and calcite + $SiO_2 \rightarrow$ garnet + fibrous fassaite + CO_2 represents a stage in the decarbonation process and the outward migration of CO_2. The titanium-rich fassaite (anal. 16) described by Dixon and Kennedy (1933) occurs as a segregation within a sphene-rich plagioclase–diopside hornfels xenolith in the Haddo norite, Aberdeenshire.

Fassaites, in the eclogite inclusions of the basic breccia nephelinite pipes at Delegate, New South Wales, have been described by Lovering and White (1969).

The fassaites (anal. 4) are extremely rich in Al, ranging from 9·77 to 14·38 weight per cent Al_2O_3, corresponding with 0·43 to 0·61 Al on the basis of six oxygens. Comparable compositions have been reported for the fassaite from the kimberlite pipe at Udachnaya, Siberia (Bobrievich *et al.*, 1959).

Fassaite-bearing hypersthene eclogite nodules occur in the tuff of Salt Lake Crater, Oahu (Yoder and Tilley, 1962). The fassaitic pyroxene (anal. 10) and the hypersthene form an intimate intergrowth and the texture is suggestive of exsolution. Fassaite (anal. 15) is a common constituent of the phlogopitic, feldspathic and scapolite pyroxenites of southern Madagascar (Majmundar, 1971), and has been reported as a constituent of the secondary igneous assemblage in the kyanite eclogite from the Roberts Victor mine, South Africa (Switzer and Melson, 1969). The latter is richer in sodium (Na_2O 1·4 per cent) than typical fassaite but has a much lower content than that of the primary omphacite (Na_2O 6·4 per cent) in the eclogite. Bloxam and Allen (1960) have described a fassaite (anal. 5) from the eclogite at Knockormal, Ayrshire.

Fassaite (anal. 13) is the main constituent of the clinopyroxenite fragments in the alkali basalts and nepheline basanites of the Hocheifel area, West Germany (Huckenholz, 1973). The green fassaite is rimmed by different zones of colourless clinopyroxenes, the outer zones of which merge with the augite of the groundmass of the surrounding alkali basalts. In addition to the fassaite in the pyroxenite fragments, the mineral (anal. 14) also occurs as single anhedral crystals in the alkali basalts and basanites. Both fassaites are rich in ferric iron, and although the content of sodium is relatively high the $CaFe^{3+}AlSiO_6$ component (ferri-aluminium-calcium Tschermak's molecule) constitutes over 30 per cent of the composition compared with 8 per cent of the acmitic component.

The occurrence of fassaite (anal. 3) as phenocrysts in a nepheline jacupirangite from Hessereau Hill, Oka, Quebec, has been reported by Peacor (1967) and fassaite, $Ca_{0\cdot96}(Mg_{0\cdot60}Fe^{3+}_{0\cdot16}Ti_{0\cdot07}Al_{0\cdot18})(Si_{1\cdot53}Al_{0\cdot47})O_6$, associated with ferroan monticellite and spinel in the olivine of the Sharps chondrite, Richmond County, Virginia, has been described by Dodd (1971).

An unusual occurrence of fassaite (Table 39, anal. 7) coexisting with diopside in a metasomatized dolerite at Vilyuy, Yakutia, has been described by Ginzburg (1969). The fassaite occurs in large grains, 0·5 to 1·0 cm in length, with a brown core, richer in Ti and Al than the green margin, and as smaller bright-green crystals without the brown core. The fassaite and diopside (Al_2O_3 4·90, Fe_2O_3 2·09 weight per cent) are not intergrown and are considered to be comagmatic. The compositional break is regarded as due to the lack of complete solid solution between fassaite and diopside dating from the time of their simultaneous metasomatic crystallization.

The presence of fassaite in calcium-aluminium-rich inclusions in the Allende meteorite has been described by Blander and Fuchs (1975). The pyroxene occurs in the following assemblages: fassaite (TiO_2 9 per cent, Al_2O_3 18 per cent)–spinel–melilite–anorthite, fassaite (TiO_2 4–6 per cent, Al_2O_3 15–21 per cent)–olivine–spinel–anorthite–melilite, and fassaite (TiO_2 8 per cent, Al_2O_3 20 per cent)–melilite–anorthite–perovskite–grossular–spinel.

References

Bell, P. M. and **Mao, H. K.**, 1976. Crystal-field spectra of fassaite from the Angra dos Reis meteorite. *Carnegie Inst. Washington, Ann. Rept. Dir. Geophys. Lab.*, 1975–76, 701–705.

Blander, M. and **Fuchs, L. H.**, 1975. Calcium-aluminium-rich inclusions in the Allende meteorite: evidence for a liquid origin. *Geochim. Cosmochim. Acta*, **39**, 1605–1619.

Bloxam, T. W. and **Allen, J. B.**, 1960. Glaucophane-schist, eclogite and associated rocks from Knockormal in the Girvan–Ballantrae complex, south Ayrshire. *Trans. Roy. Soc. Edinburgh*, **64**, 1–27.

Bobrievich, A. P., Bondarenko, M. N., Gunashev, M. A., Krasor, A. M. and **Smirnov, G. I.**, 1959. [The diamond deposits of Yakutia.] *Gosgeoltekhizdat*, Moscow (in Russian).

Boyd, F. R., 1970. The system $CaSiO_3$–$MgSiO_3$–Al_2O_3. *Carnegie Inst. Washington, Ann. Rept. Dir. Geophys. Lab.*, 1968–69, 214–221.

Burns, R. G. and **Huggins, F. E.**, 1973. Visible-region absorption spectra of a Ti^{3+} fassaite from the Allende meteorite: a discussion. *Amer. Min.*, **58**, 955–961.

Clark, S. P., Schairer, J. F. and **DeNeufville, J.**, 1962. Phase relations in the system $CaMgSi_2O_6$–$CaAl_2SiO_6$–SiO_2 at low and high pressure. *Carnegie Inst. Washington, Ann. Rept. Dir. Geophys. Lab.*, 1961–62, 59–68.

Coleman, L. C., 1962. Effect of ionic substitution on the unit-cell dimensions of synthetic diopside. *Geol. Soc. America* (Buddington vol.), 429–446.

Desnoyers, C., 1975. Exsolutions d'amphibole, de grenat et de spinelle dans les pyroxènes de roches ultrabasiques: péridotite et pyroxénolites. *Bull. Soc. franç. Min. Crist.*, **98**, 65–77.

Dixon, B. E. and **Kennedy, W. Q.**, 1933. Optically uniaxial titanaugite from Aberdeenshire. *Z. Krist.*, **86**, 112–120.

Dodd, R. T., 1971. Calc-aluminous insets in olivine of the Sharps chondrite. *Min. Mag.*, **38**, 451–458.

Dowty, E. and **Clark, J. R.**, 1973. Crystal structure refinement and optical properties of a Ti^{3+} fassaite from the Allende meteorite. *Amer. Min.*, **58**, 230–242.

Eckermann, H. von, 1922. The rocks and contact minerals of the Mansjö Mountain. *Geol. För. Förh. Stockholm*, **44**, no. 349.

Ginzburg, I. V., 1969. Immiscibility of the natural pyroxenes diopside and fassaite and the criterion for it. *Dokl. Acad. Sci. U.S.S.R., Earth Sci. Sect.*, **186**, 106–109.

Griffin, W. L., 1971. Mineral reactions at a peridotite-gneiss contact Jotunheimen, Norway. *Min. Mag.*, **38**, 435–445.

Hallimond, A. F., 1947. Pyroxenes, amphibole and mica from the Tiree marble. *Min. Mag.*, **28**, 230–243.

Hays, J. F., 1966. Stability and properties of the synthetic pyroxene $CaAl_2SiO_6$. *Amer. Min.*, **51**, 1524–1529.

Hieke, O., 1945. I giacimenti di contatto del Monte Costone (Adamello meridionale). *Mem. Inst. Geol. Univ. Padova*, **15**, 1–46 (M.A. **11**-149).

Hijikata, K., 1968. Unit-cell dimensions of the clinopyroxenes along the join $CaMgSi_2O_6$–$CaFe^{3+}AlSiO_6$. *J. Fac. Sci. Hokkaido Univ., ser. 4, Geol. Min.*, **4**, 149–157.

Hijikata, K. and **Onuma, K.**, 1969. Phase equilibria of the system $CaMgSi_2O_6$–$CaFe^{3+}AlSiO_6$ in air. *J. Japanese Assoc. Min. Petr. Econ. Geol.*, **62**, 209–217.

Huckenholz, H. G., 1973. The origin of fassaite augite in the alkali basalt suite of the Hocheifel area, West Germany. *Contr. Min. Petr.*, **40**, 315–326.

Huckenholz, H. G., Lindhuber, W. and **Springer, J.**, 1974. The join $CaSiO_3$–Al_2O_3–Fe_2O_3 of the CaO–Al_2O_5–Fe_2O_3–SiO_2 quarternary system and its bearing on the formation of granditic garnets and fassaite pyroxenes. *Neues Jahrb. Min. Abhdl.*, **121**, 160–207.

Hytönen, K. and **Schairer, J. F.**, 1961. The plane enstatite–anorthite–diopside and its relation to basalts. *Carnegie Inst. Washington, Ann. Rept. Dir. Geophys. Lab.*, 1960–61, 125–141.

Ivanova-Panayotova, V., 1972. [On the mineralogical characterization of the magnesian-skarn deposits in south-east Bulgaria.] *Bull. Geol. Inst. Bulg. Acad. Sci., ser. Geochim. Min. Petr.*, **21**, 125–138 (in Bulgarian, with Russian and English summaries).

Knopf, A. and **Lee, D. E.**, 1957. Fassaite from near Helena, Montana. *Amer. Min.*, **42**, 73–77.

Kushiro, I., 1965. Clinopyroxene solid solutions at high pressures. *Carnegie Inst. Washington, Ann. Rept. Dir. Geophys. Lab.*, 1964–65, 112–117.

Kushiro, I., 1969. Clinopyroxene solid solutions formed by reactions between diopside and plagioclase at high pressures. *Min. Soc. America, Spec. Paper* **2**, 179–192.

Lewis, J. F., 1967. Unit cell dimensions of some aluminous natural clinopyroxenes. *Amer. Min.*, **52**, 42–54.

Lovering, J. F. and **White, A. J. R.**, 1969. Granulitic and eclogitic inclusions from basic pipes at Delegate, Australia. *Contr. Min. Petr.*, **21**, 9–52.

MacGregor, I. D. and **Carter, J. F.**, 1970. The chemistry of clinopyroxenes and garnets of eclogite and peridotite xenoliths from the Roberts Victor Mine, South Africa. *Phys. Earth Planet. Interiors*, **3**, 391–397.

Majmundar, H. H., 1971. Fassaite from Madagascar. *Can. Min.*, **10**, 899–903.

Mao, H. K. and **Bell, P. M.**, 1974. Crystal field effects of trivalent titanium in fassaite from the Pueblo de Allende Meteorite. *Carnegie Inst. Washington, Ann. Rept. Dir. Geophys. Lab.*, 1973–74, 488–492.

Newton, R. C., Charlu, T. V. and **Kleppa, O. J.**, 1977. Thermochemistry of high pressure garnets and

clinopyroxenes in the system CaO–MgO–Al_2O_3–SiO_2. *Geochim. Cosmochim. Acta*, **41**, 369–377.
Okamura, F. P., Ghose, S. and **Ohashi, H.**, 1974. Structure and crystal chemistry of calcium Tschermak's pyroxene, $CaAlAlSiO_6$. *Amer. Min.*, **59**, 549–557.
Peacor, D. R., 1967. Refinement of the crystal structures of a pyroxene of formula $M_I M_{II}(Si_{1\cdot5}Al_{0\cdot5})O_6$. *Amer. Min.*, **52**, 31–41.
Rao, A. T. and **Rao, M. V.**, 1970. Fassaite from a calc-silicate skarn vein near Gondivalasa, Orissa, India. *Amer. Min.*, **55**, 975–980.
Sakata, Y., 1957. Unit cell dimensions of synthetic aluminian diopsides. *Japan. J. Geol. Geogr.*, **28**, 161–168.
Segnit, E. R., 1953. Some data on synthetic aluminous and other pyroxenes. *Min. Mag.*, **30**, 218–226.
Shabynin, L. I., 1969. The fassaite nature of clinopyroxenes from magnesian skarns. *Dokl. Acad. Sci. U.S.S.R., Earth Sci. Sect.*, **187**, 169–172.
Switzer, G. and **Melson, W. G.**, 1969. Partially melted kyanite eclogite from the Roberts Victor mine, South Africa. *Smithsonian Contr. Earth Sci.*, no. 1, 1–9.
Tilley, C. E., 1938. Aluminous pyroxenes in metamorphosed limestones. *Geol. Mag.*, **75**, 81–86.
Tomasi, L., 1940. Fassaite di val di Solda e sua paragenesi. *Studi Trentini Sci. Natur.*, **21**, 85–111 (M.A. **10**-340).
Toropov, N. A. and **Khotimchenko, V. C.**, 1967. Isomorphous substitution in aluminous pyroxenes. *Geochem. Intern.*, **4**, 831–835.
Tröger, E., 1951. Über den Fassaite und über die Einteilung der Klinopyroxene. *Neues Jahrb. Min., Monat.*, 132–139.
Willemse, J. and **Bensch, J. J.**, 1964. Inclusions of original carbonate rocks in gabbro and norite of the eastern part of the Bushveld complex. *Trans. Geol. Soc. South Africa*, **67**, 1–87.
Yoder, H. S. and **Tilley, C. E.**, 1962. Origin of basalt magmas: An experimental study of natural and synthetic rock systems. *J. Petr.*, **3**, 342–532.
Zvetkov, A. I., 1945. [Synthesis of alumina pyroxenes and dependence of their optics on composition.] *Mém. Soc. Russe Min., ser. 2*, **74**, 215–222 (in Russian, English summary).

Johannsenite $Ca(MnFe^{2+})[Si_2O_6]$

Monoclinic (+)

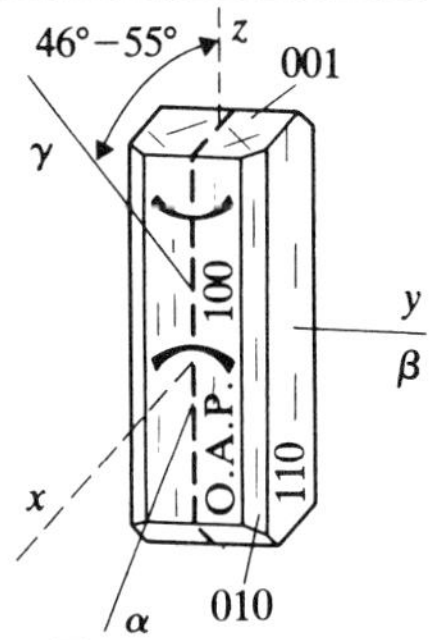

α	1·699–1·710
β	1·710–1·719
γ	1·725–1·738
δ	0·022–0·029
$2V_\gamma$	58°–72°
O.A.P.	(010); $\beta = y$
$\gamma : z$	46°–55°
Dispersion:	$r \gtrless v$, weak.
D	3·27–3·54
H	6
Cleavage:	{110} good, {100}, {010}, {001} partings. (110):(1$\bar{1}$0) $\simeq$ 87°
Twinning:	{100}, simple and lamellar, very common.
Colour:	Clove-brown, greyish, green, colourless, blue; colourless in thin section.
Unit cell:	*a* 9·83–9·98, *b* 9·04–9·16, *c* 5·26–5·29 Å, β 105°–105°29′ Z = 4. Space group *C*2/*c*.

Decomposed by heating with HCl.

Johannsenite, the manganese analogue of diopside and hedenbergite, occurs in metasomatized limestones, in association with some manganese ores, and less frequently as a vein mineral. The main variation in the chemical composition is related to the replacement of Mn by Fe^{2+}, and the iron-rich varieties are described as ferroan johannsenite. The mineral is named in honour of A. Johannsen, late Professor of Geology, Chicago University.

Structure

X-ray powder data for johannsenite and ferroan johannsenite were given respectively by Schiavinato (1953) and Momoi (1964), and Hutton (1956). The crystal structure was determined by Morimoto *et al.* (1966) for a specimen from Teragochi, Okayama, Japan with composition $Na_{0\cdot02}Ca_{0\cdot85}Mg_{0\cdot03}Mn_{0\cdot96}Fe^{2+}_{0\cdot11}Fe^{3+}_{0\cdot01}Al_{0\cdot02}(Al_{0\cdot02}Si_{1\cdot98})O_6$ and *a* 9·93, *b* 9·11, *c* 5·26 Å, β 105°11′. Another structure determination was by Freed and Peacor (1967) for a specimen from Venetia, northern Italy, with composition $Ca_{0\cdot915}Mn_{0\cdot98}Fe_{0\cdot02}Mg_{0\cdot015}Si_{2\cdot00}O_{6\cdot05}$ (normalized to 2Si), and *a* 9·978, *b* 9·156, *c* 5·293 Å, β 105°29′.

Johannsenite is isostructural with diopside (space group *C*2/*c*); all Si–O chains are equivalent and they are kinked somewhat more than in diopside (O3–O3–O3 = 163·78° compared with 166·38°). The mean Si–O (non-bridging) distance is 1·599 Å, and mean Si–O (bridging) is 1·688 Å; the latter is unusually large for a pure

Si–O tetrahedron. Mn is in almost regular octahedral co-ordination in *M*1 sites with a mean Mn–O distance of 2·17 Å, while Ca is in irregular eight-fold co-ordination in *M*2. There are four short Ca–O distances with mean 2·35 Å and four longer distances (Ca–O3) with average 2·71 Å. The co-ordination of *M*2 in johannsenite is more like 4+4 than the 6+2 of Na and Li pyroxenes, or the 4+2+2 of other Ca pyroxenes.

Freed and Peacor suggest that the large size of Mn as an *M*1 cation leads to a less stable structure than in other *C*2/*c* pyroxenes, and that the inversion to the pyroxenoid bustamite (at 830°C), and its common twinning may result from this.

Johannsenite transforms to bustamite with an orientation relationship such that an approximately close-packed framework of oxygen ions along a dense zone of metal cations and silicon atoms is preserved. The direction [012] of johannsenite is parallel to $[11\bar{1}]$ of bustamite. Migration of Si and other metal atoms must take place through the oxygen framework (Morimoto *et al.*, 1966).

Chemistry

Seven analyses of johannsenite and two of ferroan johannsenite, are shown in Tables 41 and 42 respectively; other analyses and earlier references are given by Schaller (1938). The ferroan johannsenites have compositions approximately intermediate between those of hedenbergite and johannsenite. Hutton (1956) has suggested that the distinction between johannsenite and hedenbergite should be made on the basis of the dominant molecule present (Mg being distributed equally between johannsenite and hedenbergite). Minerals having a composition closer to johannsenite than hedenbergite are described as ferroan johannsenite, and those closer to hedenbergite as manganoan hedenbergite (see anal. 5, Table 19).

Johannsenite inverts at approximately 830°C to the triclinic polymorph bustamite, and shows the same thermal relation to bustamite as hedenbergite does to the wollastonite solid solution series (Bowen *et al.*, 1933). Finely powdered and crystal fragments of johannsenite heated at 800°, 850°, 900° and 950°C for 24 hours showed that the powdered material is completely transformed to bustamite at 900°C and the crystal fragments at 950°C. The latter preserved their original morphology and the bustamite pseudomorphs are regularly oriented with respect to the original johannsenite (Morimoto *et al.*, 1966). The transformation of johannsenite to bustamite is accompanied by a volume change of 4·2 cm^3/mole ($\simeq$ 6 per cent). The ferroan johannsenite (Table 42, anal. 2) has been transformed to a manganese-iron wollastonite solid solution, or bustamite, by heating with nitrogen in a sealed tube at 1070°C for $15\frac{1}{2}$ hours. The high temperature in this experiment was used to ensure a rapid and complete transition (Hutton, 1956).

At subsolidus temperatures (830°–900°C) at 1 atm, there is a limited solid solution of up to 35% Ca-Tschermak's molecule ($CaAl_2SiO_6$) in johannsenite, and a linear relationship between d_{150} and the Ca-Tsch content (Yuquan *et al.*, 1977). On heating, the johannsenite solid solution is converted to bustamite; the inversion temperature (830°–950°C) rises as the Ca-Tsch content increases, whereas the inversion velocity decreases.

Table 41. Johannsenite Analyses

	1	2	3	4
SiO_2	47·90	47·62	48·02	48·18
TiO_2	tr.	—	—	—
Al_2O_3	tr.	0·91	0·80	0·80
Fe_2O_3	0·25	0·04	0·30	1·00
FeO	0·98	0·70	3·20	4·82
MnO	26·81	27·47	27·53	26·29
MgO	0·96	0·53	0·48	0·47
CaO	21·62	22·18	19·10	18·79
Na_2O	—	—	0·25	0·20
K_2O	—	—	0·04	0·04
H_2O^+	0·50	0·40	0·23	0·10
H_2O^-	0·26	0·09	0·00	0·00
Total	100·39	100·18	99·95	100·69
α	1·710	1·709	1·707	—
β	1·719	1·718	1·720	—
γ	1·738	1·737	1·736	—
$2V_\gamma$	70°	—	72°	—
γ:z	48°	—	55°	—
D	3·44–3·46	—	3·54	—

Numbers of ions on the basis of six O

	1	Σ	2	Σ	3	Σ	4	Σ
Si	2·014	2·01	1·971	2·00	1·984	2·00	1·977	2·00
Al	—		0·029		0·016		0·023	
Al	—	1·06	0·015	1·04	0·023	1·14	0·016	1·16
Ti	—		—		0·000		0·000	
Fe^{3+}	0·008		0·001		0·009		0·031	
Mg	0·060		0·032		0·030		0·029	
Fe^{2+}	0·034		0·024		0·111		0·165	
Mn	0·955		0·963		0·963		0·914	
Na	—	0·91	—	0·97	0·020	0·97	0·016	0·84
Ca	0·911		0·970		0·945		0·826	
K	—		—		0·002		0·002	

1 Johannsenite, limestone with rhodonite, quartz, sphalerite and manganese oxide, Monte Civillina, Recoaro, Italy (Schiavinato, 1953). Includes CO_2 1·11. Cell parameters *a* 9·83, *b* 9·04, *c* 5·27 Å, β 105°.

2 Johannsenite, Pueblo, Mexico (Schaller, 1938). Anal. Steiger (includes CO_2 0·24).

3 Cobalt blue johannsenite, manganese ore deposit, Teragōchi, Okayama Prefecture, Japan (Momoi, 1964). Anal. H. Momoi (cell parameters *a* 9·916, *b* 9·107, *c* 5·280 Å, β 105°11′).

4 Dark green johannsenite, manganese ore deposit, Teragōchi, Okayama Prefecture, Japan (Momoi, 1964). Anal. H. Momoi (cell parameters *a* 9·913, *b* 9·104, *c* 5·279 Å, β 105°12′).

Johannsenite is very susceptible to oxidation, hydration and carbonation, and is commonly altered to rhodonite. This alteration may occur in the form of irregularly shaped masses of rhodonite embedded in johannsenite, or as a pseudomorphous replacement in which compact and columnar crystals of rhodonite retain the shape of the original johannsenite (Schaller, 1938). Johannsenite may also be replaced by impure black oxides of manganese and by calcite. The alteration of johannsenite is frequently associated with the introduction, subsequent to its formation, of manganese ore-bearing solutions. The common occurrence of alteration in johannsenite results in its frequent intergrowth with secondary products, and it is evident that many of the earlier analyses were carried out on impure material.

Table 41. Johannsenite Analyses – *continued*

	5		6		7	
SiO_2	48·29		46·66		48·81	
TiO_2	—		—		0·01	
Al_2O_3	—		2·14		0·74	
Fe_2O_3	—		1·21		0·79	
FeO	tr.		1·77		1·54	
MnO	26·10		24·90		22·58	
MgO	2·84		1·11		2·29	
CaO	22·59		21·62		21·87	
Na_2O	—		—		0·07	
K_2O	—		—		0·02	
H_2O^+	—		0·78		0·32	
H_2O^-	—		0·27		0·35	
Total	99·82		100·46		99·40	
α	1·699		1·7031		1·700	
β	1·712		—		1·710	
γ	1·734		1·7256		1·7325	
$\gamma:z$	—		58°–61°		60°	
$2V_\gamma$	46°		43°–48°		47°–52°	
D	3·52		—		3·37	
Numbers of ions on the basis of six O						
Si	1·969	1·97	1·919	2·00	1·990	2·00
Al	0·000		0·081		0·010	
Al	0·000	1·08	0·023	1·06	0·026	1·02
Ti	0·000		0·000		0·000	
Fe^{3+}	0·000		0·037		0·024	
Mg	0·173		0·068		0·139	
Fe^{2+}	0·000		0·061		0·053	
Mn	0·902		0·868		0·780	
Ca	0·000	0·99	0·000	0·95	0·006	0·96
Na	0·987		0·953		0·955	
K	0·000		0·000		0·001	

5 Johannsenite, with nasonite and calcite in breccia interstices, Franklin, New Jersey (Frondel, 1965). Anal. L. H. Bauer.

6 Johannsenite, associated with rhodonite, metasomatized limestone, Madan district, Bulgaria (Padera *et al.*, 1964). Cell parameters *a* 9·90, *b* 9·10, *c* 5·27 Å, β 105·5°.

7 Johannsenite, irregular mass replacing limestone, Aravaipa mining district, Arizona (Simons and Munson, 1963). (includes P_2O_5 0·01) Semiquantitative analysis for B, Ba, Co, Cr, Cu, Ni, Pb, Sn, Sr, V, Y; also reported to contain 0·18 per cent Zn.

Table 42 Ferroan Johannsenite Analyses

	1.	2.	Numbers of ions on the basis of six O	1.		2.	
SiO_2	48·98	48·39	Si	1·989	2·00	1·979	2.00
TiO_2	0·14	0·00	Al	0·011		0·021	
Al_2O_3	0·84	0·58	Al	0·029	1·07	0·007	1·09
Fe_2O_3	1·07	0·00	Ti	0·004		—	
FeO	11·30	13·44	Fe^{3+}	0·032		—	
MnO	14·13	14·14	Mg	0·132		0·133	
MgO	2·19	2·19	Fe^{2+}	0·384		0·460	
CaO	20·64	20·79	Mn	0·486		0·490	
Na_2O	0·10	0·09	Na	0·008	0·91	0·006	0·92
K_2O	0·05	0·05	Ca	0·898		0·911	
H_2O^+	0·44	0·08	K	0·002		0·002	
H_2O^-	—	0·12					
Total	99·88	99·87					

	1.	2.
α	1·703	1·716
β	1·711	1·728
γ	1·732	1·745
$2V_\gamma$	70°	64°
$\gamma:z$	48°	48°
D	—	3·55

1. Feroan johannsenite, metasomatized limestone, Star Mine, Vanadium, New Mexico (Allen and Fahey, 1953). Anal. J. J. Fahey.
2. Ferroan johannsenite, associated with manganpyrosmalite and bustamite, Broken Hill, New South Wales, Australia (Hutton, 1956). Anal. C. O. Hutton.

Optical and Physical Properties

The optical properties of johannsenite and ferroan johannsenite are similar to those of hedenbergite and demonstrate that Mn^{2+} and Fe^{2+} do not differ greatly in their effect on the optical properties of pyroxenes of the diopside–hedenbergite series. The relation between the optical properties and composition of the manganese-rich pyroxenes of the johannsenite–diopside–hedenbergite series has been investigated

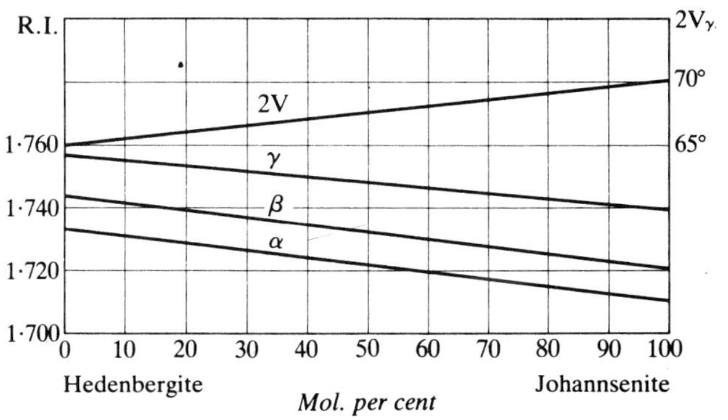

Fig. 195. The relationship between the optical properties and chemical composition in the hedenbergite–johannsenite minerals (after Zharikov and Vlasova, 1955).

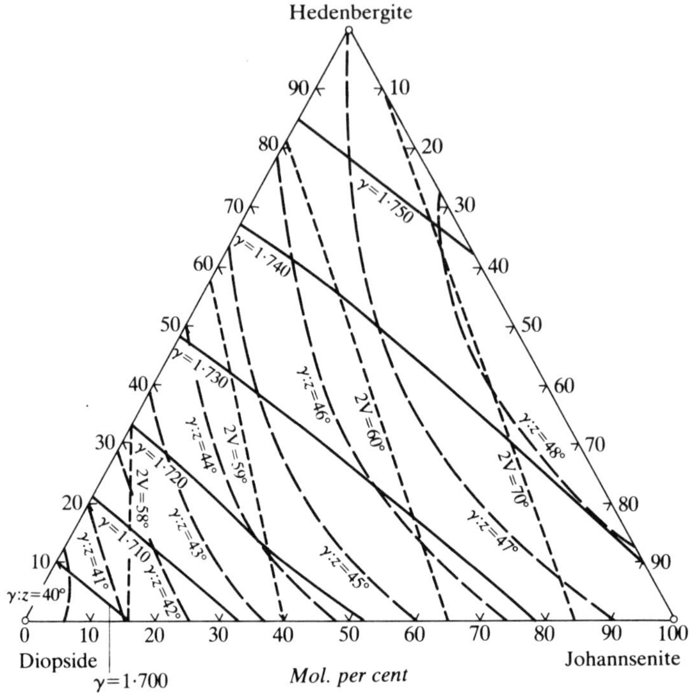

Fig. 196. The relationship between the optical properties and chemical composition in the diopside–hedenbergite–johannsenite minerals (after Zharikov and Vlasova, 1955).

Table 43 Optical properties of diopside–hedenbergite–johannsenite pyroxenes (after Zharikov and Vlasova, 1955)

Composition mol. per cent									
Di	**Hd**	**Jh**	**α**	**β**	**γ**	**δ**	**γ:z**	**$2V_\gamma$**	**Reference**
91	9	0	1·672	1·679	1·701	0·029	39·5°	56·5°	Adirondacks (Hess, 1949)
81	4	15	1·669	1·677	1·701	0·032	41°	58°	Altin Topkan, central Asia
70	30	0	1·692	1·698	1·719	0·027	45°(?)	59°	Clifton, New York (Hess, 1949)
53	21	26	1·695	1·702	1·724	0·029	44°	58°	Altin Topkan, central Asia
37	27	36	1·702	1·709	1·730	0·028	46°	59°	Altin Topkan, central Asia
31	33	36	1·705	1·712	1·733	0·028	47°	60°	Altin Topkan, central Asia
1	2	97	1·710	1·719	1·738	0·028	48°	70°	Italy (Schiavinato, 1953)
3	2	95	1·709	1·718	1·737	0·028	48°	70°	Pueblo, Mexico (Schaller, 1938)
3	16	81	1·712	1·721	1·740	0·028	48°	72°	New Mexico (Spencer and Page, 1935)
14	14	72	1·708	1·718	1·737	0·028	47°	70°	Venedia, Italy (Schaller, 1938)
0	56	44	1·710	1·715	1·735	0·025	44°(?)	54·5°(?)	Adirondacks (Hess, 1949)
16	26	58	1·708	1·718	1·736	0·028	47°	—	Vanadium, New Mexico (Allen and Fahey, 1953)
22	55	23	1·716	1·722	1·741	0·025	47°	60°	Kurusa, central Asia
8	70	22	1·724	1·730	1·748	0·024	47°	70°	Tetuke, Far Eastern Russia
7	81	12	1·734	1·741	1·754	0·020	47°	64°(?)	Tetuke, Far Eastern Russia
8	91	1	1·726	1·732	1·755	0·029	48°	62·5°	New Hampshire (Hess, 1949)

by Zharikov and Vlasova (1955) and some of the relevant data are given in Table 43. The variation in the refractive indices and optic axial angles in the johannsenite–hedenbergite series is shown in Fig. 195 and the variation in the optical properties of the diopside–hedenbergite–johannsenite pyroxenes in Fig. 196. The refractive indices and densities of ferroan johannsenite are usually slightly higher than the values for johannsenite given at the beginning of the section. Blue johannsenite zoned by green johannsenite (Table 41, anals. 3 and 4), has been described by Momoi (1964) and the colour change ascribed to the difference in iron content of the two varieties.

Distinguishing Features

Johannsenite may be distinguished from diopside by its higher refractive indices, optic axial angle and extinction angle, and from hedenbergite by lower refractive indices and high optic axial angle. The refractive indices of iron-rich johannsenite overlap with those of some intermediate members of the diopside–hedenbergite series, and the somewhat higher optic axial angle is the main optical characterization. The very common presence of black surface staining by manganese oxide is a most useful diagnostic feature of johannsenite, and usually serves to distinguish it from other members of the pyroxene group. The usual fibrous and spherulitic habit is also characteristic. Johannsenite is distinguished from bustamite by its larger optic axial angle, weaker dispersion and higher refractive indices, from rhodonite by its higher birefringence and straight extinction in the [010] zone, and from pyroxmangite by its larger optic axial angle.

Paragenesis

Johannsenite is commonly associated with bustamite and rhodonite in metasomatized limestones adjacent to acid and intermediate igneous rocks, and in the Aravaipa mining district, Arizona, forms large tabular and irregular masses replacing limestone (Simons and Munson, 1963). It has also been reported, with bustamite and manganoan hedenbergite in skarns from Nakatatsu mine, Fukui Prefecture, Japan (Tokunaga, 1965), with nasonite and calcite, filling interstices of an andradite-manganophyllite breccia at Franklin, New Jersey (Frondel, 1965), as radiating aggregates, associated with rhodonite, in metasomatized marble at Madan, Bulgaria (Padera *et al.*, 1964), and in quartz and calcite veins in rhyolite (Schaller, 1938). Johannsenite, much altered to manganese oxide and opaline silica, is the chief component of the manganese ore deposits at Teragōchi, Okayama Prefecture, Japan (Momoi, 1964). Johannsenite is associated particularly with the formation of copper, zinc and lead ores in metasomatized carbonate rocks, and is thus often found with chalcopyrite, sphalerite, galena, pyrite and magnetite. At Pueblo, Mexico, johannsenite (Table 41, anal. 2) occurs in calcite veins in rhyolite (Schaller, 1938), and is altered to rhodonite and xonotlite:

$$\underset{\text{johannsenite}}{6CaMnSi_2O_6} + H_2O \rightarrow \underset{\text{rhodonite}}{6MnSiO_3} + \underset{\text{xonotlite}}{Ca_6Si_6O_{17}(OH)_2}$$

The alteration of johannsenite to bustamite with increasing grade of metamorphism has been reported from a number of localities, e.g. Vanadium, New Mexico (Allen and Fahey, 1953). The ferroan johannsenite (Table 42, anal. 2) described by Hutton (1956) is associated with manganpyrosmalite and bustamite.

References

Allen, V. T. and **Fahey, J. J.**, 1953. Rhodonite, johannsenite and ferroan johannsenite at Vanadium, New Mexico. *Amer. Min.*, **38**, 883–890.

Bowen, N. L., **Schairer, J. F.** and **Posnjak, E.**, 1933. The system CaO–FeO–SiO_2. *Amer. J. Sci.*, ser. 5, **26**, 193–284.

Freed, R. L. and **Peacor, D. R.**, 1967. Refinement of the crystal structure of johannsenite. *Amer. Min.*, **52**, 709–720.

Frondel, C., 1965. Johannsenite and manganoan hortonolite from Franklin, N.J. *Amer. Min.*, **50**, 780–782.

Hess, H. H., 1949. Chemical compositions and optical properties of common pyroxenes. Part I. *Amer. Min.*, **34**, 234–248.

Hutton, C. O., 1956. Manganpyrosmalite, bustamite and ferroan johannsenite from Broken Hill, New South Wales, Australia. *Amer. Min.*, **41**, 581–591.

Momoi, H., 1964. Johannsenite from Teragōchi, Okayama Prefecture, Japan. *Mem. Fac. Sci., Kyushu Univ., ser. D., Geol.*, **15**, 65–72.

Morimoto, N., **Koto, K.** and **Shinohara, T.**, 1966. Oriented transformations of johannsenite to bustamite. *Min. J. Japan*, **5**, 44–64.

Padera, K., **Minčeva-Stefanova, I.** and **Kirov, G. K.**, 1964. [Johannsenite from the Borieva deposit, Madan ore district.] *Bull. Geol. Inst. Bulg. Acad. Sci.*, **13**, 5–13. (Bulgarian with German and Russian summaries.) (M.A. **17**-392.)

Schaller, W. T., 1938. Johannsenite, a new manganese pyroxene. *Amer. Min.*, **23**, 575–582.

Schiavinato, G., 1953. Sulla johannsenite dei giacimenti a silicati manganesi-feri del Monte Civillina presso Recoaro (Vicenza). *Rend. Soc. Min., Ital.*, **9**, 210–218.

Simons, F. S. and **Munson, E.**, 1963. Johannsenite from the Aravaipa mining district, Arizona. *Amer. Min.*, **48**, 1154–1158.

Spencer, A. C. and **Page, S.**, 1935. Geology of the Santa Rita Mining area, New Mexico. *U.S. Geol. Surv., Bull.*, **859**.

Tokunaga, M., 1965. On the zoned skarn including bustamite, ferroan johannsenite and manganoan hedenbergite from Nakatatsu mine, Fukui Prefecture, Japan. *Sci. Rept. Tokyo Univ.*, **9**, 67–87.

Yuquan, S., **Danian, Y.** and **Jingxiong, G.**, 1977 [Experimental studies of $CaMnSi_2O_6$–$CaAlSiAlO_6$ system.] *Sci. Geol. Sinica*, 343–354 (Chinese with English abstract).

Zharikov, V. A. and **Vlasova, D. K.**, 1955. [The diagram: composition–properties for pyroxenes of the isomorphous series diopside–hedenbergite–johannsenite.] *Dokl. Acad. Sci. U.S.S.R.*, **105**, 814–817 (in Russian).

Calcium-Sodium Pyroxenes

Omphacite $(Ca,Na)(Mg,Fe^{2+},Fe^{3+},Al)[Si_2O_6]$
(Aegirine-augite)

Omphacite $(Ca,Na)(Mg,Fe^{2+},Fe^{3+},Al)[Si_2O_6]$

Monoclinic (+)

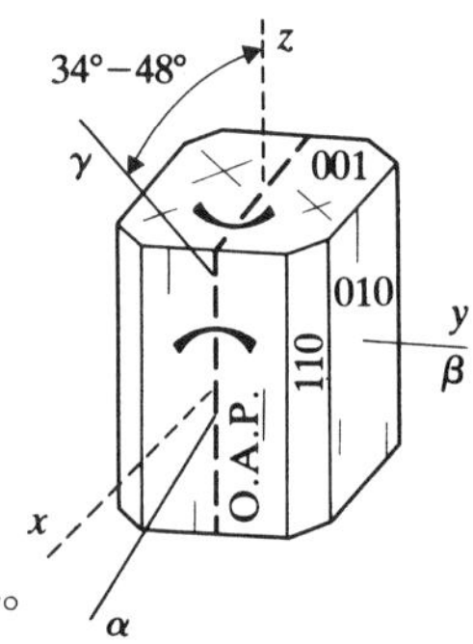

α	1·662–1·701
β	1·670–1·712
γ	1·685–1·723
δ	0·012–0·028
$2V_\gamma$	56°–84°
γ:z	34°–48°
O.A.P.	(010)
Dispersion:	$r > v$, moderate.
D	3·16–3·43
H	5–6
Cleavage:	{110} good, {100} parting; (110):(1$\bar{1}$0) $\simeq$ 87°
Twinning:	{100} simple, lamellar, common.
Colour:	Green, dark green; colourless to pale green in thin section.
Pleochroism:	Feeble, with α colourless, β very pale green, γ very pale green, blue-green (darker in ferroan variety).
Unit cell:	a 9·45–9·68, b 8·57–8·90, c 5·23–5·28 Å, β 105°–108° Z = 4. Space group $C2/c$ or $P2/n$.

Insoluble in HCl.

Omphacite is the characteristic clinopyroxene occurring in eclogites, but is also found in rocks without garnet, e.g. glaucophane schists and amphibolites. The mineral resembles diopside but is distinguished chemically by considerable replacement of Ca by Na and Mg by Al. Omphacite is named from the Greek, *omphax*, an unripe grape, in allusion to its characteristic colour. The derivation of the name is discussed more fully by Clark and Papike (1968).

Structure

The crystal structures of omphacites are in general similar to those of other pyroxenes, but while some have been reported to have the normal $C2/c$ space group others were assigned to $P2$ (Clark and Papike, 1968; Clark *et al.*, 1969). The latter authors refined the crystal structure of an omphacite from the Tiburon Peninsula, California, with formula $Na_{0{\cdot}51}Ca_{0{\cdot}48}Mg_{0{\cdot}44}Fe_{0{\cdot}10}Fe^{3+}_{0{\cdot}10}Al_{0{\cdot}39}(Si_{1{\cdot}96}Al_{0{\cdot}04})O_6$, and unit cell a 9·596, b 8·771, c 5·265 Å, β 106°56′, V 423·9 Å^3, $D_{calc.}$ 3·34 g/cm^3. With the space group $P2$ there are two crystallographically distinct Si–O chains (A and C) and within each chain two distinct Si–O tetrahedra (Si1 and Si2). Also there are four distinct $M1$ and four $M2$ sites. Matsumoto *et al.* (1975) have reinvestigated the Californian omphacite and also one from Bessi, Japan, and for both they show that the true space group is $P2/n$ (see also, Matsumoto and Banno, 1970). The Bessi omphacite has the formula $Na_{0{\cdot}484}Ca_{0{\cdot}516}Mg_{0{\cdot}392}Fe^{2+}_{0{\cdot}077}Fe^{3+}_{0{\cdot}137}Al_{0{\cdot}398}Ti_{0{\cdot}005}$-

$(Si_{1\cdot918}Al_{0\cdot082})O_6$, and unit cell *a* 9·585, *b* 8·776, *c* 5·260 Å, *β* 106·85°, *V* 423·5 Å^3, $D_{calc.}$ 3·39 g/cm^3. The earlier structure determination had assigned the space group *P*2 on the basis of some weak X-ray reflections which the later workers attributed to a multiple reflection effect and therefore discounted. The *P*2/*n* symmetry is also supported by electron diffraction (Phakey and Ghose, 1973).

With the space group *P*2/*n* all Si–O chains are equivalent but each chain contains two kinds of Si site, and there are two distinct *M*1 and *M*2 sites. The structure determination shows that Mg and Al atoms are strictly ordered between the two kinds of *M*1 site, and that Na and Ca are partially ordered between the two *M*2 sites. Mg and Al alternate in successive octahedra parallel to the *z* direction. The *P*2/*n* space group is no doubt a consequence of the difference in sizes of the two main *M*1 cations Al and Mg, and their presence in approximately equal numbers. The partial ordering of Ca and Na, with Ca-rich and Na-rich *M*2 sites alternating parallel to *z* is such as to achieve maximum proximity of Na polyhedra to Al octahedra and Ca polyhedra to Mg octahedra (Fig. 197). An ideal formula for an ordered omphacite

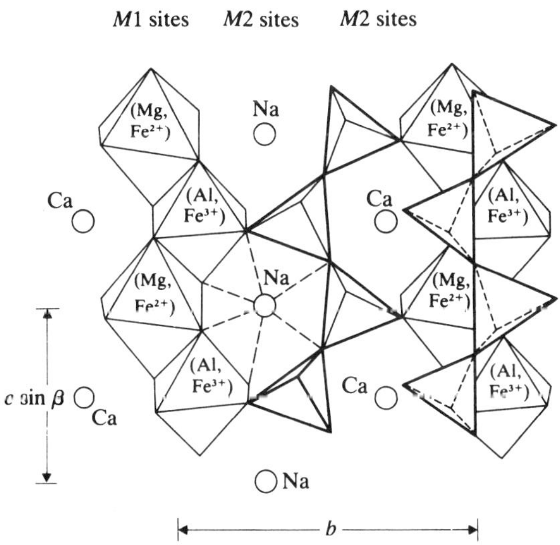

Fig. 197. View along *x* of selected portion of the omphacite structure (after Clark and Papike, 1968).

would be $Na_{0\cdot5}Ca_{0\cdot5}Mg_{0\cdot5}Al_{0\cdot5}Si_2O_6$ (i.e. $Jd_{50}Di_{50}$), but a range of compositions between $Jd_{64}Di_{36}$ and $Jd_{35}Di_{65}$ has been reported. It is assumed that for an ordered omphacite there must be sufficient Al in octahedral co-ordination to initiate the ordering process. In the aegirine-augite series where Fe^{3+} occurs instead of Al, ordering may not ensue since the size of Fe^{3+} is more similar to that of Mg; this may also be the reason why omphacites with higher Fe^{3+} content are not ordered. There is also the possibility, however, that some omphacites may have retained disorder and the *C*2/*c* space group from their high-temperature origin, and that there is an order/disorder transformation for omphacite which can be quenched in. An antiphase domain structure for a *P*2/*n* omphacite ($Jd_{31}Di_{52}Ac_{17}$) has been observed by electron microscopy (Champness, 1973) and has been attributed to this order/disorder transition. The domains are relatively large (0·2 μm), and appear to have nucleated on dislocations and subgrain boundaries, and the order/disorder transition is considered to involve a nucleation and growth mechanism. Champness (1973) suggested a narrow solvus for intermediate composition in the jadeite–diopside join, and that metastable crystallization of *C*-lattice omphacites was

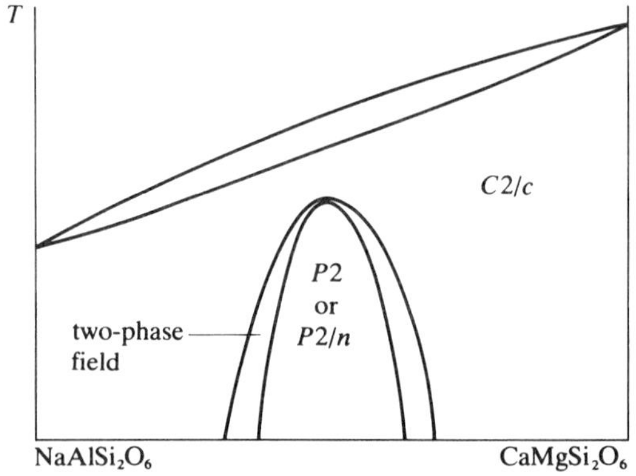

Fig. 198. Tentative phase diagram of the join jadeite–diopside showing a single-phase ordered and a disordered region separated by a two-phase field (after Champness, 1973).

followed by the $C \rightarrow P$ transition. Yokoyama *et al.* (1976) however, assume stable crystallization in the conditions of regional metamorphism with a rather low temperature $C \rightarrow P$ transition (Fig. 198).

An omphacite $(Ca_{0\cdot47}Na_{0\cdot53})(Mg_{0\cdot31}Fe^{2+}_{0\cdot15}Fe^{3+}_{0\cdot07}Al_{0\cdot47})(Si_{1\cdot98}Al_{0\cdot02})O_6$, from pyroxenite nodules in the eclogitic mica schist formation of Seisa-Lanzo, Piedmont, Italy, shows some X-ray reflections not consistent with *C2/c* symmetry (Ogniben, 1968). The cell parameters are *a* 9·575, *b* 8·770, *c* 5·263 Å, *β* 107°2′ and D 3·378 g/cm^3.

An unusual ferrous iron and Ti-rich omphacite (*a* 9·622, *b* 8·826, *c* 5·279 Å, *β* 106·92°) has a similar *P2/n* structure to those described above (Curtis *et al.*, 1975). It seems likely that other omphacites described as having the space group *P2* (see, for example, Heritsch, 1970; Morgan, 1970; Black, 1970) are in fact *P2/n*. Mössbauer spectra from omphacites (Bancroft *et al.*, 1969; Matsui *et al.*, 1970) show characteristics which are consistent with the *P2/n* space groups. It should be noted that the distinction between *P2/n* and *C2/c* cannot be made from X-ray powder data, but requires single-crystal diffraction techniques.

Details of the structure refinements of omphacites are all similar. The difference between the two kinds of tetrahedra in a single chain is slight and the Si–O bonds of bridging oxygens are longer than non-bridging, as in other pyroxenes. The *M*1

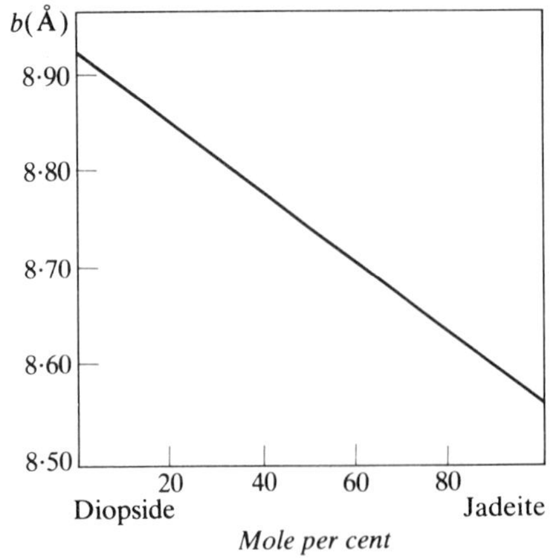

Fig. 199. Relationship of the *b*-parameter and mol. per cent $NaAlSi_2O_6$ along the join diopside–jadeite (after Edgar *et al.*, 1969).

cations are in octahedral and the *M*2 are in eight-fold co-ordination (six at about 2·4 Å and two at about 2·7 Å). The *P*2/*n* omphacites appear to be associated with low-temperature and high-pressure environments. Consideration of the compositional range of *P*2/*n* omphacites suggests that the jadeite–diopside binary contains single-phase ordered and disordered regions separated by a narrow two-phase field (Fig. 198).

Edgar *et al.* (1969) have determined the cell dimensions by X-ray powder diffraction for fifty-five omphacites and related pyroxenes, and have demonstrated that by comparing the measured *b* parameters with the theoretical *b* parameters of the diopside–jadeite solid solutions, the ratio of these two end-members can be determined to within ±5 mol. per cent (Fig. 199).

Chemistry

The composition of most omphacites can be expressed in terms of the four components, $CaMgSi_2O_6$, $CaFeSi_2O_6$, $NaAlSi_2O_6$ and $NaFe^{3+}Si_2O_6$. The substitution of Si by Al is thus generally small (≃ 3 atomic per cent). Two of the analyses in Tables 44–46, however, show higher values of tetrahedrally co-ordinated Al, and it is 13 atomic per cent in the omphacite from a corundum-bearing eclogite (Table 44, anal. 2). Significant amounts of tetrahedrally co-ordinated Al have also been reported in the omphacites from the kyanite-bearing eclogites of the Fay de Bretagne region, France (Velde, 1966). The compositional representation of such pyroxenes requires a fifth component, $CaAl_2SiO_6$, and Clark and Papike (1968) define omphacites within the system Di–Hd–Jd–Ac–CaTs as lying between the boundary planes $0{\cdot}2 \leq Na/(Na+Ca) \geq 0{\cdot}8$, and limited to the ratio, $[Al]^6/[Al]^6+Fe^{3+}) \geq 0{\cdot}5$ (see Fig. 2, p. 4). Some authors restrict the name omphacite to a smaller chemical range than shown in Fig. 2, and use the term chloromelanite to describe those clinopyroxenes which contain substantial amounts of the acmite as well as the augite and jadeite components (Table 47, Fig. 200; see also Tröger, 1962).

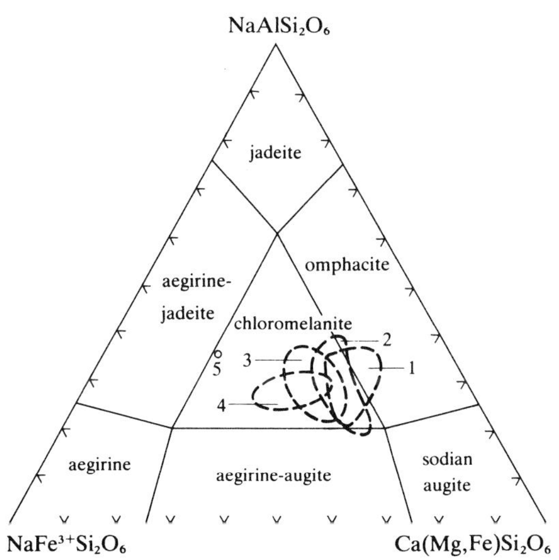

Fig. 200. Compositional fields of chloromelanites from (1) Basic schists, Île de Groix (Makanjuola and Howie, 1972); (2) and (3) veins and groundmass respectively of glaucophane schist facies, Franciscan formation, California (Essene and Fyfe, 1967); (4) Lawsonite, lawsonite–epidote and epidote zone metabasalts, Ouégua district, New Caledonia (Black, 1974); (5) Lawsonite schist, Suberi-dani, Japan (Iwasaki, 1960).

Table 44. Omphacite Analyses (eclogite nodules in kimberlite)

	1	2	3	4
SiO_2	54·49	48·99	54·29	55·06
TiO_2	0·41	0·13	0·34	0·51
Al_2O_3	5·48	17·78	4·40	7·67
Fe_2O_3	1·77	0·72	2·92	1·68
FeO	1·77	0·74	5·29	4·01
MnO	0·07	0·02	0·28	0·07
MgO	13·08	9·73	13·66	11·97
CaO	18·94	18·23	14·20	14·22
Na_2O	2·72	3·30	2·99	4·70
K_2O	0·12	0·06	0·20	0·08
H_2O^+	—	0·36	0·77	—
H_2O^-	—	—	0·25	—
Total	100·00	100·11	99·70	99·97
α	1·676	1·674 (α')	—	1·672
β	—	—	1·685	1·679
γ	1·704	1·697 (γ')	—	1·699
$\gamma:z$	40°	40°	—	36°
$2V_\gamma$	66°	59°	—	64°
D	3·33	—	—	—

Numbers of ions on the basis of six O

	1	Σ	2	Σ	3	Σ	4	Σ
Si	1·961	2·00	1·74	2·00	1·991	2·00	1·973	2·00
Al	0·039		0·26		0·009		0·027	
Al	0·193		0·50		0·181		0·297	
Ti	0·011		—		0·009		0·014	
Fe^{3+}	0·048		0·01		0·080		0·045	
Mg	0·702	1·04[a]	0·52	1·05[b]	0·751	1·19[c]	0·639	1·12
Fe^{2+}	0·053		0·02		0·162		0·120	
Mn	0·002		0·00		0·009		0·002	
Na	0·190		0·22		0·212		0·327	
Ca	0·730	0·93	0·69	0·91	0·558	0·78	0·546	0·88
K	0·006		0·00		0·009		0·004	
$\frac{100\,Na}{(Na+Ca)}$	20·7		24·2		27·5		37·5	

[a] Includes Cr^{3+} 0·033.
[b] Includes Cr^{3+} 0·002

1 Chrome omphacite, eclogite nodule. Obnazhennaya pipe. Olenek basin (Milashev, 1960). Anal T. M. Sablina (includes Cr_2O_3 1.15; anal. recalc. on water-free basis).
2 Omphacite, corundum eclogite, Obnazhennaya pipe, Yakutia (Sobolev and Kuznetsova, 1965). Anal. I. K. Kuznetsova (includes Cr_2O_3 0·05, H_2O^+ quoted as volatile).
3 Omphacite, eclogite, kimberlite pipe, Roberts Victor mine, Kimberley (Kushiro and Aoki, 1968). Anal. K. Aoki (includes Cr_2O_3 0·11).
4 Omphacite, eclogite nodule in kimberlite pipe, Kaalvalle, Basutoland (Nixon *et al.*, 1963). Anal. O. von Knorring (trace element data includes Cr^{3+} 1300, Ni^{2+} 1000 p.p.m.).

5		6		7		8		
55·25		54·73		55·50		55·75		SiO_2
0·23		0·41		0·18		1·62		TiO_2
16·17		14·10		15·63		14·66		Al_2O_3
1·64		1·17		1·94		3·28		Fe_2O_3
1·36		1·74		0·96		1·95		FeO
0·04		0·04		0·02		0·02		MnO
7·23		8·84		7·17		4·99		MgO
12·34		11·62		10·52		7·63		CaO
6·19		6·61		7·61		9·70		Na_2O
0·11		0·24		0·08		0·01		K_2O
—		0·38		—		0·19		H_2O^+
—		0·17		—		0·00		H_2O^-
100·60		100·15		99·65		100·00		Total
1·673		—		—		1·671		α
—		1·676		—		1·678		β
1·700		—		—		1·690		γ
37°		—		—		47°		$\gamma:z$
71°		—		—		79°–82°		$2V_\gamma$
—		—		—		—		D
1·921	2·00	1·930	2·00	1·944	2·00	1·962	2·00	Si
0·079		0·070		0·056		0·038		Al
0·584		0·516		0·589		0·570		Al
0·006		0·011		0·005		0·043		Ti
0·043	$1·05^c$	0·031	$1·08^b$	0·050	$1·05^c$	0·087	$1·02^d$	Fe^{3+}
0·375		0·465		0·374		0·261		Mg
0·040		0·051		0·028		0·057		Fe^{2+}
0·001		0·001		0·001		0·001		Mn
0·417		0·452		0·516		0·661		Na
0·460	0·88	0·439	0·89	0·395	0·92	0·287	$0·95^e$	Ca
0·005		0·001		0·004		0·000		K
47·5		50·7		56·6		69·7		100 Na/(Na + Ca)

[c] Includes Cr^{3+} 0·001.
[d] Includes V^{5+} 0·003, Ni 0·001, Cr^{3+} 0·001.
[e] Includes Li 0·003.

5 Omphacite, corundum eclogite, Yakutia (Sobolev, 1968a) (includes Cr_2O_3 0·04).
6 Omphacite, eclogite kimberlite pipe, Roberts Victor mine, Kimberley (Kushiro and Aoki, 1968). Anal. K. Aoki (includes Cr_2O_3 0·10).
7 Omphacite, kyanite eclogite, kimberlite pipe, Zagadochnaya, Yakutia (Sobolev *et al.*, 1968) (includes Cr_2O_3 0·04).
8 Sodium-rich omphacite, kimberlite pipe, Grnet Ridge, Arizona (Watson and Morton, 1969). Anal. C. O. Ingamells (includes Cr_2O_3 0·03, V_2O_3 0·11, NiO 0·02, Sr 0·00, BaO 0·02, Li_2O 0·02; analysis corrected for 1·1 per cent impurities).

Table 45. Omphacite Analyses (eclogites from gneiss terrains)

	1		2		3		4		5	
SiO_2	50·78		53·00		53·80		52·65		52·31	
TiO_2	0·23		0·03		0·02		1·34		0·43	
Al_2O_3	7·24		4·85		4·93		10·26		9·91	
Fe_2O_3	2·96		2·79		2·95		2·98		1·13	
FeO	3·35		0·71		0·85		4·08		5·43	
MnO	0·03		tr.		0·00		0·04		0·02	
MgO	11·92		14·86		14·01		9·13		10·64	
CaO	20·12		20·35		19·55		15·32		16·60	
Na_2O	2·84		3·08		3·64		3·68		4·00	
K_2O	0·06		0·04		0·03		0·59		0·16	
H_2O^+	0·66		0·18		0·18		0·03		0·02	
H_2O^-	0·00		0·06		0·06		—		0·04	
Total	100·21		'99·55'		100·02		100·10		100·75	
α	1·677		1·6687		1·6684		—		—	
β	—		1·6756		1·6749		—		—	
γ	1·705		1·6956		1·6927		—		—	
$\gamma:z$	48°		—		—		38°		43°	
$2V_\gamma$	62°		64°		62°		71°		75°	
D	—		3·3017		3·2938		3·33		—	
	Numbers of ions on the basis of six O									
Si	1·869	2·00	1·923	2·00	1·948	2·00	1·901	2·00	1·83	2·00
Al	0·131		0·077		0·052		0·099		0·17	
Al	0·183		0·131		0·149		0·337		0·24	
Ti	0·006		0·001		—		0·036		0·01	
Fe^{3+}	0·082		0·076		0·082		0·082		0·03	
Mg	0·654	1·03	0·804	1·03	0·775	1·03	0·491	1·07	0·56	1·00
Fe^{2+}	0·103		0·022		0·026		0·123		0·16	
Mn	0·001		0·000		0·000		0·001		—	
Na	0·202		0·216		0·256		0·257		0·27	
Ca	0·794	1·00	0·791	1·01	0·757	1·02	0·592	0·88	0·62	0·90
K	0·002		0·002		0·002		0·028		0·01	
$\frac{100\,Na}{(Na+Ca)}$	20·3		21·4		25·3		30·3		30·3	

1 Omphacite, hornblende-bearing eclogite, Gongen-yama, Bessi district, Japan (Shidô, 1959). Anal. H. Haramura (includes P_2O_5 0·02).
2 Omphacite, zoisite eclogite, Silberbach, Fichtelgebirge, Germany (Wolff, 1942). Anal. E. Eberius.
3 Omphacite, zoisite eclogite, Silberbach, Fichtelgebirge, Germany (Wolff, 1942). Anal. E. Eberius.
4 Omphacite, eclogite, Gertrush, Kärnten, Austria (Angel and Schaider, 1950).
5 Omphacite, eclogite, Sauviat-sur-Vige, central France (Coffrant, 1974). Anal. D. Coffrant (includes P_2O_5 0·06).

6		7		8		9		10		
54·08		54·74		55·73		55·72		54·88		**SiO_2**
0·14		0·09		0·27		0·16		0·39		**TiO_2**
9·20		5·61		12·76		11·23		10·44		**Al_2O_3**
0·83		4·56		0·07		1·18		5·80		**Fe_2O_3**
2·18		3·83		3·23		1·64		3·31		**FeO**
tr.		0·03		0·04		0·02		—		**MnO**
11·51		10·83		9·07		9·97		6·41		**MgO**
17·50		15·11		14·07		14·80		12·94		**CaO**
4·20		4·56		4·40		5·65		5·31		**Na_2O**
0·04		0·01		tr.		0·05		0·34		**K_2O**
0·46		0·50		0·02		0·17		0·23		**H_2O^+**
0·00		0·20		tr.		0·00		—		**H_2O^-**
100·14		'100·23'		99·98		100·61		100·05		**Total**
1·665		1·688		1·665		1·673		—		α
1·674		1·693		1·673		1·767		1·67		β
1·687		1·709		1·688		1·691		—		γ
37°		—		37°–40°		—		36°		$\gamma:z$
64°		56°		71°–72°		—		—		**$2V_\gamma$**
—		3·394		—		—		3·31		**D**
1·926	2·00	1·973	2·00	1·966	2·00	1·961	2·00	1·973	2·00	**Si**
0·074		0·027		0·034		0·039		0·027		**Al**
0·312		0·221		0·496		0·427		0·415		**Al**
0·004		0·002		0·007		0·004		0·010		**Ti**
0·022	1·01	0·123	1·05	0·002	1·08[b]	0·031	1·03	0·156	1·02	**Fe^{3+}**
0·611		0·582		0·477		0·523		0·343		**Mg**
0·065		0·116		0·095		0·048		0·100		**Fe^{2+}**
0·000		0·001		0·001		0·001		0·000		**Mn**
0·290		0·325		0·301		0·385		0·370		**Na**
0·668	0·96	0·583	0·91[a]	0·532	0·83	0·558	0·95	0·497	0·88	**Ca**
0·002		0·000		0·000		0·002		0·016		**K**
30·3		35·9		36·1		40·8		42·7		**$\frac{100\,Na}{(Na+Ca)}$**

[a] Includes Sr 0·001.
[b] Includes Cr 0·003.

6 Omphacite, type locality, kyanite eclogite, Kupperbrunn, Saualpe, Austria (Mottana *et al.*, 1968). Anal A. Mottana (*a* 9·671, *b* 8·842, *c* 5·212 Å, *β* 106·77°).
7 Omphacite, eclogite, Eiksundsdal, Sanmøre, Norway (Warner, 1964). Anal. J. Ito (includes Cr_2O_3 0·01, NiO 0·01, Sr 0·05. Cell parameters *a* 9·662, *b* 8·819, *c* 5·228 Å, *β* 106·55°).
8 Omphacite, eclogite, Bielice, Sudeten Mountains (Smulikowski, 1964). Anal. S. Rossol (includes Cr_2O_3 0·17, V_2O_3 0·03, P_2O_5 0·12).
9 Omphacite, eclogite, Silberbach, Fichtelgebirge (Banno, 1967a). Anal. H. Haramura (includes P_2O_5 0·02).
10 Omphacite, eclogite, Fay de Bretagne, France (Brière, 1920). Anal. M. Raoult.

Table 45. Omphacite Analyses (eclogites from gneiss terrains) – *continued*

	11	12	13
SiO_2	53·32	54·14	54·03
TiO_2	1·12	0·25	0·54
Al_2O_3	8·69	8·06	11·54
Fe_2O_3	7·27	6·29	5·62
FeO	3·95	3·26	4·09
MnO	0·05	0·00	0·05
MgO	7·82	8·09	5·13
CaO	11·73	12·98	11·82
Na_2O	5·36	6·90	6·81
K_2O	0·19	0·03	0·20
H_2O^+	—	0·11	—
H_2O^-	0·60	0·02	0·29
Total	100·29	100·13	100·12
α	—	1·690	—
β	—	1·698	1·697
γ	—	1·708	—
$\gamma:z$	—	40°	41°
$2V_\gamma$	—	80°	82°40′
D	—	3·39	3·365
	Numbers of ions on the basis of six O		
Si	1·940 } 2·00	1·964 } 2·00	1·951 } 2·00
Al	0·060	0·036	0·049
Al	0·312 } 1·09[c]	0·309 } 1·02	0·443 } 1·01
Ti	0·028	0·007	0·015
Fe^{3+}	0·201	0·172	0·152
Mg	0·422	0·437	0·276
Fe^{2+}	0·120	0·099	0·123
Mn	0·002	0·000	0·001
Na	0·381 } 0·85	0·485 } 0·99	0·476 } 0·94
Ca	0·457	0·504	0·457
K	0·009	0·001	0·010
$\frac{100\,Na}{(Na+Ca)}$	45·5	49·0	51·0

[c] Includes Cr 0·004.

11 Omphacite, eclogite, Aktyuz complex, U.S.S.R. (Dobretsov and Sobolev, 1970). Anal. I. K. Kuznetsova (includes Cr_2O_3 0·19).
12 Omphacite, barroisite eclogite, Naustdal, Norway (Binns, 1967). Anal. R. A. Binns.
13 Omphacite, eclogite, Vanelvsdalen, Norway (Eskola, 1921). Anal. Thomassen.

Table 46. Omphacite Analyses (eclogites from glaucophane schist facies)

	1		2		3		4		5	
SiO_2	53·21		53·12		53·31		55·80		53·14	
TiO_2	0·54		0·31		0·26		0·33		0·89	
Al_2O_3	12·86		10·31		10·52		11·05		9·21	
Fe_2O_3	1·76		4·55		4·11		2·24		5·61	
FeO	2·79		3·00		2·84		2·36		3·20	
MnO	0·03		0·08		0·05		0·04		0·01	
MgO	8·59		7·54		8·42		8·68		7·66	
CaO	14·69		15·03		14·50		13·50		12·89	
Na_2O	4·51		5·81		5·90		6·40		6·23	
K_2O	0·23		0·03		0·05		0·00		0·17	
H_2O^+	0·11		0·13		0·16		—		1·12	
H_2O^-	0·25		0·06		—		—		0·08	
Total	99·63		99·97		100·12		100·40		100·26	
α	1·664		—		1·673		1·668		1·682	
β	1·670		—		1·679		1·676		—	
γ	1·685		—		1·691		1·691		1·694	
$\gamma:z$	35°		—		39°		43°		—	
$2V_\gamma$	66°		—		60°		72·5°		76°	
D	3·401		3·30		3·34		3·28		—	
	Numbers of ions on the basis of six O									
Si	1·907	2·00	1·924	2·00	1·919	2·00	1·972	2·00	1·934	2·00
Al	0·093		0·076		0·081		0·028		0·066	
Al	0·450		0·364		0·365		0·431		0·333	
Ti	0·015		0·008		0·007		0·009		0·024	
Fe^{3+}	0·047	1·06	0·124	1·00	0·111	1·05	0·059	1·03	0·154	1·03
Mg	0·459		0·407		0·478		0·460		0·418	
Fe^{2+}	0·084		0·091		0·085		0·070		0·098	
Mn	0·001		0·002		0·002		0·001		0·000	
Na	0·313		0·408		0·412		0·438		0·444	
Ca	0·564	0·89	0·583	0·99	0·540	0·95	0·511	0·95	0·505	0·96
K	0·011		0·001		0·002		—		0·007	
$\frac{100\,Na}{(Na+Ca)}$	35·7		41·2		43·2		46·2		46·8	

1 Omphacite, eclogite in greenschist facies, Lyell Highway, Tasmania (Spry, 1963). Anal. H. Asari (includes P_2O_5 0·06).
2 Omphacite (acmitic diopside–jadeite), eclogite, Valley Ford, California (Bloxam, 1959). Anal. T. W. Bloxam.
3 Omphacite, eclogite in glaucophane schist, Healdsburg, Sonoma County, California (Switzer, 1945). Anal. F. A. Gonyer.
4 Omphacite, eclogite, nodule in serpentinite, Motagua fault zone, Guatemala (McBirney *et al.*, 1967). Anal. K. Aoki (*a* 9·56, *b* 8·78, *c* 5·25 Å, β 106°47′, *V* 421·6 Å³).
5 Omphacite, glaucophanite, Île de Groix, France (Triboulet, 1974). Anal. N. Vassard (includes P_2O_5 0·05).

Table 46. Omphacite Analyses (eclogites from glaucophane schist facies) – *continued*

	6	7	8	9	10
SiO_2	55·36	54·4	56·02	56·42	52·8
TiO_2	0·21	0·24	0·38	0·40	0·88
Al_2O_3	10·32	10·6	12·74	11·84	11·5
Fe_2O_3	2·86	3·91	0·88	0·88	4·98
FeO	2·05	2·5	1·64	2·15	10·5
MnO	0·007	—	0·00	tr.	0·64
MgO	8·73	8·4	8·01	8·50	0·84
CaO	13·49	12·70	12·45	12·79	11·2
Na_2O	6·56	6·6	7·05	6·82	6·99
K_2O	0·028	0·05	0·40	0·12	0·05
H_2O^+	—	0·4	—	} 0·34	—
H_2O^-	0·03	0·00	—		—
Total	99·65	99·8	99·57	100·26	100·4
α	1·672 (min)	1·680	—	—	—
β	1·679 (aver.)	1·685	1·675	—	—
γ	1·697 (max)	1·700	—	—	—
$\gamma:z$	48°–60°	36°	—	—	—
$2V_\gamma$	68°–70°	72°	72°	—	—
D	—	—	3·32	—	—
	Numbers of ions on the basis of six O				
Si	1·977 } 2·00	1·96 } 2·00	1·985 } 2·00	1·988 } 2·00	1·973 } 2·00
Al	0·023	0·04	0·015	0·012	0·027
Al	0·411	0·41	0·517	0·479	0·480
Ti	0·005	0·01	0·010	0·011	0·025
Fe^{3+}	0·077 } 1·02	0·11 } 1·06	0·023 } 1·00	0·025 } 1·03	0·140 } 1·04
Mg	0·465	0·45	0·423	0·447	0·047
Fe^{2+}	0·061	0·08	0·031	0·064	0·332
Mn	0·000	0·00	0·000	0·000	0·020
Na	0·454	0·46	0·484	0·483	0·507
Ca	0·516 } 0·97	0·49 } 0·95	0·473 } 0·98	0·466 } 0·95	0·448 } 0·96
K	0·001	0·00	0·018	0·004	0·002
$\frac{100\ Na}{(Na+Ca)}$	46·8	48·4	50·6	50·9	53·1

6 Omphacite, eclogite, Guajira Peninsula, Colombia, South America (Green *et al.*, 1968) (includes Cr_2O_3 0·005; cell parameters *a* 9·596, *b* 8·783, *c* 5·266, β 106°54′).

7 Omphacite, omphacitite, Monviso Massif, Cottian Alps (Mottana, 1971). Anal. A. Mottana (*a* 9·599, *b* 8·785, *c* 5·26 Å, β 106·87°, *V* 424·6 Å³).

8 Omphacite, eclogite associated with epidote-amphibolite facies, Puerto Cabello, Venezuela (Morgan, 1970). Anal. Technical Services Lab., Toronto, Canada (*P*2 symmetry, *a* 9·551, *b* 8·751, *c* 5·254 Å, β 106°52′, *V* 420·2 Å³).

9 Omphacite, eclogite, Maksyutov complex, U.S.S.R. (Dobretsov and Sobolev, 1970). Anal. I. K. Kuznetsova.

10 Ferroan omphacite, regionally metamorphosed peralkaline rocks, Red Wine alkalic province, central Labrador (Curtis *et al.*, 1975). Microprobe anal., Fe^{3+}/Fe^{2+} ratio by wet chemical methods (*a* 9·622, *b* 8·826, *c* 5·279 Å, β 106·92°, *P*2/*n* space group).

11		12		13		
54·75		52·9		52·5		SiO_2
1·95		0·05		2·60		TiO_2
9·40		11·1		10·2		Al_2O_3
4·00		3·0		5·05		Fe_2O_3
4·64		13·1		10·8		FeO
0·01		0·5		0·57		MnO
6·47		0·7		1·04		MgO
10·95		10·9		10·7		CaO
6·92		7·1		7·60		Na_2O
tr.		—		0·05		K_2O
0·14		—		—		H_2O^+
—		—		—		H_2O^-
99·63		99·35		100·1		Total
1·662		1·696		1·701		α
1·670		1·703		1·712		β
1·680		1·720		1·723		γ
64°		46°		74°		$\gamma:z$
84°		65°		84°		$2V_\gamma$
3·315		3·43		3·42		D
1·983	2·00	1·985	2·00	1·952	2·00	Si
0·017		0·015		0·048		Al
0·383		0·476		0·399		Al
0·052		0·001		0·073		Ti
0·109	1·04	0·085	1·03	0·141	1·02	Fe^{3+}
0·352		0·039		0·058		Mg
0·139		0·411		0·334		Fe^{2+}
0·000		0·016		0·018		Mn
0·484		0·517		0·548		Na
0·424	0·91	0·438	0·96	0·426	0·98	Ca
—		—		0·002		K
53·3		54·1		56·3		$\frac{100\,Na}{(Na+Ca)}$

11 Omphacite, eclogite in glaucophane schist, Pian della Mussa, Alpes piémontaises (Nicolas, 1966) (includes P_2O_5 0·10, ign. loss 0·30).

12 Ferroan omphacite, metamorphosed acid volcanic rock, Bouehndep, New Caledonia (Black, 1970). Microprobe anal., Fe^{3+}/Fe^{2+} ratio by wet chemical methods (*a* 9·594, *b* 8·820, *c* 5·272 Å, β 106°55′, *P*2 space group).

13 Titanian ferroan omphacite, regionally metamorphosed peralkaline rocks, Red Wine alkalic province, central Labrador (Curtis *et al.*, 1975). Microprobe anal., Fe^{3+}/Fe^{2+} ratio by wet chemical methods; pleochroism α dark blue, β azure blue, γ colourless.

Table 47. Chloromelanite Analyses

	1		2		3	
SiO_2	54·54		53·47		52·41	
TiO_2	0·23		0·42		1·56	
Al_2O_3	8·89		9·93		9·47	
Fe_2O_3	7·07		8·96		15·54	
FeO	2·89		2·19		1·43	
MnO	0·03		0·03		0·12	
MgO	6·48		5·94		2·96	
CaO	12·38		11·66		5·06	
Na_2O	7·20		7·28		9·97	
K_2O	0·06		0·14		0·10	
H_2O^+	—		—		1·10	
H_2O^-	—		—		0·08	
Total	99·77		100·02		99·8	
α	1·685		1·688		1·726	
β	1·697		1·698		—	
γ	1·702		1·704		1·750	
$\gamma:z$	—		—		—	
$2V_\gamma$	76°		78°		74°–80°	
D	—		—		—	
	Numbers of ions on the basis of six O					
Si	1·977	2·00	1·937	2·00	1·937	2·00
Al	0·023		0·063		0·063	
Al	0·357		0·361		0·350	
Ti	0·006		0·012		0·043	
Fe^{3+}	0·193	1·00	0·244	1·01	0·433	1·04
Mg	0·350		0·321		0·163	
Fe^{2+}	0·088		0·066		0·044	
Mn	0·001		0·001		0·004	
Na	0·506		0·512		0·714	
Ca	0·481	0·99	0·452	0·97	0·201	0·92
K	0·003		0·007		0·005	
$\frac{100\ Na}{(Na+Ca)}$	51·3		53·1		78·0	

1 Chloromelanite, basic schist, Île de Groix (Makanjuola and Howie, 1972).
2 Chloromelanite, basic schist, Île de Groix (Makanjuola and Howie, 1972).
3 Chloromelanite, lawsonite schist, Suberi-dani, Tokushima Prefecture, Japan (Iwasaki, 1960). Anal. H. Haramura (includes P_2O_5 0·06; *a* 9·56, *b* 8·77, *c* 5·24 Å, β 107°47′).

Table 48 Average omphacite compositions for rocks of deep seated origin (after Sobolev, 1968b). **Numbers of cations on the basis of six oxygens**

	$[Al]^4$	$[Al]^6$	Fe^{2+}	Fe^{3+}	Mg	Ca	Na
Corundum eclogites	0·260	0·500	0·010	0·020	0·520	0·690	0·230
Grosspydites and kyanite eclogites	0·091	0·561	0·038	0·025	0·419	0·491	0·398
Diamond eclogites	0·040	0·335	0·079	0·069	0·619	0·515	0·384

A number of surveys have been made of the compositions of omphacitic pyroxenes in relation to their paragenesis. The compositions of clinopyroxenes from nodules in kimberlites, rocks of the granulite facies and the glaucophane schist facies (Coleman *et al.*, 1965) range from augitic omphacite to omphacite to jadeitic omphacite. Sobolev and Kuznetsova (1965) have tabulated the percentages of the main components of omphacite from a variety of rocks, Smulikowski (1968) has detailed the maximum, minimum and average weight per cent of the main oxides in omphacitic pyroxenes from garnet peridotites, common eclogites and ophiolitic eclogites, and a more general review of the variation in the composition of sixty-four clinopyroxenes from eclogites and rocks of the granulite facies has been presented by White (1964). An investigation of omphacites from different environments by Mottana *et al.* (1971) demonstrated that the compositions of the pyroxenes are controlled essentially by diopside–jadeite solid solution, and to a lesser extent by substitution of the acmite component. A survey of the omphacites from the kimberlite pipes of Yakutia (Sobolev, 1968a) has shown that the highest content of aluminium in tetrahedral + octahedral co-ordination and lowest sodium content occurs in the pyroxenes of the corundum-bearing eclogites; the highest content of octahedrally co-ordinated aluminium is found in the grospydites and kyanite eclogites (Table 48). The omphacites of the diamond-bearing eclogites have the smallest content of both tetrahedrally and octahedrally co-ordinated aluminium and approximate most closely to diopside–jadeite compositions. Other studies of the range of omphacite compositions include those of Essene and Fyfe (1967) and Matthes *et al.*, (1970). The former gave the compositions of eleven matrix and vein omphacites from the glaucophane schist facies of the Franciscan formation, California. An unusually ferrous iron-rich omphacite (Table 46, anal. 12), intermediate in composition between hedenbergite and jadeite has been described by Black (1970).

Omphacites rich in ferrous iron and titanium have been described by Curtis *et al.* (1975) and may be represented by the simplified formulae, $(Na_{0\cdot51}Ca_{0\cdot45})(Fe_{0\cdot47}Ti_{0\cdot03}Al_{0\cdot48}Si_2O_6$ and $(Na_{0\cdot55}Ca_{0\cdot43})(Fe_{0\cdot47}Ti_{0\cdot07}Al_{0\cdot40})Si_2O_6$. Both omphacites (Table 46, anals. 10 and 13) have an exceptionally low content of Mg (0·64 and 1·04 weight per cent MgO). Major, minor and some trace element (Ni, Co, V, Sr and Cr) data for more than a hundred omphacites from the eclogites and eclogite amphibolites in the Muenchberger gneiss massif, north-east Bavaria have been collated by Matthes *et al.* (1970). The variation in the cation content, on the basis of six oxygens, for seventy-one omphacites is Na 0·19–0·57, K 0·00–0·01, Ca 0·39–0·73, Mg 0·27–0·69, Fe^{2+} 0·02–0·13, Fe^{3+} 0·01–0·11, $[Al]^6$ 0·30–0·62, $[Al]^4$ 0·00–0·07, Si 1·93–2·01.

The distribution of Fe and Mg between the clinopyroxenes and garnets of eclogites has been studied by Banno (1970). The distribution coefficient corresponds to the exchange reaction:

$$CaMgSi_2O_6 + \tfrac{1}{3}Fe_3Al_2Si_3O_{12} \rightleftharpoons CaFeSi_2O_6 + \tfrac{1}{3}Mg_3Al_2Si_3O_{12}$$

for which $\Delta V = -1{\cdot}33\ \mathrm{cm}^3$. As the above reaction moves to the left with increase in pressure the apparent distribution coefficient:

$$K' = \left(\frac{X_{Fe}^{Gar}}{X_{Mg}^{Gar}}\right) \Bigg/ \left(\frac{X_{Fe}^{Cpx}}{X_{Mg}^{Cpx}}\right)$$

also increases with pressure. By plotting the Fe^{2+}/Mg ratios of the coexisting clinopyroxenes and garnets of eclogites from low-, intermediate- and high-pressure environment (Fig. 201), Banno has demonstrated that K' decreases with increasing temperature.

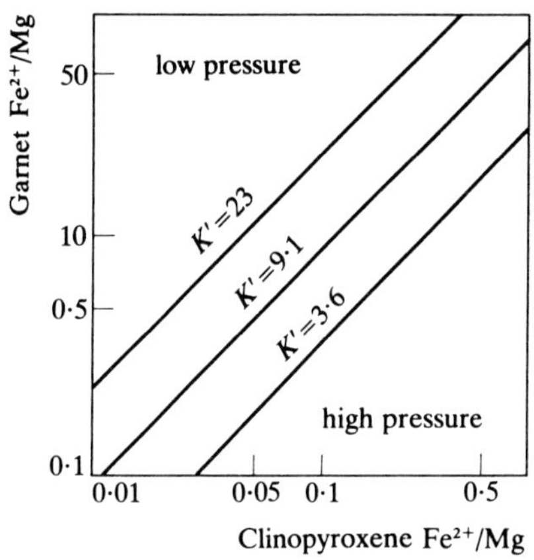

Fig. 201. Fe–Mg distribution between clinopyroxene and garnet for low-temperature eclogites, amphibolite facies terrain eclogites, and eclogite inclusions in kimberlite (after Banno, 1970).

Alteration. Omphacite is particularly susceptible to alteration during the process of amphibolitization of eclogites in gneiss terrains in which the breakdown of omphacite may begin with the formation of a border zone of green hornblende, and the development of homoaxial fibres of the amphibole throughout the pyroxene. The exsolution of the jadeite component of omphacites to yield diopside–plagioclase symplectites, first described in detail by Eskola (1921), is also a common feature of eclogites. Alderman (1936) and Davidson (1943) have described the replacement of omphacite by fine-grained symplectites of diopside and plagioclase, the symplectites being subsequently replaced by amphibole. Analyses of unaltered and altered omphacite in the kyanite eclogite from the Roberts Victor mine, South Africa are given by Switzer and Melson (1969). Plagioclase and a clinopyroxene are the main constituents of the altered omphacite, and electron microprobe analyses of the alteration products indicate that the reaction was essentially isochemical. The breakdown of omphacite to pyroxene–plagioclase symplectites is, however, not in all cases a strictly isochemical process but commonly involves the oxidation of iron in the pyroxenes. The Fe^{3+}/Fe^{2+} ratio of the symplectite is higher than that of the original omphacite, and Mysen and Griffin (1973) have suggested the possible operation of the reaction:

ferrous omphacite + O_2 → ferric augite + plagioclase

due to the introduction of an aqueous phase facilitating the oxidation (see also Vogel, 1966).

A detailed electron microprobe investigation of the symplectization of the omphacites from the eclogites of the Nordfjord area, Norway (Wikström, 1970b) has shown that during the early stages of alteration the omphacite unmixes to pyroxene and plagioclase, and that the secondary pyroxene is replaced subsequently by hornblende. The process is considered to operate not only in response to changing pressure–temperature conditions, but is also influenced by migratory water from the country rocks and higher oxygen fugacity. Scanning electron microscope photographs of the alteration of omphacite are given by Wikström (1971).

Two chemically and texturally distinct symplectic clinopyroxenes derived from the omphacite in the Hareidland eclogite, western Norway have been described by Mysen (1972). In one the original omphacite composition is modified by the exsolution of parallel rods of plagioclase, and the symplectic clinopyroxene, apart from an overall lower jadeite and Ca-Ts content, displays internal chemical variation in which the rims are poorer than the core in both these components. The plagioclase of the symplectite is albite and the exsolution of the jadeite molecule requires the addition of silica:

$$NaAlSi_2O_6 + SiO_2 \rightarrow NaAlSi_3O_8$$

(Jd-component) (Ab-component of plagioclase)

In the second variety of symplectic pyroxene the plagioclase is concentrated along the margins of the parental omphacite, and represents a later stage in the process. The plagioclase is more calcium-rich, demonstrating that the CaTs component of the omphacite, as well as the jadeite component have been exsolved.

$$CaAl_2SiO_6 + SiO_2 \rightarrow CaAl_2Si_2O_8$$

(CaTs component)(An-component of plagioclase)

The replacement of omphacite may also occur as a result of reaction between the pyroxene and garnet which leads to the formation of hornblende–plagioclase kelyphitic aggregates. These develop initially around the garnet before gradually extending into the omphacite which is progressively replaced. The omphacite of eclogites associated with the metamorphic rocks of Alpine-type orogenic zones shows a different mode of alteration. Thus Bloxam (1959) has described the marginal replacement of omphacite, and pseudomorphism by glaucophane in the form of stout prisms, commonly zoned by crossite and a deep green hornblende, in the eclogites associated with glaucophane schists near Valley Ford, California.

Experimental

Omphacite is essentially a solid solution of jadeite and diopside, but acmite ($NaFe^{3+}Si_2O_6$), commonly present in amounts up to 10 weight per cent, and Tschermak's molecule ($CaAl_2SiO_6$) are also important constituents. The binary system jadeite–diopside has been studied at pressures of 30 and 40 kbar by Bell and Davis (1969). The system is binary throughout the whole compositional range, with a small melting interval particularly for jadeite-rich compositions (Fig. 202). The heats of melting, ΔH, at various temperatures have been computed by Bell and Davis; ΔH decreases for both components towards a central minimum demonstrating the non-ideality of the jadeite–diopside solid solutions. The unit cell volumes display an excess volume of mixing compared with the ideal linear relationship, an indication of disorder in the intermediate synthetic composition that is not equalled by natural omphacites, and may be due to quenching from high temperatures of the synthetic products.

The hydrothermal synthesis of omphacitic pyroxenes, ($\simeq Di_{85}Jd_{15}$–$Di_{75}Jd_{25}$), plagioclase and a sodium-rich silicate phase, from starting materials of composition

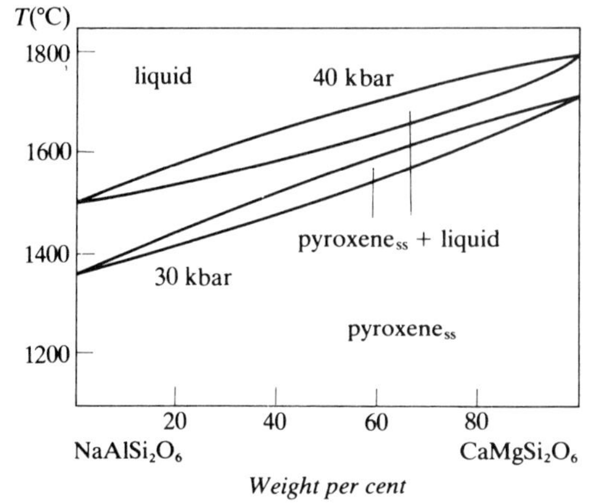

Fig. 202. Temperature–composition diagram for the jadeite–diopside join at 30 and 40 kbar (after Bell and Davis, 1969).

$Di_{70}Jd_{30}$, $Di_{60}Jd_{40}$ and $Di_{50}Jd_{50}$ at temperatures from 450° to 650°C and pressures between 1000 and 5000 atm had been reported by Wikström (1970a). The jadeitic component of the pyroxene increases as the pressure increases and temperature decreases; dP/dT gradients for samples with the same content of jadeite are approximately 20 ± 6 bar/°C.

The melting interval at 1 atm of a natural omphacite from the Roberts Victor mine has been determined by Switzer and Melson (1969); the beginning of melting occurs at $\simeq$ 1030°C and is complete at $\simeq$ 1230°C. Omphacite melts incongruently at high pressures, and O'Hara and Yoder (1967) have shown that the mineral begins melting at $\simeq$ 1570°C and is complete at 1600°C at a pressure of 30 kbar (Fig. 203). The early liquids are highly undersaturated and probably contain appreciable contents of the nepheline and albite components.

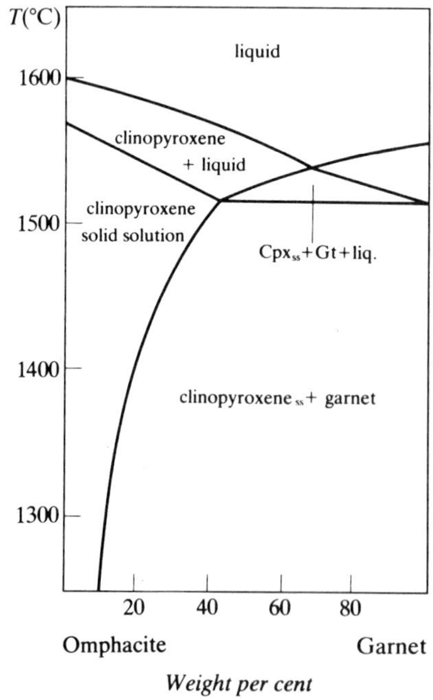

Fig. 203. Omphacite–garnet relationship from kimberlite eclogite at 30 kbar (after O'Hara and Yoder, 1967).

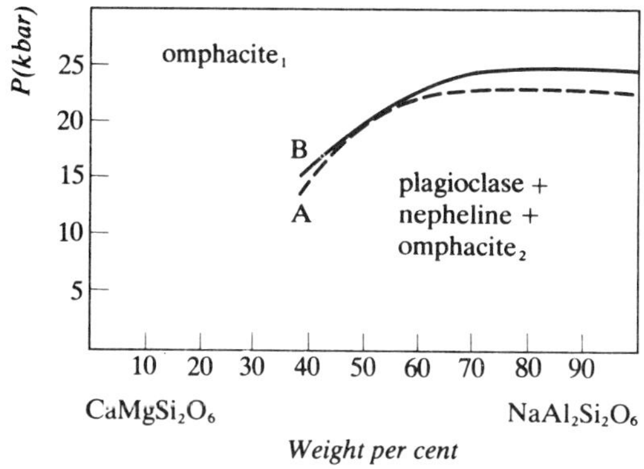

Fig. 204. Pressure–composition section for omphacite at 1150°C (curve A) and 1225°C (curve B) (after Bell and Kalb, 1969).

Pure jadeite breaks down according to the univariant reaction, 2 jadeite = albite + nepheline (see jadeite section). The effect on the reaction of the addition of the diopside component to the pyroxene has been evaluated by Bell and Kalb (1969) in their investigation of the reaction:

omphacite ⇌ plagioclase + nepheline + diopsidic pyroxene

for the compositions $Jd_{50}Di_{50}$ and $Jd_{90}Di_{10}$. The pressure–composition reactions for omphacite at 1150° and 1225°C are shown in Fig. 204 and indicate that the breakdown is only affected in those pyroxenes in which the diopside content in solid solution exceeds 30 weight per cent.

The stability of omphacitic pyroxenes in equilibrium with albite and quartz may be estimated from the free energies of the two reactions:

(a) albite ⇌ jadeite + quartz
(b) jadeite + diopside + acmite + hedenbergite ⇌ omphacite

The transition pressure for the first reaction is 6 kbar at 25°C (Birch and Lecombe, 1961). Data for the second reaction are not available, but the equilibrium pressure, based on ideal solution models has been estimated by Essene and Fyfe (1967). Their pressure–composition diagram is shown in Fig. 205 from which it can be seen that for omphacites with between 30 and 50 mol. per cent of the jadeite component, and containing little of the acmite molecule, the equilibrium pressures are in the range between 4 and 7 kbar at 200°C.

The solidus relationships for three compositions, $Di_{55}Ab_{45}$, $Di_{40}Ab_{60}$ and $Di_{25}Ab_{75}$ (weight per cent), on the join diopside–albite have been investigated in the pressure range 25 to 34·5 kbar at 1250°C (Kushiro, 1969a). The boundary for the omphacite + plagioclase + quartz and the omphacite + quartz assemblages for the composition $Di_{55}Ab_{45}$ is 26·5 kbar at 1250°C (Fig. 206).

The effect of pressure on the composition of the clinopyroxene solid solution in the system $CaMgSi_2O_6$–$NaAlSi_3O_8$–$CaAl_2Si_2O_8$ at 15, 20, 28 and 35 kbar and a temperature of 1150°C (Kushiro, 1969b), is shown in four isobaric and isothermal projections (Fig. 207). As the pressure increases the pyroxene solid solution + quartz field expands towards albite composition, and at 28 kbar extends to $Di_{40}Ab_{60}$ (pyroxene composition $Di_{45}Jd_{55}$ mol. per cent), and is accompanied by a contraction of the clinopyroxene + plagioclase + quartz volume. At 35 kbar albite is unstable and the clinopyroxene solid solution + quartz field extends from the

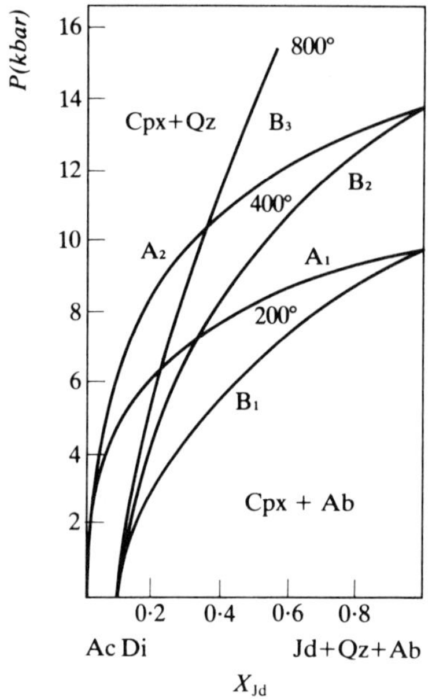

Fig. 205. Ideal solid solution model for diopside–jadeite–acmite. Ac–Jd, curve A_1 at 200°C, A_2 at 400°C: Di–Jd, curve B_1 at 200°C, B_2 at 400°C and curve B_3 at 800°C (after Essene and Fyfe, 1967).

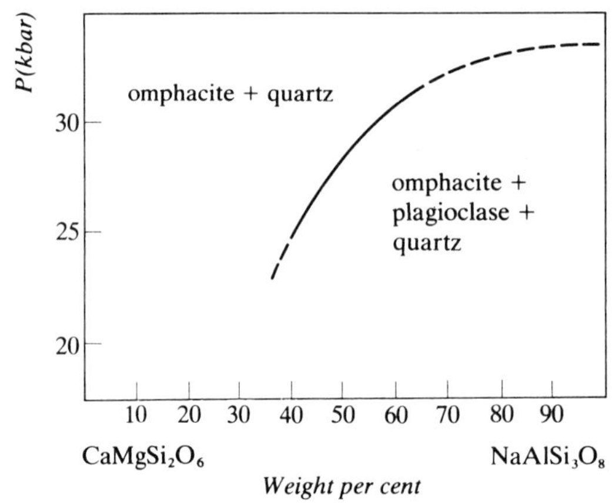

Fig. 206. Subsolidus phase relations along the join diopside–albite at 1250°C (after Kushiro, 1969b).

diopside to albite. The clinopyroxene + quartz field does not expand towards the diopside–anorthite join due to the appearance of a garnet solid solution phase and the enlargement of the field of clinopyroxene + garnet + plagioclase + quartz. The change in volume and location of the primary phase fields as the pressure increases shows that pyroxene compositions may extend to jadeite, but solid solution of anorthite in diopside is limited in the presence of quartz at a temperature of 1150°C. The changing composition of the clinopyroxene solid solution with increasing pressure may be illustrated by considering the bulk composition Di_{40} $Plag(An_{75})_{60}$ (A in Fig. 207). At 15 kbar this composition lies within the clinopyroxene + plagioclase + quartz field, and the projection of the clinopyroxene to crystallize from this composition onto the plane diopside–albite–anorthite is $\simeq Di_{76 \cdot 9}Jd_{5 \cdot 3}$Ca-$Tsch_{17 \cdot 8}$ mol. per cent (point B). At 20 kbar and 28 kbar the composition lies in the clinopyroxene + garnet + plagioclase + quartz volume and the estimated compositions of the clinopyroxene solid solution are $Di_{75 \cdot 1}Jd_{7 \cdot 9}$Ca-$Ts_{17 \cdot 0}$ and $Di_{61 \cdot 1}$

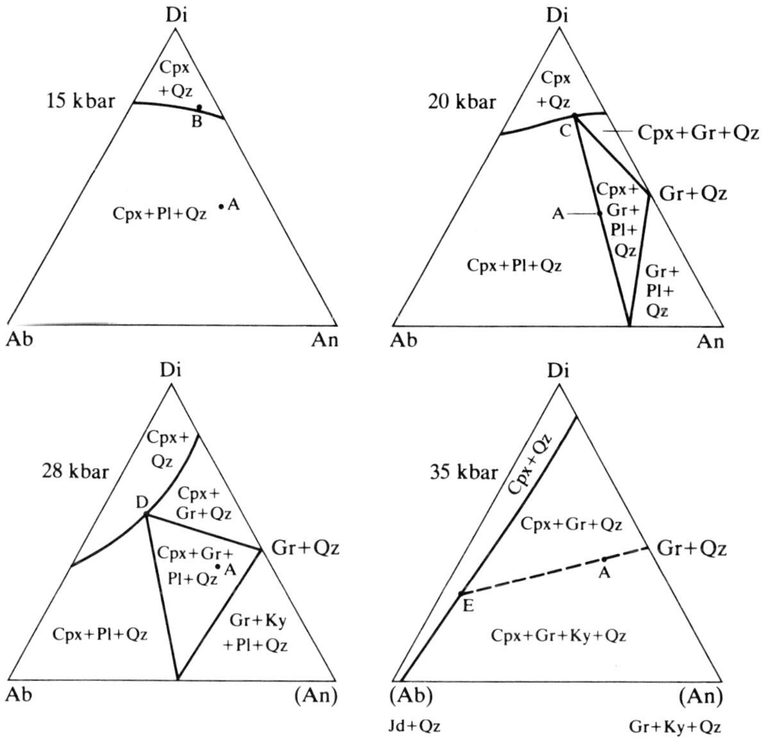

Fig. 207. **Diagrammatic isobaric and isothermal projections in the system $CaMgSi_2O_6$–$NaAlSi_3O_8$–$CaAl_2Si_2O_8$ at 15, 20, 28 and 35 kbar at 1150°C. Cpx, clinopyroxene solid solution; Pl, plagioclase; Qz, quartz; Gr, garnet solid solution; Ky, kyanite (after Kushiro, 1969b).**

$Jd_{27}Ca$-$Ts_{11\cdot9}$ respectively (points C and D). At 35 kbar the composition lies in the clinopyroxene+garnet+kyanite+quartz, or possibly in the clinopyroxene+garnet+quartz field, and the composition of the clinopyroxene solid solution becomes richer in the jadeite and less rich in the $CaAl_2SiO_6$ component.

The phase relations in the subsolidus of dry synthetic mixtures on the join diopside–pyrope, $Mg_3Al_2Si_3O_{12}$, have been determined (O'Hara and Yoder, 1967). The diopside–pyrope compositions and those of mixtures of natural omphacite ($Na_2O = 2{\cdot}22$) and garnet ($CaO = 10{\cdot}63$ weight per cent), from a garnet-peridotite nodule in kimberlite (see Fig. 203), show that at 30 kbar the clinopyroxene solid solution extends to approximately 40 weight per cent garnet.

The formation of garnet and an increase in the acmitic component of the clinopyroxene, accompanied by a decrease in feldspar and magnetite is observed in natural assemblages. Banno and Green (1968) have investigated the reactions between clinopyroxene solid solutions, albite, garnet and magnetite at pressures up to 36 kbar and at temperatures between 1100° and 1300°C (Fig. 208). The low-pressure assemblage of clinopyroxene, magnetite and plagioclase is replaced at higher pressure by clinopyroxene, garnet and magnetite according to the reactions of the type:

$$\underset{\text{albite}}{2NaAlSi_3O_8} + \underset{\text{magnetite}}{Fe_3O_4} + \underset{\text{diopside}}{CaMgSi_2O_6} \rightleftharpoons$$

$$\underset{\text{acmite}}{2NaFe^{3+}Si_2O_6} + \underset{\text{garnet}}{CaMgFe^{2+}Al_2Si_3O_{12}} + SiO_2 \qquad (1)$$

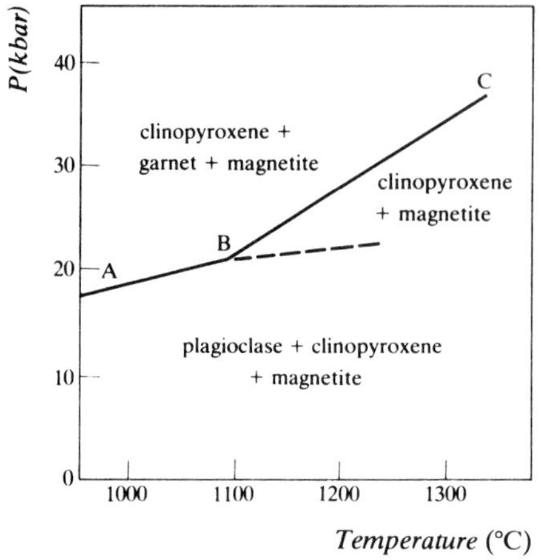

Fig. 208. Pressure–temperature relationships for the composition, diopside + albite + magnetite (after Banno and Green, 1968).

$$\underset{(\text{jadeite}+\text{diopside})_{ss}}{2NaAlSi_2O_6 + CaMgSi_2O_6} + \underset{\text{magnetite}}{Fe_3O_4} \rightleftharpoons \underset{\text{acmite}}{2NaFe^{3+}Si_2O_6} + \underset{\text{garnet}}{CaMgFe^{2+}Al_2Si_3O_{12}} \quad (2)$$

The break in the slope of the line ABC in Fig. 208 is interpreted by Banno and Green as an effect of the reaction:

albite + pyroxene $\rightleftharpoons$ (jadeite$_{ss}$ in pyroxene) + quartz

$CaAl_2SiO_6$ is an important component of omphacite and is a stable phase in the system $CaO–Al_2O_3–SiO_2$ (Hays, 1966a, b), but a pyroxene of such a composition has not been recognized in igneous or metamorphic rocks in which it is present solely as the CaTs component ($CaAl_2SiO_6$) in solid solution with other pyroxene components. In both natural rocks and in synthetic products the excess SiO_2 released during the breakdown of $CaAl_2Si_2O_8 \rightarrow CaAl_2SiO_6 + SiO_2$ has not been observed, and it has generally been assumed that the excess SiO_2 is present in the resulting diopside–$CaAl_2SiO_6$ solid solution. This has been confirmed by the X-ray microprobe analysis of a pyroxene, $Jd_{65}An_{35}$ in composition, synthesized at 40 kbar (Bell and Mao, 1971).

The stability field of omphacite and the subsolidus assemblages at high pressure for high-alumina basalt, kyanite eclogite and grosspydite compositions has been investigated by Green (1967); see also Green and Ringwood (1967, 1972). This synthesis of kyanite eclogite and grosspydite was found to require higher pressures than those needed to synthesize eclogite assemblages from normal basalt compositions, and showed that there is a direct relation between the Jd/Di ratio of the pyroxenes and the Gro/(Alm + Pyr) ratio of the coexisting garnets. The syntheses showed also that at pressures below 10 kbar the stable mineral assemblage is dominated by plagioclase + clinopyroxene. In the pressure range 20·3–24·8 kbar at 1100°C the assemblage is eclogitic, garnet + clinopyroxene ± quartz ± kyanite, while between pressures of 27–36 kbar there is a significant increase in the garnet/clinopyroxene ratio with increasing pressures as illustrated by the two equations:

$$\underset{\text{olivine}}{2(Mg,Fe)_2SiO_4} + \underset{\text{anorthite}}{2CaAl_2Si_2O_8} \rightleftharpoons$$

$$\rightleftharpoons \underset{\text{Al-clinopyroxene}}{Ca(Mg,Fe)Si_2O_6.CaAl_2SiO_6} + \underset{\text{Al-orthopyroxene}}{2(Mg,Fe)SiO_3.(Mg,Fe)Al_2SiO_6} \quad (1)$$

for which $\Delta V = -37\ cm^3$.

$$mCa(Mg,Fe)Si_2O_6 + CaAl_2Si_2O_8 \rightleftharpoons mCa(Mg,Fe)Si_2O_6.CaAl_2SiO_6 + SiO_2 \quad (2)$$

diopside, anorthite, Al-clinopyroxene, quartz

for which $\Delta V \simeq -14\ cm^3$.

The above reactions involve the anorthite content of the plagioclase solid solution, and the plagioclase probably becomes more sodium-rich until at a pressure of approximately 22·5 kbar the albite component is accommodated in the jadeite content of the clinopyroxene solid solution thereby releasing free SiO_2. Prior to the breakdown to albite, the clinopyroxene has a high content of $CaAl_2SiO_6$, but at pressures greater than 22·5 kbar the increase in the jadeite component, combined with the breakdown of the aluminium clinopyroxene by the reaction, leads to a clinopyroxene solid solution with a high content of jadeite and a lower content of the CaTs components, thus

$$mCa(Mg,Fe)Si_2O_6.CaAl_2SiO_6 + 2(Mg,Fe)SiO_3 \rightleftharpoons Ca(Mg,Fe)_2Al_2Si_3O_{12} + mCa(Mg,Fe)Si_2O_6$$

Al-clinopyroxene, orthopyroxene, grossular, diopside

for which $\Delta V \simeq -7\ cm^3$, and

$$NaAlSi_3O_8 + mCa(Mg,Fe)Si_2O_6 \rightleftharpoons mCa(Mg,Fe)Si_2O_6.NaAlSi_2O_6 + SiO_2$$

albite, diopside, omphacite, quartz

for which reaction $\Delta V \simeq -17\ cm^3$

The above reversible reactions, and particularly those involving plagioclase which are not controlled by the amount of the mineral present nor take account of the Ab:An ratio of the feldspar, do not represent the complete picture of the processes taking place, and it is likely that more complex reactions are involved in attaining equilibrium between jadeite and the $CaAl_2SiO_6$ components in solid solution in the pyroxene.

The high-pressure assemblages of undersaturated basalt compositions have been investigated by Green and Ringwood (1967) and may be represented by the reaction:

$$2Mg_2SiO_4 + 2CaAl_2Si_2O_8 \rightleftharpoons CaMgSi_2O_6.CaAl_2SiO_6 + 2MgSiO_3.MgAl_2SiO_6$$

olivine, anorthite, Al-diopside$_{ss}$, Al-enstatite$_{ss}$

In rocks with high Fe^{3+}/Na or magnetite/albite ratios it is possible the following reaction may occur:

$$2NaAlSi_3O_8 + Fe_3O_4 + nCaMgSi_2O_6 = nCaMgSi_2O_6.FeAl_2SiO_6.2NaFe^{3+}Si_2O_6 + SiO_2$$

albite, magnetite, diopside, acmitic omphacite

An experimental study of the temperature and pressure dependence of the Fe–Mg partition coefficient for coexisting omphacitic pyroxene [$100 \times 2Na/(Mg+Fe^{2+}+Ca+2Na)$ 31·5 to 35·3] and garnet, using a mineral mix, glass of typical tholeiite composition, and glasses of tholeiitic compositions with $6{\cdot}2 < 100\,Mg/(Mg+Fe^{2+}) < 93$, has been made by Råheim and Green (1974). The investigation showed that as the temperature is increased at constant pressure the Mg content of the pyroxene decreases, the jadeite component is not significantly affected provided there is no change in the mineral association (e.g. the presence of phengite at low temperatures), the amounts of the $CaAl_2SiO_6$ and $CaTiAl_2O_6$ molecules increase, and the pyroxene becomes more subcalcic. The variation in both the jadeite and $CaAl_2SiO_6$ and $CaTiAl_2O_6$ (Ca-Ti-Tschermak's molecule) components takes place in steps related to the mineral parageneses, from the low-temperature association, clinopyroxene + garnet + phengite + quartz, through clinopyroxene + garnet + quartz to the high-temperature association, clinopyroxene + garnet + melt. With increasing pressure at constant temperature (1100°C) the jadeite component increases and those of $CaAl_2SiO_6$ and $CaTiAl_2O_6$ decrease. Under conditions of constant temperature and pressure but different Mg-values (Mg < 69·4) of the starting material, the jadeite content increases and $CaAl_2SiO_6$ and $CaTiAl_2O_6$ decrease, as the Mg-values decline.

The garnet–clinopyroxene distribution coefficient, K_D, decreases with increasing temperature, and in a plot of $\ln K_D$ v. $1/T$°K a single straight line is defined, for compositions > $Mg_{6\cdot2}$ < Mg_{85}, by the linear relationship:

$$\ln K_D = \frac{4\,639}{T(°\mathrm{K})} - 2{\cdot}418 \text{ (at } P = 30 \text{ kbar)}$$

which gives the temperature of equilibration for an eclogite assemblage crystallized at 30 kbar. K_D increases with increasing pressure at constant temperature. Changes in the Mg-value of the bulk chemical system appear to have little, if any, effect on K_D, and the straight line relationship

$$\ln K = \frac{\Delta V}{RT}(P - P_0) + \ln K_0 \; (T \text{ const.})$$

where K and K_0 are the distribution constants at pressure $P = P$ and $P = P_0$ at given temperature respectively, can be used to calculate a pressure coefficient $p^*(= 2{\cdot}357)$, corresponding with ΔV in the above equation, for the exchange reactions for the Fe and Mg end-members of the clinopyroxene and garnet solid solutions in tholeiitic compositions. The effects of pressure, temperature and chemical composition on K_D can be expressed by the general relationship:

$$\ln K_\mathrm{D} = \frac{\mathrm{a}}{\mathrm{T}} + \frac{p^*}{RT}(P - P_0) + b$$

where a and b are constants. Using the relationship,

$$T(°\mathrm{K}) = \frac{\mathrm{a} + \frac{p^*}{R}(P - P_0)}{\ln K_D - \mathrm{b}} = \frac{3686 + 28{\cdot}35 \times P\,(\mathrm{kbar})}{\ln K_D + 2{\cdot}33}$$

or, more simply, Fig. 209, the temperature of equilibration of the omphacitic pyroxene–garnet assemblage may be obtained provided the K_D-value is known, and a pressure estimate can be provided.

Optical and Physical Properties

There is a relatively wide variation in the refractive indices and optic axial angle of omphacites, but the data do not show any clear relationship with the chemical composition. For most of the minerals detailed in Tables 44–46 the optic axial angle

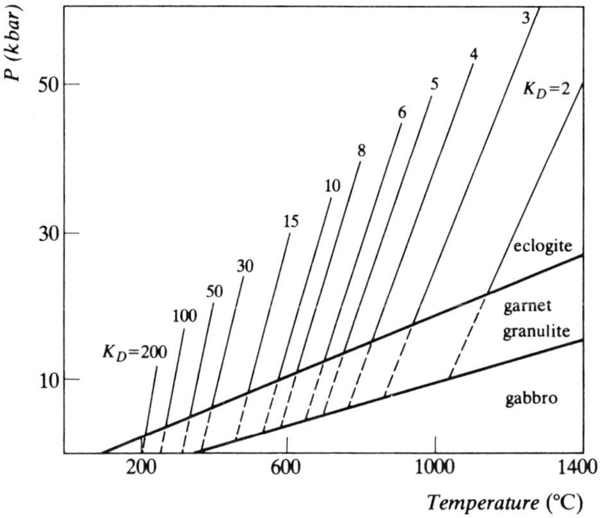

Fig. 209. K_D (garnet–clinopyroxene) as a function of pressure and temperature. Eclogite, garnet granulite and gabbro fields from Green and Ringwood, 1972 (after Raheim and Green, 1974).

varies between 60° and 75°, but although the Vanelvsdalen omphacite and the mineral from the Motagua fault zone have high contents of $[Al]^6 + Fe^{3+}$ and the largest optic axial angle, the other minerals in the tables display no obvious relationship between this property and the chemical composition. A ferrous omphacite (Table 46, anal. 12) has α 1·696, β 1·703, γ 1·720, γ:z 46°, $2V_\gamma$ 65°–73° and density 3·43 g/cm³. The mineral is zoned from a light green core to a deeper green at the margin, the ferrous iron increasing (FeO from 14·0 to 17·8 per cent) and aluminium decreasing (Al_2O_3 from 13·0 to 8·7 weight per cent).

An unusual pleochroism, with α dark blue, β azure blue and γ colourless and high γ:z extinction angle, has been reported in iron-rich omphacites (Table 46, anal. 13) from the metamorphosed alkaline complex in central Labrador (Curtis *et al.*, 1975). The omphacite shows some colour zoning, the darker blue zone having higher titanium and lower Ca and Al contents than the paler blue zones.

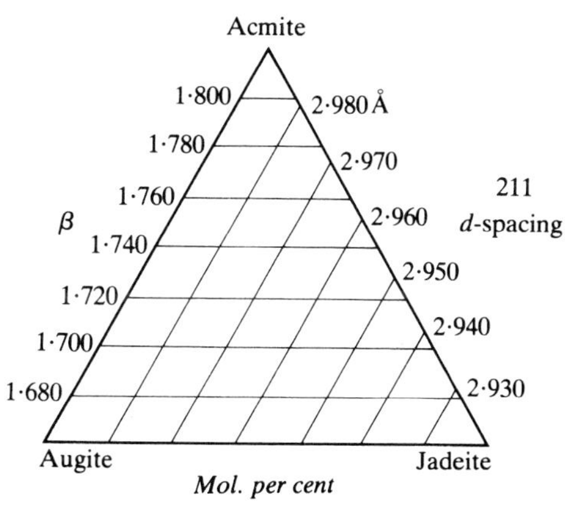

Fig. 210. Variation of β refractive index and 211 *d*-spacing for sodic pyroxenes of glaucophane schists (after Essene and Fyfe, 1967).

Patchy and undulose extinction, due to a high density of stacking faults, is not uncommon in omphacites of eclogites associated with low-temperature blueschist facies metamorphism (Champness *et al.*, 1974). The density of omphacite in relation to its chemical composition is high in accord with its crystallization in high-pressure environments, and increases with increasing iron content by approximately 0·01 for each per cent by weight $FeO + Fe_2O_3$ (Matthes *et al.*, 1970).

Small variations in the refractive indices and densities of omphacites from the same rock have been reported by Wolff (1942). Analyses of two density fractions of omphacites from the zoisite eclogite, Silberbach, Fichtelgebirge, are detailed in Table 45, (anals. 1 and 2) and show that the fraction less rich in sodium has somewhat higher refractive indices, optic axial angle and density.

The approximate composition of omphacitic pyroxenes from low-temperature eclogites may be determined from the measurement of the β refractive index and 211 *d*-spacing (Fig. 210).

Distinguishing Features

Omphacite is distinguished from fassaite by its larger optic axial angle, lower refractive indices, and greater density, from diopside and diopsidic augite by larger optic axial angle and from jadeite by its stronger colour, higher refractive indices, birefringence and extinction angle. Aegirine-augites are distinguished from omphacite by their higher refractive indices, smaller extinction angles and stronger pleochroism. In rocks in which omphacite is associated with hornblende it is distinguished from the amphibole by its paler colour and weaker pleochroism as well as by the pyroxene cleavage.

Paragenesis

Omphacite is restricted in its occurrence to eclogites and closely related rocks. Eclogites are generally regarded as a product of a high-pressure regime of either an igneous or metamorphic environment. This view, originally supported mainly by the high density of the two main constituents, omphacite and garnet, has been substantiated by more recent experimental evidence. It is, nevertheless, equally obvious that eclogites have formed in a number of different environments, and in recent years various schemes grouping eclogites on a paragenetical basis have been formulated (e.g. Smulikowski, 1964; Coleman *et al.*, 1965; Sobolev, N. R., 1968a; Banno, 1970). In that adopted here (Coleman *et al.*, 1965), eclogites are considered in three groups. The first, group A, embraces its occurrence in kimberlite pipes, basalts and layers in metamorphic rocks of very deep-seated origin; group B, eclogites found in migmatitic gneiss terrains in which they are commonly associated, as schlieren and lenses, with garnet-bearing anorthosites, granulites, amphibolites and granite gneiss; group C, eclogites occurring in metamorphic rocks of Alpine-type orogenic zones (Fig. 211).

The chemical composition of many eclogites is closely comparable with that of basic igneous rocks, and the formation of omphacite and garnet, due to crystallization or recrystallization of pre-existing igneous rocks under conditions of

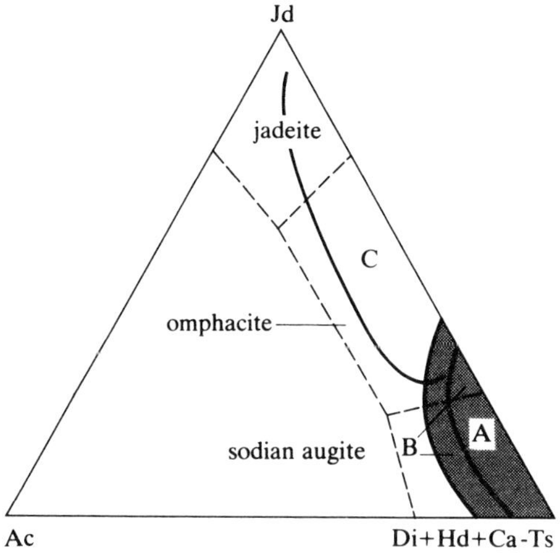

Fig. 211. Compositions of pyroxenes from group A eclogites, group B eclogites, group C eclogites from the glaucophane schist facies (after Coleman *et al.*, 1965). Compositional fields of jadeite, omphacite and sodian augite after Essene and Fyfe (1967).

high temperature and high pressure may be expressed (Alderman, 1936) by the equation:

$$\underset{\text{augite}}{\overset{Mg_2Fe_2^{2+}Si_4O_{12}}{Ca_2Mg_2Fe^{2+}Fe^{3+}AlSi_5O_{18}}} + \underset{\text{plagioclase}}{\overset{NaAlSi_3O_8}{Ca_2Al_4Si_4O_{16}}} + \underset{\text{ilmenite}}{2FeTiO_3} \rightarrow$$

$$\underset{\text{omphacite}}{NaCa_2MgFe^{2+}AlSi_6O_{18}} + \underset{\text{garnet}}{Ca_2Mg_3Fe_4^{2+}Fe^{3+}Al_5Si_9O_{36}} + \underset{\text{quartz}}{SiO_2} + \underset{\text{rutile}}{2TiO_2}$$

or, if the eclogite is considered in terms of its equivalence with olivine gabbro, by the equation:

$$\underset{\text{labradorite}}{3CaAl_2Si_2O_8 + 2NaAlSi_3O_8} + \underset{\text{olivine}}{3Mg_2SiO_4} + \underset{\text{diopside}}{nCaMgSi_2O_6} \rightarrow$$

$$\underset{\text{omphacite}}{2NaAlSi_2O_6 + nCaMgSi_2O_6} + \underset{\text{garnet}}{3CaMg_2Al_2Si_3O_{12}} + \underset{\text{quartz}}{2SiO_2}$$

The presence of eclogite nodules in the kimberlite pipes of South Africa is well known from the work of Wagner (1928) and Holmes (1936). In a recent study of the eclogites from the Roberts Victor mine, Kimberley, Kushiro and Aoki (1968) have shown that the eclogite compositions are undersaturated and correspond with olivine and picrite basalt. The jadeite component of the omphacites varies between 9·9 and 40·8 mol. per cent, and is accompanied by a corresponding decrease in the $(Mg,Fe)SiO_3$ component. Most of the omphacites (Table 44, anals. 3 and 6) are homogeneous but some, although the X-ray diffraction patterns do not display the presence of another mineral, show very thin exsolution lamellae, probably of an orthopyroxene. The partition coefficients of $Mg–Fe^{2+}$ distribution between the omphacites and garnets

$$K(T,P) = \left(\frac{Fe/(Mg+Fe)}{1-Fe/(Mg+Fe)}\right)\text{Gar}\left(\frac{1-Fe/(Mg+Fe)}{Fe/(Mg+Fe)}\right)\text{Omph}$$

lie between $K = 3$ and 5 (see also MacGregor and Carter, 1970) and, together with the general fields of the partition coefficient of pyroxene–garnet pairs of group B and C eclogites, are shown in Fig. 212. The variation of the Mg/(Mg+Ca) ratio in the omphacites, as well as in coexisting garnets, changes systematically indicating that equilibration occurred over a range of pressure (see chemistry section). An eclogitic xenolith, showing a rim consisting of omphacite and garnet, from the Roberts Victor mine kimberlite pipe has been described by Chinner and Cornell (1974). The original omphacite, containing some 60 per cent $NaAlSi_2O_6$ (8–9 per cent Na_2O), in the cores has been largely replaced by a quench product of melted omphacite consisting of augite and albite. The omphacite of the rim is less altered and has a smaller jadeitic content ($\simeq$ 4·5 per cent Na_2O). The high-level melting occurred later than the formation of the rim, and the different response of the core and rim omphacite to melting is attributed to the lower jadeite component of the latter which rendered it more refractory under the low-pressure conditions of melt formation.

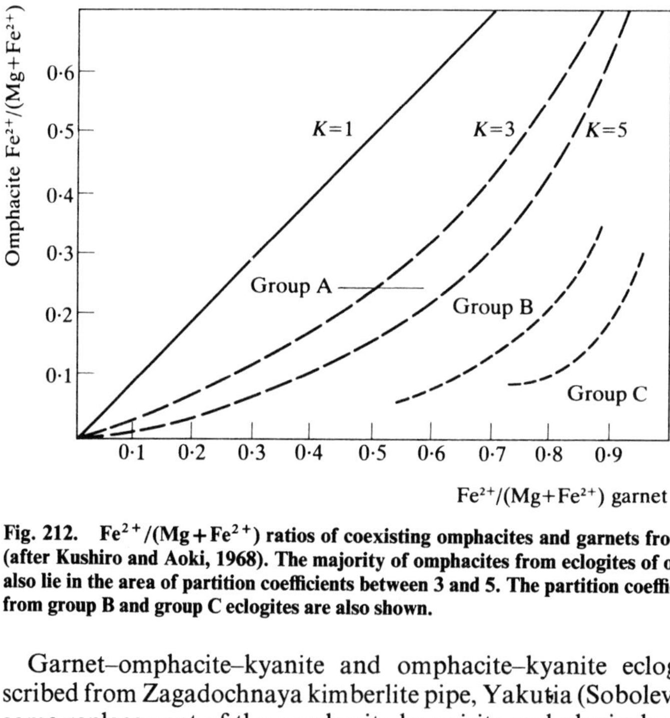

Fig. 212. $Fe^{2+}/(Mg+Fe^{2+})$ ratios of coexisting omphacites and garnets from the Roberts Victor mine (after Kushiro and Aoki, 1968). The majority of omphacites from eclogites of other nodules in kimberlites also lie in the area of partition coefficients between 3 and 5. The partition coefficients fields for omphacites from group B and group C eclogites are also shown.

Garnet–omphacite–kyanite and omphacite–kyanite eclogites have been described from Zagadochnaya kimberlite pipe, Yakutia (Sobolev *et al.*, 1968). There is some replacement of the omphacite by zoisite and plagioclase, and the pyroxenes have a higher content of Al_2O_3 than those of typical eclogites. Most of the aluminium is located in octahedral sites and this co-ordination may be due to the isomorphous admixture of an aluminium-rich molecule, possibly aluminosilicate, in view of the solution of kyanite in an eclogitic clinopyroxene synthesized at 30 kbar and 1500°C reported by O'Hara and Yoder (1967). An unusually aluminium-rich omphacite, containing 17·78 weight per cent Al_2O_3, equivalent to 26·5 per cent of the $CaAl_2SiO_6$ component (Table 44, anal. 2) occurs in the corundum-bearing eclogites of the Obnazhennaya pipe (Sobolev and Kuznetsova, 1965). Omphacites from the diamond-bearing eclogites in the Mir pipe, Yakutia (Sobolev and Kuznetsova, 1966) are more typical in composition, ($Na_{0\cdot42}K_{0\cdot01}$-

$Ca_{0\cdot49})(Mg_{0\cdot50}Fe^{2+}_{0\cdot10}Fe^{3+}_{0\cdot11}Al_{0\cdot37})(Al_{0\cdot03}Ti_{0\cdot02}Si_{1\cdot95})O_6$ and $(Na_{0\cdot31}K_{0\cdot01}Ca_{0\cdot51})$ $(Mg_{0\cdot66}Fe^{2+}_{0\cdot13}Fe^{3+}_{0\cdot05}Ti_{0\cdot02}Al_{0\cdot39})(Al_{0\cdot03}Si_{1\cdot97})O_6$ corresponding to $Di_{46}Jd_{31}$ and $Di_{50}Jd_{26}$ respectively.

Omphacites with high contents of the jadeite component (25–60 mol. per cent Jd) have been described from a number of South African diamondiferous eclogites (Reid *et al.*, 1976). More aluminium-rich clinopyroxenes are present in some of the eclogites, and are considered to be of secondary origin consequent on the breakdown of omphacite to a more fassaitic pyroxene and albite. The Fe–Mg distribution coefficients, $K_D = (FeO/MgO)_{gar}/(FeO/MgO)_{cpx}$, of a number of coexisting omphacite-garnet pairs in these rocks are also given by Reid *et al.*

Group A eclogites have recently been described from two localities in the United States. The mineralogy of the eclogite nodules in the kimberlite pipes, Riley County, Kansas, is similar to that of the eclogites of South Africa and Yakutia. The omphacite $\simeq Di_{52}Jd_{24}$ is associated with phlogopite, and orthopyroxene; as the latter increases modally there is a corresponding decrease in jadeite and acmitic components of the omphacite (Meyer and Brookins, 1971). The omphacites (Table 44, anal. 8) in the eclogite inclusions in the kimberlite pipes at Garnet Ridge, north-eastern Arizona, are usually rich in sodium, the jadeite component varying between 58 and 73 per cent (Watson and Morton, 1969), and the possibility that the eclogites are xenoliths derived from the basement rocks can be discounted.

At Salt Lake Crater, Oahu, Hawaii, eclogite nodules have been found in nephelinite tuff. The proportion of jadeite to the $CaAl_2SiO_6$ component in the pyroxene, however, is not typical of the omphacites in inclusions in kimberlite, and the mineral shows extensive exsolution of orthopyroxene as plates and mantling overgrowths and, on the basis of the chemical composition and petrography, Green (1966) has suggested that the aluminium-rich clinopyroxenes crystallized and accumulated from a magma of basaltic composition at pressures between 13 and 18 kbar and a temperature of about 1000°C.

The omphacite (Table 45, anal 6) from the type locality, Kupperbrunn, Austria, occurs in a kyanite-bearing eclogite typical of group B. Similar occurrences have also been described from Kärnten (Angel and Schaider, 1950), Bielice (Smulikowski, 1964), Bohemia (Dudek and Fediuková, 1974), Silberbach (Wolff, 1942; Banno, 1967a), Fay de Bretagne and northern Vendée, France (Brière, 1920; Velde and Sabatier, 1972), and from Vanelvsdalen and Naustdal (Eskola, 1921; Binns, 1967). In his classic paper on the eclogites of Norway, Eskola considered them to be igneous in origin, and their mineralogy the result of crystallization of basic magmas under conditions of high pressure. Subsequently most workers have tacitly assumed a purely metamorphic origin for eclogites of this type. More recently, in his account of the eclogite from Naustdal, Binns has tentatively suggested that the omphacite-garnet assemblage here has been derived by primary crystallization at high pressure followed, as the magma ascended to higher crystal levels, by the later crystallization from the residual liquid of additional omphacite and a barroisitic amphibole.

Conditions for the formation of eclogites, and the processes by which a rock of basaltic composition may be converted to an eclogitic assemblage, with particular reference to the Kristiansund area of western Norway, have been examined by Griffin and Råheim (1973). The region consists of an eclogite–amphibolite series, and includes late or post-tectonic olivine tholeiite sills. The eclogite pyroxenes contain thin spindles of quartz parallel to the cleavage, and commonly display exsolution of plagioclase; the feldspar occurs as thin rod-shaped bodies parallel to the cleavage, and also as rims between the pyroxene and garnet. In areas where the amount of exsolution is greater the rods of plagioclase are lath-shaped and zoned from more sodium-rich centres to more calcium-rich margins, and the pyroxene has recrystallized. Remote from the plagioclase exsolution the pyroxene is an omphacite containing approximately 30 per cent of the jadeite molecule, e.g. $Na_{0\cdot23}Ca_{0\cdot70}Mg_{0\cdot70}Fe^{2+}_{0\cdot13}Al_{0\cdot23}(Si_{1\cdot98}Al_{0\cdot02})O_6$, while the rims are sodian augites in composition and show a higher content of the $CaAl_2SiO_6$ molecule. The margins of some of the tholeiite sills are comparable in appearance with the eclogites, and the original clinopyroxene, plagioclase, iron oxide assemblages with or without olivine and orthopyroxene, have been transformed to omphacite, garnet, rutile, plagioclase and quartz. This eclogitic assemblage was subsequently affected by a period of retrograde metamorphism, and in conformity with the changes in eclogite, the omphacitic pyroxene exsolved plagioclase and became poorer in the jadeite component (Fig. 213).

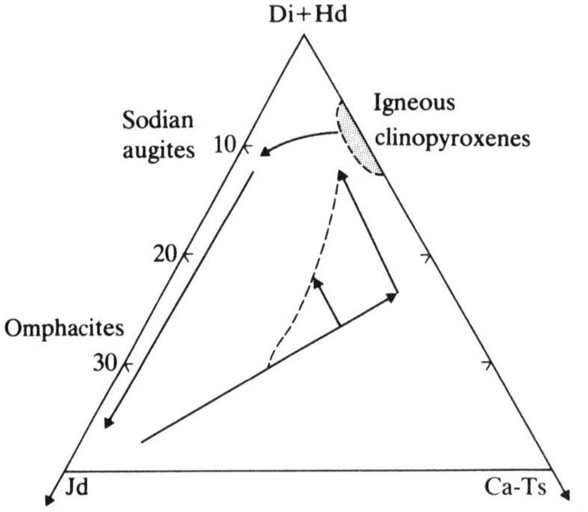

Fig. 213. Evolution of eclogitic pyroxenes in basic rocks (after Griffin and Råheim, 1973).

An *in situ* origin of eclogite, interleaved with thin layers of gneiss, has also been described from Weisenstein, north-east Bavaria (Matthes *et al.*, 1974). Here the eclogite assemblages (omphacite $\simeq (DiHd)_{70}(JdAc)_{30}$–$(DiHd)_{48}(JdAc)_{52}$) have been derived from basalt, the metamorphic transformation of which has taken place in a water-deficient environment at pressures and temperatures between 5–6 kbar and 500°–600°C respectively.

The omphacitic pyroxene (Table 45, anal. 1) from a hornblende-bearing eclogite at Gongen-yama, Bessi, Japan (Shidô, 1959) is poorer in sodium than typical omphacite, and its composition is considered to be due to crystallization under *P–T* conditions comparable with the highest grade of the garnet-amphibolite facies. Knorring and Kennedy (1958) have described an omphacite from the garnet–hornblende–pyroxene–scapolite gneisses of the Mampong inselberg, Ghana. The jadeite component of this pyroxene is also lower than is usual in omphacite. In common with other eclogites of group B, it is considered to represent an original basic rock and, like the Bessi eclogite, its present mineralogical composition is the result of recrystallization under high grade garnet-amphibolite metamorphic conditions. The eclogites in the gneiss terrains of the eclogite–kyanite gneiss, formations of the Kokchetav Ulutau-Makbal-Aktyuz massifs, U.S.S.R., and including a series of eclogitized rocks ranging from weakly eclogitized gabbro–diabase to typical eclogites, have been described by Dobretsov and Sobolev (1970).

Small areas of amphibole eclogites, i.e. predominantly omphacite-, garnet- and amphibole-bearing rocks, occur in the Puy des Ferrières, Uzerche, central France

(Velde *et al.*, 1970). The eclogites are now mainly transformed to symplectized or amphibolitized assemblages, and the original jadeitic omphacite (4·65 per cent Na_2O) is partially represented by augite (0·49 per cent Na_2O).

Omphacite is not an uncommon constituent in a variety of rocks of Alpine-type orogenic zones. In the southern Hohe Tauern area, Tyrol, Austria (Abraham *et al.*, 1974) the eclogite assemblages represent the culmination of the progressive metamorphic sequence, prasinite, amphibolite, eclogite. The development of the garnet-omphacite association may be represented schematically, where the low grade rocks were either relatively poor or rich in calcite by the reactions:

$$\underset{\text{amphibole}}{NaCa_2Fe_2Mg_3AlSi_7O_{22}(OH)_2} + \underset{\text{epidote}}{2Ca_2Al_3Si_3O_{12}(OH)} + \underset{\text{chlorite}}{Mg_5Al_2Si_3O_{10}(OH)_8} + \quad (1)$$

$$\underset{\text{albite}}{3NaAlSi_3O_8} + \underset{\text{ilmenite}}{4FeTiO_3} + \underset{\text{quartz}}{3SiO_2} \rightarrow \underset{\text{garnet}}{4Ca_{0.5}MgFe_{1.5}Al_2Si_3O_{12}} +$$

$$\underset{\text{omphacite}}{8Na_{0.5}Ca_{0.5}Mg_{0.5}Al_{0.5}Si_2O_6} + \underset{\text{rutile}}{4TiO_2} + 6H_2O$$

$$\underset{\text{amphibole}}{NaCa_2Fe_2Mg_3AlSi_7O_{22}(OH)_2} + \underset{\text{chlorite}}{Mg_5Al_2Si_3O_{10}(OH)_8} + \underset{\text{paragonite}}{3NaAl_3Si_3O_{10}(OH)_2} + \quad (2)$$

$$\underset{\text{ilmenite}}{4FeTiO_3} + \underset{\text{quartz}}{9SiO_2} + \underset{\text{calcite}}{4CaCO_3} \rightarrow \underset{\text{garnet}}{4Ca_{0.5}MgFe_{1.5}Al_2Si_3O_{12}} +$$

$$\underset{\text{omphacite}}{8Na_{0.5}Ca_{0.5}Mg_{0.5}Al_{0.5}Si_2O_6} + \underset{\text{rutile}}{4TiO_2} + 8H_2O + 4CO_2$$

Omphacites with an intermediate jadeitic (29–36 mol. per cent) and low to intermediate acmitic content (Table 46, anal. 7) are characteristic in the ophiolites of the Piedmont zone of the Alpine orogenic belt, in the Monviso Massif, Cottian Alps (Bearth, 1965; Mottana, 1971), the Zermatt-Saas Fee complex (Bearth, 1973), and the Pennidic belt, Liguria, Italy (Bocchio and Mottana, 1974).

Evidence for the derivation of the eclogite assemblage, omphacite + garnet, from original gabbroic rocks of the Oetztal Alps, Tyrol, has been presented by Miller (1970). The diopsidic augite of the gabbro first altering to omphacite along grain boundaries and fissures and subsequently to an aggregate of fine-grained omphacite, the jadeitic component of which is derived from the albite molecule of the original plagioclase:

$$\underset{\text{diopside}}{CaMgSi_2O_6} + \underset{\text{albite}}{xNaAlSi_3O_8} \rightleftharpoons \underset{\text{omphacite}}{CaMgSi_2O_6.xNaAlSi_2O_6} + SiO_2$$

and a smaller content of the $CaAl_2SiO_6$ molecule from the anorthite component of the plagioclase. The change from diopsidic augite to omphacite is accompanied by an increase in the Jd/Ts ratio of 0–0·13 for the igneous, to 5·9 for the metamorphic pyroxene (see also Richter, 1973).

Omphacitic pyroxenes, chloromelanites (Table 47, anals. 1 and 2) with 14–25 mol. per cent $NaFe^{3+}Si_2O_6$ occur in the basic schists of the Île de Groix, Brittany (Makanjuola and Howie, 1972).

Omphacite is a particularly common mineral in the Franciscan formation, California (Switzer, 1945; Borg, 1956; Bloxam, 1959; Ernst *et al.*, 1970), where it is associated with lawsonite, pumpellyite, glaucophane, epidote, clinozoisite, hornblende, chlorite, muscovite and, less frequently, paragonite, quartz and albite. Here omphacite also occurs in late veins cutting glaucophane schists and amphibolites. The veins are often monomineralic but may also contain sphene, apatite and muscovite. The pyroxene occurs in columnar growths normal to the vein walls, and in radiating varioles and fan-like aggregates. From its occurrence in veins it is clear that localized stress is not essential for the formation of omphacite, and its presence in radial growths and drusy cavities is indicative of crystallization in a hydrostatic environment. In this paragenesis omphacite is stable at low temperatures and high water pressure, and its association with lawsonite and the veining of the rocks by aragonite, imposes a lower limit for the formation of omphacite + quartz at 6–8 kbar and at temperatures between 200° and 300°C (Essene and Fyfe, 1967). Additional evidence of the lower temperatures of crystallization of the eclogite assemblage in the Franciscan formation is provided by the lower value (0·033–0·052) for the partition coefficient:

$$K_D = (Mg/Fe^{2+})_{garnet}/(Mg/Fe^{2+})_{Cpx}$$

compared with the values for the garnet–clinopyroxene pairs in amphibolites (0·076–0·11), in granulites (0·11–0·13) and in kimberlite pipes (0·16–0·34).

Omphacites ($\simeq$ Di + Hd)$_{50}$Jd$_{30-40}$Ac$_{20-10}$, similar in composition to those of the Franciscan formation occur in eclogite blocks in the Otter Point formation of south-western Oregon (Ghent and Coleman, 1973). The pyroxenes are zoned with Fe generally enriched in the rim relative to core, and have a maximum range of the jadeite component of 7 mol. per cent within single grains.

Omphacite is present in many of the rocks of the Maksyutova and Ufaley metamorphic complexes in the southern Urals, and there occurs in eclogites, amphibolites, glaucophane schists, as well as in monomineralic and Alpine-type veins (Kazak, 1970; see also Dobretsov and Sobolev, 1970). Typical assemblages of the lenticular eclogite and amphibolite bodies within the schists, include garnet–omphacite, garnet–omphacite–hastingsite and garnet–glaucophane–omphacite. Lenses of omphacitite, 30 to 50 cm in thickness, occur in the eclogite. The omphacite of these monomineralic layers is bright green in colour and the composition is similar to that of the bluish green omphacite in the enclosing rutile-bearing eclogite, $(Ca_{0·47}Na_{0·44})(Mg_{0·36}Fe^{2+}_{0·10}Fe^{3+}_{0·12}Ti_{0·01}Al_{0·42})(Si_{1·98}Al_{0·02})O_6$. Green omphacite assemblages also occur in the schists and include garnet–chloritoid–glaucophane–muscovite–quartz, and glaucophane–muscovite–quartz. Colourless omphacite, $(Ca_{0·46}Na_{0·48})(Mg_{0·33}Fe^{2+}_{0·09}Fe^{3+}_{0·07}Ti_{0·01}Al_{0·49})(Si_{1·97}Al_{0·03})O_6$, occurs in the ec-logitized basic rocks assemblages, omphacite–garnet–glaucophane–rutile, and omphacite–garnet–muscovite–hastingsite–rutile, and in glaucophane–muscovite–quartz and phengite–quartz–rutile veins. The omphacite of the veins forms crystals 2 to 5 cm in length and is $(Ca_{0·39}Na_{0·42})(Mg_{0·35}Fe^{2+}_{0·08}Fe^{3+}_{0·27}Ti_{0·01}Al_{0·35})(Si_{1·95}Al_{0·05})O_6$ in composition.

The occurrence of Alpine-type eclogite, similar to those of the Franciscan, and the Pennine area of the Central Alps, has been reported from Attunga, in the Great Serpentine belt of New South Wales (Shaw and Flood, 1974), and the Sanbagawa metamorphic belt, Japan (Iwasaki, 1960, Ernst *et al.*, 1970).

Omphacite eclogites and omphacite–glaucophane–lawsonite–garnet rocks have been described from the serpentinite of the Motagua fault zone, Guatemala (McBirney *et al.*, 1967). The omphacite (Table 46, anal. 4) displays a pronounced elongation and occurs in foliation clusters around garnet porphyroblasts. Morgan (1970) has described the occurrence of omphacite (Table 46, anal. 8) in the eclogites and eclogite-amphibolites associated with rocks of the epidote-amphibolite facies along the northern flank of the Cordillera de la Costa, Venezuela.

In the Ouégoa district, northern New Caledonia, omphacitic pyroxenes occur in the interbedded metabasalts, metamorphosed acid volcanics and quartzofeldspathic schists (Black, 1974). In this blueschist facies terrain, the pyroxene first developed in the metabasalts of the upper grade lawsonite zone as rims and veins around relicts of original augite which, in the higher grade epidote zone, is completely replaced by an omphacitic pyroxene. In the lawsonite zone the pyroxene composition may be either chloromelanite or omphacite; in the rocks of the epidote zone it is an omphacite. Thus with increasing metamorphic grade the pyroxene in the metabasalts is enriched in Al (jadeite) and Ca (diopside). A ferroan omphacite, associated with albite, potassium feldspar, quartz, phengite, sphene and pyrite, occurs in some of the acid volcanic rocks, of the area associated with pelitic schists of the epidote subzone (Black, 1970). The occurrence of hedenbergitic rather than a more acmitic omphacite is considered to be due to the low oxygen fugacity in the metamorphosed acid volcanic rock.

Blue omphacites (Table 46, anal. 10), rich in iron, occur in the alkaline gneisses and metamorphosed syenites and malignites of the regionally metamorphosed peralkaline (agpaitic) rocks of the Red Wine alkalic province of central Labrador. The omphacites occur both as the single pyroxene phase, or coexisting with aluminous aegirines (chloromelanite) either as single crystals or more rarely within a single grain, and are associated with alkali feldspar, arfvedsonite, and nepheline (Curtis *et al.*, 1975, see also Curtis and Gittins, 1978, and Table 46, anal. 13). The development of omphacitic pyroxenes in metasomatized ultrabasic rocks of the Urushten–Markopidzh fault zone in the Caucasus has been reported by Afanas'yev (1969). The omphacite occurs in an amphibole–pyroxene–apatite rock, and is associated with pargasite–hastingsite, delessite, carbonates, albite, allanite, sphene and rutile.

References

Abraham, K., **Hörmann, P. D.** and **Raith, M.**, 1974. Progressive metamorphism of basic rocks from the southern Hohe Tauern area, Tyrol (Austria). *Neues Jahrb. Min., Abhdl.*, **122**, 1–35.

Afanas'yev, G. D., 1969. Omphacite from metasomatites of the front range of the Caucasus. *Dokl. Acad. Sci., U.S.S.R., Earth Sci. Sect.*, **188**, 133–136.

Alderman, A. R., 1936. Eclogites from the neighbourhood of Glenelg, Inverness-shire. *Quart. J. Geol. Soc.*, **92**, 488–528.

Angel, F. and **Schaider, F.**, 1950. Granat und Omphazit aus dem Eklogit des Gertrusk (Saualpe, Kärnten). Carinthia II. *Naturwiss. Beitr. Kärntens*, Jahrg. 58–60, 33–36 (M.A. **11**-391).

Bancroft, G. M., **Williams, G. L.** and **Essene, E. J.**, 1969. Mössbauer spectra of omphacites. *Min. Soc. America, Spec. Paper* **2**, 59–65.

Banno, S., 1967a. Mineralogy of two eclogites from Fichtelgebirge and their mineralogical characteristics in relation to other eclogite occurrences. *Neues Jahrb. Min., Mh.*, 116–124.

Banno, S., 1967b. Effects of jadeite component on the paragenesis of eclogite rocks. *Earth Planet. Sci. Letters*, **2**, 249–254.

Banno, S., 1970. Classification of eclogites in terms of physical conditions of their origin. *Phys. Earth Planet. Interiors*, **3**, 405–421.

Banno, S. and **Green, D. H.**, 1968. Experimental studies on eclogites: the roles of magnetite and acmite in eclogitic assemblages. *Chem. Geol.*, **3**, 21–32.

Bearth, P., 1965. Zur Entstehung alpinotyper Eklogite. *Schweiz. Min. Petr. Mitt.*, **45**, 179–188.

Bearth, P., 1973. Gesteins-und mineralparageneten aus den Ophiolithen von Zermatt. *Schweiz. Min. Petr. Mitt.*, **53**, 299–334.

Bell, P. M. and **Davis, B. T. C.**, 1969. Melting relations in the system jadeite–diopside at 30 and 40 kilobars. *Amer. J. Sci.* **267A** (Schairer vol.), 17–32.

Bell, P. M. and **Kalb, K.**, 1969. Stability of omphacite in the absence of excess silica. *Carnegie Inst. Washington, Ann. Rept. Dir. Geophys. Lab.*, 1967–1968, 97–98.

Bell, P. M. and **Mao, H. K.**, 1971. Composition of clinopyroxene in the system $NiAlSi_2O_6$–$CaAl_2Si_2O_8$. *Carnegie Inst. Washington, Ann. Rept. Dir. Geophys. Lab.*, 1970–1971, 131.

Binns, R. A., 1967. Barroisite-bearing eclogite from Naustdal, Sognog Fjordane, Norway. *J. Petr.*, **8**, 349–371.

Birch, F. and **Lecombe, P.**, 1961. Temperature pressure plane for albite composition. *Amer. J. Sci.*, **258**, 209–217.

Black, P. M., 1970. *P*2 omphacite, intermediate in composition between jadeite and hedenbergite from metamorphosed acid volcanics, Bouehndep, New Caledonia. *Amer. Min.*, **55**, 512–514.

Black, P. M., 1974. Mineralogy of New Caledonian metamorphic rocks. III. Pyroxenes and major element partitioning between coexisting pyroxenes, amphiboles and garnets from the Ouégoa district. *Contr. Min. Petr.*, **45**, 281–288.

Bloxam, T. W., 1959. Glaucophane-schists and associated rocks near Valley Ford, California. *Amer. J. Sci.*, **257**, 95–112.

Bocchio, R. and **Mottana, A.**, 1974. Le eclogiti amfiboliche in serpentina di Vara (Gruppo di Voltri). *Rend. Soc. Italiana Min. Petr.*, **30**, 855–891.

Borg, I., 1956. Glaucophane schists and eclogites near Healdsburg, California. *Bull. Geol. Soc. America*, **67**, 1563–1584.

Brière, Y., 1920. Les éclogites françaises–leur composition, minéralogique et chimique; leur origine. *Bull. Soc. franç. Min. Crist.*, **42**, 72–222.

Champness, P., 1973. Speculation on an order–disorder transformation in omphacite. *Amer. Min.*, **58**, 540–542.

Champness, P., **Fyfe, W. S.** and **Lorimer, G. W.**, 1974. Dislocations and voids in pyroxene from a low-temperature eclogite: mechanism of eclogite formation. *Contr. Min. Petr.*, **43**, 91–98.

Chinner, G. A. and **Cornell, D. H.**, 1974. Evidence of kimberlite–grospydite reaction. *Contr. Min. Petr.*, **45**, 153–160.

Clark, J. R., **Appleman, D. F.** and **Papike, J. J.**, 1969. Crystal-chemical characterization of clinopyroxenes based on eight new structure refinements. *Min. Soc. America, Spec. Paper* **2**, 31–50.

Clark, J. R. and **Papike, J. J.**, 1968. Crystal-chemical characterization of omphacites. *Amer. Min.*, **53**, 840–868.

Coffrant, D., 1974. Les éclogites et les roches basiques et ultrabasiques associées du massif de Sauviat-sur-Vige, Massif central français. *Bull. Soc. franç. Min. Crist.*, **97**, 70–78.

Coleman, R. G. and **Clarke, J. R.**, 1968. Pyroxenes in the blue schist facies of California. *Amer. J. Sci.*, **266**, 42–59.

Coleman, R. G., **Lee, D. E.**, **Beatty, L. B.** and **Brannock, W. W.**, 1965. Eclogites and eclogites: their differences and similarities. *Bull. Geol. Soc. America*, **76**, 483–508.

Curtis, L. W. and **Gittins, J.**, 1978. Aluminous and titaniferous clinopyroxenes from regionally metamorphosed agpaitic rocks in central Labrador. *J. Petr.* (in press).

Curtis, L., **Gittins, J.**, **Kocman, V.**, **Rucklidge, J. D.**, **Hawthorne, F. E.** and **Ferguson, R. B.**, 1975. Two crystal structure refinements of a *P*2/*n* titanium ferro-omphacite. *Can. Min.*, **13**, 62–67.

Davidson, C. F., 1943. The Archaean rocks of the Rodil district, South Harris, Outer Hebrides. *Trans. Roy. Soc. Edinburgh.*, **61**, 71–112.

Dobretsov, N. L. and **Sobolev, N. V.**, 1970. Eclogites from metamorphic complexes of the U.S.S.R. *Phys. Earth Planet. Interiors*, **3**, 462–470.

Dudek, A. and **Fediukova, E.**, 1974. Eclogites of the Bohemian Moldanubicum. *Neues Jahrb. Min., Abhdl.*, **121**, 127–159.

Edgar, A. D., **Mottana, A.** and **MacRea, N. D.**, 1969. The chemistry and cell parameters of omphacites and related pyroxenes. *Min. Mag.*, **37**, 61–74.

Ernst, W. G., **Seki, Y.**, **Onuki, H.** and **Gilbert, M. C.**, 1970. Comparative study of low-grade metamorphism in the California Coast Ranges and the outer metamorphic belt of Japan. *Mem. Geol. Soc. America*, **124**, 1–276.

Eskola, P., 1921. On the eclogites of Norway. *Vidensk. Skr. I, Mat.-Naturv., kl.*, No. 8.

Essene, E. J. and **Fyfe, W. S.**, 1967. Omphacite in Californian metamorphic rocks. *Contr. Min. Petr.*, **15**, 1–23.

Ghent, E. D. and **Coleman, R. G.**, 1973. Eclogites from south-western Oregon. *Bull. Geol. Soc. America*, **84**, 2471–2488.

Green, D. H., 1966. The origin of the 'eclogites' from Salt Lake Crater, Hawaii. *Earth Planet. Sci. Letters*, **1**, 414–420.

Green, D. H., **Lockwood, J. P.** and **Kiss, E.**, 1968. Eclogite and almandine–jadeite–quartz rock from the Guajira Peninsula, Colombia, South America. *Amer. Min.*, **53**, 1320–1335.

Green, D. H. and **Ringwood, A. E.**, 1967. An experimental investigation of the gabbro to eclogite transformation and its petrological applications. *Geochim. Cosmochim. Acta*, **31**, 767–833.

Green, D. H. and **Ringwood, A. E.**, 1972. A comparison of recent experimental data on the gabbro–garnet granulite–eclogite transition. *J. Geol.*, **80**, 277–288.

Green, T. H., 1967. An experimental investigation of sub-solidus assemblages formed at high pressure in high alumina basalt, kyanite eclogite and grosspydite compositions. *Contr. Min. Petr.*, **16**, 84–114.
Griffin, W. L. and **Heier, K. S.**, 1973. Petrological implications of some corona structures. *Lithos*, **6**, 315–335.
Griffin, W. L. and **Råheim, A.**, 1973. Convergent metamorphism of eclogites and dolerites, Kristiansund area, Norway. *Lithos*, **6**, 21–40.
Hays, J. F., 1966a. Lime–alumina–silica. *Carnegie Inst. Washington, Ann. Rept. Dir. Geophys. Lab.*, 1964–65, 234–236.
Hays, J. F., 1966b. Stability and properties of the synthetic pyroxene $CaAl_2SiO_6$. *Amer. Min.*, **51**, 1524–1528.
Heritsch, H., 1970. Ueber Omphazite der Kiralpe, Steiermark. *Oesterr. Akad. Wiss., Math-Naturw, Kl.*, **107**, 10–11.
Hermes, O. D., 1973. Paragenetic relationships in an amphibolite tectonic block in the Franciscan terrain, Panoche Pass, California. *J. Petr.*, **14**, 1–32.
Holmes, A., 1936. Contributions to the petrology of kimberlites and its inclusions. *Trans. Geol. Soc. South Africa*, **39**, 379–428.
Iwasaki, M., 1960. Clinopyroxene intermediate between jadeite and aegirine from Suberi-dani, Tokushima Prefecture, Japan. *J. Geol. Soc. Japan*, **66**, 334–340.
Kazak, A. P., 1970. Omphacite species from glaucophane schists, amphibolite and eclogite in the southern Urals. *Dokl. Acad. Sci. U.S.S.R., Earth Sci. Sect.*, **190**, 122–125.
Knorring, O. von and **Kennedy, W. Q.**, 1958. The mineral paragenesis and metamorphic status of garnet–hornblende–pyroxene–scapolite gneiss from Ghana (Gold Coast). *Min. Mag.*, **31**, 846–859.
Kushiro, I., 1965. Clinopyroxene solid solutions at high pressure. *Carnegie Inst. Washington, Ann. Rept. Dir. Geophys. Lab.*, 1964–1965, 112–117.
Kushiro, I., 1969a. Stability of omphacite in the presence of excess silica. *Carnegie Inst. Washington, Ann. Rept. Dir. Geophys. Lab.*, 1967–1968, 98–100.
Kushiro, I., 1969b. Clinopyroxene solid solutions formed by reactions between diopside and plagioclase at high pressures. *Min. Soc. America, Spec. Paper* **2**, 179–191.
Kushiro, I. and **Aoki, K.-I.**, 1968. Origin of some eclogite inclusions in kimberlite. *Amer. Min.*, **53**, 1347–1365.
Kushiro, I., **Syono, Y.** and **Akimoto, S.**, 1967. Effect of pressure on garnet pyroxene equilibrium in the system $MgSiO_3$–$CaSiO_3$–Al_2O_3. *Earth Planet. Sci. Letters*, **2**, 460–464.
Lorimer, G. W., **Champness, P. E.** and **Spooner, E. T. C.**, 1972. Dislocation distributions in naturally deformed omphacite and albite. *Nature (Phys. Sci.)*, **239**, 108–109.
MacGregor, I. D. and **Carter, J. L.**, 1970. The chemistry of clinopyroxenes and garnets of eclogite and peridotite xenoliths from the Roberts Victor Mine, South Africa. *Phys. Earth Planet. Interiors*, **3**, 391–397.
Makanjuola, A. A. and **Howie, R. A.**, 1972. The mineralogy of the glaucophane schists and associated rocks from the Île de Groix, Brittany. *Contr. Min. Petr.*, **35**, 83–118.
Matsui, Y., **Maeda, Y.** and **Syono, Y.**, 1970. Mössbauer study of synthetic calcium-rich pyroxenes. *Geochem. J.*, **4**, 15–26.
Matsumoto, T. and **Banno, S.**, 1970. A natural pyroxene with the space group C^4_{2h}–$P2/n$. *Proc. Japan. Acad.*, **46**, 173–175.
Matsumoto, T., **Tokonami, M.** and **Morimoto, N.**, 1975. The crystal structure of omphacite. *Amer. Min.*, **60**, 634–641.
Matthes, S., **Richter, P.** and **Schmidt, K.**, 1970. Die Eklogitvorkommen des kristallinen Grundgebirges in NE-Bayern. II. Der Klinopyroxen der Eklogite und Eklogitamphibolite des Munchberger Gneisgebietes. *Neues. Jahrb. Min., Abhdl.*, **112**, 1–46.
Matthes, S., **Richter, P.** and **Schmidt, K.**, 1974. Die Eklogitvorkommen des kristallinen Grundgebirges in NE-Bayern. VII. Ergebnisse aus einer Kernbohrung durch den Eklogitkorper des Weisensteins. *Neues Jahrb. Min., Abhdl.*, **120**, 270–314.
McBirney, A., **Aoki, K.-I.** and **Bass, M. N.**, 1967. Eclogites and jadeite from Motagua fault zone, Guatemala. *Amer. Min.*, **52**, 908–918.
Meyer, H. O. and **Brookins, D. G.**, 1971. Eclogite xenoliths from Stockdale kimberlite, Kansas. *Contr. Min. Petr.*, **34**, 60–72.
Milashev, V. A., 1960. [Cognate inclusions in the kimberlite pipe, 'Obnazhennaya' (Olenek basin).] *Mem. All-Union Min. Soc.*, **89**, 284–299 (in Russian).
Miller, C., 1970. Petrology of some eclogites and metagabbros of the Oetztal Alps, Tyrol, Austria. *Contr. Min. Petr.*, **28**, 42–56.
Miyashiro, A. and **Seki, Y.**, 1958. Mineral assemblages and subfacies of the glaucophane-schist facies. *Jap. J. Geol. Geogr.*, **29**, 199–208.
Morgan, B. A., 1970. Petrology and mineralogy of eclogite and garnet amphibolite from Puerto Cabello, Venezuela. *J. Petr.*, **11**, 101–146.

Mottana, A., 1971. Pyroxenes in the ophiolitic metamorphism of the Cottian Alps. *Min. Soc. Japan, Spec. Paper* **1** *(Proc. IMA–IAGOD Meetings 1970 IMA* vol.*)*, 140–146.

Mottana, A. and **Bocchio, R.**, 1975. Superferric eclogites of the Voltri group (Pennidic belt, Apennines). *Contr. Min. Petr.*, **49**, 201–210.

Mottana, A., Church, W. R. and **Edgar, A. D.**, 1968. Chemistry, mineralogy and petrology of an eclogite from the type locality (Saualpe Austria). *Contr. Min. Petr.*, **18**, 338–346.

Mottana, A., Sutterlin, P. G. and **May, R. W.**, 1971. Factor analysis of omphacites and garnets: a contribution to the geochemical classification of eclogites. *Contr. Min. Petr.*, **31**, 238–250.

Mysen, B. O., 1972. Five clinopyroxenes in the Hareidland eclogite, western Norway. *Contr. Min. Petr.*, **34**, 315–325.

Mysen, B. O. and **Griffin, W. L.**, 1973. Pyroxene stoichiometry and the breakdown of omphacite. *Amer. Min.*, **58**, 60–63.

Nicolas, A., 1966. Étude pétrochimique des roches vertes et de leurs minéraux entre Dora Maira et Grand Paradis (Alpes piémontaises). *Nantes, Fac. Sci.*, 1–299.

Nixon, P. A., von Knorring, O. and **Rooke, J. M.**, 1963. Kimberlite and associated inclusions of Basutoland: a mineralogical and geochemical study. *Amer. Min.*, **48**, 1090–1132.

Ogniben, G., 1968. Il pirosseno omfacitico di Cesnola–Tavagnosco (Piemonte). *Periodico Min., Roma*, **37**, 1–12.

O'Hara, M. J. and **Yoder, H. S.**, 1967. Formation and fractionation of basic magmas at high pressures. *Scott. J. Geol.*, **3**, 67–117.

Parker, R. B., 1961. Rapid determination of the approximate composition of amphiboles and pyroxenes. *Amer. Min.*, **46**, 892–900.

Phakey, P. P. and **Ghose, G.**, 1973. Direct observation of domain structure in omphacite. *Contr. Min. Petr.*, **39**, 239–245.

Råheim, A. and **Green, D. H.**, 1974. Experimental determination of the temperature and pressure dependence of the Fe–Mg partition coefficient for coexisting garnet and clinopyroxene. *Contr. Min. Petr.*, **48**, 179–203.

Reid, A. M., Brown, R. W., Dawson, J. B., Whitfield, G. G. and **Siebert, J. C.**, 1976. Garnet and pyroxene compositions in some diamondiferous eclogites. *Contr. Min. Petr.*, **58**, 203–220.

Richter, W., 1973. Vergleichende Untersuchungen an ostalpinen Eclogiten. *Tschermaks. Min. Petr. Mitt.*, **19**, 1–50.

Sahlstein, T. G., 1935. Petrographie der Eklogiteinschlüsse in den Gneisen des südwestlichen Liverpool-Landes in Öst-Gronland. *Medd. om Grønland*, **95**, 5.

Shaw, S. E. and **Flood, R. H.**, 1974. Eclogite from serpentinite near Attunga New South Wales. *J. Geol. Soc. Australia*, **21**, 357–385.

Shidô, F., 1959. Notes on rock-forming minerals (9). Hornblende-bearing eclogite from Gongen-yama of Higasi-Akaisi in the Bessi district, Sikoku. *J. Geol. Soc. Japan.*, **65**, 701–703.

Smulikowski, K., 1964. Le problème des eclogites. *Geol. Sudetica, Warsaw*, **1**, 13–52.

Smulikowski, K., 1965. Chemical differentiation of garnets and clinopyroxenes in eclogites. *Bull. Acad. Polonaise Sci., Sér Sci. géol. géogr.*, **13**, 11–18.

Smulikowski, K., 1968. Differentiation of eclogites and its possible causes. *Lithos*, **1**, 89–101.

Sobolev, N. V., 1968a. Eclogite clinopyroxenes from the kimberlite pipes of Yakutia. *Lithos*, **1**, 54–57.

Sobolev, N. V., 1968b. Eklogit-xenolithe in den Kimberlitpipes von Jakutien. *Chemie der Erde*, **27**, 164–177.

Sobolev, N. V. and **Kuznetsova, I. K.**, 1965. More facts on the mineralogy of eclogite from Yakutian kimberlite pipes. *Dokl. Acad. Sci. U.S.S.R. Earth Sci. Sect.*, **163**, 137–140.

Sobolev, N. V. and **Kuznetsova, I. K.**, 1966. The mineralogy of diamond-bearing eclogites. *Dokl. Acad. Sci. U.S.S.R., Earth Sci. Sect.*, **167**, 112–115.

Sobolev, N. V., Kuznetsova, I. K. and **Zyuzin, N. I.**, 1968. The petrology of grosspydite xenoliths from Zagadochnaya kimberlite pipe in Yakutia. *J. Petr.*, **9**, 253–280.

Spry, A., 1963. The occurrence of eclogite on the Lyell Highway, Tasmania. *Min. Mag.*, **33**, 589–593.

Switzer, G., 1945. Eclogite from the California glaucophane schists. *Amer. J. Sci.*, **243**, 1–8.

Switzer, G. and **Melson, W. G.**, 1969. Partially melted kyanite eclogite from the Roberts Victor Mine, South Africa. *Smithsonian Contr. Earth Sci.*, No. 1, 1–9.

Triboulet, C., 1974. Les glaucophanites et roches associées de L'ile de Groix (Morbihan, France): étude minéralogique et pétrogenetique. *Contr. Min. Petr.*, **45**, 65–90.

Tröger, W. C., 1962. Zur Systematik und Optik der Chloromelanit-Reihe. *Tschermaks Min. Petr. Mitt.*, ser. 3, **8**, 24–35.

Velde, B., 1966. Étude minéralogique d'une éclogite de Fay-de-Bretagne (Loire-Atlantique). *Bull. Soc. franç. Min. Crist.*, **89**, 385–393.

Velde, B., Hervé, F. and **Kornprobst, J.**, 1970. The eclogite amphibole transition at 650°C and 6·5 kbar pressure, as exemplified by basic rocks of the Uzerche area, central France. *Amer. Min.*, **55**, 953–974.

Velde, B. and **Sabatier, H.,** 1972. Eclogites from northern Vendée, France. *Bull. Soc. franç. Min. Crist.*, **95**, 397–400.
Vogel, D. E., 1966. Nature and chemistry of the formation of clinopyroxene–plagioclase symplectite from omphacite. *Neues Jahrb. Min., Mh.*, 185–189.
Wagner, P. A., 1928. The evidence of kimberlite pipes on the constitution of the outer part of the earth. *South African J. Sci.*, **25**, 127–148.
Warner, J., 1964. X-ray crystallography of omphacite. *Amer. Min.*, **49**, 1461–1467.
Watson, K. D. and **Morton, D. M.,** 1969. Eclogite inclusions in kimberlite pipes at Garnet Ridge, north-eastern Arizona. *Amer. Min.*, **54**, 267–285.
Wetzel, R., 1972. Zur Petrographie und Mineralogie der Furgg-Zone (Mont Rosa-Decke). *Schweiz. Min. Petr. Mitt.*, **52**, 161–236.
White, A. J. R., 1964. Clinopyroxenes from eclogites and basic granulites. *Amer. Min.*, **49**, 883–888.
Wikström, A., 1970a. Hydrothermal experiments in the system jadeite–diopside. *Norsk. Geol. Tidsskr.*, **50**, 1–14.
Wikström, A., 1970b. Electron micro-probe studies of the alteration of omphacite in eclogites from the Nordfjord area, Norway. *Norsk. Geol. Tidsskr.*, **50**, 137–155.
Wikström, A., 1971. Scanning electron microscope photographs of the alteration of omphacite. *Norsk. Geol. Tidsskr.*, **51**, 191–192.
Wolff, T. von, 1942. Methodisches zur quantitativen Gesteins- und Mineral-Untersuchung mit Hilfe der Phaseanalyse (Am Beispiel der mafischen Komponenten des Eklogits von Silberbach). *Min. Petr. Mitt. (Tschermak)*, **54**, 1–122.
Yoder, H. S., 1950. The jadeite problem. *Amer. J. Sci.*, **248**, 225–248.
Yokoyama, K., Banno, S. and **Matsumoto, T.,** 1976. Composition range of *P*2/*n* omphacite from the eclogitic rocks of central Shikoku, Japan. *Min. Mag.*, **40**, 773–779.

Sodium Pyroxenes

Jadeite $NaAl[Si_2O_6]$
Aegirine $NaFe^{3+}[Si_2O_6]$
Aegirine-augite $(Na, Ca)(Fe^{3+}, Fe^{2+}, Mg, Al)[Si_2O_6]$
Ureyite $NaCr[Si_2O_6]$

Jadeite $NaAl[Si_2O_6]$

Monoclinic (+)

α	1·640–1·681
β	1·645–1·684
γ	1·652–1·692
δ	0·006–0·021
$2V_\gamma$	60°–96°
$\gamma:z$	32°–55°
O.A.P.	(010); $\beta = y$
Dispersion:	$r > v$ moderate to strong.
D	3·24–3·43
H	6
Cleavage:	{110} good, (110):(1$\bar{1}$0) $\simeq$ 87°
Twinning:	{100}, {001} simple, lamellar.
Colour:	Colourless, white, green, greenish blue, more rarely blue and violet; colourless in thin section.
Unit cell:[1]	*a* 9·418, *b* 8·562, *c* 5·219, β 107·58° Z = 4. Space group *C*2/*c*.

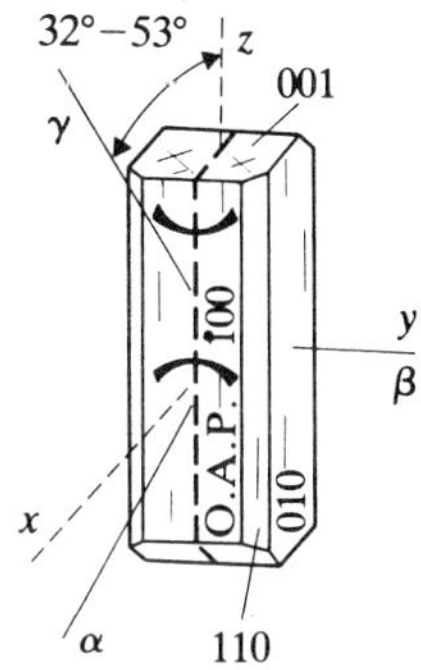

Insoluble in HCl.

Jadeite, originally considered to be an uncommon pyroxene has, during the last decade, been reported from an increasing number of localities, especially in Alpine-type metamorphic terrains. The paragenesis of jadeite is of particular interest because, although it has generally been regarded as a high pressure mineral, it occurs as a stable phase in rocks of low metamorphic grade. Jadeite composition is also an important component of the omphacite pyroxenes that in some instances formed under the *P–T* conditions of the upper mantle. Jadeite is one of the two varieties of jade, and has been used for many centuries in jewellery and other artefacts. The other variety, nephrite, is an amphibole belonging to the tremolite-actinolite series. Most of the jadeite for ornamental artefacts comes from upper Burma, California, Mexico, Guatemala and Omi-Kutaki, central Japan. A summary of the early references to jade is given by Foshag (1955) who showed that the Aztec gem stone, chalchihuitl, is jadeite or one of the related minerals, diopside-jadeite or chloromelanite. The various types of jadeite used in the manufacture of artefacts are also described by Foshag. An account of jade axes from the British Isles and Europe is given by Smith (1963, 1965), and Schmidt and Štelcl (1971) have described jadeite axes from the Moravian neolithic period. Jadeite is named from the Spanish *Piedra de jada*, colic stone, from the belief that it would prevent nephritic colic.

[1] Values for nearly pure jadeite (Prewitt and Burnham, 1966).

Structure

Wyckoff *et al.* (1925) first demonstrated that the X-ray diffraction pattern of jadeite was essentially the same as that of diopside. The structure of a jadeite (Fig. 214) from Santa Rita Peak, California $(Na_{0\cdot98}Ca_{0\cdot02})(Al_{0\cdot99}Mg_{0\cdot01})(Si_{1\cdot99}Fe^{3+}_{0\cdot01})O_6$ with *a* 9·418, *b* 8·562, *c* 5·219 Å, *β* 107·58°, space group *C*2/*c*, has been determined by Prewitt and Burnham (1966).

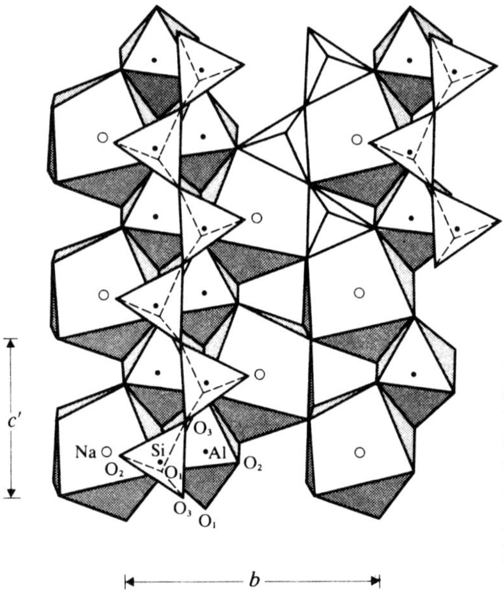

Fig. 214. Projection of the jadeite structure along a direction about half way between *a* and *a, showing chains of Si–O tetrahedra and Al–O octahedra. The latter, together with Na–O polyhedra form bands parallel to *z* which are linked by shared edges of polyhedra to form sheets parallel to (100) (after Prewitt and Burnham, 1966).**

As in all *C*2/*c* pyroxenes, the Si–O chains are all equivalent. The mean Si–O bridging distance is 1·632 Å and non-bridging 1·613 Å. Al is in very nearly regular octahedral co-ordination in *M*1 sites with a mean Al–O distance of 1·928 Å, while Na is in eight-fold co-ordination in *M*2 sites with six nearly equal shorter distances (average 2·378 Å) and two longer distances (2·741 Å). In diopside, with Ca in *M*2, the two longest distances are similar but two out of the remaining six are larger than the other four. The different environment of *M*2 is accompanied by a less kinked Si–O chain in jadeite, where the O3–O3–O3 angle is 174·7° compared with 166·38° in diopside. This straighter Si–O chain is characteristic of Na and Li pyroxenes.

Re-determination of the structure of the same jadeite at room temperature by Cameron *et al.* (1973) gave similar results to those described above. They also determined the structure at 400°, 600° and 800°C. Jadeite and other Na and Li pyroxenes show lower rates of thermal expansion than do Ca pyroxenes. Whereas for calcium pyroxenes the *β* angle increases with increasing temperature, *β* for jadeite remains constant (see also pp. 424, 521).

Coleman (1961) gave X-ray powder data for jadeites with varying substitution of Ac, Di and Hd components, which serve to increase the *d*-spacings, and for two jadeites with higher acmite and diopside components noted an extra reflection with *d* = 21·0 Å possibly indicating a super-cell. McBirney *et al.* (1967) gave for a jadeite, $(Na_{0\cdot95}Ca_{0\cdot05})(Al_{0\cdot95}Mg_{0\cdot05})Si_2O_6$, from Guatemala, *a* 9·439, *b* 8·5846, *c* 5·226 Å, *β* 107°27·5′, confirming the increase of cell parameters with diopside substitution. Similar relationships for cell parameters are shown by those measured by Seki and

Onuki (1967) for natural clinopyroxenes with compositions between jadeite and diopside–hedenbergite, but with some acmite component. Determination of cell dimensions for a number of jadeites from Omi-Kotaki, central Japan (Chihara, 1971) showed a range of *a* 9·388–9·504, *b* 8·583–8·605, *c* 5·216–5·252 Å, *β* 108°40′–106°18′.

Diffractometer patterns of a Burmese jadeite and of a synthetic jadeite crystallized from a jadeite glass at 800°C and 20 kbar were given by Robertson *et al.* (1957), and *d*-spacings from a number of jadeites were given by Yoder (1950).

For synthetic jadeite–acmite pyroxenes, the cell parameters *a*, *b*, *c* all increase with increasing Fe^{3+} (Gilbert, 1968). The molar volumes show no significant deviation from a linear relationship between the two end-members, and indicate random distribution of Al and Fe^{3+} in the *M*1 six-fold sites. Unit cell parameters for jadeite–anorthite solid solutions increase with increasing contents of the CaAl $SiAlO_6$ molecule and are accompanied by a decrease in *β* (Mao, 1971).

Chemistry

The compositions of recently analysed jadeites do not depart greatly from the ideal $NaAlSi_2O_6$, and in most natural jadeites more than 80 per cent of the *M*1 and *M*2 sites are occupied by Al and Na respectively. Yoder (1950) plotted the Na_2O and Al_2O_3 content of sixty-five pyroxenes and showed that there was almost continuous variation from Na- and Al- free pyroxenes to the composition of the pure end-member. Most of the analyses considered by Yoder, however, are old; there can be little doubt that some of the analyses were made on impure samples. There is no replacement of Si by Al in tetrahedral co-ordination, particularly in the normal range of composition, in accord with its high-pressure paragenesis. Substitution of Al by Fe^{3+} occurs to a limited extent, e.g. in the zoned jadeites (Table 49, anal. 5) of the jadeite-bearing metamorphosed arkoses of Celebes (de Roever, 1955), the cores being more iron-rich than the margins. A statistical survey of the compositions of 119 jadeitic pyroxenes from eclogites, pyrope peridotites, jadeite rocks and glaucophane schists has been made by Dobretsov (1964).

Experimental

The synthesis of jadeite by Coes was reported in 1953 (see Roy and Tuttle, 1956). Many earlier attempts to prepare jadeite both from anhydrous melts, and in the presence of water vapour at various temperatures and pressures have been summarized by Yoder (1950). Jadeite has a composition intermediate between nepheline and albite, but jadeite does not crystallize at low pressures in the binary system $NaAlSiO_4$–$NaAlSi_3O_8$ (Greig and Barth, 1938), nor in the ternary system Na_2O–Al_2O_3–SiO_2 (Schairer and Bowen, 1956). Jadeite is unstable at liquidus temperatures below approximately 25 kbar and melts incongruently to albite + liquid. At pressures above ≃29 kbar it melts congruently. The binary system $NaAlSiO_4$–$NaAlSi_3O_8$ at 1 bar and at 32 and 40 kbar is shown in Fig. 215.

Natural jadeite heated to 800°C first forms some glass and later recrystallizes to a mixture of nepheline and albite; glasses of jadeite composition crystallize between 900° and 1000°C to a mixture of the same two minerals. Both nepheline and albite have framework structures and densities of approximately 2·6, whereas jadeite has a

Table 49. Jadeite Analyses

	1	2	3	4	5
SiO_2	59·06	58·97	59·51	59·67	58·51
TiO	0·08	0·00	0·01	—	—
Al_2O_3	24·62	23·77	24·31	23·61	22·00
Fe_2O_3	0·41	0·32	0·35	0·33	3·31
FeO	0·18	0·11	0·03	0·16	0·98
MnO	0·03	0·02	0·01	0·16	tr.
MgO	0·17	0·97	0·58	0·47	0·50
CaO	0·35	1·43	0·77	0·82	0·71
Na_2O	14·95	14·42	14·37	14·24	13·74
K_2O	0·01	0·00	0·02	0·71	0·26
H_2O^+	0·07	—	0·06	0·13	—
H_2O^-	0·03	—	—	0·00	—
Total	99·96	100·01	100·03	100·30	100·01
α	1·654	1·653	1·658	1·656	—
β	1·656	1·658	1·663	1·660	1·67–1·68
γ	1·666	1·666	1·673	1·668	—
$\gamma:z$	35°	55°	33°	37°	40°–50°
$2V_\gamma$	70°	74°	—	68°–72°	70°–75°
D	—	—	—	3·30	3·26–3·30

Numbers of ions on the basis of six O

	1		2		3		4		5	
Si	1·997	2·00	1·996	2·00	2·005		2·016		2·001	
Al	0·003		0·004		—		—		—	
Al	0·977		0·942		0·966		0·940		0·886	
Ti	0·002		—		—		0·000		0·000	
Fe^{3+}	0·010	1·00	0·008	1·00	0·008	1·00	0·008	0·98	0·084	1·02
Mg	0·008		0·049		0·029		0·024		0·025	
Fe^{2+}	0·005		0·003		0·001		0·005		0·028	
Mn	0·001		0·001		—		0·005		0·000	
Na	0·978		0·945		0·940		0·933		0·910	
Ca	0·012	0·99	0·051	1·00	0·028	0·97	0·030	0·99	0·026	0·95
K	0·000		0·000		0·001		0·031		0·012	

1 White jadeite, vein cutting albite-crossite-acmite schist, New Idria district, California (Coleman, 1961). Anal. E. H. Oslund.
2 Jadeite, inclusions in serpentinite, Manganal, Guatemala (McBirney *et al.*, 1967). Anal. K. Aoki (analysis recalculated after subtraction of 2 weight per cent albite, An_5).
3 Jadeite, Burma (Yoder, 1950). Anal. E. G. Zies (includes Cr_2O_3 0·01).
4 Jadeite, vein in metagabbro, Sizuoka Prefecture, Japan (Seki *et al.*, 1960). Anal. C. Kato.
5 Jadeite, blastopsammitic quartzite, Koesek river, Celebes (de Roever, 1955). Anal. J. H. Scoon (anal. recalc. after subtraction of TiO_2 0·44, H_2O^+ 0·15 and quartz 5 per cent).

6	7	8	9	10	
60·03	58·12	58·02	61·66	59·35	SiO_2
—	0·31	0·04	0·05	0·18	TiO_2
22·78	20·32	22·96	21·81	22·18	Al_2O_3
0·32	2·49	0·77	0·32	1·15	Fe_2O_3
—	0·77	0·18	0·24	0·32	FeO
0·11	0·07	0·01	0·05	0·01	MnO
1·00	2·16	1·70	0·98	1·77	MgO
1·14	3·13	1·58	1·38	2·57	CaO
13·77	12·43	12·38	12·27	12·20	Na_2O
0·22	0·10	0·16	0·57	0·20	K_2O
—	0·16	0·87	0·44	0·20	H_2O^+
—	—	0·61	0·10	—	H_2O^-
99·37	100·07	99·28	99·87	'100·15'	Total
1·665	—	1·658	1·640	—	α
1·659	—	1·663	1·645	—	β
1·669	—	1·673	1·652	—	γ
35°	—	33°	40°	—	$\gamma{:}z$
69°–71°	—	—	67°	—	$2V_\gamma$
—	—	—	3·245	—	D
2·035	1·994 ⎫ 2·00	2·000 ⎫ 2·00	2·081	2·011	Si
—	0·006 ⎭	0·000 ⎭	0·000	0·000	Al
0·910 ⎫	0·816 ⎫	0·932 ⎫	0·868 ⎫	0·886 ⎫	Al
0·000	0·008	0·001	0·001	0·004	Ti
0·008 ⎬ 0·97	0·064 ⎬ 1·02	0·020 ⎬ 1·05	0·008 ⎬ 0·93	0·029 ⎬ 1·02	Fe^{3+}
0·051	0·110	0·087	0·049	0·089	Mg
0·000	0·022	0·005	0·007	0·009	Fe^{2+}
0·003 ⎭	0·002 ⎭	0·000 ⎭	0·001 ⎭	0·000 ⎭	Mn
0·905 ⎫	0·827 ⎫	0·828 ⎫	0·802 ⎫	0·801 ⎫	Na
0·041 ⎬ 0·96	0·115 ⎬ 0·95	0·058 ⎬ 0·89	0·050 ⎬ 0·88	0·093 ⎬ 0·90	Ca
0·010 ⎭	0·004 ⎭	0·006 ⎭	0·024 ⎭	0·008 ⎭	K

6 Jadeite, albitite in serpentinite, Kanto Mountains, Japan (Seki, 1961). Anal. Y. Seki.
7 Jadeite, pea-green coloured celt, Guatemala (Foshag, 1955). Anal. J. Fahey (includes Cr_2O_3 0·01).
8 Green jadeite, serpentinite, Kotaki, Japan (Yoder, 1950), Anal. Y. Kawano.
9 Colourless jadeite, with serpentine, calcite and quartz, in veinlets of glaucophane rock, Cloverdale, California (Wolfe, 1955). Anal. Univ. Minnesota Laboratory.
10 'Blue jade', Mexico (Foshag, 1955). Anal. J. Fahey.

Table 49. Jadeite Analyses – *continued*

	11	12	13	14	15
SiO_2	56·7	56·54	57·39	56·35	56·28
TiO_2	0·61	0·44	0·44	0·32	0·03
Al_2O_3	18·4	18·38	18·93	18·15	12·18
Fe_2O_3	5·8	5·67	4·45	5·22	0·85
FeO	0·62	1·05	0·81	0·75	1·28
MnO	0·12	0·10	0·09	0·03	0·13
MgO	1·8	1·44	1·92	2·83	9·02
CaO	2·4	2·69	2·74	4·23	12·60
Na_2O	13·2	13·00	12·46	12·11	5·94
K_2O	0·23	0·03	0·11	0·016	0·25
H_2O^+	0·12	0·20	0·54	—	0·30
H_2O^-	0·05	0·05	—	—	—
Total	100·05	99·60	99·88	100·01	'99·80'
α	1·681	1·679	—	1·671	—
β	1·684	1·681	—	1·676	—
γ	1·692	1·685	—	1·686	—
$\gamma:z$	30°–46°	38°	—	49°	—
$2V_\gamma$	88°–96°	64°	—	73°	—
D	>3·31	—	—	—	—

Numbers of ions on the basis of six O

	11		12		13		14		15	
Si	1·97	2·00	1·981	2·00	1·995	2·00	1·962	2·00	1·985	2·00
Al	0·03		0·019		0·005		0·038		0·015	
Al	0·72	0·99	0·740	1·00	0·771	1·02	0·707	1·02	0·491	1·03
Ti	0·02		0·011		0·011		0·008		0·001	
Fe^{3+}	0·15		0·146		0·116		0·137		0·022	
Mg	0·09		0·074		0·099		0·147		0·474	
Fe^{2+}	0·01		0·030		0·024		0·022		0·038	
Mn	0·00		0·003		0·003		0·001		0·004	
Na	0·89	0·99	0·882	0·98	0·840	0·95	0·818	0·98	0·474	0·96
Ca	0·09		0·101		0·103		0·158		0·476	
K	0·01		0·001		0·005		0·001		0·011	

11 Jadeitic pyroxene, albite–jadeite–pyroxene–glaucophane schist, Valley Ford, California (Keith and Coleman, 1968). Anal. R. E. Mays and C. Heropoulos (spectrographic data for Cu, Zn, Co, Ni, Cr, V, Ga, Sc, Y, Yb, Zr, Ba, Mo, Pb, Sr. High content of TiO_2 possibly due to inclusions of sphene; cell parameters *a* 9·48, *b* 8·63, *c* 5·24 Å, β 107°26′).

12 Green jadeite, Clear Creek, New Idria district, California (Coleman, 1961). Anal. E. H. Oslund (includes Li_2O 0·01).

13 Chloromelanite, greyish green celt, Guatemala (Foshag, 1955). Anal. J. Fahey.

14 Jadeite, 'eclogite' Guajira Peninsula, Colombia (Green *et al.*, 1968). Analysis corrected for admixture of 7·8 per cent quartz, 0·2 per cent apatite; *a* 9·494, *b* 8·623, *c* 5·223 Å, β 107°26′; includes Cr_2O_3 0·002.

15 Diopside-jadeite, fragment from tomb, Kaminaljuva, Guatemala (Foshag 1955). Anal. J. Fahey.

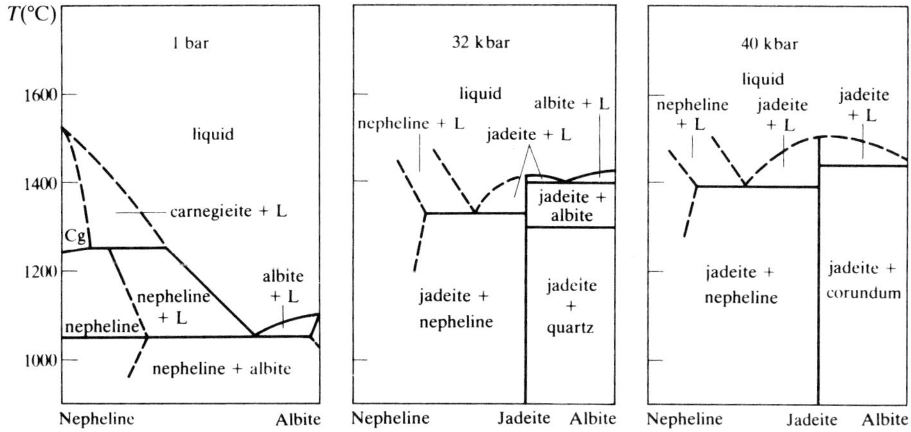

Fig. 215. The binary system $NaAlSiO_4$–$NaAlSi_3O_8$ at 1 bar (Greig and Barth, 1938), and at 32 kbar and 40 kbar (Bell and Roseboom, 1969).

pyroxene structure and a density of about 3·3 g/cm^3. The reductions in volume for the formation of jadeite for the reactions:

nepheline + albite = 2 jadeite
albite = jadeite + quartz
analcite = jadeite + water

are $\Delta V = -34{\cdot}9$, $-17{\cdot}8$ and $-18{\cdot}8$ cm^3/mole respectively.

The melting relations of jadeite and albite compositions have been studied experimentally by Bell and Roseboom (1969); see also Williams and Kennedy (1970). The jadeite composition melts congruently at pressure above 29 kbar, melts incongruently to albite + liquid between 25 and 29 kbar, and below 25 kbar is replaced by albite and nepheline (Fig. 216). The melting of jadeite between 30 and 60 kbar has been examined by Williams and Kennedy (1970). The data show some discrepancy with the earlier work of Bell and Roseboom at pressures greater than 40 kbar. For albite compositions jadeite coexists with either quartz or coesite at

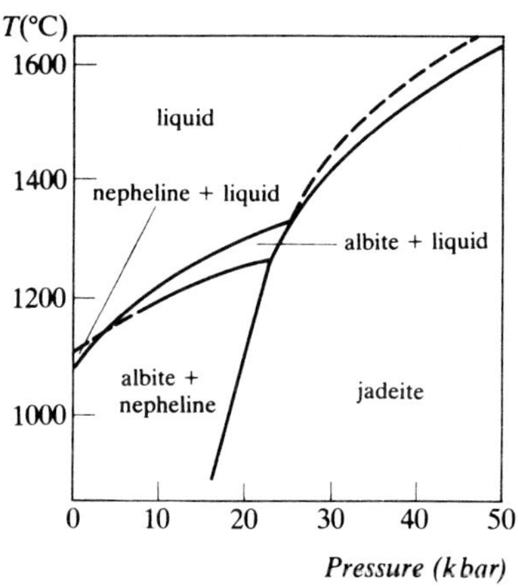

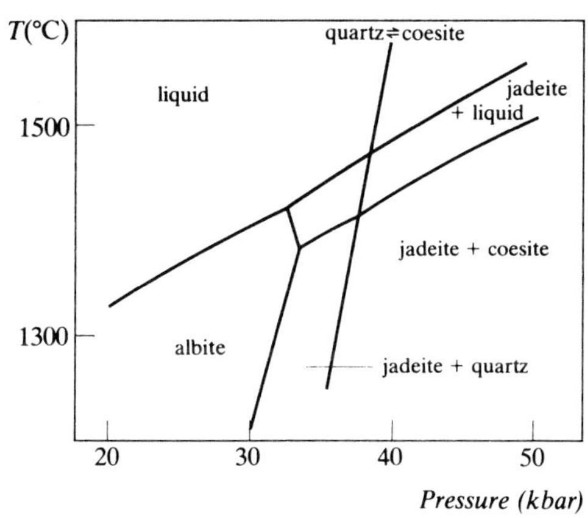

Fig. 216. Temperature–pressure relations for jadeite composition (after Bell and Roseboom, 1969).

Fig. 217. Temperature-pressure relations for albite composition (after Bell and Roseboom, 1969).

pressures between 30 and 34 kbar and temperatures of 1200° and 1380°C respectively (Fig. 217).

Robertson *et al.* (1957) determined the stability fields of jadeite and albite + nepheline at pressures between 10 and 25 kbar and temperatures between 600° and 1200°C, and showed that on heating in the albite + nepheline stability field natural jadeite breaks down to these two minerals; similarly on heating in the jadeite stability field a mixture of albite and nepheline begins to form jadeite:

$$\underset{\text{nepheline}}{NaAlSiO_4} + \underset{\text{albite}}{NaAlSi_3O_8} \rightleftharpoons \underset{\text{jadeite}}{2NaAlSi_2O_6}$$

Robertson *et al.*, using synthetic glass of jadeite composition as the starting material, determined the equilibrium curve for the reaction albite + nepheline ⇌ 2 jadeite and showed that it falls close to the line $P(\text{bars}) = 1000 + 18{\cdot}5T(°\text{C})$, comparable to the calculated curve of Kelly *et al.* (1953). The equilibrium curve for the reaction has been determined more recently by Newton and Kennedy (1968) as $P\ (\text{bars}) = 25\ T(°\text{C}) - 4000$; and the slope of the curve has been given by Bell and Boyd (1968) as $dT/dP = 54°\text{C/kbar}$. The results of these investigations are shown in Fig. 218.

The reaction nepheline + high albite ⇌ jadeite and high albite ⇌ jadeite + quartz, in the presence of high-pressure vapour and silicate melt, in the system $NaAlSiO_4$–SiO_2–H_2O and at temperatures between 600°–800°C, has been studied by Boettcher and Wyllie (1968). The fluxing action of high-pressure H_2O-rich vapours on the

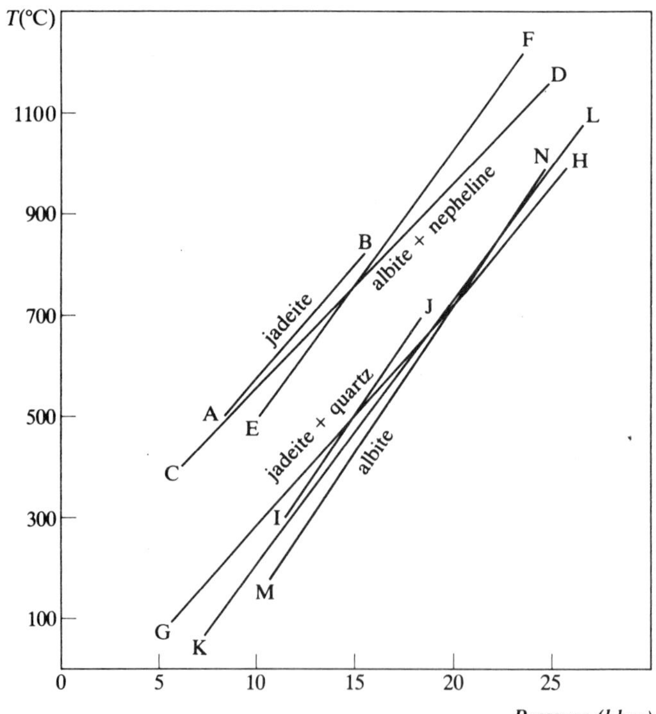

Fig. 218. Equilibrium curves for the reaction albite + nepheline ⇌ 2 jadeite. A–B, calculated curve of Kelly *et al.* (1953); C–D, Newton and Kennedy (1968); E–F, Robertson *et al.* (1957), and albite ⇌ jadeite + quartz, G–H, Newton and Smith (1967); I–J, Popp and Gilbert (1972); K–L, Birch and LeComte (1960); M–N, Newton and Kennedy (1968).

growth of the jadeite crystal was observed indicating that the influence of solutions during metamorphism may result in the formation of jadeitic pyroxenes at pressures lower than indicated by the univariant reactions of the synthetic experimental data. The influence of shear on the reaction Ab + Ne ⇌ Jd is demonstrated by the 60 per cent conversion to jadeite at 350°C and approximately 14 kbar (Fig. 219), conditions under which less than 10 per cent jadeite was produced in static runs (Dachille and Roy, 1961).

The heat of reaction of albite + nepheline ⇌ 2 jadeite (Table 50) has been determined by Kracek *et al.* (1951). The slope of the equilibrium curve is given by $dP/dT = \Delta S/\Delta V$, where ΔS is the difference in entropy between the products and the reactants, ΔV the difference in volume. From the entropy and volume data Kelly *et al.* (1953) calculated the slope of the equilibrium curve albite + nepheline ⇌ 2 jadeite as 20 bars/°C, a value which is in moderately good agreement with the experimentally determined values.

Table 50 Heat of reaction, $NaAlSi_3O_8 + NaAlSiO_4 \rightleftharpoons 2NaAlSi_2O_6$ at 25°C and 1 bar (Kracek *et al.*, 1951)

Albite	**Nepheline**	**Jadeite**	**ΔH kcal/mole**
Varutrask	Synthetic	Burma	−5·5
Varutrask	Synthetic	Japan, ground in agate mortar	−6·7
Varutrask	Synthetic	Japan, ground in mullite mortar	−3·6
Amelia	Synthetic	Burma	−7·2
Amelia	Synthetic	Japan, ground in agate mortar	−8·4
Amelia	Synthetic	Japan, ground in mullite mortar	−5·3

The reaction albite ⇌ jadeite + quartz, at pressures and temperatures between 15 and 25 kbar and 600° and 1000°C, respectively, has been studied by Birch and LeComte (1960). The slope of the equilibrium line is 50°/kbar, and the equation of the equilibrium $P\,(\text{bars}) = 20 \pm 2\; T(°\text{C}) + 000\; (\pm 500)$. Later results by Newton and Smith (1967), Newton and Kennedy (1968), Popp and Gilbert (1972) are shown in Fig. 218.

Thermochemical data, from oxide melt solution calorimetry at 691°C for the reactions low albite ⇌ jadeite + quartz and low albite + nepheline ⇌ 2 jadeite, have been presented by Hlabse and Kleppa (1968). The need to take account of the enthalpy and entropy changes associated with the low ⇌ high-albite transformation has been emphasized by Hlabse and Kleppa and by Boettcher and Wyllie (1968). This leads to lower pressures than those previously accepted for the reactions albite ⇌ jadeite + quartz and albite + nepheline = 2 jadeite. Figure 220 shows that at a pressure of about 7 kbar jadeite + quartz is stable below 140°C, whereas low albite is stable above those temperatures, and that jadeite + quartz is stable with respect to high albite at all temperatures below about 220°C; it is possible that the reaction albite = jadeite + quartz may be facilitated by high albite acting as a metastable intermediary.

In natural jadeite containing some replacement of Al by Fe^{3+}, the pressure of the reaction albite ⇌ jadeite + quartz is lower than that for pure jadeite, and is approximately 1 kbar less at 200°C for the composition $Jd_{82}Ac_{14}Di_4$ (Fig. 221). The uncertainties associated with the determination of the albite ⇌ jadeite + quartz equilibrium are discussed by Hays and Bell (1973) who give the tentative value of 16·4 ± 0·5 kbar for the equilibrium pressure at 600°C.

The reaction analcite ⇌ jadeite + quartz has been studied by Griggs and Kennedy

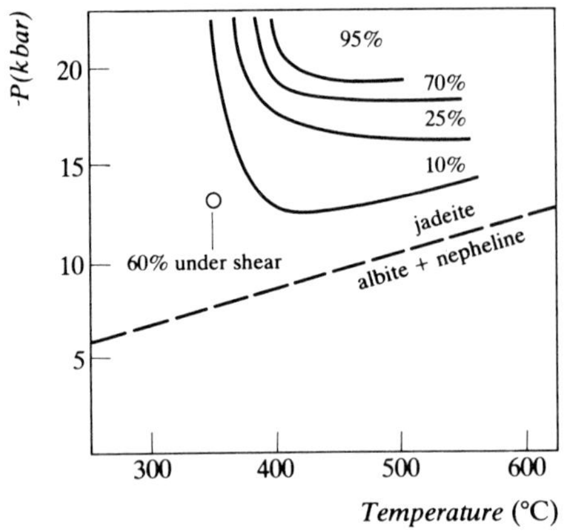

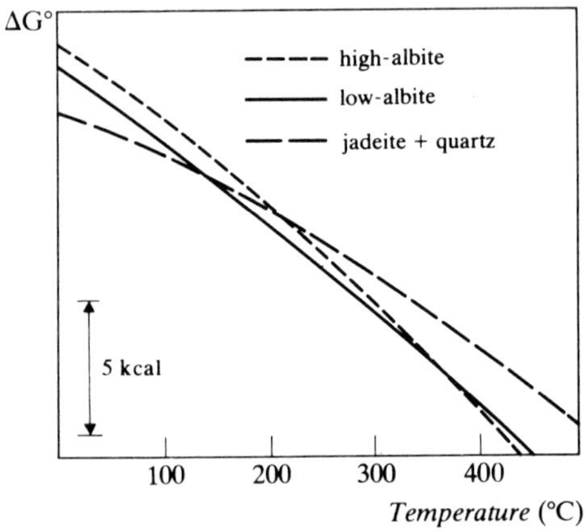

Fig. 219. Reaction albite + nepheline → jadeite. Contours for 10, 25, 70 and 95 per cent conversion in 40 hours of reaction. Addition of shearing increases the conversion to 60 per cent in region of very low yield (after Dachille and Roy, 1961).

Fig. 220 Free-energy–temperature diagram for low-albite, high-albite and jadeite + quartz at approximately 7 kbar. Note that in temperature range 140° to 220°C, jadeite + quartz is stable with respect to high-albite, but unstable with respect to low albite (after Hlabse and Kleppa, 1968).

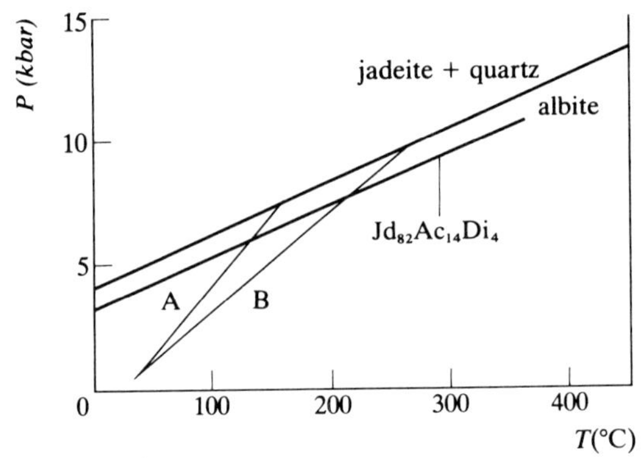

Fig. 221. The jadeite + quartz ⇌ albite transition curve in relation to thermal gradients (A) 6·2°C/km, (B) 8·6°C/km. The stability curve for $Jd_{82}Ac_{14}Di_4$ + quartz is also shown (after Newton and Smith, 1967).

(1956) and Fyfe and Valpy (1959). The equilibrium equation for the reaction is P (bars) = 9·5 T(°C) + 6000 (Newton and Kennedy, 1968). Campbell and Fyfe (1965) have given the values for $\Delta G^\circ_{298} = +1540$ cals, and $\Delta S^\circ_{294} = +7{\cdot}42$ e.u. A tentative phase diagram for the system $NaAlSi_3O_8$–H_2O (Campbell and Fyfe, 1965) showing the stability field for jadeite + quartz is given in Fig. 222. An investigation of the stability field of jadeite, albite and analcite at pressures and temperatures of 8 to 16 kbar and 500° to 800°C respectively has been made by Newton and Kennedy (1968), and the boundary curves for the reactions nepheline + albite ⇌ 2 jadeite, and analcite ⇌ jadeite + water are shown in Fig. 223 (see also Thompson, 1971).

Jadeite–diopside form a binary system at 30 and 40 kbar, and a melting interval extends over the whole compositional range (Bell and Davis, 1965, 1967, 1969). Thus at high pressure the jadeite–diopside system is analogous to albite–anorthite at atmospheric pressure although the solidus and liquidus temperature range is small throughout the series (see Fig. 202, p. 440). Jadeite–$CaAl_2SiO_6$ clinopyroxene solid solutions have been synthesized by Mao (1971) from glasses of compositions $Jd_{95}An_5$, $Jd_{85}An_{15}$, $Jd_{75}An_{25}$ and $Jd_{65}An_{35}$ at 1300°C and 40 kbar. The clinopyroxenes dissolve up to 7·5 weight per cent excess SiO_2 and the reaction

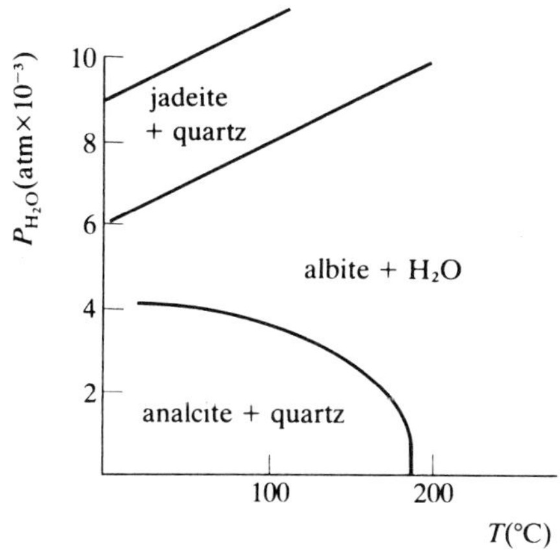

Fig. 222. Phase diagram for the system $NaAlSi_3O_8$–H_2O (after Campbell and Fyfe, 1965).

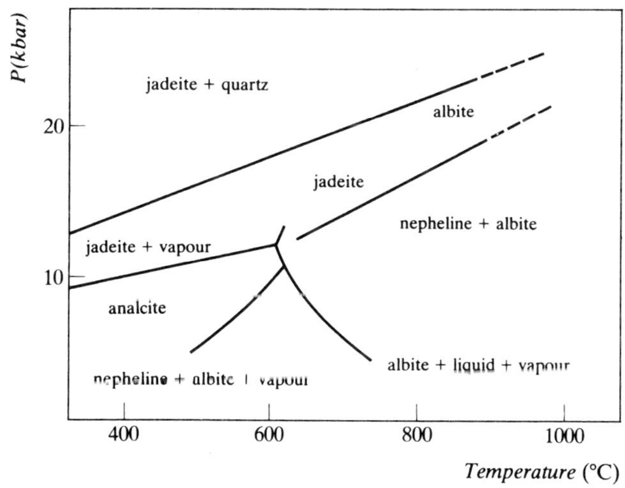

Fig. 223. Equilibrium curves for the reactions, albite ⇌ jadeite + quartz, nepheline + albite ⇌ 2 jadeite, and analcite ⇌ jadeite + water (after Newton and Kennedy, 1968). The triple point at which the fields of jadeite + vapour, analcite and albite + liquid + vapour meet occurs at approximately 11·8 kbar and 610°C. The second triple point where analcite, nepheline + albite + vapour and albite + liquid + vapour meet is at 11 kbar and 620°C.

$NaAlSi_2O_6 + xCaAl_2Si_2O_8 = (NaAlSi_2O_6 + xCaAl_2SiO_6) + xSiO_2$ does not occur. In these pyroxenes containing excess SiO_2 the total cation–oxygen ratio is less than the ideal 2:3 pyroxene ratio, and the calculated formula for the composition $Jd_{65}An_{35}$ is $(Na_{0\cdot66}Ca_{0\cdot26})(Al_{1\cdot00})(Al_{0\cdot17}Si_{0\cdot83})O_6$. *P–T* phase diagrams for compositions $Jd_{95}An_5$ and $Jd_{65}An_{35}$, at pressures between 0 and 40 kbar, are given by Mao (1971) and show that the stability field of the jadeitic pyroxenes increases with decreasing pressure. Increasing amounts of excess silica are dissolved in the pyroxene as the pressure increases and as the silica solubility in a Jd–CaTs pyroxene is a function of pressure and temperature, the *P* and *T* at the time of formation are indicated by the chemical composition, but until the effects of diopside, hedenbergite and acmite usually present in natural clinopyroxenes are known the chemical composition cannot be used as a precise geothermometer or geobarometer. The solid solution of jadeite in diopside increases with pressure in the experimental system $CaMgSi_2O_6$–$NaAlSi_3O_8$–$CaAl_2Si_2O_8$ (Kushiro, 1969; see also omphacite section and Fig. 207), and at 35 kbar pressure may extend to the jadeite composition. The system $NaAlSi_2O_6$–$NaCrSi_2O_6$ in the pressure range 1 bar to 25 kbar at 800°C has been investigated by Abs-Wurmbach and Neuhaus (1976) (see p. 521 and Fig. 241).

Orthoclase (K_2O 11·17, Na_2O 4·40 per cent) has been transformed to sanidine + jadeite + quartz, sanidine + jadeite + coesite and, with increasing water pressure, to $KAlSi_3O_8.H_2O$ + jadeite + coesite. The jadeite produced in these reactions does not contain any appreciable amount of potassium at pressure and temperature up to 62 kbar and 1000°C, respectively (Seki and Kennedy, 1964).

Waldbaum and Thompson (1969) have examined the thermodynamic relations of the jadeite–feldspar–quartz equilibria in the context of the system $NaAlO_2$–$KAlO_2$–SiO_2, and calculated the isobaric temperature–composition phase relations at 16, 20, 26 and 32 kbar, as well as the effect of the $KAlSi_3O_8$ component on the equilibria between alkali feldspars, jadeite and quartz for temperatures up to 1300°C.

The transformation of nepheline to jadeite + $NaAlO_2$ and of $NaAlGe_2O_6$ to $NaAlGeO_4$ (calcium ferrite structure) + GeO_2 (rutile structure) at high pressures and temperatures between 1000° and 1100°C has been reported by Ringwood *et al.* (1967). The synthesis of a pyroxene containing trivalent titanium ($NaTiSi_2O_6$) has been reported by Prewitt *et al.* (1972).

Optical and Physical Properties

The small but significant substitution of Al by Fe^{3+} that occurs in some jadeites is accompanied by the higher refractive indices of such samples compared with jadeites having a more ideal composition. Substitution by other components in general gives rise to higher refractive indices and extinction angles, and lower refringence, optical axial angle and density. In some jadeites, however, the small chemical differences do not appear to be of sufficient magnitudes to produce the observed wide variations in refractive indices (e.g. anals. 8 and 9, Table 49). Some of the observed variation may be associated with incipient alteration and the existence of some vacant sites in the structure (e.g. anal. 6).

Coleman (1961) has reported very strong dispersion ($r > v$) and low birefringence (0·000–0·008) in jadeitic pyroxenes containing between 10 and 30 weight per cent of the diopside + hedenbergite + acmite components. There are other data indicating that jadeities containing appreciable amounts of Fe^{3+} can be distinguished by their anomalous interference colours and by the very strong dispersion of the optic axes ($r > v$). Higher values of the optic axial angle and extinction than shown in the data at the beginning of the section, and averaging 86° and 53° respectively, have been given by Bloxam (1956); de Roever (1955) also gives a higher range of optic axial angles, extinction angles and β refractive indices for some zoned jadeites from Celebes. Accurate measurements of 2V and $\gamma{:}z$ are difficult for jadeites showing very high dispersion.

The optical spectrum of a natural green jadeite from Burma has been determined by Rossman (1974). The infrared absorption spectrum of jadeite is given by Kolesova (1959), and Kuznetsova and Moskaleva (1968) have shown that there is a close linear relationship between the frequency of the absorption band in the 674–748 cm^{-1} range and jadeite content for members of jadeite–diopside solid solutions.

The volume thermal expansion, α_v, of jadeite is given by the equation:

$$\alpha_v = 19{\cdot}96 \times 10^{-6} + 1{\cdot}166 \times 10^{-8}\,(T-20)$$

where T is the temperature in °C, and α_v the coefficient of volume thermal expansion. Compressibility data are also given by Yoder and Weir (1951). The thermal diffusivity K, of a gem quality jadeite, in the temperature range 27° to 827°C has been measured by Kanamori *et al.* (1968); $1/K$ increases almost linearly with temperature to 427°C: above 627°C, $1/K$ decreases, probably due to radiative heat transfer within the crystal.

Jadeite has a high fracture surface energy γ_f (121000 ergs/cm^2) and fracture toughness, K_c ($7{\cdot}1 \times 10^8$ dynes/cm$^{-3/2}$). The resistance of jadeite to breakage is not directly related to the atomic bonding but probably due to crack propagation taking place by transgranular and not by intergranular fracture. The fracture surface energy of jadeite is approximately 50 per cent of the value for nephrite, but is two orders of magnitude greater than that of most ceramic materials, e.g. for quartz $(10\bar{1}0)$ $\gamma_f = 1030$, and for corundum $(10\bar{1}1)$ $\gamma_f = 600$ ergs/cm^2 (Bradt *et al.*, 1973).

Distinguishing Features

The refractive indices of jadeite are lower than those of all other pyroxenes except spodumene. Jadeite is, however, distinguished from spodumene by its lower birefringence and higher extinction angle. Jadeite is distinguished from omphacite and fassaite by its lower refractive indices and birefringence and from aegirine by the absence of colour in thin section, higher extinction angle and lower refractive indices and birefringence. It is distinguished from actinolite by its pyroxene cleavage, higher extinction angle and lack of colour. In some jadeites the erratic variation of optic axial and extinction angles, strong dispersion, anomalous interference colours and low birefringence may lead to its confusion with zoisite but the latter shows straight extinction and higher refractive indices.

Paragenesis

Jadeite is found only in metamorphic rocks and has long been considered a typical high-pressure mineral, partly on account of the presence of the jadeite molecule in the composition of the eclogite pyroxene, omphacite, and also because of its smaller volume compared with that of its chemical equivalents: this assessment was strengthened by the many early failures to effect its synthesis. In almost all known parageneses of jadeite it is associated with albite; other mineral associates include quartz, analcite, lawsonite, glaucophane, garnet, pectolite, chlorite, carbonates, zoisite, sphene, actinolite, stilpnomelane and mica, associates which, in general, indicate formation in environments of relatively low metamorphic grade.

Jadeite and jadeitic pyroxenes are relatively common constituents in the metagreywackes and related rocks of regional metamorphic belts. First reported by de Roever (1955) in eastern Celebes, they have subsequently been described by Bloxam (1956), Coleman (1961) and others from California, from various Alpine localities by Lovenzoni (1963), Lefèvre and Michard (1965) and Bearth (1966), and from the Sanbagawa metamorphic belt, Japan, by Seki and Shidô (1959).

Both jadeitic and omphacitic pyroxenes are common constituents in the rocks of the glaucophane schist facies of the Franciscan formation, California. These metamorphic rocks are derived from Franciscan sediments and lavas and developed in the folded geosynclinal prism (Essene *et al.*, 1965). Here jadeite occurs in two main environments. Pure jadeite is found in monomineralic pods and veins and, with albite and analcite, in cross-cutting veins in metasomatized tectonic inclusions within sheared serpentinites. Jadeitic pyroxenes (Jd 70–90 per cent) coexist with quartz in the greywackes, and with omphacitic or acmitic pyroxenes in the metabasalts.

Jadeite, associated with serpentine and calcite, in small veins in glaucophane rocks has been described by Wolfe (1955), and in other areas it is found with albite, analcite and other zeolites in small veins cutting albite–glaucophane–aegirine schists (Coleman, 1961). The jadeite of San Benito County occurs in a lens-shaped body surrounded by serpentinite (Yoder and Chesterman, 1951). A 60 cm wide zone, consisting of grossular, lawsonite, pumpellyite and a green amphibole, separates the jadeite and serpentinite, and pods of natrolite, pectolite and albite are enclosed in the jadeite mass. Coleman (1961) has also described lens-like inclusions in serpentinite which consist of a monomineralic core of green jadeite (composition jadeite 75, acmite 15, diopside 10 per cent) surrounded by a zone consisting largely of pectolite and hydrogrossular. Jadeite ($Jd_{98}Ac_1Di_1$), with minor amounts of albite, occurs in veins which show a transgressive relation to inclusions of albite–crossite schist and, with acmite, is present also in the schist. The albite–crossite schist has a high content of Na_2O (9·8 per cent) and Coleman has suggested that the formation of jadeite may be the result, at least in part, of the unusual composition of this rock. Locally both the albite and jadeite are replaced by analcite. Jadeite-bearing rocks in tectonic inclusions occur within serpentinite in the New Idria district, California, and here jadeite is found as green fibrous crystals within the schists, and as veins cross-cutting albite–crossite–aegirine schists (Coleman, 1961).

One of the most interesting occurrences of jadeite in California is its formation in quartz–jadeite rocks, or metagreywackes (Bloxam, 1956). The jadeite-bearing metamorphic rocks, and the indurated greywackes from which they are derived, show no significant differences in chemistry; clastic textures and original bedding planes are preserved in the quartz–jadeite rocks which are associated with glaucophane schists and antigoritized and actinolitized serpentinites. The constituents of the metagreywackes are jadeite, glaucophane, lawsonite, sericite, chlorite and albite; the latter, however, is not abundant and has been largely replaced by jadeite. The jadeite contains 12·5 weight per cent quartz inclusions, and Bloxam concluded that it is a product of the reaction

$$\underset{\text{albite}}{NaAlSi_3O_8} \rightarrow \underset{\text{jadeite}}{NaAlSi_2O_6} + \underset{\text{quartz}}{SiO_2}$$

Inclusions of lawsonite also occur in the jadeite, and may represent the anorthitic component of the original plagioclase.

In some of the metagreywackes, glaucophane and lawsonite are present in greater amounts, and the glaucophane is derived in part from the original chloritic material and in part also from jadeite, the latter showing marginal alteration to glaucophane (Bloxam, 1959). In some cases the jadeite develops a pale green colour and a higher birefringence indicating a change in chemical composition possibly to a more Fe^{3+}-rich variety.

The Franciscan metagreywackes from the Diablo Range, central California Coast Ranges have been described by Ernst (1971). Here the most thoroughly reconstituted sediments consist of jadeitic pyroxene + lawsonite ± aragonite. Acicular prisms of the pyroxene replacing albite are displayed in some examples, but generally the former presence of the feldspar is marked by sheaves and stellate prismatic groups of the jadeitic pyroxene that retain the original clastic outlines of the plagioclase. In other rocks there is no evidence of replacement and the pyroxene and albite occur in adjacent grains (see also Kerrick and Cotton, 1971). Jadeite has also been described by McKee (1958, 1962) from the metasedimentary and igneous rocks of the Franciscan group, south-east of San Francisco. Jadeite replaces albite in the clastic sediments, and both albite and augite in the greenstones. In the metasediments the jadeite develops initially as needles parallel to one or more cleavages of the albite and finally completely replaces the feldspars, while in the meta-igneous rocks the replacement of augite is incomplete. At this locality the jadeite crystallization occurred subsequently to the formation of lawsonite which persists as relict grains in the jadeite:

$$\underset{\text{plagioclase}}{NaCaAl_3Si_5O_{16}} + 2H_2O \rightarrow \underset{\text{jadeite}}{NaAlSi_2O_6} + \underset{\text{lawsonite}}{CaAl_2Si_2O_7(OH)_2.H_2O} + \underset{\text{quartz}}{SiO_2}$$

Jadeite, associated with lawsonite and glaucophane also occurs in the pebbles of the glaucophane-lawsonite facies metamorphic rocks in the conglomerates of the Franciscan formation (Fyfe and Zardini, 1967).

Jadeite-bearing metabasalts, spilitic and keratophyric in composition, occur in the low to intermediate, but jadeite is replaced by omphacite in the higher grade glaucophane schists, and Coleman (1965) has suggested that the derivation of the jadeite–glaucophane assemblage may be due to reactions of the type:

$$\underset{\text{albite}}{8NaAlSi_3O_8} + \underset{\text{montmorillonite}}{2Ca_{0\cdot5}(Mg_{3\cdot5}Al_{0\cdot5})Si_8O_{20}(OH)_4.nH_2O} \rightarrow \quad (1)$$

$$\underset{\text{jadeitic pyroxene}}{5(Na_{0\cdot8}Ca_{0\cdot2})(Mg_{0\cdot2}Al_{0\cdot8})Si_2O_6} +$$

$$+ \underset{\text{glaucophane}}{2Na_2(Mg_3)Al_2(Al_{0\cdot5}Si_{7\cdot5})O_{22}(OH)_2} + 15SiO_2 + 6H_2O$$

$$\underset{\text{albite}}{8NaAlSi_3O_8} + \underset{\text{chlorite}}{(Mg_{4\cdot0}Fe_{2\cdot0})(Al_{1\cdot0}Si_{3\cdot0})O_{10}(OH)_8} + CaCO_3 \rightarrow \quad (2)$$

$$\underset{\text{jadeitic pyroxene}}{5(Na_{0\cdot8}Ca_{0\cdot2})(Fe_{0\cdot2}Al_{0\cdot8})Si_2O_6} +$$

$$+ \underset{\text{glaucophane}}{2Na_2(Mg_{2\cdot0}Fe_{1\cdot0})Al_2(Al_{0\cdot5}Si_{7\cdot5})O_{22}(OH)_2} + 2SiO_2 + CO_2 + 2H_2O$$

The occurrence of jadeite (Jd_{70-80}), associated with glaucophane, in the original varioles of pillow lava has also been reported by Coleman and Lee (1963). With increasing metamorphic grade the jadeitic component of the pyroxene decreases and the mineral becomes omphacitic in composition. The replacement of plagioclase by jadeite in a diabase sill, Angel Island, San Francisco, has been described by Bloxam (1960).

Keith and Coleman (1968) have reported an unusual association of jadeite, glaucophane and albite with minor amounts of analcite, in which the formation of the albite post-dates the ferromagnesian minerals, and have explained this relationship as resulting from several periods of blueschist facies metamorphism.

Single-crystal X-ray study of the pyroxene (Table 49, anal. 11) shows the presence of two mineral phases, the predominant jadeite and a minor phase an amphibole presumably due to the incipient alteration of the jadeite with the changing metamorphic environment.

Coleman and Clark (1968) have made a survey of the pyroxenes in the blueschist facies of California and their chemical relationships are shown in Fig. 224 from

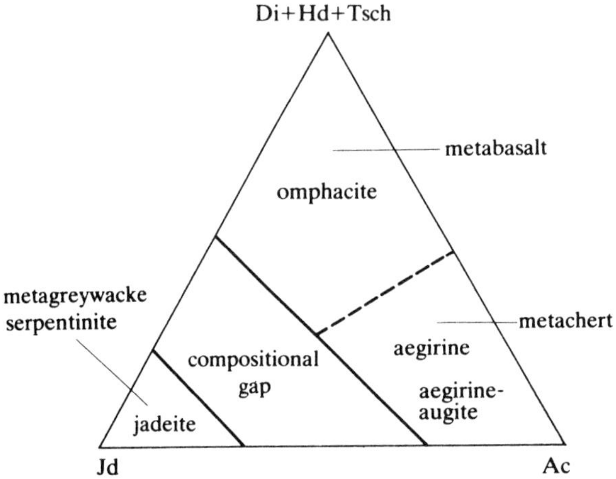

Fig. 224. Diagram showing the variation of pyroxenes in the blueschist facies of California in terms of the composition, jadeite, acmite and diopside-hedenbergite + Tschermak's molecule in relation to the host rock (after Coleman and Clark, 1968).

which it appears that there is a compositional gap, along the jadeite–diopside join, between 75 and 45 mol. per cent jadeite. Earlier, Dobretsov (1964) had concluded that jadeites and omphacites are separated by an immiscibility region, between 80 and 60 per cent jadeite (but see Fig. 211). The paragenesis of jadeite, jadeitic pyroxene, omphacite and aegirine-augite, and the occurrence of jadeite-bearing rocks in the U.S.S.R., Japan and the Alps have been discussed by Coleman and Clark. The field, petrographic and petrogenetic relationships of the jadeite-bearing rocks of the Franciscan formation have been reviewed by Ernst (1965).

Occurrences of jadeite from Alpine localities include the description of jadeite pseudomorphing potassium feldspar in schists and gneisses derived from a rhyolite complex, Cottian Alps, Italy, by Lefèvre and Michard (1965), jadeite-bearing metagreywackes, western Alps (Bearth, 1966), metagranites of the Mount Mucrone area (Compagnoni and Maffeo, 1973) and the formation of jadeite and quartz from the breakdown of albite in the jadeite–glaucophane metagreywackes of the Ambin massif (Lovenzoni, 1963). Sodium-rich pyroxenes with an intermediate jadeite and low to medium acmite content, are the characteristic pyroxenes of the Monviso massif (Mottana, 1971), and here are restricted to highly silicic schists, metagreywackes and metarhyolites. They display a compositional gap between the jadeitic and omphacitic pyroxenes similar to that observed by Coleman and Clark (1968) for the jadeitic pyroxenes of California. Pyroxenes, rich in the jadeite component and associated with quartz, have been reported from schists in Corsica (Autran, 1964), and Essene (1969) has described a relatively pure jadeite (Jd_{94}) in a siliceous gneiss from this area.

Jadeite-bearing metamorphosed arkoses have been described from Celebes (de Roever, 1947, 1955) and here the mineral is associated with lawsonite, albite and sericitic mica. The albite is considered to have crystallized at a later stage in the history of the rock, and the jadeite at an earlier period according to the reaction albite → jadeite + quartz. In these rocks the association of jadeite, quartz and albite is considered by de Roever to be a disequilibrium assemblage, in which the jadeite persists as an unstable relic in the physical environment responsible for the later formation of the albite.

Jadeite-bearing rocks (e.g. metagabbro consisting chiefly of jadeite and pumpellyite) have been described from the Sanbagawa and Kamuikotan metamorphic belts, Japan (Seki, 1960; Seki and Shidô, 1959). At Sanbagawa the jadeite rocks occur in pelitic and psammitic sediments and mafic volcanic rocks in the lawsonite–pumpellyite–epidote–glaucophane–actinolite zone. The jadeite has replaced augite and the *c* axes of the two pyroxenes are completely or nearly parallel. Jadeite-bearing metagabbros within serpentinites, associated with the crystalline schists of the lawsonite–pumpellyite–epidote–glaucophane subfacies, also occur in the Sibukawa district, and here the pyroxene is associated with pumpellyite, vesuvianite, grossular and lawsonite; veins composed chiefly of jadeite (Table 49, anal. 4) also occur in the metagabbros (Seki *et al.*, 1960). A description of the jadeitic rock, jadeite–albite and jadeitic veins in serpentinite, located along fault boundaries at Omi-Kataki, central Japan, is given by Chihara (1971). Jadeite, enclosed in serpentine, from a jadeite-bearing albitite at Kotaki has been reported by Iwao (1953) and Shidô (1958). The jadeite albitite forms a narrow 20 cm zone around an albitite core and is itself enclosed by an outer rim of an amphibole rock containing both jadeite and albite.

Green *et al.* (1968) have described the association of a jadeite-rich pyroxene (Table 49, anal. 14) with almandine garnet and quartz. The rock, of sodic trachyte composition, occurs together with other eclogite-type assemblages as boulders in a Tertiary conglomerate in the Guajira Peninsula, Colombia. Minor phases include scapolite, calcite, clinozoisite and paragonite, the latter two minerals are intimately associated and Green *et al.* have suggested that they may have been derived from lawsonite by the reaction

$$\underset{\text{lawsonite}}{4CaAl_2Si_2O_8.2H_2O} + \underset{\text{jadeite}}{NaAlSi_2O_6} \rightleftharpoons \underset{\text{clinozoisite}}{2Ca_2Al_3Si_3O_{12}(OH)} + \underset{\text{paragonite}}{NaAl_3Si_3O_{10}(OH)_2} + SiO_2 + 6H_2O$$

A second rock, of alkali olivine basalt hawaiite composition, consists of primary almandine and omphacite. The Guajîra mineral extends the compositional range of jadeite-bearing pyroxenes occurring in eclogites and its unique paragenesis is probably related to the bulk rock composition. Jadeite-rich inclusions of jadeite–albite rocks have been described in serpentinite from the Motagua fault zone, Guatemala, and here occur as small grains in optical continuity and are apparently relics of larger crystals that have been partially replaced by the surrounding albite (McBirney *et al.*, 1967).

At Tawmaw, Burma, jadeite occurs associated with albite, actinolite and glaucophane in a dyke-like mass in serpentine. Descriptions and discussion of the paragenesis of the Burma occurrences are given by Yoder (1950) and de Roever (1955). Jadeite rock has been described from the Pay-Yer Mountain in the Polar Urals (Dobretsov and Ponomareva, 1965) and jadeite-bearing amygdales have been recorded in regionally metamorphosed mafic extrusive rocks in north-western Kamchatka, and are considered by Ponomareva and Dobretsov (1966) to have developed from analcite amygdales during the metamorphism. The unique occurrence of jadeite with nepheline in a rock from 'Thibet' originally described by Bauer (1896), and later by Lacroix (1930), was re-examined by Tilley (1956). The jadeite is traversed by thin veins of nepheline; the latter mineral has been derived from the breakdown of the jadeite, residual grains of which are enclosed within the nepheline. The replacement of jadeite by analcite in this rock was also described by Tilley.

The pressure required for the development of the jadeite–quartz association of the glaucophane schist facies has, on the basis of experimental data, been shown to be lower than had earlier been considered necessary. The data presented by Newton and Smith (1967) indicate that a minimum pressure of 6·8 kbar is necessary at a temperature of 150°C, and 7·9 kbar at 200°C. The association of aragonite with jadeite assemblages favours the lower temperature–pressure conditions. A pressure of 7 kbar requires some 25 km of burial in the geosynclinal column, a thickness of over-burden that at some localities is not greatly at variance with the stratigraphical evidence. The experimental data refer to the pure jadeite composition, and lower pressures than are indicated by the experimentally determined univariant reactions may be envisaged due to the replacement of Al by Fe^{3+} in natural jadeites of the glaucophane-schist facies.

References

Abs-Wurmbach, I. and **Neuhaus, A.**, 1976. Das System $NaAlSi_2O_6$ (Jadeit)–$NaCrSi_2O_6$ (Kosmochlor) im Druckbereich von 1 bar bis 25 kbar bei 800°C. *Neues Jahrb. Min., Abhdl.*, **127**, 213–241.

Adams, L. H., 1953. A note on the stability of jadeite. *Amer. J. Sci.*, **251**, 299–308.

Autran, M. A., 1964. Description de l'association jadeite + quartz et des paragénèses minerales associées, dans les schistes lustres de Sant Andrea di Cotone (Corse). *Bull. Soc. franç. Min. Crist.*, **87**.

Bauer, M., 1896. Jadeit von Tibet. *Neues Jahrb. Min.*, **1**, 89–95.

Bearth, P., 1966. Zur Mineralfaziellen Stellung der Glaukophangesteine der Westalpen. *Schweiz. Min. Petr. Mitt.*, **46**, 13–23.

Bell, P. M. and **Boyd, F. R.**, 1968. Phase equilibrium data bearing on the pressure and temperature of shock metamorphism. In *Shock Metamorphism of Natural Material*, French and Short (eds.). Baltimore (Mono Book Corp.), 43–50.

Bell, P. M. and **Davis, B. T. C.**, 1965. Temperature–composition section for jadeite–diopside. *Carnegie Inst. Washington, Ann. Rept. Dir. Geophys. Lab.*, 1964–65, 120–123.

Bell, P. M. and **Davis, B. T. C.**, 1967. Investigation of a solvus in the system jadeite–diopside. *Carnegie Inst. Washington, Ann. Rept. Dir. Geophys. Lab.*, 1965–66, 239–241.

Bell, P. M. and **Davis, B. T. C.**, 1969. Melting relations in the system jadeite–diopside at 30 and 40 kilobars. *Amer. J. Sci.*, **267-A** (Schairer vol.), 17–32.

Bell, P. M. and **Roseboom, E. H.**, *Jr.*, 1969. Melting relations of jadeite and albite to 45 kilobars with comments on melting diagrams of binary systems at high pressures. *Min. Soc. America, Spec. Paper* **2**, 151–161.

Birch, F. and **LeComte, P.**, 1960. Temperature–pressure plane for albite composition. *Amer. J. Sci.*, **258**, 209–217.

Bloxam, T. W., 1956. Jadeite-bearing metagraywackes in California. *Amer. Min.*, **41**, 488–496.

Bloxam, T. W., 1959. Glaucophane schists and associated rocks near Valley Ford, California. *Amer. J. Sci.*, **257**, 95–112.

Bloxam, T. W., 1960. Jadeite-rocks and glaucophane schists from Angel Island, San Francisco Bay. California. *Amer. J. Sci.*, **258**, 555–573.

Boettcher, A. L. and **Wyllie, P. J.**, 1968. Jadeite stability measured in the presence of silicate liquids in the system $NaAlSiO_4$–SiO_2–H_2O. *Geochim. Cosmochim. Acta*, **32**, 999–1012.

Bradt, R. C., **Newnham, R. E.** and **Biggers, J. V.**, 1973. The toughness of jade. *Amer. Min.*, **58**, 727–732.

Cameron, M., **Sueno, S.**, **Prewitt, C. T.** and **Papike, J. J.**, 1973. High-temperature crystal chemistry of acmite, diopside, hedenbergite, jadeite, spodumene and ureyite. *Amer. Min.*, **58**, 594–618.

Campbell, A. S. and **Fyfe, W. S.**, 1965. Analcite–albite equilibria. *Amer. J. Sci.*, **263**, 807–816.

Chihara, K., 1971. Mineralogy and paragenesis of jadeites from the Omi-Kotaki area, central Japan. *Min. Soc. Japan, Spec. Paper* **1** *(Proc. IMA–IAGOD Mtgs 1970, IMA vol.)*, 147–156.

Coes, L., 1955. High-pressure minerals. *J. Amer. Ceram. Soc.*, **38**, 298.

Coleman, R. G., 1961. Jadeite deposits of the Clear Creek area, New Idria district, San Benito County, California. *J. Petr.*, **2**, 209–247.

Coleman, R. G., 1965. Composition of jadeitic pyroxene from the California metagraywackes. *Prof. Paper U.S. Geol. Surv.*, **525-C**, 25–34.

Coleman, R. G. and **Clark, J. R.**, 1968. Pyroxenes in the blueschist facies of California. *Amer. J. Sci.*, **266**, 43–59.

Coleman, R. G. and **Lee, D. E.**, 1963. Glaucophane-bearing metamorphic rock types of the Cazadero area, California. *J. Petr.*, **4**, 260–301.

Compagnoni, R. and **Maffeo, B.**, 1973. Jadeite-bearing metagranites l.s. and related rocks in the Mount Mucrone area (Sesia-Lanzo zone, western Italian Alps). *Schweiz. Min. Petr. Mitt.*, **53**, 355–378.

Dachille, F. and **Roy, R.**, 1961. Influence of displacive-shearing stresses on the kinetics of reconstructive transformations affected by pressure in the range of 0–100,000 bars. *Proc. 4th Intern. Symposium on Reactivity of Solids* (Elsevier Publ. Co., Amsterdam), 502–511.

Dobretsov, N. L., 1964. Miscibility limits and mean compositions of jadeite pyroxenes. *Dokl. Acad. Sci. U.S.S.R., Earth Sci. Sect.*, **146**, 118–120.

Dobretsov, N. L. and **Ponomareva, L. G.**, 1965. [Comparative characteristics of the Polar Ural and Balkhash region jadeite and its associated rocks.] *Akad. Nauk S.S.S.R., Sibirskoe otd-ie, Inst. Geol. Geofiz., Trudy*, no. 31, 178–244 (in Russian with English summary).

Ernst, W. G., 1965. Mineral parageneses in Franciscan metamorphic rocks, Panoche Pass, California. *Bull. Geol. Soc. America*, **76**, 879–914.

Ernst, W. G., 1971. Petrologic reconnaissance of Franciscan metagraywackes from the Diablo Range, central California coast ranges. *J. Petr.*, **12**, 413–437.

Essene, E. J., 1969. Relatively pure jadeite from a siliceous Corsican gneiss. *Earth Planet. Sci. Letters*, **5**, 270–272.

Essene, E. J., Fyfe, W. S. and **Turner, F. J.**, 1965. Petrogenesis of Franciscan glaucophane schists and associated metamorphic rocks, California. *Beitr. Min. Petr.*, **11**, 695–704.

Foshag, W. F., 1955. Chalchihuitl – a study in jade. *Amer. Min.*, **40**, 1062–1070.

Fyfe, W. S. and **Valpy, G. W.**, 1959. The analcime–jadeite phase boundary: some indirect deductions. *Amer. J. Sci.*, **257**, 316–320.

Fyfe, W. S. and **Zardini, R.**, 1967. Metaconglomerate in the Franciscan formation near Pacheco Pass, California. *Amer. J. Sci.*, **265**, 819–830.

Gilbert, M. C., 1968. X-ray properties of jadeite acmite pyroxenes. *Carnegie Inst. Washington, Ann. Rept. Dir. Geophys. Lab.*, 1966–67, 374–375.

Green, D. H. and **Lambert, I. B.**, 1965. Experimental crystallization of anhydrous granite at high pressures and temperatures. *J. Geophys. Res.*, **70**, 5259–5268.

Green, D. H., Lockwood, J. P. and **Kiss, E.**, 1968. Eclogite and almandine–jadeite–quartz rock from the Guajira Peninsula, Colombia, South America. *Amer. Min.*, **53**, 1320–1335.

Green, D. H. and **Ringwood, A. E.**, 1967. An experimental investigation of the gabbro to eclogite transformation. *Geochim. Cosmochim. Acta*, **31**, 767–833.

Green, T. M., 1967. An experimental investigation of sub-solidus assemblages formed at high pressure in high alumina basalt, kyanite eclogite and grosspydite compositions. *Contr. Min. Petr.*, **16**, 84–114.

Greenwood, H. J., 1961. The system $NaAlSi_2O_6$–H_2O–argon: total pressure and water pressure in metamorphism. *J. Geophys. Res.*, **66**, 3923–3946.

Greig, J. W. and **Barth, T. F. W.**, 1938. The system $Na_2O.Al_2O_3.2SiO_2$ (nepheline, carnegieite)–$Na_2O.Al_2O_3.6SiO_2$ (albite). *Amer. J. Sci.*, 5th ser., **35-A**, 93–112.

Griggs, D. T. and **Kennedy, G. C.**, 1956. A simple apparatus for high pressures and temperatures. *Amer. J. Sci.*, **254**, 722–735.

Hays, J. F. and **Bell, P. M.**, 1973. Albite–jadeite–quartz equilibrium: a hydrostatic determination. *Carnegie Inst. Washington, Ann. Rept. Dir. Geophys. Lab.*, 1972–73, 706–708.

Hlabse, T. and **Kleppa, O. J.**, 1968. The thermochemistry of jadeite. *Amer. Min.*, **53**, 1281–1292.

Iwao, S., 1953. Albitite and associated jadeite rock from Kotaki district, Japan; a study in ceramic raw material. *Geol. Surv. Japan, Rept.*, **153**, 25 pp (M.A. **12**-381).

Kanamori, H., Fujii, N. and **Mizutani, H.**, 1968. Thermal diffusivity measurement of rock-forming minerals from 300° to 1100°K. *J. Geophys. Res.*, **73**, 595–605.

Keith, T. E. C. and **Coleman, R. G.**, 1968. Albite–pyroxene–glaucophane schist from Valley Ford, California. *Prof. Paper U.S. Geol. Surv.*, **600-C**, 13–17.

Kelly, K. K., Todd, S. S., Orr, R. L., King, E. G. and **Bonnickson, K. R.**, 1953. Thermodynamic properties of sodium-aluminium and potassium-aluminium silicates. *U.S. Bur. Mines, Rept. Invest.* 4955.

Kerrick, D. M. and **Cotton, W. R.**, 1971. Stability relations of jadeite pyroxene in Franciscan metagraywackes near San Jose, California. *Amer. J. Sci.*, **271**, 350–369.

Kolesova, V. A., 1959. Infrared absorption spectra of the silicates containing aluminium and of certain crystalline aluminates. *Optics and Spectroscopy*, **6**, 20.

Kozlowski, K., 1965. [The granulitic complex of Stary Gieraltów, east Sudetes.] *Polska Akad. Nauk., Arch. Min.*, 1961 (publ. 1965), **25**, 5–108 (in Polish; English summary).

Kracek, F. C., Neuvonen, K. J. and **Burley, G.**, 1951. Thermochemistry of mineral substances. I. A thermodynamic study of the stability of jadeite. *J. Washington Acad. Sci.*, **41**, 373–383.

Kushiro, I., 1965. Coexistence of nepheline and enstatite at high pressures. *Carnegie Inst. Washington, Ann. Rept. Dir. Geophys. Lab.*, 1964–1965, 109–112.

Kushiro, I., 1965. Clinopyroxene solid solutions at high pressures. *Carnegie Inst. Washington, Ann. Rept. Dir. Geophys. Lab.*, 1964–65, 112–117.

Kushiro, I., 1969. Clinopyroxene solid solutions formed by reactions between diopside and plagioclase at high pressures. *Min. Soc. America, Spec. Paper* **2**, 179–191.

Kuznetsova, L. G. and **Moskaleva, V. N.**, 1968. [Infrared absorption spectra of pyroxenes of the diopside–jadeite isomorphous series.] *Zap. Vses. Min. Obshch.*, **97**, 715–718 (in Russian).

Lacroix, A., 1930. La jadéite de Birmanie: les roches qu'elle constitue ou qui l'accompagnent. Composition et origine. *Bull. Soc. franç. Min.*, **53**, 216–264.

Lefèvre, R. and **Michard, A.**, 1965. La jadéite dans le métamorphisme alpin à propos des gisements de type nouveau, de la bande d'Acceglio (Alpes cottiennes, Italie). *Bull. Soc. franç. Min. Crist.*, **88**, 664–677.

Lovenzoni, S., 1963. Metagrovache in facies epimetamorfica a giadeite e gastaldite, affioranti nel gruppo montuoso d'Ambin (Alpi Cozie). *Ricerca Sci., Roma, Rend. A.* **3**, ser. 2, 1059–1066.

Mao, H. K., 1971. The system jadeite ($NaAlSi_2O_6$)–anorthite ($CaAl_2Si_2O_8$) at high pressures. *Carnegie Inst. Washington, Ann. Rept. Dir. Geophys, Lab.*, 1969–70, 163–168.

McBirney, A., Aoki, K.-I. and **Bass, M. N.**, 1967. Eclogites and jadeite from the Motagua fault zone, Guatemala. *Amer. Min.*, **52**, 908–918.

McKee, B., 1958. Jadeite alteration of sedimentary and igneous rocks. *Bull. Geol. Soc. America*, **69**, 1612 (abstract).

McKee, B., 1962. Widespread occurrence of jadeite, lawsonite, and glaucophane in central California. *Amer. J. Sci.*, **260**, 596–610.

Miyashiro, A. and **Banno, S.**, 1958. Nature of glaucophanitic metamorphism. *Amer. J. Sci.*, **256**, 97–110.

Mottana, A., 1971. Pyroxenes in the ophiolitic metamorphism of the Cottian Alps. *Min. Soc. Japan, Spec. Paper* **1** *(Proc. IMA–IAGOD Mtgs '70, IMA vol.)*, 140–146.

Newton, M. S. and **Kennedy, G. C.**, 1968. Jadeite, analcite, nepheline and albite at high temperatures and pressures. *Amer. J. Sci.*, **266**, 728–735.

Newton, R. C. and **Smith, J. V.**, 1967. Investigations concerning the breakdown of albite at depth in the earth. *J. Geol.*, **75**, 268–286.

Ponomareva, L. G. and **Dobretsov, N. L.**, 1965. Jadeite-bearing and other amygdules in meta-extrusives of north-western Kamchatka. *Dokl. Acad. Sci. U.S.S.R., Earth Sci. Sect.*, **167**, 87–91.

Popp, R. K. and **Gilbert, M. C.**, 1972. Stability of acmite–jadeite pyroxenes at low pressure. *Amer. Min.*, **57**, 1210–1231.

Prewitt, C. T. and **Burnham, C. W.**, 1966. The crystal structure of jadeite, $NaAlSi_2O_6$. *Amer. Min.*, **51**, 956–975.

Prewitt, C. T., Shannon, R. D. and **White, W. B.**, 1972. Synthesis of a pyroxene containing trivalent titanium. *Contr. Min. Petr.*, **35**, 77–82.

Ringwood, A. E., Reid, A. F. and **Wadsley, A. D.**, 1967. A high pressure transformation of alkali aluminosilicates and aluminogermanates. *Earth and Planet. Sci. Letters*, **3**, 38–40.

Robertson, E. C., Birch, F. and **MacDonald, G. J. F.**, 1957. Experimental determination of jadeite stability relations to 25,000 bars. *Amer. J. Sci.*, **255**, 115–137.

Roever, W. P. de, 1947. *Igneous and Metamorphic Rocks in Eastern Central Celebes. Geological Explorations in the Island of Celebes.* Amsterdam.

Roever, W. P. de, 1955. Genesis of jadeite by low grade metamorphism. *Amer. J. Sci.*, **253**, 283–298.

Rossman, G. R., 1974. Lavender jade. The optical spectrum of Fe^{3+} and $Fe^{2+} \rightarrow Fe^{3+}$ intervalence charge transfer in jadeite from Burma. *Amer. Min.*, **59**, 868–870.

Roy, R. and **Tuttle, O. F.**, 1956. Investigation under hydrothermal conditions. *Phys. Chem. Earth*, **1**, 138–180.

Schairer, J. F. and **Bowen, N. L.**, 1956. The system Na_2O–Al_2O_3–SiO_2. *Amer. J. Sci.*, **254**, 129–195.

Schmidt, J. and **Štelcl, J.**, 1971. Jadeites from Moravian neolithic period. *Acta Univ. Carolinae, Geol.*, 141–152.

Schüller, A., 1960. Das Jadeitproblem vom petrogenetischen und mineralfaziellen Standpunkt. *Neues Jahrb. Min., Abhdl.*, **94**, 1295–1308.

Seki, Y., 1960. Jadeite in Sanbagawa crystalline schists of central Japan. *Amer. J. Sci.*, **258**, 705–715.

Seki, Y., 1961. Jadeite from Kanasaki of the Kanto Mountains, central Japan. *J. Geol. Soc. Japan*, **67**, 101–104.

Seki, Y., **Aiba, M.** and **Kato, C.**, 1960. Jadeite and associated minerals of metagabbroic rocks in the Sibukawa district, central Japan. *Amer. Min.*, **45**, 668–679.

Seki, Y. and **Kennedy, G. C.**, 1964. The breakdown of potassium feldspar, $KAlSi_3O_8$, at high temperatures and pressures. *Amer. Min.*, **49**, 1688–1706.

Seki, Y. and **Onuki, H.**, 1967. Variation of unit-cell dimensions in natural jadeite–diopside mineral series. *J. Japan. Assoc. Min. Petr. Econ. Geol.*, **58**, 233–237.

Seki, Y. and **Shidô, F.**, 1959. Finding of jadeite from the Sanbagawa and Kamuikotan metamorphic belts, Japan. *Proc. Japan. Acad.*, **35**, 137–138.

Shidô, F., 1958. Calciferous amphibole rich in sodium from jadeite-bearing albite of Kotaki, Niigata Prefecture. *J. Geol. Soc. Japan*, **64**, 595–600.

Shidô, F. and **Seki, Y.**, 1959. Notes on rock-forming minerals (11). Jadeite and hornblende from the Kamuikotan metamorphic belt. *J. Geol. Soc. Japan*, **65**, 673–677.

Smith, W. C., 1963. Jade axes from sites in the British Isles. *Proc. Prehist. Soc.*, **29**, 133–172.

Smith, W. C., 1965. The distribution of jade axes in Europe with a supplement to the catalogue of those from the British Isles. *Proc. Prehist. Soc.*, **31**, 25–33.

Thompson, A. B., 1971. Analcite–albite equilibria at low temperatures. *Amer. J. Sci.*, **271**, 79–92.

Tilley, C. E., 1956. Nepheline associations. *Kon. Ned. Geol-Mijn., Gen., Geol. Sci.* (Brouwer vol.), 403–414.

Waldbaum, D. R. and **Thompson, J. B.**, *Jr.*, 1969. Mixing properties of sanidine crystalline solution: IV. Phase diagrams from equations of state. *Amer. Min.*, **54**, 1274–1298.

Wikström, A., 1970. Hydrothermal experiments in the system jadeite–diopside. *Norsk. Geol. Tidsskr.*, **50**, 1–14.

Williams, D. W. and **Kennedy, G. C.**, 1970. The melting of jadeite to 60 kilobars. *Amer. J. Sci.*, **269**, 481–488.

Wolfe, C. W., 1955. Crystallography of jadeite crystals from near Cloverdale, California. *Amer. Min.*, **40**, 248–260.

Wyckoff, R. W. G., **Merwin, H. E.** and **Washington, H. S.**, 1925. X-ray diffraction measurements upon the pyroxenes. *Amer. J. Sci.*, 5th ser., **10**, 383–397.

Yoder, H. S., 1950. The jadeite problem, Parts I and II. *Amer. J. Sci.*, **248**, 225–248 and 312–334.

Yoder, H. S. and **Chesterman, C. E.**, 1951. Jadeite of San Benito County, California. *Spec. Rept., Div. Mines, California*, no. 10-c, 1.

Yoder, H. S. and **Weir, C. E.**, 1951. Change of free energy with pressure of the reaction nepheline + albite = 2 jadeite. *Amer. J. Sci.*, **249**, 683–694.

Aegirine $NaFe^{3+}[Si_2O_6]$
Aegirine-augite $(Na,Ca)(Fe^{3+},Fe^{2+},Mg,Al)[Si_2O_6]$

Monoclinic

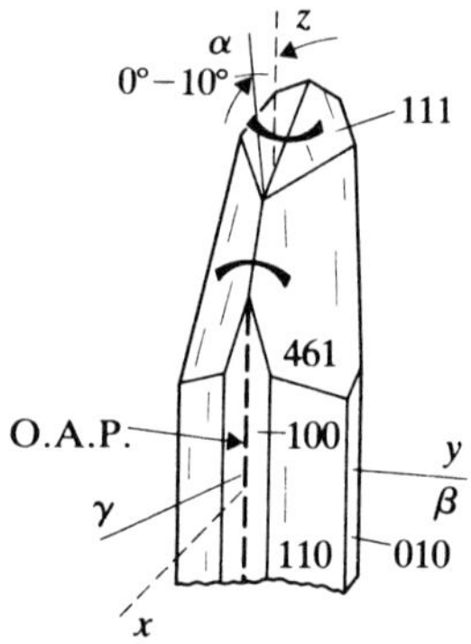

	Aegirine (−)	**Aegirine-Augite (−) (+)**
α	1·750–1·776	1·700–1·760
β	1·780–1·820	1·710–1·800
γ	1·795–1·836	1·730–1·813
δ	0·040–0·060	0·028–0·050
$2V_\alpha$	60°–70°	70°–110°
$\alpha:z$	0°–10° (β obtuse) $\beta = y$, O.A.P. (010)	0°–20° (β acute) $\beta = y$, O.A.P. (010)
Dispersion:	$r > v$, moderate to strong.	$r > v$, moderate to strong.
D	3·50–3·60	3·40–3·60
H	6	6
Cleavage:	{110} good, {100} parting. (110):(1$\bar{1}$0) ≃ 87°	{110} good, {100} parting. (110):(1$\bar{1}$0) ≃ 87°
Twinning:	{100} simple and lamellar, common.	{100} simple and lamellar, common
Colour:	Aegirine, dark green to greenish black; pale to dark green and yellowish green in thin section. Acmite, reddish brown, dark green to black; light brown to yellow or greenish yellow in thin section.	Dark green to black, green, yellow-green or brown; pale green, green and yellowish green in thin section.
Pleochroism:	α emerald-green, deep green.	bright green, deep green, brownish green.
	β grass green, deep green, yellow.	yellowish green.
	γ brownish green, green, yellowish brown, yellow.	yellow, greenish brown, brownish green.

Unit cell:[1]	a(Å) 9·658	9·68–9·74
	b(Å) 8·795	8·79–8·935
	c(Å) 5·294	5·26–5·30
	β 107·42°	105°–106·85°

Z = 4. Space group $C2/c$.

Insoluble in HCl.

Aegirine and aegirine-augite are characteristic minerals of the alkaline igneous rocks, and occur commonly in syenites and syenite pegmatites. They also occur in alkali granites and equivalent dyke rocks as well as in some crystalline schists, metamorphosed iron-rich sediments, and hydrothermal, metasomatic and diagenetic environments.

Although the term acmite has generally been adopted to describe the $NaFe^{3+}Si_2O_6$ 'molecule', both aegirine and acmite have been used for pyroxenes of approximately this composition. Aegirine is generally restricted to the green to black crystals of $NaFe^{3+}Si_2O_6$ which are characteristically bluntly terminated by faces of the {111} form, and acmite to the brown variety showing pointed terminations of the forms {221} and {661}. Aegirine is strongly pleochroic in thin section, acmite only weakly so, but zoned crystals showing both types of pleochroism are not uncommon. The name aegirine-augite, first introduced by Rosenbusch, is used to describe the green strongly pleochroic pyroxenes, with high $\gamma{:}z$ extinction angles, which are intermediate in composition between aegirine and augite. Acmite is named from the Greek *akme*, a point, in reference to its pointed habit, and aegirine after Aegir, the Scandinavian sea-god, the mineral being first reported from Norway.

There is no general agreement on the nomenclature of aegirine pyroxenes; Washington and Merwin (1927) restricted aegirine to minerals containing 80 per cent or more $NaFe^{3+}Si_2O_6$ and Tröger (1952) put the division between aegirine and aegirine-augite at 75 mol. per cent aegirine. Sabine (1950) suggested a nomenclature based on the change in the optic sign which occurs at about $Na_{0\cdot45}Fe^{3+}_{0\cdot45}$: the more (Na,Fe^{3+})-rich varieties having a negative sign he referred to as aegirine, and for the less (Na,Fe^{3+})-rich varieties having a positive sign he suggested the name aegirine-augite. In the aegirine–aegirine-augite series precise correlation of optical properties with the chemical composition is difficult due to the variable ionic replacement in the augitic component of this series, and the change in optic sign in natural minerals appears to take place over a range of some 10 mol. per cent of the aegirine component. The extinction angle $\alpha{:}z$ appears to be a more reliable guide to the chemical composition, particularly in those members of the series having high (Na,Fe^{3+}) contents. In minerals containing more than $Fe^{3+}_{0\cdot7-0\cdot8}$ the α vibration direction lies in the obtuse angle β, and in those less rich in Fe^{3+} this direction lies in the acute angle β. It is here proposed to restrict the name aegirine to those members of the series containing between $Fe^{3+}_{0\cdot8}$ and $Fe^{3+}_{1\cdot0}$. The division between aegirine-augite and sodian augite is taken at $Fe^{3+}_{0\cdot2}$. in conformity with the nomenclature proposed by Clark and Papike (1968) (see Fig. 2, p. 4), i.e. the name aegirine-augite is used for those members of $Ca(Mg,Fe)Si_2O_6$–$NaFe^{3+}Si_2O_6$ series containing between $Fe^{3+}_{0\cdot2}$ and $Fe^{3+}_{0\cdot8}$.

[1] For aegirine, values are for synthetic end-member (Nolan, 1969).

Structure

Single-crystal X-ray diffraction patterns showed that the structure of aegirine is similar to that of diopside, with space group *C2/c* (Warren and Biscoe, 1931). Synthetic acmite has the unit cell *a* 9·658, *b* 8·795, *c* 5·294 Å, *β* 107·42°, *V* 429·1 Å^3, $D_{calc.}$ 3·577 g/cm^3.

A full structure determination has been carried out by Clark *et al.* (1969) for a nearly pure acmite, $Na_{0.95}(Fe^{3+}_{0.99}Mg_{0.01}Al_{0.01})Si_2O_6$, $D_{calc.}$ 3·577 g/cm^3, from the Green River formation, Wyoming. Detail of the structure is very similar to that of jadeite. All of the Si–O chains are equivalent and they are almost as unkinked as those in jadeite; O3–O3–O3 is 174·0° in aegirine and 174·7° in jadeite as compared with 166·38° in diopside. The mean Si–O (bridging) distance is 1·642 Å and non-bridging 1·614 Å. The Fe^{3+} cations are in almost regular octahedral co-ordination in *M*1 sites with a mean Fe–O distance of 2·025 Å, whereas Na is in eight-fold co-ordination in *M*2 with six shorter Na–O distances (mean 2·414 Å) and two longer (2·831 Å).

The structure of the same acmite has been determined also at 400°, 600° and 800°C by Cameron *et al.* (1973). The cell parameters *a*, *b* and *c* increase linearly with temperature and at 800°C are *a* 9·711, *b* 8·876, *c* 5·312 Å. Whereas for the Ca pyroxenes *β* increases with temperature, for aegirine (and ureyite, $NaCr^{3+}Si_2O_6$) *β* decreases. The mean thermal expansion coefficients are lower for Na- and Li- than for Ca-pyroxenes.

The structures of a synthetic ureyite (*a* 9·550, *b* 8·712, *c* 5·273 Å, *β* 107·44°, *V* 418·6 Å^3, $D_{calc.}$ 3·603 g/cm^3, *C2/c*) and a synthetic NaIn pyroxene (*a* 9·916, *b* 9·132, *c* 5·371 Å, *β* 107·0°, *V* 465·1 Å^3, $D_{calc.}$ 4·13 g/cm^3, *C2/c*) were determined by Clark *et al.* (1969) and Christensen and Hazell (1967), respectively, and were found to have features very similar to those of aegirine.

Among other Na-pyroxenes which have been synthesized and which are isostructural with aegirine, are the following:
$NaFe^{3+}Ge_2O_6$ with *a* 10·01, *b* 8·94, *c* 5·52 Å, *β* 108° (Soloveva and Bakakin, 1968), $NaScSi_2O_6$ with *a* 9·83, *b* 9·06, *c* 5·37 Å, *β* 107·2° and $NaSc_{0.25}Fe^{3+}_{0.75}Si_2O_6$ with *a* 9·69, *b* 8·84, *c* 5·32 Å (Ito and Frondel, 1968). Unit cell dimensions for the solid solution series $NaFe^{3+}Si_2O_6$–$NaInSi_2O_6$ have been determined by Ito (1968).

Synthetic $NaTi^{3+}Si_2O_6$, has *a* 9·7114, *b* 8·8956, *c* 5·3089 Å, *β* 106·779°, *V* 439·10 Å^3, and is assumed to have the space group *C2/c* (Prewitt *et al.*, 1972). The latter authors show that the cell parameters *a*, *b*, *c* of sodium pyroxenes follow an approximately linear increasing trend in the series NaAl, NaCr, $NaTi^{3+}$, NaSc. The $NaInSi_2O_6$ cell is still larger, but not to the extent that it should be if this trend were continued. This effect has also been noted by Hawthorne and Grundy (1974), who refined the crystal structure of $NaInSi_2O_6$ and gave its cell parameters as *a* 9·9023, *b* 9·1307, *c* 5·3589 Å, *β* 107·20°. The same authors (1973) refined the crystal structure of $NaScSi_2O_6$.

Cell dimensions for pyroxenes in the Ac–Di–Hd system have been determined by Nolan (1969) (Fig. 225 and Table 51). In the Di–Ac system, with increasing ($NaFe^{3+}$) content, *a* and *b* decrease while *c* and *β* increase (Nolan and Edgar, 1963). Nolan also measured the cell parameters of thirteen aegirine-augites from Uganda (Tyler and King, 1967), together with an aegirine from the aegirine granite of Rockall (Sabine, 1960). The values for the natural alkali pyroxenes are in good agreement with those for the synthetic compositions (Fig. 225). The variation of the *b* parameters and *α* refractive indices are shown in Fig. 226.

The cell parameters and molar volumes for a number of compositions of synthetic

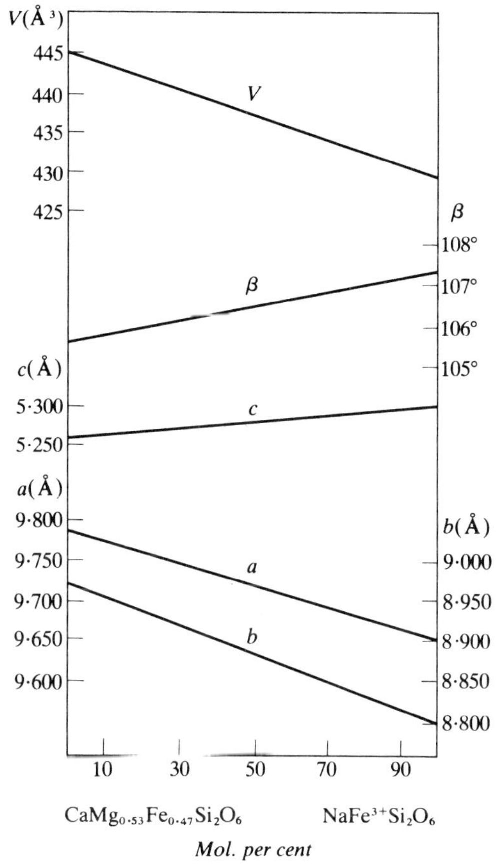

Fig. 225. Unit cell parameters of synthetic and natural pyroxenes of the acmite (aegirine)–aegirine-augite and augite series (after Nolan, 1969).

acmite–jadeite solid solutions, $Ac_{53}Jd_{47}$ to Ac_0Jd_{100}, have been measured by Gilbert (1968a). The variation of the molar volumes shows no significant departure from a straight-line relationship between the end-member values indicating a random distribution of Fe^{3+} and Al in the octahedrally co-ordinated *M*1 site. The relationship between synthetic acmite–jadeite pyroxenes and the diffraction angle of the (221) plane has been determined by Newton and Smith (1967).

Chemistry

The chemical composition of eight aegirines and eighteen aegirine-augites, together with the numbers of cations on the basis of six oxygens per mole are shown in Tables 52 and 53: for earlier analyses of aegirine and aegirine-augite see Washington and Merwin (1927) and Niggli (1943). Aegirines show a very wide range of chemical composition but in most minerals the main replacement is $NaFe^{3+} \rightleftharpoons Ca(Mg,Fe^{2+})$. Sabine (1950) has shown that sodium and potassium together vary directly with ferric iron, titanium, six-fold co-ordinated aluminium and vanadium; magnesium and ferrous iron together vary directly with calcium; sodium and potassium together vary inversely with calcium. The content of Al in the minerals of the series is small, and in aegirines in particular the replacement of Si by

Table 51 Unit cell parameters for some synthetic pyroxenes in the system aegirine–diopside–hedenbergite (Nolan, 1969)

Ac	Di	Hd	*a*(Å)	*b*(Å)	*c*(Å)	*β*	*V*(Å³)
100	—	—	9·658	8·795	5·294	107·42°	429·1
51·78	—	48·22	9·737	8·904	5·278	106·21°	439·41
—	—	100	9·841	9·027	5·247	104·80°	450·69
80·02	10·67	9·31	9·679	8·831	5·286	106·96°	432·20
40·03	32·02	27·95	9·733	8·899	5·269	106·19°	438·25
20·02	42·70	37·28	9·761	8·937	5·261	105·82°	441·53
—	100	—	9·748	8·924	5·251	105·79°	439·48

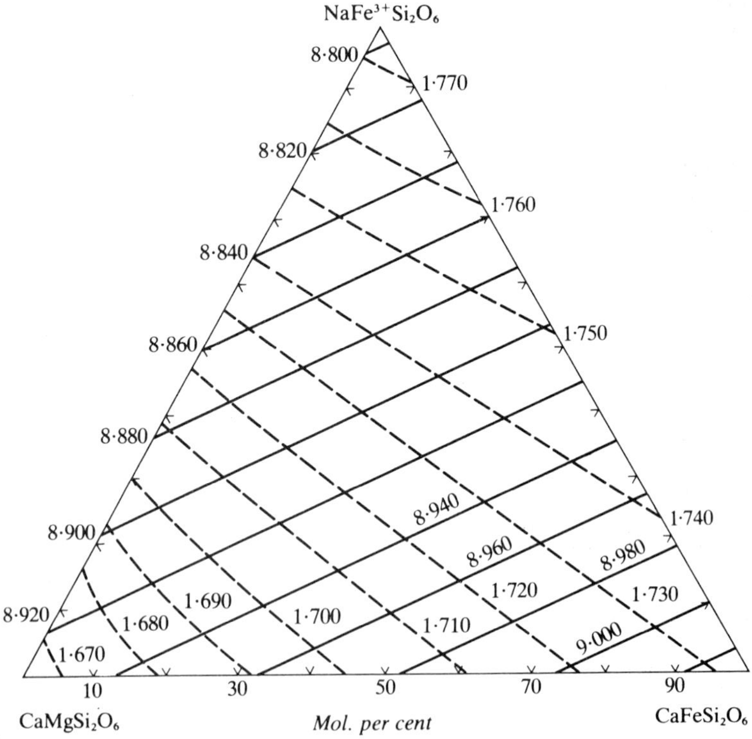

Fig. 226. Relationship between composition and α refractive indices and *b* unit cell parameters for synthetic and natural pyroxenes in the system aegirine–diopside–hedenbergite (after Nolan, 1969).

Al is commonly negligible and the minor amount of Al is mainly in octahedral co-ordination.

The titanium content of most aegirines is small, and generally does not exceed 1·0 weight per cent TiO_2. Larger amounts occur more frequently in aegirine-augites, and up to 2·5 per cent TiO_2 is not uncommon. Much higher values, however, have recently been reported (Schmincke, 1969, 1974; Flower, 1974; Curtis and Gittins, 1978, and Ferguson, 1977), particularly in late stage aegirines of alkaline rocks. The high titanium content in such aegirines is accompanied by lower values of Fe^{3+} and higher (Mg,Fe^{2+}), and their composition may be regarded to result from solid solution between $NaFe^{3+}Si_2O_6$ and an end-member $Na(Mg,Fe)_{0\cdot5}Ti_{0\cdot5}Si_2O_6$

Table 52. Aegirine Analyses

	1		2	3	4	5
SiO_2	51·64		52·10	51·64	51·31	51·92
TiO_2	0·43		0·64	0·04	0·06	0·77
Al_2O_3	0·14		0·29	1·05	1·15	1·85
Fe_2O_3	33·47		33·10	32·98	32·08	31·44
FeO	0·70		0·64	0·14	0·09	0·75
MnO	0·18		—	0·26	0·92	—
MgO	0·09		tr.	0·38	0·38	—
CaO	0·31		0·32	0·60	0·59	—
Na_2O	13·00		12·90	12·21	11·70	12·86
K_2O	tr.		0·22	0·06	0·18	0·19
H_2O^+	0·23		—	0·41	0·35	0·17
H_2O^-	—		—	0·04	0·01	—
Total	100·25		100·21	99·82	'99·86'	99·95
α	1·766	(α')	1·767	1·76 ± 0·02	1·76 ± 0·02	1·770
β	1·802		1·794	—	—	1·812
γ	1·817	(γ')	1·814	1·80 ± 0·02	1·81 ± 0·02	1·830
$\alpha:z$	—		5°	6°	6°	9°
$2V_\alpha$	—		61°	80° ± 10°	80° ± 10°	—
D	—		3·518	3·50	3·526	—

Numbers of ions on the basis of six O

	1		2		3		4		5	
Si	1·989	2·00	1·997	2·00	1·988	2·00	1·992	2·00	1·986	2·00
Al	0·006		0·003		0·012		0·008		0·014	
Al	0·000	1·02	0·003	1·00	0·036	1·03	0·044	1·04[a]	0·069	1·02
Ti	0·012		0·018		0·001		0·002		0·022	
Fe^{3+}	0·970		0·955		0·954		0·938		0·905	
Mg	0·005		—		0·022		0·022		—	
Fe^{2+}	0·023		0·021		0·004		0·003		0·024	
Mn	0·006		—		0·008		0·030		—	
Na	0·971	0·98	0·959	0·98	0·912	0·94	0·880	0·91	0·953	0·96
Ca	0·013		0·013		0·025		0·024		—	
K	0·000		0·011		0·002		0·008		0·008	

[a] Includes V^{3+} 0·001.

1 Aegirine, quartz–albite–microline–riebeckite–aegirine–lepidomelane–astrophyllite gneiss, Vigo, Spain (Floor, 1966). Anal. H. M. I. Bult-Bik (includes P_2O_5 0·06; partial analyses ZrO_2 0·135, Nb_2O_5 0·036, Ta_2O_3 < 0·02 not included in total).
2 Aegirine, segregation in aegirine–riebeckite schist, Gout Creek, south Westland, New Zealand (White, 1962). Anal. A. J. R. White.
3 Aegirine, vein, with quartz and manganosiderite, cutting iron formation, Cuyana Range, Minnesota (Grout, 1946). Anal. L. C. Peck (includes V_2O_3 0·01).
4 Acmite, taconite, iron formation, Cuyuna Range, Minnesota (Grout, 1946). Anal. L. C. Peck (includes V_2O_3 0·04).
5 Aegirine, riebeckite–albite granite, Kigom Hills, Nigeria (Greenwood, 1951). Anal. F. A. Gonyer.

Table 52. Aegirine Analyses – *continued*

	6	7	8	9
SiO_2	52·48	51·35	50·12	51·7
TiO_2	0·57	1·10	1·14	9·4
Al_2O_3	0·96	2·15	2·57	0·72
Fe_2O_3	31·74	28·66	28·74	17·1
FeO	0·93	2·24	1·57	3·4
MnO	0·10	tr.	0·60	0·08
MgO	0·15	0·10	0·59	2·67
CaO	0·28	1·25	1·30	2·07
Na_2O	12·05	12·66	12·24	12·8
K_2O	0·35	0·15	0·19	—
H_2O^+	—	0·12	—	—
H_2O^-	—	0·17	—	—
Total	100·50	99·95	99·73	99·94
α	1·767	—	—	—
β	1·806	1·797	—	—
γ	1·823	—	1·795	—
γ:z	6°	—	—	—
$2V_\alpha$	60°	64°	65°	—
D	3·587	—	—	—

Numbers of ions on the basis of six O

	6		7		8		9	
Si	2·004	2·00	1·975	2·00	1·932	2·00	1·944	2·00[e]
Al	—		0·025		0·068		0·032	
Al	0·043	1·02[b]	0·072	1·01	0·049	1·03[d]	0·000	0·98
Ti	0·016		0·032		0·033		0·265	
Fe^{3+}	0·912		0·826		0·834		0·458	
Mg	0·008		0·006		0·034		0·150	
Fe^{2+}	0·030		0·072		0·051		0·107	
Mn	0·003		—		0·020		0·003	
Na	0·892	0·93[c]	0·941	0·97	0·915	0·98	0·933	1·02
Ca	0·011		0·022		0·054		0·083	
K	0·017		0·007		0·009		—	

[b] Includes Zr 0·007.
[c] Includes Ce 0·006.
[d] Includes Zr 0·012; V 0·001.
[e] Includes Fe^{3+} 0·024.

6 Aegirine, Quincy, Massachusetts (Washington and Merwin, 1927). Anal. H. S. Washington (includes ZrO_2 0·41, $(Ce,Y)_2O_3$ 0·48).

7 Aegirine, cancrinite-albite syenite (mariupolite), Itapirapuã, São Paulo, Brazil (Gomes *et al.*, 1970). Anal. S. L. Moro (trace element data).

8 Aegirine, vein in transitional pulaskite, Kangerdlugssuaq, east Greenland (Kempe and Deer, 1970). Anal. D. R. C. Kempe (includes ZrO_2 0·65, V_2O_3 0·01, BaO 0·01; trace element data; cell parameters *a* 9·645, *b* 8·789, *c* 5·287 Å, β 106·99°).

9 Titanian aegirine, phonolitic segregation in melanocratic leucitite, New South Wales, Australia (Ferguson, 1977). Anal. A. K. Ferguson.

Table 53. Aegirine-augite Analyses

	1	2	3	4	5
SiO_2	49·78	48·82	49·73	52·25	49·97
TiO_2	0·75	0·43	0·82	0·92	0·05
Al_2O_3	2·97	2·60	4·25	2·86	1·84
Fe_2O_3	6·44	10·29	11·95	12·69	12·69
FeO	7·78	9·91	9·87	7·82	11·03
MnO	0·76	2·93	0·38	0·52	0·63
MgO	8·90	4·08	3·82	4·55	3·64
CaO	20·80	16·48	13 30	14·20	15·47
Na_2O	1·76	3·92	5·45	3·76	4·80
K_2O	0·22	0·16	0·40	0·37	—
H_2O^+	0·55	—	0·09	0·06	—
H_2O^-	—	0·25	—	0·03	—
Total	100·71	100·11	100·11	100·03	100·12
α	1·705	—	1·730	1·722	—
β	1·715	—	1·748	1·742	—
γ	1·735	1·750–1·755	1·768	1·760	—
$\gamma:z$	78°	—	76°	—	
$2V_\gamma$	72°	—	90°	80°	—
D	3·35	—	3·516	—	—
	Numbers of ions on the basis of six O				
Si	1·893 } 2·00	1·913 } 2·00	1·915 } 2·00	1·983 } 2·00	1·947 } 2·00
Al	0·107	0·087	0·085	0·017	0·053
Al	0·025 } 1·01	0·033 } $1·01^a$	0·108 } 1·03	0·111 } 1·02	0·031 } 1·00
Ti	0·021	0·013	0·024	0·026	0·001
Fe^{3+}	0·184	0·303	0·346	0·363	0·372
Mg	0·504	0·238	0·219	0·257	0·211
Fe^{2+}	0·247	0 325	0·318	0·248	0·359
Mn	0·024	0·097	0·012	0·017	0·021
Na	0·130 } 0·99	0·298 } 1·00	0·406 } $0·98^b$	0·277 } 0·87	0·363 } 1·01
Ca	0·848	0·692	0·549	0·577	0·646
K	0·010	0·008	0·020	0·018	0·000

[a] Includes Zr 0·004, V 0·001.
[b] Includes Ba 0·001.

1 Aegirine-augite, ijolite pegmatite, Homa Bay area, Kenya (Pulfrey, 1950). Anal. W. P. Horne.
2 Aegirine-augite, vein in foyaite, Kangerdlugssuaq, east Greenland (Kempe and Deer, 1970). Anal. D. R. C. Kempe (includes ZrO_2 0·19, V_2O_3 0·03, SrO 0·01, BaO 0·01; cell parameters *a* 9·741, *b* 8·934, *c* 5·264 Å, *β* 105·69°: trace element data.
3 Aegirine-augite, banded syenite, Iron Hill, Colorado (Larsen, 1942). Anal. I. A. Gonyer (includes BaO 0·05).
4 Aegirine-augite, pyroxene nepheline, syenodiorite, Khamman district, Andhra Pradesh, India (Subbarao, 1971). Anal. K. V. Subbarao.
5 Aegirine-augite, rim around salite, ijolite, Seabrook Lake, Ontario, Canada (Mitchell, 1972). Anal. R. H. Mitchell (microprobe analysis).

Table 53. Aegirine-augite Analyses – *continued*

	6	7	8	9	10
SiO_2	50·44	53·98	51·80	49·75	50·96
TiO_2	0·38	0·00	0·06	0·47	2·70
Al_2O_3	2·20	1·13	3·61	3·07	2·52
Fe_2O_3	13·99	14·82	16·49	17·77	18·53
FeO	7·49	1·15	2·03	5·37	7·56
MnO	0·61	2·90	0·96	1·03	1·01
MgO	5·31	8·79	6·98	3·50	0·31
CaO	13·40	10·51	10·86	10·29	3·46
Na_2O	5·34	6·23	7·01	7·39	10·34
K_2O	0·39	0·02	0·13	0·39	0·51
H_2O^+	0·47	0·00	0·43	0·97	0·36
H_2O^-	0·24	0·24	0·08	0·17	0·04
Total	'100·30'	99·77	100·46	100·17	99·84
α	1·720	1·728	—	1·750	1·737
β	1·740	1·746	1·735	1·777	1·769
γ	1·757	1·756	—	1·794	1·787
$\alpha:z$	14°–15°	[13°]	10°–20°	8°	≃0°
$2V_\alpha$	82°–88°	—	70°–80°	77°	74°
D	—	3·40	—	3·589	3·517

Numbers of ions on the basis of six O

	6		7		8		9		10	
Si	1·944	2·00	2·018	2·02	1·941	2·00	1·935	2·00	1·951	2·00[a]
Al	0·056		0·000		0·059		0·065		0·031	
Al	0·044		0·050		0·100		0·076		0·083	
Ti	0·011		0·000		0·002		0·015		0·078	
Fe^{3+}	0·406	1·03	0·417	1·09	0·465	1·00	0·540	1·04	0·534	1·01[b]
Mg	0·305		0·490		0·340		0·205		0·018	
Fe^{2+}	0·242		0·036		0·064		0·174		0·242	
Mn	0·020		0·092		0·030		0·034		0·033	
Na	0·399		0·421		0·569		0·554		0·768	
Ca	0·553	0·97	0·451	0·87	0·436	1·01	0·427	1·00	0·142	0·94
K	0·018		0·001		0·006		0·019		0·025	

[a] Includes Be 0·018.
[b] Includes Nb 0·018.

6 Aegirine-augite, syenite pegmatite, Ilmen Mountains, U.S.S.R. (Zavaritsky, 1946).

7 Aegirine-augite, specularite–augite–riebeckite–tremolite–calcite–quartz rock, Upper Wabush, Labrador (Klein, 1966). Anal. C. Klein (cell parameters, *a* 9·694, *b* 8·855, *c* 5·260 Å, β 106·85°, *V* 432·22 Å³).

8 Aegirine-augite, alkali–amphibole–garnet–apatite–hematite–albite–quartz schist, Hodono, Bessi district, Japan (Banno, 1959). Anal. H. Haramura (includes P_2O_3 < 0·02).

9 Aegirine-augite, quartz syenitic fenite, Alnö, Sweden (Eckermann, 1974). Anal. R. Blix.

10 Aegirine-augite, banded paragneiss, Seal Lake, Labrador (Nickel and Mark, 1965). Anal. E. Mark (includes Nb_2O_5 1·04, BeO 0·20, F 0·51, O≡F 0·21; cell parameters, *a* 9·681, *b* 8·793, *c* 5·303 Å, β 105°06′).

11		12		13		14		15		
51·68		53·11		52·10		52·04		51·35		SiO_2
1·59		2·57		1·62		0·78		0·65		TiO_2
1·28		1·25		0·57		0·45		2·88		Al_2O_3
19·10		21·73		25·14		24·88		25·43		Fe_2O_3
5·17		1·57		4·29		3·47		3·70		FeO
0·45		0·44		0·33		0·17		0·18		MnO
4·29		3·97		1·05		1·21		0·58		MgO
9·56		5·15		3·62		3·73		3·29		CaO
7·25		10·22		11·12		11·56		11·65		Na_2O
—		0·18		0·01		0·18		0·14		K_2O
0·05		0·27		—		0·62		0·24		H_2O^+
0·05				—		0·28		0·09		H_2O^-
100·47		100·46		99·85		99·37		100·20		Total
1·741		1·742		1·751		1·755		1·751		α
1·767		1·768		1·786		—		1·786		β
1·789		1·787		1·802		1·813		1·800		γ
12°		2°		—		4°		1°		$\alpha:z$
85°		81°		—		64°		65°–73°		$2V_\alpha$
3·561		3·52		—		—		—		D
1·963	2·00	1·992	2·00	2·002	2·00	2·024	2·02	1·968	2·00	Si
0·037		0·008		—		0·000		0·032		Al
0·020		0·047		0·026		0·021		0·098		Al
0·045		0·073		0·047		0·023		0·019		Ti
0·546	1·03	0·613	1·02	0·727	1·01	0·728	0·96	0·734	1·01	Fe^{3+}
0·243		0·222		0·060		0·070		0·033		Mg
0·164		0·049		0·138		0·113		0·119		Fe^{2+}
0·014		0·014		0·011		0·006		0·006		Mn
0·534		0·743		0·829		0·872		0·866		Na
0·389	0·92	0·207	0·96	0·149	0·98	0·155	1·04	0·135	1·01	Ca
—		0·008		0·000		0·009		0·007		K

11 Aegirine-augite, aegirine granulite, Glen Lui, Aberdeenshire (McLachlan, 1951). Anal. G. R. McLachlan.
12 Aegirine-augite, syenite pegmatite dyke, Libby, Montana (Goranson, 1927). Anal. H. E. Vassar.
13 Aegirine-augite vein in shonkinite, Shonkin Sag, Montana (Nash and Wilkinson, 1970).
14 Aegirine-augite, quartz-aegirine pegmatite, Synnirskii pluton, northern Near-Baikal (Kostyuk *et al.*, 1970). Anal. E. S. Guletskaya.
15 Aegirine-augite, nepheline-syenite pegmatite dyke, Assynt district, Scotland (Sabine, 1950). Anal. Geochemical Laboratory, Wembley (includes V_2O_3 0·02).

Table 53. Aegirine-augite Analyses – *continued*

	16		17		18		19	
SiO_2	52·03		52·42		51·72		53·33	
TiO_2	0·81		0·60		1·32		7·94	
Al_2O_3	1·59		0·85		1·56		2·21	
Fe_2O_3	25·64		25·77		26·14		6·71	
FeO	3·01		1·95		2·38		13·21	
MnO	0·27		—		0·21		0·68	
MgO	0·61		2·47		1·41		1·23	
CaO	4·12		2·98		2·56		3·70	
Na_2O	10·96		10·99		11·28		10·74	
K_2O	0·33		0·86		0·34		0·00	
H_2O^+	0·30		2·23		0·50		—	
H_2O^-					—		—	
Total	99·67		'100·12'		99·66		99·75	
α	1·748		1·760		1·757		—	
β	1·776		1·800		1·786		—	
γ	1·795		1·812		1·797		—	
$\alpha:z$	—		—		—		—	
$2V_\alpha$	68°–75°		65°		62°		—	
D	—		3·57		—		—	
Numbers of ions on the basis of six O								
Si	2·002	2·00	2·016	2·02	1·988	2·00	2·026	2·03
Al	0·000		—		0·012		0·000	
Al	0·072		0·039		0·059		0·098	
Ti	0·023		0·017		0·038		0·227	
Fe^{3+}	0·742	0·98	0·746	1·01	0·756	1·02[a]	0·192	1·03
Mg	0·035		0·142		0·081		0·070	
Fe^{2+}	0·097		0·063		0·076		0·420	
Mn	0·009		—		0·007		0·021	
Na	0·818		0·819		0·841		0·794	
Ca	0·170	1·00	0·123	0·98	0·105	0·96	0·152	0·95
K	0·016		0·042		0·016		0·000	

[a] Includes Zr 0·004.

16 Aegirine-aguite, tinguaite dyke, Harohalli, Bangalore district, Mysore State (Ikramuddin and Sadashivaiah, 1966).
17 Aegirine-augite, with authigenic albite, hematite, carbonates, and sulphates, voids in Permian marl, Kumolinsk, Kazakhstan (Beĭseev, 1966).
18 Fibrous aegirine-augite, pegmatite dike in shonkinite, Bearpaw Mountains, Montana (Pecora, 1942). Anal. F. A. Gonyer (includes ZrO_2 0·24).
19 Titanian aegirine-augite, hybrid syenite gneiss, Red River alkaline province, central Labrador, Canada (Curtis and Gittins, 1978).

(neptunite). Thus the composition of anal. 9 (Table 52) can be expressed as $(NaFe^{3+}Si_2O_6)_{49}(Na(Mg,Fe)_{0\cdot5}Ti_{0\cdot5}Si_2O_6)_{51}$.

Acmite and aegirine not infrequently contain small amounts of Zr. The aegirines from the main units and veins of the Kangerdlugssuaq alkaline intrusion, east Greenland (Kempe and Deer, 1970) are relatively rich in zirconium (ZrO_2 ranging from 0·25 to 0·65 weight per cent). The ZrO_2 and HfO_2 contents of three aegirines from the Lovozero massif vary between 0·16 and 0·48 weight per cent, and have a ZrO_2/HfO_2 ratio of 40 (Gerasimovskii *et al.*, 1962). Ce and other rare earth ions are also commonly present in aegirine in trace amounts.

Acmite not infrequently contains small amounts of Zr, and also Ce and other rare earth ions. This feature is less common in aegirine but the suggestion that the brown colour of acmite is due to the presence of Zr and rare earth ions has not been substantiated. Acmites with a high vanadium content have been reported, and two vanadium-rich aegirine-augites from Libby, Montana, have been described (e.g. anal. 7, Table 54). It is unlikely, however, that the brown colour of acmite is due to the presence of vanadium, and Grout (1946) has described a brown acmite and a green aegirine (Table 52, anals. 3, 4) from Cuyuna in which the contents of V_2O_3 are respectively 0·01 and 0·04 per cent. Aegirines with higher vanadium contents have been reported, one containing 8·18 weight per cent V_2O_3 (the trivalency of vanadium was determined by X-ray photoelectron spectroscopy) from the Tanohata mine, Iwate Prefecture, Japan (Nakai *et al.*, 1976), and others (Table 54, anal. 8) containing between 11·76 and 13·84 weight per cent V_2O_5 from an aegirine–sphene–graphite–pyrite quartzite, Mendocino County, California (Wood, 1977). A high manganese content has also been suggested as the cause of the difference in colour between acmite and aegirine, thus the manganoan aegirine-augite, blanfordite (Table 54, anal. 3), from Chikla (Bilgrami, 1956), is brown to deep lavender in colour, and the Cuyuna acmite contains appreciably more manganese than the aegirine from the same locality.

Compositional zoning is very common in aegirine and aegirine-augite; the earlier formed crystals are generally richer in the augitic component and the rims richer in the $NaFe^{3+}Si_2O_6$ molecule. Some aegirines and aegirine-augites show a reaction relationship with the residual magmatic liquids as evidenced by the frequent mantling and, in some examples, the complete replacement of the pyroxene by a sodium- and iron-rich amphibole.

In the syenite of the Morotu district, Japan, the clinopyroxenes are strongly zoned and consist of a pale-coloured core of sodian augite (Fe_2O_3 4·60, Na_2O 1·91) mantled by a deep greenish aegirine-augite (Fe_2O_3 12·75, Na_2O 4·92 weight per cent).

The scandian analogue of aegirine, $NaScSi_2O_6$, and compositions between $NaScSi_2O_6$ and $NaFeSi_2O_6$ with ratios $(Sc_{0\cdot75}Fe_{0\cdot25})$, $(Sc_{0\cdot50}Fe_{0\cdot50})$ and $(Sc_{0\cdot25}Fe_{0\cdot75})$ have also been synthesized (Ito and Frondel, 1968). Scandium in appreciable amounts, however, is not generally present in natural sodium-rich pyroxenes. Neumann (1961) reports 20 p.p.m. in an acmite from Rundemyr, Norway, while six aegirines from the Kangerdlugssuaq alkaline intrusion have scandium contents between 5 and 60 p.p.m.; values for Ga, Li, Nb, Ni, Cu, V (71 to 415 p.p.m.), Y, Yb, Sr, La, Ba and Rb are also given for these pyroxenes. Cadmium-, cobalt-, nickel-, zinc- and zirconium-bearing acmites (Schüller, 1958), indium aegirine end-members of the $NaFe^{3+}Si_2O_6$–$NaInSi_2O_6$ solid solution series (Christensen and Hazell, 1967), and the sodium ferric iron metagermanate, $NaFe^{3+}Ge_2O_6$ (Soloveva and Bakakin, 1968) have been synthesized.

The common alteration products of aegirine, and aegirine-augite include chlorite,

Table 54. Manganoan-, zincian- and vanadian-rich aegirine and aegirine-augite analyses

	1	2	3	4
SiO_2	52·51	50·85	52·52	51·35
TiO_2	0·00	—	0·59	—
Al_2O_3	0·25	0·20	5·88	4·29
Fe_2O_3	22·28	29·46	18·15	18·29
V_2O_3	—	—	—	—
FeO	1·01	1·98	1·11	—
MnO	6·02	4·30	4·72	3·00
ZnO	—	—	—	—
MgO	3·85	0·81	3·26	4·74
CaO	4·79	1·73	4·38	8·41
Na_2O	8·99	10·38	9·75	9·28
K_2O	0·02	0·07	0·00	—
H_2O^+	—	—	—	0·82
H_2O^-	0·16	0·09	—	
Total	99·88	99·87	100·36	100·18
α	1·749	1·768	1·732	1·760
β	1·770	1·797	1·756	1·768
γ	1·781	1·815	1·770	1·780
$\alpha:z$	1°–2°	0°–1°	9°	—
$2V_\alpha$	—	—	80°	—
D	3·51	3·59	3·28	—
α	yellow	dark brown	rose-pink	pink
β	yellow-brown	brownish yellow	pale blue	pale pink
γ	pale yellow-brown	pale yellow-brown	lilac-blue	blue

Numbers of ions on the basis of six O

	1		2		3		4	
Si	2·018	2·02	1·984	1·99	1·960	2·00	1·927	2·00
Al	0·000		0·009		0·040		0·073	
Al	0·011		0·000		0·218		0·111	
Ti	0·000		0·000		0·017		—	
Fe^{3+}	0·644		0·865		0·510		0·526	
V^{3+}	0·000		0·000		—		—	
Mg	0·220	1·10	0·047	1·12	0·181	1·11	0·270	1·00
Fe^{2+}	0·032		0·065		0·034		—	
Zn	0·000		0·000		—		—	
Mn	0·196		0·142		0·149		0·097	
Na	0·670		0·785		0·706		0·689	
Ca	0·197	0·87	0·072	0·86	0·175	0·89	0·345	1·03
K	0·001		0·003		—		—	

1 Manganoan aegirine-augite, rhodonite–aegirine-augite–specularite–calcian rhodochrosite assemblage, Wabush district, Labrador (Klein, 1966). Anal. J. Ito (*a* 9·689, *b* 8·843, *c* 5·286 Å, β 107·26°, *V* 432·51 Å^3).

2 Manganoan aegirine, rhodonite–rhodochrosite–specularite–calderite assemblage, Wabush District, Labrador (Klein, 1966). Anal. J. Ito (*a* 9·676, *b* 8·818, *c* 5·285 Å, β 107·37°, *V* 430·44 Å^3).

3 Manganoan aegirine-augite (blanfordite), pegmatite contaminated by manganese ore, Chikla, Bhandara district, India (Bilgrami, 1956). Anal. R. K. Phillips.

4 Manganoan aegirine-augite (blandfordite), pegmatite, Ponia, India (Kilpady, 1960).

5	6	7	8	
52·86	46·38	51·91	54·87	SiO_2
0·75	0·10	0·91	—	TiO_2
1·43	1·36	0·38	3·62	Al_2O_3
27·61	12·90	21·79	—	Fe_2O_3
—	—	3·98	14·45[a]	V_2O_3
—	1·35	1·48	12·27	FeO
2·00	7·18	0·58	—	MnO
—	8·77	—	—	ZnO
1·47	1·18	3·08	—	MgO
1·57	14·86	5·53	1·05	CaO
11·94	4·99	10·46	13·44	Na_2O
0·15	0·18	0·22	—	K_2O
0·07	0·55	0·06 (H₂O⁺ + H₂O⁻)	—	H_2O^+
0·02	0·25		—	H_2O^-
99·91	100·05	100·51	99·89	**Total**
1·748	1·741	1·745	—	α
1·773	1·751	1·770	—	β
1·788	1·774	1·782	0·026[c]	γ
6°	≃ 60° (γ)	1–4°	8°	$\alpha:z$
61°	—	69°	—	$2V_\alpha$
3·516	3·59	3·55	—	D
rose-pink	brown	—	pale-yellow-green	α
violet	—	—	orange	β
blue	yellow-brown	—	red-brown	γ

5	Σ	6	Σ	7	Σ	8	Σ	
1·992	2·00	1·904	1·97	1·974	1·99	2·02	2·02	Si
0·008		0·066		0·016		—		Al
0·057	0·99[b]	0·000	1·04	0·000	1·01	0·16	0·97[d]	Al
0·022		0·003		0·026		—		Ti
0·778		0·398		0·624		0·06		Fe^{3+}
—		—		0·121		0·43		V^{3+}
0·085		0·072		0·175		—		Mg
—		0·046		0·047		0·32		Fe^{2+}
—		0·266		—		—		Zn
0·002		0·250		0·019		—		Mn
0·880	0·95	0·397	1·06	0·772	1·01	0·96	1·00	Na
0·062		0·654		0·225		0·04		Ca
0·007		0·009		0·010		—		K

[a] V_2O_5.
[b] Includes Mn^{3+} 0·050.
[c] Birefringence.
[d] Includes Cr 0·004.

5 Manganoan aegirine, banded granular rock with microcline, manganarfvedsonite (jadeite), manganchlorite and braunite, north Tirodi mine, Goldongri, India (Nayak and Neuvonen, 1963). Anal. P. Väänänen. (Manganese given as Mn_2O_5 1·91, MnO 0·09, includes P_2O_5 0·04: *a* 9·64, *b* 8·78, *c* 5·28 Å, β 106°57′, *V* 427·5 Å³).
6 Zincian and manganoan aegirine-augite, skarn, Franklin, New Jersey (Frondel and Ito, 1966). Anal. J. Ito.
7 Vanadian aegirine-augite, quartz vein, Rainy Creek, Libby, Montana (Larsen and Hunt, 1914). Includes S 0·13.
8 Vanadian aegirine, aegirine–sphene–graphite–pyrite quartzite, Laytonville quarry, Mendocino County, California (Wood, 1977). Anal. R. M. Wood (microprobe anal. includes Cr_2O_3 0·19).

epidote, hematite and limonite. The alteration of the manganoan aegirine, blanfordite, to limonite has been reported by Bilgrami (1956).

Experimental

Aegirine is readily synthesized from the appropriate molecular proportions of SiO_2, Fe_2O_3 and $Na_2CO_3.H_2O$ fused with NaCl (Washington and Merwin, 1927). Bowen and Schairer (1929) showed that at one atmosphere pressure, aegirine melts incongruently at 990°C with the formation of hematite and liquid. Subsequently Bailey and Schairer (1966) have demonstrated that there is a small melting interval of 975°–988° ± 5°C. The effect of pressure on the incongruent melting of aegirine and the liquidus for $NaFe^{3+}Si_2O_6$ composition has been investigated by Gilbert (1967, 1969). The incongruent melting of aegirine to hematite (+magnetite) and liquid persists to pressures of at least 45 kbar according to the equation:

$$T(°C) = 988 + 20{\cdot}87P(\text{kbars}) - 0{\cdot}155\,P^2$$

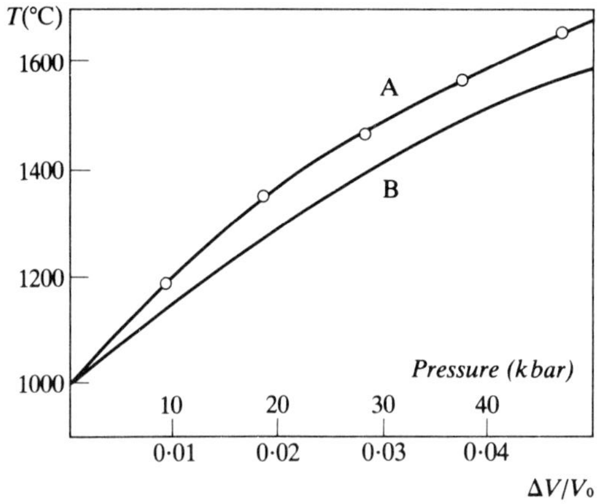

Fig. 227. Melting temperature of acmite versus compression at room temperature (curve A). Aegirine incongruent melting curve, B (after Gilbert, 1968b, 1969).

The initial slope of the $NaFe^{3+}Si_2O_6$ melting curve is 20°C/kbar (Fig. 227), under conditions of f_{O_2} controlled by hematite + magnetite buffer, and is steeper than that of other silicates so far investigated; from the Clapeyron relation, $dT/dP = \Delta V/\Delta S$, either ΔV is unusually large, or ΔS unusually small. The former is unlikely as aegirine melts to relatively dense oxides and liquid, and a small ΔS is favoured by Gilbert who considers that it may be due to the low entropy of the oxides, or to a similar co-ordination of the ions in the liquid and the acmite. The melting curve is very sensitive to oxygen pressure, and at low P_{H_2O} the temperature of the incongruent melting is lowered by 250 to 300°C over the range to 30 kbar. Gilbert (1968b, 1969) has investigated the temperature–compression melting relations of acmite and shown that $\Delta V/V_0$ and melting temperature relationships are non-linear but similar in form to the fusion curve (Fig. 227).

The stability of aegirine in the presence of water has been investigated under conditions of controlled partial pressures of oxygen provided by magnetite–hematite and quartz–fayalite–magnetite buffers at fluid (= total) pressures between 2 and 5 kbar (Bailey, 1969). The incongruent melting, as shown by Gilbert, is very sensitive to the oxygen fugacity but relatively insensitive to total pressure. The melting curves have low dT/dP values, under hematite–magnetite conditions

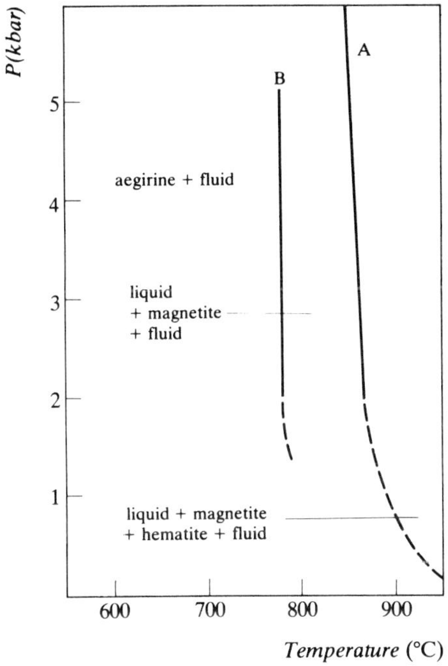

Fig. 228. $P_{(\text{Total})}$**–*T*–diagram for aegirine + excess water. A, magnetite–hematite buffer, B, quartz–fayalite–magnetite buffer (after Bailey, 1969).**

−5°C/kbar (Fig. 228, curve A). With lower partial pressures of oxygen the temperature of melting is even less dependent on total pressure, and under conditions controlled by the quartz–fayalite–magnetite buffer dT/dP is $\simeq$ 0°C/kbar (curve B). Reduced partial pressure of oxygen, however, leads to a marked lowering of the high temperature stability of aegirine, and the reduction of the melting temperature between the conditions of the magnetite–hematite and quartz–fayalite–magnetite buffers is approximately 100°C. Under oxidation conditions controlled by iron–wüstite and wüstite–magnetite buffers, aegirine ceases to be stable and is replaced by an arfvedsonite–riebeckite solid solution:

$$\underset{\text{aegirine}}{20NaFe^{3+}Si_2O_6} + 4H_2O = \underset{\text{riebeckite}}{4Na_2Fe_3^{2+}Fe_2^{3+}Si_8O_{22}(OH)_2} + \\ + 4(Na_2O.SiO_2] + 2(Na_2O.2SiO_2) + 3O_2 \quad (1)$$

$$\underset{\text{aegirine}}{10NaFe^{3+}Si_2O_6} + 2H_2O = \underset{\text{arfvedsonite}}{2Na_2Fe_4^{2+}Fe^{3+}Si_8O_{22}(OH)_2} + \\ + 2(Na_2O.2SiO_2) + 2O_2 \quad (2)$$

the sodium metasilicate and sodium disilicate forming a glass.

Aegirine has been reported as a by-product of quartz synthesis in the system SiO_2–NaOH–Na_2CO_3–H_2O at pressures between 400 and 1000 atm and temperatures of 250° to 350°C (Nosyrev *et al.*, 1969). In one sample the pyroxene contained large quantities of nickel (10·22 weight per cent NiO) and chromium (5·46 per cent Cr_2O_3) and originated by reaction of the melt with the autoclave metal. Inclusions of aegirine in the pyramidal faces of synthetic quartz have been described by Tsyganov and Novozhilova (1966).

Newton and Smith (1967) have shown that aegirine forms a nearly ideal solid solution series with jadeite. A hydrothermal investigation of the system $NaFeSi_3O_8$ (acmite + quartz)–$NaAlSi_3O_8$ (jadeite + quartz) by Popp and Gilbert (1972) has defined the stability fields of clinopyroxene + quartz and clinopyroxene + albite + quartz. The solubility of jadeite in the Fe^{3+}Al-clinopyroxene decreases with pressure and at 600°C and 4 kbar is between 4 and 5 mol. per cent jadeite, while at 500°C and the same pressure the solubility is increased to between 5 and 6 per cent. Curves for the ideal system $NaFeSi_3O_8$–$NaAlSi_3O_8$ at 600°, 500°, 400° and 300°C, assuming no solution of iron in albite, and the curve at 600°C for 5 per cent substitution of Fe^{3+} for Al in albite assuming ideal mixing in all four tetrahedral

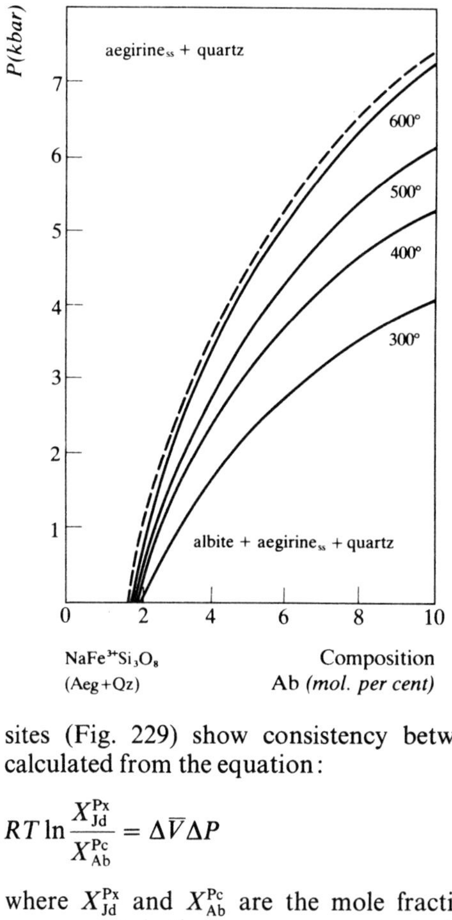

Fig. 229. The iron-rich portion of the join $NaFeSi_3O_8$–$NaAlSi_3O_8$ at 300°, 400°, 500° and 600°C. Broken line, ideal phase relations at 600°C, assuming 5 mol. per cent substitution of Fe^{3+} for Al in albite, $X_{Ab}^{Pc} = 0{\cdot}95$ (after Popp and Gilbert, 1972).

sites (Fig. 229) show consistency between the experimental data and values calculated from the equation:

$$RT \ln \frac{X_{Jd}^{Px}}{X_{Ab}^{Pc}} = \Delta \bar{V} \Delta P$$

where X_{Jd}^{Px} and X_{Ab}^{Pc} are the mole fractions of the $NaAlSi_2O_6$ and $NaAlSi_3O_8$ components in the pyroxene and plagioclase respectively, $\Delta \bar{V}$ the molar volume of the pure jadeite–pure albite reaction assuming pressure independence, ΔP the pressure difference between pure jadeite–pure albite equilibrium and the given solid solution reaction.

On the basis of a study of the clinopyroxenes from the alkaline rocks of Japan and Pantelleria, Aoki (1964) considered that the calcium- and sodium-rich clinopyroxenes are separated by a region of immiscibility. This conclusion, however, is not substantiated by the experimental data or by later investigations of natural aegirines and aegirine-augites (King, 1965; Yagi, 1966; Perchuk, 1966; Klein, 1966; see also section on petrogenesis).

$NaFe^{3+}Si_2O_6$–$CaMgSi_2O_6$ solid solutions melt incongruently at one atmosphere, and the incongruent melting range shows a gradual decrease as the diopside content increases. A number of aegirine–diopsides have been synthesized by Ostrovsky (1946) and Yagi (1958) and the latter author (1966) determined the equilibrium relations in the acmite–diopside system at 1 atm. The melting relation for the compositions, $\simeq Ac_{89}$, Ac_{75}, Ac_{67}, Ac_{57} and Ac_{33} at 10 kbar has been investigated by Cassie (1971), and at this pressure shows that the range of incongruent melting is considerably smaller (Fig. 230). The melting behaviour of $Ac_{67}Di_{33}$, at pressures up to 40 kbar, is shown in Fig. 231. Between 30 and 40 kbar the

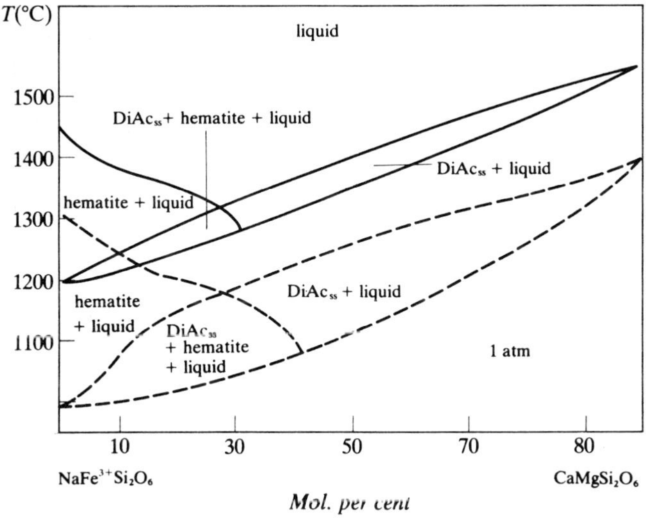

Fig. 230. Melting curves for $NaFeSi_2O_6$–$CaMgSi_2O_6$, broken lines at 1 atm; full lines 10 kbar (after Cassie, 1971).

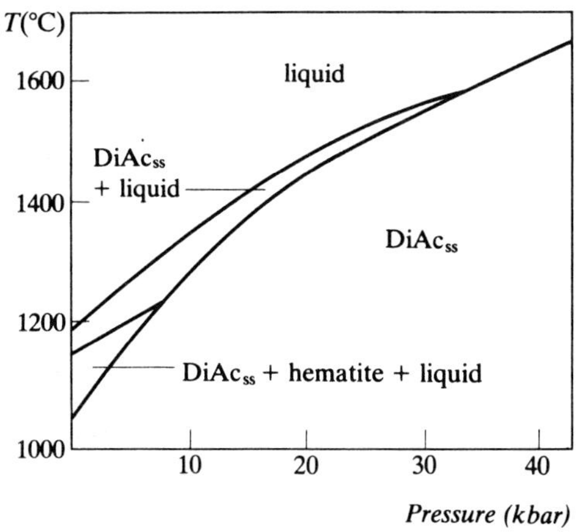

Fig. 231. Pressure–temperature plane for diopside–aegirine solid solution, $Ac_{67}Di_{33}$ (after Cassie, 1971).

incongruent melting-interval is either very small or melting becomes congruent. X-ray examination of these pyroxenes shows some discrepancies from the determinative curves of Nolan and Edgar (1963) which are possibly due to some departure from strictly $NaFe^{3+}Si_2O_6$–$CaMgSi_2O_6$ solid solution. The pyroxenes were found to contain small amounts of glass filaments that may consist of $CaFe_2^{3+}SiO_6$ and enstatite components in the pyroxene solid solutions as a product of a subsolidus reaction:

$$\underbrace{4NaFe^{3+}Si_2O_6 + 2CaMgSi_2O_6}_{Ac_{67}Di_{33}} \rightarrow$$

$$\underbrace{2NaFe^{3+}Si_2O_6 + CaMgSi_2O_6 + CaFe_2^{3+}SiO_6 + MgSiO_3}_{\text{pyroxene}_{ss}} + Na_2Si_2O_5 + 2SiO_2$$

The phase relations on the pseudobinary joins, $NaFe^{3+}Si_2O_6$–$NaTiFe^{3+}SiO_6$ and $NaFe^{3+}Si_2O_6$–$NaTiAlSiO_6$ at 1 kbar P_{H_2O} and oxygen fugacities controlled by the Mn_2O_3–Mn_3O_4 buffer have been determined by Flower (1974). The solid solutions with maximum contents of $NaTiFe^{3+}SiO_6$ (28 mol. per cent at 700°C) and $NaTiAlSiO_6$ (15 mol. per cent at 700°C) melt incongruently over intervals of $\simeq$ 100° and 130°C with melting beginning at 775°, and 745°C respectively. The data also show that the entry of Ti in aegirine is favoured by high temperature, is affected by the activity of silica, and is particularly influenced by high oxygen fugacity.

Nolan and Edgar (1963) synthesized a series of aegirine–diopside solid solutions at 1000 kg/cm^2 water vapour pressure, and Nolan (1966) has determined the phase relations for the systems $NaFeSi_2O_6$–$NaAlSiO_4$–H_2O, $NaFeSi_2O_6$–$NaAlSi_3O_8$–H_2O, $NaFeSi_2O_6$–$NaAlSiO_4$–$NaAlSi_3O_8$–H_2O, and for pyroxene compositional planes within the system $NaFeSi_2O_6$–$CaMgSi_2O_6$–$NaAlSiO_4$–$NaAlSi_3O_8$–H_2O at the same water vapour pressure (Fig. 232). The low melting point in the system, P, is

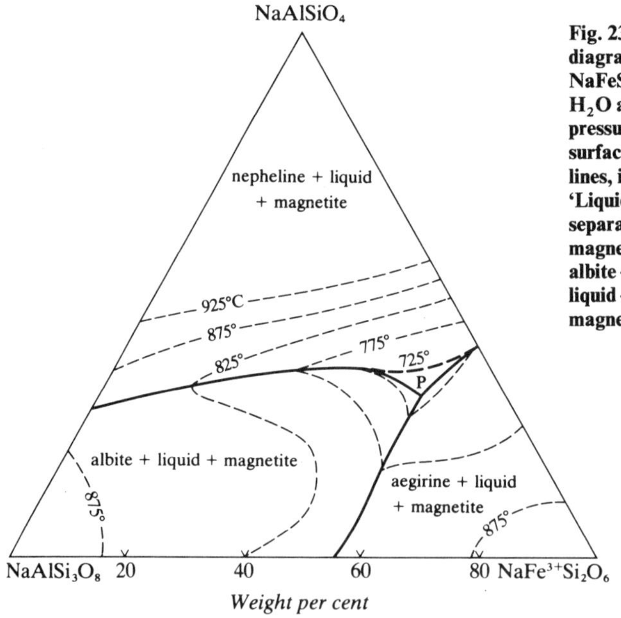

Fig. 232. Isobaric equilibrium diagram for part of the system $NaFeSi_2O_6$–$NaAlSi_3O_8$–$NaAlSiO_4$–H_2O at 1000 kg/cm^2 water vapour pressure. Full lines, traces of boundary surfaces and phase volumes. Dashed lines, isotherms in the 'liquidus' surface. 'Liquidus' denotes the boundary surface separating the phase volume liquid + magnetite from the phase volumes albite + liquid + magnetite, nepheline + liquid + magnetite and aegirine + magnetite + liquid (after Nolan, 1966).

located close to the composition $Ac_{55}Ne_{30}Ab_{15}$ at a temperature of 715°C (this is not a ternary invariant point but a piercing-point of a univariant line, with the five-component system meeting the plane $NaFeSi_2O_6$–$NaAlSiO_4$–$NaAlSi_3O_8$–H_2O where the crystalline phases, aegirine, nepheline, albite and magnetite, and a liquid and gas phase coexist in equilibrium). The introduction of small amounts of $CaMgSi_2O_6$ into the system produced a marked effect on the phase volumes. Thus in the compositional plane $Ac_{95}Di_5$ the boundary curve between the pyroxene and feldspar shifts by about 20 weight per cent feldspar in the direction of the albite–nepheline join. With the introduction of further increments of $CaMgSi_2O_6$ the boundary curve continues to move towards the albite–nepheline sideline but by

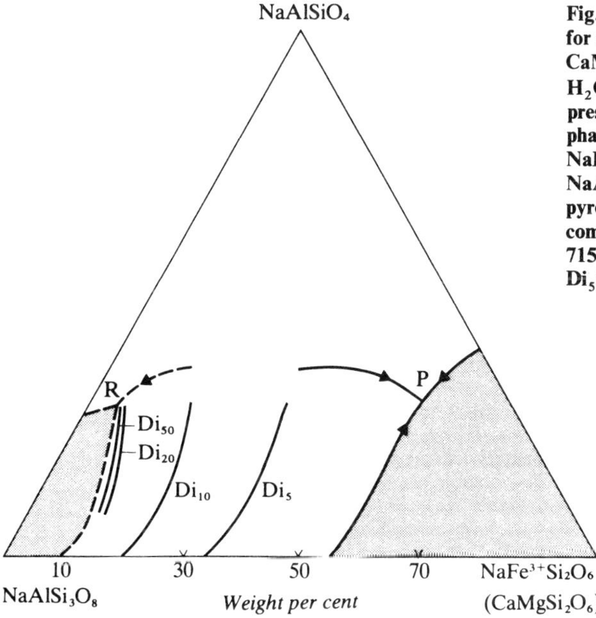

Fig. 233. Composite phase diagram for a portion of the system $NaFeSi_2O_6$–$CaMgSi_2O_6$–$NaAlSiO_4$–$NaAlSi_3O_8$–H_2O at 1000 kg/cm^2 water vapour pressure. Stippled area (right): plagioclase phase volume determined in the system $NaFeSi_2O_6$–$CaMgSi_2O_6$–$NaAlSiO_4$–$NaAlSi_3O_8$–H_2O. Stippled area (left), pyroxene phase volume. P piercing point, composition $Ac_{55}Ne_{30}Ab_{15}$, temperature 715°C. R, reaction and piercing point, $Di_5Ne_{29}Ab_{66}$ (after Nolan, 1966).

diminishing amounts (Fig. 233).

The effect of adding 4·5 and 8·3 per cent $NaFe^{3+}Si_2O_6$ to the system $NaAlSi_3O_8$–$KAlSi_3O_8$–SiO_2–H_2O at a P_{H_2O} of 1000 kg/cm^2 was investigated by Carmichael and MacKenzie (1963). Aegirine does not crystallize from the system or enter into solid solution with the other phases but the addition of $NaFe^{3+}Si_2O_6$ and sodium metasilicate leads to a progressive shift in the minima of the liquidus surface towards the $KAlSi_3O_8$–SiO_2 join on the $NaAlSi_3O_8$–$KAlSi_3O_8$–SiO_2 plane.

Aegirine is one of the ternary compounds in the system $Na_2O.SiO_2$–Fe_2O_3–SiO_2 (Fig. 234), the low temperature region of which is rich in silica and alkali, and

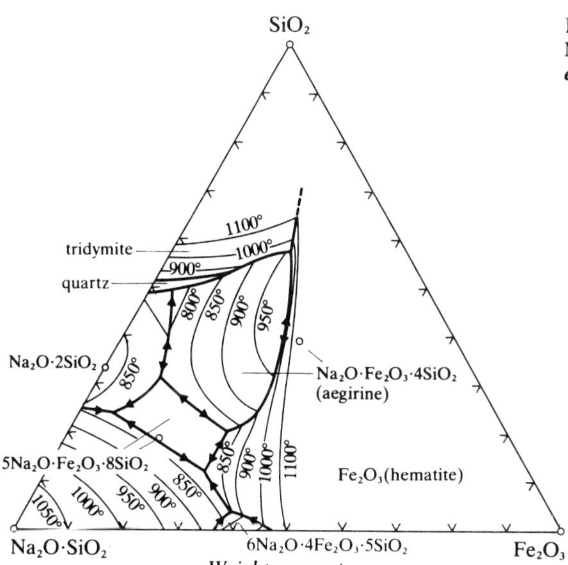

Fig. 234. Phase diagram of the system $Na_2O.SiO_2$–Fe_2O_3–SiO_2 (after Bowen *et al.*, 1930).

relatively poor in iron (Bowen *et al.*, 1930). Bailey and Schairer (1966) have studied the system Na_2O–Al_2O_3–Fe_2O_3–SiO_2 at one atmosphere, and have presented liquidus diagrams for the following joins in the quarternary system: aegirine–jadeite $Na_2O.4SiO_2$, aegirine–nepheline–silica, aegirine–nepheline–sodium disilicate (Fig. 235), aegirine–albite–sodium disilicate, aegirine–nepheline–

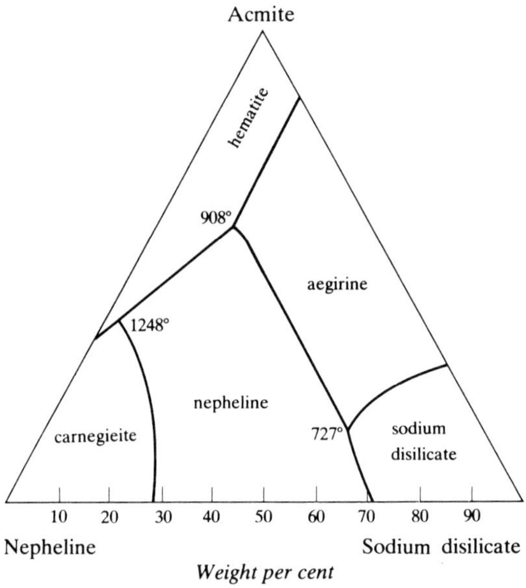

Fig. 235. Liquidus diagram for the join, aegirine–nepheline–sodium disilicate in the system Na_2O–Al_2O_3–Fe_2O_3–SiO_2 (after Bailey and Schairer, 1966).

$5Na_2O.Fe_2O_3.8SiO_2$ and aegirine–nepheline–($Na_2O.4SiO_2$). The outstanding feature of the system is that aegirine crystallization occurs in equilibrium with a sodium silicate normative liquid in association with both nepheline + albite and albite + quartz. Aegirine does not appear on the liquidus of the join aegirine–nepheline–silica, and the relationships generally are dominated by the incongruent melting behaviour of the pyroxene. A large primary hematite phase region is a characteristic feature of the joins in the Na_2O–Al_2O_3–Fe_2O_3–SiO_2 system.

The formation of aegirine, rather than arfvedsonite, is favoured by higher partial oxygen pressures, lower P_{H_2O} and higher temperature: e.g.

$$Na_3Fe_4^{2+}Fe^{3+}Si_8O_{22}(OH)_2 + Na_2O + 2SiO_2 + O_2 = 5NaFe^{3+}Si_2O_6 + H_2O$$

or

$$Na_3Fe_4^{2+}Fe^{3+}Si_8O_{22}(OH)_2 + Na_2O + 2SiO_2 + H_2O = 5NaFe^{3+}Si_2O_6 + 2H_2$$

An explanation of the crystallization trend in the system aegirine–hedenbergite–diopside, based on the effect of the acid-base interaction of the components assuming that the relative basicity of the components is Ac > Di > Hd has been put forward by Perchuk (1964). Increased alkalinity causes an increase in the activities of all the bases but primarily of CaO and MgO, and favours a shift to the right in the reaction $Fe^{2+} \rightarrow Fe^{3+} + e^-$, or using the equation for the melt:

$$[2Fe^{2+} + 2O^{2-}] + [2Na^+ + O^{2-}] = [2Fe^{3+} + 3O^{2-}] + [2Na^+ + 2e^-]$$

As the alkalinity rises the activity of MgO, being a stronger base, increases more rapidly than that of FeO, and the relationship for the solution and melt respectively may be expressed:

$$\frac{\partial a_{MgO}}{\partial pH} > \frac{\partial a_{FeO}}{\partial pH}; \quad \frac{\partial a_{MgO}}{\partial a_{O^{2-}}} > \frac{\partial a_{FeO}}{\partial a_{O^{2-}}}$$

Thus the crystallization of the diopside component will be enhanced, the Fe^{2+} concentration increased in the melt, and the composition of the solid phases displaced towards the aegirine–hedenbergite join in the system aegirine–hedenbergite–diopside. The stability fields of aegirine and aenigmatite in relation to temperature and oxygen fugacity, with reference to their occurrence in the Granitberg fayaite ring complex, South-West Africa, are discussed by Marsh (1975).

Optical and Physical Properties

The correlation between the optical properties and chemical composition of aegirine and aegirine-augite has been investigated by Larsen (1942), Sabine (1950), Winchell (1951), Tröger (1952) and Kostyuk (1965). The optical properties of synthetic aegirine given by Washington and Merwin (1927) are α 1·776, β 1·819, γ 1·836, $2V_\alpha$ 61°, $\alpha{:}z$ 8°–10°. The optical properties of synthetic aegirine-diopside solid solutions have been studied by Ostrovsky (1946), Yagi (1958, 1966) and Nolan (1969). The latter author has also presented data for the aegirine–hedenbergite and aegirine–diopside–hedenbergite systems (see also Perchuk, 1966).

The negative optic axial angle of $NaFe^{3+}Si_2O_6$ is approximately 60°; with replacement of Na by Ca and Fe^{3+} by (Mg,Fe^{2+}) the angle becomes larger and in the compositional range $Fe^{3+}_{0\cdot35}$ to $Fe^{3+}_{0\cdot40}$ $2V \simeq 90°$. With further replacement of (Na,Fe^{3+}) by (Ca,Mg,Fe^{2+}) the γ direction becomes the acute bisectrix, and the more augite-rich members of the series are optically positive. The extinction angle $\alpha{:}z$ (in obtuse angle β) is 9° for the pure $NaFe^{3+}Si_2O_6$ end-member, the angle decreases to zero at a composition of approximately $Fe^{3+}_{0\cdot75}$, and with further replacement of Fe^{3+} by (Mg,Fe^{2+}) the extinction angle (in acute angle β) increases to 20° at a composition of about $Fe^{3+}_{0\cdot2}$. The precise determination of $\alpha{:}[001]$ and optic axial angle in aegirine-augites may be difficult because of the high dispersion and strong absorption. An alternative procedure has been suggested by King (1962) and involves the measurement of the angle between an optic axis and [001]. A further method, involving the measurement of the extinction angle $\alpha' \wedge (110)$, using a three-axis universal stage, has been described by Santos (1973).

The refractive indices, optic axial angles and extinction angles of the minerals in Tables 52 and 53 are plotted in Fig. 236 against the number of Fe^{3+} ions on the basis of six oxygens. The refractive indices are shown as varying linearly with the number of Fe^{3+} ions. The departure of individual minerals from this relationship may be due in part to the variable $Mg{:}Fe^{2+}$ ratios of the natural pyroxenes, as well as to the non-linearity of the relationship between composition and optical properties shown by members of the synthetic aegirine–diopside–hedenbergite pyroxenes (see below). Sabine (1950) has expressed the relationship between composition and optical properties on a triangular diagram using $(Na+K) = 2$, $(Fe^{3+}+V+Ti+Al^{[6]}) = 2$, and $Mg+Fe^{2+}+Zr+Mn+Ca) = 2$ and a diagram showing the relationship between chemical composition and α refractive index, optic axial angle (Fig. 237) and extinction angle for pyroxenes consisting essentially of diopside, hedenbergite and aegirine components has been constructed by Kostyuk (1965).

Yagi (1966) has determined the refractive indices of glasses for ten weight per cent

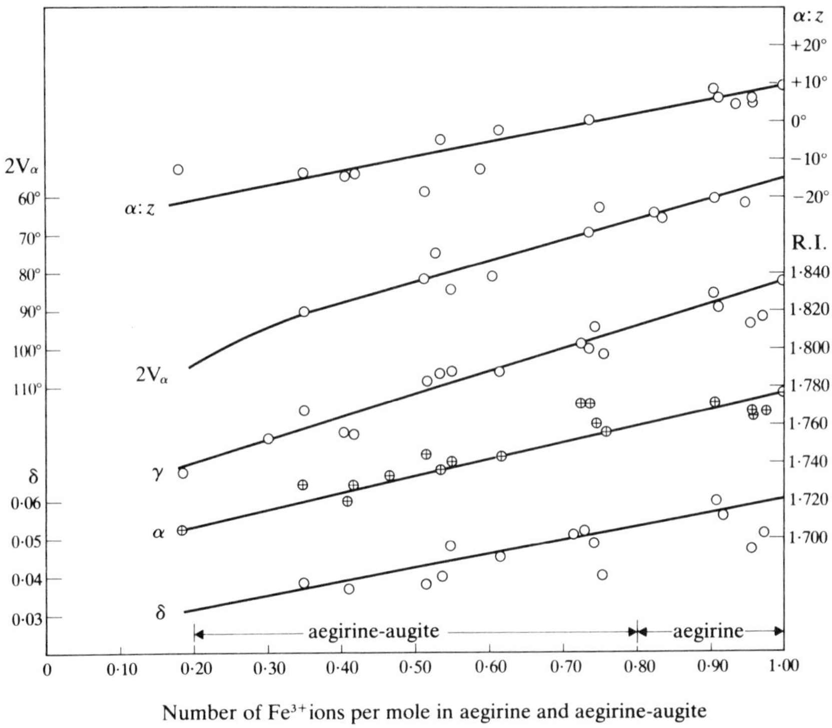

Fig. 236. The relation between the optical properties and the number of Fe^{3+} ions per mole in aegirine and aegirine augite.

composition intervals between the two end-members in the system aegirine (*n* 1·643)–diopside (*n* 1·607). The variation is linear for compositions between $Ac_{100}Di_0$ and $Ac_{50}Di_{50}$ and between $Ac_{30}Di_{70}$ and Ac_0Di_{100}. The optical properties of synthetic aegirine–diopside solid solutions for the same compositional intervals have also been measured by Yagi, and show only small differences from the earlier values presented by Ostrovsky (1946) for four compositions, ≃ 6·5, 13·6, 20·9 and 45·7 mol. per cent acmite. The α refractive index for a number of synthetic pyroxenes (Table 55) on the aegirine–diopside and aegirine–hedenbergite joins, and for twenty-two compositions in the ternary system aegirine–diopside–hedenbergites are given by Nolan (1969). Yagi's data show that the variations of the α and γ indices, extinction angle and optic axial angle are not linear, the non-linearity being most

Table 55 Values of α refractive index for some synthetic pyroxenes on the joins aegirine–hedenbergite, aegirine–diopside and in the system aegirine–diopside–hedenbergite (after Nolan, 1969)

Composition	mol. per cent	α	Composition	mol. per cent		α
Ac_{100}	Hd_0	1·776	$Ac_{80·02}$	$Di_{10·67}$	$Hd_{9·31}$	1·759
$Ac_{90·63}$	$Hd_{9·37}$	1·768	$Ac_{70·02}$	$Di_{16·01}$	$Hd_{13·97}$	1·751
$Ac_{51·76}$	$Hd_{48·24}$	1·749	$Ac_{60·03}$	$Di_{21·34}$	$Hd_{18·63}$	1·747
$Ac_{31·52}$	$Hd_{68·48}$	1·742	$Ac_{59·97}$	$Di_{26·77}$	$Hd_{23·26}$	1·737
$Ac_{10·66}$	$Hd_{89·34}$	1·736	$Ac_{39·49}$	$Di_{42·12}$	$Hd_{18·39}$	1·723
$Ac_{78·95}$	$Di_{21·05}$	1·754	$Ac_{29·04}$	$Di_{61·95}$	$Hd_{9·01}$	1·705
$Ac_{58·44}$	$Di_{41·56}$	1·732	$Ac_{20·57}$	$Di_{21·95}$	$Hd_{57·48}$	1·728
$Ac_{28·06}$	$Di_{71·34}$	1·697	$Ac_{20·02}$	$Di_{42·70}$	$Hd_{37·28}$	1·716
$Ac_{9·43}$	$Di_{90·57}$	1·676	$Ac_{14·47}$	$Di_{72·05}$	$Hd_{13·48}$	1·694

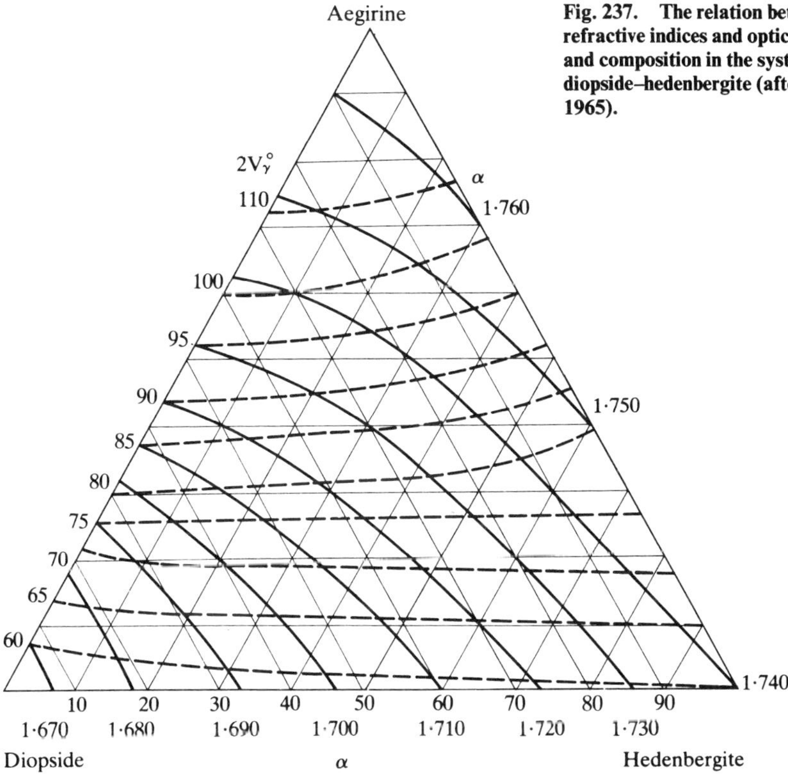

Fig. 237. The relation between α refractive indices and optic axial angles and composition in the system aegirine–diopside–hedenbergite (after Kostyuk, 1965).

pronounced for compositions between ~ Ac_{50}–Ac_{30}, and a similar feature is shown by the α indices given by Nolan. Although less marked the variation of the α index with composition for the aegirine–hedenbergite synthetic pyroxenes also shows some departure from linearity.

The variation of the α index and *b* cell parameters for synthetic pyroxenes in the system aegirine–diopside–hedenbergite (Nolan, 1969) are given in Fig. 226, p. 486. Reasonable estimates of the compositions of natural pyroxenes with low contents of aluminium and titanium may be obtained by the use of this diagram.

Many aegirines and aegirine-augites show conspicuous colour zoning (Frontispiece Fig. E) and variable pleochroism and extinction angles, e.g. the aegirine-augites in the ijolites and melteigites of Homa Bay, Kenya (Pulfrey, 1950), show simple and multiple colour zoning, and between the lighter and darker components of the crystals there are differences in extinction angles of up to 17°. The zoning, which may be of hourglass pattern, is normally from a more augite-rich core to a more aegirine-rich margin, but reversed and oscillatory zoning also occur (Dawson, 1962). The growth of secondary pale greenish brown acmite on aegirine-augite, and of pale brown to colourless acmite around aegirine has been reported by Stephenson (1972). Anomalous rose-violet and indigo blue interference colours in sections perpendicular to $\alpha\beta$ in aegirine-augite have been reported by Shenderova and Sokolova (1958). Aegirine often occurs in well formed stout prismatic crystals, sometimes in slender needles elongated parallel to the *z* axis and more rarely in bundle-like or felted aggregates of minute fibres. Large crystals of aegirine-augite, up to 100 × 35 × 20 cm in size, have been recorded in nepheline-syenite adjacent to aegirinite, and in coarse-grained ijolite-urtite pegmatite (Polkanov, 1939).

The pleochroism of the manganoan aegirines and aegirine-augite shows considerable variation (Table 54). Many are pleochroic from α rose-pink to γ blue, but others show α in yellows and browns and γ yellow-browns. The manganoan and zincian aegirine-augite from Franklin is brown to yellowish brown in transmitted light and moderately pleochroic in shades of brown and yellow-brown with absorption $\alpha > \gamma$. The optical properties of the synthetic scandian analogue, $NaScSi_2O_6$, are α 1·683, β 1·715, γ 1·724, $\alpha:z$ 8°, optically negative.

The infrared absorption spectra, in the frequency range 4000 cm^{-1} to 400 cm^{-1}, of pyroxenes formed by substitutions involving the cation pairs Na–Al, Na–Fe, Na–Cr, Ca–Mg, Ca–Fe, Ca–Co and Ca–Ni have been investigated by Ohashi and Yagi (1968). The frequency of two absorption bands of solid solution in the aegirine–diopside and aegirine–hedenbergite vary linearly with composition (see also Lazarev, and Tenisheva, 1961).

Distinguishing Features

Aegirine is distinguished from other pyroxenes by its small $\alpha:z$ extinction angle, high refractive indices, high birefringence, negative optic axial angle and strong pleochroism, and aegirine-augites by their high $\gamma:z$ extinction angles, large optic axial angles and characteristic strong pleochroism. Aegirine and aegirine-augite resemble some alkali amphiboles; in basal and near basal sections the pyroxene cleavage angle is diagnostic and in prismatic sections the pleochroism is sufficiently characteristic to distinguish them from arfvedsonite and riebeckite. The length-fast optical character of aegirine and aegirine-augite distinguishes them from other amphiboles. Aegirine and aegirine-augite have refractive indices, birefringence, $\alpha:z$ extinction angles, optical axial angles, colour and pleochroism comparable with those of iron-rich epidotes: they have stronger absorption colours, however, which reduce the apparent birefringence, and can generally be distinguished from epidote by the higher polarization colours of the latter mineral. In sections showing one cleavage and small extinction angles the optic axial plane is parallel to the cleavage in aegirine and aegirine-augite and normal to the cleavage in epidote. Aegirine and aegirine-augite are distinguished from tourmaline by their biaxial character, higher refringence and birefringence, and better cleavage.

Paragenesis

Igneous rocks

Aegirine and aegirine-augite are pre-eminently the products of the crystallization of alkaline magmas, and many peralkaline rocks, $(Na_2O+K_2O) > Al_2O_3$, consist essentially of aegirinitic pyroxene (+sodium amphiboles) and alkali feldspar with either nepheline or quartz. The sodium-rich pyroxenes are thus common constituents of alkali granites, quartz syenites, syenites and nepheline-syenites and their associated pegmatites and veins, in which they are often associated with arfvedsonitic amphiboles, magnesioriebeckite, melanite garnet, astrophyllite, aenigmatite, catapleiite, eucolite and låvenite. In most of these rocks aegirine forms late in the crystallization sequence, and commonly mantles earlier pyroxenes or sodian pyroxenes and is not a product of reaction between these earlier phases and liquid.

Aegirine is one of the principal minerals of the Lovozero massif (Vlasov *et al.*, 1966), and in the aegirine lujavrite makes up between 20 and 35 per cent of the rock. Acmite and aegirine-augite also occur but are usually present in only small amounts. Three aegirine generations are distinguished in the various units and pegmatites of the complex. Aegirine, Fe_2O_3 21, FeO 8 per cent, crystallized earlier than the associated nepheline, sodalite, hydrosodalite and potassium feldspars. A more ferric iron-rich aegirine, Fe_2O_3 22·5–25·5, FeO 1·3–3·5 per cent, is the characteristic pyroxene to crystallize subsequent to the formation of nepheline, the early potassium feldspars and the high temperature dark mineral components. The most iron-rich aegirine, Fe_2O_3 28·5, FeO 1·3 per cent, occurs on the walls of leached cavities and in mariolitic cavities and crystallized at a late hydrothermal stage. Replacement by arfvedsonite is common, and in the late hydrothermal and hypergene stages the aegirine may be replaced by chlorite, celadonite and nontronite. Rocks extremely rich in aegirine, nepheline aegirinites, and almost monomineralic aegirinites have been described from the Gremiakha–Vyrmes complex, Kola Peninsula, by Polkanov (1940). Aegirine-augite and aegirine are the commonest ferromagnesian constituents of the alkali gabbros, quartz syenites, syenites and nepheline-syenites of the alkali complexes of the central Turkestan–Alai region (Perchuk, 1964).

In the Kangerdlugssuaq alkaline intrusion, east Greenland (Kempe and Deer, 1970), aegirine (Table 52, anal. 8) and aegirine-augite (Table 53, anal. 2) are the main ferromagnesian minerals. The pyroxene of the majority of the main units, nordmarkites, transitional pulaskite, and pulaskite, is aegirine, $\simeq Ac_{80}$, and similar compositions also occur in most of the veins. The foyaite and its veins and pegmatitic segregations contain an aegirine-augite, Ac_{30}. This compositional trend is in contrast to that shown in the alkali pyroxenes of the sheets, laccoliths and dykes of the Morotu district, Sakhalin (Yagi, 1953), and other alkaline complexes. In the Kangerdlugssuaq intrusion, however, the trend is related primarily to the mineralogy of the two rock types, both of which crystallized from an alkaline magma and not, as is the case in the Morotu district, from a wide range of liquids. At Kangerdluggsuaq the major change from Ac_{80} to Ac_{30} is associated with the composition of the residual foyaitic fraction in which the content of Ca increased and that of Fe^{2+} decreased. Thus with Na being increasingly removed from the magma by the crystallization of nepheline, the residual liquids became enriched in Ca, Mg and Fe^{3+} from which aegirine-augite and melanite garnet were precipitated. Compared with the sodium-rich pyroxenes of the Lovozero massive the Kangerdlugssuaq minerals have a high content of manganese (MnO 0·59–2·93), notably higher zirconium and lower titanium. Aegirine, in association with arfvedsonite, occurs in many of the rocks of the Ilímaussaq intrusion, south-west Greenland (Sørensen, 1962; Engell *et al.*, 1971; Larsen, 1976). In veins in the lujavrite the aegirine is partially replaced by arfvedsonite, and less frequently by acmite, while in other veins the arfvedsonite is pseudomorphed by aggregates of irregularly oriented grains of acmite.

Aegirine-augite is the dominant ferromagnesian constituent of the foyaite, augite syenite and minor intrusions of the South Qôroq centre, south Greenland. The pyroxenes show continuous solid solution from $Di_{69}Hd_{25}Ac_6$ to $Di_6Hd_{17}Ac_{77}$, and the range is further extended, from $Di_4H_{10}Ac_{86}$ to Ac_{100}, by the green aegirine and brown to colourless acmite in the pegmatites (Stephenson, 1972). The pyroxenes of the individual intrusions show a restricted compositional range, and are commonly rimmed by alkali amphibole. The formation of the latter, and the termination of pyroxene crystallization, is probably due to the instability of the aegirine-augite at the high water pressures developed during the late stages of the crystallization of the successive intrusions.

In the alkaline district of Alnö Island (Eckermann, 1948, 1966, 1974) sodium-rich pyroxenes occur in the sövite, sövite pegmatite, fenite, juvite, malignite, vibetoite, jacupirangite, ijolite and melteigite, and in the last named may amount to as much as 70 per cent of the rock. Sodium-rich pyroxenes also occur in the dyke rocks, phonolite, tinguaite, alvikite, and in the foyaite, nephelinite and melteigite porphyries. Analyses of sixteen pyroxenes are presented (Eckermann, 1974) all of which are described as aegirine-augite although some have relatively high Al and Fe^{2+} contents and are more correctly classified as sodian augites. Aegirine-augite is also present in the micaceous alkaline kimberlite portion of a composite carbonatitic kimberlite dyke associated with the Alnö complex (Eckermann, 1964). Phenocrysts of acmitic pyroxenes are present in the damkjernite of the Fen alkaline complex (Griffin and Taylor, 1975). The phenocrysts are commonly zoned to more iron-rich margins; the trend is, however, generally reversed in the outermost rims, and the pyroxenes of the groundmass are also more magnesium-rich than the phenocrysts. Some of the phenocryst cores are partially resorbed and overgrown by magnesium titanaugite, probably as the consequence of a reduction in the oxygen fugacity of the crystallizing magma.

At Alnö the formation of aegirine-augite accompanies the fenitization of the adjacent gneissose granite in which the metasomatic growth of the pyroxenes begins as narrow rims around quartz grains. With increasing intensity of fenitization the amount of aegirine-augite increases, due in part to the replacement of the original biotite and hornblende. At the highest temperature developed during the fenitization process aegirine is replaced by garnet.

$$\underset{\text{aegirine}}{2NaFe^{3+}Si_2O_6} + 3CaO \rightarrow \underset{\text{andradite}}{Ca_3Fe_2^{3+}Si_3O_{12}} + SiO_2 + Na_2O$$

At Fen, southern Norway, the early stage of the fenitization of the granite gneiss is marked by the development of aegirine-augite at the expense of the original biotite, the silicon and sodium required for the conversion being supplied by the solution of quartz and the metasomatic solutions respectively (Saether, 1957). A detailed account of the formation of aegirine in the country granite gneiss during the process of fenitization associated with the Oldonyo Dili carbonatite rock, Tanganyika has been given by McKie (1966). Here aegirine, as fine needles, commonly intergrown with magnesioarfvedsonite, appears initially in veins. As the intensity of the fenitization increases aegirine first forms radiating groups of needles disseminated throughout the rock and subsequently abundant, larger and stouter crystals. Aegirine is a common constituent in veins cutting the fenites of the Tur'il Peninsula, Murmansk region (Kulakov *et al.*, 1974).

Aegirine-augite is an essential constituent of the melteigite–ijolite–urtite–carbonatite series, and the nepheline-syenites, syenites and fenites of Uganda (King, 1965; Tyler and King, 1967). The sequence of the pyroxene crystallization begins with sodian augites in the melteigites, and is followed by aegirine-augite in the ijolites and nepheline-syenites, and by aegirine in the alkali syenites. On the evidence of the petrography, King has suggested that the enrichment of the pyroxene in the acmite component is dependent, in part, on a reaction between nepheline and diopside, and is associated with a progressive decrease in the temperature of crystallization. Aegirine-augite is also present in the metanephelinites and nephelinites of Moroto Mountain, eastern Uganda (Varne, 1968).

In the Shonkin Sag laccolith, aegirine-augite occurs only in the most differentiated rocks represented by the soda syenite, soda syenite veins in the shonkinite and the soda syenite final residuum. Crystallization of the pyroxene, along the margins of diopside–hedenbergite cores and as discrete crystals, taking place after the olivine ceased to form, does not follow a single compositional path, and the pyroxenes formed under conditions specific to each environment (Fig. 238(a)). In common with many highly alkaline suites the sodium-rich pyroxenes are associated with arfvedsonite both as cores and rims to the amphibole. It has been suggested (Bailey, 1969) that these fluctuations are due to variations in the oxygen fugacity and not to changes in magmatic composition, but Nash and Wilkinson (1970) consider it more likely to be due to the fact that aegirine and arfvedsonite are in equilibrium over a finite but limited f_{O_2}–T interval. The compositional trend of the aegirine-augite at the Shonkin Sag is very similar to the differentiation sequence at Morotu, Sakhalin (Yagi, 1953, 1966) and that of the eastern Uganda alkaline complexes (Tyler and King, 1967). Similar textural relations between salite cores and aegirine-augite rims that occur at Shonkin Sag are also exhibited by the pyroxenes in the ijolite of the Seabrook Lake complex, Ontario (Mitchell, 1972). The same general compositional trend is shown by the aegirine-augites and aegirine (Table 52, anal. 7) of the nepheline-syenite complex, Itapirapuã, São Paulo, Brazil (Gomes *et al.*, 1970). The fractionation initially involved the substitution of CaMg by $NaFe^{3+}$ and some $CaFe^{2+}$. The pyroxenes form a continuous series within the diopside–hedenbergite–acmite series and vary from sodian augite through aegirine-augite to aegirine as the temperature decreases and the partial pressure of oxygen increases (Fig. 238(b)).

Aegirine-augite is present in both the miaskitic nepheline-syenites and in the agpaitic zone of the Granitberg foyaite ring complex, South-West Africa. In the

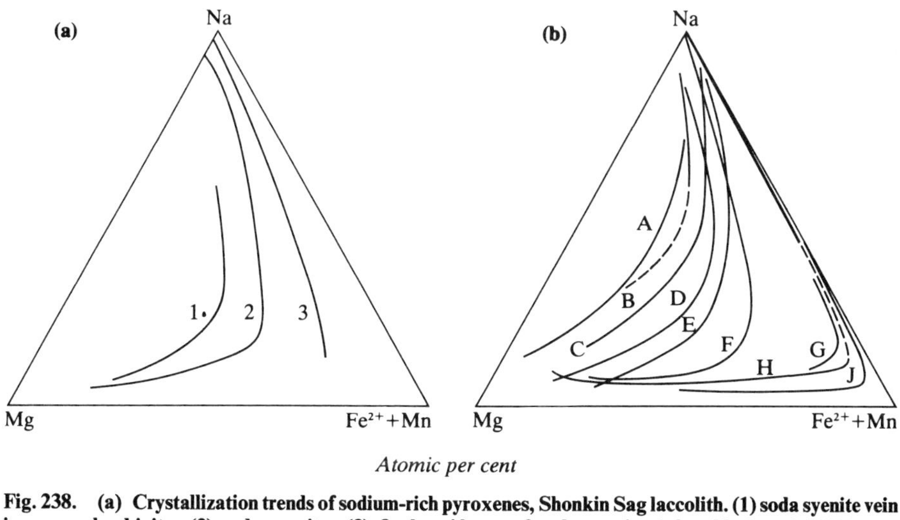

Fig. 238. (a) Crystallization trends of sodium-rich pyroxenes, Shonkin Sag laccolith. (1) soda syenite vein in upper shonkinite; (2) soda syenite; (3) final residuum of soda syenite (after Nash and Wilkinson, 1970). (b) Alkali pyroxene trends. Undersaturated: (A), Auvergne, France (Varet, 1969); (B), Lovozero, U.S.S.R. (Bussen and Sakharov, 1972); (C), Itapirapuã (Gomes *et al.*, 1970); (D), Uganda (Tyler and King, 1967); (E), Morotu, Sakhalin (Yagi, 1966); (F), South Qôroq centre, south Greenland (Stephenson, 1972); (J), Ilímaussaq, south Greenland (Larsen, 1976). Oversaturated: (G), Pantellerite trend (Nicholls and Carmichael, 1969); (H), Nandewar volcano, Australia (Abbott, 1969), (after Larsen, 1976).

rocks of the agpaitic zone the pyroxene, zoned with the acmitic component varying from Ac_{16} to Ac_{46}, is accompanied by aenigmatite, and the stability fields of the two phases with regard to temperature and oxygen fugacity are considered by Marsh (1975).

Aegirine-augite (Table 53, anal. 4) has been described from nepheline syenodiorite, ferrohastingsite–nepheline–sodalite–cancrinite syenite and barkevikite–nepheline syenite, Khamman district, Andhra Pradesh, India (Subbarao, 1971), from the nepheline–barkevikite microsyenite of the Mundwara magmatic suite, Sirohi district, Rajasthan (Bose and Das Gupta, 1973), and aegirine-augite (Table 53, anal. 16) from a tinguaite dyke, Harohalli, Bangalore, Mysore (Ikramuddin and Sadashivaiah, 1966). Aegirine, associated with ankerite, hematite, magnetite, apatite, quartz and baryte occurs in lenticular carbonatite veins in the Mundwara complex, Rajasthan (Subrahmanyam and Rao, 1972).

Aegirine and aegirine-augite occur commonly in syenite pegmatites where their formation is sometimes associated with late replacement processes due to the activity of hydrothermal solutions; for example, the aegirine-augite (Table 53, anal. 12), described by Goranson (1927), from a coarse syenite pegmatite, occurs in fracture zones in the syenite, and has crystallized as a replacement of the earlier formed microline. Aegirine-augite (Table 53, anal. 6) in graphic intergrowth with microperthite, has been reported by Zavaritsky (1946) from a syenite pegmatite vein.

Aegirine occurs in some alkali granites and their hypabyssal equivalents, e.g. in the riebeckite-biotite granite and microgranite of the Buji complex, northern Nigeria (Greenwood, 1951; Beer, 1952; Jacobson *et al.*, 1958). In these rocks aegirine is commonly intergrown with riebeckite in spongy forms which show no regular intergrowth pattern. Both aegirine and riebeckite are late crystallizing constituents; aegirine occurs both as cores within riebeckite, and as marginal rims around riebeckite, and Jacobson *et al.* suggested that the variable relationship between the two minerals may be due to fluctuating chemical conditions in the magma causing the partial transformation of one mineral into the other. Aegirine granites are present in some, notably the Yllymakh, ring complexes of central Aldan (Maksimov, 1970). Aegirine does not occur commonly in acid volcanic rocks but has been reported in rhyolite at Hadjer-el-Hamis, French Sudan (Gentil and Freydenberg, 1908).

Aegirine-augite has been described from cognate xenoliths in some Japanese andesites (Kuno and Taneda, 1940). The xenoliths consist essentially of hornblende with smaller amounts of plagioclase; the margins of the former are dark in colour and consist of augite, hypersthene, magnetite, feldspar and biotite. Aggregates of these minerals, together with aegirine-augite and anorthoclase, occur between the hornblende crystals, and the aegirine-augite is continuous with the augite of the opacitized rims of the hornblende. In some of the andesites, aegirine-augite occurs in vein-like areas and in small patches in the groundmass. In this paragenesis the aegirine-augite is not considered to be a product of alteration of hydrothermal solutions, but its crystallization is believed to have occurred during the deuteric stage of consolidation, and the sodium to have been derived from the hornblende at the time the opacitized rims were formed. The formation of aegirine-augite from augite adjacent to veins of natrolite in analcite–olivine dolerite has been described by Suzuki (1938).

Aegirine is present in some alkali volcanic rocks and has been reported in rhyolite from Oki Island (Tomita, 1935), from an aegirine–arfvedsonite comendite from the Oga Peninsula, Japan (Yagi and Chihara, 1963), and titanian aegirine (Table 52, anal. 9) occurs in a segregation vein in a leucitite from New South Wales, Australia (Ferguson, 1977). Carmichael (1962) noted that aegirine does not crystallize from liquids of pantelleritic composition, and the general absence of aegirine and aegirine-augite in volcanic rocks is most probably due to the low partial pressure of oxygen in the volcanic environment under which conditions fayalite, sodic hedenbergite or ferropigeonite are the stable phases. Aegirine-augite, associated with alkali feldspar, haüyne, and sodium-rich amphibole, in the alkaline lavas (tahitites) of Gran Canaria has been described by Hernández–Pacheco (1960).

An unusual occurrence of aegirine-augite and aegirine within pegmatoid zones of basalt flows at Picture Gorge, Spray, Oregon, has been described by Lindsley *et al.* (1971). The sodium-rich pyroxenes, are associated with alkali feldspar (Ab_{65-56} Or_{35-44}), fayalitic olivine (Fa_{84-95}), ferroaugite ($Ca_{43}Mg_{18}Fe_{39}$–$Ca_{40}Mg_{10}Fe_{50}$), aenigmatite, titanomagnetite and ilmenite. The aegirine-augite occurs in highly differentiated pockets as independent crystals and as alteration rims or overgrowths on the ferroaugite. The aegirine forms overgrowths on the aegirine-augite projecting into cavities indicating crystallization at a lower temperature than the aegirine-augite and possibly from a gas phase after solidification of the residual magma. In pegmatites in the syenites of the Pajarito Mountain intrusion, Otero County, New Mexico, aegirine has replaced hornblende and riebeckite (Kelley, 1968). Here the late development of aegirine and its replacement of the amphiboles is attributed to the release of water possibly due to the late fracturing of the cooling intrusion.

Metamorphic rocks

The occurrence of aegirine-augite in regionally metamorphosed rocks has been reported in crystalline schists from Hokkaidô (Suzuki, 1931, 1933; Suzuki and Suzuki, 1959) and from the Bessi–Shirataki district, Japan (Kojima and Hide, 1957). In the former occurrence the aegirine-augite is associated with glaucophane and riebeckite-bearing schists, and in the latter area occurs in garnet–crossite–quartz and garnet–muscovite–amphibole–quartz schists. In the Bessi–Shirataki area the (010) planes of the aegirine-augite are oriented parallel with the bedding schistocity plane (*a b* fabric plane), and the *z* axis parallel with the *b* fabric axis of the schist. In the Sanbagawa schists of the Bessi district (Banno, 1959), aegirine-augite is present in three assemblages, alkali amphibole–garnet–apatite–hematite–albite–quartz, magnesioriebeckite–hematite–albite–muscovite–garnet–quartz, and albite–calcite–garnet–muscovite–alkali amphibole. The sodium-rich pyroxene does not occur in the lower grade, chlorite–epidote-bearing schists where its absence may be due to a sufficiently high P_{CO_2} for the following reactions:

$$\underset{\text{acmite}}{4NaFe^{3+}Si_2O_6} + \underset{\text{epidote}}{2Ca_2Al_2Fe^{3+}Si_3O_{12}(OH)} + 4CO_2 \rightarrow$$

$$\underset{\text{albite}}{4NaAlSi_3O_8} + \underset{\text{hematite}}{3Fe_2O_3} + \underset{\text{quartz}}{2SiO_2} + \underset{\text{calcite}}{4CaCO_3} + H_2O \qquad (1)$$

$$\underset{\text{antigorite}}{Mg_3Si_2O_5(OH)_4} + \underset{\text{acmite}}{2NaFe^{3+}Si_2O_6} + \underset{\text{quartz}}{2SiO_2} \rightarrow$$

$$\underset{\text{magnesioriebeckite}}{Na_2Mg_3Fe_2^{3+}Si_8O_{22}(OH)_2} + H_2O \qquad (2)$$

The aegirine-augite assemblage described by Suzuki (1934) from the Kamuikotan schists indicates that here the P_{CO_2} was lower than that required for the equilibrium of the above reactions.

Acmitic pyroxenes have been reported from blueschists within the high pressure blueschist belts of Japan (Iwasaki, 1960) and California (Coleman and Clark, 1968; Onuki and Ernst, 1969). In the Franciscan formation acmitic pyroxenes are common in the metabasalts in which the pyroxene is a sodium-rich omphacite containing as much as 33 per cent of the acmite component, and may be due to the reactions:

$$\underset{\text{albite}}{2NaAlSi_3O_8} + \underset{\text{diopside}}{CaMgSi_2O_6} + \underset{\text{magnetite}}{Fe_3O_4} \rightleftharpoons \underset{\text{acmite}}{2NaFe^{3+}Si_2O_6} + \underset{\text{garnet}}{CaMgFe^{2+}Al_2Si_3O_{12}} + \underset{\text{quartz}}{SiO_2} \quad (1)$$

for which $\Delta V \simeq -43\ cm^3$ at normal temperature and pressure (Banno and Green, 1968),

$$\underset{\text{jadeite}}{2NaAlSi_2O_6} + \underset{\text{diopside}}{CaMgSi_2O_6} + \underset{\text{magnetite}}{Fe_3O_4} + SiO_2 \rightleftharpoons \underset{\text{acmite}}{2NaFeSi_2O_6} + \underset{\text{Ca–Tsch}}{CaAl_2SiO_6} + \underset{\text{hypersthene}}{MgFeSi_2O_6} \quad (2)$$

$\Delta V \simeq +1\ cm^3$.

Aegirine-augite, commonly associated with riebeckitic amphibole, occurs in the siliceous metasediments within the lawsonite zone of the Ouégoa–Cor d'Arama district, New Caledonia (Black, 1974). Shenderova and Sokolova (1958) have described aegirine-augite from the ferruginous quartzites of Krivoy Rog. Aegirine–riebeckite–magnetite quartz schists have been reported from Ishigaki-shima, Japan, by Hashimoto (1974), and an account of the aegirine–riebeckite schists from south Westland, New Zealand, has been given by White (1962).

In the aegirine granulites of Glen Lui, Aberdeenshire, aegirine-augite (Table 53, anal. 11) is associated with orthoclase, albite, almandine, andradite and sodic amphibole. The formation of the sodium-rich pyroxene in these rocks is ascribed by McLachlan (1951) to the introduction of sodium into the mica and garnet–mica schists from the adjacent Cairngorm granite. Sodium metasomatism is also considered to have been the cause of the development of bands of aegirine-augite in some metamorphosed quartzose rocks from Hokkaidô (Suzuki, 1934), and the formation of aegirine–diopside in hornfels at Langesundfjord has been attributed by Goldschmidt (1914) to the introduction of sodium from larvikite and nepheline-syenite. Nesterenko and Kornakov (1965) have described the formation of aegirine-augite from pigeonite during the metasomatic alteration of the gabbro–dolerites of the Bolshaya Botnobiya River area, Siberia. In the fenitized quartzites from the Borolan complex, Scotland, aegirine-augite occurs in banded quartz–alkali feldspar–pyroxene and quartz–alkali feldspar–pyroxene–amphibole rocks, as randomly distributed clusters and lines defining fine veins (Woolley *et al.*, 1972).

The replacement of wollastonite by aegirine-augite has been reported by Cann (1965) in the high grade thermally metamorphosed amygdales in the basalt at 'S Airde Beinn, Mull, Scotland. With increasing metamorphic grade the aegirine-augite is unstable and breaks down marginally to a vermicular intergrowth of augite, iron oxide and plagioclase.

Aegirine, titanian aegirine-augite (Table 53, anal. 19) and aegirine–jadeite have been described from the amphibolite facies metamorphic agpaitic alkalic rocks of the Red Wine alkaline province, central Labrador, Canada (Curtis and Gittins, 1978). Aegirine has also been reported in the quartz–microcline–riebeckite–lepidomelane–astrophyllite gneisses at Galiñeiro, Spain (Floor, 1966), and both aegirine and aegirine-augite occur in the alkali syenite orthogneiss, Ouguela, Portugal (Teixeira and Assunção, 1958). A niobium-bearing aegirine-augite (Table 53, anal. 10), associated with arfvedsonite, is found in an area of beryllium and niobium mineralization in banded paragneiss at Seal Lake, Labrador (Nickel and Mark, 1965). Aegirine, associated with quartz, microcline, oligoclase-andesine, sodic amphibole, magnetite and zircon, occurs within a sixty mile zone in the paragneisses of central Malawi and the assemblage is considered to have been derived from the regional metamorphism of glauconite-rich feldspathic sandstones (Bloomfield, 1967).

Acmite (Table 52, anal. 4) occurs in the altered iron-bearing shales and cherts of the Cuyuna district, Minnesota (Grout, 1946), and aegirine (Table 52, anal. 3), associated with quartz, adularia and carbonates, in high temperature veins of the same region. The acmite is considered by Grout to have formed as a result of the action of hot sodium-rich waters on hematite cherts. Manganese-rich aegirine-augite and aegirine are common constituents of the metamorphosed banded Wabush Iron Formation, south-western Labrador. The mineralogy of the original sediments, chert, hematite, magnetite, siderite, ferrodolomite, and probably chamosite or greenalite, manganese oxide and rhodochrosite has been replaced during the regional metamorphism (kyanite zone on the basis of the pelitic schists underlying the iron formation) by three manganoan pyroxene associations: rhodonite–specularite–calcian rhodochrosite, rhodonite–specularite–calderite–rhodochrosite, and specularite–riebeckite–tremolite–quartz–manganoan calcite. The manganese content of the aegirine and aegirine-augites in association with rhodonite is probably close to the maximum in such pyroxene formed under conditions of the kyanite zone of regional metamorphism (Klein, 1966). Aegirine, associated with magnesioriebeckite, also occurs in the iron formation of Bababudan Hills, Mysore, India (Kutty *et al.*, 1974).

The formation of aegirine during the sodium metasomatism of the Arunta feldspathic gneiss adjacent to the metamorphosed carbonatite in the Strangways Range, north-east of Alice Springs, Northern Territory, Australia, has been reported by Moore (1973), and the aegirinization of the pyroxenites of the Maĭmecha-Kotnĭ province, east Siberia, was described by Landa (1966).

Manganoan aegirine-augite (blanfordite) occurs in the regionally metamorphosed manganiferous sediments of the gondite and kodurite series and a number of other areas in India (Roy, 1966). In the gondite, the pyroxene commonly occurs as large porphyroblasts and may be replaced by tirodite, winchite or piemontite, and at some localities by pyrolusite and cryptomelane. In the Gowari Wadhona Chindwara district, blanfordite, in association with juddite, is a common constituent of the rocks of the greenschist and almandine-amphibolite facies (Roy and Purkait, 1968). At the Tirodi mine, manganoan aegirine is associated with manganarfvedsonite, manganchlorite, microcline and braunite (Nayak and Neuvonen, 1963). Manganiferous pyroxenes containing an unusually high content of aluminium have been described from Madhya Pradesh and Maharashtra, central India, by Roy and Mitra (1964).

Pyroxenes containing variable and sometimes large amounts of trivalent iron, sodium, manganese and zinc occur in the skarn zones at Franklin and Sterling Hill, New Jersey. Their composition is intermediate between aegirine-augite and sodian and ferrian augite and the pyroxenes were originally described as jeffersonite. The name, however, lacks species significance and should be discarded (Frondel and Ito, 1966). Manganese-rich aegirine-augite (MnO 2·5–5·9 weight per cent) has been described from the metamorphosed radiolarites of the Bernina region, Alps (Trommsdorff *et al.*, 1970).

Authigenic aegirine has been reported from the Green River formation in Colorado, Wyoming and Utah. The aegirine is commonly associated with shortite, $Na_2Ca_2(CO_3)_3$, and is considered to have formed at or close to room temperature (Milton and Eugster, 1959). Authigenic aegirine, with albite, hematite, carbonates and sulphates, is present in voids in the Permian marls of the Kumolinsk syncline, Kazakhstan (Beĭseev, 1966).

References

Abbott, M. J., 1969. Petrology of the Nandewar volcano, N.S.W., Australia. *Contr. Min. Petr.*, **20**, 115–134.

Allen, E. T. and **White, W. P.**, 1909. Diopside and its relations to calcium and magnesium metasilicates: with optical study by F. E. Wright and E. S. Larsen. *Amer. J. Sci.*, ser. 4, **27**, 1–47.

Aoki, K.-I., 1964. Clinopyroxenes from alkaline rocks of Japan. *Amer. Min.*, **49**, 1199–1223.

Bailey, D. K., 1969. The stability of acmite in the presence of H_2O. *Amer. J. Sci.*, **267-A** (Schairer vol.), 1–16.

Bailey, D. K. and **Schairer, J. F.**, 1966. The system Na_2O–Al_2O_3–Fe_2O_3–SiO_2 at 1 atmosphere, and the petrogenesis of alkaline rocks. *J. Petr.*, **7**, 114–170.

Banno, S., 1959. Aegirinaugites from crystalline schists in Sokoku. *J. Geol. Soc. Japan*, **65**, 652–657.

Banno, S. and **Green, D. H.**, 1968. Experimental studies on eclogites: the roles of magnetite and acmite in eclogitic assemblages. *Chem. Geol.*, **3**, 21–32.

Beer, K. E., 1952. The petrography of some of the riebeckite-granites of Nigeria. *Rept. Geol. Surv., Atomic Energy Divn.*, 1–38.

Beĭseev, O. B., 1966. [Authigenic aegirine from the Kumolinsk syncline (Kazakhstan).] *Zap. Vses. Min. Obshch.*, **95**, 483–488 (in Russian) (M.A. **18**-192).

Bilgrami, S. A., 1956. Manganese silicate minerals from Chikla, Bhandara district, India. *Min. Mag.*, **31**, 236–244.

Black, P. M., 1974. Mineralogy of New Caledonian metamorphic rocks. III. Pyroxenes, and major element partitioning between coexisting pyroxenes, amphiboles and garnets from the Ouégoa District. *Contr. Min. Petr.*, **45**, 281–288.

Bloomfield, K., 1967. Aegirine-gneisses in central Malawi. *Quart. J. Geol. Soc.*, **123**, 93–98.

Bose, M. K. and **Das Gupta, D. K.**, 1973. Petrology of the alkali syenites of the Mundwara magmatic suite, Sirohi, Rajasthan, India. *Geol. Mag.*, **110**, 457–466.

Bowen, N. L. and **Schairer, J. F.**, 1929. The fusion relations of acmite. *Amer. J. Sci.*, 5th ser., **18**, 365–374.

Bowen, N. L., **Schairer, J. F.** and **Willems, H. W. V.**, 1930. The ternary system: Na_2SiO_3–Fe_2O_3–SiO_2. *Amer. J. Sci.*, 5th ser., **20**, 405–455.

Bussen, I. V. and **Sakharov, A. S.**, 1972. [*Petrology of the Lovozero alkaline massif.*] Nauka, Leningrad (in Russian).

Cameron, M., **Shigeho, S.**, **Prewitt, C. T.** and **Papike, J. J.**, 1973. High-temperature crystal chemistry of acmite, diopside, hedenbergite, jadeite, spodumene and ureyite. *Amer. Min.*, **58**, 594–618.

Cann, J. R., 1965. The metamorphism of amygdales at 'S Airde Beinn, northern Mull. *Min. Mag.*, **34** (Tilley vol.), 92–106.

Carmichael, I. S. E., 1962. Pantelleritic liquids and their phenocrysts. *Min. Mag.*, **33**, 86–113.

Carmichael, I. S. E. and **MacKenzie, W. S.**, 1963. Feldspar–liquid equilibria in pantellerites: an experimental study. *Amer. J. Sci.*, **261**, 382–396.

Cassie, R. M., 1971. The melting behaviour of diopside–acmite pyroxenes at high pressure. *Carnegie Inst. Washington, Ann. Rept. Dir. Geophys. Lab.*, 1969–70, 170–175.

Christensen, A. N. and **Hazell, R. C.**, 1967. The crystal structure of $NaIn(SiO_3)_2$. *Acta Chim. Scand.*, **21**, 1425–1429.

Clark, J. R., **Appleman, D. E.** and **Papike, J. J.**, 1969. Crystal-chemical characterization of clinopyroxenes based on eight new structure refinements. *Min. Soc. America, Spec. Paper* **2**, 31–50.

Clark, J. R. and **Papike, J. J.**, 1968. Crystal-chemical characterization of omphacites. *Amer. Min.*, **53**, 840–868.

Coleman, R. G. and **Clark, J. R.**, 1968. Pyroxenes in the blueschist facies of California. *Amer. J. Sci.*, **266**, 43–59.

Curtis, L. W. and **Gittins, J.**, 1978. Aluminous and titaniferous clinopyroxenes from regionally metamorphosed agpaitic rocks in central Labrador. *J. Petr.*, in press.

Dawson, J. B., 1962. The geology of Oldoinyo Lengai. *Bull. Volcanol.*, **24**, 349–387.

De Villiers, P. R., 1970. The geology and mineralogy of the Kalahari manganese-field north of Sishen, Cape Province, with information on the Sishen–Postmasburg area. *Mem. Dept. Mines, South Africa*, **59**, 1–84.

Eckermann, H. von., 1948. The alkali district of Alnö Island. *Sveriges Geol. Undersok., Ser. Ca*, no. 36.

Eckermann, H. von., 1964. Contributions to the knowledge of the alkaline dykes of the Alnö region. XI–XII. *Arkiv. Min. Geol. Stockholm*, **3**, 521–535.

Eckermann, H. von., 1966. The pyroxenes of the Alnö carbonatite (Søvite) and of the surrounding families. *Min. Soc. India, I.M.A. vol. (I.M.A. Papers & Proc. 4th Gen. Meeting, New Delhi, 1964)*, 125–139.

Eckermann, H. von., 1974. The chemistry and optical properties of some minerals of the Alnö alkaline rocks. *Arkiv. Min. Geol.*, **5**, 93–210.

Edgar, A. D. and **Nolan, J.**, 1966. Phase relations in the system $NaAlSi_3O_8$ (albite)–$NaAlSiO_4$ (nepheline)–$NaFeSi_2O_6$ (acmite)–$CaMgSi_2O_6$ (diopside)–H_2O and its importance in the genesis of alkaline undersaturated rocks. *Min. Soc. India, I.M.A. vol. (I.M.A. Papers and Proc., 4th Gen. Meeting, New Delhi 1964)*, 176–181.

Engell, J., Hansen, J., Jenson, M., Kunzendorff, H. and **Lovborg, L.**, 1971. Beryllium mineralization in the Ilímaussaq intrusion, south Greenland, with description of a field beryllometer and chemical methods. *Rept. Grønlands Geol. Unders.*, **33**, 1–40.

Ferguson, A. K., 1977. The natural occurrence of aegirine–neptunite solid solution. *Contr. Min. Petr.*, **60**, 247–253.

Floor, P., 1966. Petrology of an aegirine–riebeckite gneiss-bearing part of the Hesperian massif: the Galiñeiro and surrounding areas, Vigo, Spain. *Leidse Geol. Mededelingen*, **36**, 1–204.

Frondel, C. and **Ito, J.**, 1966. Zincian aegirine–augite and jeffersonite from Franklin, New Jersey. *Amer. Min.*, **51**, 1406–1413.

Flower, M. F. J., 1974. Phase relations of titan-acmite in the system Na_2O–Fe_2O–Al_2O_3–TiO_2–SiO_2 at 1000 bars total water pressure. *Amer. Min.*, **59**, 536–548.

Gentil, L. and **Freydenberg**, 1908. *Bull. Geol. Soc. franç.*, sér. IV.

Gerasimovskii, V. I., Tuzova, A. M. and **Shevaleyevskiĭ, I. D.**, 1962. Zirconia-hafnia ratio in minerals and rocks of the Lovozero massif. *Geochem.*, no. 6, 585–592. Transl. from *Geokhimiya Publ. Acad. Sci. U.S.S.R.*

Gilbert, M. C., 1967. Acmite. *Carnegie Inst. Washington, Ann. Rept. Dir. Geophys. Lab.*, 1965–66, 241–244.

Gilbert, M. C., 1968a. X-ray properties of jadeite–acmite pyroxenes. *Carnegie Inst. Washington, Ann. Rept. Dir. Geophys. Lab.*, 1966–67, 374–375.

Gilbert, M. C., 1968b. The temperature–compression melting relationship. *Carnegie Inst. Washington, Ann. Rept. Dir. Geophys. Lab.*, 1967–68, 100–101.

Gilbert, M. C., 1969. High-pressure stability of acmite. *Amer. J. Sci.*, **267-A** (Schairer vol.), 145–159.

Goldschmidt, V. M., 1914. Ueber ein Fall von Natronzufuhr bei Kontakt-metamorphose. *Neues Jahrb. Min.*, **39**, 193–224.

Gomes, C. de B., Moro, S. L. and **Dutra, C. B.**, 1970. Pyroxenes from the alkaline rocks of Itapirapuã, São Paulo, Brazil. *Amer. Min.*, **55**, 224–230.

Goranson, R. W., 1927. Aegirine from Libby, Montana. *Amer. Min.*, **12**, 37–39.

Greenwood, R., 1951. Younger intrusive rocks of Plateau Province, Nigeria, compared with the alkalic rocks of New England. *Bull. Geol. Soc. America*, **62**, 1151–1178.

Griffin, W. L. and **Taylor, P. N.**, 1975. The Fen damkjernite: petrology of a 'central complex kimberlite'. *Phys. Chem. Earth*, **9**, 163–177.

Grout, F. F., 1946. Acmite occurrences in the Cuyuna Range, Minnesota. *Amer. Min.*, **31**, 125–130.

Hashimoto, M., 1974. Riebeckite–aegirine–quartz–schist of Ishigaki-shima. *Mem. Nat. Sci. Mus., Tokyo*, (7), 19–24.

Hawthorne, F. C. and **Grundy, H. D.**, 1973. Refinement of the crystal structure of $NaScSi_2O_6$. *Acta Cryst.*, **B29**, 2615–2616.

Hawthorne, F. C. and **Grundy, H. D.**, 1974. Refinement of the crystal structure of $NaInSi_2O_6$. *Acta Cryst.*, **B30**, 1882–1884.

Hernández-Pacheco, A., 1970 (for 1964). The tahitites of Gran Canaria and haüynitization of their inclusions. *Bull. Volc.*, **33**, 701–728.

Ikramuddin, M. and **Sadashivaiah, M. S.**, 1966. Aegirine from a tinguaite dyke near Harohalli, Bangalore District, Mysore State. *Proc. Indian Acad. Sci.*, **64B**, 169–175.

Ishibashi, K., 1964. [Aegirine from the Iwagishima, Ehime, Japan.] *J. Min. Soc. Japan*, **6**, 361–367 (in Japanese) (M.A. **17**-702).

Ito, J., 1968. Synthetic indium silicate and indium hydrogarnet. *Amer. Min.*, **53**, 1663–1673.

Ito, J. and **Frondel, C.**, 1968. Syntheses of the scandium analogues of aegirine, spodumene, andradite and melanotekite. *Amer. Min.*, **53**, 1276–1280.

Iwasaki, M., 1960. Clinopyroxene intermediate between jadeite and aegirine from Suberi-dani, Tokusima Prefecture, Japan. *J. Geol. Soc. Japan*, **66**, 334–340.

Jacobson, R. R. E., Macleod, W. N. and **Black, R.**, 1958. Ring-complexes in the younger granite province of northern Nigeria. *Mem. Geol. Soc.*, **1**.

Kelley, V. C., 1968. Geology of the Precambrian rocks at Pajarito Mountain, Otero County, New Mexico. *Bull. Geol. Soc. America*, **79**, 1565–1572.

Kempe, D. R. C. and **Deer, W. A.**, 1970. Geological investigations in east Greenland: Pt. IX. The mineralogy of the Kangerdlugssuaq alkaline intrusion, east Greenland. *Medd. om Grønland*, **190**, No. 3, 1–95.

Kilpady, S. R., 1960. An X-ray study and re-examination of blanfordite. *Proc. Nat. Inst. Sci. India*, **26**, 250–259.

King, B. C., 1962. Optical determination of aegirine-augite with the universal stage. *Min. Mag.*, **33**, 132–137.

King, B. C., 1965. Petrogenesis of the alkaline igneous rock suites of the volcanic and intrusive centres of eastern Uganda. *J. Petr.*, **6**, 67–100.

Klein, C., *Jr.*, 1966. Mineralogy and petrology of the metamorphosed Wabush Iron formation, south-western Labrador. *J. Petr.*, **7**, 246–305.

Kojima, G. and **Hide, K.**, 1957. On new occurrence of aegirine-augite–amphibole–quartz–schists in the Sambagawa crystalline schists of the Besshi–Shirataki District, with special reference to the preferred orientation of aegirine-augite and amphibole. *J. Sci., Hiroshima Univ., ser. C*, **2**, 1–21.

Kostyuk, V. P., 1965. Additional data on the diopside–hedenbergite–aegirine diagram. *Dokl. Acad. Sci., U.S.S.R., Earth Sci. Sect.*, **156**, 110–113.

Kostyuk, V. P., Panina, L. I. and **Guletskaya, E. C.**, 1970. [The mineralogy of high potassic alkaline rocks of Synnirskii pluton (northern Near-Baikal).] 181–197. Editor, Dobretsov, N. L. [*Problems of petrology and genetic mineralogy*] **2**, Moscow (Publ. House Nauka) (in Russian).

Kulakov, A. N., Evdokimov, M. D. and **Bulakh, A. G.**, 1974. [Mineral veins in fenites of the Tur'il Peninsula in the Murmansk region.] *Zap. Vses Min. Obshch.*, **103**, 179–191 (in Russian).

Kuno, H. and **Taneda, S.**, 1940. Occurrence of aegirine-augite in some andesites from Japan. *J. Geol. Soc. Japan*, **47**, 62.

Kutty, T. R. N., Iyer, G. V. A. and **Ramakrishnan, M.**, 1974. Coexisting aegirine and magnesioriebeckite from Bababudan Hills, Mysore State. *Current Sci., India*, **43**, 1–3.

Landa, E. A., 1966. [The genesis of nepheline–pyroxene rocks in alkaline ultrabasic massifs.] *Zap. Vses. Min. Obshch.*, **95**, 652–664 (in Russian).

Larsen, E. S., 1942. Alkali rocks of Iron Hill, Gunnison County, Colorado. *U.S. Geol. Surv., Prof. Paper*, **197A**.

Larsen, E. S. and **Hunt, F. W.**, 1914. Zwei vanadinhaltige Aegirine von Libby, Montana. *Z. Krist.*, **53**, 209–218.

Larsen, L. M., 1976. Clinopyroxenes and coexisting mafic minerals from the alkaline Ilímaussaq intrusion, south Greenland. *J. Petr.*, **17**, 258–290.

Lazarev, A. N. and **Tenisheva, T. F.**, 1961. The vibrational spectra of silicate. II. Infra-red absorption spectra of silicates and germanates with chain anions. *Optics and Spectroscopy*, **10**, 37–40.

Lindsley, D. G., Smith, D. and **Haggerty, S. E.**, 1971. Petrography and mineral chemistry of a differentiated flow of Picture Gorge basalt, near Spray, Oregon. *Carnegie Inst. Washington, Ann. Rept. Dir. Geophys. Lab.*, 1969–70, 264–285.

Maksimov, Y. P., 1970. New data on the geology of volcano–plutonic ring complexes of the central Aldan. *Dokl. Acad. Sci., U.S.S.R., Earth Sci. Sect.*, **190**, 51–54.

McLachlan, G. R., 1951. The aegirine–granulites of Glen Lui, Braemar, Aberdeenshire. *Min. Mag.*, **29**, 476–495.

McKie, D., 1966. Fenitization in carbonatites. In *Carbonatites*. O. F. Tuttle and J. Gittins (eds.). New York and London (Wiley), 267–294.

Marsh, J. S., 1975. Aenigmatite stability in silica-undersaturated rocks. *Contr. Min. Petr.*, **50**, 135–144.

Milton, C. and **Eugster, H. P.**, 1959. Mineral assemblages of the Green River formation. In *Researches in Geochemistry*. New York (John Wiley).

Mitchell, R. H., 1972. Composition of nepheline, pyroxene and biotite in ijolite from the Seabrook Lake complex, Ontario, Canada. *Neues Jahrb. Min., Mh.*, 415–432.

Moore, A. C., 1973. Carbonatites and kimberlites in Australia: a review of the evidence. *Min. Sci. Eng.*, **5**, 81–91.

Nakai, I., Hideo, O., Sugitani, Y. and **Niwa, Y.**, 1976. X-ray photoelectron spectroscopic study of vanadium-bearing aegirines. *Min. J. Japan*, **8**, 129–134.

Nash, W. P. and **Wilkinson, J. F. G.**, 1970. Shonkin Sag laccolith, Montana. I. Mafic minerals and estimates of temperature, pressure, oxygen fugacity and silica activity. *Contr. Min. Petr*, **25**, 241–269.

Nayak, V. K. and **Neuvonen, K. J.**, 1963. Some manganese minerals from India. *Bull. Comm. géol. Finlande*, No. 212, 27–36.

Nesterenko, G. V. and **Kornakov, Yu. N.**, 1965. Subalkalic traps of the Vilyuy River as a metasomatic product. *Dokl. Acad. Sci., U.S.S.R. Earth Sci. Sect.*, **153**, 181–182.

Neumann, H., 1961. The scandium content of some Norwegian minerals and the formation of thortveitite, a reconnaissance survey. *Norsk Geol. Tidsskr.*, **41**, 197–210.

Newton, R. C. and **Smith, J. V.**, 1967. Investigations concerning the breakdown of albite at depth in the earth. *J. Geol.*, **75**, 268–286.

Nicholls, J. and **Carmichael, I. S. E.**, 1969. Peralkaline acid liquids: a petrological study. *Contr. Min. Petr.*, **20**, 268–294.

Nickel, E. H. and **Mark, E.**, 1965. Arfvedsonite and aegirine-augite from Seal Lake, Labrador. *Canad. Min.*, **8**, 185–197.

Niggli, P., 1943. Gesteinschemismus und Mineralchemismus II. Die Pyroxene der magmatischen Erstarrung. *Schweiz. Min. Petr. Mitt.*, **23**, 538–607.

Nolan, J., 1966. Melting relations in the system $NaAlSi_3O_8$–$NaAlSiO_4$–$NaFeSi_2O_6$–$CaMgSi_2O_6$–H_2O and their bearing on the genesis of alkaline undersaturated rocks. *Quart. J. Geol. Soc.*, **122**, 119–157.

Nolan, J., 1969. Physical properties of synthetic and natural pyroxenes in the system diopside–hedenbergite–acmite. *Min. Mag.*, **37**, 216–229.

Nolan, J. and **Edgar, A. D.**, 1963. An X-ray investigation of synthetic pyroxenes in the system acmite–diopside–water at 1000 kg/cm² water-vapour pressure. *Min. Mag.*, **33**, 625–634.

Nosyrev, I. V., Shaposhnikov, A. A., Smirnov, A. A., Sipavina, L. V. and **Khadzhi, I. P.**, 1969. Some properties of hydrothermally synthesized clinopyroxene. *Dokl. Acad. Sci. U.S.S.R., Earth Sci. Sect.*, **187**, 109–112.

Ohashi, H. and **Yagi, K.**, 1968. Infrared absorption spectra of Na-pyroxenes and Ca-pyroxenes. *J. Min. Soc. Japan*, **9**, 99–103 (in Japanese). (Abstract in *Min. Journ.*, 1970, **6**, 138.)

Onuki, H. and **Ernst, W. G.**, 1969. Coexisting sodic amphiboles and sodic pyroxenes from blueschist facies metamorphic rocks. *Min. Soc. America, Spec. Paper*, **2**, 241–250.

Ostrovsky, I. A., 1946. Optical properties of synthetic aegirine–diopsides. *Acad. Sci. U.S.S.R., Belyankin vol.*, 505–506.

Pecora, W. T., 1942. Nepheline syenite pegmatites, Rocky Boy Stock, Bearpaw Mountain, Montana. *Amer. Min.*, **27**, 397–424.

Pederson, A. K., Engell, J. and **Rønsbo, J. G.**, 1975. Early Tertiary volcanism in the Skagerrak: new chemical evidence from ash-layers in the mo-clay of northern Denmark. *Lithos*, **8**, 255–268.

Perchuk, L. L., 1964. Effect of the acid-base interaction of the components in the system aegirine–hedenbergite–diopside. *Dokl. Acad. Sci. U.S.S.R., Earth Sci. Sect.*, **147**, 219–222.

Perchuk, L. L., 1966. [Composition property diagram for the aegirine–hedenbergite–diopside system and possible limited miscibility in the subsolidus of the system.] *Zap. Vses. Min. Obshch.*, **95**, 619–626 (in Russian).

Polkanov, A. A., 1939. [On the gigantic aegirine-augite crystals from the plutonic rocks of Gremiakha–Vyrmes (Kola Peninsula).] *Dokl. Acad. Sci. U.S.S.R.*, **24**, 935–937.

Polkanov, A. A., 1940. [The aegirinites of the Gremiakha-Vyrmes pluton on the Kola peninsula.] *Mém. Soc. Russe Min.*, ser. 2, **69**, 303–308 (in Russian, English summary).

Popp, R. K. and **Gilbert, M. C.**, 1972. Stability of acmite–jadeite pyroxenes at low pressure. *Amer. Min.*, **57**, 1210–1231.

Prewitt, C. T., Shannon, R. D. and **White, W. B.**, 1972. Synthesis of a pyroxene containing trivalent titanium. *Contr. Min. Petr.*, **35**, 77–82.

Pulfrey, W., 1950. Ijolitic rocks near Homa Bay, western Kenya. *Quart. J. Geol. Soc.*, **105**, 425–460.

Roy, S., 1966. *Syngenetic manganese formations of India*. Jadarpur University, Calcutta, 1–219.

Roy, S., 1970. Manganese-bearing silicate minerals from metamorphosed manganese formations of India. I. Juddite. *Min. Mag.*, **37**, 708–716.

Roy, S. and **Mitra, F. N.**, 1964. Mineralogy and genesis of the gondites associated with metamorphic manganese ore bodies of Madhya Pradesh and Maharashtra, India. *Proc. Nat. Inst. Sci. India*, **30**, 395–438.

Roy, S. and **Purkait, P. K.**, 1968. Mineralogy and genesis of the metamorphosed manganese silicate rocks (gondite) of Gowari Wadhona, Madhya Pradesh, India. *Contr. Min. Petr.*, **20**, 86–114.

Sabine, P. A., 1950. The optical properties and composition of acmitic pyroxenes. *Min. Mag.*, **29**, 113–125.

Sabine, P. A., 1960. The geology of Rockall, North Atlantic. *Bull. Geol. Surv. Gr. Brit.*, no. 16, 156–178.

Saether, E., 1957. The alkaline rock province of the Fen area in southern Norway. *Norsk. Vidensk. Sersk. Skrifter*, no. 1.
Santos, A. R. dos., 1973. Extinction curves method for optical study of the alkaline pyroxene series. *Bol. Soc. Geol. Portugal*, **18**, 179–198.
Schminche, H.-U., 1969. Ignimbrite sequence of Gran Canaria. *Bull. Volcanol.*, **33**, 1199–1218.
Schminche, H.-U., 1974. Volcanological aspects of peralkaline silicic welded ash-flow tuffs. *Bull. Volcanol.*, **38**, 594–636.
Schüller, K.-H., 1958. Das Problem Akmit–Aegirin. *Beitr. Min. Petr.*, **6**, 112–138.
Shenderova, A. G. and **Sokolova, E. P.**, 1958. [Aegirine–diopside from the central Dnieper region (Ukrainian S.S.R.).] *Min. Mag. Lvov. Geol. Soc.*, **12**, 306–316 (in Russian, English summary).
Singh, S. K. and **Bonardi, M.**, 1972. Mössbauer resonance of arfvedsonite and aegirine-augite from the Joan Lake agpaitic complex, Labrador, *Lithos*, **5**, 217–225.
Soloveva, L. P. and **Bakakin, V. V.**, 1968. X-ray study of the sodium-rich metagermanate, $NaFeGe_2O_6$. *Soviet Physics – Crystallography*, **12**, 517–520.
Sørensen, H., 1962. On the occurrence of steenstrupine in the Ilímaussaq Massif, south-west Greenland. *Medd. om Grønland*, **167**, 25 p.
Stephenson, D., 1972. Alkali clinopyroxenes from nepheline syenites of the South Qôroq centre, south Greenland. *Lithos*, **5**, 187–201.
Subbarao, K. V., 1971. The Kunavaram series – A group of alkaline rocks, Khammam district, Andhra Pradesh, India. *J. Petr.*, **12**, 621–391.
Subrahmanyam, N. P. and **Rao, G. V. U.**, 1972. Carbonatite veins of Mundwara igneous complex, Rajasthan. *J. Geol. Soc. India*, **13**, 388–391.
Suzuki, J., 1931. Aegirine augite–glaucophane–quartz schist from the province of Teshio, Hokkaidô. *Proc. Imp. Acad. Japan*, **7**, 283–286.
Suzuki, J., 1933. Aegirine augite-bearing riebeckite–quartz–schist from Kamuikotan and some other localities in Hokkaidô. *Proc. Imp. Acad. Japan*, **9**, 617–620.
Suzuki, J., 1934. On some soda-pyroxene and amphibole-bearing quartz schists from Hokkaidô. *J. Fac. Sci. Hokkaidô Univ.*, ser. 4, **2**, 339–353.
Suzuki, J., 1938. On the occurrence of aegirine-augite in natrolite veins in the dolerite from Nemiro, Hokkaidô. *J. Fac. Sci. Hokkaidô Univ.*, **4**, 183–191.
Suzuki, J. and **Suzuki, Y.**, 1959. Petrological studies of the Kamuikotan metamorphic complex in Hokkaidô, Japan. *J. Fac. Sci. Hokkaidô Univ.*, ser. 4, **10**, 349–446.
Teixeira, C. and **Assunção, C. T. de.**, 1958. Sur la géologie et la pétrographic des gneiss à riebeckite et aegyrine et des syénites à nepheline et sodalite de Devadais, près d'Ouguela (Campo Maior), Portugal. *Com. Serv. Geol. Portugal*, **62**, 31–56.
Tomita, T., 1935. On the chemical compositions of the Cenozoic alkaline suite of the circum-Japan Sea region. *J. Shanghai Sci. Inst.*, sect. 2, **1**, 227–306.
Tröger, W. E., 1952. *Tabellen zur optischen Bestimmung der gesteinsbildenden Minerale.* Stuttgart (Schweizerbart'sche Verlag).
Trommsdorff, V., **Schwander, H.** and **Peters, T. J.**, 1970. Mangansilikate der alpinen Metamorphose in Radiolariten des Julier-Bernina-Gebietes. *Schweiz. Min. Petr. Mitt.*, **50**, 439–545.
Tsyganov, E. M. and **Novozhilova, Zh. V.**, 1966. [Acmite inclusions in synthetic quartz crystals.] *Zap. Vses, Min. Obshch.*, **95**, 329–333 (in Russian).
Tyler, R. C. and **King, B. C.**, 1967. The pyroxenes of the alkaline igneous complexes of eastern Uganda. *Min. Mag.*, **36**, 5–21.
Varet, J., 1969. Les pyroxènes des phonolites du Cantal (Auvergne, France). *Neues Jahrb. Min., Mh.*, **4**, 174–178.
Varne, R., 1968. The petrology of Moroto Mountain, eastern Uganda, and origin of nephelinites. *J. Petr.*, **9**, 167–190.
Vlasov, K. A., **Kuz'menko, M. Z.** and **Es'kova, E. M.**, 1966. *The Lovozero Alkali Massif.* Edinburgh (Oliver and Boyd).
Warren, B. E. and **Biscoe, J.**, 1931. The crystal structure of the monoclinic pyroxenes. *Z. Krist.*, **80**, 391–401.
Washington, H. S. and **Merwin, H. E.**, 1927. The acmitic pyroxenes. *Amer. Min.*, **12**, 233–252.
Winchell, A. N., 1951. *Elements of Optical Mineralogy. Part II. Descriptions of Minerals.* New York (Wiley).
White, A. J. R., 1962. Aegirine–riebeckite schists from south Westland, New Zealand. *J. Petr.*, **3**, 38–48.
Wood, R. M., 1977. Iron-rich sediments in blueschist facies metamorphism. *Ph.D. thesis, Univ. Cambridge.*
Woolley, A. R., **Symes, R. F.** and **Elliott, C. J.**, 1972. Metasomatized quartzites from Borralan complex, Scotland. *Min. Mag.*, **38**, 819–836.
Yagi, K., 1953. Petrochemical studies of the alkalic rocks of the Morotu district, Sakhalin. *Bull. Geol. Soc. America*, **64**, 769–810.

Yagi, K., 1958. [Synthetic pyroxenes of the acmite–diopside system.] *J. Min. Soc. Japan*, **3**, 763–769 (in Japanese, English summary).

Yagi, K., 1962. A reconnaissance of the systems acmite–diopside and acmite–nepheline. *Carnegie Inst. Washington, Ann. Rept. Dir. Geophys. Lab.*, 1961–62, 98–99.

Yagi, K., 1966. The system acmite–diopside and its bearing on the stability relations of natural pyroxenes of the acmite–hedenbergite–diopside series. *Amer. Min.*, **51**, 976–1000.

Yagi, K. and **Chihara, K.**, 1963. Occurrence of arfvedsonite comendite from Oga Peninsula, Japan. *J. Japanese Assoc. Min. Petr. Econ. Geol.*, **49**, 22–28.

Zavaritsky, A. N., 1946. [An interesting example of a syenite–pegmatite from Ilmen Mountains.] *Acad. Sci. U.S.S.R.*, Fersman memorial vol., 319–325 (in Russian) (M.A. **10**-433).

Zwaan, P. C. and **Plas, L. van der**, 1958. Optical and X-ray investigations of some pyroxenes and amphiboles from Nagpur, Central Provinces, India. *Proc. Koninkl. Nederl. Akad. Westensch.*, **61B**, 265–277.

Ureyite $NaCr[Si_2O_6]$

Monoclinic (−)

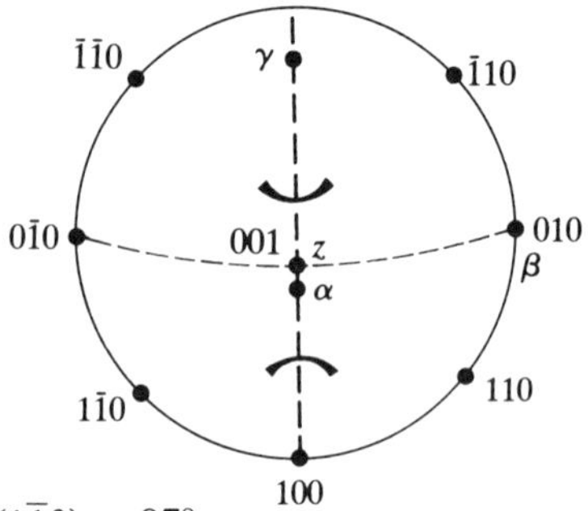

α	1·740–1·766
β	1·756–1·778
γ	1·762–1·781
δ	0·015–0·022
$2V_\gamma$	53°
α:z	8°–22°
β = y	O.A.P. (010)
D	3·60
H	≃ 6
Cleavage:	{110} good; {001} parting. (110):(11̄0) ≃ 87°
Colour:	Dark emerald green; yellow or green in thin section.
Pleochroism:	α yellowish green. β grass green. γ emerald green.
Unit cell:	*a* 9·54–9·58 Å *b* 8·69–8·73 Å *c* 5·26–5·28 Å β 107·3°–107·5° Z = 4. Space group *C2/c*.

Insoluble in HCl, HF.

Ureyite is a rare accessory constituent of some iron meteorites, and was first described by Laspeyres (1897) from the Toluca octahedrite, and later from the hexahedrites of the Coahuila and Hex River Mountains, Mexico, by Frondel and Klein (1965). The precise chemical composition of the Toluca mineral, named kosmochlor by Laspeyres, was not determined, but the later investigation by Frondel and Klein demonstrated that the three minerals have the general formula $NaCrSi_2O_6$, to which composition these authors gave the name of ureyite after Professor H. C. Urey.

Structure

The structure of an almost pure synthetic ureyite was determined by Clark *et al.* (1969). It has *a* 9·550, *b* 8·712, *c* 5·273 Å, β 107·44°, *V* 418·6 $Å^3$, space group *C2/c*, $D_{calc.}$ 3·603 g/cm^3, and its formula was given as $Na_{1\cdot00}Ca_{0\cdot05}Mg_{0\cdot04}Cr_{0\cdot97}Mn_{0\cdot01}Si_{1\cdot97}O_6$. The structure is typical of *C2/c* pyroxenes and in detail is very similar to that of aegirine. All Si–O chains are equivalent; the difference between bridging (1·642 Å) and non-bridging (1·606 Å) Si–O distances is slightly greater than in aegirine, and the Si–O chains are a little more kinked, O3–O3–O3, = 172·1° compared with 174·0° for aegirine.

Chromium is in nearly regular octahedral co-ordination in *M*1 sites with a mean Cr–O distance of 1·998 Å and Na is in eight-fold co-ordination in *M*2 with six shorter Na–O distances (mean 2·397 Å) and two longer (2·764 Å).

The structure of the same ureyite at 400° and 600°C was determined by Cameron *et al.* (1973), who also gave the mean thermal expansion coefficients for the cell parameters, cell volume, and the sizes of the Na and Cr polyhedra. Whereas for the Ca pyroxenes β increases with temperature, for ureyite and aegirine β decreases. The cell edges a, b and c and volume all increase with temperature. Certain structural parameters of ureyite are compared with those of other pyroxenes in Fig. 10, p. 12 and Fig. 242, p. 528.

The cell parameters of members of the pseudobinary system $NaAlSi_2O_6$–$NaCrSi_2O_6$ have been determined by Abs-Wurmbach and Neuhaus (1976).

Chemistry

Analyses of ureyite from the Coahuila and Toluca meteorites, and of ureyite synthesized at 700°C and P_{H_2O} 20 kbar are shown in Table 56. Both natural minerals show some replacement of NaCr by CaMg but neither contains aluminium although the Toluca mineral is associated with sodium-rich feldspar.

Early syntheses of ureyite have been reported by Weyberg (1908), Schüller (1958), and Coes (1955). Subsequent syntheses include those at 1 atm using non-stoichiometric mixtures by Frondel and Klein (1965), and at high pressures under anhydrous conditions reported by Neuhaus (1967). Ureyite has not been synthesized from its own composition at 1 atm, but has been successfully grown hydrothermally at 2 kbar and temperatures between 500° and 700°C (Yoder and Kullerud, 1971). The latter investigation also demonstrated the presence of complete solid solution between diopside and ureyite.

The system $CaMgSi_2O_6$–$NaCrSi_2O_6$ has been investigated at 1 atm (Fig. 239) by Ikeda and Yagi (1972), and at 15 kbar by Ikeda and Ohashi (1974). The latter investigation showed that at this pressure at 1100°C the amount of ureyite in solid solution in diopside is 19·5 weight per cent, and of diopside in ureyite is about 55 weight per cent. The diopside-rich part of the diopside–ureyite system, Di_{100}–Di_{70} at 1 atm (anhydrous), and 1, 5 and 20 kbar (hydrous) has been studied by Vredeboogd and Forbes (1975). The phase diagram of the system with excess water at 20 kbar (Fig. 240) shows a maximum content of $NaCrSi_2O_6$ in diopside of 13 weight per cent (4·6 per cent Cr_2O_3). The incongruent melting of ureyite to eskolaite, Cr_2O_3, and liquid occurs at $\simeq$ 1150°C at 1 atm, the temperature decreasing to 850°C at 20 kbar.

The system $NaAlSi_2O_6$–$NaCrSi_2O_6$ has been studied at 1 atm (Malinovskii *et al.*, 1971), and at pressures up to 25 kbar (Abs-Wurmbach and Neuhaus, 1976). In the latter (hydrous) investigation the formation of single phase mixed crystals shows that there is complete solid solution between jadeite and ureyite at 800°C and a P_{H_2O} greater than 18 kbar. With decreasing pressure the solid solutions, except for those containing more than 86 mol. per cent $NaCrSi_2O_6$ break down to albite, nepheline and a jadeite–ureyite solid solution richer in a ureyite component than the original, according to the reaction

$$Na(Cr_nAl_{1-n})Si_2O_6 \rightarrow Na_{1-x}(Cr_nAl_{1-x-n})(Si_2O_6)1-x + 0{\cdot}5x NaAlSiO_4 + 0{\cdot}5x NaAlSi_3O_8$$

Table 56. Ureyite Analyses

	1	2	3
SiO_2	55·5	56·0	52·44
Fe_2O_3	0·2	0·4	—
Cr_2O_3	30·6	22·6	32·46
MgO	0·8	5·4	0·50
CaO	1·7	3·7	0·52
Na_2O	11·6	11·6	13·63
Total	100·4	99·7	99·55
α	1·748	1·740	1·766
β	1·756	1·756	—
γ	1·765	1·762	1·781
δ	0·017	0·022	0·015
$\alpha:z$	14°	22°	8°
2V	—	—	—
D	—	—	3·60
	Numbers of ions on the basis of six O		
Si	2·058	2·065	2·004
Fe^{3+}	0·006	0·010	—
Cr	0·897	0·660	0·981
Mg	0·044 } 1·85	0·297 } 1·94	0·029 } 1·97
Ca	0·068	0·146	0·021
Na	0·834	0·830	0·936

1 Ureyite, Coahuila hexahedrite, Mexico (Frondel and Klein, 1965).
2 Ureyite, Toluca octahedrite, Mexico (Frondel and Klein, 1965).
3 Ureyite, synthesized at 700°C and 20 kbar (Vredevoogd and Forbes, 1975).

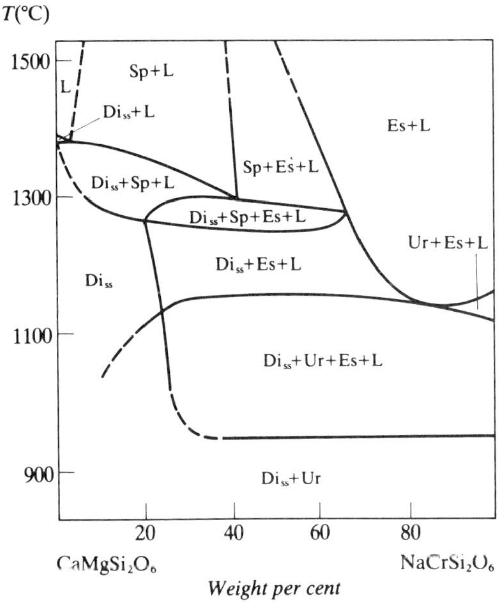

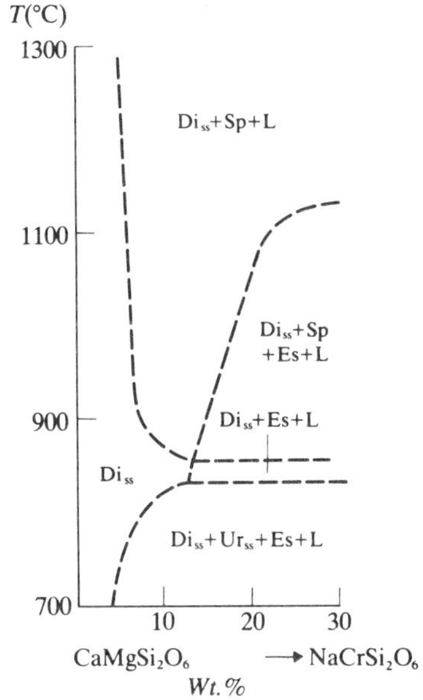

Fig. 239. Phase equilibrium diagram of the join $CaMgSi_2O_6$–$NaCrSi_2O_6$ at 1 atm (after Ikeda and Yagi, 1972). Di_{ss} diopside solid solution. Ur ureyite. Sp magnesio-chromite. Es eskolaite. L liquid.
Fig. 240. Diopside-rich portion of the $CaMgSi_2O_6$–$NaCrSi_2O_6$ system with excess water at 20 kbar (after Vredeboogd and Forbes, 1975).

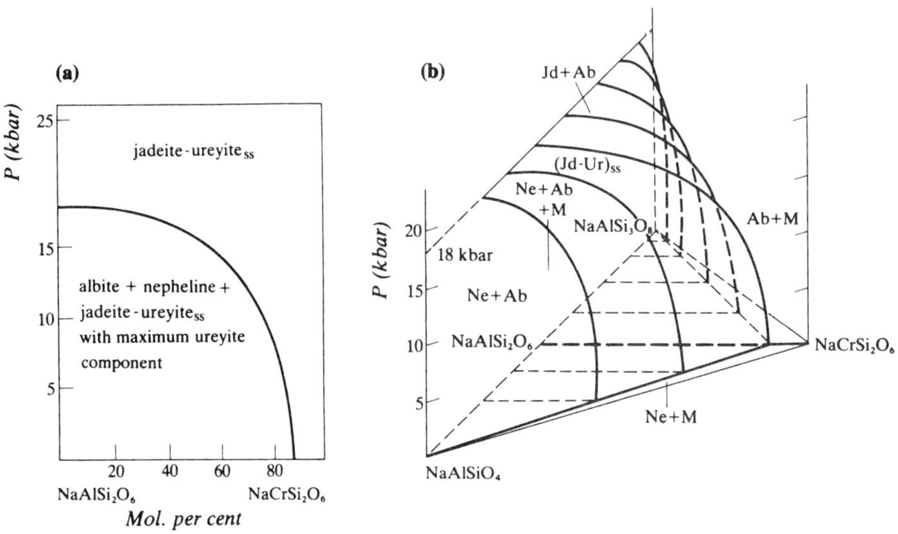

Fig. 241. (a) Relationship of pressure and composition in the system $NaAlSi_2O_6$–$NaCrSi_2O_6$ (after Abs-Wurmbach and Neuhaus, 1976). (b) Hypothetical pressure–composition relationships in the system $NaAlSiO_4$–$NaAlSi_3O_8$–$NaCrSi_2O_6$ at 800°C (after Abs-Wurmbach and Neuhaus, 1976). Ne nepheline, Ab albite. $(Jd–Ur)_{ss}$ $NaAlSi_2O_6$–$NaCrSi_2O_6$ solid solution. M jadeite–ureyite solid solution with maximum ureyite component – pressure controlled. Diopside–ureyite solid solution stable phase above domed surface. Nepheline, albite, diopside–ureyite with maximum ureyite component stable assemblage below domed surface.

where x is the pressure dependent content of the jadeite component in the solid solution. The breakdown curve at 800°C for the reaction varies from 18 kbar for pure jadeite to 1 bar at the composition $Jd_{14}Ur_{86}$ (Fig. 241(a)). The hypothetical pressure–composition relationships in the system nepheline–albite–ureyite at 800°C are shown in Fig. 241(b). Abs-Wurmbach and Neuhaus have also presented isothermal sections at 1 bar, 9, 18 and 22 kbar for the subsystem jadeite–ureyite–quartz.

Optical and Physical Properties

Ureyite is strongly pleochroic and has high refractive indices. Those for natural minerals are somewhat lower than for synthetic $NaCrSi_2O_6$, and are in conformity with the small replacements of NaCr by CaMg, comparable with the effects of the $NaFe^{3+} \rightleftharpoons CaMg$ replacement in the aegirine–aegirine-augites.

The absorption spectra of ureyite show two strong bands, at 15600 and 22000 cm^{-1}, characteristic of Cr^{3+} in six-fold co-ordination (White *et al.*, 1971). The mean thermal expansion coefficients of the unit cell parameters, in the temperature range 24–600°C have been determined by Cameron *et al.* (1973).

Distinguishing Features

Ureyite is distinguished from other members of the pyroxene group, except aegirine and aegirine-augite by its negative optic axial angle. The extinction angle is not diagnostic but the birefringence of ureyite is markedly lower than that of aegirine and aegirine-augite and serves to distinguish it from members of that series.

Paragenesis

Ureyite is unknown in terrestrial rocks, and its absence in the ultramafics may be due to the low sodium content of the liquids from which these rocks have formed, combined with the entry of chromium into diopside resulting from solid solution between diopside and Cr-Tschermak's molecule, $CaCrAlSiO_6$ (Dickey *et al.*, 1971).

In the hexahedrite meteorites of Coahuila and Hex River Mountains, ureyite occurs as polycrystalline aggregates embedded in nodular daubréelite, $FeCr_2S_4$, and in the Toluca medium octahedrite is enclosed in the cliftonite (a form of graphite) rims of troilite nodules that also contain a chromian diopside ($\simeq$ 1·0 weight per cent Cr_2O_3), feldspar, zircon and quartz. The coexistence in the latter meteorite of ureyite and a chromian diopside suggests that $CaMgSi_2O_6$–$NaCrSi_2O_6$ do not form a complete solid solution series under natural conditions. The absence of aluminium in meteoritic ureyites indicates that they formed under conditions of very low pressure as primary crystallization products of a silicate melt.

References

Abs-Wurmbach, I. and **Neuhaus, A.,** 1976. Das System $NaAlSi_2O_6$ (Jadeit)–$NaCrSi_2O_6$ (Kosmochlor) im Druckbereich von 1 bar bis 25 kbar bei 800°C. *Neues Jahrb. Min., Abhdl.*, **127**, 213–241.

Cameron, M., Sueno, S., Prewitt, C. T. and **Papike, J. J.,** 1973. High-temperature crystal chemistry of acmite, diopside, hedenbergite, jadeite, spodumene and ureyite. *Amer. Min.*, **58**, 594–618.
Clark, J. R., Appleman, D. E. and **Papike, J. J.,** 1969. Crystal-chemical characterization of clinopyroxenes based on eight new structure refinements. *Min. Soc. America, Spec. Paper* **2**, 31–50.
Coes, L., 1955. High pressure minerals. *J. Amer. Ceram. Soc.*, **38**, 298–301.
Dickey, J. S., Yoder, H. S. and **Schairer, J. F.,** 1971. Chromium in silicate–oxide systems. *Carnegie Inst. Washington, Ann. Rept. Dir. Geophys. Lab.*, 1970–71, 118–122.
Frondel, C. and **Klein, C.,** 1965. Ureyite, $NaCrSi_2O_6$: a new meteoritic pyroxene. *Science*, **149**, 742–744.
Ikeda, K. and **Ohashi, H.,** 1974. Crystal field spectra of diopside–kosmochlor solid solutions formed at 15 kb pressure. *J. Japan Assoc. Min. Petr. and Econ. Geol.*, **69**, 103–109.
Ikeda, K. and **Yagi, K.,** 1972. Synthesis of kosmochlor and phase equilibria in the join $CaMgSi_2O_6$–$NaCrSi_2O_6$. *Contr. Min. Petr.*, **36**, 63–72.
Laspeyres, A., 1897. Mittheilungen aus dem mineralogischen Museum der Universität Bonn. VIII. Theil. *Z. Krist.*, **27**, 586–600.
Malinovskii, J. Yu, Doroshev, A. M. and **Mikhailov, M. Yu.,** 1971. [The system $NaAlSi_2O_6$ (jadeite)–$NaCrSi_2O_6$ (ureyite) at 1 atm.] *Sibir. Inst. Geol. Geophys. Expt. Min.*, 65–71 (in Russian).
Neuhaus, A., 1967. Über Kosmochlor (Ureyit). *Naturw.*, **54**, 440–441.
Neuhaus, A. and **Abs-Wurmbach, I.,** 1968. Über Kosmochlor (Ureyit). *Min. Soc., I.M.A. vol. (I.M.A. Papers and Proc., 5th Gen. Meeting, Cambridge 1966)*, 329–333.
Schüller, K. H., 1958. Das Problem Akmit–Aegirin. *Beitr. Min. Petr.*, **6**, 112–138.
Vredevoogd, J. J. and **Forbes, W. C.,** 1975. The system diopside-ureyite at 20 kbar. *Contr. Min. Petr.*, **52**, 147–156.
Weyberg, A., 1908. Über die Natriumchromsilikate. *Zbl. Min. Geol. Paläontol.*, 519–523.
White, W. B., McCarthy, G. J. and **Scheetz, B. E.,** 1971. Optical spectra of chromium nickel and cobalt-containing pyroxenes. *Amer. Min.*, **56**, 72–89.
Yoder, H. S. and **Kullerud, G.,** 1971. Kosmochlor and the chromite–plagioclase association. *Carnegie Inst. Washington, Ann. Rept. Dir. Geophys. Lab.*, 1969–70, 155–157.

Lithium Pyroxenes

Spodumene $LiAlSi_2O_6$

Spodumene $LiAl[Si_2O_6]$

Monoclinic (+)

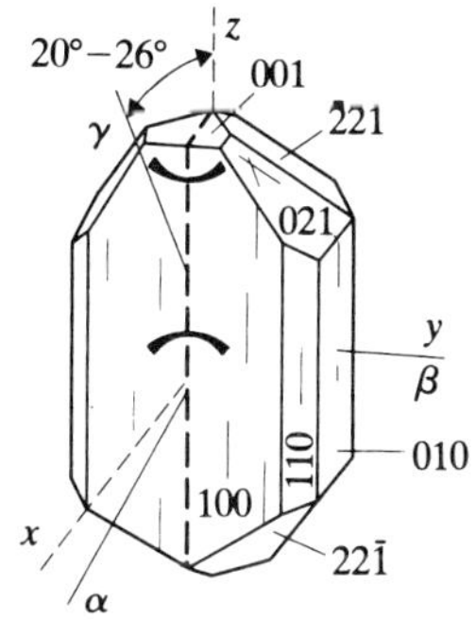

α	1·648–1·663
β	1·655–1·669
γ	1·662–1·679
δ	0·014–0·027
$2V_\gamma$	58°–68°
γ:z	20°–26°
β = y; O.A.P.	(010)
Dispersion:	r < v
D	3·03–3·23
H	$6\frac{1}{2}$–7
Cleavage:	{110} good, {100}, {010} partings. (110):(1$\bar{1}$0) ≃ 87°
Twinning:	{100} common.
Colour:	Colourless, greyish white, pale amethyst, pale green, yellowish emerald-green (hiddenite), lilac (kunzite); usually colourless in thin section.
Pleochroism:	kunzite α purple, γ colourless; hiddenite α green, γ colourless.
Unit cell:	*a* 9·45, *b* 8·39, *c* 5·215, β 110° Z = 4. Space group *C*2/*c*.

Insoluble in acids.

Spodumene is a characteristic mineral of the lithium-bearing granite pegmatites in which it occasionally forms crystals of great size, e.g. 2 m thick and 13 m long at Etta, South Dakota. Spodumene also occurs in some aplites and has been reported in a few gneisses. The clear emerald-green and lilac coloured crystals of gem quality are known as hiddenite and kunzite respectively. Spodumene is one of the principal sources of lithium and plays an important role in the crystallization processes associated with the formation of the complex lithium-rich pegmatites. There are three polymorphic forms of the compound $LiAlSi_2O_6$ but the low-temperature α-spodumene is the only naturally occurring form. The name spodumene is from the Greek *spodoumenos*, reduced to ashes, in allusion to its common greyish white colour.

Structure

Although the axial ratios of spodumene differ significantly from those of the other monoclinic pyroxenes, a preliminary investigation by Warren and Biscoc (1931) showed the structure to be similar to that of diopside. The *c* parameter of spodumene is similar to that of diopside but both *a* and *b* are appreciably smaller.

The smaller cell volume of spodumene is in keeping with the replacement of the larger Ca and Mg ions of diopside by Li and Al, which results in a closer packing of the chains of Si–O tetrahedra.

The structure of a spodumene from Newry, Maine, was determined by Clark *et al.* (1969). It has *a* 9·449, *b* 8·386, *c* 5·215 Å, β 110·10°, *V* 388·1 Å^3, $D_{calc.}$ 3·184 g/cm^3, and its formula is $Li_{1·00}Na_{0·01}Al_{1·00}Si_{1·99}O_6$. They also determined the structure of a synthetic Fe^{3+} analogue.

The main difference between the structure and that of other *C*2/*c* pyroxenes is that the *M*2 cation is in an irregular six-fold rather than eight-fold co-ordination. In comparison with the Mg,Fe pyroxenes, however, because Li is singly charged, the *M*2–O2 distance is longer and *M*2–O3 is shorter, and the Si–O chains remain equivalent and do not become very kinked. In spodumene the angle O3–O3–O3 is 170·5° (but it should be noted that the tetrahedra are rotated in the opposite sense (*S*) from those of other *C*2/*c* pyroxenes) and in $LiFe^{3+}Si_2O_6$ the angle is 180°. The Si–O bridging distance is greater than non-bridging, but the difference is rather small.

The structure of another spodumene with almost ideal composition (from Mt. Marion, Western Australia) was determined by Graham (1975), with results very similar to those obtained by Clark *et al.* (1969). Although the latter authors refined the crystal structure in space group *C*2/*c*, they noted some weak extra reflections and concluded that the true space group was *C*2. With this space group, all Si–O chains remain equivalent but there would be two distinct tetrahedra in each chain. The extra reflections vary as reported by different investigators and are sometimes not observed at all. Graham suggests therefore that they are artefacts, although not due to double diffractions.

Graham reports also a 'rotten spodumene' which had transformed to muscovite in three orientations, with the silicate sheets parallel respectively to (110) and (100) of the original spodumene.

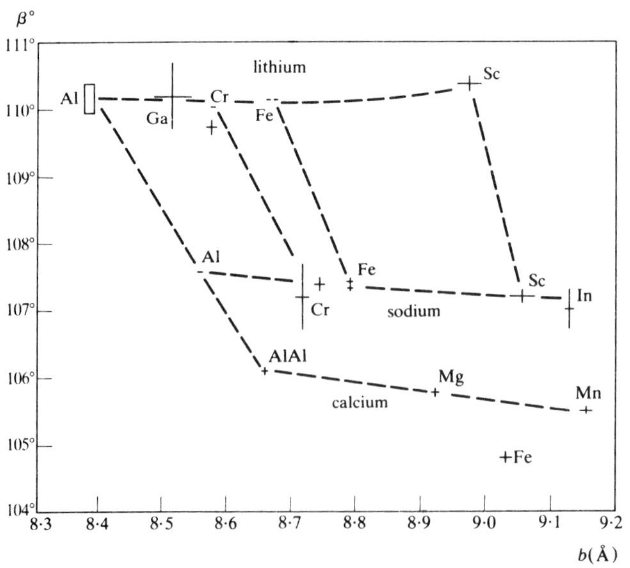

Fig. 242. Plot of *b* against β for lithium, sodium and calcium pyroxenes. Crosses indicate the probable errors in the lattice parameters. The open rectangle for spodumene contains six such crosses (Brown, 1971). Data from Brown (1971) and earlier literature.

The structures of spodumene at 24°, 300°, 460° and 760°C were determined by Cameron *et al.* (1973). Mean thermal expansion coefficients are given for cell parameters, cell volume and the sizes of the Li and Al polyhedra. The cell edges *a*, *b* and *c* increase, while β decreases linearly with increasing temperature. (See also p. 462, 484, 521.)

Cell parameters for a number of synthetic $LiM^{3+}Si_2O_6$ pyroxenes were given by Brown (1971). That they do not follow a linear trend with change in atomic number of the *M* cation is attributed to crystal field effects. It did not prove possible to synthesize a LiMn pyroxene. Figure 242 shows a β–b plot for some Li, Na and Ca pyroxenes.

$LiAlGe_2O_6$ and $LiGaGe_2O_6$ synthesized under normal pressure but within a narrow high temperature range have the pyroxene structure (Hahn and Behruzi, 1968); the former has *a* 9·888, *b* 8·399, *c* 5·398 Å, β 110·61°, and a few weak extra reflections placing it in space group $P2_1/n$ rather than $C2/c$. The latter compound has *a* 9·790, *b* 8·706, *c* 5·347 Å, β 108·88° and space group $P2_1/c$. There is almost complete solid solution between the two germanates.

At high temperatures $LiAlSi_2O_6$ can exist in the form of 'stuffed silica derivatives'. β-Spodumene is tetragonal $P4_32_12$ or $P4_12_12$ with *a* 7·541, *c* 9·156 Å, $D_{obs.}$ 2·365 g/cm^3 and has the keatite structure (Li and Peacor, 1968a; Clarke and Spink, 1969), while γ-spodumene is hexagonal $P6_222$ or $P6_422$ with *a* 5·217, *c* 5·464 Å, $D_{obs.}$ 2·395 g/cm^3 and has the high-quartz structure (Li and Peacor, 1968b; Ioffe and Zonn, 1970).

Chemistry

The chemical composition of spodumene does not show any large variation from the ideal formula $LiAlSi_2O_6$; there is little or no replacement of Si by Al, and only one minor substitution of Al by Fe^{3+} (see Table 57). Some analyses show marginally less than two Si and correspondingly more than one Al on the basis of six oxygens; this may be related to very small amounts of micaceous impurities. Other spodumene analyses show a small amount of Si in excess of two atoms per formula unit that may represent small amounts of SiO_2 in solid solution, or result from slight contamination by quartz or other impurities. Thus Graham (1975) reports spodumene crystals, including gem-quality hiddenite and kunzite, containing extensive veining and microinclusions. Mica, albite and plagioclase are the most frequent but pollucite, quartz, mangan-apatite, lithiophilite, prehnite and cassiterite may also occur as inclusions.

Some of the earlier analyses of spodumene show a higher content of Fe^{3+} and Mn than is indicated by the analyses in Table 57, but it is probable that these higher values are due to impurities, or to the presence of films of iron and manganese oxides coating the spodumene. The spectrochemical analysis of eleven carefully purified specimens (Gabriel *et al.*, 1942) gave compositional ranges of Fe_2O_3 0·02 to 0·78 weight per cent, MnO 0·01 to 0·08 per cent (one specimen 0·54 per cent) and $TiO_2 \leq 0{\cdot}01$ per cent. The substitution of Li by Na, in spite of the difference in ionic radius, appears to be significant though some Na may be present in solution in brines in fluid inclusions (Heinrich, 1975). Minor and trace elements (including Cr, V, Ga, Mn, Co, Ni, Cu, Sn, Ge) for ten spodumenes have been determined by Claffy (1953), and Rb, Be, Ga, Zn, Ta, Nb and Sc for spodumenes of the Sayan Mountains, Siberia, pegmatites by Slepnev (1964). The average content of BeO in spodumenes

Table 57. Spodumene Analyses

	1	2	3	4	5
SiO_2	62·91	63·45	64·03	63·60	63·30
Al_2O_3	28·42	27·40	27·69	27·48	27·19
Fe_2O_3	0·53	0·053	0·02	0·04	0·45
FeO	—	—	—	—	0·10
MnO	—	—	0·02	0·14	0·19
MgO	0·13	0·012	<0·02	<0·02	0·20
CaO	0·11	0·16	0·04	<0·02	0·06
Na_2O	0·46	0·114	0·11	0·22	0·13
K_2O	0·69	0·038	<0·02	—	0·10
Li_2O	6·78	7·87	7·87	7·79	7·76
H_2O^+	0·28	0·30	0·17	0·17	0·64
H_2O^-	—	0·11	—	—	0·04
Total	100·31	99·53	99·95	99·44	100·21
α	1·660	1·660	1·658	1·660	1·659
β	1·667	1·666	1·667	1·665	1·664
γ	1·672	1·675	1·678	1·678	1·676
γ:z	—	20°	24°	26°	25°
$2V_\gamma$	—	—	—	—	65°
D	3·167	3·153	3·109	3·123	3·155

Numbers of ions on the basis of six O[a]

	1		2		3		4		5	
Si	1·927	2·00	1·987	2·00	1·986	2·00	1·989	2·00	1·984	2·00
Al	0·073		0·013		0·014		0·011		0·016	
Al	0·982	0·99	0·998	1·00	1·001	1·00	1·003	1·00	0·988	1·00
Fe^{3+}	0·012		0·001		0·001		0·001		0·011	
Li	1·010		0·991		0·984		0·980		0·978	
Mg	0·006		0·001		0·000		0·000		0·009	
Fe^{2+}	—		—		—		—		0·003	
Mn	—	1·08	—	1·01	0·001	0·99	0·004	1·00	0·005	1·01
Na	0·028		0·007		0·007		0·013		0·008	
Ca	0·004		0·005		0·002		0·000		0·002	
K	0·028		0·002		0·000		0·000		0·004	
H	—		—		—		—		—	

[a] Bikitaite calculated on basis of seven (O).

1 Spodumene, showing slight alteration, pegmatite, Etta mine, Black Hills, South Dakota (Schwartz and Leonard, 1926).
2 White spodumene, Tanco pegmatite, Bernic Lake, Manitoba (Černý and Ferguson, 1972). Anal. Ramlal and Hill (includes Rb_2O 0·002, Cs_2O 0·001, P_2O_5 0·02).
3 Spodumene, pegmatite, Keystone, South Dakota (Edgar, 1968). Anal. J. Esson.
4 Pale lilac spodumene (kunzite), Pala, San Diego, California (Edgar, 1968). Anal. J. Esson.
5 Spodumene, Haapaluoma, Finland (Haapala, 1966). Anal. P. Ojanperä (includes P_2O_5 0·05; *a* 9·52, *b* 8·32, *c* 5·25 Å, *β* 110°20′).

6	7	8	9	10	
62·61	64·3	64·89	64·34	63·82	SiO_2
27·20	26·0	26·74	27·01	28·18	Al_2O_3
1·03	0·57	0·57	0·24	0·62	Fe_2O_3
0·29	0·43	0·04	0·00	0·36	FeO
0·13	—	0·01	0·01	—	MnO
0·39	0·01	0·00	0·00	tr.	MgO
0·45	0·06	0·00	0·00	tr.	CaO
0·23	0·08	0·05	1·04	0·37	Na_2O
0·09	0·03	0·16	0·08	0·10	K_2O
7·60	7·55	7·12	7·00	6·60	Li_2O
} 0·11	—	0·48	0·24	0·20	H_2O^+
	—	0·06	0·07	—	H_2O^-
100·26	99·03	100·12	100·03	100·25	Total
—	—	1·661	1·661	—	α
—	—	1·666	1·666	—	β
—	—	1·676	1·676	—	γ
—	—	—	—		γ:z
—	—	—	—	—	$2V_\gamma$
—	—	3·163	3·130	—	D
1·961 } 2·00	2·020 } 2·02	2·026 } 2·03	2·012 } 2·01	1·992 } 2·00	Si
0·039	0·000	0·000	0·000	0·008	Al
0·966 } 0·99[b]	0·966 } 0·98	0·984 } 1·00	0·996 } 1·00	1·029 } 1·04	Al
0·024	0·013	0·014	0·006	0·014	Fe^{3+}
0·957 }	0·955 }	0·894 }	0·880 }	0·829 }	Li
0·018	0·001	0·000	0·000	0·000	Mg
0·007	0·011	0·001	0·001	0·009	Fe^{2+}
0·003 1·02	— 0·98	0·000 0·90	0·000 0·95	0·000 0·86	Mn
0·014	0·005	0·002	0·063	0·022	Na
0·015	0·002	0·000	0·000	0·000	Ca
0·004	0·001	0·006	0·003	0·004	K
—	—	—	—	—	H

[b] Includes Ti 0·002.

6 Spodumene, pegmatite, China (Chang, 1974). Includes P_2O_5 0·04, TiO_2 0·09.
7 Spodumene, pegmatite, Mt. Marion, Western Australia (Graham, 1975). Anal. C. E. S. Davis.
8 Wine-yellow spodumene, pegmatite, Varuträsk, Sweden (Quensel, 1938). Anal. T. Berggren.
9 Spodumene, pegmatite, Varuträsk, Sweden (Quensel, 1938). Anal. J. Berggren.
10 Spodumene, pegmatite, Katumba, Congo (Thoreau, 1953).

Table 57. Spodumene Analyses – *continued*

	11	12	13	14	15
SiO_2	65·05	63·90	63·21	64·16	55·79
Al_2O_3	26·70	26·82	26·92	27·74	26·68
Fe_2O_3	0·04	0·41	1·58	1·03	0·07
FeO	—	0·70	—	—	—
MnO	—	0·24	tr.	0·32	—
MgO	—	—	tr.	—	0·33
CaO	0·09	—	tr.	—	—
Na_2O	1·68	1·14	1·16	1·03	0·10
K_2O	0·21	—	0·52	—	0·17
Li_2O	6·35	6·11	5·83	5·80	6·51
H_2O^+	—	0·81	0·77	0·52	9·82
H_2O^-	0·17	—	—	—	—
Total	100·29	100·13	99·99	100·60	99·47
α	1·653	—	1·663	1·653	1·510
β	1·659	—	1·668	1·667	1·521
γ	1·677	—	1·679	1·678	1·523
γ:z	25°	—	26°	26°	28°
$2V_\gamma$	66°	—	67°30′	58°	45°[a]
D	3·140	—	3·216	—	2·34

Numbers of ions on the basis of six O

	11		12		13		14		15	
Si	2·033	2·03	2·019	2·02	2·007	2·01	2·011	2·01	1·901	2·00
Al	0·000		0·000		0·000		0·000		0·099	
Al	0·984	0·99	1·000	1·01	1·008	1·05	1·025	1·05	0·973	0·98
Fe^{3+}	0·001		0·010		0·038		0·024		0·002	
Li	0·798		0·777		0·744		0·731		0·891	
Mg	0·000		—		0·000		—		0·016	
Fe^{2+}	—		0·018		—		—		—	
Mn	0·000	0·91	0·006	0·87	0·000	0·84	0·008	0·80	—	0·92
Na	0·102		0·070		0·072		0·062		0·007	
Ca	0·003		—		0·000		—		—	
K	0·008		—		0·020		—		0·007	
H	—		—		—		—		2·232	

[a] $2V_\alpha$.

11 Colourless spodumene, pegmatite, Kluntarna Island, Sweden (Grip, 1941).
12 Emerald green spodumene (hiddenite), mica pegmatite, Kabbur, Hassan district, Mysore, India (Babu, 1975). Anal. S. K. Babu (trace element data including Cr 1200, Cu 150, Be 50, B 50 p.p.m.).
13 Spodumene, pegmatite, St. Radegund, Graz, Austria (Angel, 1933). Anal. K. Schoklitsch.
14 Purplish grey to greenish spodumene, pegmatite, Ooregum mine, Kolar gold field, India (Rao and Rao, 1939). Anal. E. R. Tirumalachar.
15 Bikitaite, ($LiAlSi_2O_6.H_2O$) pegmatite, Bikita, Southern Rhodesia (Hurlbut, 1957). Anal. J. Ito (*a* 8·63, *b* 4·95, *c* 7·64 Å, β 114°34′, space group $P2_1/m$).

(eleven specimens) of the Rhodesian pegmatites is 15 p.p.m., and ranges between < 5 and 40 p.p.m. (Gallagher, 1975).

The absolute abundance of the lithium isotopes in spodumene has been determined by Svec and Anderson (1965) who gave the mean absolute $^6Li/^7Li$ ratio for thirteen spodumenes as 0·08182 ± 0·00033 (range 0·08117–0·08233).

There are three polymorphs of $LiAlSi_2O_6$; α-spodumene is the low-temperature form with a pyroxene structure, and the only naturally occurring phase. β-spodumene is a high-temperature tetragonal form, isostructural with the silica polymorph keatite. Solid solutions between $LiAlSi_2O_6$ and SiO_2, based on the keatite structure, can be synthesized, and are termed β-spodumene$_{ss}$. The third phase, referred to as β-quartz$_{ss}$ has hexagonal symmetry and is believed to be isostructural with β-quartz (high-temperature quartz). The term[1] is used for both hexagonal $LiAlSi_2O_6$ and solid solutions along the join $Li_2O.Al_2O_3$–SiO_2. The X-ray powder pattern of β-quartz$_{ss}$ ($LiAlSi_2O_6$) is essentially the same as that of β-eucryptite$_{ss}$, shown to be metastable at 1 atm but stable at high pressure (Munoz, 1969).

The occurrence of α-spodumene as a typical constituent of the Li-bearing pegmatites suggests that crystallization occurs at temperatures of some 500°C or below, temperatures that are not in disagreement with those extrapolated from the experimentally determined α–β-spodumene inversion temperatures. Some α-spodumene may, however, have crystallized at lower temperatures, evidence of which is provided by the homogenization temperatures, 440° to 200°C, of inclusions in spodumenes of the quartz–cleavelandite–spodumene zones of rare-metal pegmatites of the U.S.S.R. (Bazarov and Motorina, 1969). The pressure existing in the inclusions at the time of homogenization is estimated to be 1·4 kbar at 300°C (see also Sheshulin, 1961).

Experimental

α-Spodumene inverts to the high-temperature β-spodumene phase at about 900°C, but the transformation can be achieved at approximately 700°C by fine grinding and prolonged heating (Roy and Osborn, 1949). The transformation appears to be monotropic and α-spodumene has not been obtained from the high-temperature form. Later work by Grubb (1973) has, however, shown that the $\alpha \rightarrow \beta$ transformation may begin at a temperature as low as 720°C, and only becomes irreversible at temperatures above 900°C. Thus where heating cycles are terminated prior to the inversion being completed, α-spodumene may subsequently reinvert from the β-phase. The transformation temperature is pressure sensitive, and at $P_{H_2O} \simeq 0{\cdot}7$ kbar, α-spodumene was reported by Roy *et al.* (1950) to invert to β-spodumene at a maximum temperature of 500°C. A number of failures to synthesize α-spodumene from gels of $LiAlSi_2O_6$ composition have been recorded including one at pressures up to 4 kbar (Isaacs and Roy, 1958). In this investigation, α-eucryptite ($LiAlSiO_4$) and petalite ($LiAlSi_4O_{10}$) were obtained throughout the *P–T* range used in the experimental procedure.

Roy *et al.* (1950), however, synthesized a modification of $LiAlSi_2O_6$ which corresponds with the low-temperature form, α-spodumene, except that the crystals have slightly higher refractive indices and slightly different *d*-values compared to

[1] Compositions richer in SiO_2 than the range of β-eucryptite solid solution shown in Fig. 245, p. 536 and based on the β-eucryptite structure occur among synthetic products in the system Li_2O–Al_2O_3–SiO_2 (Henglein, 1956), for which Roy (1959) suggested the name 'O' series, and the term silica O for a family of phases, with end-members β-eucryptite ($LiAlSiO_4$) and silica O (SiO_2).

natural spodumene, due possibly to the substitution of small amounts of Al by Fe^{3+} (derived from the steel vessels in which the synthesis was carried out). This low-temperature form was synthesized by heating Li_2CO_3, Al_2O_3 and H_2SiO_3 at 450°C and 0·8 kbar for two weeks in steel vessels. During their investigation of the system $LiAlSi_2O_6$–SiO_2 Roy *et al.* synthesized this α-spodumene at temperatures between 375° and 500°C. At temperatures below about 375°C and at moderate and low pressures hydrated phases crystallized.

Barrer and White (1951) have also reported the crystallization of α-spodumene at a temperature of approximately 360°C from gels of Li_2O, Al_2O_3, 4–8 SiO_2 composition. Considerable contamination of this material by ferric iron is, however, indicated by its high refractive index (≃ 1·76). A green iron-bearing α-spodumene (*a* 9·58, *b* 8·63, *c* 5·28 Å, *β* 110°10′) has been synthesized by Šćavničar and Sabatier (1957) by heating albite with LiCl at temperatures between 550° and 600°C, and at pressures up to 1 kbar. The relatively high iron content (derived from the steel autoclave) of this mineral is indicated by its refractive index, 1·72 and its density, 3·48 g/cm³.

The formation of β-spodumene by solid state reaction of pyrophyllite, $Al_4[Si_8O_{20}](OH)_4$, and lithium carbonate has been studied by X-ray powder diffraction and single crystal X-ray techniques (Udagawa *et al.*, 1973). Decomposition of the pyrophyllite–Li_2CO_3 mixture begins at about 460°C, followed at about 500°C by the formation of β-quartz$_{ss}$ (silica O) and at about 850°C by a β-spodumene solid solution.

The α-→ β-spodumene transition has been examined by Edgar (1968), using three natural spodumenes of high purity ($Li_{0\cdot978-0\cdot982}Al_{1\cdot012-1\cdot032}Si_{1\cdot982-1\cdot984}O_6$), in the P_{H_2O} range 0·3 to 3 kbar. The P_{H_2O}–T curves for the three specimens gave dT/dP values of 0·025, 0·028 and 0·034°C kg^{-1} cm^2 which, extrapolated to atmospheric pressure gave the inversion temperature between 530° and 550°C (Fig. 243). The transition was not sharp, probably due to the sluggish kinetics of the transitions, but was completed over a 5° to 25°C interval. Values of ΔH for the α–β-spodumene transition have been calculated by Edgar from the equation $\Delta H = T\Delta V \mathrm{d}P/\mathrm{d}T$, where ΔV is the difference in molar volume of β-spodumene and α-spodumene in cm³, T the absolute temperature of the transition (°K), and dT/dP the experimentally determined slope of the transition curve in °K kg^{-1} cm^2. The ΔH values for the three specimens vary between 10250 and 14201 ± 3000 cals/mole, and are high compared to those of other silicate reconstructive transformations.

An investigation of the stability of $LiAlSi_2O_6$ composition in the temperature and pressure range 900° to 1700°C and 5 to 45 kbar respectively (Munoz, 1969) shows that under these conditions the stable assemblages include α-spodumene, β-spodumene, β-quartz$_{ss}$ ($LiAlSi_2O_6$), β-spodumene$_{ss}$ + β-quartz$_{ss}$ and β-quartz$_{ss}$ + liquid (Fig. 244).

β-Spodumene melts congruently at about 1430°C and 1 atm and this melting behaviour persists to 1460°C at 8·5 kbar when β-spodumene melts incongruently to β-quartz$_{ss}$ + liquid. The solidus curve β-quartz$_{ss}$ ⇌ β-quartz$_{ss}$ + liquid terminates at the univariant point, spodumene + β-quartz$_{ss}$ + liquid at 1510°C and 26·5 kbar. At higher temperatures and pressures β-spodumene melts incongruently to β-quartz$_{ss}$ + liquid.

Extrapolation of the inversion curve α-spodumene ⇌ β-quartz$_{ss}$ ($LiAlSi_2O_6$) leads to a value of between 520° and 375°C (Fig. 243) for the upper stability of α-spodumene at 1 atm, compared with the range 530° to 550°C obtained by Edgar.

A hydrothermal investigation at 1 kbar and temperatures between 350° and 750°C, using spodumene, spodumene + quartz mixtures, petalite and gels of

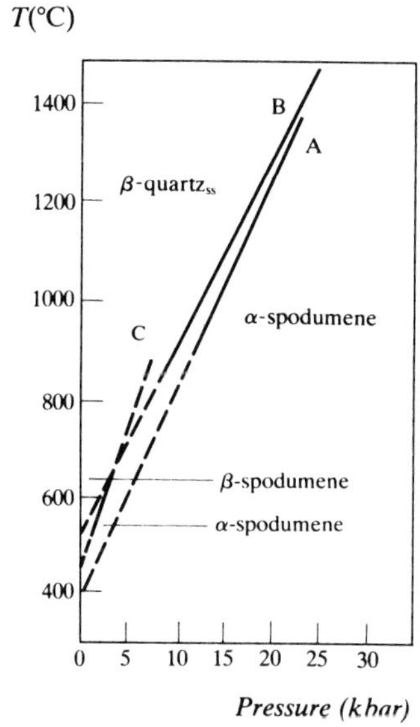

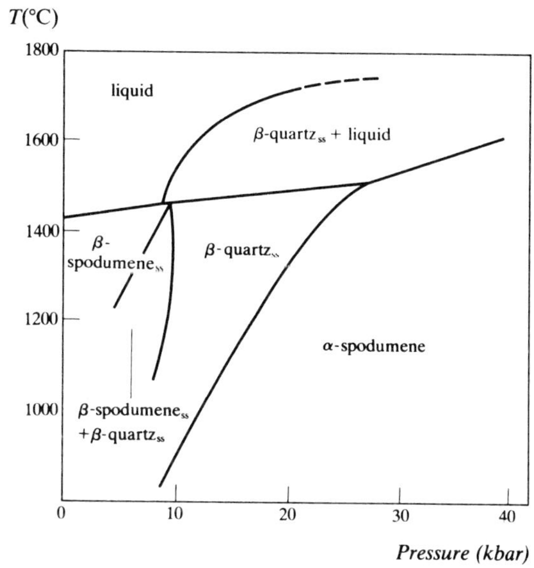

Fig. 243. ***P–T*** **projection for the inversion curves of α-spodumene ⇌ β-quartz$_{ss}$ ($LiAlSi_2O_6$). A piston-in run, B piston-out run (Munoz, 1969). C α-spodumene ⇌ β-spodumene solid solution (Edgar, 1968).**
Fig. 244. ***P–T*** **stability relations for the bulk composition $LiAlSi_2O_6$ (after Munoz, 1969).**

spodumene composition with and without the addition of LiOH, LiF, NaF, NaCl, KF, KCl and $Na_2B_4O_7.H_2O$ as starting materials, has been described by Grubb (1973). The investigation demonstrated that under purely aqueous hydrothermal conditions the α-⇌β-spodumene inversion occurs at temperatures between 550° and 600°C at 1 kbar, and under alkaline and especially lithium-rich conditions at temperatures below 550°C. The α-⇌β-spodumene inversion thus provides an upper temperature limit of 600°C for the formation of the α-phase. Above this temperature petalite, which replaces β-spodumene, is the stable phase. Although temperature is probably the main factor affecting the formation of either spodumene or petalite the availability of Li^+ is also significant and the replacement of petalite by α-spodumene is enhanced by an increased lithium content in the nutrient fluid. Very high Li^+ concentrations (and correspondingly high pH), however, lead to the crystallization of eucryptite or cookeite. Although eucryptite is a common derivative of α-spodumene in purely aqueous hydrothermal runs at temperatures above 600°C, its formation is apparently less dependent on temperature under alkaline conditions with high Li^+ concentrations, and it may be derived from either spodumene or petalite irrespective of temperature. In addition eucryptite was found to form more readily from spodumene than from either a spodumene–quartz mixture or petalite. In the system $LiAlSiO_4$–$NaAlSi_3O_8$–H_2O at 2 kbar H_2O pressure eucryptite and albite form a eutectic at 725°C (Stewart, 1960).

The phase equilibrium relations at high temperatures in the system $LiAlSiO_4$–SiO_2 (Fig. 245) have been investigated by Hatch (1943) and Roy *et al.* (1950). β-Spodumene forms a series of solid solutions with SiO_2, the limit of solid solution occurring at 79 per cent SiO_2. There are two series of solid solutions between high-temperature spodumene and high-temperature eucryptite, a β-spodumene solid solution series and a β-eucryptite series; the diversion between the two series occurs at 58 per cent SiO_2 (β-spodumene solid solutions are uniaxial positive, β-eucryptite

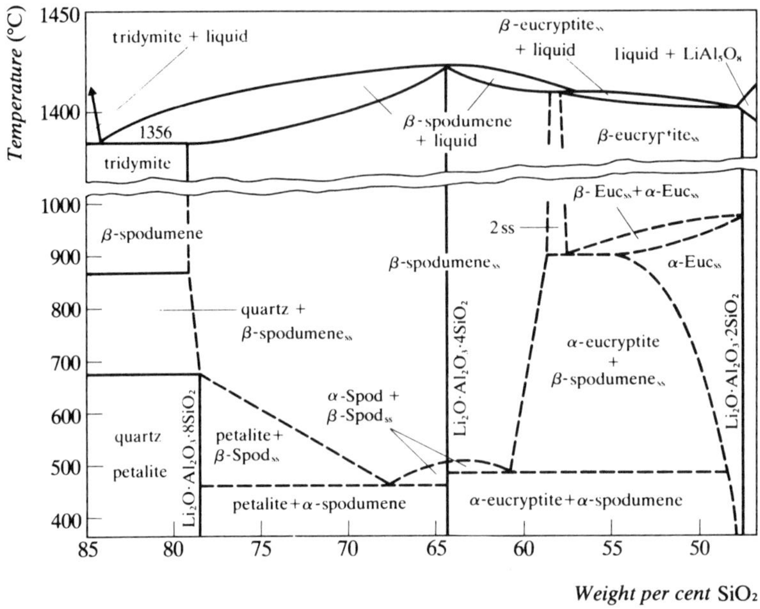

Fig. 245. The phase relations along the join eucryptite ($LiAlSiO_4$)–silica. The higher temperature part of the diagram is after Hatch (1943), slightly modified. The lower part of the diagram shows the probable phase relations at about 10000 lb/in^2 pressure (after Roy *et al.*, 1950).

solid solutions are uniaxial negative). The phase relationships at subsolidus temperatures (Roy *et al.*, 1950) are illustrated in the lower part of Fig. 245, which shows the probable field boundaries at approximately 0·7 kbar.

The effect of pressure on the $LiAlSi_2O_6$–SiO_2 system has not been studied but from the application of high pressure data on the stability of $LiAlSi_2O_6$ composition, Munoz (1969) concludes that the solubility of Li and Al in β-quartz solid solution greatly increases as the pressure increases, and that a complete series of solid solutions are stable at pressures > 10 kbar.

The anhydrous system Li_2SiO_3–$LiAlSi_2O_6$–SiO_2 (Roy and Osborn, 1949) contains two eutectics and a maximum on the boundary curve joining them. At the first eutectic (A in Fig. 246) lithium disilicate ($Li_2O.2SiO_2$), tridymite and a β-spodumene solid solution (containing 33 per cent SiO_2) are in equilibrium, and at the second invariant point (B in Fig. 246) $Li_2O.SiO_2$, $Li_2O.2SiO_2$ and a β-spodumene solid solution (containing 24 per cent SiO_2) are in equilibrium at 975°C.

The low thermal expansion coefficients of glasses in the Li_2O–Al_2O_3–SiO_2 system, and particularly of β-spodumene solid solutions are of particular relevance to glass ceramic technology. The mechanism of crystallization and the role of nucleating additive TiO_2 in glasses with compositions within the system have been studied by a number of investigators (e.g. Blinov, 1969; Barry *et al.*, 1969, 1970).

Pyroxenes with the composition $LiM^{3+}Si_2O_6$, where M^{3+} is Sc, V, Cr and Fe have been synthesized dry at 1 atm and about 1000°C (Brown, 1971; Appleman and Stewart, 1968). Mixes with Ga^{3+} under the same conditions, however, gave β-gallium spodumene, β-$LiGaSi_2O_6$, together with a minor amount of a phase with the same structure as $LiAlSiO_4$, α-eucryptite. The failure to synthesize Ga-, as well as Al-, pyroxenes under these conditions is probably related to the small radii of the Ga^{3+} and Al^{3+} ions, and their consequent preference for tetrahedrally co-

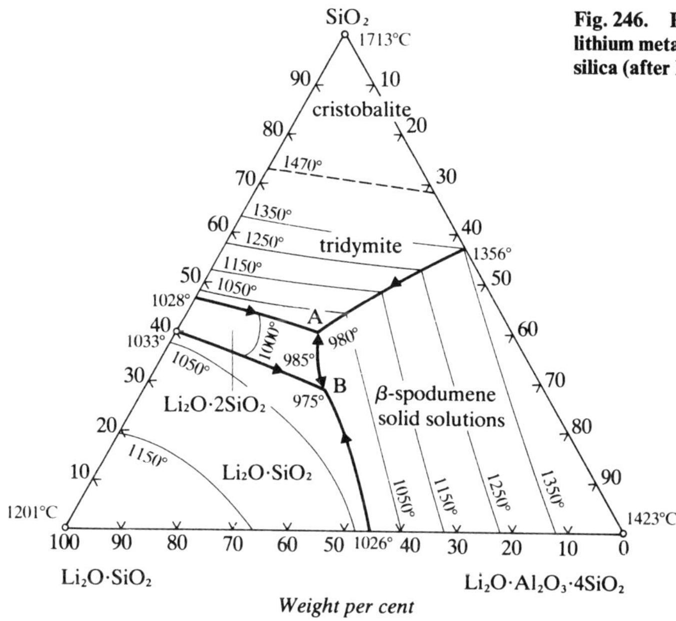

Fig. 246. Phase diagram of the system lithium metasilicate–β-spodumene–silica (after Roy and Osborn, 1949).

ordinated sites in framework structures. The gallium analogue, $LiGaSi_2O_6$, of α-spodumene, as well as germanium α- and β-spodumene, $LiAlGe_2O_6$, have been obtained from glasses of appropriate composition subjected to a pressure of 60000 atm (Blinov, 1969).

The synthesis of α-Cr, V, In and Fe spodumenes from gels of appropriate composition in the temperature range 400°–700°C at 2 kbar has been reported by Drysdale (1975). Under these conditions $LiAlSi_2O_6$ gave β-spodumene at temperatures above 400°C and zeolites + bikitaite, $LiAlSi_2O_6.H_2O$, below 350°C.

Scandium spodumene, $LiScSi_2O_6$, grown from a mixture of Sc_2O_3 and SiO_2, using a lithium molybdate flux has been reported by Ito and Frondel (1968) (see also Toropov and Tsu-Hsiang, 1961).

β-Eucryptite has been synthesized from a melt using V_2O_5 as a mineralizer (Tscherry and Schmid, 1971), and its breakdown at 1000°C and pressures between 30 and 55 kbar, $2\alpha[Li]^4[Al]^4[Si]^4O_4 \rightarrow \alpha[Li]^6[Al]^6[Si]^4_2O_6 + \alpha[Li]^6[Al]^6O_2$, has been described by Neuhaus and Meyer (1965).

Alteration. Spodumene alters readily, particularly as the result of late alkaline hydrothermal activity, and a number of investigations of unaltered to completely altered spodumene have been presented (e.g. Schwartz, 1937; Ginsburg, 1959).

A common alteration product consists of a mixture of eucryptite, hexagonal $LiAlSiO_4$ and albite; the eucryptite is commonly subsequently altered to a lithium-bearing mica, and the resulting mixture is known as cymatolite.

A 'rotten spodumene' from Ravensthorpe, Western Australia, has been shown by Graham (1975) to consist of muscovite, the silicate sheets of which are parallel to {110} and {100} of the original spodumene. Thus transformation has probably taken place by a material exchange in solution by an ideal reaction such as:

$$3LiAlSi_2O_6 + 2H_2O + K^+ \rightarrow KAl_2(Si_3Al)O_{10}(OH)_2 + 3Li^+ + 3SiO_2 + 2(OH)^-$$

Analysis of the Ravensthorpe muscovite shows some replacement of Al by Mg, and the mica could have been derived from a pure spodumene by the reaction:

$$2{\cdot}37LiAlSi_2O_6 + 0{\cdot}24Mg^{2+} + 0{\cdot}85K^+ + 1{\cdot}52H_2O \rightarrow$$
$$K_{0{\cdot}85}Mg_{0{\cdot}24}Al_{1{\cdot}76}(Si_{3{\cdot}39}Al_{0{\cdot}61})O_{10}(OH)_2 + 2{\cdot}37Li^+ + 1{\cdot}35SiO_2 + 4(OH)^-$$

a reaction that involves only a 2 per cent decrease in volume. The replacement of spodumene by aggregates of muscovite and lithium-bearing mica has also been described by Jahns (1953).

Alteration products of the spodumene–quartz aggregates in the intermediate zones of the Tanco pegmatite, Bernic Lake, Manitoba (Čěrný, 1972), include caesian analcite, caesian beryl, lithiophosphate, cookeite, albite, carbonate apatite and montmorillonite–illite. The wide range of secondary mineral association and of the chemical composition of some species in both different and single assemblages indicates that the alteration resulted from the activity of local residual solutions, a conclusion put forward by Wright (1963) for the replacement bodies of the Mountgary pegmatite in the same area.

In the Varuträsk pegmatite (Quensel, 1938) spodumene is locally replaced by cookeite; other alteration products consist of a mixture of clay minerals. Illite and halloysite pseudomorphs after spodumene have been described by Rao (1961), and kaolinite by Schwartz (1937). The combined hydrothermal and supergene alteration of spodumene in the Namivu pegmatite, Mozambique, to dioctahedral chlorite and interlayered dioctahedral chlorite–montmorillonites is reported by Figueiredo Gomes (1967).

The marked decrease in the content of LiO_2 and the disappearance of free quartz above 600°C in spodumene and spodumene–quartz mixtures leached by distilled water at a water vapour pressure of 1 kbar is attributed to the formation of β-spodumene$_{ss}$ and petalite (Grubb, 1973).

Optical and Physical Properties

The refractive indices, optic axial angles and extinction angles $\gamma{:}z$ of spodumenes, in agreement with their relatively small differences in chemical composition, do not show much variation. Spodumenes in which some Li is replaced by Na have a lower α refractive index than those minerals closer to $LiAlSi_2O_6$ in composition; the γ index is not affected by this substitution and the more sodium-rich spodumenes have in consequence a higher birefringence. Scully and Walker (1914) reported a Namaqualand spodumene with a birefringence of 0·028 (α 1·651, β 1·670, γ 1·679) but the sodium content of this mineral is not known. Considerably higher refractive indices are reported for synthetic spodumene, e.g. 1·72 (Šćavničar and Sabatier, 1957), and 1·76 (Barrer and White, 1951), but it is most probable in both cases that the minerals have much higher iron contents than natural spodumenes.

β-Spodumene is tetragonal; the refractive indices given by Roy *et al.* (1950) are ε 1·522, ω 1·516 and by Hatch (1943) ε 1·523, ω 1·518. The optical properties of the synthetic scandium spodumene are α 1·702, β 1·710, γ 1·716, $\alpha{:}z$ 37°, 2V negative, $y = b$, twinning on (100). The natural hydrated lithium aluminium silicate, bikitaite, $LiAlSi_2O_6.H_2O$ (Table 57, anal. 15), is colourless to white, with {001} and {100} cleavage, H = 6, D 2·34 and α 1·510, β 1·521, γ 1·523 $\alpha{:}z$ 38°, $2V_\alpha$ 45° (Hurlbut, 1957).

The colour of the kunzite variety of spodumene is due to the presence of relatively high concentrations of manganese and especially to a low Fe:Mn ratio (Claffy,

1953). The green colour of hiddenite has been attributed to the presence of small amounts of Cr^{3+}, but Leckebusch *et al.* (1974) have shown that it may also be related to the content of Fe^{3+}. Violet-coloured kunzite develops a pink colour on heating, and on irradiation becomes green in colour similar to hiddenite. Kunzite can, however, be distinguished from hiddenite by a luminescence which is not displayed by hiddenite (Bank, 1939). The relationship between luminescence and the contents of Cr, Mn and Fe in ten spodumenes from Afghanistan has been examined by Claffy (1953). The same author has reviewed the literature on the luminescence and tenebrescence of spodumene.

The infrared spectra of the compounds and some solid solutions in the system Li_2O–Al_2O_3–SiO_2 are given by Murthy and Kirby (1962), and of the synthetic α- and β-forms of both spodumene and eucryptite by Blinov (1969). The reflectivity of infrared active modes in spodumene have been investigated by Gervais *et al.* (1973).

The thermal expansion of β-spodumene–silica solid solutions, having the general composition $Li_2O.Al_2O_3.nSiO_2$ ($4 \leq n \leq 11$), have been determined by Ostertag *et al.* (1968). The expansion is anisotropic in the temperature range 0° to 1200°C, with contraction of the *a* and increase of the *c* dimension. The volume expansion becomes increasingly smaller as the SiO_2 content of the solid solution increases. The coefficients of thermal expansion of eucryptite, parallel to *y* and *z* are $8{\cdot}21 \times 10^{-6}$ and $17{\cdot}6 \times 10^{-6}$ respectively.

A petrofabric study of spodumene in some east Siberian pegmatites has shown that the optic orientations of the pyroxenes are related to the fracture system within the pegmatite (Kochnev, 1964).

Distinguishing Features

Spodumene is distinguished from other monoclinic pyroxenes, except clinoenstatite and some aegirine-augites, by the small extinction angle. It is distinguished from clinoenstatite by the position of the optic axial plane, and from aegirine-augite by the pleochroic scheme and higher refractive indices of the latter.

Paragenesis

Spodumene is a characteristic mineral of the lithium-rich granitic pegmatites in which it is usually restricted to the more central parts of the pegmatite, and commonly associated with quartz, petalite, albite, lepidolite and beryl. In the Varuträsk pegmatite (Quensel, 1938, 1955), spodumene (Table 57, anals. 8, 9) occurs in the intermediate zone between the wall zone and the core. During the pegmatite period of crystallization the spodumene, developed in thick tabular masses, is associated chiefly with quartz, the two minerals commonly occurring together in approximately equal amounts. A second generation of spodumene crystallized during a later pneumatolytic period of mineralization. This spodumene occurs in semitranslucent slender laths, and is commonly altered to a mixture of clay minerals.

The Rhodesian lithium-bearing pegmatites consist of two main varieties, the Kamativi and Bikita types (Grubb, 1973). The former is characterized by the presence of coarse-grained spodumene and the absence of primary petalite indicating that crystallization of the pegmatite liquid began at or below 600°C, and was completed in an essentially closed system. The Bikita type contains primary petalite and secondary fine-grained aggregates of spodumenes and quartz, an association that is probably related to a period of secondary boiling in a closed system that was then followed by fracturing of the wall rock. The consequent escape of the gaseous phase led to supersaturation in the liquid and the rapid crystallization of a metastable spodumene–quartz assemblage in place of petalite. That this phase in the formation of the pegmatites was followed by late stage metasomatism is shown by the large scale replacement of petalite by fine acicular spodumene and quartz, and the common alteration of spodumene to lepidolite (Hornung and von Knorring, 1962). The Augustus pegmatite near Salisbury is estimated to contain up to 12 per cent spodumene (Gallagher, 1975), and higher values are reported for the pegmatites of the Kings Mountain district, North Carolina (Griffiths, 1954).

Spodumene is the main lithium mineral in the pegmatites of the Saltpond area, Ghana (Amaoko-Mensah, 1972). Lepidolite is absent and only small amounts of muscovite, beryl and tourmaline are present in these pegmatites, presumably the result of an unusually low content of volatiles such as water, boron and fluorine in the magmatic fluids from which they were derived. The pegmatites, however, show a remarkable series of secondary phosphates including frondelite, $(Mn,Fe^{2+})Fe_4^{3+}(OH)_5(PO_4)_3$, manganoan lipscombite, $(Fe^{2+},Mn)Fe_2^{3+}(OH)_2(PO_4)_2$ and heterosite, $(Fe,Mn)PO_4$.

The complex Tanco pegmtite at Bernic Lake, Manitoba, is similar to the Varuträsk and the Bikita pegmatites. The spodumene, occurring as tabular aggregates of waterclear fibrous crystals intergrown with quartz, is located in the lower and upper intermediate zones in which it is associated with microcline-perthite + albite + quartz (+ amblygonite), and quartz + amblygonite (+ petalite) respectively (Crouse and Černý, 1972). In some parts of the spodumene-bearing zones, particularly those rich in quartz, it occurs in white lath-shaped crystals (Table 57, anal. 2), and in the central intermediate zone the pyroxene forms fibrous and columnar crystals, waterclear to pale greenish in colour, that are usually embedded in quartz. Elsewhere veinlets of fibrous spodumene, intergrown with quartz, are developed along the petalite cleavages. The fibre axes, parallel to z, are oriented along both cleavages and the γ direction of petalite, and are similar in appearance to the fibrous tabular masses of spodumene and quartz that commonly show the characteristic wedge-shaped habit of spodumene present in other parts of the zones. This textural evidence of the formation of spodumene + quartz by the breakdown of petalite is supported by the almost identical chemical composition of the spodumene + quartz aggregates and petalite (Černý and Ferguson, 1972). In the central intermediate zone, however, spodumene appears to have a primary origin, a conclusion based on the absence of relict petalite morphology, as well as its somewhat different physical properties including a higher content of iron.

One of the most abundant developments of spodumene is probably that of the North Carolina spodumene pegmatite belt for which Hess (1946) has estimated a reserve of some 1500000 tonnes of spodumene. The pegmatites consist mainly of microcline, albite, quartz, muscovite and spodumene; smaller amounts of cassiterite, tourmaline, apatite and amblygonite also occur. The spodumene crystals vary considerably in size and the largest are some 80 cm in length, but there are no very large crystals as are found in the pegmatites of the Black Hills, South Dakota. In the pegmatites of North Carolina, spodumene never occurs associated with much muscovite, and Hess has suggested that the spodumene is in part derived by replacement of the mica. The replacement of microcline by spodumene also occurs, and partly and wholly replaced microcline crystals are commonly observed.

Tantalum and caesium-bearing minerals are not infrequently associated with the lithium-rich pegmatites, and the accessory minerals of the Siberian pegmatite province spodumene–microcline–albite bodies include microlite $(Ca,Na)_2(Ta,Nb)_2O_6(O,OH,F)$, tantalite $(Fe,Mn)(Ta,Nb)_2O_6$, and pollucite $(Cs,Na)(AlSi_2O_6).H_2O$, as well as montebrasite, beryl and cassiterite (Filippova, 1970). Rare metal granitic pegmatites are also reported from the Sayan Mountains, Siberia, the spodumene–microcline–albite–quartz–muscovite pegmatites of which contain montebrasite, triphylite, pollucite and lepidolite as accessory constituents. Here the spodumene was formed during an early stage of pegmatitization, and was later altered by a subsequent period of sodium metasomatism during which large amounts of lithium were removed from the pegmatites (Slepnev, 1964). The association of cassiterite with spodumene is not uncommon, and has been reported from the Doade pegmatites, Galicia, Spain (Hensen, 1967) and the Katumba pegmatites, Congo (Varlamoff, 1961). Graphic intergrowths of spodumene and quartz, the replacement of K-feldspar by spodumene, and its development in symplectitic rims around albite occur in the Galician pegmatites.

The spodumene (Table 57, anal. 14) from the Ooregum mine described by Rao and Rao (1939) is associated with lithiophilite, apatite, tourmaline, beryl, sillimanite, kyanite and native arsenic. The paragenesis of spodumene in pegmatites as well as the formation of other pegmatite minerals has been discussed by Nikitin (1957).

A number of other occurrences of spodumene-bearing pegmatites, particularly in relation to their economic importance, are described by Heinrich (1975).

References

Amaoko-Mensah, A., 1972. Geochemical aspects of spodumene pegmatites of the Saltpond area, Ghana. *16th Ann. Rept. Res. Inst. African Geol., Univ. Leeds*, 58–61.

Angel, F., 1933. Spodumen und Beryll aus dem Pegmatiten von St. Radegund bei Graz. *Min. Petr. Mitt., Tschermak*, **43**, 441–446.

Appleman, D. E. and **Stewart, D. B.**, 1968. Crystal chemistry of spodumene-type pyroxenes. *Geol. Soc. America Spec. Paper* (Abs), **101**, 5–6.

Babu, S. K., 1975. On an occurrence of hiddenite from a pegmatite near Kabbur, Hassan district, Mysore State. *Current Sci., India*, **44**, 387–388.

Bank, H., 1939. Spodumene, seine Varietäten. Hiddenit und Kunzite. *Z. Deutsch. Gemmol. Gesell.*, **18**, 176–180.

Barrer, R. M. and **White, E. A. D.**, 1951. The hydrothermal chemistry of silicates, Pt. I. Synthetic lithium aluminosilicates. *J. Chem. Soc.*, 1267–1278.

Barry, T. I., Clinton, D., Lay, L. A., Mercer, R. A. and **Miller, R. P.**, 1969. The crystallisation of glasses based on eutectic compositions in the system $Li_2O–Al_2O_3–SiO_2$. Pt. I. Lithium metasilicate–β-spodumene. *J. Materials Sci.*, **4**, 596–612.

Barry, T. I., Clinton, D., Lay, L. A., Mercer, R. A. and **Miller, R. P.**, 1970. The crystallisation of glasses based on eutectic compositions in the system $Li_2O–Al_2O_3–SiO_2$. Pt. II. Lithium metasilicate–β-eycryptite. *J. Materials Sci.*, **5**, 117–126.

Bazarov, L. Sh. and **Motorina, I. V.**, 1969. Physicochemical conditions during formation of rare-metal pegmatite of the sodium–lithium type. *Dokl. Acad. Sci. U.S.S.R., Earth Sci. Sect.*, **188**, 124–126.

Blinov, V. A., 1969. The mechanism of nucleated crystallisation of glasses in lithia–alumina–silica and cordierite systems. *J. Materials Sci.*, **4**, 461–468.

Brown, W. L., 1971. On lithium and sodium trivalent-metal pyroxenes and crystal-field effects. *Min. Mag.*, **38**, 43–48.

Cameron, M., Sueno, S., Prewitt, C. T. and **Papike, J. J.**, 1973. High-temperature crystal chemistry of acmite, diopside, hedenbergite, jadeite, spodumene and ureyite. *Amer. Min.*, **58**, 594–618.

Černý, P., 1972. The Tanco pegmatite at Bernic Lake, Manitoba. VIII. Secondary minerals from the spodumene-rich zones. *Can. Min.*, **11**, 714–726.

Černý, P. and **Ferguson, R. B.**, 1972. The Tanco pegmatite at Bernic Lake, Manitoba. IV. Petalite and spodumene relations. *Can. Min.*, **11**, 660–678.

Chang, J.-P., 1974. [Preliminary study of a spodumene-pegmatite in China.] *Geochimica*, 182–191 (Chinese with English abstract).

Claffy, E. W., 1953. Composition, tenebrescence and luminescence of spodumene minerals. *Amer. Min.*, **38**, 919–931.

Clark, J. R., **Appleman, D. E.** and **Papike, J. J.**, 1969. Crystal chemical characterization of clinopyroxenes based on eight new structure refinements. *Min. Soc. America, Spec. Paper* **2**, 31–50.

Clarke, P. T. and **Spink, J. M.**, 1969. The crystal structure of β-spodumene $LiAlSi_2O_6$–II. *Z. Krist.*, **130**, 420–426.

Crouse, R. A. and **Černý, P.**, 1972. The Tanco pegmatite at Bernic Lake, Manitoba. I. Geology and paragenesis. *Can. Min.*, **11**, 591–608.

Drysdale, D. J., 1975. Hydrothermal synthesis of various spodumenes. *Amer. Min.*, **60**, 105–110.

Edgar, A. D., 1968. The α–β $LiAlSi_2O_6$ (spodumene) transition from 5000 to 45000 lb/in^2 P_{H_2O}. *Intern. Min. Assoc. Papers Proc. Fifth General Meeting, Cambridge, 1966*, 222–231.

Eppler, R. A., 1963. Glass formation and recrystallization in the lithium metasilicate region of the system Li_2O–Al_2O_3–SiO_2. *J. Amer. Ceramic Soc.*, **46**, 97–101.

Figueiredo Gomes, C. S., 1967. Alteration of spodumene and lepidolite with formation of dioctahedral chlorite, plus dioctahedral chlorite–dioctahedral montmorillonite interstratifications. *Mem. Noticias, Mus. Lab. Min. Geol. Univ. Coimbra*, **64**, 32–57 (M.A. 75-2419).

Filippova, Yu. I., 1970. A new paragenetic type of tantalum- and cesium-bearing pegmatite. *Dokl. Acad. Sci. U.S.S.R., Earth Sci. Sect.*, **192**, 123–126.

Gabriel, A., **Slavin, M.** and **Carl, H. F.**, 1942. Minor constituents in spodumene. *Econ. Geol.*, **37**, 116–125.

Gallagher, M. J., 1975. Composition of some Rhodesian lithium–beryllium pegmatites. *Trans. Geol. Soc. South Africa*, **78**, 35–41.

Gervais, F., **Piriou, B.** and **Servoin, J.-L.**, 1973. Étude par réflexion infrarouge des modes internes et externes de quelques silicates. *Bull. Soc. franç. Min. Crist.*, **96**, 81–90.

Gillery, F. H. and **Bush, E. A.**, 1959. Thermal contraction of β-eucryptite ($Li_2O.Al_2O_3.2SiO_2$) by X-ray and dilatometer methods. *J. Amer. Ceram. Soc.*, **42**, 175–177.

Ginsburg, A. I., 1959. Spodumene and the processes of its alteration. *Trans. Min. Mus. Acad. Sci. U.S.S.R.*, **9**, 19–52.

Gordienko, V. V. and **Kalenchuk, G. E.**, 1966. [The chemical nature of spodumene.] *Zap. Vses. Min. Obshch.*, **95**, 169–180.

Graham, J., 1975. Some notes on α-spodumene, $LiAlSi_2O_6$. *Amer. Min.*, **60**, 919–923.

Griffiths, W. R., 1954. Beryllium resources of the tin-spodumene belt. North Carolina. *U.S. Geol. Surv., Circ.*, **309**, 1–12.

Grip, E., 1941. A lithium pegmatite on Kluntarna in the archipelago of Piteå. *Geol. För. Förh. Stockholm*, **62**, 380–390.

Grubb, P. L. C., 1973. Paragenesis of spodumene and other lithium minerals in some Rhodesian pegmatites. *Spec. Publ. Geol. Soc. South Africa*, **3**, 201–216.

Haapala, I., 1966. On the granitic pegmatites in the Peräseinäjoki–Alavus area, South Pohjanmaa, Finland. *Bull. Comm. géol. Finlande*, **224**, 1–98.

Hahn, Th. and **Behruzi, M.**, 1968. New germanates with chain structures. *Z. Krist.*, **127**, 160–162.

Hatch, R. A., 1943. Phase equilibrium of the system Li_2O–Al_2O_3–SiO_2. *Amer. Min.*, **28**, 471–496.

Henglein, E., 1956. Zur Kenntnis der Hochtemperatur-Modifikationen von Lithium-Aluminium-Silikaten. *Fortschr. Min.*, **34**, 40–43.

Heinrich, E. W., 1975. Economic geology and mineralogy of petalite and spodumene pegmatites. *Indian J. Earth Sci.*, **2**, 18–29.

Hensen, B. J., 1967. Mineralogy and petrography of some tin, lithium and beryllium bearing albite-pegmatites near Doade, Galicia, Spain. *Leidse Geol. Mededel.*, **39**, 249–259.

Hess, F. L., 1946. The spodumene pegmatites of North Carolina. *Econ. Geol.*, **35**, 942–966.

Hornung, G. and **Knorring, O. von**, 1962. The pegmatites of the North Mtoko region, Southern Rhodesia. *Trans. Geol. Soc. South Africa*, **65**, 153–180.

Hurlbut, C. S., *Jr.*, 1957. Bikitaite, $LiAlSi_2O_6.H_2O$, a new mineral from Southern Rhodesia. *Amer. Min.*, **42**, 792–797.

Ioffe, V. A. and **Zonn, Z. N.**, 1970. Growth of single crystals of β-eucryptite and β- and γ-spodumene. *Soviet Physics – Crystallography*, **15**, 342–343.

Isaacs, T. and **Roy, R.**, 1958. The α–β inversions of eucryptite and spodumene. *Geochim. Cosmochim. Acta*, **15**, 213–217.

Ito, J. and **Frondel, C.**, 1968. Syntheses of the scandium analogues of aegirine, spodumene, andradite and melanotekite. *Amer. Min.*, **53**, 1276–1280.

Jahns, R. H., 1953. The genesis of pegmatites. II. Quantitative analysis of lithium-bearing pegmatite, Mona County, New Mexico. *Amer. Min.*, **38**, 1078–1112.

Kochnev, A. P., 1964. [The orientation of spodumene in East Siberian pegmatites.] *Zap. Vses. Min. Obshch.* (Mem. All-Union Min. Soc.), **93**, 46–53 (in Russian).

Leckebusch, R., **Recker, K.** and **Triché, C.**, 1974. Relation entre couleur et luminescence des spodumènes d'Afghanistan et teneurs en chrome, manganese et fer. *C.R. Acad. Sci., Paris*, **278**, sér. D, 1541–1544.

Li, C. T. and **Peacor, D. R.**, 1968a. The crystal structure of $LiAlSi_2O_6$–II (β-spodumene). *Z. Krist.*, **126**, 46–65.

Li, C. T. and **Peacor, D. R.**, 1968b. The crystal structure of $LiAlSi_2O_6$–III (high quartz solid solution). *Z. Krist.*, **127**, 327–348.

Murthy, M. K. and **Kirby, E. M.**, 1962. Infrared study of compounds and solid solutions in the system lithia–alumina–silica. *J. Amer. Ceram. Soc.*, **45**, 324–329.

Munoz, J. L., 1969. Stability relations of $LiAlSi_2O_6$ at high pressures. *Min. Soc. America, Spec. Paper* **2**, 203–209.

Neuhaus, A. and **Meyer, H. J.**, 1965. Hochdruckverhalten des Eukryptits, (α-$LiAlSiO_4$) und anderer oxidischer Phasen. *Naturw.*, **52**, 639–640.

Nikitin, V. D., 1957. [Features of the rare metals mineralization in pegmatite veins.] *Mem. Soc. Russe Min.*, **86**, 27 (in Russian) (M.A. **13**-366).

Ostertag, W., **Fischer, G. R.** and **Williams, J. P.**, 1968. Thermal expansion of synthetic β-spodumene and β-spodumene–silica solid solutions. *J. Amer. Ceram. Soc.*, **51**, 651–654.

Pye, E. G., 1965. Geology and lithium deposits of Georgia Lake area. *Ontario Dept. Mines Geol., Rept.*, **31**, 1–113.

Quensel, P., 1938. Minerals of the Varuträsk pegmatite, X. Spodumene and its alterations products. *Geol. För. Förh. Stockholm*, **60**, 201–215.

Quensel, P., 1955. The paragenesis of the Varuträsk pegmatite, including a review of its mineral assemblage. *Arkiv. Min. Geol.*, **2**, 9–125.

Rao, A. B., 1961. Pseudomorfos de espodumênio na Borborema. *Noticias de Perquisas, J. Clube Min. Recife, Brazil*, **2**, 28 (M.A. **16**-104)..

Rao, B. R. and **Rao, M. D. R.**, 1939. Spodumene and its associated minerals from the Ooregum mine, Kolar gold field. *Rec. Mysore Geol. Dept.*, **37**, 38–42 (M.A. **8**-86).

Roy, R., 1959. Silica O, a new common form of silica. *Zeit. Krist*, **111**, 185–189.

Roy, R. and **Osborn, E. F.**, 1949. The system lithium metasilicate–spodumene–silica. *J. Amer. Chem. Soc.*, **71**, 2086–2095.

Roy, R., **Roy, D.** and **Osborn, E. F.**, 1950. Compositional and stability relationships among the lithium alumino-silicates, eucryptite, spodumene and petalite. *J. Amer. Ceram. Soc.*, **33**, 152–159.

Ščavničar, S. and **Sabatier, G.**, 1957. Action du chlorure du lithium sur les feldspaths alcalins. Données nouvelles sur le feldspath-Li, le spodumène-Fe et l'α-eucryptite. *Bull. Soc. franç. Min. Crist.*, **80**, 308–317.

Schwartz, G. M., 1937. Alteration of spodumene to kaolinite in the Etta mine. *Amer. J. Sci.*, 5th ser., **33**, 303–307.

Schwartz, G. M. and **Leonard, R. J.**, 1926. Alteration of spodumene in the Etta mine, Black Hills, S.D. *Amer. J. Sci.*, 5th ser., **11**, 257–264.

Scully, G. C. and **Walker, A. R. E.**, 1914. Note on spodumene from Namaqualand. *Trans. Roy. Soc. S. Africa*, **4**, 65–67.

Sheshulin, G. I., 1961. Composition of gas-liquid inclusions in the minerals of spodumene pegmatites. In *New Data on Rare Element Mineralogy*. Ginsburg (ed.). New York (Consultants Bureau), 47–85.

Slepnev, Yu. S., 1964. Geochemical characteristics of the rare metal granitic pegmatites of the Sayan Mountains. *Geochem. Intern.*, **1**, 221–228.

Stewart, D. B., 1960. The system $LiAlSiO_4$–$NaAlSi_3O_8$–H_2O at 2000 bars. *Intern. Geol. Congr., 21 Sess.*, pt. **17**, 15–30.

Svec, H. J. and **Anderson, A. R.**, *Jr.*, 1965. The absolute abundance of the lithium isotopes in natural sources. *Geochim. Cosmichim. Acta*, **29**, 633–641.

Thoreau, J., 1953. Minéraux lithiques de pegmatites du Congo belge et du Ruanda. *Bull. Cl. Sci. Acad. Roy. Belgique*, 5th Ser., **39**, 684–687.

Toropov, N. R. and **Tsu-Hsiang, L. N.**, 1961. The system $LiAlSi_2O_6$–$LiGaSi_2O_6$. *J. Inorg. Chem. (U.S.S.R.)*, **6**, 928–932.

Tscherry, V. and **Schmid, R.**, 1971. Zuchtung und optische Eigenschaften von β-Eukryptit-Einkristallen, $LiAlSiO_4$. *Z. Krist.*, **133**, 110–113.

Udagawa, S., **Ikawa, H.** and **Urabe, K.**, 1973. Formation of β-spodumene by solid state reaction between pyrophyllite and lithium carbonate. *Proc. Intern. Clay Conf. (Madrid)*, 141–147.

Varlamoff, N., 1961. Matériaux pour l'étude des pegmatites du Congo et du Ruanda. Quatrième note: Pegmatites à amblygonite et à spodumène et pegmatites fortement albitisées à spodumène et à cassiterite de la région de Katumba (Ruanda). *Ann. (Bull.) Soc. géol. Belgique*, **84**, 257–278.

Warren, B. E. and **Biscoe, J.,** 1931. The crystal structure of the monoclinic pyroxenes. *Z. Krist.*, **80**, 394–401.

Wright, C. M., 1963. Geology and origin of the pollucite-bearing Mountgary pegmatite, Manitoba. *Bull. Geol. Soc. America*, **74**, 919–946.

Single-Chain Silicates: Non-Pyroxenes

Wollastonite $CaSiO_3$
Pectolite $Ca_2NaH[SiO_3]_3$
Bustamite $(Mn,Ca,Fe)SiO_3$
Rhodonite $(Mn,Fe,Ca)SiO_3$
Pyroxmangite $(Mn,Fe)SiO_3$
Sapphirine $(Mg,Fe^{2+},Fe^{3+},Al)_8O_2[(Al,Si)_6O_{18}]$
Aenigmatite $Na_2Fe_5^{2+}TiO_2[Si_6O_{18}]$
Rhönite $Ca_2(Mg,Fe^{2+},Fe^{3+})_5TiO_2[Si,Al)_6O_{18}]$
Serendibite $Ca_2(Mg,Al)_6O_2[(Si,Al,B)_6O_{18}]$

Wollastonite $Ca[SiO_3]$

Triclinic (−)

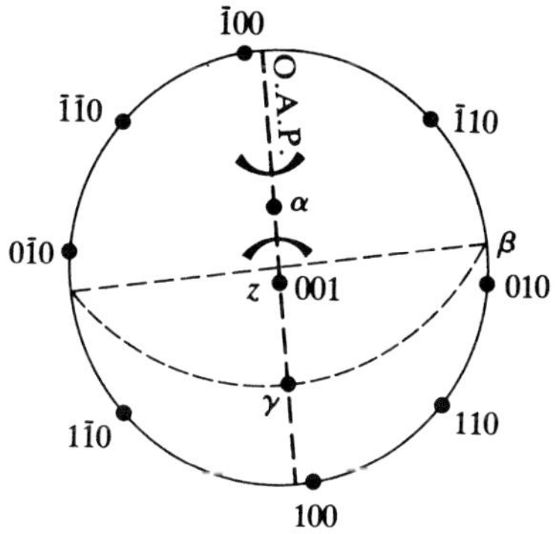

α	1·616–1·640
β	1·628–1·650
γ	1·631–1·653
δ	0·013–0·014
$2V_\alpha$	36°–60°
$\alpha:z = 30°–44°$, $\beta:y = 0°–5°$	
O.A.P.	approx. (010)
Dispersion:	$r > v$
D	2·86–3·09
H	$4\frac{1}{2}$–5
Cleavage:	{100} perfect, {001} and {$\bar{1}$02} good; on (010) sections (100):(001) = $84\frac{1}{2}$°, (100):($\bar{1}$02) = 70°
Twinning:	Common; twin axis [010], composition plane (100).
Colour:	Usually white, sometimes colourless, grey or very pale green; colourless in thin section.
Unit cell:	*a* 7·94, *b* 7·32, *c* 7·07 Å α 90°02′, β 95°22′, γ 103°26′ Z = 6. Space group $P\bar{1}$.

Decomposed by concentrated HCl.

There are three normal modifications of calcium metasilicate: pseudowollastonite (β-$CaSiO_3$),[1] the high-temperature form, is triclinic (pseudo-orthorhombic), and is known in nature only in pyrometamorphosed rocks. Wollastonite-Tc and wollastonite-2M are both referred to as the low-temperature form (α-$CaSiO_3$): their properties are closely related, the triclinic form being distinguished from the monoclinic form (previously known as parawollastonite) by the slight inclination of the optic axial plane to the plane normal to *y*, or by asymmetry in the single-crystal X-ray pattern. Wollastonite is a common mineral of metamorphosed limestones and similar assemblages. The name is after W. H. Wollaston, British chemist and mineralogist.

[1] The high-temperature form pseudowollastonite is termed β-$CaSiO_3$ in accordance with accepted mineralogical usage for the high-temperature phases of a compound. Cement research workers have, however, following Rankin and Wright (1915), exactly reversed this convention and term pseudowollastonite α-$CaSiO_3$. The use of the terms high-temperature or low-temperature is thus to be preferred. The I.M.A. Commission on New Minerals and Mineral Names approved the use of the names wollastonite (low temperature form) and pseudowollastonite (high-temperature form), but recommended that the term parawollastonite be dropped in favour of the use of suffixes (-Tc, -2M, etc.) to distinguish stacking polymorphs of $CaSiO_3$ (Amorós, 1963); parawollastonite thus becomes wollastonite-2M.

Structure

Wollastonite was at one time classed structurally with the pyroxene group, but Warren and Biscoe (1931) showed that its unit cell has no relation to that of diopside: the triclinic nature of the wollastonite structure was confirmed by Peacock (1935), and the cell parameters of the monoclinic form parawollastonite were obtained by Barnick (1935). Further detailed work on wollastonite by Dornberger-Schiff *et al.* (1954), Buerger (1956) and by Mamedov and Belov (1956) indicated that it has a different type of infinite-chain silicate structure. Three-dimensional X-ray single-crystal work has confirmed that the monoclinic phase, wollastonite-2M (Table 58, anal. 1), which is related to the triclinic wollastonite by a simple stacking modification (Ito, 1950), also has this type of structure (Trojer, 1968). Buerger and Prewitt (1961) have shown that the crystal structure proposed by Mamedov and Belov (1956) for wollastonite is correct (see also Prewitt and Buerger, 1963; Peacor and Prewitt, 1963). Basically the structure consists of infinite chains containing three tetrahedra per unit cell running parallel to *y*; this repeat unit can be considered as consisting of a pair of tetrahedra joined apex to apex as in the $[Si_2O_7]$ group, alternating with a single tetrahedron with one edge parallel to the chain direction (Fig. 247(a)). The arrangement of the oxygen atoms approximates close packing in a crude way with an obvious layering parallel to (101). Layers of Ca atoms in octahedral co-ordination alternate with layers composed of Si atoms between the sheets of oxygen atoms, as can be seen from a *y*-axis projection (Fig. 251(b), p. 566).

Unusual features in the structure of wollastonite include the presence of a

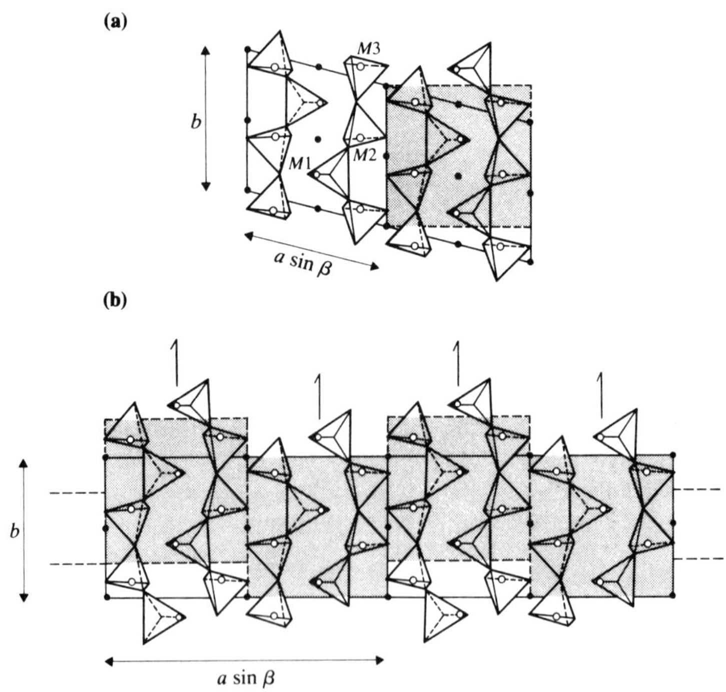

Fig. 247. (a) Projection of the structure of wollastonite-Tc along the *z* axis showing chains of Si–O tetrahedra and Ca atoms (o). Shaded area is that described by Ito (1950) as having pseudomonoclinic symmetry (after Trojer, 1968). (b) Stacking of pseudomonclinic units (see above) with $\pm b/4$ displacements to produce the structure of wollastonite-2M (after Trojer, 1968).

substructure along y with a period of $b/2$. Ito (1950) suggested that triclinic wollastonite could be constructed by starting with a hypothetical monoclinic cell and shifting successive cells along (100) in increments of $\pm\frac{1}{4}b$. Prewitt and Buerger (1963) expanded this idea and proposed the space group $P2_1/a$ for wollastonite-2M with additional local centres of symmetry not required by this space group. Trojer (1968) verified this relationship and showed (Fig. 247(b)) how a cell regarded as having pseudomonoclinic symmetry can be selected from the structure of wollastonite. When these units are stacked together a rough model of the structure of wollastonite-2M (parawollastonite) is obtained.

A fibrous wollastonite from Devon was considered by Jeffery (1953) to consist of complex intergrowths of both triclinic and monoclinic forms with packing mistakes, or alternatively with a twinning relationship. Coe (1970) suggested that the parawollastonite–wollastonite transition may be analogous with that of orthoenstatite to clinoenstatite, and be associated with shear stress.

Wollastonite with a well-ordered superstructure with $a \simeq 32$ Å was reported by Wenk (1969) who found that polymorphism of wollastonite is very common from various parageneses. Single-crystal precession photographs show structural differences; the overall symmetry is monoclinic but in detail, regarding intensities, at least some of the structures are clearly triclinic. If A is the normal wollastonite 1T unit cell and B is a unit cell related to the original one by a mirror reflection on (100), or by rotation or a screw axis around [010], or by translation or glide parallel to [010], all with (100) as composition plane, then for the different polymorphs the following stacking sequences of A and B could occur:

1T: A A A A
2M: A B A B
4T: A A A B A A A B
4M: A A B B A A B B
∞T: A B A A B B B

As the units might be distorted in the packing, a slightly triclinic symmetry can be expected also for the 2M and 4M arrangement (Wenk, 1969).

It has been suggested (Jefferson and Bown, 1973) that a crystal of natural wollastonite reported to give X-ray diffraction patterns of 4M polytype with triclinic distortion may represent a fine intergrowth of 1T and 2M forms. Further electron microscopic evidence of polytypism in wollastonite has been presented by Wenk, Müller, Liddell, and Phakey (1976) and by Hutchison and McLaren (1976) who confirmed that simple stacking faults and twinning which also involves displacement in the composition plane both occur. A wide domain of parawollastonite may be produced by stacking faults in every second triclinic unit cell, whereas a twin lamella may be produced by faults in every adjacent triclinic unit cell (Hutchison and McLaren, 1977). In wollastonites in which there is partial substitution of iron or manganese for calcium (still with the wollastonite as opposed to the bustamite structure) there may be some initial degree of ordering of Fe and Mn in two of the three Ca sites (*M*1, *M*2 and *M*3 of Fig. 247(a)). With the entry of Mn and Fe into the structure there is a distinct decrease in cell volume at least as far as 10 mol. per cent $(Fe,Mn,Mg)SiO_3$ where there may be a structural break (Rutstein, 1971; Matsueda, 1973), though for Mn in synthetic material, Rutstein and White (1971) have shown that the characteristic wollastonite infrared spectrum remains intact to about $Ca_{0.75}Mn_{0.25}SiO_3$ and then changes over a very narrow compositional range to the bustamite pattern. The wollastonite spectrum has two distinctive clusters, each with three bands, in the high frequency region at around

$1000\,cm^{-1}$. With the addition of Fe, this characteristic wollastonite pattern persists only to $Ca_{0.88}Fe_{0.12}SiO_3$ and then gives way to that of bustamite. It thus appears that wollastonite will accept only limited amounts of either Fe or Mn before cation ordering takes place. Detailed structural refinement of the occupancies in the octahedral sites of two wollastonites containing approximately 4 mol. per cent $MnSiO_3$ or $FeSiO_3$ (Ohashi and Finger, 1976) has shown that Mn or Fe atoms are distributed over the three sites *M*1, *M*2 and *M*3 with the *M*3 site in iron-wollastonite being only slightly richer in Fe. This nearly disordered distribution thus places a limit on how far the smaller cations Mn, Fe or Mg can substitute for Ca in the wollastonite structure.

Pseudowollastonite has been investigated by Clark (1946) and more fully by Jeffery and Heller (1953). The latter authors derived cell dimensions on synthetic material and showed that the cell is triclinic but with planes of pseudosymmetry perpendicular to all three crystallographic axes so that it is thus pseudohexagonal: a larger pseudo-orthorhombic cell may also be chosen. Pseudowollastonite has been shown to be isostructural with $SrGeO_3$ (Hilmer, 1958) and has a similar basic structure to that originally proposed by Barnick (1935) for low-temperature wollastonite.

The cell parameters for $CaSiO_3$ may thus be tabulated:

	a(Å)	*b*(Å)	*c*(Å)	α	β	γ	Space group	*Z*
Wollastonite-Tc	7·94	7·32	7·07	90°02′	95°22′	103°26′	$P\bar{1}$	6
Wollastonite-2M	15·43	7·32	7·07	90°	95°24′	90°	$P2_1/a$	12
Pseudowollastonite	6·90	11·78	19·65	90°	90°48′	90°	$P1$ or $P\bar{1}$	24
	(pseudohexagonal cell *a* 6·82, *c* 19·65 Å)							

Powder X-ray data for these various forms have been listed by Heller and Taylor (1956) who give in detail the distinguishing features of the X-ray powder diffraction patterns of wollastonite-Tc and wollastonite-2M. It is claimed that the distinction can be made by noting the separation between the strong 3·83 Å line and the one adjacent on the low-angle side; this has a *d*-spacing of 4·05 Å in wollastonite and 4·37 Å in wollastonite-2M. See also Ueda and Tomita (1968).

Chemistry

Although normally fairly close in composition to $CaSiO_3$, the wollastonite structure can accept considerable amounts of Fe and Mn, and lesser amounts of Mg, all replacing Ca (Table 58). Some material earlier reported as ferroan wollastonite has been found to have the bustamite structure, but the wollastonite structure appears to be retained up to about 10 mol. per cent (Rutstein, 1971) and it would appear that around double this amount of Mn^{2+} can be accommodated (e.g. Table 58, anal. 12). Matsueda (1973) reports a break in the linear variation in cell parameters at near 10 mol. per cent (Fe,Mn,Mg) but the infrared spectra for synthetic material maintain the wollastonite pattern to $Ca_{0.75}Mn_{0.25}SiO_3$ (Rutstein and White, 1971). Analyses 9 and 11 (Table 58) are considered by Mason (1973) to represent bustamite on the basis of their refractive indices, though Nambu *et al.* (1971) have reported that manganoan wollastonite of composition $(Ca_{4.18}Mn_{1.08}Fe^{2+}_{0.59}Mg_{0.16})(Si_{5.98}Fe^{3+}_{0.06}Al_{0.04})O_{18}$ still shows the wollastonite pattern on a Weissenberg X-ray photograph.

Table 58. Wollastonite Analyses

	1		2		3		4		5		6	
SiO_2	51·56		50·82		51·22		50·15		52·02		50·46	
Al_2O_3	0·15		—		0·22		0·14		—		0·08	
Fe_2O_3	0·21		—		0·08		0·04		—		0·72	
FeO	0·08		0·18		0·62		0·60		1·94		0·46	
MnO	0·06		0·03		0·04		0·96		0·32		0·61	
MgO	0·26		0·22		0·08		0·00		1·81		0·26	
CaO	47·73		48·16		47·61		46·97		43·89		46·69	
Na_2O	0·02		0·12		0·04		0·35		—		—	
K_2O	0·00		0·07		0·00		0·05		—		—	
H_2O^+	0·03		0·08		0·05		0·38		—		0·48	
H_2O^-	0·02		0·00		0·11		—		—		—	
Total	100·12		99·68		100·07		100·09		99·98		99·76	
α	1·618		1·619		1·616		1·618		—		1·623	
β	1·628		1·630		1·630		1·629		1·629		1·632	
γ	1·631		1·632		1·632		1·633		—		—	
δ	0·013		0·013		0·016		0·015		—		—	
$2V_\gamma$	—		39°		42°		40°		37°–38°		35°–36°	
D	2·922		—		2·882		2·859		2·95		—	
	Numbers of ions on the basis of eighteen O											
Si	5·976	6.00 (Si+Al)	5·947		5·962	6·00 (Si+Al+Fe³⁺)	5·939	5·96 (Si+Al+Fe³⁺)	6·015		5·929	6·00 (Si+Al+Fe³⁺)
Al	0·021		—		0·030		0·020		—		0·011	
Fe^{3+}	0·018	6·01 (Fe³⁺–K)	—	6·13 (Fe³⁺–K)	0·007		0·004		—		0·064	
Mg	0·045		0·037		0·014	6·025 (Mg–K)	—	6·13 (Mg–K)	0·312	5·97 (Mg–K)	0·015	6·03 (Mg–K)
Fe^{2+}	0·007		0·018		0·060		0·059		0·188		0·045	
Mn	0·006		0·003		0·004		0·096		0·031		0·061	
Na	0·004		0·009		0·009		0·080		—		—	
Ca	5·928		6·031		5·938		5·888		5·438		5·878	
K	—		0·027		—		0·007		—		—	

[a] Includes Ti 0·006, Sr 0·007.
[b] Includes Ti 0·021.

1 White coarsely crystalline parawollastonite (wollastonite-2M), associated with vesuvianite, blue calcite and diopside, Sky Blue Hill, Crestmore, California (Tolliday, 1959). Anal. R. A. Howie.
2 Wollastonite, Remonmaki, Finland (Simonen, 1953). Anal. H. B. Wiik.
3 Wollastonite, sövite, Smedsgarden Farm, Alnö alkaline complex, Sweden (Eckermann, 1974). Anal. N. Sahlbom.
4 Wollastonite, wollastonite-melanite melteigite, Oka carbonatite and alkaline complex, Oka, Quebec, Canada (Gold, 1966). Anal. H. Ulk (includes CO_2 0·45, subtracted with equivalent CaO before recalculation).
5 Wollastonite, metamorphosed sediments, Monti Peloritani, Sicily (Atzori, 1969); β: [010] = 3°–4°.
6 Fibrous wollastonite, contact metamorphic zone between marble and granodiorite, north-west of Xanthi, northern Greece (Sapountzis, 1973); $\beta{:}y$ = 4°–5°.

Table 58. Wollastonite Analyses – *continued*

	7	8	9	10	11	12
SiO_2	51·9	50·78	49·92	50·64	50·24	49·37
Al_2O_3	0·1	0·54	0·87	0·37	0·46	0·20
Fe_2O_3	0·3	0·13	0·90	0·37	tr.	0·42
FeO	1·8	0·72	0·08	5·04	5·54	5·32
MnO	0·1	0·53	1·93	5·06	8·16	9·18
MgO	1·0	0·11	0·16	tr.	0·07	0·85
CaO	44·8	46·62	45·02	38·46	35·93	33·93
Na_2O	—	0·18	tr.	—	—	—
K_2O	—	0·04	tr.	—	—	—
H_2O^+	0·5	0·05	0·66	0·09	0·00	0·37
H_2O^-	0·1	0·02	0·16	0·08	0·00	0·13
Total	100·6	99·89	100·14	100·11	100·54	99·77
α	1·623	1·619	1·640	1·639	1·644	—
β	1·632	1·632	1·641	—	1·654	—
γ	1·638	1·634	1·655	1·650	1·657	—
δ	0·015	0·015	0·015	0·011	0·013	—
$2V_\gamma$	—	41°	—	—	—	—
D	—	2·922	2·80	—	—	—

Numbers of ions on the basis of eighteen O

	7		8		9		10		11		12	
Si	6·012		5·928		5·876		5·994		5·972		5·948	
Al	0·014	6·05	0·074	6·01	0·120	6·08	0·052	5·98	0·064	6·04	0·028	6·01
F^{3+}	0·026		0·011		0·080		0·033		—		0·038	
Mg	0·173		0·019		0·028		—		0·012		0·153	
Fe^{2+}	0·174		0·070		0·008		0·499		0·551		0·536	
Mn	0·010	5·92	0·053	6·03[a]	0·192	5·93[b]	0·507	5·88	0·821	5·96	0·937	6·01
Na	—		0·040		—		—		—		—	
Ca	5·560		5·831		5·678		4·878		4·576		4·380	
K	—		0·006		—		—		—		—	

[a] Includes Ti 0·006, Sr 0·007.
[b] Includes Ti 0·021.

7 Wollastonite, hybrid pyroxenite, dolerite plug cutting chalk and flint, Carneal, Co. Antrim, Northern Ireland (Sabine, 1975a). Anal. G. A. Sergeant.

8 Wollastonite, wollastonite ijolite, Oldoinyo Lengai, Tanganyika (Dawson and Sahama, 1963); $\beta{:}y$, $\alpha{:}z$ = 32 in acute angle β; inclined dispersion with $r > v$. Anal. H. B. Wiik (includes SrO 0·10, TiO_2 0·07 per cent).

9 Wollastonite, section above 1170 ft. level, South mine, Broken Hill, New South Wales, Australia (Stilwell, 1959). Anal. G. C. Carlos (includes TiO_2 0·24, P_2O_5 0·14, SO_3 0·02, Cl_2 0·04 per cent).

10 White acicular, manganoan ferroan wollastonite (Mel'nitskaya, 1967).

11 Manganoan ferroan wollastonite, 7th floor above 2300 ft. level, Section 28, North mine, Broken Hill, New South Wales, Australia (Stilwell, 1959). Anal. Avery and Anderson (includes S 0·14 per cent).

12 Manganoan wollastonite, calcareous zone of hornfels in Mn ore deposit, Hijikuzu mine, Japan (Nambu *et al.*, 1971).

The relatively few analyses reported for wollastonites from igneous rocks appear to show a greater variety of minor amounts of other elements (anals. 3, 4 and 8). Further data on the compositional limits of wollastonite and bustamite are given by Hodgson (1975) and Mason (1975).

Experimental

Wollastonite can be synthesized readily from its component oxides, or from hydrous gels via xonotlite, $Ca_6Si_6O_{17}(OH)_2$, which breaks down on heating to yield wollastonite. The wollastonite–pseudowollastonite inversion temperature is 1125° ± 10°C for pure $CaSiO_3$ (Osborn and Schairer, 1941) rising to 1610°C at 23 kbar (Kushiro, 1964); at atmospheric pressure the melting point of pseudowollastonite is 1544°C. The heat of inversion at 25°C is 1·56 kcal/mole. The inversion temperature is raised very considerably by the solid solution of 21 per cent of diopside, when the inversion temperature is 1368°C (Schairer and Bowen, 1942): there is no solid solution of diopside in pseudowollastonite. Only limited solid solution of åkermanite or gehlenite in wollastonite or pseudowollastonite takes place (Juan, 1950; but see also Yoder, 1964); it is possible, however, that pseudowollastonite may contain appreciable Mg replacing Ca. Recently Shinno (1970) has shown that a new phase of diopside composition but with the wollastonite structure appears at the initial stage of crystallization from a glass or from a chemical mixture; wollastonites of this type were found to appear in the whole range $CaSiO_3$–$CaMgSi_2O_6$. The X-ray powder and infrared absorption data for these solid solutions are described by Shinno (1974). Wollastonite of this series between Wo_{100} and Wo_{85} can be synthesized stably, but that richer in the diopside component is metastable at all temperature ranges.

Bowen *et al.* (1933) showed that wollastonite forms a series of solid solutions in the system $CaSiO_3$–$FeSiO_3$, but it is now known that compositions in this series containing more than about 10 per cent $FeSiO_3$ crystallize with the bustamite or pyroxmangite structures. The $CaSiO_3$–$CaFeSi_2O_6$ join has been re-examined by Rutstein (1971) who has interpreted the experimental results schematically (Fig. 248). In pure wollastonite-Tc, the Ca atoms are distributed over three general positions (Fig. 247(a)) but the retention of the structure with the addition of Fe requires that the Fe and Ca be distributed randomly, at least in the *M*1 and *M*2 sites. On the other hand, in the pure bustamite structure the Fe and Ca must be ordered; but for a composition in the range $Ca_{0\cdot88}Fe_{0\cdot12}SiO_3$ to $Ca_{0\cdot5}Fe_{0\cdot5}SiO_3$ there is a probability of a 'disordered' bustamite. It is considered that the distribution of Fe in low-iron wollastonite solid solutions proceeds in a random fashion, but that at a certain temperature-composition boundary the substitution of Fe becomes ordered with the development of the bustamite-type phase which over a compositional interval coexists with the wollastonite structural type. With increasing temperature, a given composition can interact to produce a variety of ordered and disordered phases related to both structural types (Rutstein, 1971).

Hydrothermal synthesis of phases in the system $MnSiO_3$–$CaSiO_3$ at a total pressure of 2 kbar produced wollastonites (as opposed to bustamites or pyroxmangites) with up to 38 mol. per cent $MnSiO_3$ (Albrecht and Peters, 1975) where there is an abrupt structural break; within the wollastonite series a gradual decrease in *d*-spacings with increasing Mn content was noted.

Wollastonite often occurs in nature as a result of the reaction of quartz and calcite in metamorphosed limestones. A tentative *P–T* curve for the reaction $CaCO_3 + SiO_2 \rightleftharpoons CaSiO_3 + CO_2$ was derived by Goldschmidt (1912), and Danielsson (1950)

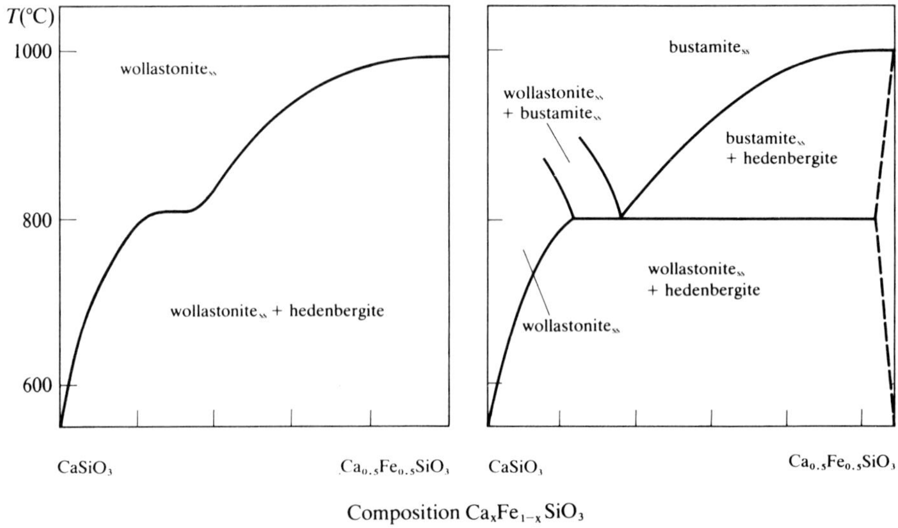

Fig. 248. (a) Solvus for wollastonite solid solutions and hedenbergite at 1 kbar P_{H_2O} and f_{O_2} defined by QFM buffer. (b) Possible equilibrium relations to explain inflected solvus. Details of phase boundaries are schematic (after Rutstein 1971).

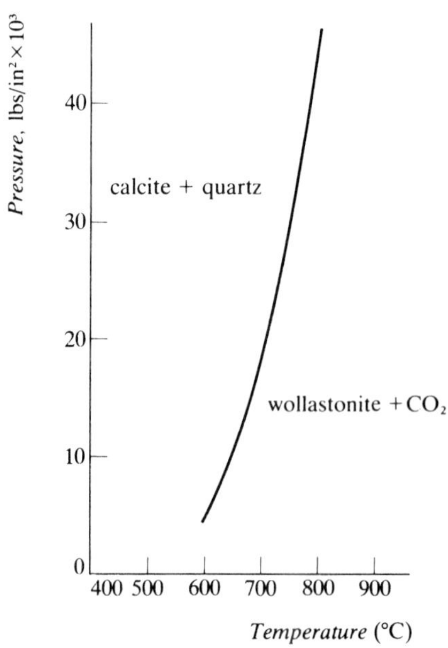

Fig. 249. The experimentally determined univariant P_{CO_2}–T curve for the reaction $CaCO_3 + SiO_2 \rightleftharpoons CaSiO_3 + CO_2$ (after Harker and Tuttle, 1956a).

calculated the equilibrium range to be 250° to 850°C, while the calculated curve was further checked by Ellis and Fyfe (1956) using later thermodynamic data for CO_2. Further work has shown that for this the activation energy is in the range 25·5 to 29·0 kcal/mole, and that the mechanism determining the rate of reaction involves the solid state diffusion of calcium, silicon, and probably oxygen through a product layer of wollastonite (Kridelbaugh, 1973). Harker and Tuttle (1956a) determined experimentally the univariant P_{CO_2}–T curve for this reaction (Fig. 249), for pressures greater than 5000 lb/in^2 and estimated the approximate position of the curve at pressures less than 5000 lb/in^2 using the Nernst approximation formula and Van't Hoff's equation: the curves thus derived indicate that at atmospheric pressure the reaction takes place at, or slightly below, 400°C. The synthesis of wollastonite at temperatures as low as 350°C over periods of 2 to 10 days at pressures of 150–1700 atm. was reported by Kalinin (1967). The formation of wollastonite has been studied by Greenwood (1967) who determined its stability in supercritical mixtures of H_2O and CO_2 at pressures of 1 and 2 kbar, and showed that ideal mixing occurs in the pressure range below 2 kbar at temperatures above about 500°C. The kinetics of the dissolution of wollastonite in H_2O–CO_2 and buffered systems at 25°C has been studied by Bailey and Reesman (1971) who showed that hydrogen ion attack is mainly responsible, the release of calcium being controlled by diffusion through an aqueous surface film; the calcium release decreases with increasing pH values until pH and the rate of calcium release become constant. As noted by Harker and Tuttle the effect of the additional components, such as MgO and FeO, most likely to be encountered in the solid phases in natural limestones remains unknown, but their addition, if they actually enter the structure of the phases involved, would result in the reaction no longer remaining strictly univariant.

An investigation of the system CaO–TiO_2–SiO_2 by De Vries *et al.* (1955) has shown that there is a 3 per cent substitution of Ti for Si between $CaSiO_3$ and $CaTiO_3$, while Buckner *et al.* (1960) have shown from a study of the system $CaSiO_3$–$SrSiO_3$–H_2O that the wollastonite structure will only admit small amounts of strontium (< 5 per cent Sr^{+2}); experimental work by Moir and Glasser (1974) showed that extensive solid solutions occur in the ternary system Na_2SiO_3–$CaSiO_3$–$SrSiO_3$ including the substitution $2Na^+ \rightleftharpoons$ (Ca,Sr). The system α-$CaSiO_3$–$SrSiO_3$ has been shown to have a characteristic discontinuous change in symmetry (Moir *et al.*, 1975) with a two-polytype region in the range 50–60 mol. per cent $CaSiO_3$; the pseudohexagonal polytype of $SrSiO_3$ is analogous to pseudowollastonite.

Buckner *et al.* (1960) also determined the *P–T* curve for the decomposition of xonotlite to triclinic wollastonite and vapour: the equilibrium temperature ranges from 400°C at 5000 lb/in^2 to 420°C at 20000 lb/in^2. The reaction wollastonite + calcite $\rightleftharpoons$ spurrite + carbon dioxide has been investigated experimentally by Tuttle and Harker (1957): at 5000 lb/in^2 CO_2 pressure the reaction takes place at about 1000°C. Similarly for the reaction wollastonite + monticellite $\rightleftharpoons$ åkermanite, the *P–T* curve lies between 700° and 750°C in the pressure range 30000 to 60000 lb/in^2, i.e. the *P–T* curve of this reaction is nearly parallel to the pressure axis and thus intersects other *P–T* curves for reactions involving volatile phases, such as that for quartz + calcite $\rightleftharpoons$ wollastonite + CO_2 (Harker and Tuttle, 1956b). Although wollastonite is normally regarded in metamorphism as a low-temperature mineral compared to spurrite, Wyllie and Haas (1965, 1966) have shown that in the system CaO–SiO_2–CO_2–H_2O it has no stability field on the vapour-saturated liquidus surface below 1000°C at 1 kbar, whereas there is a large primary phase volume for spurrite extending down to 677°C at 1 kbar. The dehydration of prehnite to give wollastonite + anorthite + H_2O occurs at 440° ± 5°C at 1 kbar (Liou, 1971), rising to 550° ± 5°C at 5 kbar.

Wollastonite transforms to a high-pressure polymorph, wollastonite-II, at pressures above 25 kbar (Essene, 1974). It is reported to be triclinic, with Si_3O_9 rings between layers of Ca atoms (Trojer, 1969). It is about 5 per cent denser than wollastonite, and an initial experiment indicated it to have a mean refractive index considerably below 1·70 (Ringwood and Major, 1967) and to be stable to well over 100 kbar. Phase transitions between pseudowollastonite, wollastonite and wollastonite-II have been reversed and the melting curve for these three polymorphs determined between 10 and 35 kbar by Huang and Wyllie (1975); the pseudowollastonite melting curve rises from 1544°C at 1 bar with an unusually steep slope to a triple point at 1588°C and 23 kbar, where the solidus meets the pseudowollastonite–wollastonite transition curve (Fig. 250). The rapid reactivity of this transition over a range of at least 1000°C, its insensitivity to temperature and its location at around 30 kbar indicate that it has potential as a pressure calibration curve at high temperature.

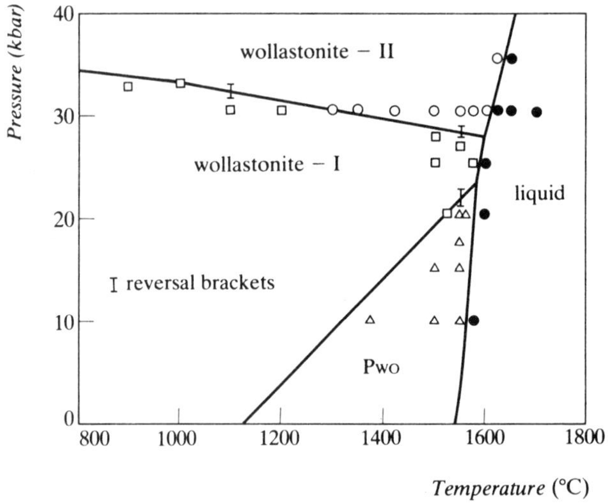

Fig. 250. Experimental results showing the subsolidus and melting relations for $CaSiO_3$. Solid circles denote runs above liquidus; open symbols denote subsolidus runs. The definite reversal brackets on the subsolidus transitions are indicated (after Huang and Wyllie, 1975).

Synthetic wollastonite and $CaSiO_3$ glass on compression to loading pressures above 160 kbar and after heating to 1500°C by a laser in a diamond-anvil cell gave (after cooling but with the sample maintained at high pressure) an X-ray diffraction pattern of a cubic perovskite-type polymorph with *a* 3·485 Å (Liu and Ringwood, 1975). After release of pressure the same sample showed a mixture of glass plus a few weak lines corresponding with ϵ-$CaSiO_3$. The density of the perovskite-type polymorph of $CaSiO_3$ is 9·2 per cent greater than that of an isochemical mixture of $CaO + SiO_2$ (stishovite) at about 160 kbar.

The reaction $Ca_3Al_2Si_3O_{12}$ (grossular) + SiO_2 = $CaAl_2Si_2O_8$ (anorthite) + $2CaSiO_3$ (wollastonite) was investigated at 1100°C to determine the shift of equilibrium pressure as a function of plagioclase composition at *P–T* conditions representative of the Earth's lower crust and upper mantle (Windom and Boettcher, 1976); at this temperature, the equilibrium pressure was found to be $14{\cdot}8 \pm 0{\cdot}3$ kbar.

Detectable reactions on the DTA curve can be obtained by heating wollastonite with Li_2WO_4 as a flux: Zuberi and Kopp (1976) obtained a weak reaction at 740° and a strong reaction at 980° on heating and 930°C on cooling, the stronger reaction being tentatively attributed to the formation of $CaWO_4$.

The hydrothermal alteration of wollastonite to stevensite and talc has been recorded (De Rudder and Beck, 1963). It may also be replaced by grossular, hydrogrossular and pectolite.

Optical and Physical Properties

Typical values for pure $CaSiO_3$ of the various structural modifications are:

	α	β	γ	δ	$\alpha:z$	$\beta:y$	Sign
Wollastonite-Tc	1·618	1·630	1·632	0·014	39°	4°	(−)
Wollastonite-2M	1·618	1·630	1·632	0·014	38°	0°	(−)
Pseudowollastonite	1·610	1·611	1·654	0·044	9°	—	(+)

A wollastonite with particularly low refractive indices (α'1·609) but with a normal optic axial angle has been reported (Belyankin and Petrov, 1939): it is suggested that the entry of about 0·5 per cent water into the structure would have this effect.

The introduction of iron increases the refractive indices and the optic axial angle (Bowen *et al.*, 1933; see also Sabine, 1975b). A series of determinations of the optics, specific gravity and thermal expansion properties of the phases in the CaO–FeO–SiO_2 system have also been given by Rigby *et al.* (1945). It is now recognized, however, that there is a break in the $CaSiO_3$–$FeSiO_3$ series at around $Ca_{0\cdot90}Fe_{0\cdot10}SiO_3$, the limit of the wollastonite structure type, more iron-rich compositions having a bustamite-type structure. After the break the *a*, *b*, and *c* cell dimensions decrease even more rapidly with increasing iron and the optic axial angle increases more rapidly (Matsueda, 1973). The effect of the entry of manganese is in general similar to that of iron, up to the structural discontinuity at around $Ca_{0\cdot75}Mn_{0\cdot25}SiO_3$.

The infrared spectrum of wollastonite has been studied in some detail (Klimov *et al.*, 1960; Lazarev and Tenisheva, 1961; Ryall and Threadgold, 1966; Rutstein and White, 1971): there are two distinct clusters each with three bands in the high frequency region (around 1000 cm^{-1}). A blue fluorescence is excited at 365 μm (Przibram, 1962); a bright green cathodo-luminescence was recorded in wollastonite by Long and Agrell (1965) the activator probably being manganese. The natural thermoluminescence of wollastonite shows two peaks, both of which are shifted upwards by pressure (Angino, 1964). The compressibility of wollastonite to 45 kbar has been studied by Vaidya *et al.* (1973).

For fibrous wollastonite, the thermal expansion coefficients over the range 25°–650°C for directions parallel to the *y*-axis fibres [010] and at right angles (010) were found by Weston and Rogers (1976) to be [010] $6\cdot23 \times 10^{-6}$/°C and (010) $7\cdot77 \times 10^{-6}$/°C. The coefficient is constant for [010] but varies for (010).

Twinning occurs on {100}, the common twin law having its twin axis [010]. Köppen (1948) has deduced a further twin law with twin axis [001] for a triclinic iron-bearing "wollastonite" with α 1·705, β 1·718, γ 1·725, $2V_\alpha 72\frac{1}{2}°$ (composition $\simeq$ 64–65 per cent $FeSiO_3$) from a glassy slag on steel ingots: he reported that with increasing amount of iron the differences in optic orientation and twinning become more evident and suggested that monoclinic parawollastonite is really triclinic with multiple lamellar twinning.

Distinguishing Features

The distinction between wollastonite-Tc and wollastonite-2M is based on the extinction angle $\beta:y$, which is 3°–5° in wollastonite-Tc and 0° in wollastonite-2M. These two forms may also be distinguished with certainty by single-crystal X-ray photographs or, it is claimed, by their X-ray powder patterns (see p. 550). Pseudowollastonite is distinct, with much higher birefringence and small 2V(+). Wollastonite differs from tremolite and pectolite in its weaker birefringence and its variable sign of elongation ($\beta \parallel y$), and it may be distinguished from zoisite and clinozoisite by its lower relief and normally by its lower 2V. Diopside, which often occurs in association with wollastonite, has higher relief, higher 2V, and is optically positive. The three cleavage directions of wollastonite are also distinctive. Minerals of wollastonite structural type can best be distinguished from the bustamite type by their distinctive infrared spectrum.

Paragenesis

Wollastonite is a common constituent of thermally metamorphosed impure limestone, and may occur also in contact altered calcareous sediments where the silicon is due to metasomatism, and in the invading igneous rock where it is due to contamination: in most of these occurrences it is the result of the reaction $CaCO_3 + SiO_2 \rightarrow CaSiO_3 + CO_2$. The geological significance of the *P–T* curve for this reaction (Fig. 249) has been discussed by Harker and Tuttle (1956a): in certain circumstances the CO_2 pressure may be effectively reduced, either by a dilution by some other volatile component or by the escape of CO_2 through fissures, and under these conditions wollastonite may form at somewhat lower temperatures than those indicated. The reported sequence of reactions in the progressive metamorphism of siliceous dolomites (Bowen, 1940; Tilley, 1951) is modified in view of the experimental evidence, which indicates that it is unlikely that periclase is ever stable at lower temperatures than wollastonite (Harker and Tuttle, 1956a; Weeks, 1956): the early stages of the sequence are thus talc–tremolite–diopside–forsterite–wollastonite–periclase–monticellite. The development of wollastonite in a diffusion reaction skarn has been recorded by Kennedy (1959): in this occurrence limestone concretions within sandstone have been metamorphosed by a dolerite dyke, giving rise to wollastonite in the sandstone as a result of outward migration of Ca from the limestone. Aggregates of stellate wollastonite, set in a calcite matrix and often around quartz granules, have been recorded from calc-silicate rocks some distance from the contact of a porphyritic biotite granite with interfolded gneiss and crystalline limestone (Sathe and Choudhary, 1967). Amygdales in basalts in northern Mull, originally containing gyrolite, were metamorphosed by a volcanic plug first to reyerite and then to wollastonite, the wollastonite-filled amygdales being enveloped in a rim of aegirine-augite due to reaction between the wollastonite and the basalt (Cann, 1965). Wollastonite also occurs in regionally metamorphosed impure calcareous sediments but is less commonly developed there than as a result of contact metamorphism.

Wollastonite-2M is rarer than the normal triclinic wollastonite but may occur in association with it, though the more massive deposits are apparently formed of triclinic wollastonite (Peacock, 1935). Wollastonite-2M, identified by single-crystal X-ray photographs, has been recorded from Monte Somma, Vesuvius (Peacock, 1935), from Crestmore, California (Table 58, anal. 1), from Csiklova, Romania (Barnick, 1935), and from several localities in metamorphic rocks in the Sierra Nevada, California. An investigation of the various polymorphic superstructures of wollastonite (2M, 4M, 4T, etc.) has shown that their occurrence is independent of any mineralogical isograds, but they do appear to be restricted to rocks with well developed lineation and strong preferred orientation (Wenk, 1969); it has been suggested that consecutive annealing of strained disordered wollastonite might cause periodic stacking sequences of [SiO_3]-chains along the *a** axis.

Wollastonite is known from certain alkaline igneous rocks and carbonatites as at Oka, Quebec (Table 58, anal. 4) and Alnö, Sweden (anal. 3; see also Eckermann, 1948, 1967), in the urtite of the Prairie Lake complex, Ontario (Watkinson, 1973), and in the ijolitic alkaline rocks of Kenya (Pulfrey, 1949) and Tanzania (Table 58, anal. 8), and also occurs in wollastonite phonolites and related rocks (Erdmansdörffer, 1935). It has been found in xenoliths of impure sediments in igneous rocks; Isshiki (1954) has reported a xenolith of augite–hypersthene andesite in volcanic lava, consisting of anorthite with a rim of zoned iron-wollastonite (mean α 1·632, β 1·644, γ 1·647, $2V_\alpha$ 54°, composition $\simeq$ 11 mol. per cent $FeSiO_3$) and a further outer rim of clinopyroxene. The assemblage calcic clinopyroxene–wollastonite–plagioclase is commonly developed in thermally metamorphosed and metasomatized fragments of basement calc-silicate rocks found as inclusions in agglomerates and recent lavas of Santorini volcano, Greece (Nicholls, 1971). In one such inclusion a primary melilite–wollastonite–magnetite assemblage has given rise to a secondary melilite–wollastonite–andradite association during its inclusion by the dacite.

A cordierite-granulite subfacies of metamorphism has been proposed (Hapuarachchi, 1968) which would include regionally developed wollastonite–scapolite–diopside–feldspar assemblages. The stability of wollastonite in the granulite facies has been considered in detail by Mukherjee and Rege (1972) who conclude that very high values of both temperature and pressure, together with low P_{CO_2}, are indicated. The assemblage wollastonite–anorthite has been recognized as a stable association in calc-silicate rocks in the amphibolite facies (Misch, 1964).

Acicular needles of wollastonite-Tc about 100 μm long have been found in diopside-lined cavities in the Allende carbonaceous chondrite (Fuchs, 1971); electron microprobe analysis has shown its composition to be close to pure $CaSiO_3$. This is the first recorded occurrence of the mineral in a meteorite.

Pseudowollastonite was first reported as a natural occurrence, in pyrometamorphosed rocks in south-west Persia (McLintock, 1932), where sediments have been baked by the burning of hydrocarbons in prehistoric times. It occurs frequently in synthetic preparations as the high-temperature form of $CaSiO_3$, as well as in slags and glasses. A second natural occurrence is reported to have been identified in material from the dumps of the graphite mines near Hauzenberg, Bavaria (Hochleitner, 1972).

Wollastonite is becoming increasingly important as an industrial mineral. Wall tile manufacture is the major use but the mineral is finding increasing use in ceramics generally, due to its fluxing properties, freedom from volatile constituents, whiteness, and acicular particle shape. World resources have been reviewed by Andrews (1970).

References

Albrecht, J. and **Peters, Tj.**, 1975. Hydrothermal synthesis of pyroxenoids in the system $MnSiO_3$–$CaSiO_3$ at *Pf* = 2 kbar. *Contr. Min. Pet.*, **50**, 241–246.

Allen, E. T., White, W. P. and **Wright, F. E.**, 1906. On wollastonite and pseudowollastonite, polymorphic forms of calcium metasilicate. *Amer. J. Sci.*, ser. 4, **21**, 89–108.

Amorós, J. L., 1963. Proceedings of the International Mineralogical Association Third General Business Meeting of Delegates, Washington, D.C., April 18 and 20, 1962. *Min. Soc. America, Spec. Paper* **1**, 315–325.

Andrews, R. W., 1970. *Wollastonite.* London (Min. Resources Divn., Inst. Geol. Sci.), 114 pp.

Angino, E. E., 1964. Some effects of pressure on the thermoluminescence of amblygonite, pectolite, orthoclase, scapolite and wollastonite. *Amer. Min.*, **49**, 386–394.

Atzori, P., 1969. Metamorfiti a pirosseni e wollastonite nel cristallino dei Monti Peloritani (Sicilia). *Atti Accad. Gioenia Sci. Natur.*, ser. 6, **20** (*Suppl. Sci. Geol.*), 163–172.

Bailey, A. and **Reesman, A. L.**, 1971. A survey study of the kinetics of wollastonite dissolution in H_2O–CO_2 and buffered systems at 25°C. *Amer. J. Sci.*, **271**, 464–472.

Barnick, M., 1935. Strukturuntersuchung des natürlichen Wollastonits. *Naturwiss.*, **35**, 770–771.

Belyankin, D. S. and **Petrov, V. P.**, 1939. Hibschite in Georgia. *Dokl. Acad. Sci. U.S.S.R.*, **24**, 349–352.

Bowen, N. L., 1940. Progressive metamorphism of siliceous limestone and dolomite. *J. Geol.*, **48**, 225–274.

Bowen, N. L., **Schairer, J. F.** and **Posnjak, E.**, 1933. The system CaO–FeO–SiO_2. *Amer. J. Sci.*, ser. 5, **26**, 193–283.

Buckner, D. A., **Roy, D. M.** and **Roy, R.**, 1960. Studies in the system CaO–Al_2O_3–SiO_2–H_2O. II. The system $CaSiO_3$–H_2O. *Amer. J. Sci.*, **258**, 132–147.

Buerger, M. J., 1956. The arrangement of atoms in crystals of the wollastonite group of metasilicates. *Proc. Nat. Acad. Sci. U.S.A.*, **42**, 113–116.

Buerger, M. J. and **Prewitt, C. T.**, 1961. The crystal structure of wollastonite and pectolite. *Proc. Nat. Acad. Sci.*, **47**, 1883–1888.

Cann, J. R., 1965. The metamorphism of amygdales at 'S Airde Beinn, northern Mull. *Min. Mag.*, **34** (Tilley vol.), 92–106.

Clark, C. B., 1946. X-ray diffraction data for compounds in the system CaO–MgO–SiO_2. *J. Amer. Ceram. Soc.*, **29**, 25–30.

Coe, R. S., 1970. The thermodynamic effect of shear stress on the ortho–clino inversion in enstatite and other coherent phase transitions characterized by a finite simple shear. *Contr. Min. Petr.*, **26**, 247–264.

Danielsson, A., 1950. Das Calcit-Wollastonitgleichgewicht. *Geochim. Cosmochim. Acta*, **1**, 55–69.

Dawson, J. B. and **Sahama, Th. G.**, 1963. A note on parawollastonite from Oldoinyo Lengai, Tanganyika. *Schweiz Min. Petr. Mitt.*, **43**, 131–133.

De Rudder, R. D. and **Beck, C. W.**, 1963. Stevensite and talc – hydrothermal alteration products of wollastonite. *Clays and Clay Minerals, Proc. 11th Conf. Clays and Clay Min.*, Pergamon, 188–199.

De Vries, R. C., **Roy, R.** and **Osborn, E. F.**, 1955. Phase equilibria in the system CaO–TiO_2–SiO_2. *J. Amer. Ceram. Soc.*, **38**, 158–171.

Dornberger-Schiff, K., **Liebau, F.** and **Thilo, E.**, 1954. Über die Kristallstruktur des $(NaAsO_3)_x$, des Maddrellschen Salzes und des β-Wollastonits. *Naturwiss.*, **41**, 551.

Eckermann, H. von, 1948. The alkaline district of Alnö Island. *Sveriges Geol. Undersok., Ser. Ca.*, no. 36, 176 pp.

Eckermann, H. von, 1967. Wollastonite in carbonatite rocks. *Geochim. Cosmochim. Acta*, **31**, 2253–2254.

Eckermann, H. von, 1974. The chemistry and optical properties of some minerals of the Alnö alkaline rocks. *Arkiv. Min. Geol.*, **5**, 93–210.

Ellis, A. J. and **Fyfe, W. S.**, 1956. A note on the calcite–wollastonite equilibrium. *Amer. Min.*, **41**, 805.

Erdmannsdörffer, O. H., 1935. Über Wollastonit Urtit und die Entstehungsweisen von Alkaligesteinen. *Akad. Wiss. Heidelberg-Sitzungsber.*, pp. 1–23.

Essene, E., 1974. High-pressure transformations in $CaSiO_3$. *Contr. Min. Petr.*, **45**, 247–250.

Fuchs, L. H., 1971. Occurrence of wollastonite, rhönite, and andradite in the Allende meteorite. *Amer. Min.*, **56**, 2053–2068.

Gold, D. P., 1966. The minerals of the Oka carbonatite and alkaline complex, Oka, Quebec. *Papers and Proc. I.M.A. 4th Gen. Meeting, New Delhi* (Min. Soc. India), 109–125.

Goldschmidt, V. M., 1912. Die Gesetze der Gesteinsmetamorphose mit Beispielen aus der Geologie des Südlichen Norwegens. *Vidensk. Skrift.* 1, *Math. Natur. Kl.*, **22**, 238.

Greenwood, H. J., 1967. Wollastonite: stability in H_2O–CO_2 mixtures and occurrence in a contact-metamorphic aureole near Salmo, British Columbia. *Amer. Min.*, **52**, 1669–1680.

Hapuarachchi, D. J. A. C., 1968. Cordierite and wollastonite-bearing rocks of south-western Ceylon. *Geol. Mag.*, **105**, 317–324.

Harker, R. I. and **Tuttle, O. F.**, 1956a. Experimental data on the P_{CO_2}–T curve for the reaction: calcite + quartz $\rightleftharpoons$ wollastonite + carbon dioxide. *Amer. J. Sci.*, **254**, 239–256.

Harker, R. I and **Tuttle, O. F.**, 1956b. The lower limit of stability of åkermanite ($Ca_2MgSi_2O_7$). *Amer. J. Sci.*, **254**, 468–478.

Heller, L. and **Taylor, H. F. W.**, 1956. *Crystallographic data for the calcium silicates.* London (H.M.S.O.).

Hilmer, W., 1958. Zur Struktur bestimmungen von Strontiumgermanat. *Naturwiss.*, **95**, 238.
Hochleitner, R., 1972. Cyclowollastonit von den Halden des Graphitabbaus in der Nähe von Hauzenberg/Bayerischer Wald. *Der Aufschluss*, **23**, 340–341.
340–341.
Hodgson, C. J., 1975. The geology and geological development of the Broken Hill lode in the New Broken Hill Consolidated mine, Australia. Part II: Mineralogy. *J. Geol. Soc. Australia*, **22**, 33–50.
Huang, W.-L. and **Wyllie, P. J.,** 1975. Melting and subsolidus phase relations for $CaSiO_3$ to 35 kilobars pressure. *Amer. Min.*, **60**, 213–217.
Hutchison, J. L. and **McLaren, A. C.,** 1976. Two-dimensional lattice images of stacking disorder in wollastonite. *Contr. Min. Petr.*, **55**, 303–309.
Hutchison, J. L. and **McLaren, A. C.,** 1977. Stacking disorder in wollastonite and its relationship to twinning and the structure of parawollastonite. *Contr. Min. Petr.*, **61**, 11–13.
Isshiki, N., 1954. On iron-wollastonite from Kanpû volcano, Japan. *Proc. Japan Acad.*, **30**, 869–872 (M.A. **12**-533).
Ito, T., 1950. *X-ray studies on polymorphism.* Tokyo (Maruzen).
Jefferson, D. A. and **Bown, M. G.,** 1973. Polytypism and stacking disorder in wollastonite. *Nature (Phys. Sci.)*, **245**, 43–44.
Jeffery, J. W., 1953. Unusual X-ray effects from a crystal of wollastonite. *Acta. Cryst.*, **6**, 821–825.
Jeffery, J. W. and **Heller, L.,** 1953. Preliminary X-ray investigation of pseudowollastonite. *Acta Cryst.*, **6**, 807–808.
Juan, V. C., 1950. The system $CaSiO_3$–$Ca_2Al_2SiO_7$–$NaAlSiO_4$. *J. Geol.*, **58**, 1–15.
Kalinin, D. V., 1967. Lower temperature boundaries for the formation of tremolite, diopside and wollastonite under hydrothermal conditions; experimental data. *Geochem. Intern.*, **4**, 836–839.
Kennedy, W. Q., 1959. The formation of a diffusion reaction skarn by pure thermal metamorphism. *Min. Mag.*, **32**, 26–31.
Klimov, V. V., Kagarlitskaya, N. V. and **Shcherbov, D. P.,** 1960. [Infra-red spectrometry of inorganic substances. IV. Absorption spectra of some silicate minerals at wavelengths from 2 to 15 μ.] *Trudy Kazakh, Nauch.-Issled. Inst. Min. Syr'ya*, 312–317.
Köppen, N. M., 1948. Die Zwillingsgesetze des eisenhaltigen Wollastonits. *Neues Jahrb. Monat., Abt. A*, 136–143.
Kridelbaugh, S. J., 1973. The kinetics of the reaction calcite + quartz = wollastonite + carbon dioxide at elevated temperatures and pressures. *Amer. J. Sci.*, **273**, 757–777.
Kushiro, I., 1964. Wollastonite–pseudowollastonite inversion. *Carnegie Inst. Washington, Ann. Rept. Dir. Geophys. Lab.*, 1963–1964, 83–84.
Lazarev, A. N. and **Tenisheva, T. F.,** 1961. Vibrational spectra of silicates III. Infrared spectra of the pyroxenoids and other chain metasilicates. *Optics and Spectroscopy*, **11**, 316–317.
Liou, J. G., 1971. Synthesis and stability relations of prehnite, $Ca_2Al_2Si_3O_{10}(OH)_2$. *Amer. Min.*, **56**, 507–531.
Liu, L.-G. and **Ringwood, A. E.,** 1975. Synthesis of a perovskite-type polymorph of $CaSiO_3$. *Earth Planet. Sci. Letters*, **28**, 209–211.
Long, J. V. P. and **Agrell, S. O.,** 1965. The cathode-luminescence of minerals in thin section. *Min. Mag.*, **34** (Tilley vol.), 318–326.
McLintock, H. F. P., 1932. On the metamorphism produced by the combustion of hydrocarbons in the Tertiary sediments of south-west Persia. *Min. Mag.*, **23**, 207–226.
Mamedov, Kh. S. and **Belov, N. V.,** 1956. The crystal structure of wollastonite. *Dokl. Acad. Sci. U.S.S.R.*, **107**, 463–466.
Mason, B., 1973. Manganese silicate minerals from Broken Hill, New South Wales. *J. Geol. Soc. Australia*, **20**, 397–404.
Mason, B., 1975. Compositional limits of wollastonite and bustamite. *Amer. Min.*, **60**, 209–212.
Matsueda, H., 1973. Iron-wollastonite from the Sampo mine showing properties distinct from those of wollastonite. *Min. J., Japan*, **7**, 180–201.
Mel'nitskaya, E. F., 1967. [Manganiferrous wollastonite and its alteration.] *Zap. Vses. Min. Obshch.*, **96**, 297–305 (in Russian).
Misch, P., 1964. Stable association wollastonite–anorthite, and other calc-silicate assemblages in amphibolite-facies crystalline schists of Nanga Parbat, north-west Himalayas. *Beitr. Min. Petr.*, **10**, 315–356.
Moir, G. K., Gard, J. A. and **Glasser, F. P.,** 1975. Crystal chemistry and solid solutions amongst the pseudowollastonite-like polytypes of $CaSiO_3$, $SrSiO_3$ and $BaSiO_3$. *Z. Krist.*, **141**, 437–450.
Moir, G. K. and **Glasser, F. P.,** 1974. Solid solutions and phase equilibrium in the systems Na_2SiO_3–$SrSiO_3$ and Na_2SiO_3–$CaSiO_3$–$SrSiO_3$. *Trans. J. Brit. Ceram. Soc.*, **73**, 199–206.
Mukherjee, A. and **Rege, S. M.,** 1972. Stability of wollastonite in the granulite facies: some evidence from the Eastern Ghats, India. *Neues Jahrb. Min., Abhdl.*, **118**, 23–42.

Nambu, M., Tanida, K. and **Kitamura, T.,** 1971. Res. Inst. Min. Dress. Met. Rept., 552, p. 123 (quoted from H. Matsueda), 1973. Iron-wollastonite from the Sampo mine. *Min. J. Japan*, **7**, 180–201.)
Nicholls, I. A., 1971. Calcareous inclusions in lavas and agglomerates of Santorini volcano. *Contr. Min. Petr.*, **30**, 261–276.
Ohashi, Y. and **Finger, L. W.,** 1976. Stepwise cation ordering in bustamite and disordering in wollastonite. *Carnegie Inst. Washington, Ann. Rept. Dir. Geophys. Lab.*, 1975–76, 746–753.
Osborn, E. F. and **Schairer, J. F.,** 1941. The ternary system pseudowollastonite–åkermanite–gehlenite. *Amer. J. Sci.*, **239**, 715–763.
Peacock, M. A., 1935. On wollastonite and parawollastonite. *Amer. J. Sci.*, ser. 5, **30**, 495–529.
Peacor, D. R. and **Prewitt, C. T.,** 1963. Comparison of the crystal structures of bustamite and wollastonite. *Amer. Min.*, **48**, 588–596.
Prewitt, C. T. and **Buerger, M. J.,** 1963. Comparison of the crystal structure of wollastonite and pectolite. *Min. Soc. America, Spec. Paper* **1**, 293–302.
Przibram, K., 1962. Über die Fluoreszenz organischer Spuren in anorganischen Stoffen und ihre Verbreitung in der Natur. *Geochim. Cosmochim. Acta*, **26**, 1045–1054.
Pulfrey, W., 1949. Ijolitic rocks near Homa Bay, western Kenya. *Quart. J. Geol. Soc.*, **105**, 425–459.
Rankin, G. A. and **Wright, F. E.,** 1915. The ternary system $CaO–Al_2O_3–SiO_2$. *Amer. J. Sci.*, ser. 4, **39**, 1–79.
Rigby, G. R., Lovell, G. H. B. and **Green, A. T.,** 1945. The reversible thermal expansion and other properties of some calcium ferrous silicates. *Trans. Brit. Ceram. Soc.*, **44**, 37–52.
Ringwood, A. E. and **Major, A.,** 1967. Some high-pressure transformations of geophysical importance. *Earth Planet. Sci. Letters*, **2**, 106–110.
Rutstein, M. S., 1971. Re-examination of the wollastonite–hedenbergite ($CaSiO_3–CaFeSi_2O_6$) equilibria. *Amer. Min.*, **56**, 2040–2052.
Rutstein, M. S. and **White, W. B.,** 1971. Vibrational spectra of high-calcium pyroxenes and pyroxenoids. *Amer. Min.*, **56**, 877–887.
Ryall, W. R. and **Threadgold, I. M.,** 1966. Evidence for $[(SiO_3)_5]_\infty$ type chains in inesite as shown by X-ray and infrared absorption studies. *Amer. Min.*, **51**, 754–761.
Sabine, P. A., 1975a. Metamorphic processes at high temperature and low pressure: the petrogenesis of the metasomatized and assimilated rocks of Carneal, Co. Antrim. *Phil. Trans. Roy. Soc. A.*, **280**, 225–269 (no. 1294).
Sabine, P. A., 1975b. Refringence of iron-rich wollastonite. *Bull. Geol. Surv. Gt. Britain*, no. 52, 65–67.
Sapountzis, E., 1973. About the occurrence of wollastonite in the area of Xanthi (north Greece). *Neues Jahrb. Min., Abhdl.*, **120**, 98–107.
Sathe, R. V. and **Choudhary, P. D.,** 1967. Stellate wollastonite from calc silicate skarns of Jothwad Hill, Panchmahal district, Gujrat, India. *Min. Mag.*, **36**, 616–618.
Schairer, J. F. and **Bowen, N. L.,** 1942. The binary system $CaSiO_3$–diopside and the relations between $CaSiO_3$ and åkermanite. *Amer. J. Sci.*, **240**, 725–742.
Shinno, I., 1970. [A consideration on the crystallization process of diopside.] *J. Jap. Assoc. Min. Petr., Econ. Geol.*, **63**, 146–159 (Japanese with English abstract).
Shinno, I., 1974. Unit cell dimensions and infra-red absorption spectra of Mg-wollastonite in the system $CaSiO_3–CaMgSi_2O_6$. *Min. J. Japan*, **7**, 456–471.
Simonen, A., 1953. Mineralogy of the wollastonites found in Finland. *Bull. Comm. géol. Finlande*, **27**, No. 159, 9–18.
Stilwell, F. L., 1959. Petrology of the Broken Hill lode and its bearing on ore genesis. *Proc. Austral. Inst. Min. Metall.*, no. 190, 1–84.
Tilley, C. E., 1951. A note on the progressive metamorphism of siliceous limestones and dolomites. *Geol. Mag.*, **88**, 175–178.
Tolliday, J. M., 1959. The crystal structure of parawollastonite and wollastonite. *Ph.D. Thesis*, University of London.
Trojer, F. J., 1968. The crystal structure of parawollastonite. *Z. Krist.*, **127**, 291–308.
Trojer, F. J., 1969. Crystal structure of a high-pressure form of $CaSiO_3$. *Z. Krist.*, **130**, 185–206.
Tuttle, O. F. and **Harker, R. I.,** 1957. Synthesis of spurrite and the reaction wollastonite + calcite = spurrite + carbon dioxide. *Amer. J. Sci.*, **255**, 226–274.
Ueda, T. and **Tomita, K.,** 1968. Wollastonite and parawollastonite from Japan. *Mem. Fac. Sci., Kyoto Univ., Ser. Geol. and Min.*, **34**, 75–82 (M.A. 69–1047).
Vaidya, S. N., Bailey, S., Pasternack, T. and **Kennedy, G. C.,** 1973. Compressibility of fifteen minerals to 45 kilobars. *J. Geophys. Res.*, **78**, 6893–6898.
Warren, B. E. and **Biscoe, J.,** 1931. The crystal structure of the monoclinic pyroxenes. *Z. Krist.*, **80**, 391–401.
Watkinson, D. H., 1973. Pseudoleucite from plutonic alkalic rock-carbonatite complexes. *Can. Min.*, **12**, 129–134.

Weeks, W. F., 1956. A thermochemical study of equilibrium relations during metamorphism of siliceous carbonate rocks. *J. Geol.*, **64**, 245–270.
Wenk, H.-R., 1969. Polymorphism of wollastonite. *Contr. Min. Petr.*, **22**, 238–247.
Wenk, H.-R., Müller, W. F., Liddell, N. A. and **Phakey, P. P.**, 1976. Polytypism in wollastonite. In *Electron Microscopy in Mineralogy* (H.-R. Wenk (ed.)). Berlin, Heidelberg and New York (Springer-Verlag), 324–331.
Weston, R. M. and **Rogers, P. S.**, 1976. Anisotropic thermal expansion characteristics of wollastonite. *Min. Mag.*, **40**, 549–551.
Windom, K. E. and **Boettcher, A. L.**, 1976. The effect of reduced activity of anorthite on the reaction grossular + quartz = anorthite + wollastonite: a model for plagioclase in the Earth's lower crust and upper mantle. *Amer. Min.*, **61**, 889–896.
Wyllie, P. J. and **Haas, J. L.**, *Jr.*, 1965. The system $CaO–SiO_2–CO_2–H_2O$: I. Melting relationships with excess vapor at 1 kilobar pressure. *Geochim. Cosmochim. Acta*, **29**, 871–892.
Wyllie, P. J. and **Hass, J. L.**, *Jr.*, 1966. The system $CaO–SiO_2–CO_2–H_2O$. II. The petrogenetic model. *Geochim. Cosmochim. Acta*, **30**, 525–543.
Yoder, H. S., *Jr.*, 1964. Soda melilite. *Carnegie Inst. Washington, Ann. Rept. Dir. Geophys. Lab.*, 1963–64, 86–89.
Zuberi, Z. H. and **Kopp, O. C.**, 1976. Differential reaction analysis (DRA) – a technique for obtaining differential thermal analysis data from inert substances. *Amer. Min.*, **61**, 281–286.

Pectolite $Ca_2NaH[SiO_3]_3$

Triclinic (+)

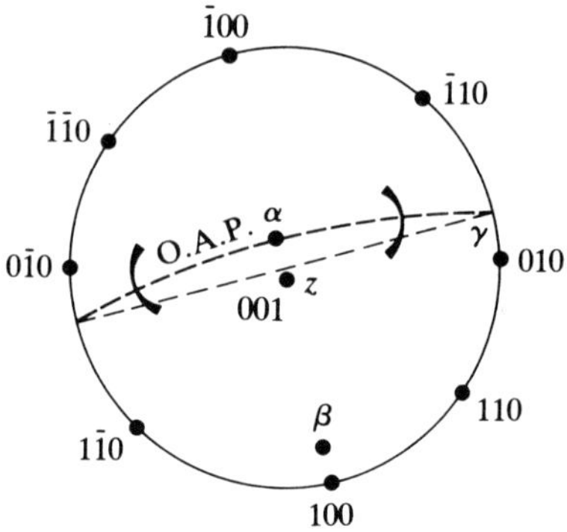

α 1·592–1·610[1]
β 1·603–1·615
γ 1·630–1·645
δ 0·030–0·038
$2V_\gamma$ 50°–63°
$\alpha:z = 10°–19°$ $\beta:x$ 10°–16°
$\gamma:y = 2°$ O.A.P. near (100)
Dispersion: $r > v$, weak.
D 2·84–2·90
H $4\frac{1}{2}$–5
Cleavage: {1̄00} and {001}, perfect.
Twinning: Rare, twin axis y, composition plane approximately (100).
Colour: Colourless or white; colourless in thin section.
Unit cell: a 7·99, b 7·04, c 7·02 Å
α 90°31′, β 96°11′, γ 102°28′
Z = 2. Space group $P\bar{1}$

Partly decomposed and gelatinized by HCl.

Pectolite generally occurs as whitish radiating aggregates of acicular crystals and is found in cavities in basaltic rocks, often in association with zeolites: it is less commonly found in lime-rich metamorphic rocks. Its name is derived from the Greek *pektos*, congealed, in allusion to its translucent appearance.

Structure

Pectolite was formerly regarded as being monoclinic with a pyroxene-type structure, but Warren and Biscoe (1931) determined the unit cell and showed that the structure was not like that of the pyroxenes but was probably triclinic. Nor is pectolite a zeolite, with which mineral group it often occurs, as it loses its structural water less readily than the zeolite minerals. Peacock (1935) confirmed the triclinic symmetry from a detailed morphological study, and suggested also that pectolite is isomorphous with wollastonite and bustamite. The structure of pectolite was first determined in detail by Buerger (1956) and further refinements were made by Buerger and Prewitt (1961) and by Prewitt (1967). Detailed comparisons between the structure of pectolite and wollastonite have been made by Prewitt and Buerger (1963) and between pectolite and the other 'Dreierketten' silicates by Prewitt and

[1] Highly manganoan pectolite has properties outside these ranges.

Peacor (1964); the Dreierketten silicates, with repeat units of three tetrahedra (Liebau, 1962), include wollastonite, parawollastonite, bustamite, and pectolite (Fig. 263, p. 601). The structure is based upon single silicon–oxygen chains containing a sequence of alternate single and double tetrahedral groups. The Ca atoms are co-ordinated by oxygen octahedra which share edges to form a lath-like strip parallel to the *y* axis in the plane parallel to (101) (Figs. 251(a), 252). The Na atoms, however, have an unorthodox environment, and show considerable anisotropic thermal motion. Each Na is close to only three of the six oxygen atoms which roughly constitute its co-ordination; the Na polyhedron has been described by Prewitt and Peacor as a distorted square antiprism. The Ca octahedral bands share edges with the Na polyhedron but not with the silicate tetrahedra; no less than five tetrahedral edges are shared with the Na polyhedron. Buerger (1956) pointed out that although the structures of wollastonite and pectolite are very similar, pectolite has a considerably smaller *b* dimension; this was taken to imply the presence of a hydrogen bond between certain oxygen atoms of neighbouring silicon tetrahedra, as was confirmed by Prewitt (1967). Although pectolite and wollastonite contain similar Si–O chains, these are in different relative orientations, and the large cations between them are in different locations (compare Figs. 251(a) and (b)). The infrared absorption spectra of pectolite have been recorded by Lazarev and Tenisheva (1962) and by Ryall and Threadgold (1966) and show a possible OH bending mode at 1395 cm^{-1}.

Peacock (1935) found that the twin law in pectolite is such that *y* is the twin axis with (100) as the composition plane. This is consistent with a 180° rotation of one of the pseudomonoclinic units in the structure around *y* which would be approximately equivalent to translation by *b*/2 along a neighbouring cell (possible only if $\alpha = 90°$). X-ray photographs of twinned pectolites confirm this interpretation (Prewitt and Buerger, 1963). The pectolite structure cannot, however, form a monoclinic parawollastonite analogue as it is not possible to form a pectolite octahedral band with the Ca at centres of symmetry (see, however, Müller, 1976).

The structure of serandite has been refined by Takeuchi *et al.* (1976) on a crystal from the Tanohata mine, Japan, with *a* 7·683, *b* 6·889, *c* 6·747 Å, α 90·53°, β 94·12°, γ 102·75°, space group $P\bar{1}$, Z = 2 [$(Mn_{1\cdot88}Ca_{0\cdot17}Mg_{0\cdot01})NaHSi_{2\cdot79}O_9$]. The Si–O and Na–O distances are all smaller than the corresponding distances in pectolite. Takeuchi *et al.* have suggested that the shrinkage of octahedral bands due to substitution of Mn for Ca gives rise to the distortion of individual silicate tetrahedra rather than to a change in the configuration of the silicate chains. Ca atoms are preferentially located in the *M*1 position (Fig. 253) corresponding with Ca(1) in pectolite; the occupancy at *M*1 is 0·84 Mn, 0·16 Ca. There thus appears to be a virtually direct relationship between the Ca/Mn ratio and the cell parameters, with all three dimensions *a*, *b* and *c* and the volume decreasing as Ca is replaced by Mn (see also Semenov *et al.*, 1976, who in addition give the infrared spectra for fourteen analysed members of the pectolite–serandite series).

Chemistry

Although pectolite is often found in a very pure state, bivalent manganese can replace calcium in the structure. Similarities in morphology and in X-ray diffraction pattern indicate that pectolite, manganoan pectolite (schizolite), and serandite (the Mn analogue of pectolite) form an isostructural series (Machatschki, 1932;

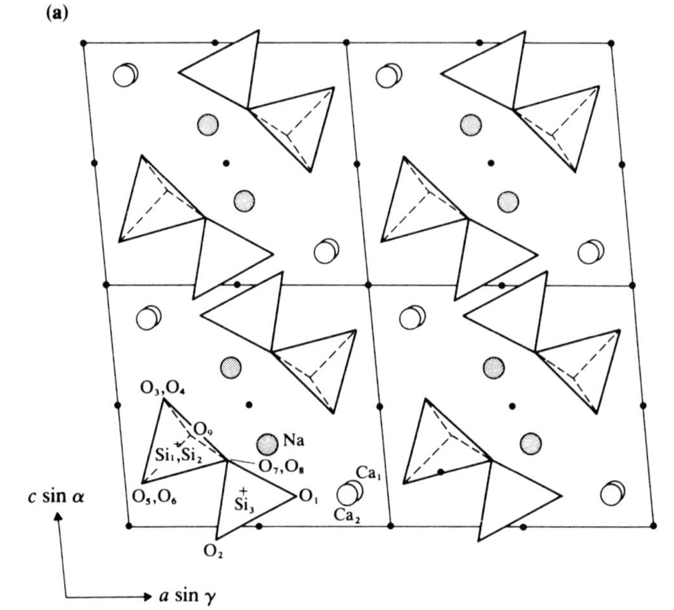

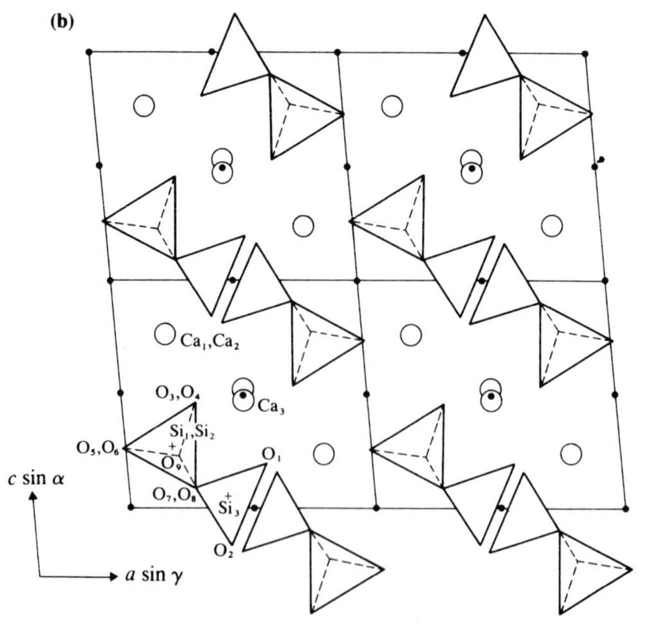

Fig. 251. (a) Projection of the pectolite structure along *y*. Although the Si_1 and Si_2 tetrahedra should not be exactly superimposed, they are presented this way to simplify the drawing. (b) Projection of the wollastonite structure along *y* (after Prewitt and Buerger (1963)).

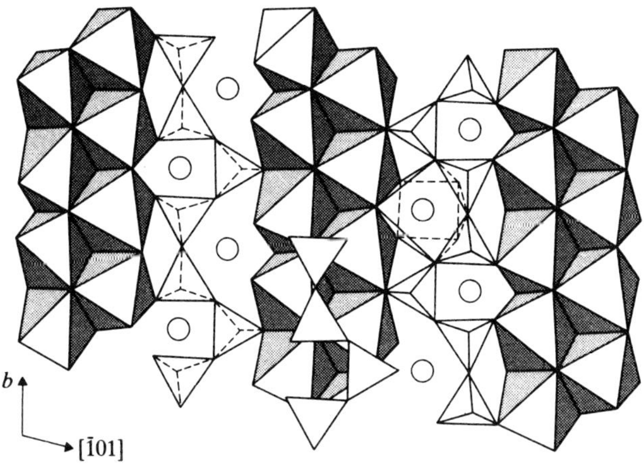

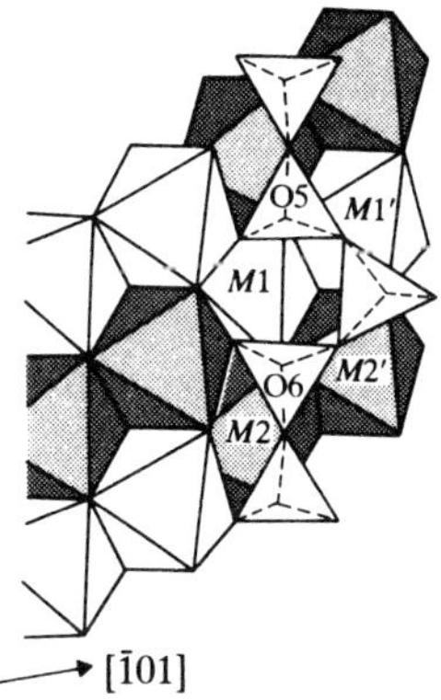

Fig. 252. Projection of a portion of an octahedral layer of the pectolite structure on to (101). The circles represent Na ions, the octahedra contain Ca, and the tetrahedra contain Si ions (after Prewitt, 1967).

Fig. 253. Relations in serandite between cation ordering in octahedral bands and locations of Si_3O_9 chains attached to the bands. The octahedra fully occupied by Mn are shaded (after Takeuchi *et al.*, 1976).

Schaller, 1955), though for a long time the only recorded analysis of serandite was of a calcium-rich variety (Lacroix, 1931). Schizolite was shown by Schaller to be a manganoan pectolite and not a distinct species.

Recently, however, Semenov *et al.* (1976) have described a complete range of specimens ranging from pectolite through manganoan pectolite and calcian serandite to serandite (Table 59, anals. 1, 13 and 14) and the structure of serandite has been determined on crystals from the Tanohata mine, Japan, for which microprobe analysis gave MnO 37·33 per cent (Takeuchi *et al.*, 1976). The more Mn-rich members of the series typically contain small amounts of rare earths; the rare-earth contents of eleven specimens are reported in detail by Semenov *et al.* who show that La and Ce are common in the Ca–Mn varieties but that in serandites Y is dominant.

Magnesium and iron may also enter the structure in small amounts, and some aluminium may enter the Z group along with silicon. The almost pure calcium end-member is, however, by far the most abundant variety. In a review of many pectolite analyses it is noted that the content of structural water reported is often high, giving more than 2(OH) ions on recalculating the analyses on the basis of 18(O,OH). This might be ascribed to adsorbed water due to the fibrous nature of most specimens, were it not for the fact that in such analyses the Z group is low. Three such analyses are shown in Table 59 (anals. 9, 10 and 11) and in several similar cases the excess water would appear to fit the formula $Na_2Ca_4Si_{6-m}O_{18}H_{2+4m}$ (cf. hydrogrossular).

Although it has been suggested that yuksporite is identical with pectolite, Minguzzi (1942) has shown that its refractive indices are higher; the X-ray powder photograph is slightly different and the reported compositions yield different formulae.

Experimental

Pectolite has been synthesized by heating SiO_2, CaO, H_2O and NaOH in a closed vessel for $3\frac{1}{2}$ days at 180°C (Clark and Bunn, 1940), and it is also known to occur in highly siliceous boiler scales. It may undergo alteration to stevensite,

Table 59. Pectolite Analyses

	1	2	3	4	5
SiO_2	51·49	54·18	52·89	54·64	53·80
Al_2O_3	0·66	—	0·21	0·07	0·00
Fe_2O_3	—	0·18	—	tr.	0·00
FeO	0·33	0·42	0·14	0·14	1·00
MnO	0·01	—	tr.	0·09	0·12
MgO	0·23	—	tr.	0·02	0·00
CaO	34·00	33·36	33·85	32·79	33·20
SrO	—	—	—	—	—
Na_2O	9·15	8·72	8·81	8·62	9·01
K_2O	0·11	0·88	0·19	0·00	0·00
H_2O^+	3·02	2·74	3·66	3·03	2·94
H_2O^-	0·98	—	—	0·31	—
Total	100·02	100·48	99·75	99·71	100·07
α	1·592	—	1·594	—	1·600
β	—	—	1·603	—	1·605
γ	1·630	—	1·631	—	1·636
δ	0·038	—	0·037	—	0·036
$2V_\gamma$	—	—	—	—	50°
D	2·77	2·834	2·86	—	—

Numbers of ions on the basis of eighteen O, OH

	1	Σ	2	Σ	3	Σ	4	Σ	5	Σ
Si	5·782		5·989		5·826		6·025		5·954	
Al	0·088	5·87	—	6·00	0·028	5·85	0·009	6·03	—	
Fe^{3+}	—		0·015		—		—		—	
Mg	0·038		—		—		0·003		—	
Fe^{2+}	0·031	4·16	0·039	3·99	0·013	4·01	0·013	3·90	0·092	4·04
Mn	0·001		—		—		0·009		0·011	
Ca	4·091		3·953		3·995		3·875		3·937	
Na	1·992	2·01	1·867	1·99	1·881	1·91	1·891		1·934	
K	0·016		0·124		0·026		—		—	
OH	2·262		2·020		2·690		2·229		2·170	

1 Pectolite, Zhelekhov, Severnaya Chekhiya (Semenov *et al.*, 1976). Includes TiO_2 0·04.
2 Glassy pectolite in granitic rock in serpentinite, Thetford mines, Quebec (Parsons, 1924). Anal. E. W. Todd.
3 White, fibrous pectolite, serpentinite, Horokanai, Isikari Province, Japan (Harada, 1934). Includes CuO 0·004.
4 Pectolite, cavity filling, Prospect alkaline diabase-picrite intrusion, New South Wales, Australia (Wilshire, 1967). Anal. M. Chiba.
5 White bladed crystals of pectolite, Paterson, New Jersey (Peacock, 1935). Anal. F. A. Gonyer.

6		7		8		9		10		
53·28		53·84		54·02		53·28		52·04		SiO_2
0·16		0·16		—		0·03		0·92		Al_2O_3
0·48		—		—		—		—		Fe_2O_3
—		0·07		—		0·07		1·29		FeO
0·33		0·04		1·53		0·09		2·31		MnO
0·26		0·09		—		0·11		0·05		MgO
33·41		33·74		32·20		32·80		31·15		CaO
—		0·002		—		—		0·12		SrO
9·14		8·82		8·88		9·98		7·97		Na_2O
0·25		0·02		0·36		0·55		0·90		K_2O
2·70		2·96		3·00		3·12		3·07		H_2O^+
—		0·10		—		0·04		—		H_2O^-
100·01		99·85		99·99		100·07		100·21		Total
1·601		1·598		1·600		1·600		1·604		α
1·604		1·606		—		—		1·610		β
1·633		1·623		1·638		1·630		1·636		γ
0·032		0·026		0·038		0·030		0·032		δ
—		35°		—		58°		—		$2V_\gamma$
2·857		2·86		—		2·863		—		D
5·921		5·954		5 977		5·902		5·816		Si
0·021	5·98	0·021	5·98	—		0·004	5·91	0·121	5·94	Al
0·040		—		—		—		—		Fe^{3+}
0·043		0·015		—		0·018		0·008		Mg
—	4·06	0·006	4·02	—	3·96	0·007	3·93	0·120	4·11[a]	Fe^{2+}
0·031		0·004		0·143		0·008		0·219		Mn
3·983		3·998		3·818		3·893		3·731		Ca
1·972	2·01	1·891	1·89	1·905	1·96	2·143	2·22	1·727	1·86	Na
0·035		0·003		0·050		0·078		0·129		K
2·004		2·184		2·214		2·306		2·289		OH

[a] Includes Sr 0·008, Ba 0·006, Zn 0·021.

6 White bladed crystals of pectolite, crevice in Keweenawan traps, Lake Nipigon, Ontario (Walker and Parsons, 1926). Anal. H. C. Rickaby.

7 Pectolite, tectonic inclusion in serpentinite, Myrtle Creek, Oregon (Coleman, 1961). Includes CuO 0·004.

8 Pectolite, nepheline-syenite, Yuksporlak, Kola Peninsula, Russia (Belyankin and Iwanova, 1933).

9 Pectolite, associated with laumontite and calcite in globular limburgite, Georgia, U.S.S.R. (Skhirtladze, 1966).

10 Transparent, colourless pectolite, Parker shaft, Franklin, New Jersey (Palache, 1937). Anal. R. B. Gage (inludes BaO 0·13, ZnO 0·26).

Table 59 Pectolite Analyses – *continued*

	11	12	13	14
SiO_2	52·36	51·44	50·61	48·96
Al_2O_3	1·36	—	0·54	0·65
Fe_2O_3	0·70	—	—	—
FeO	1·45	2·01	0·71	0·58
MnO	0·02	11·69	26·40	34·35
MgO	0·80	0·13	0·21	—
CaO	31·55	20·53	9·94	3·68
SrO	—	—	—	—
Na_2O	6·20	9·50	7·73	8·50
K_2O	2·20	—	0·24	0·56
H_2O^+	3·68	2·25	3·10	2·20
H_2O^-	—	—	—	—
Total	100·32	99·95	99·96	'99·89'
α	1·596	—	1·660	1·680
β	—	—	—	—
γ	1·630–1·632	—	1·690	1·705
δ	0·034–0·036	—	0·030	0·025
$2V_\gamma$	—	—	—	33°
D	—	—	3·22	3·33

Numbers of ions on the basis of eighteen O, OH

	11		12		13		14	
Si	5·753		6·007		5·870		5·924	
Al	0·176	5·99	—		0·074	5·95[c]	0·093	6·02[e]
Fe^{3+}	0·058		—		—		—	
Mg	0·131		0·022		0·123		—	
Fe^{2+}	0·133	3·98	0·195	4·06[b]	0·069	4·04[d]	0·059	4·07[f]
Mn	0·002		1·148		2·594		3·520	
Ca	3·714		2·550		1·235		0·477	
Na	1·321	1·63	2·135		1·738	1·77	1·994	2·08
K	0·308		—		0·036		0·086	
OH	2·696		1·740		2·398		1·776	

[b] Includes Y 0·148.
[c] Includes Ti 0·005.
[d] Includes *TR* 0·022.
[e] Includes Ti 0·003.
[f] Includes *TR* 0·017.

11 Pectolite in veins associated with quartz dolerite cutting Voĭkaro–Syninsk ultrabasic massif, polar Urals, U.S.S.R. (Maslov and Savel'eva, 1965).

12 Yellowish manganoan pectolite (schizolite), nepheline-syenite, Naujakasik, Julianehaab district, Greenland (Winther, 1901). Anal. Chr. Christensen (includes Y_2O_3 2·40).

13 Calcian serandite, Karnasurt, Lovozero (Semenov *et al.*, 1976). Includes TiO_2 0·06, rare earths 0·42.

14 Serandite, St. Hilaire, Canada (Semenov *et al.*, 1976). Includes TiO_2 0·03, rare earths 0·31.

$Mg_3Si_4O_{10}(OH)_2$, by reaction with $MgH_2(CO)_3$ and water (Glenn, 1916; Faust and Murata, 1953). Glenn, and also Thilo and Funk (1950), reported that pectolite lost water above 400°C and was changed to a mixture of wollastonite, SiO_2 and $Ca_2NaHSi_3O_9$: the latter altered to pectolite when heated with water at 200°C. McLaughlin (1957), however, recorded the DTA curve for pectolite from Bergen Hill, New Jersey, and demonstrated that the water was lost fairly sharply at a higher temperature, giving an endothermic peak at 780° and finishing at approximately 845°C; melting started at about 1000°C. See also Skhirtladze (1966). DTA, DTG and weight-loss curves for pectolite and serandite (Semenov *et al.*, 1976) all show the loss of water and a sharp endothermic peak at around 740°C for pectolite (Table 59, anal. 1) and slightly lower at 640°–725°C for calcian serandite (Table 59, anal. 13): see also Todor (1972) who gives the maximum for the endothermic peak of pectolite as 780°C. Juan *et al.* (1968) have shown experimentally that the reaction 2 pectolite + $5Al_2O_3$ + $8SiO_2$ = 2 labradorite + water is in equilibrium near 224°C at 890 bar and 230°C at 2680 bar water pressure, and have inferred that the curve for the reaction 2 pectolite + 5 prehnite + 10 CO_2 = 2 labradorite + 10 calcite + 7 quartz + 6 water lies at similar temperatures. Cadmium pectolite has been synthesized hydrothermally by Belokoneva *et al.* (1974); it has a similar structure to pectolite and has *a* 7·847, *b* 6·980, *c* 6·920 Å, α 90°24′, β 94°36′, γ 102°48′; D 4·0 g/cm³.

Material from West Paterson, New Jersey, originally called 'pink pectolite' has been shown to represent pseudomorphs of stevensite after pectolite, formed by alteration of the latter (Rothstein, 1971).

Optical and Physical Properties

The entry of iron and manganese increases the refractive indices and the specific gravity, and brings about a slight decrease in birefringence and optic axial angle. Schaller (1955) showed that in the isomorphous series pectolite–manganoan pectolite (schizolite)–serandite there is a continuous variation in physical properties: this has been confirmed by Semenov *et al.* (1976). The end-members have the following optical properties (2V values are calculated):

	α	β	γ	δ	$2V_\gamma$	D
Pectolite	1·592	1·603	1·630	0·038	62°	2·86
Serandite (by extrapolation)	1·680	1·682	1·705	0·025	33°	3·32

The replacement of calcium by magnesium is reported to lower the refractive indices and specific gravity.

The accurate optical orientation has been given by Peacock (1935) with co-ordinates as follows: α, $\phi = -115°$, $\rho = 11°$; for β, $\phi = 78°$, $\rho = 80°$; for γ, $\phi = -13°$, $\rho = 88°$. Some crystals may show a small amount of luminescence and thermoluminescence: Greg and Lettsom (1858) reported that a small fragment of pectolite from the Edinburgh Castle dolerite on being broken in the dark emitted a brilliant flash of light. Angino (1964) has reported that pectolite shows a moderate natural thermoluminescence and that the natural glow curve peaks at 275° and 350°C are both shifted upwards by 40°C on application of a pressure of 2530 bar for two minutes applied uniaxaially and producing shear stress of unknown magnitude.

Distinguishing Features

Pectolite is distinguished from wollastonite by its considerably greater birefringence, and by its length-slow character: the latter property also distinguishes it from rosenbuschite, $(Ca,Na)_3(Zr,Ti)Si_2O_8F$, which has $\alpha = z$ (parallel to fibres).

Paragenesis

Pectolite typically occurs as a hydrothermal mineral in cavities and on joint faces in basic igneous rocks: it is chiefly associated with basalt and dolerite (Table 59, anals. 4, 5, 6 and 11). Its formation is reportedly favoured by a silica-saturated environment (Juan *et al.*, 1968) and it may have a close genetic association with prehnite and be produced at the last stage of the cooling history of basic igneous rocks. It also occurs in serpentinites (anals. 2, 3 and 7) and has been recorded from mica peridotite (Franks, 1959), where it makes up 25 per cent of the rock and is the major component of the groundmass. Milk-white veins of xonotlite and pectolite occur in the border zone of a pyroxenite within a serpentinized mass in the Zlatibor Mountains, Yugoslavia and have been attributed to the alteration of calcium-bearing ultramafic rocks (Majer and Barić, 1971); its occurrence in rodingite has been noted (Bloxam, 1954). A similar assemblage occurs in fracture veins cutting tuff, pillow lava, and olivine basalt in west Greenland (Karup-Møller, 1969); although zeolites also occur in these rocks, the xonotlite-pectolite suite is not associated with zeolites and is assumed to have formed under higher temperature conditions at around 300°C ± 50° and pressures less than 300 bar. Pectolite occurs also as a primary mineral in some alkaline igneous rocks such as tinguaite and microfoyaite (Shand, 1928), phonolite (Lacroix, 1932), and various types of nepheline-syenite (Table 59, anal. 8; see also Kostyleva, 1925; Adamson, 1944; Jérémine, 1950). The manganoan variety, schizolite, is also known mainly from alkaline rocks (Table 59, anal. 12; see also Gerasimovskiĭ and Belyayev, 1963; Kapustin, 1971); intermediate varieties containing smaller amounts of manganese also occur, e.g. anal. 10 (2·31 per cent MnO) and anal. 8 (1·53 per cent MnO), and pectolites from fenite of the Turi peninsula (Kulakov *et al.*, 1974) with 1·82 and 0·60 per cent MnO. A member of the series for which microprobe analyses gave MnO values ranging from 16·30 to 23·90 per cent within the same crystal has been recorded from a phonolite dyke in south-eastern Queensland (Carr *et al.*, 1976), which also contains eudialyte, aegirine, arfvedsonite, nepheline, natrolite, and analcite.

Pectolite has also been described from some calcium-rich metamorphic rocks and skarns (e.g. Francis, 1958), as pseudomorphs after quartz in vein material (Glenn, 1917), and in siliceous boiler scales.

References

Adamson, O. J., 1944. The petrology of the Norra Kärr district. An occurrence of alkaline rocks in southern Sweden. *Geol. För. Förh.*, **66**, 113–255 (M.A. **9**–87).

Angino, E. E., 1964. Some effects of pressure on the thermoluminescence of amblygonite, pectolite, orthoclase, scapolite and wollastonite. *Amer. Min.*, **49**, 387–394.

Belokoneva, E. L., Sandomirskii, P. A., Simonov, M. A. and **Belov, N. V.**, 1974. Crystal structure of cadmium pectolite. *Soviet Physics – Doklady*, **18**, 629–630.

Belyankin, D. S. and **Iwanova, W. P.**, 1933. Untersuchung der chemischen Konstitution des Pektoliths. *Centr. Min., Abt. A*, 327–339.
Bloxam, T. W., 1954. Rodingite from the Girvan–Ballantrae complex, Ayrshire. *Min. Mag.*, **30**, 525–528.
Buerger, M. J., 1956. The determination of the crystal structure of pectolite, $Ca_2NaHSi_3O_9$. *Z. Krist.*, **108**, 248–261.
Buerger, M. J. and **Prewitt, C. T.**, 1961. The crystal structure of wollastonite and pectolite. *Proc. Nat. Acad. Sci.*, **47**, 1884–1888.
Carr, G. R., Phillips, E. R. and **Williams, P. R.**, 1976. An occurrence of eudialyte and manganoan pectolite in a phonolite dyke from south-eastern Queensland. *Min. Mag.*, **40**, 853–856.
Clark, L. M. and **Bunn, C. W.**, 1940. The scaling of boilers. Pt. IV. Identification of phases in calcium silicate scales. *J. Soc. Chem. Ind.*, **59**, 155–158.
Coleman, R. G., 1961. Jadeite deposits of the Clear Creek area, New Idria district, San Benito County, California. *J. Petr.*, **2**, 209–247.
Faust, G. T. and **Murata, K. J.**, 1953. Stevensite, redefined as a member of the montmorillonite group. *Amer. Min.*, **38**, 973–987.
Francis, G. H., 1958. Petrological studies in Glen Urquhart, Inverness-shire. *Bull. Brit. Mus. (Nat. Hist.), Min.*, **1**, 123–162.
Franks, P. C., 1959. Pectolite in mica peridotite, Woodson County, Kansas. *Amer. Min.*, **44**, 1082–1086.
Gerasimovskiĭ, V. I. and **Belyayev, Yu. I.**, 1963. Manganese, barium and strontium in the alkalic rocks of the Kola Peninsula. *Geochemistry*, **12**, 1161–1174.
Glenn, M. L., 1916. A new occurrence of stevensite, a magnesium-bearing alteration product of pectolite. *Amer. Min.*, **1**, 44–46.
Glenn, M. L., 1917. Pectolite pseudomorphous after quartz from West Paterson, N.J. *Amer. Min.*, **2**, 43–45.
Greg, R. P. and **Lettsom, W. G.**, 1858. *Manual of the Mineralogy of Great Britain and Ireland.* London.
Harada, Z., 1934. Über einen neuen Pektolith-fund in Japan. *J. Fac. Sci. Hokkaido Univ.*, ser. 4, **2**, 355–359 (M.A. **6**–92).
Jérémine, E., 1950. Sur quelques minéraux des syénites néphéliniques de Bou Agrao, Haut Atlas (Maroc). *Compt. Rend. Acad. Sci. Paris*, **250**, 110–111.
Juan, V. C., Youh, C.-C. and **Lo, H.-J.**, 1968. A synthetic study with natural pectolite and its bearing on the hydrothermal alteration of basic igneous rocks. *Proc. Geol. Soc. China* [Formosa], **11**, 99–108.
Kapustin, Yu. I., 1971. [*The mineralogy of carbonatites.*] Moscow ('Nauka' Press), 288 pp. (M.A. 72–1734).
Karup-Møller, S., 1969. Xonotlite-, pectolite- and natrolite-bearing fracture veins in volcanic rocks from Nûgssuaq, west Greenland. *Medd. om Grønland*, **186**, no. 2, 20 pp.
Kostyleva, E. E., 1925. [Pectolite from Khibinsky tundra.] *Bull. Acad. Sci. Russie*, ser. 6, **19**, 383–404 (in Russian) (M.A. **3**–111).
Kulakov, A. N., Evdokimov, M. D. and **Bulakh, A. G.**, 1974. [Mineral veins in fenites of the Tur'ii peninsula in the Murmansk region.] *Zap. Vses. Min. Obshch.*, **103**, 179–191 (in Russian).
Lacroix, A., 1931. Les pegmatites de la syénite sodalitique de l'île Rouma (Archipel de Los, Guinée française). Description d'un nouveau minéral (sérandite) qu'elles renferment. *Compt. Rend. Acad. Sci. Paris*, **192**, 187–194.
Lacroix, A., 1932. Les roches intrusives et filoniennes de la région granitique et sédimentaire du nord du Tibesti. *Compt. Rend. Acad. Sci. Paris*, **194**, 670–674.
Lazarev, A. N. and **Tenjsheva, T. F.**, 1962. Vibrational spectra of silicates. V. Silicates with anions in ribbon form. *Optics and spectroscopy (U.S.S.R.)*, **12**, 115–117.
Liebau, F., 1962. Die Systematik der Silikate. *Naturwiss.*, **49**, 481–491.
Machatschki, F., 1932. Serandit–Pektolith–Wollastonit. *Centr. Min., A.*, 69–73.
McLaughlin, R. J. W., 1957. Other minerals. In: *The Differential Thermal Investigation of Clays.* London (Min. Soc), pp. 364–388.
Majer, V. and **Barić, L.**, 1971. Xonotlit und Pektolith aus basischen Gesteinen des Peridotitgabbrokomplexes im Zlatibor-Gebirge, Jugoslawien. *Tschermaks Min. Petr. Mitt.*, ser. 3, **15**, 43–55.
Maslov, M. A. and **Savel'eva, G. H.**, 1965. [The discovery of pectolite in the polar Urals.] In: *Materialy po geol. i polezn. iskopaemym Severo-Vostoka Evrop. chasti SSSR, Skÿtÿvkar*, no. 5, 105–107 (in Russian) (M.A. **17**–599).
Minguzzi, C., 1942. Sulla non identità della juxporite con la pectolite. *Z. Krist.*, **104**, 417–424.
Müller, W. F., 1976. On stacking disorder and polytypism in pectolite and serandite. *Z. Krist.*, **144**, 401–408.
Palache, C., 1937. The minerals of Franklin and Sterling Hill, Sussex County, New Jersey. *Prof. Paper U.S. Geol. Surv.*, **180**.
Parsons, A. L., 1924. Pectolite and apophyllite from Thetford Mines, Quebec. *Univ. Toronto Studs., Geol. Ser.*, No. 17, 55–57

Peacock, M. A., 1935. On pectolite. *Z. Krist.*, **90**, 97–111.
Prewitt, C. T., 1967. Refinement of the structure of pectolite, $Ca_2NaHSi_3O_9$. *Z. Krist.*, **125**, 298–316.
Prewitt, C. T. and **Buerger, M. J.,** 1963. Comparison of the crystal structures of wollastonite and pectolite. *Min. Soc. America, Spec. Paper* **1**, 293–302.
Prewitt, C. T. and **Peacor, D. R.,** 1964. Crystal chemistry of the pyroxenes and pyroxenoids. *Amer. Min.*, **49**, 1527–1542.
Rothstein, J., 1971. 'Stevensite': a review. *Min. Record*, **2**, 30–31.
Ryall, W. R. and **Threadgold, I. M.,** 1966. Evidence for $[(SiO_3)_5]_\infty$ type chains in inesite as shown by X-ray and infra-red absorption studies. *Amer. Min.*, **51**, 754–761.
Schaller, W. T., 1955. The pectolite–schizolite–serandite series. *Amer. Min.*, **40**, 1022–1031.
Semenov, E. I., Maksimyuk, I. E. and **Arkangelskaya, V. N.,** 1976. [On the minerals of the pectolite–serandite group.] *Zap. Vses. Min. Obshch.*, **104**, 154–163 (in Russian).
Shand, S. J., 1928. The geology of Pilansberg in the western Transvaal: a study of alkaline rocks and ring-intrusions. *Trans. Geol. Soc. South Africa*, **31**, 97–156.
Skhirtladze, N. I., 1966. Pectolite found for the first time in Georgia. *Dokl. Acad. Sci. U.S.S.R., Earth Sci. Sect.*, **169**, 155–157.
Takeuchi, Y., Kudoh, Y. and **Yamanaka, T.,** 1976. Crystal chemistry of the serandite–pectolite series and related minerals. *Amer. Min.*, **61**, 229–237.
Thilo, E. and **Funk, H.,** 1950. Über einige chemische Eigenschaften des Pektoliths, $Ca_2Na(HSi_3O_9)$, und seine Synthese. *Z. anorg. Chem.*, **262**, 185–191.
Todor, D. N., 1972. *Analiza termica a mineralelor*. Bucharest (Edit. Tehnică), p. 220.
Walker, T. L. and **Parsons, A. L.,** 1926. Zeolites and related minerals from Lake Nipigon, Ontario. *Univ. Toronto Studs., Geol. Ser.*, no. 22, 17–18.
Warren, B. E. and **Biscoe, J.,** 1931. The crystal structure of the monoclinic pyroxenes. *Z. Krist.*, **80**, 391–401.
Wilshire, H. G., 1967. The Prospect alkaline diabase–picrite intrusion, New South Wales, Australia. *J. Petr.*, **8**, 97–163.
Winther, Chr., 1901. Schizolite, a new mineral. *Medd. om Grønland*, **24**, 196–203.

Bustamite $(Mn,Ca,Fe)[SiO_3]$

Triclinic (−)

α 1·640–1·695
β 1·651–1·708
γ 1·653–1·710
δ 0·013–0·017
$2V_\alpha$ 34°–60°
O.A.P. and α approx. ⊥ (100).[1]
$\alpha{:}x \simeq 15°$, $\beta{:}y \simeq 35°$, $\gamma{:}z \simeq 30°–35°$
Dispersion: $r < v$, weak to strong.
D 3·32–3·43
H $5\frac{1}{2}$–$6\frac{1}{2}$
Cleavage: {100} perfect, {110} and {1$\bar{1}$0} good, {010} poor.[1] (110):(1$\bar{1}$0) = 95°
Twinning: Not common, simple twins with composition plane (110).
Colour: Pale pink to brownish red; colourless to yellowish pink in thin section.
Pleochroism: Weak; in thick sections $\alpha = \gamma$ orange, β rose.
Unit cell:[1] a 7·736, b 7·157, c 13·824 Å
α 90°31′, β 94°35′, γ 103°52′
Z = 12. Space group $A\bar{1}$.

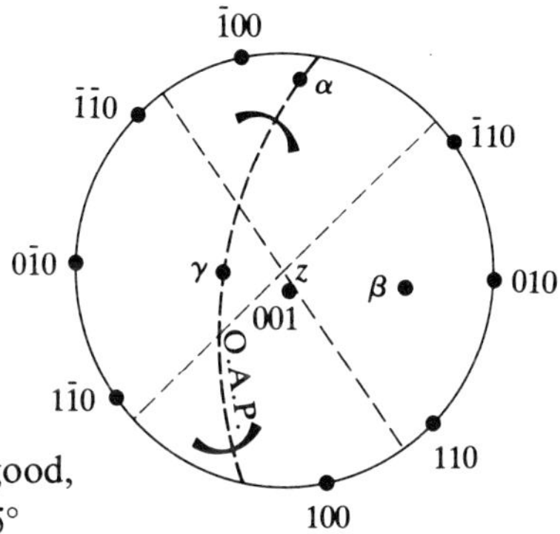

Partly soluble in HCl. Manganese reaction with fluxes.

Bustamite occurs chiefly in manganese orebodies and normally has resulted from metamorphism with accompanying metasomatism. It is named after M. Bustamente, the discoverer.

Structure

The cell parameters were first determined from X-ray measurements on a cleavage fragment by Berman and Gonyer (1937), when bustamite was shown to have a triclinic structure differing from that of rhodonite but closely approximating to that of wollastonite, as had been previously indicated by X-ray powder patterns (Bowen *et al.*, 1933). It was thus placed in the wollastonite group of the pyroxenoid family (Berman, 1937).

Later work by Buerger (1956) and by Liebau *et al.* (1958) showed that the unit cells of bustamite and wollastonite are different, though closely related. The structure was determined in detail by Peacor and Buerger (1962) using specimens

[1] Optical data and cleavages are given according to the old (morphological) choice of unit cell; cell parameters refer to later choice of axes (see p. 576).

from Franklin (Table 60, anal. 4). The arrangement of oxygen atoms crudely approximates close packing; Ca and Mn atoms and Si atoms alternate in layers between sheets of oxygen atoms, with Ca and Mn in octahedral co-ordination and Si in tetrahedral co-ordination. The SiO_4 tetrahedra share two O atoms with other tetrahedra to form a chain with a three-tetrahedra repeat (Dreierketten) and which is oriented along the *y* axis (or the *z* axis in the orientation chosen for the other Mn chain silicates, Fig. 263, p. 601). The Ca and Mn octahedra share edges (Fig. 254) to

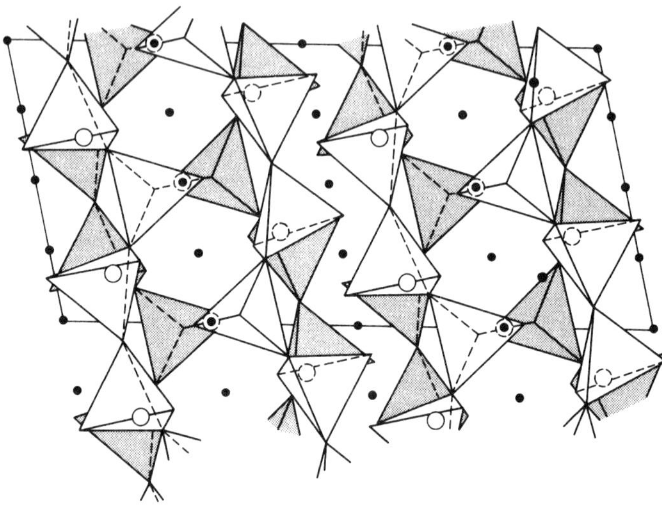

Fig. 254. Projection along the *z* axis of the structure of bustamite (after Peacor and Buerger, 1962).

form a band three octahedra wide, extending infinitely parallel to *y*: individual bands are separated by a column of unoccupied octahedrally co-ordinated voids. Since the pyroxenoids are triclinic, there is considerable freedom of choice of the unit cell. In comparing the crystal structures of bustamite and wollastonite, Peacor and Prewitt (1963) used an *A*-centred cell for bustamite which showed the structural relationships more clearly than the *F*-cell adopted previously by Peacor and Buerger (1962). The primary difference between the structures of the two phases lies in the relative arrangement of the silicate chains with respect to the planes of Ca and Mn octahedra; in bustamite there is a shift of magnitude $b/2$ of successive pairs of chains of tetrahedra within the sheets parallel to (101) (*A*-cell). Bustamite has a pseudomonoclinic cell with space group $A2/m$ (wollastonite $P2_1/m$). The Ca and Mn are ordered in bustamite, compared with the distribution of Ca in wollastonite over three general positions, with one Ca and one Mn on inversion centres and two Ca and two Mn in general positions (there is not complete equivalence of positions; for details see Peacor and Prewitt, 1963). With substitution of Mn for Ca, the characteristic wollastonite infrared spectra pattern on synthetic materials remains intact to about $Ca_{0.75}Mn_{0.25}SiO_3$ (Rutstein and White, 1971) and then changes over a very narrow compositional range to the bustamite pattern; similarly with the addition of iron to $CaSiO_3$, there is a boundary beyond which the iron becomes ordered with the development of the bustamite structure, giving the bustamite infrared pattern (Rutstein, 1971).

Although the high-temperature CaFe metasilicate phase found in the CaO–FeO–SiO_2 ternary system was earlier referred to as 'wollastonite solid solution' (Bowen *et al.*, 1933), more recent work has shown that this phase has the bustamite type of structure and it has been identified from its infrared spectrum as the iron analogue of bustamite (Rutstein and White, 1971). The structures of a synthetic $Wo_{50}Fs_{50}$ phase and of natural ferrobustamite (Table 60, anal. 1) were determined by Rapoport and Burnham (1973) who confirmed that they have approximately the same cell parameters and structures as bustamite; in particular, the natural ferrobustamite $Ca_{0\cdot79}Fe_{0\cdot19}Mn_{0\cdot02}SiO_3$ (originally described as iron-wollastonite; Tilley, 1948) has *a* 7·83, *b* 7·23, *c* 13·925 Å, α 90°01′, β 95°24′, γ 103°21′, thus indicating a slight enlargement of the unit cell on the replacement of Mn by Fe. The clear distinction between the X-ray powder diffraction pattern of bustamite and those of rhodonite and pyroxmangite has been illustrated by Momoi (1964) and the break between the cell parameters of wollastonites and ferrobustamites (iron-wollastonites) has been demonstrated by Matsueda (1973). Rapoport and Burnham (1973) have shown that the synthetic crystal of $Ca_{0\cdot5}Fe_{0\cdot5}SiO_3$ has Fe and Ca ordered over two general and two special positions; (see also Burnham, 1976); in the Mössbauer spectrum of this phase, the inner doublet has been assigned to iron in the more distorted *M*1 and *M*2 sites (Dowty and Lindsley, 1974). Refinements of the cation occupancies of the octahedral sites in three bustamites with $MnSiO_3$ 60, 37 and 12 mol. per cent has shown that cation substitution in bustamite is a two-step process (Ohashi and Finger, 1976). The smaller cations such as Mn or Fe occupy the *M*3 site and then part of the *M*1 site. If the smaller cations completely fill *M*1, i.e. their total exceeds that of Ca, then the *M*2 site begins to accommodate them. Analysis of the geometrical relationships between the tetrahedra and octahedra indicates that *M*3 and *M*4 are quite different in bustamite probably resulting in the occupancy being essentially fixed.

Chemistry

Bustamite analyses showing varying calcium and iron contents are given in Table 60, where they have also been recalculated on the basis of eighteen oxygen ions. The original bustamite from Puebla, Mexico, has been shown to be a mixture of johannsenite and rhodonite (Schaller, 1938). The magnesium content is generally low, and zinc is only found in material from a zincian environment as at Franklin, New Jersey. Ferrous iron is normally low but rises to just over 9 per cent in the fibrous bustamite from Broken Hill (Table 60, anal. 7), see also the nine analyses of bustamites from this locality reported by Mason (1973) which have FeO 4·17–9·18 per cent. When rhodonite and bustamite occur together, their compositional fields can be seen to be quite distinct, bustamite being much richer in calcium (e.g. Howie, 1965; Binns, 1968; Mason, 1973; Hodgson, 1975). Detailed work by Schaller on the bustamites from Broken Hill, Australia, showed that their maximum $MnSiO_3$ component is about 67 per cent, corresponding with the formula $CaMn_2Si_3O_9$; the maximum content of $CaSiO_3$ is around 20 per cent (Mason, 1973, 1975). From the recalculations of the analyses of Table 60 it is clear that Fe substitutes mainly for Mn, and as the so-called iron-wollastonites have the bustamite structure, members of the bustamite family with Fe > Mn have been termed ferrobustamite (e.g. anal. 1). A slag mineral, vogtite, named in honour of J. H. L. Vogt by Hlawatsch (1907), and hitherto considered to be related to rhodonite, may also more correctly be referred to ferrobustamite; an analysis of vogtite and its properties is given in Table 60, anal. 12. On the basis of infrared spectral studies (Rutstein and White, 1971), the limits of solid solution of Fe and Mn in wollastonite have been set at approximately

Table 60. Bustamite Analyses

	1	2	3	4	5	6
SiO_2	50·00	48·84	49·72	48·44	48·23	45·82
TiO_2	—	—	—	—	—	0·30
Al_2O_3	—	1·29	0·17	—	0·59	1·84
Fe_2O_3	0·00	0·07	0·00	—	0·00	4·31
FeO	9·29	4·65	5·55	0·27	5·01	0·43
MnO	1·22	7·10	11·08	25·20	26·48	28·32
MgO	—	0·16	n.d.	0·65	n.d.	1·94
CaO	38·86	37·43	33·39	25·20	19·80	17·53
Na_2O	—	—	—	—	—	—
K_2O	—	—	—	—	—	—
H_2O^+	—	—	—	—	—	0·34
H_2O^-	—	—	—	—	—	0·06
Total	99·82	99·54	99·91	100·63	100·11	100·89
α	1·640	—	—	1·662	—	—
β	—	1·653	1·663	1·674	1·693	—
γ	1·653	—	—	1·676	—	—
δ	0·013	—	—	0·014	—	—
$2V_\alpha$	60°	—	—	44°	—	35°
D	3·09	3·10	3·133	—	3·380	—

Numbers of ions on the basis of eighteen O

	1		2		3		4		5		6	
Si	5·985		5·850	6·03	5·985	6·01	5·940		5·973	6·06	5·640	5·91
Al	—		0·182		0·024		—		0·086		0·267	
Fe^{3+}	—		0·006		—		—		—		0·399	
Mg	—		0·029		—		0·119		—		0·356	
Ti	—		—		—		—		—		0·028	
Fe^{2+}	0·930	6·03	0·466	6·02	0·559	5·99	0·026	6·12[a]	0·519	5·92	0·044	6·09
Mn	0·123		0·720		1·130		2·618		2·778		2·953	
Ca	4·981		4·804		4·306		3·311		2·627		2·312	
Na	—		—		—		—		—		—	
K	—		—		—		—		—		—	

[a] Includes Zn 0·048.

1 White ferrobustamite ('iron-wollastonite'), around chert nodules in dolomite metamorphosed by granite, Camas Malag, Skye (Tilley, 1948). Anal. H. G. C. Vincent (includes 0·45 insol. in HCl).
2 Bustamite, Broken Hill, New South Wales, Australia (Mason, 1973). Anal. G. Steiger.
3 Bustamite, Broken Hill, New South Wales, Australia (Mason, 1973). Anal. M. L. Lindberg.
4 Pink prismatic bustamite, Franklin, New Jersey (Larsen and Shannon, 1922). Includes ZnO 0·53, ign. loss 0·34 per cent.
5 Bustamite, Broken Hill, New South Wales, Australia (Mason, 1973). Anal. M. L. Lindberg.
6 Bustamite, manganiferous koduritic assemblage, Ravinella di Sotto, near Forno, Strona valley, Novara, Italy (Bertolani, 1967).

7	8	9	10	11	12	
47·65	46·32	48·31	47·24	46·25	47·4	SiO_2
0·01	0·03	—	0·00	0·02	0·10	TiO_2
0·02	1·02	tr.	1·91	0·31	0·15	Al_2O_3
0·40	0·00	—	1·24	0·02	2·7	Fe_2O_3
9·07	8·63	1·87	6·89	0·75	15·95	FeO
27·10	26·50	33·04	30·37	38·09	12·95	MnO
0·07	0·88	1·90	0·10	1·07	5·26	MgO
15·32	14·98	14·93	12·75	12·24	15·1	CaO
0·15	—	—	0·02	0·17	—	Na_2O
0·05	—	—	0·01	0·49	—	K_2O
0·05	1·05	—	—	0·00	—	H_2O^+
0·01	0·24	—	0·08	0·13	—	H_2O^-
99·91	99·65	100·05	100·61	100·08	99·61	**Total**
1·693	1·692	1·687	1·695	—	—	α
1·708	1·705	1·701	1·708	—	—	β
1·710	1·707	1·703	1·710	—	1·701	γ
0·017	0·015	0·016	0·015	—	(0·018)	δ
37°	35°	36·2°	34°	—	65·5°	$2V_\alpha$
3·44	3·425	3·410	3·46	—	3·39	D
5·997 } 6·00	5·892 } 6·00	5·999	5·881 } 6·00	5·895 } 5·94	5·823 } 5·85	Si
0·003	0·153	—	0·280	0·046	0·022	Al
0·038	—	—	0·116	0·001	0·250	Fe^{3+}
0·013	0·166	0·352	0·019	0·203	0·961	Mg
0·001	0·003	—	—	0·002	0·009	Ti
0·955 } 6·01	0·918 } 6·03	0·195 } 6·01	0·718 } 5·93	0·080 } 6·16[b]	1·636 } 6·19	Fe^{2+}
2·888	2·856	3·475	3·204	4·113	1·346	Mn
2·066	2·042	1·986	1·701	1·671	1·985	Ca
0·036	—	—	0·004	0·041	—	Na
0·008	—	—	0·002	0·040	—	K

[b] Includes Ba 0·005.

7 Flesh-pink asbestiform bustamite, lining cavity in lead ore, Broken Hill, New South Wales, Australia (Binns, 1968). Anal. G. I. Z. Kalocsai (includes P_2O_5 0·01 per cent).

8 Coarse, fibrous, flesh-pink bustamite, Treburland manganese mine, Altarnum, Cornwall (Tilley, 1946). Anal. Geochemical Laboratories.

9 Bustamite, Långban, Sweden (Sundius, 1931). Anal. A. Bygdén.

10 Coarse, radiating, light brown bustamite, contact metamorphosed calc-silicate hornfels, Meldon Railway Quarry, Okehampton, Devonshire (Howie, 1965). Anal. R. A. Howie.

11 Bustamite, Nodatamagawa mine, Iwate Prefecture, Japan (Watanabe and Kato, 1956; quoted from Takeuchi *et al.*, 1961). Includes BaO 0·10, CO_2 0·44.

12 Amber-yellow ferrobustamite ('vogtite'), acid steel furnace slag Hallimond, 1919). Anal. J. H. Whiteley.

12 and 25 atomic per cent respectively; at above these concentrations the bustamite structure becomes the stable phase. Solid solution of bustamite with diopside is also limited.

Experimental

The phase equilibria in the system $CaSiO_3$–$MnSiO_3$ have been investigated experimentally by Voos (1935) and by Glasser (1962); the phase diagram (Fig. 258, p. 591) shows an extensive series of solutions between bustamite and $MnSiO_3$ (rhodonite) but there is some uncertainty as to the structural state of the 'bustamites'. Further experimental work on the $CaSiO_3$–$MnSiO_3$ system, at 2 kbar (Fig. 255), showed that d_{220} for bustamite increases with increasing $CaSiO_3$, until around Wo_{63} when the structural break to the wollastonite structure occurs (Fig.

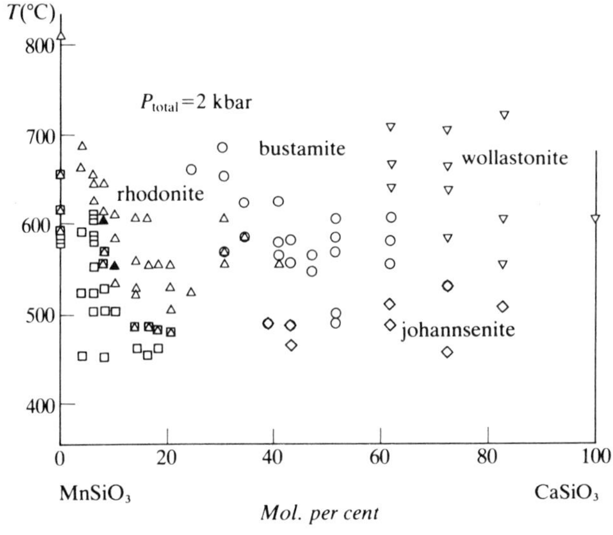

Fig. 255. Results of experimental work in the system $MnSiO_3$–$CaSiO_3$ at P_{total} = 2 kbar. Squares = pyroxmangite, triangles = rhodonite, circles = bustamite, diamonds = johannsenite, inverted triangles = wollastonite. Duration of runs 14 to 60 days; X_{CO_2} varies 0·3–0·6 (after Abrecht and Peters, 1975).

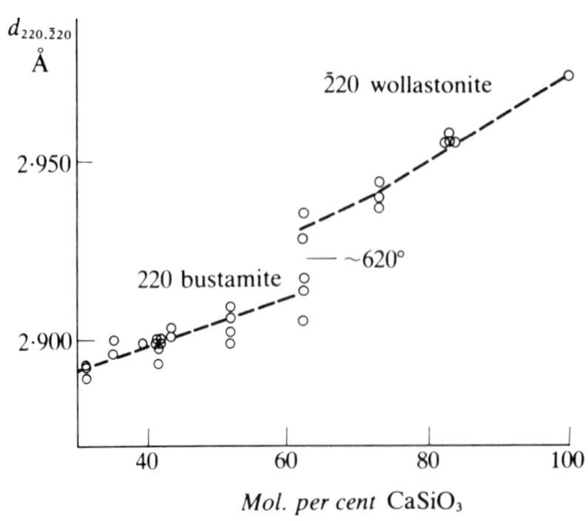

Fig. 256. Variation in the 220 or $\bar{2}20$ X-ray spacing for compositions having either the bustamite or the wollastonite structure in the series $MnSiO_3$–$CaSiO_3$ (after Abrecht and Peters, 1975).

256). Abrecht and Peters (1975) also showed that it was possible to convert non-reversibly the pyroxenoids rhodonite → bustamite, bustamite → wollastonite, and johannsenite → bustamite or wollastonite with increase of temperature. Investigation of the series $CaSiO_3$–$CaFeSi_2O_6$ (Bowen *et al.*, 1933; Rutstein, 1971) led Rutstein to suggest a relationship as shown in Fig. 248(b) (p. 554), where, over a range of composition approximately $Ca_{0\cdot90}Fe_{0\cdot10}SiO_3$ to $Ca_{0\cdot83}Fe_{0\cdot17}SiO_2$, phases with the bustamite and wollastonite structures coexist at around 850°C; with increasingly Fe-rich bulk compositions, the development of an ordered structure predominates and the bustamite variant is stable alone. Further work on Fe (and Mg) calcium metasilicate systems is needed with careful characterization of the exact structural type of the phases involved. (See also note on p. 585.)

Bowen *et al.* (1933) showed that johannsenite inverts to bustamite as its high-temperature form; the lowest temperature for the inversion of johannsenite to bustamite was recorded as 830°C (Schaller, 1938). The topotactic nature of the johannsenite to bustamite inversion has been studied in detail by Morimoto *et al.* (1966). See also Yuquan *et al.*, (1977).

Bustamite fairly rapidly undergoes alteration on exposure, a brownish oxide of manganese forming on the surface of the grains, obscuring the pink colour.

Optical and Physical Properties

When fresh, bustamite has a brownish or flesh-pink colour, paler than rhodonite, and it often has a fibrous character. Its detailed optical properties have been investigated by Hey (1929) and by Sundius (1931); the variation of refractive indices and density in the bustamite–wollastonite series has been plotted by Mason (1975) and the β index has been similarly plotted by Hodgson (1975) against both mol. per cent $(Ca,Mg)SiO_3$ and mol. per cent $FeSiO_3$. A plot of the β refractive index, the optic axial angle, and the specific gravity against molecular percentage of (MnO + FeO) in the monoxides is given in Fig. 257. The refractive indices and specific gravity increase steadily with increasing Mn and Fe, while the birefringence varies only slightly and the optic axial angle appears to decrease: Tilley (1946) gave values for 2V of from 30° to 35° for iron-bearing bustamites from Cornwall, and reported that lower values have been recorded. Ferrobustamite, however, appears to have a markedly higher 2V. Dispersion of the optic axes may be fairly strong $v > r$: for a Långban specimen, in red light (656·3 μm) $\beta = 1{\cdot}68280$ and $2V_\alpha$ calc. = 40°28′, and in green light (485·15 μm) $\beta = 1{\cdot}69648$ and $2V_\alpha$ calc. = 43°12′ (Sundius, *loc. cit*). The extinction angles in sections perpendicular to the zone of cleavages are distinctive; thus for the Cornish bustamite (Table 60, anal. 8) Tilley (1946) records α':(100) = 47°; sections perpendicular to the acute bisectrix show γ at 29° to 30° to the trace of the cleavage.

The optical absorption spectra of the manganese silicates were studied by Manning (1968) who attributed the brownish colour of bustamite to the super-imposition of the octahedrally bonded Mn(II) bands on strong background absorption; there is also a band at around 5000 cm^{-1} which is characteristic of Fe(II) in tetrahedral sites. Lazarev and Tenisheva (1961) and Ryall and Threadgold (1966) suggested that the number of sharp weak bands in the infrared spectra of chain silicates near 700 cm^{-1} is related to the number of tetrahedra in the chain repeat unit; this was confirmed by Rutstein and White (1971). Thus diopside has two sharp peaks in this region whereas wollastonite and bustamite (both

Dreierketten) have three, and rhodonite (Fünferketten) has five, though the spectra for wollastonite and bustamite are quite unlike each other. The frequencies of these groups of peaks change very little with composition (unlike the infrared spectra of most solid solutions) and instead the patterns shift abruptly at phase boundaries. The bustamite pattern is present for compositions from $Ca_{0\cdot75}Mn_{0\cdot25}SiO_3$ to $Ca_{0\cdot10}Mn_{0\cdot90}SiO_3$, and is replaced at the calcic end by the wollastonite pattern and the manganese-rich end by the rhodonite pattern. These results agree in general with those of X-ray structural work and led Rutstein and White to propose solid solution and phase boundaries in the system $CaSiO_3$–$MgSiO_3$–$MnSiO_3$–$FeSiO_3$ as shown in Fig. 262, p. 595.

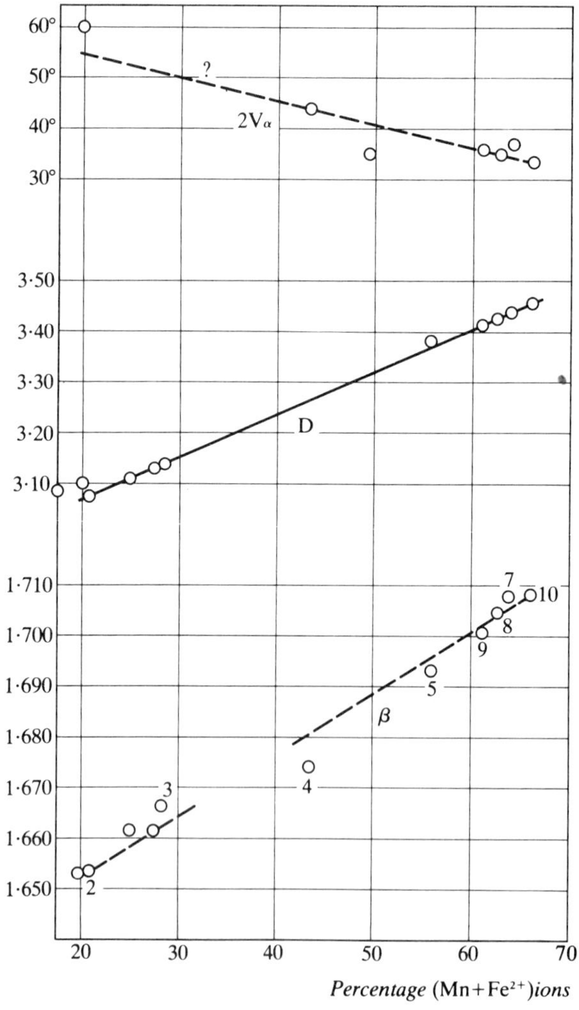

Fig. 257. Variation in optical and physical properties against molecular percentage of $(Mn+Fe^{2+})$ for bustamite.

Distinguishing Features

The pink colour, manganese reaction, triclinic optics and relatively weak birefringence help to place it in the group of triclinic manganese silicates. Bustamite differs from rhodonite and pyroxmangite in its negative optic sign, lower refractive indices and smaller specific gravity: in addition the optic axial angle is less than that of rhodonite and the birefringence less than that of pyroxmangite.

Paragenesis

Bustamite is a mineral typical of manganese orebodies, usually resulting from metamorphism with associated metasomatism, and often occurs in skarn deposits. It is a moderately high-temperature mineral, its existence at ordinary temperatures showing that the bustamite–clinopyroxene inversion may be a sluggish one. Typical bustamite assemblages (Tilley, 1946) are bustamite–rhodonite–tephroite and bustamite–tephroite–calcite (manganocalcite): in a higher grade of metamorphism, involving increasing decarbonation, the last assemblage reacts to form glaucochroite, as at Franklin where glaucochroite–bustamite–calcite and glaucochroite–tephroite–bustamite assemblages occur. Bustamite from the manganese skarns at Långban, Sweden (Table 60, anal. 9) was described by Sundius (1931), and bustamite similar in properties to that of anal. 5 has been reported associated with manganpyrosmalite and johannsenite at Broken Hill, New South Wales (Hutton, 1956). Bustamite, with wollastonite, diopside and grossular, been reported from the Ōhori mine, Japan (Takeuchi *et al.*, 1961). At both Nakatatsu and Chichibu skarn deposits, Japan, an instability of the assemblage hedenbergite–calcite is indicated by enrichment of the pyroxene in Mn towards the contact with marble, followed by zones of johannsenite and bustamite at the marble contact (Burt, 1972). Rosettes of slender needles of bustamite occur in association with rhodochrosite and rhodonite in metamorphosed banded siliceous tuffites of Culm age in the Harz Mountains (Haage, 1964), and an unusually fibrous variety with a silky lustre has been reported as part of the lining of a cavity several centimetres wide between rhodonite-bearing ore and high-grade calcitic lead ore at Broken Hill (Table 60, anal. 7). Experimental work on the $CaSiO_3$–$CaFeSi_2O_6$ join (Fig. 248(b), p. 554) indicates that ferrobustamite is stable only above about 800°C, a temperature likely to be reached only in contact metamorphic skarn zones or in slags.

References

Abrecht, J. and **Peters, Tj.**, 1975. Hydrothermal synthesis of pyroxenoids in the system $MnSiO_3$–$CaSiO_3$ at $Pf = 2$ kbar. *Contr. Min. Petr.*, **50**, 241–246.

Berman, H., 1937. Constitution and classification of the natural silicates. *Amer. Min.*, **22**, 342–408.

Berman, H. and **Gonyer, F. A.**, 1937. The structural lattice and classification of bustamite. *Amer. Min.*, **22**, 215–216.

Bertolani, M., 1967. Rocce manganesifere tra le granuliti della valle Strona (Novara). *Periodico Min.*, **36**, 1011–1032.

Binns, R. A., 1968. Asbestiform bustamite from a cavity lining within the Broken Hill lode, New South Wales. *J. Geol. Soc. Australia*, **15**, 1–8.

Bowen, N. L., **Schairer, J. F.** and **Posnjak, E.**, 1933. The system CaO–FeO–SiO_2. *Amer. J. Sci.*, ser. 5, **26**, 193–284.

Buerger, M. J., 1956. The arrangement of atoms in crystals of the wollastonite group of metasilicates. *Proc. Nat. Acad. Sci. U.S.A.*, **42**, 113–116.

Burnham, C. W., 1976. Ferrobustamite: the crystal structure of two Ca,Fe bustamite-type pyroxenoids. *Z. Krist.*, **142**, 450–462.

Burt, D. M., 1972. The facies of some Ca–Fe–Si skarns in Japan. *Rept. 24th Intern. Geol. Congr.*, Montreal, Sect. 2, 284–288.

Dowty, E. and **Lindsley, D. H.**, 1974. Mössbauer spectra of synthetic Ca–Fe pyroxenoids and lunar pyroxferroite. *Contr. Min. Petr.*, **48**, 229–232.

Glasser, F. P., 1962. The ternary system CaO–MnO–SiO_2. *J. Amer. Ceram. Soc.*, **45**, 242–249.

Haage, R., 1964. Beitrag zur Petrographie des Kieselschiefer-Mangankieselvorkommens in Schävenholz bei Elbingerode (Harz). *Wiss. Zeits. Martin-Luther Univ.*, Halle-Wittenberg, **13**, 213–225 (M.A. **17**-367).

Hallimond, A. F., 1919. The crystallography of vogtite, an anorthic, metasilicate of iron, calcium, manganese, and magnesium from acid steel-furnace slags. *Min. Mag.*, **18**, 368–372.

Hlawatsch, C., 1907. Eine trikline, rhodonitähnliche Schlacke. *Z. Krist.*, **42**, 590–593.

Harada, K., **Sekino, H.**, **Nagashima, K.**, **Watanabe, T.** and **Momoi, K.**, 1974. High-iron bustamite and fluorapatite from the Broken Hill mine, New South Wales, Australia. *Min. Mag.*, **39**, 601–604.

Hey, M. H., 1929. The variation of optical properties with chemical composition in the rhodonite–bustamite series. *Min. Mag.*, **22**, 193–205.

Hodgson, D. J., 1975. The geology and geological development of the Broken Hill lode, in the New Broken Hill Consolidated mine, Australia. Part II: Mineralogy. *J. Geol. Soc. Australia*, **22**, 33–50.

Howie, R. A., 1965. Bustamite, rhodonite, spessartine, and tephroite from Meldon, Okehampton, Devonshire. *Min. Mag.*, **34** (Tilley vol.), 249–255.

Hutton, C. O., 1956. Manganpyrosmalite, bustamite and ferroan johannsenite from Broken Hill, New South Wales, Australia. *Amer. Min.*, **41**, 581–591.

Larsen, E. S. and **Shannon, E. V.**, 1922. Bustamite from Franklin Furnace, New Jersey. *Amer. Min.*, **7**, 95–100.

Lazarev, A. N. and **Tenisheva, T. F.**, 1961. Vibrational spectra of silicates, III. Infrared spectra of pyroxenoids and other chain metasilicates. *Optics and Spectroscopy* (U.S.S.R.), **11**, 316–317.

Liebau, F., **Sprung, M.** and **Thilo, E.**, 1958. Über das System $MnSiO_3$–$CaMn(SiO_3)_2$. *Z. Anorg. Chem.*, **297**, 213–225.

Manning, P. G., 1968. Absorption spectra of the manganese-bearing chain silicates pyroxmangite, rhodonite, bustamite and serandite. *Can. Min.*, **9**, 348–357.

Mason, B., 1973. Manganese silicate minerals from Broken Hill, New South Wales. *J. Geol. Soc. Australia*, **20**, 397–404.

Mason, B., 1975. Compositional limits of wollastonite and bustamite. *Amer. Min.*, **60**, 209–212.

Matsueda, H., 1973. Iron-wollastonite from the Sampo mine showing properties distinct from those of wollastonite. *Min. J., Japan*, **7**, 180–201.

Momoi, H., 1964. Mineralogical study of rhodonites in Japan, with special reference to contact metamorphism. *Mem. Fac. Sci., Kyushu Univ.*, ser. D, Geol., **15**, 39–63.

Morimoto, N., **Koto, K.** and **Shinohara, T.**, 1966. Oriented transformation of johannsenite to bustamite. *Min. J., Japan*, **5**, 44–64.

Ohashi, Y. and **Finger, L. W.**, 1976. Stepwise cation ordering in bustamite and disordering in wollastonite. *Carnegie Inst. Washington, Ann. Rept. Dir. Geophys. Lab.*, 1975–76, 746–753.

Palache, C., 1937. The minerals of Franklin and Sterling Hill, Sussex County, New Jersey. *U.S. Geol. Surv., Prof. Paper* **180**.

Peacor, D. R. and **Buerger, M. J.**, 1962. Determination and refinment of the crystal structure of bustamite, $CaMnSi_2O_6$. *Z. Krist.*, **117**, 331–343.

Peacor, D. R. and **Prewitt, C. T.**, 1963. Comparison of the crystal structures of bustamite and wollastonite. *Amer. Min.*, **48**, 588–596.

Rapoport, P. A. and **Burnham, C. W.**, 1973. Ferrobustamite: the crystal structures of two Ca,Fe bustamite-type pyroxenoids. *Z. Krist.*, **138**, 419–438.

Rutstein, M. S., 1971. Re-examination of the wollastonite–hedenbergite ($CaSiO_3$–$CaFeSi_2O_6$) equilibria. *Amer. Min.*, **56**, 2040–2052.

Rutstein, M. S. and **White, W. B.**, 1971. Vibrational spectra of high-calcium pyroxenes and pyroxenoids. *Amer. Min.*, **56**, 877–887.

Ryall, W. R. and **Threadgold, I. M.**, 1966. Evidence for $[(SiO_3)_5]_\infty$ type chains in inesite as shown by X-ray and infrared absorption studies. *Amer. Min.*, **51**, 754–761.

Schaller, W. T., 1938. Johannsenite, a new manganese pyroxene. *Amer. Min.*, **23**, 575–582.

Sundius, N., 1931. On the triclinic manganiferous pyroxenes. *Amer. Min.*, **16**, 411–429 and 488–518.

Takeuchi, T., **Sugaki, A.**, **Suzuki, T.** and **Abe, H.**, 1961. Ōhori mine, a Tertiary pyrometasomatic deposit, Yamagata Prefecture, Japan. *Sci. Rept. Tohoku Univ.*, ser. 3, **7**, p. 153.

Tilley, C. E., 1946. Bustamite from Treburland manganese mine, Cornwall, and its paragenesis. *Min. Mag.*, **27**, 236–241.
Tilley, C. E., 1948. On iron-wollastonite in contact skarns: an example from Skye. *Amer. Min.*, **33**, 736–738.
Voos, E., 1935. Untersuchung des Schnittes $CaO.SiO_2$–$MnO.SiO_2$ imternären System SiO_2–CaO–MnO. *Z. anorg. Chem.*, **222**, 201–224.
Yuquan, S., Danian, Y. and **Jingxiong, G.,** 1977. [Experimental studies of $CaMnSi_2O_6$–$CaAlSiAlO_6$ system.] *Sci. Geol. Sinica*, 343–354 (Chinese with English abstract).

Note added in proof: A natural sample with composition $Wo_{82}Fs_{18}$ has been described from Japan and is close to the Fe-poor end of the ferrobustamite solid solution. It has *a* 7·862, *b* 7·253, *c* 13·967 Å, α 89°44′· β 95°28′, γ 103°29′, space group $A\bar{1}$, $Z = 12$ (**Yamanaka** *et al.*, 1977. Amer. Min., **62**, 1216–1224).

Rhodonite (Mn,Fe,Ca)[SiO_3]

Triclinic (+)

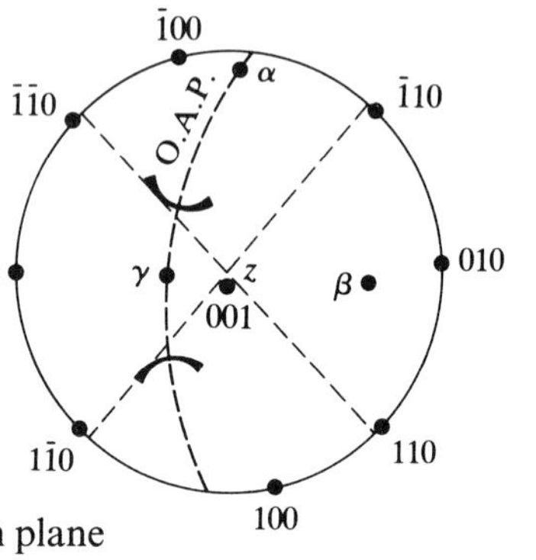

α	1·711–1·734
β	1·716–1·739
γ	1·724–1·748
δ	0·011–0·017
$2V_\gamma$	63°–87°
$\alpha:x \simeq 5°, \beta:y \simeq 20°, \gamma:z \simeq 25°$	
Dispersion:	$r < v$
D	3·57–3·76
H	$5\frac{1}{2}$–$6\frac{1}{2}$
Cleavage:	{110} and {1$\bar{1}$0} perfect, {001} good; (110):(1$\bar{1}$0) = $92\frac{1}{2}°$
Twinning:	Lamellar twinning with composition plane (010), not common.
Colour:	Rose-pink to brownish red; colourless to faint pink in thin section.
Pleochroism:	Weak; in thick sections α yellowish red, β pinkish red, γ pale yellowish red.
Unit cell:	*a* 9·758, *b* 10·499, *c* 12·205 Å α 108·58°, β 102·92°, γ 82·52° *V* 1152·9 Å^3. Z = 20. Space group $C\bar{1}$.

Slightly attacked by HCl. Gives manganese reaction with fluxes.

Rhodonite occurs in manganese-bearing orebodies of varying origin: the name is derived from the Greek, *rhodon*, a rose, in allusion to the characteristic colour of the mineral.

Structure

Rhodonite was first investigated structurally by Gossner and Brückl (1928) who determined the cell dimensions of material from Franklin, New Jersey, and who demonstrated that rhodonite was not isostructural with diopside. Berman (1937) classified the so-called 'pyroxenoids' into two series, rhodonite and several related manganese silicates falling in one series, with wollastonite and some other calcium metasilicates including bustamite in the other series. The structure of rhodonite from Franklin, with chemical composition approximately $(Mn_4Ca)(SiO_3)_5$, was determined by Liebau *et al.* (1959) and that of a rhodonite from Switzerland was studied by Mamedov (1958). The two structures proposed, although differing in detail, both contain the same basic features in that Mn and Ca are octahedrally co-ordinated, with octahedra occurring in discontinuous sheets, and chains of SiO_4 tetrahedra, of composition $(SiO_3)_n$, with a repeat unit which contains five tetrahedra

(Fünferketten). Thus the structure is seen to have similarities with that of wollastonite and bustamite (three tetrahedra periodicity or Dreierketten) and pyroxmangite (seven tetrahedra periodicity or Siebenerketten). The structure was confirmed and refined by Peacor and Niizeki (1963), using a crystal from Pajsberg, Sweden (cf. Table 61, anal. 5), who showed that planes of Mn, Ca, Mg and Fe ions in octahedral co-ordination alternate between planes of oxygen ions with planes of Si in tetrahedral co-ordination.

From chemical and structural considerations Liebau *et al.* (1959) suggested that not more than one in five of the cations which link the (SiO_3) chain can be calcium. A linear relation was found by Momoi (1964) between the Ca content and the spacings of three doublets in the X-ray powder diffraction pattern of rhodonites; in particular, 2θ 200–2θ 020 increases linearly from 2·09° to 2·25° for 3·5 to 14·5 mol. per cent Ca, i.e. the *a* dimension increases with Ca (Momoi, 1969), which is in line with the general trend in these chain silicates.

The planes of approximately close-packed oxygens are similar to those which occur in other pyroxenoids and in pyroxenes. The triclinic nature of pyroxenoids allows a varied choice of unit cells so that this relationship and the designation of the chain axis (*z*) for each mineral can be emphasized. Earlier X-ray work tended to choose the reduced cell ($P\bar{1}$), but suitable transformations can relate this to the $C\bar{1}$ cell which best shows the close relationship, and is also in accord with the morphological and optical descriptions. Some choices of unit cell are:

	*a*Å	*b*Å	*c*Å	Z	α	β	γ	Reference
$P\bar{1}$	6·71	7·08	12·23	10	111·54	85·25	93·95	Liebau, 1959; Prewitt and Peacor, 1964
$P\bar{1}$	7·66	12·27	6·68	10	86·0	93·2	111·1	Liebau *et al.*, 1959
$P\bar{1}$	7·68	11·82	6·71	10	92·35	93·95	105·67	Peacor and Niizeki, 1963
$C\bar{1}$	9·758	10·499	12·205	20	108·58	102·92	82·52	Ohashi and Finger, 1975

A structural refinement of a rhodonite of composition $Mn_{0·81}Fe_{0·07}Mg_{0·06}Ca_{0·05}SiO_3$ from Taguchi mine, Japan, has been reported by Ohashi and Finger (1975) and compared with that of pyroxmangite from the same mine and with similar composition, see Fig. 264, p. 602. The *M*1, *M*2 and *M*3 sites in rhodonite are essentially occupied by Mn; in comparison with the more Ca-rich rhodonite studied by Peacor and Niizeki (1963), these three sites are similar in size but the *M*–O distance in *M*5, the site partially occupied by Ca, is greater in Ca-rich rhodonite than in Ca-poor rhodonite. The *M*4 site is very distorted and possibly can accommodate Ca as well as Fe and Mg; Mössbauer study indicates that in an iron-rich rhodonite from Broken Hill Fe atoms prefer the *M*4 to the other sites (Dickson, 1975).

Unit cell dimensions for synthetic rhodonites in the range $MnSiO_3$–$(Mn_{0·60}Mg_{0·40})SiO_3$ are given by Ito (1972). The polymorphism of $MnSiO_3$ is discussed by Liebau *et al.* (1958): see also Akimoto and Syono (1972), Maresch and Mottana (1976) and Narita *et al.* (1977).

The absorption spectra of an iron-rich calcium-rich rhodonite and of an iron-poor rhodonite have been studied (Marshall and Runciman, 1975); the bands between 330 and 470 cm^{-1} are due to vibrations involving the *M*1 to *M*5 ions whereas those between 490 and 1100 cm^{-1} arise from silicate groups. The comparison of low-iron and high-iron spectra has permitted most non-vibrational bands, including the entire visible range, to be assigned to Fe^{2+} or Mn^{2+}.

Chemistry

Rhodonite is never pure $MnSiO_3$ but always contains a certain amount of CaO: anal. 1 (Table 61) with only 2·25 per cent CaO is one of the most manganese-rich examples known. There is not a continuous isomorphous series between rhodonite and bustamite, and from structural and optical considerations it seems likely that rhodonite cannot contain more than about 20 per cent $CaSiO_3$ in solid solution: thus Liebau *et al.* (1959) gave its formula $(Mn,Mg)_{1-x}Ca_xSiO_3$, with $0 < x < 0{\cdot}2$. The chemistry of rhodonites, particularly with respect to their calcium contents, was investigated in detail by Momoi (1964) with sixteen new analyses and a review of published data. In comparison to pyroxmangite, he demonstrated that rhodonite had a more restricted manganese content (rhodonite $MnSiO_3$ 52–95 mol. per cent, pyroxmangite $MnSiO_3$ 36–96 mol. per cent), but that in each mineral Mn decreases with increasing Ca. In both minerals magnesium also increases with increasing calcium but rhodonite tends to have a higher Ca/Mg ratio. In a plot of over fifty analysed rhodonites their mean composition appears to be near $Mn_{80}Ca_{10}(Fe,Mg)_{10}$. Electron microprobe analyses of seven Japanese rhodonites, checked by X-ray diffraction pattern, have been given by Ohashi *et al.* (1975), and microprobe analyses of several Broken Hill rhodonites were reported by Hodgson (1975) who also considered the compositions of coexisting rhodonites and bustamites, the latter being somewhat poorer in Fe as well as richer in Ca. Microprobe analyses of thirteen samples of rhodonite from Plainfield, Massachusetts are tabulated by Dunn (1976).

The ferric iron content is usually low, but ferrous iron may replace manganese to a considerable extent: iron-rhodonite with 14·51 per cent FeO has been reported by Sundius (1930), from Tuna Hästberg, Sweden (Table 61, anal. 10) and this would appear to be about the limit of substitution of Fe^{2+} for Mn^{2+} (at near 25 mol. per cent $FeSiO_3$). The slag 'iron-rhodonites' of Whiteley and Hallimond (1919) are pyroxferroite. Most rhodonites contain small amounts of magnesium, generally < 3 per cent MgO; however, a distinct magnesian variety with 14 per cent MgO has been recorded by Klein (1966) from a magnesioriebeckite–manganoan cummingtonite–rhodonite–hematite assemblage (anal. 11). Zinc may enter the rhodonite structure if it is available during crystallization, and appears to replace ferrous iron: the zinc-rich variety has been named fowlerite (after Dr. Samuel Fowler) and occurs at Franklin, New Jersey (Table 61, anal. 12); a microprobe analysis (ZnO 5·72 per cent) is given by Gibbons *et al.* (1974).

Experimental

The binary system MnO–SiO_2 has been investigated by Glaser (1926), Herty (1930), White *et al.* (1934), and by Glasser (1958). Glasser synthesized material in this system from MnO_2 and SiO_2 in an atmosphere of low partial pressure of oxygen so that manganese remained in the divalent state. He found that rhodonite melts incongruently at 1291°C to tridymite plus liquid, and that the rhodonite-tephroite eutectic is at 1251°C and 38·3 weight per cent SiO_2. Rhodonite quenched from temperatures just beneath 1291°C appeared to have properties identical to rhodonite heat-treated at low temperatures, and to natural rhodonite. The phase equilibria in the system manganese oxide–silica in air were investigated by Muan (1959a, b) who found that the $MnO.SiO_2$ (solid solution) phase persisted down to a temperature of 1048°C before oxidation to Mn_2O_3 (solid solution) and tridymite took place. The free energy of formation of rhodonite and activity/composition

Table 61. Rhodonite Analyses

	1	2	3	4	5	6
SiO_2	45·46	46·42	46·72	46·58	46·33	46·41
TiO_2	—	—	—	—	—	0·00
Al_2O_3	0·27	0·07	0·49	0·25	0·26	0·24
Fe_2O_3	0·00	0·11	—	tr.	0·83	0·66
FeO	0·96	1·49	0·29	1·66	—	3·03
MnO	50·54	47·62	46·33	44·89	44·28	42·66
MgO	0·55	0·92	2·27	1·52	0·04	0·08
ZnO	—	—	—	0·21	0·07	tr.
CaO	2·25	3·26	3·30	4·46	8·02	7·25
Na_2O	—	—	0·02	—	—	0·01
K_2O	—	—	—	—	—	0·01
H_2O^+	0·00	—	0·51	0·58 (H₂O⁺ + H₂O⁻)	—	—
H_2O^-	0·00	0·18	—		—	0·02
Total	100·03	100·07	99·93	100·15	100·05	100·37
α	1·725	1·723	1·722	1·726	1·720	—
β	1·728	1·729	1·730	1·731	1·725	1·734
γ	1·736	1·737	1·739	1·739	1·733	1·742
δ	0·011	0·014	0·017	0·013	0·013	—
$2V_\gamma$	74°	70°–73°	87°	64°	75°	72°
D	3·66	3·60	3·60	3·57	3·615	3·62

Numbers of ions on the basis of eighteen O

	1		2		3		4		5		6	
Si	5·920	5·96	5·988	6·00	5·976	6·00	5·979	6·00	5·946	5·99	5·942	5·98
Al	0·042		0·011		0·024		0·021		0·040		0·037	
Al	—	6·10	—	6·00	0·050	5·99	0·017	6·00	—	6·01	—	6·03[a]
Fe^{3+}	—		0·011		—		—		0·080		0·063	
Mg	0·106		0·177		0·433		0·291		0·008		0·015	
Fe^{2+}	0·104		0·160		0·031		0·178		—		0·325	
Zn	—		—		—		0·020		0·006		—	
Mn	5·576		5·204		5·020		4·882		4·814		4·627	
Na	—		—		0·004		—		—		0·002	
Ca	0·314		0·451		0·452		0·613		1·104		0·994	

[a] Includes K 0·001.

1 Rose-pink rhodonite, pegmatite cutting manganese ore bands, Chikla, Bhandara district, India (Bilgrami, 1956). Anal. R. K. Phillips.
2 Rhodonite, rhodonite-spessartine-rhodochrosite schist, Arrow Valley, Crown Range, western Otago, New Zealand (Hutton, 1957). Anal. C. O. Hutton.
3 Rhodonite, regionally metamorphosed Precambrian Wabush Iron Formation, south-western Labrador (Klein, 1966).
4 Rhodonite, X mine, Honshu, Japan (Lee, 1955). Anal. W. H. Herdsman.
5 Bright red rhodonite, Hartsig mine, Pajsberg, Sweden (Hey, 1929). Anal. M. H. Hey (includes loss on ign. 0·22).
6 Pink massive rhodonite, B.R. Quarry, Meldon, Okehampton, Devonshire (Howie, 1965). Anal. R. A. Howie.

Table 61. Rhodonite Analyses – *continued*

	7	8	9	10	11	12
SiO_2	46·20	46·84	46·45	47·78	51·2	46·87
TiO_2	—	—	—	—	—	—
Al_2O_3	0·00	—	—	0·08	—	—
Fe_2O_3	0·00	—	0·12	0·11	—	—
FeO	9·03	7·33	12·65	14·51	4·6	tr.
MnO	39·72	38·92	35·10	29·20	25·7	38·22
MgO	0·12	2·83	0·40	1·93	14·0	0·79
ZnO	0·30	—	—	—	—	6·38
CaO	4·66	3·44	5·45	6·55	1·6	7·61
Na_2O	—	—	—	—	—	—
K_2O	—	—	—	—	—	—
H_2O+	—	0·26	0·27	0·09	—	—
H_2O-	—	—	—	—	—	—
Total	100·03	99·62	100·44	100·25	97·1	99·87
α	—	1·729	1·726	1·725	—	—
β	1·731	1·734	1·730	1·728	—	—
γ	—	1·741	1·739	1·737	1·730	—
δ	—	0·012	0·013	0·012	—	—
$2V_\gamma$	—	68°–73°	74°	70°	—	—
D	3·71	3·64	3·68	3·653	—	—
	Numbers of ions on the basis of eighteen O					
Si	5·983	5·994	5·982	6·034	6·120	6·019
Al	—	—	—	—	—	—
Al	—	—	—	0·012	—	—
Fe^{3+}	—	—	0·010	0·009	—	—
Mg	0·023	0·540	0·076	0·363	2·494	0·151
Fe^{2+}	0·978	0·784	1·362	1·533	0·460	—
Zn	0·029	—	—	—	—	0·605
Mn	4·357	4·218	3·830	3·124	2·602	4·159
Na	—	—	—	—	—	—
Ca	0·647	0·471	0·752	0·886	0·205	1·048
	6·03	6·01	6·03	5·93	5·76	5·96

7 Rhodonite, Broken Hill, New South Wales, Australia (Mason, 1973). Anal. J. G. Fairchild.

8 Brownish red rhodonite, quartzite, Simsiö, Lapua, south-west Finland (Hietanen, 1938, no. 27c). Anal. A. Hietanen.

9 Iron-rich rhodonite, Broken Hill, New South Wales, Australia (Henderson and Glass, 1936). Anal. E. P. Henderson.

10 Pink iron rhodonite, knebelite skarn ore, Tuna Hästberg mine, Sweden (Sundius, 1930). Anal. A. Bygdén.

11 Magnesian rhodonite, regionally metamorphosed Precambrian Wabush Iron Formation, Labrador (Klein, 1966). Electron-probe analysis.

12 Fowlerite (zincian rhodonite), skarn associated with pegmatite, Franklin, New Jersey (Palache, 1935). Anal. D. Jenkins and F. H. Bauer.

relations in the system $MnSiO_3$–$FeSiO_3$ were derived by Muan (1967). An experimental study of the system $MnSiO_3$–$CaSiO_3$ was made by Voos (1935) who claimed that complete miscibility exists between wollastonite, bustamite and rhodonite, though this was at variance with the break observable in the optical properties of the natural minerals, e.g. Sundius (1931). Liebau *et al.* (1958), however, resolved this apparent contradiction by a consideration of the polymorphism of $MnSiO_3$: they considered that the low-temperature form, rhodonite, is limited to 20 per cent $CaSiO_3$ in solid solution, but that the higher temperature polymorphs with the bustamite or pseudowollastonite type structures may show complete solid solution between $MnSiO_3$ and $CaSiO_3$ (though, as noted above, Glasser was not able to confirm the existence of distinct temperature-dependent structural polymorphs, while Muan deliberately left open the question as to the exact crystal structure of the phases obtained and refers to them as $MnO.SiO_2$ solid solution). The work of Voos was repeated by Glasser (1962) who produced a revised diagram (Fig. 258) showing that rhodonite solid solutions extend over only a small

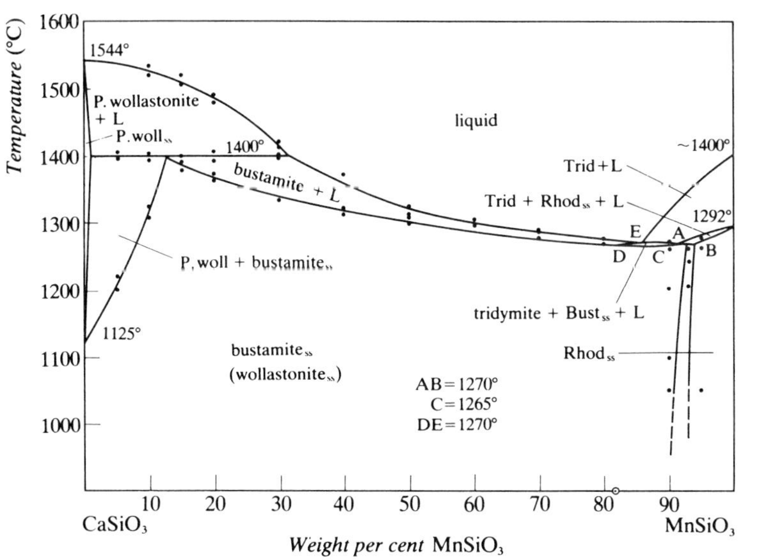

Fig. 258. Phase equilibrium diagram for the system $CaSiO_3$–$MnSiO_3$ (after Glasser, 1962).

compositional range, e.g. from 0 to 6 per cent $CaSiO_3$ at 1270°C; at lower temperatures this range is likely to be only slightly greater. Glasser and Osborn (1960) reported that, in the system MgO–MnO–SiO_2, rhodonite solid solutions extend from $MnSiO_3$ to 94·5 per cent $MgSiO_3$ at solidus temperatures, with the extent of solid solution decreasing to 78 per cent $MgSiO_3$ at 1300°C: a very narrow two-phase region (high-temperature enstatite solid solution plus rhodonite solid solution) was also found. More recently a rhodonite–pyroxmangite peritectic was found on the liquidus at a composition near $(Mn_{0\cdot67}Mg_{0\cdot33})SiO_3$ at 1350° ± 5°C in air (Ito, 1972). Single crystals of rhodonite were grown to 5 × 4 × 2 mm for the composition $(Mn_{0\cdot80}Mg_{0\cdot20})SiO_3$ by fairly rapid cooling (in air at 10°C/hour) of melt. The crystals are dark red but turn bright pink when heated at 1250°C for several hours, attributed to reduction of the trace amount of higher valency manganese. Likewise the loss of the pink colour when zincian rhodonite is shock-loaded to 496 kbar is attributed to the reduction of Mn^{3+} to Mn^{2+}, possibly by water shown to be present in the sample (Gibbons *et al.*, 1974), though this has been challenged by Faye (1975) who considered that the important spectral changes could be explained more readily by pressure-induced intensification or broadening of the $O^{2-} \rightarrow (Mn^{2+},Fe^{2+})$ charge-transfer absorption centred in the ultraviolet (see also Gibbons *et al.*, 1975, who find this unacceptable). The phase transformation at fixed *P* and *T* for $(Mn,Mg)SiO_3$ appears to be controlled by the mean cationic radii for the variable Mn and Mg contents; the presence of wide-ranging solid-solution series for rhodonites and pyroxmangites in the join $MnSiO_3$–$MgSiO_3$ (Fig. 266, p. 608) probably indicates that Mg^{2+} ions must be distributed over several sites in the structure of these phases; they cannot be restricted to a single site.

The system MnO–Al_2O_3–SiO_2 was investigated by Snow (1943), and the *P–T* equilibrium relations for the reactions $MnCO_3 + SiO_2 \rightleftharpoons MnSiO_3 + CO_2$ and $MnCO_3 + MnSiO_3 \rightleftharpoons Mn_2SiO_4 + CO_2$ have been calculated by Yoshinga (1958). Hori (1962) considered that the formation of rhodonite by the reaction of rhodochrosite + quartz in burial or load metamorphism at a rock pressure of 1500 bars (= CO_2 pressure) would occur at 460°C.

Hydrothermal syntheses of rhodonites and their conversion to bustamite or production from pyroxmangite were reported by Abrecht and Peters (1975), in the system $MnSiO_3$–$CaSiO_3$ at 2 kbar total pressure: the type of structure obtained was dependent mainly on the Mn/Ca ratio but also to a lesser extent on temperature. They found that from pure $MnSiO_3$ to 17 mol. per cent $CaSiO_3$ there is a gradual increase in d_{220} but that this spacing is constant from 17 to 42 per cent $CaSiO_3$ (Fig. 259). The rhodonites synthesized in the latter range were generally accompanied by

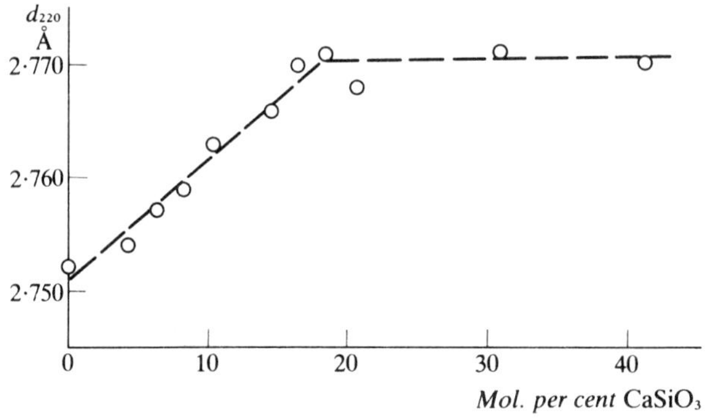

Fig. 259. Variation in the d_{220} spacing for compositions in the series $MnSiO_3$–$CaSiO_3$ giving synthetic rhodonite (after Abrecht and Peters, 1975).

a carbonate more Ca-rich than the starting (Ca,Mn) carbonate; this is taken to indicate that the rhodonite structure cannot accommodate more than 17–18 mol. per cent $CaSiO_3$ in the range 500°–550°C. The Gibbs free energy of formation of rhodonite is −297·390 kcal/mole (Robie and Waldbaum, 1968).

In further work on the pyroxmangite – rhodonite polymorphic transformation (Maresch and Mottana, 1976) the transformation of pure $MnSiO_3$ has been reversibly bracketed in the presence of water at 3 kbar (between 425° and 450°C), 6 kbar (between 475° and 525°C), 20 kbar (between 500° and 900°C), 25 kbar (between 800° and 900°C), and 30 kbar (between 900° and 1000°C). Rhodonite of $MnSiO_3$ composition is shown to be the low-pressure high-temperature polymorph with respect to pyroxmangite of the same composition (Fig. 260).

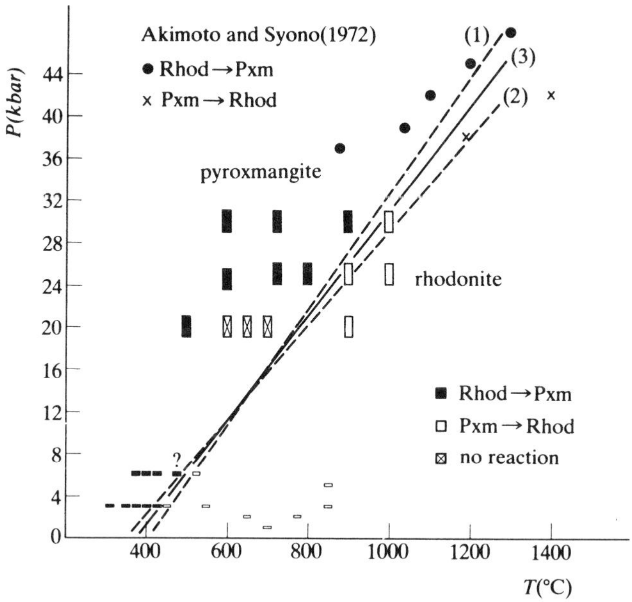

Fig. 260. Experimental results for the system $MnSiO_3$. Curve (3), the mean of several runs, is given by the equation $T°(C) = 378 + 20{\cdot}2\, P$ (kbars). (After Maresch and Mottana, 1976).

Nambulite, $LiNaMn_8Si_{10}O_{28}(OH)_2$, has been considered to be an alkaline analogue of rhodonite (Yoshi *et al.*, 1972) whereas the phase $LiMn_4Si_5O_{14}OH$ has been termed Li-hydrorhodonite (Ito, 1972); both phases are triclinic. Li-hydrorhodonite (similar to material earlier termed hydrorhodonite) dehydrates reversibly at 850°C and 1·5 kbar water pressure to rhodonite + quartz.

Alteration products of rhodonite include pyrolusite and rhodochrosite, intermediate stages probably being represented by such ill-defined material as marceline, dyssnite, allagite, photicite, etc. Hydration may also occur to penwithite, neotocite, or to a manganese-bearing serpentine as at Franklin (Palache, 1935), where the altered material in the mass is locally called 'hydro-rhodonite', a name which has also been used for hydrated rhodonite from Långban, Sweden.

Optical and Physical Properties

The entry of Ca (and Mg) into the rhodonite structure decreases the refractive indices and specific gravity and tends to increase the optic axial angle (Fig. 261). The effect of the replacement of MnO by FeO appears to be slight. Detailed optical properties are listed by Hey (1929) and Sundius (1931), while Voos (1935) gave values for synthetic material which differ slightly from those for natural crystals: see also Tilley (1937). Additional data by Dilaktorsky (1934) for artificial rhodonite (containing 3–8 per cent tephroite and SiO_2) include a melting point of 1310°C, D 3·766, β 1·739, $2V_\gamma$ 56°.

Because of the triclinic symmetry of rhodonite, the choice of unit cell is somewhat arbitrary: two orientations have been in use in recent years. In the one accepted by Dana (6th Edit.) and by Sundius (1931) the planes $(\bar{1}\bar{1}0)$, $(00\bar{1})$ and $(1\bar{1}0)$ correspond with (100), (010) and (001) respectively in the orientation used by Gossner and

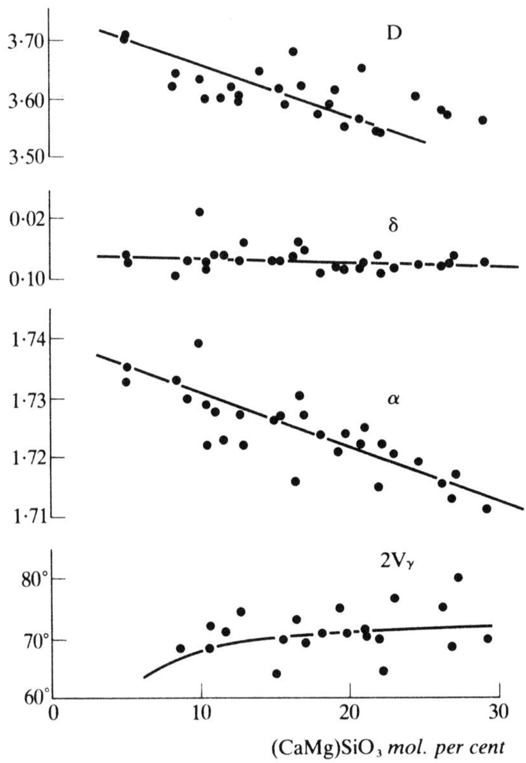

Fig. 261. Relation of optical properties and specific gravity to chemical composition for rhodonite (after Momoi, 1964).

Brückl (1928) and by Perutz (1937). The older orientation was chosen with the zone of the chief cleavages as the *z*-axis, and the cleavages indexed as (110) and (1$\bar{1}$0) in a similar fashion to the pyroxenes, while the newer orientation was based on early single-crystal X-ray studies. The orientation recently used by structural crystallographers to express the unit cell parameters again differs as it takes the chain direction of the Fünferketten repeat as the *z*- axis. Still more recently, the *P* unit cell has been transformed to a *C*-cell which conforms with the early morphological description (see p. 587; also Koto *et al.*, 1976).

Optical absorption spectra of rhodonite have been studied by Ryall and Threadgold (1966), Manning (1968), Keester and White (1968), Rutstein and White (1971), Lakshman and Reddy (1973), and by Hunt *et al.* (1973). The infrared spectrum between 550 and 750 cm^{-1} shows five peaks (Ryall and Threadgold, 1966; Rutstein and White, 1971), confirming the claim by Lazarev and Tenisheva (1961) that in general for these chain silicates the number of peaks in this region provide information as to the number of tetrahedra in the repeat unit of the structure. The study of infrared spectra has enabled Rutstein and White (1971) to establish the composition limits of the various structural types in the system $CaSiO_3$–$MgSiO_3$–$MnSiO_3$–$FeSiO_3$ with some degree of accuracy (Fig. 262).

Rhodonite is normally pale pink to red in colour, though yellow and grey varieties have been reported. Manning (1968) attributed the pink colour of rhodonite primarily to octahedrally bonded Mn^{2+}, though as noted above, it was found that the colour could be reduced in intensity by heating (Ito, 1972) or by shock (Gibbons *et al.*, 1974), which was attributed to the reduction of trace amounts of higher

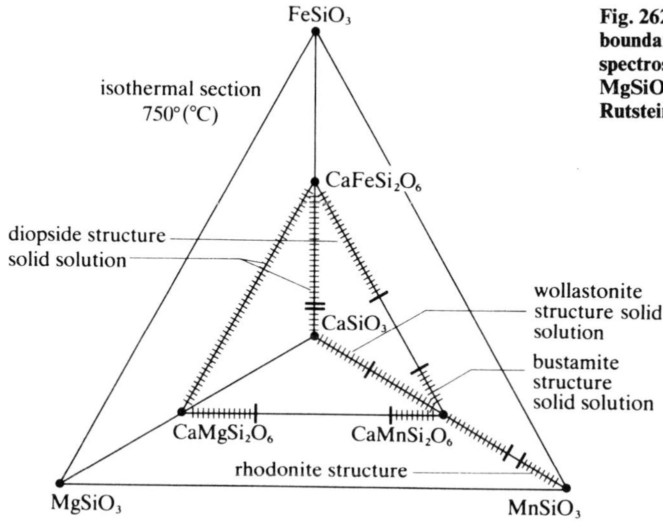

Fig. 262. Solid solution and phase boundaries found by infrared spectroscopy in the system $CaSiO_3$–$MgSiO_3$–$MnSiO_3$–$FeSiO_3$ (after Rutstein and White, 1971).

valency manganese and the consequent disappearance of the 540 nm Mn^{3+} absorption band. The descriptions of material as being brown or black are due to the very frequent association of oxidation products on the surface of the crystals. So-called 'green rhodonite' from New South Wales has been shown to be manganhedenbergite (Smith, 1926). The fine-grained rose-pink massive variety may be cut and polished as an ornamental stone: a 40-tonne block has been mined at Ekaterinburg in the Urals. The Plainfield, Massachusetts, rhodonite varies from a deep reddish pink to a very light pink; rare zones of orange material, which in some massive rhodonites from elsewhere are associated with spessartine, are shown to be orange-coloured rhodonite (Dunn, 1976). Transparent reddish rhodonite from Broken Hill, New South Wales, has been noted by Bank *et al.* (1974) who give details of its distinction from pyroxmangite (higher 2V and band in absorption spectrum at 18650–19100 cm^{-1} compared with pyroxmangite at 18200 cm^{-1}). As noted by Sundius (1931) rhodonite forms cleavage fragments readily, but as all the chief optical directions are oblique to them, determination of the refractive indices by the immersion method is difficult unless a stage-goniometer is used: the acute bisectrix forms the smallest angle with [001], the axis of the best cleavages, thus the γ value is the most easily obtained.

Distinguishing Features

Its pink colour (sometimes masked by dark brown or black alteration products), distinct cleavages, inclined extinction and manganese reactions help to distinguish rhodonite as one of the triclinic manganese silicates. It differs from bustamite in its positive optic sign, and from pyroxmangite in having a larger 2V and smaller birefringence. The low birefringence and also the inclined extinction in the [010] zone distinguish it from the monoclinic pyroxenes: kyanite has more distinct cleavages.

Paragenesis

Rhodonite occurs in many manganese orebodies and is usually associated with metasomatic activity. Fermor (1909) considered it to be a normal igneous mineral, but when it is found in pegmatites etc., it is more likely to be the result of assimilation of manganese-rich country rock (e.g. Bilgrami, 1956). It commonly results also from the contact metamorphism of rhodochrosite, where the metasomatic introduction of SiO_2 has occurred, and its production purely by load metamorphism, by the reaction of rhodochrosite + quartz in buried geosynclinal sediments has been postulated (Hori, 1962). In some cases it is possible that the manganese is introduced by circulating solutions, as in the South African manganese deposits at Postmasburg where argillaceous material has been replaced giving rhodonite in association with psilomelane and pyrolusite (Hall, 1926). It is also found in limestone in ore pockets associated with sub-marine extrusions of spilite and pillow lavas (Park, 1946). The occurrence of rhodonite in the manganese deposits of Långban, Sweden, has been described by Magnusson (1924) and Palache (1929), and rhodonite from contact metamorphosed calc-silicate rocks of Devon and Cornwall has been reported by Russell (1946) and by Howie (1965), where it is sometimes associated with bustamite and tephroite or with bementite (Tilley, 1946). The Japanese occurrences have been discussed by Yosimura (1939) and by Lee (1955), see also Momoi (1964), and Hietanen (1938) has described rhodonites from lenticles in quartzite (Table 61, anal. 8) in Finland.

Rhodonite occurs in the manganese silicate assemblage at Broken Hill, New South Wales, where it is found (anal. 9) associated with pyroxmangite, bustamite, hedenbergite, spessartine, galena, and sphalerite (see also Stilwell, 1959; Binns, 1968; Mason, 1973); a similar assemblage of Mn silicates has been reported from near Forno, Italy, where it has been attributed to the invasion of limestones recrystallized by deep tectonic metamorphism and then invaded by pegmatites (Bertolani, 1967). Although experimental work on pure $MnSiO_3$ has shown that, relative to pyroxmangite, rhodonite is the low-pressure, high-temperature polymorph, the application of this result to natural occurrences is premature until the influence of CaO, FeO and MgO on the relative stabilities of Mn minerals is better known (Maresch and Mottana, 1976). The well-known locality of Franklin, New Jersey, which has yielded rhodonite and fowlerite (anal. 12) has a similar complex geological history (Palache, 1935). Segnit (1962) has recorded rhodonite as a primary phase with tephroite or knebelite in silicified Palaeozoic strata in the general vicinity of granitic batholiths in the Tamworth area of New South Wales. Nodules of rhodonite in gondite have been reported by Sinharay (1966) to have a core of rhodonite fringed by braunite, surrounded by rhodonite and black quartz. Geochemically, manganese frequently accompanies iron and thus in regionally metamorphosed iron-rich siliceous sediments rhodonite may accompany a suite of iron-rich silicates, e.g. in the metamorphosed Wabush Iron Formation of the Labrador Trough (Table 61, anal. 3), from which a magnesian rhodonite (anal. 11) has also been reported (Klein, 1966). The occurrence of α-$MnSiO_3$, the X-ray powder pattern of which shows a similarity to that of pseudowollastonite, has been reported from silicified manganese ores in Liassic crinoidal limestone in the Tatra Mountains (Korczyńska-Oszacka, 1975).

References

Abrecht, J. and **Peters, Tj.**, 1975. Hydrothermal synthesis of pyroxenoids in the system $MnSiO_3$–$CaSiO_3$ at *Pf* = 2 kb. *Contr. Min. Petr.*, **50**, 241–246.

Akimoto, S. and **Syono, Y.**, 1972. High pressure transformation in $MnSiO_3$. *Amer. Min.*, **57**, 76–84.

Bank, H., Berdesinski, W., Ottemann, J. and **Schmetzer, K.**, 1974. Durchsichtiger rötlicher eisenreicher Rhodonit aus Australien. *Z. Deutsch. Gem. Gesell.*, **23**, 180–188.

Berman, H., 1937. Constitution and classification of the natural silicates. *Amer. Min.*, **22**, 342–408.

Bertolani, M., 1967. Rocce manganesifere tra le granuliti della valle Strona (Novara). *Periodico Min.*, **36**, 1011–1032.

Bilgrami, S. A., 1956. Manganese silicate minerals from Chikla, Bhandara district, India. *Min. Mag.*, **31**, 236–244.

Binns, R. A., 1968. Asbestiform bustamite from a cavity lining within the Broken Hill lode, New South Wales. *J. Geol. Soc. Australia*, **15**, 1–8.

Dickson, B. L., 1975. The iron distribution in rhodonite. *Amer. Min.*, **60**, 98–104.

Dilaktorsky, N. L., 1934. [On the artificial metasilicates of manganese.] *Trav. Inst. Pétrogr. Acad. Sci. U.R.S.S.*, **6**, 369–379 (in Russian) (M.A. **7**–142).

Dunn, P. J., 1976. On gem rhodonite from Massachusetts, U.S.A. *J. Gemmology*, **15**, 76–80.

Faye, G. H., 1975. Spectra of shock-affected rhodonite: a discussion. *Amer. Min.*, **60**, 939–941.

Fermor, L. L., 1909. The manganese-ore deposits of India. *Mem. Geol. Surv. India*, **37**, pts. 1–3.

Gibbons, R. V., Ahrens, T. J. and **Rossman, G. R.**, 1974. A spectrographic interpretation of the shock-produced color change in rhodonite ($MnSiO_3$): the shock-induced reduction of Mn(III) to Mn(II). *Amer. Min.*, **59**, 177–182.

Gibbons, R. V., Ahrens, T. J. and **Rossman, G. R.**, 1975. Spectra of shock-affected rhodonite: a reply. *Amer. Min.*, **60**, 942–943.

Glaser, O., 1926. Thermische und Mikroskopische Untersuchungen an den für die Kupolofenschlacke Bedeutsamen System: MnO–Al_2O_3–SiO_2, MnS–$MnSiO_3$, CaS–$CaSiO_3$. *Centr. Min. Abt. A*, 81–96.

Glasser, F. P., 1958. The system MnO–SiO_2. *Amer. J. Sci.*, **256**, 398–412.

Glasser, F. P., 1962. The ternary system CaO–MnO–SiO_2. *J. Amer. Ceram. Soc.*, **45**, 242–249.

Glasser, F. P. and **Osborn, F. F.**, 1960. The ternary system MgO–MnO–SiO_2. *J. Amer. Ceram. Soc.*, **43**, 132–140.

Gossner, B. and **Brückl, K.**, 1928. Über strukturelle Beziehungen von Rhodonit zu anderen Silikaten. *Centr. Min.*, Abt. A, 316–322.

Hall, A. L., 1926. The manganese deposits near Postmasburg, west of Kimberley. *Trans. Geol. Soc. South Africa*, **29**, 17–46.

Hallimond, A. F., 1919. The crystallography of vogtite, an anorthic, metasilicate of iron, calcium, manganese and magnesium from acid steel-furnace slags. *Min. Mag.*, **18**, 368–372.

Henderson, E. P. and **Glass, J. J.**, 1936. Pyroxmangite, new locality: identity of sobralite and pyroxmangite. *Amer. Min.*, **21**, 273–294.

Herty, C., 1930. Fundamental and applied research on the physical chemistry of steel-making. *Metals and Alloys*, **1**, 883–889.

Hey, M. H., 1929. The variation of optical properties with chemical composition in the rhodonite-bustamite series. *Min. Mag.*, **22**, 193–205.

Hietanen, A., 1938. On the petrology of Finnish quartzites. *Bull. Comm. géol. Finlande*, **21**, no. 122, 1–119.

Hodgson, C. J., 1975. The geology and geological development of the Broken Hill lode, in the New Broken Hill Consolidated mine, Australia. Part II: Mineralogy. *J. Geol. Soc. Australia*, **22**, 33–50.

Hori, F., 1962. On the load metamorphic formation of rhodonite, tephroite and manganosite. *Sci. Papers Coll. Gen. Educ. Univ. Tokyo*, **12**, 117–142.

Howie, R. A., 1965. Bustamite, rhodonite, spessartine, and tephroite from Meldon, Okehampton, Devonshire. *Min. Mag.*, **34** (Tilley vol.), 249–255.

Hunt, G. R., Salisbury, J. W. and **Lenhoff, C. J.**, 1973. Visible and near infrared spectra of minerals and rocks: VI. Additional silicates. *Modern Geol.*, **4**, 85–106.

Hutton, C. O., 1957. Contributions to the mineralogy of New Zealand, Part IV. *Trans. Roy. Soc. New Zealand*, **84**, 791–803.

Ito, J., 1972. Rhodonite–pyroxmangite peritectic along the join $MnSiO_3$–$MgSiO_3$ in air. *Amer. Min.*, **57**, 865–876.

Keester, K. L. and **White, W. B.**, 1968. Crystal-field spectra and chemical bonding in manganese minerals. Min. Soc. (London), *I.M.A. vol. (I.M.A., Papers and Proc. 5th Gen. Meeting, Cambridge 1966)*, 22–35.

Klein, C., *Jr.*, 1966. Mineralogy and petrology of the metamorphosed Wabush Iron Formation, southwestern Labrador. *J. Petr.*, **7**, 246–305.

Korczyńska-Oszacka, B., 1975. The occurrence of α-$MnSiO_3$ in the manganese-bearing rocks of the Tatra Mountains. *Min. Polonica*, **6**, 75–81 (publ. 1976).

Koto, K., Morimoto, N. and **Narita, H.,** 1976. Crystallographic relationships of the pyroxenes and pyroxenoids. *J. Japanese Assoc. Min. Petr., Econ. Geol.*, **71**, 248–254.

Lakshman, S. V. J. and **Reddy, B. J.,** 1973. Optical absorption spectrum of Mn^{2+} in rhodonite. *Physica*, **66**, 601–610.

Lazarev, A. N. and **Tenisheva, T. F.,** 1961. Vibrational spectra of silicates. III. Infrared spectra of the pyroxenoids and other chain metasilicates. *Optics and Spectroscopy (U.S.S.R.)*, **11**, 316–317.

Lee, D. E., 1955. Mineralogy of some Japanese manganese ores. *Stanford Univ. Publ., Geol. Sci.*, **5**, 1–64.

Liebau, F., 1959. Über die Kristallstruktur des Pyroxmangits $(Mn,Fe,Ca,Mg)SiO_3$. *Acta Cryst.*, **12**, 177–181.

Liebau, F., Hilmer, W. and **Lindemann, G.,** 1959. Über die Kristallstruktur des Rhodonits $(Mn,Ca)SiO_3$. *Acta Cryst.*, **12**, 182–187.

Liebau, F., Sprung, M. and **Thilo, E.,** 1958. Chemische Untersuchungen von Silikaten. XXIII. Über das System $MnSiO_3$–$CaMn(SiO_3)_2$. *Z. anorg. Chem.*, **297**, 213–225.

Magnusson, N. H., 1924. The Långban minerals from a geological point of view. *Geol. För. Förh., Stockholm*, **46**, 284–300.

Mamedov, K. S., 1958. The crystal structure of rhodonite. *Dokl. Akad. Nauk Azerb. S.S.R.*, **16**, 445–450.

Manning, P. G., 1968. Absorption spectra of the manganese-bearing chain silicates pyroxmangite, rhodonite, bustamite and serandite. *Can. Min.*, **9**, 348–357.

Maresch, W. V. and **Mottana, A.,** 1976. The pyroxmangite–rhodonite transformation for the $MnSiO_3$ composition. *Contr. Min. Petr.*, **55**, 69–79.

Marshall, M. and **Runciman, W. A.,** 1975. The absorption spectrum of rhodonite. *Amer. Min.*, **60**, 88–97.

Mason, B., 1973. Manganese silicate minerals from Broken Hill, New South Wales. *J. Geol. Soc. Australia*, **20**, 397–404.

Momoi, H., 1964. Mineralogical study of rhodonites in Japan, with special reference to contact metamorphism. *Mem. Fac. Sci., Kyushu Univ., ser. D, Geol.*, **15**, 39–63.

Momoi, H., 1969. [Determination of lattice constants by X-ray diffractometer, with an example applied to rhodonite.] *Sci. Rept., Fac. Sci., Kyushu Univ., Geol.*, **9**, 59–65 (Japanese with English summary).

Momoi, H., 1974. Hydrothermal crystallization of $MnSiO_3$ polymorphs. *Min. J., Japan*, **7**, 359–373.

Morimoto, N., Koto, K. and **Shinohara, T.,** 1966. Oriented transformation of johannsenite to bustamite. *Min. J., Japan*, **5**, 44–64.

Muan, A., 1959a. Phase equilibria in the system manganese oxide–SiO_2 in air. *Amer. J. Sci.*, **257**, 297–315.

Muan, A., 1959b. Stability relations among some manganese minerals. *Amer. Min.*, **44**, 946–960.

Muan, A., 1967. Stabilities of oxide compounds and activity/composition relations in oxide solution systems. *Proc. Brit. Ceram. Soc.*, **8**, 103–113.

Narita, H., Koto, K. and **Morimoto, N.,** 1977. The crystal structure of $MnSiO_3$ polymorphs (rhodonite- and pyroxmangite-type). *Min. J., Japan*, **8**, 329–342.

Ohashi, Y. and **Finger, L. W.,** 1975. Pyroxenoids: a comparison of refined structures of rhodonite and pyroxmangite. *Carnegie Inst. Washington, Ann. Rept. Dir. Geophys. Lab.*, 1974–75, 565–569.

Ohashi, Y., Kato, A. and **Matsubara, S.,** 1975. Pyroxenoids: a variation in chemistry of natural rhodonites and pyroxmangites. *Carnegie Inst. Washington, Ann. Rept. Dir. Geophys. Lab.*, 1974–75, 561–564.

Palache, C., 1929. A comparison of the ore deposits of Långban, Sweden, with those of Franklin, New Jersey. *Amer. Min.*, **14**, 43–47.

Palache, C., 1935. The minerals of Franklin and Sterling Hill, Sussex County, New Jersey. *U.S. Geol. Surv., Prof. Paper*, **180**.

Park, C. F., *Jr.*, 1946. The spilite and manganese problems of the Olympic peninsula, Washington. *Amer. J. Sci.*, **244**, 305–323.

Peacor, D. R. and **Niizeki, N.,** 1963. The redetermination and refinement of the crystal structure of rhodonite, $(Mn,Ca)SiO_3$. *Z. Krist.*, **119**, 98–116.

Perutz, M., 1937. Iron-rhodonite (from slag) and pyroxmangite and their relation to rhodonite. *Min. Mag.*, **24**, 573–576.

Prewitt, C. T. and **Peacor, D. R.,** 1964. Crystal chemistry of the pyroxenes and pyroxenoids. *Amer. Min.*, **49**, 1527–1542.

Robie, R. A. and **Waldbaum, D. R.,** 1968. Thermodynamic properties of minerals and related substances at 298·15°K (25°C) and one atmosphere (1·013 bars) pressure and at higher temperatures. *Bull. U.S. Geol. Surv.*, **1259**.

Russell, Sir Arthur, 1946. On rhodonite and tephroite from Treburland manganese mine, Altarnun,

Cornwall: and on rhodonite from other localities in Cornwall and Devonshire. *Min. Mag.*, **27**, 221–235.

Rutstein, M. S. and **White, W. B.**, 1971. Vibrational spectra of high-calcium pyroxenes and pyroxenoids. *Amer. Min.*, **56**, 877–887.

Ryall, W. R. and **Threadgold, I. M.**, 1966. Evidence for $[(SiO_3)_5]_\infty$ type chains in inesite as shown by X-ray and infrared absorption studies. *Amer. Min.*, **51**, 754–761.

Segnit, E. R., 1962. Manganese deposits in the neighbourhood of Tamworth, New South Wales. *Proc. Austral. Inst. Mining Metall.*, no. 202, 47–61.

Sinharay, S., 1966. Manganese nodules in gondites from Chikla mines, Bhandara district, Mahrashtra. *Quart. J. Geol. Mining Metall. Soc. India*, **38**, 115–116.

Smith, T. H., 1926. Mineralogical notes, no. 2. *Rec. Australian Museum*, **15**, 69–78.

Snow, R. B., 1943. Equilibrium relationships on the liquidus surface in part of the $MnO–Al_2O_3–SiO_2$ system. *J. Amer. Ceram. Soc.*, **26**, 11–20.

Stilwell, F. L., 1959. Petrology of the Broken Hill lode and its bearing on ore genesis. *Proc. Austral. Inst. Mining Metall.*, no. 190, 1–84.

Sundius, N., 1930. Iron-rhodonite from Tuna-Hästberg. *Geol. För. Förh. Stockholm*, **52**, 403–406.

Sundius, N., 1931. On the triclinic manganiferous pyroxenes. *Amer. Min.*, **16**, 411–429 and 488–518.

Tilley, C. E., 1937. Pyroxmangite from Inverness-shire, Scotland. *Amer. Min.*, **22**, 720–727.

Tilley, C. E., 1946. Bustamite from Treburland manganese mine, Cornwall, and its paragenesis. *Min. Mag.*, **27**, 236–241.

Voos, E., 1935. Untersuchung des Schnittes $CaO.SiO_2–MnO.SiO_2$ im ternären System $SiO_2–CaO–MnO$. *Z. anorg. Chem.*, **222**, 201–224.

White, J., Howat, D. D. and **Hay, R.**, 1934. The binary system $MnO–SiO_2$. *J. Roy. Tech. Coll. Glasgow*, **3**, 231–240.

Whiteley, J. H. and **Hallimond, A. F.**, 1919. The acid hearth and slag. *J. Iron Steel Inst.*, **99**, 199–242.

Yoshi, M., Aoki, Y. and **Maeda, K.**, 1972. Nambulite, a new lithium- and sodium-bearing manganese silicate from the Funakozawa mine, north-eastern Japan. *Min. J.*, **7**, 29–44.

Yoshinga, M., 1958. A thermodynamic study of equilibrium relations among some manganese minerals introduced during the epoch of thermal metamorphism. *J. Min. Soc. Japan*, **3**, 406–417.

Yosimura, T., 1939. Studies on the minerals from the manganese deposit of the Kaso mine, Japan. *J. Fac. Sci. Hokkaido Univ.*, ser. 4, **4**, 313–451.

Pyroxmangite (Mn,Fe)[SiO_3]

Triclinic (+)

α	1·728–1·748
β	1·730–1·742
γ	1·746–1·758
δ	0·016–0·019
$2V_\gamma$	37°–46°
O.A.P.	approximately $\perp$ (1$\bar{1}$0)
Dispersion:	$r > v$, moderate.
D	3·61–3·80
H	$5\frac{1}{2}$–6
Cleavage:	{110} and {1$\bar{1}$0} perfect, {010} and {001} poor. (110):(1$\bar{1}$0) = 92°, (010):(110) = 45°, (001):z = 64°
Twinning:	Lamellar on {010}, simple on {001}; not common.
Colour:	Pink or red, but normally covered with brown or black oxidation products; colourless to faint lilac in thin section, ranging to yellow for pyroxferroite.
Unit cell:	a 9·690, b 10·505, c 17·391 Å. α 112·17°, β 102·85°, γ 82·93°. V 1596·7 Å^3. Z = 28. Space group $C\bar{1}$.

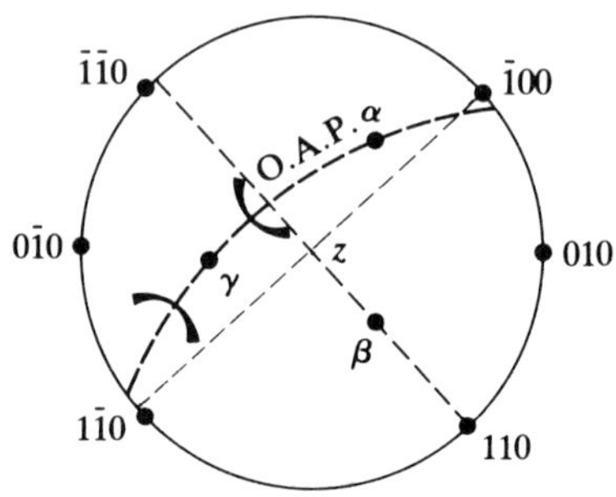

Insoluble in HCl. Manganese reaction with fluxes.

Pyroxmangite is a mineral typical of manganiferous metasomatic or metamorphic rocks. It was described originally from Iva, South Carolina, by Ford and Bradley (1913), who gave it this name thinking it to be a manganese member of the pyroxene group. Palmgren (1917) gave the name sobralite to a similar manganese-iron silicate from Sweden, which apparently differed in optical orientation, naming it after his former teacher, Dr. José M. Sobral of Buenos Aires: the original description of pyroxmangite was, however, in error. Pyroxferroite, the iron-rich analogue of pyroxmangite, has been found in microgabbros from the Sea of Tranquillity, on the Moon (Chao *et al.*, 1970), and both the Iva material and the 'sobralite' with an Fe/Mn ratio of > 1 are now known to be pyroxferroite rather than pyroxmangite.

Structure

That pyroxmangite is structurally very similar to rhodonite was suggested by Tunnell (1936) from examination of the respective powder photographs, although an appreciable difference was noted. Perutz (1937), using single-crystal X-ray techniques, confirmed the fundamental differences and concluded that although rhodonite and pyroxmangite had features in common, they did not belong to the same solid solution series. He also demonstrated that the 'iron-rhodonite' of Whiteley and Hallimond (1919) was structurally similar to pyroxmangite. The structure of pyroxmangite from Iva, South Carolina, was determined by Liebau (1959), and shown to contain chains of linked SiO_4 tetrahedra, of composition $(SiO_3)_n$, parallel to the *y* axis, with seven tetrahedra in the identity period of the chain. In pyroxenes the chain repeats after every two tetrahedra (Zweierketten); in wollastonite and bustamite after every three (Dreierketten); in rhodonite after every five (Fünferketten); and in pyroxmangite after every seven (Siebenerketten); a synthetic $MgSiO_3$ phase, ferrosilite III, has been shown to have a Neunerketten configuration (Burnham, 1966). Liebau (1956) noted a general correlation between size of the octahedral cation and the frequency of offsets in the silicate chain; as the cation size increases the chain type progresses from Zweierketten with no offsets through Neunerketten, Siebenerketten, Fünferketten, to Dreierketten (Fig. 263); even longer repeat lengths may yet be found in phases with pyroxene composition.

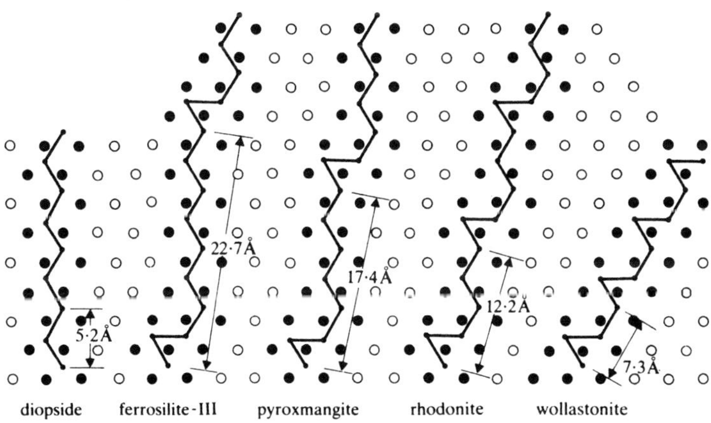

Fig. 263. Schematic diagram of silicate chains and octahedral cation distribution in pyroxenoids and diopside. Solid circles indicate occupied octahedral sites, and lines connect Si atoms in the single chains. The *z* axes, whose lengths are indicated, are parallel to the chain direction in all cases (after Burnham, 1966).

The discovery that the lunar phase pyroxferroite was the iron analogue of pyroxmangite (Chao *et al.*, 1970) has led to the refinement of the pyroxferroite structure by Burnham (1971). The Siebenerketten structure was confirmed, with Si tetrahedra forming single chains as in pyroxenes but with offsets after every seventh tetrahedron (Fig. 263). The structure can be visualized in terms of approximate close packing of oxygen atoms with metal atoms occupying some octahedral sites between octahedral layers. Layers containing metal atoms alternate with layers containing Si tetrahedral chains. The oxygen layers are parallel to (110) when the unit cell is chosen as $P\bar{1}$ with *a* 6·671, *b* 7·557, *c* 17·45 Å, α 113·7°, β 84·0°, γ 94·3°, *V* 800·6 A^3, Z = 14 (Liebau, 1959; for a calcian manganoan pyroxmangite from Iva, South Carolina). As for rhodonite, however, an alternative cell can be chosen which relates the pyroxenoids to each other and to pyroxenes and has (100) as the plane of the oxygen layer. Such a cell is given by Ohashi and Finger (1975) for a pyroxmangite from Taguchi mine, Japan, with *a* 9·690, *b* 10·505, *c* 17·391 Å, α 112·17°, β 102·85°, γ 82·93°, *V* 1596·7 $Å^3$, Z = 28, and space group $C\bar{1}$. Other cell parameters in the $P\bar{1}$ space group are those for synthetic pyroxmangite, *a* 6·621, *b* 7·552, *c* 17·23 Å, α 114°14′, β 82°53′, γ 94°73′, (Peters, 1971) and *a* 6·727, *b* 7·603, *c* 17·455 Å, α 113°10′, β 82°16′, γ 94°08′, *V* 808 $Å^3$ (Akimoto and Syono, 1972). For a lunar pyroxferroite with composition near $(Fe_{0\cdot83}Ca_{0\cdot13}Mg_{0\cdot02}Mn_{0\cdot02})SiO_3$, Burnham (1971) gave *a* 6·6213, *b* 7·5506, *c* 17·3806 Å, α 114·267°, β 82·684°, γ 94·756°, *V* 785·3 Å^3. For the pyroxenoids generally, it is preferable to label the chain axis direction '*z*', as in the pyroxenes.

Within the metal–oxygen layer in pyroxmangite, co-ordination polyhedra share edges to form continuous bands parallel to the *x* axis, but with each band consisting of an offset sequence of polyhedral chains running at an angle to the *z* axis. It seems probable that Ca is distributed between three of the seven *M* sites in the pyroxferroite structure, *M*5, *M*6 and *M*7; both Mn and Fe probably occur in all seven *M* sites in the pyroxmangite–pyroxferroite series. Although there is only a slight decrease in cell size from the Mn to the Fe end-member, there are quite large differences of intensity between the X-ray powder patterns for these two phases (Chao *et al.*, 1970). The Mössbauer spectra of a lunar pyroxferroite and of synthetic $Ca_{0\cdot15}Fe_{0\cdot85}SiO_3$ have been studied by Dowty and Lindsley (1974) who tentatively assigned the outer doublet in their spectra to *M*2 and *M*3, which are completely occupied by Fe and are apparently the least distorted of the seven cation sites, and the inner doublet to Fe in the five remaining sites. The structural relationships of pyroxmangite and rhodonite have been studied by Ohashi and Finger (1975) who refined the structures of specimens of both minerals, of virtually identical composition (pyroxmangite $Mn_{0\cdot82}Fe_{0\cdot07}Mg_{0\cdot09}Ca_{0\cdot02}SiO_3$, rhodonite $Mn_{0\cdot81}Fe_{0\cdot07}Mg_{0\cdot06}Ca_{0\cdot05}SiO_3$), both from the Taguchi mine, Japan. There is a close correspondence in the cation polyhedral configuration and to a first approximation

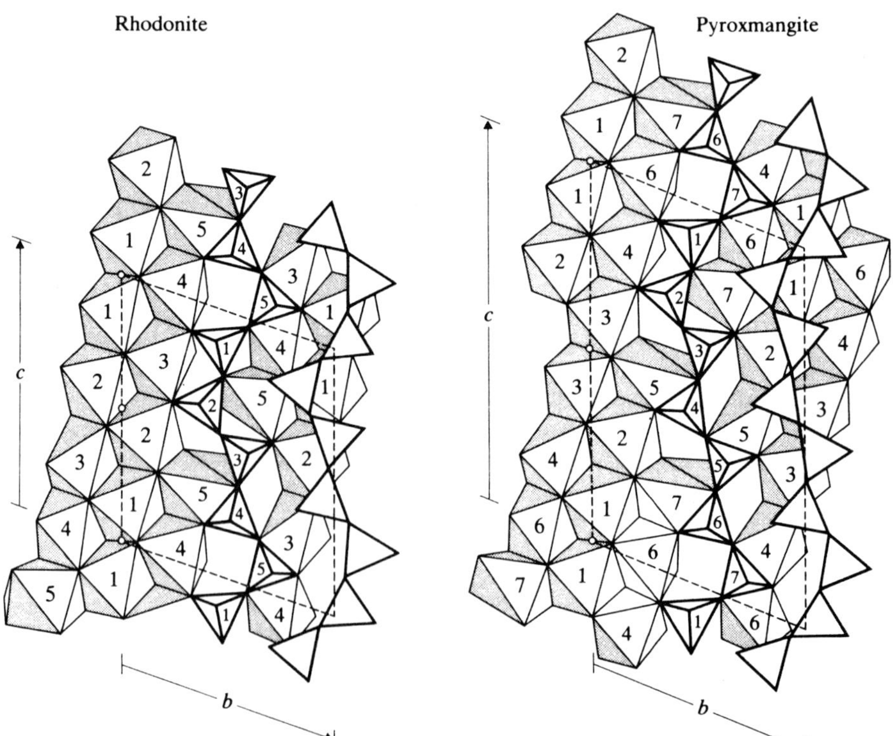

Fig. 264. Portions of the rhodonite and pyroxmangite structures projected onto the plane parallel to the oxygen closest-packed layer (after Ohashi and Finger, 1975; Ohashi *et al.*, 1975).

the pyroxmangite structure can be regarded as the rhodonite structure plus the band consisting of *M*3, *M*5, Si4, and Si5 (Fig. 264); see also Narita *et al.* (1977). The designation of cation sites in pyroxenoids has been further discussed by Takéuchi (1977).

Chemistry

Thirteen analyses are given in Table 62, together with their structural formulae calculated on the basis of eighteen oxygen ions. They represent fairly pure $(Mn,Fe)SiO_3$, with only small amounts of Fe_2O_3 being reported. CaO and MgO are often present in considerable proportions, though CaO is generally less abundant than in rhodonite: this agrees with the conclusion of Liebau (1959) who suggested that while possibly one in five of the cation positions could be filled by Ca in the rhodonite structure, for pyroxmangite this may be limited to one in seven. The chemical differences between pyroxmangite and rhodonite were examined in detail by Momoi (1964). Pyroxmangite has a wider range of Mn content; the Mn contents of both minerals decrease with increasing Ca. The Fe content is much more variable in pyroxmangite although it increases with increasing Ca in both minerals. Consideration of their Ca/Mg ratios shows that pyroxmangite tends to contain more Mg and rhodonite more Ca. On a plot of mol. per cent Mn–Ca–(Fe,Mg), typical rhodonite compositions cluster in an area centred around $Mn_{80}Ca_{10}(Fe,Mg)_{10}$ whereas the pyroxmangites have a field ranging from $Mn_{95}Ca_1(Fe,Mg)_4$ to around $Mn_{40}Ca_{14}(Fe,Mg)_{46}$, which would obviously extend virtually to the Ca–(Fe,Mg) side-line for pyroxferroite. As noted by Henderson and Glass (1936), usually neither MnO nor FeO is entirely dominant, though some pyroxmangites (Table 62, anals. 1, 2, 3) are manganese-rich, with the material of anal. 1 containing approximately 96 per cent $MnSiO_3$. Further Japanese occurrences of pyroxmangite relatively rich in Mn are reported by Yoshimura *et al.* (1958), and by Momoi (1964), and microprobe analyses of seven Japanese pyroxmangites with MnO 51·27–30·17 per cent are given by Ohashi *et al.* (1975) who also discuss the chemical variations in pyroxmangites and rhodonites. Following the discovery of the natural occurrence of pyroxferroite (Table 62, anal. 13) which approaches the iron end-member composition, Chao *et al.* (1970) reviewed the chemistry of these Siebenerketten silicates and established that there is a series between the two end-members. The members of the series with Mn > Fe are classified as pyroxmangites whereas those with Mn < Fe (e.g. Table 62, anals. 10, 11) are pyroxferroites. Further material which must now be classed as pyroxferroite includes the specimens from the Tango Peninsula, Japan (Tatekawa, 1964), with 25·10 and 26·10 per cent FeO. Available data appear to indicate a tendency for CaO to increase in amount towards the pyroxferroite end of the series. Although pyroxferroite was not expected to contain appreciable Mg, by analogy with the limited substitution shown by the more Ca-rich pyroxenoids in igneous rocks, Boyd and Smith (1971) have shown that the central part of a relatively large and homogeneous lunar crystal had a composition $Ca_{0.13}Mg_{0.12}Fe_{0.75}SiO_3$ (Table 62, anal. 12) and the pyroxferroite structure (see also Agrell *et al.*, 1970; Smith and Finger, 1971).

Experimental

The experimental determination of the curve of the reaction rhodochrosite + quartz $\rightleftharpoons$ pyroxmangite was reported by Peters (1971) using gas mixtures with

Table 62. Pyroxmangite

	1	2	3	4	5
SiO_2	45·74	47·4	47·56	45·60	47·04
Al_2O_3	tr.	—	0·37	0·00	0·00
Fe_2O_3	tr.	—	0·29	0·00	0·66
FeO	0·39	0·7	1·28	16·68	12·35
MnO	52·42	48·0	45·53	35·84	33·37
MgO	0·68	1·0	3·84	0·20	3·48
CaO	0·46	3·4	0·68	1·72	2·88
Na_2O	0·05 (Na_2O + K_2O)	—	—	—	—
K_2O		—	—	—	—
H_2O^+	0·13	—	0·49 (H_2O^+ + H_2O^-)	—	0·57
H_2O^-	0·19	—		—	0·08
Total	100·06	100·5	100·27	100·04	100·43
α	1·732	—	1·728	—	1·731
β	1·736	—	1·732	1·745	1·734
γ	1·751	—	1·746	—	1·749
δ	0·019	—	0·018	—	0·018
$2V_\gamma$	43°–46°	—	41°–42°	—	40°
D	3·69	—	3·61	3·79	3·68

Numbers of ions on the basis of eighteen O

	1		2		3		4		5	
Si	5·976		6·043		6·013		5·963		5·975	
Al	—		—		—		—		—	
Al	—	6·05[a]	—	5·91	0·055	5·93[b]	—	6·07	—	6·02
Fe^{3+}	—		—		0·027		—		0·062	
Mg	0·132		0·190		0·724		0·039		0·659	
Fe^{2+}	0·042		0·075		0·135		1·824		1·312	
Mn	5·803		5·184		4·877		3·970		3·591	
Ca	0·064		0·465		0·092		0·241		0·392	

[a] Includes (Na + K) 0·011.
[b] Includes 0·021 Zn.

1 Rose-pink pyroxmangite, Ajiro mine, Honshu, Japan (Lee, 1955). Anal. W. H. Herdsman.
2 Pyroxmangite, Bernina area, Rhaetic Alps, Switzerland (Peters *et al.*, 1973). Microprobe analysis (cell parameters *a* 7·558, *b* 17·187, *c* 6·616 Å, α 82·61°, β 94·78°, γ 113·94°; *P*-cell).
3 Purplish pink pyroxmangite, Kinko mine, Honshu, Japan (Lee, 1955). Anal. W. H. Herdsman (includes ZnO 0·23).
4 Pyroxmangite, Broken Hill, New South Wales, Australia (Mason, 1973). Anal. J. G. Fairchild.
5 Brown porphyroblastic pyroxmangite, quartzite, Simsiö, Lapua, south-west Finland (Hietanen, 1938, no. 27a). Anal. E. Ståhlberg.

6	7	8	9	10	
46·51	47·44	45·47	45·78	46·48	SiO_2
—	0·66	—	0·30	0·00	Al_2O_3
—	1·45	1·50	tr.	2·37	Fe_2O_3
19·12	15·02	20·91	21·69	22·32	FeO
29·34	28·25	27·06	26·80	21·09	MnO
1·96	4·56	2·14	0·94	3·11	MgO
2·94	3·00	2·62	3·66	4·64	CaO
—	—	—	tr.	—	Na_2O
—	—	—	0·31	—	K_2O
0·25	—	0·32	0·30	0·45	H_2O^+
—	—	—	0·15	0·20	H_2O^-
100·12	100·38	100·02	100·49	100·66	**Total**
1·738	1·732	1·737	1·739	1·734	α
1·742	1·735	1·740	1·743	1·737	β
1·754	1·750	1·754	1·756	1·751	γ
0·016	0·018	0·017	0·017	0·017	δ
39·5°	41°	39°	43°	37·5°	$2V_\gamma$
3·75	3·63	3·66	—	3·72	D
5·981	5·923 } 6·00	5·888	5·936 } 5·98	5·901	Si
—	0·077	—	0·046	—	Al
—	0·022	—	—	—	Al
—	0·136	0·146	—	0·226	Fe^{3+}
0·376 } 6·04	0·849 } 5·97	0·412 } 6·15	0·181 } 6·06[c]	0·590 } 6·09	Mg
2·056	1·568	2·264	2·352	2·370	Fe^{2+}
3·198	2·988	2·968	2·944	2·268	Mn
0·405	0·402	0·364	0·508	0·631	Ca

[c] Includes Ti 0·021, Zn 0·005, K 0·052.

6 Brown pyroxmangite, quartzite, Simsiö, Lapua, south-west Finland (Hietanen, 1938, no. 2). Anal. A. Hietanen.
7 Pink pyroxmangite, grunerite-garnet schist, Glen Beag, Glenelg, Scotland (Tilley, 1937). Anal. H. Bennett.
8 Pale pink pyroxmangite, Idaho (Henderson and Glass, 1936). Anal. E. P. Henderson.
9 Pyroxmangite, 230 ft. level, Zinc Corporation mine, Broken Hill, New South Wales, Australia (Stilwell, 1959). Anal. Avery and Anderson (includes TiO_2 0·22, Zn 0·04, S 0·30 per cent).
10 Dark brown 'pyroxmangite' (= calcian manganoan pyroxferroite), quartzite, Simsiö, Lapua, south-west Finland (Hietanen, 1938, no. 5). Anal. E. Ståhlberg.

Table 62. Pyroxmangite Analyses – *continued*

	11		12		13	
SiO_2	46·53		47·0		46·8	
Al_2O_3	0·21		0·26		0·3	
Fe_2O_3	0·85		—		—	
FeO	24·69		41·7		44·6	
MnO	20·50		0·76		0·8	
MgO	1·39		3·8		0·8	
CaO	5·46		5·4		6·0	
Na_2O	—		<0·1		tr.	
K_2O	—		—		—	
H_2O^+	0·39		—		—	
H_2O^-	—		—		—	
Total	100·10		99·3		99·8	
α	1·738		—		1·753–1·756	
β	1·740		—		1·755–1·758	
γ	1·755		—		1·766–1·767	
δ	0·017		—		0·013	
$2V_\gamma$	42°		—		35°–40°	
D	—		—		3·76	
	Numbers of ions on the basis of eighteen O					
Si	5·967	6·00	5·957	6·00	5·997	
Al	0·032		0·039		—	
Al	—	5·98[d]	—	5·99[e]	0·045	5·94[f]
Fe^{3+}	0·082		—		—	
Mg	0·265		0·718		0·152	
Fe^{2+}	2·648		4·420		4·779	
Mn	2·228		0·081		0·087	
Ca	0·750		0·733		0·824	

[d] Includes 0·004 Ba.
[e] Includes Ti 0·035, Cr 0·004.
[f] Includes Ti 0·048.

11 Yellowish red-brown 'sobralite' (= calcian manganoan pyroxferroite), Vester Silvberg, Dalecarlia, central Sweden (Sundius, 1931). Anal. A. Bygdén (includes BaO 0·08).

12 Magnesian calcian pyroxferroite, coarse-grained lunar basalt 12021, Oceanus Procellarum (Boyd and Smith, 1971). Microprobe analysis (includes TiO_2 0·37, Cr_2O_3 0·04).

13 Calcian pyroxferroite, basalt, Tranquillity Base, on the Moon (Chao *et al.*, 1970). Microprobe analysis (includes TiO_2 0·5; cell parameters *a* 6·62, *b* 7·54, *c* 17·35 Å, α 114·4°, β 82·7°, γ 94·5°; *P*-cell).

different CO_2/H_2O ratios at a total pressure of 2 kbar; in pure CO_2, the equilibrium temperature is 508 ± 2°C, with a heat of reaction of 51·7 kcal/mole. The stability relations of $MnSiO_3$ at pressures up to 130 kbar were studied by Akimoto and Syono (1972) who identified four polymorphs (see also Ringwood and Major, 1967); $MnSiO_3$-II with a pyroxmangite structure has an equilibrium phase boundary with $MnSiO_3$-I with a rhodonite structure determined as P (kbars) $= 10 + 0{\cdot}026\,T$(°C), and is transformed to $MnSiO_3$-III with a clinopyroxene structure on the boundary curve represented by P (kbars) $= 19 + 0{\cdot}057\,T$(°C) (see Fig. 265). The transformations between rhodonite and pyroxmangite and between

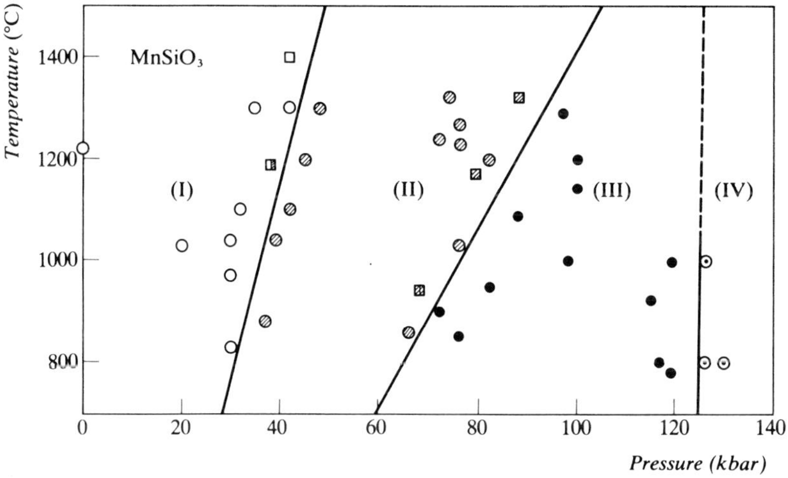

Fig. 265. Experimentally determined P–T relations for the $MnSiO_3$ system. Starting materials were $MnSiO_3$(1), rhodonite, except for reverse reaction runs indicated by squares (after Akimoto and Syono, 1972).

the clinopyroxene $MnSiO_3$-III phase and pyroxmangite are completely reversible; the rhodonite-pyroxmangite boundary lies at approximately 30 kbar at 800°C to 45 kbar at 1400°C, whereas the pyroxmangite $MnSiO_3$-III boundary runs from approximately 70 kbar at 900°C to 97 kbar at 1300°C. Further work at 2 kbar in the system MnO–SiO_2–CO_2–H_2O (Peters *et al.*, 1973) resulted in $\log K = \log f_{CO_2} = -11765/T + 18{\cdot}618$ for the reaction rhodochrosite + quartz = pyroxmangite + CO_2 and $\log f_{CO_2} = -7083/T + 11{\cdot}870$ for the reaction rhodochrosite + pyroxmangite = tephroite + CO_2; starting from rhodochrosite and quartz at a very low CO_2/H_2O ratio tephroite is obtained below 420°C and pyroxmangite is only formed above 430°C. The rhodonite–pyroxmangite peritectic along the join $MnSiO_3$–$MgSiO_3$ in air was located at approximately $(Mn_{0{\cdot}67}Mg_{0{\cdot}33})SiO_3$ at 1350°C by Ito (1972a) who was able to grow single crystals of pyroxmangite to a maximum size of 5 × 3 × 2 mm by slow cooling of the melts. The phase transformation at fixed P *and* T for $(Mn,Mg)SiO_3$ appears to be controlled by the mean cationic radii for the variable Mn and Mg contents (Fig. 266). A narrow but distinct region where rhodonite and pyroxmangite coexist was found below the solidus around $(Mn_{0{\cdot}50}Mg_{0{\cdot}50})SiO_3$–$(Mn_{0{\cdot}55}Mg_{0{\cdot}45})SiO_3$. The isomorphous pyroxmangite series extends all the way from $MgSiO_3$ to $MnSiO_3$ under hydrothermal conditions.

In syntheses on the join $MnSiO_3$–$CaSiO_3$ at 2 kbar, Abrecht and Peters (1975)

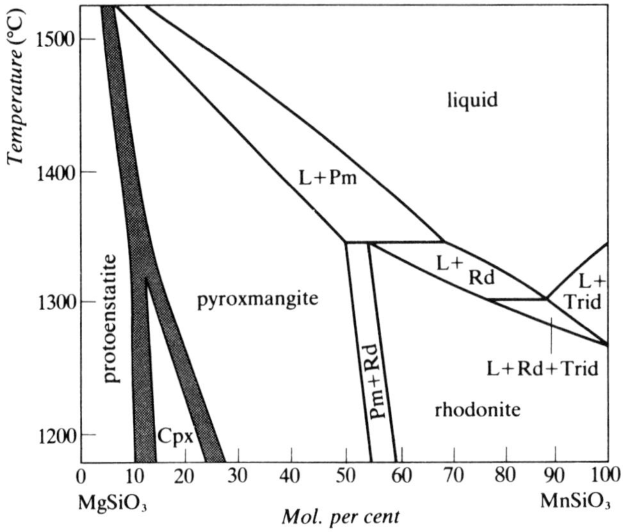

Fig. 266. Phase diagram for the join $MnSiO_3$–$MgSiO_3$. Pm = pyroxmangite, Rd = rhodonite, Trid = tridymite, Cpx = clinopyroxene and Pr = protoenstatite. The shaded area represents the two-phase region (after Ito, 1972a).

found that the type of structure obtained is mainly dependent on the Mn/Ca ratio, but also to a lesser extent on temperature. Pyroxmangite could be converted to rhodonite with increase of temperature but not the reverse (see also Fig. 260, p. 593); a schematic phase diagram for this system at 1 kbar was presented by Momoi (1968). The enthalpies of formation of the rhodonite and pyroxmangite $MnSiO_3$ polymorphs were studied by Navrotsky and Coons (1976) who obtained enthalpies of solution in a flux ($2PbO.B_2O_3$) at 713°C of $+6{\cdot}98 \pm 0{\cdot}26$ and $+6{\cdot}92 \pm 0{\cdot}19$ kcal/mole, respectively, giving $\Delta H^{\circ}_{986} = +0{\cdot}06 \pm 0{\cdot}33$ kcal/mole, leading to the comment that with such comparable energies, the effects of surface energies, coherent intergrowths or possible 'mistakes' in stacking sequences may affect the observed equilibria to an observable degree. The equilibrium constants for the reactions rhodochrosite + quartz = pyroxmangite + CO_2 and rhodochrosite + pyroxmangite = tephroite + CO_2 at 500 bars have been determined by Candia *et al.* (1975); the mean value obtained for the Gibbs free energy of formation of pyroxmangite was $-298{\cdot}005$ kcal/mole. Experimentally determined phase relations representing substitution of Mn in $MnSiO_3$ by Mg, Fe^{2+}, and Ca were presented by Ito (1972b), see Fig. 267.

Fe–Ca metasilicates approaching the composition of pyroxferroite were synthesized by Bowen *et al.* (1933) under the definition of 'wollastonite solid solution'. Subsequently it was shown by Lindsley and Munoz (1969) that such phases richer in iron than approximately $Ca_{0{\cdot}37}Fe_{0{\cdot}63}SiO_3$ are not stable at below 2 kbar, but with increasing pressure Lindsley (1967) synthesized pyroxferroite of composition $Ca_{0{\cdot}15}Fe_{0{\cdot}85}SiO_3$ at pressures from 10 to 17·5 kbar and temperatures from 1130° to 1250°C. More recently, phase equilibria studies have indicated a triangular stability field for pyroxferroite of that composition, lying between invariant points at approximately 9·5 kbar, 1190°C; 10 kbar, 1040°C; and 17·5 kbar, 1270°C (Lindsley and Burnham, 1970). At lower pressures, pyroxferroite decomposes to a Ca-enriched metasilicate phase, fayalitic olivine, and a SiO_2 phase, whereas at lower temperatures it transforms to a clinopyroxene.

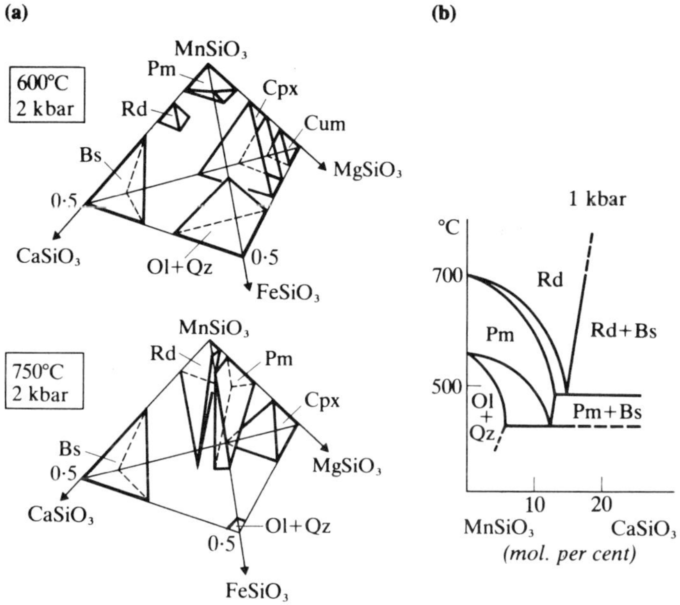

Fig. 267. Synthesis diagrams (a) for the system $MnSiO_3$–$Ca_{0\cdot5}Mn_{0\cdot5}SiO_3$–$Fe_{0\cdot5}Mn_{0\cdot5}SiO_3$–$Mg_{0\cdot5}Mn_{0\cdot5}SiO_3$ (after Ito, 1972b); and (b) for the join $MnSiO_3$–$CaSiO_3$ (after Momoi, 1968). Bs = bustamite, Cpx = clinopyroxene, Cum = cummingtonite, Ol = olivine, Pm = pyroxmangite, Qz = quartz, Rd = rhodonite.

Alteration products are nearly always present on the surfaces and cleavages of pyroxmangite as collected: these alteration products consist of manganese oxide and hydroxide with a lesser amount of iron oxide. Ford and Bradley (1913) named such alteration material on the original South Carolina specimen as skemmatite. The $Fe_2O_3:MnO_2$ ratio appears to vary in alteration products from different pyroxmangite localities though it is not known whether this simply reflects varying FeO:MnO ratios in the fresh pyroxmangites concerned or whether a variable product is involved. Palache *et al.* (1944) prefer to describe skemmatite as a ferrian wad. This coating of alteration material may be removed with HCl or H_2SO_3.

Optical and Physical Properties

The optical properties of analysed pyroxmangites are included in Table 62. The variation in optics with change in MnO:Fe ratio is not very pronounced but it can be shown that the refractive indices rise with increasing amounts of Fe^{2+} + Mn, and that the effect of Fe^{2+} alone is somewhat greater than that of Mn. The optic axial angle appears to decrease with increasing substitution of Fe^{2+} for Mn (Table 62) though, as noted by Lee (1955), data are still too few to warrant curves being drawn (see, however, Momoi, 1964): the birefringence shows a slight tendency to increase with higher Mn:Fe ratios. The 'iron-rhodonite' of Whiteley and Hallimond (1919), which has a Siebenerketten structure, has α 1·750, β 1·754, γ 1·767, $2V_\gamma$ 27° (Tilley, 1937) for a composition near Fe 63·0, Mn 26·3, Ca + Mg 10·7 mol. per cent (Perutz, 1937); it is thus a pyroxferroite.

The colour of pyroxmangite is pale pink or lilac to red, and the mineral is normally colourless in thin section. Specimens as collected are dark brown to black, due to the usual presence of a film of alteration material, and when cleaved form short nearly rectangular prisms. Lee (1955) reports that the colour is brighter pink with a higher manganese content, and suggests that for the Kinko mine pyroxmangite (Table 62, anal. 3) the purplish hue is due to the appreciable magnesium content. The optical absorption spectrum of pyroxmangite shows a set of five absorption bands in the visible and near ultraviolet regions which have been assigned to octahedrally bonded Mn^{2+} (Manning, 1968) to which the pink colour is primarily attributed; there are also prominent absorptions due to octahedrally bonded Fe^{2+}. The infrared spectrum between 550 and 750 cm^{-1} shows seven peaks (Ryall and Threadgold, 1966), confirming the claim by Lazarev and Tenisheva (1961) that, in general, the number of peaks in this region provide information as to the number of tetrahedra in the repeat unit of a single-chain silicate. On clean fresh surfaces the lustre is pearly to vitreous. The iron-rich species pyroxferroite is distinctly yellow and may appear pale yellow to yellow-orange in thin section; it may show faint pleochroism. Lamellar twinning may occur parallel to (010) and simple twins on (001) have been reported; see also Lee (1955, Fig. 5) and Diehl and Berdesinski (1970).

As for rhodonite, more than one orientation is in common use (see p. 593): in the orientation used by Sundius (1931), Henderson and Glass (1936) and Tilley (1937) the two best cleavages are taken to be (110) and ($1\bar{1}0$), whereas Perutz (1937), following the orientation adopted by Gossner and Brückl (1928) for rhodonite, takes the first principal cleavage as being parallel to (100), with the second principal cleavage parallel to (001). The various orientations have been correlated by Diehl and Berdesinski (1970), who recommend a setting in which the twin plane becomes ($\bar{1}01$); see also Koto *et al.* (1976).

The identification of sobralite with the pyroxmangite series was demonstrated by Henderson and Glass (1936). The similarity had been noted previously by Palmgren (1917), Sobral (1921) and Sundius (1931), but the optical properties and orientation given by Ford and Bradley (1913) for the original pyroxmangite were too incomplete to establish identity. The latter authors reported that the optic axial plane was normal to the plane of parting (010). Gilluly (1936) re-examined the type material from South Carolina and found that this orientation was incorrect, but that the angle between the optic axial plane and the (110) cleavage is 47°, and that between the optic axial plane and the ($1\bar{1}0$) cleavage is 78°, values similar to those obtained for sobralite, thus reconciling the only apparent difference between the two phases. The later discovery of the existence of an iron end-member pyroxferroite has led to a change in nomenclature, and the type sobralite as a member of the pyroxmangite–pyroxferroite series with Fe > Mn is now termed pyroxferroite.

Distinguished Features

The distinct cleavages, pinkish colour in fresh hand specimen, inclined extinction and manganese reaction help to distinguish pyroxmangite as one of the triclinic manganese silicates. The brownish black alteration product often present on the surface of hand specimens is also characteristic. The iron-rich species pyroxferroite has a yellowish tinge in thin section. Pyroxmangite may be distinguished from rhodonite by its smaller 2V and higher birefringence, and from bustamite by its positive optic sign.

Paragenesis

Pyroxmangite is a mineral of metamorphic or metasomatic rocks, being found typically in manganese-rich assemblages in association with spessartine garnet, dannemorite, tephroite, alleghanyite, hausmannite, pyrophanite, alabandite or rhodochrosite. It has been recorded from a grunerite-garnet schist in Lewisian paragneiss (Tilley, 1937), where it occurs associated with rhodonite and iron-rich pyroxenes, and is also known from impure quartzites (Hietanen, 1938). Pyroxferroite was first described from lunar microgabbros and breccias from Tranquillity Base (Chao *et al.*, 1970; Agrell *et al.*, 1970). Lee (1955) reported that the environment of the pyroxmangite of Table 62, anals. 1 and 3, is rich in manganese and magnesium and exceptionally poor in iron and calcium. The country rocks of the Bernina Alps in which the rhodochrosite + quartz + pyroxmangite (Table 62, anal. 2) assemblage occurs show moderate greenschist facies metamorphism (Peters, 1971); rather similar assemblages are reported from Italy (Bertolani, 1967; Sabatino, 1967). Both pyroxmangite and pyroxferroite are also known from pegmatites, particularly in Japan, e.g. Omori and Hasegawa (1956) described greyish pyroxmangite occurring with a manganoan allanite in pegmatite at Iwaizumi, Iwate Prefecture, and Tatekawa (1964) described amber or greenish amber 'pyroxmangite' (now pyroxferroite) from pegmatites of the Tango Peninsula, Japan.

Although the compositional ranges of pyroxmangite and rhodonite are different, in the manganese-rich compositions these minerals have a polymorphic relationship, pyroxmangite being stable at lower temperature and higher pressure than rhodonite (Akimoto and Syono, 1972; Abrecht and Peters, 1975). This is in line with the general polymorphic relationships that exist in the pyroxenes and pyroxenoids, i.e. the phase with a greater frequency of occurrence of tetrahedral horizontal offset in a repeat unit of a single chain tends to occur with decreasing pressure or with rising temperature (Momoi, 1974; Ohashi and Finger, 1975), rhodonite having $n = 5$. Recent work by Maresch and Mottana (1976), however, led to confirmation that pyroxmangite is the high-pressure, low-temperature polymorph with respect to rhodonite of the same $MnSiO_3$ end-member composition, and that it is the stable phase at atmospheric pressure below 350°–405°C (see Fig. 260, p. 593; Fig. 265, p. 607); this result is based on reversible runs across the transformation at 3–30 kbar and 425°–1000°C. In any event the application of experimental work to natural occurrences is premature until the influence of CaO, FeO and MgO on the stability of the $MnSiO_3$ phases is better known.

The coexistence of rhodonite and pyroxmangite (Table 62, anal. 7) has been recorded from the Glenelg grunerite schist (Tilley, 1937), and they have been reported to occur in intimate intergrowth in specimens from Broken Hill, New South Wales (Burrell, 1942); this is in line with the experimental evidence and does not necessarily mean that one of the minerals is present as a metastable phase.

References

Abrecht, J. and **Peters, Tj.,** 1975. Hydrothermal synthesis of pyroxenoids in the system $MnSiO_3$–$CaSiO_3$ at Pf = 2 kb. *Contr. Min. Petr.*, **50**, 241–246.

Agrell, S. O., Scoon, J. H., Muir, I. D., Long, J. V. P., McConnell, J. D. C. and **Peckett, A.,** 1970. Observations on the chemistry, mineralogy and petrology of some Apollo 11 lunar samples. *Proc. Apollo 11 Lunar Sci. Conf.* (Suppl. to *Geochim. Cosmochim. Acta*, **34**), vol. 1, 93–128.

Akimoto, S. I. and **Syono, Y.,** 1972. High pressure transformations in $MnSiO_3$. *Amer. Min.*, **57**, 76–84.

Bertolani, M., 1967. Rocce manganisifere tra le granuliti della valle Strona (Novara). *Periodico Min.*, **36**, 1011–1032.

Bowen, N. L., Schairer, J. F. and **Posnjak, E.,** 1933. The system CaO–FeO–SiO_2. *Amer. J. Sci.*, ser. 5, **26**, 193–284.

Boyd, F. R. and **Smith, D.,** 1971. Compositional zoning in pyroxenes from lunar rock 12021, Oceanus Procellarum. *J. Petr.*, **12**, 439–464.

Burnham, C. W., 1966. Ferrosilite III: a triclinic pyroxenoid-type polymorph of ferrous metasilicate. *Science*, **154**, 513–516.

Burnham, C. W., 1971. The crystal structure of pyroxferroite from Mare Tranquillitatis. *Proc. Second Lunar Sci. Conf.* (Suppl. to *Geochim. Cosmochim. Acta*, **2**), **1**, 47–57.

Burrell, H. C., 1942. A statistical and laboratory investigation of ore types at Broken Hill, Australia. *Harvard Univ. Ph.D. Thesis* (quoted from Lee, 1955).

Candia, M. A. F., Peters, Tj. and **Valarelli, J. V.,** 1975. The experimental investigation of the reactions $MnCO_3+SiO_2 = MnSiO_3+CO_2$ and $MnSiO_3+MnCO_3 = Mn_2SiO_4+CO_2$ in CO_2/H_2O gas mixtures at a total pressure of 500 bars. *Contr. Min. Petr.*, **52**, 261–266.

Chao, E. C. T., Minkin, J. A., Frondel, C., Klein, C., *Jr.*, **Drake, J. C., Fuchs, L., Tani, B., Smith, J. V., Anderson, A. T., Moore, P. B., Zechman, G. R.,** *Jr.*, **Trail, R. J., Plant, A. G., Douglas, J. A. V.** and **Dence, M. R.,** 1970. Pyroxferroite, a new calcium-bearing silicate from Tranquillity Base. *Proc. Apollo 11 Lunar Sci. Conf. (Suppl. to Geochim. Cosmochim. Acta,* **34***)*, **1**, 65–79.

Diehl, R. and **Berdesinski, W.,** 1970. Zwillingsbildung am Pyroxmangit der North Mine von Broken Hill, New South Wales, Australia. *Neues Jahrb. Min., Monat.*, 348–362.

Dowty, E. and **Lindsley, D. H.,** 1974. Mössbauer spectra of synthetic Ca-Fe pyroxenoids and lunar pyroxferroite. *Contr. Min. Petr.*, **48**, 229–232.

Ford, W. E. and **Bradley, W. M.,** 1913. Pyroxmangite, a new member of the pyroxene group, and its alteration product, skemmatite. *Amer. J. Sci.*, 4th ser., **36**, 169–174.

Gilluly, J., 1936. In: Henderson, E. P. and Glass, J. J. (*q.v.*).

Gossner, B. and **Brückl, K.,** 1928. Über strukturelle Beziehungen von Rhodonit zu anderen Silikaten. *Centr. Min.*, Abt. A, 316–322.

Henderson, E. P. and **Glass, J. J.,** 1936. Pyroxmangite, a new locality: identity of sobralite and pyroxmangite. *Amer. Min.*, **21**, 273–294.

Hietanen, A., 1938. On the petrology of Finnish quartzites. *Bull. Comm. géol. Finlande*, **21**, no. 122, 1–119.

Ito, J., 1972a. Rhodonite–pyroxmangite peritectic along the join $MnSiO_3$–$MgSiO_3$ in air. *Amer. Min.*, **57**, 865–876.

Ito, J., 1972b. Synthesis and crystal chemistry of Li-hydro-pyroxenoids. *Min. J., Japan*, **7**, 45–65.

Koto, K., Morimoto, N. and **Narita, H.,** 1976. Crystallographic relationships of the pyroxenes and pyroxenoids. *J. Japanese Assoc. Min. Petr. Econ. Geol.*, **71**, 248–254.

Lazarev, A. N. and **Tenisheva, T. F.,** 1961. Vibrational spectra of silicates. III. Infrared spectra of the pyroxenoids and other chain metasilicates. *Optics and Spectroscopy* (U.S.S.R.), **11**, 316–317.

Lee, D. E., 1955. Mineralogy of some Japanese manganese ores. *Stanford Univ. Publ., Univ. Ser., Geol. Sci.*, **5**, 1–64.

Liebau, F., 1956. Bemerkungen zur Systematik der Kristallstrukturen von Silikaten mit hochkondensierten Anione. *Z. Phys. Chem.*, **206**, 73–92.

Liebau, F., 1959. Über die Kristallstruktur des Pyroxmangits (Mn,Fe,Ca,Mg)SiO_3. *Acta Cryst.*, **12**, 177–181.

Lindsley, D. H., 1967. The join hedenbergite–ferrosilite at high pressures and temperatures. *Carnegie Inst. Washington, Ann. Rept. Dir. Geophys. Lab.*, Yearbook 65, 230–232.

Lindsley, D. H. and **Burnham, C. W.,** 1970. Pyroxferroite: stability and X-ray crystallography of synthetic $Ca_{0\cdot15}Fe_{0\cdot85}SiO_3$ pyroxenoid. *Science*, **168**, 364–367.

Lindsley, D. H. and **Munoz, J. L.,** 1969. Subsolidus relations along the join hedenbergite–ferrosilite. *Amer. J. Sci.* (Schairer vol.), 295–324.

Manning, P. G., 1968. Absorption spectra of the manganese-bearing chain silicates pyroxmangite, rhodonite, bustamite and serandite. *Canad. Min.*, **9**, 348–357.

Maresch, W. V. and **Mottana, A.**, 1976. The pyroxmangite–rhodonite transformation for the $MnSiO_3$ composition. *Contr. Min. Petr.*, **55**, 69–79.
Mason, B., 1973. Manganese silicate minerals from Broken Hill, New South Wales. *J. Geol. Soc. Australia*, **20**, 397–404.
Momoi, H., 1964. Mineralogical study of rhodonites in Japan, with special reference to contact metamorphism. *Mem. Fac. Sci., Kyushu Univ., ser. D, Geol.*, **15**, 39–63.
Momoi, H., 1968. Some manganese pyroxenoids. *J. Min. Soc. Japan*, **8**, Spec. Issue 2, 1–5 (in Japanese).
Momoi, H., 1974. Hydrothermal crystallization of $MnSiO_3$ polymorphs. *Min. J. Japan*, **7**, 359–373.
Narita, H., **Koto, K.** and **Morimoto, N.**, 1977. The crystal structures of $MnSiO_3$ polymorphs (rhodonite- and pyroxmangite-type). *Min. J., Japan*, **8**, 329–342.
Navrotsky, A. and **Coons, W. E.**, 1976. Thermochemistry of some pyroxenes and related compounds. *Geochim. Cosmochim. Acta*, **40**, 1281–1288.
Ohashi, Y. and **Finger, L. W.**, 1975. Pyroxenoids: a comparison of refined structures of rhodonite and pyroxmangite. *Carnegie Inst. Washington, Ann. Rept. Dir. Geophys. Lab.*, 1974–75, 565–569.
Ohashi, Y., **Kato, A.** and **Matsubara, S.**, 1975. Pyroxenoids: a variation in chemistry of natural rhodonites and pyroxmangites. *Carnegie Inst. Washington, Ann. Rept. Dir. Geophys. Lab.*, 1974–75, 561–564.
Omori, K. and **Hasegawa, S.**, 1956. Chemical compositions of perthite, ilmenite, allanite, and pyroxmangite that occurred in pegmatites of a vicinity of Iwaizumi town, Iwate Prefecture. *Sci. Rept. Tohoku Univ.*, ser. 3, **5**, 129–137.
Palache, C., **Berman, H.** and **Frondel, C.**, 1944. *Dana's System of Mineralogy*. 7th edition, vol. 1. New York (John Wiley).
Palmgren, J., 1917. Die Eulysit von Sodermanland (Sobralit). *Bull. Geol. Inst. Upsala*, **14**, 109–228.
Peacor, D. R. and **Niizeki, N.**, 1963. The redetermination and refinement of the crystal structure of rhodonite, $(Mn,Ca)SiO_3$. *Z. Krist.*, **119**, 98–116.
Perutz, M., 1937. Iron-rhodonite (from slag) and pyroxmangite and their relation to rhodonite. *Min. Mag.*, **24**, 573–576.
Peters, Tj., 1971. Pyroxmangite: stability in H_2O–CO_2 mixtures at a total pressure of 2000 bars. *Contr. Min. Petr.*, **32**, 267–273.
Peters, Tj., **Schwander, H.** and **Tromsdorff, V.**, 1973. Assemblages among tephroite, pyroxmangite, rhodochrosite, quartz: experimental data and occurrences in the Rhetic Alps. *Contr. Min. Petr.*, **42**, 325–332.
Ringwood, A. E. and **Major, A.**, 1967. Some high-pressure transformations of geophysical importance. *Earth Planet. Sci. Letters*, **2**, 106–110.
Ryall, W. R. and **Threadgold, I. M.**, 1966. Evidence for $[(SiO_3)_5]_\infty$ type chains in inesite as shown by X-ray and infrared absorption studies. *Amer. Min.*, **51**, 755–761.
Sabatino, B. D., 1967. Su una paragenesi del giacimento manganesifero di Scortico (Alpi Apuane). *Periodico Min.*, **36**, 965–992.
Smith, D. and **Finger, L. W.**, 1971. Magnesian pyroxferroite in lunar rock 12021. *Carnegie Inst. Washington, Ann. Rept. Dir. Geophys. Lab.*, 1970–71, 133–134.
Sobral, J. M., 1921. Optical investigation of the new pyroxene sobralite. *Bull. Geol. Inst. Upsala*, **18**, 57–53.
Stilwell, F. L., 1959. Petrology of the Broken Hill lode and its bearing on ore genesis. *Proc. Austral. Inst. Min. Metall.*, no. 190, 1–84.
Sundius, N., 1931. On the triclinic manganiferous pyroxenes. *Amer. Min.*, **16**, 411–429 and 488–518.
Takéuchi, Y., 1977. Designation of cation sites in pyroxenoids. *Min. J., Japan*, **8**, 431–438.
Tatekawa, M., 1964. [Manganese minerals from the pegmatites in the neighbourhood of the Tango Peninsula, Japan.] *Kobutsugaku Zasshi*, **6**, 324–329 (in Japanese).
Tilley, C. E., 1937. Pyroxmangite from Inverness-shire, Scotland. *Amer. Min.*, **22**, 720–727.
Tunnell, G., 1936. In: Henderson, E. P. and Glass, J. J. (*q.v.*).
Whiteley, J. H. and **Hallimond, A. F.**, 1919. The acid hearth and slag. *J. Iron Steel Inst.*, **99**, 199–242.
Yoshimura, T., **Shirozu, H.** and **Hirowatari, F.**, 1958. Bementite and pyroxmangite from the Ichinomata mine, Kumamoto Prefecture. *J. Min. Soc. Japan*, **3**, 457–467.

Sapphirine $(Mg,Fe^{2+},Fe^{3+}Al)_8O_2[(Al,Si)_6O_{18}]$

Monoclinic (−) or (+)

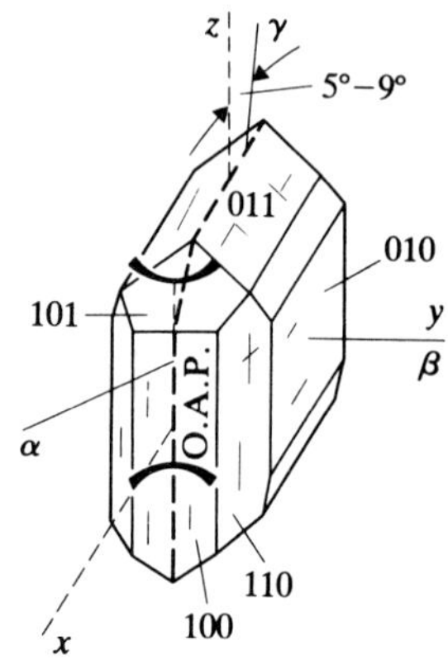

α 1·701–1·726[1]
β 1·703–1·728
γ 1·705–1·734
δ 0·005–0·007
$2V_\alpha$ 47°–114°[2]
γ:z 5°–9°[3]
$\beta = y$:O.A.P. (010)
Dispersion: $v > r$ strong.
D 3·40–3·58
H $7\frac{1}{2}$
Cleavage: {010} moderate, {001}, {100} poor.
Twinning: {010} repeated; not common.
Colour: Light blue or green, sometimes grey, pinkish grey, pale red; colourless to blue (sometimes pale pink) in thin section.
Pleochroism: α colourless, pale reddish or pink, yellowish green, pale yellow, greenish grey, greyish green.
β sky-blue, pale blue, lavender-blue, grey-blue.
γ blue, sapphire-blue, dark blue, greenish blue, blue-green, pale green.
Unit cell: $a \simeq 9·8$, $b \simeq 14·4$, $c \simeq 9·9$ Å, $\beta \simeq 110°$
Z = 4. Space group $P2_1/n$.
or $a \simeq 11·3$, $b \simeq 14·4$, $c \simeq 9·9$ Å, $\beta \simeq 125°$
Z = 4. Space group $P2_1/a$.

Insoluble in acids, decomposed by fusion in Na_2CO_3 or $KHSO_4$.

Sapphirine, once considered to be a rare mineral, has in recent years been reported from an increasing number of localities mainly of high metamorphic grade. Sapphirine occurs characteristically in aluminium-rich, silicon-poor rocks of the granulite and hornblende granulite facies, but is also found in contact assemblages in which metasomatic processes have been involved. It is invariably associated with other aluminium- and/or magnesium-rich minerals such as spinel, cordierite, corundum, sillimanite, kyanite, phlogopite, gedrite, orthopyroxene, calcium-rich plagioclase, chlorite, kornerupine, mullite and talc. Sapphirine is named on account of its common sapphire-blue colour.

[1] Lower values, α' 1·690, γ' 1·696 are given by Chekirda and Entin (1969).
[2] Smaller values of $2V_\alpha$ have been reported, see optics section.
[3] Larger values are reported by Clifford *et al.* (1975) and Monchoux (1972).

Structure

Before the structure of sapphirine was determined (in 1969) it had generally been regarded as an orthosilicate, perhaps related to the spinels, and its formula was written as $(Mg,Fe)_4Al_8O_{12}[SiO_4]_2$. The structure determination by Moore (1968, 1969) showed, however, that sapphirine is a chain silicate and a more appropriate formula is $(Mg,Fe,Al)_8O_2[(Al,Si)_6O_{18}]$.

The structure of sapphirine is based upon cubic close-packed arrays of oxygen atoms in layers parallel to (100)[1]. Bands of (Mg,Al,Fe) atoms in octahedra run along the *z* axis direction and lie parallel to (100). The bands are themselves made up of rows of four and of three octahedra as shown in Fig. 268(a). The octahedral bands are separated from each other in the *y* direction but are connected by chains of tetrahedra of the kind shown in Fig. 268(b). These chains can be likened to those of a pyroxene but they contain additional 'wings' of corner-sharing tetrahedra. The repeat unit of the chain is T_6O_{18} where T is Al or Si. An additional octahedron connects the bands of octahedra in the direction normal to (100), sharing three of its edges with the bands on each side of it. It also forms a lateral link for the tetrahedral chains. The structural formula is thus best written as $M_7(M)O_2[T_6O_{18}]$ where M and T are octahedra and tetrahedra and (M) is the additional octahedron.

(a)

(b)

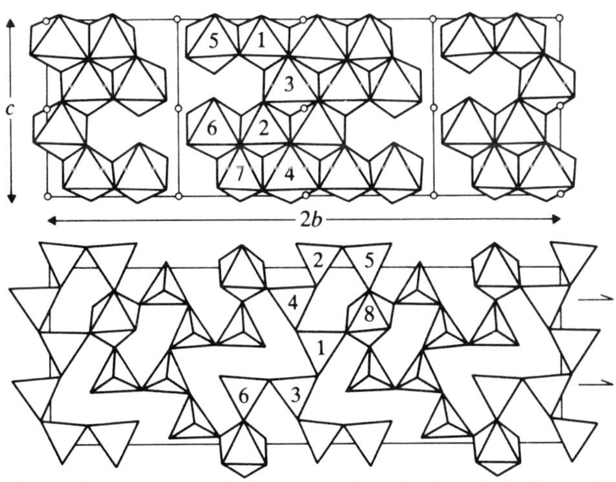

Fig. 268. (a) Idealized polyhedral diagram of bands of octahedra in sapphirine, lying in the (100) plane and running parallel to *z*. (b) Idealized polyhedral diagram of chains of tetrahedra in sapphirine, running parallel to *z* and linked laterally by single octahedra (after Cannillo *et al.*, 1971).

For the sapphirine specimen he studied, Moore (1969) showed that there was a high degree of cation ordering. In the octahedral bands three of the octahedra are purely Mg, three Al and one is $Mg_{0\cdot5}Al_{0\cdot5}$. The bridging octahedron is purely Al. In the T_6O_{18} chains, half the tetrahedra are Al, one is $Al_{0\cdot5}Si_{0\cdot5}$, one $Al_{0\cdot25}Si_{0\cdot7}$ and one $Al_{0\cdot75}Al_{0\cdot25}$ (Fig. 269).

Sapphirine also exhibits stacking disorder. Merlino (1973) describes the structure in terms of layers of monoclinic symmetry with *a* 9·783, *c* 9·929 Å, β 100°17′ and of thickness $b/2$, where $b = 14{\cdot}401$ Å (parameters from Moore, 1969). Such layers are related by displacements $a/2+b/2-c/4$ (t_1) or $a/2+b/2+c/4$ (t_2) and can give rise to two ordered stacking sequences, $t_1t_1t_1\ldots$ (or $t_2t_2t_2\ldots$) and $t_1t_2t_1t_2\ldots$

Most sapphirines have the latter two-layered monoclinic (2M) cell, but Merlino

[1] Structure described in terms of $P2_1$ *a* cell of Moore (1969).

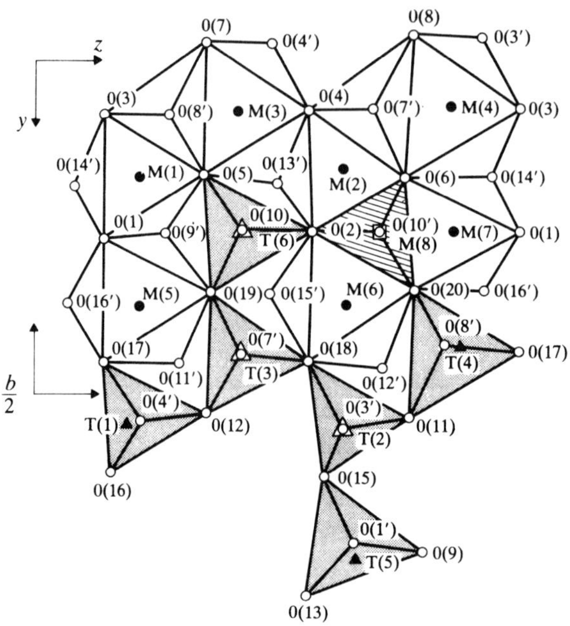

Fig. 269. The asymmetric unit of the structure of sapphirine as viewed perpendicular to (100). Octahedra in the bands are unshaded and tetrahedra are stippled. The octahedron between the bands lies over the shaded region (after Moore, 1969).

found for a sapphirine in granulites from Wilson Lake Labrador the one-layered triclinic (1 Tc) sequence. He suggests that this polymorph (polytype?) is the stable form at the high temperatures appropriate to the (sapphirine + quartz) stability field. For the 1 Tc structure there is a pseudomonoclinic cell, with doubled *b* axis, with *a* 9·87, *b* 29·08, *c* 10·04 Å, *β* 110°38′, but the true triclinic cell has *a* 10·04 Å, *b* 10·38, *c* 8·65 Å, *α* 107°33′, *β* 95°07′, *γ* 123°55′.

A yellow sapphirine from Mautia Hill, Tanzania, with disordered stacking was described by McKie (1963). It has composition $Mg_{3\cdot67}Mn_{0\cdot04}Fe_{0\cdot17}Ti_{0\cdot01}Fe^{3+}_{0\cdot33}Al_{8\cdot07}Si_{1\cdot75}P_{20}$ and *a* 9·85, *b* 28·6, *c* 9·96 Å, *β* 110½°.

Dornberger-Schiff and Merlino (1974) showed that sapphirine, Mautia Hill sapphirine and aenigmatite are members of an isomorphous series of order-disorder structures in which sapphirine and aenigmatite are ordered and Mautia Hill sapphirine is disordered.

The unit cell of the more common 2M sapphirine has been described in an alternative setting, which has space group $P2_1/a$ as compared with $P2_1/n$ for that given above. Thus Kuzel (1961) gave *a* 11·26, *b* 14·42, *c* 9·93 Å, *β* 125·33°, $P2_1/a$, Z = 4 (for twenty oxygens). This setting was used also by Moore (1969) for his structure determination of a sapphirine from Fiskenaesset, Greenland, with composition $Mg_{3\cdot5}Al_{9\cdot0}Si_{1\cdot5}O_{20}$ (Bøggild, 1953), while $P2_1/n$ was implied by the choice of cell made by Gossner and Mussgnug (1928), and Fleet (1967). The various cell parameters are shown in Table 63. The transformation matrix to change from $P2_1/a$ to $P2_1/n$ is $[101/0\bar{1}0/00\bar{1}]$ and for the pseudomonoclinic to triclinic transformation $[00\bar{1}/\frac{1}{2}\frac{1}{4}\frac{1}{4}/\frac{1}{2}\frac{\bar{1}}{4}\frac{1}{4}]$.

The cell edges of sapphirine increase approximately linearly with increasing iron content, as shown in Fig. 270, from Sahama *et al.* (1974).

Table 63 Sapphirine cell parameters and orientations

	Fleet (1967)	Merlino (1973)		McKie (1963)		Moore (1969)		Kuzel (1961)[a]	
	2M	pseudo-monoclinic	triclinic 1Tc	$P2_1/n$ 2M	$P2_1/a$ 2M	$P2_1/n$ 2M	$P2_1/a$ 2M	$P2_1/n$ 2M	$P2_1/a$ 2M
*a*Å	9·77	9·87	10·04	9·85	11·29	9·783	11·266	9·73	11·26
*b*Å	14·54	29·08	10·38	28·6	28·6	14·401	14·401	14·42	14·42
*c*Å	10·06	10·04	8·65	9·96	9·96	9·929	9·929	9·93	9·93
α			107°33′						
β	110°20′	110°38′	95°07′	110°30′	125°15′	110°17′	125°27′	109°50′	125°20′
γ			123°55′						
Z	4	8	2	8	8	4	4	4	4

[a] Synthetic $Mg_{3\cdot5}Al_{9\cdot0}Si_{1\cdot5}O_{20}$. Kuzel gave $P2_1/c$, an error for $P2_1/a$ according to Moore (1969).

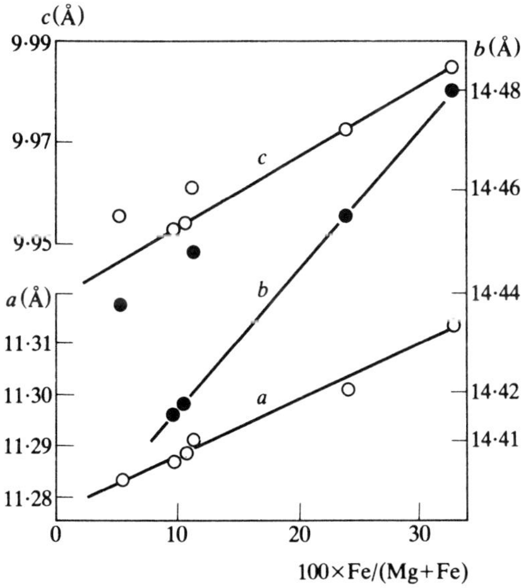

Fig. 270. Unit cell edges of sapphirine plotted against iron content (after Sahama *et al.*, 1974).

Chemistry

The composition of sapphirine has been the subject of various investigations and a number of different formulae have been proposed: $Mg_2Al_4SiO_{10}$ (Gossner and Mussgnug, 1928), $Mg_4Al_{10}Si_2O_{23}$ (Foster, 1950), $Mg_{15-n}Al_{34+2n}Si_{7-n}O_{80}$, $n = 0–1{\cdot}0$ (Palache *et al.*, 1944), $Mg_{16-n}Al_{32+2n}Si_{8-n}O_{80}$, $n = 0–2{\cdot}5$ (Vogt, 1947), $Mg_7Al_{18}Si_3O_{40}$ (Kuzel, 1961; Moore, 1969), variations that are essentially the result of the coupled substitution Mg,Si $\rightleftharpoons$ 2Al. Replacement of Mg by Fe^{2+} and Al by Fe^{3+} also occur, and the three substitutions together account for most of the chemical variations displayed by the sapphirine minerals.

The crystallochemical formula of the sapphirine, $Mg_{3\cdot5}Al_{9\cdot0}Si_{1\cdot5}O_{20}$ (Table 64, anal. 1), used by Moore (1969) to determine the crystal structure may be expressed in a condensed form as $[Mg_{3\cdot5}Al_{4\cdot5}]^6[Al_{4\cdot5}Si_{1\cdot5}]^4O_{20}$, or using an extended form to illustrate site occupancy as:

$$[Mg_{3\cdot0}(Mg_{0\cdot5}Al_{0\cdot5})Al_{4\cdot0}]^6[Al_{3\cdot0}Al_{0\cdot25}Si_{0\cdot75})(Al_{0\cdot5}Si_{0\cdot5})(Al_{0\cdot75}Si_{0\cdot25})]^4O_{20}$$

Table 64. Sapphirine Analyses

	1	2	3	4	5
SiO_2	12·83	15·30	16·71	11·2	13·84
TiO_2	—	0·07	0·67	—	—
Al_2O_3	65·29	61·34	54·77	67·5	62·09
Fe_2O_3	0·93	0·57	⎫ 3·16	—	2·06
FeO	0·65	1·50	⎭	2·9	1·40
MnO	—	tr.	—	0·10	—
MgO	19·78	20·80	19·89	17·6	19·54
CaO	—	0·30	1·13	—	—
Na_2O	—	tr.	0·20	—	—
K_2O	—	tr.	0·95	—	—
H_2O^+	—	0·33	—	—	1·20
H_2O^-	—	0·05	0·21	—	0·34
Total	99·79	100·26	99·59	100·0	100·63
α	1·7055	1·704	1·703	—	—
β	1·7088	1·707	—	—	—
γ	1·7112	1·710	1·710	—	—
$2V_\alpha$	—	85°	≃100°	—	—
$\gamma:z$	—	11°	—	—	—
D	3·486	—	—	—	—

Numbers of ions on the basis of twenty O

	1		2		3		4		5	
Si	1·490	6·00	1·776	6·00	2·018	6·00	1·302	6·00	1·620	6·00
Al	4·510		4·224		3·982		4·698		4·380	
Al	4·430		4·168		3·816		4·552		4·194	
Ti	—		0·006		0·062		—		—	
Fe^{3+}	0·080		0·048		0·140		—		0·182	
Mg	3·424		3·598		3·580		3·048		3·356	
Fe^{2+}	0·064	8·00	0·146	8·00	0·178	8·11	0·282	7·95[a]	0·136	7·92[b]
Mn	—		—		—		0·010		—	
Na	—		—		0·046		—		—	
Ca	—		0·036		0·146		—		—	
K	—		—		0·146		—		—	
fe*	4·0		5·1		8·2		8·5		8·7	

[a] Includes Cr^{3+} 0·046, Ni^{2+} 0·014.
[b] Includes Be^{3+} 0·052.

* fe = $100(Fe^{2+}+Fe^{3+})/(Fe^{2+}+Fe^{3+}+Mg)$.

1 Sapphirine, Fiskenaesset, west Greenland (Bøggild, 1953). Anal. Ussing (includes loss on ignition 0·31).

2 Sapphirine, Bekily, Madagascar (Sahama *et al.*, 1974). Anal. O. von Knorring (*a* 11·284, *b* 14·438, *c* 9·955 Å, β 125·44°).

3 Sapphirine, enstatite–spinel–sapphirine rock, Mawson area, MacRobertson Land, Antarctica (Segnit, 1957). Anal. P. C. Hemingway and R. N. Lewis (includes loss on ignition 1·90).

4 Sapphirine, sillimanite-mullite–corundum–sapphirine rock, xenolith in critical zone, Bushveld complex, Burgersfoot South Africa (Cameron, 1976). Average of six analyses corrected to 100 per cent (includes Cr_2O_3 0·50, NiO 0·15). Anal. W. E. Cameron.

5 Sapphirine, sakénite, Sakena, Madagascar (Lacroix, 1940). Anal. M. Raoult (includes Be_2O_3 0·16).

6	7	8	9	10	
12·2	16·0	12·2	15·6	16·85	SiO_2
0·02	0·01	—	—	0·06	TiO_2
62·1	58·0	63·1	61·1	56·86	Al_2O_3
—	—	2·3	1·2	1·86	Fe_2O_3
3·3	3·90	1·5	2·3	2·34	FeO
0·08	0·04	—	—	0·06	MnO
18·0	21·5	19·1	18·9	19·50	MgO
0·04	0·52	—	—	0·34	CaO
0·01	—	—	—	0·14	Na_2O
0·01	—	—	—	0·08	K_2O
—	—	0·37 (H₂O⁺ + H₂O⁻)	1·3 (H₂O⁺ + H₂O⁻)	1·84 (H₂O⁺ + H₂O⁻)	H_2O^+
—	—				H_2O^-
100·7	100·27	98·61	100·4	99·96	Total
—	—	1·713	1·705	1·702	α
—	—	1·716	1·710	—	β
—	—	1·718	1·715	1·708	γ
—	—	65°	70°	—	$2V_\alpha$
—	—	—	—	8°	$\gamma:z$
—	—	3·518	—	$>3·58<3·60$	D
1·435 } 6·00	1·879 } 6·00	1·450 } 6·00	1·834 } 6·00	2·004 } 6·00	Si
4·565	4·121	4·550	4·166	3·996	Al
4·073 } 8·04[a]	3·919 } 8·10[b]	4·308 } 8·05[c]	4·284 } 7·98	3·964 } 7·94	Al
0·002	0·001	—	—	0·004	Ti
—	—	0·204	0·112	0·166	Fe^{3+}
3·166	3·695	3·382	3·330	3·480	Mg
0·326	0·383	0·150	0·254	0·232	Fe^{2+}
0·008	0·004	—	—	0·004	Mn
0·002	—	—	—	0·032	Na
0·005	0·066	—	—	0·042	Ca
0·001	—	—	—	0·012	K
9·3	9·4	9·5	9·9	10·3	fe*

[a] Includes Cr^{3+} 0·457, Zn^{2+} 0·003, Ni^{2+} 0·001.
[b] Includes Cr^{3+} 0·010. Ni^{2+} 0·019.
[c] Includes Cr^{3+} 0·004.

6 Chromian sapphirine, associated with picotite, corundum, chromian hornblende, phlogopite and plagioclase, Lower Angnertussoq, west Greenland (Herd, 1973). Includes Cr_2O_3 4·9, ZnO 0·03, NiO 0·01, V_2O_5 0·01; total iron as FeO.
7 Sapphirine, entatite-spinel-sapphirine, layer between peridotite and pyroxenite, Finero, western Italian Alps (Lensch, 1971). Includes Cr_2O_3 0·10, NiO 0·20.
8 Sapphirine, corundum–kornerupine–sapphirine assemblage, amphibolite, Kittilä, Finnish Lapland (Haapala *et al.*, 1971). Anal. J. Siivola (includes Cr_2O_3 0·04).
9 Sapphirine, cordierite–anthophyllite–kornerupine–phlogopite rock, Lherz, Ariège, France (Monchoux, 1972).
10 Sapphirine, amphibolite, Haut Allier, central France (Forestier and Lasnier, 1969). Includes P_2O_5 0·03.

Table 64. Sapphirine Analyses – *continued*

	11	12	13	14	15
SiO_2	14·79	14·1	15·19	11·2	12·95
TiO_2	0·12	—	0·25	<0·04	—
Al_2O_3	58·01	61·6	61·69	66·3	62·38
Fe_2O_3	3·74	—	—	—	1·69
FeO	1·73	4·9	4·31	4·60	3·09
MnO	0·43	—	0·12	0·01	tr.
MgO	20·87	19·8	16·23	16·7	15·22
CaO	0·00	—	0·49	<0·05	—
Na_2O	0·04	—	—	—	tr.
K_2O	0·05	—	—	—	0·10
H_2O^+	—	—	1·60	—	4·80
H_2O^-	0·00	—	0·19	—	0·05
Total	99·78	100·4	100·07	99·3	100·28
α	1·725	1·709	—	—	1·714
β	—	1·716	—	—	1·719
γ	1·732	1·724	—	—	1·720
$2V_\alpha$	114°	65°	—	—	50·5°
$\gamma{:}z$	≃7°	—	—	—	6°
D	—	—	—	—	3·398
	Numbers of ions on the basis of twenty O				
Si	1·746 } 6·00	1·654 } 6·00	1·806 } 6·00	1·332 } 6·00	1·584 } 6·00
Al	4·254	4·346	4·194	4·668	4·416
Al	3·814	4·158	4·450	4·578	4·582
Ti	0·011	—	0·022	0·000	—
Fe^{3+}	0·332	—	—	—	0·156
Mg	3·668	3·456	2·874	2·948	2·776
Fe^{2+}	0·171 } 8·05	0·478 } 8·09	0·428 } 7·85	0·456 } 8·05[a]	0·316 } 7·85
Mn	0·043	—	0·012	0·002	—
Na	0·008	—	—	—	—
Ca	—	—	0·062	0·000	—
K	0·007	—	—	—	0·016
fe*	12·1	12·2	13·0	13·4	14·5

[a] Includes Cr^{3+} 0·042.

11 Sapphirine, enstatite–hornblende–sapphirine rock, Mautia Hill, Tanganyika (McKie, 1963). Anal. J. Ḣ. Scoon (trace element data; *a* 9·85, *b* 28·6, *c* 9·96 Å, β 110°30′).

12 Sapphirine, bronzite–cordierite–phlogopite–sapphirine rock, Nababeep district, Namaqualand, South Africa (Clifford *et al.*, 1975). Anal. E. F. Stumpfl (total Fe as FeO).

13 Sapphirine, Val Codera, Italy (Cornelius and Dittler, 1929) (mean of the two analyses).

14 Sapphirine, metagabbro, xenolith in kimberlite, Stockdale, Kansas (Meyer and Brookins, 1976). Includes Cr_2O_3 0·45.

15 Sapphirine, associated with corundum and biotite in granitic rock, Blinkwater, northern Transvaal (Mountain, 1939).

16	17	18	19	20	21	
14·42	15·26	13·16	16·82	16·00	13·76	SiO_2
0·06	0·04	0·12	0·40	0·22	0·12	TiO_2
58·95	61·93	62·32	57·10	55·56	57·21	Al_2O_3
1·27	—	—	6·32	2·72	10·44	Fe_2O_3
5·43	6·61	7·95	2·84	8·35	2·97	FeO
0·02	0·21	0·02	0·02	0·02	0·35	MnO
18·75	15·61	16·18	13·83	17·16	14·73	MgO
0·00	0·00	—	0·30	—	0·00	CaO
—	0·22	—	0·25	—	—	Na_2O
—	0·02	—	0·15	—	—	K_2O
} 0·25	—	—	1·66	0·14	0·33	H_2O^+
}	—	—	0·14	—	0·08	H_2O^-
99·84	100·00	99·75	99·84	100·17	99·99	**Total**
1·720	—	—	1·726	1·717	1·731	α
1·723	1·720	—	—	—	1·741	β
1·725	—	—	1·734	1·724	1·743	γ
48°	≃ 50°	—	—	64°	40°	$2V_\alpha$
—	7°	—	—	—	—	$\gamma{:}z$
—	3·554	—	—	3·58	3·60	D
1·710 } 6·00	1·802 } 6·00	1·570 } 6·00	2·027 } 6·00	1·924 } 6·00	1·663 } 6·00	Si
4·290	4·198	4·430	3·973	4·076	4·337	Al
3·944	4·428	4·334	4·139	3·806	3·821	Al
0·006	0·004	0·011	0·036	0·020	0·011	Ti
0·114	—	—	0·572	0·246	0·950	Fe^{3+}
3·312	2·748	2·877	2·484	3·078	2·655	Mg
0·538 } 8·10[a]	0·654 } 7·91	0·793 } 8·02	0·287 } 7·64	0·840 } 7·99	0·301 } 7·77	Fe^{2+}
0·002	0·022	0·002	0·002	0·002	0·036	Mn
—	0·050	—	0·058	—	—	Na
—	—	—	0·038	—	—	Ca
—	0·002	—	0·023	—	—	K
16·4	19·2	21·6	25·7	26·1	32·0	fe*

[a] Includes Be^{2+} 0·186.

16 Beryllian sapphirine, metasomatic zone in pyroxenite, Razor Hill, Musgrave Ranges, central Australia (Wilson and Hudson, 1967). Anal. L. J. Sutherland (includes BeO 0·65, P_2O_5 0·04).

17 Sapphirine, sapphirine–biotite–feldspar schist, Dangin, West Australia (Prider, 1945). Anal. Govt. Chem. Lab. W. Australia (includes P_2O_5 0·10; recalc. after removal of loss on ignition 0·57, Fe_2O_3 1·70).

18 Sapphirine, högbomite–kornerupine–surinamite–sapphirine granulite, north-eastern Strangways Range, central Australia (Woodford and Wilson, 1976). Anal. P. J. Woodford (trace element data).

19 Sapphirine, hypersthene–cordierite–sillimanite–biotite–garnet granulite. Anabar massif, U.S.S.R. (Lutts and Kopaneva, 1968). Includes F 0·01.

20 Sapphirine, hypersthene–cordierite–spinel rock, band in hypersthene granulite, Ganguvarpatti, Madura, India (Muthuswami, 1949). Anal. J. H. Scoon.

21 Sapphirine, charnockite, Labwor Hills, Uganda (Sahama *et al.*, 1974). Anal. P. Ojanperä.

Twenty-one sapphirine analyses are presented in Table 64. The number of silicon atoms (on the basis of twenty oxygens) varies from 1·3 to 2·0, (Al,Fe^{3+}) atoms from 7·9 to 9·3, and (Mg,Fe^{2+}) atoms from 2·8 to 4·1. This compositional range is in excess of that represented by the formulae above, and indicates that the solid solution series between $Mg_4Al_8Si_2O_{20}$ (= $2MgO,2Al_2O_3.SiO_2$) and $Mg_{3\cdot5}Al_9Si_{1\cdot5}O_{20}$ (= $7MgO.9Al_2O_3.3SiO_2$) extends towards a theoretical end-member

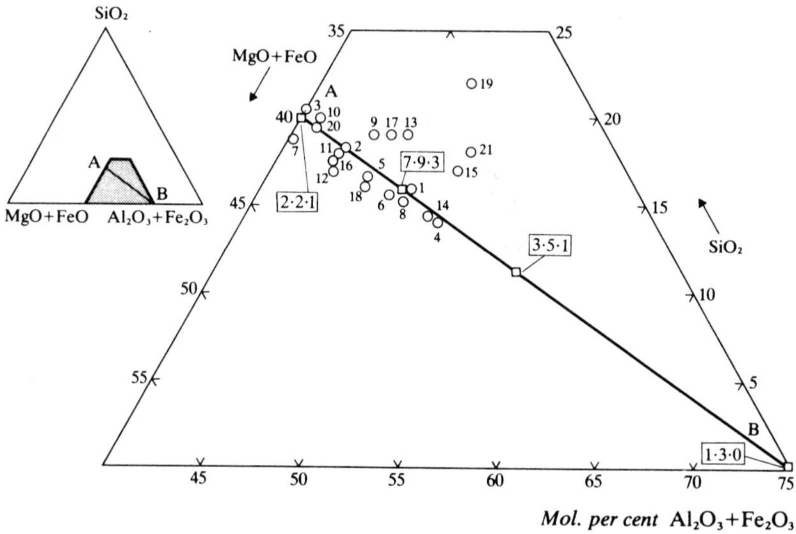

Fig. 271. Sapphirine compositions showing variation of (MgO + FeO), ($Al_2O_3 + Fe_2O_3$) and SiO_2 (mol. per cent). Numbered circles refer to analyses in Table 64. Open squares, oxide ratios of some hypothetical sapphirine compositions. Line A–B shows substitution $(Mg,Fe^{2+})Si \rightleftharpoons Al(Fe^{3+})$.

composition (Fig. 271) $Mg_3Al_{10}SiO_{20}$ (= $3MgO.5Al_2O_3.SiO_2$). Chromium is reported in many sapphirine analyses and is probably present in most sapphirines, and a chromium-rich variety (Table 64, anal. 6) has been described by Herd (1973) from a west Greenland locality. An unusual sapphirine (anal. 16) with a relatively high beryllium content is reported by Wilson and Hudson (1967). Trace element data are given by Nalivkina (1961), McKie (1963), Barker (1964a), Zotov (1966), Herd (1973) and Woodford and Wilson (1976).

Experimental

The first synthesis of sapphirine, by repeated sintering of appropriate oxide mixtures at atmospheric pressure at about 1450°C and its incongruent melting to spinel and a liquid of cordierite composition at approximately 1475°C was reported by Foster (1950). This investigation also showed that at subsolidus temperatures the stability relations of sapphirine with spinel, mullite and corundum change at about 1460°C, with spinel and mullite as the compatible phases above, and sapphirine and corundum the more stable pair below this temperature (Fig. 272). The presence in the system $MgO–Al_2O_3–SiO_2$ of a small sapphirine stability field predicted by Foster was subsequently confirmed by Keith and Schairer (1952) and later by Schreyer and Schairer (1961). These authors gave the temperatures of the ternary reaction points, sapphirine–spinel–mullite–liquid, sapphirine–mullite–cordierite–

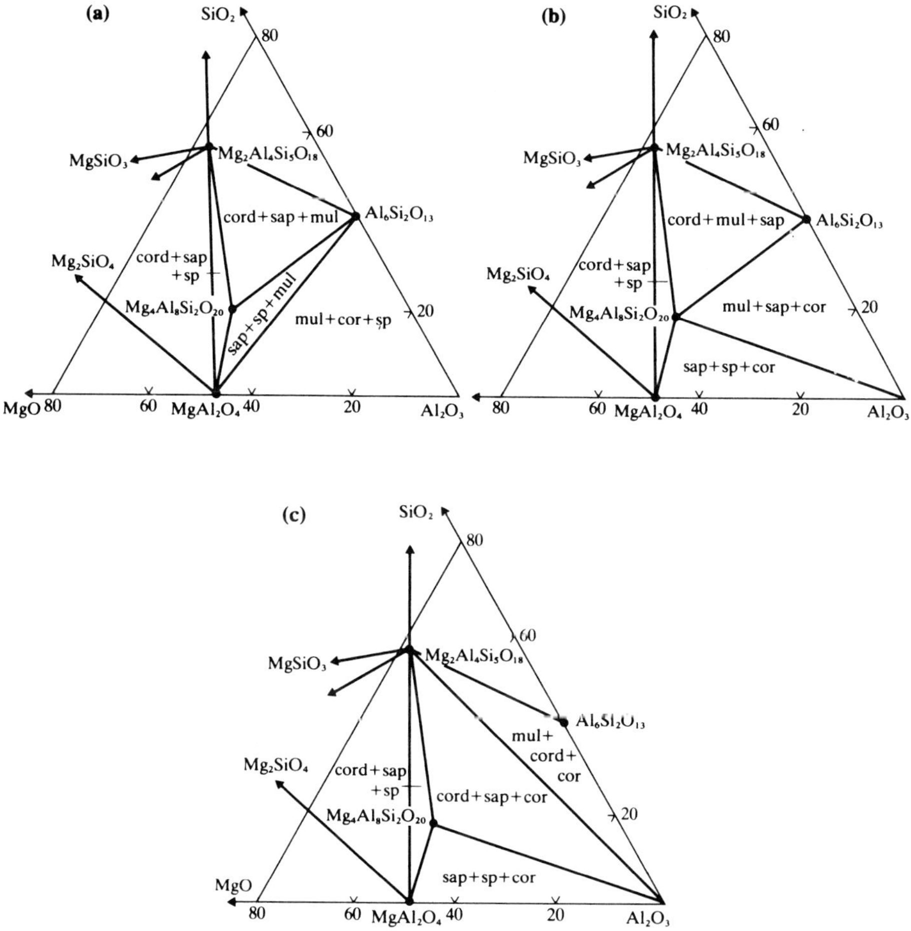

Fig. 272. Phase distributions in part of the system $MgO–Al_2O_3–SiO_2$ showing three phase regions. (a) inferred from the liquidus invariant points; mullite–spinel join stable only above 1460°C. (b) as (a) but showing a stable join between sapphirine and mullite which exists between 1386° and 1460°C. (c) below 1386°C. Solid solution of cordierite, sapphirine, mullite, and spinel omitted (after Smart and Glasser, 1976).

liquid, and sapphirine–spinel–cordierite–liquid as 1482°, 1460° and 1453°C respectively (Fig. 273). Substantial agreement with these temperatures of the liquidus invariant points was obtained by Smart and Glasser (1976). Their investigation of the system $MgO–Al_2O_3–SiO_2$ at 1 atm (Fig. 272) in addition to confirming the reversibility at 1386°C of the reaction:

$$\underset{\text{sapphirine}}{Mg_2Al_4SiO_{10}} + \underset{\text{mullite}}{2Al_6Si_2O_{13}} \rightleftharpoons \underset{\text{corundum}}{6Al_2O_3} + \underset{\text{cordierite}}{Mg_2Al_4Si_5O_{18}}$$

also showed that the addition of Na_2CO_3, $CaCO_3$ and Fe_2O_3, in amounts equivalent to 1·5 weight per cent Na_2O, CaO and Fe_2O_3, to a sinter consisting of sapphirine, cordierite and corundum and heated in the temperature range 1360° to 1410°C has little effect on the kinetics of the reaction:

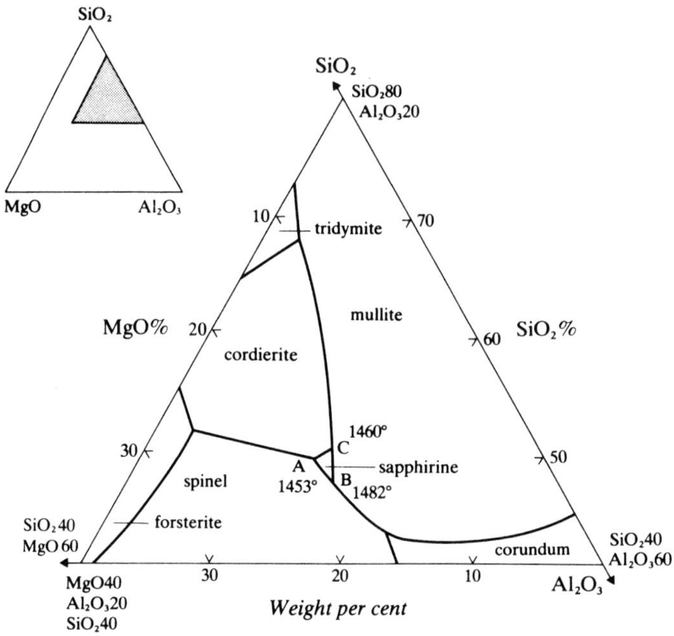

Fig. 273. Part of the system $MgO–Al_2O_3–SiO_2$ at 1 atm showing coexistence of sapphirine + spinel + cordierite + liquid at 1453°C (A), sapphirine + mullite + spinel + liquid at 1482°C (B), sapphirine + cordierite + mullite + liquid at 1460°C (C) (after Schreyer and Schairer, 1961).

$$\underset{\text{cordierite}}{2Mg_2Al_4Si_5O_{18}} + \underset{\text{corundum}}{6Al_2O_3} + \underset{\text{sapphirine}}{Mg_2Al_4SiO_{10}} \rightleftharpoons$$

$$\underset{\text{cordierite}}{Mg_2Al_4Si_5O_{18}} + \underset{\text{mullite}}{2Al_6Si_2O_{13}} + \underset{\text{sapphirine}}{2Mg_2Al_4SiO_{10}}$$

The stability of sapphirine at high temperatures and pressures has been examined by a number of workers. Earlier investigations include those of Boyd and England (1959, 1962) who showed that sapphirine + aluminous enstatite + sillimanite occur as breakdown products of pyrope, $Mg_3Al_2Si_3O_{12}$, at pressures and temperatures between 17 and 25 kbar and 1000° and 1600°C respectively (Fig. 274).

The stability relationship of sapphirine and quartz has been investigated by Hensen (1972), Newton (1972) and Chatterjee and Schreyer (1972). The invariant pressure–temperature curve for the reaction:

Al-enstatite solid solution + sillimanite ⇌ sapphirine solid solution + quartz has been defined by four reversals at pressures between 15 and 18 kbar and 1140°C and 1400°C respectively (Chatterjee and Schreyer, 1972), and at 13 kbar and 1100°C (Hensen); the average slope of the reaction in this temperature range is 0·5 bar/°C. At lower temperatures, due to the small values of ΔV and ΔS of the reaction, the curve is much flatter, and from a starting mixture of cordierite composition Newton (1972) has shown that at temperatures and pressures below about 800°C and 7 kbar both the sapphirine + quartz and enstatite + sillimanite + quartz assemblages are replaced by cordierite (Fig. 275):

$$Mg_2Al_4Si_5O_{18} \rightleftharpoons Mg_2Al_4SiO_{10} + 4SiO_2$$
$$Mg_2Al_4Si_5O_{18} \rightleftharpoons 2MgSiO_3 + 2Al_2SiO_5 + SiO_2$$

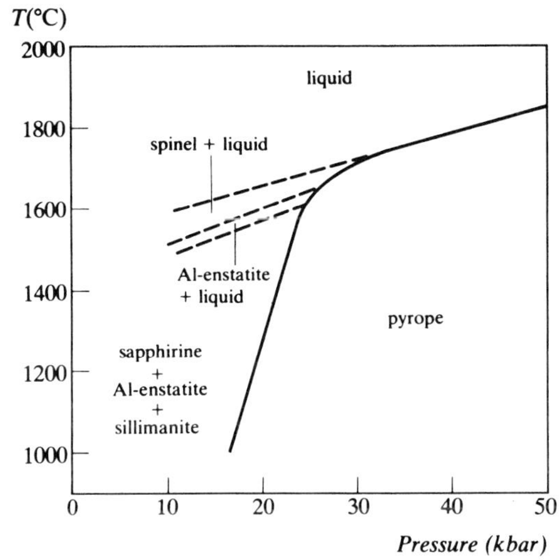

Fig. 274. Stability fields of sapphirine + Al-enstatite + sillimanite and pyrope for $Mg_3Al_2Si_3O_{12}$ composition (after Boyd and England, 1962).

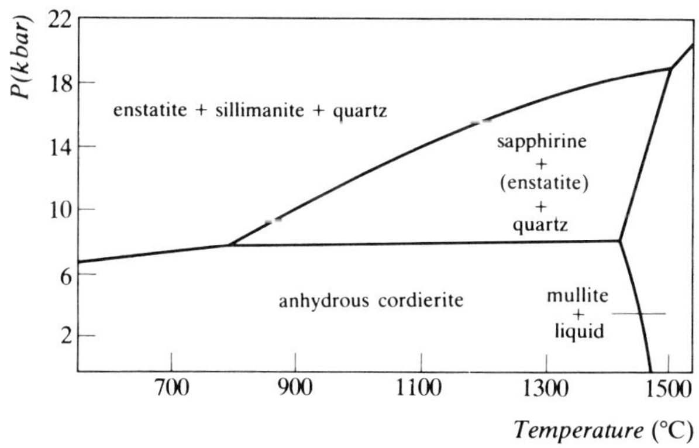

Fig. 275. Phase relations for $Mg_2Al_4Si_5O_{18}$ (cordierite) composition in the dry system (after Newton, 1972).

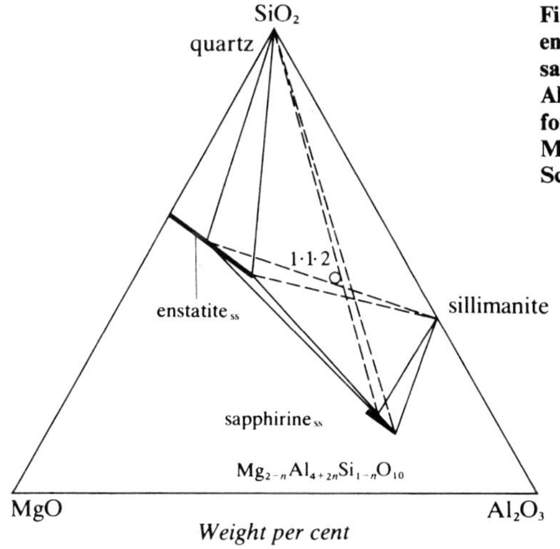

Fig. 276. Reaction relationships between enstatite solid solutions + sillimanite and sapphirine solid solutions + quartz in the $MgO–Al_2O_3–SiO_2$ system. Broken tie-lines stable only for univariant conditions. Open circle, composition $MgO.Al_2O_3.2SiO_2$ (after Chatterjee and Schreyer, 1972).

The effect of Mg,Si$\rightleftharpoons$2Al substitution in both sapphirine and enstatite on the reaction:

$$2MgSiO_3 + 2Al_2SiO_5 \rightleftharpoons Mg_2Al_4SiO_{10} + 3SiO_2$$

is illustrated in Fig. 276 and shows that as the amount of substitution increases the tie-line intersections between sapphirine and quartz and enstatite and sillimanite are displaced from the 1:1:2 ($MgO.Al_2O_3.2SiO_2$) composition.

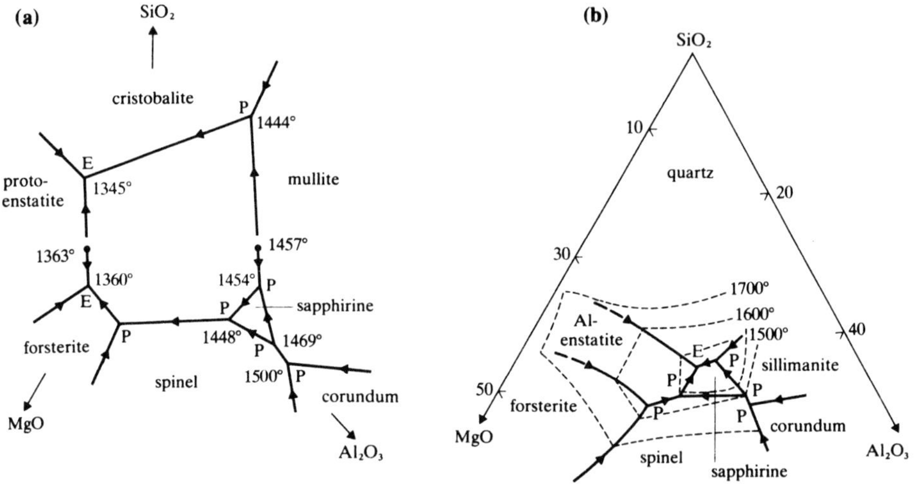

Fig. 277. (a) Schematic diagram showing liquidus invariant points bordering the primary phase fields of sapphirine and cordierite in the $MgO–Al_2O_3–SiO_2$ system (after Smart and Glasser, 1976). (b) Liquidus relations in part of the system $MgO–Al_2O_3–SiO_2$ at 15 kbar showing primary phase fields (after Taylor, 1973). E eutectic, P peritectic points.

The primary phase field in the system $MgO–Al_2O_3–SiO_2$ at 15 kbar (Fig. 277) is more extensive than at 1 atm, and the assemblage sapphirine solid solution + Al-enstatite solid solution ($\simeq$ 19 weight per cent Al_2O_3) + quartz is stable immediately below the eutectic temperature at 1430°C (Taylor, 1973).

A number of reactions relating to the stability of sapphirine at high P_{H_2O} (15–20 kbar) have been investigated by Schreyer and Yoder (1964), Schreyer (1968), Schreyer and Seifert (1969a, b), Seifert (1974) and Ackermand *et al.* (1975). The lower temperature stability at high water pressures (Fig. 278) given by the reaction:

$$\underset{\text{chlorite}}{Mg_4Al_2(Si_2Al_2)O_{10}(OH)_8} + \underset{\text{corundum}}{2Al_2O_3} \rightleftharpoons \underset{\text{sapphirine}}{2Mg_2Al_4SiO_{10}} + 4H_2O$$

is dependent on chlorite having an amesitic composition, and the more general reaction of the breakdown of sapphirine at low temperature is:

$$\underset{\text{chlorite}}{Mg_5Al(Si_3Al)O_{10}(OH)_8} + \underset{\text{corundum}}{4Al_2O_3} + \underset{\text{spinel}}{MgAl_2O_4} \rightleftharpoons \underset{\text{sapphirine}}{3Mg_2Al_4SiO_{10}} + 4H_2O$$

The upper stability of sapphirine at high pressure is defined by the reaction:

$$\underset{\text{sapphirine}}{3Mg_2Al_4SiO_{10}} \rightleftharpoons \underset{\text{pyrope}}{Mg_3Al_2Si_3O_{12}} + \underset{\text{spinel}}{3MgAl_2O_4} + \underset{\text{corundum}}{2Al_2O_3}$$

and this equilibrium has been reversed from the bulk composition $2MgO.2Al_2O_3.SiO_2.xH_2O$. The breakdown reaction has also been investigated by Doroshev and

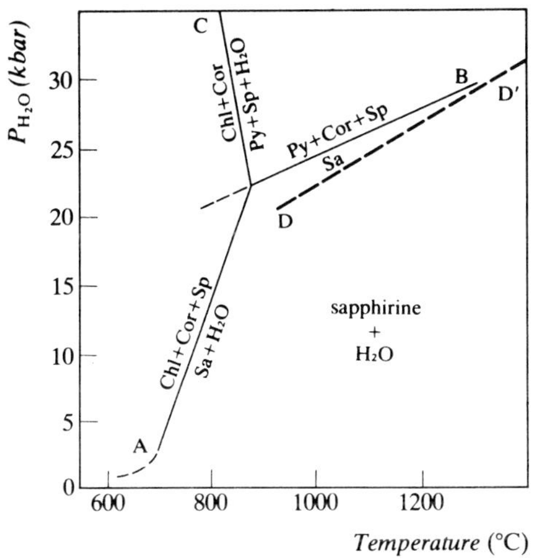

Fig. 278. P_{H_2O}–T diagram of the equilibrium curves for the reactions, A, chlorite + corundum + spinel ⇌ sapphirine + H_2O. B, pyrope + corundum + spinel ⇌ sapphirine. C, chlorite + corundum ⇌ pyrope + spinel + H_2O (after Ackermand *et al.*, 1975). Dashed part of curve B only stable in a water-deficient environment. D–D′, reaction curve pyrope + corundum + spinel ⇌ sapphirine (Doroshev and Malinovskii, 1974).

Malinovskii (1974). Their data show slightly lower pressures, in the temperature range 900° to 1250°C, at which sapphirine is stable (Fig. 278), and a *T–P* equation $T(°C) = 56{\cdot}25P$ (kbars) $- 344$. Ackermand *et al.* have also defined the curve of the reaction chlorite + corundum ⇌ pyrope + spinel + H_2O, and the invariant point chlorite + spinel + corundum + pyrope + sapphirine + H_2O is located at 880°C, 22 kbar. Although under these *P–T* conditions Mg-staurolite (Schreyer, 1968) is stable in its own compositional field, the assemblage pyroxene + corundum + H_2O was found to be more stable than the association of sapphirine + Mg-staurolite at temperatures and pressures close to the invariant point. The stability fields shown in Fig. 278 relate to $P_{H_2O} = P_{total}$; the high pressure reaction, however, is not dependent on f_{H_2O}, and the sapphirine breakdown curve in the absence of water may extend to lower temperatures than that of the invariant point with the consequent enlargement of the stability field of sapphirine.

Two assemblages, both containing sapphirine, were found to occur during the investigation by Hensen and Green (1970) of the high *P–T* phase relationships of synthetic compositions approximating to natural metasediments. The sapphirine–garnet–hypersthene–quartz assemblage first appeared at a pressure of ≃ 8 kbar and 1050°C, its stability field expanding to higher pressures with increasing temperature. The other assemblage, sapphirine–hypersthene–quartz crystallized at higher temperatures and lower pressure (Fig. 279(a)). The stability of sapphirine in association with quartz shown by their high pressure–temperature assemblage is substantiated by the increasing amount of sapphirine and decreasing content of sillimanite obtained from mixtures of garnet and sillimanite containing smaller amounts of sapphirine and hypersthene held at 9 kbar and 1100°C. Sapphirine has also been found to crystallize under comparable *P–T* conditions from natural pelitic compositions with $100Mg/(Mg+Fe^{2+})$ ratios of 70 and 30 (Hensen and Green, 1971). *P–X* diagrams of the phase relations for these natural compositions at 800°, 900°, 1000°, 1035° and 1100°C (Fig. 279(b)) show that sapphirine first appeared at a temperature of 1035°C and pressures above about 7·5 kbar, and that its stability field is enlarged as the temperature is increased.

The enthalpies of solution (in a lead borate, $2PbO.B_2O_3$, melt at 970°K) of an iron-free, $Mg_{1\cdot75}Al_{4\cdot5}Si_{0\cdot75}O_{10}$, synthetic and natural sapphirine (2·4 weight per cent FeO), and the enthalpies of formation from their oxides have been determined

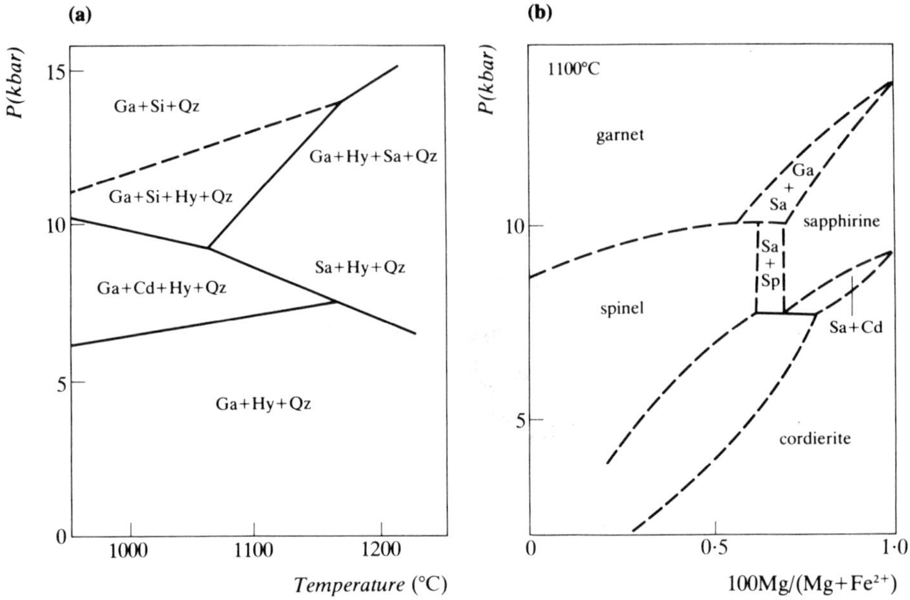

Fig. 279. (a) Phase relationships for composition SiO_2 52·82, Al_2O_3 25·43, FeO 6·83, MgO 8·95, CaO 2·42, Na_2O 1·18, K_2O 1·35 (after Hensen and Green, 1970). Sa, sapphirine, Cd, cordierite, Ga, garnet, Si, sillimanite, Hy hypersthene, Qz, quartz. (b) *P–X*, (100 Mg/Mg + Fe^{2+}), diagram based on compositions SiO_2 50·94 (52·15), Al_2O_3 32·31 (33·26), FeO, 10·13 (4·49), MgO 2·44 (5·86), CaO 1·63 (1·69) Na_2O, 1·18 (1·18), K_2O 1·35 (1·35). Phase boundaries approximate (after Hensen and Green, 1971).

by Charlu *et al.* (1975). Comparison of the enthalpies of formation of the synthetic material ($\Delta H_f = -38{\cdot}84$ kcal/mole) and natural sapphirine ($\Delta H_f = -45{\cdot}88$ kcal/mole) shows that the former is not a good stability model for the latter. The lower enthalpy of formation of the high-temperature synthetic phase is probably related to cation disordering.

Optical and Physical Properties

The variation of the optical properties with chemical composition of sapphirine is related mainly to the iron content. Thus the refractive indices increase, γ from about 1·705 to 1·745, as the value $100(Fe^{2+} + Fe^{3+})/(Mg + Fe^{2+} + Fe^{3})$ increases from 4 to 35 (Fig. 280). For those minerals with a high Fe^{3}/Fe^{2+} ratio the correlation is only moderate, possibly due in part to the highly refractory nature of sapphirine and the consequent low values of FeO returned in some analyses. The negative optic axial angle decreases as the iron content increases, but much smaller angles than those shown in Fig. 280 have been reported (e.g. Herd, 1973). Both Segnit (1957) and McKie (1963) have described sapphirines with a positive 2V, and considerable fluctuations in optic axial angle for sapphirines from a single locality are not unusual. The extinction angle, $\gamma{:}z$, generally varies between 6° and 11° for natural minerals, but values as high as 20° have been reported by Zotov (1966), and between 13° and 15° for synthetic Mg-sapphirines; there is an approximate correlation of extinction angle with iron content, $\gamma{:}z$ decreasing with increasing tenor of iron.

The usual blue colour of sapphirine is not universally present; yellow sapphirine

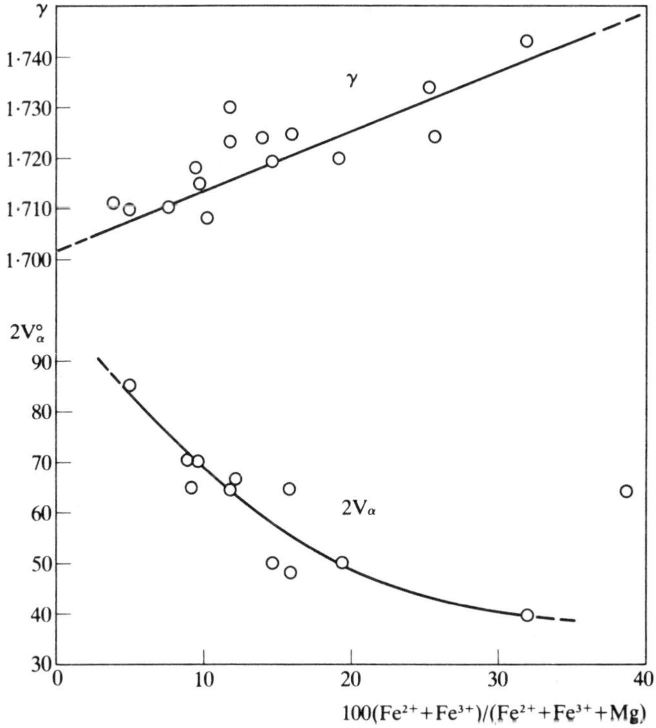

Fig. 280. Variation of γ refractive index and $2V_\alpha$ with $100(Fe^{2+}+Fe^{3+})/(Fe^{2+}+Fe^{3+}+Mg)$.

has been described by McKie (1963), and pale red sapphirine by Cameron (1976). Pleochroism is usually strong with $\gamma > \beta > \alpha$, but the pleochroic intensity may show considerable variation within the same occurrence; non-pleochroic sapphirine has been described by Zotov (1966) and Christophe-Michel-Lévy (1962). Exceptions to the common pleochroism dominated by various shades of blue are not unusual, and McKie has suggested that the absorption colours β = pale yellow, γ = pale purplish orange of the Mautia Hill mineral may be related to its high Fe^{3+}/Fe^{2+} ratio $[100Fe^{3+}/(Fe^{3+}+Fe^{2+}) = 66]$. Other sapphirines with comparable oxidation ratios, however, display the usual blue pleochroism. The Mautia sapphirine has a relatively high content of manganese (0·43 weight per cent MnO), and McKie has suggested that the chromophore group may be Fe^{3+}–O–Mn^{2+}, but the Labwor Hills blue sapphirine (Sahama *et al.*, 1974) also has a high oxidation ratio and only a slightly lower content of manganese (0·35 weight per cent MnO). The infrared absorption spectra of two sapphirines from Madagascar have been investigated by Povarennykh (1970), and are interpreted in terms of the sapphirine structure.

Inclined dispersion of the optic axial angle is generally strong with $v > r$. The dispersion of an iron-rich sapphirine (9·0 weight per cent FeO) described by Nixon *et al.* (1973) is given as $v < r$; this mineral, however, has variable $2V_\alpha$ (10°–30°), shows incomplete extinction, as well as differences in colour intensity and magnetic susceptibility within a single specimen. Abnormal interference colours, $\perp\alpha$, in a mineral showing strong dispersion, have been described by de Roever (1973). The dispersion of a sapphirine from Strangways Range (Hudson and Wilson, 1966) is $2V_\alpha$ 44° (red) and 52° (blue), and for a specimen from Val Codera (Barker, 1964a), 52° and 59° for sodium and white light respectively.

Bancroft *et al.* (1968) have investigated the Mössbauer spectra of two sapphirines, a blue-green specimen from Sukkertoppen, west Greenland, and the yellow sapphirine from Mautia Hill, Tanzania. The spectra indicate that the Fe^{2+} ions are located in two six-co-ordinated sites, and that Fe^{3+} occupies two four-co-ordinated sites, in conformity with McKie's (1963) suggestion that the Fe^{3+} ions order on tetrahedral and Fe^{2+} ions on octahedral sites.

Simple, multiple and interpenetration twins in sapphirine have been described. In simple twins the twin plane has been reported as (100) by Zotov (1966), and for multiple twins as (010) by Meng and Moore (1972). As sapphirine is sometimes developed with a tabular habit||(010) (see Mountain, 1939) the 'multiple twinning' may, however, originate from quasi-parallel tabular growth (Sahama *et al.*, 1974).

Distinguishing Features

Sapphirine is distinguished from corundum by its biaxial character and lower refractive indices, from kyanite by poorer cleavages and lower birefringence, from cordierite by higher refractive indices, and from blue alkali amphiboles and zoisite by poorer cleavage and higher specific gravity.

Paragenesis

Sapphirine is found in a wide range of assemblages, in which its common associates include sillimanite, corundum, cordierite, spinel, phlogopite, gedrite, potassium feldspar and plagioclase. The formation of sapphirine is restricted very largely to high grade regional metamorphic environments. In many sapphirine localities, metasomatism, involving the introduction of potassium and water into ultrabasic and basic rocks, has played an active role. The presence of aluminium and magnesium-rich phases, or an original bulk composition with similar characteristics of high Al and Mg, is an important factor in providing a suitable milieu for the production of sapphirine. The mineral is found rarely as a product of thermal metamorphism, and in this paragenesis is confined to xenoliths that have been in contact with high temperature basic liquid. Many sapphirine-bearing assemblages contain an unusually large number of phases, evidence of disequilibrium associations that record changing conditions of temperature and pressure, due in some cases to the development of reaction rims that have isolated intermediate and unstable products. An early summary of sapphirine localities has been given by Sørensen (1955) and a more recent review of the literature has been presented by Monchoux (1972).

Most sapphirine-bearing assemblages occur in high grade metamorphic terrains belonging to the hornblende granulite facies. In many cases their parageneses have also involved an element of metasomatism commonly resulting from the migration of potassium and silica derived from acid and intermediate gneisses adjacent to the basic and ultramafic horizons in which the sapphirine-bearing rocks developed. West Greenland is one of the principal regions in which sapphirine occurs, and many localities between Sukkertoppen and Fiskenaesset have been described. Thus sapphirine, associated with aluminium amphiboles, spinel, phlogopite and anorthite, occurring in lenticular masses of original ultramafic rocks, and enclosed in granodioritic gneisses of granulite to amphibolite facies grade, have been described from a number of localities in the Sukkertoppen area. The assemblages are generally regarded to have resulted from the transfer of K and Si from the gneisses to the ultramafic rocks (Ramberg, 1948), as illustrated by the reactions:

$$\underset{\text{anorthite}}{2{\cdot}5CaAl_2Si_2O_8} + \underset{\text{enstatite}}{10MgSiO_3} + \underset{\text{spinel}}{6MgAl_2O_4} + K_2O + 3H_2O \rightarrow$$

$$\underset{\text{Al-hornblende}}{Ca_{2\cdot5}Mg_4Al_3Si_6O_{22}(OH)_2} + \underset{\text{sapphirine}}{3Mg_2Al_4SiO_{10}} + \underset{\text{phlogopite}}{2KMg_3AlSi_3O_{10}(OH)_2}$$

$$Ca_{2\cdot5}Mg_4Al_3Si_6O_{22}(OH)_2 + 4MgAl_2O_4 + 6SiO_2 + K_2O + H_2O \rightarrow$$

$$2{\cdot}5CaAl_2Si_2O_8 + Mg_2Al_4SiO_{10} + 2KMg_3AlSi_3O_{10}(OH)_2$$

Later investigations by Herd *et al.* (1969) and Herd (1973), who have described many of the localities between Sukkertoppen and Fiskenaesset, are not at variance with Ramberg's general thesis. They concluded that many of the sapphirine-bearing rocks of the west Greenland area were formed from Mg- and Al-rich ultramafic rocks following the introduction of a hydrous potassium feldspar-rich component derived, during the mobilization accompanying the amphibolite facies event, from the neighbouring granodioritic or tonalitic gneisses.

In the Fiskenaesset area, sapphirine occurs in a layered meta-anorthosite–pyrobolite complex, at Sukkertoppen is found in meta-norites, and at Tasiussaq in chromite-rich dunites. The area has been affected by two periods of regional metamorphism, first under hornblende granulite and later by a regression to amphibolite facies conditions. The sapphirine rocks are most commonly located along discontinuous horizons of selective metasomatic activity. Four types of sapphirine-bearing assemblages have been recognized in the Fiskenaesset–Sukkertoppen region. The first development of sapphirine occurring as rims around spinel, and subsequently as individual laths, took place during a period of hornblende granulite metamorphism associated with silica metasomatism and the introduction of smaller amounts of potassium and water. A later metasomatic addition of calcium produced a further change in the original ultramafic assemblage that involved the replacement of enstatite and led to the characteristic association, sapphirine–spinel–pargasite–phlogopite. With the change in conditions to those of the cordierite amphibolite facies, gedrite became the stable amphibole phase and sapphirine was progressively altered to corundum and phlogopite until it is present solely as rims around corundum. The formation of sapphirine was not, however, solely dependent on the pressure–temperature conditions of the metamorphism and the associated metasomatism. Sapphirine is not universally present in all the ultramafic rocks of the area, and its development is restricted to those rich in both aluminium and magnesium.

A detailed account of an unusual zoned sapphirine-bearing pegmatoid body in the southern part of the Fiskenaesset complex is given by Rivalenti (1974). The complex assemblage of corundum, sillimanite, cordierite, plagioclase (An_{85}), phlogopite, anthophyllite, hornblende, spinel, talc, sapphirine, chlorite and dolomite is the result of overlapping parageneses and reactions that record changing physical conditions, particularly of P_{H_2O}, during the course of the formation of the pegmatite.

The occurrence of sapphirine in the spinel–enstatite zone of the banded skarns, Kukhilal, south-western Pamirs (Zotov, 1966) has some features in common with those of the west Greenland area. The assemblage, consisting of anthophyllite, phlogopite, enstatite, spinel and sapphirine, has developed from an original enstatite (5·5 weight per cent Al_2O_3)–spinel rock, in which the spinel, now present only as relics in sapphirine, has been largely replaced by the sapphirine which initially developed at the margins between enstatite and spinel, and which subsequently, together with enstatite, was largely replaced by anthophyllite. The formation of sapphirine is interpreted as a reaction product between spinel and Al-enstatite resulting from the introduction of silica derived from the granitization of deep-seated sialic rocks, and the consequent release of Mg to facilitate the formation of sapphirine.

A similar origin for the sapphirine-bearing assemblages in the Yuzhnyy Bug River region, has been put forward by Nalivkina (1961). Here the sapphirine occurs in zoned sapphirine–corundum–biotite–microcline (spinel), sapphirine–corundum–sillimanite–microcline (spinel), and sapphirine–sillimanite–corundum–kornerupine (var. prismatine)–microcline associations formed by the metasomatic alteration of pyroxenite during a period of regional granitization associated with the development of charnockites, in which the successive replacement of spinel by sapphirine and sillimanite, of sapphirine by prismatine, and sapphirine by microcline accompanied the metasomatic introduction of K, Si, B and possibly Al.

In their review of the sapphirine associations of the Aldan shield, Chekirda and Entin (1969) concluded that one of the main groups of sapphirine-bearing assemblages, enstatite–spinel–cordierite–phlogopite–sapphirine ($\pm$potassium feldspar), formed by the metasomatism of mafic igneous rocks of hypersthenite type. The sapphirine has replaced spinel and enstatite, and its subsequent replacement by cordierite is recorded in symplectite-like intergrowths of cordierite and sapphirine; some replacement of sapphirine by phlogopite has also taken place. A second group of sapphirine-bearing assemblages appears to be the result of a granulite facies metamorphism without metasomatic overtones that led to an assemblage consisting of cordierite, garnet, hypersthene, sillimanite and sapphirine, the latter as reaction rims around sillimanite, and also partially replaced by cordierite.

A similar mineral association is found in the granulites of the Kola Peninsula (Bondarenko, 1972). The sapphirine-bearing assemblage here, however, appears to have been derived by way of an initial isochemical metamorphic reaction of almandine–pyrope to orthopyroxene + spinel + cordierite from which the assemblage sapphirine + Al-hypersthene + cordierite + spinel + sillimanite + kyanite + plagioclase + Mg-biotite was formed during a later stage of metamorphic differentiation that involved an increase in alkalinity and water pressure.

The formation of a sapphirine-bearing assemblage, interpreted by Clifford *et al.* (1975), as the result of high pressure regional metamorphism of argillaceous sediments after the removal of some 80 per cent anatectic granitic liquid, has been described from the Nababeep district, South Africa. The sapphirine occurs mainly within granular aggregates of cordierite, and contains inclusions of phlogopite, cordierite and orthopyroxene. The rock is believed to show evidence of the earlier presence of a cordierite–phlogopite, that was later overprinted by a sapphirine–bronzite, assemblage as illustrated by the reaction:

$$4\ \text{phlogopite} + 2\ \text{cordierite} \rightarrow 2\ \text{sapphirine} + 3\ \text{bronzite} + \text{rutile} + \text{MgO}$$

The *P–T* environment in which the assemblage developed is estimated from the Al content of the orthopyroxene (7·1 weight per cent Al_2O_3) to have been $\simeq$ 1000°C, at a pressure, based on the association of sapphirine and cordierite rather than orthopyroxene and sillimanite, of somewhat less than 10 to 11 kbar.

Lenticular bodies of sapphirine-bearing mica schist in the granitic and granodioritic migmatites at Snaresund, southern Norway, have been described by Touret and Roche (1971). The rock, in addition to sapphirine, contains cordierite, spinel, gedrite, orthopyroxene and biotite, has high contents of alumina, magnesia and potash and is probably derived from an argillaceous sediment.

Sapphirine-bearing granulites, derived from silica-deficient ferruginous sediments, have been described from the granulite and amphibolite facies terrain of northern Uganda (Nixon *et al.*, 1973). The assemblages include orthopyroxene, sillimanite, cordierite, garnet, corundum, ilmenite and potassium feldspar. The Fe-rich (total iron as FeO = 9·0 per cent)–low silicon sapphirine appears to have formed mainly at the expense of spinel and garnet but, in view of the high Al/Si ratio of this mineral, sillimanite is also likely to have been involved in the reaction:

$$\underset{\text{spinel}}{7(Mg,Fe)Al_2O_4} + \underset{\text{sillimanite}}{2Al_2SiO_5} + SiO_2 \rightarrow \underset{\text{sapphirine}}{4(Mg,Fe)_{1\cdot 75}Al_{4\cdot 5}Si_{0\cdot 75}O_{10}}$$

the silica being provided by small amounts of orthopyroxene and garnet now pseudomorphed by sapphirine.

Sapphirine occurs as a constituent of the aluminium-rich granulitic gneisses, and in lenses of oxide minerals, titaniferous hematite, magnetite and spinel, enclosed in the gneisses, at Wilson Lake, Labrador (Morse and Talley, 1971; Moore and Meng, 1971; Meng and Moore, 1972). The textural relationships of the sapphirine are varied; the mineral sometimes occurs enclosing orthopyroxene and spinel, as rims between orthopyroxene and spinel or orthopyroxene and hematite, as inclusions in orthopyroxene and sillimanite, as well as forming symplectitic intergrowths with hematite. Alteration to boehmite and a chlorite group mineral is locally evident. The assemblage is interpreted by Morse and Talley as a retrograde one but as Meng and Moore have pointed out, interlayering of incompatible assemblages, as exemplified by the presence of the high and low pressure phases of the reaction orthopyroxene + quartz $\rightleftharpoons$ cordierite + sillimanite, occurs in the gneisses. The textural relationships between the principal constituents, particularly those involving sapphirine, could be explained by a combination of fluctuating physical conditions and arrested reactions. The association of a highly aluminous suite of silicates and spinel with hematite is unusual, and Meng and Moore consider that the Wilson Lake sapphirine-bearing and related rocks have been derived from aluminous and ferruginous sediments possibly of lateritic origin.

Sapphirine, högbomite, $\simeq Mg_{6\cdot 8}Fe_{1\cdot 5}Ti_{1\cdot 5}Al_{19}O_{40}$, kornerupine and surinamite, $(Mg_{1\cdot 25}Fe^{3+}_{0\cdot 40}Al_{1\cdot 35})(Al_{0\cdot 50}Si_{1\cdot 50})O_{7\cdot 24}(OH)_{0\cdot 76}$, in assemblages from the north-eastern Strangways Range, central Australia, have been described by Woodford and Wilson (1976). The sapphirine, högbomite and surinamite are considered to be derived from a high-alumina laterite or bauxite that has undergone granulite facies metamorphism; the kornerupine is confined to potassium-rich metasomatic zones that cut the granulite. Beryllium sapphirine (Table 64, anal. 16), in segregations with taaffeite, $BeMgAl_4O_8$, spinel, apatite and phlogopite, occurring in metasomatic zones, resulting from the introduction of K, F, P and Be into pyroxenite in the Musgrave Ranges, central Australia, is reported by Wilson and Hudson (1967).

Lacroix (1929, 1940, 1941, see also Boulanger, 1959; Christophe-Michel-Lévy, 1962; Foissy *et al.*, 1966) has described anorthite–pyroxene–sapphirine (Table 64, anal. 5) rocks (sapphirine sakénites), sapphirine–spinel sakénites, sapphirinites, spinel–sapphirine pyroxenites and sapphirine amphibolites (edenite and sapphirine) associated with sillimanite–cordierite–almandine paragneisses from Sakena, Madagascar. These sapphirine-bearing rocks are associated with horizons of meta-anorthosites containing chromite-bearing ultra-basics, and are similar in many respects to the anorthite–corundum–spinel rocks from west Greenland.

The sapphirine rocks at Madura, India (Muthuswami, 1949) have also been ascribed to metasomatic metamorphism. Here sapphirine is associated with cordierite, spinel, garnet, biotite and anthophyllite, the assemblage occurring as bands in hypersthene granulite. The spinel, in symplectite intergrowths with hypersthene around garnet porphyroblasts, is partially replaced by sapphirine which also occurs in intergrowths with cordierite.

Sapphirine, earlier probably mistaken as corundum, has been identified in the well known Waldheim kornerupine rock from the garnet–kyanite (sillimanite) granulites of Saxony (Schreyer, *et al.*, 1976). Two generations of sapphirine, showing only small differences in chemical composition, are present. The earlier phase is brown and is sometimes surrounded by a later colourless variety. The latter occurs in aggregates of very small elongated grains, the identical optic orientation of which may be due to the skeletal growth of larger crystals. Other phases in the granulite include plagioclase $\simeq Ab_{85}$, dumortierite, tourmaline and Mg-biotite.

Other sapphirine-bearing assemblages in granulite–charnockite terrains include those of the Quairading area, Western Australia (Wilson, 1962, 1965), Labwor Hills, Uganda (Sahama *et al.*, 1974), McRobertson Land (Segnit, 1957) and Enderby Land, Antarctica (Dallwitz, 1968). In the latter area, sapphirine, quartz and cordierite form up to 98 per cent of the rock, which includes sillimanite, garnet, hypersthene and rutile among the accessory minerals. The sapphirine is in direct contact with quartz and in strings of small grains in intergrowths of cordierite and quartz. Sapphirine has also been reported in ultrabasic rocks in the granulites of Bahia, Brazil (Fujimori and Allard, 1966), and in the mesoperthite gneiss, Bakhuis Mountains, Surinam (de Roever, 1973), where it also coexists with quartz.

The presence of small amounts of sapphirine in the Sittampundi anorthosite intrusion, Salen district, Tamil Nadu, India has been reported by Janardhanan and Leake (1974), see also Leake *et al.* (1976). Chrome spinel, gedrite and a clinoamphibole, as well as corundum and sillimanite are also present in the anorthosite (plagioclase $\simeq An_{73}$). The petrogenesis of this accessory association is uncertain, but it has been suggested that the sapphirine may have formed by incongruent melting of plagioclase at high water pressure with a high Mg^{2+} activity in the vapour phase. This explanation is based on the observations (Yardley and Blacic, 1976) that sapphirine formed as a product of the incongruent melting of plagioclase during the experimental uniaxial deformation of an anorthosite at 850°C and 10 kbar confining pressure.

Sapphirine is a not uncommon constituent in ultrabasic rocks that have not been metamorphosed at a higher grade than the amphibolite facies as at Grønøy, north Norway, where it is associated with kyanite and staurolite in lenses of foliated ultramafic rocks (Østergaard, 1969). Sapphirine is also associated with kyanite and staurolite, as well as together with corundum, in a pargasite–anorthite amphibolite, Allier valley, Massif Central (Forestier and Lasnier, 1969), and with corundum and kornerupine in amphibolite at Kittila, Finnish Lapland (Haapala *et al.*, 1971). The sapphirine-bearing rocks at Val Codera, occur in the root zone of the Alps, close to the Bergell granite. Here the assemblage, sapphirine–biotite–cordierite–hypersthene, has been derived from garnet and spinel following the metasomatic introduction of potassium into an ultrabasite (Cornelius and Dittler, 1929; Barker, 1964a).

The formation of a sapphirine–anthophyllite–phlogopite–kornerupine–cordierite rock at the contact of lherzolite and amphibolite, Lherz, France, at which the primary hornblende–spinel is partially replaced by the sapphirine–bearing assemblage has been described by Monchoux (1972). At Finero, western Italian Alps, sapphirine occurs, in association with enstatite and spinel and with enstatite, pyrope, pargasite, Al-diopside and bytownite, in layers between peridotite and pyroxenite (Lensch, 1971).

An unusual sapphirine association has been described from the Mount Painter area, South Australia (Oliver and Jones, 1965). In addition to sapphirine the rock consists of corundum, boehmite (γ-AlO.OH), cordierite and an aluminium-rich chlorite (corundophilite). Experimental data relating to the stability of Al-rich chlorite suggest that the sapphirine, cordierite, corundum and corundophilite were probably in equilibrium at the time of their formation and crystallized at a temperature between 500° and 700°C and a pressure of not less than 1·5 kbar, the boehmite forming subsequently during retrograde metamorphism at temperatures between 120° and 280°C.

The main constituents of the sapphirine-bearing rock, Mautia Hill, Tanzania (McKie, 1963) are enstatite En_{98} (4·15 weight per cent Al_2O_3), hornblende and sapphirine (Table 64, anal. 11); accessory minerals include pseudobrookite hematite and dolomite. The lack of evidence of any previous presence of either spinel or corundum, as well as the mutual relationships of the main constituents, indicate that the sapphirine crystallized in equilibrium with enstatite and hornblende. In some specimens oriented relics of sapphirine, enstatite and hornblende are enclosed in a matrix of talc, and have apparently been derived by a local retrogressive metasomatism. The physical conditions under which the main assemblage crystallized is uncertain. Sapphirine, however, is present in small amounts in the neighbouring enstatite–hornblende–chlorite–högbomite–dolomite rocks the overall compositions of which are consistent with their derivation from reaction between dolomite-marble and adjacent yoderite-bearing schist.

The occurrence of highly aluminous (67·9–68·6 weight per cent Al_2O_3) and iron-free sapphirine in a kyanite–gedrite–talc schist in the talc-rich horizon of the Sar e Sang lapis lazuli deposit, Afghanistan is described by Schreyer and Abraham (1975). Cordierite and corundum are also present in the schist and are interpreted as products of a later lower pressure reaction:

$$\underset{\text{talc}}{Mg_6Si_8O_{22}(OH)_4} + \underset{\text{kyanite}}{7Al_2SiO_5} \rightleftharpoons \underset{\text{cordierite}}{3Mg_2Al_4Si_5O_{18}} + \underset{\text{corundum}}{Al_2O_3} + 2H_2O$$

Narrow channels of cordierite within the schist are separated from kyanite by fine-grained felty crystals of sapphirine and corundum, the latter enclosing and apparently postdating the formation of the sapphirine. Like cordierite and corundum the sapphirine developed from the assemblage kyanite–gedrite or kyanite–talc. Its formation, however, is at variance with the phase compatibilities in the system $MgO–Al_2O_3–SiO_2–H_2O$ (Schreyer and Seifert, 1969a) which implies that the formation of either corundum and cordierite or cordierite and yoderite should occur. In addition, with regard to the original mineral constitution of the schist, the coexistence of both sapphirine and corundum with cordierite violates the phase rule. The textural evidence indicating that the formation of sapphirine occurred prior to corundum suggests that the sapphirine is probably an intermediate metastable product of the reaction:

aluminous anthophyllite + kyanite → cordierite + sapphirine

that preceded the formation of the stable association through the reaction:

kyanite + sapphirine → cordierite + corundum.

The occurrence of sapphirine in a zoned xenolith from the critical zone of the Bushveld complex has been noted by Willemse and Viljoen (1970). The mineral is present in both the inner and outer parts of the xenolith, the former consisting of sillimanite and mullite, with smaller amounts of sapphirine, corundum and rutile; the outer zone assemblage includes two varieties of spinel, pleonaste and a more hercynite-rich variety, sillimanite, plagioclase and sapphirine. The sapphirine (Table 64, anal. 4) of the inner zone is pale red in colour, and is one of the most aluminium-rich varieties yet described. It appears to have crystallized directly from the substance of the xenolith and to be stable with sillimanite and mullite. The sapphirine of the outer zone is enclosed in bytownite, is blue in colour and has a similar content of aluminium to the red variety; but is richer in iron (5·9 per cent total iron as FeO) and chromium (0·89 weight per cent Cr_2O_3). The temperature and pressure of the magma at the time of formation of the critical zone is estimated to have been $\simeq$ 1200°C, 4·5 kbar (Cameron, 1976).

Aluminium-rich sapphirine-bearing pyroxenite and metagabbro granulite xenoliths occur in the Stockdale kimberlite, Kansas (Brookins and Woods, 1970; Meyer and Brookins, 1976). In both types of xenoliths sapphirine has formed from spinel ($\simeq Mg_{0\cdot5}Fe_{0\cdot5}Al_2O_4$), the silica required for the formation of sapphirine probably being derived from the reaction plagioclase + pyroxene → garnet in the case of the metagabbro xenoliths, and of plagioclase to fassaitic pyroxene and sillimanite in the pyroxenite xenoliths. The disappearance of plagioclase suggests that both the metagabbro and pyroxenite re-equilibrated at pressures and temperatures between 10 and 14 kbar and 800 to 1000°C.

Sapphirine with cordierite–sillimanite, pleonaste–sillimanite, corundum–sillimanite, and corundum–pleonaste occurs in the hornfelsed pelitic schists' xenoliths in the Cortlandt complex, Peekskill, New York (Friedman, 1952, 1956; Barker, 1964b). The formation of these assemblages is interpreted as a result of the migration of SiO_2, H_2O and K_2O from the schists and its consequent enrichment in Al_2O_3, Fe-oxides and MgO.

The sapphirine from Blinkwater, Transvaal (Mountain, 1939) occurs in a granitic rock and is associated with corundum. Its paragenesis is not known but it is presumably xenolithic in origin

References

Ackermand, D., Seifert, F. and **Schreyer, W.,** 1975. Instability of sapphirine at high pressures. *Contr. Min. Petr.*, **50**, 79–92.

Bancroft, G. M., Burns, R. G. and **Stone, A. J.,** 1968. Applications of the Mössbauer effect to silicate mineralogy – II. Iron silicates of unknown and complex crystal structures. *Geochim. Cosmochim. Acta*, **32**, 547–559.

Barker, F., 1964a. Sapphirine-bearing rock, Val Codera, Italy. *Amer. Min.*, **49**, 146–152.

Barker, F., 1964b. Reaction between mafic magmas and pelitic schist, Cortlandt, New York. *Amer. J. Sci.*, **262**, 614–634.

Bishop, F. C. and **Newton, R. C.,** 1975. The composition of low-pressure synthetic sapphirine. *J. Geol.*, **83**, 511–517.

Bøggild, O. B., 1953. The mineralogy of Greenland. *Medd. om Grønland*, **149**, no. 3, 192.

Bondarenko, L. P., 1972. Hypersthene–kyanite association in garnet–sapphirine granulites; thermodynamic conditions for their formation. *Internat. Geol. Rev.*, **14**, 466–472.

Boulanger, J., 1959. Les anorthosites de Madagascar. *Annal. géol. de Madagascar*, **26**, 1–71.

Boyd, F. R. and **England, J. L.,** 1959. Pyrope. *Carnegie Inst. Washington, Ann. Rept. Dir. Geophys. Lab.*, 1958–59, 83–87.

Boyd, F. R. and **England, J. L.**, 1962. Effect of pressure on the melting of pyrope. *Carnegie Inst. Washington, Ann. Rept. Dir. Geophys. Lab.*, 1961–62, 109–112.
Brookins, D. G. and **Woods, M. J.**, 1970. High pressure mineral reactions in a pyroxenite granulite nodule from the Stockdale kimberlite, Riley County, Kansas. *Bull. Kansas Geol. Surv.*, **199**, Pt. 3, 1–6.
Cameron, W. E., 1976. Coexisting sillimanite and mullite. *Geol. Mag.*, **113**, 497–514.
Cannillo, E., Mazzi, F., Fang, J. H., Robinson, P. D. and **Ohya, Y.**, 1971. The crystal structure of aenigmatite. *Amer. Min.*, **56**, 427–446.
Charlu, T. V., Newton, R. C. and **Kleppa, O. J.**, 1975. Enthalpies of formation at 970 K of compounds in the system MgO–Al_2O_3–SiO_2 from high temperature solution calorimetry. *Geochim. Cosmochim. Acta*, **39**, 1487–1497.
Chatterjee, N. and **Schreyer, W.**, 1970. Stabilitäte beziehungen der Paragenese Sapphirin + Quartz. *Fortschr. Min.*, **47**, 9–10.
Chatterjee, N. and **Schreyer, W.**, 1972. The reaction enstatite$_{ss}$ + sillimanite $\rightleftharpoons$ sapphirine$_{ss}$ + quartz in the system MgO–Al_2O_3–SiO_2. *Contr. Min. Petr.*, **36**, 49–62.
Chekirda, A. I. and **Entin, A. R.**, 1969. New data on the sapphirine associations of the Aldan Shield. *Dokl. Acad. Sci. U.S.S.R., Earth Sci. Sect.*, **186**, 131–134.
Christophe-Michel-Lévy, M., 1962. Quelques remarques sur la saphirine. *C.R. 86th Congr. nation. Soc. Savantes Montpellier*, 1961. *Sect. Sci.*, 383–385.
Clifford, T. N., Stumpfl, E. F. and **McIver, J. R.**, 1975. A sapphirine–cordierite–bronzite–phlogopite paragenesis from Namaqualand, South Africa. *Min. Mag.*, **40**, 347–356.
Cornelius, H. P. and **Dittler, E.**, 1929. Zur Kenntniss des Sapphirinvorkommens von Alpe Brasciadega in Val Codera (Italien, Prov. Sondrio). *Neues Jahrb. Min., Abt. A.*, **59**, 27–64.
Dallwitz, W. B., 1968. Co-existing sapphirine and quartz in granulite from Enderby Land, Antarctica. *Nature*, **219**, 476–477.
Dornberger-Schiff, K. and **Merlino, S.**, 1974. Order-disorder in sapphirine, aenigmatite and aenigmatite-like minerals. *Acta Cryst.*, **A30**, 168–173.
Doroshev, A. M. and **Malinovskii, I.**, 1974. [Upper pressure boundary of sapphirine stability.] *Dokl. Akad. Sci. U.S.S.R.*, **219**, 959–961.
Fleet, S. G., 1967. Non-space group absences in sapphirine. *Min. Mag.*, **36**, 449–450.
Foissy, B., Kleiber, J. and **Picot, P.**, 1966. Note sur la présence de saphirine dans des ultrabasites d'Andriamena (centre nord de Madagascar). *C.R. Semaine Géol.*, 1965, Madagascar, 103–104.
Forestier, F. H. and **Lasnier, B.**, 1969. Découverte de niveaux d'amphibolites à pargasite, anorthite, corindon et saphirine dans les schistes cristallins de la vallée du Haut-Allier. Existence du facies granulite dans le Massif Central français. *Contr. Min. Petr.*, **23**, 194–235.
Foster, W. R., 1950. Synthetic sapphirine and its stability relations in the system MgO–Al_2O_3–SiO_2. *J. Geol.*, **58**, 135–151.
Friedman, G. M., 1952. Sapphirine occurrence of Cortlandt, New York. *Amer. Min.*, **37**, 244–249.
Friedman, G. M., 1956. The origin of spinel-emery deposits with particular reference to those of the Cortlandt complex, New York. *New York State Museum Bull.*, **351**, 1–66.
Fujimori, S. and **Allard, G. O.**, 1966. Ocorrencia de safirina em Salvador, Bahia. *Bol. Soc. Brasil Geol.*, **15**, 67–81.
Gossner, B. and **Mussgnug, F.**, 1928. Vergleichende röntgenographische Untersuchung von Magnesium-silikaten. *Neues Jahrb. Min., Abt. A*, **58**, 213–252.
Haapala, I., Shvola, J., Ojanperä, P. and **Yletyinen, V.**, 1971. Red corundum, sapphirine and kornerupine from Kittilä, Finnish Lapland. *Bull. Geol. Soc. Finland*, **43**, 221–231.
Hensen, B. J., 1972. Phase relations involving pyrope, enstatite$_{ss}$ and sapphirine$_{ss}$ in the system MgO–Al_2O_3–SiO_2. *Carnegie Inst. Washington, Ann. Rept. Dir. Geophys. Lab.*, 1971–72, 421–426.
Hensen, B. J. and **Green, D. H.**, 1970. Experimental data on coexisting cordierite and garnet under high grade metamorphic conditions. *Phys. Earth Planet. Interiors*, **3**, 431–440.
Hensen, B. J., and **Green, D. H.**, 1971. Experimental study of the stability of cordierite and garnet in pelitic compositions at high pressures and temperatures. 1. Compositions with excess alumino-silicate. *Contr. Min. Petr.*, **33**, 309–330.
Herd, R. K., 1973. Sapphirine and kornerupine occurrences within the Fiskenaesset complex. *Rapp. Grφnlands Geol. Unders.*, **51**, 65–71.
Herd, R. K., Windley, B. F. and **Ghisler, M.**, 1969. The mode of occurrence and petrogenesis of the sapphirine-bearing and associated rocks of west Greenland. *Rapp. Grφnlands Geol. Unders.*, **24**, 1–44.
Hudson, D. R. and **Wilson, A. F.**, 1966. A new occurrence of sapphirine and related anthophyllite from central Australia. *Geol. Mag.*, **103**, 293–298.
Janardhanan, A. S. and **Leake, B. E.**, 1974. Sapphirine in the Sittampundi complex, India. *Min. Mag.*, **39**, 901–902.
Keith, M. L. and **Schairer, J. F.**, 1952. The stability field of sapphirine in the system MgO–Al_2O_3–SiO_2. *J. Geol.*, **60**, 181–186.

Kuzel, H.-J., 1961. Über Formel und Elementarzelle des Sapphirin. *Neues Jahrb. Min., Mh.*, 68–71.
Lacroix, A., 1929. Sur un schist cristallin à saphirine de Madagascar et sur les roches à saphirine en général. *Bull. Soc. franç. Min.*, **52**, 76–84.
Lacroix, A., 1940. Les roches dépourvues de feldspath du cortège des sakénites (Madagascar); composition chimique de cet ensemble. *Compt. Rend. Acad. Sci. Paris*, **210**, 193–196.
Lacroix, A., 1941. Les gisements de phlogopite de Madagascar et les pyroxénites qui les renferment. *Annal. géol. du Serv. Mines*, **11**, 1–113.
Leake, B. E., **Janardhanan, A. S.** and **Kemp, A.**, 1976. High P_{H_2O} and hornblende in the Sittampundi complex, India. *Min. Mag.*, **40**, 525–526.
Lensch, G., 1971. Das Vorkommen von Sapphirin im Peridotitkörper von Finero (Zone von Ivrea, Italienische Westalpen). *Contr. Min. Petr.*, **31**, 145–152.
Lutts, B. G. and **Kopaneva, L. N.**, 1968. A pyrope–sapphirine rock from the Anabar massif and its conditions of metamorphism. *Dokl. Acad. Sci. U.S.S.R., Earth Sci. Sect.*, **179**, 161–163.
McKie, D., 1963. Order-disorder in sapphirine. *Min. Mag.*, **33**, 635–645.
Meng, K. L. and **Moore, J. M.**, *Jr.*, 1972. Sapphirine-bearing rocks from Wilson Lake, Labrador. *Can. Min.*, **11**, 777–790.
Merlino, S., 1973. Polymorphism in sapphirine. *Contr. Min. Petr.*, **41**, 23–29.
Meyer, H. O. A. and **Brookins, D. G.**, 1976. Sapphirine, sillimanite and garnet in granulite xenoliths from Stockdale kimberlite, Kansas. *Amer. Min.*, **61**, 1194–1202.
Monchoux, P., 1972. Roches à sapphirine au contact des lherzolites pyrénéennes. *Contr. Min. Petr.*, **37**, 47–64.
Moore, J. M. and **Meng, K. L.**, 1971. Comments on 'sapphirine reactions in deep-seated granulites near Wilson Lake, central Labrador, Canada'. *Earth Planet. Sci. Letters*, **12**, 355–356.
Moore, P. B., 1968. Crystal structure of sapphirine. *Nature*, **218**, 81–82.
Moore, P. B., 1969. Crystal structure of sapphirine. *Amer. Min.*, **54**, 31–49.
Morse, S. A. and **Talley, J. H.**, 1971. Sapphirine reactions in deep-seated granulites near Wilson Lake, central Labrador, Canada. *Earth Planet. Sci. Letters*, **10**, 325–328.
Mountain, E. D., 1939. Sapphirine crystals from Blinkwater, Transvaal. *Min. Mag.*, **25**, 277–282.
Muthuswami, T. N., 1949. Sapphirine – (Madura). *Proc. Indian Acad. Sci., Sect. A*, **30**, 295–301.
Nalivkina, E. G., 1961. Metasomatic zonality and genesis of sapphirine-bearing rocks in the Bug region. *Intern. Geol. Rev.*, **3**, 337–349.
Newton, R. C., 1972. An experimental determination of the high-pressure stability limits of magnesian cordierite under wet and dry conditions. *J. Geol.*, **80**, 398–420.
Nixon, P. H., **Reedman, H. J.** and **Burns, L. K.**, 1973. Sapphirine-bearing granulites from Labwor, Uganda. *Min. Mag.*, **39**, 420–428.
Oliver, R. L. and **Jones, J. B.**, 1965. A chlorite–corundum rock from Mount Painter, South Australia. *Min. Mag.*, **35**, 140–145.
Østergaard, T. V., 1969. Sapphirine in a kyanite and staurolite-bearing rock: a new occurrence of sapphirine in Norway. *Norges Geol. Unders.*, **258**, 62–65.
Palache, C., **Berman, H.** and **Frondel, C.**, 1944. *Dana's System of Mineralogy*, **1**, 724. New York (John Wiley & Sons).
Povarennykh, A. S., 1970. Spectres infrarouges de certain minéraux de Madagascar. *Bull. Soc. franç. Min. Crist.*, **93**, 224–234.
Prider, R. T., 1945. Sapphirine from Dangin, Western Australia. *Geol. Mag.*, **82**, 49–54.
Ramberg, H., 1948. On sapphirine-bearing rocks in the vicinity of Sukkertoppen (west Greenland). *Medd. om Grønland*, **142**, no. 5, 1–32.
Richardson, S. W., **Gilbert, M. C.** and **Bell, P. M.**, 1969. Experimental determination of kyanite–andalusite and andalusite–sillimanite equilibria; the aluminium silicate triple point. *Amer. J. Sci.*, **267**, 259–272.
Rivalenti, G., 1974. A ruby corundum pegmatoid in an area near Fiskenaesset, south-west Greenland. *Bull. Soc. Geol. It.*, **93**, 23–32.
Roever, de E. W. F., 1973. Preliminary note on coexisting sapphirine and quartz in a mesoperthite gneiss from the Bakhuis Mountains (Surinam). *Contr. Geol. Surinam*, **3** *(Geol. Mijnb. Dienst. Sur., Mededel.* **22***)*, 67–70 (M.A. 75–455).
Sahama, Th. G., **Lehtinen, M.** and **Rehtijarvi, P.**, 1974. Properties of sapphirine. *Ann. Acad. Sci. Fennicae*, Ser. A. III, **114**, 1–23.
Schairer, J. F., 1954. The system K_2O–MgO–Al_2O_3–SiO_2. 1. Results of quenching experiments on four joins in the tetrahedron cordierite–forsterite–leucite–silica and on the join cordierite–mullite–potash felspar. *J. Amer. Ceram. Soc.*, **37**, 501–533.
Schreyer, W., 1968. Stability of sapphirine. *Carnegie Inst. Washington, Ann. Rept. Dir. Geophys. Lab.*, 1966–67, 389–392.
Schreyer, W. and **Abraham, K.**, 1975. Peraluminous sapphirine as a metastable reaction product in

kyanite–gedrite–talc schist from Sar e Sang, Afghanistan. *Min. Mag.*, **40**, 171–180.

Schreyer, W., Abraham, K. and **Behr, H. J.,** 1976. Sapphirine and associated minerals from the kornerupine rock of Waldheim, Saxony. *Neues Jahrb. Min., Abh.*, **126**, 1–27.

Schreyer, W. and **Schairer, J. F.,** 1961. Compositions and structural states of anhydrous Mg-cordierites: a re-investigation of the central part of the system $MgO–Al_2O_3–SiO_2$. *J. Petr.*, **2**, 324–406.

Schreyer, W. and **Seifert, F.,** 1969a. Compatibility relations of aluminium silicates in the systems $MgO–Al_2O_3–SiO_2–H_2O$ and $K_2O–MgO–Al_2O_3–SiO_2–H_2O$ at high pressures. *Amer. J. Sci.*, **267**, 371–388.

Schreyer, W. and **Seifert, F.,** 1969b. High pressure phases in the system $MgO–Al_2O_3–SiO_2–H_2O$. *Amer. J. Sci.*, **367-A** (Schairer vol.), 407–443.

Schreyer, W. and **Yoder, H. S.,** 1964. The system Mg-cordierite–H_2O and related rocks. *Neues Jahrb. Min., Abhdl.*, **101**, 270–342.

Segnit, E. R., 1957. Sapphirine-bearing rocks from McRobertson Land, Antarctica. *Min. Mag.*, **31**, 690–697.

Seifert, F., 1974. Stability of sapphirine: a study of the aluminous part of the system $MgO–Al_2O_3–SiO_2–H_2O$. *J. Geol.*, **82**, 173–204.

Shen, B.-M., Zheng, X.-Z. and **Li, D.-Z.,** 1974. [Studies on the crystallization of some glasses in the system $MgO–Al_2O_3–SiO_2$. I. A new metastable phase of cordierite and the characteristic in formation of sapphirine.] *Sci. Geol. Sinica*, 171–181. (Chinese with English abstract.)

Smart, R. M. and **Glasser, F. P.,** 1976. Phase relations of cordierite and sapphirine in the system $MgO–Al_2O_3–SiO_2$. *J. Mater. Sci.*, **11**, 1459–1464.

Sørensen, H., 1955. On sapphirine from west Greenland. *Medd. om Grønland*, **137**, 18–30.

Taylor, H. C. J., 1973. Melting relations in the system $MgO–Al_2O_3–SiO_2$ at 15 kb. *Bull. Geol. Soc. America*, **84**, 1335–1348.

Touret, M. J., 1970. Le facies granulite, métamorphisme en milieu carbonique. *Compt. Rend. Acad. Sci. Paris*, **271**, 2228–2231.

Touret, M. J. and **Roche, de la H.,** 1971. Saphirine à Snaresund près de Tvedestrand (Norvège Méridionale). *Norsk. Geol. Tidsskr.*, **51**, 169–175.

Vogt, T., 1947. Mineral assemblages with sapphirine and kornerupine. *Bull. Comm. géol. Finlande*, **140**, 15–23.

Waldmann, L., 1931. Der sapphirinführende Gabbro von Stallek. *Vorh. Geol. Bundesanstalt*, Wien, 79–84.

Willemse, J. and **Viljoen, E. A.,** 1970. The fate of argillaceous material in the gabbroic magma of the Bushveld complex. *Geol. Soc. S. Africa, Spec. Publ.*, **1**, 336–366.

Wilson, A. F., 1962. A new occurrence of sapphirine and the significance of sapphirine in the granulite terrains of Western Australia. *Amer. Min.*, **47**, 207–208.

Wilson, A. F., 1965. The petrological features and structural setting of Australia granulites and charnockites. *Intern. Geol. Congr., XXII Session, India*, Part **13**, 21–42.

Wilson, A. F. and **Hudson, D. R.,** 1967. The discovery of beryllium-bearing sapphirine in the granulites of the Musgrave Ranges (central Australia). *Chem. Geol.*, **2**, 209–215.

Woodford, P. J. and **Wilson, A. F.,** 1976. Sapphirine, högbomite, kornerupine and surinamite from aluminous granulites, north-eastern Strangways Range, central Australia. *Neues Jahrb. Min., Mh.*, 15–35.

Yardley, B. W. D. and **Blacic, J. D.,** 1976. Sapphirine in the Sittampundi complex, India: a discussion. *Min. Mag.*, **40**, 523–524.

Yoder, H. S., *Jr.*, 1952. The $MgO–Al_2O_3–SiO_2–H_2O$ system and the related metamorphic facies. *Amer. J. Sci.* (Bowen vol.), 569–627.

Zotov, I. A., 1966. Sapphirine found in magnesian skarns of the south-western Pamirs. *Dokl. Acad. Sci. U.S.S.R., Earth Sci. Sect.*, **170**, 146–148.

Aenigmatite $Na_2Fe_5^{2+}TiO_2[Si_6O_{18}]$

Triclinic (+)

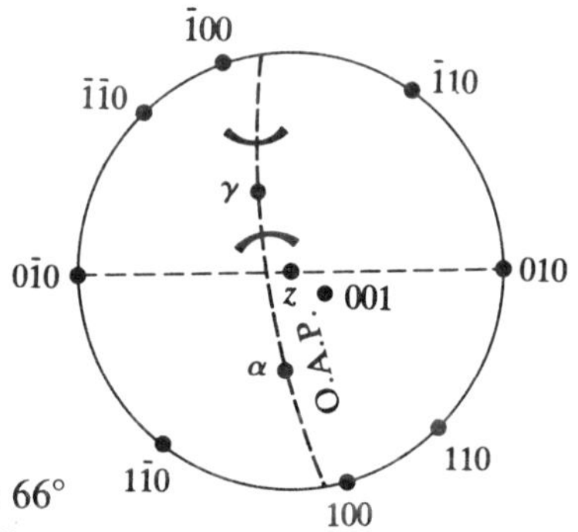

α	1·790[1]–1·81
β	1·805–1·826
γ	1·87–1·90
δ	0·07–0·08
2V	27°–55°
γ:z	40°–45°
O.A.P.	≃ ‖(010)[2]
Dispersion:	$r < v$ very strong.
D	3·74–3·86
H	5·5–6
Cleavage:	(010), (001), good; (010):(001) ≃ 66°
Twinning:	simple, multiple on (01̄1) common.
Colour:	Black, brown, reddish brown to black in thin section.
Pleochroism:	α light yellowish brown, reddish brown, brownish red. β reddish brown, dark brown. γ deep reddish brown, brownish black, opaque.
Unit cell:	*a* 10·406, *b* 10·813, *c* 8·926 Å α 104°56′, β 96°52′, γ 125°19′ Z = 2. Space group $P\bar{1}$.

Soluble in HF.

Aenigmatite is a relatively common constituent of sodium-rich alkaline rocks, and occurs in nepheline- and sodalite-syenites, foyaites, nordmarkites and occasionally in alkali granites. Aenigmatite also occurs as phenocrysts and in the groundmass of alkaline lavas, such as pantellerite, comendite, phonolite, trachyte and peralkaline rhyolite. Included among its common associates are aegirine, aegirine-augite, riebeckite–arfvedsonite, ferrorichterite, hedenbergite, and fayalite. Although rhönite is isostructural with aenigmatite, solid solution between the two compositions is very limited. Aenigmatite is so named on account of the earlier uncertainty regarding its composition. The term cossyrite, derived from its occurrence in the lavas of the island of Pantellaria the ancient name for which was *Cossyra*, has been used synonymously with aenigmatite, but is now rarely used.

Structure

The crystal morphology of aenigmatite was studied by Palache (1933) and by earlier workers. Palache determined a triclinic cell with $a:b:c = 1{\cdot}0050:1:0{\cdot}5862$, α

[1] A lower value, 1·774 has been reported by Yagi (1953).
[2] O.A.P., cleavage and twinning refer to Kelsey and McKie (1964) triclinic cell (see structure section, Table 65).

96°59·5′, β 96°49·5′, γ 112°28′. Kelsey and McKie (1964), from X-ray studies, found the unit cell described in Table 65, and gave the matrix for transforming face indices for the Palache cell to their own cell as $[001/0\frac{1}{2}\frac{\bar{1}}{2}/\frac{1}{2}00]$. They also gave some indication of the likely structure of aenigmatite as one containing pyroxene-like chains, and deduced from many chemical analyses the presently accepted structural formula.

Merlino (1970) and Cannillo *et al.* (1971) showed that the structure of aenigmatite is similar to that of sapphirine. In sapphirine (p. 615) there are bands of (Mg,Al,Fe) octahedra parallel to (100) and running along z, interconnected by chains of tetrahedra ($\|z$) and by additional Al octahedra. In aenigmatite some octahedra are replaced by Na polyhedra (distorted square antiprisms) and these, along with (Fe,Ti) octahedra, form continuous (100) sheets (Fig. 281(a)) instead of bands. These sheets are interconnected by $[Si_6O_{18}]$ chains and individual Fe octahedra (Fig. 281(b)). The tetrahedral chains are, as in sapphirine, like pyroxene chains but with additional 'wings' of tetrahedra, though the chains are somewhat less kinked in aenigmatite than in sapphirine. Whereas in sapphirine the oxygens form a close-packed spinel-like array, those in aenigmatite are more loosely arranged.

The site occupations indicated in Fig. 281 are idealized. One of the octahedra is preferentially but not wholly occupied by Ti and therefore contains some Fe^{2+}, and small proportions of the Fe^{2+} sites are occupied by Ti. The small amount of Fe^{3+} present substitutes for Si, mostly in T(3) (Fig. 281(b)).

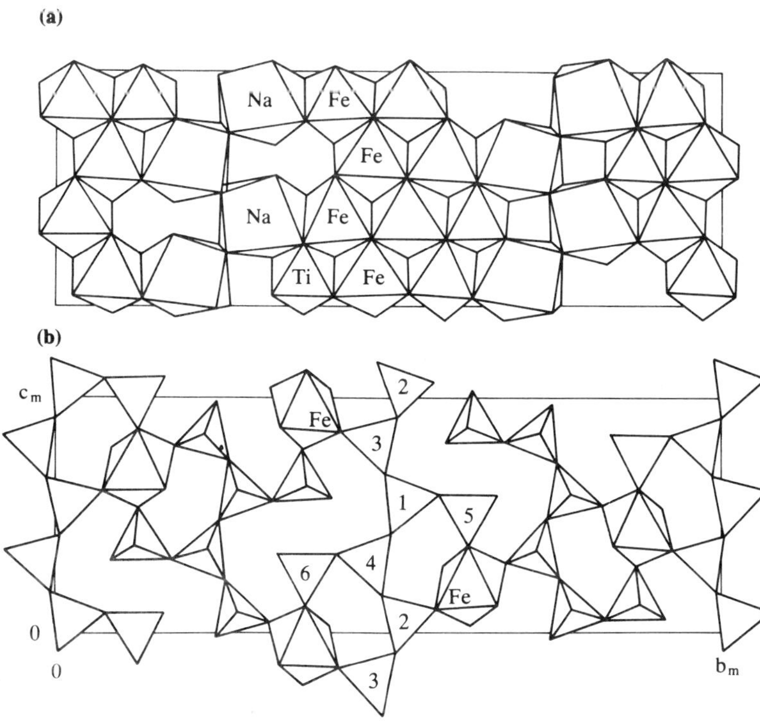

Fig. 281. (a) Idealized polyhedral diagram of bands of (Fe,Ti) octahedra in aenigmatite lying in the (100) plane and running parallel to z (pseudomonoclinic cell). The bands are linked by further octahedra (Na) to form (100) sheets. (b) Idealized polyhedral diagram of tetrahedra in aenigmatite running parallel to z and linked laterally by single octahedra (Cannillo *et al.*, 1971).

The relationship between aenigmatite and sapphirine is shown by the cell parameters listed in Table 65, where for both minerals a triclinic and an alternative pseudomonoclinic cell are shown.

Table 65 Cell parameters of aenigmatite, sapphirine and related minerals

	Aenigmatite		**Sapphirine**		**Rhönite**	**Serendibite**	**Krinovite**
	Cannillo *et al.* (1971)	**Kelsey and McKie (1964)**	**Merlino (1973)**		**Walenta (1969)**	**Machin and Süsse (1974)**	**Merlino (1972)**
	pseudo-monoclinic	**triclinic[b]**	**pseudo-monoclinic[a]**	**triclinic**	**triclinic**	**triclinic**	**triclinic**
*a*Å	12·120	10·406	11·33	10·04	10·415	10·019	10·22
*b*Å	29·63	10·813	29·08	10·38	11·207	10·393	10·67
*c*Å	10·406	8·926	10·04	8·65	9·018	8·630	8·80
α	90°04′	104°56′	90°	107°33′	102°03′	106°21′	105°8′
β	127°9′	96°52′	125°23′	95°07′	100°31′	96°4′	96°36′
γ	89°44′	125°19′	90°	123°55′	127°42′	124°21′	125°1′
Z	8	2	8	2	2	2	2

[a] The parameters given by Merlino (1973), *a* 9·87, *b* 29·08, *c* 10·04 Å, β 110°38′ have been transformed to another monoclinic cell for comparison with aenigmatite.

[b] Aenigmatite ($Na_2Fe_5TiSi_6O_{20}$) from Kola, U.S.S.R. The matrix for transforming the triclinic to pseudomonoclinic cell is [011/$\bar{1}$22/100].

The twinning commonly exhibited by aenigmatite is explained by Cannillo *et al.* (1971) in terms of the pseudomonoclinic cell, as involving an *a*/2 glide relationship on (010). Kelsey and McKie (1964) had suggested that aenigmatite in volcanic rocks crystallized first as truly monoclinic and then inverted to repeated twins of the triclinic structure. Cannillo *et al.* (1971) regarded the twinned anigmatite as having a primary origin by rapid crystallization at high temperature, slower cooling under plutonic conditions producing untwinned crystals.

As for sapphirine, different stacking relationships parallel to (010) for segments of the structure could produce different polymorphs and disordered structures as well as twins. Dornberger-Schiff and Merlino (1974) showed that aenigmatite and sapphirine are ordered members of an isostructural series, with the sapphirine from Mautia Hill (see p. 616) being a disordered member.

Other minerals which are isostructural with aenigmatite (Table 65) are rhönite $Ca_2(Mg,Fe)_4Fe^{3+}TiO_2[Al_3Si_3O_{18}]$ (Walenta, 1969), serendibite $Ca_2(Mg,Al)_6O_2[(Si,Al,B)_6O_{18}]$ (Machin and Süsse, 1974) and krinovite $Na_2Mg_4Cr_2[Si_6O_{18}]$ (Merlino, 1972).

Chemistry

The original belief that aenigmatite belonged to a group of triclinic amphiboles was first shown to be in error by Gossner and Spielberger (1929) and Kostyleva (1930). Later Fleischer (1936) suggested that the composition of aenigmatite could be represented by the formula, $X_4Y_{13}(Si_2O_7)_6$ in which X = Na,Ca,K and Y = Fe^{2+},Ti,Fe^{3+}Al. The present formula, $Na_2Fe_5^{2+}TiSi_6O_{20}$, proposed by Kelsey and McKie (1964), has since been confirmed by a full investigation of the structure (Cannillo *et al.*, 1971).

Although aenigmatite occurs in a wide variety of both plutonic and volcanic rocks, and in a number of ferromagnesian mineral associations, its chemical variability is relatively restricted. The main variations involve Fe^{2+},Ti and Fe^{3+}. In the fourteen analyses detailed in Table 66, the number of titanium ions, on the basis of twenty oxygens, varies approximately from 0·75 to 1·05 (average 0·93). The number of Fe^{3+} cations is more variable; in some aenigmatites Fe^{3+} occupies a small percentage of the tetrahedral sites (e.g. anals. 3, 9). A reciprocal relationship between Fe^{3+} and (Fe^{2+} + Ti) cations in the aenigmatites of the sodic rhyolites and trachytes of the Mount Edziza volcanic complex was noted by Yagi and Souther (1974) and considered probably to be due to solid solution between 'ideal' aenigmatite and Ti-free aenigmatite, $Na_2Fe_4^{2+}Fe_2^{3+}Si_6O_{20}$ (see experimental section). A similar conclusion was reached by Hodges and Barker (1973) on the basis of the compositional variation of the aenigmatites (titanium content varying from 6·8 to 1·7 weight per cent TiO_2) of the Sierra Prieta nepheline–analcite syenite and other Diablo Plateau intrusions. In these aenigmatites the amount of Fe^{3+} present is in excess of that required to replace Ti^{4+}, and results from a $Fe^{2+}Si \rightleftharpoons AlFe^{3+}$, in addition to the $Fe^{2+}Ti \rightleftharpoons 2Fe^{3+}$, replacement. The reason for this large substitution of Ti by Fe^{3+} is uncertain. Although oxygen fugacity may have been a contributory factor the rocks in which they occur have a very low titanium content (0·04–0·11 weight per cent TiO_2), and the activity of titanium in this instance may have been the prime cause.

The occurrence of an additional coupled replacement CaAl (core) $\rightleftharpoons$ NaSi (margin) is well illustrated by the discontinuously zoned aenigmatite (Table 66, anals. 6, 14) from the Ilímaussaq intrusion, south-west Greenland (Larsen, 1976). Most aenigmatites contain an appreciable amount of manganese, and an unusually high Mn content (4·36 weight per cent MnO) has been reported for the mineral in the foyaite of the Granitberg complex (Table 66, anal. 4). Little data pertaining to the trace element content of aenigmatites are available, but Sr, Zn, Be, V, Cu, Ga and Pb are known to be present in the minerals from the alkaline rocks of the Lovozero massif.

Experimental. The observation of Nicholls and Carmichael (1969) that there is a general antipathetic relationship between aenigmatite and Fe–Ti oxides in peralkaline volcanic rocks led to their suggestion that the two reactions:

$$\underset{\beta\text{-phase}}{4Fe_3O_4} + \underset{\text{glass}}{6Na_2Si_2O_5} + \underset{\text{quartz}}{12SiO_2} + O_2 = \underset{\text{aegirine}}{12NaFe^{3+}Si_2O_6} \qquad (1)$$

$$\underset{\text{aenigmatite}}{Na_2Fe_5TiSi_6O_{20}} + \underset{\text{glass}}{Na_2Si_2O_5} + O_2 = \underset{\text{aegirine}}{4NaFe^{3+}Si_2O_6} + \underset{\alpha\text{-phase}}{FeTiO_3} \qquad (2)$$

may be regarded as representative of the occurrence of aenigmatite and aegirine in siliceous sodic peralkaline rocks. On this basis, and making allowance for solid solution in the oxide phases and aegirine, and for the presence of $Na_2Si_2O_5$ as a component of the liquid, the oxygen fugacity can be calculated for the two reactions:

$$\log f_{O_2} = \frac{\Delta G^\circ}{2{\cdot}303RT} - 4\log a^{\beta\text{-phase}}_{Fe_3O_4} - 6\log a^{\text{liq}}_{Na_2Si_2O_5} + 12\log a^{\text{Pyx}}_{NaFeSi_2O_6} \qquad (3)$$

$$\log f_{O_2} = \frac{\Delta G^\circ}{2{\cdot}303RT} - \log a^{\text{liq}}_{Na_2Si_2O_5} + \log a^{\text{Pyx}}_{NaFeSi_2O_6} + \log a^{\alpha\text{-phase}}_{FeTiO_3} \qquad (4)$$

where a is the activity of the subscript component in the superscript phase.

Table 66. Aenigmatite Analyses

	1	2	3	4	5
SiO_2	39·66	40·86	38·59	41·61	39·62
TiO_2	9·53	8·16	10·11	8·25	9·66
Al_2O_3	2·35	1·87	1·20	1·13	0·64
Fe_2O_3	3·73	6·28	9·97	—	4·64
FeO	32·97	31·50	29·32	35·80	33·92
MnO	1·52	2·42	2·26	4·36	2·46
MgO	1·98	1·31	2·23	0·90	1·65
CaO	0·43	0·48	0·74	0·27	0·44
Na_2O	6·83	6·58	6·07	7·41	7·20
K_2O	0·17	0·62	0·16	0·01	0·04
H_2O^+	0·55	0·21	—	—	0·05
H_2O^-	0·26	0·04	—	—	0·00
Total	100·10	100·47	100·65	99·74	100·34
α	—	—	—	—	—
β	—	—	—	—	—
γ	—	—	—	—	—
$2V_\gamma$	—	—	—	—	—
$\gamma:z$	—	—	—	—	—
D	3·77	—	3·758	—	—

Numbers of ions on the basis of twenty O

	1		2		3		4		5	
Si	5·620		5·753		5·420		5·936		5·631	
Al	0·380	6·00	0·247	6·00	0·199	6·00	0·064	6·00	0·107	6·00
Fe^{3+}	0·000		0·000		0·381		0·000		0·262	
Al	0·013		0·064		0·000		0·126		0·000	
Fe^{3+}	0·399		0·665		0·674		0·000		0·234	
Ti	1·017		0·865		1·070		0·886		1·038	
Mg	0·418	6·00[a]	0·274	6·00	0·467	6·00	0·191	6·00	0·349	6·00
Fe^{2+}	3·909		3·710		3·449		4·274		4·032	
Mn	0·182		0·288		0·269		0·527		0·296	
Ca	0·057		0·134		0·071		0·000		0·051	
Ca	0·008		0·010		0·041		0·042		0·016	
Na	1·877	1·92	1·798	1·92	1·655	1·73	2·052	2·09	1·985	2·01
K	0·031		0·112		0·029		0·000		0·007	

[a] Includes Ba^{2+} 0·005.

1 Aenigmatite, alkali pegmatite, Khibina tundras, Kola Peninsula, U.S.S.R. (Shlyukova, 1963). Anal. V. G. Zaginaichenko (includes BaO 0·09; P_2O_5 0·03).
2 Aenigmatite, alkali pegmatite, Lovozero alkali massif, Kola Peninsula (Vlasov *et al.*, 1966). Includes S 0·14; Cl trace.
3 Aenigmatite, pegmatite, Chasnachorr, Khibina tundra, Kola Peninsula, U.S.S.R. (Kostyleva, 1937). Anal. I. Borneman-Starynkevich.
4 Aenigmatite, agpaitic zone, foyaite unit, Granitberg, South-West Africa (Marsh, 1975).
5 Aenigmatite, khibinite, Kirovsk, Kola Peninsula, U.S.S.R. (Kelsey and McKie, 1964). Anal. J. H. Scoon (includes Cl 0·02, F nil).

6	7	8	9	10	
41·78	40·54	41·30	39·71	41·02	**SiO_2**
9·04	6·80	7·43	10·23	8·92	**TiO_2**
0·07	1·54	0·67	0·13	0·94	**Al_2O_3**
1·16	5·79	3·75	3·27	1·31	**Fe_2O_3**
37·87	35·10	36·52	37·03	38·84	**FeO**
2·51	0·57	1·01	1·20	1·16	**MnO**
0·00	0·41	1·27	1·49	0·07	**MgO**
0·00	3·72	0·32	0·20	0·45	**CaO**
7·41	5·62	7·39	6·9	7·36	**Na_2O**
0·00	0·04	0·08	0·24	0·06	**K_2O**
—	0·06	—	—	—	**H_2O^+**
—	—	—	—	—	**H_2O^-**
99·84	100·19	99·74	100·40	100·13	**Total**
—	⩽1·806	1·81±0·015	—	1·795	α
—	1·816	1·82	—	1·805	β
—	1·873	1·90	—	1·87	γ
—	38°	—	—	—	**$2V_\gamma$**
—	—	—	—	—	$\gamma:z$
—	3·808	—	—	—	**D**

6		7		8		9		10		
5·987		5·786		5·898		5·671		5·866		**Si**
0·012	6·00	0·214	6·00	0·102	6·00	0·024	6·00	0·134	6·00	**Al**
0·001		0·000		0·000		0·305		0·000		**Fe^{3+}**
0·000		0·045		0·011		0·000		0·024		**Al**
0·124		0·623		0·403		0·045		0·141		**Fe^{3+}**
0·974		0·730		0·798		1·100		0·961		**Ti**
0·000	5·94	0·087	6·00	0·270	6·00	0·317	6·03	0·015	6·00	**Mg**
4·539		4·190		4·362		4·424		4·647		**Fe^{2+}**
0·305		0·069		0·122		0·145		0·141		**Mn**
0·000		0·256		0·034		0·000		0·068		**Ca**
0·000		0·314		0·015		0·031		0·000		**Ca**
2·059	2·06	1·554	1·88	2·046	2·08	1·910	1·98	2·042	2·05	**Na**
0·000		0·007		0·015		0·043		0·010		**K**

6 Aenigmatite, margin of zoned crystal, sodalite foyaite, Ilímaussaq intrusion, south-west Greenland (Larsen, 1976). **See anal. 14.**

7 Aenigmatite, alkalic syenite, Khusha-Gol, eastern Sayan Mountains, U.S.S.R. (Mitrofanov and Afanas'yeva, 1966).

8 Aenigmatite, peralkaline trachyte, Nandewar Mountains, New South Wales, Australia (Abbott, 1967). Anal. A. J. Easton and M. J. Abbott.

9 Aenigmatite, aegirine–riebeckite granite, Liruei complex, Nigeria (Borley, 1976).

10 Aenigmatite, obsidian, Pantelleria (Carmichael, 1962). Anal. I.S.E. Carmichael.

Table 66. Aenigmatite Analyses – *continued*

	11		12		13		14	
SiO_2	40·7		40·7		40·93		37·28	
TiO_2	8·7		8·3		8·71		7·55	
Al_2O_3	0·29		0·72		0·24		2·68	
Fe_2O_3	—		4·0		2·60		7·38	
FeO	41·5		37·5		38·45		35·29	
MnO	0·44		0·85		1·15		0·70	
MgO	0·55		0·11		0·49		0·05	
CaO	0·31		0·50		0·55		1·90	
Na_2O	6·8		7·2		6·93		6·31	
K_2O	—		0·03		0·04		0·00	
H_2O^+	—		—		0·02		—	
H_2O^-	—		—		0·00		—	
Total	99·3		99·91		100·11		99·14	
α	—		1·800		—		—	
β	—		1·813		—		—	
γ	—		1·88		—		—	
$2V_\gamma$	—		52°		—		—	
$\gamma:z$	—		43°		—		—	
D	—		—		3·813–3·817		—	
	Numbers of ions on the basis of twenty O							
Si	5·962		5·835		5·953		5·430	
Al	0·037	6·00	0·122	6·00	0·041	6·00	0·460	6·00
Fe^{3+}	—		0·043		0·006		0·110	
Al	0·014		0·000		0·000		0·000	
Fe^{3+}	—		0·388		0·275		0·699	
Ti	0·959		0·896		0·937		0·827	
Mg	0·120	6·05	0·023	5·99	0·104	6·07	0·011	6·00
Fe^{2+}	4·901		4·498		4·610		4·299	
Mn	0·055		0·103		0·140		0·087	
Ca	0·000		0·077		0·000		0·077	
Ca	0·050		0·000		0·084		0·220	
Na	1·932	1·98	2·002	2·01	1·935	2·03	1·782	2·00
K	—		0·005		0·006		0·000	

11 Aenigmatite, overgrowth on ilmenite, pegmatoid zone, Picture Gorge basalt, Spray, Oregon (Lindsley *et al.*, 1971). Total as FeO, distribution of Fe^{2+} and Fe^{3+} in structural formula taken as $Fe^{3+} = 1-(Si+Al)$.

12 Aenigmatite, phenocryst, comendite, Mt. Edziza, British Columbia, Canada (Yagi and Souther, 1974). Total Fe as FeO, Fe_2O_3 calculated by method of Finger, (1972).

13 Aenigmatite, Pantellerite, Pantelleria (Zies, 1966). Anal. E. G. Zies.

14 Aenigmatite, core of zoned crystal, sodalite foyaite, Ilímaussaq intrusion, south-west Greenland (Larsen, 1976). Total iron as FeO, Fe_2O_3 calculated assuming stoichiometry (see anal. 6).

The 'no-oxide' field, in which aenigmatite and aegirine coexist (Fig. 282(a)), is defined in f_{O_2}–T space by the field below the intersection of the curves of equations 3 and 4. The position of the 'no-oxide' field is, however, subject to considerable error, and in addition to the effects of solid solution in the components of the reactions, is also affected by their activity; in the figure the effect of different activities of $Na_2Si_2O_5$ (the activities of other components = 1·0) is shown by the broken curve.

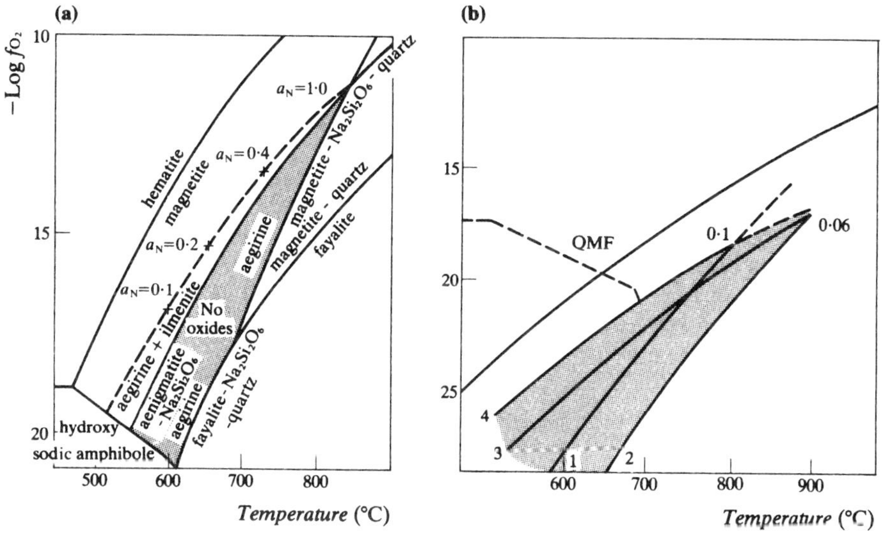

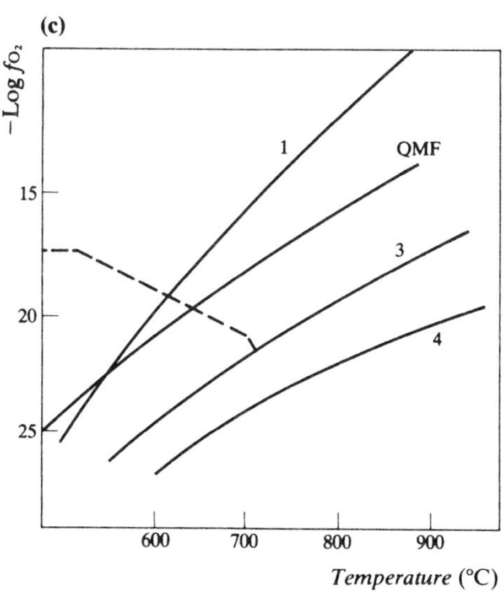

Fig. 282. (a) Diagram showing the 'no-oxide' field, in which aenigmatite and aegirine coexist, in f_{O_2}–T space. The broken curve is the locus of intersections of the two curves, aegirine–sodium disilicate–magnetite–quartz and aenigmatite–sodium disilicate–aegirine–ilmenite for varying activities (0·1, 0·2, 0·4 and 1·0) of sodium disilicate at constant unit activity of aegirine, magnetite and ilmenite (after Nicholls and Carmichael, 1969). The magnetite–hematite and quartz–fayalite–magnetite buffer curves from Eugster and Wones (1962) and Wones and Gilbert (1968) respectively. The upper stability limit of the hydroxy-riebeckite–arfvedsonite amphiboles from Ernst (1962). (b) f_{O_2}–T variation for reactions: $4Fe_3O_4 + 6Na_2Si_2O_5 + 12SiO_2 + O_2 \rightleftharpoons 12NaFeSi_2O_6$ (curves 1 and 2), $\frac{3}{2}Na_2Fe_5TiSi_6O_{20} + O_2 \rightleftharpoons \frac{3}{2}Fe_2TiO_4 + \frac{1}{2}Fe_3O_4 + 3SiO_2 + 3NaFeSi_2O_6$ (curve 3), and $2Na_2Fe_5TiSi_6O_{20} + O_2 \rightleftharpoons 2(Fe_3O_4 + Fe_2TiO_4) + 2Na_2Si_2O_5 + 8SiO_2$ (curve 4), assuming aegirine activity of 0·06 and 0·1 and sodium disilicate activity of 1·0. 'No-oxide' fields shaded. (c) f_{O_2}–T variation for above reactions assuming aegirine activity of 0·1 and sodium disilicate activity of $10^{-1\cdot5}$ (after Marsh, 1975). QFM, synthetic quartz–fayalite–magnetite buffer curve. The upper stability limit of the hydroxy-riebeckite–arfvedsonite amphiboles shown by broken line.

In a subsequent investigation of the 'no-oxide' field in f_{O_2}–T space, Marsh (1975) expressed the relationship between aenigmatite and a β-phase by the equations:

$$\underset{\text{aenigmatite}}{2Na_2Fe_5TiSi_6O_{20}} + \underset{\text{gas}}{O_2} \rightleftharpoons \underset{\beta\text{-phase solid soln.}}{2Fe_3O_4 + 2Fe_2TiO_4} + \underset{\text{glass}}{2Na_2Si_2O_5} + \underset{\text{glass}}{8SiO_2} \quad (1)$$

$$\underset{\text{aenigmatite}}{\tfrac{3}{2}Na_2Fe_5TiSi_6O_{20}} + \underset{\text{gas}}{O_2} \rightleftharpoons \underset{\beta\text{-phase solid soln.}}{\tfrac{3}{2}Fe_2TiO_4 + \tfrac{1}{2}Fe_3O_4} + \underset{\text{quartz}}{3SiO_2} + \underset{\text{aegirine}}{3NaFeSi_2O_6} \quad (2)$$

The first reaction is supported by recent observational data in accord, for example, with the mineral relationships displayed in the Nandewar Mountain trachyte (Abbott, 1967; see p. 652). The second equation marks the experimentally determined upper stability limit of aenigmatite (Lindsley, 1971; see p. 651).

The uncertainties pertaining to the use of equations of the Nicholls and Carmichael form in calculating the f_{O_2}–T curves for these reactions (Fig. 282(b), (c)) are discussed by Marsh, particularly in relation to the aegirine free energy value used, as well as to variations in the solid solutions and activities of the components. Using a range of realistic values for these parameters, Marsh concluded that the existence of a 'no-oxide' field is probable in liquids of phonolitic composition, but that the evidence for the existence of such a field in undersaturated liquids is not conclusive.

Aenigmatite with the theoretical composition, $Na_2Fe_5^{2+}TiSi_6O_{20}$, was first synthesized by Thompson and Chisholm (1969) at 700°C, 1 kbar P_{H_2O} and with oxygen fugacity controlled by iron–wüstite buffer. Earlier, Ernst (1962) had synthesized a titanium-free 'aenigmatite' (probable composition $Na_2Fe_4^{2+}Fe_2^{3+}Si_6O_{20}$). The latter synthesis was effected using a bulk composition $Na_2O.5FeO_x.8SiO_2$ plus excess water at oxygen fugacities defined by the magnetite–wüstite and wüstite–iron buffers, and showed that the aenigmatite stability field is enlarged at f_{O_2} of the wüstite–iron, compared with that of the magnetite–wüstite

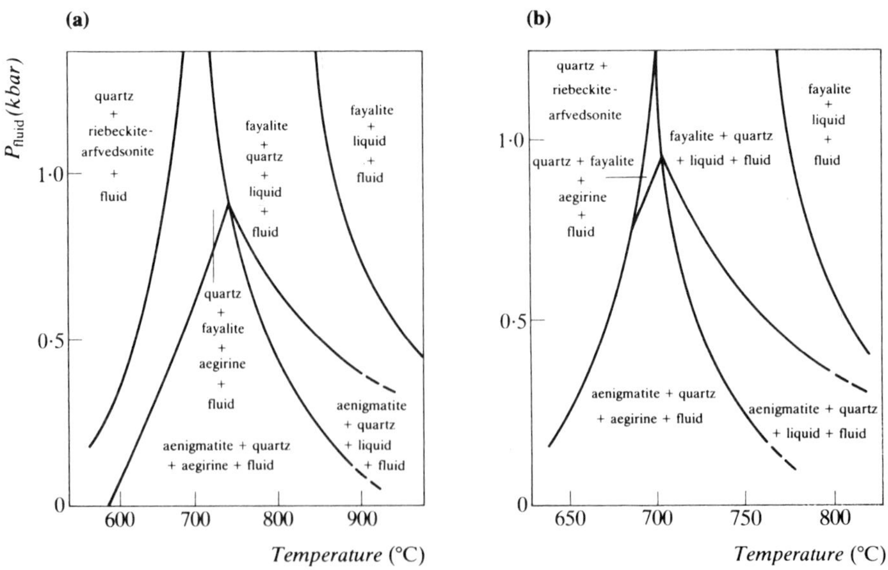

Fig. 283. P_{fluid}–T **phase relations for the bulk composition** $Na_2O.5FeO_x.8SiO_2$ **+ excess water. (a) oxygen fugacity defined by magnetite–wüstite buffer. (b) oxygen fugacity defined by wüstite–iron buffer (after Ernst, 1962).**

buffer (Fig. 283). Under these conditions titanium-free 'aenigmatite' is not stable at P_{H_2O} above $\simeq$ 0·9 kbar, and its pressure field is thus more restricted than that of 'normal', $Na_2Fe_5^{2+}TiSi_6O_{22}$, aenigmatite. At oxygen fugacities defined by the nickel–bunsenite buffer, Ti-free 'aenigmatite' does not crystallize.

Aenigmatite, $Na_2Fe_5^{2+}TiSi_6O_{22}$ composition, melts incongruently *in vacuo* at 880° to 900°C. As water lowers the melting temperature the maximum thermal stability of aenigmatite, crystallized from its own composition, is thus 900°C. From an estimated free energy of formation of aenigmatite, Nicholls and Carmichael (1969) calculated that the mineral may be stable at higher oxygen fugacities than those of the fayalite–magnetite–quartz buffer. This conclusion, however, is not supported by the experimental data of Lindsley (1971) who found that the aenigmatite stability curve, at 750°C and 0·5 kbar, lies between the Ni–NiO and FMQ buffer curves. Thus aenigmatite, previously synthesized at oxygen fugacities of the Ni–NiO buffer, was found to recrystallize, in runs of short duration and at a variety of temperatures, to aenigmatite + aegirine + Fe–Ti oxide(s), and to break down completely to aegirine + Ti-magnetite + quartz in runs over a longer period, the aenigmatite apparently persisting metastably rather than being stabilized by ferric iron. Additional experiments, involving the reaction of the breakdown assemblage, at f_{O_2} of the Ni–NiO and FMQ buffers, showed that the growth of aenigmatite took place at the expense of the oxidized assemblage only at the lower oxygen fugacity, and that f_{O_2}, rather than total pressure, has the greater influence on the stability of aenigmatite.

Optical and Physical Properties

The refractive indices and birefringence of aenigmatite are very high. The mineral shows very strong absorption, $\gamma > \beta > \alpha$, and the birefringence cannot be observed in thin sections of normal thickness. Precise measurement of the refractive indices is difficult and their determination is subject to more than the usual degree of error. Most refractive indices in the literature are quoted $\pm$0·005 to 0·010. In view of the relative similarity of aenigmatite compositions, the rather wide range of values of the optic axial angle is unexpected and has not been explained, although it may in part be related to the large inclined and horizontal dispersion displayed by the mineral.

Simple and lamellar twinning is a characteristic feature of aenigmatite, simple twins occurring mainly in the aenigmatites of plutonic rocks, multiple twins in those of the volcanic rocks. The possibility that the fine, lamellar twinning is due to the inversion of a quenched high-temperature monoclinic form has been suggested by Kelsey and McKie (1964), a view, however, not accepted on structural grounds by Cannillo *et al.* (1971). Moreover, lamellar twinning is a common feature of serendibite, a pneumatolytic skarn mineral with an aenigmatite-type structure.

Distinguishing Features

Aenigmatite and rhönite have similar optical properties but the two minerals can be distinguished by the higher γ refractive index and birefringence of aenigmatite. In some rhönites the pleochroic scheme displays a greenish component in the α

vibration direction, in contrast to the yellow-brown and brownish red colour of aenigmatites. Aenigmatite is distinguished from kaersutite, barkevikite, ferrorichterite and basaltic hornblende by higher refractive indices, smaller optic axial angle, stronger pleochroism and larger cleavage angle.

Paragenesis

Aenigmatite is a relatively common constituent of nepheline and sodalite syenites and their associated pegamtites, and much of the earlier data relating to its paragenesis come from the Ilímaussaq intrusion, south-west Greenland (e.g. Lorenzen, 1882), the alkaline complexes of Khibina and Lovozero, Kola Peninsula (Fersman, 1923; Fersman and Bonshtedt, 1937; Gerasimovsky, 1936; Kostyleva, 1930, 1937; Bussen and Dudkin, 1962; Kucharenko *et al.*, 1965), and the pegmatites of Langesundsfjord, Norway (Brøgger, 1890). Although, except for pegmatites, the occurrence of aenigmatite is generally as an accessory mineral it may be present in substantial amounts; thus in the aegirine- and aegirine-augite syenites of the Khusha-Gol intrusion, eastern Sayan Mountains (Mitrofanov and Afanas'yeva, 1966) it is a major constituent making up some 12 per cent of the rocks. In this intrusion it is associated with aegirine, riebeckite and hastingsite, and in some members of the complex aenigmatite has been replaced by albite and riebeckite during a late period of albitization; replacement by astrophyllite and, less commonly, by biotite has also occurred.

An interesting occurrence of aenigmatite has been described from a microkakortokite dyke associated with the Ilímaussaq intrusion (Larsen and Steenfelt, 1974). The dyke displays two facies. A high alkali facies in which aenigmatite is associated with aegirine, arfvedsonite and fayalite, with magnetite either very scarce or absent altogether, and a low alkali facies without aenigmatite, and in which an aegirine–hedenbergite is accompanied by magnetite, presumably the result of changing oxygen fugacity in the later stages of the crystallization of the dyke.

Aenigmatite, associated with alkali feldspar, $Or_{35}Ab_{65}$, nepheline, $Ne_{78}Ks_{22}$, aegirine-augite zoned from $Di_{38}Hd_{46}Ac_{16}$ to $Di_{10}Hd_{44}Ac_{46}$, and accessory arfvedsonite and astrophyllite, occurs in the outer agpaitic foyaite unit of the Granitberg intrusion (Marsh, 1975). Titaniferous magnetite is present as rare skeletal grains enclosed in the aenigmatite indicative of the early crystallization of the oxide and its later reaction with the liquid to form aenigmatite. In the inner miaskitic unit of the intrusion titaniferous magnetite is a common constituent, and its presence here, and its absence in the agpaitic unit, is probably related to changing oxygen fugacity during the development of the ring complex. The occurrence of astrophyllite in the miaskitic unit in place of aenigmatite, and its replacement relationship elsewhere, suggests that it may be considered as the hydrated equivalent of aenigmatite, a suggestion that is supported by the common association of this mineral with riebeckite–arfvedsonite in many quartz-bearing alkaline rocks. The aenigmatites, occurring as accessory minerals, in the nepheline–analcite syenite of the Sierra Prieta intrusion, Diablo Plateau (Hodges and Barker, 1973) have unusually low contents of titanium, and show a wide range of solid solutions towards the titanium-free composition, $Na_2Fe_4^{2+}Fe_2^{3+}Si_6O_{20}$, titanium contents ranging from 6·8 to 1·7 weight per cent TiO_2.

Aenigmatite also occurs in more silica-rich plutonic rocks, and has been reported in the nordmarkite of the York area, Maine (Woodward, 1957), and from one of the younger granite complexes of Nigeria (Borley, 1976). The aenigmatite (Table 66, anal. 9) of the Liruei complex occurs in a peralkaline aegirine–riebeckite granite. The mineral has an unusually high titanium content (10·2 weight per cent TiO_2) in which it is relatively enriched in comparison with the associated aegirine (1·58 per cent TiO_2) and arfvedsonite–riebeckite (1·61 per cent TiO_2).

Aenigmatite (anal. 11) occurs in pegmatoid zones of the Picture Gorge basalt, Spray, Oregon (Lindsley *et al.*, 1971). The pegmatoids have been derived from the residual liquids of the basalt and are enriched in SiO_2, TiO_2 and alkalis relative to the host rock. The minerals of the pegmatoids include strongly zoned augite, $Ca_{42}Mg_{38}Fe_{20}$ to $Ca_{40}Mg_{10}Fe_{50}$, fayalitic olivine, Fa_{92}, plagioclase, alkali feldspar $\simeq Or_{38\cdot5}Ab_{61\cdot5}$, ilmenite, ilmenomagnetite and small amounts of aegirine-augite (12–45 mol. per cent $NaFe^{3+}Si_2O_6$). The aenigmatite occurs as discrete primary grains that crystallized simultaneously with the aegirine-augite, as overgrowths on ilmenite and ilmenomagnetite, and replacing selectively the (111) ilmenite lamellae of the latter indicating the late stage formation of aenigmatite, and a maximum thermal stability below 900°C consistent with the experimental data on synthetic aenigmatite. The aenigmatite does not occur in contact with plagioclase or alkali feldspar less potassium-rich than Or_{35}. This incompatibility of aenigmatite with the anorthitic component of plagioclase suggests a possible reaction:

$$\underset{\text{aenigmatite}}{Na_2Fe_5TiSi_6O_{20}} + \underset{\text{in plagioclase}}{CaAl_2Si_2O_8} + \tfrac{1}{2}O_2 =$$

$$= \underset{\text{in pyroxene}}{CaFeSi_2O_6} + \underset{\text{(in alkali feldspar)}}{2NaAlSi_3O_8} + \underset{\text{(in ilmenite}_{ss})}{FeTiO_3} + \underset{\text{(in magnetite}_{ss})}{Fe_3O_4}$$

The temperature and oxygen fugacity operating during the formation of the aenigmatite is estimated by Lindsley *et al.*, on the basis of the coexisting ilmenite and ilmenomagnetite compositions, and of the fayalitic olivine, to have been 900 ± 50°C and f_{O_2} $10^{-12\cdot60}$ to $10^{-13\cdot7}$ atm, in fair agreement with the experimental data on 'normal' and Ti-free aenigmatite.

Zoned aenigmatites, with rhönite (see p. 658) cores and aenigmatite margins, separated by a sharp compositional discontinuity, are present in the monzonites and syenites at Morotu River, Sakhalin Island (Yagi, 1953). Aenigmatite, rimmed successively by aegirine-augite–arfvedsonite and aegirine also occurs in these rocks in accord with the magmatic differentiation towards lower temperature liquids enriched in ferric iron and alkalis. The occurrence of a somewhat similar relationship has been described by Tomita (1934) from a teschenite from Oki, Japan. In this rock the earlier formed kaersutitic amphibole shows peripheral alteration, and displays an inner zone of rhönite that passes outwards through a zone of dendritic aenigmatite–rhönite to a margin consisting of aenigmatite, augite and magnetite. The extent, if any, to which the aenigmatite minerals of the intermediate zone exhibit more than the very limited solid solution between aenigmatite and rhönite is not known, and a re-examination of the Oki teschenite is desirable.

In many alkali lavas, such as pantellerite, comendite, phonolite and trachyte, aenigmatite occurs commonly as a constituent of the groundmass, and less frequently as a phenocrystal component. Many of the earlier references to accounts of such occurrence have been listed by Kelsey and McKie (1964). The lavas of Pantelleria have been described by numerous workers and a more recent account is given by Carmichael (1962), and those of Major Island, New Zealand (Marshall, 1936) have been supplemented by Ewart *et al.* (1968). Here aenigmatite, together with anorthoclase, aegirine–hedenbergite and quartz, occurs as small phenocrysts, and the mineral is also present, with tuhualite, in the groundmass of the lava.

Two ferromagnesian assemblages occur in the peralkaline trachytes from the continental alkaline volcanic province in the Nandewar Mountains of New South Wales, Australia (Abbott, 1967). Aenigmatite (Table 66, anal. 8) is present as a groundmass constituent of both assemblages, with sodic ferrohedenbergite $Di_6Hd_{84}Ac_{10}$, a riebeckite–arfvedsonite solid solution and titanomagnetite, and with hedenbergite–aegirine, $Hd_{50}Ac_{50}$, and titanomagnetite. The titanomagnetite is rimmed by aenigmatite indicating a reaction of the oxide with the peralkaline sodium silicate liquid and the crystallization of aenigmatite at low oxygen fugacity, and the two assemblages may be related by a reaction of the type:

$$\underset{\text{titanomagnetite}}{6(TiFeO_3.Fe_3O_4)} + 12SiO_2 + \underset{\text{aegirine–hedenbergite}_{ss}}{12(NaFe^{3+}Si_2O_6 + CaFe^{2+}Si_2O_6)} \rightleftharpoons$$

$$\underset{\text{aenigmatite}}{6(Na_2Fe_5^{2+}TiSi_6O_{20})} + \underset{\text{hedenbergite}}{12CaFe^{2+}Si_2O_6} + \underset{\text{magnetite}}{2Fe_3O_4} + 5O_2$$

Aenigmatite occurs in variable amounts and habits in the sodic rhyolite and trachyte ash flows, ash falls and lava domes of the Mt. Edziza volcanic complex, British Columbia (Yagi and Souther, 1974). Universally present as a major constituent, up to 15 per cent, it occurs only rarely as phenocrysts and is mainly confined to the groundmass where it is intergrown with aegirine and arfvedsonite. Although the three ferromagnesian phases began crystallizing simultaneously, their order of cessation, deduced from the textural relationships of partially mantled crystals, is the same as the crystallization sequence in the Morotu River monzonites and syenites, viz. aenigmatite, aegirine, riebeckite. In the glassy trachytes it is present as small microlites and in spherulites consisting of radiating acicular crystals with aegirine and feldspar. In contrast, aenigmatite is only a minor constituent of the sodic rhyolites in which it occurs as sparse phenocrysts. Both aenigmatite-bearing rocks are considered to represent late fractionation products of a parent alkali olivine basalt, and their mineralogy is not unlike the pegmatoids in the Picture Gorge basalt (Lindsley *et al.*, 1971). Aenigmatite, together with sodian hedenbergite and anorthoclase, also occurs as phenocrysts in the Fantale ash-flow tuff in the northern part of the main Ethiopian rift (Gibson, 1970). Other aenigmatite-bearing alkaline lavas include those of the Kenya rift (Smith, 1931; Bowen, 1937; Nash *et al.*, 1969; Lippard, 1973), and the comendites of Mount Nimrud, Armenia (Prior, 1928).

References

Abbott, M. J., 1967. Aenigmatite from the groundmass of a peralkaline trachyte. *Amer. Min.*, **52**, 1895–1901.

Andrade, M. de, 1954. Contribution à l'étude des roches alcalines d'Angola (note préliminaire). *Compt. Rend. 19th Intern. Geol. Congr. Algiers*, **20**, 241–252.

Borley, G. D., 1976. Aenigmatite from an aegirine–riebeckite granite, Liruei complex, Nigeria. *Min. Mag.*, **40**, 595–598.

Bowen, N. L., 1937. A note on aenigmatite. *Amer. Min.*, **22**, 139–140.

Brøgger, W. C., 1890. Die Mineralien der Syenitpegmatitgänge der Südnorwegischen Augit- und Nephelinsyenite. 50 Ainigmatit. *Zeit. Krist.*, **16**, 423–433.

Bussen, I. V. and **Dudkin, O. B.**, 1962. New data on aenigmatite from Khibina and Lovozero alkaline massifs. *Akad. Nauk. S.S.S.R., Kol'skii Filial, Materialy po Min., Kol'skogo Poluostrova*, **2**, 96–106.

Cannillo, E., Mazzi, F., Fang, J. H., Robinson, P. D. and **Ohya, Y.**, 1971. The crystal structure of aenigmatite. *Amer. Min.*, **56**, 427–446.

Carmichael, I. S. E., 1962. Pantelleritic liquids and their phenocrysts. *Min. Mag.*, **33**, 86–113.

Dornberger-Schiff, K. and **Merlino, S.**, 1974. Order-disorder in sapphirine, aenigmatite and aenigmatite-like minerals. *Acta Cryst.*, **A30**, 168–173.

Ernst, W. G., 1962. Synthesis, stability relations and occurrence of riebeckite and riebeckite–arfvedsonite solid solutions. *J. Geol.*, **70**, 689–736.

Eugster, H. P. and **Wones, D. R.**, 1962. Stability relations of the ferruginous biotite, annite. *J. Petr.*, **3**, 82–125.

Ewart, A., Taylor, S. R. and **Capp, A. C.**, 1968. Geochemistry of the pantellerites of Major Island, New Zealand. *Contr. Min. Petr.*, **17**, 116–140.

Fersman, A. E., 1923. [Regular intergrowths of minerals in the Khibina and Lovozero tundras.] *Bull. Acad. Sci.*, ser. 6, **17**, 275–290 (in Russian).

Fersman, A. E. and **Bonshtedt, E. M.**, 1937. *Minerals of the Khibina and Lovozero tundras.* Lomonossov Inst. Acad. Sci. U.S.S.R. Acad. Sci. Press, Moscow and Leningrad.

Finger, L. W., 1972. The uncertainty in the calculated ferric iron content of a microprobe analysis. *Carnegie Inst. Washington, Ann. Rept. Dir. Geophys. Lab.*, 1971–72, 600–603.

Fleischer, M., 1936. The formula of aenigmatite. *Amer. J. Sci.*, **32**, 343–348.

Gerasimovsky, V. I., 1936. [On the mineralogy of the south-eastern part of Lujavr-urt.] *Trans. Lomonossov Inst. Geochim. Cryst. Min. Acad. Sci. U.S.S.R.* (Russian with English summary).

Gibson, I. L., 1970. A pantelleritic welded ash-flow tuff from the Ethiopian rift valley. *Contr. Min. Petr.*, **28**, 89–111.

Gossner, B. and **Spielberger, F.**, 1929. Chemische und röntgenographische Untersuchungen an Silikaten: Ein Beitrag zur Kenntnis der Hornblendegruppe. *Z. Krist.*, **72**, 111–142.

Hodges, F. N. and **Barker, D. S.**, 1973. Solid solution in aenigmatite. *Carnegie Inst. Washington, Ann. Rept. Dir. Geophys. Lab.*, 1972–73, 578–581.

Kelsey, C. H. and **McKie, D.**, 1964. The unit cell of aenigmatite. *Min. Mag.*, **33**, 986–1001.

Kostyleva, E. E., 1930. [Aenigmatit der Chibina-Tundren (Halbinsel Kola).] *Trudy Muz. Min. Akad. Sci. S.S.S.R.*, **4**, 87–107 (in Russian, German summary).

Kostyleva, E. E., 1937. Minerals of the Khibina and Lovozero tundras. *Lomonossov Inst. Acad. Sci. U.S.S.R.*, 112–114. A. E. Fersman and E. M. Bonshtedt (eds.). Moscow, Leningrad (Academy Sciences Press).

Kucharenko, A. A., Orlowa, M. P. and **Bulack, A. G.**, 1965. *Der kaledonische Komplex ultrabasischer alkalischer Gestein und Carbonatite der Halbinsel Kola in nordliches Karelien.* Moskau, 501–502.

Larsen, L. M., 1976. Clinopyroxenes and coexisting mafic minerals from the alkaline Ilimaussaq intrusion, south Greenland. *J. Petr.*, **17**, 285–290.

Larsen, L. M. and **Steenfelt, A.**, 1974. Alkali loss and retention in an iron-rich peralkaline phonolite dyke from the Gardar province, south Greenland. *Lithos*, **7**, 81–90.

Lindsley, D. H., 1971. Synthesis and preliminary results on the stability of aenigmatite ($Na_2Fe_5TiSi_6O_{20}$). *Carnegie Inst. Washington, Ann. Rept. Dir. Geophys. Lab.*, 1969–70, 188–190.

Lindsley, D. H., Smith, D. and **Haggerty, S. E.**, 1971. Petrology and mineral chemistry of a differentiated flow of Picture Gorge basalt near Spray, Oregon. *Carnegie Inst. Washington, Ann. Rept. Dir. Geophys. Lab.*, 1969–70, 264–285.

Lippard, S. J., 1973. The petrology of phonolites from the Kenya Rift. *Lithos*, **6**, 217–234.

Lorenzen, J., 1882. On some minerals from the sodalite-syenites in Julianehaab district, south Greenland. *Min. Mag.*, **5**, 49–70.

Machin, M. P. and **Süsse, P.**, 1974. Serendibite: a new member of the aenigmatite structure group. *Neues Jahrb. Min., Mh.*, 435–441.

Marsh, J. S., 1975. Aenigmatite stability in silica-understaturated rocks. *Contr. Min. Petr.*, **50**, 135–144.

Marshall, P., 1936. Geology of Major Island. *Trans. Roy. Soc. New Zealand*, **66**, 337–345.

Merlino, S., 1970. Crystal structure of aenigmatite. *Chem. Comm.*, **20**, 1288–1289.

Merlino, S., 1972. X-ray crystallography of krinovite. *Z. Krist.*, **136**, 81–88.

Merlino, S., 1973. Polymorphism in sapphirine. *Contr. Min. Petr.*, **41**, 23–29.

Mitrofanov, F. P. and **Afanas'yeva, L. I.**, 1966. Aenigmatite from alkalic syenite of the eastern Sayans. *Dokl. Acad. Sci., Earth Sci. Sect.*, **166**, 111–113.

Nash, W. P., Carmichael, I. S. E. and **Johnson, R. W.**, 1969. The mineralogy and petrology of Mount Suswa, Kenya. *J. Petr.*, **10**, 409–439.

Nicholls, J. and **Carmichael, I. S. E.**, 1969. Peralkaline acid liquids: a petrological study. *Contr. Min. Petr.*, **20**, 268–294.

Palache, C., 1933. Crystallographic notes on anapaite, aenigmatite and eudidymite. *Z. Krist.*, **86**, 280–291.

Prior, G. T., 1928. Note on the alkali-lavas of Mount Nimrud, Armenia. *Min. Mag.*, **21**, 485–488.

Shlyukova, Z. V., 1963. [Cossyrite is aenigmatite.] *Trudy Min. Muz. Akad. Nauk S.S.S.R.*, **14**, 262–264 (in Russian).

Smith, W. C., 1931. A classification of some rhyolites, trachytes, and phonolites from part of Kenya Colony, with a note on some associated basaltic rocks. *Quart. J. Geol. Soc.*, **87**, 212–258.
Thompson, R. N. and **Chisholm, J. E.**, 1969. Synthesis of aenigmatite. *Min. Mag.*, **37**, 253–255.
Tomita, T., 1934. On kaersutite from Dogo, Oki Islands, Japan, and its magmatic alteration and resorption. *J. Shanghai Sci. Inst. Sect. 2*, **1**, 99–136.
Vlasov, K. A., **Kuzmenko, M. Z.** and **Eskova, E. M.**, 1966. *The Lovozero Alkali Massif.* Edinburgh (Oliver & Boyd).
Walenta, K., 1969. Zur Kirstallographie des Rhönits. *Z. Krist.*, **130**, 214–230.
Wones, D. R. and **Gilbert, M. C.**, 1968. The stability of fayalite. *Carnegie Inst. Washington, Ann. Rept. Dir. Geophys. Lab.*, 1966–67, 402–403.
Woodward, H. H., 1957. Diffusion of chemical elements in some naturally occurring silicate inclusions. *J. Geol.*, **65**, 61–84.
Yagi, K., 1953. Petrochemical studies of the alkalic rocks of the Morotu district, Sakhalin. *Bull. Geol. Soc. America*, **64**, 769–809.
Yagi, K. and **Souther, J. G.**, 1974. Aenigmatite from Mt. Edziza, British Columbia, Canada. *Amer. Min.*, **59**, 820–829.
Zies, E. G., 1966. A new analysis of cossyrite from the island of Pantelleria. *Amer. Min.*, **51**, 200–205.

Rhönite $Ca_2(Mg,Fe^{2+},Fe^{3+})_5TiO_2[(Si,Al)_6O_{18}]$

Triclinic (−)

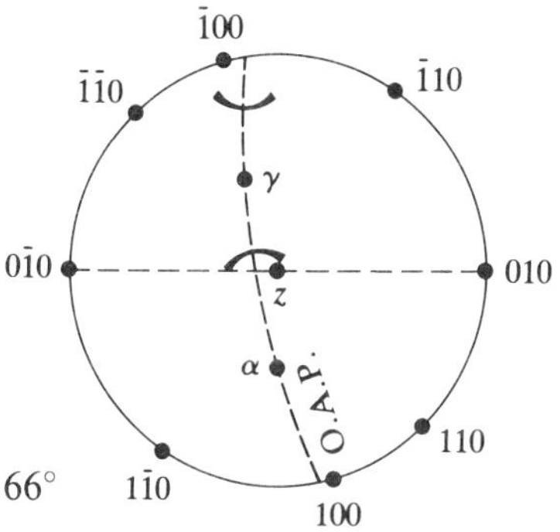

α	1·79–1·808
β	1·80–1·815
γ	1·83–1·845
δ	0·032–0·040
$2V_\alpha$	47°–66°
γ:z	≃ 40° on (010)
O.A.P.	≃ ‖(010)
Dispersion:	$r < v$ very strong inclined.
D	3·4–3·65
H	5–6
Cleavage:	(010), (001), good; (010):(001) ≃ 66°
Twinning:	$(0\bar{1}1)$ lamellar, very common.
Colour:	black; dark brown to black in thin section.
Pleochroism:	α greenish brown, dark greenish brown, red-brown. β brown, reddish brown, green-brown, brownish yellow. γ deep reddish brown, dark red, black, opaque.
Unit cell:	a 10·415, b 11 207, c 9·018 Å α 102°03′, β 100°31′, γ 127°42′ Z = 2. Space group $P\bar{1}$

Soluble with difficulty in HCl.

Rhönite is isostructural with aenigmatite (see Table 65) and its composition may be regarded as being related to that mineral by the coupled substitutions, CaAl ⇌ SiNa, TiMg ⇌ 2Al and SiMg ⇌ 2Al. Major departures from the ideal formula are uncommon and the normal compositional range is shown by analyses 1 to 7 in Table 67. The rhönite (anal. 8) of the Allende meteorite is, however, exceptionally rich in titanium and low in silicon; this is compensated by a correspondingly high value of aluminium, some 60 per cent of the tetrahedral sites being occupied by this cation. The very limited occurrence of compositions between rhönite and aenigmatite in natural minerals is generally regarded as evidence of a large miscibility gap between the two end-member compositions. Grünhagen and Seck (1972), however, have suggested that rhönite–aenigmatite solid solutions may be thermodynamically stable, and their absence is due to the lack of the necessary conditions for their formation in natural parageneses.

The optical properties of rhönite, except for the lower value of the γ refractive index and of the birefringence, have a close similarity to those of aenigmatite. Like aenigmatite, rhönite shows very strong absorption, and the pleochroic scheme can usually only be observed fully in very thin sections.

Characteristically rhönite occurs in silica-poor rocks, in which it is associated with titanaugite, olivine, plagioclase and a feldspathoid mineral. Rhönite was first described (Soellner, 1907) from the alkali basic lavas of the Rhön district, Germany.

Table 67. Rhönite Analyses

	1	2	3	4
SiO_2	29·8	23·7	24·42	23·2
TiO_2	10·2	11·9	9·46	10·0
Al_2O_3	13·6	16·8	17·25	16·55
Fe_2O_3	—	—	11·69	10·15
FeO	21·2	21·0	11·39	12·50
MnO	0·1	0·16	tr.	0·14
MgO	14·4	13·05	12·62	12·4
CaO	11·4	12·15	12·43	13·0
Na_2O	—	0·97	0·67	0·92
K_2O	—	—	0·63	0·10
H_2O^+	—	—	—	0·68
H_2O^-	—	—	—	—
Total	100·7	99·83	100·56	99·64
α	—	—	1·808	1·810
β	—	—	—	1·825
γ	—	—	1·840	1·840
$2V_\alpha$	—	—	>70°	72°
$\gamma:z$	—	—	28[b]	—
D	—	—	3·58	—

Numbers of ions on the basis of twenty O

	1		2		3		4	
Si	3·960		3·172		3·266		3·177	
Al	2·040	6·00	2·651	6·00	2·720	6·00	2·672	6·00
Fe^{3+}	—		0·177		0·014		0·151	
Al	0·090		0·000		0·000		0·000	
Fe^{3+}	—		0·676		1·163		0·894	
Ti	1·020		1·198		0·953		1·030	
Mg	2·852	6·33	2·603	6·01[a]	2·516	6·00	2·529	6·00
Fe^{2+}	2·356		1·500		1·276		1·430	
Mn	0·011		0·018		0·000		0·016	
Ca	—		0·000		0·092		0·101	
Ca	1·625		1·744		1·689		1·802	
Na	—	1·63	0·251	2·00	0·174	1·97	0·151	1·97
K	—		—		0·108		0·018	
fe*	45·2		47·5		49·4		49·5	

[a] Includes Cr^{3+} 0·006, Ni^{2+} 0·004.
[b] γ'.

* $fe = 100(Fe^{2+} + Fe^{3+})/(Fe^{2+} + Fe^{3+} + Mg)$.

1 Rhönite, cavity in pyroxenite xenolith in basalt, Monistrol d'Allier, Haute Loire, France (Babkine *et al.*, 1964).

2 Rhönite, basanite, McMurdo, Antarctica (Kyle and Price, 1975). Total iron as FeO; cations calculated assuming stoichiometry (Finger, 1972), Fe_2O_3 8·45, FeO 13·4. Includes Cr_2O_3 0·06, Ni 0·04.

3 Rhönite, volcanic breccia, Scharnhausen, Stuttgart (Soellner, 1907). Anal. M. Dittrich.

4 Rhönite, nepheline dolerite, Löbauer Berg, Saxony (Grünhagen and Seck, 1972). Anal. H. A. Seck.

5		6		7		8		
28·58		24·82		27·86		19·1		SiO_2
10·70		9·09		9·28		16·8		TiO_2
13·35		17·24		15·36		28·9		Al_2O_3
—		9·48		6·90		—		Fe_2O_3
22·49		15·98		17·63		1·9		FeO
0·17		0·26		0·21		—		MnO
12·09		10·67		9·56		15·7		MgO
10·23		11·97		9·60		17·9		CaO
2·11		0·72		1·43		—		Na_2O
0·02		0·02		0·38		—		K_2O
—		0·35		0·95		—		H_2O^+
—		0·06		—		—		H_2O^-
99·74		100·69		99·16		101·0		**Total**
—		1·795		1·805		1·79		α
—		1·806		1·815		—		β
—		1·83		1·845		1·83		γ
—		65°		50°		—		$2V_\alpha$
—		—		—		—		$\gamma:z$
—		3·64		—		3·41		D
3·831		3·365		3·823		2·390		Si
2·108	6·00	2·635	6·00	2·177	6·00	3·610	6·00	Al
0·061		0·000		0·000		—		Fe^{3+}
0·000		0·121		0·309		0 638		Al
0·554		0·964		0·712		—		Fe^{3+}
1·079		0·927		0·959		1·582		Ti
2·416	6·00	2·156	6·04	1·955	6·00	2·927	6·00[a]	Mg
1·905		1·812		2·024		0·199		Fe^{2+}
0·019		0·059		0·025		—		Mn
0·027		0·000		0·016		0·579		Ca
1·442		1·740		1·394		1·822		Ca
0·550	2·00	0·189	1·93	0·381	1·84	—	1·82	Na
0·003		0·003		0·066		—		K
51·1		56·3		58·3		—		fe*

[a] Includes V^{3+} 0·075.

5 Rhönite, nepheline hawaiite, Dunedin district, New Zealand (Kyle and Price, 1975). Total iron given as FeO; cations calc. assuming stoichiometry (Finger, 1972), Fe_2O_3 6·10, FeO 17·0.
6 Rhönite, metasyenite, Terlingua, Texas (Cameron *et al.*, 1970). Anal. H. Asari (includes P_2O_5 0·03).
7 Rhönite, metaphonolite differentiate in olivine-nephelinite layer, Puy de Saint-Sandoux (Auvergne) France (Grünhagen and Seck, 1972). Anal. H. A. Seck.
8 Rhönite, carbonaceous chondritic meteorite, Allende (Fuchs, 1971). Anal. J. E. Sanechi (includes V_2O_3 0·7).

Here the mineral occurs, both as small phenocrysts, and as a primary constituent of the groundmass, in nepheline and melilite basalts, limburgites, olivine leucitites, nepheline basanites and other related rocks. A more recent account of the parageneses of rhönite in the Rhön volcanics, including its occurrence in tephrites and pipe breccias, as well as in the basic alkaline lavas of the Swabian Alps, Saxony, has been presented by Ficke (1961). The occurrence of rhönite in the titanaugite nepheline dolerites of the Puy-de-Barneire has been described by Lacroix (1909). The mineral here occurs in large crystals showing partial resorption and is rimmed by titanomagnetite. Other examples of rhönite reaction relationships include its replacement of kaersutite in the analcite-bearing teschenite, Oki Island, Japan (Tomita, 1934), and in the nepheline hawaiite at Dunedin, New Zealand (Kyle and Price, 1975). In the alkali syenites and monzonites of Sakhalin Island (Yagi, 1953), rhönite occurs as cores surrounded by aenigmatite. The occurrence of these discontinuously zoned crystals shows that the rhönite formed at a somewhat higher temperature than aenigmatite, its crystallization leading to depletion in aluminium and calcium in the residual magma and the local development of a peralkaline liquid, thus providing a suitable milieu for the crystallization of aenigmatite.

Rhönite is a major constituent of the metasyenite unit of a mafic alkaline sill, Big Bend National Park, Texas (Cameron *et al.*, 1970). The rhönite occurs as small phenocrysts with titanaugite, olivine (Fa_{65}), plagioclase (An_{55}), and nepheline, in a fine-grained groundmass of alkali feldspar and zeolites. The ferromagnesian phenocrysts all display evidence of reaction with the liquid, and the rhönite is rimmed by an unidentified opaque material, and is also partially replaced by a very fine-grained aggregate consisting of a carbonate and possible sphene and chlorite.

The occurrence of rhönite (Table 67, anal. 1), associated with spinel and titanomagnetite, in glass-filled cavities in a pyroxenite xenolith in a basalt from Haute Loire, France, has been described by Babkine *et al.* (1964). The glass, from which the rhönite has formed during its subsequent recrystallization, is considered to have been derived from local pockets of fused pyroxene.

References

Babkine, J. F., Conquéré, J. C. and **Duong, P. K.,** 1964. Sur un nouveau gisement de rhönite (Monistrol-d'Allier, Haute Loire). *C.R. Acad. Sci., Paris*, Ser. D, **258**, 5479–5481.

Cameron, K. L., Carman, M. F. and **Butler, J. C.,** 1970. Rhönite from Big Bend National Park, Texas. *Amer. Min.*, **55**, 864–874.

Ficke, B., 1961. Petrologische Untersuchungen an Tertiaren basaltischen bis phonolithischen Vulkaniten der Rhön. *Tsch. Min. Petr. Mitt.*, Ser. 3, **7**, 337–436.

Finger, L. W., 1972. The uncertainty in the calculated ferric iron content of a microprobe analysis. *Carnegie Inst. Washington, Ann. Rept. Dir. Geophys. Lab.*, 1971–2, 600–603.

Fuchs, L. H., 1971. Occurrence of wollastonite, rhönite and andradite in the Allende meteorite. *Amer. Min.*, **56**, 2053–2068.

Grünhagen, H. and **Seck, H. A.,** 1972. Rhönit aus einen Melaphonolith von Puy de Saint-Sandoux (Auvergne). *Min. Petr. Mitt., Tschermak*, **18**, 17–38.

Kyle, P. R. and **Price, R. C.,** 1975. Occurrences of rhönite in alkalic lavas of the McMurdo Volcanic Group, Antarctica, and Dunedin Volcano, New Zealand. *Amer. Min.*, **60**, 722–725.

Lacroix, M. A., 1909. Note sur la rhönite du Puy de Barneire a Saint-Sandoux. *Bull. Soc. franç Min. Crist.*, **32**, 325–331.

Soellner, J., 1907. Üeber Rhönit, ein neues änigmatitähnliches Mineral und über die Verbreitung desselben in basaltischen Gesteinen. *Neues Jahrb. Min.*, **24**, 475–547.

Tomita, T., 1934. On kersutite from Dogo, Oki Islands, Japan, and its magmatic alteration and resorption. *J. Shanghai Sci. Inst.*, Sect. 2, **1**, 99–136.

Yagi, K., 1953. Petrochemical studies of the alkalic rocks of the Morotu district, Sakhalin. *Bull. Geol. Soc. America*, **64**, 769–809.

Serendibite $Ca_2(Mg,Al)_6O_2[(Si,Al,B)_6O_{18}]$

Triclinic (+)

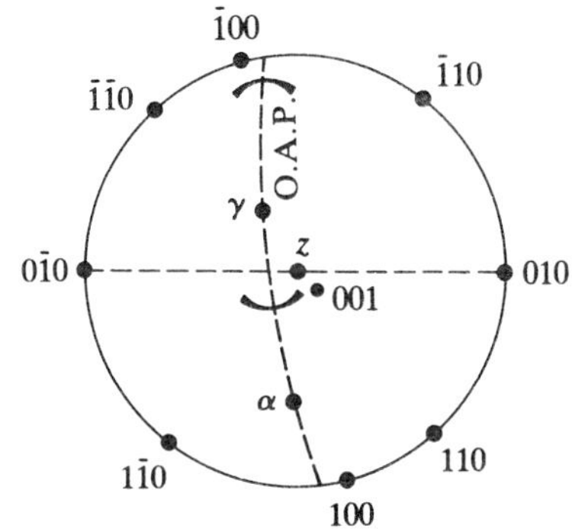

α	1·700–1·738
β	1·703–1·741
γ	1·706–1·743
δ	0·005–0·006
$2V_\gamma$	78°–90° (high negative values also reported)
γ:z	26°–40°
Cleavage:	(010), (001), good; (010):(001) ≃ 66°
Dispersion:	r > v strong, inclined, weaker horizontal.
D	3·42–3·515
Twinning:	lamellar, usually thin, 0·003–0·5 mm, on $(0\bar{1}1)$
Colour:	blue; pale yellow, pale to dark blue in thin section.
Pleochroism:	α pale greenish blue, yellowish green, blue-green, greyish blue-green, pale yellow, pale yellow-green, pale brownish yellow. β pale blue, blue-green, pale yellow (almost colourless). γ pale blue, blue, deep indigo blue, prussian blue, greenish yellow.
Absorption:	γ > β > α or γ > β = α
Unit cell:	$a \simeq 10{\cdot}02$, $b \simeq 10{\cdot}39$, $c \simeq 8{\cdot}63$ Å α 106°21′, β 96°4′, γ 124°21′ Z = 2. Space group $P\bar{1}$.

Serendibite is isostructural with aenigmatite (see Table 65). The composition of serendibite (Table 68) varies a little from the ideal formula and there is commonly some replacement of (Mg,Al) by (Fe^{2+},Fe^{3+}) in octahedral sites. Compared with aenigmatite and rhönite, titanium is present in only very small amounts.

The refractive indices of serendibite are considerably lower than those of both aenigmatite and rhönite, and it is easily distinguished from these structurally related minerals by its very low birefringence. The dispersion is strong and the mineral may display anomalous interference colours. Values of $2V_\gamma$ between 78·5° to 82° for λ 460 and 600 nm respectively are reported by Hutcheon *et al.* (1977). It is also distinguished from aenigmatite and rhönite by its absorption colours, the pleochroism showing mainly blue, green and yellow tints. Twinning is characteristically developed, and is commonly multiple with fine lamellae between 0·003 and 0·5 mm in thickness.

The paragenesis of serendibite is restricted to skarns, in which boron metasomatism has been active, at carbonate with acid rock contacts. Thus at Gangapitiya, Sri Lanka, the locality from which serendibite (Table 68, anal. 3) was first described (Prior and Coomaraswamy, 1903), it occurs in association with diopside, spinel, scapolite, apatite and plagioclase, in a contact zone between limestone and granulite, while at Johnsberg, Warren County, New York (Larsen and Schaller,

Table 68. Serendibite Analyses

	1	2	3	4	5
SiO_2	26·30	20·85	25·33	25·02	22·08
TiO_2	—	0·06	—	0·26	0·17
Al_2O_3	34·05	40·20	34·96	32·46	30·68
B_2O_3	8·37	[5·57]	4·17	6·65	6·95
Fe_2O_3	—	—	—	2·32	5·05
FeO	2·76	3·48	4·17	4·02	6·06
MnO	—	—	—	0·16	0·16
MgO	15·44	12·71	14·91	12·50	12·92
CaO	13·30	17·11	14·56	15·22	14·93
Na_2O	—	0·02	0·51	—	—
K_2O	—	—	0·22	—	—
H_2O^+	—	—	0·69	1·09	0·75
H_2O^-	—	—	—	0·26	0·25
Total	100·22	100·00	100·00	99·96	100·00
α	1·701	1·700	≃1·7	1·719	1·738
β	1·703	1·703	—	—	—
γ	1·706	1·706	—	1·725	1·743
$2V_\gamma$	83°[a]	80°	—	≃90°	≃90°
$\gamma:z$	—	—	—	—	—
D	3·376	—	3·42	3·453	3·515

Numbers of ions on the basis of twenty O

	1		2		3		4		5	
Si	3·022		2·489		3·056		3·037		2·716	
B	1·660	6·00	1·148	6·00	0·872	6·00	1·393	6·00	1·475	6·00
Al	1·318		2·363		2·072		1·570		1·809	
Al	3·294		3·297		2·922		3·081		2·641	
Fe^{3+}	—		—		—		0·212		0·467	
Ti	—		0·005		—		0·024		0·016	
Mg	2·644	6·20	2·261	6·00	2·693	6·04	2·262	6·01	2·368	6·13
Fe^{2+}	0·265		0·347		0·423		0·409		0·623	
Mn	—		—		—		0·017		0·016	
Ca	0·000		0·090		—		0·000		0·000	
Ca	1·638		2·099		1·891		1·980		1·968	
Na	—	1·64	0·004	2·10	0·120	2·04	—	1·98	—	1·97
K	—		—		0·024		—		—	

[a] $2V_\alpha$.

1 Serendibite, Grenville limestone–granite contact, Johnsburg, Warren County, New York (Larsen and Schaller, 1932). Anal. W. T. Schaller.
2 Serendibite, marble–granite contact, Melville Peninsula, Franklin district, Canada (Hutcheon *et al.*, 1977). (B_2O_3 by difference.)
3 Serendibite, contact zone between limestone and granulite, Gangapitiya, Kandy, Sri Lanka (Prior and Coomaraswamy, 1903). Anal. G. T. Prior (includes P_2O_5 0·48).
4 Serendibite, skarn, Tayezhnoye, South Yakutia, Siberia (Pertsev and Nikitina, 1959). Anal. I. B. Nikitina.
5 Serendibite, skarn, Tayezhnoye, South Yakutia, Siberia (Pertsev and Nikitina, 1959). Anal. I. B. Nikitina.

1932), serendibite (anal. 1) with diopside and phlogopite, is present at the Grenville limestone–granite contact. At Riverside, California, serendibite is developed in thin replacement bands of hydrothermal contact metamorphosed dolomitic marbles. Here the associated minerals include diopside, grossularite, vesuvianite and wollastonite (Richmond, 1939).

A number of serendibite (anals. 4, 5) occurrences in spinel–diopside skarns have been described from the Tayezhnoye iron ore localities of southern Yakutia (Shabynin and Pertsev, 1956; Pertsev and Nikitina, 1959). The skarn formations are restricted to contacts between carbonate rocks and feldspathic schists, acid gneisses and migmatites, and have involved boron mineralization resulting from post magmatic solutions derived from an alaskite granite. Serendibite is associated with sinhalite, $MgAlBO_4$, and warwickite, $(Mg,Fe,Ti)_2BO_4$, in the carbonate rocks, and with tourmaline in the siliceous rocks. In the former the assemblages of serendibite with diopside, pargasite, diopside + phlogopite (all with calcite) appear to be stable. In the siliceous rocks, however, the early formed tourmaline has been replaced by serendibite, which itself has later been partially replaced by an almost colourless second generation tourmaline, presumably at a lower temperature. The replacement of tourmaline by serendibite (anal. 2) is also reported from the contact zone between calcite marble and granite, Melville Peninsula, Canada (Hutcheon *et al.*, 1977).

Serendibite is also found with sinhalite, warwickite and tourmaline, together with chromian tremolite, forsterite, spinel, pyrrhotite and graphite in the skarns of the Handemi district, Tanzania (Bowden *et al.*, 1969).

References

Bowden, P., von Knorring, O. and **Bartholomew, R. W.**, 1969. Sinhalite and serendibite from Tanzania. *Min. Mag.*, **37**, 145–146.

Hutcheon, I., Gunter, A. E. and **Lecheminant, A. N.**, 1977. Serendibite from Penrhyn Group marble, Melville Peninsula, District of Franklin. *Can. Min.*, **15**, 108–112.

Larsen, E. S. and **Schaller, W. T.**, 1932. Serendibite from Warren County, New York, and its paragenesis. *Amer. Min.*, **17**, 457–465.

Pertsev, N. N. and **Nikitina, I. B.**, 1959. [New data on serendibite.] *Zap. Vses. Min. Obsch.*, **88**, 169–172 (in Russian).

Prior, G. T. and **Coomaraswamy, A. K.**, 1903. Serendibite, a new borosilicate from Ceylon. *Min. Mag.*, **13**, 224–227.

Richmond, G. M., 1939. Serendibite and associated minerals from the New City Quarry, Riverside, California. *Amer. Min.*, **24**, 725–726.

Shabynin, L. I. and **Pertsev, N. N.**, 1956. [Warwickite and serendibite from magnesium skarn of southern Yakutia.] *Zap. Vses. Min. Obsch.*, **85**, 515–528 (in Russian).

Note added in proof: Welshite, a new beryllium-bearing member of the aenigmatite group with composition $Ca_2Mg_4Fe^{3+}Sb^{5+}O_2[Si_4Be_2O_{18}]$, and cell parameters a 9·68, b 14·77, c 5·14Å, β 101°30′, pseudomonoclinic, has been described by Moore (*Min. Mag.*, **41**, 129–132, 1978). The mineral, from the manganese and iron mines at Långban, Sweden, is deep reddish-brown to reddish black in colour, with $\alpha = 1{\cdot}81$, γ 1·83, 2E $\simeq$ 45°, specific gravity 3·77.

Index

Mineral names in **bold** type are those described in detail; page numbers in **bold** type refer to the principal description or definition of the mineral.